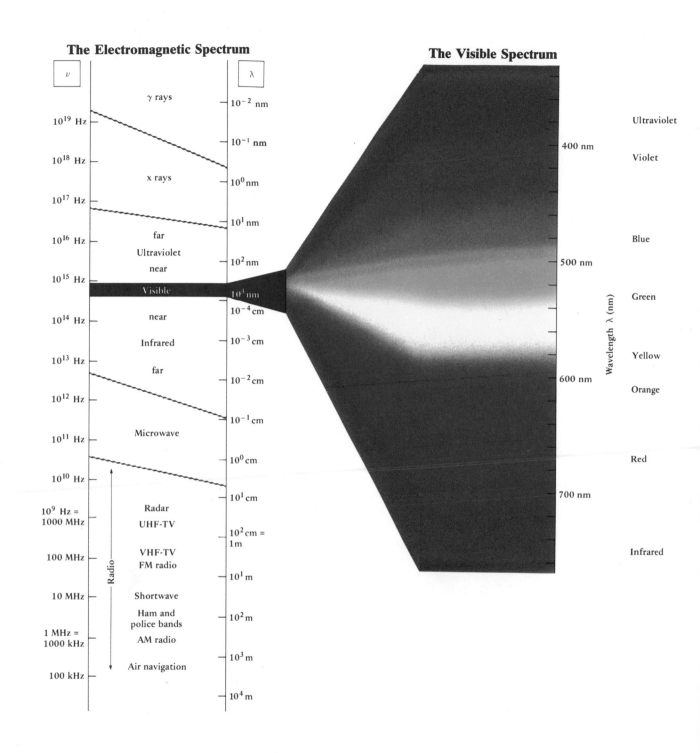

## The Electromagnetic Spectrum

| $\nu$ | | $\lambda$ |
|---|---|---|
| | $\gamma$ rays | $10^{-2}$ nm |
| $10^{19}$ Hz | | $10^{-1}$ nm |
| $10^{18}$ Hz | x rays | $10^{0}$ nm |
| $10^{17}$ Hz | | $10^{1}$ nm |
| $10^{16}$ Hz | far | |
| | Ultraviolet | $10^{2}$ nm |
| $10^{15}$ Hz | near | |
| | Visible | $10^{3}$ nm |
| $10^{14}$ Hz | near | $10^{-4}$ cm |
| | Infrared | $10^{-3}$ cm |
| $10^{13}$ Hz | far | |
| $10^{12}$ Hz | | $10^{-2}$ cm |
| | | $10^{-1}$ cm |
| $10^{11}$ Hz | Microwave | $10^{0}$ cm |
| $10^{10}$ Hz | | $10^{1}$ cm |
| $10^{9}$ Hz = 1000 MHz | Radar UHF-TV | $10^{2}$ cm = 1m |
| 100 MHz | VHF-TV FM radio | $10^{1}$ m |
| 10 MHz | Shortwave | $10^{2}$ m |
| | Ham and police bands | |
| 1 MHz = 1000 kHz | AM radio | $10^{3}$ m |
| | Air navigation | |
| 100 kHz | | $10^{4}$ m |

Radio

## The Visible Spectrum

Wavelength $\lambda$ (nm)

- Ultraviolet
- 400 nm
- Violet
- Blue
- 500 nm
- Green
- Yellow
- 600 nm
- Orange
- Red
- 700 nm
- Infrared

# FUNDAMENTALS OF

# ANALYTICAL CHEMISTRY

## SIXTH EDITION

**Douglas A. Skoog**
Stanford University

**Donald M. West**
San Jose State University

**F. James Holler**
University of Kentucky

**SAUNDERS COLLEGE PUBLISHING**
**A Harcourt Brace Jovanovich College Publisher**
Fort Worth   Philadelphia   San Diego   New York   Orlando   Austin   San Antonio
Toronto   Montreal   London   Sydney   Tokyo

Text Typeface: Times Roman
Compositor: Bi-Comp, Inc.
Acquisitions Editor: John J. Vondeling
Developmental Editor: Jennifer Bortel
Managing Editor: Carol Field
Project Editor: Margaret Mary Anderson
Copy Editor: Becca Gruliow
Manager of Art and Design: Carol Bleistine
Art Director: Doris Bruey
Art Assistant: Caroline McGowan
Cover Designer: Lawrence R. Didona
Text Artwork: J & R Technical Services
Director of EDP: Tim Frelick
Production Manager: Charlene Squibb
Marketing Manager: Marjorie Waldron


Cover Credit: Lawrence R. Didona


Printed in the United States of America

*Fundamentals of Analytical Chemistry, Sixth Edition*

*0-03-074922-0*


Library of Congress Catalog Card Number: 91-058039


234 071 98765432

THIS BOOK IS PRINTED ON **ACID-FREE, RECYCLED** PAPER

# PREFACE

The sixth edition of *Fundamentals of Analytical Chemistry*, like its predecessors, is an introductory textbook designed primarily for a one- or two-semester course for chemistry majors. Since the publication of the fifth edition, the scope of analytical chemistry has continued to expand and includes applications in biology, medicine, materials science, ecology, and other related fields. The content of courses in analytical chemistry varies from institution to institution and depends upon the available facilities, time allocated to analytical chemistry in the chemistry curriculum, and desires of individual instructors. It is with these thoughts in mind that we have designed the sixth edition of *Fundamentals of Analytical Chemistry* to form a versatile and modular basis for courses in analytical chemistry.

## Objectives

A major objective of this text is to provide a rigorous background in those chemical principles that are particularly important to analytical chemistry. A second goal is to develop an appreciation for the difficult task of judging the accuracy and precision of experimental data and to show how these judgments can be sharpened by the application of statistical methods. A third aim is to introduce a wide range of techniques of modern analytical chemistry. A final goal is to teach those laboratory skills that will give students confidence in their ability to obtain high-quality analytical data.

## Coverage

The material in this text covers both fundamental and practical aspects of chemical analysis. Following a brief introduction in Chapter 1, Chapters 2 and 3 present topics in statistics and data analysis that are important in analytical chemistry. An overview of classical gravimetric and titrimetric analysis is then presented in Chapters 4 and 5. The theory of aqueous solutions, activities, and chemical equilibria is detailed in Chapters 6–8. In Chapter 9–16, we consider the theory and practice of various titrimetric methods of analysis. Chapters 17–19 cover various electrical methods such as potentiometry and voltammetry. Chapters 20–24 on various spectroscopic methods are followed by Chapter 25, which is concerned with kinetic aspects of analytical chemistry. Analytical separations are then considered in Chapters 26–28. Practical facets of the preparation and analysis of real samples are detailed in Chapters 29–32. Finally, Chapters 33 and 34 provide discussions of the equipment and practice of chemical analysis and many detailed procedures for specific analyses.

We have included many features in the text to enhance learning and to provide a versatile teaching tool.

- **Organization**  *Fundamentals of Analytical Chemistry* provides a gradual progression from theoretical to the practical topics of analytical chemistry. Each chapter is organized as a learning unit; many chapters can stand alone, thus affording instructors the option of including or omitting topics as time and resources permit.
- **Important Equations**  Equations that we deem most important have been highlighted with a color screen for emphasis and ease of review.
- **Mathematical Level**  Most of the development of the theory of chemical analysis requires only a working knowledge of college algebra. A few concepts presented require basic differential and integral calculus.
- **Worked Examples**  A large number of worked examples serve as an aid in understanding the concepts of analytical chemistry. The examples are also models for the solution of problems found at the end of most of the chapters.
- **Problems**  An extensive set of problems and questions is included at the end of most chapters. Answers to approximately half of the problems are given at the end of the book.
- **Appendixes and Endpapers**  Included in the appendixes are important references of analytical chemistry, tables of chemical constants, designations for filtering media, a section on the use of logarithms and exponential numbers, a section on normality and equivalents (terms that are not used in the text itself), a list of compounds recommended for the preparation of standard materials, and derivations of the expressions for propagation of measurement uncertainties.

The endpapers provide a full-color chart of chemical indicators, a color display of the electromagnetic spectrum, a table of atomic numbers and weights, a table of common formula weights, and a periodic table of the elements.

## Changes in the Sixth Edition

Users of earlier editions of this text will find numerous changes not only in content but in format and style as well. We also have put increased emphasis on including units in chemical calculations and using dimensional analysis as a check on their correctness. Many changes have been made in response to suggestions by reviewers and colleagues.

## Content

- Chapter 19 (Voltammetry) has been completely rewritten to reflect a shift in emphasis and importance from classical polarography to hydrodynamic voltammetry.
- Material on molecular fluorescence spectroscopy has been combined in a new Chapter 23 to provide modularity and to shorten chapters.
- Chapter 14 (Introduction to Electrochemistry) has been largely rewritten, and many new figures are provided.
- Approximately 30 percent of the problems in the text are new or revised.

## Format

- **Style**   We have made numerous style and format changes in order to make the text more readable and student-friendly. Throughout, we have endeavored to use shorter sentences, simpler words, and the active voice. We have also shortened chapters and increased their number so that students are not so overwhelmed by the amount of material to be mastered in one sitting. Although a certain amount of vocabulary must be introduced when study of a new field is undertaken, we have attempted to avoid jargon whenever possible.
- **Marginal Notes**   Another innovation in this edition of *Fundamentals of Analytical Chemistry* is the use of marginal notes throughout the text. These notes include highlighted definitions, points of particular importance, line drawings and photographs of equipment, historical notes, and thought-provoking questions.
- **Features**   Also new to this edition is a series of boxed and highlighted "Features" that contain derivations of equations, explanations of more difficult theoretical points, and historical background notes. Instructors may make these materials optional or required reading, depending upon the rigor and scope of the course.
- **Illustrations**   Many new illustrations and photographs have been added to this edition. We have also provided full-color inserts to show in dramatic fashion various important color changes, chemical phenomena, and pieces of equipment.
- **Biographical Sketches**   New to the sixth edition is the inclusion of biographical sketches of persons whose work has contributed to the development of analytical chemistry. A photo or postage stamp containing the likeness of each person is presented with each biographical sketch. For persons interested in philately, the postage stamps are reproduced in full color in the color inserts.

## Ancillaries

- **Instructor's Manual**   The instructor's manual is a complete set of worked-out solutions to the problems in the text.
- **Overhead Transparencies**   Approximately 50 transparencies are available for overhead projection. Many of the transparencies have multiple figures from the text.

• **Computer Applications**  A supplemental book is available entitled *MathCAD Applications for Analytical Chemistry*. MathCAD® is a commercially available mathematical notebook, computational tool, and equation solver that allows students to perform many of the calculations of analytical chemistry quickly and accurately. Results can be easily plotted, printed, and formatted to enhance student understanding of the concepts presented in *Fundamentals of Analytical Chemistry*. MathCAD is available in versions for both the IBM-PC and Macintosh computers, and an inexpensive student edition for the IBM-PC. *MathCAD Applications for Analytical Chemistry* introduces students to the command structure of MathCAD gradually, introducing program syntax as needed to perform statistical calculations, solve systems of equilibrium equations, carry out least squares analysis, analyze multi-component mixtures using multiple linear regression, and accomplish many other computational and graphical tasks.

## Acknowledgments

We wish to acknowledge with thanks the comments and suggestions of the following who have reviewed the manuscript at various stages in its production: Professor John P. Walters, St. Olaf College; Professor Frank Guthrie, Rose-Hulman Institute of Technology; and Professor Larry Sveum, New Mexico Highlands University.

Our thanks also to Professor C. Marvin Lang and Gary Shulfer of the University of Wisconsin at Stevens Point for providing color slides of the postage stamps that appear in the color plates and the marginal notes of the book.

Finally, we want to acknowledge with thanks the various people at Saunders College Publishing for their friendly assistance in completing this project in record time. Among these are Senior Project Editor Margaret Mary Anderson, Copy Editor Becca Gruliow, Art Director Doris Bruey, and Production Manager Charlene Squibb. Our thanks also to Publisher John Vondeling and Developmental Editor Jennifer Bortel.

**Douglas A. Skoog**
**Donald M. West**
**F. James Holler**
December 1991

# CONTENTS OVERVIEW

# CONTENTS

xii    Contents

## CHAPTER 1

# INTRODUCTION

Analytical chemistry involves separating, identifying, and determining the relative amounts of the components in a sample of matter. Qualitative analysis reveals the chemical identity of the species in the sample. Quantitative analysis establishes the relative amount of one or more of these species, or *analytes,* in numerical terms. Qualitative information is required before a quantitative analysis can be undertaken. A separation step is usually a necessary part of both a qualitative and a quantitative analysis.

In this text, we are concerned principally with quantitative methods of analysis and methods of analytical separations. However, we refer occasionally to qualitative methods.

The components of a sample that are to be determined are often referred to as *analytes*.

## 1A ROLE OF ANALYTICAL CHEMISTRY IN THE SCIENCES

Analytical chemistry has played a vital role in the development of science. For example, in 1894 Wilhelm Ostwald wrote,

> Analytical chemistry, or the art of recognizing different substances and determining their constituents, takes a prominent position among the applications of science, since the questions which it enables us to answer arise wherever chemical processes are employed for scientific or technical purposes. Its supreme importance has caused it to be assiduously cultivated from a very early period in the history of chemistry, and its records comprise a large part of the quantitative work which is spread over the whole domain of science.

Since Ostwald's time, analytical chemistry has evolved from an art into a science with applications throughout industry, medicine, and all the sciences. To illustrate, consider a few examples. The amounts of hydrocarbons, nitrogen oxides, and carbon monoxide present in automobile exhaust gases must be measured in order to determine the effectiveness of smog-control devices. Quantitative measurements for ionized calcium in blood serum help diagnose parathyroid disease in human patients. Quantitative determination of nitrogen in foods establishes their protein content and thus their nutritional value. Analysis of steel during its production permits adjustment in the concentrations of such elements as carbon, nickel, and chromium to achieve a desired strength, hardness, corrosion resistance, and ductility. The mercaptan content of household gas supplies is monitored continually to assure that the gas has a sufficiently obnoxious odor to warn of dangerous leaks. Modern farmers tailor fertilization and irrigation schedules to meet changing plant needs during the growing season. They gauge these needs from quantitative analyses of the plants and the soil in which they grow.

Quantitative analytical measurements also play a vital role in many research areas in chemistry, biochemistry, biology, geology, and the other sciences. For example, chemists unravel the mechanisms of chemical reactions through reaction rate studies. The rate of consumption of reactants or formation of products in a chemical reaction can be calculated from quantitative measurements made at equal time intervals. Quantitative measurement of potassium, calcium, and sodium ions in the body fluids of animals permit physiologists to study the role these ions play in nerve-signal conduction and muscle contraction and relaxation. Materials scientists rely heavily upon quantitative analyses of crystalline germanium and silicon in their studies of the behavior of semiconductor devices. Impurities in these devices are in the concentration range of $1 \times 10^{-6}$ to $1 \times 10^{-10}$ percent. Archaeologists identify the source of volcanic glasses (obsidian) by measuring concentrations of minor elements in samples taken from various locations. This knowledge in turn makes it possible to trace prehistoric trade routes for tools and weapons fashioned from obsidian.

Many chemists and biochemists devote a significant part of their time in the laboratory to gathering quantitative information about systems of interest to them. Analytical chemistry serves as an important tool for the scholarly efforts of such investigators.

## 1B    CLASSIFICATION OF QUANTITATIVE METHODS OF ANALYSIS

We compute the results of a typical quantitative analysis from two measurements. One is the weight or volume of sample to be analyzed. The second is the measurement of some quantity that is proportional to the amount of analyte in that sample. This second step normally completes the analysis. Chemists classify analytical methods according to the nature of this final measurement. In a *gravimetric* method, the mass of the analyte or some compound chemically related to it is determined. In a *volumetric* method, the volume of a solution containing sufficient reagent to react completely with the analyte is measured. *Electroanalytical* methods involve the measurement of such electrical properties as potential, current, resistance, and quantity of electricity. *Spectroscopic* methods are based upon measurement of the interaction between electromagnetic radiation and analyte atoms or molecules or upon the production of such radiation by analytes. Finally, a group of miscellaneous methods should be mentioned. These include the measurement of such properties as

*Electromagnetic radiation* includes X-ray, ultraviolet, visible, infrared, microwave, and radio-frequency radiation.

mass-to-charge ratio (*mass spectrometry*), rate of radioactive decay, heat of reaction, rate of reaction, thermal conductivity, optical activity, and refractive index.

## 1C   STEPS IN A TYPICAL QUANTITATIVE ANALYSIS

A typical quantitative analysis involves the sequence of steps shown in Figure 1–1. In some instances, we can leave out one or more of these steps. Ordinarily, though, all of them play an important role in the success of an analysis.

The first 23 chapters of this textbook focus on the last three steps in Figure 1–1. In the measurement step we determine one of the physical properties mentioned in Section 1B. In the calculation step we compute the relative amount of the analyte present in the samples. In the final step we evaluate the quality of the results and estimate their reliability.

We now describe each of these steps to give you an overview of a quantitative analysis.

### 1C–1 Selecting a Method of Analysis

A vital first step in any quantitative analysis is the selection of a method. The choice is sometimes difficult, requiring experience as well as intuition on the part of the chemist. An important consideration in selection is the level of accuracy required. Unfortunately, high reliability nearly always requires a large investment of time. The method that is selected ordinarily represents a compromise between accuracy and economics.

A second consideration related to economic factors is the number of samples to be analyzed. If there are many samples, we can afford to spend a good deal of time in such preliminary operations as assembling and calibrating instruments and equipment and preparing standard solutions. If we have only a single sample or a few samples at most, we may find it more expedient to select a procedure that avoids or minimizes such preliminary steps.

Finally, the choice of method is always governed by the complexity of the sample as well as the number of components in the sample. We discuss the choice of method in detail in Section 29A.

### 1C–2 Sampling

To produce meaningful information, an analysis must be performed on a sample whose composition faithfully represents that of the bulk of material from which it was taken. Where the bulk is large and inhomogeneous, great effort is required to get a representative sample. Consider, for example, a railroad car containing 25 tons of silver ore. Buyer and seller must agree on a price based primarily upon the silver content of the shipment. The ore itself is inherently heterogeneous, consisting of many lumps that vary in size as well as in silver content. The assay of this shipment will be performed upon a sample that weighs about one gram. For the analysis to

Figure 1–1
Steps in a quantitative analysis.

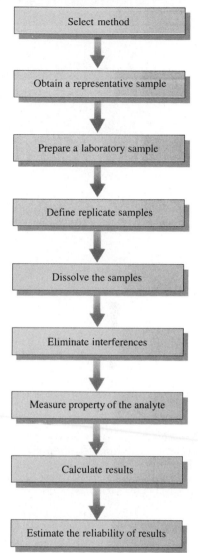

An *assay* is the process of determining how much of a given sample is the material indicated by its name. For example, a zinc alloy is assayed for its zinc content.

have significance, this small sample must have a composition that is representative of the 25 tons (or approximately 22,700,000 g) of ore in the shipment. Isolation of 1 g of material that accurately reflects the average composition of the nearly 23,000,000 g of bulk sample is a difficult undertaking that requires a careful, systematic manipulation of the entire shipment.

Many sampling problems are easier to solve than the one just described. Whether sampling is simple or complex, however, the chemist must have some assurance that the laboratory sample is representative of the whole before proceeding with an analysis. A detailed description of how various types of materials are sampled is presented in Section 30A.

## 1C–3 Preparing a Laboratory Sample

A solid laboratory sample is ground to decrease particle size, mixed to assure homogeneity, and stored for various lengths of time before the analysis begins. Absorption or desorption of water may occur during each of these steps, depending upon the humidity of the environment. Because any loss or gain of water changes the chemical composition of solids, it is a good idea to dry samples just before starting an analysis. Alternatively, the moisture content of the sample can be determined at the time of the analysis in a separate analytical procedure. Further information about preparing samples for analysis and removing moisture content is given in Chapter 30.

## 1C–4 Defining Replicate Samples

*Replicate samples* are portions of a material of approximately the same size that are carried through an analytical procedure at the same time and in the same way.

Most chemical analyses are performed on replicate samples whose weights or volumes have been determined by careful measurements with an analytical balance or with a precise volumetric device. Replication improves the quality of the results and provides a measure of their reliability.

Detailed instructions on techniques for measuring the weight or volume of samples appear in Sections 33D and 33H.

## 1C–5 Preparing Solutions of the Sample

Most analyses are performed on solutions of the sample. Ideally, the solvent should dissolve the entire sample (not just the analyte) rapidly and completely. The conditions of dissolution should be sufficiently mild so that loss of the analyte cannot occur. Unfortunately, many materials that must be analyzed are insoluble in common solvents. Examples include silicate minerals, high-molecular-weight polymers, or specimens of animal tissue. Conversion of the analyte in such materials into a soluble form can be a difficult and time-consuming task. Methods for decomposing and dissolving samples appear in various parts of Chapter 34.

## 1C–6 Eliminating Interferences

Few chemical or physical properties of importance in chemical analysis are unique to a single chemical species. Instead, the reactions used and the properties measured are characteristic of a group of elements or compounds. Species other than the analyte that affect the final measurement are called *interferences*. A scheme must be devised to isolate the analytes from interferences before the final measurement is made. No hard and fast rules can be given for eliminating interferences; indeed, resolution of this problem can be the most demanding aspect of an analysis. Chapters 26 through 28 describe separation methods.

Techniques or reactions that work for only one analyte are said to be *specific*. Techniques or reactions that work for only a few analytes are *selective*.

## 1C–7 Calibration and Measurement

All analytical results depend on a final measurement $X$ of a physical property of the analyte. This property must vary in a known and reproducible way with the concentration $c_A$ of the analyte. Ideally, the measurement of the physical property is directly proportional to the concentration. That is,

$$c_A = kX$$

where $k$ is a proportionality constant. With two exceptions, analytical methods require the empirical determination of $k$ with chemical standards for which $c_A$ is known. The exceptions are gravimetric and coulometric methods discussed in Chapters 4 and 18, respectively. The process of determining $k$ is thus an important step in most analyses and is termed a *calibration*.

## 1C–8 Calculating Results

Computing analyte concentrations from experimental data is ordinarily a simple and straightforward task, particularly with modern calculators or computers. Such computations are based on the raw experimental data collected in the measurement step, the stoichiometry of the chemical reaction upon which the analysis is based, and instrumental factors. These calculations appear throughout this book.

## 1C–9 Evaluating Results and Estimating Their Reliability

Analytical results are incomplete without an estimate of their reliability. The experimenter must provide some measure of the uncertainties associated with computed results if the data are to have any value. Chapters 2 and 3 present detailed methods for carrying out this important final step in the analytical process.

## CHAPTER 2

# ERRORS IN CHEMICAL ANALYSIS

It is impossible to perform a chemical analysis in such a way that the results are totally free of errors or uncertainties. Our goal is to keep these errors at a tolerable level and to estimate their size with acceptable accuracy. In this chapter we explore the nature of experimental errors and their effects on analytical results.

The effect of errors in analytical data is illustrated in Figure 2–1, which shows results for the quantitative determination of iron(III). Six equal portions of an aqueous solution known to contain exactly 20.00 ppm of iron(III) were analyzed in exactly the same way. Note that the results range from a low of 19.4 ppm to a high of 20.3 ppm of iron(III). The average $\bar{x}$ of the data is 19.8 ppm.

Every measurement is influenced by many uncertainties, which combine to produce a scatter of results like that in Figure 2–1. Measurement uncertainties can never be completely eliminated, so the true value for any quantity is generally unknown. The probable magnitude of the error in a measurement can be evaluated, however. It is then possible to define limits within which the true value of a measured quantity lies at a given level of probability.

It is seldom easy to estimate the reliability of experimental data. Nevertheless, we must make such estimates *because data of unknown reliability are worthless*. Moreover, results that are not especially accurate may be of considerable value if the limits of uncertainty are known.

Unfortunately, there is no simple and widely applicable method for determining the reliability of data with absolute certainty. It often requires as much effort to estimate the quality of experimental results as it requires to collect them. Reliability can be assessed in a number of different ways. Experiments designed to reveal the presence of errors can be performed. Standards of known composition can be analyzed and the results compared with the known composition. A few minutes in the library to consult the literature of analytical chemistry can be profitable. Calibrating equipment enhances the quality of data. Finally, statistical tests can be applied to the data. None of these options is perfect, so in the end we have to make *judgments* as to the probable accuracy of our

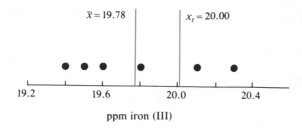

ppm iron (III)

Figure 2–1
Results from six replicate determinations for iron in aqueous samples of a standard solution containing 20.00 ppm of iron(III).

results. These judgments tend to become harsher and less optimistic with experience.

One of the first questions to answer before beginning an analysis is, "What is the maximum error I can tolerate in the result?" The answer to this question determines how much time you will spend on the analysis. For example, a tenfold increase in reliability may take hours, days, or even weeks of added labor. *No one can afford to waste time generating data that are more reliable than are needed.*

In this chapter we consider the types of errors encountered in chemical analysis, methods for recognizing errors, and techniques for estimating and reporting their size.[1]

## 2A   DEFINITION OF TERMS

Chemists carry two to five portions (*replicates*) of a sample through an entire analytical procedure. Individual results from a set of measurements are seldom the same (Figure 2–1), so a central or "best" value is used for the set. We justify the extra effort required to analyze several samples in two ways. First, the central value of a set should be more reliable than any of the individual results. Second, variation in the data should provide a measure of the uncertainty associated with the central result. Either the *mean* or the *median* may serve as the central value for a set of replicated measurements.

### 2A–1  The Mean and Median

*Mean, arithmetic mean,* and *average* ($\bar{x}$) are synonyms for the quantity obtained by dividing the sum of the results of replicate measurements by the number of measurements in the set:

$$\bar{x} = \frac{\sum\limits_{i=1}^{N} x_i}{N} \qquad (2\text{–}1)$$

The symbol $\Sigma x_i$ means to add all of the values $x_i$ for the replicates.

[1]For detailed discussions of error analysis, see A. Currie, in *Treatise on Analytical Chemistry,* 2nd ed., I. M. Kolthoff and P. J. Elving, Eds., Part I, Vol. 1, Chapter 4. New York: Wiley, 1978; and J. Mandel, *ibid.,* Chapter 5.

The *median* is the middle value in a set of data that has been arranged in increasing or decreasing order.

where $x_i$ represents the individual values of $x$ making up a set of $N$ replicate measurements.

The *median* of a set of data is the middle result when the data are arranged by size. Equal numbers of data are larger and smaller than the median. For an odd number of results, the median can be evaluated directly. For an even number, the mean of the middle pair is used.

---

Example 2–1

Calculate the mean and median for the data shown in Figure 2–1.

$$\text{mean} = \bar{x} = \frac{19.4 + 19.5 + 19.6 + 19.8 + 20.1 + 20.3}{6}$$

$$= 19.78 \approx 19.8 \text{ ppm Fe}$$

Because the set contains an even number of measurements, the median is the average of the central pair:

$$\text{median} = \frac{19.6 + 19.8}{2} = 19.7 \text{ ppm Fe}$$

Ideally, the mean and the median are identical. Frequently they are not, particularly when the number of measurements in the set is small.

---

## 2A–2 Precision

*Precision* is the closeness of a measurement to other measurements made in the same way.

*Precision* describes the agreement between two or more measurements that have been made *in exactly the same way*. There are several ways to express precision.

### Standard Deviation (*s*)

The standard deviation is a statistical term scientists and engineers use as a measure of precision. For small sets of data, we calculate the sample standard deviation $s$ using the following equation:

Equation 2–2 applies to small sets of data. It says to find the deviations of the $x_i$'s from the mean $\bar{x}$, square them, sum them, divide the sum by $N - 1$, and take the square root. The quantity $N - 1$ is called *the number of degrees of freedom*.

$$s = \sqrt{\frac{\sum_{i=1}^{N}(x_i - \bar{x})^2}{N - 1}} \qquad (2\text{–}2)$$

where $x_i - \bar{x}$ is the *deviation from the mean* of the $i$th measurement.

### Variance (*s*²)

The variance is simply the square of the standard deviation:

$$s^2 = \frac{\sum_{i=1}^{N}(x_i - \bar{x})^2}{N - 1} \qquad (2-3)$$

Table 2–1 illustrates how to obtain the standard deviation and the variance for the data in Figure 2–1. The standard deviation is 0.35 ppm Fe, the variance is 0.13 (ppm Fe)$^2$, and the number of degrees of freedom ($N - 1$) is five. The significance of these three terms in statistics is discussed in Section 3A.

Note that the standard deviation has the same units as the data, while the variance has the units of the data squared. Scientists and engineers tend to use standard deviation rather than variance as a measure of precision because it is easier to relate the precision of a measurement to the measurement itself if they both have the same units. On the other hand, we will see that variances have the advantage of being additive.

Many scientific calculators and computer software packages have the standard deviation function built in. If you use built-in functions to calculate the standard deviation of small sets of data, be sure that the built-in function uses the number of degrees of freedom rather than $N$.

The deviation from the mean is obtained by subtracting the mean of a set of measurements from an individual measurement.

### Table 2–1
### METHODS OF EXPRESSING PRECISION AND ACCURACY*

| Fe Concentration, ppm $x_i$ | Deviation from Mean $\lvert x_i - \bar{x} \rvert$ | $(x_i - \bar{x})^2$ |
|---|---|---|
| $x_1$ 19.4 | 0.38 | 0.1444 |
| $x_2$ 19.5 | 0.28 | 0.0784 |
| $x_3$ 19.6 | 0.18 | 0.0324 |
| $x_4$ 19.8 | 0.02 | 0.0004 |
| $x_5$ 20.1 | 0.32 | 0.1024 |
| $x_6$ 20.3 | 0.52 | 0.2704 |
| $\Sigma x_i = 118.7$ | | $\Sigma(x_i - \bar{x})^2 = 0.6284$ |

Mean $= \bar{x} = 118.7/6 = 19.78 \approx 19.8$ ppm Fe

Standard deviation $= s = \sqrt{\dfrac{\Sigma(x_i - \bar{x})^2}{N - 1}} = \sqrt{\dfrac{0.6284}{6 - 1}}$

$\qquad = \sqrt{0.1257} = 0.354 \approx 0.35$ ppm Fe

Variance $= s^2 = \dfrac{\Sigma(x_i - \bar{x})^2}{N - 1} = \dfrac{0.6284}{6 - 1} = 0.1257 \approx 0.13$ (ppm Fe)$^2$

Spread, or range $= w = 20.3 - 19.4 = 0.9$ ppm Fe

Relative standard deviation $= \dfrac{s}{\bar{x}} \times 1000$ ppt $= \dfrac{0.354}{19.78} \times 1000 = 17.9 \approx 18$ ppt

Coefficient of variation $= CV = \dfrac{s}{\bar{x}} \times 100\% = \dfrac{0.354}{19.78} \times 100\% \approx 1.8\%$

Absolute error† $= 19.78 - 20.00 = -0.22 = -0.2$ ppm Fe

Relative error $= \dfrac{19.78 - 20.00}{20.00} \times 100\% = -1.1\%$

*For source of data, see Figure 2–1.

†Sample known to contain 20.00 ppm Fe.

### An Alternative Expression for Standard Deviation

To calculate $s$ with a calculator that does not have a standard deviation key, the following rearrangement of Equation 2–2 is easier to use:

$$s = \sqrt{\frac{\sum\limits_{i=1}^{N} x_i^2 - \frac{\left(\sum\limits_{i=1}^{N} x_i\right)^2}{N}}{N-1}} \qquad (2\text{–}4)$$

---

**Example 2–2**

The following data were collected by finding the absorbance ($x_i$) of replicate solutions of a colored complex of iron (see Chapter 22). Determine the mean and standard deviation for the five data.

To apply Equation 2–4, we calculate $\Sigma x_i^2$ and $(\Sigma x_i)^2/N$.

| Solution | $x_i$ | $x_i^2$ |
|---|---|---|
| 1 | 0.752 | 0.565504 |
| 2 | 0.756 | 0.571536 |
| 3 | 0.752 | 0.565504 |
| 4 | 0.751 | 0.564001 |
| 5 | 0.760 | 0.577600 |
| | $\Sigma x_i = 3.771$ | $\Sigma x_i^2 = 2.844145$ |

$$\bar{x} = 3.771/5 = 0.7542 \approx 0.754$$

$$\frac{(\Sigma x_i)^2}{N} = \frac{(3.771)^2}{5} = 2.8440882$$

$$s = \sqrt{\frac{2.844145 - 2.8440882}{5-1}}$$

$$= \sqrt{\frac{0.0000568}{4}} = 0.00377 \approx 0.0038$$

---

This is a general problem. Anytime we compute the difference between two large numbers that are approximately equal to each other, the result will have a relatively large uncertainty.

Note that the difference between $\Sigma x_i^2$ and $(\Sigma x_i)^2/N$ in Example 2–2 is very small. If we had rounded these numbers before subtracting them, a serious error would have appeared in the computed value of $s$. To avoid this source of error, never use Equation 2–4 to calculate the standard deviation of numbers containing five or more digits. Use Equation 2–2 instead.[2] Note also that handheld calculators and small computers with a standard deviation function usually employ a version of Equation 2–4. Therefore, expect large errors in $s$ when these devices are used to calcu-

---

[2]The first two or three digits in a set of data are ordinarily identical to each other. As an alternative to using Equation 2–2, these identical digits can be dropped and the remaining digits used with Equation 2–4. For example, the standard deviation for the data in Example 2–2 could be based on 0.052, 0.056, 0.052, and so forth or even on 52, 56, 52, and so forth.

late the standard deviation of data that have five or more significant figures.

### Relative Standard Deviation (RSD)

Chemists frequently quote standard deviations in relative rather than absolute terms. We calculate the relative standard deviation by dividing the standard deviation by the mean of the set of data.

Relative standard deviation is often expressed in parts per thousand (ppt) or in percent. For example,

$$\text{RSD} = (s/\bar{x}) \times 1000 \text{ ppt}$$

When the relative standard deviation is multiplied by 100%, it is called the *coefficient of variation* (CV).

$$\text{CV} = (s/\bar{x}) \times 100\% \qquad (2\text{--}5)$$

The relative standard deviation for the data in Table 2–1 is 18 ppt. The coefficient of variation for the set is 1.8%.

Relative standard deviations often give a clearer picture of data quality than do absolute standard deviations. As an example, suppose that a sample contains about 50 mg of copper and that the standard deviation of a copper determination is 2 mg. The CV for this sample is 4%. For a sample containing only 10 mg of copper, the CV is 20%.

### Spread or Range (*w*)

The *spread,* or *range,* of a set of data is the difference between the largest value in the set and the smallest. Thus, the spread of the data in Table 2–1 is 0.9 ppm Fe.

### 2A–3 Accuracy

Figure 2–2 illustrates the basic difference between accuracy and precision. *Accuracy* indicates the closeness of a measurement to its true or accepted value and is expressed by the *error*. Accuracy measures agreement between a result and its true value. *Precision* describes the agreement among several results measured in the same way. Precision is determined by simply replicating a measurement. In contrast, accuracy can never be determined exactly because the true value of a quantity can never be known exactly. An accepted value is used instead.

### Absolute Error

The *absolute error E* in the measurement of a quantity $x_i$ is given by the equation

Low accuracy, low precision

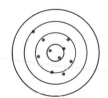

High accuracy, low precision

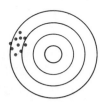

Low accuracy, high precision

High accuracy, high precision

**Figure 2–2**
**Accuracy and precision.**

The term "absolute" has a different meaning here than it does in mathematics. The absolute value in mathematics means the magnitude of a number, *ignoring its sign*. As we shall use it, the absolute error is the difference between *an experimental result and its accepted value, including its sign*.

$$E = x_i - x_t \qquad (2\text{-}6)$$

where $x_t$ is the true, or accepted, value of the quantity. Returning to our example in Table 2–1, we see that the absolute error is −0.2 ppm Fe. Note that with absolute error we retain the sign (in contrast to measures of precision). Thus, the negative sign in the example shows that the experimental result is smaller than the accepted value.

### Relative Error

The *relative error $E_r$* frequently is a more useful quantity than the absolute error. Percent relative error is given by the expression

$$E_r = \frac{x_i - x_t}{x_t} \times 100\% \qquad (2\text{-}7)$$

Relative error is also expressed in parts per thousand (ppt). Thus, the relative error for the mean of the data in Table 2–1 is reported either as −1.1% or as −11 ppt.

### 2A–4  Types of Errors in Experimental Data

The precision of a measurement is readily determined by comparing data from carefully replicated experiments. Unfortunately, an estimate of the accuracy is not so easy to obtain. To determine accuracy, we have to know the true value, and this is exactly what we are looking for.

It is tempting to assume that if we know the answer precisely, then we also know it accurately. The danger of this assumption is illustrated in Figure 2–3, which summarizes the results for the determination of nitrogen in two pure compounds. The dots show the absolute errors of replicate results obtained by four analysts. Note that analyst 1 obtained relatively high precision and high accuracy. Analyst 2 had poor precision but good accuracy. The results of analyst 3 are surprisingly common: the precision is excellent, but the numerical average for the data is quite inaccurate. Both precision and accuracy are poor for the results of analyst 4.

Indeterminate errors are errors that affect the precision of measurements.

Figures 2–1 and 2–3 show that chemical analyses are affected by at least two types of errors. One type, called *indeterminate* (or *random*) *error*, causes data to be scattered more or less symmetrically around a mean value. Refer again to Figure 2–3, and notice that the scatter in the data for analysts 1 and 3 is significantly less than that for analysts 2 and 4.

The precision of the data reflects the indeterminate errors in an analysis. For example, the relative standard deviation of 18 ppt Fe computed in Table 2–1 is a valid measure of the indeterminate errors associated with this method for determining iron.

A second type of error, called *determinate* (or *systematic*) *error*, causes the mean of a set of data to differ from the accepted value. For example,

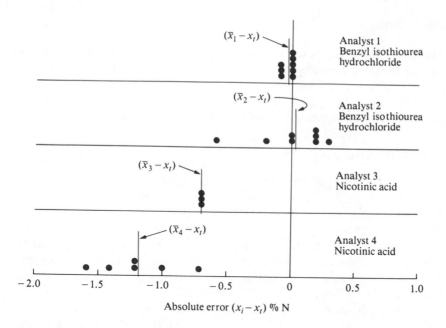

Figure 2–3
Absolute error in the micro-
Kjeldahl determination of nitrogen.
Each dot represents the error associated with a single determination.
Each vertical line labeled $(\bar{x}_i - x_t)$
is the absolute average deviation of
the set from the true value. (Data
from C. O. Willits and C. L. Ogg,
*J. Assoc. Offic. Anal. Chem.*,
**1949**, *32*, 561. With permission.)

the data in Figure 2–1 have a determinate error of about $-0.22$ ppm Fe.
The results of analysts 1 and 2 in Figure 2–3 have little determinate error,
but the data of analysts 3 and 4 show determinate errors of about $-0.7$ and
$-1.2\%$ nitrogen.

A third type of error is *gross error*. Gross errors differ from indeterminate and determinate errors in that they usually occur only occasionally,
are often large, and may cause a result to be either high or low. Gross
errors lead to *outliers*–results that differ markedly from all other data in a
set of replicate measurements. There is no evidence of a gross error in
Figures 2–1 and 2–3.

Gross errors usually affect only a
single result in a set of replicate
data, causing it to differ significantly
from the remaining results for that
set.

## 2B   DETERMINATE ERRORS

Determinate errors have a definite source that usually can be identified.
They cause all the results from replicate measurements to be either high
or low. Because they are unidirectional, determinate errors are also called
*systematic errors*.

Determinate, or systematic, errors
affect the accuracy of results.

For example, the last two sets of data in Figure 2–3 reveal a negative
determinate error. A probable source of this systematic error can be
traced to the chemical nature of the sample, nicotinic acid. The analytical
method used involves decomposition of the samples with hot concentrated sulfuric acid, which converts the nitrogen to ammonium sulfate.
The amount of ammonia in the ammonium sulfate is then determined in
the measurement step. Experiments have shown that compounds containing a pyridine ring (such as nicotinic acid) are incompletely decomposed
by sulfuric acid unless special precautions are taken. Without these precautions, low results are obtained. It is very likely that the negative errors, $(\bar{x}_3 - x_t)$ and $(\bar{x}_4 - x_t)$ in Figure 2–3 are determinate errors that can
be blamed on incomplete decomposition.

### 2B-1 Sources of Determinate Errors

There are three types of determinate error. (1) *Instrument errors* are caused by imperfections in measuring devices and instabilities in their power supplies. (2) *Method errors* arise from nonideal chemical or physical behavior of analytical systems. (3) *Personal errors* result from the carelessness, inattention, or personal limitations of the experimenter.

### Instrument Errors

All measuring devices are sources of determinate errors. For example, pipets, burets, and volumetric flasks may have volumes slightly different from those indicated by their graduations. These differences may result from using glassware at a temperature that differs significantly from the calibration temperature, from distortions in container walls due to heating while drying, from errors in the original calibration, or from contaminants on the inner surfaces of the containers. Calibration eliminates most determinate errors of this type.

Electronic instruments are subject to determinate errors, which have many sources. For example, errors emerge as the voltage of a battery-operated power supply decreases with use. Errors result from increased resistance in circuits because of dirty electrical contacts. Temperature changes cause variation in resistors and standard potential sources. Currents induced from 110-V power lines affect electronic instruments. Errors from these and other sources are detectable and correctable.

### Method Errors

The nonideal chemical or physical behavior of the reagents and reactions upon which an analysis is based can introduce determinate method errors. Such sources of nonideality include the slowness of some reactions, the incompleteness of others, the instability of some species, the nonspecificity of most reagents, and the possible occurrence of side reactions that interfere with the measurement process. For example, a common method error in volumetric methods results from the small excess of reagent required to cause an indicator to undergo the color change that signals completion of the reaction. The accuracy of such an analysis is thus limited by the very phenomenon that makes the titration possible.

Another example of method error was described earlier in connection with the data of analysts 3 and 4 in Figure 2–3. Here, the source was the incompleteness of the reaction between the pyridine nitrogen in nicotinic acid and the sulfuric acid.

Errors inherent in a method are difficult to detect and are thus the most serious of the three types of determinate error.

### Personal Errors

Many measurements require personal judgments. Examples include estimating the position of a pointer between two scale divisions, the color of a solution at the end point in a titration, or the level of a liquid with respect to a graduation in a pipet or buret. Judgments of this type are often subject to systematic, unidirectional errors. For example, one person may read a pointer consistently high, another may be slightly slow in activating a timer, and a third may be less sensitive to color changes. An analyst who is insensitive to color changes tends to use excess reagent in a volumetric analysis. Physical handicaps are often sources of personal determinate errors.

A universal source of personal error is prejudice, or personal bias. Most of us, no matter how honest, have a natural tendency to estimate scale readings in a direction that improves the precision in a set of results. Or we may have a preconceived notion of the true value for the measurement. We then subconsciously cause the results to fall close to this value. Number bias is another source of personal error that varies considerably from person to person. The most common bias encountered in estimating the position of a needle on a scale involves a preference for the digits 0 and 5. Also prevalent is a prejudice favoring small digits over large and even numbers over odd.

Digital readouts on pH meters, laboratory balances, and other electronic instruments eliminate personal bias because no judgment is involved in taking a reading.

> Color blindness is a good example of a handicap that amplifies personal errors in volumetric analysis. A famous color-blind analytical chemist enlisted his wife to come to the laboratory to help him detect color changes at titration end points.

> Persons who make measurements must guard against personal bias to preserve the integrity of the collected data.

## 2B–2  The Effect of Determinate Errors Upon Analytical Results

Determinate errors may be either *constant* or *proportional*. The magnitude of a constant error does not depend on the size of the quantity measured. Proportional errors increase or decrease in proportion to the size of the sample taken for analysis.

### Constant Errors

Constant errors become more serious as the size of the quantity measured decreases. The effect of solubility losses on the results of a gravimetric analysis illustrates this behavior.

---

Example 2–3

Suppose that 0.50 mg of precipitate is lost as a result of being washed with 200 mL of wash liquid. If the precipitate weighs 500 mg, the relative error due to solubility loss is $-(0.50/500) \times 100\% = -0.1\%$. Loss of the same quantity from 50 mg of precipitate results in a relative error of $-1.0\%$.

---

The excess of reagent required to bring about a color change during titration is another example of constant error. This volume, usually small, remains the same regardless of the total volume of reagent required for the titration. Again, the relative error from this source becomes more serious as the total volume decreases. One way of minimizing the effect of constant error is to use as large a sample as possible.

## Proportional Errors

A common cause of proportional errors is the presence of interfering contaminants in the sample. For example, a widely used method for the determination of copper is based upon the reaction of copper(II) ion with potassium iodide to give iodine. The quantity of iodine is then measured and is proportional to the amount of copper. Iron(III), if present, also liberates iodine from potassium iodide. Unless steps are taken to prevent this interference, high results are observed for the percentage of copper because the iodine produced will be a measure of the copper(II) *and* iron(III) in the sample. The size of this error is determined by the *fraction* of iron contamination, which is independent of the size of sample taken. If the sample size is doubled, for example, the amount of iodine liberated by both the copper and the iron contaminant is also doubled. Thus, the magnitude of the reported percentage of copper is independent of sample size.

## 2B–3 Detection of Determinate Instrument and Personal Errors

Determinate instrument errors are usually found and corrected by calibration. Periodic calibration of equipment is always desirable because the response of most instruments will change with time as a result of wear, corrosion, or mistreatment.

Most personal errors can be minimized by care and self-discipline. It is a good habit to check instrument readings, notebook entries, and calculations systematically. Errors that result from a known physical handicap can usually be avoided by a careful choice of method.

## 2B–4 Detection of Determinate Method Errors

The determinate error associated with an analytical method is often called its *bias*.

Determinate method errors are particularly difficult to detect. We may take one or more of the following steps to recognize and adjust for systematic errors of this type.

### Analysis of Standard Samples

*Standard reference materials* (SRM) for detecting determinate errors in analytical methods are materials containing one or more species in reliably known concentration.

The best way of estimating the determinate error of an analytical method is by the analysis of *standard reference materials* (SRM)—materials that contain one or more analytes at reliably known concentration levels. Standard reference materials are obtained in several ways.

Standard materials can sometimes be prepared by synthesis. Carefully measured quantities of the pure components of a material are measured out and mixed to produce a homogeneous sample whose composition is known from the quantities taken. The overall composition of a synthetic standard material must approximate closely the composition of the samples to be analyzed. Great care must be taken to ensure that the concentration of analyte is known reliably. Unfortunately, the synthesis of such standard samples may either be impossible or so difficult and time-consuming that this approach may not be practical.

Standard reference materials can be purchased from a number of governmental and industrial sources. For example, the National Institute of Standards and Technology (formerly the National Bureau of Standards) offers over 900 standard reference materials, including rocks and minerals, gas mixtures, glasses, hydrocarbon mixtures, polymers, urban dusts, rainwaters, and river sediments.[3] The concentration of one or more of the components in these materials has been determined in one of three ways: (1) by analysis by a previously validated reference method of analysis, (2) by analysis by two or more independent, reliable measurement methods, or (3) by analysis by a network of cooperating analysts who are technically competent and thoroughly knowledgeable about the material being tested.

Several commercial supply houses also offer analyzed materials for method testing.[4]

Standard reference materials from NIST. (Photo courtesy of the National Institute of Standards and Technology.)

### Independent Analysis

If standard samples are not available, a second independent and reliable analytical method can be used in parallel with the method being evaluated. The independent method should differ as much as possible from the one under study to minimize the possibility that some common factor in the sample has the same effect on both methods.

### Blank Determinations

Blank determinations are useful for detecting certain types of constant errors. In a blank determination, or *blank*, all steps of the analysis are performed in the absence of a sample. For instance, if directions call for 25 mL of hydrochloric acid to dissolve an ore sample prior to titration, an appropriate blank would be 25 mL of hydrochloric acid. Any reagent needed to titrate the blank would then be applied as a correction to the volumes needed to titrate samples of the analyte.

Blank determinations reveal errors due to interfering contaminants from the reagents and vessels employed in analysis. Blanks also allow the analyst to correct titration data for the volume of reagent needed to cause an indicator to change color at the end point.

A *blank* is a solution that contains the solvent and all the reagents used in an analysis but no sample.

### Variation in Sample Size

Example 2–3 demonstrates that as the size of a measurement increases, the effect of a constant error decreases. Thus, constant errors can often be detected by varying the sample size.

---

[3]See U.S. Department of Commerce, *NIST Standard Reference Materials Catalog,* 1990–91 ed., NIST Special Publication 260. Washington: Government Printing Office, 1990. For a description of the reference material programs of the NIST (formerly NBS), see R. A. Alvarez, S. D. Rasberry, and F. A. Uriano, *Anal. Chem.,* **1982,** *54,* 1226A; and F. A. Uriano, *ASTM Standardization News,* **1979,** *7,* 8.

[4]For sources of biological and environmental reference materials containing various elements, see C. Veillon, *Anal. Chem.,* **1986,** *58,* 851A.

## 2C    GROSS ERRORS

Most gross errors are personal and are attributable to carelessness, laziness, or ineptitude. Gross errors can be random but occur so infrequently that they are generally not considered as indeterminate errors. Sources of gross errors include arithmetic mistakes, transposition of numbers in recording data, reading a scale backward, reversing a sign, using a wrong scale, spilling a solution, or just bad luck. Some gross errors affect only a single result. Others, such as using the wrong scale of an instrument, affect an entire set of replicate measurements. Gross errors are also encountered as a result of momentary interruptions in power or water supplies and other unexpected events.

Gross errors can be eliminated through self-discipline. Many scientists follow the practice of re-reading an instrument after a first reading has been recorded and then making sure the two agree.

## 2D    INDETERMINATE ERRORS

*Indeterminate,* or *random, errors* arise when a system of measurement is extended to its maximum sensitivity. They are caused by the many uncontrollable variables that are an inevitable part of every physical or chemical measurement.

There are many sources of indeterminate error, but none can be positively identified or measured because most are so small that they are undetectable. The cumulative effect of the individual indeterminate errors, however, causes the data from a set of replicate measurements to fluctuate randomly around the mean of the set. For example, the data in Figures 2–1 and 2–3 show scatter as a direct result of the accumulation of small indeterminate errors. Notice that in Figure 2–3 the indeterminate errors in the results of analysts 2 and 4 are greater than in those of analysts 1 and 3.

### 2D–1    Sources of Indeterminate Error

We can get a qualitative idea of the way small errors produce an overall uncertainty in the following way. Imagine a situation in which just four small random errors combine to give an overall error. We will assume that each error has an equal probability of occurring and that each can cause the final result to be high or low by a fixed amount $\pm U$.

Table 2–2 shows all the possible ways the four errors can combine to give the indicated deviations from the mean value. Note that only one combination of errors leads to a deviation of $+4U$, four combinations give a deviation of $+2U$, and six give a deviation of $0U$. The negative errors have the same relationship. This ratio of $1:4:6:4:1$ is a measure of the probability of a deviation of each magnitude. Therefore, if we make a sufficiently large number of measurements, we can expect a frequency distribution like that shown in Figure 2–4a. Note that the ordinate in the plot is the relative frequency of occurrence of the five possible combinations.

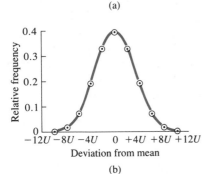

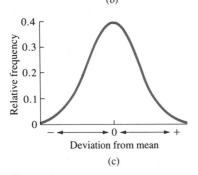

**Figure 2–4**
Frequency distribution for measurements containing (a) four indeterminate uncertainties; (b) ten indeterminate uncertainties; (c) a very large number of indeterminate uncertainties.

**Table 2–2**
**POSSIBLE COMBINATIONS OF FOUR EQUAL-SIZED UNCERTAINTIES**

| Combinations of Uncertainties | Magnitude of Indeterminate Error | Number of Combinations | Relative Frequency |
|---|---|---|---|
| $+U_1 + U_2 + U_3 + U_4$ | $+4U$ | 1 | $1/16 = 0.0625$ |
| $-U_1 + U_2 + U_3 + U_4$<br>$+U_1 - U_2 + U_3 + U_4$<br>$+U_1 + U_2 - U_3 + U_4$<br>$+U_1 + U_2 + U_3 - U_4$ | $+2U$ | 4 | $4/16 = 0.250$ |
| $-U_1 - U_2 + U_3 + U_4$<br>$+U_1 + U_2 - U_3 - U_4$<br>$+U_1 - U_2 + U_3 - U_4$<br>$-U_1 + U_2 - U_3 + U_4$<br>$-U_1 + U_2 + U_3 - U_4$<br>$+U_1 - U_2 - U_3 + U_4$ | $0$ | 6 | $6/16 = 0.375$ |
| $+U_1 - U_2 - U_3 - U_4$<br>$-U_1 + U_2 - U_3 - U_4$<br>$-U_1 - U_2 + U_3 - U_4$<br>$-U_1 - U_2 - U_3 + U_4$ | $-2U$ | 4 | $4/16 = 0.250$ |
| $-U_1 - U_2 - U_3 - U_4$ | $-4U$ | 1 | $1/16 = 0.0625$ |

Figure 2–4b shows the theoretical distribution for ten equal-sized uncertainties. Again we see that the most frequent occurrence is zero deviation from the mean. At the other extreme, a maximum deviation of $10U$ occurs only about once in 500 measurements.

When the same procedure is applied to a very large number of individual errors, a curve like that shown in Figure 2–4c results. This bell-shaped curve is called a *Gaussian curve* or a *normal error curve*.

In our example, all the uncertainties have the same magnitude. This restriction is not necessary to derive the equation for a Gaussian curve.

## 2D–2 Distribution of Experimental Data

We find empirically that the distribution of replicate data from most quantitative analytical experiments approaches that of the Gaussian curve shown in Figure 2–4c. As an example, consider the data in Table 2–3 for the calibration of a 10-mL pipet. In this experiment a small flask and stopper are weighed, and a 10-mL portion of water is transferred to the flask with the pipet, and the flask is stoppered. Finally, the flask, the stopper, and the water are weighed again. The temperature of the water is also measured to establish its density. The weight of the water is then calculated by taking the difference between the two weights; this difference is divided by the density of the water to find the volume delivered by the pipet. The experiment was performed 50 times.

The data in Table 2–3 are typical of those obtained by an experienced worker weighing to the nearest milligram (which corresponds to 0.001 mL) on a top-loading balance and making every effort to avoid determinate error. Even so, the standard deviation of the 50 measurements is 0.0056 mL and the spread is 0.025 mL. This distribution of data about the mean is caused by the many indeterminate errors in the experiment.

Table 2–3
REPLICATE DATA ON THE CALIBRATION OF A 10-mL PIPET*

| Trial | Volume, mL | Trial | Volume, mL | Trial | Volume, mL |
|-------|-----------|-------|-----------|-------|-----------|
| 1 | 9.988 | 18 | 9.975 | 35 | 9.976 |
| 2 | 9.973 | 19 | 9.980 | 36 | 9.990 |
| 3 | 9.986 | 20 | 9.994† | 37 | 9.988 |
| 4 | 9.980 | 21 | 9.992 | 38 | 9.971 |
| 5 | 9.975 | 22 | 9.984 | 39 | 9.986 |
| 6 | 9.982 | 23 | 9.981 | 40 | 9.978 |
| 7 | 9.986 | 24 | 9.987 | 41 | 9.986 |
| 8 | 9.982 | 25 | 9.978 | 42 | 9.982 |
| 9 | 9.981 | 26 | 9.983 | 43 | 9.977 |
| 10 | 9.990 | 27 | 9.982 | 44 | 9.977 |
| 11 | 9.980 | 28 | 9.991 | 45 | 9.986 |
| 12 | 9.989 | 29 | 9.981 | 46 | 9.978 |
| 13 | 9.978 | 30 | 9.969‡ | 47 | 9.983 |
| 14 | 9.971 | 31 | 9.985 | 48 | 9.980 |
| 15 | 9.982 | 32 | 9.977 | 49 | 9.983 |
| 16 | 9.983 | 33 | 9.976 | 50 | 9.979 |
| 17 | 9.988 | 34 | 9.983 | | |

Mean volume = 9.982 mL

Median volume = 9.982 mL

Spread = 0.025 mL

Standard deviation = 0.0056 mL

*Data listed in the order obtained.

†Maximum value.

‡Minimum value.

Table 2–4

FREQUENCY DISTRIBUTION OF DATA FROM TABLE 2–3

| Volume Range, mL | Number in Range | % in Range |
|------------------|-----------------|------------|
| 9.969 to 9.971 | 3 | 6 |
| 9.972 to 9.974 | 1 | 2 |
| 9.975 to 9.977 | 7 | 14 |
| 9.978 to 9.980 | 9 | 18 |
| 9.981 to 9.983 | 13 | 26 |
| 9.984 to 9.986 | 7 | 14 |
| 9.987 to 9.989 | 5 | 10 |
| 9.990 to 9.992 | 4 | 8 |
| 9.993 to 9.995 | 1 | 2 |

A histogram is a bar graph such as that shown in Figure 2-5.

The information in Table 2–3 is easier to visualize when the data are rearranged into frequency distribution groups, as in Table 2–4. Here, we tabulate the number of data falling into a series of adjacent 0.003-mL *cells* and calculate the percentage of measurements falling into each cell. Note that 26% of the data reside in the cell containing the mean and median value of 9.982 mL and that more than half the data are within ±0.004 mL of this mean. Note also that 72% of the data are within ±0.0056 mL, or one standard deviation, of the mean. As we will show in Chapter 3, the theoretical percentage in this range is 68.

The frequency distribution data in Table 2–4 are plotted as a bar graph, or *histogram*, in Figure 2–5. We also show in this figure a theoretical Gaussian curve derived for an infinite set of data. The data for this curve have the same mean (9.982 mL), the same standard deviation (0.0056 mL), and the same area under the curve as the data of the histogram. As the number of calibration experiments increases and the cell size decreases, the shape of the histogram approaches the shape of the continuous curve.

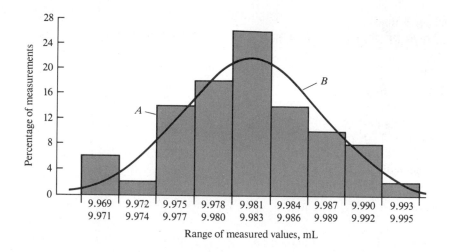

Figure 2–5
A histogram (*A*) showing distribution of the 50 results in Table 2–3 and a Gaussian curve (*B*) for data having the same mean and same standard deviation as the data in the histogram.

Variations in replicate results such as those in Table 2–3 result from numerous small and individually undetectable instrument, method, and personal indeterminate errors. These errors are attributable to uncontrolled variables in the experiment, and the net effect of such errors is also indeterminate. Ordinarily, the small errors tend to cancel one another and thus have a minimal effect. Occasionally, however, they occur in the same direction to produce a large positive or negative error.

Sources of uncertainty in the calibration of a pipet include such visual judgments as the level of the water with respect to the marking on the pipet and the mercury level in the thermometer (both personal indeterminate errors). Other sources are variation in the drainage time and in the angle of the pipet as it drains (both method errors). Instrument errors attributable to temperature fluctuations affect (1) the volume of the pipet, (2) the viscosity of the liquid, and (3) the performance of the balance. Other sources of instrument error are vibrations and drafts that cause small variations in balance readings. It is clear that many small and uncontrollable variables affect even as simple a process as calibrating a pipet. It is very difficult to determine the influence of any one of the several indeterminate errors, but their cumulative effect is responsible for the scatter of data around the mean.[5]

## 2E   THE STANDARD DEVIATION OF COMPUTED RESULTS

We often need to estimate the standard deviation of a result that has been computed from two or more experimental data, each of which has a known standard deviation. As shown in Table 2–5, the way such estimates are made depends upon the type of arithmetic that is involved.

The relationships of Table 2–5 are derived in Appendix 12.

[5]Sources of error can be separated through the statistical procedure known as analysis of variance (ANOVA). See R. L. Anderson, *Practical Statistics for Analytical Chemists*, p. 107. New York: Van Nostrand–Reinhold, 1987.

**Table 2–5**
**ERROR PROPAGATION IN ARITHMETIC CALCULATIONS**

| Type of Calculation | Example* | Standard Deviation of y | |
|---|---|---|---|
| Addition or Subtraction | $y = a + b - c$ | $s_y = \sqrt{s_a^2 + s_b^2 + s_c^2}$ | (1) |
| Multiplication or Division | $y = a \cdot b/c$ | $\dfrac{s_y}{y} = \sqrt{\left(\dfrac{s_a}{a}\right)^2 + \left(\dfrac{s_b}{b}\right)^2 + \left(\dfrac{s_c}{c}\right)^2}$ | (2) |
| Exponentiation | $y = a^x$ | $\dfrac{s_y}{y} = x\dfrac{s_a}{a}$ | (3) |
| Logarithm | $y = \log_{10} a$ | $s_y = 0.434 \dfrac{s_a}{a}$ | (4) |
| Antilogarithm | $y = \text{antilog}_{10}\, a$ | $\dfrac{s_y}{y} = 2.303\, s_a$ | (5) |

*$a$, $b$, and $c$ are experimental variables whose standard deviatives are $s_a$, $s_b$, and $s_c$, respectively.

## 2E–1 The Standard Deviation of Sums and Differences

Consider the summation:

$$
\begin{array}{ll}
+0.50 & (\pm 0.02) \\
+4.10 & (\pm 0.03) \\
-1.97 & (\pm 0.05) \\
\hline
2.63 &
\end{array}
$$

where the numbers in parentheses are absolute standard deviations. If the signs of the three individual standard deviations happen by chance to have the same sign, the standard deviation of the sum could be as large as $+0.02 + 0.03 + 0.05 = +0.10$ or $-0.02 - 0.03 - 0.05 = -0.10$. On the other hand, it is possible that the three could combine to give an accumulated value of zero: $-0.02 - 0.03 + 0.05 = 0$ or $+0.02 + 0.03 - 0.05 = 0$. More likely, however, the sum will lie between these two extremes. As shown in Table 2–5, the standard deviation of a sum or difference can be found by taking the square root of the sum of the squares of the individual absolute standard deviations. So, for the computation

$$y = a(\pm s_a) + b(\pm s_b) - c(\pm s_c)$$

the standard deviation of the result $s_y$ is given by

$$s_y = \sqrt{s_a^2 + s_b^2 + s_c^2} \tag{2-8}$$

where $s_a$, $s_b$, and $s_c$ are the standard deviations of the three terms in the sum. Substituting the standard deviations from the example gives

$$s_y = \sqrt{(\pm 0.02)^2 + (\pm 0.03)^2 + (\pm 0.05)^2} = \pm 0.06$$

and the sum should be reported as 2.63 ($\pm 0.06$).

For a sum or a difference the *absolute standard deviation of the answer* is the square root of the sum of the squares of the *absolute standard deviations* of the numbers used to calculate the sum or difference.

## 2E–2 Standard Deviations of Products and Quotients

Consider the following computation, where the numbers in parentheses are again absolute standard deviations:

$$\frac{4.10(\pm 0.02) \times 0.0050(\pm 0.0001)}{1.97(\pm 0.04)} = 0.010406(\pm ?)$$

In this situation, the standard deviations of two of the numbers in the calculation are larger than the result itself. Evidently, the approach we use to calculate standard deviation in multiplication and division cannot be the same as the one we used for addition and subtraction. As shown in Table 2–5, the relative standard deviation of the product or quotient is determined by the *relative standard deviations* of the numbers forming the computed result. For the computation

$$y = \frac{a \times b}{c} \qquad (2\text{–}9)$$

we obtain the relative standard deviation $s_y/y$ of the result $y$ by summing the squares of the relative standard deviations of $a$, $b$, and $c$ and then extracting the square root of the sum:

For multiplication or division, the *relative standard deviation of the answer* is the square root of the sum of the squares of the *relative standard deviations* of the numbers that are multiplied or divided.

$$\frac{s_y}{y} = \sqrt{\left(\frac{s_a}{a}\right)^2 + \left(\frac{s_b}{b}\right)^2 + \left(\frac{s_c}{c}\right)^2} \qquad (2\text{–}10)$$

Applying this equation to the numerical example gives

$$\frac{s_y}{y} = \sqrt{\left(\frac{\pm 0.02}{4.10}\right)^2 + \left(\frac{\pm 0.0001}{0.005}\right)^2 + \left(\frac{\pm 0.04}{1.97}\right)^2}$$

$$= \sqrt{(0.0049)^2 + (0.0200)^2 + (0.0203)^2} = \pm 0.0289$$

In order to complete the calculation, we must find the absolute standard deviation of the result,

To find the absolute standard deviation in a product or quotient, first find the relative standard deviation of the result and then multiply it by the result.

$$s_y = y \times (\pm 0.0289) = 0.0104 \times (\pm 0.0289) = \pm 0.000301$$

and we can write the answer and its uncertainty as 0.0104 ($\pm 0.0003$).

The following example demonstrates the calculation of the standard deviation of the result for a more complex calculation.

Example 2–4

Calculate the standard deviation of the result of

$$\frac{[14.3(\pm 0.2) - 11.6(\pm 0.2)] \times 0.050(\pm 0.001)}{[820(\pm 10) + 1030(\pm 5)] \times 42.3(\pm 0.4)} = 1.725(\pm ?) \times 10^{-6}$$

First, we must calculate the standard deviation of the sum and the difference. For the difference in the numerator,

$$s_a = \sqrt{(\pm 0.2)^2 + (\pm 0.2)^2} = \pm 0.283$$

and for the sum in the denominator,

$$s_b = \sqrt{(\pm 10)^2 + (\pm 5)^2} = 11.2$$

We may then rewrite the equation as

$$\frac{2.7(\pm 0.283) \times 0.050(\pm 0.001)}{1850(\pm 11.2) \times 42.3(\pm 0.4)} = 1.725 \times 10^{-6}$$

The equation now contains only products and quotients, and Equation 2–10 applies. Thus,

$$\frac{s_y}{y} = \sqrt{\left(\pm \frac{0.283}{2.7}\right)^2 + \left(\pm \frac{0.001}{0.050}\right)^2 + \left(\pm \frac{11.2}{1850}\right)^2 + \left(\pm \frac{0.4}{42.3}\right)^2} = 0.107$$

To obtain the absolute standard deviation, we write

$$s_y = y \times 0.107 = 1.725 \times 10^{-6} \times (\pm 0.107) = \pm 0.185 \times 10^{-6}$$

and round the answer to $1.7(\pm 0.2) \times 10^{-6}$.

## 2E–3 The Standard Deviation in Exponential Calculations

Consider the relationship

$$y = a^x$$

where the exponent $x$ can be considered to be free of uncertainty. As shown in Table 2–5 and Appendix 12, the relative standard deviation in $y$ resulting from the uncertainty in $a$ is

$$\frac{s_y}{y} = x \left(\frac{s_a}{a}\right) \qquad (2\text{–}11)$$

Thus, the relative standard deviation of the square of a number is twice the relative standard deviation of the number, the relative standard deviation in the cube root of a number is simply one third that of the number, and so forth.

---

Example 2–5

The standard deviation in measuring the diameter $d$ of a sphere is $\pm0.02$ cm. What is the standard deviation in the calculated volume $V$ of the sphere if $d = 2.15$ cm?

From the equation for the volume of a cube, we have

$$V = \frac{4}{3}\pi \left(\frac{d}{2}\right)^3 = \frac{4}{3}\pi \left(\frac{2.15}{2}\right)^3 = 5.20 \text{ cm}^3$$

Here we may write

$$\frac{s_V}{V} = 3 \times \frac{s_d}{d} = 3 \times \frac{0.02}{2.15} = 0.0279$$

The absolute standard deviation in $V$ is then

$$s_V = 5.20 \times 0.0279 = 0.145$$

Thus,

$$V = 5.2(\pm0.1) \text{ cm}^3$$

---

Example 2–6

The solubility product $K_{sp}$ for the silver salt AgX is $4.0(\pm0.4) \times 10^{-8}$. The solubility of AgX in water is

$$\text{Solubility} = (K_{sp})^{1/2} = (4.0 \times 10^{-8})^{1/2} = 2.0 \times 10^{-4}$$

What is the uncertainty in the calculated solubility of AgX in water? Substituting $y$ = solubility, $a = K_{sp}$, and $x = 1/2$ into Equation 2–11 gives

$$\frac{s_a}{a} = \frac{0.4 \times 10^{-8}}{4.0 \times 10^{-8}}$$

$$\frac{s_y}{y} = \frac{1}{2} \times \frac{0.4}{4.0} = 0.05$$

$$s_y = 2.0 \times 10^{-4} \times 0.05 = 0.1 \times 10^{-4}$$

$$\text{Solubility} = 2.0(\pm0.1) \times 10^{-4} \text{ M}$$

---

It is important to note that the error propagation in taking a number to a power is different from the error propagation in multiplication. For example, consider the uncertainty in the square of 4.0 ($\pm 0.2$). Here, the relative error in the result (16.0) is given by Equation 2–11:

$$\frac{s_y}{y} = 2 \times (0.2/4) = 0.1 \quad \text{or} \quad 10\%$$

Consider now the situation where $y$ is the product of *two independently measured* numbers that by chance happen to have values of $a_1 = 4.0(\pm 0.2)$ and $a_2 = 4.0(\pm 0.2)$. Here, the relative error of the product $a_1 a_2 = 16.0$ is given by Equation 2–10:

$$\frac{s_y}{y} = \sqrt{(0.2/4)^2 + (0.2/4)^2} = 0.07 \quad \text{or} \quad 7\%$$

The relative standard deviation of $y = a^3$ is *not* the same as the relative standard deviation of the product $y = abc$ where $a = b = c$.

The reason for this apparent anomaly is that, with measurements that are independent of one another, the sign associated with one error can be the same as or different from that of the other error. If they happen to be the same, the error is identical to that encountered in the first case, where the signs *must* be the same. On the other hand, if one sign is positive and the other negative, the relative errors tend to cancel. Thus, the probable error lies somewhere between the maximum (10%) and zero.

## 2E–4 The Standard Deviation of Logarithms and Antilogarithms

The last two entries in Table 2–5 show that for $y = \log a$,

$$s_y = 0.434 \frac{s_a}{a} \qquad (2\text{–}12)$$

and for $y = \text{antilog } a$,

$$\frac{s_y}{y} = 2.303\, s_a \qquad (2\text{–}13)$$

Thus, the *absolute* standard deviation of the logarithm of a number is determined by the *relative* standard deviation of the number; conversely, the *relative* standard deviation of the antilogarithm of a number is determined by the *absolute* standard deviation of the number.

---

Example 2–7

Calculate the absolute standard deviations of the results of the following computations. The absolute standard deviation for each quantity is given in parentheses.

(a)   $y = \log [2.00(\pm 0.02) \times 10^{-4}] = -3.6990 \pm ?$

(b)   $y = \text{antilog } [1.200(\pm 0.003)] = 15.849 \pm ?$

(c)   $y = \text{antilog } [45.4(\pm 0.3)] = 2.5119 \times 10^{45} \pm ?$

(a) Referring to Equation 2–12, we see that we must multiply the *relative* standard deviation by 0.434:

$$s_y = \pm 0.434 \times \frac{0.02 \times 10^{-4}}{2.00 \times 10^{-4}} = \pm 0.004$$

Thus,

$$\log [2.00(\pm 0.02) \times 10^{-4}] = -3.699(\pm 0.004)$$

(b) Applying Equation 2–13, we have

$$\frac{s_y}{y} = 2.303 \times (\pm 0.003) = \pm 0.0069$$

$$s_y = \pm 0.0069y = \pm 0.0069 \times 15.849 = \pm 0.11$$

Thus,

$$\text{antilog } [1.200(\pm 0.003)] = 15.8 \pm 0.1$$

(c)      $\dfrac{s_y}{y} = 2.303 \times (\pm 0.3) = \pm 0.69$

$$s_y = +0.69y = +0.69 \times 2.5119 \times 10^{45} = +1.7 \times 10^{45}$$

Thus,

$$\text{antilog } [45.4(\pm 0.3)] = 2.5(\pm 1.7) \times 10^{45}$$

---

Example 2–7c demonstrates that a large absolute error is associated with the antilogarithm of a number with few digits beyond the decimal point. This large uncertainty is due to the fact that the numbers to the left of the decimal (the characteristic) serve only to locate the decimal point. The large error in the antilogarithm results from the relatively large uncertainty in the *mantissa* of the number (that is, $0.4 \pm 0.3$).

## 2F   METHODS FOR REPORTING ANALYTICAL DATA

Because a numerical result is worthless unless something is known about its accuracy, you must always attempt to indicate the reliability of your data. An excellent approach is to determine the confidence limit at the 90 or 95% confidence level; see Section 3B–1. Another method is to report the absolute standard deviation or the coefficient of variation of the data. Here, it is a good idea to indicate the number of data that were used to obtain the standard deviation. A less satisfactory but more common indicator of the data quality is the *significant figure convention*.

## 2F–1 The Significant Figure Convention

A simple way of indicating the probable uncertainty associated with an experimental measurement is to round the result so that it contains only *significant figures*. By definition, the significant figures in a number are all of the certain digits *and the first uncertain digit*. For example, when you read a 50-mL buret that has graduations every 0.1 mL, you can easily tell that the liquid level is greater than 30.2 mL and less than 30.3 mL. You can also estimate the position of the liquid between the graduations to about ±0.02 mL (see Figure 2–6). So, according to the significant figure convention, you should report the volume delivered as, say, 30.26 mL, which is four significant figures. The first three digits here are certain, and the last digit (4) is uncertain.

A zero may or may not be significant, depending upon its location in a number. A zero that is surrounded by other digits is always significant (such as in 30.26 mL) because it is read directly and with certainty from a scale or instrument readout. On the other hand, zeros that only locate the decimal point for us are not significant. If we write 30.26 mL as 0.03026 L, the number of significant figures remains the same. The only function of the zero before the 3 is to locate the decimal point, so this zero is not significant. Terminal zeros may or may not be significant. For example, if the volume of a beaker is expressed as 2.0 L, the presence of the 0 tells us that the volume is known to a few tenths of a liter, and so, both the 2 and the 0 are significant figures. If this same volume is reported as 2000 mL, the situation becomes confused. The last two zeros are not significant if the uncertainty is still a few tenths of a liter or a few hundred milliliters. In order to follow the significant figure convention here, use scientific notation and report the volume as $2.0 \times 10^3$ mL.

An obvious limitation of the significant figure convention as an indicator of reliability of data is its ambiguity. In the absence of other information, the uncertainty in a result is normally assumed to be half the smallest division on the instrument used to make the measurement. Thus, the uncertainty in 61.5 could range from a high of 0.5 to a low of 0.05.

## 2F–2 Significant Figures in Numerical Computations

Care is required to determine the appropriate number of significant figures in the result of an arithmetic combination of two or more numbers.[7]

### Sums and Differences

For addition and subtraction, the number of significant figures can be found by visual inspection. For example, in the expression

> The number of significant figures is all of the certain digits plus the first uncertain digit.

> Express data in scientific notation to avoid confusion in determining whether terminal zeros are significant.

> Rules for significant figures:
>
> 1. Disregard all initial zeros.
> 2. Disregard all final zeros *unless they follow a decimal point.*
> 3. All remaining digits including zeros between nonzero digits are significant.

**Figure 2–6**
Buret section showing the liquid level and meniscus.

---

[7]For an extensive discussion of propagation of significant figures, see L. M. Schwartz, *J. Chem. Educ.*, **1985, 62**, 693.

$$3.4 + 0.020 + 7.31 = 10.73 = 10.7$$

the second and third decimal places in the answer cannot be significant because 3.4 is uncertain in the first decimal place. Note that the result contains three significant digits even though two of the numbers involved have only two significant figures.

## Products and Quotients

A rule of thumb sometimes suggested for multiplication and division is that the answer should be rounded so that it contains the same number of significant digits as the original number with the smallest number of significant digits. Unfortunately, this procedure can lead to incorrect rounding. For example, consider the two calculations

$$\frac{24 \times 4.52}{100.0} = 1.08 \quad \text{and} \quad \frac{24 \times 4.02}{100.0} = 0.965$$

By the rule just described, the first answer would be rounded to 1.1 and the second to 0.96. However, if the last digit of each number making up the first quotient is uncertain by 1, the relative uncertainties associated with each of these numbers are 1/24, 1/452, and 1/1000. Because the first relative uncertainty is much larger than the other two, the relative uncertainty in the result is also 1/24; the absolute uncertainty is then

$$1.08 \times 1/24 = 0.045 = 0.04$$

By the same argument, the absolute uncertainty of the second answer is given by

$$0.965 \times 1/24 = 0.040 = 0.04$$

Therefore, the first result should be rounded to three significant figures, or 1.08, but the second should be rounded to only two; that is, 0.96.

## Logarithms and Antilogarithms

Be especially careful in rounding the results of calculations involving logarithms. The following rules apply to most situations:[8]

1. In a logarithm of a number, keep as many digits to the right of the decimal point as there are significant figures in the original number.
2. In an antilogarithm of a number, keep as many digits as there are digits to the right of the decimal point in the original number.

---

[8]D. E. Jones, *J. Chem. Educ.*, **1971**, *49*, 753.

You have probably heard it said that a chain is only as strong as its weakest link. For addition and subtraction, the weak link is the number of decimal places in the number with the *smallest* number of decimal places.

Before you add or subtract numbers in scientific notation, be sure to express the numbers to the same power of ten.

The weak link for multiplication and division is the number of *significant figures* in the number with the smallest number of significant figures. *Use this rule of thumb with caution.*

The *mantissa* of a logarithm consists of the digits to the right of the decimal point. The *characteristic* of a logarithm consists of the digits to the left of the decimal point.

The number of significant figures in the *mantissa*, or the digits to the right of the decimal point of a logarithm, is the same as the number of significant figures in the original number.

$$\log 9.57 \times 10^4 = 4.981$$

Example 2–8

Round the following answers so that only significant digits are retained. (a) log $4.000 \times 10^{-5} = -4.3979400$ and (b) antilog $12.5 = 3.162277 \times 10^{12}$.

(a) Following rule 1, we retain four digits to the right of the decimal point:

$$\log \mathbf{4.000} \times 10^{-5} = -4.3979$$

(b) Following rule 2, we may retain only 1 digit

$$\text{antilog } 12 \; \mathbf{5} = 3 \times 10^{12}$$

## 2F–3 Rounding Data

In rounding a number ending in 5, always round so that the result ends with an even number.

Always round the computed results of a chemical analysis in an appropriate way. For example, consider the replicate results: 61.60, 61.46, 61.55, and 61.61. The mean of these data is 61.555, and the standard deviation is 0.069. When we round the mean, do we take 61.55 or 61.56? A good guide to follow when rounding a 5 is always to round to the nearest even number. In this way, we eliminate any tendency to round in a set direction. In other words, there is an equal likelihood that the nearest even number will be the higher or the lower in any given situation. Accordingly, we might choose to report the result as $61.56 \pm 0.07$. If we had reason to doubt the reliability of the estimated standard deviation, we might report the result as $61.6 \pm 0.1$.

## 2F–4 Rounding the Results from Chemical Computations

Do not round until calculations are complete.

Throughout this text and others, the reader is asked to perform calculations with data whose precision is indicated only by the significant figure convention. In these circumstances, common sense assumptions must be made as to the uncertainty in each number. The uncertainty of the result is then estimated using the techniques presented in Section 2E. Finally, the result is rounded so that it contains only significant digits. *It is especially important to postpone rounding until the calculation has been completed.* At least one extra digit beyond the significant digits should be carried through all of the computations in order to avoid a rounding error. This extra digit is sometimes called a "guard" digit. Modern calculators generally retain several extra digits that are not significant, and the user must be careful to round final results properly so that only significant figures are included. The following example illustrates this procedure.

Example 2–9

A 3.4842-g sample of a solid mixture containing benzoic acid, $C_6H_5COOH$ (fw = 122.1247) was dissolved and titrated with 41.36 mL of 0.2328 M

NaOH to a phenolphthalein end point. Calculate the percent benzoic acid (HBz) in the sample.

As will be shown in Section 5C–3, the computation takes the following form:

$$\%HBz = \frac{41.36\,mL \times \dfrac{0.2328\,mmol\,NaOH}{mL\,NaOH} \times \dfrac{1\,mmol\,HBz}{mmol\,NaOH} \times \dfrac{122.125\,g\,HBz}{1000\,mmol\,HBz}}{3.4842\,g\,sample}$$

$$\times\,100\%$$

$$= 33.749\%$$

Since all operations are either multiplication or division, the relative uncertainty of the answer is determined by the relative uncertainties of the experimental data. Let us estimate what these uncertainties are.

(a) The position of the liquid level in a buret can be estimated to ±0.02 mL. Initial and final readings must be made, however, so that the standard deviation of the volume will be

$$\sqrt{(0.02)^2 + (0.02)^2} = \pm 0.028\ mL \quad \text{(Equation 2–10)}$$

The relative uncertainty is then

$$\frac{\pm 0.028}{41.36} \times 1000\ ppt = \pm 0.68\ ppt$$

(b) The absolute uncertainty of a weight obtained with an analytical balance will be on the order of ±0.0001 g. Thus the relative uncertainty of the denominator is

$$\frac{0.0001}{3.4842} \times 1000\ ppt = 0.029\ ppt$$

(c) Usually we can assume that the absolute uncertainty molarity of a reagent solution is 0.0001, and so

$$\frac{0.0001}{0.2328} \times 1000\ ppt = 0.43\ ppt$$

(d) The relative uncertainty in the formula weight of HBz is several orders of magnitude smaller than that of the three experimental data and is of no consequence. Note, however, that we should retain enough digits in the calculation so that the formula weight is given to at least one more digit (the guard digit) than any of the experimental data. Thus, in the calculation, we use 122.125 for the formula weight (here we are carrying two extra digits).

(e) No uncertainty is associated with 100% and the 1000 mmol HBz, since these are exact numbers.

It is common practice to carry at least one extra digit throughout a series of calculations to prevent roundoff error. The extra digit is called a *guard digit*. With calculators and computers, we often carry many guard digits.

It is ordinarily sufficient to round the answer so that its relative uncertainty is of the *same order of magnitude* as the relative uncertainty of the number with the largest relative uncertainty. In practice, this means that the answer is rounded so that its relative uncertainty lies between 0.2 and 2 times the largest relative uncertainty of the input data.

We see that the largest relative uncertainty of the three input data is 0.68 ppt. The answer should then be rounded to the same order of magnitude as 0.68, or about 0.7 ppt. If the answer is rounded to 33.7, the suggested relative uncertainty is $(0.1/33.7) \times 1000$ ppt = 3 ppt, which is well over $2 \times 0.7$ ppt = 1.4 ppt. Rounding to 33.75 implies a relative uncertainty of $(0.01/33.75) \times 1000$ ppt = 0.3 ppt, which lies between $0.2 \times 1$ ppt and $2 \times 1$ ppt. Thus the answer is written as % HBz = 33.75.

With a little practice, you can make rounding decisions, such as that shown in Example 2–9, in your head. For example, looking again at the first equation in this example, we see that the relative uncertainty of the volume measurement is somewhat less than 4 parts in 4000 or 1 part in 1000 (actually 2.8 parts in 4136). Similarly, the uncertainty in the molarity is roughly 1 part in 2000, and the uncertainty in the denominator is somewhat less than 1 in 34,000. Therefore the uncertainty in the result is determined by the uncertainty in the volumetric measurement, which, as we have said, is roughly 1 part in 1000. The uncertainty in the computed result is then 1/1000th of 33.7%, or about 0.03%. Therefore we round to 33.75%.

It is important to remember that rounding decisions are an important part of *every calculation* and that such decisions *cannot* be based on the number of digits displayed on the readout of a calculator.

There is no relationship between the number of digits displayed on a calculator and the true number of significant figures.

## 2G   QUESTIONS AND PROBLEMS

2–1. Explain the difference between
   *(a) accuracy and precision.
   (b) indeterminate and determinate error.
   *(c) mean and median.
   (d) absolute and relative error.
   *(e) constant and proportional error.
   (f) variance and standard deviation.
2–2. Define
   *(a) range.
   (b) coefficient of variation.
   *(c) histogram.
   (d) Gaussian distribution.
   *(e) personal bias.
   (f) significant figures.
2–3. Name three types of determinate errors.
2–4. How are determinate method errors detected?
2–5. What kind of determinate errors are detected by varying the sample size?

2–6. Suggest some sources of indeterminate error in measuring the width of a 3-m table with a 1-m metal rule.
2–7. Consider the following sets of replicate measurements:

| *A | B | *C | D | *E | F |
|---|---|---|---|---|---|
| 16.40 | 0.888 | 35.64 | 0.1006 | 41.37 | 0.0765 |
| 16.36 | 0.889 | 35.80 | 0.0991 | 41.33 | 0.0761 |
| 16.26 | 0.904 | 35.59 | 0.1010 | 41.37 | 0.0747 |
| 16.34 | | 35.57 | 0.0997 | 41.20 | 0.0754 |
| | | | 0.0999 | 41.60 | |

For each set, calculate the (a) mean, (b) median, (c) spread, or range, (d) standard deviation, and (e) coefficient of variation.
2–8. The accepted values for the sets of data in Problem 2–7 are: *A, 16.28; B, 0.0891; *C, 35.70; D, 0.1009; *E, 41.40; F, 0.0766. For each set, calculate (a) the

*Answers to problems or parts of problems with an asterisk are found at the end of the book.

absolute error and **(b)** the relative error in parts per thousand.

**2–9.** The color change of a chemical indicator requires an overtitration of 0.03 mL. Calculate the percent relative error if the total volume of titrant is
*(a) 5.00 mL.     *(c) 25.0 mL.
(b) 10.0 mL.     (d) 40.0 mL.

**2–10.** A loss of 0.4 mg of Zn occurs in the course of an analysis for that element. Calculate the percent relative error due to this loss if the weight of Zn in the sample is
*(a) 40 mg.  (b) 175 mg.  *(c) 400 mg.  (d) 600 mg.

**2–11.** The method in Problem 2–10 is to be used for the analysis of ore samples that assay about 20% Zn. What minimum sample weight should be taken if the relative error from this source is to be less than
*(a) −1.0%.  (b) −0.6%.  *(c) −0.2%.  (d) −0.1%.

**2–12.** Estimate the absolute deviation and the coefficient of variation for the results of the following calculations. Round each result so that it contains only significant digits. The numbers in parentheses are absolute standard deviations.

**(a)** $y = 6.75(\pm0.03) + 0.843(\pm0.001) - 7.021(\pm0.001) = 0.572$

**(b)** $y = 19.97(\pm0.04) + 0.0030(\pm0.0001) + 1.29(\pm0.08) = 21.263$

**(c)** $y = 67.1(\pm0.3) \times 1.03(\pm0.02) \times 10^{-17} = 6.9113 \times 10^{-16}$

**(d)** $y = 243(\pm1) \times \dfrac{760(\pm2)}{1.006(\pm0.006)} = 183578.5$

**(e)** $y = \dfrac{143(\pm6) - 64(\pm3)}{1249(\pm1) + 77(\pm8)} = 5.9578 \times 10^{-2}$

**(f)** $y = \dfrac{1.97(\pm0.01)}{243(\pm3)} = 8.106996 \times 10^{-3}$

**2–13.** Estimate the absolute standard deviation and the coefficient of variation for the results of the following calculations. Round each result to include only significant figures. The numbers in parentheses are absolute standard deviations.

**(a)** $y = -1.02(\pm0.02) \times 10^{-7} - 3.54(\pm0.2) \times 10^{-8} = -1.374 \times 10^{-7}$

**(b)** $y = 100.20(\pm0.08) - 99.62(\pm0.06) + 0.200(\pm0.004) = 0.780$

**(c)** $y = 0.0010(\pm0.0005) \times 18.10(\pm0.02) \times 200(\pm1) = 3.62$

**(d)** $y = \dfrac{1.73(\pm0.03) \times 10^{-14}}{1.63(\pm0.04) \times 10^{-16}} = 106.1349693$

**(e)** $y = \dfrac{100(\pm1)}{2(\pm1)} = 50$

**(f)** $y = \dfrac{1.43(\pm0.02) \times 10^{-2} - 4.76(\pm0.06) \times 10^{-3}}{24.3(\pm0.7) + 8.06(\pm0.08)} = 2.948 \times 10^{-4}$

**2–14.** Round each of the following results to include only significant figures.
*(a) $y = \log 1.73 = 0.238046$
(b) $y = \log 0.0432 = -1.364516$
*(c) $y = \log 6.022 \times 10^{23} = 23.77960$
(d) $y = \log 4.213 \times 10^{-21} = -20.375409$
*(e) $y = \text{antilog}(-3.47) = 3.38844 \times 10^{-4}$
(f) $y = \text{antilog}\, 5.7 = 5.01187 \times 10^{5}$
*(g) $y = \text{antilog}\, 0.99 = 9.77237$
(h) $y = \text{antilog}(-27.2424) = 5.722687 \times 10^{28}$

## CHAPTER 3

# STATISTICAL EVALUATION OF DATA

We have shown that the magnitude of the indeterminate error associated with an individual measurement is determined by a chance combination of tiny individual errors, each of which may be positive or negative. Because chance is involved in this type of error, we can use the laws of statistics to extract information from experimental data.[1] In this chapter we describe several important statistical procedures and show how they are used to estimate the magnitude of the indeterminate error in an analysis.

## 3A THE STATISTICAL TREATMENT OF INDETERMINATE ERRORS

Statistics allows us to look at our data in different ways and make objective and intelligent decisions regarding their quality and use.

Statistics is the mathematical science that deals with chance variations. We must emphasize at the outset that statistics only reveals information that is already present in a data set. That is, *no new information is created* by statistics. Statistical treatment of a data set does, however, allow us to make objective judgments concerning the validity of results that are otherwise difficult to make.

### 3A–1 The Population and the Sample

In order to use statistics to treat our data, we must assume that the few replicate experimental results gathered in the laboratory are a tiny but representative fraction of an infinite number of results that could be collected if we had infinite time. Statisticians call this small set of data a *sample* and view it as a subset of a *population,* or a *universe,* of data that

---

[1]References on statistical applications include R. L. Anderson, *Practical Statistics for Analytical Chemists.* New York: Van Nostrand–Reinhold, 1987; R. Calcutt and R. Boddy, *Statistics for Analytical Chemists.* New York: Chapman and Hall, 1983; J. Mandel, in *Treatise on Analytical Chemistry,* 2nd ed., I. M. Kolthoff and P. J. Elving, Eds., Part I, Vol. 1, Chapter 5. New York: Wiley, 1978.

in principle exists. For example, the data in Table 2–3 make up a statistical sample of an infinite population of pipet-calibration measurements that can be imagined (but not performed).

The laws of statistics apply strictly to a population of data only. To use these laws, we must assume that the handful of data that make up the typical sample truly represents the infinite population of results. Unfortunately, there is no guarantee that this assumption is valid. As a result, statistical estimates about the magnitude of indeterminate errors are themselves subject to uncertainty and therefore can only be made in terms of probabilities.

### The Population Mean ($\mu$) and the Sample Mean ($\bar{x}$)

We will find it useful to differentiate between the *sample mean* and the *population mean*. The sample mean is the mean of a limited sample drawn from a population of data. It is defined by Equation 2–1, when $N$ is a small number. The population mean, in contrast, is the true mean for the population. It is also defined by Equation 2–1 when $N$ approaches infinity. If the data are free of determinate error, the population mean is also the true value. To emphasize the difference between the two means, the sample mean is symbolized by $\bar{x}$ and the population mean by $\mu$. More often than not, particularly when $N$ is small, $\bar{x}$ differs from $\mu$ because a small sample of data does not exactly represent its population.

### The Sample Standard Deviation ($s$) and the Population Standard Deviation ($\sigma$)

We must also differentiate between the *sample standard deviation* and the *population standard deviation*. The sample standard deviation $s$ was defined in Equation 2–2; that is,

$$s = \sqrt{\frac{\sum_{i=1}^{N}(x_i - \bar{x})^2}{N - 1}} \qquad (3\text{–}1)$$

In contrast, the population standard deviation $\sigma$, which is the true standard deviation, is given by

$$\sigma = \sqrt{\frac{\sum_{i=1}^{N}(x_i - \mu)^2}{N}} \qquad (3\text{–}2)$$

Note that this equation differs from Equation 3–1 in two ways. First, the population mean $\mu$ appears in the numerator of Equation 3–2 in place of the sample mean, $\bar{x}$. Second, $N$ replaces the number of degrees of freedom ($N - 1$) that appears in Equation 3–1.

The reason the number of degrees of freedom must be used when $N$ is small is as follows. When $\sigma$ is unknown, two quantities must be extracted from a set of data: $\bar{x}$ and $s$. One degree of freedom is used to establish $\bar{x}$

Do not confuse the *statistical sample* with the *analytical sample*. Four analytical samples analyzed in the laboratory represent a single statistical sample. This is an unfortunate duplication of the term "sample" that should cause no trouble once you are aware of its two meanings.

Sample mean = $\bar{x}$, where

$$\bar{x} = \frac{\sum_{i=1}^{N} x_i}{N} \qquad \text{when } N \text{ is small.}$$

Population mean = $\mu$, where

$$\mu = \frac{\sum_{i=1}^{N} x_i}{N} \qquad \text{when } N \to \infty.$$

In the absence of determinate error, the population mean $\mu$ is the true value of a measured quantity.

When $N \to \infty$, $\bar{x} \to \mu$ and $s \to \sigma$.

because, with their signs retained, the sum of the individual deviations must add up to zero. Thus, when $N - 1$ deviations have been computed, the final one is known. Consequently, only $N - 1$ deviations provide an *independent* measure of the precision of the set.

### 3A–2 Properties of the Normal Error Curve

Figure 3–1a shows two Gaussian curves in which the relative frequency of occurrence of various deviations from the mean is plotted as a function of deviation from the mean $(x - \mu)$. The two curves are for two populations of data that differ only in standard deviation. The standard deviation for the population yielding the broader but lower curve $(B)$ is twice that for the population yielding curve $A$.

Figure 3–1b shows another type of normal error curve in which the abscissa is now a new variable, $z$, which is defined as

$$z = \frac{(x - \mu)}{\sigma} \qquad (3-3)$$

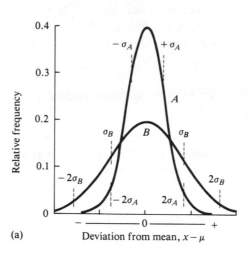

(a)

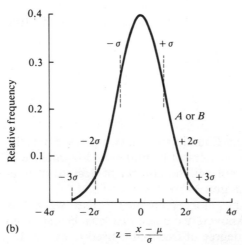

(b)

Figure 3–1

Normal error curves. The standard deviation for curve $B$ is twice that for curve $A$, that is, $\sigma_B = 2\sigma_A$. (a) The abscissa is the deviation from the mean in the units of measurement. (b) The abscissa is the deviation from the mean in units of $\sigma$. Thus, the two curves $A$ and $B$ are identical here.

**Feature 3–1**
WHY USE THE NUMBER OF DEGREES OF FREEDOM INSTEAD OF *N*?

The effect of using the number of degrees of freedom for calculating the standard deviation can be demonstrated by dividing the data in Table 2–3 into 10 samples of 5 data each, 5 samples of 10 data each, and 2 samples of 25 data each. When $\sigma$ and $s$ are calculated for each sample using Equations 3–1 and 3–2, the results are

| Number of Samples and Size | Mean $\sigma$ of Samples (Equation 3–2) | Mean $s$ of Samples (Equation 3–1) |
|---|---|---|
| 10 samples of 5 | 0.0053 | 0.0059 |
| 5 samples of 10 | 0.0058 | 0.0061 |
| 2 samples of 25 | 0.0059 | 0.0061 |
| 1 sample of 50 | 0.0060 | 0.0060 |

A negative bias accompanies application of Equation 3–2 to small sets of data; this bias is reflected in the data of column 2. Note (in column 3) that the bias disappears when $s$ is calculated using Equation 3–1.

Note that $z$ is the standard deviation from the mean of the data expressed *in units of standard deviation*. That is, when $x - \mu = \sigma$, $z$ is equal to one standard deviation; when $x - \mu = 2\sigma$, $z$ is equal to two standard deviations; and so forth. Since $z$ is the deviation of the mean in standard deviation units, a plot of relative frequency versus this parameter yields a single Gaussian curve that describes all populations of data regardless of standard deviation. Thus, Figure 3–1b is the normal error curve for both sets of data used to plot curves $A$ and $B$ in Figure 3–1a.

The normal error curve has several general properties. (1) The mean occurs at the central point of maximum frequency. (2) There is a symmetrical distribution of positive and negative deviations about the maximum. (3) There is an exponential decrease in frequency as the magnitude of the deviations increases. Thus, small indeterminate uncertainties are observed much more often than very large ones.

### Areas Under a Normal Error Curve

It can be shown that 68.3% of the area beneath any normal error curve lies within one standard deviation ($\pm 1\sigma$) of the mean $\mu$. Thus, 68.3% of the data making up the population lie within these bounds. Furthermore, approximately 95.5% of all data are within $\pm 2\sigma$ of the mean and 99.7% within $\pm 3\sigma$. The vertical dashed lines show the areas bounded by $\pm 1\sigma$, $\pm 2\sigma$, and $\pm 3\sigma$ in Figure 3–1b.

Because of area relationships such as these, the standard deviation of a population of data is a useful predictive tool. For example, we can say that the chances are 68.3 in 100 that the indeterminate uncertainty of any single measurement in a normal distribution is no more than $\pm 1\sigma$. Similarly, the chances are 95.5 in 100 that the error is less than $\pm 2\sigma$, and so forth.

### Standard Error of a Mean

The figures on percentage distribution just quoted refer to the probable error for a *single* measurement. If a series of samples, each containing $N$ data, are taken randomly from a population of data, the mean of each set will show less and less scatter as $N$ increases. The standard deviation of each mean is known as the *standard error* of the mean and is given the symbol $\sigma_m$. It can be shown that the standard error is inversely proportional to the square root of the number of data $N$ used to calculate the mean:

$$\sigma_m = \sigma/\sqrt{N} \qquad (3\text{--}4)$$

where sigma is defined by Equation 3–2. An analogous equation can be written for a sample standard deviation:

$$s_m = s/\sqrt{N} \qquad (3\text{--}5)$$

### 3A–3  Properties of the Standard Deviation

#### Effect of N on the Reliability of s

When $N > 20$, $s \cong \sigma$.

Uncertainty in the calculated value of $s$ decreases as $N$ in Equation 3–1 increases. Figure 3–2 shows the error in $\sigma$ as a function of $N$. When $N$ is greater than about 20, $s$ and $\sigma$ can be assumed to be identical for all practical purposes. For example, if the 50 measurements in Table 2–3 are divided into ten subgroups of 5 measurements each, the value of $s$ varies widely from one subgroup to another (0.0023 to 0.0079 mL) even though the average of the computed values of $s$ is that of the entire set (0.0056

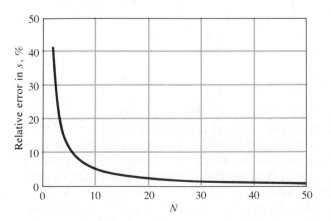

Figure 3–2

Relative error in $\sigma$ as a function of $N$.

mL). In contrast, the computed values of $s$ for two subsets of 25 measurements each are nearly identical (0.0054 and 0.0058 mL).

The rapid improvement in the reliability of $s$ as $N$ increases makes it feasible to obtain a good approximation of $\sigma$ when the method of measurement is not excessively time-consuming and when an adequate supply of sample is available. For example, if the pH of numerous solutions is to be measured in the course of an investigation, it is useful to evaluate $s$ in a series of preliminary experiments. This measurement is simple, requiring only that a pair of rinsed and dried electrodes be immersed in the test solution. The voltage between the electrodes is proportional to pH. To determine $s$, 20 to 30 portions of a buffer solution of fixed pH can be measured with all steps of the procedure being followed exactly. Normally, it is safe to assume that the indeterminate error in this test is the same as that in subsequent measurements. The value of $s$ calculated from Equation 3–1 is thus a valid and accurate measure of the theoretical $\sigma$.

## Pooling Data to Improve the Reliability of $s$

The foregoing procedure is not always practical for analyses that are time-consuming. In this situation, data from a series of samples accumulated over time can be pooled to provide an estimate of $s$ that is superior to the value for any individual subset. Again, we must assume the same sources of indeterminate error in all the samples. This assumption is usually valid if the samples have similar compositions and have been analyzed in exactly the same way.

To obtain a pooled estimate of the standard deviation, $s_{pooled}$, deviations from the mean for each subset are squared; the squares of all subsets are then summed and divided by an appropriate number of degrees of freedom, as shown in Equation 3–6. The pooled $s$ is obtained by extracting the square root of the quotient. One degree of freedom is lost for each subset. Thus, the number of degrees of freedom for the pooled $s$ is equal to the total number of measurements minus the number of subsets.

$$s_{pooled} = \sqrt{\frac{\sum_{i=1}^{N_1}(x_i - \bar{x}_1)^2 + \sum_{j=1}^{N_2}(x_j - \bar{x}_2)^2 + \sum_{k=1}^{N_3}(x_k - \bar{x}_3)^2 + \cdots}{N_1 + N_2 + N_3 + \cdots - N_s}} \qquad (3\text{–}6)$$

where $N_1$ is the number of data in set 1, $N_2$ is the number in set 2, and so forth. The term $N_s$ is the number of data sets that are being pooled.

---

Example 3–1

The mercury content in samples of seven fish taken from the Sacramento River was determined by a method based upon the absorption of radiation by gaseous elemental mercury. Calculate a pooled estimate of the standard deviation for the method, based upon the first three columns of data:

| Specimen | Number of Samples Measured | Hg Content, ppm | Mean, ppm Hg | Sum of Square of Deviations from Mean |
|----------|----------|-----------------|----------|----------|
| 1 | 3 | 1.80, 1.58, 1.64 | 1.673 | 0.0258 |
| 2 | 4 | 0.96, 0.98, 1.02, 1.10 | 1.015 | 0.0115 |
| 3 | 2 | 3.13, 3.35 | 3.240 | 0.0242 |
| 4 | 6 | 2.06, 1.93, 2.12, 2.16, 1.89, 1.95 | 2.018 | 0.0611 |
| 5 | 4 | 0.57, 0.58, 0.64, 0.49 | 0.570 | 0.0114 |
| 6 | 5 | 2.35, 2.44, 2.70, 2.48, 2.44 | 2.482 | 0.0685 |
| 7 | 4 | 1.11, 1.15, 1.22, 1.04 | 1.130 | 0.0170 |
| | $N = 28$ | | Sum of squares = | 0.2196 |

The values in the last two columns for sample 1 were computed as follows:

| $x_i$ | $|(x_i - \bar{x})|$ | $(x_i - \bar{x})^2$ |
|-------|---------------------|---------------------|
| 1.80 | 0.127 | 0.0161 |
| 1.58 | 0.093 | 0.0086 |
| 1.64 | 0.033 | 0.0011 |
| 5.02 | Sum of squares = | 0.0258 |

$$\bar{x} = \frac{5.02}{3} = 1.673$$

The other data in columns 4 and 5 were obtained similarly. Then

$$s_{pooled} = \sqrt{\frac{0.0258 + 0.0115 + 0.0242 + 0.0611 + 0.0114 + 0.0685 + 0.0170}{28 - 7}}$$

$$= 0.10 \text{ ppm Hg}$$

Note that one degree of freedom is lost for each of the seven samples. Because more than 20 degrees of freedom remain, however, the computed value of $s$ can be considered a good approximation of $\sigma$; that is, $s \rightarrow \sigma = 0.10$ ppm Hg.

## 3B  THE USES OF STATISTICS

Experimentalists use statistical calculations to sharpen their judgment concerning the effects of indeterminate errors. The most common applications of statistics to analytical chemistry include:

1. Defining the interval around the mean of a set within which the population mean can be expected to be found with a given probability.
2. Determining the number of replicate measurements required to assure

(at a given probability) that an experimental mean falls within a predetermined interval around the population mean.

3. Deciding whether an outlying value in a set of replicate results should be retained or rejected in calculating the mean for the set.
4. Estimating the probability that two samples analyzed by the same method are significantly different in composition, that is, whether a difference in experimental results is likely to be a consequence of indeterminate error or a real composition difference.
5. Estimating the probability that there is a difference in precision between two sets of data obtained by different workers or by different methods.
6. Defining and estimating detection limits.
7. Treating calibration data.

We will examine each of these applications in the sections that follow.

### 3B–1 Confidence Limits

The exact value of the mean $\mu$ for a population of data can never be determined exactly because such a determination requires an infinite number of measurements. Statistical theory allows us to set limits around an experimentally determined mean $\bar{x}$, however, and the true mean $\mu$ lies within these limits with a given degree of probability. These limits are called *confidence limits,* and the interval they define is known as the *confidence interval.*

> Confidence limits define an interval around $\bar{x}$ that probably contains $\mu$.

The size of the confidence interval, which is derived from the sample standard deviation, depends on the certainty with which $s$ is known. If there is reason to believe that $s$ is a good approximation of $\sigma$, then the confidence interval can be significantly narrower than if the estimate of $s$ is based upon only two or three measurements.

### The Confidence Interval When s Is a Good Approximation of $\sigma$

Figure 3–3 shows a series of five normal error curves. In each, the relative frequency is plotted as a function of the quantity $z$ (Equation 3–3), which is the deviation from the mean *in units of the population standard deviation.* The shaded area in each plot lies between the values of $-z$ and $+z$ that are indicated to the left and right of the curves, respectively. The number within the shaded area is the percentage of the total area under the curve that is included within the $z$ values. For example, as shown in the top curve, 50% of the area under any Gaussian curve is located between $-0.67\sigma$ and $+0.67\sigma$. Proceeding downward, we see that 80% of the total area lies between $-1.29\sigma$ and $+1.29\sigma$, and 90% lies between $-1.64\sigma$ and $+1.64\sigma$. Relationships such as these allow us to define a range of values around a measurement within which the true mean is likely to lie with a certain probability. For example, we may assume that 90 times out of 100, the true mean, $\mu$, will be within $\pm1.64\sigma$ of any measurement that we make. Here, the *confidence level* is 90% and the *confidence interval* is $\pm z\sigma = \pm1.64\sigma$.

> The confidence level is the probability expressed as a percent.

The confidence limits are the values above and below a measurement that bound its confidence interval.

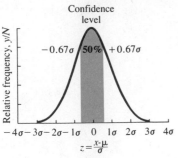

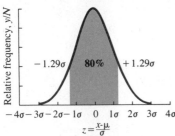

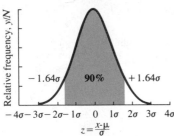

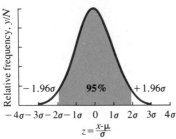

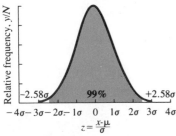

**Figure 3–3**
Areas under a Gaussian curve for various values of $\pm z$.

We find a general expression for the confidence limits (CL) of a single measurement by rearranging Equation 3–3. Remember that $z$ can take positive or negative values. Thus,

$$\text{CL for } \mu = x \pm z\sigma \qquad (3\text{–}7)$$

Values for $z$ at various confidence levels are found in Table 3–1.

---

**Example 3–2**

Calculate the 50% and 95% confidence limits for the first entry (1.80 ppm Hg) in Example 3–1.

In that example, we calculated $s$ to be 0.10 ppm Hg and had sufficient data to assume $s \rightarrow \sigma$. From Table 3–1, we see that $z = 0.67$ and 1.96 for the two confidence levels. Thus, from Equation 3–7

$$50\% \text{ CL for } \mu = 1.80 \pm 0.67 \times 0.10 = 1.80 \pm 0.07$$
$$95\% \text{ CL for } \mu = 1.80 \pm 1.96 \times 0.10 = 1.80 \pm 0.20$$

From these calculations, we conclude that the chances are 50 in 100 that $\mu$, the population mean (and, *in the absence of determinate error*, the true value), lies in the interval between 1.73 and 1.87 ppm Hg. Furthermore, there is a 95% chance that $\mu$ lies in the interval between 1.60 and 2.00 ppm Hg.

---

Equation 3–7 applies to the result of a *single measurement*. Application of Equation 3–4 shows that the confidence interval is decreased by $\sqrt{N}$ for the average of $N$ replicate measurements. Thus, a more general form of Equation 3–7 is

$$\text{CL for } \mu = \bar{x} \pm \frac{z\sigma}{\sqrt{N}} \qquad (3\text{–}8)$$

---

**Example 3–3**

Calculate the 50% and 95% confidence limits for the mean value (1.67 ppm Hg) for specimen 1 in Example 3–1. Again, $s \rightarrow \sigma = 0.10$.

For the three measurements,

$$50\% \text{ CL} = 1.67 \pm \frac{0.67 \times 0.10}{\sqrt{3}} = 1.67 \pm 0.04$$

$$95\% \text{ CL} = 1.67 \pm \frac{1.96 \times 0.10}{\sqrt{3}} = 1.67 \pm 0.11$$

Thus, the chances are 50 in 100 that the population mean is located in the interval between 1.63 and 1.71 ppm Hg and 95 in 100 that it lies between 1.56 and 1.78 ppm.

---

Example 3–4

How many replicate measurements of specimen 1 in Example 3–1 are needed to decrease the 95% confidence interval to ±0.07 ppm Hg?

The pooled value is a good estimate of $\sigma$. For a confidence interval of ±0.07 ppm Hg, then, substitution into Equation 3–8 leads to

$$0.07 = \pm \frac{zs}{\sqrt{N}} = \pm \frac{1.96 \times 0.10}{\sqrt{N}}$$

$$\sqrt{N} = \pm \frac{1.96 \times 0.10}{0.07} = \pm 2.80$$

$$N = (\pm 2.8)^2 = 7.8$$

We conclude that eight measurements would provide a slightly better than 95% chance of the population mean lying within ±0.07 ppm of the experimental mean.

---

Equation 3–8 tells us that the confidence interval for an analysis can be halved by carrying out four measurements. Sixteen measurements will narrow the interval by a factor of 4, and so on. We rapidly reach a point of diminishing returns in acquiring additional data. Ordinarily we take advantage of the relatively large gain attained by averaging two to four measurements but can seldom afford the time required for additional increases in confidence.

It is essential to keep in mind at all times that confidence intervals based on Equation 3–8 apply only *in the absence of determinate errors and only if we can assume that s ≈ σ.*

## The Confidence Limits When $\sigma$ Is Unknown

Often we are faced with limitations in time or amount of available sample that prevent us from accurately estimating $\sigma$. Here, a single set of replicate measurements must provide not only a mean but also an estimate of precision. As indicated earlier, s calculated from a small set of data may be quite uncertain. Thus, confidence limits are necessarily broader when a good estimate of $\sigma$ is not available.

To account for the variability of s, we use the important statistical parameter t, which is defined in exactly the same way as z (Equation 3–3) except that here s is substituted for $\sigma$:

$$t = \frac{x - \mu}{s} \qquad (3\text{–}9)$$

Like z in Equation 3–3, t depends on the desired confidence level. It also depends on the number of degrees of freedom in the calculation of s.

Table 3–1

CONFIDENCE LEVELS FOR VARIOUS VALUES OF z

| Confidence Levels, % | z |
|---|---|
| 50 | 0.67 |
| 68 | 1.00 |
| 80 | 1.29 |
| 90 | 1.64 |
| 95 | 1.96 |
| 96 | 2.00 |
| 99 | 2.58 |
| 99.7 | 3.00 |
| 99.9 | 3.29 |

| Number of Measurements Averaged, N | Relative Size of Confidence Interval |
|---|---|
| 1 | 1.00 |
| 2 | 0.71 |
| 3 | 0.58 |
| 4 | 0.50 |
| 5 | 0.45 |
| 6 | 0.41 |
| 10 | 0.32 |

The t statistic is often called *Student's t.* Student was the name used by W. S. Gossett when he wrote the classic paper on t that appeared in *Biometrika,* **1908,** 6, 1. Gossett was employed by the Guinness Brewery to analyze statistically the results of determinations of the alcohol content of their products. As a result of this work, he discovered the now-famous statistical treatment of small sets of data. To avoid the disclosure of any trade secrets of his employer, Gossett published the paper under the name Student.

As $N - 1 \to \infty$, $t \to z$.

### Table 3–2
### VALUES OF $t$ FOR VARIOUS LEVELS OF PROBABILITY

| Degrees of Freedom | Factor for Confidence Interval | | | | |
|---|---|---|---|---|---|
| | 80% | 90% | 95% | 99% | 99.9% |
| 1 | 3.08 | 6.31 | 12.7 | 63.7 | 637 |
| 2 | 1.89 | 2.92 | 4.30 | 9.92 | 31.6 |
| 3 | 1.64 | 2.35 | 3.18 | 5.84 | 12.9 |
| 4 | 1.53 | 2.13 | 2.78 | 4.60 | 8.60 |
| 5 | 1.48 | 2.02 | 2.57 | 4.03 | 6.86 |
| 6 | 1.44 | 1.94 | 2.45 | 3.71 | 5.96 |
| 7 | 1.42 | 1.90 | 2.36 | 3.50 | 5.40 |
| 8 | 1.40 | 1.86 | 2.31 | 3.36 | 5.04 |
| 9 | 1.38 | 1.83 | 2.26 | 3.25 | 4.78 |
| 10 | 1.37 | 1.81 | 2.23 | 3.17 | 4.59 |
| 11 | 1.36 | 1.80 | 2.20 | 3.11 | 4.44 |
| 12 | 1.36 | 1.78 | 2.18 | 3.06 | 4.32 |
| 13 | 1.35 | 1.77 | 2.16 | 3.01 | 4.22 |
| 14 | 1.34 | 1.76 | 2.14 | 2.98 | 4.14 |
| $\infty$ | 1.29 | 1.64 | 1.96 | 2.58 | 3.29 |

Table 3–2 provides values for $t$ for a few degrees of freedom. More extensive tables are found in various mathematical and statistical handbooks. Note that $t \to z$ (Table 3–1) as the number of degrees of freedom becomes infinite.

The confidence limits for the mean $\bar{x}$ of $N$ replicate measurements can be derived from $t$ by an equation similar to Equation 3–8:

$$\text{CL for } \mu = \bar{x} \pm \frac{ts}{\sqrt{N}} \tag{3–10}$$

---

### Example 3–5

A chemist obtained the following data for the alcohol content of a sample of blood: % $C_2H_5OH$: 0.084, 0.089, and 0.079. Calculate the 95% confidence limits for the mean assuming (a) no additional knowledge about the precision of the method and (b) it is known that $s \to \sigma = 0.005\%$ $C_2H_5OH$ on the basis of previous experience.

(a) $\Sigma x_i = 0.084 + 0.089 + 0.079 = 0.252$

$\Sigma x_i^2 = 0.007056 + 0.007921 + 0.006241 = 0.021218$

$$s = \sqrt{\frac{0.021218 - (0.252)^2/3}{3 - 1}} = 0.0050\% \ C_2H_5OH$$

Here, $\bar{x} = 0.252/3 = 0.084$. Table 3–2 indicates that $t = 4.30$ for two degrees of freedom and 95% confidence. Thus,

$$95\% \ \text{CL} = \bar{x} \pm \frac{ts}{\sqrt{N}} = 0.084 \pm \frac{4.30 \times 0.0050}{\sqrt{3}} = 0.084 \pm 0.012\% \ C_2H_5OH$$

(b) Because a good value of $\sigma$ is available,

$$95\% \text{ CL} = \bar{x} \pm \frac{z\sigma}{\sqrt{N}} = 0.084 \pm \frac{1.96 \times 0.0050}{\sqrt{3}} = 0.084 \pm 0.006\% \text{ C}_2\text{H}_5\text{OH}$$

Note that a sure knowledge of $\sigma$ significantly decreases the confidence interval.

---

## 3B–2 Rejection of Outliers

When a set of data contains an outlying result that appears to differ excessively from the average, a decision must be made whether to retain or reject the result.[2] The choice of criterion for the rejection of a suspected result has its perils. If we set a stringent standard that makes the rejection of a questionable measurement difficult, we run the risk of retaining results that are spurious and have an inordinate effect on the average of the data. If we set lenient limits on precision and thereby make the rejection of a result easy, we are likely to discard measurements that rightfully belong in the set and thus introduce a bias to the data. It is an unfortunate fact that no universal rule can be invoked to settle the question of retention or rejection.

> Outliers are the result of gross errors (Section 2C).

### Statistical Tests

Several statistical procedures such as the $Q$ test and the $T_n$ test have been developed to provide criteria for rejection or retention of outliers. Such tests assume that the distribution of the population data is normal, or Gaussian. Unfortunately, this condition cannot be proved or disproved for samples consisting of many fewer than 50 results. Consequently, statistical rules, which are perfectly reliable for normal distributions of data, should be *used with extreme caution* when applied to samples containing only a few data. J. Mandel, in discussing the treatment of small sets of data, writes, "Those who believe that they can discard observations with statistical sanction by using statistical rules for the rejection of outliers are simply deluding themselves."[3] Thus, statistical tests for rejection should be used only as aids to common sense when small samples are involved.

### The Q Test

The $Q$ test is a simple, widely used statistical test.[4] In this test the absolute value of the difference between the questionable result $x_q$ and its

---

[2]J. Mandel, in *Treatise on Analytical Chemistry,* 2nd ed., I. M. Kolthoff and P. J. Elving, Eds., Part I, Vol. 1, pp. 282–289. New York: Wiley, 1978.

[3]J. Mandel, in *Treatise on Analytical Chemistry,* 2nd ed., I. M. Kolthoff and P. J. Elving, Eds., Part I, Vol. 1, p. 282. New York: Wiley, 1978.

[4]R. B. Dean and W. J. Dixon, *Anal. Chem.,* **1951,** *23,* 636.

**Table 3–3**

**CRITICAL VALUES FOR REJECTION QUOTIENT Q***

| Number of Observations | $Q_{crit}$ (Reject if $Q_{exp} > Q_{crit}$) | | |
|---|---|---|---|
| | 90% Confidence | 95% Confidence | 99% Confidence |
| 3 | 0.941 | 0.970 | 0.994 |
| 4 | 0.765 | 0.829 | 0.926 |
| 5 | 0.642 | 0.710 | 0.821 |
| 6 | 0.560 | 0.625 | 0.740 |
| 7 | 0.507 | 0.568 | 0.680 |
| 8 | 0.468 | 0.526 | 0.634 |
| 9 | 0.437 | 0.493 | 0.598 |
| 10 | 0.412 | 0.466 | 0.568 |

*Reproduced from D. B. Rorabacher, *Anal. Chem.,* **1991,** *63,* 139. By courtesy of the American Chemical Society.

nearest neighbor $x_n$ is divided by the spread $w$ of the entire set to give the quantity $Q_{exp}$:

$$Q_{exp} = |x_q - x_n|/w \qquad (3\text{–}11)$$

This ratio is then compared with rejection values $Q_{crit}$ found in Table 3–3. If $Q_{exp}$ is greater than $Q_{crit}$, the questionable result can be rejected with the indicated degree of confidence.

---

Example 3–6

The analysis of a calcite sample yielded CaO percentages of 55.95, 56.00, 56.04, 56.08, and 56.23. The last value appears anomalous; should it be retained or rejected?

The difference between 56.23 and 56.08 is 0.15%. The spread (56.23 − 55.95) is 0.28%. Thus,

$$Q_{exp} = \frac{0.15}{0.28} = 0.54$$

For five measurements, $Q_{crit}$ at the 90% confidence level is 0.642. Because $0.536 < 0.642$, we must retain the value 56.23.

---

The $T_n$ Test

In the American Society for Testing Materials (ASTM) $T_n$ *test,* the quantity $T_n$ serves as the rejection criterion,[5] where

$$T_n = \frac{|x_q - \bar{x}|}{s}$$

---

[5]For further discussion of this test, see J. Mandel, in *Treatise on Analytical Chemistry,* 2nd ed., I. M. Kolthoff and P. J. Elving, Eds., Part I, Vol. 1, pp. 283–285. New York: Wiley, 1978.

Here, $x_q$ is the questionable result, and $\bar{x}$ and $s$ are the mean and standard deviations of the entire set *including the questionable result*. Rejection is indicated if the calculated $T_n$ is greater than the critical values found in Table 3–4.

---

Example 3–7

Apply the $T_n$ test to the data in Example 3–6.

$\Sigma x_i = 55.95 + 56.00 + 56.04 + 56.08 + 56.23 = 280.3$

$\Sigma x_i^2 = (55.95)^2 + (56.00)^2 + (56.04)^2 + (56.08)^2 + (56.23)^2 = 15713.6634$

$\bar{x} = 280.3/5 = 56.06$

$s = \sqrt{\dfrac{15713.6634 - (280.3)^2/5}{5 - 1}} = 0.107$

$T_n = |56.23 - 56.06|/0.107 = 1.59$

Table 3–4 indicates that the critical value of $T_n$ for five measurements is greater than the experimental value at all confidence levels. Therefore, retention is also indicated by this test.

---

Recommendations for the Treatment of Outliers

Recommendations for the treatment of a small set of results that contains a suspect value follow:

1. Reexamine carefully all data relating to the outlying result to see if a gross error could have affected its value. This recommendation demands *a properly kept laboratory notebook containing careful notations of all observations*.
2. If possible, estimate the precision that can be reasonably expected from the procedure to be sure that the outlying result actually is questionable.

Table 3–4
CRITICAL VALUES FOR REJECTION QUOTIENT $T_n$*

Use caution when rejecting data for any reason.

| Number of Observations | $T_n$ | | |
|---|---|---|---|
| | 95% Confidence | 97.5% Confidence | 99% Confidence |
| 3 | 1.15 | 1.15 | 1.15 |
| 4 | 1.46 | 1.48 | 1.49 |
| 5 | 1.67 | 1.71 | 1.75 |
| 6 | 1.82 | 1.89 | 1.94 |
| 7 | 1.94 | 2.02 | 2.10 |
| 8 | 2.03 | 2.13 | 2.22 |
| 9 | 2.11 | 2.21 | 2.52 |
| 10 | 2.18 | 2.29 | 2.41 |

*Adapted from J. Mandel, in *Treatise on Analytical Chemistry*, 2nd ed., I. M. Kolthoff and P. J. Elving, Eds., Part I, Vol. 1, p. 284. New York: Wiley, 1978. With permission of John Wiley & Sons, Inc.

3. Repeat the analysis if sufficient sample and time are available. Agreement between the newly acquired data and those data of the original set that appear to be valid will lend weight to the notion that the outlying result should be rejected. Furthermore, if retention is still indicated, the questionable result will have a relatively small effect on the mean of the larger set of data.
4. If more data cannot be secured, apply the $Q$ test or the $T_n$ test to the existing set to see if the doubtful result should be retained or rejected on statistical grounds.
5. If the statistical test indicates retention, consider reporting the median of the set rather than the mean. The median has the great virtue of allowing inclusion of all data in a set without undue influence from an outlying value. In addition, the median of a normally distributed set containing three measurements provides a better estimate of the correct value than the mean of the set after the outlying value has been discarded.

The blind application of statistical tests to retain or reject a suspect measurement in a small set of data is not likely to be much more fruitful than an arbitrary decision. The application of good judgment based on broad experience with an analytical method is usually a sounder approach. In the end, the only valid reason for rejecting a result from a small set of data is the sure knowledge that a mistake was made in the measurement process. Without this knowledge, *a cautious approach to rejection of an outlier is wise.*

## 3B–3 Statistical Aids to Hypothesis Testing

Much of scientific and engineering endeavor is based upon hypothesis testing. Thus, in order to explain an observation, a hypothetical model is advanced and tested experimentally to determine its validity. If the results from these experiments do not support the model, we reject it and seek a new hypothesis. If agreement is found, the hypothetical model serves as the basis for further experiments. When the hypothesis is supported by sufficient experimental data, it becomes recognized as a useful theory until such time as data are obtained that refute it.

Experimental results seldom agree exactly with those predicted from a theoretical model. Consequently, scientists and engineers frequently must judge whether a numerical difference is a manifestation of the indeterminate errors inevitable in all measurements. Certain statistical tests are useful in sharpening these judgments.

Tests of this kind make use of a *null hypothesis*, which assumes that the numerical quantities being compared are, in fact, the same. The probability of the observed differences appearing as a result of indeterminate error is then computed from statistical theory. Usually, if the observed difference is greater than or equal to the difference that would occur 5 times in 100 (the 5% probability level), the null hypothesis is considered questionable and the difference is judged to be significant. Other probability levels, such as 1 in 100 or 10 in 100, may also be adopted, depending upon the certainty desired in the judgment.

In statistics a *null hypothesis* postulates that two observed quantities are the same.

The notation to indicate the 5% probability level is $P > 0.05$.

The kinds of testing that chemists use most often include the comparison of (1) the mean from an analysis $\bar{x}$ with what is believed to be the true value $\mu$; (2) the means $\bar{x}_1$ and $\bar{x}_2$ from two sets of analyses; (3) the standard deviations $s_1$ and $s_2$ or $\sigma_1$ and $\sigma_2$ from two sets of measurements; and (4) the standard deviation $s$ of a small set of data with the standard deviation $\sigma$ of a larger set of measurements. The sections that follow consider some of the methods for making these comparisons.

## Comparison of an Experimental Mean with a True Value

A common way of testing for determinate errors in an analytical method is to use the method to analyze a sample whose composition is accurately known. It is likely that the experimental mean $\bar{x}$ will differ from the accepted value $\mu$; the judgment must then be made whether this difference is the consequence of indeterminate error or determinate error.

In treating this type of problem statistically, the difference $\bar{x} - \mu$ is compared with the difference that could be caused by indeterminate error. If the observed difference is less than that computed for a chosen probability level, the null hypothesis that $\bar{x}$ and $\mu$ are the same cannot be rejected; that is, no significant determinate error has been demonstrated. It is important to realize, however, that this statement does not say that there is no determinate error; it says only that whatever determinate error is present is so small that it cannot be detected. If $\bar{x} - \mu$ is significantly larger than either the expected or the critical value, we may assume that the difference is real and that the determinate error is significant.

The critical value for the rejection of the null hypothesis is calculated by rewriting Equation 3–10 in the form

$$\bar{x} - \mu = \pm \frac{ts}{\sqrt{N}} \qquad (3\text{--}12)$$

where $N$ is the number of replicate measurements employed in the test. If a good estimate of $\sigma$ is available, Equation 3–12 can be modified by replacing $t$ with $z$ and $s$ with $\sigma$.

---

Example 3–8

A new procedure for the rapid determination of sulfur in kerosenes was tested on a sample known from its method of preparation to contain 0.123% S. The results were % S = 0.112, 0.118, 0.115, and 0.119. Do the data indicate that there is a determinate error in the method?

$$\Sigma\, x_i = 0.112 + 0.118 + 0.115 + 0.119 = 0.464$$
$$\bar{x} = 0.464/4 = 0.116\%\ S$$
$$\bar{x} - \mu = 0.116 - 0.123 = -0.007\%\ S$$
$$\Sigma\, x_i^2 = 0.012544 + 0.013924 + 0.013225 + 0.014161 = 0.053854$$
$$s = \sqrt{\frac{0.053854 - (0.464)^2/4}{4 - 1}} = \sqrt{\frac{0.000030}{3}} = 0.0032$$

From Table 3–2, we find that at the 95% confidence level, $t$ has a value of 3.18 for three degrees of freedom. Thus,

$$\frac{ts}{\sqrt{4}} = \frac{3.18 \times 0.0032}{\sqrt{4}} = \pm 0.0051$$

An experimental mean can be expected to deviate by $\pm 0.0051$ or greater no more frequently than 5 times in 100. Thus, if we conclude that $\bar{x} - \mu = -0.007$ is a significant difference and that a determinate error is present, we will, on the average, be wrong fewer than 5 times in 100.

> If it was confirmed by further experiments that the method always gave low results, we would say that the method had a *negative bias*.

If we make a similar calculation employing the value for $t$ at the 99% confidence level, $ts/\sqrt{N}$ assumes a value of 0.0093. Thus, if we insist upon being wrong no more often than 1 time in 100, we must conclude that no difference between the results has been *demonstrated*. Note that this statement is different from saying that there is no determinate error.

> Even if a mean value is shown to be equal to the true value at a given confidence level, we cannot conclude that there is *no* determinate error in the data.

### Comparison of Two Experimental Means

The results of chemical analyses are frequently used to determine whether two materials are identical. Here, the chemist must judge whether a difference in the means of two sets of measurements is real and constitutes evidence that the samples are different or whether the discrepancy is simply a consequence of indeterminate errors in the two sets. To illustrate, let us assume that $N_1$ replicate analyses of material 1 yielded a mean value of $\bar{x}_1$, and $N_2$ analyses of material 2 obtained by the same method gave a mean of $\bar{x}_2$. If the data were collected in an identical way, it is usually safe to assume that the standard deviations of the two sets of measurements are the same and modify Equation 3–12 to take into account that one set of results is being compared with a second rather than with the true mean of the data, $\mu$.

In this case, as with the previous one, we invoke the null hypothesis that the samples are identical and that the observed difference in the results, $(\bar{x}_1 - \bar{x}_2)$, is the result of indeterminate errors. To test this hypothesis statistically, we modify Equation 3–12 in the following way. First, we substitute $\bar{x}_2$ for $\mu$, thus making the left side of the equation the numerical difference between the two means $\bar{x}_1 - \bar{x}_2$. Since we know from Equation 3–5 that the standard deviation of the mean $\bar{x}_1$ is

$$s_{m1} = \frac{s_1}{\sqrt{N_1}} \quad \text{and likewise for } \bar{x}_2, \quad s_{m2} = \frac{s_2}{\sqrt{N_2}}$$

Thus the variance $s_d^2$ of the difference $(d = x_1 - x_2)$ between the means is given by

$$s_d^2 = s_{m1}^2 + s_{m2}^2$$

By substituting the values for $s_d$, $s_{m1}$, and $s_{m2}$ into this equation, we have

$$\left(\frac{s_d}{\sqrt{N}}\right)^2 = \left(\frac{s_{m1}}{\sqrt{N_1}}\right)^2 + \left(\frac{s_{m2}}{\sqrt{N_2}}\right)^2$$

If we then assume that the pooled standard deviation $s_{pooled}$ is a good estimate of both $s_{m1}$ and $s_{m2}$, then

$$\left(\frac{s_d}{\sqrt{N}}\right)^2 = \left(\frac{s_{pooled}}{\sqrt{N_1}}\right)^2 + \left(\frac{s_{pooled}}{\sqrt{N_2}}\right)^2 = s_{pooled}^2 \left(\frac{N_1 + N_2}{N_1 N_2}\right)$$

and

$$\frac{s_d}{\sqrt{N}} = s_{pooled} \sqrt{\frac{N_1 + N_2}{N_1 N_2}}$$

Substituting this equation into Equation 3–12 (and also $\bar{x}_2$ for $\mu$), we find that

$$\bar{x}_1 - \bar{x}_2 = \pm t s_{pooled} \sqrt{\frac{N_1 + N_2}{N_1 N_2}} \qquad (3\text{–}13)$$

The numerical value for the term on the right is computed using $t$ for the particular confidence level desired. The number of degrees of freedom for finding $t$ in Table 3–2 is $N_1 + N_2 - 2$. If the experimental difference $\bar{x}_1 - \bar{x}_2$ is smaller than the computed value, the null hypothesis is not rejected and no significant difference between the means has been demonstrated. An experimental difference greater than the value computed from $t$ indicates that there is a significant difference between the means.

If a good estimate of $\sigma$ is available, Equation 3–12 can be modified by inserting $z$ for $t$ and $\sigma$ for $s$.

---

Example 3–9

The composition of a flake of paint found on the clothes of the victim of a hit-and-run accident was compared with that of paint from the car suspected of causing the accident. Do the following data for the spectroscopic determination of titanium in the paint suggest a difference in composition between the two materials? From previous experience, the standard deviation for the method is known to be 0.35% Ti; that is, $s \rightarrow \sigma = 0.35\%$ Ti.

$$\text{Paint from clothes} \quad \% \text{ Ti} = 4.0, 4.6$$
$$\text{Paint from car} \quad \% \text{ Ti} = 4.5, 5.3, 5.5, 5.0, 4.9$$
$$\bar{x}_1 = \frac{4.6 + 4.0}{2} = 4.3$$
$$\bar{x}_2 = \frac{4.5 + 5.3 + 5.5 + 5.0 + 4.9}{5} = 5.0$$
$$\bar{x}_1 - \bar{x}_2 = 4.3 - 5.0 = -0.7\% \text{ Ti}$$

Modifying Equation 3–13 to take into account our knowledge that $s \to \sigma$ and taking values of $z$ from Table 3–2, we calculate for the 95% confidence level

$$\pm z\sigma \sqrt{\frac{N_1 + N_2}{N_1 N_2}} = \pm 1.96 \times 0.35 \sqrt{\frac{2 + 5}{2 \times 5}} = \pm 0.57$$

and for the 99% confidence level

$$\pm z\sigma \sqrt{\frac{N_1 + N_2}{N_1 N_2}} = \pm 2.58 \times 0.35 \sqrt{\frac{2 + 5}{2 \times 5}} = \pm 0.76$$

Only 5 out of 100 data should differ by 0.57% Ti or greater, and only 1 out of 100 should differ by as much as 0.76% Ti. Thus, it seems reasonably probable (between 95% and somewhat less than 99% certain) that the observed difference of −0.7% does not arise from indeterminate error but in fact is caused, at least in part, by a real difference between the two paint samples. Hence, we conclude that the suspected vehicle was probably not involved.

---

Example 3–10

Two barrels of wine were analyzed for their alcohol content in order to determine whether they were from different sources. On the basis of six analyses, the average content of the first barrel was established to be 12.61% ethanol. Four analyses of the second barrel gave a mean of 12.53% alcohol. The ten analyses yielded a pooled value of $s = 0.070\%$. Do the data indicate a difference between the wines?

Here we employ Equation 3–13, using $t$ for eight degrees of freedom $(10 - 2)$. At the 95% confidence level,

$$\pm ts \sqrt{\frac{N_1 + N_2}{N_1 N_2}} = \pm 2.31 \times 0.070 \sqrt{\frac{6 + 4}{6 \times 4}} = \pm 0.10\%$$

The observed difference is

$$\bar{x}_1 - \bar{x}_2 = 12.61 - 12.53 = 0.08\%$$

As often as 5 times in 100, indeterminate error will be responsible for a difference as great as 0.10%. At the 95% confidence level, then, no difference in the alcohol content of the wine has been established.

---

In Example 3–10, no significant difference between the two wines was detected at the 95% probability level. Note that this statement is not equivalent to saying that $\bar{x}_1$ is equal to $\bar{x}_2$; nor do the tests prove that the wines come from the same source. Indeed, it is conceivable that one wine is a red and the other is a white. To establish with a reasonable probability that the two wines are from the same source would require extensive testing of other characteristics, such as taste, color, odor, and refractive index, as well as tartaric acid, sugar, and trace element content. If no significant differences are revealed by all of these tests and by others, then it might be possible to judge the two wines as having a common origin. In contrast, the finding of *one* significant difference in any test would clearly show that the two wines are different. Thus, the establishment of a significant difference by a single test is much more revealing than the establishment of an absence of difference.

### Estimation of Detection Limits

Equation 3–13 is useful for estimating the detection limit for a measurement. Here, the standard deviation from several blank determinations is computed. The minimum detectable quantity $\Delta x_{min}$ is

$$\Delta x_{min} = \bar{x}_1 - \bar{x}_b > t s_b \frac{N_1 + N_2}{N_1 N_2} \tag{3-14}$$

where the subscript $b$ refers to the blank determination.

---

Example 3–11

A method for the analysis of DDT gave the following results when applied to pesticide-free foliage samples: $\mu$g DDT $= 0.2, -0.5, -0.2, 1.0, 0.8, -0.6, 0.4, 1.2$. Calculate the DDT-detection limit (at the 99% confidence level) of the method for (a) a single analysis and (b) the mean of five analyses.

Here we find

$$\Sigma x_i = 0.2 - 0.5 - 0.2 + 1.0 + 0.8 - 0.6 + 0.4 + 1.2 = 2.3$$

$$\Sigma x_i^2 = 0.04 + 0.25 + 0.04 + 1.0 + 0.64 + 0.36 + 0.16 + 1.44 = 3.93$$

$$s_b = \sqrt{\frac{3.93 - (2.3)^2/8}{8 - 1}} = \sqrt{\frac{3.26875}{7}} = 0.68 \ \mu g$$

(a) For a single analysis, $N_1 = 1$ and the number of degrees of freedom is $1 + 8 - 2 = 7$. From Table 3–2, we find $t = 3.50$, and so

$$\Delta x_{min} > 3.50 \times 0.68 \sqrt{\frac{1 + 8}{1 \times 8}} > 2.5 \ \mu g \ DDT$$

Thus, 99 times out of 100, a result greater than 2.5 $\mu$g DDT indicates the presence of the pesticide on the plant.

(b) Here, $N_1 = 5$, and the number of degrees of freedom is 11. Therefore, $t = 3.11$, and

$$\Delta x_{min} > 3.11 \times 0.68 \sqrt{\frac{5 + 8}{5 \times 8}} > 1.2 \ \mu g \ DDT$$

## Comparison of the Precision of Measurements

The $F$ test provides a simple method for comparing the precision of two sets of measurements. The sets do not necessarily have to be obtained from the same sample as long as the samples are sufficiently alike that the sources of indeterminate error can be assumed to be the same. The $F$ test is also based upon the null hypothesis and thus assumes that the precisions are identical. The quantity $F$, which is defined as the ratio of the variances of the two measurements, is computed and compared with the maximum values of $F$ expected (at a certain probability level) if there was no difference in precision between the two sets of measurements. If the experimental $F$ exceeds the critical value found in probability tables, there is a statistical basis for questioning the null hypothesis that the two standard deviations are the same. Table 3–5 provides critical values for $F$ at the 5% probability level. These values will be exceeded only 5 times in 100 if the standard deviations of the two measurements are the same. You will find more extensive tables of $F$ values at various probability levels in most mathematics handbooks.

The $F$ test may be used to provide insights into either of two questions: (1) whether method A is more precise than method B and (2) whether there is a difference in the precision of the two methods. For the first of these applications, the variance of the supposedly more precise procedure is always placed in the denominator and that of the less precise procedure in the numerator. For the second application, the larger variance always appears in the numerator. This arbitrary placement of the larger variance in the numerator makes the outcome of the test less certain; thus, the uncertainty level of the $F$ values in Table 3–5 is doubled from 5% to 10%.

---

### Example 3–12

A standard method for the determination of carbon monoxide in gaseous mixtures is known from many hundreds of measurements to have a standard deviation of 0.21 ppm CO. A modification of the method yields a value for $s$ of 0.15 ppm CO for a pooled data set with 12 degrees of freedom. A second modification, also based on 12 degrees of freedom, has

### Table 3–5
### CRITICAL VALUES FOR $F$ AT THE 5% LEVEL

| Degrees of Freedom (Denominator) | Degrees of Freedom (Numerator) | | | | | | | |
|---|---|---|---|---|---|---|---|---|
| | **2** | **3** | **4** | **5** | **6** | **12** | **20** | **∞** |
| 2 | 19.00 | 19.16 | 19.25 | 19.30 | 19.33 | 19.41 | 19.45 | 19.50 |
| 3 | 9.55 | 9.28 | 9.12 | 9.01 | 8.94 | 8.74 | 8.66 | 8.53 |
| 4 | 6.94 | 6.59 | 6.39 | 6.26 | 6.16 | 5.91 | 5.80 | 5.63 |
| 5 | 5.79 | 5.41 | 5.19 | 5.05 | 4.95 | 4.68 | 4.56 | 4.36 |
| 6 | 5.14 | 4.76 | 4.53 | 4.39 | 4.28 | 4.00 | 3.87 | 3.67 |
| 12 | 3.89 | 3.49 | 3.26 | 3.11 | 3.00 | 2.69 | 2.54 | 2.30 |
| 20 | 3.49 | 3.10 | 2.87 | 2.71 | 2.60 | 2.28 | 2.12 | 1.84 |
| ∞ | 3.00 | 2.60 | 2.37 | 2.21 | 2.10 | 1.75 | 1.57 | 1.00 |

a standard deviation of 0.12 ppm CO. Is either modification significantly more precise than the original?

Because an improvement is claimed, the variances of the modifications are placed in the denominator. For the first,

$$F_1 = \frac{s_{std}^2}{s_1^2} = \frac{(0.21)^2}{(0.15)^2} = 1.96$$

and for the second,

$$F_1 = \frac{(0.21)^2}{(0.12)^2} = 3.06$$

For the standard procedure, $s \rightarrow \sigma$ and the number of degrees of freedom for the numerator can be taken as infinite. The critical value of $F$ is thus 2.30 (Table 3–5).

The value of $F$ for the first modification is less than 2.30, and so the null hypothesis has not been disproved at the 95% probability level. The second modification, however, does appear to have significantly greater precision.

It is interesting to note that if we ask whether the precision of the second modification is significantly better than that of the first, the $F$ test indicates that such a difference has not been demonstrated. That is,

$$F = \frac{s_1^2}{s_2^2} = \frac{(0.15)^2}{(0.12)^2} = 1.56$$

In this instance, the critical value for $F$ is 2.69.

---

### 3B–3  The Least-Squares Method for Deriving Calibration Plots

Most analytical methods are based on a calibration curve in which a measured quantity $y$ is plotted as a function of the known concentration $x$ of a series of standards. The typical calibration curve shown in Figure 3–4 was derived for the determination of isooctane in a hydrocarbon sample. The ordinate (the dependent variable) is the area under the chromatographic peak for isooctane, and the abscissa (the independent variable) is the mole percent of isooctane. As is typical (and desirable), the plot approximates a straight line. Note, however, that because of the indeterminate errors in the measuring process not all the data fall exactly on the line. Thus, the investigator must try to derive a "best" straight line from the points. A statistical technique called *regression analysis* provides the means for objectively obtaining such a line and also for specifying the uncertainties associated with its subsequent use. We consider here only the simplest regression procedure: the *method of least squares*.

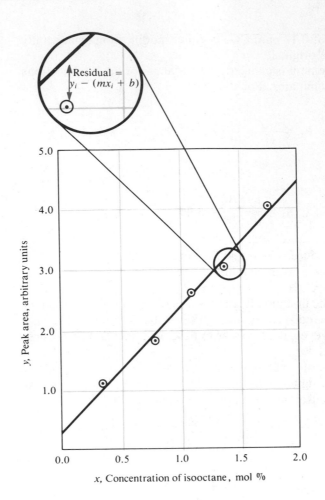

**Figure 3–4**

Calibration curve for the determination of isooctane in a hydrocarbon mixture.

## Assumptions

Linear least-squares analysis gives you the equation for the best straight line among a set of $x$, $y$ data pairs.

When the method of least squares is used to generate a calibration curve, two assumptions are required. The first is that there is actually a linear relationship between the measured variable ($y$) and the analyte concentration ($x$). This relationship is stated mathematically as

$$y = mx + b$$

where $b$ is the $y$-intercept (the value of $y$ when $x$ is zero) and $m$ is the slope of the line (see Figure 3–5). We also assume that any deviation of individual points from the straight line results from error in the *measurement*. That is, we must assume that there is no error in the $x$ values of the points. For the example of a calibration curve, we assume that exact concentrations of the standards are known. Both of these assumptions are appropriate for most analytical methods.

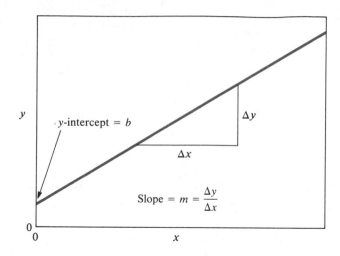

Figure 3–5
The slope-intercept form of a
straight line.

### The Derivation of a Least-Squares Line

As illustrated in Figure 3–4, the vertical deviation of each point from the straight line is called a *residual*. The line generated by the least-squares method is the one that minimizes the sum of the squares of the residuals from all of the points. In addition to providing the best fit between the experimental points and the straight line, the method gives the standard deviations for $m$ and $b$.[6]

For convenience, we define three quantities $S_{xx}$, $S_{yy}$, and $S_{xy}$ as follows:

$$S_{xx} = \Sigma\,(x_i - \bar{x})^2 = \Sigma\,x_i^2 - \frac{(\Sigma\,x_i)^2}{N} \qquad (3-15)$$

$$S_{yy} = \Sigma\,(y_i - \bar{y})^2 = \Sigma\,y_i^2 - \frac{(\Sigma\,y_i)^2}{N} \qquad (3-16)$$

$$S_{xy} = \Sigma\,(x_i - \bar{x},\,y_i - \bar{y}) = \Sigma\,x_i y_i - \frac{\Sigma\,x_i\,\Sigma\,y_i}{N} \qquad (3-17)$$

where $x_i$ and $y_i$ are individual pairs of data for $x$ and $y$, $N$ is the number of pairs of data used in preparing the calibration curve, and $\bar{x}$ and $\bar{y}$ are the average values for the variables, that is,

$$\bar{x} = \Sigma\,x_i/N \qquad \text{and} \qquad \bar{y} = \Sigma\,y_i/N$$

Note that $S_{xx}$ and $S_{yy}$ are the sum of the squares of the deviations from the mean for the individual values of $x$ and $y$. The equivalent expressions shown to the far right in Equations 3–15 to 3–17 are more convenient when a hand-held calculator is being used.

Six useful quantities can be derived from $S_{xx}$, $S_{yy}$, and $S_{xy}$.

The terms $S_{xx}$ and $S_{yy}$ are similar in form to the numerator in the equation for standard deviation.

[6]R. L. Anderson, *Practical Statistics for Analytical Chemists*, pp. 89–121. New York: Van Nostrand–Reinhold, 1987.

1. The slope of the line $m$:

$$m = S_{xy}/S_{xx} \qquad (3\text{--}18)$$

2. The intercept $b$:

$$b = \bar{y} - m\bar{x} \qquad (3\text{--}19)$$

3. The standard deviation about regression $s_r$:

$$s_r = \sqrt{\frac{S_{yy} - m^2 S_{xx}}{N - 2}} \qquad (3\text{--}20)$$

4. The standard deviation of the slope $s_m$:

$$s_m = \sqrt{s_r^2/S_{xx}} \qquad (3\text{--}21)$$

5. The standard deviation of the intercept $s_b$:

$$s_b = s_r \sqrt{\frac{\sum x_i^2}{N \sum x_i^2 - (\sum x_i)^2}} = s_r \sqrt{\frac{1}{N - (\sum x_i)^2/\sum x_i^2}} \qquad (3\text{--}22)$$

6. The standard deviation for results obtained from the calibration curve $s_c$:

$$s_c = \frac{s_r}{m} \sqrt{\frac{1}{M} + \frac{1}{N} + \frac{(\bar{y}_c - \bar{y})^2}{m^2 S_{xx}}} \qquad (3\text{--}23)$$

Equation 3–23 gives us a way to calculate the standard deviation from the mean $\bar{y}_c$ of a set of $M$ replicate analyses of unknowns when a calibration curve that contains $N$ points is used; recall that $\bar{y}$ is the mean value of $y$ for the $N$ calibration data.

The standard deviation about regression $s_r$ (Equation 3–20) is the standard deviation for $y$ when the deviations are measured not from the mean of $y$ (as is usually the case) but from the derived straight line:

$$s_r = \sqrt{\frac{\sum_{i=1}^{N} [y_i - (b + mx_i)]^2}{N - 2}} \qquad (3\text{--}24)$$

In this equation, the number of degrees of freedom is $N - 2$ since one degree is lost in calculating $m$ and one in determining $b$.

---

Example 3–13

Carry out a least-squares analysis of the experimental data provided in the first two columns in Table 3–6 and plotted in Figure 3–4.

Columns 3, 4, and 5 of the table contain computed values for $x_i^2$, $y_i^2$, and $x_i y_i$, with their sums appearing as the last entry in each column. Note that

The standard deviation about regression is analogous to the standard deviation for one-dimensional data.

**Table 3–6**

**CALIBRATION DATA FOR A CHROMATOGRAPHIC METHOD FOR THE DETERMINATION OF ISOOCTANE IN A HYDROCARBON MIXTURE**

| Mole Percent Isooctane, $x_i$ | Peak Area, $y_i$ | $x_i^2$ | $y_i^2$ | $x_iy_i$ |
|---|---|---|---|---|
| 0.352 | 1.09 | 0.12390 | 1.1881 | 0.38368 |
| 0.803 | 1.78 | 0.64481 | 3.1684 | 1.42934 |
| 1.08 | 2.60 | 1.16640 | 6.7600 | 2.80800 |
| 1.38 | 3.03 | 1.90440 | 9.1809 | 4.18140 |
| 1.75 | 4.01 | 3.06250 | 16.0801 | 7.01750 |
| 5.365 | 12.51 | 6.90201 | 36.3775 | 15.81992 |

the number of digits carried in the computed values should be the *maximum allowed by the calculator or computer; that is, rounding off should not be performed until the calculation is complete*.

We now substitute into Equations 3–15, 3–16, and 3–17 and obtain

$$S_{xx} = \Sigma x_i^2 - (\Sigma x_i)^2/N = 6.90201 - (5.365)^2/5 = 1.14537$$
$$S_{yy} = \Sigma y_i^2 - (\Sigma y_i)^2/N = 36.3775 - (12.51)^2/5 = 5.07748$$
$$S_{xy} = \Sigma x_iy_i - \Sigma x_i \Sigma y_i/N = 15.81992 - 5.365 \times 12.51/5 = 2.39669$$

Substitution of these quantities into Equations 3–18 and 3–19 yields

$$m = 2.39669/1.14537 = 2.0925 = 2.09$$
$$b = \frac{12.51}{5} - 2.0925 \times \frac{5.365}{5} = 0.2567 = 0.26$$

Thus, the equation for the least-squares line is

$$y = 2.09x + 0.26$$

Substitution into Equation 3–20 yields the standard deviation about regression:

$$s_r = \sqrt{\frac{S_{yy} - m^2S_{xx}}{N - 2}} = \sqrt{\frac{5.07748 - (2.0925)^2 \times 1.14537}{5 - 2}} = 0.144 = 0.14$$

and substitution into Equation 3–21 gives the standard deviation of the slope:

$$s_m = \sqrt{s_r^2/S_{xx}} = \sqrt{(0.144)^2/1.14537} = 0.13$$

Finally, we find the standard deviation of the intercept from Equation 3–22.

$$s_b = 0.144 \sqrt{\frac{1}{5 - (5.365)^2/6.90201}} = 0.16$$

Example 3–14

The calibration curve derived in Example 3–13 was used for the chromatographic determination of isooctane in a hydrocarbon mixture. A peak area of 2.65 was obtained. Calculate the mole percent of isooctane and the standard deviation for the result if the area was (a) the result of a single measurement and (b) the mean of four measurements.

In either case,

$$x = \frac{y - 0.26}{2.09} = \frac{2.65 - 0.26}{2.09} = 1.14 \text{ mol } \%$$

(a) Substituting into Equation 3–22, we obtain

$$s_c = \frac{0.14}{2.09} \sqrt{\frac{1}{1} + \frac{1}{5} + \frac{(2.65 - 12.51/5)^2}{(2.09)^2 \times 1.145}} = 0.074 \text{ mol } \%$$

(b) For the mean of four measurements,

$$s_c = \frac{0.14}{2.09} \sqrt{\frac{1}{4} + \frac{1}{5} + \frac{(2.65 - 12.51/5)^2}{(2.09)^2 \times 1.145}} = 0.046 \text{ mol } \%$$

## 3C QUESTIONS AND PROBLEMS

3–1. Distinguish between
   (a) the meaning of the word "sample" as it is used in a chemical and in a statistical sense.
   (b) the sample mean and the population mean.
3–2. How do the sample standard deviation and the population standard deviation resemble one another? How do they differ?
3–3. Briefly define or describe
   (a) confidence interval.
   (b) null hypothesis.
   (c) detection limit.
   (d) residuals.
*3–4. Analysis of several plant-food preparations for potassium ion yielded the following data:

| Sample | Mean Percent $K^+$ | Number of Observations | Deviation of Individual Results from Mean |
|--------|--------------------|------------------------|-------------------------------------------|
| 1 | 4.80 | 5 | 0.13, 0.09, 0.07, 0.05, 0.06 |
| 2 | 8.04 | 3 | 0.09, 0.08, 0.12 |
| 3 | 3.77 | 4 | 0.02, 0.15, 0.07, 0.10 |
| 4 | 4.07 | 4 | 0.12, 0.06, 0.05, 0.11 |
| 5 | 6.84 | 5 | 0.06, 0.07, 0.13, 0.10, 0.09 |

   (a) Evaluate the standard deviation $s$ for each sample.
   (b) Obtain a pooled estimate for $s$.

| Bottle | Percent (w/v) Residual Sugar | Number of Observations | Deviation of Individual Results from Mean |
|--------|------------------------------|------------------------|-------------------------------------------|
| 1 | 0.94 | 3 | 0.050, 0.10, 0.08 |
| 2 | 1.08 | 4 | 0.060, 0.050, 0.090, 0.060 |
| 3 | 1.20 | 5 | 0.05, 0.12, 0.07, 0.00, 0.08 |
| 4 | 0.67 | 4 | 0.05, 0.10, 0.06, 0.09 |
| 5 | 0.83 | 3 | 0.07, 0.09, 0.10 |
| 6 | 0.76 | 4 | 0.06, 0.12, 0.04, 0.03 |

**3–5.** Six bottles of wine were analyzed for residual sugar, with the results shown above:
  (a) Evaluate the standard deviation $s$ for each set of data.
  (b) Pool the data to establish an absolute standard deviation for the method.

**\*3–6.** Nine samples of illicit heroin preparations were analyzed in duplicate by a gas chromatographic technique. Pool the following data to establish an absolute standard deviation for the procedure:

| Sample | Heroin, % | Sample | Heroin, % |
|--------|-----------|--------|-----------|
| 1 | 2.24, 2.27 | 6 | 1.07, 1.02 |
| 2 | 8.4, 8.7 | 7 | 14.4, 14.8 |
| 3 | 7.6, 7.5 | 8 | 21.9, 21.1 |
| 4 | 11.9, 12.6 | 9 | 8.8, 8.4 |
| 5 | 4.3, 4.2 | | |

**3–7.** Calculate the 90% confidence interval for each of the accompanying sets of data, assuming that these are the only measurements available.

| *A | B | *C | D | *E | F |
|------|--------|-------|------|-------|-------|
| 16.35 | 0.1018 | 2.796 | 9.25 | 55.98 | 12.47 |
| 16.25 | 0.1006 | 2.814 | 9.55 | 56.05 | 12.40 |
| 16.06 | 0.0997 | 2.712 | 9.21 | 55.70 | 12.49 |
| 16.21 | 0.1004 | | 9.17 | 56.06 | 12.30 |
| | 0.1009 | | 9.24 | 55.95 | 12.40 |
| | | | | | 12.44 |

**3–8.** Calculate the 90% confidence interval for the sets of data in Problem 3–7 if $s \rightarrow \sigma$, with values of: A, 0.15; B, 0.0010; C, 0.05; D, 0.14; E. 0.21; F, 0.07.

**\*3–9.** An atomic absorption method for the determination of the amount of iron present in used jet engine oil was found, from pooling 30 triplicate analyses, to have a standard deviation $s \rightarrow \sigma = 2.4\ \mu g/mL$. Calculate the 80 and 95% confidence intervals for the result, 18.5 $\mu g$ Fe/mL, if it was based upon (a) a single analysis, (b) the mean of two analyses, (c) the mean of four analyses.

**3–10.** An atomic absorption method for determination of copper in fuels yielded a pooled standard deviation of $s \rightarrow \sigma = 0.32\ \mu g$ Cu/mL. The analysis of an oil from a reciprocating aircraft engine showed a copper content of 8.53 $\mu g$ Cu/mL. Calculate the 80 and

99% confidence interval for the result if it was based upon
  (a) a single analysis.
  (b) the mean of 4 analyses.
  (c) the mean of 16 analyses.

**\*3–11.** How many replicate measurements are needed to decrease the 95 and 99% confidence limits for the analysis described in Problem 3–9 to ±1.5 $\mu g$ Fe/mL?

**3–12.** How many replicate measurements are necessary to decrease the 95 and 99% confidence limits for the analysis described in Problem 3–10 to ±0.2 $\mu g$ Cu/mL?

**\*3–13.** A chemist obtained the following data for percent lindane in the triplicate analysis of an insecticide preparation: 7.47, 6.98, 7.27. Calculate the 90% confidence interval for the mean of the three data, assuming that
  (a) the only information about the precision of the method is the precision for the three data.
  (b) on the basis of long experience with the method, it is believed that $s \rightarrow \sigma = 0.28\%$ lindane.

**3–14.** Each set in Problem 3–7 has a possible outlying result. Apply the $Q$ test to see if rejection (95% confidence) is warranted.

**3–15.** Apply the $T_n$ test to the data sets in Problem 3–7 to determine whether rejection (95% confidence level) of the outlying result is statistically justified.

**\*3–16.** A new method for the analysis of copper was tested with a sample known to contain 16.68% Cu.

| Sample | % Cu Found |
|--------|------------|
| 1 | 16.54 |
| 2 | 16.64 |
| 3 | 16.30 |
| 4 | 16.67 |
| 5 | 16.70 |

  (a) Evaluate the mean and the median percentages of copper for these data.
  (b) Apply the $Q$ test (90% confidence level) to the outlying result.
  (c) Which value—the mean or the median—do you prefer as the "best" value for this analysis? Defend your answer in a sentence or two.

**\*3–17.** A spark-source mass-spectrometric method for the determination of various elements in steel was

| Element | Number of Analyses | Mean, % (w/w) | Relative Standard Deviation, ppt | NIST Result, % (w/w) |
|---|---|---|---|---|
| (a) V | 8 | 0.090 | 97 | 0.096 |
| (b) Ni | 5 | 0.36 | 55 | 0.39 |
| (c) Cu | 7 | 0.55 | 76 | 0.52 |

tested by analyzing several National Institute of Standards and Technology samples. The results from three of the analyses are given above. Assume that the NIST analyses are correct and determine whether or not a determinate error in any of the analyses is indicated at the 95% confidence level.

3–18. A prosecuting attorney in a criminal case presented as principal evidence small fragments of glass found imbedded in the coat of the accused. The attorney claimed that the fragments were identical in composition to a rare Belgian stained glass window broken during the crime. The averages of triplicate analyses for five elements in the glass are shown below. On the basis of these data, does the defendant have grounds for claiming reasonable doubt as to guilt? Employ the 99% confidence level as a criterion for doubt.

| Element | Concentration, ppm From Clothes | From Window | Standard Deviation $s \rightarrow \sigma$ |
|---|---|---|---|
| As | 129 | 119 | 9.5 |
| Co | 0.53 | 0.60 | 0.025 |
| La | 3.92 | 3.52 | 0.20 |
| Sb | 2.75 | 2.71 | 0.25 |
| Th | 0.61 | 0.73 | 0.043 |

*3–19. The homogeneity of a standard chloride sample was tested by analyzing portions of the material from the top and the bottom of the container, with the following results:

| % Chloride | |
|---|---|
| Top | Bottom |
| 26.32 | 26.28 |
| 26.33 | 26.25 |
| 26.38 | 26.38 |
| 26.39 | |

(a) Is nonhomogeneity indicated at the 95% confidence level?
(b) Is nonhomogeneity indicated at the 95% level if it is known that $s \rightarrow \sigma = 0.03\%$ Cl?

3–20. A method for the analysis of codeine in prescription drugs yielded the following results when applied to a codeine-free blank: 0.1, −0.2, 0.3, 0.2, 0.0, −0.1 mg codeine. Calculate the detection limit (in terms of milligrams of codeine) at the 99% confidence level, based upon the mean of (a) two analyses, (b) four analyses, (c) six analyses.

*3–21. A single alloy specimen was used to compare the results of two testing laboratories. The standard deviation $s$ and degrees of freedom DF in pooled sets for four analyses are:

| Element | Laboratory A s | DF | Laboratory B s | DF |
|---|---|---|---|---|
| *(a) Fe | 0.10 | 6 | 0.12 | 12 |
| (b) Ni | 0.07 | 12 | 0.04 | 20 |
| *(c) Cr | 0.05 | 20 | 0.07 | 6 |
| (d) Mn | 0.020 | 20 | 0.035 | 6 |

Use the $F$ test to determine whether the results from one laboratory are statistically more precise than those from the other.

3–22. Calculate the minimum detection limits at the 99% confidence level for the mean of $N_1$ measurements, given the accompanying data:

| Individual Blank Determinations | $N_1$ |
|---|---|
| *(a) 0.4, 0.1, 0.6, 0.3, 0.2 | 8 |
| (b) 1.3, 1.7, 0.9, 1.5 | 6 |
| *(c) 0.8, 1.1, 0.6, 1.4, 1.2, 1.0 | 5 |
| (d) 0.67, 0.82, 0.75, 0.77, 0.69 | 7 |

*3–23. An established method for the determination of manganese in alloy samples has a standard deviation ($\sigma$) of 0.11. Determine whether a modification to the method appears to yield improved precision if the standard deviation ($s$) is 0.07, based upon a pooled set with
*(a) 6 degrees of freedom.
(b) 12 degrees of freedom.
*(c) 20 degrees of freedom.

*3–24. The sulfate ion concentration in natural water can be determined by measuring the turbidity that results when an excess of $BaCl_2$ is added to a measured quantity of the sample. A turbidimeter, the instrument used for this analysis, was calibrated with a series of standard $Na_2SO_4$ solutions. The following data were obtained in the calibration:

| mg $SO_4^{2-}$/L, $c_x$ | Turbidimeter Reading, $R$ |
|---|---|
| 0.00 | 0.06 |
| 5.00 | 1.48 |
| 10.00 | 2.28 |
| 15.0 | 3.98 |
| 20.0 | 4.61 |

| pCa | $E$, mV |
|---|---|
| 5.00 | −53.8 |
| 4.00 | −27.7 |
| 3.00 | + 2.7 |
| 2.00 | +31.9 |
| 1.00 | +65.1 |

Assume that there is a linear relationship between the instrument reading and concentration.

(a) Plot the data and draw a straight line through the points by eye.

(b) Derive a least-squares equation for the relationship between the variables.

(c) Compare the straight line from the relationship derived in (b) with that in (a).

(d) Calculate the standard deviation for the slope and the intercept of the least-squares line.

(e) Calculate the concentration of sulfate in a sample yielding a turbidimeter reading of 3.67. Calculate the absolute standard deviation of the result and the coefficient of variation.

(f) Repeat the calculations in (e) assuming that 3.67 was a mean of six turbidimeter readings.

3–25. The following data were obtained in calibrating a calcium ion electrode for the determination of pCa. A linear relationship between the potential $E$ and pCa is known to exist.

(a) Plot the data and draw a line through the points by eye.

(b) Derive a least-squares expression for the best straight line through the points. Plot this line.

(c) Calculate the standard deviation for the slope of the least-squares line.

(d) Calculate the standard deviation of the intercept of the least-squares line.

(e) Calculate the pCa of a serum solution in which the electrode potential was 20.3 mV. Calculate the absolute and relative standard deviations for pCa if the result was from a single voltage measurement.

(f) Calculate the absolute and relative standard deviations for pCa if the millivolt reading in (e) was the mean of two replicate measurements. Repeat the calculation based upon the mean of eight measurements.

(g) Calculate the molar calcium ion concentration for the sample described in (e).

(h) Calculate the absolute and relative standard deviations in the calcium ion concentration if the measurement was performed as described in (f).

# GRAVIMETRIC METHODS OF ANALYSIS

Gravimetric methods of analysis are based upon mass measurements with an analytical balance, an instrument that yields highly accurate and precise data. In fact, if you perform a gravimetric chloride analysis in the laboratory, you may make some of the most accurate and precise measurements of your life.

$G$ravimetric methods of analysis, which are based upon the measurement of mass, are of two major types.[1] In *precipitation methods,* the analyte is converted to a sparingly soluble precipitate that is filtered, washed free of impurities, and converted to a product of known composition by suitable heat treatment. This product is then weighed. For example, in a precipitation method for determining calcium in natural water recommended by the Association of Official Analytical Chemists, an excess of oxalic acid, $H_2C_2O_4$, is added to a carefully measured volume of the sample. The addition of ammonia then causes essentially all the calcium in the sample to precipitate as calcium oxalate. The reaction is

$$Ca^{2+}(aq) + C_2O_4^{2-}(aq) \rightarrow CaC_2O_4(s)$$

The precipitate is collected in a weighed filtering crucible, dried, and then ignited at red heat. This process converts the precipitate entirely to calcium oxide:

$$CaC_2O_4(s) \rightarrow CaO(s) + CO(g) + CO_2(g)$$

The crucible and precipitate are cooled, weighed, and the weight of calcium oxide is determined by subtraction of the known weight of the crucible. The calcium content of the sample is then computed from the stoichiometry of the process as shown in the various examples in Section 4B.

In *volatilization methods,* the analyte or its decomposition products are volatilized at a suitable temperature. The volatile product is then collected and weighed, or alternatively the weight of the product is determined indirectly from the weight loss of the sample. An example of a gravimetric volatilization procedure is the determination of the sodium

---

[1]For an extensive treatment of gravimetric methods, see C. L. Rulfs, in *Treatise on Analytical Chemistry,* I. M. Kolthoff and P. J. Elving, Eds., Part I, Vol. 11, Chapter 13. New York: Wiley, 1975.

hydrogen carbonate content of antacid tablets. Here, a weighed quantity of the finely ground sample is treated with dilute sulfuric acid to convert the sodium hydrogen carbonate to carbon dioxide:

$$NaHCO_3(aq) + H_2SO_4(aq) \rightarrow CO_2(g) + H_2O(l) + NaHSO_4(aq)$$

This reaction is carried out in a flask that is connected to a weighed absorption tube containing an absorbent that retains the carbon dioxide selectively as it is evolved as the solution is heated.[2] The difference in weight of the tube before and after absorption is used to calculate the amount of sodium hydrogen carbonate.

## 4A   A REVIEW OF CHEMICAL STOICHIOMETRY

*Stoichiometry* is defined as the weight relationships among reacting chemical species. This section provides a brief review of stoichiometry.

Although as a matter of habit we speak of weighing and weights, we determine *mass* in the laboratory and use mass in chemical calculations.

### 4A–1 Empirical Formulas and Chemical Formulas

An *empirical formula* gives the simplest whole-number ratio of atoms in a chemical compound. In contrast, a *chemical formula* specifies the number of atoms in a molecule. Two or more substances may have the same empirical formula but different chemical formulas. For example, $CH_2O$ is both the empirical and the chemical formula for formaldehyde; it is also the empirical formula for such diverse substances as acetic acid, $C_2H_4O_2$, glyceraldehyde, $C_3H_6O_3$, and glucose, $C_6H_{12}O_6$, as well as more than 50 other substances containing 6 or fewer carbon atoms. The empirical formula is obtained from the percent composition of a compound. The chemical formula requires, in addition, a knowledge of the molecular weight of the species.

A *molecular formula* also provides structural information. For example, the chemically different ethanol and dimethyl ether share the same chemical formula, $C_2H_6O$. Their molecular formulas, $C_2H_5OH$ and $CH_3OCH_3$, reveal structural differences between these compounds that are not discernible in the chemical formulas that they share.

### 4A–2 The Mole

The *mole* (mol) is the SI unit for the amount of a chemical species.[3] It is always associated with a chemical formula and represents Avogadro's number ($6.022 \times 10^{23}$) of the species represented by that formula. The

One mole of a chemical species is $6.022 \times 10^{23}$ atoms, molecules, ions, electrons, ion pairs, or subatomic particles.

---

[2]See Section 4E–5 for a description of the absorbent.

[3]In the International System (SI) of Units, proposed by the International Bureau of Weights and Measures, the only chemical unit for amount of substance is the mole, defined as that quantity of a material that contains as many elementary entities (these may be atoms, ions, electrons, ion-pairs, or molecules and must be explicitly defined) as there are atoms of carbon in exactly 0.012 kg of carbon-12 (that is, Avogadro's number).

*formula weight* (fw) of a substance is the weight in grams of 1 mol of that substance. Formula weights are calculated by summing the atomic weights of all the atoms appearing in a chemical formula. For example, the formula weight of formaldehyde, $CH_2O$, is

There is a table of formula weights on the inside back cover of this book.

$$fw\ CH_2O = \frac{1\ mol\ C}{mol\ CH_2O} \times \frac{12.0\ g}{mol\ C} + \frac{2\ mol\ H}{mol\ CH_2O} \times \frac{1.0\ g}{mol\ H}$$

$$+ \frac{1\ mol\ O}{mol\ CH_2O} \times \frac{16.0\ g}{mol\ O}$$

$$= 30.0\ g/mol\ CH_2O$$

and that of glucose is

$$fw\ C_6H_{12}O_6 = \frac{6\ mol\ C}{mol\ C_6H_{12}O_6} \times \frac{12.0\ g}{mol\ C} + \frac{12\ mol\ H}{mol\ C_6H_{12}O_6} \times \frac{1.0\ g}{mol\ H}$$

$$+ \frac{6\ mol\ O}{mol\ C_6H_{12}O_6} \times \frac{16.0\ g}{mol\ O}$$

$$= 180.0\ g/mol\ C_6H_{12}O_6$$

Thus, 1 mol of formaldehyde weighs 30.0 g, and 1 mol of glucose weighs 180.0 g.

### 4A–3 The Millimole

1 mmol = $10^{-3}$ mol

1 mfw = $10^{-3}$ fw

Sometimes it is convenient to calculate with millimoles (mmol) rather than moles, where the millimole is 1/1000 of a mole. The weight in grams of one millimole, an amount called the milliformula weight (mfw), is likewise 1/1000 of the formula weight.

### 4A–4 Calculating the Amount of a Substance in Moles or Millimoles

The two examples that follow illustrate how the number of moles or millimoles of a species can be derived from its weight in grams or from the weight of a chemically related species.

---

**Example 4–1**

The number of moles of a species A is given by

$$no.\ mol\ A = \frac{wt\ A}{fw\ A}$$

The number of millimoles is given by

$$no.\ mmol\ A = \frac{wt\ A}{(fw\ A)/1000} = \frac{wt\ A}{mfw\ A}$$

How many moles and millimoles of benzoic acid (fw = 122.1 g) are contained in 2.00 g of the pure acid?

If we use HBz to represent benzoic acid, we can write

$$1\ mol\ HBz = 122.1\ g\ HBz$$

$$amount\ HBz = 2.00\ g\ HBz \times \frac{1\ mol\ HBz}{122.1\ g\ HBz} = 0.0164\ mol\ HBz$$

To obtain the number of millimoles, we divide by the milliformula weight:

$$\text{amount HBz} = 2.00 \text{ g HBz} \times \frac{1 \text{ mmol HBz}}{0.1221 \text{ g HBz}} = 16.4 \text{ mmol HBz}$$

Remember that the chemical units for the amount of a species are moles and millimoles.

Example 4–2

What weight of $Na^+$ (fw = 23.00 g) is contained in 25.00 g of $Na_2SO_4$ (fw = 142.0 g)?

   We first calculate the number of moles of $Na_2SO_4$:

$$\text{amount Na}_2\text{SO}_4 = 25.00 \text{ g Na}_2\text{SO}_4 \times \frac{1 \text{ mol Na}_2\text{SO}_4}{142.0 \text{ g Na}_2\text{SO}_4}$$

$$= 0.17606 \text{ mol Na}_2\text{SO}_4$$

Since 1 mol of $Na_2SO_4$ contains 2 mol of $Na^+$,

$$\text{amount Na}^+ = \frac{2 \text{ mol Na}^+}{\text{mol Na}_2\text{SO}_4} \times 0.17606 \text{ mol Na}_2\text{SO}_4$$

$$= 0.35211 \text{ mol Na}^+$$

$$\text{wt Na}^+ = 0.35211 \text{ mol Na}^+ \times \frac{23.00 \text{ g Na}^+}{\text{mol Na}^+} = 8.099 \text{ g Na}^+$$

In doing calculations of this kind, you should always include all units, as we do throughout this chapter. This practice often reveals errors in setting up equations.

## 4A–5 Stoichiometric Calculations

A balanced chemical equation is a statement of the combining ratios, or stoichiometry (in units of moles), that exist between reacting substances and their products. Thus, the equation

$$2 \text{ NaI(aq)} + \text{Pb(NO}_3)_2\text{(aq)} \rightarrow \text{PbI}_2\text{(s)} + 2 \text{ NaNO}_3\text{(aq)}$$

indicates that 2 mol of aqueous sodium iodide combine with 1 mol of aqueous lead nitrate to produce 1 mol of solid lead iodide and 2 mol of aqueous sodium nitrate.[4]

   Example 4–3 demonstrates how the weight of reactants in a chemical reaction is related to the weight of the products. A calculation of this type is a three-step process involving (1) transformation of the known weight of a substance in grams to the corresponding number of moles, (2) multiplication by a factor that accounts for the stoichiometry, and (3) reconversion of the data in moles back to the metric units called for in the answer (see Figure 4–1).

The stoichiometry of a reaction is the relationship among the number of moles of reactants and products as shown by a balanced equation.

The physical state of substances in chemical equations is often indicated by the letters (g), (l), (s), and (aq), which refer to gaseous, liquid, solid, and aqueous solution states, respectively.

The relationship among the various metric units of weight is

$$10^{-3} \text{ kg} = 1 \text{ g}$$
$$= 10^3 \text{ mg}$$
$$= 10^6 \text{ μg (microgram)}$$
$$= 10^9 \text{ ng (nanogram)}$$
$$= 10^{12} \text{ pg (picogram)}$$

---

[4]Here it is advantageous to depict the reaction in terms of chemical compounds. If we wish to focus on reacting species, the *net ionic equation* is preferable:

$$2 \text{ I}^-\text{(aq)} + \text{Pb}^{2+}\text{(aq)} = \text{PbI}_2\text{(s)}$$

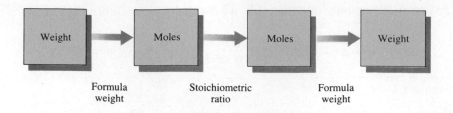

Figure 4–1

Flow diagram for making stoichiometric calculations.

---

Example 4–3

(a) What weight of $AgNO_3$ (fw = 169.9 g) is needed to convert 2.33 g of $Na_2CO_3$ (fw = 106.0 g) to $Ag_2CO_3$? (b) What weight of $Ag_2CO_3$ (fw = 275.7 g) is formed?

(a) $Na_2CO_3(aq) + 2\ AgNO_3(aq) \rightarrow Ag_2CO_3(s) + 2\ NaNO_3(aq)$

Step 1: amount $Na_2CO_3$ = 2.33 g $Na_2CO_3$ × $\dfrac{1\ \text{mol } Na_2CO_3}{106.0\ \text{g } Na_2CO_3}$

$= 0.02198\ \text{mol } Na_2CO_3$

Step 2: The balanced equation reveals that

amount $AgNO_3$ = 0.02198 mol $Na_2CO_3$ × $\dfrac{2\ \text{mol } AgNO_3}{1\ \text{mol } Na_2CO_3}$

$= 0.04396\ \text{mol } AgNO_3$

Step 3: wt $AgNO_3$ = 0.04396 mol $AgNO_3$ × $\dfrac{169.9\ \text{g } AgNO_3}{\text{mol } AgNO_3}$

$= 7.47\ \text{g } AgNO_3$

(b) no. mol $Ag_2CO_3$ = no. mol $Na_2CO_3$ = 0.02198 mol

wt $Ag_2CO_3$ = 0.02198 mol $Ag_2CO_3$ × $\dfrac{275.7\ \text{g } Ag_2CO_3}{\text{mol } Ag_2CO_3}$

$= 6.06\ \text{g } Ag_2CO_3$

---

## 4B CALCULATION OF RESULTS FROM GRAVIMETRIC DATA

The results of a gravimetric analysis are generally computed from two experimental measurements: the weight of sample and the weight of a product of known composition. When the product is the analyte, the concentration is given by the equation

$$\text{percent A} = \frac{\text{weight of A}}{\text{weight of sample}} \times 100\% \qquad (4\text{–}1)$$

where A represents the analyte. When the product is not the analyte, the numerator in Equation 4–1 is obtained by means of a calculation similar to that shown in Example 4–3.

---

Example 4–4

A 0.3516-g sample of a commercial phosphate detergent was ignited at a red heat to destroy the organic matter. The residue was then taken up in hot HCl, which converted the P to $H_3PO_4$. The phosphate was precipitated as $MgNH_4PO_4 \cdot 6H_2O$ by addition of $Mg^{2+}$ followed by aqueous $NH_3$. After being filtered and washed, the precipitate was converted to $Mg_2P_2O_7$ (fw = 222.57) by ignition at 1000°C. This residue weighed 0.2161 g. Calculate the percent P (fw = 30.974) in the sample.

The weight of analyte is

$$\text{wt P} = \text{g } Mg_2P_2O_7 \times \underbrace{\frac{1 \text{ mol } Mg_2P_2O_7}{222.57 \text{ g } Mg_2P_2O_7}}_{\substack{\text{Conversion}\\\text{factor}}} \times \underbrace{\frac{2 \text{ mol P}}{1 \text{ mol } Mg_2P_2O_7}}_{\substack{\text{Stoichiometric}\\\text{factor}}} \times \underbrace{\frac{30.974 \text{ g P}}{\text{mol P}}}_{\substack{\text{Conversion}\\\text{factor}}}$$

Metric quantity

$$= 0.2161 \times \frac{2 \times 30.974}{222.57} = 0.060147 \simeq 0.06015 \text{ g P}$$

Substituting into Equation 4–1 yields the percent analyte:

$$\text{percent P} = \frac{0.2161 \times \dfrac{2 \times 30.974}{222.57}}{0.3516} \times 100\% = 17.107 \simeq 17.11\%$$

Note that we carry guard digits for the formula weights of phosphorus and $Mg_2P_2O_7$, and then round at the end.

---

Example 4–5

An iron ore was analyzed by dissolving a 1.1324-g sample in concentrated HCl. The resulting solution was diluted with water, and the iron(III) was precipitated as the hydrous oxide $Fe_2O_3 \cdot xH_2O$ by the addition of $NH_3$. After filtration and washing, the residue was ignited at a high temperature to give 0.5394 g of pure $Fe_2O_3$ (fw = 159.69). Calculate (a) the percent Fe (fw = 55.847) and (b) the percent $Fe_3O_4$ (fw = 231.54) in the sample.

(a) $\text{wt Fe} = \text{g } Fe_2O_3 \times \dfrac{1 \text{ mol } Fe_2O_3}{159.69 \text{ g } Fe_2O_3} \times \dfrac{2 \text{ mol Fe}}{1 \text{ mol } Fe_2O_3} \times \dfrac{55.847 \text{ g Fe}}{\text{mol Fe}}$

$$= 0.5394 \times \frac{2 \times 55.847}{159.69} \text{ g}$$

$$\text{percent Fe} = \frac{0.5394 \times \dfrac{2 \times 55.847}{159.69} \text{ g}}{1.1324 \text{ g}} \times 100\% = 33.317 \simeq 33.32\%$$

(b) In this calculation, we assume that 3 mol of $Fe_2O_3$ is equivalent to 2 mol of $Fe_3O_4$:

$$3 \text{ Fe}_2O_3 \rightarrow 2 \text{ Fe}_3O_4 + \tfrac{1}{2} O_2$$

Then

$$\text{wt Fe}_3O_4 = \text{g Fe}_2O_3 \times \frac{1 \text{ mol Fe}_2O_3}{159.69 \text{ g Fe}_2O_3} \times \frac{2 \text{ mol Fe}_3O_4}{3 \text{ mol Fe}_2O_3} \times \frac{231.54 \text{ g Fe}_3O_4}{\text{mol Fe}_3O_4}$$

Note again that we carried five figures throughout the calculation and then rounded the answer to four.

$$\text{percent Fe}_3O_4 = \frac{0.5394 \times \dfrac{2 \times 231.54}{3 \times 159.69}}{1.1324} \times 100\% = 46.043 \approx 46.04\%$$

Note that the calculations in Examples 4–4 and 4–5 are similar in that the weight of analyte was obtained by multiplying the weight of the final product by a constant made up of the two conversion factors and the stoichiometric relationship between the analyte and the product weighed. This constant is sometimes called the *gravimetric factor* (GF). Thus, the gravimetric factor in Example 4–4 is

$$\text{GF} = \frac{2 \times \text{fw P}}{\text{fw Mg}_2P_2O_7} = \frac{2 \times 30.974}{222.57} = 0.27833$$

Similarly, the two gravimetric factors in Example 4–5 are

(a) $\text{GF} = \dfrac{2 \times \text{fw Fe}}{\text{fw Fe}_2O_3} = \dfrac{2 \times 55.847}{159.69} = 0.69944$

(b) $\text{GF} = \dfrac{2 \times \text{fw Fe}_3O_4}{3 \times \text{fw Fe}_2O_3} = \dfrac{2 \times 231.54}{3 \times 159.69} = 0.96662$

A general definition for the gravimetric factor is

$$\text{GF} = \frac{a}{b} \times \frac{\text{fw of substance sought}}{\text{fw of substance weighed}}$$

where $a$ and $b$ are small whole numbers that have values such that the number of formula weights in the numerator and the denominator are chemically equivalent.

A general equation for calculating the results of a gravimetric analysis is

$$\text{percent A} = \frac{\text{wt product} \times \text{GF}}{\text{wt sample}} \times 100\% \qquad (4-2)$$

Example 4–6

At elevated temperatures, $NaHCO_3$ is converted quantitatively to $Na_2CO_3$:

$$2 \text{ NaHCO}_3(s) = \text{Na}_2CO_3(s) + CO_2(g) + H_2O(g)$$

Ignition of a 0.3592-g sample containing $NaHCO_3$ and nonvolatile impurities yielded a residue weighing 0.2362 g. Calculate the percent purity of the sample.

The difference in weight before and after ignition represents the amount of $CO_2$ and $H_2O$ evolved from the $NaHCO_3$ in the sample. The equation for the reaction indicates that

$$2 \text{ mol } NaHCO_3 \equiv 1 \text{ mol } CO_2 + 1 \text{ mol } H_2O$$

Thus,

$$\text{percent } NaHCO_3 = \frac{(0.3592 - 0.2362) \times \dfrac{2 \times \text{fw } NaHCO_3}{\text{fw } CO_2 + \text{fw } H_2O}}{0.3592} \times 100\%$$

$$= \frac{0.1230 \times \dfrac{2 \times 84.01}{44.01 + 18.015}}{0.3592} \times 100\% = 92.76 \simeq 92.8\%$$

Note (1) that the denominator of the gravimetric factor in this example is the sum of the gram formula weights of the two volatile products and (2) that the combined weight of these products forms the basis for this analysis.

---

Example 4–7

A 0.2795-g sample of an organic mixture containing only $C_6H_6Cl_6$ (fw = 290.83) and $C_{14}H_9Cl_5$ (fw = 354.49) was burned in a stream of oxygen in a quartz tube. The products ($CO_2$, $H_2O$, and HCl) were passed through a solution of $NaHCO_3$. After acidification, the chloride in this solution yielded 0.7161 g of AgCl (fw = 143.32). Calculate the percent of each halogen compound in the sample.

There are two unknowns, and we must therefore develop two independent equations that can be solved simultaneously. One equation is

$$\text{wt } C_6H_6Cl_6 + \text{wt } C_{14}H_9Cl_5 = 0.2795 \text{ g}$$

A second equation is

$$\text{wt AgCl from } C_6H_6Cl_6 + \text{wt AgCl from } C_{14}H_9Cl_5 = 0.7161 \text{ g}$$

After we insert the appropriate gravimetric factors, the second equation becomes

$$\text{wt } C_6H_6Cl_6 \times \frac{6 \times \text{fw AgCl}}{\text{fw } C_6H_6Cl_6} + \text{wt } C_{14}H_9Cl_5 \times \frac{5 \times \text{fw AgCl}}{\text{fw } C_{14}H_9Cl_5} = 0.7161 \text{ g}$$

Substituting numerical values for the formula weights leads to

$$\text{wt } C_6H_6Cl_6 \times 2.9568 + \text{wt } C_{14}H_9Cl_5 \times 2.0215 = 0.7161$$

The first equation can be rewritten as

$$\text{wt } C_{14}H_9Cl_6 = 0.2795 - \text{wt } C_6H_6Cl_6$$

Substitution of this relationship into the previous equation leads to

$$2.9568 \times \text{wt } C_6H_6Cl_6 + 2.0215(0.2795 - \text{wt } C_6H_6Cl_6) = 0.7161$$

Thus,

$$\text{wt } C_6H_6Cl_6 = 0.16154 \text{ g}$$

$$\text{percent } C_6H_6Cl_6 = \frac{0.16154}{0.2795} \times 100\% = 57.80\%$$

$$\text{percent } C_{14}H_9Cl_5 = 100\% - 57.80\% = 42.20\%$$

---

## 4C    PROPERTIES OF PRECIPITATES AND PRECIPITATING REAGENTS

Ideally, a gravimetric precipitating agent (the *precipitant*) should react either *specifically* or, if not that, at least *selectively* with the analyte. Specific reagents, which are rare, react only with a single chemical species. Selective reagents, which are more common, react with a limited number of species. In addition to being either specific or selective, the ideal precipitating reagent should react with the analyte to give a product that is

1. readily filtered and washed free of contaminants.
2. of sufficiently low solubility so that no significant loss of the analyte occurs during filtration and washing.
3. unreactive with constituents of the atmosphere.
4. of known composition after it is dried or, if necessary, ignited.

Few if any reagents produce precipitates that possess all these desirable properties.

Variables that influence solubility are discussed in Chapters 6, 7, and 8. In this section we consider methods for obtaining pure and easily filtered solids of known composition.[5]

### 4C–1 Particle Size and Filterability of Precipitates

Precipitates that consist of large particles are generally desirable in gravimetric work because large particles are easy to filter and wash free of impurities. In addition, such precipitates are usually purer than finely divided precipitates.

#### Factors That Determine the Particle Size of Precipitates

The particle size of solids formed by precipitation varies enormously. At one extreme are *colloidal suspensions,* whose tiny particles are invisible to the naked eye ($10^{-7}$ to $10^{-4}$ cm in diameter). Colloidal particles show no

Dimethylglyoxime (Section 4E–3) precipitates only $Ni^{2+}$ and is thus a specific reagent. Silver nitrate, in contrast, is selective. The only common ions it precipitates from acidic solution are $Cl^-$, $Br^-$, $I^-$, and $SCN^-$.

In diffuse light, colloidal suspensions may be perfectly clear and appear to contain no solid. The presence of the second phase can be detected, however, by shining the beam of a flashlight into the solution. Because particles of colloidal dimensions scatter visible radiation, the path of the beam through the solution can be seen by the eye. This phenomenon is called the *Tyndall effect.*

---

[5]For a more detailed treatment of precipitates, see H. A. Laitinen and W. E. Harris, *Chemical Analysis,* 2nd ed., Chapters 8 and 9. New York: McGraw-Hill, 1975; A. E. Nielsen, in *Treatise on Analytical Chemistry,* 2nd ed., I. M. Kolthoff and P. J. Elving, Eds., Part I, Vol. 3, Chapter 27. New York: Wiley, 1983.

tendency to settle from solution, nor are they easily filtered. At the other extreme are particles with dimensions on the order of tenths of a millimeter or greater. The temporary dispersion of such particles in a liquid phase is called a *crystalline suspension*. The particles of a crystalline suspension tend to settle spontaneously and are readily filtered.

Scientists have studied precipitate formation for many years, but the mechanism of the process is still not fully understood. It is certain, however, that particle size is influenced by such experimental variables as precipitate solubility, temperature, reactant concentrations, and rate at which reactants are mixed. The net effect of these variables can be accounted for, at least qualitatively, by assuming that the particle size is related to a single property of the system called its *relative supersaturation,* where

The particles of colloidal suspensions are not easily filtered. To trap these particles, the pore size of the filtering medium must be so small that filtrations take too long. With suitable treatment, however, the individual colloidal particles can be made to stick together and thus give a filterable mass.

$$\text{relative supersaturation} = \frac{Q - S}{S} \qquad (4\text{--}3)$$

In this equation, $Q$ is the concentration of the solute at any instant and $S$ is its equilibrium solubility.

Precipitation reactions are generally slow, so that even when a precipitating reagent is added drop by drop to a solution of an analyte, some supersaturation is likely. Experimental evidence indicates that the particle size of a precipitate varies inversely with the average relative supersaturation during the time the reagent is being introduced. Thus, when $(Q - S)/S$ is large, the precipitate tends to be colloidal; when $(Q - S)/S$ is small, a crystalline solid is more likely.

Equation 4–3 is known as the Von Weimarn equation in recognition of the scientist who proposed it in 1925.

To increase the particle size of a precipitate, minimize the relative supersaturation while the precipitate is being formed.

## Mechanism of Precipitate Formation

The effect of relative supersaturation on particle size can be explained if we assume that precipitates form in two ways: by *nucleation* and by *particle growth*. The particle size of a freshly formed precipitate is determined by which of these two precipitation routes is the faster.

In nucleation, a few ions, atoms, or molecules (perhaps as few as four or five) come together to form a stable solid. These nuclei often form on the surface of suspended solid contaminants, such as dust particles. Further precipitation then involves a competition between additional nucleation and growth on existing nuclei (particle growth). If nucleation predominates, the precipitate is made up of a large number of small particles; if growth predominates, the precipitate consists of a smaller number of larger particles.

The rate of nucleation is believed to increase exponentially with increasing relative supersaturation. In contrast, the rate of particle growth is only moderately enhanced by high relative supersaturation. Thus, when the relative supersaturation is high, nucleation is the major precipitation mechanism, and a large number of small particles are formed. When the relative supersaturation is low, the rate of particle growth tends to predominate, and deposition of solid on existing particles occurs to the exclusion of further nucleation; a crystalline suspension results.

Nucleation is a process in which a minimum number of atoms, ions, or molecules join together to give a stable solid.

Precipitates form by nucleation and by particle growth. When nucleation predominates, the result is a large number of very fine particles; when particle growth predominates, the result is a smaller number of larger particles.

## Experimental Control of Particle Size

Experimental variables that minimize supersaturation and thus lead to crystalline precipitates include elevated temperatures to increase the solubility of the precipitate ($S$ in Equation 4–3), dilute solutions (to minimize $Q$), and slow addition of the precipitating agent with good stirring. The last two measures also minimize the concentration of the solute ($Q$) at any given instant.

Larger particles can also be obtained by pH control, provided the solubility of the precipitate depends upon pH. For example, large, easily filtered crystals of calcium oxalate are obtained by forming the bulk of the precipitate in a mildly acidic environment, in which the salt is moderately soluble. The precipitation is then completed by slowly adding aqueous ammonia until the acidity is sufficiently low that essentially all the calcium oxalate is precipitated. The additional precipitate produced during this step forms on the solid particles generated in the first step.

Unfortunately, many precipitates cannot be formed as crystals under practical laboratory conditions. A colloidal solid is generally encountered when a precipitate has such a low solubility that $S$ in Equation 4–3 always remains negligible relative to $Q$. The relative supersaturation thus remains enormous throughout precipitate formation, and a colloidal suspension results. For example, under conditions feasible for an analysis, the hydrous oxides of iron(III), aluminum, and chromium(III) and the sulfides of most heavy-metal ions form only as colloids because of their very low solubilities.[6]

Precipitates with very low solubilities, such as many sulfides and hydrous oxides, generally form as colloids.

## 4C–2 Colloidal Precipitates

Colloidal suspensions are often stable for indefinite periods and are not usable for gravimetric analysis because their particles are too small to be readily filtered. Fortunately, the stability of most such suspensions can be decreased by heating, by stirring, and by adding an electrolyte. These measures cause the individual colloidal particles to bind together to form an amorphous mass that settles out of solution and is filterable. The process of converting a colloidal suspension into a filterable solid is called *coagulation* or *agglomeration*.

### Coagulation of Colloids

Colloidal suspensions are stable because the particles are either all positively charged or all negatively charged and thus repel one another. This charge results from cations or anions that are bound to the surface of the particles. The process by which ions are retained *on the surface of a solid* is known as *adsorption*. We can readily demonstrate that colloidal parti-

Adsorption is a process in which a substance (gas, liquid, or solid) is held *on the surface* of a solid. In contrast, absorption involves retention of a substance *within* the pores of a solid.

[6]Silver chloride illustrates that the relative-supersaturation concept is imperfect. This compound ordinarily forms as a colloid, and yet its molar solubility is not significantly different from that of other compounds, such as $BaSO_4$, which generally form as crystals.

cles are charged by observing their migration when placed in an electrical field.

The tendency of ions to be adsorbed on an ionic solid surface originates in the normal bonding forces that are responsible for crystal growth. For example, a silver ion at the surface of a silver chloride particle has a partially unsatisfied bonding capacity for anions because of its surface location. Negative ions are attracted to this site by the same forces that hold chloride ions in the silver chloride lattice. In the same way, chloride ions at the surface of the solid attract cations dissolved in the solvent.

The kind and number of ions retained on the surface of a colloidal particle depend, in a complex way, on several variables. For a suspension produced in the course of a gravimetric analysis, however, the species adsorbed, and hence the charge on the particles, is readily predicted because lattice ions are generally more strongly held than any other ions. For example, when sodium chloride is first added to a solution containing silver nitrate, the colloidal particles of silver chloride formed are positively charged, as shown in Figure 4–2. This charge is due to adsorption of some of the excess silver ions from the solution. The charge on the particles becomes negative, however, when enough sodium chloride has been added to provide an excess of chloride ions. Now, the adsorbed species is primarily chloride ions. The surface charge is at a minimum when the supernatant liquid contains an excess of neither common ion.

The extent of adsorption, and thus the charge on a given particle, increase rapidly as the concentration of the lattice ion becomes greater. Eventually, however, the surface of the particles becomes covered with the adsorbed ions, and the charge then becomes constant and independent of concentration.

Figure 4–2 shows a colloidal silver chloride particle in a solution that contains an excess of silver nitrate. Attached directly to the solid surface is the *primary adsorption layer,* which consists mainly of adsorbed silver ions. Surrounding the charged particle is a layer of solution, called *the counter-ion layer,* which contains an excess of negative ions (principally nitrate). The primarily adsorbed silver ions and the negative counter-ion layer constitute an *electric double layer* that stabilizes a colloidal suspension by preventing individual particles from coming close enough to one another to agglomerate.

In Figure 4–3a, the effective charge of a colloidal silver chloride particle in a silver nitrate solution is plotted as a function of distance from its surface. The effective charge can be thought of as a measure of the repulsive force the particle exerts on like particles in the solution. Note that the effective charge falls off rapidly as the distance from the surface increases and approaches zero at the point $d_1$. This decrease in effective charge (in this case positive) is caused by the negative charge of the excess counter ions in the double layer. At point $d_1$, the number of counter ions in the layer is approximately equal to the number of primarily adsorbed ions on the surface of the particle so that the effective charge of the particle approaches zero.

> The charge on a colloidal particle formed in a precipitation is determined by the charge of the lattice ion that is in excess when the precipitation is complete.

> The electric double layer of a colloid consists of a layer of charge adsorbed on the surface of the particles and a layer of opposite charge in the solution surrounding the particles.

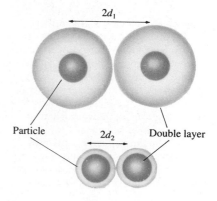

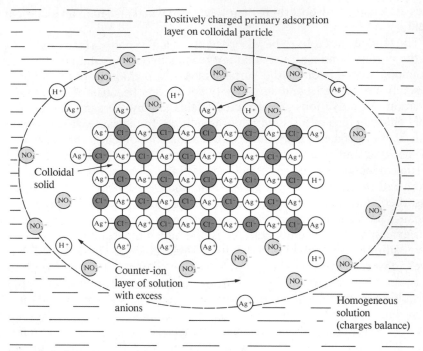

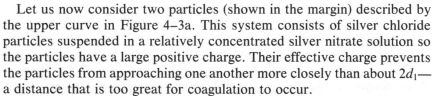

Figure 4–2

A colloidal silver chloride particle suspended in a solution of silver nitrate

Theodore W. Richards (1868–1928) and his graduate students at Harvard University developed or refined many of the techniques of gravimetric analysis involving silver and chlorine. These techniques were used to determine the atomic weights of 25 of the elements by preparing pure samples of the chlorides of the elements, decomposing known weights of these compounds, and determining their chloride content by gravimetric methods. For this work, Richards became the first American to receive the Nobel Prize in Chemistry (1914).

Let us now consider two particles (shown in the margin) described by the upper curve in Figure 4–3a. This system consists of silver chloride particles suspended in a relatively concentrated silver nitrate solution so the particles have a large positive charge. Their effective charge prevents the particles from approaching one another more closely than about $2d_1$— a distance that is too great for coagulation to occur.

The addition of more sodium chloride to this solution decreases the silver ion concentration and also the number of silver ions adsorbed on each particle. As shown in the lower of the two curves, the effective charge is now much less and extends a shorter distance into the solution. Now the particles can approach within $2d_2$ of one another. With further additions of chloride ion, this distance becomes small enough so that the forces of agglomeration can take over, and a coagulated precipitate begins to appear.

Coagulation of a colloidal suspension can often be brought about by a short period of heating, particularly if accompanied by stirring. Heating decreases the number of adsorbed ions and thus the thickness $d_1$ of the double layer. The particles may also gain enough kinetic energy at the higher temperature to overcome the barrier to close approach posed by the double layer.

An even more effective way to coagulate a colloid is to increase the electrolyte concentration of the solution. If we add a suitable ionic compound to a suspension, the concentration of counter ions increases in the vicinity of each particle. As a result, the volume of solution that contains

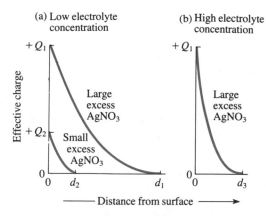

**Figure 4–3**
Effect of $AgNO_3$ and electrolyte concentration on the thickness of the double layer surrounding a colloidal AgCl particle in a solution containing excesses of $AgNO_3$.

sufficient counter ions to balance the charge of the primary adsorption layer decreases. The net effect of adding an electrolyte is thus a shrinkage of the counter-ion layer as shown in Figure 4–3b. The particles can then approach one another more closely and agglomerate.

Colloidal suspensions can often be coagulated by heating, by stirring, and by adding an electrolyte.

## Peptization of Colloids

*Peptization* refers to the process by which a coagulated colloid reverts to its original dispersed state. When a coagulated colloid is washed, some of the electrolyte responsible for its coagulation is leached from the internal liquid in contact with the solid. Removal of this electrolyte has the effect of increasing the volume of the counter-ion layer. The repulsive forces responsible for the original colloidal state are then reestablished, and particles detach themselves from the coagulated mass. The washings become cloudy as the freshly dispersed particles pass through the filter.

Peptization is a process by which a coagulated colloid is reconverted to a colloidal suspension.

The chemist is thus faced with a dilemma in working with coagulated colloids. On the one hand, washing is needed to minimize contamination; on the other, there is the risk of losses resulting from peptization if pure water is used as the wash liquid. The problem is often resolved by washing with a solution containing an electrolyte that volatilizes during the subsequent drying or ignition step. For example, silver chloride is ordinarily washed with a dilute solution of nitric acid. The precipitate undoubtedly becomes contaminated with the acid. No harm results, however, since the nitric acid is volatilized as the solid is dried.

Peptization is prevented by washing precipitates with a solution of a volatile electrolyte.

## Practical Treatment of Colloidal Precipitates

Colloids are best precipitated from hot, stirred solutions containing sufficient electrolyte to ensure coagulation. The filterability of a coagulated colloid frequently improves if the solid is allowed to stand for an hour or more in contact with the hot solution from which it was formed. During this process, which is known as *digestion*, weakly bound water appears to be lost from the precipitate; the result is a denser mass that is easier to filter.

Digestion is a process in which a precipitate is heated for an hour or more in the solution from which it was formed (the *mother liquor*).

## 4C–3 Crystalline Precipitates

Crystalline precipitates are generally more easily filtered and purified than are coagulated colloids. In addition, the size of individual crystalline particles, and thus their filterability, can be controlled to a degree.

### Methods of Improving Particle Size and Filterability

The particle size of crystalline solids can often be improved significantly by minimizing $Q$ and/or maximizing $S$ in Equation 4–3. Minimization of $Q$ is generally accomplished by using dilute solutions and adding the precipitating reagent slowly and with good mixing. Often $S$ is increased by precipitating from hot solution or by adjusting the pH of the precipitation medium.

Digestion of crystalline precipitates (without stirring) for some time after formation frequently yields a purer, more filterable product. The improvement in filterability undoubtedly results from the dissolution and recrystallization that occur continuously and at an enhanced rate at elevated temperatures. Recrystallization apparently results in bridging between adjacent particles, a process that yields larger and more easily filtered crystalline aggregates. This view is supported by the observation that little improvement in filtering characteristics occurs if the mixture is stirred during digestion.

*Digestion improves the purity and filterability of both colloidal and crystalline precipitates.*

## 4C–4 Coprecipitation

*Coprecipitation* is a phenomenon in which *otherwise soluble* compounds are removed from solution during precipitate formation. It is important to understand that contamination of a precipitate by a second substance whose solubility product has been exceeded *does not constitute coprecipitation*.

There are four types of coprecipitation: *surface adsorption, mixed-crystal formation, occlusion,* and *mechanical entrapment*.[7] Surface adsorption and mixed-crystal formation are equilibrium processes, whereas occlusion and mechanical entrapment result from the kinetics of crystal growth.

*Coprecipitation is a process whereby normally soluble compounds are carried out of solution by a precipitate.*

### Surface Adsorption

Adsorption is a common source of coprecipitation that is likely to cause significant contamination of precipitates with large specific surface areas—that is, coagulated colloids (see Feature 4–1 for the definition of

*Adsorption is often the major source of contamination in coagulated colloids but of no significance in crystalline precipitates.*

---

[7]Several classification systems have been suggested for coprecipitation phenomena. We follow the simple system proposed by A. E. Nielsen, in *Treatise on Analytical Chemistry*, 2nd ed., I. M. Kolthoff and P. J. Elving, Eds., Part I, Vol. 3, p. 333. New York: Wiley, 1983.

**Feature 4–1**

**THE SPECIFIC SURFACE AREA OF COLLOIDS**

*Specific surface area* is defined as the surface area per unit mass of solid and is ordinarily expressed in units of square centimeters per gram.

For a given mass of solid, specific surface area increases dramatically as particle size decreases, and therefore is enormous for colloids. For example, the solid cube shown in Figure 4–4 is 1 cm on a side and so has a surface area of 6 cm². If this cube weighs 2 g, its specific surface area is 6 cm²/2 g = 3 cm²/g. As shown in the lower part of Figure 4–4, this cube can be divided into 1000 cubes, each being 0.1 cm on a side. The surface area of each face of these cubes is now 0.1 cm × 0.1 cm = 0.01 cm², and the total surface area for the six faces of one of these 0.1-cm cubes is 0.06 cm². Because there are 1000 of these cubes, the total surface area for the 2 g of solid is now 60 cm²; the specific surface area is therefore 30 cm²/g.

Continuing in this way, we find that the specific surface area becomes 300 cm²/g when we have $10^6$ cubes that are 0.01 cm on a side. The particle size of a typical crystalline suspension lies in the region between 0.1 and 0.01 cm, and so a typical precipitate has a specific surface area between 30 and 300 cm²/g. Contrast these figures with that for 2 g of a colloid made up of $10^{18}$ particles with dimensions of $10^{-6}$ cm. Here, the specific surface area is $3 \times 10^6$ cm²/g, which converts to something over 3000 ft²/g. Thus 1 g of a colloidal suspension has a surface area that is equivalent to the floor area of a good-sized home.

Figure 4–4

Increase in surface area per unit mass with decrease in particle size.

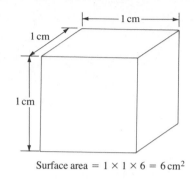

Surface area = 1 × 1 × 6 = 6 cm²

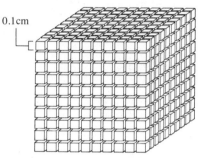

1000 cubes each 0.1 cm on a side
Surface area = 1000 × 0.1 × 0.1 × 6 = 60 cm²

specific surface area). Although adsorption does occur on crystalline solids, its effects on purity are usually undetectable because of the relatively small specific surface area of these solids.

Coagulation of a colloid does not significantly decrease the extent of adsorption because the coagulated solid still contains large internal surface areas that remain exposed to the solvent (Figure 4–5). The coprecipitated contaminant on the coagulated colloid consists of the lattice ion originally adsorbed on the surface before coagulation and the counter ion of opposite charge held in the film of solution immediately adjacent to the particle. *The net effect of surface adsorption is therefore the carrying down of an otherwise soluble compound as a surface contaminant.* For example, the coagulated silver chloride formed in the gravimetric determination of chloride ion is contaminated with primarily adsorbed silver ions and with nitrate or other anions in the counter-ion layer. As a consequence, silver nitrate, a normally soluble compound, is coprecipitated with the silver chloride.

**Methods for Minimizing Adsorbed Impurities on Colloids.** The purity of many coagulated colloids is improved by digestion. During this process,

> The specific surface area of a solid is the ratio of its surface area (cm²) to its mass (g).

> In adsorption, a normally soluble compound is carried out of solution on the surface of a coagulated colloid. This compound consists of the primarily adsorbed ion and an ion of opposite charge from the counter-ion layer.

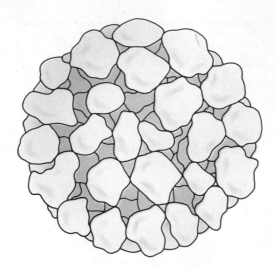

Figure 4–5
A coagulated colloid. Note the extensive *internal* surface area exposed to solvent.

Surface adsorption does not usually affect the purity of a crystalline precipitate because the specific surface area of the solid is small.

Calcium oxalate is often freed from coprecipitated magnesium oxalate by reprecipitation. Here the precipitated calcium oxalate is dissolved in dilute sulfuric acid. Ammonium oxalate is added to the acid solution, followed by ammonia, until a second precipitate of calcium oxalate forms. This precipitate is largely free of magnesium oxalate.

water is expelled from the solid to give a denser mass that has a smaller specific surface area for adsorption.

Washing a coagulated colloid will not remove many primarily adsorbed ions because attraction between these ions and the solid surface is too strong. Exchange does occur between existing *counter ions* and ions of like charge in the wash solution, however. It is thus advantageous to select a volatile electrolyte for this solution. For example, in the determination of silver by precipitation with chloride ion, the primarily adsorbed species is chloride. Washing with an acidic solution converts the counter-ion layer largely to hydrogen ions so that both chloride and hydrogen ions are retained by the solid and its counter-ion layer. Volatile HCl is then given off when the precipitate is dried.

Regardless of the method of treatment, a coagulated colloid is always contaminated to some degree, even after extensive washing. The error introduced into the analysis from this source can be as low as 1 to 2 ppt, as in the coprecipitation of silver nitrate on silver chloride. The coprecipitation of heavy-metal hydroxides on the hydrous oxides of trivalent iron or aluminum, on the other hand, may cause errors amounting to several percent, which is frequently intolerable.

**Reprecipitation.**   A drastic but very effective way to minimize the effects of adsorption is *reprecipitation,* or *double precipitation.* Here, the filtered solid is redissolved and reprecipitated. The first precipitate ordinarily carries down only a fraction of the contaminant present in the dissolved sample. Thus, the solution containing the redissolved precipitate has a significantly lower contaminant concentration than the original, and even less adsorption occurs during the second precipitation. Reprecipitation adds substantially to the time required for an analysis but is often necessary for such precipitates as the hydrous oxides of iron(III) and aluminum, which have extraordinary tendencies to adsorb the hydroxides of heavy-metal cations, such as zinc, cadmium, and manganese.

## Mixed-Crystal Formation

In mixed-crystal formation, one of the ions in the crystal lattice of a solid is replaced by an ion of another element. For this exchange to occur, it is necessary that the two ions have the same charge and that their sizes differ by no more than about 5%. Furthermore, the two salts must belong to the same crystal class. Thus, for example, barium sulfate formed by adding barium chloride to a solution containing sulfate, lead, and acetate ions is found to be severely contaminated by lead sulfate even though acetate ions should prevent precipitation of lead sulfate by complexing the lead. Here, lead ions replace some of the barium ions in the barium sulfate crystals. Other examples of coprecipitation by mixed-crystal formation include $MgKPO_4$ in $MgNH_4PO_4$, $SrSO_4$ in $BaSO_4$, and MnS in CdS.

The extent of mixed-crystal contamination is governed by the law of mass action and increases as the ratio of contaminant concentration to analyte concentration increases. Mixed-crystal formation is a particularly troublesome type of coprecipitation because little can be done about it when certain combinations of ions are present in a sample matrix. This problem is encountered with both colloidal suspensions and crystalline precipitates. When mixed-crystal formation occurs, separation of the interfering ion may have to be carried out before the final precipitation step. Alternatively, a different precipitating reagent that does not give mixed crystals with the ions in question may be used.

Mixed crystal formation is a type of coprecipitation in which a contaminant ion replaces the analyte ion in the lattice of a crystalline precipitate.

Mixed-crystal formation may occur in both colloidal and crystalline precipitates, whereas occlusion and mechanical entrapment are confined to crystalline precipitates.

## Occlusion and Mechanical Entrapment

When a crystal is growing rapidly during precipitate formation, foreign ions in the counter-ion layer may become trapped, or *occluded,* within the growing crystal. The amount of occluded material is greatest in that part of a crystal that forms first because supersaturation, and thus growth rate, decrease as a precipitation progresses.

Mechanical entrapment occurs when crystals lie close together during growth. Here, several crystals grow together and in so doing trap a portion of the solution in a tiny pocket.

Both occlusion and mechanical entrapment are at a minimum when the rate of precipitate formation is low—that is, under conditions of low supersaturation. In addition, digestion is often remarkably helpful in decreasing these types of coprecipitation. Undoubtedly, the rapid solution and reprecipitation that goes on at the elevated temperature of digestion open up the pockets and allow the impurities to escape into the solution.

Occlusion is a type of coprecipitation in which an ion and its counter ion are trapped in a pocket of a rapidly growing crystalline precipitate.

## Coprecipitation Errors

Coprecipitated impurities may cause either negative or positive errors in an analysis. If the contaminant is not a compound of the analyte ion, positive errors always result. Thus, a positive error is observed whenever colloidal silver chloride adsorbs silver nitrate during a chloride analysis.

In contrast, if the contaminant does contain the ion being determined, either positive or negative errors may occur. For example, in the determi-

Coprecipitation can cause either negative or positive errors in a gravimetric analysis.

nation of barium by precipitation as barium sulfate, occlusion of other barium salts occurs. If the occluded contaminant is barium nitrate, a positive error is observed because this compound has a larger formula weight than the barium sulfate that would have formed had no coprecipitation taken place. If barium chloride is the contaminant, the error is negative because its formula weight is less than that of the sulfate salt.

## 4C–5 Precipitation from Homogeneous Solution

In a homogeneous precipitation, a precipitating agent is generated throughout a solution by a slow chemical reaction.

Precipitation from homogeneous solution is a technique in which a precipitating agent is generated in a solution of the analyte by a slow chemical reaction.[8] Local reagent excesses do not occur because the precipitating agent appears gradually and homogeneously throughout the solution and reacts immediately with the analyte. As a result, the relative supersaturation is kept low during the entire precipitation. In general, homogeneously formed precipitates, both colloidal and crystalline, are better suited to analysis than are solids formed by direct addition of a precipitating reagent.

Urea is often used for the homogeneous generation of hydroxide ion. The reaction can be expressed by the equation

$$(H_2N)_2CO + 3\ H_2O \rightarrow CO_2 + 2\ NH_4^+ + 2\ OH^-$$

Solids formed by homogeneous precipitation are generally purer and more easily filtered than precipitates generated by direct addition of a reagent to the analyte solution.

This reaction proceeds slowly just below 100°C, and a 1- to 2-hour heating period is needed to complete a typical precipitation. Urea is particularly valuable for the precipitation of hydrous oxides or basic salts. For example, hydrous oxides of iron(III) and aluminum formed by direct addition of base are bulky and gelatinous masses that are heavily contaminated and difficult to filter. In contrast, when these same products are produced by homogeneous generation of hydroxide ion, they are dense, readily filtered, and have considerably higher purity. Figure 4–6 shows hydrous oxide precipitates of aluminum formed by direct addition of base and by homogeneous precipitation with urea. Similar pictures for hydrous oxides of iron(III) are shown in the color plates. Homogeneous precipitation of crystalline precipitates also results in marked increases in crystal size as well as improvements in purity.

Representative methods based upon precipitation by homogeneously generated reagents are given in Table 4–1.

## 4C–6 Drying and Ignition of Precipitates

After filtration, a gravimetric precipitate is heated until its weight becomes constant. Heating removes the solvent and any volatile species carried down with the precipitate. Some precipitates must be ignited to decompose the solid and form a compound of known composition. This new compound is often called the *weighing form*.

---

[8]For a general reference on this technique, see L. Gordon, M. L. Salutsky, and H. H. Willard, *Precipitation from Homogeneous Solution*. New York: Wiley, 1959.

Figure 4–6

Hydrous aluminum oxide formed by (a) homogeneous generation of hydroxide ion and (b) direct addition of base.

Table 4–1

METHODS FOR THE HOMOGENEOUS GENERATION OF PRECIPITATING AGENTS

| Precipitating Agent | Reagent | Generation Reaction | Elements Precipitated |
|---|---|---|---|
| $OH^-$ | Urea | $(NH_2)_2CO + 3\ H_2O \rightarrow$ $CO_2 + 2\ NH_4^+ + 2\ OH^-$ | Al, Ga, Th, Bi, Fe, Sn |
| $PO_4^{3-}$ | Trimethyl phosphate | $(CH_3O)_3PO + 3\ H_2O \rightarrow$ $3\ CH_3OH + H_3PO_4$ | Zr, Hf |
| $C_2O_4^{2-}$ | Ethyl oxalate | $(C_2H_5)_2C_2O_4 + 2\ H_2O \rightarrow$ $2\ C_2H_5OH + H_2C_2O_4$ | Mg, Zn, Ca |
| $SO_4^{2-}$ | Dimethyl sulfate | $(CH_3O)_2SO_2 + 4\ H_2O \rightarrow$ $2\ CH_3OH + SO_4^{2-} + 2\ H_3O^+$ | Ba, Ca, Sr, Pb |
| $CO_3^{2-}$ | Trichloroacetic acid | $Cl_3CCOOH + 2\ OH^- \rightarrow$ $CHCl_3 + CO_3^{2-} + H_2O$ | La, Ba, Ra |
| $H_2S$ | Thioacetamide* | $CH_3CSNH_2 + H_2O \rightarrow$ $CH_3CONH_2 + H_2S$ | Sb, Mo, Cu, Cd |
| DMG† | Biacetyl + hydroxylamine | $CH_3COCOCH_3 +$ $2\ H_2NOH \rightarrow DMG + 2\ H_2O$ | Ni |
| HOQ‡ | 8-Acetoxyquinoline§ | $CH_3COOQ + H_2O \rightarrow$ $CH_3COOH + HOQ$ | Al, U, Mg, Zn |

*CH₃—C(=S)—NH₂

†DMG = Dimethylglyoxime = CH₃—C(=N—OH)—C(=N—OH)—CH₃

‡HOQ = 8-Hydroxyquinoline =

§CH₃—C(=O)—O—(quinoline)

Jöns Jacob Berzelius (1779–1848), considered the leading chemist of his time, developed much of the apparatus and many of the techniques of 19th-century analytical chemistry. Examples include the use of ashless filter paper in gravimetry, the use of hydrofluoric acid to decompose silicates, and the use of the metric system in weight determinations. He performed thousands of analyses of pure compounds to determine the atomic weights of most of the elements known then. Berzelius also developed our present system of symbols for elements and compounds.

The temperature required to dehydrate a precipitate completely may be as low as 100°C or as high as 1000°C.

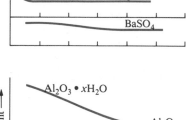

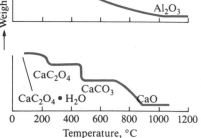

Figure 4–7

Effect of temperature on precipitate weight.

The temperature required to produce a suitable weighing form varies from precipitate to precipitate. Figure 4–7 shows weight loss as a function of temperature for several common analytical precipitates. These data were obtained with an automatic thermobalance,[9] an instrument that records the weight of a substance continuously as the temperature is increased at a constant rate in a furnace (Figure 4–8). Heating three of the precipitates—silver chloride, barium sulfate, and aluminum oxide—simply causes removal of water and perhaps volatile electrolytes. Note, however, the vastly different temperatures required to produce an anhydrous precipitate of constant weight. Moisture is completely removed from silver chloride above 110°C, but dehydration of aluminum oxide is not complete until a temperature greater than 1000°C is achieved. It is of interest to note that aluminum oxide formed homogeneously with urea can be completely dehydrated at about 650°C.

The thermal curve for calcium oxalate is considerably more complex than the others shown in Figure 4–7. Below about 135°C, unbound water is eliminated to give the monohydrate $CaC_2O_4 \cdot H_2O$. This compound is then converted to the anhydrous oxalate $CaC_2O_4$ at 225°C. The abrupt change in weight at about 450°C signals the decomposition of calcium oxalate to calcium carbonate and carbon monoxide. The final step in the curve depicts the conversion of the calcium carbonate to calcium oxide and carbon dioxide. It is evident that the compound finally weighed in a gravimetric calcium determination based upon precipitation as oxalate is highly dependent upon the ignition temperature.

## 4D    A CRITIQUE OF THE GRAVIMETRIC METHOD

Some chemists are inclined to discount the present-day value of gravimetric methods on the grounds that they are inefficient and obsolete. We, on the other hand, believe that the gravimetric approach to an analytical problem, like all others, has strengths and weaknesses and that ample situations exist where it represents the best possible choice for the resolution of an analytical problem.

### 4D–1 Time Required for a Gravimetric Analysis

Gravimetric methods are purported to be slower than other analytical procedures. This assertion is usually true if it is based upon the difference in clock time between the start of an analysis and its end. On the other hand, in terms of operator time this difference often disappears because much of the elapsed time in a gravimetric procedure involves operations such as drying, igniting, digesting, and evaporating, that require little or no attention of the operator. Based on operator time, then, the gravimetric approach for a particular analysis may prove to be the most efficient, especially where only one or two samples are to be analyzed, because

---

[9]For descriptions of thermobalances, see W. W. Wendlandt, *Thermal Methods of Analysis*, 3rd ed. New York: Wiley, 1986.

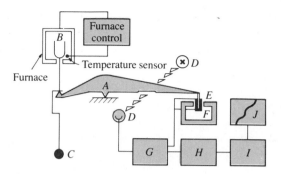

Figure 4–8

Schematic of a thermobalance: *A*, beam; *B*, sample cup and holder; *C*, counterweight; *D*, lamp and photodiodes; *E*, coil; *F*, magnet; *G*, control amplifier; *H*, tare calculator; *I*, amplifier; and *J*, recorder. (Courtesy of Mettler Instrument Corp., Hightstown, NJ.)

there is no need for calibration or standardization. All other analytical methods, with the exception of coulometry (Section 18D), require preparation of standard solutions that are then employed to either derive an empirical calibration curve or titrate a solution of the analyte. By contrast, a gravimetric method requires only the calculation of a gravimetric factor from data in a table of atomic weights. Thus, when only a few samples are to be analyzed, a gravimetric method, with its freedom from calibration or standardization, often requires less operator time, on a per sample basis, than procedures that require calibration.

## 4D–2 Sensitivity and Accuracy of Gravimetric Methods

The sensitivity and accuracy of many methods are limited by the device used for the analytical measurement. Such a limitation seldom, if ever, affects gravimetric analyses. Thus, with a suitable balance it is perfectly feasible to obtain the weight of a few micrograms of material to within a few parts per thousand of its true value; for larger masses, the weighing uncertainty can be decreased to a few parts per million. Few other analytical instruments exhibit such high precision.

   The sensitivity and accuracy of a gravimetric method is usually limited by solubility losses, coprecipitation errors, and mechanical losses of precipitate, particularly when a small amount of precipitate must be isolated from a relatively large volume of solution that contains high concentrations of other constituents from the sample. Because of these problems, the chemist is wise to rule out a gravimetric method for a constituent if its concentration is likely to be below 0.1%; for simple samples containing more than 1% of the analyte, however, gravimetric procedures are seldom surpassed in accuracy. Here, errors may often be decreased to 1 or 2 ppt. With increasing sample complexity, larger errors are inevitable unless a great deal of time is expended in circumventing them. With this type of sample, the accuracy of a gravimetric method may be no better than, and may sometimes be poorer than, that of other analytical methods.

## 4D–3 Specificity of Gravimetric Methods

Gravimetric reagents are seldom specific but are instead *selective* in the sense that they tend to form precipitates with groups of ions. Each ion

Gravimetric methods do not require a calibration or standardization step (as do all other analytical procedures except coulometry) because the results are calculated directly from the experimental data and atomic weights. Thus, when only one or two samples are to be analyzed, a gravimetric approach may be the method of choice because it requires less time and effort than a procedure that requires preparation of standards and calibration.

within a group interferes with the determination of any other ion in the group unless a preliminary separation is performed. In general, gravimetric procedures are less specific than some methods to be considered in later chapters.

### 4D–4 Equipment for Gravimetric Methods

In contrast to many analytical methods, the equipment required for a gravimetric analysis is simple, relatively inexpensive, reliable, and easy to maintain.

## 4E    APPLICATIONS OF GRAVIMETRIC METHODS

Gravimetric methods have been developed for most inorganic anions and cations, as well as for such neutral species as water, sulfur dioxide, carbon dioxide, and iodine. A variety of organic substances can also be readily determined gravimetrically. Examples include lactose in milk products, salicylates in drug preparations, phenolphthalein in laxatives, nicotine in pesticides, cholesterol in cereals, and benzaldehyde in almond extracts. Indeed, gravimetric methods are among the most widely applicable of all analytical procedures.

### 4E–1 Inorganic Precipitating Agents

Table 4–2 lists common inorganic precipitating agents. These reagents typically form slightly soluble salts or hydrous oxides with the analyte. As you can see from the many entries for each reagent, most inorganic precipitants are not very selective.

### 4E–2 Reducing Agents

Table 4–3 lists several reagents that convert an analyte to its elemental form for weighing.

### 4E–3 Organic Precipitating Agents

Numerous organic reagents have been developed for the gravimetric determination of inorganic species. Some of these reagents are significantly more selective in their reactions than are most of the inorganic reagents listed in Table 4–2.

We encounter two types of organic reagents. One forms slightly soluble, nonionic products called *coordination compounds;* the other forms products in which the bonding between the inorganic species and the reagent is largely ionic.

Organic reagents that yield sparingly soluble coordination compounds typically contain at least two functional groups. Each of these groups is

| Table 4–3 SOME REDUCING AGENTS EMPLOYED IN GRAVIMETRIC METHODS | |
| --- | --- |
| Reducing Agent | Analyte |
| $SO_2$ | Se, Au |
| $SO_2 + H_2NOH$ | Te |
| $H_2NOH$ | Se |
| $H_2C_2O_4$ | Au |
| $H_2$ | Re, Ir |
| HCOOH | Pt |
| $NaNO_2$ | Au |
| $SnCl_2$ | Hg |
| Electrolytic reduction | Co, Ni, Cu, Zn, Ag, In, Sn, Sb, Cd, Re, Bi |

Table 4–2
## SOME INORGANIC PRECIPITATING AGENTS†

| Precipitating Agent | Element Precipitated‡ |
|---|---|
| $NH_3(aq)$ | **Be** ($BeO$), **Al** ($Al_2O_3$), **Sc** ($Sc_2O_3$), Cr ($Cr_2O_3$),* **Fe** ($Fe_2O_3$), Ga ($Ga_2O_3$), Zr ($ZrO_2$), **In** ($In_2O_3$), Sn ($SnO_2$), U ($U_3O_8$) |
| $H_2S$ | Cu ($CuO$),* **Zn** ($ZnO$, or $ZnSO_4$), **Ge** ($GeO_2$), As ($\underline{As_2O_3}$, or $As_2O_5$), Mo ($MoO_3$), Sn ($SnO_2$),* Sb ($\underline{Sb_2O_3}$, or $\underline{Sb_2O_5}$), Bi ($Bi_2S_3$) |
| $(NH_4)_2S$ | Hg ($\underline{HgS}$), Co ($Co_3O_4$) |
| $(NH_4)_2HPO_4$ | **Mg** ($Mg_2P_2O_7$), Al ($AlPO_4$), Mn ($Mn_2P_2O_7$), Zn ($Zn_2P_2O_7$), Zr ($ZrP_2O_7$), Cd ($Cd_2P_2O_7$), Bi ($BiPO_4$) |
| $H_2SO_4$ | Li, Mn, **Sr, Cd, Pb, Ba** (all as sulfates) |
| $H_2PtCl_6$ | K ($K_2PtCl_6$, or Pt), Rb ($\underline{Rb_2PtCl_6}$), Cs ($\underline{Cs_2PtCl_6}$) |
| $H_2C_2O_4$ | Ca ($CaO$), Sr ($SrO$), **Th** ($\underline{ThO_2}$) |
| $(NH_4)_2MoO_4$ | Cd ($CdMoO_4$),* Pb ($\underline{PbMoO_4}$) |
| HCl | **Ag** ($AgCl$), Hg ($Hg_2Cl_2$), Na (as $NaCl$ from butyl alcohol), Si ($SiO_2$) |
| $AgNO_3$ | **Cl** ($\underline{AgCl}$), Br ($\underline{AgBr}$), $\underline{I(AgI)}$ |
| $(NH_4)_2CO_3$ | **Bi** ($\underline{Bi_2O_3}$) |
| $NH_4SCN$ | Cu [$Cu_2(SCN)_2$] |
| $NaHCO_3$ | Ru, Os, Ir (precipitated as hydrous oxides; reduced with $H_2$ to metallic state) |
| $HNO_3$ | Sn ($SnO_2$) |
| $H_5IO_6$ | Hg [$Hg_5(IO_6)_2$] |
| NaCl, $Pb(NO_3)_2$ | F ($PbClF$) |
| $BaCl_2$ | $SO_4^{2-}$ ($BaSO_4$) |
| $MgCl_2$, $NH_4Cl$ | $PO_4^{3-}$ ($Mg_2P_2O_7$) |

†From W. F. Hillebrand, G. E. F. Lundell, H. A. Bright, and J. I. Hoffman, *Applied Inorganic Analysis*. New York: Wiley, 1953. By permission of John Wiley & Sons, Inc.

‡Boldface type indicates that gravimetric analysis is the preferred method for the element or ion. The weighed form is indicated in parentheses. An asterisk indicates that the gravimetric method is seldom used. An underscore indicates the most reliable gravimetric method.

capable of bonding with a cation by sharing a pair of electrons. The functional groups are located in the molecule such that a five- or six-membered ring results from the reaction. Reagents that form compounds of this type are called *chelating agents,* and their products are called *chelates.*

Metal chelates are relatively nonpolar and, as a consequence, have solubilities that are low in water but high in organic liquids. Usually these compounds possess low densities and are often intensely colored. Because they are not wetted by water, coordination compounds are readily freed of moisture at low temperatures. Two widely used chelating reagents are described in the paragraphs that follow.

## 8-Hydroxyquinoline

Approximately two dozen cations form sparingly soluble chelates with 8-hydroxyquinoline. The structure of magnesium 8-hydroxyquinolate is

Chelates are cyclical metal-organic compounds in which the metal is a part of one or more five- or six-membered rings. The chelate pictured below is heme, which is a part of hemoglobin, the oxygen-carrying molecule in human blood.

Notice the four six-membered rings that are formed with $Fe^{2+}$.

typical of these chelates:

8-Hydroxyquinoline is sometimes called *oxine*.

$$2 \text{(oxine)} + Mg^{2+} \rightarrow \text{(chelate)} + 2H^+$$

The solubilities of metal 8-hydroxyquinolates vary widely from cation to cation and are pH-dependent because 8-hydroxyquinoline is always deprotonated during chelation. Therefore, we can achieve a considerable degree of selectivity in the use of 8-hydroxyquinoline by controlling pH.

### Dimethylglyoxime

Nickel dimethylglyoxime is spectacular in appearance. It has a beautiful and vivid red color.

Dimethylglyoxime is an organic precipitating agent of unparalleled specificity. Only nickel(II) is precipitated from a weakly alkaline solution. The reaction is

$$2 \text{(dimethylglyoxime)} + Ni^{2+} \rightarrow \text{(chelate)} + 2H^+$$

Creeping is the process by which a precipitate (usually a metal-organic chelate) moves up the sides of a wetted surface of a glass container or a filter paper.

This precipitate is so bulky that only small amounts of nickel can be handled conveniently. It also has an exasperating tendency to creep up the sides of the container as it is filtered and washed. The solid is readily dried at 110°C and has the composition indicated by its formula.

### Sodium Tetraphenylboron

Potassium tetraphenyl boron is a salt rather than a chelate.

Sodium tetraphenylboron, $(C_6H_5)_4B^-Na^+$, is an important example of an organic precipitating reagent that forms salt-like precipitates. In cold mineral-acid solutions, it is a near-specific precipitating agent for potassium and ammonium ions. The composition of the precipitates is stoichiometric and contains one mole of potassium or ammonium ion for each mole of tetraphenylboron ion; these ionic compounds are readily filtered and can be brought to constant weight at 105 to 120°C. Only mercury(II), rubidium, and cesium interfere and must be removed by prior treatment.

## 4E–4 Organic Functional-Group Analysis

Several reagents react selectively with certain organic functional groups and thus can be used for the determination of most compounds containing these groups. A list of gravimetric functional-group reagents is given in Table 4–4. Many of the reactions shown can also be used for volumetric and spectrophotometric determinations.

## 4E–5 Volatilization Methods

The two most common gravimetric methods based on volatilization are those for water and carbon dioxide.

Water is quantitatively eliminated from many inorganic samples by ignition. In its direct determination, water is collected on any of several solid desiccants, and its mass is then determined from the weight gain of the desiccant.

The indirect method, in which the amount of water is determined by the loss of weight of the sample during heating, is less satisfactory because it must be assumed that water is the only component volatilized. This as-

**Table 4–4**
**GRAVIMETRIC METHODS FOR ORGANIC FUNCTIONAL GROUPS**

| Functional Group | Basis for Method | Reaction and Product Weighed* |
|---|---|---|
| Carbonyl | Weight of precipitate with 2,4-dinitrophenyl-hydrazine | $RCHO + H_2NNHC_6H_3(NO_2)_2 \rightarrow$ <br> $\underline{R\text{—}CH = NNHC_6H_3(NO_2)_2}(s) + H_2O$ (RCOR′ reacts similarly) |
| Aromatic carbonyl | Weight of $CO_2$ formed at 230°C in quinoline; $CO_2$ distilled, absorbed, and weighed | $ArCHO \xrightarrow[CuCO_3]{230°C} Ar + \underline{CO_2}(g)$ |
| Methoxyl and ethoxyl | Weight of AgI formed after distillation and decomposition of $CH_3I$ or $C_2H_5I$ | $ROCH_3\ \ \ + HI \rightarrow ROH\ \ \ \ + CH_3I$ <br> $RCOOCH_3 + HI \rightarrow RCOOH + CH_3I$ $\Big\}$ $CH_3I + Ag^+ + H_2O \rightarrow$ <br> $ROC_2H_5\ \ \ + HI \rightarrow ROH\ \ \ \ + C_2H_5I$  $\underline{AgI}(s) + CH_3OH$ |
| Aromatic nitro | Weight loss of Sn | $RNO_2 + \frac{3}{2} Sn(s) + 6 H^+ \rightarrow RNH_2 + \frac{3}{2} Sn^{4+} + 2 H_2O$ |
| Azo | Weight loss of Cu | $RN = NR' + \underline{2\ Cu}(s) + 4 H^+ \rightarrow RNH_2 + R'NH_2 + 2 Cu^{2+}$ |
| Phosphate | Weight of Ba salt | $ROP(OH)_2 + Ba^{2+} \rightarrow ROPO_2Ba(s) + 2 H^+$ (with O double bonds on P) |
| Sulfamic acid | Weight of $BaSO_4$ after oxidation with $HNO_2$ | $RNHSO_3H + HNO_2 + Ba^{2+} \rightarrow ROH + \underline{BaSO_4}(s) + N_2 + 2 H^+$ |
| Sulfinic acid | Weight of $Fe_2O_3$ after ignition of Fe(III) sulfinate | $3\ ROSOH + Fe^{3+} \rightarrow (ROSO)_3Fe(s) + 3 H^+$ <br> $(ROSO)_3Fe \xrightarrow{O_2} CO_2 + H_2O + SO_2 + \underline{Fe_2O_3}(s)$ |

*The substance weighed is underlined.

sumption is frequently unjustified because heating of many substances results in their decomposition and a consequent change in weight, irrespective of the presence of water. Nevertheless, the indirect method has found wide use for the determination of water in items of commerce. For example, a semiautomated instrument is available for the determination of moisture in cereal grains. It consists of a platform balance upon which a 10-g sample is heated with an infrared lamp. The percent residue is read directly.

Carbonates are ordinarily decomposed by acids to give carbon dioxide, which is readily evolved from solution by heat. As in the direct analysis for water, the weight of carbon dioxide is established from the increase in the weight of a solid absorbent. Ascarite II,[10] which consists of sodium hydroxide on a nonfibrous silicate, retains carbon dioxide by the reaction

$$2\ NaOH + CO_2 \rightarrow Na_2CO_3 + H_2O$$

The absorption tube must also contain a desiccant to prevent loss of the evolved water.

Sulfides and sulfites can also be determined by volatilization. Hydrogen sulfide or sulfur dioxide evolved from the sample after treatment with acid is collected in a suitable absorbent.

Finally, the classical method for the determination of carbon and hydrogen in organic compounds is a gravimetric procedure in which the combustion products ($H_2O$ and $CO_2$) are collected selectively on weighed absorbents. The increase in weight serves as the analytical parameter.

---

[10]®Thomas Scientific, Swedesboro, NJ.

---

## 4F　QUESTIONS AND PROBLEMS

4-1. Explain the difference between
　*(a) a colloidal and a crystalline precipitate.
　(b) specific and selective precipitating reagents.
　*(c) precipitation and coprecipitation.
　(d) peptization and coagulation.
　*(e) occlusion and mixed-crystal formation.
　(f) nucleation and particle growth.

4-2. Define
　*(a) digestion.
　(b) adsorption.
　*(c) reprecipitation.
　(d) precipitation from homogeneous solution.
　*(e) counter-ion layer.
　(f) mother liquor.
　*(g) relative supersaturation.

*4-3. What is peptization and how is it avoided?

4-4. How can the relative supersaturation be controlled during precipitate formation?

*4-5. After an excess of $AgNO_3$ has been added to an aqueous solution containing $NaNO_3$ and KSCN
　(a) what is the charge on the surface of the coagulated AgSCN?
　(b) what is the source of this charge?

　(c) what ions can be expected to predominate in the counter-ion layer?

4-6. Suggest a gravimetric method for the separation of $K^+$ from $Na^+$ and $Li^+$.

4-7. What are the structural characteristics of a chelating reagent?

4-8. Calculate the number of moles in 12.2 g of
　*(a) $H_2O$.
　(b) $Hg_2Cl_2$.
　*(c) $K_2SO_4$.
　(d) $Fe(NH_4)_2(SO_4)_2 \cdot 6H_2O$.
　*(e) $NaC_2H_3O_2$.

4-9. Calculate the number of grams in 4.63 mmol of
　*(a) FeO.　　　　　　　(d) $Fe_3O_4$.
　(b) $Fe(NH_4)_2(SO_4)_2 \cdot 6H_2O$.　*(e) $K_4Fe(CN)_6$.
　*(c) $Fe_2O_3$.

4-10. How many grams of iron are contained in the substances in Problem 4-9?

*4-11. A 0.8378-g sample of calcium oxalate is heated to 1000°C:

$$CaC_2O_4(s) \rightarrow CaO(s) + CO(g) + CO_2(g)$$

Calculate
(a) the moles of CaO remaining after ignition.
(b) the mmoles of CO evolved.
(c) the weight of $CO_2$ produced.

4–12. The Rammelsberg reaction involves the thermal decomposition of an alkaline-earth iodate to the corresponding paraperiodate. With strontium iodate, for example, the reaction is

$$5\ Sr(IO_3)_2(s) \rightarrow Sr_5(IO_6)_2(s) + 4\ I_2(g) + 9\ O_2(g)$$

When a 0.6961-g sample of $Sr(IO_3)_2$ (fw = 437.4) undergoes this reaction
(a) what weight of $Sr_5(IO_6)_2$ (fw = 883.9) is produced?
(b) how many millimoles of $O_2$ are produced?
(c) how many moles of $I_2$ are produced?

*4–13. What weight of $BaCl_2 \cdot 2\ H_2O$ is needed to react with 0.1503 g of
(a) $AgNO_3$? (product AgCl)
(b) KF? (product AgF)
(c) $MgSO_4$? (product $BaSO_4$)
(d) $H_3PO_4$? (product $Ba_3(PO_4)_2$)

*4–14. What weight of solid is produced in the reactions of Problem 4–13?

4–15. What minimum weight of reagent (see Table 4–1 for reactions) is needed for the homogeneous precipitation of
*(a) 0.283 g of Al(III) as $Al(OH)_3$, with urea (fw = 60.06)?
(b) 0.214 g of Zn(II) as $ZnC_2O_4$, with ethyl oxalate (fw = 146.14)?
*(c) 0.401 g of Cd(II) as CdS, with thioacetamide (fw = 75.13)?
(d) 0.364 g of Cu(II) as $Cu_2S$, with thioacetamide (fw = 75.13)?
*(e) 0.297 g of Ag(I) as $Ag_2CO_3$, with trichloroacetic acid (fw = 163.39)?
(f) 0.305 g of Zn(II) as $Zn_2P_2O_7$, with trimethyl phosphate (fw = 140.08)?

4–16. Use chemical symbols to generate the factor needed to convert the weight of the substance in the left column into the corresponding weight of the substance in the right column.

| | Weighed | Sought |
|---|---|---|
| *(a) | $CaC_2O_4$ | $CaCO_3$ |
| (b) | $BaCrO_4$ | $K_2Cr_2O_7$ |
| *(c) | $PbI_2$ | $NaIO_3$ |
| (d) | $Fe_2O_3$ | $K_3Fe(CN)_6$ |
| *(e) | $B_2O_3$ | $Na_2B_4O_7$ |
| (f) | $Mn_3O_4$ | $KMnO_4$ |
| *(g) | $Ba(IO_3)_2$ | KI |
| (h) | $Mg(C_9H_6ON)_2$ | $MgSO_4$ |
| *(i) | $Hg_2I_2$ | HgO |
| (j) | $Ba_3(PO_4)_2$ | $H_3PO_4$ |

4–17. Calculate the weight of $PbI_2$ that can be produced from

*(a) 0.3010 g of $Pb(NO_3)_2$.
(b) 0.3010 g of KI.
*(c) 0.5379 g of a sample that is 64.2% $Pb(NO_3)_2$.
(d) 0.7263 g of a sample that assays 44.9% $Pb(NO_3)_2$.

*4–18. Calculate the weight of silver chloride produced when 0.364 g of AgI is heated in a stream of chlorine. Reaction:

$$2\ AgI(s) + Cl_2(g) \rightarrow 2\ AgCl(s) + I_2(g)$$

4–19. Calculate the weight of calcium oxide that is produced when 3.164 g of $CaC_2O_4$ are strongly ignited. Reaction:

$$CaC_2O_4(s) \rightarrow CaO(s) + CO(g) + CO_2(g)$$

*4–20. Treatment of a 0.4000-g sample of impure potassium chloride with an excess of $AgNO_3$ resulted in the formation of 0.7332 g of AgCl. Calculate the percentage of KCl in the sample.

4–21. After appropriate preliminary treatment, the thallium in a 10.20-g pesticide sample was precipitated as TlI and was weighed as such. Calculate the percentage of $Tl_2SO_4$ (fw = 504.80) in the sample if 0.1964 g of TlI (fw = 331.27) was recovered.

4–22. The aluminum in a 1.200-g sample of impure ammonium aluminum sulfate was precipitated with aqueous ammonia as the hydrous $Al_2O_3 \cdot xH_2O$. The precipitate was filtered and ignited at 1000°C to give anhydrous $Al_2O_3$ which weighed 0.1798 g. Express the result of this analysis in terms of
*(a) % $NH_4Al(SO_4)_2$.
(b) % $Al_2O_3$.
(c) % Al.

4–23. A 0.7406-g sample of impure magnesite, $MgCO_3$, was decomposed with HCl; the liberated $CO_2$ was collected on calcium oxide and found to weigh 0.1881 g. Calculate the percentage of magnesium in the sample.

*4–24. The hydrogen sulfide in a 50.0-g sample of crude petroleum was removed by distillation and collected in a solution of $CdCl_2$. The precipitated CdS was then filtered, washed, and ignited to $CdSO_4$. Calculate the percentage of $H_2S$ in the sample if 0.108 g of $CdSO_4$ was recovered.

4–25. A 5.000-g sample of a pesticide was decomposed with metallic sodium in alcohol, and the liberated chloride ion was precipitated as AgCl. Express the results of this analysis in terms of the percentage of 1,1,1-trichloro-2,2-bis-(parachlorophenyl)ethane DDT ($C_{14}H_9Cl_5$) (fw = 354.5) based upon the recovery of 0.1606 g of AgCl.

*4–26. The mercury in a 0.7152-g sample was precipitated with an excess of paraperiodic acid, $H_5IO_6$:

$$5\ Hg^{2+} + 2\ H_5IO_6 \rightarrow Hg_5(IO_6)_2(s) + 10\ H^+$$

The precipitate was filtered, washed free of precipitating agent, dried, and found to weigh 0.3408 g. Calculate the percentage of $Hg_2Cl_2$ in the sample.

4-27. Precipitates used in the gravimetric determination of uranium include $Na_2U_2O_7$ (fw = 634.0 g), $(UO_2)_2P_2O_7$ (fw = 714.0 g), and $V_2O_5 \cdot 2\ UO_3$ (fw = 753.9 g). Which of these weighing forms provides the smallest weight of precipitate from a given quantity of uranium?

*4-28. W. W. White and P. J. Murphy [*Anal. Chem.,* **51,** 1864 (1979)] report that the anionic complex between silver(I) and thiosulfate is quantitatively precipitated with hexaminecobalt(III) trichloride; the reaction is

$$Co(NH_3)_6^{3+} + Ag(S_2O_3)_2^{3-} \rightarrow$$
$$[Co(NH_3)_6][Ag(S_2O_3)_2](s)$$

The product (fw = 493.2) is dried to constant weight at 95°C. A 25.00-mL portion of a photographic fixer solution yielded 0.4161 g of precipitate when analyzed by this method. Calculate the grams of silver contained in each liter of this solution, assuming an excess of $S_2O_3^{2-}$ was present in the solution.

4-29. The nitrobenzene (fw = 123) in a 0.739-g sample was determined by dilution with an HCl/methanol mixture, followed by the introduction of 0.465 g of pure tin. The mixture was refluxed for an hour, during which time the nitrobenzene was reduced to aniline:

$$2\ C_6H_5NO_2 + 3\ Sn(s) + 12\ H^+ \rightarrow$$
$$2\ C_6H_5NH_2 + 4\ H_2O + 3\ Sn^{4+}$$

When the reaction was complete, the unused tin was isolated by filtration, dried, and found to weigh 0.128 g. Calculate the percentage of nitrobenzene in the sample.

*4-30. A series of sulfate samples is to be analyzed by precipitation as $BaSO_4$. If it is known that the sulfate content in these samples ranges between 20 and 55%, what minimum sample weight should be taken to ensure that a precipitate weight no smaller than 0.300 g is produced? What is the maximum precipitate weight to be expected if this quantity of sample is taken?

4-31. The addition of dimethylglyoxime, $H_2C_4H_6O_2N_2$, to a solution containing nickel(II) ion gives rise to a precipitate:

$$Ni^{2+} + 2\ H_2C_4H_6O_2N_2 \rightarrow$$
$$2\ H^+ + Ni(HC_4H_6O_2N_2)_2(s)$$

Nickel dimethylglyoxime is a bulky precipitate that is inconvenient to manipulate in amounts greater than 175 mg. The amount of nickel in a type of permanent-magnet alloy ranges between 24 and 35%. Calculate the sample size that should not be exceeded when analyzing this alloy for its nickel content.

4-32. The success of a particular catalyst is highly dependent upon its zirconium content. The starting material for this preparation is received in batches that assay between 68 and 84% $ZrCl_4$. Because it has been established that there are no sources of chloride ion other than the $ZrCl_4$ in the material, routine analysis based upon precipitation of AgCl is feasible.

(a) What sample weight should be taken to ensure a AgCl precipitate that weighs at least 0.400 g?

(b) If this sample weight is used, what is the maximum weight of AgCl that can be expected in this analysis?

(c) To simplify calculations, what sample weight should be taken in order to have the percentage of $ZrCl_4$ exceed the weight of AgCl produced by a factor of 100?

*4-33. Addition of an excess of $AgNO_3$ to a 0.5012-g sample yielded a mixture of AgCl and AgI that weighed 0.4715 g. The precipitate was then heated in a stream of $Cl_2$ to convert the AgI to AgCl

$$2\ AgI(s) + Cl_2(g) \rightarrow 2\ AgCl(s) + I_2(g)$$

The precipitate was found to weigh 0.3922 g after this treatment. Calculate the percentages of KI and $NH_4Cl$ in the sample.

4-34. A 1.461-g sample containing $K_2SO_4$, $NH_4NO_3$, and inert materials was dissolved in sufficient water to give exactly 250 mL of solution. A 25.0-mL aliquot of this solution yielded 0.2999 g of precipitate that consisted of $(C_6H_5)_4BK$ (fw = 358) and $(C_6H_5)_4BNH_4$ (fw = 337) upon being treated with an excess of sodium tetraphenylboron $(C_6H_5)_4BNa$. A 50.0-mL aliquot of the sample was made alkaline and warmed to drive off the ammonia

$$NH_4^+ + OH^- \rightarrow NH_3(g) + H_2O$$

following which treatment with $(C_6H_5)_4BNa$ gave 0.3230 of $(C_5H_5)_4BK$. Calculate the percentages of $K_2SO_4$ and $NH_4NO_3$ in the sample.

*4-35. A 1.008-g sample that contained $NH_4NO_3$, $(NH_4)_2SO_4$, and inert materials was dissolved in sufficient water to give exactly 500 mL of solution. A 50.0-mL aliquot of this solution yielded 0.3819 g of $(C_6H_5)_4BNH_4$ (fw = 337) when treated with an excess of sodium tetraphylboron $(C_6H_5)_4BNa$. A second 50.0-mL aliquot was made alkaline and heated with Devarda's alloy (50% Cu, 45% Al, 5% Zn) to convert the $NO_3^-$ to $NH_3$:

$$NO_3^- + 6\ H_2O + 8e^- \rightarrow NH_3 + 9\ OH^-$$

The $NH_3$ produced in this reaction, as well as that derived from $NH_4^+$, was distilled and collected in dilute acid. Treatment of the distillate with $(C_6H_5)_4BNa$ yielded 0.5996 g of $(C_6H_5)_4BNH_4$. Calculate the percentages of $(NH_4)_2SO_4$ and $NH_4NO_3$ in the sample.

**4-36.** Several alloys that contained only Ag and Cu were analyzed by dissolving weighed quantities in $HNO_3$, introducing an excess of $IO_3^-$, and bringing the filtered mixture of $AgIO_3$ and $Cu(IO_3)_2$ to constant weight. Use the accompanying data to calculate the percentage composition of the alloys.

|        | Wt of Sample, g | Wt of Precipitate, g |
|--------|-----------------|----------------------|
| *(a)   | 0.2175          | 0.7391               |
| (b)    | 0.1948          | 0.7225               |
| *(c)   | 0.2473          | 0.7443               |
| (d)    | 0.2386          | 0.9962               |
| *(e)   | 0.1864          | 0.8506               |

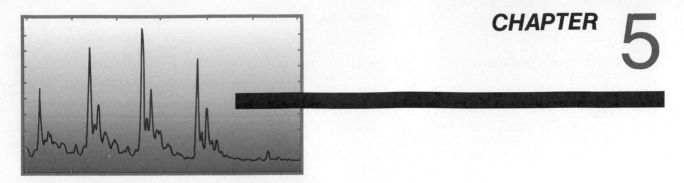

## CHAPTER 5

# TITRIMETRIC METHODS OF ANALYSIS

The three types of quantitative titrimetry are: volumetric, gravimetric, and coulometric. Volumetric is by far the most widely used.

Titrimetric methods are analytical procedures in which the amount of analyte is determined from the amount of a standard reagent required to react with the analyte completely.

Titrimetric methods constitute a large and powerful group of quantitative procedures that find widespread use in analytical chemistry. Volumetric methods represent one of three types of *titrimetry* in which analyses are based upon measuring the amount of a reagent of known concentration that is consumed by the analyte. The other two types are *weight* (or *gravimetric*) *titrimetry* and *coulometric titrimetry*. In volumetric titrimetry, the volume of a solution of known concentration that is needed to react essentially completely with the analyte is determined. Gravimetric titrimetry differs only in that the weight of the reagent is measured instead of its volume. In coulometric titrimetry, the "reagent" is a constant direct electrical current of known magnitude that reacts with the analyte; here, the time required to complete the electrochemical reaction is measured.

This chapter provides an introduction to volumetric methods. Chapters 9 through 16 cover the theory and applications of various types of volumetric titrimetry, although much of the material also applies to weight titrimetry. Coulometric titrimetry is considered in Section 18D–5. All these titrimetric methods are used for routine analyses because they are generally rapid, convenient, accurate, and readily automated.

## 5A SOME GENERAL ASPECTS OF VOLUMETRIC TITRIMETRY[1]

### 5A–1 Definition of Some Terms

A *standard solution* (or *standard titrant*) is a reagent of known concentration that is used to carry out a volumetric analysis. A *titration* is performed by slowly adding a standard solution from a buret or other volumetric measuring device to a solution of the analyte until the reaction

---

[1]For a detailed discussion of volumetric methods, see J. I. Watters, in *Treatise on Analytical Chemistry*, I. M. Kolthoff and P. J. Elving, Eds., Part I, Vol. 11, Chapter 114. New York: Wiley, 1975.

between the two is complete. The volume needed to complete the titration is determined from the difference between the initial and final buret readings.

The *equivalence point* in a titration is reached when the amount of added titrant is chemically equivalent to the amount of analyte in the sample. For example, the equivalence point in the titration of sodium chloride with silver nitrate occurs after exactly one mole of silver ion has been added for each mole of chloride ion in the sample. The equivalence point in the titration of sulfuric acid with sodium hydroxide is reached after the introduction of two moles of base for each mole of acid.

It is sometimes necessary to add an excess of the standard titrant and then determine the excess by *back-titration* with a second standard titrant. Here, the equivalence point corresponds to the point where the amount of initial titrant is chemically equivalent to the amount of analyte plus the amount of back-titrant.

> The equivalence point is the point in a titration when the amount of added standard reagent exactly equals the amount of analyte.

> Back-titrations are often required when the rate of reaction between the analyte and reagent is slow or when the reagent lacks stability.

## 5A–2 Equivalence Points and End Points

The equivalence point of a titration is a theoretical point that cannot be determined experimentally. Instead, we can only estimate it by observing some physical change associated with the condition of equivalence. This change is called the *end point* for the titration. Every effort is made to ensure that any volume difference between the equivalence point and the end point is small. Such differences do exist, however, as a result of inadequacies in the physical changes and in our ability to observe them. The difference in volume between the equivalence point and the end point is the *titration error*.

An *indicator* is often added to the analyte solution in order to give an observable physical change (the end point) at or near the equivalence point. We shall see that large changes in the relative concentrations of analyte and titrant occur in the equivalence-point region. These concentration changes cause the indicator to change in appearance. Typical indicator changes are the appearance or disappearance of a color, a change in color, and the appearance or disappearance of turbidity.

We often use instruments to detect end points. These instruments respond to certain properties of the solution that change in a characteristic way during the titration. Among such instruments are voltmeters, ammeters, and ohmmeters; colorimeters; temperature recorders; and refractometers.

> The end point is the point in a titration when a physical change occurs that is associated with the condition of chemical equivalence.

> All volumetric methods are based upon a primary standard whose chemical composition and purity are known exactly.

In volumetric methods, the titration error $E_t$ is given by

$$E_t = V_{ep} - V_{eq}$$

where $V_{ep}$ is the actual volume used to arrive at the end point, and $V_{eq}$ is the theoretical volume of reagent required to reach the equivalence point.

## 5A–3 Primary Standards

A *primary standard* is a highly purified compound that serves as a reference material in all volumetric titrimetric methods. The accuracy of such methods is critically dependent on the properties of this compound. Important requirements for a primary standard are:

1. High purity. Established methods for confirming purity should be available.

2. Stability in air.
3. Absence of hydrate water so that the composition does not change with variations in relative humidity.
4. Ready availability at modest cost.
5. Reasonable solubility in the titration medium.
6. Reasonably large formula weight so that the relative error associated with weighing is minimized.

The number of compounds that meet or even approach these criteria is small; only a limited number of primary standard substances are available to the chemist. As a consequence, less pure compounds must sometimes be employed in lieu of a primary standard. The purity of such a *secondary standard* must be established by careful analysis.

## 5B    STANDARD SOLUTIONS

Standard solutions play a central role in all volumetric methods of analysis.

### 5B–1  Desirable Properties of Standard Solutions

The ideal standard solution for a volumetric method will

1. be sufficiently stable so that it is necessary to determine its concentration only once.
2. react rapidly with the analyte so that the time required between additions of titrant is minimized.
3. react more or less completely with the analyte so that satisfactory end points are realized.
4. undergo a selective reaction with the analyte that can be described by a simple balanced equation.

Relatively few reagents meet all these ideals perfectly.

### 5B–2  Methods for Establishing the Concentration of Standard Solutions

The accuracy of a volumetric method can be no better than the accuracy of the concentration of the standard solution used in the titration. Two basic methods are used to establish the concentration of standard solutions. In the *direct method,* a carefully weighed quantity of a primary standard is dissolved and diluted to an exactly known volume in a volumetric flask. In the second, the solution is *standardized* by titrating (1) a weighed quantity of a primary standard, (2) a weighed quantity of a secondary standard, or (3) a measured volume of another standard solution. A titrant that is standardized against a secondary standard or against another standard solution is sometimes referred to as a *secondary standard solution.* A secondary standard solution is less desirable than a

> Standardization is a process in which the concentration of a solution is determined by using the solution to titrate a known amount of another reagent.

primary standard solution because the concentration of the former is subject to greater uncertainty. The best standard solutions are those prepared by the direct method. When the reagent does not possess the properties required for a primary standard, the less satisfactory indirect approach must be used.

### 5B–3 Methods for Expressing the Concentration of Standard Solutions

The concentrations of standard solutions are generally expressed in units of either *molarity c* or *normality* $c_N$. The first gives the number of moles of reagent contained in 1 L of solution, and the second gives the number of equivalents of reagent in the same volume.

Many chemists believe that the terms "normality" and "equivalent" offer so few real advantages that they can be abandoned without serious loss. We are sympathetic to this view and base volumetric calculations in this text on molarities. We also recognize, however, that the use of normalities and equivalents abounds in the chemical and biochemical literature of the last century and are also still found in various monographs and compilations of analytical methods of considerable value to present-day chemists.[2] For this reason, we have included a section in Appendix 10 that demonstrates how volumetric computations are performed with normalities and equivalents. You may find this appendix useful when consulting the analytical literature.

## 5C   VOLUMETRIC CALCULATIONS

### 5C–1 Concentration of Solutions

Chemists express the concentration of solutes in solution in several ways. The most important of these are described in this section.

### Molar Concentration

The molar concentration ($c_A$) of a solution of a chemical species A is the number of moles of that species contained in one liter of the solution (*not in one liter of the solvent*). The unit of molar concentration is *molarity*, M, which has the dimensions of mol $L^{-1}$. Molarity also expresses the number of millimoles of a solute per milliliter of solution:

$$c_A = \frac{\text{no. mol solute}}{\text{no. L solution}} = \frac{\text{no. mmol solute}}{\text{no. mL solution}} \qquad (5\text{--}1)$$

> The molarity of a solution is the number of moles of a reagent contained in one liter of the solution.

> *Analytical Chemistry,* a leading scientific journal of analytical chemistry, no longer permits the use of either "normality" or "titer" in the papers it accepts for publication.

---

[2]For example, The Association of Official Analytical Chemists employs normality in its publications, which are useful sources of tested methods for the determination of a wide variety of species in materials of agriculture and commerce. See *Official Methods of Analysis,* 15th ed., Washington, D.C.: Association of Official Analytical Chemists, 1990.

Example 5–1

Calculate the molar concentration of ethanol in an aqueous solution that contains 2.30 g of $C_2H_5OH$ (fw = 46.07 g) in 3.50 L.

We first calculate the number of moles contained in 2.30 g of ethanol:

$$2.30 \text{ g } C_2H_5OH \times \frac{1 \text{ mol } C_2H_5OH}{46.07 \text{ g } C_2H_5OH} = 0.04992 \text{ mol } C_2H_5OH$$

To obtain the molar concentration, $c_{C_2H_5OH}$, we divide the number of moles of ethanol by the volume:

$$c_{C_2H_5OH} = \frac{0.04992 \text{ mol } C_2H_5OH}{3.50 \text{ L}} = 0.01426 \text{ mol } C_2H_5OH/L = 0.0143 \text{ M}$$

Analytical molarity describes how a solution has been prepared. It gives the total moles of a solute in one liter of solution.

Equilibrium, or species, molarity gives the number of moles of a particular species in one liter of solution.

Some chemists prefer to distinguish between species and analytical concentrations in a different way. They use molar concentration for species concentration and formal concentration (F) for analytical concentration. Applying this convention to our example, we can say that the formal concentration of $H_2SO_4$ is 1.00 F, whereas its molar concentration is 0.00 M.

In this example the analytical molarity of $H_2SO_4$ is given by $c_{H_2SO_4} = [SO_4^{2-}] + [HSO_4^-]$ because these are the only two sulfate-containing species in the solution.

**Analytical Molarity.** The *analytical molarity* of a solution gives the *total* number of moles of a solute in one liter of the solution (or the total number of millimoles in one milliliter). That is, the analytical molarity specifies a recipe by which the solution can be prepared. For example, a sulfuric acid solution that has an analytical concentration of 1.0 M can be prepared by dissolving 1.0 mol, or 98 g, of $H_2SO_4$ in water and diluting to exactly 1.0 L.

**Equilibrium, or Species, Molarity.** The *equilibrium, or species, molarity* expresses the molar concentration of a particular species in a solution at equilibrium. In order to state the species molarity, it is necessary to know what happens to the solute when it is dissolved in a solvent. For example, the species molarity of $H_2SO_4$ in a solution with an analytical concentration of 1.00 M is 0.00 M because the sulfuric acid is entirely dissociated into a mixture of $H_3O^+$, $HSO_4^-$, and $SO_4^{2-}$ ions; essentially no $H_2SO_4$ molecules are present in this solution. The equilibrium concentrations and thus the species molarity of these three ions are 1.01, 0.99, and 0.01 M, respectively.

Equilibrium molar concentrations are often symbolized by placing square brackets around the chemical formula for the species. Thus, for our solution of $H_2SO_4$ with an analytical concentration of 1.00 M, we can write

$$[H_2SO_4] = 0.00 \text{ M} \qquad [H_3O^+] = 1.01 \text{ M}$$
$$[HSO_4^-] = 0.99 \text{ M} \qquad [SO_4^{2-}] = 0.01 \text{ M}$$

Example 5–2

Calculate the analytical and equilibrium molar concentrations of the solute species in an aqueous solution that contains 285 mg of trichloroacetic acid ($Cl_3CCOOH$, fw = 163.4 g) in 10.0 mL (the acid is 73% ionized in water).

Employing HA as the symbol for $Cl_3CCOOH$, we write

$$\text{amount HA} = 285 \text{ mg HA} \times \frac{1 \text{ g HA}}{1000 \text{ mg HA}} \times \frac{1 \text{ mol HA}}{163.4 \text{ g HA}}$$
$$= 1.744 \times 10^{-3} \text{ mol HA}$$

From the definition of analytical molar concentration,

$$c_{HA} = \frac{1.744 \times 10^{-3} \text{ mol HA}}{10.0 \text{ mL}} \times \frac{1000 \text{ mL}}{1 \text{ L}} = 0.174 \frac{\text{mol HA}}{\text{L}} = 0.174 \text{ M}$$

Because all but 27% of the acid is dissociated into $H_3O^+$ and $A^-$, the species concentration of HA is

$$[HA] = 0.174 \frac{\text{mol}}{\text{L}} \times 0.27 = 0.047 \frac{\text{mol}}{\text{L}} = 0.047 \text{ M}$$

The molarity of $H^+$ as well as that of $A^-$ is equal to the analytical concentration of the acid minus the species concentration of undissociated acid:

$$[H_3O^+] = [A^-] = c_{HA} - [HA] = 0.174 - 0.047$$
$$= 0.127 \text{ mol/L} = 0.127 \text{ M}$$

Note that the analytical concentration of HA is the sum of the species concentrations of HA and $A^-$:

$$c_{HA} = [HA] + [A^-]$$

---

Example 5–3

Describe the preparation of 2.00 L of 0.108 M $BaCl_2$ from $BaCl_2 \cdot 2\ H_2O$ (fw = 244 g).

We see that 1 mol of the dihydrate yields 1 mol of $BaCl_2$. Therefore, to produce this solution, we need

$$2.00 \text{ L} \times \frac{0.108 \text{ mol } BaCl_2 \cdot 2\ H_2O}{\text{L}} = 0.216 \text{ mol } BaCl_2 \cdot 2\ H_2O$$

The weight of $BaCl_2 \cdot 2\ H_2O$ is then

$$0.216 \text{ mol } BaCl_2 \cdot 2\ H_2O \times \frac{244 \text{ g}}{\text{mol } BaCl_2 \cdot 2\ H_2O} = 52.7 \text{ g } BaCl_2 \cdot 2\ H_2O$$

Therefore we would dissolve 52.7 g of $BaCl_2 \cdot 2\ H_2O$ in water and dilute to 2.00 L to prepare the 0.108 M solution.

---

Example 5–4

Describe the preparation of 500 mL of 0.0740 M $Cl^-$ solution from solid $BaCl_2 \cdot 2\ H_2O$ (fw = 244 g).

The number of moles of the species A in a solution of A is given by

$$\text{no. mol A} = c_A \times V_A$$

where $V_A$ is the volume of the solution in liters.

The number of moles of $Cl^-$ in the solution is given by

$$\text{amount Cl}^- = \frac{0.0740 \text{ mol Cl}^-}{\cancel{L}} \times \frac{1 \cancel{L}}{1000 \cancel{mL}} \times 500 \cancel{mL} = 0.0370 \text{ mol}$$

Because 1 mol of the salt contains 2 mol $Cl^-$, we need only half the number of moles, or

$$\text{amount BaCl}_2 \cdot 2 \text{ H}_2\text{O} = 0.0370 \cancel{\text{mol Cl}^-} \times \frac{1 \text{ mol BaCl}_2 \cdot 2 \text{ H}_2\text{O}}{2 \cancel{\text{mol Cl}^-}}$$

$$= 0.0185 \text{ mol BaCl}_2 \cdot 2 \text{ H}_2\text{O}$$

$$\text{wt BaCl}_2 \cdot 2 \text{ H}_2\text{O} = 0.0185 \cancel{\text{mol BaCl}_2 \cdot 2 \text{ H}_2\text{O}} \times \frac{244 \text{ g BaCl}_2 \cdot 2 \text{ H}_2\text{O}}{\cancel{\text{mol BaCl}_2 \cdot 2 \text{ H}_2\text{O}}}$$

$$= 4.51 \text{ g BaCl}_2 \cdot 2 \text{ H}_2\text{O}$$

Therefore we would dissolve 4.51 g of $BaCl_2 \cdot 2 \text{ H}_2\text{O}$ in water and dilute to 500 mL to prepare the 0.0740 M solution of $Cl^-$.

### Percent Concentration

Chemists frequently express concentrations in terms of percent (parts per hundred). Unfortunately, this practice can be a source of ambiguity because a solution's percent composition can be expressed in several ways. Three common methods are

$$\text{weight percent (w/w)} = \frac{\text{weight solute}}{\text{weight soln}} \times 100\%$$

$$\text{volume percent (v/v)} = \frac{\text{volume solute}}{\text{volume soln}} \times 100\%$$

$$\text{weight/volume percent (w/v)} = \frac{\text{weight solute, g}}{\text{volume soln, mL}} \times 100\%$$

Note that the denominator in each of these expressions refers to the *solution* rather than to the solvent. Note also that the first two expressions do not depend on the units employed (provided, of course, that there is consistency between numerator and denominator). In the third expression, units must be defined because the numerator and denominator have units that do not cancel. Of the three expressions, only weight percent has the virtue of being temperature independent.

Weight percent is frequently employed to express the concentration of commercial aqueous reagents. For example, hydrochloric acid is sold as a 37% solution, which means that the reagent contains 37 g of HCl per 100 g of solution (Figure 5–1).

Volume percent is commonly used to specify the concentration of a solution that was prepared by diluting a pure liquid compound with another liquid. For example, a 5% aqueous solution of methanol *usually* means a solution prepared by diluting 5.0 mL of pure methanol with enough water to give 100 mL.

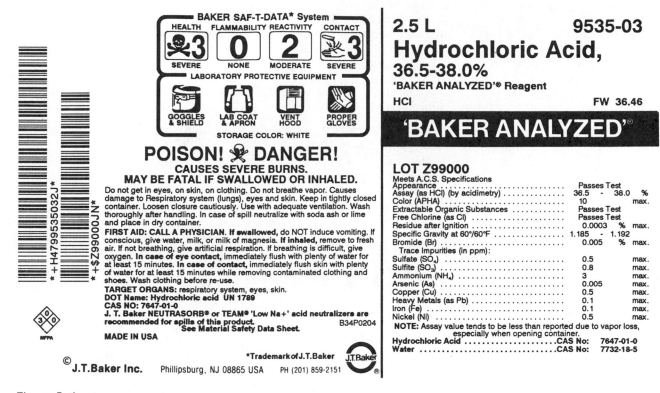

**Figure 5–1**
Label from a bottle of reagent-grade hydrochloric acid.

Weight/volume percent is often employed to indicate the composition of dilute aqueous solutions of solid reagents. For example, 5% aqueous silver nitrate *often* refers to a solution prepared by dissolving 5 g of silver nitrate in sufficient water to give 100 mL of solution.

To avoid uncertainty, always specify explicitly the type of percent composition being discussed. If this information is missing, the user must decide intuitively which of the several types is involved. The potential error resulting from a wrong choice is considerable. For example, commercial 50% (w/w) sodium hydroxide contains 763 g of the reagent per liter, which corresponds to 76.3% (w/v) sodium hydroxide.

> You should always specify the type of percent when reporting concentrations in this way.

### Parts Per Million and Parts Per Billion

For very dilute solutions, parts per million (ppm) is a convenient way to express concentration:

$$c_{ppm} = \frac{\text{weight solute}}{\text{weight solution}} \times 10^6 \text{ ppm}$$

where $c_{ppm}$ is the concentration in parts per million. Note that the units of weight in the numerator and denominator must agree.

A handy rule in calculating parts per million is to remember that, for dilute aqueous solutions whose densities are approximately 1.00 g/mL, 1 ppm = 1.00 mg/L. That is,

$$c_{ppm} = \frac{\text{weight solute (mg)}}{\text{volume soln (L)}} \qquad (5\text{–}2)$$

$$c_{ppb} = \frac{\text{weight solute g}}{\text{weight solution g}} \times 10^9 \text{ ppb}$$

For even more dilute solutions, $10^9$ ppb rather than $10^6$ ppm is used in the equation above to give the results in parts per billion (ppb).

The term "parts per thousand" (ppt) is also encountered, especially in oceanography.

---

### Example 5–5

What is the molarity of $K^+$ in a solution that contains 63.3 ppm of $K_3Fe(CN)_6$ (fw = 329.3 g)?

Because the solution is so dilute, it is reasonable to assume that its density is 1.00 g/mL. Therefore, according to Equation 5–2 (see margin note),

$$63.3 \text{ ppm } K_3Fe(CN)_6 = 63.3 \text{ mg } K_3Fe(CN)_6/L$$

$$\frac{\text{no. mol } K_3Fe(CN)_6}{L} = \frac{63.3 \text{ mg } K_3Fe(CN)_6}{L} \times \frac{1 \text{ g}}{1000 \text{ mg}}$$

$$\times \frac{1 \text{ mol } K_3Fe(CN)_6}{329.3 \text{ g } K_3Fe(CN)_6}$$

$$= 1.922 \times 10^{-4} \frac{\text{mol}}{L} = 1.922 \times 10^{-4} \text{ M}$$

$$[K^+] = \frac{1.922 \times 10^{-4} \text{ mol } K_3Fe(CN)_6}{L} \times \frac{3 \text{ mol } K^+}{1 \text{ mol } K_3Fe(CN)_6}$$

$$= 5.77 \times 10^{-4} \frac{\text{mol } K^+}{L} = 5.77 \times 10^{-4} \text{ M}$$

---

### Solution–Diluent Volume Ratios

The composition of a dilute solution is sometimes specified in terms of the volume of a more concentrated solution and the volume of the solvent used to do the diluting. The volume of the former is separated from that of the latter by a colon. Thus, a 1:4 HCl solution contains four volumes of water for each volume of concentrated hydrochloric acid.

This method of notation is frequently ambiguous in that the concentration of the original solution is not always obvious to the reader. Unfortunately, sometimes 1:4 is interpreted to mean dilute one volume with three volumes. Because of such uncertainties, you should avoid using solution–diluent ratios.

### p-Functions

For the chemical species X, $pX = -\log [X]$.

Scientists frequently express the concentration of a species in terms of its *p-function,* or *p-value.* The p-value is the negative logarithm (to base 10) of the molar concentration of a species. So, for the species X,

The best known p-function is pH, where pH = $-\log [H^+]$.

$$pX = -\log [X] \qquad (5\text{–}3)$$

As shown by the following examples, p-values offer the advantage of allowing concentrations that vary over ten or more orders of magnitude to be expressed in terms of small positive numbers.

---

**Example 5–6**

Calculate the p-value for each ion in a solution that is $2.00 \times 10^{-3}$ M in NaCl and $5.4 \times 10^{-4}$ M in HCl.

$$pH = -\log [H^+] = -\log (5.4 \times 10^{-4})$$
$$= -\log 5.4 - \log 10^{-4} = -0.73 - (-4) = 3.27$$

To obtain pNa, we write

$$pNa = -\log (2.00 \times 10^{-3}) = -\log 2.00 - \log 10^{-3}$$
$$= -0.301 - (-3.000) = 2.699$$

The total $Cl^-$ concentration is given by the sum of the concentrations of the two solutes:

$$[Cl^-] = 2.00 \times 10^{-3} \text{ M} + 5.4 \times 10^{-4} \text{ M}$$
$$= 2.00 \times 10^{-3} \text{ M} + 0.54 \times 10^{-3} \text{ M} = 2.54 \times 10^{-3} \text{ M}$$
$$pCl = -\log 2.54 \times 10^{-3} = 2.595$$

> Logarithmic functions such as p-functions are often used for measurements that have a wide *dynamic range*, that is, those that range over many orders of magnitude.

---

Note that in Example 5–6 and in the one that follows, the results are rounded according to the rules listed in Section 2F–2.

---

**Example 5–7**

Calculate the molar concentration of $Ag^+$ in a solution that has a pAg of 6.372.

$$pAg = -\log [Ag^+] = 6.372$$
$$\log [Ag^+] = -6.372 = 0.628 - 7.000$$
$$[Ag^+] = \text{antilog} (0.628) \times \text{antilog} (-7.000)$$
$$= 4.246 \times 10^{-7} = 4.25 \times 10^{-7}$$

---

## 5C–2 Density and Specific Gravity of Solutions

> In SI units, density is expressed in units of $kg/m^3$.

> Specific gravity is the ratio of the mass of a substance to the mass of an equal volume of water.

"Density" and "specific gravity" are terms often encountered in the analytical literature. The *density* of a substance is its mass per unit volume, whereas its *specific gravity* is the ratio of its mass to the mass of an equal volume of water at 4°C. Density has units of kilograms per liter or grams per milliliter in the metric system. Specific gravity is dimensionless and so is not tied to any particular system of units. For this reason, specific gravity is widely used in describing items of commerce. Since the

## SPECIFIC GRAVITIES OF CONCENTRATED ACIDS AND BASES

| Reagent | Concentration, % (w/w) | Specific Gravity |
|---|---|---|
| Acetic acid | 99.5 | 1.05 |
| Ammonia | 27 | 0.90 |
| Hydrochloric acid | 37 | 1.18 |
| Hydrobromic acid | 48 | 1.51 |
| Nitric acid | 70 | 1.42 |
| Perchloric acid | 70 | 1.66 |
| Phosphoric acid | 85 | 1.69 |
| Sulfuric acid | 95.5 | 1.83 |

Densities of liquids and solids are usually expressed in units of g/cm³.

density of water is approximately 1.00 g/mL and since we use the metric system throughout this text, density and specific gravity are used interchangeably.

### Example 5–8

Calculate the molar concentration of $HNO_3$ (fw = 63.0 g) in a solution that has a specific gravity of 1.42 and is 70% $HNO_3$ (w/w).

Let us first calculate the grams of acid per liter of concentrated solution:

$$\frac{1.42 \text{ kg reagent}}{\text{L reagent}} \times \frac{10^3 \text{ g reagent}}{\text{kg reagent}} \times \frac{70 \text{ g } HNO_3}{100 \text{ g reagent}} = \frac{994 \text{ g } HNO_3}{\text{L reagent}}$$

Then

$$c_{HNO_3} = \frac{994 \text{ g } HNO_3}{\text{L reagent}} \times \frac{1 \text{ mol } HNO_3}{63.0 \text{ g } HNO_3} = \frac{15.8 \text{ mol } HNO_3}{\text{L reagent}} = 16 \text{ M}$$

### Example 5–9

Describe the preparation of 100 mL of 6.0 M HCl from a concentrated solution that has a specific gravity of 1.18 and is 37% (w/w) HCl (fw = 36.5 g).

Proceeding as in Example 5–8, we calculate the molarity of the concentrated reagent:

$$c_{HCl} = \frac{1.18 \times 10^3 \text{ g reagent}}{\text{L reagent}} \times \frac{37 \text{ g HCl}}{100 \text{ g reagent}} \times \frac{1 \text{ mol HCl}}{36.5 \text{ g HCl}} = 12.0 \text{ M}$$

The number of moles of HCl required is

$$\text{amount HCl} = 100 \text{ mL} \times \frac{1 \text{ L}}{1000 \text{ mL}} \times \frac{6.0 \text{ mol HCl}}{\text{L}} = 0.600 \text{ mol}$$

$$\text{vol concd reagent} = 0.600 \text{ mol HCl} \times \frac{1 \text{ L reagent}}{12.0 \text{ mol HCl}}$$

$$= 0.0500 \text{ L or } 50.0 \text{ mL}$$

Thus we would dilute 50 mL of the concentrated reagent to 600 mL.

The solution to Example 5–9 is based upon the following useful relationship, which we will be using countless times:

$$V_{concd} \times c_{concd} = V_{dil} \times c_{dil} \qquad (5–4)$$

where the two terms on the left are the volume and molar concentration of a concentrated solution that is being used to prepare a diluted solution that has the volume and concentration given by the corresponding terms on the right. This equation is based upon the fact that the number of moles of solute in the diluted solution must equal the number of moles in the concentrated reagent. Note that the volumes can be in milliliters or liters as long as the same units are used for both solutions.

### 5C–3 Some Useful Algebraic Relationships

Most volumetric calculations are based on two pairs of simple equations that are derived from the definitions of millimole, mole, and molar concentration. For the chemical species A, we may write

$$\text{amount A} = \text{no. mmol A} = \frac{\text{wt A (g)}}{\text{mfw A (g/mmol)}} \qquad (5\text{–}5)$$

$$\text{amount A} = \text{no. mol A} = \frac{\text{wt A (g)}}{\text{fw A (g/mol)}} \qquad (5\text{–}6)$$

The second pair are derived from the definition of molar concentration:

$$\text{amount A} = \text{no. mmol A} = V(\text{mL}) \times c_A(\text{mmol A/mL}) \qquad (5\text{–}7)$$

$$\text{amount A} = \text{no. mol A} = V(\text{L}) \times c_A(\text{mol A/L}) \qquad (5\text{–}8)$$

where $V$ is the volume of the solution.

You should use Equations 5–5 and 5–7 when volumes are measured in milliliters and amounts are measured in millimoles. Use Equations 5–6 and 5–8 when the units are liters and moles.

---

**Feature 5–1**
**UNITS IN USING EQUATIONS 5–5 THROUGH 5–8**

It is useful to know that any combination of grams, moles, and liters can be replaced with any analogous combination expressed in milligrams, millimoles, and milliliters. For example, a 0.1 M solution contains 0.1 mol of a species per liter or 0.1 mmol per milliliter. Similarly, the number of moles of a compound is equal to either the weight in grams of that compound divided by its formula weight in grams or the weight in milligrams divided by its milliformula weight in milligrams.

---

### 5C–4 Calculation of the Molarity of Standard Solutions

The following four examples illustrate how volumetric reagents are prepared.

Equation 5–4 can be used with liters and moles per liter or with milliliters and millimoles per milliliter:

$$L_{concd} \times \frac{mol_{concd}}{L_{concd}} = L_{dil} \times \frac{mol_{dil}}{L_{dil}}$$

$$mL_{concd} \times \frac{mmol_{concd}}{mL_{concd}} = mL_{dil} \times \frac{mmol_{dil}}{mL_{dil}}$$

The units for the amount of a chemical species are millimoles and moles.

Example 5-10

Describe the preparation of 5.000 L of 0.1000 M $Na_2CO_3$ (fw = 105.99 g) from the primary-standard solid.

Since the volume is in liters, we base our calculations on the mole rather than the millimole. Thus to obtain the amount of $Na_2CO_3$ needed, we write

$$\text{amount } Na_2CO_3 = V_{soln}(L) \times c_{Na_2CO_3}(mol/L)$$
$$= 5.000 \text{ L} \times \frac{0.1000 \text{ mol } Na_2CO_3}{L} = 0.5000 \text{ mol } Na_2CO_3$$

To obtain the weight of $Na_2CO_3$, we rearrange Equation 5-6 to give

$$\text{wt } Na_2CO_3 = 0.5000 \text{ mol } Na_2CO_3 \times \frac{105.99 \text{ g } Na_2CO_3}{\text{mol } Na_2CO_3} = 53.00 \text{ g } Na_2CO_3$$

Therefore the solution is prepared by dissolving 53.00 g of $Na_2CO_3$ in water and diluting to exactly 5.000 L.

Example 5-11

A standard 0.0100 M solution of $Na^+$ is required to calibrate a flame-photometric method for determining the element. Describe how 500 mL of this solution can be prepared from primary-standard $Na_2CO_3$.

We wish to compute the weight of reagent required to give a species molarity of 0.0100. Here, we will use millimoles, since the volume is in milliliters. Because $Na_2CO_3$ dissociates to give two $Na^+$ ions, we can write that the number of millimoles of $Na_2CO_3$ needed is

$$\text{amount } Na_2CO_3 = 500 \text{ mL} \times \frac{0.0100 \text{ mmol } Na^+}{mL} \times \frac{1 \text{ mmol } Na_2CO_3}{2 \text{ mmol } Na^+}$$
$$= 2.50 \text{ mmol } Na_2CO_3$$

From the definition of millimole, we write

$$\text{wt } Na_2CO_3 = 2.50 \text{ mmol } Na_2CO_3 \times 0.10599 \frac{\text{g } Na_2CO_3}{\text{mmol } Na_2CO_3}$$
$$= 0.265 \text{ g } Na_2CO_3$$

The solution is therefore prepared by dissolving 0.265 g of $Na_2CO_3$ in water and diluting to 500 mL.

Example 5-12

How would you prepare 50.0-mL portions of standard solutions that are 0.00500 M, 0.00200 M, and 0.00100 M in $Na^+$ from the solution in Example 5-11?

The amount of $Na^+$ taken from the concentrated solution must equal the amount in the diluted solutions. Thus,

$$\text{no. mol Na}^+ \text{ from concd soln} = \text{no. mmol Na}^+ \text{ in dil soln}$$
$$V_{concd} \times c_{concd} = V_{dil} \times c_{dil} \quad \text{(Equation 5--4)}$$

and

$$V_{concd} = \frac{V_{dil} \times c_{dil}}{c_{concd}} = \frac{50.0 \text{ mL} \times 0.00500 \text{ mmol Na}^+/\text{mL}}{0.0100 \text{ mmol Na}^+/\text{mL}} = 25.0 \text{ mL}$$

Thus, to produce 50.0 mL of 0.00500 M $Na^+$, 25.0 mL of the concentrated solution should be diluted to exactly 50.0 mL.

Repeat the calculation for the other two molarities to confirm that diluting 10.0 and 5.00 mL of the concentrated solution to 50.0 mL produces the desired solutions.

Recall the relationship $V_{concd} \times c_{concd} = V_{dil} \times c_{dil}$.

## 5C–5   Treatment of Titration Data

In this section, we describe two types of volumetric calculations. The first involves computing the molarity of solutions that have been standardized against either a primary standard or another standard solution. The second involves calculating the amount of analyte in a sample from titration data. Both types are based on three algebraic relationships. Two of these are Equations 5–5 and 5–7, both of which are based on millimoles and milliliters. The third relationship is the stoichiometric ratio of the number of millimoles of analyte to the number of millimoles of titrant.

### Calculation of Molarities from Standardization Data

Examples 5–13 and 5–14 illustrate how standardization data are treated.

### Example 5–13

Exactly 50.00 mL of an HCl solution required 29.71 mL of 0.01963 M $Ba(OH)_2$ to reach an end point with bromocresol green indicator. Calculate the molarity of the HCl.

In the titration, 1 mmol of $Ba(OH)_2$ reacts with 2 mmol of HCl, and thus the stoichiometric ratio is

$$\text{stoichiometric ratio} = \frac{2 \text{ mmol HCl}}{1 \text{ mmol Ba(OH)}_2}$$

The number of millimoles of the standard is obtained by substituting into Equation 5–7:

$$\text{amount Ba(OH)}_2 = 29.71 \text{ mL Ba(OH)}_2 \times 0.01963 \frac{\text{mmol Ba(OH)}_2}{\text{mL Ba(OH)}_2}$$

To obtain the number of millimoles of HCl, we multiply this result by the stoichiometric ratio:

$$\text{amount HCl} = (29.71 \times 0.01963) \; \text{mmol Ba(OH)}_2 \times \frac{2 \text{ mmol HCl}}{1 \text{ mmol Ba(OH)}_2}$$

To obtain the number of millimoles of HCl per milliliter, we divide by the volume of acid:

$$c_{\text{HCl}} = \frac{(29.71 \times 0.01963 \times 2) \text{ mmol HCl}}{50.00 \text{ mL HCl}}$$

$$= 0.023328 \; \frac{\text{mmol HCl}}{\text{mL HCl}} = 0.02333 \text{ M}$$

---

**Example 5–14**

Titration of 0.2121 g of pure $Na_2C_2O_4$ (fw = 134.00 g) required 43.31 mL of $KMnO_4$. What is the molarity of the $KMnO_4$ solution? The chemical reaction is

$$2 \text{ MnO}_4^- + 5 \text{ C}_2\text{O}_4^{2-} + 16 \text{ H}^+ \rightarrow 2 \text{ Mn}^{2+} + 10 \text{ CO}_2 + 8 \text{ H}_2\text{O}$$

From this equation, we see that the stoichiometric ratio is

$$\text{stoichiometric ratio} = \frac{2 \text{ mmol KMnO}_4}{5 \text{ mmol Na}_2\text{C}_2\text{O}_4}$$

The amount of primary-standard $Na_2C_2O_4$ is given by Equation 5–5:

$$\text{amount Na}_2\text{C}_2\text{O}_4 = 0.2121 \; \text{g Na}_2\text{C}_2\text{O}_4 \times \frac{1 \text{ mmol Na}_2\text{C}_2\text{O}_4}{0.13400 \text{ g Na}_2\text{C}_2\text{O}_4}$$

To obtain the number of millimoles of $KMnO_4$, we multiply this result by the stoichiometric factor:

$$\text{amount KMnO}_4 = \frac{0.2121}{0.13400} \; \text{mmol Na}_2\text{C}_2\text{O}_4 \times \frac{2 \text{ mmol KMnO}_4}{5 \text{ mmol Na}_2\text{C}_2\text{O}_4}$$

The molarity is then obtained by dividing by the volume of $KMnO_4$ consumed:

$$c_{\text{KMnO}_4} = \frac{\left(\dfrac{0.2121}{0.13400} \times \dfrac{2}{5}\right) \text{ mmol KMnO}_4}{43.31 \text{ mL KMnO}_4} = 0.01462 \text{ M}$$

---

Note that units are carried through all calculations as a check on the correctness of the relationships used in Examples 5–13 and 5–14.

## Calculation of Quantity of Analyte from Titration Data

As shown by the examples that follow, the same general approach is also used to compute analyte concentrations from titration data.

---

### Example 5–15

A 0.8040-g sample of an iron ore is dissolved in acid. The iron is then reduced to $Fe^{2+}$ and titrated with 47.22 mL of 0.02242 M $KMnO_4$ solution. Calculate the results of this analysis in terms of (a) percent Fe (fw = 55.847 g) and (b) percent $Fe_3O_4$ (fw = 231.54 g). The reaction of the analyte with the reagent is described by the equation

$$MnO_4^- + 5\ Fe^{2+} + 8\ H^+ \rightarrow Mn^{2+} + 5\ Fe^{3+} + 4\ H_2O$$

(a) $\text{stoichiometric ratio} = \dfrac{5\ \text{mmol } Fe^{2+}}{1\ \text{mmol } KMnO_4}$

$\text{amount } KMnO_4 = 47.22\ \text{mL } KMnO_4 \times \dfrac{0.02242\ \text{mmol } KMnO_4}{\text{mL } KMnO_4}$

$\text{amount } Fe^{2+} = (47.22 \times 0.02242)\ \text{mmol } KMnO_4 \times \dfrac{5\ \text{mmol } Fe^{2+}}{1\ \text{mmol } KMnO_4}$

The weight of $Fe^{2+}$ is then given by

$\text{wt } Fe^{2+} = (47.22 \times 0.02242 \times 5)\ \text{mmol } Fe^{2+} \times 0.055847\ \dfrac{\text{g } Fe^{2+}}{\text{mmol } Fe^{2+}}$

The percent $Fe^{2+}$ is

$\text{percent } Fe^{2+} = \dfrac{(47.22 \times 0.02242 \times 5 \times 0.055847)\ \text{g } Fe^{2+}}{0.8040\ \text{g sample}} \times 100\%$

$\qquad\qquad = 36.77\%$

(b) In order to derive a stoichiometric ratio, we note that

$$5\ Fe^{2+} \equiv 1\ MnO_4^-$$

The symbol $\equiv$ means chemically equivalent to.

Therefore,

$$5\ \text{mmol } Fe_3O_4 \equiv 15\ \text{mmol } Fe^{2+} \equiv 3\ \text{mmol } MnO_4^-$$

and

$$\text{stoichiometric ratio} = \dfrac{5\ \text{mmol } Fe_3O_4}{3\ \text{mmol } KMnO_4}$$

As in part (a),

$$\text{amount } KMnO_4 = 47.22\ \text{mL } KMnO_4 \times \dfrac{0.02242\ \text{mmol } KMnO_4}{\text{mL } KMnO_4}$$

$$\text{amount Fe}_3\text{O}_4 = (47.22 \times 0.02242) \text{ mmol KMnO}_4^- \times \frac{5 \text{ mmol Fe}_3\text{O}_4}{3 \text{ mmol KMnO}_4^-}$$

$$\text{wt Fe}_3\text{O}_4 = (47.22 \times 0.02242 \times \tfrac{5}{3}) \text{ mmol Fe}_3\text{O}_4^- \times 0.23154 \frac{\text{g Fe}_3\text{O}_4}{\text{mmol Fe}_3\text{O}_4^-}$$

$$\text{percent Fe}_3\text{O}_4 = \frac{(47.22 \times 0.02242 \times \tfrac{5}{3}) \times 0.23154 \text{ g Fe}_3\text{O}_4}{0.8040 \text{ g sample}} \times 100\%$$

$$= 50.81\%$$

---

**Feature 5–2**
**ANOTHER APPROACH TO EXAMPLE 5–15a**

Some people find it easier to write out the solution to a problem in such a way that the units in the denominator of each term cancel the units in the numerator of the preceding term until the units of the answer are obtained. For example, the solution to part (a) of Example 5–15 can be written

$$47.22 \text{ mL KMnO}_4^- \times \frac{0.02242 \text{ mmol KMnO}_4^-}{\text{mL KMnO}_4^-} \times \frac{5 \text{ mmol Fe}^-}{1 \text{ mmol KMnO}_4^-}$$

$$\times \frac{0.055847 \text{ g Fe}}{\text{mmol Fe}^-} \times \frac{1}{0.8040 \text{ g sample}} \times 100\%$$

$$= 36.77\% \text{ Fe}$$

---

Example 5–16

The organic matter in a 3.776-g sample of a mercuric ointment is decomposed with $HNO_3$. After dilution, the $Hg^{2+}$ is titrated with 21.30 mL of a 0.1144 M solution of $NH_4SCN$. Calculate the percent Hg (fw = 200.59 g) in the ointment. This titration involves the formation of a stable neutral complex, $Hg(SCN)_2$:

$$Hg^{2+} + 2 \ SCN^- \rightarrow Hg(SCN)_2(aq)$$

At the equivalence point,

$$\text{stoichiometric ratio} = \frac{1 \text{ mmol Hg}^{2+}}{2 \text{ mmol NH}_4\text{SCN}}$$

$$\text{amount NH}_4\text{SCN} = 21.30 \text{ mL NH}_4\text{SCN} \times 0.1144 \frac{\text{mmol NH}_4\text{SCN}}{\text{mL NH}_4\text{SCN}}$$

$$\text{amount Hg}^{2+} = (21.30 \times 0.1144) \text{ mmol NH}_4\text{SCN} \times \frac{1 \text{ mmol Hg}^{2+}}{2 \text{ mmol NH}_4\text{SCN}}$$

$$\text{wt Hg}^{2+} = (21.30 \times 0.1144 \times \tfrac{1}{2}) \text{ mmol Hg}^{2+} \times \frac{0.20059 \text{ g Hg}^{2+}}{\text{mmol Hg}^{2+}}$$

$$\text{percent Hg} = \frac{(21.30 \times 0.1144 \times \frac{1}{2} \times 0.20059) \text{ g Hg}^{2+}}{3.776 \text{ g sample}} \times 100\%$$

$$= 6.472\% = 6.47\%$$

---

**Feature 5–3**
ROUNDING THE ANSWER IN EXAMPLE 5–16

You should note that the input data for Example 5–16 all contain either four or five significant figures, but the answer is rounded to three. Why is this?

Let us proceed as we did on page 32 and make the rounding decision by doing a couple of rough calculations in our heads. We will assume that the input data are uncertain to one part in the last significant figure. The largest *relative* error will then be associated with the molarity of the reagent. Here, the relative uncertainty is 0.0001/0.1144. But we really do not need to know the error this accurately, and we can simply figure that the uncertainty is about 1 part in 1000 (compared with about 1 part in 2000 for the volume and 1 part in 3800 for the weight). We then assume the calculated result is uncertain to about the same amount as the least accurate measurement, or 1 part in 1000. The absolute uncertainty of the final result is then $6.472\% \times 1/1000 = 0.0065 = 0.01\%$, and we round to the second figure to the right of the decimal point. Thus, we report 6.47%.

You should practice making this rough type of rounding decision whenever you make a computation in this course and others.

---

Example 5–17

A 0.4755-g sample containing $(NH_4)_2C_2O_4$ and inert materials was dissolved in water and made strongly alkaline with KOH, which converted $NH_4^+$ to $NH_3$. The liberated $NH_3$ was distilled into exactly 50.00 mL of 0.05035 M $H_2SO_4$. The excess $H_2SO_4$ was back-titrated with 11.13 mL of 0.1214 M NaOH. Calculate (a) the percent N (fw = 14.007 g) and (b) the percent $(NH_4)_2C_2O_4$ (fw = 124.10 g) in the sample.

(a) The $H_2SO_4$ reacts with both $NH_3$ and NaOH, and so two stoichiometric ratios are needed:

$$\frac{2 \text{ mmol NH}_3}{1 \text{ mmol H}_2SO_4} \quad \text{and} \quad \frac{1 \text{ mmol H}_2SO_4}{2 \text{ mmol NaOH}}$$

$$\text{total amount H}_2SO_4 = 50.00 \text{ mL H}_2SO_4 \times 0.05035 \frac{\text{mmol H}_2SO_4}{\text{mL H}_2SO_4}$$

$$= 2.5175 \text{ mmol H}_2SO_4$$

The amount of $H_2SO_4$ consumed by the NaOH in the back-titration is

$$\text{amount } H_2SO_4 = (11.13 \times 0.1214) \text{ mmol NaOH} \times \frac{1 \text{ mmol } H_2SO_4}{2 \text{ mmol NaOH}}$$

$$= 0.6756 \text{ mmol } H_2SO_4$$

The amount of $H_2SO_4$ that reacted with $NH_3$ is then

$$\text{amount } H_2SO_4 = (2.5175 - 0.6756) \text{ mmol } H_2SO_4$$

$$= 1.8419 \text{ mmol } H_2SO_4$$

The amount of $NH_3$, which is equal to the amount of N, is

$$\text{amount N} = \text{amount } NH_3 = 1.8419 \text{ mmol } H_2SO_4 \times \frac{2 \text{ mmol N}}{1 \text{ mmol } H_2SO_4}$$

$$= 3.6838 \text{ mmol N}$$

$$\text{percent N} = \frac{3.6838 \text{ mmol N} \times 0.014007 \text{ g N/mmol N}}{0.4755 \text{ g sample}} \times 100\%$$

$$= 10.85\%$$

(b) Since each millimole of $(NH_4)_2C_2O_4$ produces 2 mmol of $NH_3$, which reacts with 1 mmol of $H_2SO_4$,

$$\text{stoichiometric ratio} = \frac{1 \text{ mmol } (NH_4)_2C_2O_4}{1 \text{ mmol } H_2SO_4}$$

$$\text{amount } (NH_4)_2C_2O_4 = 1.8419 \text{ mmol } H_2SO_4 \times \frac{1 \text{ mmol } (NH_4)_2C_2O_4}{1 \text{ mmol } H_2SO_4}$$

$$\text{wt } (NH_4)_2C_2O_4 = 1.8419 \text{ mmol } (NH_4)_2C_2O_4 \times \frac{0.12410 \text{ g } (NH_4)_2C_2O_4}{\text{mmol } (NH_4)_2C_2O_4}$$

$$= 0.22858 \text{ g}$$

$$\text{percent } (NH_4)_2C_2O_4 = \frac{0.22858 \text{ g } (NH_4)_2C_2O_4}{0.4755 \text{ g sample}} \times 100\% = 48.07\%$$

---

### Example 5–18

The CO in a 20.3-L sample of gas was converted to $CO_2$ by passing the gas over iodine pentoxide heated to 150°C:

$$I_2O_5(s) + 5 \ CO(g) \rightarrow 5 \ CO_2(g) + I_2(g)$$

The iodine distilled at this temperature and was collected in an absorber containing 8.25 mL of 0.01101 M $Na_2S_2O_3$:

$$I_2(aq) + 2 \ S_2O_3^{2-}(aq) \rightarrow 2 \ I^-(aq) + S_4O_6^{2-}(aq)$$

The excess $Na_2S_2O_3$ was back-titrated with 2.16 mL of 0.00947 M $I_2$ solution. Calculate the number of milligrams of CO (fw = 28.01 g) per liter of sample.

Based on the two reactions, the stoichiometric ratios are

$$\frac{5 \text{ mmol CO}}{1 \text{ mmol I}_2} \quad \text{and} \quad \frac{2 \text{ mmol Na}_2\text{S}_2\text{O}_3}{1 \text{ mmol I}_2}$$

We divide the first ratio by the second to get a third useful ratio:

$$\frac{5 \text{ mmol CO}}{2 \text{ mmol Na}_2\text{S}_2\text{O}_3}$$

This relationship reveals that 5 mmol of CO is responsible for the consumption of 2 mmol of $Na_2S_2O_3$. The total amount of $Na_2S_2O_3$ is

$$\text{amount Na}_2\text{S}_2\text{O}_3 = 8.25 \; \cancel{\text{mL Na}_2\text{S}_2\text{O}_3} \times 0.01101 \; \frac{\text{mmol Na}_2\text{S}_2\text{O}_3}{\cancel{\text{mL Na}_2\text{S}_2\text{O}_3}}$$

$$= 0.09083 \text{ mmol Na}_2\text{S}_2\text{O}_3$$

The amount of $Na_2S_2O_3$ consumed in the back-titration is

$$\text{amount Na}_2\text{S}_2\text{O}_3 = 2.16 \; \cancel{\text{mL I}_2} \times 0.00947 \; \frac{\cancel{\text{mmol I}_2}}{\cancel{\text{mL I}_2}} \times \frac{2 \text{ mmol Na}_2\text{S}_2\text{O}_3}{\cancel{\text{mmol I}_2}}$$

$$= 0.04091 \text{ mmol Na}_2\text{S}_2\text{O}_3$$

The amount of CO can then be obtained by employing the third stoichiometric ratio:

$$\text{amount CO} = (0.09083 - 0.04091) \; \cancel{\text{mmol Na}_2\text{S}_2\text{O}_3} \times \frac{5 \text{ mmol CO}}{2 \; \cancel{\text{mmol Na}_2\text{S}_2\text{O}_3}}$$

$$= 0.1248 \text{ mmol CO}$$

$$\text{wt CO} = 0.1248 \; \cancel{\text{mmol CO}} \times \frac{28.01 \text{ mg CO}}{\cancel{\text{mmol CO}}} = 3.4958 \text{ mg}$$

$$\frac{\text{wt CO}}{\text{vol sample}} = \frac{3.4958 \text{ mg CO}}{20.3 \text{ L sample}} = 0.172 \; \frac{\text{mg CO}}{\text{L}}$$

## 5D  WEIGHT TITRIMETRY

*Weight* or *gravimetric titrimetry* differs from its volumetric counterpart in that the *mass* of titrant is measured rather than the volume. Thus, in a weight titration, a balance and a solution dispenser are substituted for a buret and its markings. Weight titrimetry actually predates volumetric titrimetry by more than 50 years.[3] With the advent of reliable burets, however, weight titrations were largely supplanted by volumetric meth-

---

[3]For a brief history of gravimetric and volumetric titrimetry, see B. Kratochvil and C. Maitra, *Amer. Lab.*, **1983** (1), 22.

ods because the former required relatively elaborate equipment and were tedious and time-consuming. The recent advent of sensitive top-loading balances and convenient plastic solution dispensers has changed this situation completely, and weight titrations can now be performed more easily and more rapidly than volumetric titrations.

### 5D–1  Calculations Associated with Weight Titrations

The most convenient unit of concentration for weight titrations is *weight molarity* $M_w$, which is the number of moles of a reagent in one kilogram of solution or the number of millimoles in one gram of solution. Thus aqueous 0.1 $M_w$ NaCl contains 0.1 mol of the salt in 1 kg of solution or 0.1 mmol in 1 g of the solution.

The weight molarity $c_{w(A)}$ of a solution of a solute A is computed by means of either of two equations that are analogous to Equation 5–1:

$$c_{w(A)} = \frac{\text{no. mol A}}{\text{no. kg solution}} = \frac{\text{no. mmol A}}{\text{no. g solution}} \qquad (5-9)$$

Weight titration data can then be treated by the methods illustrated in Sections 5C–4 and 5C–5 after substitution of weight molarity for molarity and grams and kilograms for milliliters and liters.

### 5D–2  Advantages of Weight Titrations

In addition to greater speed and convenience, weight titrations offer certain other advantages over their volumetric counterparts:

1. Calibration of glassware and tedious cleaning to ensure proper drainage is avoided.
2. Temperature corrections are unnecessary because weight molarity does not change with temperature in contrast to volume molarity. This advantage is particularly important in nonaqueous titrations because of the high coefficients of expansion of most organic liquids (about ten times that of water).
3. Weight measurements can be made with considerably greater precision and accuracy than can volume measurements. For example, 50 or 100 g of an aqueous solution can be readily measured to ±1 mg, which corresponds to ±0.001 mL. This greater sensitivity makes it possible to choose sample sizes that lead to significantly smaller consumption of standard reagents.
4. Weight titrations are more easily automated than are volumetric titrations.

# 5E QUESTIONS AND PROBLEMS

Unless otherwise noted, analytical molar concentrations (see Section 5C–1) are used throughout these problems.

**5–1.** Define
  *(a) millimole.
  (b) titration.
  *(c) stoichiometric factor.
  (d) formula weight.

**5–2.** Distinguish between
  *(a) the equivalence point and the end point of a titration.
  (b) the density and the specific gravity of a solution.
  *(c) analytical molarity and equilibrium molarity.
  (d) a primary standard and a secondary standard.

**5–3.** How do volumetric titrimetry and gravimetric titrimetry resemble one another? How do they differ?

**\*5–4.** Briefly explain why milligrams of solute per liter and parts per million can be used interchangeably to describe the concentration of a dilute aqueous solution.

**5–5.** Calculate the number of moles and millimoles in
  *(a) 7.772 g of $NH_4Cl$.
  (b) 0.4417 g of $Na_2SO_4$.
  *(c) 3.428 g of $HgCl_2$.
  (d) 84.33 g of $Ba(OH)_2$.
  *(e) 8.769 g of $(NH_4)_2C_2O_4 \cdot H_2O$.

**5–6.** Calculate the weight of solute (in grams) contained in
  *(a) 40.0 mL of 0.1265 M KCl.
  (b) 12.50 mL of 0.06724 M $KMnO_4$.
  *(c) 39.9 mL of 0.1718 M $MgCl_2$.
  (d) 42.21 mL of 0.5740 M $H_2SO_4$.
  *(e) 49.76 mL of 0.2569 M $CuSO_4$.

**5–7.** Refer to Problem 5–6 and calculate the volumes needed to prepare 250.0 mL of a 0.0400 M solution.

**5–8.** What volume is needed to provide 2.213 mmol of $K^+$ if a solution is 0.0696 M in
  *(a) KCl?
  (b) $K_2SO_4$?
  *(c) $K_3PO_4$?
  (d) $K_3Fe(CN)_6$?
  *(e) $K_2Cr_2O_7$?
  (f) $K_4Fe(CN)_6$?

**5–9.** What volume of each solution in Problem 5–8 will contain 183.7 mg of $K^+$?

**5–10.** What volume of each solution in Problem 5–8 will be needed to produce 250.0 mL of a solution that is 0.0503 M in $K^+$?

**5–11.** A solution is prepared by dissolving 0.1164 g of $(NH_4)_2Ce(NO_3)_6$ (fw = 548.3) in 2.500 L of water. Calculate the ppm of
  *(a) $NH_4^+$.
  (b) $Ce^{4+}$.
  (c) $NO_3^-$.

**5–12.** An average sample of sea water has a density of 1.020 and contains 1080 ppm of $Na^+$ and 270 ppm of

$SO_4^{2-}$. Calculate the molar concentration of the two ions.

**5–13.** Calculate the molar analytical concentration of solute in an aqueous solution that is
  *(a) 11.00% (w/w) $NH_3$ and has a density of 0.9538.
  (b) 18.00% (w/w) KBr and has a density of 1.149.
  *(c) 28.00% (w/w) ethylene glycol (fw = 62.07) and has a density of 1.0350.
  (d) 15.00% (w/w) sucrose (fw = 342.5) and has a density of 1.0592.

**\*5–14.** How would you prepare
  (a) 500 mL of 16% (w/v) aqueous ethanol (fw = 46.1)?
  (b) 500 mL of 16% (v/v) aqueous ethanol?
  (c) 500 g of 16% (w/w) aqueous ethanol?

**5–15.** Describe the preparation of
  (a) 250 mL of 20% (w/v) aqueous acetone (fw = 58.05).
  (b) 250 mL of 20% (v/v) aqueous acetone.
  (c) 250 mL of 20% (w/w) aqueous acetone.

**5–16.** Calculate the p-values for each ion in a solution that is
  *(a) 0.0100 M in NaBr.
  (b) 0.0100 M in $BaBr_2$.
  *(c) $3.5 \times 10^{-3}$ M in $Ba(OH)_2$.
  (d) 0.040 M in HCl and 0.020 M in NaCl.
  *(e) $5.2 \times 10^{-3}$ M in $CaCl_2$ and $3.6 \times 10^{-3}$ M in $BaCl_2$.
  (f) $4.8 \times 10^{-8}$ M in $Zn(NO_3)_2$ and $5.6 \times 10^{-7}$ M in $Cd(NO_3)_2$.

**5–17.** Calculate the p-value for each ion:
  *(a) $Na^+$, $Cl^-$, and $OH^-$ in a solution that is 0.116 M in NaCl and 0.125 M in NaOH.
  (b) $Ba^{2+}$, $Mn^{2+}$, and $Cl^-$ in a solution that is $3.80 \times 10^{-3}$ M in $BaCl_2$ and 2.22 M in $MnCl_2$.
  *(c) $H^+$, $Cl^-$, and $Zn^{2+}$ in a solution that is 1.50 M in HCl and 0.120 M in $ZnCl_2$.
  (d) $Cu^{2+}$, $Zn^{2+}$, and $NO_3^-$ in a solution that is $4.32 \times 10^{-2}$ M in $Cu(NO_3)_2$ and 0.101 M in $Zn(NO_3)_2$.
  *(e) $K^+$, $OH^-$, and $Fe(CN)_6^{4-}$ in a solution that is $3.79 \times 10^{-6}$ M in $K_4Fe(CN)_6$ and $4.12 \times 10^{-5}$ M in KOH.
  (f) $H^+$, $Ba^{2+}$, and $ClO_4^-$ in a solution that is $2.75 \times 10^{-4}$ M in $Ba(ClO_4)_2$ and $4.44 \times 10^{-4}$ M in $HClO_4$.

**5–18.** Convert the following p-values to molar concentrations:
  *(a) pH = 8.67.
  (b) pOH = 0.125.
  *(c) pBr = 0.034.
  (d) pCa = 12.35.
  *(e) pLi = −0.321.
  (f) $pNO_3$ = 7.77.
  *(g) $pMnO_4$ = 0.0025.
  (h) pCl = 1.020.

5–19. Calculate the molar $H_3O^+$ concentration of a solution that has a pH of
*(a)  9.21.
 (b)  4.58.
*(c)  0.45.
 (d)  14.12.
*(e)  7.32.
 (f)  6.76.
*(g)  −0.21.
 (h)  −0.52.

*5–20. Calculate the molarity of a dilute HCl solution if a 50.00-mL aliquot yielded 0.8620 g of AgCl upon being treated with an excess of $AgNO_3$.

5–21. Calculate the molarity of an $H_2SO_4$ solution if 0.3084 g of $BaSO_4$ was obtained by treatment of a 25.00-mL aliquot with an excess of $BaCl_2$.

*5–22. A 0.3367-g sample of primary-standard-grade $Na_2CO_3$ required 28.66 mL of a $H_2SO_4$ solution to reach the end point in the reaction

$$CO_3^{2-} + 2\,H^+ \rightarrow H_2O + CO_2(g)$$

What is the molarity of the $H_2SO_4$?

5–23. A solution of $HClO_4$ was standardized by dissolving 0.3745 g of primary-standard-grade HgO in a solution of KBr:

$$HgO(s) + 4\,Br^- + H_2O \rightarrow HgBr_4^{2-} + 2\,OH^-$$

The liberated $OH^-$ was neutralized with 37.79 mL of the acid. Calculate the molarity of the $HClO_4$.

*5–24. Titration of 50.00 mL of 0.05251 M $Na_2C_2O_4$ required 38.71 mL of a potassium permanganate solution:

$$2\,MnO_4^- + 5\,H_2C_2O_4 + 6\,H^+ \rightarrow$$
$$2\,Mn^{2+} + 10\,CO_2(g) + 8\,H_2O$$

Calculate the molarity of the $KMnO_4$ solution.

5–25. Titration of the $I_2$ produced from 0.1238 g of primary-standard $KIO_3$ required 41.27 mL of sodium thiosulfate:

$$IO_3^- + 5\,I^- + 6\,H^+ \rightarrow 3\,I_2 + 3\,H_2O$$
$$I_2 + 2\,S_2O_3^{2-} \rightarrow 2\,I^- + S_4O_6^{2-}$$

Calculate the concentration of the $Na_2S_2O_3$ solution.

5–26. A 25.00-mL aliquot of 0.08370 M $HClO_4$ required 47.21 mL of $Ba(OH)_2$. Calculate the molar concentration of the base.

*5–27. A 0.1884-g sample of impure $Na_2CO_3$ required 31.56 mL of 0.1056 M HCl;

$$CO_3^{2-} + 2\,H^+ \rightarrow H_2O + CO_2(g)$$

Calculate the percent purity of the $Na_2CO_3$.

5–28. The sulfur in a 0.5073-g organic sample was burned in a stream of $O_2$; the combustion products were bubbled through $H_2O_2$ to convert $SO_2$ to $H_2SO_4$:

$$SO_2(g) + H_2O_2 \rightarrow H_2SO_4$$

Titration of the $H_2SO_4$ required 33.29 mL of 0.1115 M NaOH. Calculate the percent S in the sample.

*5–29. The iron in a 100.0-mL sample of spring water was reduced to the +2 state, and treated with 25.00 mL of 0.002107 M $K_2Cr_2O_7$:

$$6\,Fe^{2+} + Cr_2O_7^{2-} + 14\,H^+ \rightarrow$$
$$6\,Fe^{3+} + 2\,Cr^{3+} + 7\,H_2O$$

The excess $K_2Cr_2O_7$ was back-titrated with 7.47 mL of 0.00979 M $Fe^{2+}$. Calculate the parts per million of Fe in the sample.

5–30. Exactly 40.00 mL of a solution of $HClO_4$ was added to a solution containing 0.4793 g of primary-standard-grade $Na_2CO_3$. The solution was boiled to remove $CO_2$, and the excess $HClO_4$ was back-titrated with 8.70 mL of a NaOH solution. In a separate experiment, 25.00 mL of the NaOH neutralized 27.43 mL of $HClO_4$. Calculate the molarity of the $HClO_4$ and that of the NaOH.

5–31. A 0.3396-g sample that assayed 96.4% $Na_2SO_4$ was titrated with a solution of $BaCl_2$:

$$Ba^{2+} + SO_4^{2-} \rightarrow BaSO_4(s)$$

What is the molarity of the $BaCl_2$ solution if the end point was observed when 35.70 mL of the reagent was added?

*5–32. The ethyl acetate concentration in an alcoholic solution was determined by diluting a 10.00-mL sample to exactly 100 mL. A 20.00-mL portion of the diluted solution was refluxed with 40.00 mL of 0.04672 M KOH:

$$CH_3COOC_2H_5 + OH^- \rightarrow CH_3COO^- + C_2H_5OH$$

After cooling, the excess $OH^-$ was back-titrated with 3.41 mL of 0.05042 M $H_2SO_4$. Calculate the number of grams of ethyl acetate (fw = 76.10) per 100 mL of the original sample.

5–33. The thiourea in a 1.455-g sample of organic material was extracted into a dilute $H_2SO_4$ solution and titrated with 37.31 mL of 0.009372 M $Hg^{2+}$ via the reaction

$$4\,(NH_2)_2CS + Hg^{2+} \rightarrow [(NH_2)_2CS]_4Hg^{2+}$$

Calculate the percent $(NH_2)_2CS$ in the sample.

*5–34. (a) A 0.3147-g sample of primary standard $Na_2C_2O_4$ was dissolved in dilute $H_2SO_4$ and titrated with 31.672 g of dilute $KMnO_4$:

$$2\,MnO_4^- + 5\,C_2O_4^{2-} + 16\,H^+ \rightarrow$$
$$2\,Mn^{2+} + 10\,CO_2(g) + 8\,H_2O$$

Calculate the weight molarity of the $KMnO_4$ solution.

**(b)** The iron in a 0.6656-g ore sample was reduced quantitatively to the +2 state and then titrated with 26.753 g of the $KMnO_4$ solution. Calculate the percent $Fe_2O_3$ in the sample.

**5–35. (a)** A 0.1752-g sample of primary standard $AgNO_3$ was dissolved in 502.3 g of distilled water. Calculate the weight molarity of $Ag^+$ in this solution.

**(b)** The standard solution described in part (a) was used to titrate a 25.171-g sample of a KSCN solution. An end point was obtained after adding 23.765 g of the $AgNO_3$ solution. Calculate the weight molarity of the KSCN solution.

**(c)** The solutions described in parts (a) and (b) were used to determine the $BaCl_2 \cdot 2 H_2O$ in a 0.7120-g sample. A 20.102-g portion of the $AgNO_3$ was added to a solution of the sample and the excess $AgNO_3$ was back-titrated with 7.543 g of the KSCN solution. Calculate the percent $BaCl_2 \cdot 2 H_2O$ in the sample.

**\*5–36.** A solution of $Ba(OH)_2$ was standardized against 0.1016 g of primary-standard benzoic acid, $C_6H_5COOH$ (fw = 122.12 g). An end point was observed after addition of 44.42 mL of base.

**(a)** Calculate the molarity of the base.

**(b)** Calculate the standard deviation of the molarity if the standard deviation for weighing was ±0.2 mg and that for the volume measurement was ±0.03 mL.

**(c)** Assuming an error of −0.3 mg in the weighing, calculate the absolute and relative determinate error in the molarity.

**5–37.** A 0.1475 M solution of $Ba(OH)_2$ was used to titrate the acetic acid (fw = 60.05 g) in a dilute aqueous solution. The following results were obtained.

| Sample | Sample Volume, mL | $Ba(OH)_2$ Volume, mL |
|--------|-------------------|-----------------------|
| 1 | 50.00 | 43.17 |
| 2 | 49.50 | 42.68 |
| 3 | 25.00 | 21.47 |
| 4 | 50.00 | 43.33 |

**(a)** Calculate the mean w/v percentage of acetic acid in the sample.

**(b)** Calculate the standard deviation of the results.

**(c)** At the 90% confidence level, could any of the results be discarded?

**(d)** Assume that the buret used to measure out the acetic acid had a determinate error of −0.05 mL at all volumes delivered. Calculate the determinate error in the mean result.

# AQUEOUS-SOLUTION CHEMISTRY

This chapter provides a review of aqueous-solution chemistry, including chemical equilibrium and simple equilibrium-constant calculations. The material is a prerequisite to understanding the contents of the chapters that follow.

## 6A THE CHEMICAL COMPOSITION OF AQUEOUS SOLUTIONS

Water is the most plentiful solvent available on earth and finds widespread use as a medium for carrying out chemical analyses.

### 6A–1 Solutions of Electrolytes

A salt is the product formed in the reaction of an acid with a base. Examples include NaCl, $Na_2SO_4$, and $CH_3COONa$ (sodium acetate).

Most of the solutes we will discuss are *electrolytes*. Electrolytes are substances that form ions when dissolved in water or other solvents and thus produce solutions that conduct electricity. *Strong electrolytes* ionize essentially completely in a solvent, whereas *weak electrolytes* ionize only partially. Therefore, weak electrolytes impart less conductivity to a solvent than do strong electrolytes. Table 6–1 is a compilation of solutes that act as strong and weak electrolytes in water. Among the strong electrolytes listed are acids, bases, and salts.

### 6A–2 Acids and Bases

An acid is a substance that donates protons. A base is a substance that accepts protons.

An acid donates protons only in the presence of a proton acceptor (a base). Likewise, a base accepts protons only in the presence of a proton donor (an acid).

In 1923, two chemists, J. N. Brønsted in Denmark and J. M. Lowry in England, independently proposed a theory of acid/base behavior that is particularly useful in analytical chemistry.[1] According to the Brønsted-Lowry theory, *an acid is a proton donor* and *a base is a proton acceptor*. In order for a species to behave as an acid, a proton acceptor (or base) must be present. The reverse is also true.

---

[1]For a thorough treatment of the various acid/base concepts, see I. M. Kolthoff, in *Treatise on Analytical Chemistry,* 2nd ed., I. M. Kolthoff and P. J. Elving, Eds., Part I, Vol. 2, Chapter 17. New York: Wiley, 1979.

| Table 6-1 CLASSIFICATION OF ELECTROLYTES | |
|---|---|
| **Strong** | **Weak** |
| 1. The inorganic acids $HNO_3$, $HClO_4$, $H_2SO_4$*, $HCl$, $HI$, $HBr$, $HClO_3$, $HBrO_3$<br>2. Alkali and alkaline-earth hydroxides<br>3. Most salts | 1. Many inorganic acids, including $H_2CO_3$, $H_3BO_3$, $H_3PO_4$, $H_2S$, $H_2SO_3$<br>2. Most organic acids<br>3. Ammonia and most organic bases<br>4. Halides, cyanides, and thiocyanates of Hg, Zn, and Cd |

*$H_2SO_4$ is completely dissociated into $HSO_4^-$ and $H_3O^+$ ions and for this reason is classified as a strong electrolyte. However, it should be noted that the $HSO_4^-$ ion is a weak electrolyte, being only partially dissociated.

Svante Arrhenius (1859–1927), Swedish chemist, formulated many of the early ideas regarding ionic dissociation in solution. His ideas were not accepted at first; in fact, he was given the lowest possible passing grade for his Ph.D. examination. In 1903 Arrhenius was awarded the Nobel Prize in Chemistry for these revolutionary ideas. He was one of the first scientists to suggest the relationship between the amount of carbon dioxide in the atmosphere and global temperature, a phenomenon that has come to be known as the greenhouse effect.

### Conjugate Acids and Bases

An important feature of the Brønsted-Lowry concept is that when an acid gives up a proton, a *conjugate base* is formed that is capable of accepting a proton. For example, when the species Acid₁ gives up a proton, the species Base₁ is formed, as shown by the reaction

$$Acid_1 \rightleftarrows Base_1 + Proton$$

Here, Acid₁ and Base₁ are a conjugate acid/base pair.

Similarly, every base produces its *conjugate acid* as a result of accepting a proton. That is,

$$Base_2 + Proton \rightleftarrows Acid_2$$

When these two processes are combined, the result is an acid/base, or neutralization, reaction:

$$Acid_1 + Base_2 \rightleftarrows Base_1 + Acid_2$$

The extent to which this reaction proceeds depends upon the relative tendencies of the two bases to accept a proton (or of the two acids to donate a proton). Examples of conjugate acid/base relationships are shown in Equations 6–1 through 6–4.

Many solvents are proton donors or proton acceptors and can thus induce basic or acidic behavior in solutes dissolved in them. For example, in an aqueous solution of ammonia, water donates a proton and thus acts as an acid with respect to the solute:

$$\underset{\text{Acid}_2}{H_2O} + \underset{\text{Base}_1}{NH_3} \rightleftarrows \underset{\substack{\text{Conjugate} \\ \text{Base}_2}}{OH^-} + \underset{\substack{\text{Conjugate} \\ \text{Acid}_1}}{NH_4^+} \qquad (6\text{--}1)$$

In this reaction, ammonia (Base₁) reacts with water, which is labeled Acid₂, to give the conjugate acid ammonium ion (Acid₁) and hydroxide ion, which is the conjugate base (Base₂) of the acid water. In contrast,

> A conjugate base is the species formed when an acid loses a proton.

For example, acetate ion is the conjugate base of acetic acid, and ammonium ion is the conjugate acid of ammonia.

> A conjugate acid is the species formed when a base accepts a proton.

Water can act as either an acid or a base.

---

**Feature 6–1**
AMPHIPROTIC SPECIES

Some compounds behave as both an acid and a base. An example is dihydrogen phosphate ion ($H_2PO_4^-$), which behaves as a base in the presence of a proton donor, such as $H_3O^+$:

$$H_2PO_4^- + H_3O^+ \rightleftarrows H_3PO_4 + H_2O$$

Base₁      Acid₂      Acid₁      Base₂

Here, $H_3PO_4$ is the conjugate acid of the original base. In the presence of a proton acceptor, such as water, however, $H_2PO_4^-$ behaves as an acid and forms the conjugate base $HPO_4^{2-}$:

$$H_2PO_4^- + H_2O \rightleftarrows HPO_4^{2-} + H_3O^+$$

Acid₁      Base₂      Base₁      Acid₂

---

water acts as a proton acceptor, or base, in an aqueous solution of nitrous acid:

$$HNO_2 + H_2O \rightleftarrows NO_2^- + H_3O^+ \tag{6–2}$$

Acid₂      Base₁      Conjugate Base₂      Conjugate Acid₁

Nitrite ion is the conjugate base of the acid $HNO_2$; $H_3O^+$ is the conjugate acid of the base $H_2O$. Neither $NH_3$ nor $HNO_2$ reacts completely with $H_2O$; therefore, ammonia and nitrous acid are both termed weak electrolytes.

### Amphiprotic Solvents

Water is the classic example of an *amphiprotic* solvent—that is, a solvent that can act either as an acid (Equation 6–1) or as a base (Equation 6–2), depending upon the solute. Other common amphiprotic solvents are methanol, ethanol, and anhydrous acetic acid. In methanol, for example, the equilibria analogous to those shown in Equations 6–1 and 6–2 are

$$NH_3 + CH_3OH \rightleftarrows NH_4^+ + CH_3O^- \tag{6–3}$$

$$CH_3OH + HNO_2 \rightleftarrows CH_3OH_2^+ + NO_2^- \tag{6–4}$$

Base₁      Acid₂      Conjugate Acid₁      Conjugate Base₂

It is important to emphasize that an acid that has donated a proton becomes a conjugate base capable of accepting a proton to reform the original acid; the converse holds equally well. Thus, nitrite ion, the species produced by the loss of a proton from nitrous acid, is a potential acceptor of a proton from a suitable donor. It is this proton-accepting

*Amphiprotic solvents behave as acids in the presence of basic solutes and as bases in the presence of acidic solutes.*

reaction that causes an aqueous solution of sodium nitrite to be slightly basic:

$$\underset{\text{Base}_1}{NO_2^-} + \underset{\text{Acid}_2}{H_2O} \rightleftarrows \underset{\substack{\text{Conjugate} \\ \text{Acid}_1}}{HNO_2} + \underset{\substack{\text{Conjugate} \\ \text{Base}_2}}{OH^-}$$

## 6A-3 Autoprotolysis

Amphiprotic solvents undergo self-ionization, or *autoprotolysis*, to form a pair of ionic species. Autoprotolysis is yet another example of acid/base behavior, as illustrated by the following equations:

| Base$_1$ | + | Acid$_2$ | $\rightleftarrows$ | Acid$_1$ | + | Base$_2$ |
|---|---|---|---|---|---|---|
| $H_2O$ | + | $H_2O$ | $\rightleftarrows$ | $H_3O^+$ | + | $OH^-$ |
| $CH_3OH$ | + | $CH_3OH$ | $\rightleftarrows$ | $CH_3OH_2^+$ | + | $CH_3O^-$ |
| $HCOOH$ | + | $HCOOH$ | $\rightleftarrows$ | $HCOOH_2^+$ | + | $HCOO^-$ |
| $NH_3$ | + | $NH_3$ | $\rightleftarrows$ | $NH_4^+$ | + | $NH_2^-$ |

> The hydronium ion $H_3O^+$ is the reaction product of water with an acid.

The product $H_3O^+$ formed by the autoprotolysis of water is called the *hydronium ion* and consists of a proton covalently bonded to a water molecule by one of the unshared electron pairs of the oxygen. Higher hydrates, such as $H_5O_2^+$ and $H_9O_4^+$, also exist, but they are orders of magnitude less stable than $H_3O^+$. Essentially no unhydrated protons exist in aqueous solutions.[2]

To emphasize the extraordinary stability of the singly hydrated proton, many chemists use the notation $H_3O^+$ when writing equations for aqueous solutions containing the proton. Others use $H^+$ because this notation simplifies the balancing of equations in which the proton is a participant. Ordinarily, we use $H_3O^+$ in acid/base equilibrium calculations but the simpler $H^+$ notation otherwise.

> In this text we use the symbol $H_3O^+$ in those chapters that deal with acid/base equilibria and acid/base equilibrium calculations. In the remaining chapters we simplify to the more convenient $H^+$, it being understood that this symbol represents the hydronium ion.

The extent to which water undergoes autoprotolysis is slight at room temperature. Thus, the hydronium and hydroxide ion concentrations in pure water are only about $10^{-7}$ M. Nevertheless, this dissociation reaction is of utmost importance in understanding the behavior of aqueous solutions.

## 6A-4 Strengths of Acids and Bases

Figure 6-1 shows the dissociation reactions of a few common acids in water. The first two are *strong acids* because reaction with the solvent is sufficiently complete as to leave no undissociated solute molecules in aqueous solution. The remainder are *weak acids*, which react incompletely with water to give solutions that contain significant quantities of both the parent acid and its conjugate base. Note that acids can be cationic, anionic, or electrically neutral.

> The common strong acids are HCl, $HClO_4$, $HNO_3$, the first hydrogen in $H_2SO_4$, HBr, HI, and the organic sulfonic acids ($RSO_3H$).

---

[2]See P. A. Giguere, *J. Chem. Educ.*, **1979**, *56*, 571.

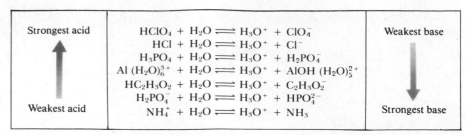

**Figure 6–1**

Dissociation reactions and relative strengths of some common acids and their conjugate bases.

The common strong bases are NaOH, KOH, $Ba(OH)_2$, and the quaternary ammonium hydroxides ($R_4NOH$, where R is an alkyl group such as $CH_3$ or $C_2H_5$).

The acids in Figure 6–1 become progressively weaker from top to bottom. Perchloric acid and hydrochloric acid are completely dissociated. In contrast, only about 1% of acetic acid ($HC_2H_3O_2$) is dissociated. Ammonium ion is an even weaker acid; only about 0.01% of this ion is dissociated into hydronium ions and ammonia molecules. Another generality illustrated in Figure 6–1 is that the weakest acid forms the strongest conjugate base; that is, ammonia has a much stronger affinity for protons than any base above it. Perchlorate and chloride ions have no affinity for protons in water.

The tendency of a solvent to accept or donate protons determines the strength of a solute acid or base dissolved in it. For example, perchloric and hydrochloric acids are strong acids in water. If anhydrous acetic acid, a poorer proton acceptor than water, is substituted *as the solvent,* neither perchloric nor hydrochloric acid undergoes complete dissociation; instead, equilibria such as the following are established:

Of all the acids listed in Figure 6–1, only perchloric acid is a strong acid in methanol and ethanol. Thus, these two alcohols are also differentiating solvents.

$$CH_3COOH + HClO_4 \rightleftharpoons CH_3COOH_2^+ + ClO_4^-$$

Base₁        Acid₂        Acid₁        Base₂

In a differentiating solvent, various acids dissociate to different degrees and are thus of different strengths. In a leveling solvent, several acids are completely dissociated and are thus of the same strength.

Perchloric acid is, however, considerably stronger than hydrochloric acid in this solvent; its dissociation is about 5000 times greater than that of the hydrochloric acid. Acetic acid thus acts as a *differentiating* solvent toward these two acids in the sense that its use reveals inherent differences in their acidities. Water, on the other hand, is a *leveling* solvent for perchloric, hydrochloric, nitric, and sulfuric acids in that all are completely ionized in it and thus exhibit no differences in strength.

## 6B  CHEMICAL EQUILIBRIUM

The reactions used in analytical chemistry are seldom complete. Instead, they proceed to a state of *chemical equilibrium* in which the ratio of concentrations of reactants and products is constant. *Equilibrium-constant expressions* are *algebraic* equations that describe the concentration relationships among reactants and products at chemical equilibrium. Such relationships permit calculation of the quantity of analyte that remains unreacted when a steady state has been reached. From these data, the error resulting from incompleteness of the reaction upon which an analysis is based can be computed.

The discussion that follows demonstrates the use of equilibrium-constant expressions to gain information about analytical systems in which no more than one or two equilibria are important. Chapter 8 extends these methods to systems containing several simultaneous equilibria. Such complex systems are often encountered in analytical chemistry.

## 6B–1  The Equilibrium State

Consider the chemical equilibrium

$$H_3AsO_4 + 3\,I^- + 2\,H^+ \rightleftharpoons H_3AsO_3 + I_3^- + H_2O \qquad (6-5)$$

This reaction is shown in the color plates.

The rate of this reaction and the extent to which it proceeds to the right can be readily judged by observing the orange-red color of the triiodide ion ($I_3^-$) since all the other participants in the reaction are colorless. If, for example, 1 mmol of arsenic acid ($H_3AsO_4$) is added to 100 mL of a solution containing 3 mmol of potassium iodide, the red color of the triiodide ion appears almost immediately, and within a few seconds the intensity of the color becomes constant, which shows that the triiodide concentration has become constant.

A solution of identical color intensity (and hence identical triiodide concentration) can also be produced by adding 1 mmol of arsenous acid ($H_3AsO_3$) to 100 mL of a solution containing 1 mmol of triiodide ion. Here, the color intensity is initially greater than in the first solution but rapidly decreases as a result of the reaction

$$H_3AsO_3 + I_3^- + H_2O \rightleftharpoons H_3AsO_4 + 3\,I^- + 2\,H^+$$

Ultimately the color of the two solutions is identical. Many other combinations of the four reactants can be employed to yield solutions that are indistinguishable from the two just described.

The foregoing observations illustrate that the concentration relationship at chemical equilibrium (that is, the *position of equilibrium*) is independent of the route by which the equilibrium state is achieved. Moreover, a system that is in equilibrium will not spontaneously depart from this condition unless a stress is applied to the system. Such stresses include changes in temperature, in pressure (if one of the reactants or products is a gas), or in total concentration of a reactant or a product. These effects can be predicted qualitatively from the *principle of Le Châtelier,* which states that the position of chemical equilibrium always shifts in the direction that tends to relieve the effect of an applied stress. Thus, an increase in temperature alters the concentration relationship in the direction that tends to absorb heat, and an increase in pressure favors those participants that occupy a smaller total volume.

In an analysis, the effect of introducing an additional amount of a participating species to the reaction mixture is particularly important. Here, the resulting stress is relieved by a shift in equilibrium in the direction that partially uses up the added substance. Thus, for the equilibrium we have been considering (Equation 6–5), the addition of either arsenic acid ($H_3AsO_4$) or hydrogen ions causes an increase in color as more triiodide ion and arsenous acid are formed; the addition of arsenous acid has the

The position of a chemical equilibrium is independent of the route by which equilibrium is reached.

The Le Châtelier principle states that the position of an equilibrium always shifts in such a direction as to relieve any stress applied to the system.

The mass action effect is a shift in the position of an equilibrium caused by adding one of the reactants or products to a system.

Chemical reactions do not cease at equilibrium. Instead, the amounts of reactants and products appear to be constant because the rates of the forward and reverse processes are identical.

Thermodynamics is a branch of chemical science that deals with heat or energy flow in chemical reactions. The position of a chemical equilibrium can be related to the heat generated or dissipated in a chemical reaction.

Equilibrium-constant expressions provide *no* information as to whether a chemical reaction is fast enough to be used for an analysis.

Cato Guldberg (1836–1902) and Peter Waage (1833–1900) were Norwegian chemists whose primary interests were in the field of thermodynamics. In 1864, these workers were the first to propose the law of mass action, which is expressed in Equation 6–7.

The term $[Z]^z$ in Equation 6–7 is replaced with

1. $p_z$ in atmospheres if Z is a gas.
2. 1.00 if Z is a pure solid present in excess.
3. 1.00 if Z is a pure liquid present in excess.
4. 1.00 if Z is the solvent $H_2O$.

reverse effect. An equilibrium shift brought about by changing the amount of one of the participating species is called a *mass-action effect*.

If it were possible to examine the system under discussion at the molecular level, we would find that interactions among the participating species continue unabated even after equilibrium is achieved. The constant concentration ratio of reactants and products results from the equality in the rates of the forward and reverse reactions. In other words, chemical equilibrium is a dynamic state in which the rates of the forward and reverse reactions are identical.

## 6B–2 Equilibrium-Constant Expressions

The influence of concentration (or pressure if the species are gases) on the position of a chemical equilibrium is conveniently described in quantitative terms by means of an equilibrium-constant expression. Such expressions are readily derived from thermodynamic theory. They are of great practical importance because they permit the chemist to predict the direction and completeness of a chemical reaction. We must emphasize, however, that an equilibrium-constant expression yields no information concerning the *rate* at which equilibrium is approached. In fact, we sometimes encounter reactions that have highly favorable equilibrium constants but are of little analytical use because their rates are low. This limitation can often be overcome by the use of a catalyst, which speeds the attainment of equilibrium without changing its position.

Let us consider a generalized equation for a chemical equilibrium:

$$wW + xX \rightleftharpoons yY + zZ \qquad (6-6)$$

where the capital letters represent the formulas of participating chemical species and the lower case italic letters are the small whole numbers required to balance the equation. Thus, the equation states that $w$ mol of W reacts with $x$ mol of X to form $y$ mol of Y and $z$ mol of Z. The equilibrium-constant expression for this reaction is

$$\frac{[Y]^y[Z]^z}{[W]^w[X]^x} = K \qquad (6-7)$$

where the square-bracketed terms have the following meanings:

1. Molar concentration if the species is a dissolved solute.
2. Partial pressure in atmospheres if the species is a gas; in fact, we will often replace the square-bracketed term (say [Z] in Equation 6–7) with the symbol $p_z$, which stands for the partial pressure of the gas Z in atmospheres.
3. Unity if the species is (a) a pure liquid, (b) a pure solid, or (c) the solvent in a dilute solution.

The rationale for these substitutions is explained in the description of the various types of equilibrium constants found in Section 6B–3.

The constant $K$ in Equation 6–7 is a temperature-dependent numerical quantity called the *equilibrium constant*. By convention, the concentrations of the products, *as the equation is written,* are always placed in the numerator and the concentrations of the reactants in the denominator.

Equation 6–7 is only an approximate form of a thermodynamic equilibrium-constant expression. The exact form is given by Equation 7–3. Generally we will use the approximate form of this equation because it is less tedious and time-consuming to use. In Section 7B, we show when the use of Equation 6–7 is likely to lead to serious errors in equilibrium calculations and how this equation can be modified to minimize such errors.

## 6B–3 Types of Equilibrium Constants Encountered in Analytical Chemistry

Table 6–2 summarizes the types of chemical equilibria and equilibrium constants important in analytical chemistry. Applications of some of these constants are illustrated in the paragraphs that follow.

### The Ion-Product Constant for Water

Aqueous solutions contain small amounts of hydronium and hydroxide ions as a consequence of the dissociation reaction

$$2 H_2O \rightleftharpoons H_3O^+ + OH^- \tag{6–8}$$

Note that assuming $[H_2O]$ to be constant and including this concentration in the equilibrium constant is simply another way of stating the convention that the concentration term for a solvent in a dilute solution is 1.00.

**Table 6–2**
**EQUILIBRIA AND EQUILIBRIUM CONSTANTS OF IMPORTANCE TO ANALYTICAL CHEMISTRY**

| Type of Equilibrium | Name and Symbol of Equilibrium Constant | Typical Example | Equilibrium-Constant Expression |
|---|---|---|---|
| Dissociation of water | Ion-product constant, $K_w$ | $2 H_2O \rightleftharpoons H_3O^+ + OH^-$ | $K_w = [H_3O^+][OH^-]$ |
| Heterogeneous equilibrium between a slightly soluble substance and its ions in a saturated solution | Solubility product, $K_{sp}$ | $BaSO_4(s) \rightleftharpoons Ba^{2+} + SO_4^{2-}$ | $K_{sp} = [Ba^{2+}][SO_4^{2-}]$ |
| Dissociation of a weak acid or base | Dissociation constant, $K_a$ or $K_b$ | $CH_3COOH + H_2O \rightleftharpoons H_3O^+ + CH_3COO^-$ $CH_3COO^- + H_2O \rightleftharpoons OH^- + CH_3COOH$ | $K_a = \dfrac{[H_3O^+][CH_3COO^-]}{[CH_3COOH]}$ $K_b = \dfrac{[OH^-][CH_3COOH]}{[CH_3COO^-]}$ |
| Formation of a complex ion | Formation constant, $\beta_n$ | $Ni^{2+} + 4 CN^- \rightleftharpoons Ni(CN)_4^{2-}$ | $\beta_4 = \dfrac{[Ni(CN)_4^{2-}]}{[Ni^{2+}][CN^-]^4}$ |
| Oxidation/reduction equilibrium | $K_{redox}$ | $MnO_4^- + 5 Fe^{2+} + 8 H^+ \rightleftharpoons Mn^{2+} + 5 Fe^{3+} + 4 H_2O$ | $K_{redox} = \dfrac{[Mn^{2+}][Fe^{3+}]^5}{[MnO_4^-][Fe^{2+}]^5[H^+]^8}$ |
| Distribution equilibrium between immiscible solvents | $K_d$ | $I_2(aq) \rightleftharpoons I_2(org)$ | $K_d = \dfrac{[I_2]_{org}}{[I_2]_{aq}}$ |

**Feature 6–2**
WHY [H₂O] DOES NOT APPEAR IN EQUILIBRIUM-CONSTANT EXPRESSIONS FOR AQUEOUS SOLUTIONS

In a dilute aqueous solution, the molar concentration of water is

$$[H_2O] = \frac{1000 \text{ g } H_2O}{\text{L } H_2O} \times \frac{1 \text{ mol } H_2O}{18.0 \text{ g } H_2O} = 55.6 \text{ M}$$

Let us suppose we have 0.1 mol of HCl in 1 L of water. The presence of this acid will shift the equilibrium shown in Equation 6–8 to the left by the mass-action effect. Originally there was only $10^{-7}$ mol/L of OH⁻ in the water to consume the added protons. Thus, even if all the OH⁻ ions are converted to H₂O, the water concentration will increase only to

$$[H_2O] = 55.6 \frac{\text{mol } H_2O}{\text{L } H_2O} + 1 \times 10^{-7} \frac{\text{mol } OH^-}{\text{L } H_2O} \times \frac{1 \text{ mol } H_2O}{\text{mol } OH^-}$$

$$= 55.6 \text{ M}$$

The percent change in water concentration is

$$\frac{10^{-7} \text{ M}}{55.6 \text{ M}} \times 100\% = 2 \times 10^{-7}\%$$

which is certainly negligible. Thus, $K[H_2O]^2$ in Equation 6–9 is, for all practical purposes, a constant. That is,

$$K(55.6)^2 = K_w = 1.00 \times 10^{-14}$$

An equilibrium constant for this reaction can be formulated as shown in Equation 6–7:

$$\frac{[H_3O^+][OH^-]}{[H_2O]^2} = K \qquad (6\text{–}9)$$

The concentration of water in dilute aqueous solutions is enormous, however, when compared with the concentration of hydrogen and hydroxide ions. As a consequence, [H₂O] in Equation 6–9 can be taken as constant, and we write

$$K[H_2O]^2 = K_w = \boxed{[H_3O^+][OH^-]} \qquad (6\text{–}10)$$

where the new constant $K_w$ is given a special name, the *ion-product constant for water.*

A useful relationship is obtained by taking the negative logarithm of Equation 6–10

$$-\log K_w = -\log[H_3O^+] - \log[OH^-]$$

By definition of p-functions,

$$pK_w = pH + pOH$$

At 25°C, $pK_w = 14.00$.

At 25°C, the ion-product constant for water is $1.008 \times 10^{-14}$. For convenience, we will use the approximation that $K_w \approx 1.00 \times 10^{-14}$. Table 6–3 shows the dependence of this constant upon temperature.

The ion-product constant for water allows us to calculate the hydronium and hydroxide ion concentrations of aqueous solutions.

| Table 6–3 VARIATION OF $K_w$ WITH TEMPERATURE | |
| --- | --- |
| Temperature, °C | $K_w$ |
| 0 | $0.114 \times 10^{-14}$ |
| 25 | $1.01 \times 10^{-14}$ |
| 50 | $5.47 \times 10^{-14}$ |
| 100 | $49 \times 10^{-14}$ |

---

Example 6–1

Calculate the hydronium and hydroxide ion concentrations of pure water at 25 and 100°C.

Because $OH^-$ and $H_3O^+$ are formed only from the dissociation of water, their concentrations must be equal:

$$[H_3O^+] = [OH^-]$$

Substitution into Equation 6–10 gives

$$[H_3O^+]^2 = [OH^-]^2 = K_w$$
$$[H_3O^+] = [OH^-] = \sqrt{K_w}$$

At 25°C,

$$[H_3O^+] = [OH^-] = \sqrt{1.00 \times 10^{-14}} = 1.00 \times 10^{-7} \text{ M}$$

At 100°C, from Table 6–3,

$$[H_3O^+] = [OH^-] = \sqrt{49 \times 10^{-14}} = 7.0 \times 10^{-7} \text{ M}$$

---

Example 6–2

Calculate the hydronium and hydroxide ion concentrations in 0.200 M aqueous NaOH.

Sodium hydroxide is a strong electrolyte, and so its contribution to the hydroxide ion concentration in this solution is 0.200 mol/L. As in Example 6–1, hydroxide ions and hydronium ions are formed *in equal amounts* from the dissociation of water. Therefore, we write

$$[OH^-] = 0.200 + [H_3O^+]$$

where $[H_3O^+]$ accounts for the hydroxide ions contributed by the solvent. The concentration of $OH^-$ and $H_3O^+$ from the water is insignificant, however, when compared with 0.200, and so we can write

$$[OH^-] \approx 0.200$$

Equation 6–10 is then used to calculate the hydronium ion concentration:

$$[H_3O^+] = \frac{K_w}{[OH^-]} = \frac{1.00 \times 10^{-14}}{0.200} = 5.00 \times 10^{-14}$$

Note that the approximation

$$[OH^-] = 0.200 + 5.00 \times 10^{-14} \approx 0.200$$

causes no significant error.

---

## Solubility-Product Constants

When we say that a sparingly soluble salt is completely dissociated, *we do not imply* that all of the salt dissolves. The very small amount that *does* go into solution dissociates completely.

Most sparingly soluble salts are essentially completely dissociated in saturated aqueous solution. For example, when an excess of barium iodate is equilibrated with water, the dissociation process is adequately described by the equation

$$Ba(IO_3)_2(s) \rightleftarrows Ba^{2+} + 2\,IO_3^-$$

Application of Equation 6–7 leads to

$$\frac{[Ba^{2+}][IO_3^-]^2}{[Ba(IO_3)_2(s)]} = K$$

Inclusion of $[Ba(IO_3)_2]$ in $K_{sp}$ is another way of saying that this bracketed term has a numerical value of 1.00.

The denominator represents the molar concentration of $Ba(IO_3)_2$ *in the solid,* which is a second phase in contact with the saturated solution. The concentration of a compound in its solid state is, however, constant. In other words, the number of moles of $Ba(IO_3)_2$ divided by the *volume* of the $Ba(IO_3)_2$ is constant no matter how much excess solid is present. Therefore, the foregoing equation can be rewritten in the form

$$[Ba^{2+}][IO_3^-]^2 = K[Ba(IO_3)_2(s)] = K_{sp} \qquad (6\text{--}11)$$

For Equation 6–11 to apply, it is necessary only that *some solid be present. You should always keep in mind that if no Ba(IO₃)₂(s) is present, Equation 6–11 is not valid.*

where the new constant is called the *solubility-product constant* or the *solubility product*. It is important to appreciate that the position of equilibrium in Equation 6–11 is independent of the total *amount* of solid $Ba(IO_3)_2$. It is necessary only that some of the solid be in contact with its saturated solution. This is a generality that applies to all solubility equilibria.

A table of solubility products for numerous inorganic salts is found in Appendix 2. The examples that follow demonstrate some typical uses of solubility-product expressions. Further applications are considered in Chapters 7, 8, and 9.

### The Solubility of a Precipitate in Water

The solubility-product expression permits us to calculate the solubility of a sparingly soluble substance that ionizes in water.

---

Example 6–3

How many grams of $Ba(IO_3)_2$ (fw = 487 g) can be dissolved in 500 mL of water at 25°C?

The solubility-product constant for $Ba(IO_3)_2$ is $1.57 \times 10^{-9}$ (Appendix 2). The equilibrium between the solid and its ions in solution is described by the equation

$$Ba(IO_3)_2(s) \rightleftarrows Ba^{2+} + 2\ IO_3^-$$

and so

$$[Ba^{2+}][IO_3^-]^2 = 1.57 \times 10^{-9} = K_{sp}$$

The equation describing the equilibrium reveals that 1 mol of $Ba^{2+}$ is formed for each mole of $Ba(IO_3)_2$ that dissolves. Therefore,

$$\text{molar solubility of } Ba(IO_3)_2 = [Ba^{2+}]$$

The molar solubility is equal to $[Ba^+]$ or to $\frac{1}{2}[IO_3^-]$.

The iodate concentration is clearly twice that for barium ion:

$$[IO_3^-] = 2[Ba^{2+}]$$

Substituting this equality into the equilibrium-constant expression gives

$$[Ba^{2+}](2[Ba^{2+}])^2 = 4[Ba^{2+}]^3 = 1.57 \times 10^{-9}$$

$$[Ba^{2+}] = \left(\frac{1.57 \times 10^{-9}}{4}\right)^{1/3} = 7.32 \times 10^{-4} \text{ M}$$

Since 1 mol $Ba^{2+}$ is produced for every mole of $Ba(IO_3)_2$

$$\text{solubility} = 7.32 \times 10^{-4} \text{ M}$$

To compute the number of millimoles of $Ba(IO_3)_2$ dissolved in 500 mL of solution, we write

$$\text{amount } Ba(IO_3)_2 = 7.32 \times 10^{-4}\ \frac{\text{mmol } Ba(IO_3)_2}{\text{mL}} \times 500\ \text{mL}$$

The weight of $Ba(IO_3)_2$ in 500 mL is given by

$$\text{wt } Ba(IO_3)_2 = 7.32 \times 10^{-4} \times 500\ \text{mmol } Ba(IO_3)_2$$

$$\times\ 0.487\ \frac{\text{g } Ba(IO_3)_2}{\text{mmol } Ba(IO_3)_2}$$

$$= 0.178 \text{ g}$$

---

**The Effect of a Common Ion on Solubility.** The common-ion effect predicted from the Le Châtelier principle is demonstrated by the following examples.

Example 6–4

Calculate the molar solubility of $Ba(IO_3)_2$ in a solution that is 0.0200 M in $Ba(NO_3)_2$.

The solubility is no longer equal to $[Ba^{2+}]$ because $Ba(NO_3)_2$ is also a source of barium ions. We know, however, that solubility is related to $[IO_3^-]$:

$$\text{molar solubility of } Ba(IO_3)_2 = \tfrac{1}{2}[IO_3^-]$$

There are two sources of barium ions: $Ba(NO_3)_2$ and $Ba(IO_3)_2$. The contribution from the former is 0.0200 M, and that from the latter is equal to the molar solubility, or $\tfrac{1}{2}[IO_3^-]$. Thus,

$$[Ba^{2+}] = 0.0200 + \tfrac{1}{2}[IO_3^-]$$

Substitution of these quantities into the solubility-product expression yields

$$(0.0200 + \tfrac{1}{2}[IO_3^-])[IO_3^-]^2 = 1.57 \times 10^{-9}$$

Since the exact solution for $[IO_3^-]$ requires solving a cubic equation, we seek an approximation that simplifies the algebra. The small numerical value of $K_{sp}$ suggests that the solubility of $Ba(IO_3)_2$ is not large, and this is confirmed by the result obtained in Example 6–3. So let us suppose that the barium ion concentration resulting from the solubility of $Ba(IO_3)_2$ is small relative to that from the $Ba(NO_3)_2$. That is, $\tfrac{1}{2}[IO_3^-] \ll 0.0200$, and

$$[Ba^{2+}] = 0.0200 + \tfrac{1}{2}[IO_3^-] \approx 0.0200$$

The original equation then simplifies to

$$0.0200[IO_3^-]^2 = 1.57 \times 10^{-9}$$
$$[IO_3^-] = \sqrt{7.85 \times 10^{-8}} = 2.80 \times 10^{-4} \text{ M}$$

The assumption that $(0.0200 + \tfrac{1}{2} \times 2.80 \times 10^{-4}) \approx 0.0200$ does not appear to cause serious error because the second term, representing the amount of $Ba^{2+}$ from the dissociation of $Ba(IO_3)_2$, is only about 0.7% of 0.0200. Ordinarily, we consider an assumption of this type to be satisfactory if the discrepancy is less than 10%. Finally, then,

$$\text{solubility of } Ba(IO_3)_2 = \tfrac{1}{2}[IO_3^-] = \tfrac{1}{2} \times 2.80 \times 10^{-4} = 1.40 \times 10^{-4} \text{ M}$$

If we compare this result with the solubility of barium iodate in pure water (Example 6–3), we see that a small excess of the common ion has lowered the molar solubility of $Ba(IO_3)_2$ by a factor of about five.

### Example 6–5

Calculate the solubility of $Ba(IO_3)_2$ in a solution prepared by mixing 200 mL of 0.0100 M $Ba(NO_3)_2$ with 100 mL of 0.100 M $NaIO_3$.

We must first establish whether either reactant is in excess at equilibrium. The amounts taken are

$$\text{amount } Ba^{2+} = 200 \text{ mL} \times 0.0100 \text{ mmol/mL} = 2.00 \text{ mmol}$$

$$\text{amount } IO_3^- = 100 \text{ mL} \times 0.100 \text{ mmol/mL} = 10.0 \text{ mmol}$$

If formation of $Ba(IO_3)_2$ is complete,

$$\text{amount excess } NaIO_3 = 10.0 \text{ mmol} - 2 \times 2.00 \text{ mmol} = 6.0 \text{ mmol}$$

Thus,

$$[IO_3^-] = \frac{6.0 \text{ mmol}}{300 \text{ mL}} = 0.0200 \text{ M}$$

As in Example 6–3,

$$\text{molar solubility of } Ba(IO_3)_2 = [Ba^{2+}]$$

Here, however,

$$[IO_3^-] = 0.0200 + 2[Ba^{2+}]$$

where $2[Ba^{2+}]$ represents the iodate contributed by the sparingly soluble $Ba(IO_3)_2$. We can obtain a provisional answer after making the assumption that $[IO_3^-] \approx 0.0200$; thus

$$\text{solubility of } Ba(IO_3)_2 = [Ba^{2+}] = \frac{K_{sp}}{[IO_3^-]^2} = \frac{1.57 \times 10^{-9}}{(0.0200)^2}$$

$$= 3.93 \times 10^{-6} \text{ mol/L}$$

The approximation appears to be reasonable.

The uncertainty in $[IO_3^-]$ is 0.1 part in 6.0 or 1 part in 60. Thus $0.0200 \times (1/60) = 0.0003$ and we round to 0.0200 M.

Note that the results from the last two examples demonstrate that an excess of iodate ion is more effective in decreasing the solubility of $Ba(IO_3)_2$ than is the same excess of barium ion.

A 0.02 M excess of $Ba^{2+}$ lowers the solubility of $Ba(IO_3)_2$ by a factor of about five; this same excess of $IO_3^-$ lowers the solubility by roughly 200.

### Acid and Base Dissociation Constants

When a weak acid or a weak base is dissolved in water, partial dissociation occurs. Thus, for nitrous acid, we can write

$$HNO_2 + H_2O \rightleftharpoons H_3O^+ + NO_2^- \qquad \frac{[H_3O^+][NO_2^-]}{[HNO_2]} = K_a$$

where $K_a$ is the *acid dissociation constant* for nitrous acid. In an analogous way, the *base dissociation constant* for ammonia is

$$NH_3 + H_2O \rightleftarrows NH_4^+ + OH^- \qquad \frac{[NH_4^+][OH^-]}{[NH_3]} = K_b$$

Note that $[H_2O]$ does not appear in the denominator of either equation because its concentration is so large relative to the concentration of the weak acid or base that the dissociation does not alter $[H_2O]$ appreciably (see Feature 6–2). Just as in the derivation of the ion-product constant for water, $[H_2O]$ is incorporated in the equilibrium constants $K_a$ and $K_b$. Dissociation constants for weak acids and weak bases are found in Appendixes 3 and 4.

### Dissociation Constants for Conjugate Acid/Base Pairs

Consider the dissociation-constant expressions for ammonia and its conjugate acid, ammonium ion:

$$NH_3 + H_2O \rightleftarrows NH_4^+ + OH^- \qquad \frac{[NH_4^+][OH^-]}{[NH_3]} = K_b$$

$$NH_4^+ + H_2O \rightleftarrows NH_3 + H_3O^+ \qquad \frac{[NH_3][H_3O^+]}{[NH_4^+]} = K_a$$

Multiplication of one equilibrium-constant expression by the other gives

$$\frac{[NH_3][H_3O^+]}{[NH_4^+]} \times \frac{[NH_4^+][OH^-]}{[NH_3]} = [H_3O^+][OH^-] = K_a K_b$$

but

$$[H_3O^+][OH^-] = K_w$$

and therefore

$$K_a K_b = K_w \qquad\qquad (6\text{--}12)$$

**Feature 6–3**
**RELATIVE STRENGTHS OF CONJUGATE ACID/BASE PAIRS**

Equation 6–12 confirms the observation in Figure 6–1 that, as the acid of a conjugate acid/base pair becomes weaker, its conjugate base becomes stronger and vice versa. Thus the conjugate base of an acid with a dissociation constant of $10^{-2}$ has a basic dissociation constant of $10^{-12}$, whereas an acid with a dissociation constant of $10^{-9}$ has a conjugate base with a dissociation constant of $10^{-5}$.

This relationship is general for all conjugate acid/base pairs. Most tables of dissociation constants do not list both the acid and the base dissociation constant for conjugate pairs, since it is so easy to calculate one from the other with Equation 6–12.

---

Example 6–6

What is $K_b$ for the equilibrium

$$CN^- + H_2O \rightleftarrows HCN + OH^-$$

Examination of Appendix 4 (dissociation constants for bases) reveals no entry for $CN^-$. Appendix 3, however, lists a $K_a$ value of $2.1 \times 10^{-9}$ for HCN. Thus,

$$\frac{[HCN][OH^-]}{[CN^-]} = K_b = \frac{K_w}{K_{HCN}}$$

$$K_b = \frac{1.00 \times 10^{-14}}{2.1 \times 10^{-9}} = 4.8 \times 10^{-6}$$

---

Hydronium Ion Concentration in Solutions of Weak Acids

When the weak acid HA is dissolved in water, two equilibria are established that yield hydronium ions:

$$HA + H_2O \rightleftarrows H_3O^+ + A^- \qquad \frac{[H_3O^+][A^-]}{[HA]} = K_a$$

$$2\,H_2O \rightleftarrows H_3O^+ + OH^- \qquad [H_3O^+][OH^-] = K_w$$

Ordinarily, the hydronium ions produced from the weak acid suppress the dissociation of water to such an extent that the contribution of hydronium ions from the water is negligible. Under these circumstances, one $H_3O^+$ ion is formed for each $A^-$ ion, and we write

$$[A^-] \approx [H_3O^+] \tag{6–13}$$

Furthermore, the sum of the molar concentrations of the weak acid and its conjugate base must equal the analytical concentration of the acid $c_{HA}$ because the solution contains no other source of $A^-$ ions. Thus,

$$c_{HA} = [A^-] + [HA] \tag{6–14}$$

Substituting $[H_3O^+]$ for $[A^-]$ (Equation 6–13) in Equation 6–14 yields

$$c_{HA} = [H_3O^+] + [HA]$$

which rearranges to

$$[HA] = c_{HA} - [H_3O^+] \tag{6–15}$$

When $[A^-]$ and $[HA]$ are replaced by their equivalent terms from Equations 6–13 and 6–15, the equilibrium-constant expression becomes

$$\frac{[H_3O^+]^2}{c_{HA} - [H_3O^+]} = K_a \qquad (6-16)$$

which rearranges to

$$[H_3O^+]^2 + K_a[H_3O^+] - K_a c_{HA} = 0 \qquad (6-17)$$

The positive solution to this quadratic equation is

$$[H_3O^+] = \frac{-K_a + \sqrt{K_a^2 + 4 K_a c_{HA}}}{2} \qquad (6-18)$$

As an alternative to using Equation 6–18, Equation 6–17 may be solved by successive approximations as shown in Feature 6–4.

Equation 6–16 can frequently be simplified by making the assumption that dissociation does not appreciably decrease the molar concentration of HA. Thus, provided $[H_3O^+] \ll c_{HA}$, $c_{HA} - [H_3O^+] \approx c_{HA}$, and Equation 6–16 reduces to

$$\frac{[H_3O^+]^2}{c_{HA}} = K_a$$

$$[H_3O^+] = \sqrt{K_a c_{HA}} \qquad (6-19)$$

Table 6–4

ERROR INTRODUCED BY ASSUMING $H_3O^+$ CONCENTRATION IS SMALL RELATIVE TO $c_{HA}$ IN EQUATION 6–15

| $K_a$ | $c_{HA}$ | $[H_3O^+]$ Using Assumption | $[H_3O^+]$ Using More Exact Equation | Percent Error |
|---|---|---|---|---|
| $1.00 \times 10^{-2}$ | $1.00 \times 10^{-3}$ | $3.16 \times 10^{-3}$ | $0.92 \times 10^{-3}$ | 244 |
| | $1.00 \times 10^{-2}$ | $1.00 \times 10^{-2}$ | $0.62 \times 10^{-2}$ | 61 |
| | $1.00 \times 10^{-1}$ | $3.16 \times 10^{-2}$ | $2.70 \times 10^{-2}$ | 17 |
| $1.00 \times 10^{-4}$ | $1.00 \times 10^{-4}$ | $1.00 \times 10^{-4}$ | $0.62 \times 10^{-4}$ | 61 |
| | $1.00 \times 10^{-3}$ | $3.16 \times 10^{-4}$ | $2.70 \times 10^{-4}$ | 17 |
| | $1.00 \times 10^{-2}$ | $1.00 \times 10^{-3}$ | $0.95 \times 10^{-3}$ | 5.3 |
| | $1.00 \times 10^{-1}$ | $3.16 \times 10^{-3}$ | $3.11 \times 10^{-3}$ | 1.6 |
| $1.00 \times 10^{-6}$ | $1.00 \times 10^{-5}$ | $3.16 \times 10^{-6}$ | $2.70 \times 10^{-6}$ | 17 |
| | $1.00 \times 10^{-4}$ | $1.00 \times 10^{-5}$ | $0.95 \times 10^{-5}$ | 5.3 |
| | $1.00 \times 10^{-3}$ | $3.16 \times 10^{-5}$ | $3.11 \times 10^{-5}$ | 1.6 |
| | $1.00 \times 10^{-2}$ | $1.00 \times 10^{-4}$ | $9.95 \times 10^{-5}$ | 0.5 |
| | $1.00 \times 10^{-1}$ | $3.16 \times 10^{-4}$ | $3.16 \times 10^{-4}$ | 0.0 |

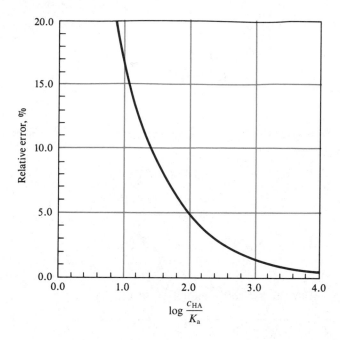

Figure 6–2
Relative error resulting from the assumption that $[H_3O^+] \ll c_{HA}$ in Equation 6–15.

The magnitude of the error introduced by the assumption that $[H_3O^+] \ll c_{HA}$ increases as the molar concentration of acid becomes smaller and as the dissociation constant for the acid becomes larger. This statement is supported by the data in Table 6–4. Note that the error introduced by the assumption is 0.5% when the ratio $c_{HA}/K_a$ is $10^4$. The error increases to 1.6% when the ratio is $10^3$, to 5.3% when it is $10^2$, and to 17% when it is 10. Figure 6–2 illustrates the effect graphically. Note that the hydronium ion concentration computed with the approximation becomes equal to or greater than the molar concentration of the acid when the ratio is unity or smaller, which is clearly a meaningless result.

In general, it is good practice to make the simplifying assumption and obtain a trial value for $[H_3O^+]$ that can be compared with $c_{HA}$ in Equation 6–15. If the trial value alters [HA] by an amount smaller than the allowable error in the calculation, the solution may be considered satisfactory. Otherwise, a better value for $[H_3O^+]$ must be obtained by solving the quadratic equation rigorously or by the method of successive approximations (Feature 6–4).

---

Example 6–7

Calculate the hydronium ion concentration in 0.120 M nitrous acid. The principal equilibrium is

$$HNO_2 + H_2O \rightleftarrows H_3O^+ + NO_2^-$$

for which (Appendix 3)

$$\frac{[H_3O^+][NO_2^-]}{[HNO_2]} = K_a = 5.1 \times 10^{-4}$$

Substitution into Equations 6–13 and 6–15 gives

$$[NO_2^-] = [H_3O^+]$$
$$[HNO_2] = 0.120 - [H_3O^+]$$

**Feature 6–4**
**THE METHOD OF SUCCESSIVE APPROXIMATIONS**

For convenience, let us write the quadratic equation in Example 6–8 in the form

$$x^2 + 2.54 \times 10^{-5}x - 5.08 \times 10^{-9} = 0$$

where $x = [H_3O^+]$.

As a first step, let us rearrange the equation to the form

$$x = \sqrt{5.08 \times 10^{-9} - 2.54 \times 10^{-5}x}$$

We then assume that $x$ on the right-hand side of the equation is zero and calculate a first value, $x_1$.

$$x_1 = \sqrt{5.08 \times 10^{-9} - 2.54 \times 10^{-5} \times 0} = 7.13 \times 10^{-5}$$

We then substitute this value into the original equation and compute a second value $x_2$:

$$x_2 = \sqrt{5.08 \times 10^{-9} - 2.54 \times 10^{-5} \times 7.13 \times 10^{-5}} = 5.72 \times 10^{-5}$$

Repeating this calculation gives

$$x_3 = \sqrt{5.08 \times 10^{-9} - 2.54 \times 10^{-5} \times 5.72 \times 10^{-5}} = 6.02 \times 10^{-5}$$

Continuing in the same way we obtain

$$x_4 = 5.96 \times 10^{-5}$$
$$x_5 = 5.97 \times 10^{-5}$$
$$x_6 = 5.97 \times 10^{-5}$$

Note that after three iterations, $x_3$ is $6.02 \times 10^{-5}$, which is within about 0.8% of the final value of $5.97 \times 10^{-5}$ M.

Occasionally, the final result will oscillate between a high and a low value. In this case, you may have to solve the quadratic equation.

The method of successive approximations is particularly useful when cubic or higher-power equations need to be solved.

When these relationships are introduced into the expression for $K_a$, we obtain

$$\frac{[H_3O^+]^2}{0.120 - [H_3O^+]} = 5.1 \times 10^{-4}$$

If we now assume $[H_3O^+] \ll 0.120$, we find

$$[H_3O^+] = \sqrt{0.120 \times 5.1 \times 10^{-4}} = 7.8 \times 10^{-3} \text{ M}$$

We now examine the assumption that $0.120 - 0.0078 \approx 0.120$ and see that the error is about 7%. The relative error in $[H_3O^+]$ is smaller than this, however, as we can see by calculating $\log (c_{HA}/K_a) = 2.4$, which, from Figure 6–2, suggests an error of about 3%. If a more accurate value is needed, solution of the quadratic equation yields $7.6 \times 10^{-3}$ M for the hydronium ion concentration.

---

Example 6–8

Calculate the hydronium ion concentration in a solution that is $2.0 \times 10^{-4}$ M in aniline hydrochloride ($C_6H_5NH_3Cl$).

In aqueous solution, dissociation of the salt to $Cl^-$ and $C_6H_5NH_3^+$ is complete. The weak acid $C_6H_5NH_3^+$ dissociates as follows:

$$C_6H_5NH_3^+ + H_2O \rightleftarrows C_6H_5NH_2 + H_3O^+ \qquad \frac{[H_3O^+][C_6H_5NH_2]}{[C_6H_5NH_3^+]} = K_a$$

Inspection of Appendix 3 reveals no entry for $C_6H_5NH_3^+$, but Appendix 4 gives a base constant for aniline ($C_6H_5NH_2$):

$$C_6H_5NH_2 + H_2O \rightleftarrows C_6H_5NH_3^+ + OH^- \qquad K_b = 3.94 \times 10^{-10}$$

The acid dissociation constant is then obtained from Equation 6–12:

$$K_a = \frac{1.00 \times 10^{-14}}{3.94 \times 10^{-10}} = 2.54 \times 10^{-5}$$

Proceeding as in Example 6–7, we have

$$[H_3O^+] = [C_6H_5NH_2]$$
$$[C_6H_5NH_3^+] = 2.0 \times 10^{-4} - [H_3O^+]$$

Let us assume that $[H_3O^+] \ll 2.0 \times 10^{-4}$ and substitute the simplified value for $[C_6H_5NH_3^+]$ into the dissociation-constant expression:

$$\frac{[H_3O^+]^2}{2.0 \times 10^{-4}} = 2.54 \times 10^{-5}$$
$$[H_3O^+] = \sqrt{5.08 \times 10^{-9}} = 7.1 \times 10^{-5} \text{ M}$$

Comparison of $7.1 \times 10^{-5}$ with $2.0 \times 10^{-4}$ suggests that a significant error has been introduced by the assumption that $[H_3O^+] \ll c_{C_6H_5NH_3^+}$ (Figure 6–2 indicates that this error is about 20%). Thus, unless only an approximate value for $[H_3O^+]$ is needed, it is necessary to use the expression

$$\frac{[H_3O^+]^2}{2.0 \times 10^{-4} - [H_3O^+]} = 2.54 \times 10^{-5}$$

which rearranges to

$$[H_3O^+]^2 + 2.54 \times 10^{-5}[H_3O^+] - 5.08 \times 10^{-9} = 0$$

$$[H_3O^+] = \frac{-2.54 \times 10^{-5} + \sqrt{(2.54 \times 10^{-5})^2 + 4 \times 5.08 \times 10^{-9}}}{2}$$

$$= 6.0 \times 10^{-5}$$

The quadratic equation can also be solved by the iterative method shown in Feature 6–4.

### Hydronium-Ion Concentration in Solutions of Weak Bases

The approaches described in previous sections are readily adapted to the calculation of the hydroxide or hydronium ion concentration in solutions of weak bases.

Aqueous ammonia is basic by virtue of the reaction

$$NH_3 + H_2O \rightleftharpoons NH_4^+ + OH^-$$

The predominant species in such solutions has been clearly demonstrated to be $NH_3$. Nevertheless, solutions of ammonia are sometimes called ammonium hydroxide because at one time chemists thought that $NH_4OH$ rather than $NH_3$ was the undissociated form of the base. Application of the mass law to the equilibrium as written yields

$$\frac{[NH_4^+][OH^-]}{[NH_3]} = K_b$$

Example 6–9

Calculate the hydronium ion concentration of a 0.075 M $NH_3$ solution.
  The predominant equilibrium is

$$NH_3 + H_2O \rightleftharpoons NH_4^+ + OH^-$$

From Appendix 4

$$\frac{[NH_4^+][OH^-]}{[NH_3]} = 1.76 \times 10^{-5} = K_b$$

The concentration of $OH^-$ from water dissociation is negligible. Therefore,

$$[NH_4^+] = [OH^-]$$

Both $NH_4^+$ and $NH_3$ come from the 0.075 M solution. Thus

$$[NH_4^+] + [NH_3] = c_{NH_3} = 0.075 \text{ M} \qquad (1)$$

If we substitute $[OH^-]$ for $[NH_4^+]$ in Equation (1) and rearrange, we find that

$$[NH_3] = 0.075 - [OH^-]$$

Substituting these quantities into the dissociation-constant expression yields

$$\frac{[OH^-]^2}{7.5 \times 10^{-2} - [OH^-]} = 1.76 \times 10^{-5}$$

Provided that $[OH^-] \ll 7.5 \times 10^{-2}$, this equation simplifies to

$$[OH^-]^2 \approx 7.5 \times 10^{-2} \times 1.76 \times 10^{-5}$$
$$[OH^-] = 1.15 \times 10^{-3} \text{ M}$$

Upon comparing the calculated value for $[OH^-]$ with $7.5 \times 10^{-2}$, we see that the error in $[OH^-]$ is less than 2%. If needed, a better value for $[OH^-]$ can be obtained by solving the quadratic equation.

Finally, then,

$$[H_3O^+] = \frac{K_w}{[OH^-]} = \frac{1.00 \times 10^{-14}}{1.15 \times 10^{-3}} = 8.7 \times 10^{-12} \text{ M}$$

---

Example 6–10

Calculate the hydronium ion concentration in a 0.010 M sodium hypochlorite solution.

The equilibrium between $OCl^-$ and water is

$$OCl^- + H_2O \rightleftarrows HOCl + OH^-$$

for which

$$\frac{[HOCl][OH^-]}{[OCl^-]} = K_b$$

Appendix 4 does not contain a value for $K_b$, but Appendix 3 reveals that the acid dissociation constant for HOCl is $3.0 \times 10^{-8}$. Therefore, we rearrange Equation 6–12 and write

$$K_b = \frac{K_w}{K_a} = \frac{1.00 \times 10^{-14}}{3.0 \times 10^{-8}} = 3.3 \times 10^{-7}$$

Proceeding as in Example 6–9, we have

$$[OH^-] = [HOCl]$$
$$[OCl^-] + [HOCl] = 0.010$$
$$[OCl^-] = 0.010 - [OH^-] \approx 0.010$$

Here we have assumed that $[OH^-] \ll 0.010$. Substitution into the equilibrium-constant expression gives

$$\frac{[OH^-]^2}{0.010} = 3.3 \times 10^{-7}$$

$$[OH^-] = 5.74 \times 10^{-5} \text{ M}$$

Note that the error resulting from the approximation is small. To obtain the hydronium ion concentration, we write

$$[H_3O^+] = \frac{1.00 \times 10^{-14}}{5.74 \times 10^{-5}} = 1.7 \times 10^{-10} \text{ M}$$

### Formation Constants for Complex Ions

An analytically important class of reactions involves the formation of soluble complex ions. Two examples are

$$Fe^{3+} + SCN^- \rightleftharpoons Fe(SCN)^{2+} \qquad \frac{[Fe(SCN)^{2+}]}{[Fe^{3+}][SCN^-]} = K_f$$

$$Zn(OH)_2(s) + 2\ OH^- \rightleftharpoons Zn(OH)_4^{2-} \qquad \frac{[Zn(OH)_4^{2-}]}{[OH^-]^2} = K_f$$

Equilibria involving complex ions are sometimes described in terms of dissociation reactions and *instability constants;* the latter, which are found in the older literature, are the reciprocals of formation constants. For example,

$$Fe(SCN)^{2+} \rightleftharpoons Fe^{3+} + SCN^-$$

$$\frac{[Fe^{3+}][SCN^-]}{[Fe(SCN)^{2+}]} = K_{inst} = \frac{1}{K_f}$$

where $K_f$ is called the *formation constant* for the complex. Note that the second constant applies only to a saturated solution that is in contact with the sparingly soluble zinc hydroxide. Note also that no concentration term for the zinc hydroxide appears in the formation-constant expression because its concentration in the solid is invariant and is included in $K_f$. In this regard, the treatment is analogous to that described for solubility-product expressions.

Formation constants for numerous complex ions appear in Appendix 5, and an application is given in the following example.

Example 6–11

A person with average eyesight can see the red color that $Fe(SCN)^{2+}$ imparts to an aqueous solution when the concentration of the complex is $6.4 \times 10^{-6}$ M or greater. What minimum concentration of KSCN is required to make it possible to detect 1 ppm of iron(III) in water from a spring if the formation constant for the complex is $1.4 \times 10^2$?

$$Fe^{3+} + SCN^- \rightleftharpoons Fe(SCN)^{2+} \qquad K_f = 1.4 \times 10^2$$

$$\frac{[Fe(SCN)^{2+}]}{[Fe^{3+}][SCN^-]} = 1.4 \times 10^2$$

To convert the minimum detectable concentration of iron(III) to molarity, we recall (page 102) that 1 ppm corresponds to 1 mg/L. Thus,

$$c_{Fe^{3+}} = \frac{1 \text{ mg}}{L} \times \frac{1 \text{ g}}{1000 \text{ mg}} \times \frac{1 \text{ mol}}{55.8 \text{ g}} = 1.8 \times 10^{-5} \text{ mol/L}$$

This detectable concentration is distributed between two species, $Fe^{3+}$ and $Fe(SCN)^{2+}$, and we may write

$$1.8 \times 10^{-5} = [Fe^{3+}] + [Fe(SCN)^{2+}]$$

Since $[Fe(SCN)^{2+}]$ must be $6.4 \times 10^{-6}$ M if this species is to be seen, we have

$$[Fe^{3+}] = 1.8 \times 10^{-5} - 6.4 \times 10^{-6} = 1.16 \times 10^{-5}$$

We now substitute these concentrations into the formation-constant expression in order to determine the $SCN^-$ concentration required:

$$\frac{6.4 \times 10^{-6}}{(1.16 \times 10^{-5})[SCN^-]} = 1.4 \times 10^2$$
$$[SCN^-] = 0.0039 \simeq 0.004$$

Thus, if the test solution is made 0.004 M or greater in KSCN, a detectable amount of $Fe(SCN)^{2+}$ will form, provided the total iron(III) concentration is 1 ppm or greater. In practice, higher concentrations of KSCN are desirable in order to convert a larger fraction of the $Fe^{3+}$ to the complex and thus enhance the sensitivity.

---

## Oxidation/Reduction Equilibrium Constants

Equilibrium constants for oxidation/reduction reactions can be formulated in the usual way. For example,

$$6\,Fe^{2+} + Cr_2O_7^{2-} + 14\,H_3O^+ \rightleftharpoons 6\,Fe^{3+} + 2\,Cr^{3+} + 21\,H_2O$$
$$\frac{[Fe^{3+}]^6[Cr^{3+}]^2}{[Fe^{2+}]^6[Cr_2O_7^{2-}][H_3O^+]^{14}} = K_{redox}$$

As in earlier examples, no term for the concentration of water is needed.

Equilibrium constants for oxidation/reduction reactions are not generally available because they are readily derived from more fundamental constants called *standard electrode potentials*. Calculations of this kind are treated in Chapter 14.

## Distribution Equilibrium Constants

Another important type of equilbrium involves the distribution of a solute between two immiscible liquid phases. For example, if an aqueous solution of iodine is shaken with an immiscible organic solvent such as hexane or chloroform, a portion of the iodine is extracted into the organic layer. Ultimately, an equilibrium is established between the two phases that can be described by

$$I_2(aq) \rightleftharpoons I_2(org) \qquad \frac{[I_2]_{org}}{[I_2]_{aq}} = K_d$$

The equilibrium constant $K_d$ for this reaction is often called a *distribution* or *partition coefficient*.

We shall see in Chapters 26 and 32 that distribution equilibria are of vital importance in many separation processes.

## Stepwise Equilibrium Constants

Many electrolytes associate or dissociate in a stepwise manner, and an equilibrium constant can be written for each step. For example, when ammonia is added to a solution containing zinc ions, the following equilibria are established:

$$Zn^{2+} + NH_3 \rightleftharpoons Zn(NH_3)^{2+} \qquad K_1 = \frac{[Zn(NH_3)^{2+}]}{[Zn^{2+}][NH_3]} = 2.5 \times 10^4$$

$$Zn(NH_3)^{2+} + NH_3 \rightleftharpoons Zn(NH_3)_2^{2+} \qquad K_2 = \frac{[Zn(NH_3)_2^{2+}]}{[Zn(NH_3)^{2+}][NH_3]} = 2.5 \times 10^4$$

$$Zn(NH_3)_2^{2+} + NH_3 \rightleftharpoons Zn(NH_3)_3^{2+} \qquad K_3 = \frac{[Zn(NH_3)_3^{2+}]}{[Zn(NH_3)_2^{2+}][NH_3]} = 3.2 \times 10^4$$

$$Zn(NH_3)_3^{2+} + NH_3 \rightleftharpoons Zn(NH_3)_4^{2+} \qquad K_4 = \frac{[Zn(NH_3)_4^{2+}]}{[Zn(NH_3)_3^{2+}][NH_3]} = 1.3 \times 10^4$$

where $K_1$, $K_2$, $K_3$, and $K_4$ are *stepwise formation constants* for the four complexes.

*Overall formation constants* are obtained by multiplying stepwise constants together. These are symbolized by $\beta_n$, where $n$ corresponds to the number of moles of complexing species that combine with 1 mole of cation. For the foregoing example,

$$\beta_2 = K_1 K_2 \qquad = 2.5 \times 10^4 \times 2.5 \times 10^4 = 6.2 \times 10^8$$

where $\beta_2$ is the overall formation constant for the reaction

$$Zn^{2+} + 2\,NH_3 \rightleftharpoons Zn(NH_3)_2^{2+} \qquad \beta_2 = \frac{[Zn(NH_3)_2^{2+}]}{[Zn^{2+}][NH_3]^2} = 6.2 \times 10^8$$

Similarly,

$$\beta_3 = K_1 K_2 K_3 \qquad = 6.2 \times 10^8 \times 3.2 \times 10^4 = 2.0 \times 10^{13}$$

where $\beta_3$ is the equilibrium constant for the reaction.

$$Zn^{2+} + 3\,NH_3 \rightleftharpoons Zn(NH_3)_2^{2+} \qquad \beta_3 = \frac{[Zn(NH_3)_3^{2+}]}{[Zn^{2+}][NH_3]^3} = 2.0 \times 10^{13}$$

Obviously $\beta_1$ and $K_1$ are identical.

Many common acids and bases undergo stepwise dissociation. For example, the following equilibria exist in an aqueous solution of phosphoric acid:

$$H_3PO_4 + H_2O \rightleftharpoons H_2PO_4^- + H_3O^+ \qquad \frac{[H_3O^+][H_2PO_4^-]}{[H_3PO_4]} = K_1 = 7.11 \times 10^{-3}$$

$$H_2PO_4^- + H_2O \rightleftharpoons HPO_4^{2-} + H_3O^+ \qquad \frac{[H_3O^+][HPO_4^{2-}]}{[H_2PO_4^-]} = K_2 = 6.34 \times 10^{-8}$$

$$HPO_4^{2-} + H_2O \rightleftharpoons PO_4^{3-} + H_3O^+ \qquad \frac{[H_3O^+][PO_4^{3-}]}{[HPO_4^{2-}]} = K_3 = 4.2 \times 10^{-13}$$

As illustrated by $H_3PO_4$, numerical values of $K_n$ for an acid or a base become smaller with each successive dissociation step.

## 6C   QUESTIONS AND PROBLEMS

Please note: Round any p-values to two places beyond the decimal.

6–1. Define
   *(a) salt.
   (b) electrolyte.
   *(c) hydronium ion.
   (d) conjugate base.
   *(e) base.
   (f) mass law.
   *(g) Le Châtelier principle.
   (h) autoprotolysis.
   *(i) amphiprotic solvent.
   (j) common-ion effect.

*6–2. Explain why water is termed a leveling solvent toward mineral acids, whereas ethanol is a differentiating solvent.

6–3. The molar analytical concentrations of saturated aqueous solutions of $Ni(OH)_2$ (fw = 92.7) and $Yb(OH)_3$ (fw = 224) are nearly identical ($5.45 \times 10^{-7}$ and $5.77 \times 10^{-7}$, respectively). Briefly account for any difference that may exist in the numerical values for their solubility-product constants.

6–4. Write balanced net ionic equations and equilibrium-constant expressions for
   *(a) the equilibrium that exists between the sparingly soluble magnesium hydroxide and its ions in solution.
   (b) the basic dissociation of methylamine, $CH_3NH_2$.
   *(c) the basic dissociation of sodium hypochlorite, NaOCl.
   (d) the acid dissociation constant for methylammonium chloride, $CH_3NH_3^+$.
   *(e) the acid dissociation constant for hypochlorous acid, HOCl.

6–5. Obtain numerical values for the constants in Problem 6–4 from Appendixes 3 and 4.

*6–6. Write balanced net-ionic equations and equilibrium-constant expressions for the acidic dissociation and also the base dissociation of sodium dihydrogen phosphate, $NaH_2PO_4$. What are the numerical values for these constants?

6–7. Write a balanced net-ionic equation and an equilibrium-constant expression for
   *(a) the reaction between magnesium hydroxide and hydrochloric acid.
   (b) the reduction of iron(III) with tin(II) [products: iron(II), Sn(IV)].
   *(c) the reaction between iodine and hydrogen sulfide [products: S(s), I$^-$].
   (d) the oxidation of metallic copper with nitric acid [products: NO(g), $Cu^{2+}$].
   *(e) the reaction between silver(I) and cyanide ion (product: $Ag(CN)_2^-$).
   (f) the reaction between zinc(II) and ammonia [product: $Zn(NH_3)_4^{2+}$].

6–8. Calculate the solubility-product constant for each of the following substances, given that the molar concentrations of their saturated solutions are as indicated:
   *(a) AgSeCN ($2.0 \times 10^{-8}$ mol/L; products are $Ag^+$ and $SeCN^-$).
   (b) $RaSO_4$ ($6.6 \times 10^{-6}$ mol/L).
   *(c) $Pb(BrO_3)_2$ ($1.7 \times 10^{-1}$ mol/L).
   (d) $PbBr_2$ ($2.1 \times 10^{-2}$ mol/L).
   *(e) $Ce(IO_3)_3$ ($1.9 \times 10^{-3}$ mol/L).
   (f) $BiI_3$ ($1.3 \times 10^{-5}$ mol/L).

6–9. Use the solubility data provided to complete the accompanying table.

| Substance (fw) | Solubility, mg/L | Molar Concentration | | $K_{sp}$ |
| --- | --- | --- | --- | --- |
| | | **Of Cation** | **Of Anion** | |
| (a) $AgI$ (234) | $2.14 \times 10^{-3}$ | | | |
| *(b) $SrSO_4$ (184) | | | | $3.20 \times 10^{-7}$ |
| (c) $Pb(IO_3)_2$ (557) | | | $8.6 \times 10^{-5}$ | |
| *(d) $Hg_2I_2$ (655) | | $2.2 \times 10^{-10}$ | $4.4 \times 10^{-10}$ | |
| *(e) $La(IO_3)_3$ (664) | | | | $6.2 \times 10^{-12}$ |
| (f) $In(OH)_3$ (166) | $2.17 \times 10^{-4}$ | | | |
| *(g) $Zn_2Fe(CN)_6$ (343) | | $9.4 \times 10^{-6}$ | | |
| (h) $Ag_3Fe(CN)_6$ (535) | | | | $6.3 \times 10^{-23}$ |
| (i) $In_4[Fe(CN)_6]_3$ (1109) | 0.18 | | | |

**6–10.** Calculate the weight (in grams) of $PbI_2$ that dissolves in 100 mL of
(a) $H_2O$.
(b) $2.00 \times 10^{-2}$ M KI.
(c) $2.00 \times 10^{-2}$ M $Pb(NO_3)_2$.

**\*6–11.** The solubility products for a series of iodides are
TlI    $K_{sp} = 6.5 \times 10^{-8}$

AgI    $K_{sp} = 8.3 \times 10^{-17}$

$PbI_2$    $K_{sp} = 7.1 \times 10^{-9}$

$BiI_3$    $K_{sp} = 8.1 \times 10^{-19}$
List these four compounds in order of decreasing molar solubility in
(a) water.
(b) 0.10 M NaI.
(c) a 0.010 M solution of the solute cation.

**6–12.** The solubility products for a series of iodates are
$AgIO_3$        $K_{sp} = 3.0 \times 10^{-8}$

$Sr(IO_3)_2$      $K_{sp} = 3.3 \times 10^{-7}$

$La(IO_3)_3$      $K_{sp} = 6.2 \times 10^{-12}$

$Ce(IO_3)_4$      $K_{sp} = 4.7 \times 10^{-17}$
List these four compounds in order of decreasing molar solubility in
(a) water.
(b) 0.10 M $NaIO_3$.
(c) a 0.10 M solution of the solute cation.

**\*6–13.** The solubility-product constant for $Ag_2CrO_4$ is $1.1 \times 10^{-12}$. What chromate-ion concentration is needed to
(a) initiate precipitation from a solution that is $4.00 \times 10^{-3}$ M in $Ag^+$?
(b) lower the silver-ion concentration to $5.0 \times 10^{-6}$ M?

**6–14.** The solubility-product constant for $Co(OH)_2$ is $2.0 \times 10^{-16}$. What hydroxide-ion concentration is needed to
(a) initiate precipitation of $Co(OH)_2$ from a solution that is $8.4 \times 10^{-4}$ M in $CoSO_4$?
(b) lower the cobalt-ion concentration to $1.00 \times 10^{-6}$ M?

**\*6–15.** The solubility-product constant for $Ce(IO_3)_3$ is $3.2 \times 10^{-10}$. What is the $Ce^{3+}$ concentration in a solution prepared by mixing 50.0 mL of 0.0500 M $Ce^{3+}$ with 50.00 mL of
(a) water?
(b) 0.50 M $IO_3^-$?
(c) 0.150 M $IO_3^-$?
(d) 0.300 M $IO_3^-$?

**6–16.** The solubility-product constant for $K_2PtCl_6$ is $1.1 \times 10^{-5}$ ($K_2PtCl_6 \rightleftarrows 2\,K^+ + PtCl_6^{2-}$). What is the $K^+$ concentration of a solution prepared by mixing 50.0 mL of 0.400 M KCl with 50.0 mL of
(a) 0.100 M $PtCl_6^{2-}$?
(b) 0.200 M $PtCl_6^{2-}$?
(c) 0.400 M $PtCl_6^{2-}$?

**6–17.** Establish whether a precipitate can be expected when the indicated volumes of solutions A and B are mixed

| | Solution A | Solution B |
| --- | --- | --- |
| *(a) | 40.0 mL of 0.0060 M HCl | 40.0 mL of 0.0400 M $TlNO_3$ |
| (b) | 60.0 mL of 0.0150 M KCl | 40.0 mL of 0.0400 M $TlNO_3$ |
| *(c) | 60.0 mL of 0.0150 M $CaCl_2$ | 40.0 mL of 0.0400 M $TlNO_3$ |
| (d) | 40.0 mL of 0.0400 M KCl | 80.0 mL of 0.0600 M $Pb(NO_3)_2$ |
| *(e) | 40.0 mL of 0.0400 M $CaCl_2$ | 80.0 mL of 0.0600 M $Pb(NO_3)_2$ |
| (f) | 40.0 mL of 0.0400 M $Pb(NO_3)_2$ | 80.0 mL of 0.0600 M KCl |

**6-18.** Establish whether a precipitate can be expected when 25.0 mL of 0.030 M $HIO_3$ are mixed with 35.0 mL of a solution that is $3.0 \times 10^{-3}$ M in
  *(a) KCl [$K_{sp}$ $KIO_3 = 5 \times 10^{-2}$].
  (b) $CuCl_2$ [$K_{sp}$ $Cu(IO_3)_2 = 7.4 \times 10^{-8}$].
  *(c) $CaCl_2$ [$K_{sp}$ $Ca(IO_3)_2 = 7.1 \times 10^{-7}$].
  (d) $CeCl_3$ [$K_{sp}$ $Ce(IO_3)_3 = 3.2 \times 10^{-10}$].

**6-19.** Potassium iodate is slowly introduced into a solution that is 0.040 M in $Ag^+$ and 0.050 M in $Ba^{2+}$.
  (a) Which iodate will precipitate first?
  (b) What will be the concentration of the ion that forms the less soluble iodate at the onset of precipitation by the more soluble iodate?

**6-20.** What is the hydronium ion concentration in water at 0°C?

**6-21.** Calculate the pH of a 0.100 M solution of
  *(a) hydrogen cyanide.
  (b) hypochlorous acid.
  *(c) acetic acid.
  (d) lactic acid.
  *(e) chloroacetic acid.
  (f) picric acid.

**6-22.** Calculate the pH of a 0.0100 M solution of
  *(a) hydrogen cyanide.
  (b) hypochlorous acid.
  *(c) acetic acid.
  (d) lactic acid.
  *(e) chloroacetic acid.
  (f) picric acid.

**6-23.** Calculate the pH of a 0.100 M solution of
  *(a) piperidine.
  (b) sodium phenolate.
  *(c) sodium cyanide.
  (d) hydroxylamine.
  *(e) hydrazine.
  (f) sodium hypochlorite.

**6-24.** Calculate the pH of a 0.0100 M solution of
  *(a) piperidine.
  (b) sodium phenolate.
  *(c) sodium cyanide.
  (d) hydroxylamine.
  *(e) hydrazine.
  (f) sodium hypochlorite.

**6-25.** Mercury(II) forms the soluble neutral complex $Hg(SCN)_2$ with $SCN^-$. The formation constant $\beta_2$ of the complex has a value of $2.0 \times 10^{17}$. Calculate the $Hg^{2+}$ and $SCN^-$ concentrations in a solution prepared by mixing 10 mL of 0.0200 M $Hg^{2+}$ with
  (a) 10 mL of 0.0200 M $SCN^-$.
  (b) 10 mL of 0.0400 M $SCN^-$.
  (c) 10 mL of 0.0600 M $SCN^-$.

**6-26.** Given that the formation constant for the reaction $CuI(s) + I^- \rightleftharpoons CuI_2^-$ is $8 \times 10^{-4}$, calculate the molar concentration of $CuI_2^-$ in a solution that is saturated with CuI and has a molar KI concentration of
  (a) 0.00100. (b) 0.0100. (c) 0.100.

**6-27.** The formation constant for $Pb(OH)_3^-$ is

$$Pb(OH)_2(s) + OH^- \rightleftharpoons Pb(OH)_3^- \qquad \beta_2 = 5 \times 10^{-2}$$

Calculate the concentration of $Pb(OH)_3^-$ in a saturated $Pb(OH)_2$ solution that is
  (a) 0.00100 M in NaOH.
  (b) 0.100 M in NaOH.

## CHAPTER 7

# ACTIVITIES AND ACTIVITY COEFFICIENTS

$W$e have noted that Equation 6–7 (page 124) is an approximate form of the thermodynamic equilibrium-constant expression derived from theory. In this chapter we describe when the use of this approximate form is likely to lead to error. We also show how the equation is modified to give a more exact description of a system at chemical equilibrium.

## 7A  THE EFFECT OF ELECTROLYTES ON CHEMICAL EQUILIBRIA

Experimentally we find that the position of most solution equilibria depends upon the electrolyte concentration of the medium. This happens even when the added electrolyte contains no ion in common with those involved in the equilibrium. For example, consider again the oxidation of iodide ion by arsenic acid:

$$H_3AsO_4 + 3\ I^- + 2\ H^+ \rightleftharpoons H_3AsO_3 + I_3^- + H_2O$$

If an electrolyte, such as barium nitrate, potassium sulfate, or sodium perchlorate, is added to this solution, the color becomes less intense. This decrease in color intensity indicates that the concentration of $I_3^-$ has decreased and that the equilibrium has been shifted to the left by the added electrolyte.

Figure 7–1 further illustrates the effect of electrolytes. Curve $A$ is a plot of the product of the molar hydronium and hydroxide ion *concentrations* ($\times\ 10^{14}$) as a function of the concentration of sodium chloride. This *concentration*-based ion product is designated $K_w'$. At low sodium chloride concentrations, $K_w'$ becomes independent of the electrolyte concentration and is equal to $1.00 \times 10^{-14}$, which is the *thermodynamic* ion-product constant for water, $K_w$. A relationship that approaches a constant value as some parameter (here, the electrolyte concentration) approaches zero is called a *limiting law*; the constant numerical value observed at this limit is referred to as a *limiting value*.

Concentration-based equilibrium constants are commonly indicated by adding a prime mark. For example: $K_w'$, $K_{sp}'$, $K_a'$.

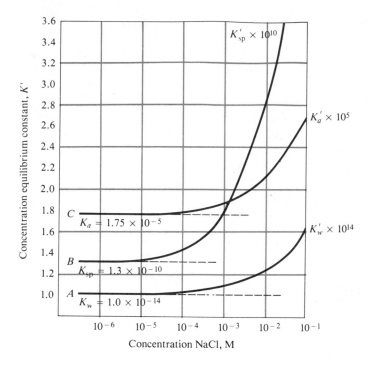

Figure 7–1

Effect of electrolyte concentration on concentration-based equilibrium constants.

The vertical axis for curve $B$ in Figure 7–1 is the product of the molar concentrations of barium and sulfate ions ($\times 10^{10}$) in saturated solutions of barium sulfate. This concentration-based solubility product is designated as $K'_{sp}$. At low electrolyte concentrations, $K'_{sp}$ has a limiting value of $1.3 \times 10^{-10}$, which is the accepted thermodynamic value of $K_{sp}$ for barium sulfate.

Curve $C$ is a plot of $K'_a$ ($\times 10^5$), the concentration quotient for the equilibrium involving the dissociation of acetic acid, as a function of electrolyte concentration. Here again, the ordinate function approaches a limiting value $K_a$, which is the thermodynamic acid dissociation constant for acetic acid.

The dashed lines in Figure 7–1 represent ideal behavior of the solutes. Note that departures from ideality can be significant. For example, the product of the molar concentrations of hydrogen and hydroxide ion increases from $1.0 \times 10^{-14}$ in pure water to about $1.7 \times 10^{-14}$ in a solution that is 0.1 M in sodium chloride. The effect is even more pronounced with barium sulfate; here, $K'_{sp}$ in 0.1 M sodium chloride is more than double that of its limiting value.

The electrolyte effect shown in Figure 7–1 is not peculiar to sodium chloride. Indeed, we would see identical curves when potassium nitrate or sodium perchlorate is substituted for sodium chloride. In each case, the effect has as its origin the electrostatic attraction between the ions of the electrolyte and the ions of reacting species of opposite charge. Since the electrostatic forces associated with all singly charged ions are approximately the same, the three salts exhibit essentially identical effects on equilibria.

We must now consider how we can take the electrolyte effect into account when we wish to make more accurate equilibrium calculations.

As the electrolyte concentration becomes very small, concentration-based equilibrium constants approach their thermodynamic values: $K_w$, $K_{sp}$, $K_a$.

Figure 7–2

Effect of electrolyte concentration on the solubility of some salts.

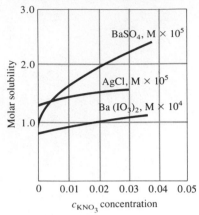

## 7A–1 The Effect of Reactant and Product Charge

Extensive studies have revealed that the magnitude of the electrolyte effect is highly dependent upon the charges of the participants in an equilibrium. When only neutral species are involved, the position of equilibrium is essentially independent of electrolyte concentration. With ionic participants, the magnitude of the electrolyte effect increases with charge. This generality is demonstrated by the three solubility curves in Figure 7–2. Note, for example, that in a 0.02 M solution of potassium nitrate, the solubility of barium sulfate, with its pair of doubly charged ions, is larger than it is in pure water by a factor of 2. This same change in electrolyte concentration increases the solubility of barium iodate by a factor of only 1.25 and that of silver chloride by 1.2. The enhanced effect due to doubly charged ions is also reflected in the greater slope of curve $B$ in Figure 7–1.

## 7A–2 The Effect of Ionic Strength

Systematic studies have shown that the effect of added electrolyte on equilibria is independent of the chemical nature of the electrolyte but depends upon a property of the solution called the *ionic strength*. This quantity is defined as

$$\text{ionic strength} = \mu = \tfrac{1}{2}([A]Z_A^2 + [B]Z_B^2 + [C]Z_C^2 + \cdots) \qquad (7\text{–}1)$$

where [A], [B], [C], . . . represent the molar concentrations of ions A, B, C, . . . and $Z_A$, $Z_B$, $Z_C$, . . . are their charges.

---

Example 7–1

Calculate the ionic strength of (a) a 0.1 M solution of $KNO_3$ and (b) a 0.1 M solution of $Na_2SO_4$.

(a) For the $KNO_3$ solution, $[K^+]$ and $[NO_3^-]$ are 0.1 M and

$$\mu = \tfrac{1}{2}(0.1 \times 1^2 + 0.1 \times 1^2) = 0.1$$

(b) For the $Na_2SO_4$ solution, $[Na^+] = 0.2$ and $[SO_4^{2-}] = 0.1$. Therefore,

$$\mu = \tfrac{1}{2}(0.2 \times 1^2 + 0.1 \times 2^2) = 0.3$$

---

Example 7–2

What is the ionic strength of a solution that is 0.05 M in $KNO_3$ and 0.1 M in $Na_2SO_4$?

$$\mu = \tfrac{1}{2}(0.05 \times 1^2 + 0.05 \times 1^2 + 0.2 \times 1^2 + 0.1 \times 2^2) = 0.35 \text{ M}$$

---

As these examples show, the ionic strength of a solution of a strong electrolyte consisting solely of singly charged ions is identical with its total molar salt concentration. The ionic strength is greater than the molar concentration if the solution contains ions with multiple charges.

For solutions with ionic strengths of 0.1 M or less, the electrolyte effect is independent of the *kind* of ions and dependent *only upon the ionic strength.* Thus, the degree of dissociation of acetic acid is the same in aqueous sodium chloride, potassium nitrate, or barium iodide, provided the concentrations of these species are such that their ionic strengths are identical. Note that this independence with respect to electrolyte species disappears at high ionic strengths.

| EFFECT OF CHARGE ON IONIC STRENGTH | | |
|---|---|---|
| Type of Electrolyte | Example | Ionic Strength* |
| 1 : 1 | NaCl | $c$ |
| 1 : 2 | $Ba(NO_3)_2$, $Na_2SO_4$ | $3c$ |
| 1 : 3 | $Al(NO_3)_3$, $Na_3PO_4$ | $6c$ |
| 2 : 2 | $MgSO_4$ | $4c$ |

*$c$ = molarity of the salt.

## 7B    ACTIVITY COEFFICIENTS

In order to describe the quantitative effect of ionic strength on equilibria, chemists use a term called *activity,* which for the species X is defined as

$$a_X = [X]f_X \qquad (7\text{-}2)$$

where $a_X$ is the activity of the species X, [X] is its molar concentration, and $f_X$ is a dimensionless quantity called the *activity coefficient.* The activity coefficient and thus the activity of X vary with ionic strength such that substitution of $a_X$ for [X] in any equilibrium-constant expression frees the numerical value of the constant from dependence on the ionic strength. To illustrate, for the generalized equilibrium we considered in Chapter 6 (page 124),

$$wW + xX \rightleftarrows yY + zZ$$

The thermodynamic equilibrium constant $K$ is defined by the equation

$$K = \frac{a_Y^y \cdot a_Z^z}{a_W^w \cdot a_X^x} \qquad (7\text{-}3)$$

Applying Equation 7–2 gives

$$K = \frac{[Y]^y[Z]^z}{[W]^w[X]^x} \times \frac{f_Y^y \cdot f_Z^z}{f_W^w \cdot f_X^x} = K' \frac{f_Y^y \cdot f_Z^z}{f_W^w \cdot f_X^x} \qquad (7\text{-}4)$$

where

$$K' = \frac{[Y]^y[Z]^z}{[W]^w[X]^x}$$

Here $K'$ is the *concentration equilibrium constant* and $K$ is the thermodynamic equilibrium constant.[1] The activity coefficients $f_W, f_X, f_Y,$ and $f_Z$

The activity of a species is a measure of its effective concentration as determined by the lowering of the freezing point of water, by electrical conductivity, and by the mass action effect.

---

[1]In the chapters that follow, we use the prime notation only when it is necessary to distinguish between thermodynamic and concentration equilibrium constants.

vary with ionic strength in such a way as to keep $K$ numerically constant and independent of ionic strength (in contrast to the concentration constants $K'$ shown in Figure 7–1).

## 7B–1 Properties of Activity Coefficients

Activity coefficients have the following properties:

1. The activity coefficient of a species is a measure of the effectiveness with which that species influences an equilibrium in which it is a participant. In very dilute solutions, where the ionic strength is minimal, this effectiveness becomes constant, and the activity coefficient is unity. Under such circumstances, the activity and the molar concentration are identical (as are the thermodynamic and concentration equilibrium constants). As the ionic strength increases an ion loses some of its effectiveness, and its activity coefficient decreases. We may summarize this behavior in terms of Equation 7–2. At moderate ionic strengths $f_X < 1$; as the solution approaches infinite dilution, however, $f_X \rightarrow 1$ and thus $a_X \rightarrow [X]$ and $K' \rightarrow K$. At high ionic strengths ($\mu > 0.1$ M), activity coefficients often increase and may even become greater than unity. Because interpretation of the behavior of solutions in this region is difficult, we shall confine our discussion to regions of low or moderate ionic strength (that is, where $\mu < 0.1$ M). The variation of typical activity coefficients as a function of ionic strength is shown in Figure 7–3.
2. In solutions that are not too concentrated, the activity coefficient for a given species is independent of the nature of the electrolyte and dependent only upon the ionic strength.
3. For a given ionic strength, the activity coefficient of an ion departs farther from unity as the charge carried by the species increases. This effect is shown in Figure 7–3. The activity coefficient of an uncharged molecule is approximately unity, regardless of ionic strength.
4. At any given ionic strength, the activity coefficients of ions of the same charge are approximately equal. The small variations that do exist can be correlated with the effective diameter of the hydrated ions.

As $\mu \rightarrow 0$, $f_X \rightarrow 1$, $a_X \rightarrow [X]$, and $K' \rightarrow K$.

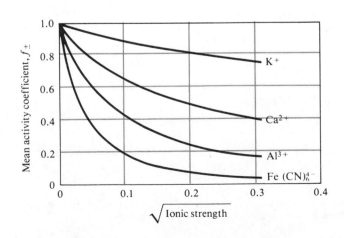

Figure 7–3
Effect of ionic strength on activity coefficients.

5. The activity coefficient of a given ion describes the ion's effective behavior in all equilibria in which it participates. For example, at a given ionic strength, a single activity coefficient for cyanide ion describes the influence of that species upon any of the following equilibria:

$$HCN + H_2O \rightleftharpoons H_3O^+ + CN^-$$

$$Ag^+ + CN^- \rightleftharpoons AgCN(s)$$

$$Ni^{2+} + 4\ CN^- \rightleftharpoons Ni(CN)_4^{2-}$$

## 7B–2  The Debye-Hückel Equation

As we have noted, the electrolyte effect results from the electrostatic attractive and repulsive forces that exist between the ions of an electrolyte and the ions involved in an equilibrium. These forces cause each ion from the dissociated reactant to be surrounded by a sheath of solution that contains a slight excess of electrolyte ions of opposite charge. For example, when a barium sulfate precipitate is equilibrated with a sodium chloride solution, each dissolved barium ion is surrounded by an ionic atmosphere that carries a net negative charge on the average due to repulsion of sodium ions and attraction of chloride ions. Similarly, each sulfate ion is surrounded by an ionic atmosphere that tends to be slightly positive. These charged layers make the barium ions somewhat less positive and the sulfate ions somewhat less negative than they would be in the absence of electrolyte. The consequence of this effect is a decrease in overall attraction between barium and sulfate ions and an increase in solubility, which becomes greater as the number of electrolyte ions in the solution becomes larger.

In 1923, P. Debye and E. Hückel used this model to derive a theoretical expression that permits the calculation of activity coefficients of ions from their charge and average size.[2] This equation, which has become known as the *Debye-Hückel equation*, takes the form

$$-\log f_X = \frac{0.51 Z_X^2 \sqrt{\mu}}{1 + 0.33\alpha_X \sqrt{\mu}} \qquad (7\text{--}5)$$

where

$f_X$ = activity coefficient of the species X

$Z_X$ = charge on the species X

$\mu$ = ionic strength of the solution

$\alpha_X$ = effective diameter of the hydrated ion X in angstrom units (1 Å = $10^{-8}$ cm)

[2]P. Debye and E. Hückel, *Physik. Z.*, **1923**, *24*, 185.

Peter Debye (1884–1966) was born and educated in Europe but became Professor of Chemistry at Cornell University in 1940. He was noted for his work in several different areas of chemistry including electrolyte solutions, x-ray diffraction, and the properties of polar molecules. He received the 1936 Nobel Prize in Chemistry.

Table 7–1

## ACTIVITY COEFFICIENTS FOR IONS AT 25°C*

| Ion | $\alpha_X$, Å | Activity Coefficient at Indicated Ionic Strength | | | | |
|---|---|---|---|---|---|---|
| | | 0.001 | 0.005 | 0.01 | 0.05 | 0.1 |
| $H_3O^+$ | 9 | 0.967 | 0.933 | 0.914 | 0.86 | 0.83 |
| $Li^+$, $C_6H_5COO^-$ | 6 | 0.965 | 0.929 | 0.907 | 0.84 | 0.80 |
| $Na^+$, $IO_3^-$, $HSO_3^-$, $HCO_3^-$, $H_2PO_4^-$, $H_2AsO_4^-$, $OAc^-$ | 4–4.5 | 0.964 | 0.928 | 0.902 | 0.82 | 0.78 |
| $OH^-$, $F^-$, $SCN^-$, $HS^-$, $ClO_3^-$, $ClO_4^-$, $BrO_3^-$, $IO_4^-$, $MnO_4^-$ | 3.5 | 0.964 | 0.926 | 0.900 | 0.81 | 0.76 |
| $K^+$, $Cl^-$, $Br^-$, $I^-$, $CN^-$, $NO_2^-$, $NO_3^-$, $HCOO^-$ | 3 | 0.964 | 0.925 | 0.899 | 0.80 | 0.76 |
| $Rb^+$, $Cs^+$, $Tl^+$, $Ag^+$, $NH_4^+$ | 2.5 | 0.964 | 0.924 | 0.898 | 0.80 | 0.75 |
| $Mg^{2+}$, $Be^{2+}$ | 8 | 0.872 | 0.755 | 0.69 | 0.52 | 0.45 |
| $Ca^{2+}$, $Cu^{2+}$, $Zn^{2+}$, $Sn^{2+}$, $Mn^{2+}$, $Fe^{2+}$, $Ni^{2+}$, $Co^{2+}$, phthalate$^{2-}$ | 6 | 0.870 | 0.749 | 0.675 | 0.48 | 0.40 |
| $Sr^{2+}$, $Ba^{2+}$, $Cd^{2+}$, $Hg^{2+}$, $S^{2-}$ | 5 | 0.868 | 0.744 | 0.67 | 0.46 | 0.38 |
| $Pb^{2+}$, $CO_3^{2-}$, $SO_3^{2-}$, $C_2O_4^{2-}$ | 4.5 | 0.868 | 0.742 | 0.665 | 0.46 | 0.37 |
| $Hg_2^{2+}$, $SO_4^{2-}$, $S_2O_3^{2-}$, $CrO_4^{2-}$, $HPO_4^{2-}$ | 4.0 | 0.867 | 0.740 | 0.660 | 0.44 | 0.36 |
| $Al^{3+}$, $Fe^{3+}$, $Cr^{3+}$, $La^{3+}$, $Ce^{3+}$ | 9 | 0.738 | 0.54 | 0.44 | 0.24 | 0.18 |
| $PO_4^{3-}$, $Fe(CN)_6^{3-}$ | 4 | 0.725 | 0.50 | 0.40 | 0.16 | 0.095 |
| $Th^{4+}$, $Zr^{4+}$, $Ce^{4+}$, $Sn^{4+}$ | 11 | 0.588 | 0.35 | 0.255 | 0.10 | 0.065 |
| $Fe(CN)_6^{4-}$ | 5 | 0.57 | 0.31 | 0.20 | 0.048 | 0.021 |

*From J. Kielland, *J. Am. Chem. Soc.*, **1937**, *59*, 1675. By courtesy of the American Chemical Society.

When $\mu$ is less than 0.01, $1 + \sqrt{\mu} \approx 1$, and Equation 7–5 becomes

$$-\log f_X = 0.51\, Z_X^2 \sqrt{\mu}$$

This equation is referred to as the Debye-Hückel Limiting Law (DHLL). Thus, in solutions of very low ionic strength, the DHLL can be used to calculate approximate activity coefficients.

The constants 0.51 and 0.33 are applicable to aqueous solutions at 25°C; other values must be employed at other temperatures.

Unfortunately, considerable uncertainty exists regarding the magnitude of $\alpha_X$ in Equation 7–5. Its value appears to be approximately 3 Å for most singly charged ions; for these species, then, the denominator of the Debye-Hückel equation simplifies to approximately $1 + \sqrt{\mu}$. For ions with higher charge, $\alpha_X$ may be as large as 10 Å. This increase in size with increase in charge makes good chemical sense. The larger the charge on an ion, the larger the number of polar water molecules that will be held in the solvation shell about the ion. It should be noted that the second term of the denominator in Equation 7–5 is small with respect to the first when the ionic strength is less than 0.01, so that at these ionic strengths, uncertainties in $\alpha_X$ are of little significance in calculating activity coefficients.

Kielland[3] has derived values of $\alpha_X$ for numerous ions from a variety of experimental data. His best values for effective diameters are given in Table 7–1. Also presented are activity coefficients calculated from Equation 7–5 using these values for the size parameter.

Experimental determination of single-ion activity coefficients such as those shown in Table 7–1 is unfortunately impossible because all experimental methods give only a mean activity coefficient for the positively and negatively charged ions in a solution. In other words, it is impossible to measure the properties of individual ions in the presence of counter ions of opposite charge and solvent molecules. It should be pointed out, however, that mean activity coefficients calculated from the data in Table 7–1 agree satisfactorily with the experimental values.

[3]J. Kielland, *J. Amer. Chem. Soc.*, **1937**, *59*, 1675.

**Example 7–3**

Use Equation 7–5 to calculate the activity coefficient for $Hg^{2+}$ in a solution that has an ionic strength of 0.085. Use 5 Å for the effective diameter of the ion. Compare the calculated value with $f_{Hg^{2+}}$ obtained by interpolation of data from Table 7–1.

$$-\log f_{Hg^{2+}} = \frac{(0.51)(2)^2 \sqrt{0.085}}{1 + (0.33)(5) \sqrt{0.085}} = 0.4016$$

$$\frac{1}{f_{Hg^{2+}}} = 2.52$$

$$f_{Hg^{2+}} = 0.397 = 0.40$$

Table 7–1 indicates that $f_{Hg^{2+}} = 0.46$ when $\mu = 0.05$ and 0.38 when $\mu = 0.1$. Thus, for $\mu = 0.085$

$$f_{Hg^{2+}} = 0.38 + \frac{(0.10 - 0.085)}{(0.10 - 0.05)} \times (0.46 - 0.38) = 0.404 = 0.40$$

The Debye-Hückel relationship and the data in Table 7–1 give satisfactory activity coefficients for ionic strengths up to about 0.1. Beyond this value, the equation fails, and experimentally determined mean activity coefficients must be used.

Values for activity coefficients at ionic strengths not listed in Table 7–1 can be approximated by interpolation as shown here.

**Feature 7–1**
**MEAN ACTIVITY COEFFICIENTS**

The mean activity coefficient of the electrolyte $A_m B_n$ is defined as

$$f_{\pm} = \text{mean activity coefficient} = (f_A^m \cdot f_B^n)^{1/(m+n)}$$

The mean activity coefficient can be measured in any of several ways, but it is impossible experimentally to resolve this term into the individual activity coefficients for $f_A$ and $f_B$. For example, if $A_m B_n$ is a precipitate, we can write

$$K_{sp} = [A]^m[B]^n \cdot f_A^m \cdot f_B^n = [A]^m[B]^n \cdot f_{\pm}^{m+n}$$

By measuring the solubility of $A_m B_n$ in a solution in which the electrolyte concentration approaches zero (that is, where both $f_A$ and $f_B \rightarrow 1$), we can obtain $K_{sp}$. A second solubility measurement at some ionic strength $\mu_1$ gives values for [A] and [B]. These data then permit the calculation of $f_A^m \cdot f_B^n = f_{\pm}^{m+n}$ for ionic strength $\mu_1$.

It is important to understand that this procedure does not provide enough experimental data to permit the calculation of the *individual* quantities $f_A$ and $f_B$ and that there appears to be no additional experimental information that would permit evaluation of these quantities. This situation is general, and the *experimental* determination of individual activity coefficients is impossible.

## 7B–3 Equilibrium Calculations Employing Activity Coefficients

Equilibrium calculations with activities yield more correct results than those obtained with molar concentrations. Unless otherwise specified, equilibrium constants found in tables are generally based upon activities and are thus thermodynamic equilibrium constants. The example that follows illustrates how activity coefficients from Table 7–1 are applied to such data.

---

Example 7–4

Use activities to calculate the solubility of $Ba(IO_3)_2$ in a 0.033 M solution of $Mg(IO_3)_2$. The thermodynamic solubility product for $Ba(IO_3)_2$ is $1.57 \times 10^{-9}$ (Appendix 2).

At the outset, we write the solubility-product expression in terms of activities:

$$a_{Ba^{2+}} \times a_{IO_3^-}^2 = K_{sp} = 1.57 \times 10^{-9}$$

where $a_{Ba^{2+}}$ and $a_{IO_3^-}$ are the activities of barium and iodate ions. Replacing activities in this equation by activity coefficients and concentrations (Equation 7–2) yields

$$[Ba^{2+}]f_{Ba^{2+}} \times [IO_3^-]^2 f_{IO_3^-}^2 = K_{sp}$$

where $f_{Ba^{2+}}$ and $f_{IO_3^-}$ are the activity coefficients for the two ions. Rearranging this expression gives

$$[Ba^{2+}][IO_3^-]^2 = \frac{K_{sp}}{f_{Ba^{2+}} \times f_{IO_3^-}^2} = K_{sp}' \qquad (7-6)$$

where $K_{sp}'$ is the *concentration-based solubility product*.

The ionic strength of the solution is obtained by substituting into Equation 7–1:

$$\mu = \tfrac{1}{2}([Mg^{2+}] \times 2^2 + [IO_3^-] \times 1^2)$$
$$= \tfrac{1}{2}(0.033 \times 4 + 0.066 \times 1) = 0.099 \approx 0.1$$

In calculating $\mu$, we have assumed that the $Ba^{2+}$ and $IO_3^-$ ions from the precipitate do not significantly affect the ionic strength of the solution. This simplification seems justified, considering the low solubility of barium iodate and the relatively high concentration of $Mg(IO_3)_2$. In situations where it is not possible to make such an assumption, the concentrations of the two ions can be approximated by solubility calculations in which activities and concentrations are assumed to be identical (as in Examples 6–3 and 6–4). These concentrations can then be introduced to give a better value for $\mu$.

Turning now to Table 7–1, we find that at an ionic strength of 0.1,

$$f_{Ba^{2+}} = 0.38 \qquad f_{IO_3^-} = 0.78$$

If the calculated ionic strength did not match that of one of the columns in the table, $f_{Ba^{2+}}$ and $f_{IO_3^-}$ could be calculated from Equation 7–5 or by interpolation.

Substituting into the thermodynamic solubility-product expression (Equation 7–6) gives

$$\frac{1.57 \times 10^{-9}}{(0.38)(0.78)^2} = 6.8 \times 10^{-9} = K_{sp}'$$

$$[Ba^{2+}][IO_3^-]^2 = 6.8 \times 10^{-9}$$

Proceeding now as in earlier solubility calculations,

$$solubility = [Ba^{2+}]$$

$$[IO_3^-] \cong 0.066$$

$$[Ba^{2+}](0.066)^2 = 6.8 \times 10^{-9}$$

$$[Ba^{2+}] = solubility = 1.56 \times 10^{-6} \ M$$

It is interesting to note that the calculated solubility, if we neglect the effects of ionic strength, is $3.6 \times 10^{-7}$ M, which is less than one fourth as large as the value obtained here.

---

### Example 7–5

Using activities, calculate the hydronium ion concentration in a 0.120 M solution of $HNO_2$ that is also 0.050 M in NaCl.

The ionic strength of this solution is

$$\mu = \tfrac{1}{2}(0.0500 \times 1^2 + 0.0500 \times 1^2) = 0.0500$$

In Table 7–1, at ionic strength 0.050, we find

$$f_{H_3O^+} = 0.86 \qquad f_{NO_2^-} = 0.80$$

Also, from rule 3 (page 150), we can write

$$f_{HNO_2} = 1.0$$

These three values for $f$ permit the calculation of a concentration-based dissociation constant from the thermodynamic constant of $5.1 \times 10^{-4}$ (Appendix 3):

$$\frac{[H_3O^+][NO_2^-]}{[NHO_2]} = \frac{K_a \times f_{HNO_2}}{f_{H_3O^+} \times f_{NO_2^-}} = \frac{5.1 \times 10^{-4} \times 1.0}{0.86 \times 0.81} = K_a' = 7.3 \times 10^{-4}$$

Proceeding as in Example 6–7, we write

$$[H_3O^+] = \sqrt{0.120 \times 7.3 \times 10^{-4}} = 9.4 \times 10^{-3}$$

Note that neglecting the activity coefficient gives $[H_3O^+] = 7.8 \times 10^{-3}$.

## 7B–4 Omission of Activity Coefficients in Equilibrium Calculations

We shall ordinarily neglect activity coefficients and simply use molar concentrations in applications of the equilibrium law. This recourse simplifies the calculations and greatly decreases the amount of data needed. For most purposes, the error introduced by the assumption of unity for the activity coefficient is not large enough to lead to false conclusions. It should be apparent from the preceding examples, however, that disregard of activity coefficients may introduce a significant numerical error in calculations of this kind. Note, for example, that an error greater than $-75\%$ was incurred in Example 7–4. The reader should be alert to the conditions under which the substitution of concentration for activity is likely to lead to the largest error. Significant discrepancies occur when the ionic strength is large (0.01 or larger) and when the ions involved have multiple charges (Table 7–1). With dilute solutions (ionic strength $< 0.01$) of non-electrolytes or of singly charged ions, the use of concentrations in a mass-law calculation often provides reasonably accurate results.

It is also important to note that the decrease in solubility resulting from the presence of an ion common to the precipitate is in part counteracted by the larger electrolyte concentration associated with the presence of the salt containing the common ion. This effect is illustrated in Example 7–4.

## 7C QUESTIONS AND PROBLEMS

7–1. What is a thermodynamic equilibrium constant?

*7–2. What is the numerical value of the activity coefficient of acetic acid ($CH_3COOH$)?

7–3. Why is the slope of Curve B in Figure 7–1 greater than that of curve A?

*7–4. What is a mean activity coefficient?

7–5. Calculate the ionic strength of a solution that is
  *(a) 0.060 M in NaBr.
  (b) 0.080 M in $K_2SO_4$.
  *(c) 0.100 M in $FeCl_2$.
  (d) 0.0100 M in $FeCl_3$.
  *(e) 0.050 M in $FeCl_2$ and 0.030 M in KBr.
  (f) 0.080 M in $K_2CrO_4$ and 0.060 M in $FeCl_3$.

7–6. Use Equation 7–5 to calculate the activity coefficient of
  *(a) $Fe^{3+}$ at $\mu = 0.075$.
  (b) $Pb^{2+}$ at $\mu = 0.012$.
  *(c) $Ce^{4+}$ at $\mu = 0.080$.
  (d) $Sn^{4+}$ at $\mu = 0.060$.

7–7. Calculate activity coefficients for the species in Problem 7–6 by linear interpolation of the data in Table 7–1.

7–8. Use the data in Table 7–1 to calculate the mean activity coefficient for each of the following compounds at an ionic strength of 0.010:
  *(a) LiOH.      *(e) $K_4Fe(CN)_6$.
  (b) $ZnSO_4$.    (f) $Ce_2(SO_4)_3$.
  *(c) $LaI_3$.      *(g) $Zn_3(PO_4)_2$.
  (d) $Co(ClO_4)_2$.  *(h) $KAl(SO_4)_2$.

7–9. Calculate $K'_{sp}$ for the following compounds in a solution that has an ionic strength of 0.050:
  *(a) AgI.        (d) $Ag_2CrO_4$.
  (b) $PbSO_4$.    *(e) $Zn(OH)_2$.
  *(c) $MgNH_4PO_4$.  (f) $La(IO_3)_2$.

7–10. Use (1) activities and (2) concentrations to calculate the solubilities of the substances in Problem 7–9 in a solution that is 0.0167 M in $Mg(ClO_4)_2$.

*7–11. Use (1) activities and (2) molar concentrations to

calculate the solubilities of the following compounds in a 0.0333 M solution of $Mg(ClO_4)_2$:
(a) AgSCN.
(b) $PbI_2$.
(c) $BaSO_4$.
(d) $Zn_2Fe(CN)_6$.

$$Zn_2Fe(CN)_6(s) \rightleftharpoons 2\,Zn^{2+} + Fe(CN)_6^{4-}$$
$$K_{sp} = 3.2 \times 10^{-17}$$

**7–12.** Use (1) activities and (2) molar concentrations to calculate the solubilities of the following compounds in a 0.0167 M solution of $Mg(NO_3)_2$:
(a) $AgIO_3$.  (c) $PbSO_4$.
(b) $Mg(OH)_2$.  (d) $La(IO_3)_3$.

**7–13.** Use (1) activities and (2) molar concentrations to calculate the hydronium ion concentration in the following solutions:
*(a) 0.0200 M $HNO_2$ that is 0.0500 M in $NaNO_3$.
(b) 0.0200 M $HNO_2$ that is 0.100 M in $NaNO_3$.
*(c) 0.0500 M $NH_3$ that is 0.0500 M in NaCl.
(d) 0.0500 M $NH_3$ that is 0.100 M in NaCl.
*(e) 0.0400 M $NH_4Cl$ that is 0.0100 M in NaCl.
(f) 0.0500 M $NH_4Cl$ that is 0.0500 M in NaCl.
*(g) 0.0400 M $NaNO_2$ that is 0.0100 M in NaCl.
(h) 0.0500 M $NaNO_2$ that is 0.0500 M in NaCl.

**7–14.** Use activities to calculate the molar solubility of $Mg(OH)_2$ in
(a) 0.0100 M $NaNO_3$.
*(b) 0.0167 M $K_2SO_4$.
(c) the solution that results when 60.0 mL of 0.0333 M $MgCl_2$ is mixed with 40.0 mL of 0.0500 M KOH.

**7–15.** The solubility-product constant for $Ce(OH)_3$ is $2.0 \times 10^{-20}$. Use activities (interpolating between values in Table 7–1 as necessary) to calculate the molar analytical concentration of $Ce(OH)_3$ in the solution that results upon mixing 40.0 mL of 0.0200 M $CeCl_3$ with 60.0 mL of
*(a) 0.0300 M KOH.
(b) 0.0200 M $Ba(OH)_2$.
(c) 0.0400 M $Ba(OH)_2$.

**7–16.** Mercury(II) forms a soluble neutral complex with $Cl^-$:

$$Hg^{2+} + 2\,Cl^- \rightleftharpoons HgCl_2 \qquad K_f = 1.6 \times 10^{13}$$

Use (1) activities and (2) molarities to calculate the $Hg^{2+}$ concentration in:
*(a) a solution prepared by dissolving 0.0100 mol of $HgCl_2$ in 1.00 L of water.
(b) a solution prepared by dissolving 0.0100 mol of $HgCl_2$ in 1.00 L of 0.0500 M $NaNO_3$.
*(c) a solution prepared by dissolving 0.0100 mol of $HgCl_2$ in 1.00 L of 0.0500 M NaCl.
(d) a solution prepared by dissolving 0.0100 mol of $HgCl_2$ in 1.00 L of 0.0333 M $Hg(NO_3)_2$.
*(e) a solution prepared by mixing 50.0 mL of a solution that is 0.0100 M in $Hg(NO_3)_2$ with 50.0 mL of a solution that is 0.0400 M in NaCl and 0.0600 M in $NaNO_3$.
(f) a solution prepared by mixing 50.0 mL of a solution that is 0.0100 M in $Hg(NO_3)_2$ with 50.0 mL of a solution that is 0.0400 M in $BaCl_2$ and 0.0500 M in $NaNO_3$.

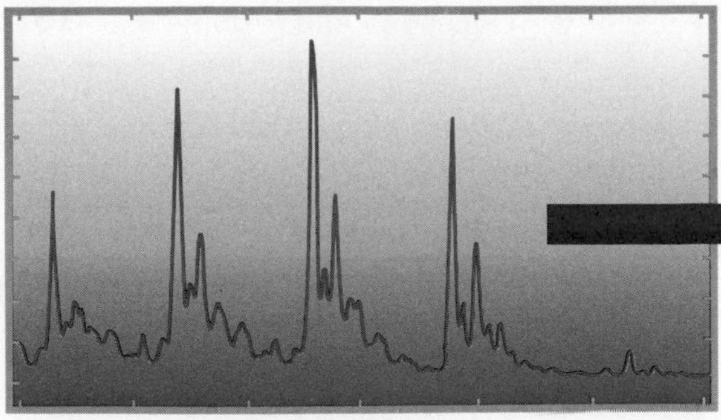

# A SYSTEMATIC METHOD FOR PERFORMING EQUILIBRIUM CALCULATIONS

Aqueous solutions encountered in the laboratory often contain several species that interact with one another and with water to yield two or more equilibria. For example, when 1.0 mmol of sodium hydrogen carbonate is dissolved in 100 mL of water, the species sodium ion concentration is 0.010 M. The species concentration of hydrogen carbonate ion is less than 0.010 M, however, because this anion reacts with water to form two products:

$$HCO_3^- + H_2O \rightleftarrows CO_3^{2-} + H_3O^+ \tag{8-1}$$

$$HCO_3^- + H_2O \rightleftarrows H_2CO_3 + OH^- \tag{8-2}$$

All equilibria involving $H_3O^+$ or $OH^-$ are necessarily influenced by the autoprotolysis of water. Therefore, in order to describe a sodium hydrogen carbonate system in quantitative terms, we must consider a third equilibrium:

$$2 H_2O \rightleftarrows H_3O^+ + OH^- \tag{8-3}$$

We see from these three equations that a 0.010 M solution of sodium hydrogen carbonate contains five species in addition to $Na^+$: $HCO_3^-$, $CO_3^{2-}$, $H_2CO_3$, $H_3O^+$, and $OH^-$. To calculate the concentration of one or more of these species from equilibrium-constant expressions, we must generate five independent algebraic expressions and solve them simultaneously. As we demonstrate in this chapter, generating such algebraic equations is relatively easy if the problem is approached systematically.

Throughout this discussion, you should bear in mind that *the validity and form of a particular equilibrium-constant expression are unaffected by other equilibria in the solution.* For example, if barium chloride is added to the sodium hydrogen carbonate solution just described, barium carbonate precipitates and a new equilibrium is established:

More often than not, developing the necessary algebraic expressions is easier than solving them unless a suitable computer program is available.

$$BaCO_3(s) \rightleftarrows Ba^{2+} + CO_3^{2-} \tag{8-4}$$

This additional equilibrium does not affect the equilibrium relationships among the other carbonate species and the hydronium and hydroxide ion. That is, the following two equilibrium-constant expressions, which describe the equilibria shown in Equations 8–1 and 8–2, are valid *regardless of whether barium chloride has been added*.

$$\frac{[H_3O^+][CO_3^{2-}]}{[HCO_3^-]} = K_a \qquad (8–5)$$

$$\frac{[OH^-][H_2CO_3]}{[HCO_3^-]} = K_b \qquad (8–6)$$

To be sure, except for the sodium ion, the concentrations of all species in the original solution are markedly altered by the added barium ion. Nevertheless, Equations 8–5 and 8–6 still describe the relationships among the species involved.

In this chapter, we first outline the steps in a systematic approach to solving multiple-equilibrium problems. We then illustrate this approach with several examples that involve calculating solubilities in the presence of species that interact with the ions of precipitates and thus enhance their solubilities. In subsequent chapters, we will use this same systematic approach for the solution of problems involving other types of equilibria.

The introduction of a new equilibrium system into a solution does not change the equilibrium constants for any existing equilibria.

## 8A    A SYSTEMATIC METHOD FOR SOLVING MULTIPLE-EQUILIBRIUM PROBLEMS

In order to solve a multiple-equilibrium problem, we need as many independent algebraic equations as there are participants in the system being studied. Thus, five such equations are required for the sodium hydrogen carbonate system just described. A sixth will be needed if we add barium ion to the solution.

Three types of algebraic equations are used in solving multiple-equilibrium problems: (1) equilibrium-constant expressions, (2) *mass-balance* equations, and (3) a single *charge-balance* equation. We have already shown how equilibrium-constant expressions are written; methods for obtaining the other two types of equations are given in the next two sections.

### 8A–1 Mass-Balance Equations

Mass-balance equations relate the equilibrium concentrations of various species in a solution to one another and to the analytical concentrations of the various solutes. They are derived from information about how the solution was prepared and from a knowledge of the kinds of equilibria established in the solution.

The term "mass-balance equation," although widely used, is misleading because such equations are really based upon balancing *concentrations* rather than *masses*.

---

### Example 8–1

Write mass-balance expressions for a solution that is 0.050 M in acetic acid (HOAc).

For a slightly soluble salt with a 1 : 1 stoichiometry, the equilibrium molarity of the cations is equal to equilibrium molarity of the anion. This equality is the mass-balance expression.

First we write chemical equations for all important equilibria. Here,

$$HOAc + H_2O \rightleftarrows H_3O^+ + OAc^-$$

$$2\,H_2O \rightleftarrows H_3O^+ + OH^-$$

Because there is only one source of HOAc and $OAc^-$, we may write that the analytical concentration of acetic acid $c_{HOAc}$ is equal to the sum of their concentrations:

$$c_{HOAc} = 0.050 = [HOAc] + [OAc^-]$$

A second mass-balance expression can be obtained by examining the two chemical equations. Here we see that one mole of $H_3O^+$ forms for each mole of $OAc^-$ and for each mole of $OH^-$. Therefore, the total concentration of $H_3O^+$ is

$$[H_3O^+] = [OAc^-] + [OH^-]$$

For slightly soluble salts with stoichiometries other than 1:1, the mass-balance expression is obtained by multiplying the concentration of one of the ions by the stoichiometric ratio. For example, the mass-balance expression for a saturated solution of $Ag_2CrO_4$ is

$$2[CrO_4^{2-}] = [Ag^+]$$

---

Example 8–2

Write mass-balance expressions for the system formed when a 0.010 M $NH_3$ solution is saturated with AgBr.

Here, equations for the pertinent equilibria are

$$AgBr(s) \rightleftarrows Ag^+ + Br^-$$

$$Ag^+ + 2\,NH_3 \rightleftarrows Ag(NH_3)_2^+$$

$$NH_3 + H_2O \rightleftarrows NH_4^+ + OH^-$$

Because the only source of $Br^-$, $Ag^+$, and $Ag(NH_3)_2^+$ is AgBr and because silver and bromide ions are present in a 1:1 ratio in the starting material, one mass-balance equation is

$$[Ag^+] + [Ag(NH_3)_2^+] = [Br^-]$$

where the bracketed terms are molar species concentrations. Also, we know that the only source of ammonia-containing species is the 0.010 M $NH_3$. Therefore, a second mass-balance expression is

$$c_{NH_3} = [NH_3] + [NH_4^+] + 2[Ag(NH_3)_2^+] = 0.010$$

---

## 8A–2 Charge-Balance Equation

We know that electrolyte solutions are electrically neutral even though they may contain many millions of charged ions. Solutions are neutral because the *molar concentration of positive charge* in an electrolyte solution always equals *the molar concentration of negative charge*. That is, for any solution containing electrolytes, we may write

$$\text{no. mol positive charge/L} = \text{no. mol negative charge/L}\quad(8-7)$$

This equation represents the charge-balance condition and is called the *charge-balance equation*. To make this equation useful for equilibrium-constant calculations, we must express both sides of the equation in terms of molar concentrations of negative and positive ions, which carry the charge in the solution.

How much charge is contributed to a solution by 1 mol of $Na^+$? Or, how about 1 mol of $Mg^{2+}$ or 1 mol of $PO_4^{3-}$? The concentration of charge contributed to a solution by an ion is equal to the molar concentration of that ion multiplied by its charge. Thus the molar concentration of positive charge in a solution due to sodium ions is

Always remember that a charge-balance equation is based upon the equality in *molar charge concentrations*. To obtain the charge concentration of an ion, you must multiply the molar concentration of the ion by its charge.

$$\text{no. mol positive charge/L} = \frac{\text{mol } Na^+}{L} \times \frac{1 \text{ mol positive charge}}{\text{mol } Na^+}$$

$$= 1 \times [Na^+]$$

The concentration of positive charge due to magnesium ions is

$$\text{no. mol positive charge/L} = \frac{\text{mol } Mg^{2+}}{L} \times \frac{2 \text{ mol positive charge}}{\text{mol } Mg^{2+}}$$

$$= 2 \times [Mg^{2+}]$$

since each mole of magnesium ion contributes 2 mol of positive charge to the solution. Similarly, we may write for phosphate ion

$$\text{no. mol negative charge/L} = \frac{\text{mol } PO_4^{3-}}{L} \times \frac{3 \text{ mol negative charge}}{\text{mol } PO_4^{3-}}$$

$$= 3 \times [PO_4^{3-}]$$

Now, consider how we would write a charge-balance equation for a 0.100 M solution of sodium chloride. Positive charges in this solution are supplied by $Na^+$ and $H_3O^+$ (from dissociation of water). Negative charges come from $Cl^-$ and $OH^-$. The molarity of positive and negative charges are

$$\text{mol positive charge/L} = [Na^+] + [H_3O^+] = 0.100 + 1 \times 10^{-7}$$
$$\text{mol negative charge/L} = [Cl^-] + [OH^-] = 0.100 + 1 \times 10^{-7}$$

We write the charge-balance equation by equating the concentrations of positive and negative charges. That is,

$$[Na^+] + [H_3O^+] = [Cl^-] + [OH^-] = 0.100 + 1 \times 10^{-7}$$

Let us now consider a solution that has an analytical concentration of magnesium chloride of 0.100 M. Here, the molarities of positive and negative charge are given by

$$\text{mol positive charge/L} = 2[Mg^{2+}] + [H_3O^+] = 2 \times 0.100 + 1 \times 10^{-7}$$

$$\text{mol negative charge/L} = [Cl^-] + [OH^-] = 2 \times 0.100 + 1 \times 10^{-7}$$

In the first equation, the molar concentration of magnesium ion is multiplied by two ($2 \times 0.100$) because 1 mol of that ion contributes 2 mol of positive charge to the solution. In the second equation the molar chloride ion concentration is twice that of the magnesium chloride, or $2 \times 0.100$. To obtain the charge-balance equation, we equate the concentration of positive charge with the concentration of negative charge to obtain

$$2[Mg^{2+}] + [H_3O^+] = [Cl^-] + [OH^-] = 0.200 + 1 \times 10^{-7}$$

For a neutral solution, $[H_3O^+]$ and $[OH^-]$ are very small and equal so that we can simplify the charge-balance equation to

$$2[Mg^{2+}] = [Cl^-] = 0.200$$

---

**Example 8–3**

Write a charge-balance equation for the system in Example 8–2.

$$[Ag^+] + [Ag(NH_3)_2^+] + [H_3O^+] + [NH_4^+] = [OH^-] + [Br^-]$$

---

**Example 8–4**

Neglecting the dissociation of water, write a charge-balance equation for a solution that contains NaCl, $Mg(NO_3)_2$, and $Al_2(SO_4)_3$.

$$[Na^+] + 2[Mg^{2+}] + 3[Al^{3+}] = [Cl^-] + [NO_3^-] + 2[SO_4^{2-}]$$

---

## 8A–3 Steps for Solving Problems Involving Several Equilibria

1. Write a set of balanced chemical equations for all pertinent equilibria.
2. Write an equation that expresses the quantity being sought in terms of the equilibrium concentrations of species in the system.
3. Write equilibrium-constant expressions for all equilibria developed in step 1, and find numerical values for the constants in tables of equilibrium constants.
4. Write mass-balance expressions for the system.
5. If possible, write a charge-balance expression for the system.
6. Count the number of unknown concentrations in the equations developed in steps 3, 4, and 5, and compare this number with the number of independent equations. If the number of equations is equal to the number of unknowns, proceed to step 7. If the number of equations is smaller than the number of unknowns, seek additional equations. If enough equations cannot be developed, try to eliminate unknowns by suitable approximations regarding the concentration of one or more of

In some systems, charge-balance equations either cannot be written or are of no use. Often the charge-balance equation is identical to one of the mass-balance expressions.

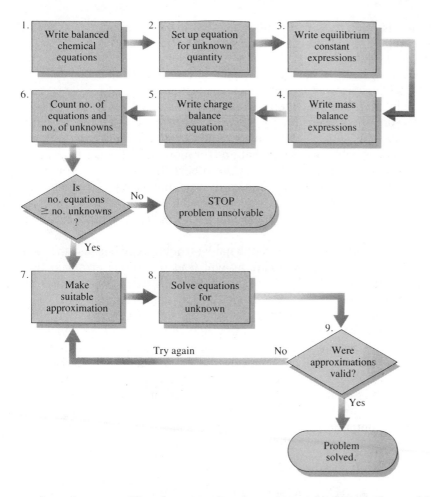

Flowchart for the systematic method.

the unknowns. If such approximations cannot be found, the problem cannot be solved.

7. Make suitable approximations to simplify the algebra.
8. Solve the algebraic equations for the equilibrium concentrations needed to give a provisional answer to the equation developed in step 2.
9. Check the validity of the approximations made in steps 6 and 7 using the provisional concentrations computed in step 8.

Step 6 is particularly important because it shows whether or not an exact solution to the problem is possible. If the number of unknowns is identical to the number of equations, the problem has been reduced to one of *algebra* alone. That is, answers can be obtained with sufficient perseverance. On the other hand, if there are not enough equations even after approximations are made, the problem should be abandoned.

## 8A–4 The Use of Approximations in Equilibrium Calculations

When step 6 of the systematic approach is complete, we have a *mathematical* problem of solving several nonlinear simultaneous equations.

Do not waste time starting the algebra in an equilibrium calculation until you are absolutely sure you have enough independent equations to allow a solution to the problem.

Approximations can be made only in charge-balance and mass-balance equations—never in equilibrium-constant expressions.

Never be afraid to make an assumption in attempting to solve an equilibrium problem. If the assumption is not valid, you will know it as soon as you have a provisional answer.

This job is often formidable, tedious, and time-consuming unless a suitable computer program is available or unless we can find approximations that decrease the number of unknowns and equations. In this section, we consider in general terms how equations describing equilibrium relationships can be simplified by suitable approximations.

Bear in mind that *only* the mass-balance and charge-balance equations can be simplified because only in these equations do the concentration terms appear as sums or differences rather than as products or quotients. It is always possible to assume that one (or more) of the terms in a sum or difference is so much smaller than the others that it can be assumed to have a value of zero without significantly affecting the equality. However, if a term in an equilibrium-constant expression is assumed to be zero, the product or quotient becomes equal to zero or infinity and is thus meaningless.

Many students find step 7 to be the most troublesome because they fear that making invalid approximations will lead to serious errors in their computed results. Such fears are groundless. Experienced scientists are often as puzzled as beginners when making an approximation that simplifies an equilibrium calculation. Nonetheless, they make such approximations without fear because they know that the effects of an invalid assumption will become obvious by the time a computation is completed. Generally, questionable assumptions should be tried at the outset and provisional answers computed. If the assumption leads to an intolerable error (which is easily recognized), a recalculation without the faulty approximation is then performed. Usually, it is more efficient to try a questionable assumption at the outset than to make a more time-consuming and tedious calculation without the assumption.

## 8B THE CALCULATION OF SOLUBILITY BY THE SYSTEMATIC METHOD

The use of the systematic method is illustrated in this section with examples involving the solubility of precipitates under various conditions. In later chapters we apply this method to other types of equilibria.

### 8B–1 Metal Hydroxides

Examples 8–5 and 8–6 involve calculating the solubilities of two metal hydroxides, the first of which is relatively soluble compared with the second. These examples illustrate the importance of making approximations and provisional calculations in solving equilibrium problems.

---

Example 8–5

Calculate the molar solubility of $Mg(OH)_2$ in water.
Step 1. Pertinent Equilibria
Two equilibria that need to be considered are

$$Mg(OH)_2(s) \rightleftharpoons Mg^{2+} + 2\ OH^-$$

$$2\ H_2O \rightleftharpoons H_3O^+ + OH^-$$

**Step 2. Definition of Unknown**

Since 1 mol of $Mg^{2+}$ is formed for each mole of $Mg(OH)_2$ dissolved,

$$\text{solubility } Mg(OH)_2 = [Mg^{2+}]$$

**Step 3. Equilibrium-Constant Expressions**

$$[Mg^{2+}][OH^-]^2 = K_{sp} = 1.8 \times 10^{-11} \qquad (8\text{--}8)$$

$$[H_3O^+][OH^-] = K_w = 1.00 \times 10^{-14} \qquad (8\text{--}9)$$

**Step 4. Mass-Balance Expression**

$$[OH^-] = 2[Mg^{2+}] + [H_3O^+] \qquad (8\text{--}10)$$

The first term on the right-hand side of Equation 8–10 represents the hydroxide ion concentration resulting from dissolved $Mg(OH)_2$, and the second term is the hydroxide ion concentration resulting from the dissociation of water.

**Step 5. Charge-Balance Expression**

$$[OH^-] = 2[Mg^{2+}] + [H_3O^+]$$

Note that this equation is identical to Equation 8–10. Often a mass-balance equation and a charge-balance equation are the same.

**Step 6. Number of Independent Equations and Unknowns**

We have developed three independent algebraic equations (Equations 8–8, 8–9, and 8–10) and have three unknowns ($[Mg^{2+}]$, $[OH^-]$, and $[H_3O^+]$). Therefore, the problem can be solved rigorously.

**Step 7. Approximations**

We can make approximations only in Equation 8–10. Since the solubility-product constant for $Mg(OH)_2$ is relatively large, the solution will be somewhat basic. Therefore, it is reasonable to assume that $[H_3O^+] \ll [Mg^{2+}]$. Equation 8–10 then simplifies to

$$2[Mg^{2+}] \approx [OH^-] \qquad (8\text{--}11)$$

**Step 8. Solution to Equations**

Substitution of Equation 8–11 into Equation 8–8 gives

$$[Mg^{2+}](2[Mg^{2+}])^2 = 1.8 \times 10^{-11}$$

$$[Mg^{2+}]^3 = \frac{1.8 \times 10^{-11}}{4} = 4.50 \times 10^{-12}$$

$$[Mg^{2+}] = \text{solubility} = 1.651 \times 10^{-4} = 1.7 \times 10^{-4} \text{ mol/L}$$

**Step 9. Check of Assumptions**

If $[OH^-]_{H_2O}$ and $[OH^-]_{Mg(OH)_2}$ are the concentrations of $OH^-$ produced from $H_2O$ and $Mg(OH)_2$, respectively, then

$$[OH^-]_{H_2O} = [H_3O^+]$$

$$[OH^-]_{Mg(OH)_2} = 2[Mg^{2+}]$$

$$[OH^-]_{total} = [OH^-]_{H_2O} + [OH^-]_{Mg(OH)_2}$$

$$= [H_3O^+] + 2[Mg^{2+}]$$

Substitution into Equation 8–11 yields

$$[OH^-] = 2 \times 1.651 \times 10^{-4} = 3.30 \times 10^{-4} = 3.3 \times 10^{-4} \text{ mol/L}$$

$$[H_3O^+] = \frac{1.00 \times 10^{-14}}{3.33 \times 10^{-4}} = 3.0 \times 10^{-11} \text{ mol/L}$$

Thus our assumption that $3.0 \times 10^{-11} \ll 1.7 \times 10^{-4}$ is certainly valid.

---

Example 8–6

Calculate the solubility of $Fe(OH)_3$ in water. Proceeding by the systematic approach used in Example 8–5, we write

Step 1.  Pertinent Equilibria

$$Fe(OH)_3(s) \rightleftarrows Fe^{3+} + 3\ OH^-$$

$$2\ H_2O \rightleftarrows H_3O^+ + OH^-$$

Step 2.  Definition of Unknown

$$\text{solubility} = [Fe^{3+}]$$

Step 3.  Equilibrium-Constant Expressions

$$[Fe^{3+}][OH^-]^3 = 4 \times 10^{-38}$$

$$[H_3O^+][OH^-] = 1.00 \times 10^{-14}$$

Steps 4 and 5.  Mass-Balance and Charge-Balance Equations
As in Example 8–5, the mass-balance and charge-balance equations are identical. That is,

$$[OH^-] = 3[Fe^{3+}] + [H_3O^+]$$

Step 6.  Number of Independent Equations and Unknowns
We see that we have three equations and three unknowns.

Step 7.  Approximations
As in Example 8–5, let us assume $[H_3O^+] \ll 3[Fe^{3+}]$, so that

$$3[Fe^{3+}] \approx [OH^-]$$

Step 8.  Solution to Equations
Substituting this equation into the solubility-product expression gives

$$[Fe^{3+}](3[Fe^{3+}])^3 = 4 \times 10^{-38}$$

$$[Fe^{3+}] = \left(\frac{4 \times 10^{-38}}{27}\right)^{1/4} = 2 \times 10^{-10}$$

$$\text{solubility} = [Fe^{3+}] = 2 \times 10^{-10} \text{ mol/L}$$

Step 9. Check of Assumptions

From the assumptions made in step 7, we can calculate a provisional value of $[OH^-]$. That is,

$$[OH^-] \approx 3[Fe^{3+}] = 3 \times 2 \times 10^{-10} = 6 \times 10^{-10}$$

Let us use this value of $[OH^-]$ to compute a *provisional* value for $[H_3O^+]$:

$$[H_3O^+] = \frac{1.00 \times 10^{-14}}{6 \times 10^{-10}} = 1.7 \times 10^{-5}$$

But $1.7 \times 10^{-5}$ is not much smaller than three times our provisional value of $[Fe^{3+}]$. This discrepancy means that our assumption was invalid and the provisional values for $[Fe^{3+}]$, $[OH^-]$, and $[H_3O^+]$ are all significantly in error. Therefore, let us go back to step 7 and assume that

$$3[Fe^{3+}] \ll [H_3O^+]$$

Now the mass-balance expression becomes

$$[H_3O^+] = [OH^-]$$

Substituting this equality into the expression for $K_w$ gives

$$[H_3O^+] = [OH^-] = 1.00 \times 10^{-7}$$

Substituting this number into the solubility-product expression developed in step 3 gives

$$[Fe^{3+}] = \frac{4 \times 10^{-38}}{(1.00 \times 10^{-7})^3} = 4 \times 10^{-17} \text{ mol/L}$$

Here we have assumed that $3[Fe^{3+}] \ll [OH^-]$ or $3 \times 4 \times 10^{-17} \ll 10^{-7}$. Clearly, the assumption is valid and we may write

$$\text{solubility} = 4 \times 10^{-17} \text{ mol/L}$$

Note the very large error introduced by the invalid assumption.

The faulty assumption in Example 8–6 was readily recognized.

Example 8–6 illustrates how readily the effects of an invalid assumption are detected. It may seem inconsistent that we used a faulty value for $[OH^-]$ to obtain $[H_3O^+]$, which we then compared with $3[Fe^{3+}]$. But the point is that the faulty value of $[OH^-]$ also arose because of the invalid assumption. Had the assumption been valid we would have had an internally consistent set of calculated concentrations. Even one internal inconsistency is a clear indication of an invalid assumption. A valid assumption leads to concentrations that satisfy all of the algebraic equalities developed in steps 3, 4, and 5.

## 8B–2 The Solubility of Precipitates in the Presence of Complexing Agents

The solubility of a precipitate often increases dramatically in the presence of reagents that form complexes with the anion or the cation of the solid. For example, fluoride ions prevent the quantitative precipitation of aluminum hydroxide even though the solubility product of this precipitate is remarkably small ($2 \times 10^{-32}$). The cause of the increase in solubility is shown by the equations

$$Al(OH)_3(s) \rightleftharpoons Al^{3+} + 3\,OH^-$$
$$+$$
$$6\,F^-$$
$$\updownarrow$$
$$AlF_6^{3-}$$

The fluoride complex is sufficiently stable to permit fluoride ions to compete successfully with hydroxide ions for aluminum ions.

*The solubility of a precipitate always increases in the presence of a complexing agent that reacts with the cation of the precipitate.*

---

### Example 8–7

The solubility product of CuI is $1.1 \times 10^{-12}$. The formation constant $K_2$ for the reaction of CuI with $I^-$ to give $CuI_2^-$ is $7.9 \times 10^{-4}$. Calculate the molar solubility of CuI in a $1.0 \times 10^{-4}$ M solution of KI.

**Step 1. Pertinent Equilibria**

From the input data, we assume that the following equilibria exist in a solution of KI that is saturated with CuI:

$$CuI(s) \rightleftharpoons Cu^+ + I^-$$
$$CuI(s) + I^- \rightleftharpoons CuI_2^-$$

Since neither $H_3O^+$ nor $OH^-$ reacts to any significant extent with any of the participants in these equilibria, we need not include the dissociation of water in our list.

**Step 2. Definition of Unknown**

From inspection of the two equations, we see that the dissolved copper(I) is present as either $Cu^+$ or $CuI_2^-$. Therefore,

$$\text{solubility} = [Cu^+] + [CuI_2^-]$$

**Step 3. Equilibrium-Constant Expressions**

$$[Cu^+][I^-] = K_{sp} = 1.1 \times 10^{-12} \tag{8–12}$$

$$\frac{[CuI_2^-]}{[I^-]} = K_2 = 7.9 \times 10^{-4} \tag{8–13}$$

**Step 4. Mass-Balance Expression**

Here, we may write

$$[I^-] = c_{KI} + [Cu^+] - [CuI_2^-] \tag{8–14}$$

The first term on the right-hand side of this equation represents the concentration of iodide from the KI, and the second term corresponds to the contribution of iodide from the dissolution of CuI. The third term gives the concentration of iodide needed to form the complex from CuI.

Step 5.  Charge-Balance Equation

$$[Cu^+] + [K^+] = [I^-] + [CuI_2^-] \tag{8-15}$$

The $K^+$ concentration is known to be $1.00 \times 10^{-4}$ M. Substituting this figure into Equation 8–15 gives, after rearranging,

$$[I^-] = 1.0 \times 10^{-4} + [Cu^+] - [CuI_2^-] \tag{8-16}$$

Note, however, that as in Example 8–5, the charge-balance and mass-balance equations are identical. Therefore, we have only three independent equations (Equations 8–12, 8–13, and 8–14).

Step 6.  Number of Independent Equations and Unknowns

We have three unknowns, $[Cu^+]$, $[I^-]$, and $[CuI_2^-]$, and as noted in step 5, three algebraic equations. Therefore, an exact solution to the equations is possible.

Step 7.  Approximations

Two approximations are possible in Equation 8–16. The first is that $[Cu^+]$ is so much smaller than $1.0 \times 10^{-4}$ that the former can be eliminated from the equation. The second is that $[CuI_2^-]$ is also small enough to be ignored. The first assumption is surely valid in light of the very small numerical value for the solubility-product constant for CuI. The validity of the second assumption is less obvious. As we have noted, however, it is usually wise to make the assumption, calculate a provisional value for the concentration, and see whether the provisional value is indeed much smaller than the number with which it is being compared. If it is not, a recalculation must be made with the questionable concentration term retained in the equation.

Following this procedure, we assume $([Cu^+] - [CuI_2^-]) \ll 1.0 \times 10^{-4}$, whereupon Equation 8–16 simplifies to

$$[I^-] = 1.0 \times 10^{-4}$$

Step 8.  Solution to Equations

Substituting for $[I^-]$ in Equation 8–12 gives, after rearranging,

$$[Cu^+] = (1.1 \times 10^{-12})/(1.0 \times 10^{-4}) = 1.1 \times 10^{-8}$$

Substituting the value for $[I^-]$ into Equation 8–13 and rearranging yield

$$[CuI_2^-] = 1.0 \times 10^{-4} \times 7.9 \times 10^{-4} = 7.9 \times 10^{-8}$$

When these concentrations are substituted into the relationship developed in step 2, we obtain

$$\text{solubility} = 1.1 \times 10^{-8} + 7.9 \times 10^{-8} = 9.0 \times 10^{-8} \text{ mol/L}$$

Step 9. Check of Assumptions

We see that $1.0 \times 10^{-4}$ is much larger than either of the calculated values for $[Cu^+]$ and $[CuI_2^-]$. The provisional values for these species are therefore acceptable.

---

Example 8–8

Calculate the solubility of AgBr in 0.0200 M $NH_3$, given that the formation constant for $Ag(NH_3)_2^+$ is $1.3 \times 10^7$, the solubility product for AgBr is $5.2 \times 10^{-13}$, and the dissociation constant for $NH_3$ is $1.76 \times 10^{-5}$.

Step 1. Pertinent Equilibria

$$AgBr \rightleftharpoons Ag^+ + Br^-$$

$$Ag^+ + 2\,NH_3 \rightleftharpoons Ag(NH_3)_2^+$$

$$NH_3 + H_2O \rightleftharpoons NH_4^+ + OH^-$$

Step 2. Definition of Unknown

It is apparent from these equations that 1 mol of AgBr produces 1 mol of $Br^-$ and 1 mol of silver-containing species. Therefore,

$$\text{solubility} = [Br^-] = [Ag^+] + [Ag(NH_3)_2^+]$$

Step 3. Equilibrium-Constant Expressions

$$[Ag^+][Br^-] = K_{sp} = 5.2 \times 10^{-13} \tag{8–17}$$

$$\frac{[Ag(NH_3)_2^+]}{[Ag^+][NH_3]^2} = \beta_2 = 1.3 \times 10^7 \tag{8–18}$$

$$\frac{[NH_4^+][OH^-]}{[NH_3]} = K_b = 1.76 \times 10^{-5} \tag{8–19}$$

Step 4. Mass-Balance Expressions

As shown in Example 8–2 and in step 2,

$$[Br^-] = [Ag^+] + [Ag(NH_3)_2^+] \tag{8–20}$$

Since the analytical concentration of $NH_3$ is 0.0200,

$$0.0200 = [NH_3] + [NH_4^+] + 2[Ag(NH_3)_2^+] \tag{8-21}$$

Furthermore, reaction of $NH_3$ with water produces one $NH_4^+$ for each $OH^-$. Therefore,

$$[OH^-] = [NH_4^+] \tag{8–22}$$

We do not include $[H_3O^+]$ in the charge balance equation because its concentration is negligible in a solution made basic with $NH_3$.

Step 5. Charge-Balance Expression

$$[Ag^+] + [NH_4^+] + [Ag(NH_3)_2^+] = [Br^-] + [OH^-] \tag{8–23}$$

Step 6. Number of Equations and Unknowns

We count seven equations (8–17 through 8–23) but only six unknowns ($[OH^-]$, $[Br^-]$, $[Ag^+]$, $[Ag(NH_3)_2^+]$, $[NH_4^+]$, and $[OH^-]$). This discrepancy suggests that the algebraic equations are not all independent. Close examination shows that Equation 8–23 is in fact the sum of Equations 8–20 and 8–22 and therefore not independent. Thus, we remove Equation 8–23 from consideration and now have six independent equations and an equal number of unknowns.

Step 7. Approximations

To locate possible approximations, we again turn to those equations in which concentrations appear as sums or differences—that is, Equations 8–20, 8–21, and 8–23.

1. Examining Equation 8–20 first, we speculate that $[Ag^+]$ may be considerably smaller than $[Ag(NH_3)_2^+]$ because the formation constant for the complex is so very large. Therefore, we assume provisionally that

$$[Ag^+] \ll [Ag(NH_3)_2^+]$$

2. The value for $K_b$ for ammonia is small, which suggests that the concentration of $NH_3$ is significantly greater than the concentration of $NH_4^+$. We therefore assume that in Equation 8–21

$$[NH_4^+] \ll [NH_3] + 2[Ag(NH_3)_2^+]$$

3. A second possible assumption regarding Equation 8–21 is that $[Ag(NH_3)_2^+]$ is also significantly smaller than $[NH_3]$. The reason for this assumption is not immediately obvious from the equations at hand. If, however, Equations 8–17 and 8–18 are multiplied together, we obtain

$$\frac{[Ag(NH_3)_2^+][Br^-]}{[NH_3]^2} = \beta_2 K_{sp} = 6.8 \times 10^{-6} \qquad (8\text{–}24)$$

This equation suggests that the numerator is significantly smaller than $[NH_3]^2$. Therefore, we shall try the additional assumption that

$$2[Ag(NH_3)_2^+] \ll [NH_3]$$

Each of these three assumptions is open to some doubt; however, no harm is done in making them because a lack of validity in any one will become obvious by the time the calculation is completed. With the assumptions, Equations 8–19 and 8–21 are no longer needed and Equations 8–20 and 8–21 simplify to

$$[Br^-] = [Ag(NH_3)_2^+]$$
$$0.0200 = [NH_3]$$
$$(8\text{–}25)$$

We now have just three equations (8–25, 8–17, and 8–18) and three unknowns ($[Ag^+]$, $[Ag(NH_3)_2^+]$, and $[Br^-]$).

Step 8.  Solution to Equations

Substituting $[NH_3] = 0.0200$ into Equation 8–18 gives, after rearranging,

$$[Ag(NH_3)_2^+] = (1.3 \times 10^7)(0.0200)^2[Ag^+] = (5.2 \times 10^3)[Ag^+]$$

Equation 8–25 then becomes

$$[Br^-] = (5.2 \times 10^3)[Ag^+]$$

Replacing $[Ag^+]$ with $K_{sp}/[Br^-]$ (Equation 8–17) gives

$$[Br^-] = \frac{5.2 \times 10^3 \times 5.2 \times 10^{-13}}{[Br^-]}$$

$$[Br^-] = \sqrt{5.2 \times 10^3 \times 5.2 \times 10^{-13}} = 5.2 \times 10^{-5}$$

Referring to step 2, we see

$$\text{solubility} = [Br^-] = 5.2 \times 10^{-5}\ \text{mol/L}$$

Step 9.  Check of Assumptions

To check assumption 1, we turn to Equation 8–25 and find

$$[Ag(NH_3)_2^+] = [Br^-] = 5.2 \times 10^{-5}$$

We assumed that this concentration is much smaller than $[NH_3]$, that is, $5.2 \times 10^{-5} \ll 0.0100$. Essentially no error is introduced by this assumption.

To check assumption 2, we substitute $[NH_3] = 0.0200$ and Equation 8–22 into Equation 8–19, which gives

$$\frac{[NH_4^+]^2}{0.0200} = 1.76 \times 10^{-5}$$

$$[NH_4^+] = \sqrt{1.76 \times 10^{-5} \times 0.0200} = 5.9 \times 10^{-4}$$

We have assumed that

$$5.9 \times 10^{-4} = 0.00059 \ll 0.0100$$

The resulting error is then less than $(0.00059/0.0100) \times 100\% = 3.0\%$, which is acceptable in most cases.

To check assumption 3, we must calculate $[Ag^+]$ by substituting $[Br^-] = 5.2 \times 10^{-5}$ into the expression for $K_{sp}$:

$$[Ag^+] = (5.2 \times 10^{-13})/(5.2 \times 10^{-5}) = 1.0 \times 10^{-8}$$

We know from Equation 8–25 that

$$[Ag(NH_3)_2^+] = [Br^-] = 5.2 \times 10^{-5}$$

and we have assumed that $1.0 \times 10^{-8} \ll 5.2 \times 10^{-5}$. The percent error introduced by this assumption is less than 0.1%.

## 8B–3  The Effect of pH on Solubility

The solubility of precipitates containing an anion with basic properties, a cation with acidic properties, or both is dependent upon pH. For example, when barium sulfate is equilibrated with a solution containing hydrochloric acid, the following equilibria are established:

$$BaSO_4(s) \rightleftharpoons Ba^{2+} + SO_4^{2-}$$

$$SO_4^{2-} + H_3O^+ \rightleftharpoons HSO_4^- + H_2O$$

If more acid is added to this system, the sulfate ion concentration decreases by the common-ion effect. This decrease causes the first equilibrium to shift to the right, thus increasing the solubility. The examples that follow illustrate how the effect of pH on solubility can be treated in quantitative terms.

> All precipitates that contain an anion that is the conjugate base of a weak acid are more soluble at low pH than at high pH.

### Solubility Calculations When the pH Is Fixed and Known

Analytical precipitations are frequently performed in solutions in which the hydronium ion concentration is fixed at some predetermined and known value. The calculation of solubility under this circumstance is a straightforward process, as illustrated by the following example.

---

**Example 8–9**

Calculate the molar solubility of calcium oxalate in a solution that has been buffered to a pH of 4.00.

Step 1. Pertinent Equilibria

$$CaC_2O_4(s) \rightleftharpoons Ca^{2+} + C_2O_4^{2-} \qquad (8-26)$$

Oxalate ions react with water to form $HC_2O_4^-$ and $H_2C_2O_4$. Thus, two other equilibria in this solution are

$$H_2C_2O_4 + H_2O \rightleftharpoons H_3O^+ + HC_2O_4^- \qquad (8-27)$$

$$HC_2O_4^- + H_2O \rightleftharpoons H_3O^+ + C_2O_4^{2-} \qquad (8-28)$$

Step 2. Definition of the Unknown
Calcium oxalate is a strong electrolyte, and so its molar analytical concentration is equal to the equilibrium calcium ion concentration:

$$solubility = [Ca^{2+}]$$

Step 3. Equilibrium-Constant Expressions

$$[Ca^{2+}][C_2O_4^{2-}] = K_{sp} = 2.3 \times 10^{-9} \qquad (8\text{--}29)$$

$$\frac{[H_3O^+][HC_2O_4^-]}{[H_2C_2O_4]} = K_1 = 5.36 \times 10^{-2} \qquad (8\text{--}30)$$

$$\frac{[H_3O^+][C_2O_4^{2-}]}{[HC_2O_4^-]} = K_2 = 5.42 \times 10^{-5} \qquad (8\text{--}31)$$

Step 4. Mass-Balance Expressions
Because $CaC_2O_4$ is the only source of $Ca^{2+}$ and of the three oxalate species,

$$[Ca^{2+}] = [C_2O_4^{2-}] + [HC_2O_4^-] + [H_2C_2O_4] \qquad (8\text{--}32)$$

Moreover, the problem states that the pH is 4.00. Thus,

$$[H_3O^+] = 1.00 \times 10^{-4} \qquad (8\text{--}33)$$

Step 5. Charge-Balance Expressions

A buffer keeps the pH of a solution constant.

A buffer is required to maintain the pH at 4.00. The buffer most likely consists of some weak acid HA and its conjugate base $A^-$ (Section 10C). The nature of the two species and their concentrations have not been specified, however, and so we do not have enough information to write a charge-balance equation.

Step 6. Number of Independent Equations and Unknowns
We have four unknowns ($[Ca^{2+}]$, $[C_2O_4^{2-}]$, $[HC_2O_4^-]$, and $[H_2C_2O_4]$) as well as four independent algebraic relationships (Equations 8–29, 8–30, 8–31, and 8–32). Therefore, an exact solution can be obtained, and the problem becomes one of algebra.

Step 7. Approximations
An exact solution is so readily obtained that we will not bother with approximations.

Step 8. Solution of the Equations
A convenient way to solve the problem is to substitute Equations 8–30 and 8–31 into 8–32 in such a way as to develop a relationship between $[Ca^{2+}]$, $[C_2O_4^{2-}]$, and $[H_3O^+]$. Thus, we rearrange Equation 8–31 to give

$$[HC_2O_4^-] = \frac{[H_3O^+][C_2O_4^{2-}]}{K_2}$$

Substituting numerical values gives

$$[HC_2O_4^-] = \frac{1.00 \times 10^{-4}[C_2O_4^{2-}]}{5.42 \times 10^{-5}} = 1.85[C_2O_4^{2-}]$$

Substituting this relationship into Equation 8–30 gives, upon rearranging,

$$[H_2C_2O_4] = \frac{[H_3O^+][C_2O_4^{2-}] \times 1.85}{K_2}$$

$$= \frac{1.85 \times 10^{-4}[C_2O_4^{2-}]}{5.36 \times 10^{-2}} = 3.45 \times 10^{-3}[C_2O_4^{2-}]$$

Substituting this equation and the earlier equation for $[HC_2O_4^-]$ into Equation 8–32 gives

$$[Ca^{2+}] = [C_2O_4^{2-}] + 1.85[C_2O_4^{2-}] + 3.45 \times 10^{-3}[C_2O_4^{2-}]$$

$$= 2.85[C_2O_4^{2-}]$$

or

$$[C_2O_4^{2-}] = \frac{[Ca^{2+}]}{2.85}$$

Substituting into Equation 8–29 gives

$$\frac{[Ca^{2+}][Ca^{2+}]}{2.85} = 2.3 \times 10^{-9}$$

$$[Ca^{2+}] = \text{solubility} = \sqrt{2.85 \times 2.3 \times 10^{-9}} = 8.1 \times 10^{-5} \text{ mol/L}$$

## Solubility Calculations When the pH Is Variable

Saturating an unbuffered solution with a sparingly soluble salt made up of a basic anion or an acidic cation causes the pH of the solution to change. For example, pure water saturated with barium carbonate is basic as a consequence of the reactions

$$BaCO_3(s) \rightleftharpoons Ba^{2+} + CO_3^{2-}$$

$$CO_3^{2-} + H_2O \rightleftharpoons HCO_3^- + OH^-$$

$$HCO_3^- + H_2O \rightleftharpoons H_2CO_3 + OH^-$$

In contrast to Example 8–9, the hydroxide ion concentration now becomes an unknown, and an additional algebraic equation must therefore be developed if the solubility of barium carbonate is to be calculated.

In many instances, the reaction of a precipitate with water cannot be neglected without introducing an appreciable error in the calculation. As shown by the data in Table 8–1, the magnitude of the error depends upon the solubility of the precipitate as well as on the base dissociation constant of the anion. The solubilities of the hypothetical precipitate MA, shown in column 4, were obtained by taking into account the reaction of $A^-$ with water. Column 5 gives the calculated results when the basic properties of $A^-$ are neglected; here, the solubility is simply the square root of the solubility product. Two solubility products, $1.0 \times 10^{-10}$ and

Table 8-1

CALCULATED SOLUBILITY OF MA FROM VARIOUS ASSUMED VALUES OF $K_{sp}$ and $K_b$

| Assumed $K_{sp}$ for MA | Assumed $K_{HA}$ | Dissociation Constant for $A^-$ $K_b = K_w/K_{HA}$ | Calculated Solubility of MA, M | Calculated Solubility of MA Neglecting Reaction of $A^-$ with Water, M |
|---|---|---|---|---|
| $1.0 \times 10^{-10}$ | $1.0 \times 10^{-6}$ | $1.0 \times 10^{-8}$ | $1.02 \times 10^{-5}$ | $1.0 \times 10^{-5}$ |
| | $1.0 \times 10^{-8}$ | $1.0 \times 10^{-6}$ | $1.2 \ \times 10^{-5}$ | $1.0 \times 10^{-5}$ |
| | $1.0 \times 10^{-10}$ | $1.0 \times 10^{-4}$ | $2.4 \ \times 10^{-5}$ | $1.0 \times 10^{-5}$ |
| | $1.0 \times 10^{-12}$ | $1.0 \times 10^{-2}$ | $10 \ \ \times 10^{-5}$ | $1.0 \times 10^{-5}$ |
| $1.0 \times 10^{-20}$ | $1.0 \times 10^{-6}$ | $1.0 \times 10^{-8}$ | $1.05 \times 10^{-10}$ | $1.0 \times 10^{-10}$ |
| | $1.0 \times 10^{-8}$ | $1.0 \times 10^{-6}$ | $3.3 \ \times 10^{-10}$ | $1.0 \times 10^{-10}$ |
| | $1.0 \times 10^{-10}$ | $1.0 \times 10^{-4}$ | $32 \ \ \times 10^{-10}$ | $1.0 \times 10^{-10}$ |
| | $1.0 \times 10^{-12}$ | $1.0 \times 10^{-2}$ | $290 \ \times 10^{-10}$ | $1.0 \times 10^{-10}$ |

$1.0 \times 10^{-20}$, have been assumed for these calculations as well as several values for $K_b$ (column 3). It is apparent that neglecting the reaction of the anions with water leads to a negative error that becomes more pronounced both as the solubility of the precipitate decreases (smaller $K_{sp}$) and as the conjugate base becomes stronger. Note that the error becomes insignificant for anions derived from acids with dissociation constants greater than about $10^{-6}$.

It is not difficult to write the algebraic relationships needed to calculate the solubility of such a precipitate. Solving the equations, however, is tedious. Fortunately, it is ordinarily possible to invoke one of two simplifying assumptions to decrease the algebraic labor:

1. The first simplification is applicable to moderately soluble compounds containing an anion that reacts extensively with water. It is assumed that sufficient hydroxide ions are formed to make it unnecessary to consider the hydronium ion concentration in the calculations. Another way of stating this assumption is to say that the hydroxide ion concentration of the solution is determined exclusively by the reaction of the anion with water and that the contribution of hydroxide ions from the dissociation of water is negligible by comparison.
2. The second simplification is applicable to precipitates of very low solubility, particularly those containing an anion that does not react extensively with water. In such a system, it can be assumed that dissolution of the precipitate does not significantly change the hydronium or hydroxide ion concentrations and that, at room temperature, these concentrations remain essentially $10^{-7}$ mol/L. The solubility calculation then follows the course shown in Example 8-9.

---

Example 8-10

Calculate the solubility of $BaCO_3$ in water.

Step 1. Pertinent Equilibria

$$BaCO_3(s) \rightleftharpoons Ba^{2+} + CO_3^{2-} \qquad (8-34)$$

$$CO_3^{2-} + H_2O \rightleftharpoons HCO_3^- + OH^- \qquad (8-35)$$

$$HCO_3^- + H_2O \rightleftharpoons H_2CO_3 + OH^- \qquad (8-36)$$

$$2\,H_2O \rightleftharpoons H_3O^+ + OH^- \qquad (8-37)$$

**Step 2.  Definition of the Unknown**

$$\text{solubility} = [Ba^{2+}] = [CO_3^{2-}] + [HCO_3^-] + [H_2CO_3]$$

**Step 3.  Equilibrium-Constant Expressions**

$$[Ba^{2+}][CO_3^{2-}] = K_{sp} = 5.1 \times 10^{-9} \qquad (8-38)$$

$$\frac{[HCO_3^-][OH^-]}{[CO_3^{2-}]} = \frac{K_w}{K_2} = \frac{1.00 \times 10^{-14}}{4.7 \times 10^{-11}} = 2.13 \times 10^{-4} \qquad (8-39)$$

$$\frac{[H_2CO_3][OH^-]}{[HCO_3^-]} = \frac{K_w}{K_1} = \frac{1.00 \times 10^{-14}}{4.45 \times 10^{-7}} = 2.25 \times 10^{-8} \qquad (8-40)$$

and

$$[H_3O^+][OH^-] = 1.00 \times 10^{-14} \qquad (8-41)$$

**Step 4.  Mass-Balance Expression**

$$[Ba^{2+}] = [CO_3^{2-}] + [HCO_3^-] + [H_2CO_3] \qquad (8-42)$$

**Step 5.  Charge-Balance Expression**

$$2[Ba^{2+}] + [H_3O^+] = 2[CO_3^{2-}] + [HCO_3^-] + [OH^-] \qquad (8-43)$$

**Step 6.  Number of Equations and Unknowns**

We have developed six equations (Equations 8-38 through 8-43), which are sufficient to solve for the six unknowns ($[Ba^{2+}]$, $[CO_3^{2-}]$, $[HCO_3^-]$, $[H_2CO_3]$, $[OH^-]$, and $[H_3O^+]$).

**Step 7.  Approximations**

We examine Equations 8-42 and 8-43 with the goal of eliminating one or more terms on the grounds that their elimination does not create a significant error. One such candidate is $[H_3O^+]$ in Equation 8-43. Because the solution is basic as a consequence of reactions 8-35 and 8-36, $[H_3O^+]$ must be smaller than $10^{-7}$. Also, $[Ba^{2+}]$ must be considerably larger than $10^{-7}$ because, if no reaction occurs between $CO_3^{2-}$ and $H_2O$, $[Ba^{2+}]$ is simply the square root of $K_{sp}$, or $7 \times 10^{-5}$ M. The fact that reaction does occur means that $[CO_3^{2-}]$ is smaller than $7 \times 10^{-5}$ and therefore $[Ba^{2+}] > 7 \times 10^{-5}$ M. Thus, the assumption that $[H_3O^+] \ll 2[Ba^{2+}]$ in Equation 8-43 appears entirely reasonable. With this assumption, we no longer need Equation 8-41.

A second possible assumption is that $[H_2CO_3]$ is so much smaller than $[HCO_3^-]$ that the former can be deleted from Equation 8-42. We base this assumption on our knowledge that the solution is basic and therefore

$[OH^-] > 10^{-7}$. If $[OH^-]$ is $10^{-6}$ M, substitution into Equation 8–40 reveals that

$$\frac{[H_2CO_3]}{[HCO_3^-]} = \frac{2.25 \times 10^{-8}}{10^{-6}} = 0.0225$$

If $[OH^-]$ is $10^{-5}$, this ratio is about 0.002. Although these calculations suggest that $[H_2CO_3] \ll [HCO_3^-]$, we cannot be sure that this assumption is valid. It is surely worth trying, however.

As a result of the first assumption, Equation 8–43 becomes

$$2[Ba^{2+}] = 2[CO_3^{2-}] + [HCO_3^-] + [OH^-] \qquad (8\text{–}44)$$

Further, because $[HCO_3^-] \gg [H_2CO_3]$, the mass-balance expression simplifies to

$$[Ba^{2+}] = [CO_3^{2-}] + [HCO_3^-] \qquad (8\text{–}45)$$

Equations 8–40 and 8–41 are now no longer needed. Thus, we have reduced both the number of equations and the number of unknowns to four.

Step 8.  Solution of the Equations

If we multiply Equation 8–45 by 2 and subtract the product from Equation 8–44, we obtain, upon rearrangement,

$$[OH^-] = [HCO_3^-] \qquad (8\text{–}46)$$

Substitution of $[HCO_3^-]$ for $[OH^-]$ in Equation 8–39 gives

$$\frac{[HCO_3^-]^2}{[CO_3^{2-}]} = \frac{K_w}{K_2}$$

$$[HCO_3^-] = \sqrt{\frac{K_w}{K_2}[CO_3^{2-}]}$$

This expression permits elimination of $[HCO_3^-]$ from Equation 8–45:

$$[Ba^{2+}] = [CO_3^{2-}] + \sqrt{\frac{K_w}{K_2}[CO_3^{2-}]} \qquad (8\text{–}47)$$

From Equation 8–38, we have

$$[CO_3^{2-}] = \frac{K_{sp}}{[Ba^{2+}]}$$

Substituting for $[CO_3^{2-}]$ in Equation 8–47 yields

$$[Ba^{2+}] = \frac{K_{sp}}{[Ba^{2+}]} + \sqrt{\frac{K_w K_{sp}}{K_2[Ba^{2+}]}}$$

It is convenient to multiply through by $[Ba^{2+}]$ and rearrange:

$$[Ba^{2+}]^2 - \sqrt{\frac{K_w}{K_2} K_{sp}[Ba^{2+}]} - K_{sp} = 0$$

Finally, after numerical values are supplied for the constants, we obtain

$$[Ba^{2+}]^2 - (1.04 \times 10^{-6})[Ba^{2+}]^{1/2} - 5.1 \times 10^{-9} = 0$$

In order to solve this equation by successive approximations (see Feature 6–4, page 136), we let $x = [Ba^{2+}]$ and rewrite the equation in the form

$$x = (1.04 \times 10^{-6} \sqrt{x} + 5.1 \times 10^{-9})^{1/2}$$

If we let $x_1 = 0$ and solve for $x_2$, we obtain

$$x_2 = (0 + 5.1 \times 10^{-9})^{1/2} = 7.14 \times 10^{-5}$$

Substituting this value for $x$ in the original equation gives

$$x_3 = (1.04 \times 10^{-6} \times \sqrt{7.14 \times 10^{-5}} + 5.1 \times 10^{-9})^{1/2} = 1.18 \times 10^{-4}$$

Further iteration leads to $x_4 = 1.28 \times 10^{-4}$, $x_5 = 1.30 \times 10^{-4}$, and $x_6 = 1.30 \times 10^{-4}$. Thus

$$\text{solubility} = [Ba^{2+}] = 1.30 \times 10^{-4} \simeq 1.3 \times 10^{-4} \text{ M}$$

### Step 9.  Check of Approximations

To check the two assumptions that were made, we must calculate the concentrations of most of the other ions in the solution. We can evaluate $[CO_3^{2-}]$ from Equation 8–38:

$$[CO_3^{2-}] = \frac{5.1 \times 10^{-9}}{1.3 \times 10^{-4}} = 3.9 \times 10^{-5}$$

From Equation 8–45,

$$[HCO_3^-] = 13.0 \times 10^{-5} - 3.9 \times 10^{-5} = 9.1 \times 10^{-5}$$

From Equation 8–46,

$$[OH^-] = [HCO_3^-] = 9.1 \times 10^{-5}$$

From Equation 8–40,

$$\frac{[H_2CO_3](9.1 \times 10^{-5})}{9.1 \times 10^{-5}} = 2.25 \times 10^{-8}$$

$$[H_2CO_3] = 2.2 \times 10^{-8}$$

Finally, from Equation 8–41,

$$[H_3O^+] = \frac{1.00 \times 10^{-14}}{9.1 \times 10^{-5}} = 1.1 \times 10^{-10}$$

We see that the two assumptions do not lead to large errors; $[HCO_3^-]$ is about 4000 times greater than $[H_2CO_3]$, and $[H_3O^+]$ is clearly much smaller than any of the species in Equation 8–43.

Failure to take into account the basic reaction of $CO_3^{2-}$ would have yielded a solubility of $7.1 \times 10^{-5}$, which is only about one half the value yielded by the more rigorous method.

---

An example of the second type of calculation referred to on page 176 follows. Here, it is assumed that the hydronium and hydroxide ion concentrations are $10^{-7}$ M after the solid has dissolved.

---

Example 8–11

Calculate the solubility of silver sulfide in pure water.

Step 1.  Pertinent Equilibria

$$Ag_2S(s) \rightleftharpoons 2\ Ag^+ + S^{2-} \tag{8–48}$$

$$S^{2-} + H_2O \rightleftharpoons HS^- + OH^- \tag{8–49}$$

$$HS^- + H_2O \rightleftharpoons H_2S + OH^- \tag{8–50}$$

$$2\ H_2O \rightleftharpoons H_3O^+ + OH^- \tag{8–51}$$

Step 2.  Definition of Unknown

$$\text{solubility} = \tfrac{1}{2}[Ag^+] = [S^{2-}] + [HS^-] + [H_2S]$$

Step 3.  Equilibrium-Constant Expressions

$$[Ag^+]^2[S^{2-}] = 6 \times 10^{-50} \tag{8–52}$$

$$\frac{[HS^-][OH^-]}{[S^{2-}]} = \frac{K_w}{K_2} = \frac{1.0 \times 10^{-14}}{1.2 \times 10^{-15}} = 8.3 \tag{8–53}$$

$$\frac{[H_2S][OH^-]}{[HS^-]} = \frac{K_w}{K_1} = \frac{1.0 \times 10^{-14}}{5.7 \times 10^{-8}} = 1.8 \times 10^{-7} \tag{8–54}$$

$$[H_3O^+][OH^-] = 1.00 \times 10^{-14}$$

Step 4.  Mass-Balance Expression

$$\tfrac{1}{2}[Ag^+] = [S^{2-}] + [HS^-] + [H_2S] \tag{8–55}$$

Step 5.  Charge-Balance Expression

$$[Ag^+] + [H_3O^+] = 2[S^{2-}] + [HS^-] + [OH^-] \tag{8–56}$$

Step 6. Comparison of Equations and Unknowns

We have six unknowns and six equations. Thus, an exact solution is feasible.

Step 7. Approximations

The solubility product for $Ag_2S$ is very small; therefore, it is probable that there is little change in hydroxide ion concentration as the precipitate dissolves. As a consequence, we can assume tentatively that

$$[OH^-] \simeq [H_3O^+] = 1.0 \times 10^{-7}$$

This assumption will be correct if, in Equation 8–56,

$$[Ag^+] \ll [H_3O^+] \quad \text{and} \quad 2[S^{2-}] + [HS^-] \ll [OH^-]$$

Step 8. Solution of Equations

We now proceed exactly as we did in Example 8–9. Substitution of $1.0 \times 10^{-7}$ for $[OH^-]$ in Equations 8–53 and 8–54 gives

$$\frac{[HS^-]}{[S^{2-}]} = \frac{8.3}{1.0 \times 10^{-7}} = 8.3 \times 10^7$$

$$[H_2S] = \frac{(1.8 \times 10^{-7})[HS^-]}{1.0 \times 10^{-7}} = 1.8[HS^-]$$

Substituting the first equation into the second yields

$$[H_2S] = (1.8)(8.3 \times 10^7)[S^{2-}] = (14.9 \times 10^7)[S^{2-}]$$

When these relationships are substituted into Equation 8–55, we obtain

$$\tfrac{1}{2}[Ag^+] = [S^{2-}] + (8.3 \times 10^7)[S^{2-}] + (14.9 \times 10^7)[S^{2-}]$$
$$[S^{2-}] = (2.3 \times 10^{-9})[Ag^+]$$

Substituting this relationship into the solubility-product expression gives

$$(2.3 \times 10^{-9})[Ag^+]^3 = 6 \times 10^{-50}$$
$$[Ag^+] = 3.0 \times 10^{-14}$$
$$\text{solubility} = \tfrac{1}{2}[Ag^+] = 1.5 \times 10^{-14} \simeq 2 \times 10^{-14} \text{ mol/L}$$

Step 9. Check of Approximations

The assumption that $[Ag^+]$ is much smaller than $[H_3O^+]$ is clearly valid. We can readily calculate a value for $2[S^{2-}] + [HS^-]$ and confirm that this sum is likewise much smaller than $[OH^-]$. Therefore, we conclude that the assumptions made are reasonable and that the approximate solution is satisfactory.

## 8B–4 The Effect of Undissociated Solute on Solubility

Thus far, we have considered only solutes that dissociate completely when dissolved in aqueous media. There are some inorganic substances, however, such as calcium sulfate and the silver halides, that act as weak electrolytes and only partially dissociate in water. For example, a saturated solution of silver chloride contains significant amounts of undissociated silver chloride molecules as well as silver and chloride ions. Here, two equilibria are required to describe the system:

$$AgCl(s) \rightleftharpoons AgCl(aq) \qquad (8-57)$$

$$AgCl(aq) \rightleftharpoons Ag^+ + Cl^- \qquad (8-58)$$

The equilibrium constant for the first reaction takes the form

$$\frac{[AgCl(aq)]}{[AgCl(s)]} = K$$

where the numerator is the concentration of the undissociated species *in the solution* and the denominator is the concentration of silver chloride *in the solid phase*. The latter term is a constant, however (page 128), and so the equation can be written

$$[AgCl(aq)] = K[AgCl(s)] = K_s \qquad (8-59)$$

where $K_s$ is the constant for the equilibrium shown in Equation 8–57. It is evident from this equation that, at a given temperature, the concentration of the undissociated silver chloride is constant and *independent* of the chloride and silver ion concentrations.

The equilibrium constant $K_d$ for the dissociation reaction (Equation 8–58) is

$$\frac{[Ag^+][Cl^-]}{[AgCl(aq)]} = K_d = 3.9 \times 10^{-4} \qquad (8-60)$$

The product of these two constants is equal to the solubility product:

$$[Ag^+][Cl^-] = K_d K_s = K_{sp}$$

As shown by Example 8–12, both reaction 8–57 and reaction 8–58, as well as two others, contribute significantly to the solubility of silver chloride.

## 8B–5 The Solubility of Silver Chloride as a Function of Chloride Ion Concentration

The reaction between silver ions and chloride ions finds widespread use in analytical chemistry for the gravimetric and volumetric determination of both species. The effect of excess chloride ion on the solubility of the silver chloride precipitate is complex, as revealed by the following set of equations that describe the chemistry of the system:

$$AgCl(s) \rightleftharpoons AgCl(aq) \qquad (8-61)$$

$$AgCl(aq) \rightleftharpoons Ag^+ + Cl^- \qquad (8-62)$$

$$AgCl(s) + Cl^- \rightleftharpoons AgCl_2^- \qquad (8-63)$$

$$AgCl_2^- + Cl^- \rightleftharpoons AgCl_3^{2-} \qquad (8-64)$$

Note that equilibrium 8–62 and thus equilibrium 8–61 shift to the left with added chloride ion, whereas equilibria 8–63 and 8–64 shift to the right under the same circumstance. The consequence of these opposing effects is that a plot of silver chloride solubility as a function of concentration of added chloride exhibits a minimum. Example 8–12 illustrates how this behavior can be described in quantitative terms.

---

Example 8–12

Derive an equation that describes the effect of the analytical concentration of KCl on the solubility of AgCl in an aqueous solution. Calculate the concentration of KCl at which the solubility is a minimum.

Step 1.  Pertinent Equilibria

Equations 8–61 through 8–64 describe the pertinent equilibria.

Step 2.  Definition of Unknown

The molar solubility $S$ of AgCl is equal to the sum of the concentrations of the silver-containing species:

$$\text{solubility} = S = [AgCl(aq)] + [Ag^+] + [AgCl_2^-] + [AgCl_3^{2-}] \quad (8-65)$$

Step 3.  Equilibrium-Constant Expressions

Equilibrium constants available in the literature include

$$[Ag^+][Cl^-] = K_{sp} = 1.82 \times 10^{-10} \qquad (8-66)$$

$$\frac{[Ag^+][Cl^-]}{[AgCl(aq)]} = K_d = 3.9 \times 10^{-4} \qquad (8-67)$$

$$\frac{[AgCl_2^-]}{[Cl^-]} = K_2 = 2.0 \times 10^{-5} \qquad (8-68)$$

$$\frac{[AgCl_3^{2-}]}{[AgCl_2^-][Cl^-]} = K_3 = 1 \qquad (8-69)$$

Step 4.  Mass-Balance Equation

$$[Cl^-] = c_{KCl} + [Ag^+] - [AgCl_2^-] - 2[AgCl_3^{2-}] \qquad (8-70)$$

The second term on the right-hand side of this equation gives the chloride ion concentration produced by the dissolution of the precipitate, and the next two terms correspond to the *decrease* in chloride ion concentration resulting from the formation of the two chloro complexes from AgCl.

Step 5. Charge-Balance Equation

As in some of the earlier examples, the charge-balance equation is identical to the mass-balance equation.

Step 6. Number of Equations and Unknowns

We have five equations (8–66 through 8–70) and five unknowns ($[Ag^+]$, $[AgCl(aq)]$, $[AgCl_2^-]$, $[AgCl_3^{2-}]$, and $[Cl^-]$).

Step 7. Assumptions

We assume that, over a considerable range of chloride ion concentration, the solubility of AgCl is so small that Equation 8–70 can be greatly simplified by the assumption that

$$[Ag^+] - [AgCl_2^-] - 2[AgCl_3^{2-}] \ll c_{KCl}$$

It is not certain that this is a valid assumption, but it is worth trying because it simplifies the problem so much. With this assumption, then, Equation 8–70 reduces to

$$[Cl^-] = c_{KCl} \tag{8–71}$$

Step 8. Solution of Equations

For convenience, we multiply Equations 8–68 and 8–69 together to give

$$\frac{[AgCl_3^{2-}]}{[Cl^-]^2} = K_2K_3 = 2.0 \times 10^{-5} \times 1 = 2.0 \times 10^{-5} \tag{8–72}$$

To calculate $[AgCl(aq)]$, we divide Equation 8–66 by Equation 8–67 and rearrange:

$$[AgCl(aq)] = \frac{K_{sp}}{K_d} = \frac{1.82 \times 10^{-10}}{3.9 \times 10^{-4}} = 4.7 \times 10^{-7} \tag{8–73}$$

Note that the concentration of this species is *constant and independent of the chloride concentration.*

Substitution of Equations 8–73, 8–66, 8–68, and 8–72 into Equation 8–65 permits us to express the solubility in terms of the chloride ion concentration and the several constants:

$$S = \frac{K_{sp}}{K_d} + \frac{K_{sp}}{[Cl^-]} + K_2[Cl^-] + K_2K_3[Cl^-]^2 \tag{8–74}$$

Substitution of Equation 8–71 yields the desired relationship between the solubility and the analytical concentration of KCl:

$$S = \frac{K_{sp}}{K_d} + \frac{K_{sp}}{c_{KCl}} + K_2c_{KCl} + K_2K_3c_{KCl}^2 \tag{8–75}$$

To find the minimum in $S$, we set the derivative of $S$ with respect to $c_{KCl}$ equal to zero:

$$\frac{dS}{dc_{KCl}} = 0 = -\frac{K_{sp}}{c_{KCl}^2} + K_2 + 2\,K_2K_3c_{KCl}$$

$$2\,K_2K_3c_{KCl}^3 + c_{KCl}^2K_2 - K_{sp} = 0$$

Substituting numerical values gives

$$(4.0 \times 10^{-5})c_{KCl}^3 + (2.0 \times 10^{-5})c_{KCl}^2 - 1.82 \times 10^{-10} = 0$$

Following the procedure shown in Feature 6–4, we can solve this equation by successive approximations to obtain

$$c_{KCl} = 0.0030 = [Cl^-]$$

In order to check the assumption made earlier, we calculate the concentration of the various species. Substitutions into Equations 8–66, 8–68, and 8–70 yield

$$[Ag^+] = (1.82 \times 10^{-10})/0.0030 = 6.1 \times 10^{-8}$$

$$[AgCl_2^-] = 2.0 \times 10^{-5} \times 0.0030 = 6.0 \times 10^{-8}$$

$$[AgCl_3^{2-}] = 2.0 \times 10^{-5} \times (0.0030)^2 = 1.8 \times 10^{-10}$$

Thus our assumption that $c_{KCl}$ is much larger than the concentrations of the ions of the precipitate is reasonable. The minimum solubility is obtained by substitution of these concentrations and [AgCl(aq)] into Equation 8–65:

$$S = 4.7 \times 10^{-7} + 6.1 \times 10^{-8} + 6.0 \times 10^{-8} + 1.8 \times 10^{-10}$$
$$= 5.9 \times 10^{-7}\ M$$

---

The solid curve in Figure 8–1 illustrates the effect of chloride ion concentration on the solubility of silver chloride; data for the curve were obtained by substituting various chloride concentrations into Equation 8–75. Note that at high concentrations of the common ion, the solubility becomes greater than that in pure water. The broken lines represent the equilibrium concentrations of the various silver-containing species as a function of $c_{KCl}$. Note that at the solubility minimum, undissociated silver chloride, AgCl(aq), is the major silver species in the solution, representing about 80% of the total dissolved silver. Its concentration is invariant, as has been demonstrated.

Unfortunately, reliable equilibrium data regarding undissociated species such as AgCl(aq) and complex species such as $AgCl_2^-$ are not abundant; consequently, solubility calculations are often, of necessity, based on solubility-product equilibria alone. Example 8–12 shows that, under some circumstances, such neglect of other equilibria can lead to serious error.

To find the minimum in a function, take the derivative of the function, set it equal to zero, and solve the resulting equation.

To verify that you have found a minimum, take the second derivative. If the second derivative is greater than zero, the solution is a minimum; if it is less than zero, the solution is a maximum. Verify that $c_{KCl} = 0.0030$ M occurs at a minimum.

In gravimetric procedures, a small excess of precipitating agent minimizes solubility losses, but a large excess often causes increased losses due to complex formation.

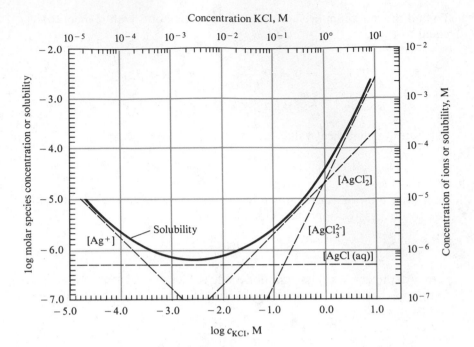

Figure 8-1
The effect of chloride ion concentration on the solubility of AgCl. The solid curve shows the total concentration of dissolved AgCl. The broken lines show the concentrations of the various silver-containing species.

## 8C   SEPARATION OF IONS BY CONTROL OF THE CONCENTRATION OF A PRECIPITATING REAGENT

When two different ions react with a reagent to form precipitates of different solubilities, the less soluble compound will form at a lower reagent concentration. If the solubilities are sufficiently different, quantitative removal of the first species from solution may be achieved without precipitation of the second. Such separations require careful control of the precipitating reagent concentration at some suitable predetermined level. Many important analytical separations, notably those involving sulfide ion, hydroxide ion, and organic reagents, are based on this concept.

### 8C-1   Calculation of the Feasibility of Separations

The following examples illustrate how solubility-product calculations are used to determine the feasibility of separations based upon solubility differences.

---

Example 8-13

Can $Fe^{3+}$ and $Mg^{2+}$ be separated quantitatively as hydroxides from a solution that is 0.10 M in each cation? If the separation is possible, what range of $OH^-$ concentrations is permissible? Solubility-product constants for the two precipitates are

$$[Fe^{3+}][OH^-]^3 = 4 \times 10^{-38}$$

$$[Mg^{2+}][OH^-]^2 = 1.8 \times 10^{-11}$$

The $K_{sp}$ for $Fe(OH)_3$ is so much smaller than that for $Mg(OH)_2$ that it appears likely that the former will precipitate at a lower $OH^-$ concentration. We can answer the questions posed in this problem by (1) calculating the $OH^-$ concentration required to achieve quantitative precipitation of $Fe^{3+}$ and (2) computing the $OH^-$ concentration at which $Mg(OH)_2$ just begins to precipitate. If (1) is smaller than (2), a separation is feasible in principle, and the range of permissible $OH^-$ concentrations is defined by the two values.

Be aware that it is only the enormous numerical difference between the $K_{sp}$'s that permits this judgment. The solubility products are not strictly comparable because the hydroxide concentration appears as a squared factor in one and as a cubed factor in the other.

To determine (1), we must first specify what constitutes a quantitative removal of $Fe^{3+}$ from the solution. The decision here is arbitrary and depends upon the purpose of the separation. In this example and the next, we shall consider a precipitation to be quantitative when all but 1 part in 1000 of the ion has been removed from the solution—that is, when $[Fe^{3+}] \ll 1 \times 10^{-4}$.

We can readily calculate the $OH^-$ concentration in equilibrium with $1 \times 10^{-1}$ M $Fe^{3+}$ by substituting directly into the solubility-product expression:

$$(1.0 \times 10^{-4})[OH^-]^3 = 4 \times 10^{-38}$$

$$[OH^-] = [(4 \times 10^{-38})/(1.0 \times 10^{-4})]^{1/3} = 7 \times 10^{-12} \text{ M}$$

Thus, if we maintain the $OH^-$ concentration at about $7 \times 10^{-12}$ mol/L, the $Fe^{3+}$ concentration will be lowered to $1 \times 10^{-4}$ mol/L. Note that quantitative precipitation of $Fe(OH)_3$ is achieved in a distinctly acidic medium.

To determine what maximum $OH^-$ concentration can exist in the solution without causing formation of $Mg(OH)_2$, we note that precipitation cannot occur until the product $[Mg^{2+}][OH^-]^2$ exceeds the solubility product, $1.8 \times 10^{-11}$. Substitution of 0.1 (the molar $Mg^{2+}$ concentration of the solution) into the solubility-product expression permits the calculation of the *maximum* $OH^-$ concentration that can be tolerated:

$$0.10[OH^-]^2 = 1.8 \times 10^{-11}$$

$$[OH^-] = 1.3 \times 10^{-5}$$

When the $OH^-$ concentration exceeds this level, the solution will be supersaturated with respect to $Mg(OH)_2$ and precipitation can begin.

From these calculations, we conclude that quantitative separation of $Fe(OH)_3$ can be achieved if the $OH^-$ concentration is greater than $7 \times 10^{-12}$ mol/L and that $Mg(OH)_2$ will not precipitate until a $OH^-$ concentration of $1.3 \times 10^{-5}$ mol/L is reached. Therefore, it is possible, in principle, to separate $Fe^{3+}$ from $Mg^{2+}$ by maintaining the $OH^-$ concentration between these levels. In practice, the concentration of $OH^-$ is kept as low as practical—often about $10^{-10}$ M.

## 8C–2 Sulfide Separations

A number of important methods for the separation of metallic ions involve controlling the concentration of the precipitating anion by regulating the

hydronium ion concentration of the solution. Such methods are particularly attractive because of the relative ease with which the hydronium ion
concentration can be maintained at some predetermined level by the use
of a suitable buffer. Perhaps the best known of these methods makes use
of hydrogen sulfide as the precipitating reagent. Hydrogen sulfide is a
weak acid, dissociating as follows:

$$H_2S + H_2O \rightleftharpoons H_3O^+ + HS^- \qquad \frac{[H_3O^+][HS^-]}{[H_2S]} = K_1 = 5.7 \times 10^{-8}$$

$$HS^- + H_2O \rightleftharpoons H_3O^+ + S^{2-} \qquad \frac{[H_3O^+][S^{2-}]}{[HS^-]} = K_2 = 1.2 \times 10^{-15}$$

These equations may be combined to give an expression for the overall
dissociation of hydrogen sulfide to sulfide ion:

$$H_2S + 2\,H_2O \rightleftharpoons 2\,H_3O^+ + S^{2-} \qquad \frac{[H_3O^+]^2[S^{2-}]}{[H_2S]} = K_1K_2 = 6.8 \times 10^{-23}$$

The constant for this overall reaction is simply the product of $K_1$ and $K_2$.

In sulfide separations, the solution is continuously saturated with hydrogen sulfide so that the molar concentration of the reagent is essentially
constant throughout the precipitation. Because hydrogen sulfide is such a
weak acid, its species concentration corresponds closely to its solubility
in water, which is about 0.1 M. It is thus permissible to assume that,
throughout any sulfide precipitation,

$$[H_2S] \approx 0.10 \text{ mol/L}$$

Substituting this value into the overall dissociation-constant expression
gives

$$\frac{[H_3O^+]^2[S^{2-}]}{0.10} = 6.8 \times 10^{-23}$$

Thus,

$$[S^{2-}] = \frac{6.8 \times 10^{-24}}{[H_3O^+]^2} \qquad (8-76)$$

Thus, we see that the sulfide ion concentration of a saturated hydrogen
sulfide solution varies inversely as the square of the hydronium ion concentration. Figure 8–2, which was obtained with Equation 8–76, reveals
that the sulfide ion concentration of an aqueous solution can be varied by
over 20 orders of magnitude by changing the pH from 1 to 11.

Substituting Equation 8–76 into the solubility-product expression gives

$$\frac{[M^{2+}] \times 6.8 \times 10^{-24}}{[H_3O^+]^2} = K_{sp}$$

$$[M^{2+}] = \text{solubility} = \frac{[H_3O^+]^2 K_{sp}}{6.8 \times 10^{-24}}$$

Thus, the solubility of a divalent metal sulfide increases as the square of the hydronium ion concentration.

___

### Example 8–14

Cadmium sulfide is less soluble than thallium(I) sulfide. Find the conditions under which $Cd^{2+}$ and $Tl^+$ can, in theory, be separated quantitatively with $H_2S$ from a solution that is 0.1 M in each cation.

The constants for the two solubility equilibria are

$$CdS(s) \rightleftarrows Cd^{2+} + S^{2-} \qquad [Cd^{2+}][S^{2-}] = 2 \times 10^{-28}$$
$$Tl_2S(s) \rightleftarrows 2\,Tl^+ + S^{2-} \qquad [Tl^+]^2[S^{2-}] = 1 \times 10^{-22}$$

Since CdS precipitates at a lower sulfide ion concentration than does $Tl_2S$, we first compute the sulfide ion concentration necessary for quantitative removal of $Cd^{2+}$ from solution. In order to make such a calculation, we must first specify what constitutes a quantitative removal. As before, this decision is arbitrary and depends upon the purpose of the separation. In this example, we again consider a separation to be quantitative when all but 1 part in 1000 of the $Cd^{2+}$ has been removed, or, in other words, when the concentration of the cation has been lowered to $1.00 \times 10^{-4}$ M. Substituting this value into the solubility-product expression gives

$$10^{-4}[S^{2-}] = 2 \times 10^{-28}$$
$$[S^{2-}] = 2 \times 10^{-24}$$

Thus, if we maintain the sulfide ion concentration at this level or greater, we may assume that the desired quantitative removal of cadmium will take place.

Next, we compute the sulfide ion concentration needed to initiate precipitation of $Tl_2S$ from a 0.1 M solution. Precipitation will begin when the solubility product is just exceeded. Since the solution is 0.1 M in $Tl^+$,

$$(0.1)^2[S^{2-}] = 1 \times 10^{-22}$$
$$[S^{2-}] = 1 \times 10^{-20}$$

These two calculations show that quantitative precipitation of $Cd^{2+}$ takes place when $[S^{2-}]$ is made greater than $2 \times 10^{-24}$ M. No precipitation of $Tl^+$ occurs, however, until $[S^{2-}]$ becomes greater than $1 \times 10^{-20}$ M.

Substituting these two values for $[S^{2-}]$ into Equation 8–76 permits calculation of the $[H_3O^+]$ range required for the separation:

$$[H_3O^+]^2 = \frac{6.8 \times 10^{-24}}{1 \times 10^{-24}} = 3.4$$
$$[H_3O^+] = 1.8 \text{ M}$$

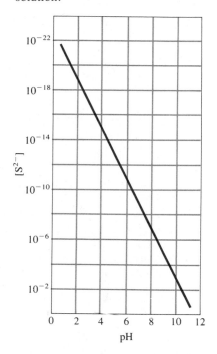

**Figure 8–2**

Sulfide ion concentration as a function of pH in a saturated $H_2S$ solution.

Ordinarily we shall assume a precipitation is quantitative if less than one part per thousand of the analyte remains in the solution.

and

$$[H_3O^+]^2 = \frac{6.8 \times 10^{-24}}{1 \times 10^{-20}} = 6.8 \times 10^{-4}$$

$$[H_3O^+] \approx 0.026 = 0.03 \text{ M}$$

By maintaining $[H_3O^+]$ between 0.03 and 1.8 M, we can in theory separate $CdS$ quantitatively from $Tl_2S$.

## 8D    QUESTIONS AND PROBLEMS

Note: Additional problems requiring use of the solubility-product constant are to be found at the end of Chapters 6 and 7. Use molar concentrations rather than activities for these problems.

*8–1. How does calculation of the molar solubility for $Fe(OH)_2$ differ from that for $Fe(OH)_3$?

8–2. Why are simplifying assumptions restricted to relationships that are sums or differences?

*8–3. Why do molar concentrations of some species appear as multiples in charge-balance equations?

8–4. Demonstrate how the sulfide-ion concentration is related to the pH of a solution that is kept saturated with hydrogen sulfide.

8–5. Briefly explain why the concentration of the unionized form of an incompletely dissociated and sparingly soluble substance is constant.

8–6. Generate the solubility-product expression for
*(a) $AgIO_3$.        (e) $CuI$.
*(b) $Ag_2SO_3$.      (f) $PbI_2$.
*(c) $Ag_3AsO_4$.     (g) $BiI_3$.
*(d) $PbClF$.         (h) $MgNH_4PO_4$.

8–7. Express the solubility-product constant for each substance in Problem 8–6 in terms of its molar solubility $S$.

8–8. Calculate the solubility-product constant for each of the following substances, given that the molar analytical concentrations of their saturated solutions are as indicated:
*(a) $BaCrO_4$        $1.2 \times 10^{-5}$ M
(b) $Cd(OH)_2$        $1.14 \times 10^{-5}$ M
*(c) $La(IO_3)_3$     $6.92 \times 10^{-4}$ M
(d) $Zn_2Fe(CN)_6$    $3.74 \times 10^{-6}$ M

*8–9. Calculate the solubility-product constant for $Fe(OH)_2$ if a saturated solution is $1.17 \times 10^{-5}$ M in $OH^-$.

*8–10. What is the $Pb^{2+}$ concentration in a saturated aqueous solution of
(a) $PbSO_4$?        (c) $Pb(OH)_2$?
(b) $PbI_2$?         (d) $PbCl_2$?

8–11. What is the $I^-$ concentration in a saturated aqueous solution of
(a) $AgI$?        (b) $PbI_2$?
(c) $BiI_3$ ($K_{sp} = 8.1 \times 10^{-19}$)?

*8–12. What $IO_3^-$ concentration is needed to
(a) initiate precipitation of $Cu(IO_3)_2$ ($K_{sp} = 7.4 \times 10^{-8}$) from a solution that is $5.0 \times 10^{-3}$ M in $Cu^{2+}$?
(b) lower the $Cu^{2+}$ concentration of a solution to $2.0 \times 10^{-6}$ M?

8–13. What is the $Hg^{2+}$ concentration of a
(a) saturated aqueous solution of $Hg(IO_3)_2$ ($K_{sp} = 3.2 \times 10^{-13}$)?
(b) 0.100 M $Hg(NO_3)_2$ solution that is saturated with $Hg(IO_3)_2$?
(c) 0.100 M $KIO_3$ that is saturated with $Hg(IO_3)_2$?

*8–14. Calculate the equilibrium concentration of each ion in the solution that results when 0.180 g of $Mg(OH)_2$ is added to 45.0 mL of (a) 0.0204 M HCl and (b) 0.204 M HCl.

*8–15. Calculate the equilibrium concentration of each ion in the solution that results after mixing 40.0 mL of 0.0450 M $MgI_2$ with 60.0 mL of
(a) 0.0400 M KOH.
(b) 0.0500 M $Ba(OH)_2$.
(c) 0.0500 M $AgNO_3$.
(d) 0.0600 M $AgNO_3$.

8–16. Calculate the equilibrium concentration of each ion in the solution that results after mixing 16.0 mL of 0.150 M $MgSO_4$ with 64.0 mL of
(a) 0.0400 M $BaCl_2$.
(b) 0.0325 M KOH.
(c) 0.0325 M $BaI_2$.
(d) 0.0375 M $Ba(OH)_2$.

8–17. Write the mass-balance expression for a solution that is
*(a) 0.10 M in $H_3PO_4$.
(b) 0.10 M in $Na_2HPO_4$.
*(c) 0.100 M in $HNO_2$ and 0.0500 M in $NaNO_2$.
(d) 0.025 M in $NaF$ and saturated with $CaF_2$.
*(e) 0.100 M in $NaOH$ and saturated with $Zn(OH)_2$ (which undergoes the reaction $Zn(OH)_2 + 2\,OH^- \rightleftharpoons Zn(OH)_4^{2-}$).
(f) saturated with $MgCO_3$.
*(g) saturated with $CaF_2$.
(h) 0.010 M in $NaF$ and saturated with $Al(OH)_3$. ($Al^{3+}$ forms a series of six F complexes having formulas of $AlF^{2+}$, $AlF_2^+$, . . . . , $AlF_6^{3-}$.)

*(i) 0.0100 M in $NH_3$ and saturated with $Cd(OH)_2$. ($Cd^{2+}$ forms a series of ammine complexes having formulas of $Cd(NH_3)^{2+}$, $Cd(NH_3)_2^{2+}$, ...., $Cd(NH_3)_6^{2+}$.)

**8–18.** Write the charge-balance equations for the solutions in Problem 8–17.

**8–19.** The molar solubility, $S$, of silver cyanide in water is sought.
  (a) Write net equations for the equilibria that affect the solubility of AgCN. Silver cyanide reacts with $CN^-$ to form the complex ion $Ag(CN)_2^-$.
  (b) How is $S$ related to the species that appear in the equilibrium expressions in (a)?
  (c) Write any mass-balance and charge-balance equations that apply to the system.
  (d) Do you have sufficient independent expressions to solve for $S$?
  (e) Make simplifying assumptions that appear reasonable, and briefly explain your reasoning behind any such assumptions.
  (f) Solve for $S$; check the validity of any assumptions that you have made.

**8–20.** Repeat the steps in Problem 8–19 to calculate the solubility of AgCN in a solution that is buffered to a pH of
  (a) 5.00.     *(b) 7.00.     (c) 9.00.

**\*8–21.** The molar solubility, $S$, of $PbC_2O_4$ in a solution buffered to a pH of 6.00 is sought. Mass-balance requires that

$$[Pb^{2+}] = [H_2C_2O_4] + [HC_2O_4^-] + [C_2O_4^{2-}]$$

  (a) Express $[H_2C_2O_4]$ and $[HC_2O_4^-]$ in terms of $[H_3O^+]$, $[C_2O_4^{2-}]$, and the dissociation constants for oxalic acid. Recall that

$$K_1K_2 = \frac{[H_3O^+]^2[C_2O_4^{2-}]}{[H_2C_2O_4]}$$

  (b) Substitute the expressions generated in (a) into the mass-balance equation.
  (c) Solve for the fraction of oxalate-containing species that exist as oxalate ion (that is, $[C_2O_4^{2-}]/S$) through rearrangement of (b).

**8–22.** Refer to Problem 8–21.
  (a) Express $[HC_2O_4^-]$ and $[C_2O_4^{2-}]$ in terms of $[H_3O^+]$, $[H_2C_2O_4]$ and the dissociation constants for oxalic acid.
  (b) Substitute the expressions generated in (a) into the mass-balance equation.
  (c) Solve for the fraction $[H_2C_2O_4]/S$ through rearrangement of (b).

**8–23.** Refer to Problem 8–21.
  (a) Express $[H_2C_2O_4]$ and $[C_2O_4^{2-}]$ in terms of $[HC_2O_4^-]$, $[H_3O^+]$, and the dissociation constants for oxalic acid.
  (b) Substitute the expressions generated in (a) into the mass-balance equation.
  (c) Solve for the fraction $[HC_2O_4^-]/S$ through rearrangement of (b).

**8–24.** The solubility products for a series of hydroxides are

| | |
|---|---|
| BiOOH | $K_{sp} = 4.0 \times 10^{-10} = [BiO^+][OH^-]$ |
| $Be(OH)_2$ | $K_{sp} = 7.0 \times 10^{-22}$ |
| $Tm(OH)_3$ | $K_{sp} = 3.0 \times 10^{-24}$ |
| $Hf(OH)_4$ | $K_{sp} = 4.0 \times 10^{-26}$ |

Which hydroxide has
  (a) the lowest molar solubility in $H_2O$?
  (b) the lowest molar solubility in a solution that is 0.10 M in NaOH?

**8–25.** Calculate the molar solubility of $Ag_2CO_3$ in a solution that has a $H_3O^+$ concentration of
  *(a) $1.0 \times 10^{-6}$ M.     *(c) $1.0 \times 10^{-9}$ M.
  (b) $1.0 \times 10^{-7}$ M.     (d) $1.0 \times 10^{-11}$ M.

**8–26.** Calculate the molar solubility of $BaSO_4$ in a solution in which $[H_3O^+]$ is
  *(a) 2.0 M.     *(c) 0.50 M.
  (b) 1.0 M.     (d) 0.10 M.

**8–27.** Calculate the molar solubility of CuS in a solution in which the $H_3O^+$ concentration is held constant at (a) $1.0 \times 10^{-1}$ M and (b) $1.0 \times 10^{-4}$ M.

**8–28.** Calculate the concentration of CdS in a solution in which the $[H_3O^+]$ is held constant at (a) $1.0 \times 10^{-1}$ M and (b) $1.0 \times 10^{-4}$ M.

**8–29.** Calculate the molar solubility of MnS in a solution with a constant $[H_3O^+]$ of
  *(a) $1.00 \times 10^{-5}$.
  (b) $1.00 \times 10^{-7}$.

**\*8–30.** Calculate the molar solubility of $PbCO_3$
  (a) in water.
  (b) in a solution buffered to a pH of 7.00.

**8–31.** Calculate the molar solubility of $Ag_2SO_3$ ($K_{sp} = 1.5 \times 10^{-14}$)
  (a) in water.
  (b) in a solution buffered to a pH of 4.00.

**\*8–32.** Calculate the molar solubility of $MgCO_3$ in
  (a) water.
  (b) a solution in which $[H_3O^+]$ is $1.0 \times 10^{-8}$ M.
  (c) 0.10 M $Na_2CO_3$.

**8–33.** The solubility-product constant for $CdCO_3$ is $2.5 \times 10^{-14}$. Calculate the equilibrium solubility of $CdCO_3$ in
  (a) water.
  (b) a solution in which the $[H_3O^+]$ is fixed at $1.0 \times 10^{-8}$ M.

**\*8–34.** Dilute NaOH is introduced into a solution that is 0.050 M in $Cu^{2+}$ and 0.040 M in $Mn^{2+}$.
  (a) Which hydroxide precipitates first?
  (b) What $OH^-$ concentration is needed to initiate precipitation of the first hydroxide?
  (c) What is the concentration of the cation forming the less soluble hydroxide when the more soluble hydroxide begins to form?

**8–35.** A solution is 0.040 M in $Na_2SO_4$ and 0.050 M in $NaIO_3$. To this is added a solution containing $Ba^{2+}$.
  (a) Which barium salt will precipitate first?

**(b)** What is the $Ba^{2+}$ concentration as the first precipitate forms?

**(c)** What is the concentration of the anion that forms the less soluble barium salt when the more soluble precipitate begins to form?

**\*8–36.** Silver ion is being considered as a reagent for separating $I^-$ from $SCN^-$ in a solution that is 0.060 M in KI and 0.070 M in NaSCN.

**(a)** What $Ag^+$ concentration is needed to lower the $I^-$ concentration to $1.0 \times 10^{-6}$ M?

**(b)** What is the $Ag^+$ concentration of the solution when AgSCN begins to precipitate?

**(c)** What is the ratio of $SCN^-$ to $I^-$ ion when AgSCN begins to precipitate?

**(d)** What is the ratio of $SCN^-$ to $I^-$ when the $Ag^+$ concentration is $1.0 \times 10^{-3}$ M?

**8–37.** Using $1.0 \times 10^{-6}$ M as the criterion for quantitative removal, determine whether it is feasible to use

**(a)** $SO_4^{2-}$ to separate $Ba^{2+}$ and $Sr^{2+}$ in a solution that is initially 0.10 M in $Sr^{2+}$ and 0.25 M in $Ba^{2+}$.

**(b)** $SO_4^{2-}$ to separate $Ba^{2+}$ and $Ag^+$ in a solution that is initially 0.040 M in each cation. For $Ag_2SO_4$, $K_{sp} = 1.6 \times 10^{-5}$.

**(c)** $OH^-$ to separate $Be^{3+}$ and $Hf^{4+}$ in a solution that is initially 0.020 M in $Be^{2+}$ and 0.010 M in $Hf^{4+}$. See Problem 8–24 for $K_{sp}$ data.

**(d)** $IO_3^-$ to separate $In^{3+}$ and $Tl^+$ in a solution that is initially 0.11 M in $In^{3+}$ and 0.060 M in $Tl^+$. For $In(IO_3)_3$, $K_{sp} = 3.3 \times 10^{-11}$; for $TlIO_3$, $K_{sp} = 3.1 \times 10^{-6}$.

**\*8–38.** What weight of AgBr dissolves in 200 mL of 0.100 M NaCN?

$$Ag^+ + 2\,CN^- \rightleftharpoons Ag(CN)_2^- ; \quad K_f = 1.3 \times 10^{21}$$

**8–39.** The equilibrium constant for formation of $CuCl_2^-$ is given by

$$Cu^+ + 2\,Cl^- \rightleftharpoons CuCl_2^-$$

$$K_f = \frac{[CuCl_2^-]}{[Cu^+][Cl^-]^2} = 7.9 \times 10^4$$

What is the solubility of CuCl in solutions having the following analytical NaCl concentrations:

**(a)** 1.0 M?          **(d)** $1.0 \times 10^{-3}$ M?

**(b)** $1.0 \times 10^{-1}$ M?     **(e)** $1.0 \times 10^{-4}$ M?

**(c)** $1.0 \times 10^{-2}$ M?

**\*8–40.** Calcium sulfate is only partially dissociated in aqueous solution:

$$CaSO_4(aq) \rightleftharpoons Ca^{2+} + SO_4^{2-} \qquad K_d = 5.2 \times 10^{-3}$$

The solubility-product constant for $CaSO_4$ is $2.6 \times 10^{-5}$. Calculate the solubility of $CaSO_4$ in (a) water and (b) 0.0100 M $Na_2SO_4$. In addition, calculate the percent of undissociated $CaSO_4$ in each solution.

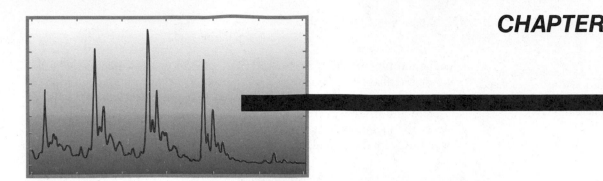

## CHAPTER 9

# PRECIPITATION TITRATIONS WITH SILVER NITRATE

$S$ilver nitrate is an important and widely used volumetric reagent. It is employed for the determination of anions that precipitate as silver salts. Among such species are the halides, several divalent anions, mercaptans, and certain fatty acids. Titrimetric methods based upon silver nitrate are sometimes termed *argentometric methods*.

With a few minor exceptions, no other precipitating agents react rapidly enough to be useful as titrimetric reagents. Thus, most precipitation titrations make use of silver nitrate.

"Argentometric" is derived from the Latin noun *argentum*, which means silver.

## 9A    TITRATION CURVES IN TITRIMETRIC METHODS

As noted in Section 5A–2, an end point is an observable physical change that occurs near the equivalence point of a titration. The two most widely used end points involve (1) a change in color due to the reagent, the analyte, or an indicator and (2) a change in potential of an electrode that responds to the concentration of one of the reactants.

To help us understand the theoretical basis of end points and the sources of titration errors, we will derive *titration curves* for the systems under consideration. Titration curves consist of a plot of reagent volume as the horizontal axis and some function of the analyte or reagent concentration as the vertical axis.

Titration curves are plots of a concentration-related variable as a function of reagent volume.

### 9A–1 Types of Titration Curves

Two general types of titration curves (and thus two general types of end points) are encountered in titrimetric methods. In the first type, called a *sigmoidal curve,* important observations are confined to a small region (typically ±0.1 to ±0.5 mL) surrounding the equivalence point. A sigmoidal curve is shown in Figure 9–1a. In the second type, called a *linear-segment curve,* measurements are made on both sides of, but well away from, the equivalence point. Measurements near equivalence are avoided. A typical linear-segment curve is shown in Figure 9–1b. The

Sigmoidal means s-shaped.

The vertical axis in a sigmoidal titration curve is either the p-function of the analyte or reagent or the potential of an analyte- or reagent-sensitive electrode.

The vertical axis in a linear-segment titration is an instrumental signal that is proportional to the concentration of the analyte or the reagent.

sigmoidal type offers the advantages of speed and convenience. The linear-segment type is advantageous for reactions that are complete only in the presence of a considerable excess of reagent or analyte.

Here and in the several chapters that follow, we will be dealing exclusively with sigmoidal titration curves. Linear-segment curves are considered in Chapters 19 and 22.

### 9A–2 Concentration Changes During Titrations

At the beginning of the titration described in Table 9–1, about 41 mL of reagent brings about a tenfold decrease in the concentration of A; only 0.1 mL is required to cause this same change at the equivalence point.

The equivalence point in a titration is characterized by major changes in the *relative* concentrations of reagent and analyte. Table 9–1 illustrates this phenomenon. The data in the second column of the table show the changes in concentration as a 50.00-mL aliquot of a 0.1000 M solution of an analyte A is titrated with a 0.1000 M solution of a reagent R. The chemical reaction produces the precipitate AR, which has a solubility product of $1.0 \times 10^{-10}$. In order to emphasize the changes in *relative* concentration that occur in the equivalence region, the volume increments selected are those required to cause tenfold decreases in the concentration of A. Thus, we see in the third column that an addition of 40.9 mL of R is needed to decrease the concentration of A by one order of magnitude, from 0.10 M to 0.010 M. A volume of only 8.1 mL is required to lower the concentration by another factor of 10, to 0.0010 M; 0.9 mL causes yet another tenfold decrease. Corresponding increases in [R] occur at the same time. End-point detection, then, is based upon this large change in *relative* concentration of analyte (or reagent) that occurs at the equivalence point for every type of titration.

The large relative-concentration changes that occur in the region of chemical equivalence are shown by plotting the negative logarithm of the analyte or reagent concentration (the p-function) against reagent volume, as has been done in Figure 9–2. The data for these plots are found in the fourth and fifth columns of Table 9–1. Titration curves for reactions in-

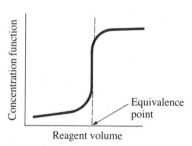

(a) Sigmoidal curve

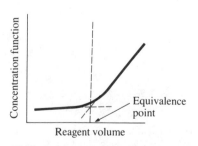

(b) Linear-segment curve

Figure 9–1
Two types of titration curves.

### Table 9–1
#### CONCENTRATION CHANGES DURING A TITRATION*

| Volume of 0.1000 M R, mL | [A], mol/L | Volume of R That Causes a Tenfold Decrease in [A], mL | pA | pR |
|---|---|---|---|---|
| 0.00 | $1.0 \times 10^{-1}$ | — | 1.00 | — |
| 40.90 | $1.0 \times 10^{-2}$ | 40.9 | 2.00 | 8.00 |
| 49.00 | $1.0 \times 10^{-3}$ | 8.1 | 3.00 | 7.00 |
| 49.90 | $1.0 \times 10^{-4}$ | 0.9 | 4.00 | 6.00 |
| 50.00 | $1.0 \times 10^{-5}$ | 0.1 | 5.00 | 5.00 |
| 50.10 | $1.0 \times 10^{-6}$ | 0.1 | 6.00 | 4.00 |
| 51.00 | $1.0 \times 10^{-7}$ | 0.9 | 7.00 | 3.00 |
| 61.10 | $1.0 \times 10^{-8}$ | 10.1 | 8.00 | 2.00 |

*Reaction: $A(aq) + R(aq) \rightarrow AR(s)$
For AR(s), $K_{sp} = 1.00 \times 10^{-10}$
Volume of 0.1000 M A = 50.00 mL

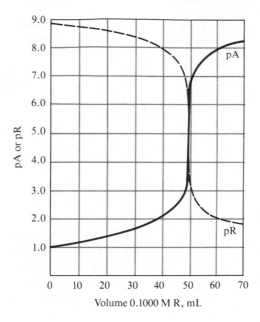

**Figure 9–2**
Precipitation titration curve for 50.00 mL of 0.1000 M A with 0.1000 M R.

volving precipitation, neutralization, complex formation, and oxidation/reduction all exhibit the same sharp change in p-function shown in Figure 9–2. Titration curves define the properties required of an indicator and allow us to estimate the error associated with titration methods.

## 9B   TITRATION CURVES FOR ARGENTOMETRIC METHODS

The derivation of a curve for an argentometric titration requires three types of calculation, each of which applies to a particular stage in the titration: (1) preequivalence points, (2) equivalence point, and (3) postequivalence points.

In the preequivalence stage, we compute the concentration of the analyte from its starting concentration and the volumetric data. The silver ion concentration is then obtained by substituting the analyte concentration into the solubility-product expression. At the equivalence point, silver ion and analyte ion exist in stoichiometric proportion, and the silver ion concentration is derived directly from the solubility-product constant. In the postequivalence stage, the analytical concentration of the excess silver nitrate is computed and assumed to be identical to its equilibrium concentration.

We cannot determine pAg initially because no $AgNO_3$ has been added.

---

Example 9–1

Derive a curve for the titration of 50.00 mL of 0.00500 M NaBr with 0.01000 M $AgNO_3$ (for AgBr, $K_{sp} = 5.2 \times 10^{-13}$).
**Initial Point**
At the outset, the solution is 0.000 M in $Ag^+$, and pAg is indeterminate.
**After Addition of 5.00 mL of Reagent**

Expressions for calculating concentrations during a titration can be derived from the charge-balance equation for the system. See Feature 9–1 for details.

The bromide ion concentration is decreased as a result of both precipitate formation and dilution. Thus, the analytical concentration of NaBr is

$$c_{NaBr} = \frac{\text{no. mmol NaBr after addition of AgNO}_3}{\text{total volume soln}}$$

$$= \frac{\text{original no. mmol NaBr} - \text{no. mmol AgNO}_3 \text{ added}}{\text{total volume soln}}$$

$$= \frac{(50.00 \text{ mL} \times 0.00500 \text{ M}) - (5.00 \text{ mL} \times 0.01000 \text{ M})}{50.00 \text{ mL} + 5.00 \text{ mL}}$$

$$= \frac{(0.2500 \text{ mmol} - 0.0500 \text{ mmol})}{55.00 \text{ mL}} = 3.64 \times 10^{-3} \text{ M}$$

The first term in the numerator of these equations is the number of millimoles of NaBr originally in the sample, and the second term is the number of millimoles of AgNO$_3$ added, which is equal to the number of millimoles of Br$^-$ that have reacted. The denominator takes into account the dilution of the solution by the added reagent.

Both the unreacted NaBr and the slightly soluble AgBr contribute to the species concentration of bromide ion. Thus, the equilibrium concentration of Br$^-$ is larger than the analytical concentration of NaBr by an amount equal to the molar solubility of the precipitate:

From the charge-balance expression,

$$[\text{Br}^-] = ([\text{Na}^+] - [\text{NO}_3^-]) + [\text{Ag}^+]$$
$$= 3.64 \times 10^{-3} + [\text{Ag}^+]$$

$$[\text{Br}^-] = 3.64 \times 10^{-3} + [\text{Ag}^+]$$

The contribution of the silver bromide to the equilibrium bromide ion concentration is equal to [Ag$^+$] because one silver ion is formed for each bromide ion from this source. Unless the concentration of NaBr is very small, this term can be neglected. That is, [Ag$^+$] $\ll 3.64 \times 10^{-3}$, and so

$$[\text{Br}^-] \approx c_{NaBr} = 3.64 \times 10^{-3}$$

The silver ion concentration is given by the relationship

$$[\text{Ag}^+] = \frac{K_{sp}}{[\text{Br}^-]}$$

To obtain an expression for pAg, we take the negative logarithm of both sides of this equation. Thus,

$$-\log[\text{Ag}^+] = -\log K_{sp} - (-\log[\text{Br}^-])$$

$$pAg + pBr = pK_{sp}$$
$$= -\log 5.2 \times 10^{-13}$$
$$= 12.28$$

From the definition of p-functions, we may write

$$pAg = pK_{sp} - pBr$$
$$= -\log 5.2 \times 10^{-13} - pBr$$
$$= 12.28 - (-\log 3.64 \times 10^{-3})$$
$$= 12.28 - 2.44 = 9.84$$

This relationship applies to any solution containing silver and bromide ions in contact with solid silver bromide. Note that the calculated pAg corresponds to $[Ag^+] = 1.4 \times 10^{-10}$, which, as we assumed at the outset, is certainly much smaller than $3.64 \times 10^{-3}$.

Other points in the region before chemical equivalence can be derived in this same way. Data for several such points are found in column 3 of Table 9–2.

Equivalence Point

At the equivalence point, neither NaBr nor $AgNO_3$ is in excess, and so the concentrations of silver and bromide ions must be equal. Substituting this equality into the solubility-product expression yields

$$[Ag^+] = [Br^-] = \sqrt{5.2 \times 10^{-13}} = 7.21 \times 10^{-7}$$
$$pAg = pBr = -\log(7.21 \times 10^{-7}) = 6.14$$

After Addition of 25.10 mL of Reagent

The solution now contains an excess of $AgNO_3$, and we can write

$$c_{AgNO_3} = \frac{\text{total no. mmol } AgNO_3 - \text{original no. mmol NaBr}}{\text{total volume of solution}}$$
$$= \frac{(25.10 \text{ mL} \times 0.01000 \text{ M}) - (50.00 \text{ mL} \times 0.00500 \text{ M})}{(50.00 + 25.10) \text{ mL}}$$
$$= 1.33 \times 10^{-5} \text{ M}$$

The equilibrium concentration of silver ion is then

$$[Ag^+] = 1.33 \times 10^{-5} + [Br^-] \approx 1.33 \times 10^{-5}$$

In this equation, $[Br^-]$ is a measure of the $Ag^+$ concentration resulting from the slight solubility of AgBr; it can ordinarily be neglected. Thus,

$$pAg = -\log(1.33 \times 10^{-5}) = 4.876 = 4.88$$

Additional points defining the titration curve beyond the equivalence point can be obtained in an analogous way and are found in Table 9–2. The data are plotted as curve B in Figure 9–3.

**Feature 9–1**
DERIVING TITRATION CURVES FROM THE CHARGE-BALANCE EQUATION

In Example 9–1, we derived a precipitation titration curve from the stoichiometry of the reaction and the volumes and concentrations of the reagents. We can show that all points on the titration curve can also be derived from the charge-balance equation.

**Before the Equivalence Point**

To calculate the bromide ion concentration, we write the charge-balance equation for the system.

$$[Na^+] + [Ag^+] = [Br^-] + [NO_3^-]$$

We may solve this equation for $[Br^-]$.

$$[Br^-] = ([Na^+] - [NO_3^-]) + [Ag^+]$$

The concentrations of $Na^+$ and $NO_3^-$ are given by

$$[Na^+] = \frac{c_{NaBr} V_{NaBr}}{V_{NaBr} + V_{AgNO_3}} \quad \text{and} \quad [NO_3^-] = \frac{c_{AgNO_3} V_{AgNO_3}}{V_{NaBr} + V_{AgNO_3}}$$

where $c_{NaBr}$ and $c_{AgNO_3}$ are the molar analytical concentrations of the sample and reagent solutions. We then substitute these expressions into the charge-balance equation to obtain

$$[Br^-] = \left( \frac{c_{NaBr} V_{NaBr}}{V_{NaBr} + V_{AgNO_3}} - \frac{c_{AgNO_3} V_{AgNO_3}}{V_{NaBr} + V_{AgNO_3}} \right) + [Ag^+]$$

$$= \left( \frac{c_{NaBr} V_{NaBr} - c_{AgNO_3} V_{AgNO_3}}{V_{NaBr} + V_{AgNO_3}} \right) + \frac{K_w}{[Br^-]}$$

This equation is a quadratic equation in $[Br^-]$ that can be solved exactly, or we can assume that the silver ion concentration is negligible as in Example 9–1. The silver ion concentration is then calculated from the $K_{sp}$

$$[Ag^+] = \frac{K_{sp}}{[Br^-]}$$

**At the Equivalence Point**

At the equivalence point $[Na^+] = [NO_3^-]$, so

$$[Ag^+] = [Br^-]$$

and

$$[Br^-] = \sqrt{K_{sp}}$$

**After the Equivalence Point**

Beyond the equivalence point, the solution contains an excess of $AgNO_3$, so we rearrange the charge-balance equation to solve for $[Ag^+]$.

$$[Ag^+] = ([NO_3^-] - [Na^+]) + [Br^-]$$

Substituting the concentrations and volumes of the reagents into this equation, we have

$$[Ag^+] = \left( \frac{c_{AgNO_3}V_{AgNO_3}}{V_{NaBr} + V_{AgNO_3}} - \frac{c_{NaBr}V_{NaBr}}{V_{NaBr} + V_{AgNO_3}} \right) + [Br^-]$$

$$= \left( \frac{c_{AgNO_3}V_{AgNO_3} - c_{NaBr}V_{NaBr}}{V_{NaBr} + V_{AgNO_3}} \right) + \frac{K_w}{[Ag^+]}$$

Once again, this equation can be solved exactly for $[Ag^+]$ using the quadratic equation, or $[Br^-]$ can be neglected as in Example 9–1.

## 9B–1  Significant Figures in Titration-Curve Calculations

The concentration data associated with the equivalence-point region of a titration curve are usually of low precision because they are based upon small differences between large numbers. For example, in the calculation of $c_{AgNO_3}$ following the introduction of 25.10 mL of 0.01000 M $AgNO_3$, the numerator ($0.2510 - 0.2500 = 0.0010$) contains only two significant figures; at best then, $c_{AgNO_3}$ is known to two significant figures. To minimize the rounding error, however, three digits ($1.33 \times 10^{-5}$) were retained in the calculation, and rounding was postponed until after pAg was computed.

In rounding the calculated value for pAg, it is important to recall (page 29) that it is the *mantissa of a logarithm (that is, the number to the right*

### Table 9–2
### CHANGES IN pAg DURING TITRATION WITH SOLUTIONS OF DIFFERENT CONCENTRATIONS

| | pAg | | |
|---|---|---|---|
| Volume AgNO₃, mL | 50.00 mL 0.0500 M Br⁻ with 0.1000 M AgNO₃ | 50.00 mL 0.00500 M Br⁻ with 0.01000 M AgNO₃ | 50.00 mL 0.000500 M Br⁻ with 0.001000 M AgNO₃ |
| 0.00 | — | — | — |
| 10.00 | 10.68 | 9.68 | 8.68 |
| 20.00 | 10.13 | 9.13 | 8.13 |
| 23.00 | 9.72 | 8.72 | 7.72 |
| 24.90 | 8.41 | 7.41 | 6.50* |
| 24.95 | 8.10 | 7.10 | 6.33* |
| 25.00 | 6.14 | 6.14 | 6.14 |
| 25.05 | 4.18 | 5.18 | 5.95† |
| 25.10 | 3.88 | 4.88 | 5.78† |
| 27.00 | 2.58 | 3.58 | 4.58 |
| 30.00 | 2.20 | 3.20 | 4.20 |

*Approximation $[Ag^+] \ll c_{NaBr}$ not valid.
†Approximation $[Br^-] \ll c_{AgNO_3}$ not valid.

pAg = 4.88 has only *two* significant figures and not three because the 4 merely sets the decimal point.

*of the decimal point) that should be rounded to include only significant figures* because the characteristic (the number to the left of the decimal point) serves merely to locate the decimal point. Thus, in Example 9–1, we rounded pAg to two figures to the right of the decimal point; that is, pAg = 4.88. Fortunately, the large changes in p-functions characteristic of most equivalence points are not obscured by the limited precision of the calculated data.

Note that the pAg values for points well away from the equivalence point contain one additional significant figure. For example, the initial pBr could be reported as 2.301. Nothing is gained by the added precision, however, because our primary interest is in the equivalence-point region. For this reason, in deriving data for titration curves, we shall generally round p-functions to two digits to the right of the decimal point.

## 9B–2 Factors Influencing End-Point Sharpness

As a rule of thumb, satisfactory end points require a change of 2 in p-function within ±0.1 mL of the equivalence point.

A sharp and easily located end point is observed when small additions of titrant cause large changes in p-function. It is therefore of interest to examine the variables that influence the magnitude of such changes during a titration.

### Reagent Concentration

End points improve as solutions become more concentrated.

Table 9–2 is a compilation of data computed by the method shown in Example 9–1. Three concentrations of titrant and analyte, each differing by a factor of 10, are used. The data are plotted in Figure 9–3. It is apparent that increases in analyte and reagent concentrations enhance the change in pAg in the equivalence-point region (an analogous effect is observed when pBr is plotted rather than pAg).

These effects have practical significance for the titration of bromide ion. If the analyte concentration is sufficient to permit the use of a silver

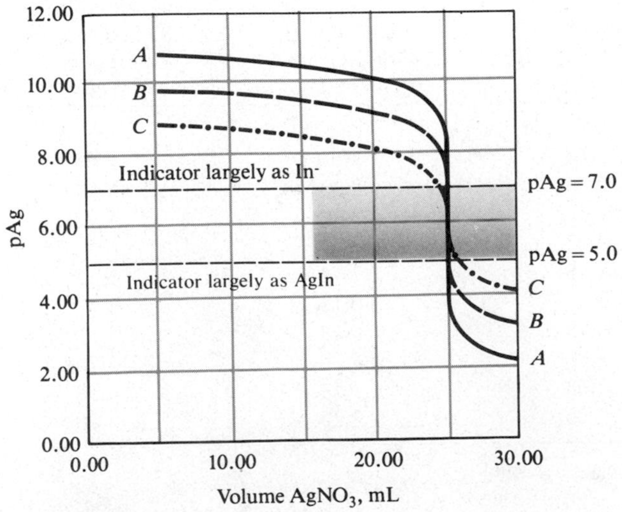

Figure 9–3

Effect of titrant concentration on titration curves: *A*, 50.00 mL of 0.0500 M NaBr with 0.1000 M AgNO$_3$; *B*, 50.00 mL of 0.00500 M NaBr with 0.01000 M AgNO$_3$; *C*, 50.00 mL of 0.000500 M NaBr with 0.001000 M AgNO$_3$.

nitrate solution that is 0.1 M or stronger, easily detected end points are observed, and the titration error is minimal. On the other hand, with standard titrant solutions that are 0.001 M or less, the change in pAg or pBr is so small that end-point detection is difficult and a large titration error must be expected. We shall see that titrant concentration also affects end-point sharpness in neutralization and complex-formation titrations.

### Reaction Completeness

Figure 9–4 demonstrates the effect of precipitate solubility on the sharpness of the end point in titrations using 0.1 M silver nitrate. Clearly, the greatest change in pAg occurs in the titration of iodide ion, which, of all the anions considered, forms the least soluble silver salt and hence reacts most completely with silver ion. The smallest change in pAg is observed for the reaction product that is most soluble—that is, in the titration of bromate ion.

> End points improve as the analytical reaction becomes more complete.

### 9B–3 Chemical Indicators for Precipitation Titrations

The end point produced by a chemical indicator usually consists of a color change or, occasionally, the appearance or disappearance of turbidity in the solution being titrated. Chemical indicators function by reacting competitively with one of the reactants of the titration. For example, consider the titration of an analyte A with a titrant R in the presence of an indicator In that reacts with R. A chemical description of this system throughout the titration is

> Indicators compete with the analyte for reagent or with the reagent for analyte.

$$A + R \rightleftarrows AR(s)$$
$$In + R \rightleftarrows InR$$

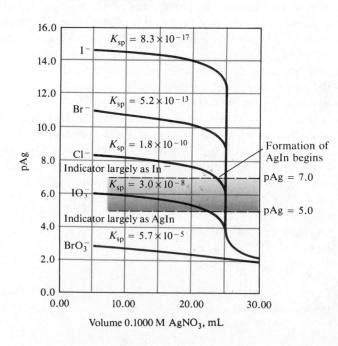

Figure 9–4

Effect of reaction completeness on titration curves. For each curve, 50.00 mL of a 0.0500 M solution of the anion was titrated with 0.1000 M AgNO$_3$.

For a color change to be seen, [InR]/[In] must change by a factor of 10 to 100.

For satisfactory indicator action, a very low concentration of InR must radically change the appearance of the titration solution. In addition, the equilibrium constant for the indicator reaction must be such that the ratio [InR]/[In] increases very quickly as a consequence of the change in [R] (or pR) that occurs in the equivalence-point region. For most indicators, the ratio [InR]/[In] must shift by one to two orders of magnitude in order to produce a color change detectable to the human eye. Such a change corresponds to a change of 1 or 2 in pR.

To illustrate, consider the application of a hypothetical indicator In$^-$ to the three bromide titrations described in Table 9–2 and Figure 9–3. We will assume that the equilibrium constant for the reaction of the indicator with silver ion to give AgIn is such that a full color change occurs when pAg shifts from 7 to 5. It is clear that each of the three titrations requires a different volume of titrant to encompass this range. For example, the data in the second column of Table 9–2 indicate that pAg changes from 8.10 to 4.18 as the volume of 0.1000 M silver nitrate is increased from 24.95 mL to 25.05 mL. Thus, an indicator with a pAg range of 7 to 5 would exhibit its color change within less than ±0.05 mL of the equivalence point at 25.00 mL. This would result in an abrupt change in color and a minimal titration error. In contrast, with 0.001 M AgNO$_3$, it can be shown that the color starts to change at about 24.5 mL and is complete at 25.8 mL. Exact location of the end point in this titration is impossible. The pAg change for the titration with 0.01 M reagent is such that somewhat less than 0.2 mL is required to bring about a color change; here, the indicator is usable, but the titration error is significant.

Consider the applicability and effectiveness of this same indicator for the titrations represented by the curves in Figure 9–4. Here, the indicator is largely AgIn throughout the titration of bromate and iodate ions, and so no color change is observed after the very first drops of silver nitrate are added. In contrast, the solubilities of silver bromide and silver iodide are small enough to prevent formation of significant amounts of AgIn until the equivalence-point region is reached. Note that pAg remains well above 7 until just before the equivalence point for the bromide titration and until just beyond the equivalence point for the iodide titration. In both cases, the color change occurs over a range of less than 0.01 mL of reagent, and titration errors are therefore negligible.

Turning to the titration curve for chloride ion in Figure 9–4, we see that an indicator with a pAg range of 5 to 7 is not satisfactory because appreciable amounts of AgIn will form approximately 1 mL short of the equivalence point and AgIn formation continues over a range of about 1 mL, making exact location of the end point impossible. In contrast, an indicator with a pAg range of 4 to 6 would be perfectly satisfactory. No chemical indicator exists for the iodate and bromate titrations because the changes in pAg in the equivalence-point region are far too small.

### The Volhard End Point

In the Volhard method, silver ions are titrated with a standard solution of thiocyanate ion:

$$Ag^+ + SCN^- \rightleftharpoons AgSCN(s)$$

Iron(III) serves as the indicator. The solution turns red with the first slight excess of thiocyanate ion:

$$Fe^{3+} + SCN^- \rightleftharpoons Fe(SCN)^{2+} \qquad K_f = 1.4 \times 10^2 = \frac{[Fe(SCN)^{2+}]}{[Fe^{3+}][SCN^-]}$$

Red

The titration must be carried out in acidic solution to prevent precipitation of iron(III) as the hydrated oxide.

---

Example 9–2

It has been found from experiment that the average observer can just detect the red color of $Fe(SCN)^{2+}$ when its concentration is $6.4 \times 10^{-6}$ M. In the titration of 50.0 mL of 0.050 M $Ag^+$ with 0.100 M KSCN, what concentration of $Fe^{3+}$ should be used to reduce the titration error to zero?

For a zero titration error, the $Fe(SCN)^{2+}$ color should appear when the concentration of $Ag^+$ remaining in the solution is identical to the sum of the two thiocyanate species. That is, at the equivalence point,

$$[Ag^+] = [SCN^-] + [Fe(SCN)^{2+}]$$
$$= [SCN^-] + 6.4 \times 10^{-6}$$

or
$$\frac{K_{sp}}{[SCN^-]} = \frac{1.1 \times 10^{-12}}{[SCN^-]} = [SCN^-] + 6.4 \times 10^{-6}$$

which rearranges to

$$[SCN^-]^2 + 6.4 \times 10^{-6}[SCN^-] - 1.1 \times 10^{-12} = 0$$
$$[SCN^-] = 1.7 \times 10^{-7}$$

The formation constant for $Fe(SCN)^{2+}$ is

$$K_f = 1.4 \times 10^2 = \frac{[Fe(SCN)^{2+}]}{[Fe^{3+}][SCN^-]}$$

If we now substitute the $[SCN^-]$ necessary to give a detectable concentration of $FeSCN^{2+}$ at the equivalence point, we obtain

$$1.4 \times 10^2 = \frac{6.4 \times 10^{-6}}{[Fe^{3+}]1.7 \times 10^{-7}}$$
$$[Fe^{3+}] = 0.27$$

---

The indicator concentration is not critical in the Volhard titration. In fact, calculations similar to those in Example 9–2 demonstrate that a titration error of one part in a thousand or less is, in theory, possible if the iron(III) concentration is held between 0.002 and 1.6 M. In practice, it is found that an indicator concentration greater than 0.2 M imparts sufficient color to the solution to make detection of the thiocyanate complex difficult. Therefore, lower concentrations (usually about 0.01 M) of iron(III) ion are employed.

The most important application of the Volhard method is for the indirect determination of halide ions. A measured excess of standard silver nitrate solution is added to the sample, and the excess silver ion is determined by back-titration with a standard thiocyanate solution. The strongly acidic environment required for the Volhard procedure represents a distinct advantage over other methods of halide analysis because such ions as carbonate, oxalate, and arsenate (which form slightly soluble silver salts in neutral media) do not interfere.

Silver chloride is more soluble than silver thiocyanate. As a consequence, in chloride determinations by the Volhard method, the reaction

$$AgCl(s) + SCN^- \rightleftharpoons AgSCN(s) + Cl^-$$

occurs to a significant extent near the end of the back-titration of the excess silver ion. This reaction causes the end point to fade and results in an overconsumption of thiocyanate ion, which in turn leads to low values for the chloride analysis. This error can be circumvented by filtering out the silver chloride before undertaking the back-titration. Filtration is not required in the determination of other halides because they all form silver salts that are less soluble than silver thiocyanate.

## Adsorption Indicators

An *adsorption indicator* is an organic compound that tends to be adsorbed onto the surface of the solid in a precipitation titration. Ideally, the adsorption (or desorption) occurs near the equivalence point and results not only in a color change but also in a transfer of color from the solution to the solid (or the reverse).

Fluorescein is a typical adsorption indicator that is useful for the titration of chloride ion with silver nitrate. In aqueous solution, fluorescein partially dissociates into hydronium ions and negatively charged fluoresceinate ions that are yellow-green. The fluoresceinate ion forms a highly colored reddish silver salt. Whenever fluorescein is used as an indicator, however, *its concentration is never large enough to precipitate as silver fluoresceinate*.

In the early stages of the titration of chloride ion with silver nitrate, the colloidal silver chloride particles are negatively charged because of adsorption of excess chloride ions (Section 4C–2). The indicator anions are repelled from this surface by electrostatic repulsion and impart a yellow-green color to the solution. Beyond the equivalence point, however, the silver chloride particles strongly adsorb silver ions and thereby acquire a positive charge. Fluoresceinate anions are now attracted *into the counterion layer* that surrounds each colloidal silver chloride particle. The net result is the appearance of the red color of silver fluoresceinate *in the surface layer of the solution surrounding the solid*. It is important to emphasize that the color change is an *adsorption* (not a precipitation) process, since the solubility product of the silver fluoresceinate is never

*The Volhard procedure requires that the analyte solution be distinctly acidic.*

*At one time a small amount of nitrobenzene was added to coat the precipitated AgCl, and thereby eliminate the need for filtration. Nitrobenzene has now been identified as a health hazard and is no longer used in this way.*

*Adsorption indicators were introduced in 1935 by a Polish chemist, K. Fajans. Titrations with these indicators are sometimes called Fajans titrations.*

exceeded. The adsorption is reversible, the indicator being desorbed upon back-titration with chloride ion.

Titrations involving adsorption indicators are rapid, accurate, and reliable, but their application is limited to the relatively few precipitation reactions in which a colloidal precipitate is formed rapidly.

## 9B–4  Other Methods of End-Point Detection

The *Mohr method* makes use of yet another chemical system to signal the end point for an argentometric titration. The indicator is a solution of potassium chromate. After reaction with the analyte is complete, the first excess of silver ion combines with chromate to form a bright red solid:

$$2\ Ag^+ + CrO_4^{2-} \rightarrow Ag_2CrO_4(s)$$

The appearance of this color signals the end point.

Potentiometric methods, which can be applied to the detection of end points in some precipitation reactions, are described in Chapter 17.

## 9B–5  Titration Curves for Mixtures

The methods developed in the previous section for deriving titration curves can be extended to mixtures that form precipitates of different solubilities. To illustrate, consider the titration of 50.00 mL of a solution that is 0.0500 M in iodide ion and 0.0800 M in chloride ion with 0.1000 M silver nitrate. The curve for the initial stages of this titration is identical to the curve shown for iodide in Figure 9–4 because silver chloride, with its much larger solubility product, does not begin to precipitate until well into the titration.

It is of interest to determine how much iodide is precipitated before appreciable amounts of silver chloride form. With the appearance of the smallest amount of solid silver chloride, the solubility-product expressions for both precipitates apply, and division of one by the other provides the useful relationship

$$\frac{[\cancel{Ag^+}][I^-]}{[\cancel{Ag^+}][Cl^-]} = \frac{8.3 \times 10^{-17}}{1.82 \times 10^{-10}} = 4.56 \times 10^{-7}$$

$$[I^-] = (4.56 \times 10^{-7})[Cl^-]$$

It is apparent from this relationship that the iodide concentration is decreased to a tiny fraction of the chloride ion concentration prior to the onset of silver chloride precipitation. Thus, for all practical purposes, formation of silver chloride will occur only after 25.00 mL of titrant have been added in this titration. At this point, the chloride ion concentration, because of dilution, is approximately

$$c_{Cl} = [Cl^-] = \frac{50.00 \times 0.0800}{50.00 + 25.00} = 0.0533\ M$$

Substituting into the previous equation yields

$$[I^-] = 4.56 \times 10^{-7} \times 0.0533 = 2.43 \times 10^{-8}$$

The percentage of iodide unprecipitated at this point can be calculated as follows:

$$\text{no. mmol } I^- \text{ remaining} = (75.00 \text{ mL})(2.43 \times 10^{-8} \text{ mmol } I^-/\text{mL}) = 1.82 \times 10^{-6}$$

$$\text{original no. mmol } I^- = (50.00 \text{ mL})(0.0500 \text{ mmol/mL}) = 2.50$$

$$I^- \text{ unprecipitated} = \frac{1.82 \times 10^{-6}}{2.50} \times 100\% = 7.3 \times 10^{-5}\%$$

Thus, to within about $7.3 \times 10^{-5}$ percent of the equivalence point for iodide, no silver chloride forms, and up to this point, the titration curve is indistinguishable from that for the iodide alone (Figure 9–4). The first part of the titration curve shown by the solid line in Figure 9–5 was derived on this basis.

As chloride ion begins to precipitate, the rapid decrease in pAg is terminated abruptly at a level that can be calculated from the solubility-product constant for silver chloride and the computed chloride concentration:

$$[Ag^+] = \frac{1.82 \times 10^{-10}}{0.0533} = 3.41 \times 10^{-9}$$

$$pAg = -\log(3.41 \times 10^{-9}) = 8.47$$

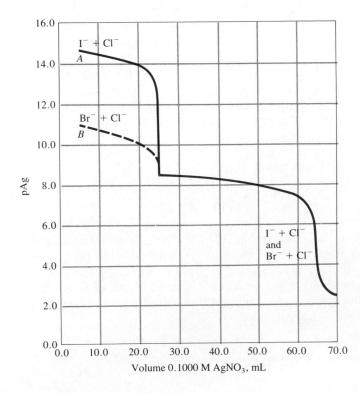

Figure 9–5
Titration curves for 50.00 mL of a solution 0.0800 M in Cl⁻ and 0.0500 M in I⁻ or Br⁻.

Further additions of silver nitrate decrease the chloride ion concentration, and the curve then becomes that for the titration of chloride by itself. For example, after 30.00 mL of titrant has been added,

$$c_{Cl} = [Cl^-] = \frac{50.00 \times 0.0800 + 50.00 \times 0.0500 - 30.00 \times 0.100}{50.00 + 30.00}$$

Here, the first two terms in the numerator give the number of millimoles of chloride and iodide, respectively, and the third is the number of millimoles of titrant added. Thus,

$$[Cl^-] = 0.0438$$

$$[Ag^+] = \frac{1.82 \times 10^{-10}}{0.0438} = 4.16 \times 10^{-9}$$

$$pAg = 8.38$$

The remainder of the curve can be derived in the same way as a curve for chloride by itself.

Curve $A$ in Figure 9–5, which is the titration curve for the chloride/iodide mixture just considered, is a composite of the individual curves for the two anionic species. Two equivalence points are evident. Curve $B$ is the titration curve for a mixture of bromide and chloride ions. Clearly, the change associated with the first equivalence point becomes less distinct as the solubilities of the two precipitates approach one another. In the bromide/chloride titration, the initial pAg values are lower than they are in the iodide/chloride titration because the solubility of silver bromide exceeds that of silver iodide. Beyond the first equivalence point, however, where chloride ion is being titrated, the two titration curves are identical.

It is possible to obtain experimental curves similar to those shown in Figure 9–5 by measuring the potential between a silver electrode and a reference electrode immersed in the solution. This technique, which is discussed in Chapter 17, permits the analysis of the individual components in certain halide-ion mixtures.

## 9C    APPLICATIONS OF STANDARD SILVER NITRATE SOLUTIONS

Table 9–3 lists some typical applications of precipitation titrations in which silver nitrate is the titrant. In most of these methods, the analyte is precipitated with a measured excess of silver nitrate and the excess determined by a Volhard back-titration with standard potassium thiocyanate.

Both silver nitrate and potassium thiocyanate are obtainable in primary-standard quality. The latter is somewhat hygroscopic, however, and so thiocyanate solutions are ordinarily standardized against silver nitrate. Both standard solutions are stable indefinitely.

**Table 9–3**
**TYPICAL ARGENTOMETRIC PRECIPITATION METHODS**

| Substance Being Determined | End Point | Remarks |
|---|---|---|
| $AsO_4^{3-}$, $Br^-$, $I^-$, $CNO^-$, $SCN^-$ | Volhard | Removal of silver salt not required |
| $CO_3^{2-}$, $CrO_4^{2-}$, $CN^-$, $Cl^-$, $C_2O_4^{2-}$, $PO_4^{3-}$, $S^{2-}$, $NCN^{2-}$ | Volhard | Removal of silver salt required before back-titration of excess $Ag^+$ |
| $BH_4^-$ | Modified Volhard | Titration of excess $Ag^+$ following $BH_4^- + 8\,Ag^+ + 8\,OH^- \rightarrow 8\,Ag(s) + H_2BO_3^- + 5\,H_2O$ |
| Epoxide | Volhard | Titration of excess $Cl^-$ following hydrohalogenation |
| $K^+$ | Modified Volhard | Precipitation of $K^+$ with known excess of $B(C_6H_5)_4^-$, addition of excess $Ag^+$ giving $AgB(C_6H_5)_4(s)$, and back-titration of the excess |
| $Br^-$, $Cl^-$ | $2\,Ag^+ + CrO_4^{2-} \rightarrow Ag_2CrO_4(s)$<br>Red | In neutral solution |
| $Br^-$, $Cl^-$, $I^-$, $SeO_3^{2-}$ | Adsorption indicator | |
| $V(OH)_4^+$, fatty acids, mercaptans | Electroanalytical | Direct titration with $Ag^+$ |
| $Zn^{2+}$ | Modified Volhard | Precipitation as $ZnHg(SCN)_4$, filtration, dissolution in acid, addition of excess $Ag^+$, back-titration of excess $Ag^+$ |
| $F^-$ | Modified Volhard | Precipitation as $PbClF$, filtration, dissolution in acid, addition of excess $Ag^+$, back-titration of excess $Ag^+$ |

## 9D   QUESTIONS AND PROBLEMS

**9–1.** Write a balanced net-ionic equation for the indicator reaction in the argentometric determinations of chloride by the
(a) Volhard method.
*(b) Fajans method.
(c) Mohr method.

**9–2.** Briefly distinguish between sigmoidal- and linear-segment titration curves. How do the two differ in the matter of data collection?

*9–3. Why are p-values useful as the ordinate of a titration curve?

**9–4.** The Volhard method can be used for the titration of bromide ion as well as chloride ion. Why does the one titration have more steps than the other?

*9–5. Briefly describe how you would determine carbonate ion by the Volhard method.

**9–6.** Write net-ionic equations for the determination of oxalate by the Volhard method. Will it be necessary to filter $Ag_2C_2O_4$ before back-titration with KSCN? Defend your answer.

*9–7. The Association of Official Analytical Chemists recommends a Volhard titration for analysis of the insecticide heptachlor, $C_{10}H_5Cl_7$. The percentage of heptachlor is given by

$$\text{percent heptachlor} = \frac{(\text{mL}_{Ag} \times c_{Ag} - \text{mL}_{SCN} \times c_{SCN}) \times 37.33}{\text{wt sample}}$$

What does this calculation reveal concerning the stoichiometry of this titration?

**9–8.** Why must the Fajans titration of $Cl^-$ be carried out in subdued light?

**9–9.** A standard solution is prepared by dissolving 8.3018 g of $AgNO_3$ in a 1.000-L volumetric flask. Calculate
*(a) the molar silver-ion concentration of this solution.
(b) the volume of 0.1079 M KSCN needed to titrate 25.00 mL of this solution.
*(c) the volume of this solution needed to titrate the $Cl^-$ in 0.1364 g of pure NaCl.
(d) the volume of this solution needed to titrate the $Cl^-$ in a 0.1948-g sample that is 85.3% $BaCl_2 \cdot 2\,H_2O$.

*(e) the mass of $COCl_2$ (fw = 98.92) with which 1.00 mL of this solution reacts. Equation:

$$COCl_2 + 2\ Ag^+ + H_2O \rightarrow$$
$$2\ AgCl(s) + CO_2 + 2\ H^+$$

9–10. A potassium thiocyanate solution was prepared by dissolving approximately 10 g of KSCN in about 1 L of water. Titration of a 50.0-mL aliquot of this solution required 41.37 mL of 0.1244 M $AgNO_3$.
  (a) What is the molar concentration of the KSCN solution?
  (b) What volume of this solution would be needed to titrate a 25.00-mL aliquot of 0.08302 M $AgNO_3$?
  (c) With how many milligrams of $Ag^+$ will 1.000 mL of this solution react?

9–11. Calculate the molar $Ag^+$ concentration of a solution if 1.000 mL reacts with 4.13 mg of
  *(a) $KIO_3$.      (e) KSCN.
  *(b) $H_2S$.       (f) $MgBr_2$.
  *(c) $LaI_3$.      (g) $CeCl_3$.
  *(d) $Al_2Cl_6$.   (h) $K_4Fe(CN)_6$.

9–12. What is the molar concentration of a silver nitrate solution if 16.35 mL react with
  *(a) 0.3017 g of $KIO_3$?
  (b) 28.75 mg of $H_2S$?
  (c) 0.4593 g of $LaI_3$?
  *(d) 69.47 mg of $Al_2Cl_6$?
  *(e) 14.86 mL of 0.1185 M KSCN?
  (f) 21.79 mL of 0.03884 M $MgBr_2$?
  *(g) 19.25 mL of 0.03176 M $CeCl_3$?
  (h) 12.33 mL of 0.02149 M $K_4Fe(CN)_6$?

9–13. What minimum volume of 0.1090 M $AgNO_3$ will be needed to assure a 5.0% excess of silver ion in the precipitation of AgCl from
  (a) 0.315 g of KCl?
  *(b) 16.8 mL of 0.126 M $MgCl_2$?
  (c) 16.8 mL of 0.126 M $FeCl_3$?
  *(d) 0.300 g of a sample that assays 71.3% $FeCl_3$?

*9–14. Standardization of a KSCN solution against 0.3341 g of primary standard $AgNO_3$ required a 21.55-mL titration. Calculate the molar concentration of the solution.

*9–15. The chloride in a 0.2720-g sample was precipitated by introducing 50.00 mL of 0.1030 M $AgNO_3$. Titration of the silver ion in excess required 8.65 mL of 0.1260 M KSCN. Express the results of this analysis in terms of percentage $MgCl_2$.

9–16. The phosphate in a 4.258-g sample of plant food was precipitated as $Ag_3PO_4$ through the addition of 50.00 mL of 0.0820 M $AgNO_3$:

$$3\ Ag^+ + HPO_4^{2-} \rightarrow Ag_3PO_4(s) + H^+$$

The solid was filtered and washed, following which the filtrate and washings were diluted to exactly 250.0 mL. Titration of a 50.00-mL aliquot of this solution required a 4.64-mL back-titration with

0.0625 M KSCN. Express the results of this analysis in terms of the percentage of $P_2O_5$.

*9–17. After pretreatment to convert the arsenic to the +5 state, a 0.821-g sample of a pesticide was treated with 25.00 mL of 0.0800 M $AgNO_3$. The $Ag_3AsO_4$ was filtered, washed free of excess silver ion, and then redissolved by treatment with nitric acid. The resulting solution required a 7.40-mL titration with 0.0865 M KSCN.
  (a) Express the results of this analysis in terms of percentage $As_2O_3$ in the sample.
  (b) If the analysis had been completed by titrating the excess silver ion in the filtrate and washings, what volume of 0.0865 M KSCN would have been used?

9–18. The theobromine ($C_7H_8N_4O_2$) in a 2.95-g sample of ground cocoa beans was converted to the sparingly soluble silver salt $C_7H_7N_4O_2Ag$ by warming in an ammoniacal solution containing 25.0 mL of 0.0100 M $AgNO_3$. After reaction was complete, all solids were removed by filtration. Calculate the percentage of theobromine (fw = 180.1) in the sample if the combined filtrate and washings required a back-titration with 7.69 mL of 0.0108 M KSCN.

*9–19. The acetylene content of a gas stream was determined by passing a 3.00-L sample through 100.0 mL of ammoniacal 0.0508 M $AgNO_3$. Reaction:

$$2\ Ag^+ + C_2H_2 \rightarrow Ag_2C_2 + 2H^+$$

Titration of the excess silver ion required 26.5 mL of 0.0845 M KSCN. Calculate the number of milligrams of acetylene contained in each liter of the gas.

*9–20. A 20-tablet sample of soluble saccharin was treated with 20.00 mL of 0.08181 M $AgNO_3$. The reaction is

After removal of the solid, titration of the filtrate and washings required 2.81 mL of 0.04124 M KSCN. Calculate the average number of milligrams of saccharin (fw = 205.17) in each tablet.

9–21. The monochloroacetic acid ($ClCH_2COOH$) preservative in 100.0 mL of a carbonated beverage was extracted into diethyl ether and then returned to aqueous solution as $ClCH_2COO^-$ by extraction with 1 M NaOH. This aqueous extract was acidified and treated with 50.00 mL of 0.04521 M $AgNO_3$. The reaction is

$$ClCH_2COOH + Ag^+ + H_2O \rightarrow$$
$$HOCH_2COOH + H^+ + AgCl(s)$$

After filtration of the AgCl, titration of the filtrate and washings required 10.43 mL of an $NH_4SCN$ solution. Titration of a blank taken through the entire process used 22.98 mL of the $NH_4SCN$. Calculate the weight (in milligrams) of $ClCH_2COOH$ in the sample.

*9–22. The iodoform in a 1.380-g sample of disinfectant was dissolved in alcohol and decomposed by treatment with concentrated $HNO_3$ and 33.60 mL of 0.0845 M $AgNO_3$:

$$CHI_3 + 3\ Ag^+ + H_2O \rightarrow 3\ AgI(s) + 3\ H^+ + CO(g)$$

When reaction was complete, the excess silver ion was titrated with 3.58 mL of 0.0950 M KSCN solution. Calculate the percentage of iodoform in the sample.

9–23. The action of an alkaline $I_2$ solution upon the rodenticide warfarin, $C_{19}H_{16}O_4$ (fw = 308.34), results in the formation of 1 mol of iodoform, $CHI_3$ (fw = 393.73), for each mole of the parent compound reacted. Analysis for warfarin can then be based upon the reaction between $CHI_3$ and $Ag^+$ (see Problem 9–22). A 13.96-g sample was treated with 25.00 mL of 0.02979 M $AgNO_3$, and the excess $Ag^+$ was then titrated with 2.85 mL of 0.05411 M KSCN. Calculate the percentage of warfarin in the sample.

*9–24. Each formula weight of acetone ($CH_3COCH_3$) yields one formula weight of iodoform ($CHI_3$) when treated with excess alkaline $I_2$. The $CHI_3$ produced from the acetone in a 100-mL urine specimen was treated with 20.00 mL of 0.0232 M $AgNO_3$ (see Problem 9–22 for reaction). Calculate the weight (mg) of acetone in the sample if 0.83 mL of 0.0209 M KSCN was needed to titrate the excess $Ag^+$.

9–25. The Fajans method is to be used in the routine analysis of solids for their chloride content. It is desired that the volume of standard $AgNO_3$ used in these titrations be numerically equal to the percent Cl when 0.2500-g samples are used. What should the molarity of the $AgNO_3$ solution be?

9–26. A Fajans titration of a 0.7908-g sample required 45.32 mL of 0.1046 M $AgNO_3$. Express the results of this analysis in terms of the percentage of
(a) $Cl^-$.
(b) $BaCl_2 \cdot 2\ H_2O$.
(c) $ZnCl_2 \cdot 2\ NH_4Cl$ (fw = 243.3).

*9–27. The sulfide in a sample of brackish water was determined by making a 100.0-mL sample ammoniacal and titrating with 7.04 mL of 0.0150 M $AgNO_3$. Reaction:

$$2\ Ag^+ + S^{2-} \rightarrow Ag_2S(s)$$

Express the results of this analysis in terms of ppm $H_2S$.

9–28. Calculate the percentage of $BaCl_2 \cdot 2\ H_2O$ in 1.504 g of an impure sample if a Fajans titration required 44.92 mL of 0.1027 M $AgNO_3$.

*9–29. A 0.1080-g pesticide sample was decomposed by the action of sodium biphenyl in toluene. The liberated $Cl^-$ was extracted with dilute $HNO_3$ and titrated with 24.31 mL of 0.04068 M $AgNO_3$ in a Fajans titration. Express the results of this analysis in terms of percent aldrin $C_{12}H_8Cl_6$ (fw = 364.92).

9–30. Titration of a 0.485-g sample by the Mohr method required 36.8 mL of standard 0.1060 M $AgNO_3$ solution. Calculate the percentage of chloride in the sample.

*9–31. A 2.00-L sample of mineral water was evaporated to a small volume, following which $K^+$ was precipitated with excess sodium tetraphenylboron:

$$K^+ + NaB(C_6H_5)_4 \rightarrow KB(C_6H_5)_4(s)$$

The precipitate was filtered, washed, and redissolved in acetone. The analysis was completed by a Mohr titration that required 43.85 mL of 0.03941 M $AgNO_3$:

$$KB(C_6H_5)_4 + Ag^+ \rightarrow AgB(C_6H_5)_4(s) + K^+$$

Calculate the $K^+$ concentration (in ppm) of the water sample.

9–32. A 1.50-L water sample was evaporated to a small volume and treated with an excess of sodium tetraphenylboron. The precipitated potassium tetraphenylboron was filtered and then redissolved in acetone. The analysis was completed by a Mohr titration, 24.90 mL of 0.0405 M $AgNO_3$ being used. See Problem 9–31. Express the results of this analysis in terms of parts per million of potassium (that is, mg K/L).

*9–33. Calculate the $CrO_4^{2-}$ concentration required to initiate formation of $Ag_2CrO_4$ at the equivalence point in a Mohr titration for $Cl^-$.

9–34. A 2.000-L water sample was evaporated to a small volume and treated with an excess of sodium tetraphenylboron, $NaB(C_6H_5)_4$. The precipitated $KB(C_6H_5)_4$ was filtered and then redissolved in acetone. The analysis was completed by a Mohr titration, with 37.90 mL of 0.03981 M $AgNO_3$ being used. See Problem 9–31. Express the results of this analysis in terms of parts per million of K (that is, mg K/L).

*9–35. A 4.269-g sample containing $NH_4Cl$, $(NH_4)_2SO_4$, and inert materials was diluted to exactly 500.0 mL. The $Cl^-$ in a 50.00-mL aliquot of this solution required 24.04 mL of 0.0682 M $AgNO_3$. The $NH_4^+$ in a 25.00-mL aliquot was converted to $NH_3$ and collected in 100.0 mL of a 0.03070 M sodium tetraphenylboron solution

$$NH_3(g) + NaB(C_6H_5)_4 + H^+ \rightarrow$$
$$NH_4B(C_6H_5)_4(s) + Na^+$$

After the solid had been removed by filtration, titration of the filtrate and washings required 7.50 mL of the $AgNO_3$ solution.

$$Ag^+ + NaB(C_6H_5)_4 \rightarrow AgB(C_6H_5)_4(s) + Na^+$$

Calculate the percentages of $NH_4Cl$ and $(NH_4)_2SO_4$ in the sample.

*9–36. A 1.998-g sample containing $Cl^-$ and $ClO_4^-$ was dissolved in sufficient water to give 250.0 mL of solution. A 50.00-mL aliquot required 13.97 mL of 0.08551 M $AgNO_3$ to titrate the $Cl^-$. A second 50.00-mL aliquot was treated with $V_2(SO_4)_3$ to reduce the $ClO_4^-$ to $Cl^-$:

$$ClO_4^- + 4\ V_2(SO_4)_3 + 4\ H_2O \rightarrow$$
$$Cl^- + 12\ SO_4^{2-} + 8\ VO^{2+} + 8\ H^+$$

Titration of the reduced sample required 40.12 mL of the $AgNO_3$ solution. Calculate the percentages of $Cl^-$ and $ClO_4^-$ in the sample.

9–37. A 3.095-g sample containing KCl, $KClO_4$, and inert materials was dissolved in sufficient water to give 250.0 mL of solution. A 50.00-mL aliquot required 38.32 mL of 0.0637 M $AgNO_3$ in a Mohr titration. A 25.00-mL aliquot was then treated with $V_2(SO_4)_3$ to reduce $ClO_4^-$ to $Cl^-$ (see Problem 9–36), following which titration required 39.63 mL of the same $AgNO_3$ solution. Calculate the respective percentages of KCl and $KClO_4$ in the sample.

*9–38. A 0.2185-g sample containing only KCl and $K_2SO_4$ yielded a precipitate of $KB(C_6H_5)_4$ that—after isolation and solution in acetone—required a Fajans titration involving 25.02 mL of 0.1126 M $AgNO_3$ for the reaction

$$KB(C_6H_5)_4 + Ag^+ \rightarrow AgB(C_6H_5)_4(s) + K^+$$

Calculate the percentages of KCl and $K_2SO_4$ in the sample.

9–39. The reaction between AgCl and $SCN^-$ is shown on page 204.
*(a) Calculate the equilibrium constant for this reaction.
(b) Assuming that the $SCN^-$ concentration needed to produce a detectable amount of $Fe(SCN)^{2+}$ is $2 \times 10^{-7}$ M, calculate the relative error resulting from the reaction between AgCl and $SCN^-$ in a titration involving the introduction of 30.0 mL of 0.100 M $AgNO_3$ to 50.0 mL of 0.0500 M NaCl, followed by a back-titration with 5.00 mL of 0.100 M KSCN.

9–40. Because the yellow color of $CrO_4^{2-}$ tends to obscure the first appearance of the red $Ag_2CrO_4$, it is common practice to hold the $CrO_4^{2-}$ concentration to about $2.5 \times 10^{-3}$ M. Calculate the relative titration error (neglecting the volume of $AgNO_3$ needed to produce a detectable amount of $Ag_2CrO_4$) in the titration of
*(a) 50.0 mL of 0.0500 M NaCl with 0.1000 M $AgNO_3$.
(b) 50.0 mL of 0.0100 M NaCl with 0.0200 M $AgNO_3$.

9–41. Calculate the silver ion concentration after the addition of 5.00*, 15.00, 25.00, 30.00, 35.00, 39.00, 40.00*, 45.00*, and 50.00 mL of 0.05000 M $AgNO_3$ to 50.0 mL of 0.0400 M KBr. Construct a titration curve from these data, plotting pAg as a function of titrant volume.

9–42. Calculate the $Ag^+$ concentration after the addition of 5.00, 20.0, 30.0, 35.0, 39.0, 40.0, 41.0, 45.0, and 50.0 mL of 0.100 M $AgNO_3$ to 50.0 mL of
*(a) 0.080 M KI.
(b) 0.080 M KSCN.
*(c) 0.080 M KCl.
(d) 0.040 M $K_2CrO_4$ (under conditions where $[HCrO_4^-]$ and $[H_2CrO_4]$ are negligibly small).

9–43. Calculate the mercury(I) concentration after the addition of 0, 10.0, 20.0, 30.0, 35.0, 39.0, 40.0, 41.0, 45.0, and 50.0 mL of 0.100 M NaCl to 80.0 mL of a solution that is 0.0250 M with respect to $Hg_2^{2+}$. For the process,

$$Hg_2Cl_2 \rightleftharpoons Hg_2^{2+} + 2\ Cl^-$$

$K_{sp}$ is equal to $2.0 \times 10^{-18}$. Construct a titration curve from these data, plotting $pHg_2$ as a function of titrant volume. *NOTE:* No evidence exists for the intermediate species $Hg_2Cl^+$.

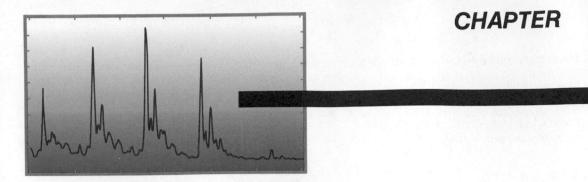

## CHAPTER 10

# THEORY OF NEUTRALIZATION TITRATIONS

$S$tandard solutions of strong acids and strong bases are used extensively for determining analytes that are themselves acids or bases or that can be converted to such species by chemical treatment. This chapter deals with the theoretical aspects of titrations with such reagents.

We begin our discussion by first describing a class of organic weak acids and bases that have indicator properties. We then show how these substances provide end points in acid/base titrations. Finally, we turn our attention to various types of pH calculations and to the derivation of titration curves for strong and weak acids and bases.[1]

In preparation for this discussion, you may find it helpful to review the material on acids and bases in Sections 6A–2 and 6B–3.

The standard reagents used in acid/base titrations are always strong acids or strong bases, most commonly HCl, $HClO_4$, $H_2SO_4$, NaOH, and KOH. Weak acids and bases are never used as standard reagents because they react incompletely with analytes.

## 10A SOLUTIONS AND INDICATORS FOR NEUTRALIZATION TITRATIONS

### 10A–1 Standard Solutions

The standard solutions employed in neutralization titrations are strong acids or strong bases because these substances react more completely with an analyte than do their weaker counterparts. Standard acids are prepared from hydrochloric, perchloric, and sulfuric acids. Nitric acid is seldom used because its property as an oxidizing agent offers the potential for undesirable side reactions. It should be noted that hot concentrated solutions of sulfuric and perchloric acids are also potent oxidizing agents and thus hazardous. Fortunately, however, dilute solutions of these reagents are relatively benign and can be used in the analytical laboratory without any special precautions other than eye protection.

---

[1]For a general reference on these topics, see D. R. Rosenthal and P. Zuman, in *Treatise on Analytical Chemistry*, 2nd ed., I. M. Kolthoff and P. J. Elving, Eds., Part I, Vol. 2, Chapter 18. New York: Wiley, 1979.

Standard basic solutions are ordinarily prepared from sodium, potassium, and occasionally barium hydroxides. Again, eye protection should always be used when handling these reagents or their solutions.

## 10A-2 The Theory of Indicator Behavior

Many substances, both naturally occurring and synthetic, display colors that depend upon the pH of the solution in which they are dissolved. Some of these substances, which have been used for centuries to indicate the acidity or alkalinity of water, find current application as acid/base indicators. Generally, acid/base indicators are weak organic acids or bases that, upon dissociation or association, undergo internal structural changes that give rise to color differences. Color changes for acid/base indicators can be ascribed to such equilibria as

For a list of common indicators and their colors, look inside the front cover of this book.

$$HIn + H_2O \rightleftharpoons H_3O^+ + In^-$$
Acid                                   Base
Color                                  Color

$$In + H_2O \rightleftharpoons InH^+ + OH^-$$
Base          Acid
Color         Color

In both cases, the color of the molecular form of the indicator is different from the color of the ionic form. The structural changes that account for these differences in the most common types of indicators are described in Section 10G-1.

Equilibrium-constant expressions for the two dissociations are

$$\frac{[H_3O^+][In^-]}{[HIn]} = K_a \qquad (10-1)$$

$$\frac{[InH^+][OH^-]}{[In]} = K_b \qquad (10-2)$$

Rearranging Equation 10-1 leads to

$$[H_3O^+] = K_a \frac{[HIn]}{[In^-]} \qquad (10-3)$$

We see then that the hydronium ion concentration determines the ratio of acid and conjugate-base form of the indicator.

The human eye is not very sensitive to color differences in a solution containing a mixture of $In^-$ and $HIn$, particularly when the ratio $[In^-]/[HIn]$ is greater than about 10 or smaller than about 0.1. To the average observer, consequently, the color of a solution of a typical indicator appears to change rapidly only within the limited concentration ratio of approximately 10 to 0.1. At greater or smaller ratios, the color becomes essentially constant to the human eye and independent of the ratio. Therefore, we can write that the indicator $HIn$ exhibits its pure acid color when

$$\frac{[HIn]}{[In^-]} \geq \frac{10}{1}$$

and its pure base color when

$$\frac{[HIn]}{[In^-]} \leq \frac{1}{10}$$

The color appears to be intermediate for ratios between these two values. These ratios vary considerably from indicator to indicator, of course. Furthermore, people differ significantly in their ability to distinguish between colors, with a color-blind person representing one extreme.

If the two concentration ratios are substituted into Equation 10–3, the range of hydronium ion concentrations needed to cause the complete indicator color change can be evaluated. Thus, for the full acid color,

$$[H_3O^+] \geq K_a \frac{10}{1}$$

and similarly for the full base color,

$$[H_3O^+] \leq K_a \frac{1}{10}$$

To obtain the indicator pH range, we take the negative logarithms of the two expressions:

$$pH \text{ (acid color)} = -\log (K_a \cdot 10) = pK_a + 1$$
$$pH \text{ (basic color)} = -\log (K_a/10) = pK_a - 1$$

The approximate pH range of most acid-type indicators is $pK_a \pm 1$.

$$\boxed{\text{indicator range} = pK_a \pm 1} \qquad (10\text{–}4)$$

Thus, the typical indicator with an acid dissociation constant of $1 \times 10^{-5}$ ($pK_a = 5$) shows a complete color change when the pH of the solution in which it is dissolved changes from 4 to 6 (Figure 10–1). A similar relationship is easily derived for a basic indicator.

## 10A–3 Some Common Indicators

The list of acid/base indicators is large and includes a number of organic structures. Indicators are available for any desired pH range. A few common indicators and their properties are listed in Table 10–1. Note that most of the transition ranges cover something less than two pH units.

## 10B  TITRATION CURVES FOR STRONG ACIDS AND STRONG BASES

When both reagent and analyte are strong electrolytes, the neutralization reaction is described by the equation

$$H_3O^+ + OH^- \rightleftarrows 2\,H_2O$$

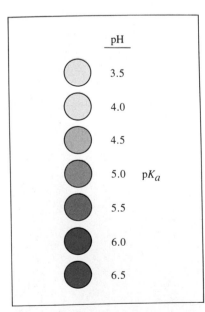

Figure 10–1
Indicator color as a function of pH ($pK_a = 5.0$).

**Table 10-1**

**SOME IMPORTANT ACID/BASE INDICATORS\***

| Common Name | Transition Range, pH | $pK_a$† | Color Change‡ | Indicator Type§ |
|---|---|---|---|---|
| Thymol blue | 1.2–2.8 | 1.65 | R–Y | 1 |
|  | 8.0–9.6 | 8.90 | Y–B |  |
| Methyl yellow | 2.9–4.0 |  | R–Y | 2 |
| Methyl orange | 3.1–4.4 | 3.46¶ | R–O | 2 |
| Bromocresol green | 3.8–5.4 | 4.66 | Y–B | 1 |
| Methyl red | 4.2–6.3 | 5.00¶ | R–Y | 2 |
| Bromocresol purple | 5.2–6.8 | 6.12 | Y–P | 1 |
| Bromothymol blue | 6.2–7.6 | 7.10 | Y–B | 1 |
| Phenol red | 6.8–8.4 | 7.81 | Y–R | 1 |
| Cresol purple | 7.6–9.2 |  | Y–P | 1 |
| Phenolphthalein | 8.3–10.0 |  | C–R | 1 |
| Thymolphthalein | 9.3–10.5 |  | C–B | 1 |
| Alizarin yellow GG | 10–12 |  | C–Y | 2 |

\*From C. A. Streuli, in *Handbook of Analytical Chemistry*, L. Meites, Ed., pp. **3**–35 and **3**–36. New York: McGraw-Hill, 1963.

†At ionic strength of 0.1.

‡B = blue; C = colorless; O = orange; P = purple; R = red; Y = yellow.

§(1) Acid type: $HIn + H_2O \rightleftharpoons H_3O^+ + In^-$

(2) Base type: $In + H_2O \rightleftharpoons InH^+ + OH^-$

¶For the reaction $InH^+ + H_2O \rightleftharpoons H_3O^+ + In$

Titration curves based on this reaction are derived by methods analogous to those shown in Section 9B with $K_w$ rather than $K_{sp}$ being used as the equilibrium constant. Normally, pH serves as the ordinate in titration curves for strong bases as well as strong acids.

## 10B-1 The Titration of a Strong Acid with a Strong Base

The hydronium ions in an aqueous solution of a strong acid have two sources: (1) the reaction of the solute with water and (2) the dissociation of water. In all but the most dilute solutions, however, the contribution from the solute far exceeds that from the solvent. Thus, for a solution of HCl with a concentration greater than about $1 \times 10^{-6}$ M, we can write

$$[H_3O^+] = c_{HCl} + [OH^-] \approx c_{HCl} \tag{10-5}$$

where $[OH^-]$ represents the contribution of water dissociation to the hydronium ion concentration.

An analogous relationship applies for a solution of a strong base, such as sodium hydroxide

$$[OH^-] = c_{NaOH} + [H_3O^+] \approx c_{NaOH} \tag{10-6}$$

As seen in the following example, the derivation of a curve for the titration of a strong acid with a strong base is simplified because the

In solutions of a strong acid that are more concentrated than about $1 \times 10^{-6}$ M, it is proper to assume that the equilibrium concentration of $[H_3O^+]$ is equal to the analytical concentration of the acid. The same is true for the hydroxide concentration in solutions of a strong base.

A useful relationship is obtained by taking the negative logarithm of each side of the ion-product constant for water:

$$-\log K_w = -\log[H_3O^+][OH^-]$$
$$= -\log[H_3O^+] - \log[OH^-]$$
$$pK_w = 14.00 = pH + pOH$$

Remember: $pH = pK_w - pOH$
$$= 14.00 - pOH$$

Before the equivalence point, we calculate the pH from the concentration of unreacted strong acid.

concentration of the hydronium ion or the hydroxide ion is obtained directly from stoichiometric calculations.

---

Example 10–1

Derive a titration curve for the reaction of 50.00 mL of 0.05000 M HCl with 0.1000 M NaOH. Round pH data to two places to the right of the decimal point.

Initial Point

The solution is $5.00 \times 10^{-2}$ M in HCl. Since HCl is completely dissociated,

$$[H_3O^+] = 5.00 \times 10^{-2} \text{ M}$$
$$pH = -\log(5.00 \times 10^{-2}) = -\log 5.00 - \log 10^{-2}$$
$$= -0.699 + 2 = 1.301 = 1.30$$

pH after Addition of 10.00 mL of NaOH

The volume of the solution is now 60.00 mL, and part of the HCl has been neutralized. Thus,

$$[H_3O^+] = \frac{50.00 \times 0.0500 - 10.00 \times 0.1000}{60.00} = 2.50 \times 10^{-2} \text{ M}$$
$$pH = 2 - \log 2.50 = 1.60$$

Additional data to define the curve in the region short of the equivalence point are calculated in the same way. The results of such calculations are given in column 2 of Table 10–2.

pH after Addition of 25.00 mL of NaOH

At the equivalence point, the solution contains neither an excess of HCl nor an excess of NaOH, and

At the equivalence point of this titration the solution is neutral.

$$[H_3O^+] = [OH^-] = \sqrt{K_w} = 1.00 \times 10^{-7}$$
$$pH = -\log 1.00 \times 10^{-7} = 7.00$$

pH after Addition of 25.10 mL of NaOH

After the equivalence point we first calculate pOH, then pH.

The concentration of excess base is

$$c_{NaOH} = \frac{25.10 \times 0.1000 - 50.00 \times 0.0500}{75.10} = 1.33 \times 10^{-4} \text{ M}$$

and so

$$[OH^-] = c_{NaOH} = 1.33 \times 10^{-4} \text{ M}$$
$$pOH = -\log(1.33 \times 10^{-4}) = 3.88$$

Thus (see marginal note),

$$pH = 14.00 - 3.88 = 10.12$$

Additional data for this titration, calculated in the same way, are given in column 2 of Table 10–2.

## The Effect of Concentration

The effects of reagent and analyte concentration on the neutralization titration curves for strong acids are shown by the two sets of data in Table 10–2 and the plots in Figure 10–2.

With 0.1 M NaOH as the titrant (curve *A*), the change in pH in the equivalence-point region is large. With 0.001 M NaOH (curve *B*), the change is markedly less but still pronounced.

## Indicator Choice

Figure 10–2 shows that the selection of an indicator is not critical when the reagent concentration is approximately 0.1 M. Here, the volume differences in titrations with the three indicators shown are of the same magnitude as the uncertainties associated with reading the buret and thus negligible. Note, however, that bromocresol green is clearly unsuited for a titration involving the 0.001 M reagent because the color begins to change after addition of about 18 mL of reagent and is complete with addition of approximately 23 mL. Phenolphthalein is equally unsatisfactory. Of the three indicators, then, only bromothymol blue provides a satisfactory end point with a minimal determinate titration error when the more dilute solutions are used.

## 10B–2  The Titration of a Strong Base With a Strong Acid

Titration curves for a strong base are derived in a way analogous to that for strong acids. Short of the equivalence point, the solution is highly basic, and the hydroxide ion concentration is equal to or a multiple of the analytical molarity of the base. The solution is neutral at the equivalence

**Table 10–2**

**CHANGES IN pH DURING THE TITRATION OF A STRONG ACID WITH A STRONG BASE**

| Volume of NaOH, mL | 50.00 mL of 0.0500 M HCl with 0.1000 M NaOH<br>pH | 50.00 mL of 0.000500 M HCl with 0.001000 M NaOH<br>pH |
|---|---|---|
| 0.00 | 1.30 | 3.30 |
| 10.00 | 1.60 | 3.60 |
| 20.00 | 2.15 | 4.15 |
| 24.00 | 2.87 | 4.87 |
| 24.90 | 3.87 | 5.87 |
| 25.00 | 7.00 | 7.00 |
| 25.10 | 10.12 | 8.12 |
| 26.00 | 11.12 | 9.12 |
| 30.00 | 11.80 | 9.80 |

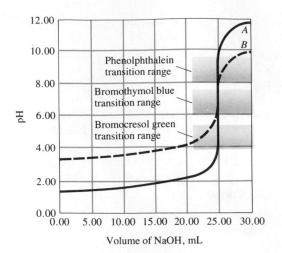

**Figure 10–2**
Titration curves for HCl with NaOH. *A*: 50.00 mL of 0.0500 M HCl with 0.1000 M NaOH. *B*: 50.00 mL of 0.000500 M HCl with 0.001000 M NaOH.

Buffers are used in all types of chemistry whenever it is desirable to maintain the pH of a solution at a constant and predetermined level.

Buffered aspirin contains buffers to help maintain stomach contents at a relatively high pH.

point. Finally, the solution is acidic in the region beyond the equivalence point; here, the hydronium ion concentration is equal to the analytical concentration of the excess strong acid. A curve for the titration of a strong base with 0.1 M hydrochloric acid is shown in Figure 10–7. Indicator selection is based upon the considerations described for the titration of a strong acid with a strong base.

## 10C  BUFFER SOLUTIONS

By definition, a *buffer solution* is one that resists changes in pH. Most buffers consist of either a weak acid and its conjugate base or a weak base and its conjugate acid. Chemists use buffers whenever they need to maintain the pH of a solution at a constant and predetermined level. You will find many references to the use of buffers throughout this text.

A buffer solution is formed whenever a weak acid is partially neutralized with a strong base or a weak base is partially neutralized with a strong acid. As a consequence of buffering, titration curves for weak acids and weak bases are significantly different from the titration curves we have thus far encountered.

### 10C–1 The Calculation of the pH of Buffer Solutions

#### Weak Acid/Conjugate Base Buffers

A solution containing a weak acid HA and its conjugate base $A^-$ may be acidic, neutral, or basic, depending upon the position of two competitive equilibria:

$$HA + H_2O \rightleftarrows H_3O^+ + A^- \qquad \frac{[H_3O^+][A^-]}{[HA]} = K_a \qquad (10\text{–}7)$$

$$A^- + H_2O \rightleftarrows OH^- + HA \qquad \frac{[OH^-][HA]}{[A^-]} = K_b = \frac{K_w}{K_a} \qquad (10\text{–}8)$$

If the first equilibrium lies farther to the right than the second, the solution is acidic. If the second equilibrium is dominant, the solution is basic. These two equilibrium-constant expressions show that the relative concentrations of the hydronium and hydroxide ions depend not only upon the magnitudes of $K_a$ and $K_b$ but also upon the ratio of the concentrations of the acid and its conjugate base.

In order to compute the pH of a solution containing both an acid HA and its salt NaA, we need to express the equilibrium concentrations of HA and NaA in terms of their analytical concentrations $c_{HA}$ and $c_{NaA}$. An examination of the two equilibria reveals that the first reaction decreases the concentration of HA by an amount equal to $[H_3O^+]$, whereas the second increases the HA concentration by an amount equal to $[OH^-]$. Thus, the species concentration of HA is related to its analytical concentration by the mass balance equation

$$[HA] = c_{HA} - [H_3O^+] + [OH^-] \tag{10-9}$$

Similarly, the first equilibrium increases the concentration of $A^-$ by an amount equal to $[H_3O^+]$, and the second decreases this concentration by the amount $[OH^-]$. Thus the equilibrium concentration is given by

$$[A^-] = c_{NaA} + [H_3O^+] - [OH^-] \tag{10-10}$$

Because of the inverse relationship between $[H_3O^+]$ and $[OH^-]$, it is *always* possible to eliminate one or the other from Equations 10–9 and 10–10. Moreover, the *difference* in concentration between these two species is often so small relative to the molar concentrations of acid and conjugate base that Equations 10–9 and 10–10 simplify to

$$[HA] \approx c_{HA} \tag{10-11}$$

$$[A^-] \approx c_{NaA} \tag{10-12}$$

Substitution of Equations 10–11 and 10–12 into the dissociation-constant expression yields, upon rearrangement,

$$[H_3O^+] = K_a \frac{c_{HA}}{c_{NaA}} \tag{10-13}$$

The assumption leading to Equations 10–11 and 10–12 sometimes breaks down with acids or bases that have dissociation constants greater than about $10^{-3}$ or when the molar concentration of either the acid or its conjugate base (or both) is very small. In these circumstances, either

**Feature 10–1**

APPLICATION OF THE SYSTEMATIC METHOD TO BUFFER CALCULATIONS

Equations 10–9 and 10–10 can also be derived from mass- and charge-balance expressions. Thus, mass-balance considerations require that

$$c_{HA} + c_{NaA} = [HA] + [A^-] \tag{1}$$

Electrical-neutrality considerations require that

$$[Na^+] + [H_3O^+] = [A^-] + [OH^-]$$

but

$$[Na^+] = c_{NaA}$$

Therefore, the charge-balance equation is

$$c_{NaA} + [H_3O^+] = [A^-] + [OH^-] \tag{2}$$

which rearranges to Equation 10–10:

$$[A^-] = c_{NaA} + [H_3O^+] - [OH^-]$$

Subtraction of Equation 2 from Equation 1 and rearrangement give

$$[HA] = c_{HA} - [H_3O^+] + [OH^-]$$

which corresponds to Equation 10–9.

The difference between the hydronium and hydroxide ion concentrations in a buffer is usually small relative to either $c_{HA}$ or $c_{NaA}$.

[OH$^-$] or [H$_3$O$^+$] must be retained in Equations 10–9 and 10–10, depending upon whether the solution is acidic or basic. In any case, Equations 10–11 and 10–12 should always be used initially. The provisional values for [H$_3$O$^+$] and [OH$^-$] can then be used to test the assumptions.

Within the limits imposed by the assumptions made in its derivation, Equation 10–13 says that the hydronium ion concentration of a solution containing a weak acid and its conjugate base is dependent only upon the *ratio* of the molar concentrations of these two solutes. Furthermore, this ratio is *independent of dilution* because the concentration of each component changes proportionately when the volume changes.

---

Example 10–2

What is the pH of a solution that is 0.400 M in formic acid and 1.00 M in sodium formate?

The equilibrium governing the hydronium ion concentration in this solution is

$$H_2O + HCOOH \rightleftarrows H_3O^+ + HCOO^-$$

for which

$$K_a = \frac{[H_3O^+][HCOO^-]}{[HCOOH]} = 1.77 \times 10^{-4}$$

$$[HCOO^-] \approx c_{HCOO^-} = 1.00$$

$$[HCOOH] \approx c_{HCOOH} = 0.400$$

Substitution into Equation 10–13 gives

$$[H_3O^+] = 1.77 \times 10^{-4} \times \frac{0.400}{1.00} = 7.08 \times 10^{-5}$$

Note that the assumption that $[H_3O^+] \ll c_{HCOOH}$ and $c_{HCOO^-}$ is valid. Thus,

$$pH = -\log (7.08 \times 10^{-5}) = 4.15$$

## Weak Base/Conjugate Acid Buffers

The calculation of the hydroxide ion concentration for a solution consisting of a weak base and its conjugate acid is entirely analogous to that developed in the preceding section.

**Feature 10–2**
### THE HENDERSON-HASSELBALCH EQUATION

The Henderson-Hasselbalch equation is an alternative form of Equation 10–13 that is encountered frequently in the biological literature and biochemical texts. It is obtained by expressing each term in the equation in the form of its negative logarithm and inverting the concentration ratio to keep all signs positive:

$$-\log [H_3O^+] = -\log K_a - \log \frac{c_{HA}}{c_{NaA}}$$

Therefore,

$$pH = pK_a + \log \frac{c_{NaA}}{c_{HA}} \qquad (10-14)$$

Example 10–3

Calculate the pH of a solution that is 0.200 M in $NH_3$ and 0.300 M in $NH_4Cl$. The base dissociation constant for $NH_3$ is $1.76 \times 10^{-5}$.

The equilibria of interest are

$$NH_3 + H_2O \rightleftharpoons NH_4^+ + OH^- \qquad K_b = 1.76 \times 10^{-5}$$

$$NH_4^+ + H_2O \rightleftharpoons NH_3 + H_3O^+ \qquad K_a = \frac{1.00 \times 10^{-14}}{1.76 \times 10^{-5}} = 5.68 \times 10^{-10}$$

Since $K_b > K_a$, the solution is alkaline. Using the arguments in the previous example, the molar concentrations of $NH_3$ and $NH_4^+$ are

$$[NH_3] = 0.200 - [OH^-] + [H_3O^+] \approx 0.200$$

$$[NH_4^+] = 0.300 + [OH^-] - [H_3O^+] \approx 0.300$$

A provisional value for $[OH^-]$ is obtained by substituting the approximate values for $[NH_3]$ and $[NH_4^+]$ into the rearranged form of the equilibrium-constant expression for $NH_3$:

$$[OH^-] = K_b \times \frac{[NH_3]}{[NH_4^+]} = 1.76 \times 10^{-5} \times \frac{0.200}{0.300} = 1.173 \times 10^{-5}$$

The approximations are clearly justified; thus,

$$pOH = 5.000 - \log 1.173 = 4.931$$

$$pH = 14.00 - 4.931 = 9.07$$

## 10C–2 Properties of Buffer Solutions

In this section we illustrate the resistance of buffers to changes in pH brought about either by dilution or by addition of strong acids or bases.

### The Effect of Dilution

The pH of a buffer solution remains essentially independent of dilution until the concentrations of the species it contains are decreased to the point where the approximations used to develop Equations 10–11 and 10–12 become invalid. Figure 10–3 contrasts the behavior of buffered and unbuffered solutions with dilution. For each, the initial solute concentration is 1.00 M. The resistance of the buffered solution to changes in pH as a result of dilution is clear.

### The Effect of Added Acids and Bases

Example 10–4 illustrates a second property of buffer solutions: their resistance to pH change after addition of small amounts of strong acids or bases.

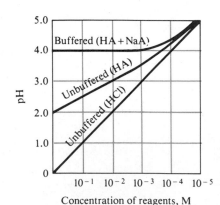

Concentration of reagents, M

Figure 10–3

The effect of dilution on the pH of buffered and unbuffered solutions. The dissociation constant for HA is $1.00 \times 10^{-4}$. Initial solute concentrations are 1.00 M.

Example 10-4

Calculate the pH change that takes place when a 100-mL portion of (a) 0.0500 M NaOH and (b) 0.0500 M HCl is added to 400 mL of the buffer solution described in Example 10-3.

(a) Addition of NaOH converts part of the $NH_4^+$ in the buffer to $NH_3$:

$$NH_4^+ + OH^- \rightarrow NH_3 + H_2O$$

The analytical concentrations of $NH_3$ and $NH_4Cl$ then become

$$c_{NH_3} = \frac{400 \times 0.200 + 100 \times 0.0500}{500} = \frac{85.0}{500} = 0.170 \text{ M}$$

$$c_{NH_4Cl} = \frac{400 \times 0.300 - 100 \times 0.0500}{500} = \frac{115}{500} = 0.230 \text{ M}$$

When substituted into the dissociation-constant expression, these values yield

$$[OH^-] = \frac{1.76 \times 10^{-5} \times 0.170}{0.230} = 1.30 \times 10^{-5}$$

$$pH = 14.00 - (-\log 1.30 \times 10^{-5}) = 9.11$$

Buffers do not maintain pH at an absolutely constant level, but changes in pH are relatively small when small amounts of acid or base are added.

In Example 10-3, we found the original pH to be 9.07. Thus, the change in pH is

$$\Delta pH = 9.11 - 9.07 = 0.04$$

(b) Addition of HCl converts part of the $NH_3$ to $NH_4^+$; thus,

$$NH_3 + H_3O^+ \rightarrow NH_4^+ + H_2O$$

$$c_{NH_3} = \frac{400 \times 0.200 - 100 \times 0.0500}{500} = \frac{75}{500} = 0.150 \text{ M}$$

$$c_{NH_4^+} = \frac{400 \times 0.300 + 100 \times 0.0500}{500} = \frac{125}{500} = 0.250 \text{ M}$$

$$[OH^-] = 1.76 \times 10^{-5} \times \frac{0.150}{0.250} = 1.06 \times 10^{-5}$$

$$pH = 14.00 - (-\log 1.06 \times 10^{-5}) = 9.02$$

$$\Delta pH = 9.02 - 9.07 = -0.05$$

It is of interest to contrast the behavior of an unbuffered solution with a pH of 9.07 to that of the buffer in Example 10-3. You can show that adding the same quantity of base to the unbuffered solution would increase the pH to 12.00—a pH change of 2.93 units. Adding the acid would decrease the pH by slightly more than 7 units.

## Buffer Capacity

Figure 10–3 and Example 10–4 demonstrate that a solution containing a conjugate acid/base pair possesses remarkable resistance to changes in pH. The ability of a buffer to prevent a significant change in pH is directly related to the total concentration of the buffering species as well as to their concentration ratio. For example, if we diluted the buffer described in Example 10–3 by a factor of 10, the pH would still be 9.07. However, if we treated the diluted buffer in the same way as the more concentrated buffer was treated in Example 10–4, the pH change would be 0.4 to 0.5 unit instead of 0.04 to 0.05.

The buffer capacity is the number of moles of strong acid or strong base that 1 L of the buffer can absorb while changing the pH by 1 unit.

The *buffer capacity* of a solution is the number of moles of a strong acid or a strong base that causes 1.00 L of the buffer to undergo a 1.00-unit change in pH. The capacity of a buffer depends not only on the total concentrations of the two buffer components but also on their concentration ratio. Buffer capacity falls off moderately rapidly as the concentration ratio of acid to conjugate base becomes larger or smaller than unity (Figure 10–4). For this reason, the $pK_a$ of the acid chosen for a given application should lie within $\pm 1$ unit of the desired pH in order for the buffer to have a reasonable resistance to added acid or base.

## The Preparation of Buffers

In principle, a buffer solution of any desired pH can be prepared by combining calculated quantities of a suitable conjugate acid/base pair. In practice, however, the pH values of buffers prepared from theoretically generated recipes differ from the predicted values (1) because of uncertainties in the numerical values of many dissociation constants and (2) because of the simplifications used in calculations. Most important is the fact that the ionic strength of a buffer is usually so high that good values for the activity coefficients of the ions in the solution cannot be obtained from the Debye-Hückel relationship. Because of these uncertainties, we prepare buffers by making up a solution of approximately the desired pH and then adjust by adding acid or conjugate base until the required pH is indicated by a pH meter.

Table 17–4 gives the components of buffers that are recommended by the National Institute of Standards and Technology.

Alternatively, empirically derived recipes for preparing buffer solutions of known pH are available in chemical handbooks and reference works.[2]

Buffers are of tremendous importance in biological and biochemical studies where a low but constant concentration of hydronium ion ($10^{-6}$ to $10^{-10}$ M) must be maintained. Several biological supply houses offer a variety of such buffers.

## 10D    TITRATION CURVES FOR WEAK ACIDS

Four distinctly different types of calculations are needed to derive a titration curve for a weak acid (or weak base):

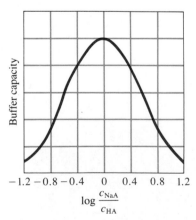

**Figure 10–4**
Buffer capacity as a function of the ratio $c_{NaA}/c_{HA}$.

[2]See, for example, L. Meites, Ed., *Handbook of Analytical Chemistry*, pp. **5**-112 and **11**-3 to **11**-8. New York: McGraw-Hill, 1963.

1. At the beginning, the solution contains only a weak acid or a weak base, and the pH is calculated from the concentration of that solute and the corresponding dissociation constant.
2. After various increments of titrant have been added (in volumes up to, but not including, the equivalence point), the solution consists of a series of buffers. The pH of each buffer is calculated from the analytical concentrations of the conjugate base or acid and the residual concentrations of the weak acid or base.
3. At the equivalence point, the solution contains only the conjugate of the weak acid or base being titrated (that is, a salt), and the pH is calculated from the concentration of this product.
4. Beyond the equivalence point, the excess of strong acid or base titrant represses the acidic or basic character of the reaction product to such an extent that the pH is governed largely by the concentration of the excess titrant.

---

Example 10–5

Derive a curve for the titration of 50.00 mL of 0.1000 M acetic acid ($K_a = 1.75 \times 10^{-5}$) with 0.1000 M sodium hydroxide.

Initial pH

Initially, we must calculate the pH of a 0.1000 M solution of HOAc using Equation 6–19:

$$[H_3O^+] = \sqrt{K_a c_{HOAc}} = \sqrt{1.75 \times 10^{-5} \times 0.1000} = 1.32 \times 10^{-3}$$
$$pH = -\log 1.32 \times 10^{-3} = 2.88$$

pH after Addition of 10.00 mL of Reagent

A buffer solution consisting of NaOAc and HOAc has now been produced. The analytical concentrations of the two constituents are

$$c_{HOAc} = \frac{50.00 \text{ mL} \times 0.1000 \text{ M} - 10.00 \text{ mL} \times 0.1000 \text{ M}}{60.00 \text{ mL}} = \frac{4.000}{60.00} \text{ M}$$

$$c_{NaOAc} = \frac{10.00 \text{ mL} \times 0.1000 \text{ M}}{60.00 \text{ mL}} = \frac{1.000}{60.00} \text{ M}$$

We substitute these concentrations into the dissociation-constant expression for acetic acid and obtain

$$\frac{[H_3O^+](1.000/60.00)}{4.000/60.00} = K_a = 1.75 \times 10^{-5}$$

$$[H_3O^+] = 7.00 \times 10^{-5}$$

$$pH = 4.16$$

Calculations similar to this provide points on the curve throughout the buffer region. Data from such calculations are given in column 2 of Table 10–3.

**Table 10–3**

**CHANGES IN pH DURING THE TITRATION OF A WEAK ACID WITH A STRONG BASE**

| Volume of NaOH, mL | pH 50.00 mL of 0.1000 M HOAc Titrated with 0.1000 M NaOH | 50.00 mL of 0.001000 M HOAc Titrated with 0.001000 M NaOH |
|---|---|---|
| 0.00 | 2.88 | 3.91 |
| 10.00 | 4.16 | 4.30 |
| 25.00 | 4.76 | 4.80 |
| 40.00 | 5.36 | 5.38 |
| 49.00 | 6.45 | 6.46 |
| 49.90 | 7.46 | 7.47 |
| 50.00 | 8.73 | 7.73 |
| 50.10 | 10.00 | 8.09 |
| 51.00 | 11.00 | 9.00 |
| 60.00 | 11.96 | 9.96 |
| 75.00 | 12.30 | 10.30 |

**Equivalence-Point pH**

At the equivalence point, all the acetic acid has been converted to sodium acetate. The solution is therefore similar to one formed by dissolving this conjugate base in water, and the pH calculation is identical to that shown in Example 6–10 for a weak base. In the present example, the NaOAc concentration is 0.0500 M. Thus,

$$OAc^- + H_2O \rightleftarrows HOAc + OH^-$$

$$[OH^-] = [HOAc]$$

$$[OAc^-] = 0.0500 - [OH^-] \approx 0.0500$$

Substituting in the base dissociation-constant expression for $OAc^-$ gives

$$\frac{[OH^-]^2}{0.0500} = \frac{K_w}{K_a} = \frac{1.00 \times 10^{-14}}{1.75 \times 10^{-5}} = 5.714 \times 10^{-10}$$

$$[OH^-] = \sqrt{0.0500 \times 5.714 \times 10^{-10}} = 5.35 \times 10^{-6}$$

$$pH = 14.00 - (-\log 5.35 \times 10^{-6}) = 8.73$$

Note that the pH at the equivalence point of this titration is greater than 7. The solution is alkaline.

**pH after Addition of 50.10 mL of Base**

After the addition of 50.10 mL of NaOH, both the excess base and the acetate ion are sources of hydroxide ion. The contribution of the latter is small, however, because the strong base represses the reaction of acetate ion with water. This fact becomes evident when we consider that the hydroxide ion concentration is only $5.34 \times 10^{-6}$ M at the equivalence point; once an excess of strong base is added, the contribution from the reaction of the acetate is even smaller. Thus,

$$[OH^-] \approx c_{NaOH} = \frac{50.10 \text{ mL} \times 0.1000 \text{ M} - 50.00 \text{ mL} \times 0.1000 \text{ M}}{100.1 \text{ mL}}$$

$$= 1.00 \times 10^{-4} \text{ M}$$

$$pH = 14.00 - (-\log 1.00 \times 10^{-4}) = 10.00$$

Thus in the region slightly beyond the equivalence point, the titration curve for a weak acid with a strong base is identical to that for a strong acid with a strong base (Figure 10–2).

Note that the analytical concentrations of acid and conjugate base are identical when an acid has been half neutralized (in Example 10–5, after the addition of exactly 25.00 mL of base). Thus, these terms cancel in the equilibrium-constant expression, and the hydronium ion concentration is numerically equal to the dissociation constant. Likewise, in the titration of a weak base, the hydroxide ion concentration is numerically equal to the dissociation constant of the base at the midpoint in the titration curve. In addition, the buffer capacity of both solutions is at a maximum at this point.

At the half neutralization point in the titration of a weak acid, $[H_3O^+] = K_a$ or $pH = pK_a$.

At the half neutralization point in the titration of a weak base, $[OH^-] = K_b$ or $pOH = pK_b$.

---

**Feature 10–3**
## DETERMINATION OF DISSOCIATION CONSTANTS FOR WEAK ACIDS AND BASES

The dissociation constants of weak acids or weak bases are often determined by monitoring the pH of the solution while the acid or base is being titrated. A pH meter and a glass electrode are used for the measurements. For an acid, the measured pH when the acid is exactly half neutralized is numerically equal to $pK_a$. For a weak base, the pH at half neutralization must be converted to pOH, which is then equal to $pK_b$.

---

## 10D–1 The Effect of Concentration

The second and third columns of Table 10–3 contain pH data for the titration of 0.1000 M and 0.001000 M acetic acid with sodium hydroxide solutions of the same two concentrations. In deriving the data for the more dilute acid, none of the approximations shown in Example 10–5 were valid, and solution of a quadratic equation was necessary throughout.

Figure 10–5 is a plot of the data in Table 10–3. Note that the initial pH values are higher and the equivalence-point pH is lower for the more dilute solution. At intermediate titrant volumes, however, the pH values of the two solutions differ from each other only slightly because of the buffering action of the acetic acid/sodium acetate system present in this region. Figure 10–5 is graphic confirmation that the pH of buffers is largely independent of dilution.

Challenge: Derive the data in the third column of Table 10–3.

The pH of a buffer is relatively independent of dilution.

## 10D–2 The Effect of Reaction Completeness

Titration curves for 0.1000 M solutions of acids with different dissociation constants are shown in Figure 10–6. Note that the pH change in the

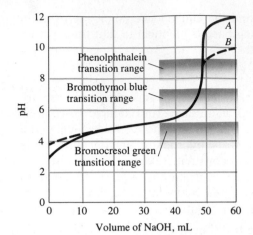

**Figure 10–5**

Curve for the titration of acetic acid with sodium hydroxide: $A$: 0.1000 M acid with 0.1000 M base. $B$: 0.001000 M acid with 0.001000 M base.

equivalence-point region becomes smaller as the acid becomes weaker—that is, as the reaction between the acid and the base becomes less complete. The effects of reactant concentration and reaction completeness illustrated by Figures 10–5 and 10–6 are analogous to these effects on precipitation titration curves (Section 9B–2).

### 10D–3 Indicator Choice; The Feasibility of Titration

Figures 10–5 and 10–6 show clearly that the choice of indicator for the titration of a weak acid is more limited than that for a strong acid. For example, from Figure 10–5, it is obvious that bromocresol green is totally unsuited for the titration of 0.1000 M acetic acid. Bromothymol blue is also unsatisfactory because its full color change occurs over a range from about 46 to 50 mL of 0.1000 M base. An indicator exhibiting a color change in the basic region, such as phenolphthalein, provides a sharp end point with a minimal titration error.

The end-point pH change associated with the titration of 0.001000 M acetic acid (curve $B$, Figure 10–5) is so small that a significant titration error is likely to be introduced regardless of indicator. However, use of an

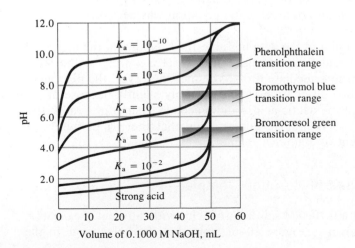

**Figure 10–6**

The effect of acid strength on titration curves. Each curve represents the titration of 50.0 mL of 0.1000 M acid with 0.1000 M NaOH.

indicator with a transition range between that of phenolphthalein and that of bromothymol blue in conjunction with a suitable color-comparison standard makes it possible to establish the end point in this titration with a reproducibility of a few percent relative.

Figure 10–6 illustrates that there are similar problems as the strength of the acid being titrated decreases. Precision on the order of $\pm 2$ ppt can be achieved in the titration of a 0.1000 M solution of an acid with a dissociation constant of $10^{-8}$, provided a suitable color-comparison standard is available. With more concentrated solutions, somewhat weaker acids can be titrated with reasonable precision.

## 10E   TITRATION CURVES FOR WEAK BASES

The derivation of a curve for the titration of a weak base is analogous to that of a weak acid.

---

Example 10–6

A 50.00-mL aliquot of 0.0500 M NaCN is titrated with 0.1000 M HCl. The reaction is

$$CN^- + H_3O^+ \rightleftharpoons HCN + H_2O$$

Calculate the pH after the addition of (a) 0.00, (b) 10.00, (c) 25.00, and (d) 26.00 mL of acid.

(a) Initial pH

The pH of a solution of NaCN can be derived by the method shown in Example 6–10:

$$CN^- + H_2O \rightleftharpoons HCN + OH^-$$

$$\frac{[OH^-][HCN]}{[CN^-]} = K_b = \frac{K_w}{K_a} = \frac{1.00 \times 10^{-14}}{2.1 \times 10^{-9}} = 4.76 \times 10^{-6} \qquad (1)$$

$$[OH^-] = [HCN]$$

$$[CN^-] = c_{NaCN} - [OH^-] \approx c_{NaCN} = 0.0500$$

Substitution into the dissociation-constant expression gives, after rearrangement,

$$[OH^-] = \sqrt{K_b c_{NaCN}} = \sqrt{4.76 \times 10^{-6} \times 0.0500} = 4.88 \times 10^{-4}$$

$$pH = 14.00 - (-\log 4.88 \times 10^{-4}) = 10.69$$

(b) pH after Addition of 10.00 mL of Titrant

Addition of acid produces a buffer with a composition given by

$$c_{NaCN} = \frac{50.00 \times 0.0500 - 10.00 \times 0.1000}{60.00} = \frac{1.500}{60.00} \text{ M}$$

$$c_{HCN} = \frac{10.00 \times 0.1000}{60.00} = \frac{1.000}{60.00} \text{ M}$$

These values are then substituted into the expression for the acid dissociation constant of HCN to give

$$[H_3O^+] = \frac{2.1 \times 10^{-9} \times (1.000/\cancel{60.00})}{1.500/\cancel{60.00}} = 1.4 \times 10^{-9}$$

$$pH = 8.85$$

The pH of the buffer can be calculated with $K_a$ for HCN, as was done here, or equally well with $K_b$ (Equation 1). We used $K_a$ because it gives $[H_3O^+]$ directly; $K_b$ gives $[OH^-]$.

(c) pH after Addition of 25.00 mL of Titrant

This volume corresponds to the equivalence point, where the principal solute species is the weak acid HCN. Thus,

$$c_{HCN} = \frac{25.00 \times 0.1000}{75.00} = 0.03333 \text{ M}$$

Applying Equation 6–19 gives

Since the principal solute species at the equivalence point is HCN, the pH is acidic.

$$[H_3O^+] = \sqrt{K_a c_{HCN}} = \sqrt{2.1 \times 10^{-9} \times 0.03333} = 8.37 \times 10^{-6}$$

$$pH = 5.08$$

(d) pH after Addition of 26.00 mL of Titrant

The excess of strong acid now present represses the dissociation of the HCN to the point where its contribution to the pH is negligible. Thus,

$$[H_3O^+] = c_{HCl} = \frac{26.00 \times 0.1000 - 50.00 \times 0.0500}{76.00} = 1.32 \times 10^{-3}$$

$$pH = 2.88$$

Figure 10–7 shows theoretical curves for a series of weak bases of different strengths. Clearly, indicators with *acidic* transition ranges must be employed for the titration of weak bases.

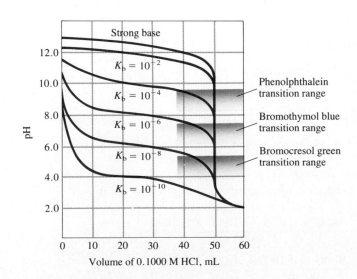

Figure 10–7

The effect of base strength on titration curves. Each curve represents the titration of 50.0 mL of 0.1000 M base with 0.1000 M HCl.

## 10F   THE COMPOSITION OF BUFFER SOLUTIONS AS A FUNCTION OF pH

The changes in composition that occur while a solution of a weak acid or a weak base is being titrated are sometimes of interest and can be visualized by plotting the *relative* concentration of the weak acid as well as the relative concentration of the conjugate base as a function of pH. These relative concentrations are called *alpha values*. For example, if we let $c_T$ be the sum of the analytical concentrations of acetic acid and sodium acetate at any point in the titration discussed in Example 10–5, we may write

$$c_T = c_{HOAc} + c_{NaOAc} \qquad (10\text{–}15)$$

We then define $\alpha_0$ as

$$\alpha_0 = \frac{[HOAc]}{c_T} \qquad (10\text{–}16)$$

and $\alpha_1$ as

$$\alpha_1 = \frac{[OAc^-]}{c_T} \qquad (10\text{–}17)$$

Alpha values are unitless ratios whose sum must equal unity. That is,

Alpha values do not depend on $c_T$.

$$\alpha_0 + \alpha_1 = 1$$

Alpha values are determined by $[H_3O^+]$ and $K_a$ alone and are independent of $c_T$. To obtain an expression for $\alpha_0$, we rearrange the dissociation-constant expression to

$$[OAc^-] = \frac{K_a[HOAc]}{[H_3O^+]} \qquad (10\text{–}18)$$

Mass balance requires that

$$c_T = [HOAc] + [OAc^-] \qquad (10\text{–}19)$$

Substituting Equation 10–18 into Equation 10–19 gives

$$c_T = [HOAc] + \frac{K_a[HOAc]}{[H_3O^+]} = [HOAc]\left(\frac{[H_3O^+] + K_a}{[H_3O^+]}\right)$$

Upon rearrangement, we obtain the expression for $\alpha_0$ as given by Equation 10–16.

$$\alpha_0 = \frac{[HOAc]}{c_T} = \frac{[H_3O^+]}{[H_3O^+] + K_a} \qquad (10\text{–}20)$$

To obtain an expression for $\alpha_1$, we rearrange the dissociation-constant expression to

$$[HOAc] = \frac{[H_3O^+][OAc^-]}{K_a}$$

and substitute into Equation 10–19

$$c_T = \frac{[H_3O^+][OAc^-]}{K_a} + [OAc^-] = [OAc^-]\left(\frac{[H_3O^+] + K_a}{K_a}\right)$$

Rearranging this equation gives $\alpha_1$ as defined by Equation 10–17

$$\alpha_1 = \frac{[OAc^-]}{c_T} = \frac{K_a}{[H_3O^+] + K_a} \qquad (10\text{–}21)$$

Note that the denominator is the same for both $\alpha_0$ and $\alpha_1$.

The solid lines labeled $\alpha_0$ and $\alpha_1$ in Figure 10–8 were derived with Equations 10–20 and 10–21 using values for $[H_3O^+]$ shown in column 2 of Table 10–3. The titration curve is shown as the curved line in Figure 10–8. Note that $\alpha_0$ is nearly 1 (0.987) at the outset of the titration, meaning that 98.7% of the acetate-containing species is present as HOAc and only 1.3% is present as OAc$^-$. At the equivalence point, $\alpha_0$ has decreased to $1.1 \times 10^{-4}$ and $\alpha_1$ approaches 1. Thus, only about 0.011% of the acetate-containing species is HOAc. Note that when the acid is half neutralized (25.00 mL), $\alpha_0$ and $\alpha_1$ are each 0.5.

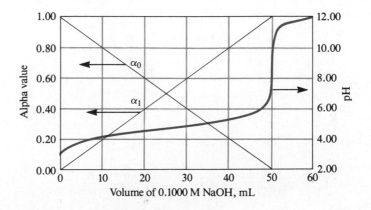

Figure 10–8
Black lines show the change in relative amounts of HOAc ($\alpha_0$) and OAc$^-$ ($\alpha_1$) during the titration of 50.00 mL of 0.1000 M acetic acid. The titration curve for the system is in color.

## 10G   COMMON TYPES OF ACID/BASE INDICATORS

Numerous organic compounds serve as indicators for neutralization titrations.

### 10G–1 Common Indicator Structures

The majority of acid/base indicators possess structural properties that permit classification into perhaps half a dozen categories.[3] Three of these classes are described in the following paragraphs.

#### Phthalein Indicators

Most phthalein indicators are colorless in moderately acidic solutions and exhibit a variety of colors in alkaline media. These colors tend to fade slowly in strongly alkaline solutions, which is an inconvenience in some applications. As a group, the phthaleins are sparingly soluble in water but readily dissolve in ethanol to give dilute solutions of the indicator.

   The best-known phthalein indicator is *phenolphthalein*, whose structures can be represented as

Note that a quinoid ring, which imparts color to most organic compounds, is formed in the second reaction. The pH at which the pink color of this quinoid structure first becomes detectable depends on the concentration of the indicator and on the visual acuity of the observer. For most people, however, the pink appears in the pH range of 8.0 to 8.2.

   The other phthalein indicators have various functional groups substituted on the phenolic rings. For example, *thymolphthalein* contains two alkyl groups on each ring. The structural alterations associated with the color change of this indicator are similar to those of phenolphthalein.

---

[3]See E. Banyai, in *Indicators*, E. Bishop, Ed., Chapter 3. New York: Pergamon, 1972.

### Sulfonphthalein Indicators

Many of the sulfonphthaleins exhibit two useful color-change ranges—one in somewhat acidic solutions and the other in neutral or moderately basic media. In contrast to the phthaleins, the basic color shows good stability toward strong alkali.

The sodium salts of the sulfonphthaleins are ordinarily used for the preparation of indicator solutions, owing to the appreciable acidity of the parent molecule. Solutions can be prepared directly from the sodium salt or indirectly by dissolving the sulfonphthalein in its acidic form in an appropriate volume of dilute aqueous sodium hydroxide.

The simplest sulfonphthalein indicator is *phenolsulfonphthalein*, known also as *phenol red*. The principal equilibria for a solution of the sodium salt of this compound are

Only the second color change, which occurs in the pH range between 6.4 and 8.0, is useful.

Substitution of halogens or alkyl groups for the hydrogens in the phenolic rings of the parent compound yields sulfonphthaleins that differ in color and pH range.

### Azo Indicators

Most azo indicators exhibit a color change from red to yellow with increasing basicity; their transition ranges are generally on the acidic side of neutrality. The most commonly encountered examples are *methyl orange* and *methyl red*. The behavior of the former is described by the equations

Methyl red contains a carboxylic acid group in place of the sulfonic acid group. Variations in the substituents on the amino nitrogen and in the rings give rise to a series of indicators with slightly different properties.

### 10G-2 Titration Errors with Acid/Base Indicators

Two types of titration errors are encountered in neutralization titrations. The first is a determinate error that occurs when the pH at which the indicator changes color differs from the pH at chemical equivalence. This type of error can usually be minimized by judicious indicator selection or by a blank correction.

The second type is an indeterminate error that originates from the limited ability of the eye to distinguish reproducibly the intermediate color of the indicator. The magnitude of this error depends on the change in pH per milliliter of reagent at the equivalence point, on the concentration of the indicator, and on the sensitivity of the eye to the two indicator colors. On the average, the visual uncertainty with an acid/base indicator is in the range of $\pm 0.5$ to $\pm 1$ pH unit. This uncertainty can often be decreased to as little as $\pm 0.1$ pH unit by matching the color of the solution being titrated with that of a reference standard containing a similar amount of indicator at the appropriate pH. These uncertainties are of course approximations that vary considerably from indicator to indicator as well as from person to person.

### 10G-3 Variables That Influence the Behavior of Indicators

The pH interval over which a given indicator exhibits a color change is influenced by temperature, by the ionic strength of the medium, and by the presence of organic solvents and colloidal particles. Some of these effects, particularly the last two, can cause the transition range to shift by one or more pH units.[4]

---

[4]For a discussion of these effects, see H. A. Laitinen and W. E. Harris, *Chemical Analysis,* 2nd ed., pp. 48–51. New York: McGraw-Hill, 1975.

---

## 10H   QUESTIONS AND PROBLEMS

*10-1. Explain the fundamental difference between the equivalence point and the end point of a titration.

10-2. Why are strong acids and strong bases used as titrants in neutralization titrations?

*10-3. Even though it is strong, nitric acid is seldom encountered as a standard solution for neutralization titrations. Briefly explain why this is so.

10-4. Why does the transition range of an acid/base indicator extend over approximately 2 pH units?

*10-5. What variables can cause the pH range of an acid/base indicator to shift?

10-6. Solutions of bromocresol green, bromothymol blue, and phenolphthalein are available (see Table

10-1). Which of these indicators would you use to detect the equivalence point in the titration of

*(a) 0.0400 M Ba(OH)$_2$ with standard 0.100 M HCl?

(b) 0.0400 M propanoic acid with 0.0500 M NaOH?

*(c) 0.0600 M ethanolamine with 0.0500 M HCl?

(d) $2.00 \times 10^{-3}$ M HCl with $2.00 \times 10^{-3}$ M NaOH?

*(e) 0.0500 M anilinium chloride (C$_6$H$_5$NH$_3$Cl) with 0.0400 M NaOH?

(f) 0.0500 M sodium phenolate with 0.0400 M HCl?

*10-7. An error no larger than $\pm 0.05$ mL can be tolerated in the titration of 50.00 mL of 0.0500 M formic acid with 0.1000 M KOH. Select an indicator from Table 10–1 that should meet this goal.

**10–8.** Select an indicator from Table 10–1 that will permit the titration of 50.00 mL of 0.1000 M ethylamine with 0.1000 M HCl with a titration error no greater than ±0.05 mL.

***10–9.** Briefly account for points of resemblance and points of difference between curves for the titration of 0.100 M HCl and 0.100 M HOCl with standard 0.100 M NaOH.

**10–10.** What is a buffer solution and what are its properties?

***10–11.** Consider a solution consisting of the weak acid HA and its conjugate base NaA. The molar equilibrium concentrations of HA and $A^-$ are given by

$$[HA] = c_{HA} - [H_3O^+] + [OH^-]$$

$$[A^-] = c_{A^-} + [H_3O^+] - [OH^-]$$

where $c$ represents the molar analytical concentration of the solute.
  (a) Why is it always possible to eliminate either $[H_3O^+]$ or $[OH^-]$ from these expressions?
  (b) Under what circumstances is it likely that *both* $[H_3O^+]$ and $[OH^-]$ can be neglected without appreciable error?

**10–12.** Establish whether solutions containing equal quantities of the following conjugate acid/base pairs will be acidic or basic:
  *(a) formic acid/sodium formate.
  (b) hydrazoic acid/sodium azide.
  *(c) phenol/sodium phenolate.
  (d) hydrogen cyanide/potassium cyanide.
  *(e) piperidine hydrochloride ($C_5H_{11}NHCl$)/piperidine.
  (f) pyridinium chloride ($C_5H_5NHCl$)/pyridine.

**10–13.** Equations 10–9 and 10–10 (see also Problem 10–3) can always be simplified by neglecting either $[H_3O^+]$ or $[OH^-]$. Which of these terms would you eliminate before calculating the pH of the solutions in Problem 10–12?

**10–14.** What is the ratio of acid to conjugate base in a solution that has a pH of 7.17 and contains
  *(a) hypochlorous acid and sodium hypochlorite?
  (b) ammonium chloride and ammonia?
  *(c) hydroxylammonium chloride and hydroxylamine?
  (d) hydrogen cyanide and potassium cyanide?
  *(e) trimethylacetic acid ($K_a = 9.33 \times 10^{-6}$) and sodium trimethylacetate?

**10–15.** Consult Appendixes 3 and 4; select a suitable conjugate acid/base pair to produce a buffer with a pH of
  *(a) 4.30.      (f) 2.50.
  (b) 3.90.      *(g) 8.12.
  *(c) 5.25.      (h) 6.30.
  (d) 7.57.      *(i) 10.10.
  *(e) 9.40.      (j) 5.00.

***10–16.** Arrange the following bases in order of decreasing end-point sharpness when 0.10 M solutions are titrated with 0.10 M HCl:

  (a) KCN.
  (b) methylamine.
  (c) KOH.
  (d) trimethylamine.
  (e) sodium hypochlorite.

**10–17.** Arrange the following acids in order of decreasing end-point sharpness when 0.10 M solutions are titrated with 0.10 M NaOH:
  (a) ethanolammonium chloride.
  (b) picric acid.
  (c) lactic acid.
  (d) anilinium chloride.
  (e) perchloric acid.

**10–18.** Calculate the pH of the solution that results upon mixing 20.0 mL of 0.200 M HCl with 25.0 mL of
  *(a) distilled water.
  (b) 0.132 M $AgNO_3$.
  *(c) 0.132 M NaOH.
  (d) 0.132 M $NH_3$.
  *(e) 0.232 M NaOH.

**10–19.** Calculate the pH of the solution that results upon mixing 0.105 g of $Mg(OH)_2$ with
  *(a) 75.0 mL of 0.0600 M HCl.
  (b) 15.0 mL of 0.600 M HCl.
  *(c) 30.0 mL of 0.0600 M HCl.
  (d) 30.0 mL of 0.0600 M $MgCl_2$.

***10–20.** Calculate the pH of a hypochlorous acid solution that is
  (a) $1.00 \times 10^{-1}$ M.
  (b) $1.00 \times 10^{-2}$ M.
  (c) $1.00 \times 10^{-4}$ M.

***10–21.** Calculate the pH of a sodium hypochlorite solution that is
  (a) $1.00 \times 10^{-1}$ M.
  (b) $1.00 \times 10^{-2}$ M.
  (c) $1.00 \times 10^{-4}$ M.

**10–22.** Calculate the pH of an ammonia solution that is
  (a) $1.00 \times 10^{-1}$ M.
  (b) $1.00 \times 10^{-2}$ M.
  (c) $1.00 \times 10^{-4}$ M.

**10–23.** Calculate the pH of an ammonium chloride solution that is
  (a) $1.00 \times 10^{-1}$ M.
  (b) $1.00 \times 10^{-2}$ M.
  (c) $1.00 \times 10^{-4}$ M.

**10–24.** Calculate the pH of a solution in which the piperidine concentration is
  *(a) $1.00 \times 10^{-1}$ M.
  (b) $1.00 \times 10^{-2}$ M.
  (c) $1.00 \times 10^{-4}$ M.

**10–25.** Calculate the pH of an iodic acid solution that is
  (a) $1.00 \times 10^{-1}$ M.
  *(b) $1.00 \times 10^{-2}$ M.
  (c) $1.00 \times 10^{-4}$ M.

**10–26.** Calculate the pH of a picric acid solution that is
  (a) $1.00 \times 10^{-1}$ M.
  (b) $1.00 \times 10^{-2}$ M.
  *(c) $1.00 \times 10^{-4}$ M.

***10–27.** Calculate the hydronium ion concentration and the pH of a solution that is 0.0500 M in HCl

(a) neglecting activities.

(b) using activities.

10–28. A solution is 0.0500 M in $NH_4Cl$ and 0.0300 M in $NH_3$. Calculate the molar $OH^-$ concentration and the pH of this solution
(a) neglecting activities.
(b) using activities.

*10–29. Calculate the pH of the solution that results upon mixing 20.0 mL of 0.200 M formic acid with 25.0 mL of
(a) water.
(b) 0.200 M NaOH.
(c) 0.160 M NaOH.
(d) 0.100 M NaOH.
(e) 0.200 M sodium formate.

10–30. Calculate the pH of the solution that results upon mixing 40.0 mL of 0.100 M $NH_3$ with 20.0 mL of
(a) water.
(b) 0.250 M HCl.
(c) 0.200 M HCl.
(d) 0.100 M HCl.
(e) 0.200 M $NH_4Cl$.

*10–31. Calculate the pH of a solution that is
(a) 0.0670 M in lactic acid and 0.0379 M in sodium lactate.
(b) 0.0460 M in HCN and 0.204 M in NaCN.
(c) 0.0963 M in piperidine and 0.148 M in its chloride salt.
(d) 0.0582 M in hydroxylamine and 0.0129 M in its chloride salt.
(e) 0.0256 M in trichloroacetic acid and 0.0348 M in its sodium salt.

10–32. Calculate the pH of a solution that is produced by dissolving
(a) 9.20 g of lactic acid and 11.15 g of sodium lactate in water and diluting to 1.00 L.
(b) 3.85 g of nitrous acid and 7.65 g of sodium nitrite in water and diluting to 500.0 mL.
(c) 2.32 g of aniline in 100.0 mL of 0.0200 M HCl and diluting to 250.0 mL.
(d) 3.30 g of $(NH_4)_2SO_4$ in water, adding 125.0 mL of 0.1011 M NaOH, and diluting to 500 mL.
(e) 4.09 g of methylamine in 180 mL of 0.344 M HCl and diluting to 250.0 mL.

10–33. Calculate the change in pH that occurs in each of the accompanying solutions as a result of a tenfold dilution with water:
*(a) $H_2O$.
(b) 0.0500 M HCl.
*(c) 0.0500 M NaOH.
(d) 0.0500 M $NH_3$.
*(e) 0.0500 M $NH_4Cl$.
(f) 0.500 M $NH_3$ + 0.500 M $NH_4Cl$.
*(g) 0.0500 M $NH_3$ + 0.0500 M $NH_4Cl$.

10–34. Calculate the change in pH that occurs when 1.00 mmol of HCl is added to 100.0 mL of the solutions listed in Problem 10–33.

*10–35. Calculate the change in pH that occurs when 1.00 mmol of NaOH is added to 100.0 mL of the solutions listed in Problem 10–33.

10–36. Calculate the change in pH that occurs when 0.500 mmol of HCl is added to 100.0 mL of
(a) 0.0800 M glycolic acid.
(b) 0.0600 M sodium glycolate.
(c) 0.0800 M glycolic acid and 0.0600 M sodium glycolate.
(d) 0.0400 M glycolic acid and 0.0300 M sodium glycolate.

*10–37. Calculate $\alpha$-values for mandelic acid in a solution buffered to a pH of
(a) 2.50.
(b) 4.00.
(c) 5.68.

10–38. Calculate $\alpha$-values for hydrogen cyanide in a solution buffered to a pH of
(a) 7.31.
(b) 9.05.
(c) 10.84.

*10–39. Calculate $\alpha$-values for ethylamine in a solution buffered to a pH of
(a) 7.46.
(b) 9.39.
(c) 11.10.

10–40. Calculate $\alpha$-values for hydroxylamine in a solution buffered to a pH of
(a) 6.57.
(b) 8.00.
(c) 8.88.

10–41. A 500-mL portion of a solution in which the molar analytical concentration of acetic acid is 0.600 M is partially neutralized with NaOH and diluted to 1.000 L. The diluted solution has a pH of 5.01. Calculate
(a) $\alpha$-values for the acetate-containing species in the resulting solution.
(b) the molar equilibrium concentrations of acetic acid and acetate ion in the resulting solution.

*10–42. What weight of sodium formate must be added to 500.0 mL of 0.800 M formic acid to produce a buffer solution with a pH of 3.50?

10–43. What weight of sodium glycolate should be added to 250.0 mL of 1.00 M glycolic acid to produce a buffer solution with a pH of 4.00?

*10–44. What volume of 2.00 M NaOH should be added to 250.0 mL of 1.00 M glycolic acid to produce a buffer solution with a pH of 4.00?

10–45. What volume of 2.00 M HCl should be added to 250.0 mL of 1.00 M sodium glycolate acid to produce a buffer solution with a pH of 4.00?

*10–46. A 25.00-mL aliquot of 0.2000 M NaOH is diluted to 50.00 mL and titrated with 0.1000 M HCl. Calculate the pH of the solution after the addition of 0.00, 10.00, 25.00, 40.00, 45.00, 49.00, 50.00, 51.00, 55.00, and 60.00 mL of the acid. Prepare a titration curve from the data.

10–47. Calculate the pH after the addition of 0.00, 5.00, 15.00, 25.00, 40.00, 45.00, 49.00, 50.00, 51.00, 55.00, and 60.00 mL of 0.1000 M NaOH in the titration of 50.00 mL of 0.1000 M

*(a) $HNO_2$.
(b) glycolic acid.
*(c) pyridinium chloride.

10–48. Calculate the pH after the addition of 0.00, 5.00, 15.00, 25.00, 40.00, 45.00, 49.00, 50.00, 51.00, 55.00, and 60.00 mL of 0.1000 M HCl in the titration of 50.00 mL of 0.1000 M
(a) ammonia.
*(b) hydroxylamine.
(c) sodium cyanide.

10–49. Calculate the pH after the addition of 0.00, 5.00, 15.00, 25.00, 40.00, 45.00, 49.00, 50.00, 51.00, 55.00, and 60.00 mL of 0.1000 M titrant in the titration of 50.00 mL of
(a) 0.1000 M anilinium chloride with 0.1000 M NaOH.
(b) 0.0100 M picric acid with 0.01000 M NaOH.
(c) 0.1000 M hypochlorous acid with 0.1000 M NaOH.
(d) 0.1000 M trimethylamine with 0.1000 M HCl.

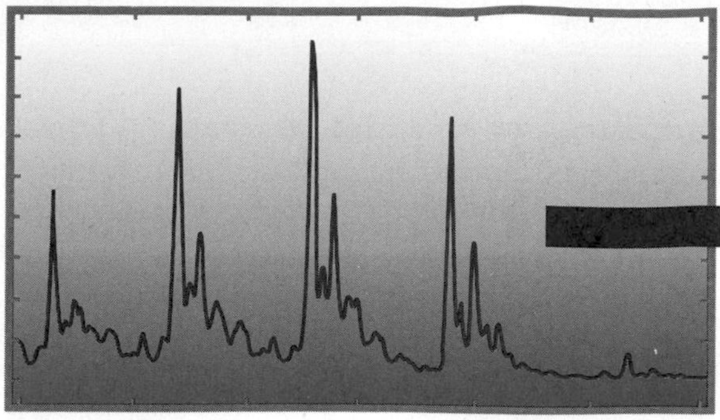

## CHAPTER 11

# TITRATION CURVES FOR COMPLEX ACID/BASE SYSTEMS

In this chapter we describe methods for deriving titration curves for complex acid/base systems. For the purpose of this discussion, complex systems are defined as solutions made up of (1) two acids or two bases of different strengths, (2) an acid or base that has two or more acidic or basic functional groups, or (3) an amphiprotic substance, which is capable of acting both as an acid and a base. Equations for more than one equilibrium are required to describe the characteristics of any of these systems.

## 11A MIXTURES OF STRONG AND WEAK ACIDS OR STRONG AND WEAK BASES

To illustrate the derivation of titration curves for mixtures of a strong and a weak acid, consider the titration of a solution containing hydrochloric acid and a weak acid HA that has a dissociation constant of $1.00 \times 10^{-4}$. The molar hydronium ion concentration in the early stages of this titration is

$$[H_3O^+] = c_{HCl} + [A^-]$$

The terms on the right account for the contributions of the two solute acids to the hydronium ion concentration of the solution. A term for hydronium ions resulting from dissociation of water is omitted because it is vanishingly small. This assumption is surely valid for any system that contains an appreciable amount of either hydrochloric acid or HA.

Example 11–1 demonstrates that hydrochloric acid represses the dissociation of the weak acid in the early stages of the titration to such an extent that we can assume that $[A^-] \ll c_{HCl}$. The hydronium ion concentration is then simply the molar concentration of the strong acid.

---

Example 11–1

Calculate the pH of a mixture that is 0.1200 M in hydrochloric acid and 0.0800 M in the weak acid HA ($K_a = 1.00 \times 10^{-4}$).

$$[H_3O^+] = 0.1200 + [A^-]$$

If we assume that $[A^-] \ll 0.1200$, then $[H_3O^+] \simeq 0.1200$ and the pH is 0.92. To check this assumption, the provisional value for $[H_3O^+]$ is substituted into the dissociation-constant expression for HA, which upon rearrangement gives

$$\frac{[A^-]}{[HA]} = \frac{K_a}{[H_3O^+]} = \frac{1.00 \times 10^{-4}}{0.1200} = 8.33 \times 10^{-4}$$

which can be rearranged to

$$[HA] = [A^-]/(8.33 \times 10^{-4})$$

From mass-balance considerations, we can write

$$0.0800 = c_{HA} = [HA] + [A^-]$$

Substituting the value of [HA] from the previous equation gives

$$0.0800 = [A^-]/(8.33 \times 10^{-4}) + [A^-] = (1.20 \times 10^3)[A^-]$$
$$[A^-] = 6.7 \times 10^{-5}$$

We see that $[A^-]$ is indeed much smaller than 0.1200 M, as assumed.

---

The approximation employed in Example 11–1 can be shown to apply until most of the hydrochloric acid has been neutralized by the titrant. Therefore, the curve in this region *is identical to the titration curve for a 0.1200 M solution of a strong acid by itself*.

As shown by Example 11–2, the presence of HA must be taken into account as the first end point in the titration is approached.

---

Example 11–2

Calculate the pH of the solution that results when 29.00 mL of 0.1000 M NaOH is added to 25.00 mL of the solution described in Example 11–1.
   Here,

$$c_{HCl} = \frac{25.00 \times 0.1200 - 29.00 \times 0.1000}{54.00} = 1.85 \times 10^{-3} \ M$$

$$c_{HA} = \frac{25.00 \times 0.0800}{54.00} = 3.70 \times 10^{-2} \ M$$

A provisional result based (as in the previous example) on the assumption that $[H_3O^+] = 1.85 \times 10^{-3}$ yields a value of $1.90 \times 10^{-3}$ for $[A^-]$. Clearly, $[A^-]$ is no longer much smaller than $[H_3O^+]$, and we must write

$$[H_3O^+] = c_{HCl} + [A^-] = 1.85 \times 10^{-3} + [A^-] \qquad (11-1)$$

In addition, from mass-balance considerations, we know that

$$[HA] + [A^-] = c_{HA} = 3.70 \times 10^{-2} \qquad (11\text{--}2)$$

We rearrange the acid dissociation constant expression for HA and obtain

$$[HA] = \frac{[H_3O^+][A^-]}{1.00 \times 10^{-4}}$$

Substitution of this expression into Equation 11–2 yields

$$\frac{[H_3O^+][A^-]}{1.00 \times 10^{-4}} + [A^-] = 3.70 \times 10^{-2}$$

$$[A^-] = \frac{3.70 \times 10^{-6}}{[H_3O^+] + 1.00 \times 10^{-4}}$$

Substitution for $[A^-]$ and $c_{HCl}$ in Equation 11–1 yields

$$[H_3O^+] = 1.85 \times 10^{-3} + \frac{3.70 \times 10^{-6}}{[H_3O^+] + 1.00 \times 10^{-4}}$$

$$[H_3O^+]^2 + (1.00 \times 10^{-4})[H_3O^+] =$$
$$(1.85 \times 10^{-3})[H_3O^+] + 1.85 \times 10^{-7} + 3.7 \times 10^{-6}$$

Collecting terms gives

$$[H_3O^+]^2 - (1.75 \times 10^{-3})[H_3O^+] - 3.885 \times 10^{-6} = 0$$

$$[H_3O^+] = 3.03 \times 10^{-3}$$

$$pH = 2.52$$

> The composition of a mixture of a strong acid and a weak acid can be determined by titration with suitable indicators provided that the weak acid has a dissociation constant that lies between $10^{-4}$ and $10^{-8}$ and provided that the concentrations of the two acids are of the same order of magnitude.

Note that the contributions to the hydronium ion concentration from HCl ($1.85 \times 10^{-3}$ M) and HA ($3.03 \times 10^{-3}$ M $- 1.85 \times 10^{-3}$ M) are of comparable magnitude.

---

When the amount of base added is equivalent to the amount of hydrochloric acid originally present, the solution is identical in all respects to one prepared by dissolving appropriate quantities of the weak acid and sodium chloride in a suitable volume of water. The sodium chloride, however, has no effect on the pH (neglecting the influence of increased ionic strength); thus, the remainder of the titration curve is identical to that for a dilute solution of HA.

The shape of the curve for a mixture of weak and strong acids, and hence the information obtainable from it, depend in large measure upon the strength of the weak acid. Figure 11–1 depicts the pH changes that occur during the titration of mixtures containing hydrochloric acid and several weak acids. Note that the rise in pH at the first equivalence point is small or essentially nonexistent when the weak acid has a relatively

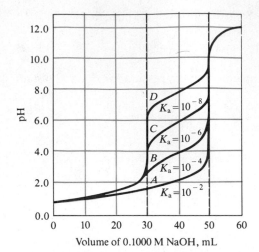

**Figure 11–1**

Curves for the titration of strong acid/weak acid mixtures with 0.1000 M NaOH. Each titration is on 25.00 mL of a solution that is 0.1200 M in HCl and 0.0800 M in HA.

large dissociation constant (curves *A* and *B*). For titrations such as these, only the total number of millimoles of weak and strong acid can be ascertained. Conversely, when the weak acid has a very small dissociation constant, only the strong acid content can be determined. For weak acids of intermediate strength ($K_a$ somewhat less than $10^{-4}$ but greater than $10^{-8}$), there are usually two useful end points.

Determination of the amount of each component in a mixture that contains a strong base and a weak base is also possible, subject to the constraints just described for the strong acid/weak acid system. The derivation of a curve for such a titration is analogous to that for a mixture of acids.

## 11B   POLYFUNCTIONAL ACIDS

Phosphoric acid is a typical polyfunctional acid. In aqueous solution it undergoes the following three dissociation reactions:

Generally, $K_1 > K_2$, often by a factor of $10^4$ to $10^5$, because of electrostatic forces. That is, the first dissociation involves separating a single positively charged proton from a singly charged anion. In the second step, a proton is separated from a doubly charged anion—a process that requires considerably more energy.

$$H_3PO_4 + H_2O \rightleftarrows H_2PO_4^- + H_3O^+$$

$$K_1 = \frac{[H_3O^+][H_2PO_4^-]}{[H_3PO_4]} = 7.11 \times 10^{-3}$$

$$H_2PO_4^- + H_2O \rightleftarrows HPO_4^{2-} + H_3O^+$$

$$K_2 = \frac{[H_3O^+][HPO_4^{2-}]}{[H_2PO_4^-]} = 6.34 \times 10^{-8}$$

$$HPO_4^{2-} + H_2O \rightleftarrows PO_4^{3-} + H_3O^+$$

$$K_3 = \frac{[H_3O^+][PO_4^{3-}]}{[HPO_4^{2-}]} = 4.2 \times 10^{-13}$$

With this acid, as with others, $K_1 > K_2 > K_3$.

**Feature 11–1**
COMBINING EQUILIBRIUM-CONSTANT EXPRESSIONS

When two adjacent stepwise equilibria are *added*, the equilibrium constant for the resulting overall reaction is the *product* of the two constants. Thus for the first two dissociation equilibria for $H_3PO_4$, we may write

$$H_3PO_4 + H_2O \rightleftarrows H_2PO_4^- + H_3O^+$$
$$\underline{H_2PO_4^- + H_2O \rightleftarrows HPO_4^- + H_3O^+}$$
$$H_3PO_4 + 2\,H_2O \rightleftarrows HPO_4^- + 2\,H_3O^+$$

and

$$\beta_2 = K_1K_2 = \frac{[H_3O^+]^2[HPO_4^-]}{[H_3PO_4]}$$
$$= 7.11 \times 10^{-3} \times 6.34 \times 10^{-8} = 4.51 \times 10^{-10}$$

Similarly,

$$\beta_3 = K_1K_2K_3 = \frac{[H_3O^+]^3[PO_4^{3-}]}{[H_3PO_4]}$$
$$= 7.11 \times 10^{-3} \times 6.34 \times 10^{-8} \times 4.2 \times 10^{-13}$$
$$= 1.9 \times 10^{-22}$$

## 11C  POLYFUNCTIONAL BASES

Polyfunctional bases are also common, an example being sodium carbonate. Carbonate ion, the conjugate base of the hydrogen carbonate ion, is involved in the stepwise equilibria

$$CO_3^{2-} + H_2O \rightleftarrows HCO_3^- + OH^-$$
$$K_{b1} = \frac{[HCO_3^-][OH^-]}{[CO_3^{2-}]} = \frac{K_w}{K_2} = \frac{1.00 \times 10^{-14}}{4.7 \times 10^{-11}} = 2.1 \times 10^{-4}$$
$$HCO_3^- + H_2O \rightleftarrows H_2CO_3 + OH^-$$
$$K_{b2} = \frac{[H_2CO_3][OH^-]}{[HCO_3^-]} = \frac{K_w}{K_1} = \frac{1.00 \times 10^{-14}}{4.45 \times 10^{-7}} = 2.25 \times 10^{-8}$$

where $K_1$ and $K_2$ are the first and second dissociation constants for carbonic acid and $K_{b1}$ and $K_{b2}$ are the first and second basic dissociation constants for carbonate ion.

The overall basic dissociation reaction of sodium carbonate is described by the equations

$$CO_3^{2-} + 2 H_2O \rightleftharpoons H_2CO_3 + 2 OH^-$$

$$K_{b1}K_{b2} = \frac{[H_2CO_3][OH^-]^2}{[CO_3^{2-}]} = 2.1 \times 10^{-4} \times 2.25 \times 10^{-8} = 4.8 \times 10^{-12}$$

The pH of polyfunctional systems, such as phosphoric acid or sodium carbonate, can be computed rigorously through use of the systematic approach to multiple-equilibrium problems described in Chapter 8. Solution of the several simultaneous equations involved is difficult and time-consuming, however, unless a simplifying assumption is made.

## 11D  BUFFER SOLUTIONS INVOLVING POLYPROTIC ACIDS

Two buffer systems can be prepared from a weak dibasic acid and its salts. The first consists of free acid $H_2A$ and its conjugate base NaHA, and the second makes use of the acid NaHA and its conjugate base $Na_2A$. The pH of the latter system is higher than that of the former because the acid dissociation constant for $HA^-$ is always less than that for $H_2A$.

Sufficient independent equations are readily written to permit a rigorous evaluation of the hydronium ion concentration for either of these systems. Ordinarily, however, it is permissible to introduce the simplifying assumption that only one of the equilibria is important in determining the hydronium ion concentration of the solution. Thus, for a buffer prepared from $H_2A$ and NaHA, the dissociation of $HA^-$ to yield $A^{2-}$ is neglected, and the calculation is based on the first dissociation only. With this simplification, the hydronium ion concentration is calculated by the method described in Section 10C–1 for a simple buffer solution. As before, it is an easy matter to check the validity of the assumption by calculating an approximate concentration of $A^{2-}$ and comparing this value with the concentrations of $H_2A$ and $HA^-$.

---

Example 11–3

Calculate the hydronium ion concentration for a buffer solution that is 2.00 M in phosphoric acid and 1.50 M in potassium dihydrogen phosphate.

The principal equilibrium in this solution is the dissociation of $H_3PO_4$:

$$H_3PO_4 + H_2O \rightleftharpoons H_3O^+ + H_2PO_4^- \qquad \frac{[H_3O^+][H_2PO_4^-]}{[H_3PO_4]} = K_1 = 7.11 \times 10^{-3}$$

The dissociation of $H_2PO_4^-$ is assumed to be negligible, that is, $[HPO_4^{2-}]$ and $[PO_4^{3-}] \ll [H_2PO_4^-]$ or $[H_3PO_4]$. Then,

$$[H_3PO_4] \simeq c_{H_3PO_4} = 2.00$$

$$[H_2PO_4^-] \simeq c_{KH_2PO_4} = 1.50$$

$$[H_3O^+] = \frac{7.11 \times 10^{-3} \times 2.00}{1.50} = 9.48 \times 10^{-3}$$

We now use the equilibrium-constant expression for $K_2$ to show that $[HPO_4^{2-}]$ can be neglected:

$$\frac{[H_3O^+][HPO_4^{2-}]}{[H_2PO_4^-]} = \frac{9.48 \times 10^{-3}[HPO_4^{2-}]}{1.50} = 6.34 \times 10^{-8}$$

$$[HPO_4^{2-}] = 1.00 \times 10^{-5}$$

and our assumption is valid. Note that $[PO_4^{3-}]$ is even smaller than $[HPO_4^{2-}]$.

---

For a buffer prepared from NaHA and Na$_2$A, the second dissociation will ordinarily predominate, and the reaction

$$HA^- + H_2O \rightleftharpoons H_2A + OH^-$$

is disregarded. The concentration of H$_2$A is negligible compared with that of HA$^-$ or A$^{2-}$; the hydronium ion concentration can then be calculated from the second dissociation constant, again employing the techniques for a simple buffer solution. To test the assumption, an estimate of the H$_2$A concentration is compared with the concentrations of HA$^-$ and A$^{2-}$.

---

Example 11–4

Calculate the hydronium ion concentration of a buffer that is 0.0500 M in potassium hydrogen phthalate (KHP) and 0.150 M in potassium phthalate (K$_2$P).

$$HP^- + H_2O \rightleftharpoons H_3O^+ + P^{2-} \qquad \frac{[H_3O^+][P^{2-}]}{[HP^-]} = K_2 = 3.91 \times 10^{-6}$$

Provided the concentration of H$_2$P in this solution is negligible,

$$[HP^-] \simeq c_{KHP} \simeq 0.0500$$

$$[P^{2-}] \simeq c_{K_2P} = 0.150$$

$$[H_3O^+] = \frac{3.91 \times 10^{-6} \times 0.0500}{0.150} = 1.30 \times 10^{-6}$$

To check the first assumption, an approximate value for [H$_2$P] is calculated by substituting numerical values for [H$_3$O$^+$] and [HP$^-$] into the expression for $K_1$:

$$\frac{(1.30 \times 10^{-6})(0.0500)}{[H_2P]} = K_1 = 1.12 \times 10^{-3}$$

$$[H_2P] = 6 \times 10^{-5}$$

This result justifies the assumption that $[H_2P] \ll [HP^-]$ and $[P^{2-}]$, that is, that the dissociation of HP$^-$ as a base can be neglected.

---

In all but a few situations, the assumption of a single principal equilibrium, as invoked in Examples 11–3 and 11–4, provides a satisfactory estimate of the pH of buffer mixtures derived from polybasic acids. Appreciable errors occur, however, when the concentration of the acid or the salt is very low or when the two dissociation constants are numerically close to one another. A more laborious and rigorous calculation is then required.

## 11E   CALCULATION OF THE pH OF NaHA SOLUTIONS

Amphiprotic salts have both acidic and basic properties.

Thus far, we have not considered how to calculate the pH of solutions of salts that have both acidic and basic properties—that is, salts that are *amphiprotic*. Such salts are formed during the neutralization titration of polyfunctional acids and bases. For example, when 1 mol of NaOH is added to a solution containing 1 mol of the acid $H_2A$, 1 mol of NaHA is formed. The pH of this solution is determined by the equilibria

$$HA^- + H_2O \rightleftarrows A^{2-} + H_3O^+$$

$$HA^- + H_2O \rightleftarrows H_2A + OH^-$$

One of these reactions produces hydronium ions and the other hydroxide ions. A solution of NaHA will be either acidic or basic, depending upon the relative magnitude of the equilibrium constants for these processes:

$$K_2 = \frac{[H_3O^+][A^{2-}]}{[HA^-]} \tag{11–3}$$

$$K_{b2} = \frac{K_w}{K_1} = \frac{[H_2A][OH^-]}{[HA^-]} \tag{11–4}$$

If $K_{b2}$ is greater than $K_2$, the solution is basic; otherwise, it is acidic.

A solution of NaHA can be described in terms of mass balance:

$$c_{NaHA} = [HA^-] + [H_2A] + [A^{2-}] \tag{11–5}$$

and charge balance:

$$[Na^+] + [H_3O^+] = [HA^-] + 2\,[A^{2-}] + [OH^-]$$

Since the sodium ion concentration is equal to the molar analytical concentration of the salt, the last equation can be rewritten as

$$c_{NaHA} + [H_3O^+] = [HA^-] + 2\,[A^{2-}] + [OH^-] \tag{11–6}$$

One additional algebraic equation is needed to solve for the five unknowns. The ion-product constant for water serves this purpose:

$$K_w = [H_3O^+][OH^-]$$

The rigorous computation of the hydronium ion concentration from these five equations is difficult. However, a reasonable approximation applicable to solutions of most amphiprotic salts can be obtained as follows.

When we subtract the mass-balance equation from the charge-balance equation, we obtain

$$c_{NaHA} + [H_3O^+] = [HA^-] + 2\,[A^{2-}] + [OH^-]$$
$$\underline{c_{NaHA} \qquad\qquad = [HA^-] + [H_2A] + [A^{2-}]}$$
$$[H_3O^+] = [A^{2-}] + [OH^-] - [H_2A]$$

We then express this last equation in terms of the principal solute species $[HA^-]$, $[H_3O^+]$, the dissociation constants for carbonic acid, and $K_w$:

$$[H_3O^+] = \frac{K_2[HA^-]}{[H_3O^+]} + \frac{K_w}{[H_3O^+]} - \frac{[H_3O^+][HA^-]}{K_1}$$

Note that

$$[H_2A] = \frac{[H_3O^+][HA^-]}{K_1}$$

Multiplication by $[H_3O^+]$ gives

$$[H_3O^+]^2 = K_2[HA^-] + K_w - \frac{[H_3O^+]^2[HA^-]}{K_1}$$

We collect terms to obtain

$$[H_3O^+]^2 \left(\frac{[HA^-]}{K_1} + 1\right) = K_2[HA^-] + K_w$$

Finally, this equation rearranges to

$$[H_3O^+] = \sqrt{\frac{K_2[HA^-] + K_w}{1 + [HA^-]/K_1}} \qquad (11-7)$$

Under most circumstances, it can be assumed that

$$[HA^-] \approx c_{NaHA} \qquad (11-8)$$

Substitution of this equality into Equation 11–7 gives

$$[H_3O^+] = \sqrt{\frac{K_2 c_{NaHA} + K_w}{1 + c_{NaHA}/K_1}} \qquad (11-9)$$

It is important to understand that the approximation shown as Equation 11–8 requires that $[HA^-]$ be much larger than any of the other equilibrium concentrations in Equations 11–5 and 11–6. This assumption is not valid for very dilute solutions of NaHA or when $K_2$ or $K_w/K_1$ is relatively large.

Frequently, the ratio $c_{NaHA}/K_1$ is much larger than unity and $K_2 c_{NaHA}$ is considerably greater than $K_w$. With these assumptions, Equation 11–9 simplifies to

$$[H_3O^+] \approx \sqrt{K_1 K_2} \qquad (11-10)$$

Make sure that you always check the assumptions that are inherent in Equation 11–10.

Note that Equation 11–10 does not contain $c_{NaHA}$, which implies that the pH of solutions of this type remains constant over a considerable range of solute concentrations.

---

Example 11–5

Calculate the hydronium ion concentration of a 0.100 M $NaHCO_3$ solution.

We first examine the assumptions leading to Equation 11–10. The dissociation constants for $H_2CO_3$ are $K_1 = 4.45 \times 10^{-7}$ and $K_2 = 4.7 \times 10^{-11}$. Clearly, $c_{NaHA}/K_1$ is much larger than unity; in addition, $K_2 c_{NaHA}$ has a value of $4.7 \times 10^{-12}$, which is substantially greater than $K_w$. Thus Equation 11–10 applies and

$$[H_3O^+] = \sqrt{4.45 \times 10^{-7} \times 4.7 \times 10^{-11}} = 4.6 \times 10^{-9}$$

---

Example 11–6

Calculate the hydronium ion concentration of a $1.0 \times 10^{-3}$ M $Na_2HPO_4$ solution.

The pertinent dissociation constants are $K_2$ and $K_3$, which both contain $[HPO_4^{2-}]$. Their values are $K_2 = 6.34 \times 10^{-8}$ and $K_3 = 4.2 \times 10^{-13}$. Considering again the assumptions that led to Equation 11–10, we find that $(1.0 \times 10^{-3})/(6.34 \times 10^{-8})$ is again large enough so that the denominator can be simplified. The product $K_2 c_{Na_2HPO_4}$ is by no means much larger than $K_w$, however. We therefore use a partially simplified version of Equation 11–9:

$$[H_3O^+] = \sqrt{\frac{4.2 \times 10^{-13} \times 1.0 \times 10^{-3} + 1.0 \times 10^{-14}}{(1.0 \times 10^{-3})/(6.34 \times 10^{-8})}} = 8.1 \times 10^{-10}$$

Use of Equation 11–10 yields a value of $1.6 \times 10^{-10}$ M.

---

Example 11–7

Find the hydronium ion concentration of a 0.0100 M $NaH_2PO_4$ solution.

The two dissociation constants of importance (those containing $[H_2PO_4^-]$) are $K_1 = 7.11 \times 10^{-3}$ and $K_2 = 6.34 \times 10^{-8}$. We see that the denominator of Equation 11–9 cannot be simplified, but the numerator reduces to $K_2 c_{NaH_2PO_4}$. Thus, Equation 11–9 becomes

$$[H_3O^+] = \sqrt{\frac{6.34 \times 10^{-8} \times 1.0 \times 10^{-2}}{1.00 + (1.0 \times 10^{-2})/(7.11 \times 10^{-3})}} = 1.6 \times 10^{-5}$$

## 11F   TITRATION CURVES FOR POLYFUNCTIONAL ACIDS

Compounds with two or more acidic functional groups yield multiple end points in a titration, provided the functional groups differ sufficiently in their strengths as acids. The computational techniques described in Chapter 10 permit derivation of reasonably accurate theoretical titration curves for polyprotic acids if the ratio $K_1/K_2$ is somewhat greater than $10^3$. If this ratio is smaller, the error, particularly in the region of the first equivalence point, becomes excessive, and a more rigorous treatment of the equilibrium relationships is required.

Figure 11–2 shows the titration curve for a diprotic acid $H_2A$ with dissociation constants of $1.00 \times 10^{-3}$ and $1.00 \times 10^{-7}$. Because $K_1/K_2$ is significantly greater than $10^3$, we can derive this curve (except for the first equivalence point) using the techniques developed in Chapter 10 for monoprotic weak acids. Thus, to obtain the initial pH (point $A$), we treat the system as if it contained a single monoprotic acid with a dissociation constant of $1.00 \times 10^{-3}$. In region $B$ we have the equivalent of a simple buffer solution consisting of the weak acid $H_2A$ and its conjugate base NaHA. That is, we assume that the concentration of $A^{2-}$ is negligible with respect to that of the other two A-containing species. At the first equivalence point ($C$), we have a solution of an amphiprotic salt and so use Equation 11–9 or 11–10 to compute the hydronium ion concentration. In the region labeled $D$, we have a second buffer, this one consisting of a weak acid $HA^-$ and its conjugate base $Na_2A$, and we calculate the pH using the second dissociation constant, $1.00 \times 10^{-7}$. At point $E$, the solu-

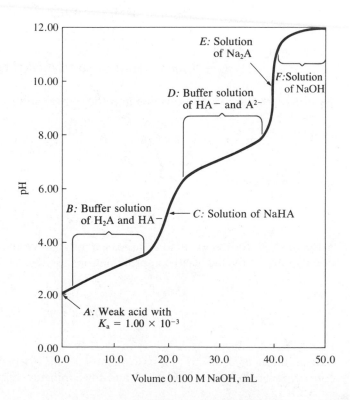

**Figure 11–2**
Titration of 20.0 mL of 0.100 M $H_2A$ with 0.100 M NaOH. For $H_2A$, $K_1 = 1.00 \times 10^{-3}$ and $K_2 = 1.00 \times 10^{-7}$. Method of pH calculation is shown for several points and regions on the titration curve.

tion contains the conjugate base of a weak acid with a dissociation constant of $1.00 \times 10^{-7}$. That is, we assume that the hydroxide ion concentration of the solution is determined solely by the reaction of $A^{2-}$ with water to form $HA^-$ and $OH^-$. Finally, in the region labeled $F$, we compute the hydroxide ion concentration from the molarity of the NaOH and find the pH from this quantity.

---

### Example 11-8

Derive a curve for the titration of 25.00 mL of 0.1000 M maleic acid, $C_2H_2(COOH)_2$, with 0.1000 M NaOH.

The two dissociation equilibria can be written as

$$H_2M + H_2O \rightleftarrows H_3O^+ + HM^- \qquad K_1 = 1.20 \times 10^{-2}$$
$$HM^- + H_2O \rightleftarrows H_3O^+ + M^{2-} \qquad K_2 = 5.96 \times 10^{-7}$$

where $H_2M$ symbolizes the free acid. Because the ratio $K_1/K_2$ is large ($2 \times 10^4$), we proceed as just described.

Initial pH

Only the first dissociation makes an appreciable contribution to $[H_3O^+]$; thus,

$$[H_3O^+] = [HM^-]$$

Mass balance requires that

$$[H_2M] + [HM^-] = 0.1000$$

or

$$[H_2M] = 0.1000 - [HM^-] = 0.1000 - [H_3O^+]$$

Substituting these relationships into the expression for $K_1$ gives

$$K_1 = 1.20 \times 10^{-2} = \frac{[H_3O^+]^2}{0.1000 - [H_3O^+]}$$

Rearranging yields

$$[H_3O^+]^2 + 1.20 \times 10^{-2}[H_3O^+] - 1.20 \times 10^{-3} = 0$$

Because $K_1$ for maleic acid is large, we must solve the quadratic equation exactly or by successive approximations. When we do so, we obtain

$$[H_3O^+] = 2.92 \times 10^{-2}$$
$$pH = 2 - \log 2.92 = 1.54$$

First Buffer Region

The addition of 5.00 mL of base results in the formation of a buffer consisting of the weak acid $H_2M$ and its conjugate base $HM^-$. To the extent that dissociation of $HM^-$ to give $M^{2-}$ is negligible, the solution can be treated as a simple buffer system. Thus applying Equations 10–11 and 10–12 gives

$$c_{NaHM} \approx [HM^-] = \frac{5.00 \times 0.1000}{30.00} = 1.67 \times 10^{-2} \text{ M}$$

$$c_{H_2M} \approx [H_2M] = \frac{25.00 \times 0.1000 - 5.00 \times 0.1000}{30.00} = 6.67 \times 10^{-2} \text{ M}$$

Substitution of these values into the equilibrium-constant expression for $K_1$ yields a provisional value of $4.8 \times 10^{-2}$ M for $[H_3O^+]$. It is clear, however, that the approximation $[H_3O^+] \ll c_{H_2M}$ or $c_{HM^-}$ is not valid; therefore Equations 10–9 and 10–10 must be used:

$$[HM^-] = 1.67 \times 10^{-2} + [H_3O^+] - [OH^-]$$
$$[H_2M] = 6.67 \times 10^{-2} - [H_3O^+] + [OH^-]$$

Because the solution is quite acidic, the approximation that $[OH^-]$ is very small is surely justified. Substitution of these last two expressions into the dissociation-constant relationship gives

$$\frac{[H_3O^+](1.67 \times 10^{-2} + [H_3O^+])}{6.67 \times 10^{-2} - [H_3O^+]} = 1.20 \times 10^{-2} = K_1$$

$$[H_3O^+]^2 + (2.87 \times 10^{-2})[H_3O^+] - 8.00 \times 10^{-4} = 0$$

$$[H_3O^+] = 1.74 \times 10^{-2}$$

$$pH = 1.76$$

Additional points in the first buffer region can be computed in a similar way.

First Equivalence Point

At the first equivalence point,

$$[HM^-] \approx c_{NaHM} = \frac{2.500}{50.00} = 5.00 \times 10^{-2}$$

Simplification of the numerator in Equation 11–9 is clearly justified. On the other hand, the second term in the denominator is not $\ll 1$. Hence,

$$[H_3O^+] = \sqrt{\frac{K_2 c_{HM^-}}{1 + c_{HM^-}/K_1}} = \sqrt{\frac{5.96 \times 10^{-7} \times 5.00 \times 10^{-2}}{1 + (5.00 \times 10^{-2})/(1.20 \times 10^{-2})}}$$

$$= 7.60 \times 10^{-5}$$

$$pH = 4.12$$

Second Buffer Region

Further additions of base to the solution create a new buffer system consisting of $HM^-$ and $M^{2-}$. When enough base has been added so that the reaction of $HM^-$ with water to give $OH^-$ can be neglected (a few tenths of a milliliter beyond the first equivalence point), the pH of the mixture is readily obtained from $K_2$. With the introduction of 25.50 mL of NaOH, for example,

$$c_{Na_2M} = \frac{(25.50 - 25.00)(0.1000)}{50.50} = \frac{0.050}{50.50} \text{ M}$$

and the molar concentration of NaHM is

$$c_{NaHM} = \frac{(25.00 \times 0.1000) - (25.50 - 25.00)(0.1000)}{50.50} = \frac{2.45}{50.50} \text{ M}$$

Substituting these values into the expression for $K_2$ gives

$$\frac{[H_3O^+](0.050/50.50)}{2.45/50.50} = 5.96 \times 10^{-7}$$

$$[H_3O^+] = 2.92 \times 10^{-5}$$

$$pH = 4.54$$

The assumption that $[H_3O^+]$ is small relative to $c_{HM^-}$ and $c_{M^{2-}}$ is valid.

Second Equivalence Point

After the addition of 50.00 mL of 0.1000 M sodium hydroxide, the solution is 0.0333 M in $Na_2M$. Reaction of the base $M^{2-}$ with water is the predominant equilibrium in the system and the only one that must be taken into account. Thus,

$$M^{2-} + H_2O \rightleftarrows OH^- + HM^-$$

$$\frac{[OH^-][HM^-]}{[M^{2-}]} = \frac{K_w}{K_2} = \frac{1.00 \times 10^{-14}}{5.96 \times 10^{-7}} = 1.68 \times 10^{-8}$$

$$[OH^-] = [HM^-]$$

$$[M^{2-}] = 0.0333 - [OH^-] \approx 0.0333$$

$$\frac{[OH^-]^2}{0.0333} = \frac{1.00 \times 10^{-14}}{5.96 \times 10^{-7}}$$

$$[OH^-] = 2.36 \times 10^{-5}$$

$$pH = 14.00 - (-\log 2.36 \times 10^{-5}) = 9.37$$

Beyond the Second Equivalence Point

Further additions of sodium hydroxide suppress the basic dissociation of $M^{2-}$. The pH is calculated from the concentration of NaOH added in excess of that required for the complete neutralization of $H_2M$. Thus after addition of 51.00 mL of NaOH, we have 1.00 mL excess of 0.1000 M NaOH, and

$$[OH^-] = \frac{1.00 \times 0.1000}{76.00} = 1.32 \times 10^{-3}$$

$$pH = 14.00 - (-\log 1.32 \times 10^{-3}) = 11.12$$

Figure 11–3 is the titration curve for 0.1000 M maleic acid derived as shown in Example 11–8. Two end points are apparent, either of which could in principle be used as a measure of the concentration of the acid. The second end point is clearly more satisfactory, however, because the pH change is more pronounced.

Figure 11–4 shows titration curves for three other polyprotic acids. These curves illustrate that a well-defined end point corresponding to the first equivalence point is observed only when the degree of dissociation of the first proton is sufficiently different from that of the second. The ratio of $K_1$ to $K_2$ for oxalic acid (curve $B$) is approximately 1000. The curve for this titration shows an inflection corresponding to the first equivalence point. However, the magnitude of the pH change is too small to permit precise location of equivalence with an indicator. The second end point, however, provides a means for the accurate determination of oxalic acid concentration.

Curve $A$ in Figure 11–4 is the theoretical titration curve for triprotic phosphoric acid. Here, the ratio $K_1/K_2$ is approximately $10^5$, as is $K_2/K_3$. This results in two well-defined end points, either of which is satisfactory for analytical purposes. An acid-range indicator will provide a color change when 1 mol of base has been introduced for 1 mol of acid; a base-range indicator will require 2 mol of base for 1 mol of acid. The third hydrogen of phosphoric acid is so slightly dissociated ($K_3 = 4.2 \times 10^{-13}$) that no practical end point is associated with its neutralization. The buffering effect of the third dissociation is noticeable, however, and causes the pH for curve $A$ to be lower than the pH for the other two curves in the region beyond the second equivalence point.

Curve $C$ is the titration curve for sulfuric acid, a substance that has one fully dissociated proton and one that is dissociated to a relatively large extent ($K_2 = 1.2 \times 10^{-2}$). Because of the similarity in strengths of the two acids, only a single end point, corresponding to the titration of both protons, is observed.

To summarize, the titration of acids or bases that have two reactive groups yields two end points of practical value only when the ratio between the two dissociation constants is at least $10^4$. If the ratio is much

Figure 11–3

Titration curve for 25.00 mL of 0.1000 M maleic acid, $H_2M$, with 0.1000 M NaOH.

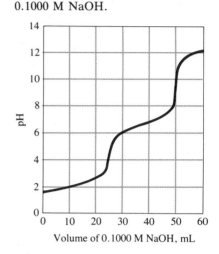

In titrating a polyprotic acid or base, two usable end points are obtained if the ratio of dissociation constants is greater than $10^4$ and if the weaker acid or base has a dissociation constant greater than $10^{-8}$.

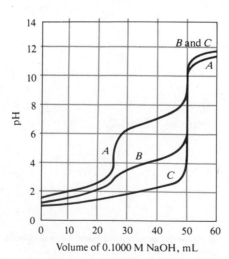

Figure 11–4

Curves for the titration of polyprotic acids. A 0.1000 M NaOH solution is used to titrate 25.00 mL of 0.1000 M $H_3PO_4$ ($A$), 0.1000 M oxalic acid ($B$), and 0.1000 M $H_2SO_4$ ($C$).

Challenge: Derive a titration curve for 50.0 mL of 0.0500 M $H_2SO_4$ with 0.100 M NaOH.

**Feature 11–2**
THE DISSOCIATION OF SULFURIC ACID

Sulfuric acid is unusual in that one of its protons behaves as a strong acid and the other as a weak acid ($K_2 = 1.20 \times 10^{-2}$). Let us see how the hydronium ion concentration of sulfuric acid solutions is computed, using as an example a solution that has an analytical concentration of 0.0400 M.

If the dissociation of $HSO_4^-$ is assumed to be negligible, it follows that

$$[H_3O^+] = [HSO_4^-] = 0.0400$$

However, an estimate of $[SO_4^{2-}]$ based upon this approximation and the expression for $K_2$ reveals that

$$\frac{\cancel{0.0400}[SO_4^{2-}]}{\cancel{0.0400}} = 1.20 \times 10^{-2}$$

Clearly, $[SO_4^{2-}]$ is *not* small relative to $[HSO_4^-]$, and so a more rigorous solution is required.

Stoichiometric considerations require that

$$[H_3O^+] = 0.0400 + [SO_4^{2-}]$$

The first term on the right is the concentration of $H_3O^+$ from dissociation of the $H_2SO_4$ to $HSO_4^-$. The second term is the contribution of the dissociation of $HSO_4^-$. Rearrangement yields

$$[SO_4^{2-}] = [H_3O^+] - 0.0400$$

Mass-balance considerations require that

$$c_{H_2SO_4} = 0.0400 = [HSO_4^-] + [SO_4^{2-}]$$

Combining the last two equations and rearranging yield

$$[HSO_4^-] = 0.0800 - [H_3O^+]$$

Introduction of these equations for $[SO_4^{2-}]$ and $[HSO_4^-]$ into the expression for $K_2$ yields

$$\frac{[H_3O^+]([H_3O^+] - 0.0400)}{0.0800 - [H_3O^+]} = 1.20 \times 10^{-2}$$
$$[H_3O^+]^2 - (0.0280)[H_3O^+] - 9.60 \times 10^{-4} = 0$$
$$[H_3O^+] = 0.0480$$

smaller than this, the pH change at the first equivalence point will prove unsatisfactory for analysis.

## 11G   TITRATION CURVES FOR POLYFUNCTIONAL BASES

The derivation of a titration curve for a polyfunctional base introduces no new principles. To illustrate, consider the titration of a sodium carbonate solution with standard hydrochloric acid. The important equilibrium constants are

$$CO_3^{2-} + H_2O \rightleftarrows OH^- + HCO_3^- \qquad K_{b1} = \frac{K_w}{K_2} = 2.13 \times 10^{-4}$$

$$HCO_3^- + H_2O \rightleftarrows OH^- + H_2CO_3 \qquad K_{b2} = \frac{K_w}{K_1} = 2.25 \times 10^{-8}$$

The reaction of carbonate ion with water governs the initial pH of the solution, which can be computed by the method shown for the second equivalence point in Example 11–8. With the first additions of acid, a carbonate/hydrogen carbonate buffer is established. In this region, the pH can be derived from *either* the hydroxide ion concentration calculated from $K_{b1}$ *or* the hydronium ion concentration calculated from $K_2$.

Sodium hydrogen carbonate is the principal solute species at the first equivalence point, and Equation 11–10 is used to compute the hydronium ion concentration. With the addition of more acid, a new buffer consisting of sodium hydrogen carbonate and carbonic acid is formed. The pH of this buffer is readily obtained from either $K_{b2}$ or $K_1$.

At the second equivalence point, the solution consists of carbonic acid and sodium chloride. The carbonic acid can be treated as a simple weak acid having a dissociation constant $K_1$. Finally, after excess hydrochloric acid has been introduced, the dissociation of the weak acid is suppressed to a point where the hydronium ion concentration is essentially that of the molar concentration of the strong acid.

Figure 11–5 illustrates that two end points are observed in the titration of sodium carbonate, the second being appreciably sharper than the first.

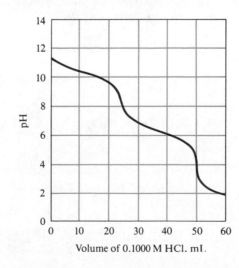

Volume of 0.1000 M HCl. mL

**Figure 11–5**
Curve for the titration of 25.00 mL of 0.1000 M $Na_2CO_3$ with 0.1000 M HCl.

It is apparent that the individual components in mixtures of sodium carbonate and sodium hydrogen carbonate can be determined by neutralization methods.

## 11H TITRATION CURVES FOR AMPHIPROTIC SPECIES

As noted earlier, an amphiprotic substance, when dissolved in a suitable solvent, behaves both as a weak acid and as a weak base. If either its acidic or its basic character predominates sufficiently, titration of the species with a strong base or a strong acid may be feasible. For example, in sodium dihydrogen phosphate solution, the following equilibria exist:

$$H_2PO_4^- + H_2O \rightleftharpoons H_3O^+ + HPO_4^{2-} \qquad K_{a2} = 6.34 \times 10^{-8}$$

$$H_2PO_4^- + H_2O \rightleftharpoons OH^- + H_3PO_4 \qquad K_{b3} = \frac{K_w}{K_{a1}} = 1.41 \times 10^{-12}$$

Note that $K_{b3}$ is much too small to permit titration of $H_2PO_4^-$ with an acid, but $K_{a2}$ is large enough for a successful titration of the ion with a standard base solution.

A different situation prevails in solutions containing disodium hydrogen phosphate, for which the analogous equilibria are

$$HPO_4^{2-} + H_2O \rightleftharpoons H_3O^+ + PO_4^{3-} \qquad K_{a3} = 4.2 \times 10^{-13}$$

$$HPO_4^{2-} + H_2O \rightleftharpoons OH^- + H_2PO_4^- \qquad K_{b2} = \frac{K_w}{K_{a2}} = 1.58 \times 10^{-7}$$

The magnitude of the constants indicates that $HPO_4^{2-}$ can be titrated with standard acid but not with standard base.

The simple amino acids are an important class of amphiprotic compounds that contain both a weak acid and a weak base functional group. In an aqueous solution of a typical amino acid, such as glycine, three important equilibria operate:

$$NH_2CH_2COOH \rightleftharpoons NH_3^+CH_2COO^- \qquad (11-11)$$

$$NH_3^+CH_2COO^- + H_2O \rightleftharpoons NH_2CH_2COO^- + H_3O^+ \qquad K_a = 2 \times 10^{-10}$$
$$(11-12)$$

$$NH_3^+CH_2COO^- + H_2O \rightleftharpoons NH_3^+CH_2COOH + OH^- \qquad K_b = 2 \times 10^{-12}$$
$$(11-13)$$

The first reaction constitutes a kind of internal acid/base reaction and is analogous to the reaction one would observe between a carboxylic acid and an amine:

$$R_1NH_2 + R_2COOH \rightleftharpoons R_1NH_3^+ + R_2COO^- \qquad (11-14)$$

Amino acids are amphiprotic.

The typical aliphatic amine has a base dissociation constant of $10^{-4}$ to $10^{-5}$ (Appendix 4), while many carboxylic acids have acid dissociation constants of about the same magnitude. The consequence is that both reaction 11–11 and reaction 11–14 proceed far to the right, with the product or products being the predominant species in the solution.

The amino acid species in Equation 11–11, bearing both a positive and a negative charge, is called a *zwitterion*. As shown by Equations 11–12 and 11–13, the zwitterion of glycine is slightly stronger as an acid than as a base. Thus, an aqueous solution of glycine is slightly acidic.

*A zwitterion is an ionic species that has both a positive and a negative charge.*

The zwitterion of an amino acid, containing as it does a positive and a negative charge, has no tendency to migrate to an electric field, whereas the singly charged anionic and cationic species are attracted to electrodes of opposite charge. No *net* migration of the amino acid occurs in an electric field when the pH of the solvent is such that the concentrations of the anionic and cationic forms are identical. The pH at which no net migration occurs is called the *isoelectric point* and is an important physical constant for characterizing amino acids. The isoelectric point is readily related to the ionization constants for the species. Thus, for glycine

*The isoelectric point is the pH at which no net migration of amino acids occurs when they are placed in an electric field.*

$$\frac{[H_3O^+][NH_2CH_2COO^-]}{[NH_3^+CH_3COO^-]} = K_a$$

$$\frac{[OH^-][NH_3^+CH_2COOH]}{[NH_3^+CH_2COO^-]} = K_b$$

At the isoelectric point,

$$[NH_2CH_2COO^-] = [NH_3^+CH_2COOH]$$

Thus, division of $K_a$ by $K_b$ gives

$$\frac{[H_3O^+][\cancel{NH_2CH_2COO^-}]}{[OH^-][\cancel{NH_3^+CH_2COOH}]} = \frac{[H_3O^+]}{[OH^-]} = \frac{K_a}{K_b}$$

Substitution of $K_w/[H_3O^+]$ for $[OH^-]$ and rearrangement yield

$$[H_3O^+] = \sqrt{\frac{K_a K_w}{K_b}}$$

The isoelectric point for glycine occurs at a pH of 6.0. That is,

$$[H_3O^+] = \left(\frac{2 \times 10^{-10}}{2 \times 10^{-12}} \times 1 \times 10^{-14}\right)^{1/2} = 1 \times 10^{-6}$$

For simple amino acids, $K_a$ and $K_b$ are generally so small that their determination by direct neutralization titration is impossible. Addition of formaldehyde removes the amine functional group, however, and leaves the carboxylic acid available for titration with a standard base. For example, with glycine,

$$NH_3^+CH_2COO^- + CH_2O \rightarrow CH_2{=}NCH_2COOH + H_2O$$

The titration curve for the product is that of a typical carboxylic acid.

## 11I    THE COMPOSITION OF POLYPROTIC-ACID SOLUTIONS AS A FUNCTION OF pH

In Section 10F, we showed how alpha values are useful in visualizing the various concentration changes that occur in a titration of a simple weak acid. Alpha values can also be derived for polyfunctional acids and bases. For example, if we let $c_T$ be the sum of the molar concentrations of maleate-containing species throughout the titration described in Example 11–8, the alpha value for the free acid is

$$\alpha_0 = \frac{[H_2M]}{c_T}$$

where

$$c_T = [H_2M] + [HM^-] + [M^{2-}] \tag{11-15}$$

The alpha values for $HM^-$ and $M^{2-}$ are given by similar equations:

$$\alpha_1 = \frac{[HM^-]}{c_T}$$

$$\alpha_2 = \frac{[M^{2-}]}{c_T}$$

As noted earlier, the sum of the alpha values for a system must equal unity:

$$\alpha_0 + \alpha_1 + \alpha_2 = 1$$

The alpha values for the maleic acid system are readily expressed in terms of $[H_3O^+]$, $K_1$, and $K_2$. To obtain such expressions, we follow the method used to derive Equations 10–20 and 10–21 in Section 10F and obtain

$$\alpha_0 = \frac{[H_3O^+]^2}{[H_3O^+]^2 + K_1[H_3O^+] + K_1K_2} \tag{11-16}$$

$$\alpha_1 = \frac{K_1[H_3O^+]}{[H_3O^+]^2 + K_1[H_3O^+] + K_1K_2} \tag{11-17}$$

$$\alpha_2 = \frac{K_1K_2}{[H_3O^+]^2 + K_1[H_3O^+] + K_1K_2} \tag{11-18}$$

Note that the denominator is the same for each expression. Note also that the fractional amount of each species is fixed at any pH and is *independent* of the total concentration $c_T$.

---

**Feature 11–3**
**A GENERAL EXPRESSION FOR ALPHA VALUES**

For the weak acid $H_nA$, the denominator in all alpha-value expressions takes the form:

$$[H_3O^+]^n + K_1[H_3O^+]^{(n-1)} + K_1K_2[H_3O^+]^{(n-2)} + \cdots + K_1K_2 \cdots K_n$$

The numerator for $\alpha_0$ is the first term in the denominator, the numerator for $\alpha_1$ is the second term, and so forth. Thus, if we let $D$ be the denominator, $\alpha_0 = [H_3O^+]^n/D$ and $\alpha_1 = K_1[H_3O^+]^{(n-1)}/D$.

  Alpha values for polyfunctional bases are generated in an analogous way, with the equations being written in terms of base dissociation constants and $[OH^-]$.

---

The three curves plotted in Figure 11–6 show the alpha value for each maleate-containing species as a function of pH. The solid curves in Figure 11–7 depict the same alpha values but now plotted as a function of sodium hydroxide volume as the acid is titrated. The titration curve is shown by the dashed line. Consideration of these curves gives a clear picture of all concentration changes that occur during the titration. For example, Figure 11–7 reveals that, before the addition of any base, $\alpha_0$ for $H_2M$ is roughly 0.7, $\alpha_1$ for $HM^-$ is approximately 0.3, and for all practical purposes, $\alpha_2$ is zero. Thus, approximately 70% of the maleic acid exists as $H_2M$ and 30% as $HM^-$. With addition of base, the pH rises, as does the

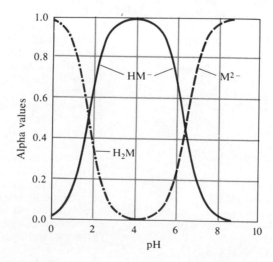

Figure 11–6

Composition of $H_2M$ solutions as a function of pH.

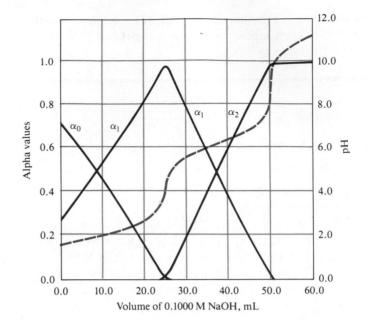

**Figure 11–7**

Titration of 25.00 mL of 0.1000 M maleic acid with 0.1000 M NaOH. The solid curves are plots of alpha values as a function of volume. The broken curve is a plot of pH as a function of volume.

fraction of $HM^-$. At the first equivalence point (pH = 4.12), essentially all of the maleate is present as $HM^-$ ($\alpha_1 \to 1$). Beyond the first equivalence point, $HM^-$ decreases and $M^{2-}$ increases. At the second equivalence point (pH = 9.37) and beyond, essentially all of the maleate exists as $M^{2-}$.

## 11J  QUESTIONS AND PROBLEMS

**11–1.** Consider solutions of $H_2A$ ($K_1 = 10^{-2}$, $K_2 = 10^{-6}$) and species derived from it. Briefly explain why
  **(a)** dissociation of the second hydrogen can be neglected in a solution that contains an appreciable amount of $H_2A$.
  **(b)** calculation of $[H_3O^+]$ for a $1.00 \times 10^{-2}$ M solution of $H_2A$ will require use of the quadratic equation.
  **(c)** solutions containing $H_2A$ and $A^{2-}$ as principal solute species cannot exist.

**11–2.** What simplification(s) to Equation 11–9 appear reasonable in calculation of $[H_3O^+]$ for a $1.00 \times 10^{-2}$ M solution of NaHA (see Problem 11–1)?

**\*11–3.** Suggest an indicator (Table 10–1) that would be suitable for the titration of
  **(a)** $Na_3PO_4$ to the $H_2PO_4^-$ end point.
  **(b)** $H_2C_2O_4$ to the $C_2O_4^{2-}$ end point.
  **(c)** $H_2NCH_2CH_2NH_2$ to the $H_3NCH_2CH_2NH_3^{2+}$ end point.
  **(d)** $Na_2CO_3$ to the $H_2CO_3$ end point.
  **(e)** HCl in a mixture of that acid and HOCl.
  **(f)** the total acidity of the mixture in (e).

**11–4.** Select an indicator (Table 10–1) for the titration of
  **(a)** $H_3PO_4$ to the $H_2PO_4^-$ end point.
  **(b)** $H_3NCH_2CH_2NH_3^{2+}$ to the $H_2NCH_2CH_2NH_2$ end point.

  **(c)** potassium hydrogen phthalate (KHP) to the phthalate ($P^{2-}$) end point.
  **(d)** phosphorous acid to the $HPO_3^{2-}$ end point.
  **(e)** NaOH in a mixture of that base and methylamine.
  **(f)** the total basicity of the mixture in (e).

**\*11–5.** Calculate the pH of a 0.0200 M solution of
  **(a)** malonic acid.
  **(b)** sulfurous acid.
  **(c)** *o*-phthalic acid.
  **(d)** carbonic acid.

**11–6.** Calculate the pH of a 0.0200 M solution of
  **(a)** fumaric acid.
  **(b)** phosphorous acid.
  **(c)** oxalic acid.
  **(d)** arsenous acid.

**\*11–7.** Calculate the pH of a 0.0500 M solution of
  **(a)** potassium hydrogen malonate.
  **(b)** sodium hydrogen sulfite.
  **(c)** potassium hydrogen phthalate.
  **(d)** sodium hydrogen carbonate.

**11–8.** Calculate the pH of a 0.0500 M solution of
  **(a)** potassium hydrogen fumarate.
  **(b)** potassium hydrogen phosphite.
  **(c)** sodium hydrogen oxalate.
  **(d)** sodium hydrogen arsenite.

**\*11–9.** Calculate the pH of a 0.0250 M solution of
  **(a)** potassium malonate.
  **(b)** sodium phosphite.
  **(c)** potassium phthalate.
  **(d)** sodium carbonate.

**11–10.** Calculate the pH of a 0.0250 M solution of
  **(a)** sodium fumarate.
  **(b)** potassium phosphite.
  **(c)** sodium oxalate.
  **(d)** sodium arsenite.

**\*11–11.** Calculate the pH for a 0.120 M solution of
  **(a)** $H_3PO_4$.
  **(b)** $NaH_2PO_4$.
  **(c)** $Na_2HPO_4$.
  **(d)** $Na_3PO_4$.

**11–12.** Calculate the pH of a 0.180 M solution of
  **(a)** $H_3AsO_4$.
  **(b)** $NaH_2AsO_4$.
  **(c)** $Na_2HAsO_4$.
  **(d)** $Na_3AsO_4$.

**\*11–13.** Identify the one (or perhaps two) principal phosphate-containing species and its (their) approximate amount(s) in the solution that results from mixing 40.0 mL of 0.0900 M $H_3PO_4$ with 60.00 mL of
  **(a)** 0.0600 M NaOH.
  **(b)** 0.0400 M $NaH_2PO_4$.
  **(c)** 0.1200 M $Na_2HPO_4$.
  **(d)** 0.0800 M $Na_3PO_4$.
  **(e)** 0.0800 M $NaH_2PO_4$.

**11–14.** Identify the one (or perhaps two) principal phosphate-containing species and its (their) approximate amount(s) in the solution that results from mixing 80.0 mL of 0.1000 M $Na_3PO_4$ with 20.00 mL of
  **(a)** 0.400 M HCl.
  **(b)** 0.400 M $Na_2HPO_4$.
  **(c)** 0.600 M $NaH_2PO_4$.
  **(d)** 0.600 M $H_3PO_4$.
  **(e)** 0.100 M $H_3PO_4$.

**\*11–15.** Calculate the pH of a solution produced by mixing 50.00-mL of 0.0800 M sulfurous acid with 40.00 mL of
  **(a)** 0.0500 M NaOH.
  **(b)** 0.1000 M NaOH.
  **(c)** 0.1500 M NaOH.
  **(d)** 0.2000 M NaOH.

**11–16.** Calculate the pH of a solution prepared by mixing 60.00 mL of 0.0500 M ethylenediamine with 40.00 mL of
  **(a)** 0.0400 M HCl.
  **(b)** 0.0750 M HCl.
  **(c)** 0.0900 M HCl.
  **(d)** 0.1500 M HCl.

**\*11–17.** Calculate the pH of a solution prepared by mixing 50.00 mL of 0.0800 M $NaH_2PO_3$ with 40.00 mL of
  **(a)** 0.0400 M HCl.
  **(b)** 0.0400 M NaOH.
  **(c)** 0.0600 M $NaHPO_3$.
  **(d)** 0.0600 M $H_3PO_3$.

**11–18.** Calculate the pH of the solution produced by mixing 40.00 mL of 0.0600 M $NaHSO_3$ with an equal volume of
  **(a)** 0.0400 M HCl.
  **(b)** 0.0400 M NaOH.
  **(c)** 0.0500 M $Na_2SO_3$.
  **(d)** 0.0500 M $H_2SO_3$.

**\*11–19.** Calculate the pH of a solution prepared by mixing 25.00 mL of 0.0600 M $NaH_2PO_4$ with 40.00 mL of
  **(a)** 0.0300 M $Na_2HPO_4$.
  **(b)** 0.0300 M $H_3PO_4$.
  **(c)** 0.0300 M $Na_3PO_4$.
  **(d)** 0.0800 M $Na_3PO_4$.

**11–20.** Calculate the pH of a solution prepared by mixing 40.00 mL of 0.0500 M $Na_2HPO_4$ with 20.00 mL of
  **(a)** 0.0800 M $NaH_2PO_4$.
  **(b)** 0.150 M $H_3PO_4$.
  **(c)** 0.0800 M $H_3PO_4$.
  **(d)** 0.0800 M $Na_3PO_4$.

**\*11–21.** What is the pH of the buffer formed by adding 250 mL of 0.150 M potassium hydrogen phthalate (KHP) to 250.0 mL of
  **(a)** 0.0800 M HCl?
  **(b)** 0.0800 M NaOH?

**11–22.** What is the pH of the buffer produced by mixing 100.0 mL of 0.200 M $Na_2HPO_4$ with an equal volume of 0.0800 M
  **(a)** HCl?
  **(b)** NaOH?

**\*11–23.** What weight of $NaH_2PO_4 \cdot 2\ H_2O$ (fw = 156.01) should be added to 500 mL of 0.120 M $Na_2HPO_4$ to produce a buffer with a pH of 7.00?

**11–24.** What weight of $Na_2HPO_4 \cdot 7\ H_2O$ (fw = 268.07) should be added to 400 mL of 0.200 M $NaH_2PO_4$ to produce a buffer with a pH of 7.50?

**\*11–25.** Your laboratory is stocked with the following regents:

3.00 M NaOH

3.00 M HCl

0.600 M $H_3PO_4$

0.600 M $NaH_2PO_4$

0.600 M $Na_2HPO_4$

0.600 M $Na_3PO_4$

How would you prepare 500 mL of a buffer with a pH of 3.00, starting with
  **(a)** $H_3PO_4$ and $NaH_2PO_4$?
  **(b)** $H_3PO_4$ and NaOH?
  **(c)** $NaH_2PO_4$ and HCl?
  **(d)** $Na_2HPO_4$ and HCl?

**11–26.** Use the data in Problem 11–25 to describe how you would prepare 500 mL of a buffer with a pH of 7.50, starting with
  **(a)** $NaH_2PO_4$ and $Na_2HPO_4$.

(b) $NaH_2PO_4$ and $NaOH$.
(c) $Na_2HPO_4$ and $HCl$.
(d) $NaH_2PO_4$ and $Na_3PO_4$.

*11–27. Calculate $\alpha$-values for species derived from
  (a) phosphorous acid in a solution buffered to a pH of 4.50.
  (b) carbonic acid in a solution buffered to a pH of 8.10.
  (c) tartaric acid in a solution buffered to a pH of 4.00.
  (d) arsenic acid in a solution buffered to a pH of 5.60.
  (e) arsenic acid in a solution buffered to a pH of 10.50.

11–28. Calculate $\alpha$-values for species derived from
  (a) oxalic acid in a solution buffered to a pH of 3.70.
  (b) hydrogen sulfide in a solution buffered to a pH of 10.00.
  (c) malic acid in a solution buffered to a pH of 5.00.
  (d) phosphoric acid in a solution buffered to a pH of 3.00.
  (e) phosphoric acid in a solution buffered to a pH of 10.00.

*11–29. Derive a curve for the titration of 50.00 mL of a solution that is 0.1000 M in $HCl$ and 0.0800 M in $HOCl$ with 0.2000 M $NaOH$. Calculate the pH after addition of 0.00, 10.00, 20.00, 24.00, 25.00, 26.00, 35.00, 44.00, 45.00, 46.00, and 50.00 mL of the base.

11–30. Derive a curve for the titration of 50.00 mL of a solution that is 0.0800 M $NaOH$ and 0.100 M in $NaOCl$ with 0.2000 M $HCl$. Calculate the pH after addition of 0.00, 10.00, 19.00, 20.00, 21.00, 30.00, 40.00, 44.00, 45.00, 46.00, and 50.00 mL of titrant.

11–31. Generate a curve for the titration of 50.00 mL of a 0.1000 M solution of the analyte in column A with a 0.2000 M solution of the titrant in column B. Calculate the pH after the addition of 0.00, 12.50, 20.00, 24.00, 25.00, 26.00, 37.50, 45.00, 49.00, 50.00, 51.00, and 60.00 mL of titrant.

| A | B |
| --- | --- |
| *(a) $Na_2CO_3$ | HCl |
| (b) ethylenediamine | HCl |
| *(c) $H_2SO_4$ | NaOH |
| (d) $H_2SO_3$ | NaOH |

## CHAPTER 12

# APPLICATIONS OF NEUTRALIZATION TITRATIONS

Neutralization titrations are widely used for determining the concentration of analytes that either are acids or bases or else are convertible to such species by suitable treatment.[1] Water is the usual solvent for neutralization titrations because it is readily available, inexpensive, and nontoxic. Its low temperature coefficient of expansion is an added virtue. Some analytes, however, are not titratable in aqueous media because their solubilities are too low or because their strengths as acids or bases are not sufficiently great to provide satisfactory end points. The concentration of such substances can often be determined by titration in a solvent other than water.[2]

## 12A  REAGENTS FOR NEUTRALIZATION REACTIONS

Standard solutions for neutralization titrations are always prepared from strong acids or strong bases because this type of reagent provides the sharpest end point.

### 12A–1 Preparation of Standard Acid Solutions

Hydrochloric acid is widely used for the titration of bases. Dilute solutions of the reagent are stable indefinitely and do not cause troublesome precipitation reactions with most cations. It is reported that 0.1 M solutions of HCl can be boiled for as long as 1 hr without loss of acid, provided that the water lost by evaporation is periodically replaced; 0.5 M solutions can be boiled for at least 10 min without significant loss.

---

[1]For a review of applications of neutralization titrations, see D. Rosenthal and P. Zuman, in *Treatise on Analytical Chemistry,* 2nd ed., I. M. Kolthoff and P. J. Elving, Eds., Part I, Vol. 2, Chapter 18. New York: Wiley, 1979.

[2]For a review of nonaqueous acid/base titrimetry, see *Treatise on Analytical Chemistry,* 2nd ed., I. M. Kolthoff and P. J. Elving, Eds., Part I, Vol. 2, Chapters 19A–19E. New York: Wiley, 1979.

Solutions of HCl, HClO$_4$, and H$_2$SO$_4$ are stable indefinitely. Restandardization is never required.

Solutions of perchloric acid and sulfuric acid are also stable and are useful for titrations where chloride ion interferes by forming precipitates. Standard solutions of nitric acid are seldom used because of their oxidizing properties.

Standard acid solutions are ordinarily prepared by diluting an approximate volume of the concentrated reagent and subsequently standardizing the diluted solution against a primary-standard base. Less frequently, the composition of the concentrated acid is established through careful density measurement; a weighed quantity is then diluted to an exact volume. (Tables relating reagent density to composition are found in most chemistry and chemical engineering handbooks.) A stock solution with an exactly known hydrochloric acid concentration can also be prepared by dilution of a quantity of the concentrated reagent with an equal volume of water followed by distillation. Under controlled conditions, the final quarter of the distillate, which is known as *constant-boiling* HCl, has a fixed and known composition, its acid content being dependent only upon atmospheric pressure. For a pressure $P$ between 670 and 780 torr, the weight in air of the distillate that contains exactly one mole of H$_3$O$^+$ is[3]

$$\frac{\text{wt constant-boiling HCl in g}}{\text{mol H}_3\text{O}^+} = 164.673 + 0.02039\,P \quad (12\text{--}1)$$

Standard solutions are prepared by diluting weighed quantities of this acid to accurately known volumes.

## 12A–2 The Standardization of Acids

### Sodium Carbonate

Acids are frequently standardized by titrating weighed quantities of sodium carbonate. Primary-standard-grade sodium carbonate is available commercially or can be prepared by heating purified sodium hydrogen carbonate at 270 to 300°C for 1 hr:

$$2\,\text{NaHCO}_3(s) \rightarrow \text{Na}_2\text{CO}_3(s) + \text{H}_2\text{O}(g) + \text{CO}_2(g)$$

Two end points are observed in the titration of sodium carbonate. The first, at about pH 8.3, corresponds to the conversion of carbonate to hydrogen carbonate; the second, at about pH 3.8, involves the formation of carbonic acid. The second end point is always used for standardization because the change in pH here is greater than at the first end point.

An even sharper end point can be achieved by boiling the solution briefly to eliminate the carbonic acid produced by the reaction. The sample is titrated to the first appearance of the acidic color of an indicator such as bromocresol green or methyl orange. At this point, the solution contains a large amount of carbonic acid and a small amount of unreacted hydrogen carbonate. Boiling effectively destroys this buffer by eliminating the carbonic acid:

---

[3]*Official Methods of Analysis of the AOAC*, 15th ed., p. 692. Washington, D.C.: Association of Official Analytical Chemists, 1990.

$$H_2CO_3(aq) \rightarrow CO_2(g) + H_2O(l)$$

The solution then becomes alkaline again due to the residual hydrogen carbonate ion, as illustrated in Figure 12–1. The titration is completed after the solution has cooled. Now, however, a substantially larger decrease in pH occurs during the final additions of acid, thus giving a more abrupt color change.

As an alternative, the acid solution can be introduced in an amount slightly in excess of that needed to convert the sodium carbonate to carbonic acid. The solution is boiled as before to remove carbon dioxide and cooled; the excess acid is then back-titrated with a dilute solution of base. Any indicator suitable for a strong acid/strong base titration is satisfactory. The volume ratio of acid to base must of course be established by an independent titration.

### Other Primary Standards for Acids

*Tris*-(hydroxymethyl)aminomethane, $(HOCH_2)_3CNH_2$, known also as TRIS or THAM, is available in primary-standard purity from commercial sources. It has the advantage of a substantially greater equivalent weight (121.1) than sodium carbonate (53.0). (See Feature 12–1.)

Sodium tetraborate decahydrate and mercury(II) oxide have also been recommended as primary standards. The reaction of an acid with the tetraborate is

$$B_4O_7^{2-} + 2\,H_3O^+ + 3\,H_2O \rightarrow 4\,H_3BO_3$$

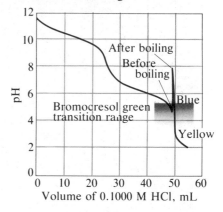

**Figure 12–1**

Titration of 25.00 mL of 0.1000 M $Na_2CO_3$ with 0.1000 M HCl. After about 49 mL of HCl have been added, the solution is boiled, causing the increase in pH shown. The change in pH on further addition of HCl is much larger.

A high formula weight is desirable in a primary standard because a large weight of reagent must be used, thus reducing the relative weighing error.

---

**Feature 12–1**
### EQUIVALENT WEIGHTS OF ACIDS AND BASES

The equivalent weight of a participant in a neutralization reaction is the weight that reacts with or supplies one mole of protons *in a particular reaction*. For example, the equivalent weight of $H_2SO_4$ is one half of its formula weight. The equivalent weight of $Na_2CO_3$ is usually one half of its formula weight because in most applications its reaction is

$$Na_2CO_3 + 2\,H_3O^+ \rightarrow H_2O + H_2CO_3 + 2\,Na^+$$

When titrated with some indicators, however, it consumes but a single proton:

$$Na_2CO_3 + H_3O^+ \rightarrow NaHCO_3 + Na^+$$

Here, the equivalent weight and the formula weight of $Na_2CO_3$ are identical. These observations demonstrate that the equivalent weight of a compound cannot be defined without having a particular reaction in mind (Appendix 10).

## 12A–3 Preparation of Standard Base Solutions

Sodium hydroxide is the most common base for preparing standard solutions, although potassium hydroxide and barium hydroxide are also encountered. None of these is obtainable in primary-standard purity, and so standardization is required after preparation.

### The Effect of Carbon Dioxide upon Standard Base Solutions

In solution as well as in the solid state, the hydroxides of sodium, potassium, and barium react avidly with atmospheric carbon dioxide to produce the corresponding carbonate:

$$CO_2(g) + 2\ OH^- \rightarrow CO_3^{2-} + H_2O$$

Although production of each carbonate ion uses up two hydroxide ions, the uptake of carbon dioxide by a solution of base does not necessarily alter its combining capacity for hydronium ions. Thus, at the end point of a titration that requires an acid-range indicator (such as bromocresol green), each carbonate ion produced from sodium or potassium hydroxide will have reacted with two hydronium ions of the analyte (Figure 11–5):

$$CO_3^{2-} + 2\ H_3O^+ \rightarrow H_2CO_3 + 2\ H_2O$$

Because the amount of hydronium ion consumed by this reaction is identical to the amount of hydroxide lost during formation of the carbonate ion, no error is incurred.

Unfortunately, most applications that use standard base require an indicator with a basic transition range (phenolphthalein, for example). Here, each carbonate ion has reacted with only one hydronium ion when the color change of the indicator is observed:

$$CO_3^{2-} + H_3O^+ \rightarrow HCO_3^- + H_2O$$

The effective concentration of the base is thus diminished by absorption of carbon dioxide, and a determinate error (called a *carbonate error*) results.

The solid reagents used to prepare standard solutions of base are always contaminated by significant amounts of carbonate ion. Thus, freshly prepared solutions of base contain substantial quantities of carbonate ion unless steps are taken to remove this species. The presence of this contaminant does not cause a carbonate error if the same indicator is used for both standardization and analysis. It does, however, lead to less sharp end points. Consequently, it is customary to remove carbonate ion before a solution of a base is standardized.

The best method for preparing carbonate-free sodium hydroxide solutions takes advantage of the very low solubility of sodium carbonate in concentrated solutions of the reagent. An approximately 50% aqueous solution of sodium hydroxide is prepared (or purchased from commercial sources). The solid sodium carbonate is allowed to settle, and then the

A negative determinate error results from the absorption of carbon dioxide by a standardized solution of sodium or potassium hydroxide in titrations that require a basic range indicator. No determinate error is incurred with an acidic range indicator.

Generally, carbonate ion in standard solutions of bases is undesirable because it decreases the sharpness of end points.

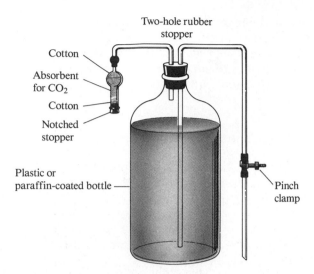

Two-hole rubber stopper

Cotton

Absorbent for $CO_2$

Cotton

Notched stopper

Plastic or paraffin-coated bottle

Pinch clamp

Figure 12–2
Arrangement for the storage of standard base solutions.

clear liquid is decanted and diluted to give the desired concentration (alternatively, the solid is removed by vacuum filtration).

Water for preparing carbonate-free solutions of base must be free of carbon dioxide. Distilled water, which is sometimes supersaturated with carbon dioxide, should be boiled briefly to eliminate the gas. The water is then allowed to cool to room temperature before the introduction of base because hot alkali solutions rapidly absorb carbon dioxide. Deionized water ordinarily does not contain significant amounts of carbon dioxide.

Standard solutions of base are reasonably stable as long as they are protected from contact with the atmosphere. Figure 12–2 shows an arrangement for preventing the uptake of atmospheric carbon dioxide during storage and when the reagent is dispensed. Air entering the vessel is passed over a solid absorbent for $CO_2$, such as soda lime or Ascarite II®.[4] The contamination that occurs as the solution is transferred from this storage bottle to the buret is ordinarily negligible.

A tightly capped polyethylene bottle usually provides sufficient short-term protection against the uptake of atmospheric carbon dioxide. Before being capped, the bottle is squeezed to minimize the interior air space. Care should also be taken to keep the bottle closed except during the brief periods when the contents are being transferred to a buret. Sodium hydroxide solutions will ultimately cause a polyethylene bottle to become brittle.

The concentration of sodium hydroxide solutions decreases slowly (by 0.1 to 0.3% per week) when the base is stored in glass bottles. The loss in strength is caused by the reaction of the base with the glass to form sodium silicates. For this reason, standard solutions of base should not be stored for extended periods (longer than one or two weeks) in glass containers. In addition, glass-stoppered containers should be avoided be-

Solutions of bases are preferably stored in polyethylene bottles. Such solutions should never be stored in glass-stoppered bottles because removal of the stopper often becomes impossible after a brief period.

[4]Thomas Scientific, Swedesboro, NJ. Ascarite II® consists of sodium hydroxide deposited on a nonfibrous silicate structure.

cause the reaction between the base and the stopper may cause the latter to "freeze" after a brief period. Finally, to avoid the same type of freezing, burets with glass stopcocks should be promptly drained and thoroughly rinsed with water after use with standard base solutions. This is not a problem with burets equipped with Teflon stopcocks.

### 12A-4  The Standardization of Bases

Standard solutions of bases cannot be prepared directly by weight and must always be standardized against a primary standard or a standard acid solution.

Several excellent primary standards are available for the standardization of bases. Most are weak organic acids that require the use of an indicator with a basic transition range.

### Potassium Hydrogen Phthalate, $KHC_8H_4O_4$

Potassium hydrogen phthalate is an ideal primary standard. It is a nonhygroscopic crystalline solid with a high equivalent weight (204.2). For most purposes, the commercial analytical-grade salt can be used without further purification. For the most exacting work, potassium hydrogen phthalate of certified purity is available from the National Institute of Standards and Technology.

### Other Primary Standards for Bases

Benzoic acid is obtainable in primary-standard purity and can be used for the standardization of bases. Because its solubility in water is limited, this reagent is ordinarily dissolved in ethanol prior to dilution with water and titration. A blank should always be carried through this standardization procedure because commercial alcohol is sometimes slightly acidic.

Potassium hydrogen iodate, $KH(IO_3)_2$, is an excellent primary standard with a high equivalent weight. It is also a strong acid that can be titrated using virtually any indicator with a transition range between pH 4 and 10.

### 12B  TYPICAL APPLICATIONS OF NEUTRALIZATION TITRATIONS

Neutralization titrations are among the most widely used analytical methods.

Neutralization titrations are used to determine the innumerable inorganic, organic, and biological species that have inherent acidic or basic properties. Of equal importance are the many applications that involve conversion of an analyte to an acid or base by suitable chemical treatment followed by titration with a standard strong base or acid.

Two major types of end points find widespread use in neutralization titrations. The first is a visual end point based on indicators such as those described in Section 10A. The second is a *potentiometric* end point, in which the potential of a glass/calomel electrode system is determined with a voltage-measuring device. The measured potential is directly proportional to pH. Potentiometric end points are described in Chapter 17.

## 12B–1 Elemental Analysis

Several important elements that occur in organic and biological systems are conveniently determined by methods that involve an acid/base titration as the final step. Generally, the elements susceptible to this type of analysis are nonmetallic and include carbon, nitrogen, sulfur, chlorine, bromine, and fluorine as well as a few other less common species. In each instance, the element is converted to an inorganic acid or base that is then titrated. A few examples follow.

### Nitrogen

Nitrogen is found in numerous substances of interest in research, industry, and agriculture. For example it is found in amino acids, proteins, synthetic drugs, fertilizers, explosives, soils, potable water supplies, and dyes. Thus analytical methods for the determination of nitrogen, particularly in organic substrates, are very important.

The most common method for determining organic nitrogen is the *Kjeldahl method,* which is based on a neutralization titration. The procedure is straightforward, requires no special equipment, and is readily adapted to the routine analysis of large numbers of samples. It is the standard means for determining the protein content of grains, meats, and other biological materials. Since most proteins contain approximately the same percentage of nitrogen, multiplication of this percentage by a suitable factor (6.25 for meats, 6.38 for dairy products, and 5.7 for cereals) gives the percentage of protein in a sample.

Hundreds of thousands of Kjeldahl nitrogen determinations are performed each year, primarily to provide a measure of the protein content of meats, grains, and animal feeds.

---

**Feature 12–2**
**OTHER METHODS FOR DETERMINING ORGANIC NITROGEN**

Two other methods are used to determine the nitrogen content of organic materials. In the *Dumas method,* the sample is mixed with powdered copper(II) oxide and ignited in a combustion tube to give carbon dioxide, water, nitrogen, and small amounts of nitrogen oxides. A stream of carbon dioxide carries these products through a packing of hot copper, which reduces any oxides of nitrogen to elemental nitrogen. The mixture is then passed into a gas buret filled with concentrated potassium hydroxide. The only component not absorbed by the base is nitrogen, and its volume is measured directly.

The newest method for determining organic nitrogen involves combusting the sample at 1100°C for a few minutes to convert the nitrogen to nitric oxide, NO. Ozone is then introduced into the gaseous mixture, which oxidizes the nitric oxide to nitrogen dioxide. This reaction gives off visible radiation (*chemiluminescence*), the intensity of which is proportional to the nitrogen content of the sample. An instrument for this procedure is available from commercial sources.

The Kjeldahl method was developed by a Danish chemist who first described it in 1883: J. Kjeldahl, *Z. Anal. Chem.* **1883,** *22,* 366.

In the Kjeldahl method, the sample is decomposed in hot, concentrated sulfuric acid to convert the bound nitrogen to ammonium ion. The resulting solution is then cooled, diluted, and made basic. The liberated ammonia is distilled, collected in an acidic solution, and determined by a neutralization titration.

The critical step in the Kjeldahl method is the decomposition with sulfuric acid, which oxidizes the carbon and hydrogen in the sample to carbon dioxide and water. The fate of the nitrogen, however, depends upon its state of combination in the original sample. Amine and amide nitrogen is quantitatively converted to ammonium ion. In contrast, nitro, azo, and azoxy groups are likely to yield the element or various nitrogen oxides, all of which are lost from the hot acidic medium. This loss can be avoided by first treating the sample with a reducing agent to form reduced products that behave as amide or amine nitrogen. In one such reduction scheme, salicylic acid and sodium thiosulfate are added to the concentrated sulfuric acid solution containing the sample. After a brief period, the digestion is performed in the usual way.

Certain aromatic heterocyclic compounds, such as pyridine and its derivatives, are particularly resistant to complete decomposition by sulfuric acid. Such compounds yield low results as a consequence (Figure 2–3) unless special precautions are taken.

The decomposition step is frequently the most time-consuming aspect of a Kjeldahl determination. Some samples may require heating periods in excess of 1 hr. Numerous modifications of the original procedure have been proposed with the aim of shortening the digestion time. In the most widely used modification, a neutral salt, such as potassium sulfate, is added to increase the boiling point of the sulfuric acid solution and thus the temperature at which the decomposition occurs.

Many substances catalyze the decomposition of organic compounds by sulfuric acid. Mercury, copper, and selenium, either combined or in the elemental state, are effective. Mercury(II), if present, must be precipitated with hydrogen sulfide prior to distillation to prevent retention of ammonia as a mercury(II) ammine complex.

Figure 12–3 illustrates typical equipment for a Kjeldahl distillation. The long-necked container, which is used for both digestion and distillation, is called a *Kjeldahl flask*. After the decomposition is judged complete, the digested mixture is cooled, diluted with water, and made basic to liberate the ammonia. The equilibria are:

$$NH_4^+ + OH^- \rightleftarrows NH_3(aq) + H_2O$$
$$NH_3(aq) \rightleftarrows NH_3(g)$$

In the apparatus shown in Figure 12–3a, the base is added slowly by partially opening the stopcock from the NaOH storage vessel; the liberated ammonia is then carried to the receiving flask by steam distillation.

In an alternative method (Figure 12–3b), a dense, concentrated sodium hydroxide solution is carefully poured down the side of the Kjeldahl flask to form a second, lower layer. The flask is then quickly connected to a spray trap and an ordinary condenser before loss of ammonia can occur. Only then are the two layers mixed by gentle swirling of the flask.

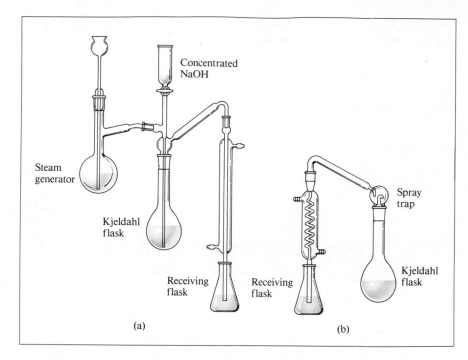

(a)

(b)

Figure 12–3
Kjeldahl distillation apparatus.

Quantitative collection of ammonia requires the tip of the condenser to extend into the liquid in the receiving flask throughout the distillation step. The tip must be removed before heating is discontinued, however. Otherwise, the liquid in the receiver will be drawn back into the apparatus.

Two methods are commonly used for collecting and determining the ammonia liberated from the sample. In one, the ammonia is distilled into a measured volume of standard acid. After the distillation is complete, the excess acid is back-titrated with standard base. An indicator with an acidic transition range is required because of the acidity of the ammonium ions present at equivalence. A convenient alternative, which requires only one standard solution, involves the collection of the ammonia in an unmeasured excess of boric acid, which retains the ammonia by the reaction

$$H_3BO_3 + NH_3 \rightarrow NH_4^+ + H_2BO_3^-$$

The dihydrogen borate ion produced is a reasonably strong base that can be titrated with a standard solution of hydrochloric acid:

$$H_2BO_3^- + H_3O^+ \rightarrow H_3BO_3 + H_2O$$

At the equivalence point, the solution contains boric acid and ammonium ions; an indicator with an acidic transition interval (such as bromocresol green) is again required.

## Sulfur

Sulfur in organic and biological materials is conveniently determined by burning the sample in a stream of oxygen. The sulfur dioxide (as well as any sulfur trioxide) formed during the oxidation is collected by distillation into a dilute solution of hydrogen peroxide:

$$SO_2(g) + H_2O_2 \rightarrow H_2SO_4$$

The sulfuric acid is then titrated with standard base.

### Other Elements

Table 12–1 lists other elements that can be determined by neutralization methods.

## 12B–2 The Determination of Inorganic Substances

Numerous inorganic species can be determined by titration with strong acids or bases. A few examples follow.

### Ammonium Salts

Ammonium salts are conveniently determined by conversion to ammonia with strong base followed by distillation in the Kjeldahl apparatus shown in Figure 12–3. The ammonia is collected and titrated as in the Kjeldahl method.

### Nitrates and Nitrites

The method just described for ammonium salts can be extended to the determination of inorganic nitrate or nitrite. These ions are first reduced to ammonium ion by Devarda's alloy (50% Cu, 45% Al, 5% Zn). Granules of the alloy are introduced into a strongly alkaline solution of the sample in a Kjeldahl flask. The ammonia is distilled after reaction is complete. Arnd's alloy (60% Cu, 40% Mg) has also been used as the reducing agent.

**Table 12–1**
**ELEMENTAL ANALYSES BASED ON NEUTRALIZATION TITRATIONS**

| Element | Converted to | Absorption or Precipitation Products | Titration |
|---|---|---|---|
| N | $NH_3$ | $NH_3(g) + H_3O^+ \rightarrow NH_4^+ + H_2O$ | Excess HCl with NaOH |
| S | $SO_2$ | $SO_2(g) + H_2O_2 \rightarrow H_2SO_4$ | NaOH |
| C | $CO_2$ | $CO_2(g) + Ba(OH)_2 \rightarrow BaCO_3(s) + H_2O$ | Excess $Ba(OH)_2$ with HCl |
| Cl (Br) | HCl | $HCl(g) + H_2O \rightarrow Cl^- + H_3O^+$ | NaOH |
| F | $SiF_4$ | $SiF_4(g) + H_2O \rightarrow H_2SiF_6$ | NaOH |
| P | $H_3PO_4$ | $12 H_2MoO_4 + 3 NH_4^+ + H_3PO_4 \rightarrow$ $(NH_4)_3PO_4 \cdot 12 MoO_3(s) + 12 H_2O + 3 H^+$ $(NH_4)_3PO_4 \cdot 12 MoO_3(s) + 26 OH^- \rightarrow$ $HPO_4^{2-} + 12 MoO_4^{2-} + 14 H_2O + 3 NH_3(g)$ | Excess NaOH with HCl |

Table 12–2
VOLUME RELATIONSHIPS IN THE ANALYSIS OF MIXTURES
CONTAINING HYDROXIDE, CARBONATE, AND HYDROGEN
CARBONATE IONS

| Constituent(s) in Sample | Relationship between $V_{phth}$ and $V_{bcg}$ in the Titration of an Equal Volume of Sample* |
|---|---|
| NaOH | $V_{phth} = V_{bcg}$ |
| $Na_2CO_3$ | $V_{phth} = \frac{1}{2}V_{bcg}$ |
| $NaHCO_3$ | $V_{phth} = 0; V_{bcg} > 0$ |
| NaOH, $Na_2CO_3$ | $V_{phth} > \frac{1}{2}V_{bcg}$ |
| $Na_2CO_3$, $NaHCO_3$ | $V_{phth} < \frac{1}{2}V_{bcg}$ |

*$V_{phth}$ = volume of acid needed for a phenolphthalein end point; $V_{bcg}$ = volume of acid needed for a bromocresol green end point.

## Carbonate and Carbonate Mixtures

The qualitative and quantitative determination of the constituents in a solution containing sodium carbonate, sodium hydrogen carbonate, and sodium hydroxide, either alone or admixed, provides interesting examples of how neutralization titrations can be employed to analyze mixtures. No more than two of these three constituents can exist in appreciable amount in any solution because reaction eliminates the third. Thus, mixing sodium hydroxide with sodium hydrogen carbonate results in the formation of sodium carbonate until one or the other (or both) of the original reactants is exhausted. If the sodium hydroxide is used up, the solution will contain sodium carbonate and sodium hydrogen carbonate; if the sodium hydrogen carbonate is depleted, sodium carbonate and sodium hydroxide will remain; if equimolar amounts of sodium hydrogen carbonate and sodium hydroxide are mixed, the principal solute species will be sodium carbonate.

The analysis of such mixtures requires two titrations; one with an alkaline-range indicator, such as phenolphthalein, and the other with an acid-range indicator, such as bromocresol green. The composition of the solution can then be deduced from the relative volumes of acid needed to titrate equal volumes of the sample (Table 12–2 and Figure 11–5). Once the composition of the solution has been established, the volume data can be used to determine the concentration of each component in the sample.

---

Example 12–1

A solution contains $NaHCO_3$, $Na_2CO_3$, and/or NaOH, either alone or in permissible combination. Titration of a 50.0-mL portion to a phenolphthalein end point requires 22.1 mL of 0.100 M HCl. A second 50.0-mL aliquot requires 48.4 mL of the HCl when titrated to a bromocresol green end point. Deduce the composition, and calculate the molar solute concentrations of the original solution.

If the solution contained only NaOH, the volume of acid required would be the same regardless of indicator (that is, $V_{phth} = V_{bcg}$). Similarly, we can rule out the presence of $Na_2CO_3$ alone because titration of this

Compatible mixtures containing two of the following can also be analyzed in a similar way: HCl, $H_3PO_4$, $NaH_2PO_4$, $Na_2HPO_4$, $Na_3PO_4$, and NaOH.

compound to a bromocresol green end point would require just twice the volume of acid required to reach the phenolphthalein end point. In fact, however, the second titration requires 48.4 mL. Because less than half of this amount is involved in the first titration, the solution must contain some $NaHCO_3$ in addition to $Na_2CO_3$. We can now calculate the concentration of the two constituents.

When the phenolphthalein end point is reached, the $CO_3^{2-}$ originally present is converted to $HCO_3^-$. Thus,

$$\text{amount } Na_2CO_3 = 22.1 \text{ mL} \times 0.100 \text{ mmol/mL} = 2.21 \text{ mmol}$$

The titration from the phenolphthalein end point to the bromocresol green end point (48.4 − 22.1 = 26.3 mL) involves both the hydrogen carbonate originally present and that formed by titration of the carbonate. Thus,

$$\text{amount } NaHCO_3 + \text{amount } Na_2CO_3 = 26.3 \times 0.100 = 2.63 \text{ mmol}$$

Hence,

$$\text{amount } NaHCO_3 = 2.63 - 2.21 = 0.42 \text{ mmol}$$

The molar concentrations are readily calculated from these data:

$$c_{Na_2CO_3} = \frac{2.21 \text{ mmol}}{50.0 \text{ mL}} = 0.0442 \text{ M}$$

$$c_{NaHCO_3} = \frac{0.42 \text{ mmol}}{50.0 \text{ mL}} = 0.0084 \text{ M}$$

How could you analyze a mixture of HCl and $H_3PO_4$? A mixture of $Na_3PO_4$ and $Na_2HPO_4$? See Figure 11–4, curve A.

The method described in Example 12–1 is not entirely satisfactory because the pH change corresponding to the hydrogen carbonate equivalence point is not sufficient to give a sharp color change with a chemical indicator (Figure 11–5). Relative errors of 1% or more must be expected as a consequence.

The accuracy of methods for analyzing solutions containing mixtures of carbonate and hydrogen carbonate ions or carbonate and hydroxide ions can be greatly improved by taking advantage of the limited solubility of barium carbonate in neutral and basic solutions. For example, in the *Winkler method* for the analysis of carbonate/hydroxide mixtures, both components are titrated with a standard acid to the end point with an acid-range indicator, such as bromocresol green (the end point is established after the solution is boiled to remove carbon dioxide). An unmeasured excess of neutral barium chloride is then added to a second aliquot of the sample solution to precipitate the carbonate ion, following which the hydroxide ion is titrated to a phenolphthalein end point. The presence of the sparingly soluble barium carbonate does not interfere as long as the concentration of barium ion is greater than 0.1 M.

Carbonate and hydrogen carbonate ions can be accurately determined in mixtures by first titrating both ions with standard acid to an end point with an acid-range indicator (with boiling to eliminate carbon dioxide). The hydrogen carbonate in a second aliquot is converted to carbonate by the addition of a known excess of standard base. After a large excess of barium chloride has been introduced, the excess base is titrated with standard acid to a phenolphthalein end point.

The presence of solid barium carbonate does not hamper end-point detection in either of these methods.

### 12B–3 The Determination of Organic Functional Groups

Neutralization titrations provide convenient methods for the direct or indirect determination of several organic functional groups. Brief descriptions of methods for the more common groups follow.

#### Carboxylic and Sulfonic Acid Groups

Carboxylic and sulfonic acid groups are the two most common structures that impart acidity to organic compounds. Most carboxylic acids have dissociation constants that range between $10^{-4}$ and $10^{-6}$, and thus these compounds are readily titrated. An indicator that changes color in the basic range, such as phenolphthalein, is required.

Many carboxylic acids are not sufficiently soluble in water to permit direct titration in this medium. Where this problem exists, the acid can be dissolved in ethanol and titrated with aqueous base. Alternatively, the acid can be dissolved in an excess of standard base followed by back-titration with standard acid.

Sulfonic acids are generally strong acids and readily dissolve in water. Their titration with a base is therefore straightforward.

Neutralization titrations are often employed to determine the equivalent weight of purified organic acids (see Feature 12–1). Equivalent weights serve as an aid in qualitative identification of organic acids.

#### Amine Groups

Aliphatic amines generally have base dissociation constants on the order of $10^{-5}$ and can thus be titrated directly with a solution of a strong acid. In contrast aromatic amines, such as aniline and its derivatives, are usually too weak for titration in aqueous medium ($K_b \approx 10^{-10}$). The same is true for cyclic amines with aromatic character, such as pyridine and its derivatives. Many saturated cyclic amines, such as piperidine, tend to resemble aliphatic amines in their acid/base behavior and thus can be titrated in aqueous media.

Many amines that are too weak to be titrated as bases in water are readily titrated in nonaqueous solvents, such as anhydrous acetic acid, which enhance their basicity. These procedures are discussed in Section 12C.

#### Ester Groups

Esters are commonly determined by *saponification* with a measured quantity of standard base:

$$R_1COOR_2 + OH^- \rightarrow R_1COO^- + HOR_2$$

The excess base is then titrated with standard acid.

The term saponification refers to breaking down an ester into its constituent acid and alcohol by treatment with strong base.

Esters vary widely in their rate of saponification. Some require several hours of heating with a base to complete the process. A few react rapidly enough to permit direct titration with the base. Typically, the ester is refluxed with standard 0.5 M KOH for 1 to 2 hr. After cooling, the excess base is determined with standard acid.

## Hydroxyl Groups

Hydroxyl groups in organic compounds can be determined by esterification with various carboxylic acid anhydrides or chlorides; the two most common reagents are acetic anhydride and phthalic anhydride. With acetic anhydride, the reaction is

$$(CH_3CO)_2O + ROH \rightarrow CH_3COOR + CH_3COOH$$

The acetylation is ordinarily carried out by mixing the sample with a carefully measured volume of acetic anhydride in pyridine. After heating, water is added to hydrolyze the unreacted anhydride:

$$(CH_3CO)_2O + H_2O \rightarrow 2\ CH_3COOH$$

The acetic acid is then titrated with a standard solution of alcoholic sodium or potassium hydroxide. A blank is carried through the analysis to establish the original amount of anhydride.

Amines, if present, are converted quantitatively to amides by acetic anhydride; a correction for this source of interference is frequently possible by a direct titration of another portion of the sample with standard acid.

## Carbonyl Groups

Many aldehydes and ketones can be determined with a solution of hydroxylamine hydrochloride. The reaction, which produces an oxime, is

$$\begin{array}{c} R_1 \\ \diagdown \\ \phantom{xx}C{=}O + NH_2OH \cdot HCl \rightarrow \\ \diagup \\ R_2 \end{array} \qquad \begin{array}{c} R_1 \\ \diagdown \\ \phantom{xx}C{=}NOH + HCl + H_2O \\ \diagup \\ R_2 \end{array}$$

where $R_2$ may be an atom of hydrogen. The liberated hydrochloric acid is titrated with base. Here again, the conditions necessary for quantitative reaction vary. Typically, 30 min suffices for aldehydes. Many ketones require refluxing with the reagent for 1 hr or more.

## 12B–4 The Determination of Salts

The total salt content of a solution can be accurately and readily determined by an acid/base titration. The salt is converted to an equivalent amount of an acid or a base by passage through a column packed with an ion-exchange resin. (This application is considered in more detail in Section 32E–3).

Standard acid or base solutions can also be prepared with ion-exchange resins. Here, a solution containing a known weight of a pure compound, such as sodium chloride, is washed through the resin column and diluted to a known volume. The salt liberates an equivalent amount of acid or base from the resin, permitting calculation of the molarity of the reagent in a straightforward way.

## 12C APPLICATION OF NEUTRALIZATION TITRATIONS IN NONAQUEOUS MEDIA

Two types of compounds that are not titratable in aqueous media can be determined by neutralization titration in suitable nonaqueous solvents. The first are high-molecular-weight organic acids and bases that have limited solubility in water. The second are inorganic and organic compounds that are such weak acids or bases ($K_a$ or $K_b < 10^{-8}$) that they do not yield satisfactory end points in aqueous solutions. Examples of the latter include aromatic amines, phenols, and the salts of a variety of inorganic and carboxylic acids. Often compounds that yield unsatisfactory end points in water exhibit sharp end points in solvents that enhance their acidic or basic character.[5]

Although nonaqueous titrations make possible the determination of species that cannot be titrated in water, several disadvantages attend their use. Generally, the solvents are expensive and in addition are often volatile and toxic. Furthermore, most have significantly larger coefficients of expansion than water and require a much closer control of reagent temperature to avoid determinate errors in the measurement of volume.

### 12C-1 Solvents for Nonaqueous Titrations

A wide variety of organic solvents have been employed for nonaqueous acid/base titrations. The most common, and the only ones considered here, are *amphiprotic solvents,* which possess both acidic and basic properties and undergo self-dissociation or *autoprotolysis* to yield an acid and a base. Although water is the most common amphiprotic solvent, many other substances exhibit analogous behavior. Three amphiprotic solvents that have found considerable use in nonaqueous titrimetry are anhydrous acetic acid, ethanol, and ethylenediamine. Their autoprotolysis equilibria are:

$$2\ CH_3COOH \rightleftharpoons CH_3COOH_2^+ + CH_3COO^- \qquad (12\text{--}2)$$

$$2\ C_2H_5OH \rightleftharpoons C_2H_5OH_2^+ + C_2H_5O^- \qquad (12\text{--}3)$$

$$2\ NH_2CH_2CH_2NH_2 \rightleftharpoons NH_2CH_2CH_2NH_3^+ + NH_2CH_2CH_2NH^- \qquad (12\text{--}4)$$

Other examples of autoprotolytic reactions are found on page 121.

---

[5]For more extensive discussions of nonaqueous neutralization titrations, see J. S. Fritz, *Titrations in Nonaqueous Solvents.* Boston: Allyn and Bacon, 1973; I. M. Kolthoff *et al., Treatise on Analytical Chemistry,* 2nd ed., I. M. Kolthoff and P. J. Elving, Eds., Part I, Vol. 2, Chapter 19. New York: Wiley, 1979.

## 12C–2 The Completeness of Acid/Base Reactions in Amphiprotic Solvents

The completeness of a neutralization reaction in an amphiprotic solvent depends not only upon the strength of the analyte as an acid or base but also upon the autoprotolysis constant, the inherent acidity or basicity, and the dielectric constant of the solvent.

### The Effect of Solvent Autoprotolysis Constant

In water, the titration of a weak base B with a strong acid can be formulated as

$$B + H_3O^+ \rightleftharpoons BH^+ + H_2O \tag{12-5}$$

The magnitude of the equilibrium constant, which provides a measure of the completeness of this reaction, is readily shown to be the ratio of base dissociation constant for B divided by the autoprotolysis constant for water:

$$K_{equil} = \frac{K_b}{K_w} = \frac{[BH^+][OH^-]}{[B][H_3O^+][OH^-]} = \frac{[BH^+]}{[B][H_3O^+]} \tag{12-6}$$

Note that the completeness of the reaction is determined in part by the magnitude of $K_w$, *the autoprotolysis constant for the solvent.*

By analogy, the completeness of the reaction between a weak acid HA and a strong base in water can be expressed by an equilibrium constant that is numerically equal to $K_a/K_w$.

Similar relationships can be derived for reactions in nonaqueous solvents. For example, when the weak base B reacts with perchloric acid in anhydrous acetic acid, the reaction is

$$B + CH_3COOH_2^+ \rightleftharpoons BH^+ + CH_3COOH \tag{12-7}$$

Here, $CH_3COOH_2^+$ is the solvated proton formed when perchloric acid is dissolved in anhydrous acetic acid:

$$HClO_4 + CH_3COOH \rightleftharpoons CH_3COOH_2^+ + ClO_4^-$$

The solvated proton $CH_3COOH_2^+$ is analogous to $H_3O^+$ in an aqueous solution. The equilibrium constant for the neutralization reaction (Equation 12–7) is given by

$$K_{equil} = \frac{[BH^+]}{[B][CH_3COOH_2^+]} = \frac{K_b'}{K_s} \tag{12-8}$$

where $K_b'$ is the dissociation constant for the base *in acetic acid;* that is,

$$B + CH_3COOH \rightleftharpoons BH^+ + CH_3COO^- \qquad K_b' = \frac{[BH^+][CH_3COO^-]}{[B]} \tag{12-9}$$

The constant $K_s$ in Equation 12–8 is the autoprotolysis constant for the equilibrium shown as Equation 12–2 and is analogous to the ion-product constant for water:

$K_s$ is analogous to $K_w$.

$$K_s = [CH_3COOH_2^+][CH_3COO^-] \qquad (12–10)$$

As with the aqueous equilibrium constants, the concentration of the solvent $CH_3COOH$ is essentially invariant and is thus included in $K_b'$ and $K_s$.

A similar set of relationships can be developed for the titration of a weak acid with a strong base in an anhydrous medium.

Equations 12–6 and 12–8 show that the completeness of an acid/base reaction is directly related to the dissociation constant of the weak acid but is inversely related *to the autoprotolysis constant of the solvent in which the titration is carried out.* These equations can be understood by considering a neutralization reaction as a competition for protons between solvent and base. Thus, for example, the extent of reaction 12–7 is governed by the success with which base molecules B compete with solvent molecules $CH_3COOH$ for a stoichiometrically limited number of hydrogen ions. The effectiveness of each participant in this competition is measured by its dissociation constant $K_b'$ and its autoprotolysis constant $K_s$, respectively.

### Effect of Acid or Base Characteristics of the Solvent

The behavior of solutes as acids or bases is strongly influenced by the strength of the solvent as an acid or a base. A number of amphiprotic solvents, including formic acid, acetic acid, and sulfuric acid, are considerably better proton donors than proton acceptors and are therefore classified as acidic solvents. In such media, the basic properties of a solute are magnified while its acidic properties are attenuated. Thus, for example, aniline, $C_6H_5NH_2$, cannot be titrated in water because its base dissociation constant is only about $10^{-10}$. In anhydrous acetic acid, however, aniline is an appreciably stronger base because the solvent gives up protons more readily than does water. Thus, the equilibrium constant, $K_b'$ for the reaction

$$C_6H_5NH_2 + CH_3COOH \rightleftharpoons C_6H_5NH_3^+ + CH_3COO^-$$

is significantly larger than $K_b$ for the analogous reaction in water:

$$C_6H_5NH_2 + H_2O \rightleftharpoons C_6H_5NH_3^+ + OH^-$$

Solvents such as ethylenediamine and liquid ammonia have a strong affinity for protons and are therefore classified as basic solvents. In these media, the acidic character of a solute is enhanced. For example, phenol, which has an acid dissociation constant of about $10^{-10}$ in water, is strong enough in ethylenediamine to be readily titrated with a standard base. Of course, the strengths of bases are diminished in solvents of this type and solutes that are strong bases in water may only partially dissociate in solvents with basic properties.

**DIELECTRIC CONSTANTS FOR SEVERAL SOLVENTS**

| Solvent | $D$ |
|---|---|
| Water | 78.5 |
| Methanol | 32.6 |
| Ethanol | 24.3 |
| Acetic acid | 6.2 |
| Ammonia | 22.4 |
| Acetone | 20.7 |
| Ethyl ether | 4.3 |
| Toluene | 2.4 |
| Benzene | 2.3 |
| Chloromethane | 12.6 |
| Methylene chloride | 9.1 |
| Chloroform | 4.8 |
| Carbon tetrachloride | 2.2 |

Glacial acetic acid derives its name from the fact that at about 16°C large ice-like crystals of acetic acid form in the medium.

### Effect of Solvent Dielectric Constant

The dielectric constant of a solvent measures its capacity to cause particles of opposite charge to separate from one another and thus behave as independent entities. For example, in a solvent with a high dielectric constant, such as water ($D_{H_2O} = 78.5$), a minimum amount of work is required to separate a positively charged ion from one with a negative charge. In contrast, a much greater amount of energy is expended in accomplishing this process in a solvent with a low dielectric constant, such as acetic acid ($D_{HOAc} = 6.2$). Methanol and ethanol, with dielectric constants of 33 and 24, respectively, are intermediate in their behavior.

Solvents with a low dielectric constant suffer a disadvantage when used as a titration medium involving acids or bases that dissociate to give two oppositely charged ions because the extent of dissociation is markedly decreased by the limited ability of the solvent to cause ion separation. For example, perchloric acid, which behaves as a strong acid in solvents with high dielectric constants, has a dissociation constant of only $1.1 \times 10^{-4}$ in *tert*-butyl alcohol, a solvent with a dielectric constant of 12.5. That is,

$$HClO_4 + C_4H_9OH \rightleftharpoons C_4H_9OH_2^+ + ClO_4^- \qquad K_a' = 1.1 \times 10^{-4}$$

It is important to note that the dielectric constant of a solvent has little effect on the degree of dissociation of an acid or a base when the dissociation does not require a separation of charged particles. For example, the dielectric effect is minimal for dissociation reactions such as

$$NH_4^+ + C_2H_5OH \rightleftharpoons C_2H_5OH_2^+ + NH_3$$

### 12C–3 Choice of Amphiprotic Solvents for Neutralization Titrations

In light of the discussion in Section 12C–2, it is possible to conclude that the best solvent for a given acid/base titration hinges upon three interrelated properties:

1. Its autoprotolysis constant; a numerically small value is desirable.
2. Its properties as a proton donor or acceptor. For the titration of a weak base, a solvent with strong proton donor tendencies (that is, an acidic solvent) is helpful; for the determination of a weak acid, a solvent that is a good proton acceptor is desirable.
3. Its dielectric constant; a high value is most useful.

In addition, of course, the solute must be reasonably soluble in the solvent.

Anhydrous acetic acid (often called glacial acetic acid) is frequently chosen as solvent for titration of very weak bases because it tends to donate protons and thus enhances the strength of a dissolved base. Also, its autoprotolysis constant ($3.6 \times 10^{-15}$) is somewhat more favorable than that of water. On the other hand, its low dielectric constant partially offsets these advantages. The two favorable properties outweigh the single disadvantage, however, and acetic acid is a generally superior solvent for the titration of weak bases; it is clearly inferior to water for the titration of weak acids because of its weakness as a proton acceptor.

Ethanol, which has been widely applied for nonaqueous titrations, is classified as a neutral solvent because its proton donor and acceptor properties do not differ markedly. The autoprotolysis constant for ethanol is far superior to that of water ($8 \times 10^{-20}$). On the other hand, its low dielectric constant compared with that of water frequently offsets the autoprotolysis advantage. For example, the dissociation constants of most uncharged acids, such as benzoic acid, are about $10^{-6}$ as great in ethanol as in water. At the same time, the ratio of autoprotolysis constants is smaller by nearly the same factor ($8 \times 10^{-6}$). Thus the ratio $K_a'/K_s$ is only slightly more favorable in ethanol than in water, and the improvement in end points gained by use of this solvent is modest. In contrast, significant gain is realized by the use of ethanol for the titration of a charged weak acid such as the ammonium ion because no charge separation is involved in the dissociation:

$$NH_4^+ + C_2H_5OH \rightleftharpoons NH_3 + C_2H_5OH_2^+$$

As a consequence, dissociation of the acid $NH_4^+$ is not significantly less in ethanol than it is in water. The reaction of $NH_4^+$ with a strong base is much more complete in ethanol, however, because of the low autoprotolysis constant of the solvent. Thus, $NH_4^+$ can be titrated successfully in ethanol but not in water.

## 12C–4 End-Point Detection in Nonaqueous Titrations

End points in nonaqueous titrations are most commonly based on potential measurements with a glass electrode. This type of end point is discussed in detail in Chapter 17.

Many of the acid/base indicators developed for aqueous titrations are also applicable in nonaqueous media. To be sure, their behavior in an aqueous environment cannot be extrapolated to predict their properties in nonaqueous solutions. The limited information available with respect to these properties in solvents other than water makes the choice of indicator largely a matter of experience and empirical observation.

## 12C–5 Applications of Nonaqueous Acid/Base Titrations

Many combinations of solvents, titrants, and end points are discussed in the literature of nonaqueous titrimetry. The choice among them is usually based upon such considerations as solubility, convenience, reagent cost, toxicity, and availability of suitable equipment.[6]

---

[6]Reviews of applications of nonaqueous titrations are found in B. Kratochvil, *Anal. Chem.*, **1982**, *54*, 105R; **1980**, *52*, 151R; **1978**, *50*, 153R.

## Titration of Bases in Glacial Acetic Acid

Glacial acetic acid is the most widely used acidic solvent. The titrant is a standard perchloric acid solution prepared in the anhydrous medium. A standard solution of sodium acetate in the same solvent serves as a base for back-titrations when needed. Solutions of the acid are generally standardized against sodium carbonate or potassium acid phthalate (in acetic acid, the hydrogen phthalate ion is a sufficiently strong base to make it an excellent primary standard for *acids*).

## Some Typical Applications

Standard solutions of perchloric acid in glacial acetic acid are useful for the determination of aromatic amines, amides, urea, and other very weak nitrogen bases. An important application of acetic acid is to the direct titration of most amino acids with a standard acid. It was noted earlier (page 257) that in aqueous media, these compounds exist largely as zwitterions which are not strong enough acids or bases for titration. In glacial acetic acid, however, the dissociation of the carboxylic acid group is essentially completely repressed, leaving the amine group available for titration with perchloric acid.

Many inorganic and organic salts that are neutral or very weak bases in water are easily determined by titration with perchloric acid in anhydrous acetic acid. For example, the sodium salts of inorganic anions such as chloride, bromide, iodide, nitrate, chlorate, and sulfate have all been titrated as bases in glacial acetic acid. The ammonium and alkali metal salts of most carboxylic acids can also be determined in this medium; typical examples include ammonium benzoate, sodium salicylate, sodium acetate, potassium tartrate, and sodium citrate.

## 12C–6 Titration of Acids

Several basic solvents are useful for determining acids that are too weak to be titrated in water. Some of these solvents are ethylenediamine, dimethylformamide, pyridine, dimethylsulfoxide, and butylamine. In addition, aliphatic alcohols, acetone, and acetonitrile find use in the determination of acids.

A variety of inorganic and organic compounds including amine salts, inorganic salts, carboxylic acids, phenols, enols, and imides are soluble in ethylenediamine and exhibit enhanced acidic characteristics therein. Tetrabutylammonium hydroxide [$(C_4H_9)_4NOH$] often serves as titrant.

## 12D  QUESTIONS AND PROBLEMS

*12–1. Carbon dioxide and hydrogen chloride are gases with similar boiling points ($-78°$ and $-85°C$, respectively). Briefly explain why a short period of boiling results in the removal of dissolved $CO_2$, while the same treatment of an aqueous HCl solution results in no loss of solute.

12–2. What is the objection to the use of $HNO_3$ as a standard acid?

*12–3. In addition to solubility of the analyte, what factors influence the choice of a solvent for a nonaqueous acid/base titration?

12–4. Potassium hydrogen phthalate functions as a primary standard for bases in aqueous solution and as a primary standard for acids in anhydrous acetic acid. Briefly explain.

*12–5. Define what is meant by the equivalent weight of an acid. What additional information do you need if you base your calculations on equivalent weight?

12–6. The absorption of carbon dioxide by a standard sodium hydroxide may or may not cause an error in the concentration of the base. Briefly explain.

*12–7. Describe the preparation of approximately 2.5 L of
(a) 0.12 M $HClO_4$ from a 0.400 M solution of the acid.
(b) 0.08 M $H_3PO_4$ from the concentrated reagent (sp. gr. 1.69, 85% $H_3PO_4$).
(c) 0.15 M $NH_3$ from the concentrated reagent (sp. gr. 0.90, 27% $NH_3$).
(d) 0.01 M $Ba(OH)_2$ from the solid reagent.

12–8. Describe preparation of 1.000 L of 0.1800 M HCl from constant-boiling HCl distilled at 740 torr.

*12–9. A concentrated solution of HCl is distilled under a pressure of 765 torr. What is the acid concentration of the constant-boiling residue (density = 1.1011 g/mL)? How many grams of the constant boiling acid will be needed to prepare 2000 mL of 0.120 M HCl?

12–10. Titration of 0.5639 g of pure sodium tetraborate decahydrate required 42.77 mL of a dilute perchloric acid solution (see Problem 12–18d). Calculate the molar concentration of the acid.

*12–11. Calculate the molar concentration of an HCl solution if a 50.00-mL portion yielded 1.005 g of AgCl when treated with an excess of $AgNO_3$.

12–12. A 0.3798-g sample of $Na_2C_2O_4$ was quantitatively converted to $Na_2CO_3$ and titrated with 27.41 mL of HCl to the bromocresol green end point. Calculate the molar concentration of the acid.

*12–13. A dilute perchloric acid solution was standardized by dissolving 0.2445 g of primary standard sodium carbonate in 50.00 mL of the acid, boiling to eliminate $CO_2$, and back-titrating with 4.13 mL of dilute NaOH. In a separate titration, a 25.00 mL portion of the acid required 26.88 mL of the base. Calculate the molar concentrations of the acid and the base.

12–14. Calculate the molar concentration of a solution prepared by dissolving 1.181 g of $Na_2CO_3 \cdot 7 H_2O$ in sufficient water to give 250.0 mL. What volume of 0.1076 M $HClO_4$ would be needed to neutralize 50.00 mL of the carbonate solution (products: $CO_2$, $H_2O$)?

*12–15. A 0.5239-g sample of primary-standard benzoic acid was dissolved in 50.00 mL of a KOH solution. Calculate the molar concentration of the base if a 1.93-mL back-titration with 0.1020 M HCl was needed.

12–16. A 0.2245-g sample of potassium hydrogen phthalate (KHP) required 37.41 mL of a sodium hydroxide solution to reach a phenolphthalein end point. Calculate the molar concentration of the base.

*12–17. Calculate the average molar concentration of a KOH solution from the accompanying titration data. In addition, calculate the standard deviation of the data.

| Weight of KHP Taken, g | Volume of NaOH Used, mL |
|---|---|
| 0.7644 | 34.56 |
| 0.5131 | 22.90 |
| 0.6985 | 31.70 |
| 0.7214 | 32.56 |

12–18. Suggest a range of sample weights appropriate for a 30- to 45-mL titration in the standardization of
(a) 0.050 M $HClO_4$ with $Na_2CO_3$ (products: $CO_2$, $H_2O$).
(b) 0.10 M NaOH with potassium hydrogen phthalate, $KHC_8H_8O_4$.
(c) 0.010 M $Ba(OH)_2$ with $KH(IO_3)_2$ [products: $Ba(IO_3)_2(s)$, $K^+$, $IO_3^-$, $H_2O$].
(d) 0.10 M HCl with borax, $Na_2B_4O_7 \cdot 10 H_2O$. Reaction:

$$B_4O_7^{2-} + 2 H_3O^+ + 3 H_2O \rightleftharpoons 4 H_3BO_3$$

(e) 0.08 M $HClO_4$ with THAM. Reaction:

$$(HOCH_2)_3CNH_2 + H_3O^+ \rightarrow (HOCH_2)_3CNH_3^+ + H_2O$$

*12–19. A 0.250-g sample of impure sodium carbonate require 36.58 mL of 0.1055 M HCl in a titration to a bromocresol green end point. Calculate the percentage of $Na_2CO_3$ in the sample.

12–20. A 1.016-g sample of impure mercury(II) oxide was dissolved in a solution containing potassium iodide. Reaction:

$$HgO + H_2O + 4 I^- \rightarrow HgI_4^{2-} + 2 OH^-$$

Calculate the percentage of HgO in the sample if titration of the liberated hydroxide ion required 43.62 mL of 0.1347 M HCl.

*12–21. A 50.00-mL sample of vinegar (density = 1.060 g/mL) was diluted to 250.0 mL in a volumetric flask. A 25.00-mL aliquot of the diluted solution required 34.60 mL of 0.0965 M NaOH. Calculate the mg of acetic acid ($CH_3COOH$, fw = 60.05) in each mL of the undiluted vinegar.

12–22. A 25.00-mL sample of white table wine was diluted to about 100 mL and titrated with 28.40 mL of 0.05412 M NaOH with phenolphthalein as indicator. Express the acidity of the wine in terms of grams tartaric acid ($H_2C_4H_4O_6$, fw = 150.09) per 100 mL; to a phenolphthalein end point, both acidic hydrogens are titrated.

*12–23. Specifications for an acid rinse tank in a plating plant call for an operating range between 36 and 48 g of $H_2SO_4$ per liter. Routine analysis for the acid content of this tank involves dilution of 50.00-

mL samples to 500.0 mL in volumetric flasks and titration of 25.00 aliquots of the diluted solutions with standard base.

(a) What should be the concentration of the standard base if the maximum titrant volume is not to exceed 45.0 mL?

(b) With this concentration of base, what will be the volume associated with titration of acid at the lowest tolerable operating limit?

(c) What is the molar concentration of a sodium hydroxide solution if titration of 0.710 g of primary standard potassium hydrogen phthalate required 30.17 mL?

(d) What volume of 50% (w/w) ($d$ = 1.5253 g/mL) NaOH is needed to prepare about 5.00 L of 0.12 M solution?

12–24. A 0.270-g sample of impure sodium carbonate required 24.12 mL of 0.1684 M HCl and a back-titration with 2.96 mL of a sodium hydroxide solution. The volume ratio of acid to base was 1.250. Calculate the percentage of $Na_2CO_3$ in the sample.

*12–25. The thorium(IV) in a mineral sample that weighed 1.95 g was precipitated as $Th(CO_3)_2$. After it was filtered and washed, the solid was redissolved in 50.00 mL of 0.1438 M HCl. The solution was heated to expel $CO_2$ and then titrated with 7.50 mL of 0.1029 M NaOH. Calculate the percentage of thorite, $ThSiO_4$ (fw = 324.1) contained in the sample.

*12–26. The carbon dioxide in a 3.00-L sample of urban air was bubbled into 50.00 mL of 0.0116 M $Ba(OH)_2$. The excess hydroxide ion was titrated with 23.6 mL of 0.0108 M HCl to a phenolphthalein end point. Express the results of this analysis in terms of ppm $CO_2$ (i.e., mL $CO_2$/$10^6$ mL air). The density of $CO_2$ is 1.98 g/mL.

12–27. The ammonia produced in a Kjeldahl decomposition of a 0.760-g sample was collected in 50.00 mL of 0.1005 M HCl. Titration of the excess acid required 2.44 mL of 0.1168 M NaOH. Express the results of this analysis in terms of percent

(a) nitrogen.

(b) cereal protein.

*12–28. A Kjeldahl analysis was performed upon a 0.0550-g sample of impure biguanide, $C_2H_7N_5$. The liberated ammonia, collected in 40 mL of 4% boric acid, was titrated with 19.51 mL of 0.1060 M HCl. Calculate the percentage of biguanide (fw = 101.1) in the sample.

*12–29. A 0.4117-g sample of a pesticide preparation was transferred to a flask containing 50.00 mL of 0.0996 M NaOH and 50.00 mL of 3% $H_2O_2$. Heating caused oxidation of formaldehyde according to the reaction

$$OH^- + HCHO + H_2O_2 \rightarrow HCOO^- + 2\ H_2O$$

The solution was cooled and the excess base was titrated with 19.66 mL of 0.0550 M $H_2SO_4$. Calculate the percentage of formaldehyde in the sample.

12–30. Air was bubbled at the rate of 3.00 L/min through a trap containing 100.0 mL of 3% $H_2O_2$. Reaction:

$$SO_2(g) + H_2O_2 \rightarrow H_2SO_4$$

After 100.0 minutes, the acid produced was titrated with 11.1 mL of 0.00204 M NaOH. Calculate the parts per million of $SO_2$ (density $SO_2$ = 2.85 g/L).

*12–31. A 0.8160-g sample containing dimethylphthalate and unreactive species was saponified by refluxing with 50.00 mL of 0.1031 M NaOH. After the reaction was complete, the excess NaOH was back-titrated with 24.27 mL of 0.1644 M HCl. Calculate the percentage of dimethylphthalate [$C_6H_4(COOCH_3)_2$ (fw = 194.19)] in the sample.

12–32. A 50.00-mL sample containing methylethyl ketone and unreactive species was treated with an excess of hydroxylamine hydrochloride $H_2HOH \cdot HCl$. After oxime formation was complete, the liberated HCl was titrated with 19.15 mL of 0.01123 M NaOH. Calculate the number of milligrams of $CH_3COC_2H_5$ (fw = 71.11) per liter of sample.

*12–33. Calculate the volume of 0.0765 M NaOH needed to titrate

(a) 25.00 mL of 0.0928 M $H_3PO_4$ to a thymolphthalein end point.

(b) 15.00 mL of 0.1477 M $H_3PO_4$ to a bromocresol green end point.

(c) 15.00 mL of a solution that is 0.0952 M in $H_3PO_4$ and 0.1073 M in $NaH_2PO_4$ to a thymolphthalein end point.

12–34. Calculate the volume of 0.1049 M HCl needed to titrate

(a) 31.6 mL of 0.08301 M $Na_3PO_4$ to a bromocresol green end point.

(b) 28.37 mL of 0.1234 M $Na_2HPO_4$ to a bromocresol green end point.

(c) 25.00 mL of a solution that is 0.0847 M in $Na_3PO_4$ and 0.04110 M in $Na_2HPO_4$ to a bromocresol green end point.

(d) the solution in (c) titrated to a thymolphthalein end point.

*12–35. A series of solutions can contain HCl, $H_3PO_4$, or $NaH_2PO_4$, alone or in any of the compatible combinations of these solutes. Data are provided for the volumes of 0.1000 M NaOH needed to titrate 25.00-mL portions of each solution to (1) a bromocresol green end point and (2) a thymolphthalein end point. Use this information to deduce the composition of each solution and calculate the number of milligrams of solute(s)/mL of solution.

|     | (1)   | (2)   |
|-----|-------|-------|
| (a) | 20.88 | 41.75 |
| (b) | 32.35 | 32.32 |
| (c) | 0.00  | 28.20 |
| (d) | 28.37 | 45.10 |
| (e) | 14.56 | 47.50 |

12–36. The accompanying volumes of 0.1083 M HCl were needed to titrate solutions containing NaOH, $Na_3PO_4$, and $Na_2HPO_4$, alone or in any compatible combination of these solutes to (1) a bromocresol green end point and (2) a thymolphthalein end point. Use these data to deduce the composition of each solution and calculate the milligrams of solute(s)/mL of solution.

|     | (1)   | (2)   |
|-----|-------|-------|
| (a) | 43.65 | 21.83 |
| (b) | 46.40 | 17.08 |
| (c) | 34.69 | 34.71 |
| (d) | 39.54 | 0.00  |
| (e) | 48.26 | 29.45 |

*12-37. A 0.5000-g sample consisting of $Na_2CO_3$, $NaHCO_3$, and inert materials was dissolved in sufficient water to give 250.0 mL of solution. A 25.00-mL aliquot of this solution was boiled with 50.00 mL of 0.01255 M HCl. Titration of the excess HCl to a phenolphthalein end point required 2.34 mL of 0.01063 M NaOH. The carbonate species in a second 25.00-mL aliquot were precipitated with an excess of $BaCl_2$ and 25.00 mL of the NaOH. Titration of the excess base required 7.63 mL of the standard acid. Calculate the percentages of $Na_2CO_3$ and $NaHCO_3$ in the sample.

12-38. A 1.219-g sample that contained $(NH_4)_2SO_4$, $NH_4NO_3$, and inert materials was diluted to 250.0 mL in a volumetric flask. A 50.00-mL aliquot was heated with strong NaOH, and the liberated $NH_3$ was distilled into 30.00 mL of 0.08421 M HCl. The excess HCl required 10.17 mL of 0.08802 M NaOH. A 25.00-mL aliquot of the sample was made basic after the addition of Devarda's alloy to reduce the $NO_3^-$ to $NH_3$. The $NH_3$ from both $NH_4^+$ and $NO_3^-$ was then distilled into 30.00 mL of the standard acid and back-titrated with 14.16 mL of the standard base. Calculate the percentages of $(NH_4)_2SO_4$ and $NH_4NO_3$ in the sample.

12-39. Several liquids and their autoprotolysis constants are listed below:

| (a) | $H_2O$     | $1.00 \times 10^{-14}$ |
|-----|------------|------------------------|
| (b) | $C_2H_5OH$ | $8 \times 10^{-20}$    |
| (c) | $H_2SO_4$  | $1.4 \times 10^{-4}$   |
| (d) | $CH_3OH$   | $2 \times 10^{-17}$    |
| (e) | $NH_3$     | $1 \times 10^{-33}$    |
| (f) | HCOOH      | $6 \times 10^{-7}$     |

Write autoprotolysis constant expressions for each liquid.

12-40. Calculate p-cation for each of the solvents in Problem 12-39.

12-41. In anhydrous methanol, the dissociation constant for acetic acid is $3.0 \times 10^{-10}$. Calculate $-\log CH_3OH_2^+$ for a solution that is
(a) 0.200 M in acetic acid.
(b) 0.100 M in acetic acid and 0.200 M in sodium acetate.
(c) 0.200 M in sodium acetate.

## CHAPTER 13

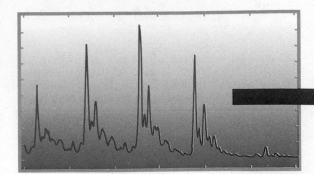

# COMPLEX-FORMATION TITRATIONS

Complex-formation reagents are widely used to titrate cations. The most useful of these reagents are organic compounds with several electron-donor groups that form multiple covalent bonds with metal ions.

## 13A  COMPLEX-FORMATION REACTIONS

Most metal ions react with electron-pair donors to form coordination compounds, or complex ions. The donor species, or *ligand,* must have at least one pair of unshared electrons available for bond formation. Water, ammonia, and halide ions are common inorganic ligands.

The number of covalent bonds a cation tends to form with electron donors is its *coordination number.* Typical values for coordination numbers are two, four, and six. The species formed as a result of coordination can be electrically positive, neutral, or negative. For example, copper(II), which has a coordination number of four, forms a cationic complex with ammonia, $Cu(NH_3)_4^{2+}$; a neutral complex with glycine, $Cu(NH_2CH_2COO)_2$; and an anionic complex with chloride ion, $CuCl_4^{2-}$.

Titrimetric methods based upon complex formation (sometimes called *complexometric methods*) have been used for more than a century. The truly remarkable growth in their analytical application began in the 1940s, however, and is based upon a particular class of coordination compounds called *chelates.* A chelate is produced when a metal ion coordinates with two (or more) donor groups of a single ligand to form a five- or six-membered heterocyclic ring. The copper complex of glycine, mentioned in the previous paragraph, is an example. Here, copper bonds to both the oxygen of the carboxyl group and the nitrogen of the amine group.

> A ligand is an ion or a molecule that forms a covalent bond with a cation by donating a pair of electrons, which are then shared by the ligand and the cation.

> The coordination number of a cation is the number of covalent bonds it tends to form with electron donor species.

> Complexometric methods are titrimetric methods that are based upon complex-formation reactions.

> A chelate is a cyclic complex formed when a cation is bonded by two or more donor groups contained in a single ligand.

> "Chelate" is pronounced kee'late.

> "Dentate" means having tooth-like projections.

$$Cu^{2+} + 2\ \underset{\underset{\displaystyle H}{|}}{\overset{\overset{\displaystyle NH_2}{|}}{H-C}} \overset{\overset{\displaystyle O}{\|}}{-C-OH} \longrightarrow \begin{matrix} O=C-O \\ | \\ H_2C-N \\ | \\ H_2 \end{matrix} \begin{matrix} \\ Cu \\ \\ \end{matrix} \begin{matrix} O-C=O \\ | \\ N-CH_2 \\ | \\ H_2 \end{matrix} + 2\ H^+$$

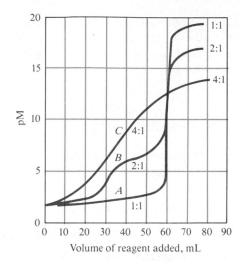

Figure 13–1
Curves for complex-formation titrations. Titration of 60.0 mL of a solution that is 0.020 M in M with (*A*) a 0.020 M solution of the tetradentate ligand D to give MD as product, (*B*) a 0.040 M solution of the bidentate ligand B to give $MB_2$, and (*C*) a 0.080 M solution of the unidentate ligand A to give $MA_4$. The overall formation constant for each product is $1.0 \times 10^{20}$.

A ligand that has a single donor group, such as ammonia, is called *unidentate* (single-toothed), whereas one such as glycine, which has two groups available for covalent bonding, is *bidentate*. *Tridentate, tetradentate, pentadentate,* and *hexadentate* chelating agents are also known.

As titrants, multidentate ligands, particularly those having four or six donor groups, have two advantages over their unidentate counterparts. First, multidentate ligands generally react more completely with cations and thus provide sharper end points. Second, they ordinarily react with metal ions in a single-step process, whereas complex formation with unidentate ligands usually involves two or more intermediate species.

The advantage of single-step reaction is illustrated by the titration curves shown in Figure 13–1. Each titration involves a reaction that has an overall equilibrium constant of $10^{20}$. Curve *A* is derived for a reaction in which a metal ion M with a coordination number of four reacts with a tetradentate ligand D to form the complex MD (here we have omitted the charges on the two reactants for convenience). Curve *B* is for the reaction of M with a hypothetical bidentate ligand B to give $MB_2$ in two steps. The formation constant for the first step is $10^{12}$, and that for the second is $10^8$. Curve *C* involves a unidentate ligand A that forms $MA_4$ in four steps with successive formation constants of $10^8$, $10^6$, $10^4$, and $10^2$. These curves demonstrate that a much sharper end point is obtained with a reaction that takes place in a single step. For this reason, polydentate ligands are ordinarily preferred for complexometric titrations.

Tetradentate and hexadentate ligands are more satisfactory as titrants than ligands with a lesser number of donor groups for two reasons. First, they react more completely with cations, and second, they tend to form 1 : 1 complexes.

## 13B  TITRATIONS WITH AMINOPOLYCARBOXYLIC ACIDS

Tertiary amines that also contain carboxylic acid groups form remarkably stable chelates with many metal ions.[1] Schwarzenbach first recognized

[1]These reagents are the subject of several excellent monographs. See, for example, R. Pribil, *Applied Complexometry*. New York: Pergamon, 1982; A. Ringbom and E. Wanninen, in *Treatise on Analytical Chemistry*, 2nd ed., I. M. Kolthoff and P. J. Elving, Eds., Part I, Vol. 2, Chapter 11. New York: Wiley, 1979.

their potential as analytical reagents in 1945. Since this original work, investigators throughout the world have described applications of these compounds to the volumetric determination of most of the metals in the periodic table.

### 13B–1 Ethylenediaminetetraacetic Acid (EDTA)

EDTA, a hexadentate ligand, is among the most important and widely used reagents in titrimetry.

Ethylenediaminetetraacetic acid, also called (ethylenedinitrilo)tetraacetic acid and commonly shortened to EDTA, has the structure

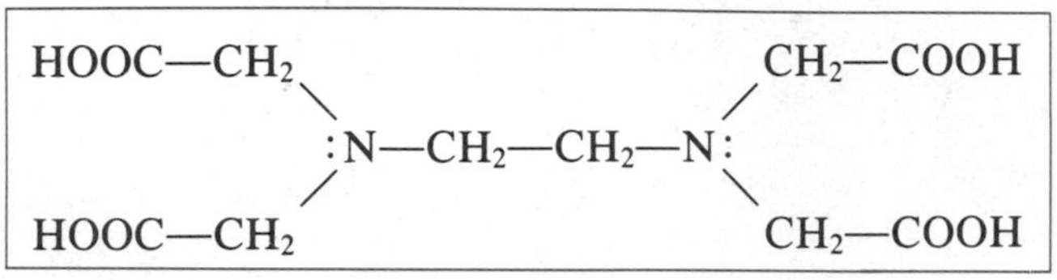

$$\text{HOOC—CH}_2 \diagdown \qquad\qquad \diagup \text{CH}_2\text{—COOH}$$
$$:\text{N—CH}_2\text{—CH}_2\text{—N}:$$
$$\text{HOOC—CH}_2 \diagup \qquad\qquad \diagdown \text{CH}_2\text{—COOH}$$

The EDTA molecule has six potential sites for bonding with a metal ion: the four carboxyl groups and the two amino groups. Thus, EDTA is a hexadentate ligand.

### Acidic Properties of EDTA

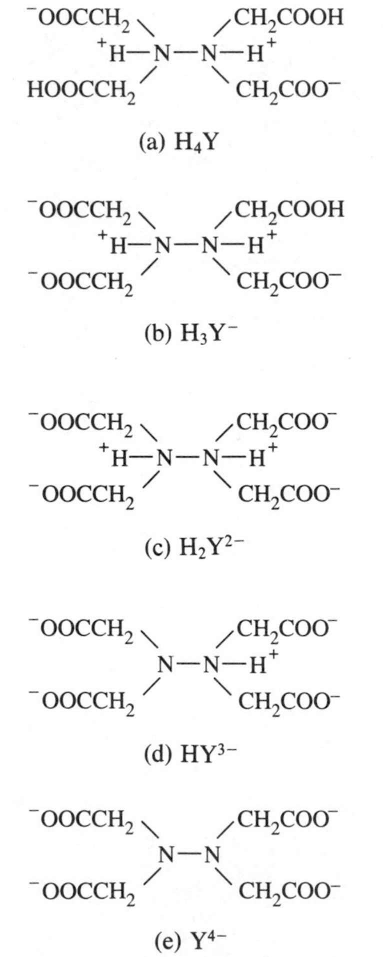

When EDTA is dissolved in water, it behaves like an amino acid, such as glycine (see Equation 11–9, Section 11G). In this case, however, a double zwitterion forms, which has the structure shown in Figure 13–2a. Note that the net charge on this species is zero and that it contains four dissociatable protons, two associated with two of the carboxyl groups and the other two with the two amine groups. For simplicity, we generally formulate the double zwitterion as $H_4Y$, where $Y^{4-}$ is the fully deprotonated ion of Figure 13–2e. The first and second steps in the dissociation process involve successive loss of protons from the two carboxylic acid groups; the third and fourth steps involve dissociation of the protonated amine groups. The structures of $H_3Y^-$, $H_2Y^{2-}$, and $HY^{3-}$ are shown in Figure 13–2b,c,d. The dissociation constants for the deprotonation of $H_4Y$ are $K_1 = 1.02 \times 10^{-2}$, $K_2 = 2.14 \times 10^{-3}$, $K_3 = 6.92 \times 10^{-7}$, and $K_4 = 5.50 \times 10^{-11}$. It is of interest that the first two constants are of the same order of magnitude, which suggests that the two protons involved dissociate from opposite ends of the rather long molecule. As a consequence of the physical separation of these protons, the negative charge created by the first dissociation does not greatly inhibit the removal of the second proton. The same cannot be said for the dissociation of the other two protons, however, which are much closer to the negatively charged carboxylate ions created by the initial dissociations.

Figure 13–3 illustrates how the relative amounts of the species derived from EDTA vary as a function of pH. It is apparent that $H_2Y^{2-}$ predominates in a moderately acidic medium (pH 3 to 6). Only at pH values greater than 10 does $Y^{4-}$ become a major component of EDTA solutions.

(a) $H_4Y$

(b) $H_3Y^-$

(c) $H_2Y^{2-}$

(d) $HY^{3-}$

(e) $Y^{4-}$

Figure 13–2
Structure of $H_4Y$ and its dissociation products.

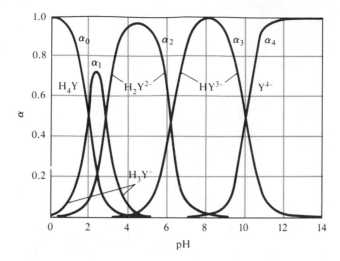

**Figure 13–3**
Composition of EDTA solutions as a function of pH.

## Reagents

The free acid of EDTA, $H_4Y$, and the dihydrate of the sodium salt, $Na_2H_2Y \cdot 2 H_2O$, are available in reagent quality. The former can serve as a primary standard after it has been dried for several hours at 130 to 145°C. It is then dissolved in the minimum amount of base required for complete solution.

Under normal atmospheric conditions, the dihydrate contains 0.3% moisture in excess of the stoichiometric amount. For all but the most exacting work, this excess is sufficiently reproducible to permit use of a corrected weight of the salt in the direct preparation of a standard solution. If necessary, the pure dihydrate can be prepared by drying at 80°C for several days in an atmosphere of 50% relative humidity.

A number of compounds related chemically to EDTA have also been investigated but do not appear to offer advantages over the original species. We shall thus confine our discussion to the properties and applications of EDTA.

Standard EDTA solutions are ordinarily prepared by dissolving weighed quantities of $Na_2H_2Y \cdot 2 H_2O$ and diluting to the mark in a volumetric flask.

Nitrilotriacetic acid (NTA) is the second most common aminopolycarboxylic acid used for titrimetry. Its structure is

$$HOOC-CH_2 \quad CH_2-COOH$$
$$N$$
$$CH_2-COOH$$

## 13B–2 Complexes of EDTA and Metal Ions

EDTA combines with metal ions in a 1:1 ratio regardless of the charge on the cation. For example, formation of the silver and aluminum complexes is described by the equations

$$Ag^+ + Y^{4-} \rightleftarrows AgY^{3-}$$
$$Al^{3+} + Y^{4-} \rightleftarrows AlY^-$$

The reagent is remarkable not only because it forms chelates with all cations but also because most of these chelates are sufficiently stable to form the basis for a titrimetric method. This great stability undoubtedly results from the several complexing sites within the molecule that give rise to a cage-like structure in which the cation is effectively surrounded

The reaction of the EDTA anion with a metal ion $M^{n+}$ is described by the equation

$$M^{n+} + Y^{4-} \rightleftarrows MY^{(n-4)+}$$

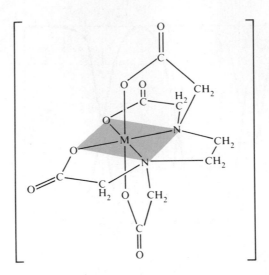

Figure 13–4
Structure of a metal/EDTA chelate.

and isolated from solvent molecules. One form of the complex is depicted in Figure 13–4. Note that all six ligand groups are involved in bonding the divalent metal ion.

Table 13–1 lists formation constants $K_{MY}$ for common EDTA complexes. Note that the constant refers to the equilibrium involving the species $Y^{4-}$ with the metal ion:

$$M^{n+} + Y^{4-} \rightleftarrows MY^{(n-4)+} \qquad \frac{[MY^{(n-4)+}]}{[M^{n+}][Y^{4-}]} = K_{MY} \qquad (13–1)$$

### 13B–3 Equilibrium Calculations Involving EDTA

A titration curve for the reaction of a cation with EDTA consists of a plot of pM as a function of reagent volume. Values for pM are readily computed in the early stage of a titration by assuming that the equilibrium

**Table 13–1**
**FORMATION CONSTANTS FOR EDTA COMPLEXES***

| Cation | $K_{MY}$ | log $K_{MY}$ | Cation | $K_{MY}$ | log $K_{MY}$ |
|---|---|---|---|---|---|
| $Ag^+$ | $2.1 \times 10^7$ | 7.32 | $Cu^{2+}$ | $6.3 \times 10^{18}$ | 18.80 |
| $Mg^{2+}$ | $4.9 \times 10^8$ | 8.69 | $Zn^{2+}$ | $3.2 \times 10^{16}$ | 16.50 |
| $Ca^{2+}$ | $5.0 \times 10^{10}$ | 10.70 | $Cd^{2+}$ | $2.9 \times 10^{16}$ | 16.46 |
| $Sr^{2+}$ | $4.3 \times 10^8$ | 8.63 | $Hg^{2+}$ | $6.3 \times 10^{21}$ | 21.80 |
| $Ba^{2+}$ | $5.8 \times 10^7$ | 7.76 | $Pb^{2+}$ | $1.1 \times 10^{18}$ | 18.04 |
| $Mn^{2+}$ | $6.2 \times 10^{13}$ | 13.79 | $Al^{3+}$ | $1.3 \times 10^{16}$ | 16.13 |
| $Fe^{2+}$ | $2.1 \times 10^{14}$ | 14.33 | $Fe^{3+}$ | $1.3 \times 10^{25}$ | 25.1 |
| $Co^{2+}$ | $2.0 \times 10^{16}$ | 16.31 | $V^{3+}$ | $7.9 \times 10^{25}$ | 25.9 |
| $Ni^{2+}$ | $4.2 \times 10^{18}$ | 18.62 | $Th^{4+}$ | $1.6 \times 10^{23}$ | 23.2 |

*Data from G. Schwarzenbach, *Complexometric Titrations*, p. 8. London: Chapman and Hall, 1957. With permission. Constants are valid at 20°C and an ionic strength of 0.1.

concentration of $M^{n+}$ is equal to its analytical concentration, which in turn is readily derived from stoichiometric data.

Calculation of $M^{n+}$ at and beyond the equivalence point requires the use of Equation 13–1. The computation in this region is troublesome and time-consuming if the pH is unknown and variable because both $[MY^{(n-4)+}]$ and $[M^{n+}]$ are pH-dependent. Fortunately, EDTA titrations are always performed in solutions that are buffered to a known pH to avoid interference by other cations and to ensure satisfactory indicator behavior. Calculating $[M^{n+}]$ in a buffered solution containing EDTA is a relatively straightforward procedure provided the pH is known. In this computation, use is made of the alpha value for $H_4Y$. Recall (Section 11I) that $\alpha_4$ for $H_4Y$ is defined as

$$\alpha_4 = \frac{[Y^{4-}]}{c_T} \qquad (13\text{–}2)$$

$[Y^{4-}] = \alpha_4 c_T$

where $c_T$ is the total molar concentration of *uncomplexed* EDTA:

$$c_T = [Y^{4-}] + [HY^{3-}] + [H_2Y^{2-}] + [H_3Y^-] + [H_4Y]$$

### Conditional Formation Constants

*Conditional,* or *effective, formation constants* are pH-dependent equilibrium constants that apply *at a single pH only*. To obtain the conditional constant for the equilibrium shown in Equation 13–1, we substitute $\alpha_4 c_T$ from Equation 13–2 for $[Y^{4-}]$ in the formation-constant expression (Equation 13–1):

Conditional formation constants are pH-dependent.

$$M^{n+} + Y^{4-} \rightleftarrows MY^{(n-4)+} \qquad \frac{[MY^{(n-4)+}]}{[M^{n+}]\alpha_4 c_T} = K_{MY} \qquad (13\text{–}3)$$

Solving this equation for the product of the two constants $K_{MY}$ and $\alpha_4$ yields a new constant:

$$\frac{[MY^{(n-4)+}]}{[M^{n+}]c_T} = \alpha_4 K_{MY} = K'_{MY} \qquad (13\text{–}4)$$

where $K'_{MY}$, the conditional formation constant, describes equilibrium relationships *only at the pH for which $\alpha_4$ is applicable*.

Conditional constants are readily computed and provide a simple means by which the equilibrium concentrations of the metal ion and the complex can be calculated at any point in a titration curve. Note that replacement of $[Y^{4-}]$ with $c_T$ in the equilibrium-constant expression greatly simplifies calculations because $c_T$ is readily determined from the reaction stoichiometry, whereas $[Y^{4-}]$ is not.

### The Computation of $\alpha_4$ Values for EDTA Solutions

An expression for calculating $\alpha_4$ at a given hydrogen ion concentration is readily derived by the method demonstrated in Section 11I. Thus, $\alpha_4$ for EDTA is

Alpha values for the other EDTA species are readily obtained in a similar way and are found to be

$$\alpha_0 = [H^+]^4/D$$
$$\alpha_1 = K_1[H^+]^3/D$$
$$\alpha_2 = K_1K_2[H^+]^2/D$$
$$\alpha_3 = K_1K_2K_3[H^+]/D$$

Only $\alpha_4$ is needed, however, in deriving titration curves.

In this chapter and those that follow, we shall revert to the use of $H^+$ as a convenient shorthand notation for $H_3O^+$; from time to time, we shall refer to the species represented by $H^+$ as the hydrogen ion.

$$\alpha_4 = \frac{K_1K_2K_3K_4}{[H^+]^4 + K_1[H^+]^3 + K_1K_2[H^+]^2 + K_1K_2K_3[H^+] + K_1K_2K_3K_4} \quad (13\text{--}5)$$

$$\alpha_4 = \frac{K_1K_2K_3K_4}{D} \quad (13\text{--}6)$$

where $K_1$, $K_2$, $K_3$, and $K_4$ are the four dissociation constants for $H_4Y$ and $D$ is the denominator of Equation 13–5.

Table 13–2 lists $\alpha_4$ at selected pH values. Note that only about $4 \times 10^{-12}\%$ of the EDTA exists as $Y^{4-}$ at pH 2.00. Example 13–1 illustrates how $[Y^{4-}]$ is calculated for a solution of known pH.

---

Example 13–1

Calculate $\alpha_4$ and the mole percent of $Y^{4-}$ in a solution of EDTA that is buffered to pH 10.20.

$$[H^+] = \text{antilog}\,(-10.20) = 6.31 \times 10^{-11} \simeq 6.3 \times 10^{-11}$$

From the values for the dissociation constants for $H_4Y$ (page 288), we obtain

$$K_1 = 1.02 \times 10^{-2} \qquad K_1K_2K_3 = 1.51 \times 10^{-11}$$
$$K_1K_2 = 2.18 \times 10^{-5} \qquad K_1K_2K_3K_4 = 8.31 \times 10^{-22}$$

Numerical values for the several terms in the denominator in Equation 13–5 are

$$
\begin{aligned}
[H^+]^4 &= (6.31 \times 10^{-11})^4 & &= 1.58 \times 10^{-41} \\
K_1[H^+]^3 &= (1.02 \times 10^{-2})(6.31 \times 10^{-11})^3 & &= 2.56 \times 10^{-33} \\
K_1K_2[H^+]^2 &= (2.18 \times 10^{-5})(6.31 \times 10^{-11})^2 & &= 8.68 \times 10^{-26} \\
K_1K_2K_3[H^+] &= (1.51 \times 10^{-11})(6.31 \times 10^{-11}) & &= 9.53 \times 10^{-22} \\
K_1K_2K_3K_4 & & &= \underline{8.31 \times 10^{-22}} \\
& & D &= 1.78 \times 10^{-21}
\end{aligned}
$$

and Equation 13–6 becomes

$$\alpha_4 = \frac{K_1K_2K_3K_4}{D} = \frac{8.31 \times 10^{-22}}{1.78 \times 10^{-21}} = 0.466 \simeq 0.47$$
$$\text{mol \% } Y^{4-} = 0.47 \times 100\% = 47\%$$

Note that only the last two terms in the denominator contribute significantly to the sum $D$ at pH 10.20. At low pH values, in contrast, only the first two or three terms are important.

---

Table 13–2

VALUES FOR $\alpha_4$ FOR EDTA AT SELECTED pH VALUES

| pH | $\alpha_4$ | pH | $\alpha_4$ |
|---|---|---|---|
| 2.0 | $3.7 \times 10^{-14}$ | 7.0 | $4.8 \times 10^{-4}$ |
| 3.0 | $2.5 \times 10^{-11}$ | 8.0 | $5.4 \times 10^{-3}$ |
| 4.0 | $3.6 \times 10^{-9}$ | 9.0 | $5.2 \times 10^{-2}$ |
| 5.0 | $3.5 \times 10^{-7}$ | 10.0 | $3.5 \times 10^{-1}$ |
| 6.0 | $2.2 \times 10^{-5}$ | 11.0 | $8.5 \times 10^{-1}$ |
| | | 12.0 | $9.8 \times 10^{-1}$ |

## The Calculation of the Cation Concentrations in EDTA Solutions

Example 13–2 demonstrates how the cation concentration in a solution of an EDTA complex is computed. Example 13–3 illustrates this calculation for a solution that contains an excess of EDTA.

---

### Example 13–2

Calculate the equilibrium concentration of $Ni^{2+}$ in a solution with an analytical $NiY^{2-}$ concentration of 0.0150 M at pH (a) 3.0 and (b) 8.0.
From Table 13–1,

$$Ni^{2+} + Y^{4-} \rightleftarrows NiY^{2-} \qquad \frac{[NiY^{2-}]}{[Ni^{2+}][Y^{4-}]} = K_{MY} = 4.2 \times 10^{18}$$

The equilibrium concentration of $NiY^{2-}$ is equal to the analytical concentration of the complex minus the concentration lost by dissociation. The latter is given by the equilibrium nickel ion concentration. Thus,

Assume:

$[NiY^{2-}] \approx c_{Ni} = 0.0150 \gg [Ni^{2+}]$

$$[NiY^{2-}] = 0.0150 - [Ni^{2+}]$$

If we assume $[Ni^{2+}] \ll 0.0150$, an assumption that is almost certainly valid in light of the large formation constant of the complex, the foregoing equation simplifies to

$$[NiY^{2-}] \approx 0.0150$$

Since the complex is the only source of both $Ni^{2+}$ and the EDTA species,

$$[Ni^{2+}] = [Y^{4-}] + [HY^{3-}] + [H_2Y^{2-}] + [H_3Y^-] + [H_4Y] = c_T$$

Substitution of this equality into Equation 13–4 gives

$$\frac{[NiY^{2-}]}{[Ni^{2+}]c_T} = \frac{[NiY^{2-}]}{[Ni^{2+}]^2} = \alpha_4 K_{MY} = K'_{MY}$$

(a) Table 13–2 indicates that $\alpha_4$ is $2.5 \times 10^{-11}$ at pH 3.0. Substitution of this value and the concentration of $NiY^{2-}$ into the equation for $K'_{MY}$ gives

$$\frac{0.0150}{[Ni^{2+}]^2} = 2.5 \times 10^{-11} \times 4.2 \times 10^{18} = 1.05 \times 10^8$$

$$[Ni^{2+}] = \sqrt{1.43 \times 10^{-10}} = 1.2 \times 10^{-5} \text{ M}$$

Note that $[Ni^{2+}] \ll 0.0150$, as assumed.

Our assumption is valid.

(b) At pH 8.0, the conditional constant is much larger. Thus,

$$K'_{MY} = 5.4 \times 10^{-3} \times 4.2 \times 10^{18} = 2.27 \times 10^{16}$$

and substitution into the equation for $K'_{MY}$ followed by rearrangement gives

$$[Ni^{2+}] = \sqrt{0.0150/(2.27 \times 10^{16})} = 8.1 \times 10^{-10} \text{ M}$$

---

### Example 13-3

Calculate the concentration of $Ni^{2+}$ in a solution prepared by mixing 50.0 mL of 0.0300 M $Ni^{2+}$ with 50.0 mL of 0.0500 M EDTA. The mixture is buffered to a pH of 3.00.

Here, the solution has an excess of EDTA, and so the analytical concentration of the complex is determined by the amount of $Ni^{2+}$ originally present. Thus,

$$c_{NiY^{2-}} = 50.0 \times \frac{0.0300}{100} = 0.0150 \text{ M}$$

$$c_{EDTA} = \frac{50.0 \times 0.0500 - 50.0 \times 0.0300}{100} = 0.0100 \text{ M}$$

Again let us assume that $[Ni^{2+}] \ll [NiY^{2-}]$, so that

$$[NiY^{2-}] = 0.0150 - [Ni^{2+}] \approx 0.0150$$

At this point, the total concentration of uncomplexed EDTA species is:

$$c_T = 0.0100 \text{ M}$$

Substitution into Equation 13-4 gives

$$\frac{0.0150}{[Ni^{2+}]0.0100} = \alpha_4 K_{MY} = K'_{MY}$$

$$= 2.5 \times 10^{-11} \times 4.2 \times 10^{18} = 1.05 \times 10^{8}$$

Again our assumption is valid.

$$[Ni^{2+}] = \frac{0.0150}{0.0100 \times 1.05 \times 10^{8}} = 1.4 \times 10^{-8} \text{ M}$$

---

## 13B-4 The Derivation of EDTA Titration Curves

Example 13-4 demonstrates how an EDTA titration curve for a metal ion is derived for a solution of fixed pH.

---

### Example 13-4

Derive a curve (pCa as a function of EDTA volume) for the titration of 50.0 mL of 0.00500 M $Ca^{2+}$ with 0.0100 M EDTA in a solution buffered to pH 10.0.

**Calculation of a Conditional Constant**

The conditional formation constant for the calcium/EDTA complex at pH

10 is obtained from the formation constant of the complex (Table 13–1) and the $\alpha_4$ value for EDTA at pH 10 (Table 13–2). Thus, substitution into Equation 13–4 gives

$$\frac{[CaY^{2-}]}{[Ca^{2+}]c_T} = \alpha_4 K_{CaY} = K'_{CaY}$$

$$= 0.35 \times 5.0 \times 10^{10} = 1.75 \times 10^{10}$$

Preequivalence-Point Values for pCa

Before the equivalence point is reached, the equilibrium concentration of $Ca^{2+}$ is equal to the sum of the contributions from the untitrated excess of the cation and from the dissociation of the complex, the latter being numerically equal to $c_T$. It is ordinarily reasonable to assume that $c_T$ is small relative to the analytical concentration of the uncomplexed calcium ion. Thus, for example, after the addition of 10.0 mL of reagent,

$$[Ca^{2+}] = \frac{50.0 \times 0.00500 - 10.0 \times 0.0100}{60.0} + c_T \approx 2.50 \times 10^{-3} \text{ M}$$

$$pCa = -\log 2.50 \times 10^{-3} = 2.60$$

Other preequivalence-point data are calculated in this same way.

The Equivalence-Point pCa

Following the method shown in Example 13–2, we first compute the analytical concentration of $CaY^{2-}$:

$$c_{CaY^{2-}} = \frac{50.0 \times 0.00500}{50.0 + 25.0} = 3.33 \times 10^{-3} \text{ M}$$

The only source of $Ca^{2+}$ ions is the dissociation of this complex. It also follows that the calcium ion concentration must be identical to the sum of the concentrations of the uncomplexed EDTA, $c_T$. Thus,

$$[Ca^{2+}] = c_T$$

$$[CaY^{2-}] = 0.00333 - [Ca^{2+}] \approx 0.00333 \text{ M}$$

Substituting into the conditional formation-constant expression gives

$$\frac{[CaY^{2-}]}{[Ca^{2+}]C_T} = \frac{0.00333}{[Ca^{2+}]^2} = 1.75 \times 10^{10}$$

$$[Ca^{2+}] = \sqrt{\frac{0.00333}{1.75 \times 10^{10}}} = 4.36 \times 10^{-7} \text{ M}$$

$$pCa = -\log 4.36 \times 10^{-7} = 6.36$$

Postequivalence-Point pCa

Beyond the equivalence point, analytical concentrations of $CaY^{2-}$ and EDTA are obtained directly from the stoichiometric data. Then a calculation similar to that in Example 13–3 is performed. Thus, after the addition of 35.0 mL of reagent,

$$c_{CaY^{2-}} = \frac{50.0 \times 0.00500}{85.0} = 2.94 \times 10^{-3} \text{ M}$$

$$c_{EDTA} = \frac{35.0 \times 0.0100 - 50.0 \times 0.00500}{85.0} = 1.18 \times 10^{-3} \text{ M}$$

As approximations, we can write

$$[CaY^{2-}] = 2.94 \times 10^{-3} - [Ca^{2+}] \approx 2.94 \times 10^{-3}$$

$$c_T = 1.18 \times 10^{-3} + [Ca^{2+}] \approx 1.18 \times 10^{-3} \text{ M}$$

and substitution into the conditional formation-constant expression gives

$$\frac{2.94 \times 10^{-3}}{[Ca^{2+}] \times 1.18 \times 10^{-3}} = 1.75 \times 10^{10} = K'_{CaY}$$

$$[Ca^{2+}] = \frac{2.94 \times 10^{-3}}{1.18 \times 10^{-3} \times 1.75 \times 10^{10}} = 1.42 \times 10^{-10}$$

$$pCa = -\log 1.42 \times 10^{-10} = 9.85$$

The approximations were clearly valid.

Other postequivalence-point data are computed in this same way.

---

Curve A in Figure 13–5 is a plot of data for the titration in Example 13–4. Curve B is the titration curve for a solution of magnesium ion under identical conditions. The formation constant for the EDTA complex of magnesium is smaller than that for the calcium complex. Consequently, the reaction of calcium ion with the EDTA is more complete, and a larger change in p-function occurs in the equivalence region. The effect here is analogous to the effects seen earlier for precipitation and neutralization titrations.

Figure 13–6 shows titration curves for calcium ion in solutions buffered to various pH levels. Recall that $\alpha_4$, and hence $K'_{CaY}$, become smaller as the pH decreases. The less favorable equilibrium constant leads to a smaller change in pCa in the equivalence-point region. It is apparent from Figure 13–6 that an adequate end point in the titration of calcium requires a pH of about 8 or greater. As shown in Figure 13–7, however, cations with larger formation constants provide good end points even in acidic media. Figure 13–8 shows the minimum permissible pH for a satisfactory end point in the titration of various metal ions in the absence of competing complexing agents. Note that a moderately acidic environment is satisfactory for many divalent heavy-metal cations and that a strongly acidic medium can be tolerated in the titration of such ions as iron(III) and indium(III).

End points for EDTA titrations become less sharp with pH decreases because complex formation is less complete under these circumstances.

Figure 13–8 shows that most cations having a +3 or +4 charge can be titrated in a distinctly acidic solution.

## 13B–5 The Effect of Other Complexing Agents on EDTA Titration Curves

Many cations form hydrous oxide precipitates when the pH is raised to the level required for their successful titration with EDTA. When this

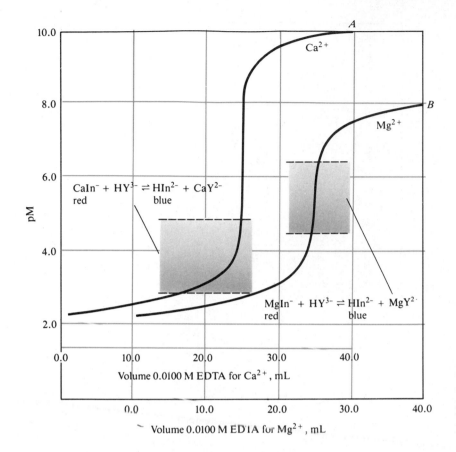

Figure 13–5

EDTA titration curves for 50.00 mL of 0.00500 M $Ca^{2+}$ ($K'$ for $CaY^{2-} = 1.75 \times 10^{10}$) and $Mg^{2+}$ ($K'$ for $MgY^{2-} = 1.72 \times 10^{8}$) at pH 10.0. The shaded areas show the transition range for Eriochrome Black T.

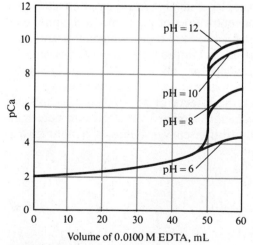

Figure 13–6

Influence of pH on the titration of 0.0100 M $Ca^{2+}$ with 0.0100 M EDTA.

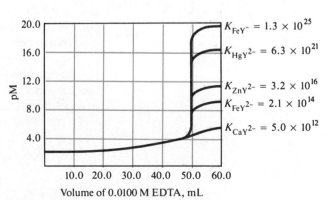

Figure 13–7

Titration curves for 50.0 mL of 0.0100 M cation solutions at pH 6.0.

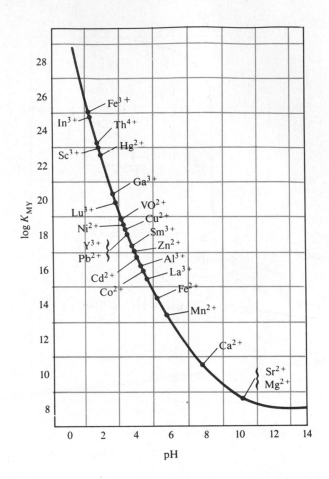

Figure 13–8

Minimum pH needed for satisfactory titration of various cations with EDTA. (Reprinted with permission, C. N. Reilley and R. W. Schmid, *Anal. Chem.*, **1958**, *30*, 947. Copyright 1958 American Chemical Society.)

Often, auxiliary complexing agents must be used in EDTA titrations to prevent precipitation of the analyte as a hydrous oxide. Such reagents cause end points to be less sharp.

problem is encountered, an auxiliary complexing agent is needed to keep the cation in solution. For example, zinc(II) is ordinarily titrated in a medium that has fairly high concentrations of ammonia and ammonium chloride. These species buffer the solution to a pH that ensures complete reaction between cation and titrant; in addition, ammonia forms ammine complexes with zinc(II) and prevents formation of the sparingly soluble zinc hydroxide, particularly in the early stages of the titration.

A quantitative description of the effects of an auxiliary complexing reagent can be derived by a procedure similar to that used to determine the influence of pH on EDTA titration curves. Here, a quantity $\alpha_M$ is defined that is analogous to $\alpha_4$:

$$\alpha_M = \frac{[M^{n+}]}{c_M} \qquad (13\text{--}7)$$

where $c_M$ is the sum of the concentrations of species containing the metal ion *exclusive* of that combined with EDTA. For solutions containing zinc(II) and ammonia, then

$$c_M = [Zn^{2+}]+[Zn(NH_3)^{2+}]+[Zn(NH_3)_2^{2+}]+[Zn(NH_3)_3^{2+}]+[Zn(NH_3)_4^{2+}]$$

$$(13\text{--}8)$$

The value of $\alpha_M$ can be expressed readily in terms of the ammonia concentration and the formation constants for the various ammine complexes. To arrive at such an expression, we write

$$K_1 = \frac{[Zn(NH_3)^{2+}]}{[Zn^{2+}][NH_3]}$$

$$[Zn(NH_3)^{2+}] = K_1[Zn^{2+}][NH_3]$$

Similarly, it is readily shown that

$$[Zn(NH_3)_2^{2+}] = K_1K_2[Zn^{2+}][NH_3]^2$$

$$[Zn(NH_3)_3^{2+}] = K_1K_2K_3[Zn^{2+}][NH_3]^3$$

$$[Zn(NH_3)_4^{2+}] = K_1K_2K_3K_4[Zn^{2+}][NH_3]^4$$

Substitution of these expressions into Equation 13–8 gives

$$c_M = [Zn^{2+}](1+K_1[NH_3]+K_1K_2[NH_3]^2+K_1K_2K_3[NH_3]^3+K_1K_2K_3K_4[NH_3]^4)$$

Substituting this expression for $c_M$ in Equation 13–7 (here, $[M^{n+}] = [Zn^{2+}]$)

$$\alpha_M = \frac{1}{1 + K_1[NH_3] + K_1K_2[NH_3]^2 + K_1K_2K_3[NH_3]^3 + K_1K_2K_3K_4[NH_3]^4}$$

$$(13\text{--}9)$$

Finally, a conditional constant for the equilibrium between EDTA and zinc(II) in an ammonia/ammonium chloride buffer is obtained by substituting Equation 13–7 into Equation 13–4 and rearranging:

$$\frac{[ZnY^{2-}]}{c_M c_T} = \alpha_4\alpha_M K_{ZnY} = K''_{ZnY} \qquad (13\text{--}10)$$

where $K''_{ZnY}$ is a new conditional constant that applies at a single pH as well as a single concentration of ammonia. The following example shows how this conditional constant is employed in the derivation of titration curves.

---

### Example 13–5

Calculate the pZn of solutions prepared by adding 20.0, 25.0, and 30.0 mL of 0.0100 M EDTA to 50.0 mL of 0.00500 M $Zn^{2+}$. Assume that both the $Zn^{2+}$ and EDTA solutions are 0.100 M in $NH_3$ and 0.176 M in $NH_4Cl$ to provide a constant pH of 9.0.

In Appendix 5, we find that the logarithms of the stepwise formation constants for the four zinc complexes with ammonia are 2.4, 2.4, 2.5, and 2.1. Thus,

$$K_1 = \text{antilog } 2.4 = 2.51 \times 10^2$$

$$K_1 K_2 = \text{antilog } (2.4 + 2.4) = 6.31 \times 10^4$$

$$K_1 K_2 K_3 = \text{antilog } (2.4 + 2.4 + 2.5) = 2.00 \times 10^7$$

$$K_1 K_2 K_3 K_4 = \text{antilog } (2.4 + 2.4 + 2.5 + 2.1) = 2.51 \times 10^9$$

Calculation of a Conditional Constant

A value for $\alpha_M$ can be obtained from Equation 13–9 by assuming that the molar and analytical concentrations of ammonia are essentially the same; thus, for $[NH_3] = 0.100$,

$$\alpha_M = \frac{1}{1 + 25 + 631 + 2.00 \times 10^4 + 2.51 \times 10^5} = 3.68 \times 10^{-6}$$

A value for $K_{ZnY}$ is found in Table 13–1, and $\alpha_4$ for pH 9.0 is given in Table 13–2. Substituting into Equation 13–10, we find

$$K''_{ZnY} = 5.2 \times 10^{-2} \times 3.68 \times 10^{-6} \times 3.2 \times 10^{16}$$

$$= 6.12 \times 10^9 \simeq 6.1 \times 10^9$$

Calculation of pZn After Addition of 20.0 mL of EDTA

At this point, only part of the zinc has been complexed by EDTA. The remainder is present as $Zn^{2+}$ and the four ammine complexes. By definition, the sum of the concentrations of these five species is $c_M$. Therefore,

$$c_M = \frac{50.0 \times 0.00500 - 20.0 \times 0.0100}{70.0} = 7.14 \times 10^{-4} \text{ M}$$

Substitution of this value into Equation 13–7 gives

$$[Zn^{2+}] = c_M \alpha_M = (7.14 \times 10^{-4})(3.68 \times 10^{-6}) = 2.63 \times 10^{-9} \text{ M}$$

$$pZn = 8.58$$

Calculation of pZn After Addition of 25.0 mL of EDTA

At the equivalence point, the analytical concentration of $ZnY^{2-}$ is

$$c_{ZnY^{2-}} = \frac{50.0 \times 0.00500}{50.0 + 25.0} = 3.33 \times 10^{-3} \text{ M}$$

The sum of the concentrations of the various zinc species not combined with EDTA equals the sum of the concentrations of the uncomplexed EDTA species:

$$c_M = c_T$$

and

$$[ZnY^{2-}] = 3.33 \times 10^{-3} - c_M \simeq 3.33 \times 10^{-3} \text{ M}$$

Substituting into Equation 13–10, we have

$$\frac{3.33 \times 10^{-3}}{c_M^2} = 6.12 \times 10^9 = K''_{ZnY}$$

$$c_M = 7.38 \times 10^{-7} \text{ M}$$

Employing Equation 13–7, we obtain

$$[Zn^{2+}] = c_M\alpha_M = (7.38 \times 10^{-7})(3.68 \times 10^{-6}) = 2.72 \times 10^{-12}$$
$$pZn = 11.57$$

Calculation of pZn After Addition of 30.0 mL of EDTA
The solution now contains an excess of EDTA; thus,

$$c_{EDTA} = c_T = \frac{30.0 \times 0.0100 - 50.0 \times 0.00500}{80.0} = 6.25 \times 10^{-4}\ M$$

and since essentially all the original $Zn^{2+}$ is now complexed,

$$c_{ZnY^{2+}} = [ZnY^{2-}] = \frac{50.0 \times 0.00500}{80.0} = 3.12 \times 10^{-3}\ M$$

Rearranging Equation 13–10 gives

$$c_M = \frac{[ZnY^{2-}]}{c_T K''_{ZnY}} = \frac{3.12 \times 10^{-3}}{(6.25 \times 10^{-4})(6.12 \times 10^9)} = 8.16 \times 10^{-10}\ M$$

and, from Equation 13–7,

$$[Zn^{2+}] = c_M\alpha_M = (8.16 \times 10^{-10})(3.68 \times 10^{-6}) = 3.00 \times 10^{-15}$$
$$pZn = 14.52$$

Figure 13–9 shows two theoretical curves for the titration of zinc(II) with EDTA at pH 9.00. The equilibrium concentration of ammonia was 0.100 M for one titration and 0.0100 M for the other. Note that the pres-

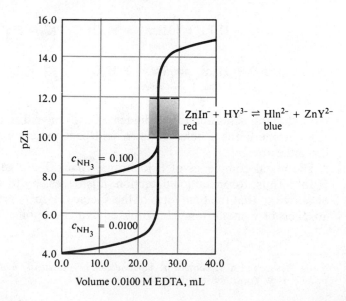

Figure 13–9
Influence of ammonia concentration on the end point for the titration of 50.0 mL of 0.00500 M $Zn^{2+}$. Solutions are buffered to pH 9.00. The shaded area shows the transition range for Eriochrome Black T.

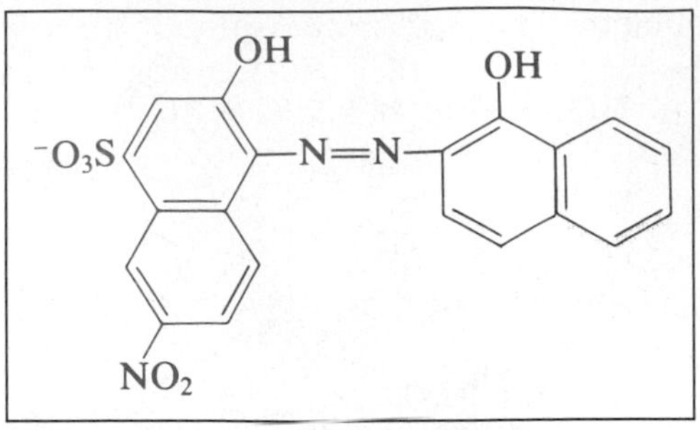

Figure 13–10
Eriochrome Black T.

ence of ammonia decreases the change in pZn near the equivalence point. For this reason, the concentration of an auxiliary complexing reagent should always be kept to the minimum required to prevent analyte precipitation. Note that the auxiliary complexing agent does not affect pZn beyond the equivalence point. Keep in mind, on the other hand, that $\alpha_4$, and thus pH, play an important role in defining this part of the titration curve (Figure 13–6).

## 13B–6 Indicators for EDTA Titrations

Reilley and Barnard[2] have listed nearly 200 organic compounds that have been suggested as indicators for metal ions in EDTA titrations. In general, these indicators are organic dyes that form colored chelates with metal ions in a pM range that is characteristic of the particular cation and dye. The complexes are often intensely colored and are discernible to the eye at concentrations in the range of $10^{-6}$ to $10^{-7}$ M.

*Eriochrome Black T* is a typical metal ion indicator that is used in the titration of several common cations. As shown in Figure 13–10, this compound contains a sulfonic acid group, which is completely dissociated in water, and two phenolic groups, which dissociate only partially. Its behavior as a weak acid is described by the equations

$$H_2O + H_2In^- \rightleftharpoons HIn^{2-} + H_3O^+ \qquad K_1 = 5 \times 10^{-7}$$
$$\text{Red} \qquad\qquad \text{Blue}$$

$$H_2O + HIn^{2-} \rightleftharpoons In^{3-} + H_3O^+ \qquad K_2 = 2.8 \times 10^{-12}$$
$$\text{Blue} \qquad\qquad \text{Orange}$$

Note that the acids and their conjugate bases have different colors. Thus, Eriochrome Black T behaves as an acid/base indicator as well as a metal ion indicator.

The metal complexes of Eriochrome Black T are generally red, as is $H_2In^-$. Thus, for metal ion detection, it is necessary to adjust the pH to 7 or above so that the blue form of the species, $HIn^{2-}$, predominates in the absence of a metal ion. Until the equivalence point in a titration, the

[2]C. N. Reilley and A. J. Barnard Jr., in *Handbook of Analytical Chemistry*, L. Meites, Ed., p. 3-77. New York: McGraw-Hill, 1963.

indicator complexes the excess metal ion, and so the solution is red. When EDTA becomes present in slight excess, the solution turns blue as a consequence of the reaction

$$MIn^- + HY^{3-} \rightleftharpoons HIn^{2-} + MY^{2-}$$

Red            Blue

Eriochrome Black T forms red complexes with more than two dozen metal ions, but the formation constants of only a few are appropriate for end-point detection. As shown in the following example, the applicability of a given indicator to an EDTA titration can be determined from the change in pM in the equivalence-point region, provided the formation constant for the metal-indicator complex is known.[3]

---

### Example 13–6

Determine the transition ranges for Eriochrome Black T in titrations of $Mg^{2+}$ and $Ca^{2+}$ at pH 10.0, given (a) that the second acid dissociation constant for the indicator is

$$HIn^{2-} + H_2O \rightleftharpoons In^{3-} + H_3O^+ \qquad K_2 = 2.8 \times 10^{-12} = \frac{[H_3O^+][In^{3-}]}{[HIn^{2-}]}$$

(b) that the formation constant for $MgIn^-$ is

$$Mg^{2+} + In^{3-} \rightleftharpoons MgIn^- \qquad K_f = 1.0 \times 10^7 = \frac{[MgIn^-]}{[Mg^{2+}][In^{3-}]}$$

and (c) that the analogous constant for $Ca^{2+}$ is $2.5 \times 10^5$.

We shall assume, as we did earlier (Section 10A–2), that a detectable color change requires a tenfold excess of one or the other of the colored species; that is, a detectable color change is observed when $[MgIn^-]/[HIn^{2-}]$ changes from 10 to 0.10.

Multiplication of $K_2$ for the indicator by $K_f$ for $MgIn^-$ gives an expression that contains the foregoing ratio:

$$\frac{[MgIn^-][H_3O^+]}{[HIn^{2-}][Mg^{2+}]} = 2.8 \times 10^{-12} \times 1.0 \times 10^7 = 2.8 \times 10^{-5}$$

which rearranges to

$$[Mg^{2+}] = \frac{[MgIn^-]}{[HIn^{2-}]} \times \frac{[H_3O^+]}{2.8 \times 10^{-5}}$$

Substitution of $1.0 \times 10^{-10}$ for $[H_3O^+]$ and 10 and 0.10 for the ratio yields the range of $[Mg^{2+}]$ over which the color change occurs:

---

[3]For a complete discussion of the principles of indicator choice in complex-formation titrations, see C. N. Reilley and R. W. Schmid, *Anal. Chem.*, **1959**, *31*, 887.

$$[Mg^{2+}] = 3.6 \times 10^{-5} \text{ to } 3.6 \times 10^{-7}$$

$$pMg = 5.4 \pm 1.0$$

Proceeding in the same way, we find the range for pCa to be $3.8 \pm 1.0$.

---

The ranges for magnesium and calcium are indicated on the titration curves in Figure 13–5. Eriochrome Black T is clearly an ideal indicator for magnesium but a totally unsatisfactory one for calcium. Note that the formation constant for $CaIn^-$ is only about one fortieth that for magnesium. As a consequence, significant conversion of $CaIn^-$ to $HIn^{2-}$ occurs well before equivalence.

A similar calculation reveals that Eriochrome Black T is also well suited for the titration of zinc with EDTA (Figure 13–9).

A limitation of Eriochrome Black T is that its solutions decompose slowly with standing. It is claimed that solutions of Calmagite, an indicator that for all practical purposes is identical in behavior to Eriochrome Black T, do not suffer this disadvantage. The structure of Calmagite is similar to that of Eriochrome Black T:

Many other metal indicators have been developed for EDTA titrations.[4] In contrast to Eriochrome Black T, some of these indicators can be employed in strongly acidic media.

### 13B–7 Titration Methods Employing EDTA

The paragraphs that follow describe several methods for the determination of cations with EDTA.

#### Direct Titration

*Direct-titration procedures with a metal ion indicator are the easiest and the most convenient to use.*

**Methods Based on Indicators for the Analyte Ion.** Reilley and Barnard[5] list 40 elements that can be determined by direct titration with EDTA using metal ion indicators. The direct method is not applicable to all

---

[4]See, for example, L. Meites, *Handbook of Analytical Chemistry,* pp. 3-101 to 3-165. New York: McGraw-Hill, 1963.

[5]C. N. Reilley and A. J. Barnard Jr., in *Handbook of Analytical Chemistry,* L. Meites, Ed., pp. 3-166 to 3-200. New York: McGraw-Hill, 1963.

cations, however, either because no good indicators have been developed or because the reaction between the metal ion and EDTA is so slow as to make titration impractical. Such cations can usually be determined by one of the several methods described in the paragraphs that follow.

**Methods Based on Indicators for an Added Metal Ion.** It is often convenient to introduce into an EDTA solution a small amount of a cation that forms an EDTA complex that is less stable than the analyte complex and for which a good indicator exists. For example, indicators for calcium ion are generally less satisfactory than those we have described for magnesium ion. Consequently, a small amount of magnesium chloride is often added to an EDTA solution that is to be used for the determination of calcium with Eriochrome Black T. In the initial stages in the titration, magnesium ions are displaced from the EDTA complex by calcium ions and are free to combine with the Eriochrome Black T, thus imparting a red color to the solution. When all of the calcium ions have been complexed, however, the liberated magnesium ions again combine with the EDTA until the end point is observed. This procedure requires standardization of the EDTA solution against primary-standard calcium carbonate.

**Potentiometric Methods.** Potential measurements can be used for endpoint detection in the EDTA titration of those metal ions for which specific ion electrodes are available. Electrodes of this type are described in Section 17D. In addition, a mercury electrode can be made sensitive to EDTA ions and used in titrations with this reagent.

## Back-Titration Methods

Back-titrations are useful for the determination of cations that form stable EDTA complexes and for which a satisfactory indicator is not available. The method is also useful for cations that react only slowly with EDTA. A measured excess of standard EDTA solution is added to the analyte solution. After the reaction is judged complete, the excess EDTA is back-titrated with a standard magnesium or zinc ion solution to an Eriochrome Black T or Calmagite end point.[6] For this procedure to be successful, it is necessary that the magnesium or zinc ions form an EDTA complex that is less stable than the corresponding analyte complex.

Back-titration is also useful for analyzing samples that contain anions that would otherwise form sparingly soluble precipitates with the analyte under the analytical conditions. Here, the excess EDTA prevents precipitate formation.

Back-titration procedures are used when no suitable indicator is available, when the reaction between analyte and EDTA is slow, or when the analyte forms precipitates at the pH required for its titration.

## Displacement Methods

In displacement titrations, an unmeasured excess of a solution containing the magnesium or zinc complex of EDTA is introduced into the analyte

---

[6]For an analysis of the back-titration procedure, see C. Macca and M. Fiorana, *J. Chem. Educ.*, **1986**, *63*, 121.

solution. If the analyte forms a more stable complex than that of magnesium or zinc, the following displacement reaction occurs:

$$MgY^{2-} + M^{2+} \rightarrow MY^{2-} + Mg^{2+}$$

where $M^{2+}$ represents the analyte cation. The liberated magnesium or zinc is then titrated with a standard EDTA solution.

Displacement methods are useful where no satisfactory indicator is available for the metal ion being determined.

*Displacement titrations are used when no indicator for an analyte is available.*

### 13B–8 The Scope of EDTA Titrations

Complexometric titrations with EDTA have been applied to the determination of virtually every metal cation with the exception of the alkali metal ions. Because of the ubiquity of EDTA chelates, the reagent might appear at first glance to be totally lacking in selectivity. In fact, however, considerable control over interferences can be realized by pH regulation. For example, trivalent cations can usually be titrated without interference from divalent species in solutions that have a pH of about 1 (Figure 13–8). At this pH, the less stable divalent chelates do not form to any significant extent, but the trivalent ions are quantitatively complexed.

Similarly, ions such as cadmium and zinc, which form more stable EDTA chelates than does magnesium, can be determined in the presence of the latter ion by buffering the mixture to pH 7 before titration. Eriochrome Black T serves as an indicator for the cadmium or zinc end points without interference from magnesium because the indicator chelate with magnesium is not formed at this pH.

*A masking agent is a complexing agent that reacts selectively with a component in a solution and in so doing prevents that component from interfering in an analysis.*

Finally, interference from a particular cation can sometimes be eliminated by adding a suitable *masking agent*, an auxiliary ligand that preferentially forms highly stable complexes with the potential interference.[7] For example, cyanide ion is often employed as a masking agent to permit the titration of magnesium and calcium ions in the presence of ions such as cadmium, cobalt, copper, nickel, zinc, and palladium. All of the latter form sufficiently stable cyanide complexes to prevent reaction with EDTA.

Specific directions for the preparation and use of EDTA solutions are given in Section 34D.

### 13B–9 The Determination of Water Hardness

*Hard water contains calcium, magnesium, and heavy-metal ions that form precipitates with soap (but not detergents).*

Historically, water "hardness" was defined in terms of the capacity of cations in the water to replace the sodium or potassium ions in soaps and

[7]For further information, see D. D. Perrin, *Masking and Demasking of Chemical Reactions*. New York: Wiley-Interscience, 1970; and C. N. Reilley and A. J. Barnard Jr., in *Handbook of Analytical Chemistry*, L. Meites, Ed., pp. 3-208 to 3-225. New York: McGraw-Hill, 1963.

form sparingly soluble products. Most multiply charged cations share this undesirable property. In natural waters, however, the concentration of calcium and magnesium ions generally far exceeds that of any other metal ion. Consequently, hardness is now expressed in terms of the concentration of calcium carbonate that is equivalent to the total concentration of all the multivalent cations in the sample.

The determination of hardness is a useful analytical test that provides a measure of the quality of water for household and industrial uses. The test is important to industry because hard water, upon being heated, precipitates calcium carbonate, which then clogs boilers and pipes.

Water hardness is ordinarily determined by an EDTA titration after the sample has been buffered to pH 10. Magnesium, which forms the least stable EDTA complex of all of the common multivalent cations in typical water samples, is not titrated until enough reagent has been added to complex all of the other cations in the sample. Therefore, a magnesium ion indicator, such as Calmagite or Eriochrome Black T, can serve as indicator in water-hardness titrations. Often, a small concentration of the EDTA complex of magnesium is incorporated in the buffer or in the titrant to ensure the presence of sufficient magnesium ions for satisfactory indicator action.

Test kits for determining the hardness of household water are available commercially. They consist of a vessel calibrated to contain a known volume of water, a measuring scoop to deliver an appropriate amount of a solid buffer mixture, an indicator solution, and a bottle of standard EDTA, which is equipped with a medicine dropper. The volume of standard reagent consumed is obtained by counting drops to the end point. The concentration of the EDTA solution is ordinarily such that one drop corresponds to one grain (about 0.065 g) of calcium carbonate per gallon of water.

## 13C   TITRATIONS WITH INORGANIC COMPLEXING AGENTS

Complexometric titrations with inorganic reagents are among the oldest volumetric methods.[8] For example, the titration of iodide ion with mercury(II) ions,

$$Hg^{2+} + 4\ I^- \rightleftharpoons HgI_4^{2-}$$

was first described in 1833. Table 13–3 lists the common inorganic complexing agents as well as some of their applications.

---

[8]For further information, see I. M. Kolthoff and V. A. Stenger, *Volumetric Analysis*, Vol. 2, pp. 282–331. New York: Interscience, 1947.

**Table 13–3**
**TYPICAL INORGANIC COMPLEX-FORMATION TITRATIONS***

| Titrant | Analyte | Remarks |
|---|---|---|
| $Hg(NO_3)_2$ | $Br^-$, $Cl^-$, $SCN^-$, $CN^-$, thiourea | Products are neutral mercury(II) complexes; various indicators used |
| $AgNO_3$ | $CN^-$ | Product is $Ag(CN)_2^-$; indicator is $I^-$; titrate to first turbidity of AgI |
| $NiSO_4$ | $CN^-$ | Product is $Ni(CN)_4^{2-}$; indicator is AgI; titrate to first turbidity of AgI |
| KCN | $Cu^{2+}$, $Hg^{2+}$, $Ni^{2+}$ | Products are $Cu(CN)_4^{2-}$, $Hg(CN)_2$, $Ni(CN)_4^{2-}$; various indicators used |

*For further applications and selected references, see L. Meites, *Handbook of Analytical Chemistry*, p. 3-226. New York: McGraw-Hill, 1963.

## 13D    QUESTIONS AND PROBLEMS

**13–1.** Define
  *(a) chelate.
  (b) tetradentate chelating agent.
  *(c) ligand.
  (d) coordination number.
  *(e) conditional formation constant.
  (f) NTA.
  *(g) water hardness.
  (h) EDTA displacement titration.

**\*13–2.** Describe three general methods for performing EDTA titrations. What are the advantages of each?

**13–3.** Propose a complexometric method for the determination of the individual components in a solution containing $In^{3+}$, $Zn^{2+}$, and $Mg^{2+}$.

**\*13–4.** Why are multidentate ligands preferable to unidentate ligands for complexometric titrations?

**13–5.** Write chemical equations and equilibrium-constant expressions for the stepwise formation of
  *(a) $Ag(S_2O_3)_2^{3-}$.    *(c) $Cd(NH_3)_4^{2+}$.
  (b) $AlF_6^{3-}$.       (d) $Ni(SCN)_3^-$.

**13–6.** Explain how stepwise and overall formation constants are related.

**\*13–7.** Why is a small amount of $MgY^{2-}$ often added to a solution that is to be titrated for hardness?

**\*13–8.** An EDTA solution was prepared by dissolving 3.853 g of purified and dried $Na_2H_2Y \cdot 2\ H_2O$ in sufficient water to give 1.000 L. Calculate the molar concentration, given that the solute contained 0.3% excess moisture (page 289).

**13–9.** An EDTA solution was prepared by dissolving about 3.0 g of $Na_2H_2Y \cdot 2\ H_2O$ in approximately 1 L of water. Calculate the molar concentration of this solution if an average of 32.22 mL was required to titrate 50.00-mL aliquots of 0.004517 M $Mg^{2+}$.

**\*13–10.** Calculate the concentration of an EDTA solution from the accompanying titration data:

| | Buret Containing | |
|---|---|---|
| | EDTA Soln | 0.01470 M $Mg^{2+}$ Soln |
| Final volume, mL | 46.39 | 31.69 |
| Initial volume, mL | 0.04 | 0.00 |

**13–11.** A 50.00-mL aliquot of a solution containing iron(II) and iron(III) required 13.73 mL of 0.01200 M EDTA when titrated at pH 2.0 and 29.62 mL when titrated at pH 6.0. Express the concentration of the solution in terms of the parts per million of each solute.

**\*13–12.** A 24-hr urine specimen was diluted to 2.000 L. After being buffered to pH 10, a 10.00-mL aliquot was titrated with 26.81 mL of 0.003474 M EDTA. The calcium in a second 10.00-mL aliquot was isolated as $CaC_2O_4(s)$, redissolved in acid, and titrated with 11.63 mL of the EDTA solution. Assuming that 15 to 300 mg of magnesium and 50 to 400 mg of calcium per day are normal, did this specimen fall within these ranges?

**13–13.** An EDTA solution was prepared by dissolving approximately 4 g of the disodium salt in approximately 1 L of water. An average of 42.35 mL of this solution was required to titrate 50.00-mL aliquots of a standard that contained 0.7682 g of $MgCO_3$ per liter. Titration of a 25.00-mL sample of mineral water at pH 10 required 18.81 mL of the EDTA solution. A 50.00-mL aliquot of the mineral water was rendered strongly alkaline to precipitate the magnesium as $Mg(OH)_2$. Titration with a calcium-specific indicator required 31.54 mL of the EDTA solution. Calculate
  (a) the molarity of the EDTA solution.

(b) the ppm of $CaCO_3$ in the mineral water.

(c) the ppm of $MgCO_3$ in the mineral water.

*13–14. The sulfate in a 1.515-g sample was homogeneously precipitated as $BaSO_4$ by adding an excess of $BaY^{2-}$ solution and slowly increasing the acid concentration to liberate $Ba^{2+}$. The precipitate was filtered and washed, and the filtrate and washings were collected in a 250-mL volumetric flask. A pH-10.00 buffer was added and the solution diluted to the mark. A 25.00-mL aliquot required a 28.73-mL titration with 0.01545 M $Mg^{2+}$ solution. Express the results of this analysis in terms of percent $Na_2SO_4 \cdot 10\ H_2O$.

13–15. Calamine, which is used for relief of skin irritations, is a mixture of zinc and iron oxides. A 1.022-g specimen of dried calamine was dissolved in acid and diluted to 250.0 mL. Potassium fluoride was added to a 10.00-mL aliquot of the diluted solution to mask the iron; after suitable adjustment of the pH, $Zn^{2+}$ consumed 38.78 mL of 0.01294 M EDTA. A second 50.00-mL aliquot was suitably buffered and titrated with 2.40 mL of 0.002727 M $ZnY^{2-}$ solution:

$$Fe^{3+} + ZnY^{2-} \rightarrow FeY^- + Zn^{2+}$$

Calculate the percentages of $ZnO$ and $Fe_2O_3$ in the sample.

*13–16. A 3.650-g sample containing bromate and bromide was dissolved in sufficient water to give 250.0 mL. After acidification, silver nitrate was introduced to a 25.00-mL aliquot to precipitate AgBr, which was filtered, washed, and then redissolved in an ammoniacal solution of potassium tetracyanonickelate(II):

$$Ni(CN)_4^{2-} + 2\ AgBr(s) \rightarrow$$
$$2\ Ag(CN)_2^- + Ni^{2+} + 2\ Br^-$$

The liberated nickel ion required 26.73 mL of 0.02089 M EDTA. The bromate in a 10.00-mL aliquot was reduced to bromide with arsenic(III) prior to the addition of silver nitrate; 21.94 mL of the EDTA solution was needed to titrate the nickel ion subsequently released. Calculate the percentages of $NaBr$ and $NaBrO_3$ in the sample.

13–17. The chromium ($d$ = 7.10 g/$cm^3$) plated on a 9.75-$cm^2$ surface was dissolved with hydrochloric acid and diluted to 100.0 mL. A 25.00-mL aliquot was buffered to pH 5, and 50.00 mL of 0.00862 M EDTA was added. Titration of the excess chelating reagent required 7.36 mL of 0.01044 M $Zn^{2+}$. Calculate the average thickness of the chromium plating.

*13–18. The silver ion in a 25.00-mL sample was converted to dicyanoargentate(I) ion by the addition of an excess of a solution containing $Ni(CN)_4^{2-}$:

$$Ni(CN)_4^{2-} + 2\ Ag^+ \rightarrow 2\ Ag(CN)_2^- + Ni^{2+}$$

The liberated nickel ion was titrated with 43.77 mL of 0.02408 M EDTA. Calculate the molar concentration of the silver solution.

13–19. The potassium ion in a 250.0-mL sample of mineral water was precipitated with sodium tetraphenylboron:

$$K^+ + B(C_6H_4)_4^- \rightarrow KB(C_6H_5)_4(s)$$

The precipitate was filtered, washed, and redissolved in an organic solvent. An excess of the mercury(II)/EDTA chelate was added:

$$4\ HgY^{2-} + B(C_6H_4)_4^- + 4\ H_2O \rightarrow$$
$$H_3BO_3 + 4\ C_6H_5Hg^+ + 4\ HY^{3-} + OH^-$$

The liberated EDTA was titrated with 29.64 mL of 0.05581 M $Mg^{2+}$. Calculate the potassium ion concentration in parts per million.

*13–20. A 0.4085-g sample containing lead, magnesium, and zinc was dissolved and treated with cyanide to complex and mask the zinc:

$$Zn^{2+} + 4\ CN^- \rightarrow Zn(CN)_4^{2-}$$

Titration of the lead and magnesium required 42.22 mL of 0.02064 M EDTA. The lead was next masked with BAL (2,3-dimercaptopropanol), and the released EDTA was titrated with 19.35 mL of a 0.007657 M magnesium solution. Finally, formaldehyde was introduced to demask the zinc:

$$Zn(CN)_4^{2-} + 4\ HCHO + 4\ H_2O \rightarrow$$
$$Zn^{2+} + 4\ HOCH_2CN + 4\ OH^-$$

which was titrated with 28.63 mL of 0.02064 M EDTA. Calculate the percentages of the three metals in the sample.

13–21. Chromel is an alloy composed of nickel, iron, and chromium. A 0.6472-g sample was dissolved and diluted to 250.0 mL. When a 50.00-mL aliquot of 0.05180 M EDTA was mixed with an equal volume of the diluted sample, all three ions were chelated, and a 5.11-mL back-titration with 0.06241 M copper(II) was required. The chromium in a second 50.0-mL aliquot was masked through the addition of hexamethylenetetramine; titration of the Fe and Ni required 36.28 mL of 0.05182 M EDTA. Iron and chromium were masked with pyrophosphate in a third 50.0-mL aliquot, and the nickel was titrated with 25.91 mL of the EDTA solution. Calculate the percentages of nickel, chromium, and iron in the alloy.

*13–22. A 0.3284-g sample of brass (containing lead, zinc, copper, and tin) was dissolved in nitric acid. The sparingly soluble $SnO_2 \cdot 4\ H_2O$ was removed by filtration, and the combined filtrate and washings were then diluted to 500.0 mL. A 10.00-mL aliquot was suitably buffered; titration of the lead, zinc, and copper in this aliquot required 37.56 mL of 0.002500 M EDTA. The copper in a 25.00-mL aliquot was masked with thiosulfate; the lead and zinc were then titrated with 27.67 mL of the EDTA solution. Cyanide ion was used to mask the copper and zinc in a 100-mL aliquot; 10.80 mL of the EDTA solution was needed to titrate the lead ion. Determine the composition of the brass sample; evaluate the percentage of tin by difference.

*13–23. Calculate conditional constants for the formation of the EDTA complex of $Mn^{2+}$ at pH (a) 6.0, (b) 8.0, and (c) 10.0.

13–24. Calculate conditional constants for the formation of the EDTA complex of $Sr^{2+}$ at pH (a) 7.0, (b) 9.0, and (c) 11.0.

*13–25. Formation constants for the successive ammine complexes of cadmium are 320, 91, 20, and 6.2. Calculate the conditional constant for the reaction of $Cd^{2+}$ with $Y^{4-}$ when
   (a) the molar concentration of $NH_3$ is 0.050 and the pH is 9.0.
   (b) the molar concentration of $NH_3$ is 0.050 and the pH is 11.0.
   (c) the molar concentration of $NH_3$ is 0.50 and the pH is 9.0.
   (d) the molar concentration of $NH_3$ is 0.50 and the pH is 11.0.

13–26. Formation constants for the successive ethylenediamine (en) complexes of nickel are $4.6 \times 10^7$, $2.5 \times 10^6$, and $3.4 \times 10^4$. Calculate the conditional constant for the reaction of $Ni^{2+}$ with $Y^{4-}$ when

   (a) the concentration of en is 0.025 M and the pH is 9.0.
   (b) the concentration of en is 0.025 M and the pH is 11.0.
   (c) the concentration of en is 0.25 M and the pH is 9.0.
   (d) the concentration of en is 0.25 M and the pH is 11.0.

*13–27. Derive a titration curve for 50.00 mL of 0.01000 M $Sr^{2+}$ with 0.02000 M EDTA in a solution buffered to pH 11.0. Calculate pSr values after the addition of 0.00, 10.00, 24.00, 24.90, 25.00, 25.10, 26.00, and 30.00 mL of titrant.

13–28. Derive a titration curve for 50.00 mL of 0.0150 M $Fe^{2+}$ with 0.0300 M EDTA in a solution buffered to pH 7.0. Calculate pFe values after the addition of 0.00, 10.00, 24.00, 24.90, 25.00, 25.10, 26.00, and 30.00 mL of titrant.

*13–29. Derive a curve for the titration of 25.00 mL of 0.04000 M $Co^{2+}$ with 0.05000 M $Na_2H_2Y$ in a solution maintained at pH 9.00 with an $NH_3/NH_4^+$ buffer. Assume that the $NH_3$ concentration is constant at 0.04000 M throughout. Calculate values for pCo after the addition of 0.00, 5.00, 10.00, 19.00, 20.00, 21.00, and 30.00 mL of reagent.

13–30. Derive a curve for the titration of 25.00 mL of 0.02000 M $Ni^{2+}$ with 0.01000 M $Na_2H_2Y$ in a solution maintained at pH 10.00 with an $NH_3/NH_4^+$ buffer. Assume that the concentration of $NH_3$ remains 0.1000 M throughout. Calculate values for pNi after the addition of 0.00, 10.00, 25.00, 40.00, 45.00, 49.00, 50.00, 51.00, 55.00, and 60.00 mL of reagent.

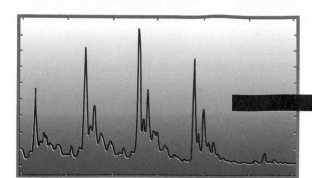

## CHAPTER 14

# AN INTRODUCTION TO ELECTROCHEMISTRY

$W$e now turn our attention to several analytical methods that are based upon oxidation/reduction reactions. These methods, which are described in Chapters 15 through 19, include oxidation/reduction titrimetry, potentiometry, coulometry, electrogravimetry, and voltammetry. This chapter deals with the fundamentals of electroanalytical chemistry that are necessary for understanding the theory of these methods.

## 14A   OXIDATION/REDUCTION REACTIONS

An *oxidation/reduction reaction* is one in which electrons are transferred from one reactant to another. An example is the oxidation of iron(II) by cerium(IV). The reaction is described by the equation

$$Ce^{4+} + Fe^{2+} \rightleftarrows Ce^{3+} + Fe^{3+} \qquad (14-1)$$

Here, $Ce^{4+}$ ion extracts an electron from $Fe^{2+}$ to form $Ce^{3+}$ and $Fe^{3+}$ ions. A substance like $Ce^{4+}$, that has a strong affinity for electrons and thus tends to remove them from other species is called an *oxidizing agent* or an *oxidant*. A *reducing agent,* or *reductant,* is a reagent that, like $Fe^{2+}$, readily donates electrons to another species. With regard to Equation 14–1, we say that $Fe^{2+}$ is *oxidized* by $Ce^{4+}$ and $Ce^{4+}$ is *reduced* by $Fe^{2+}$.

We can split any oxidation/reduction equation into two *half-reactions* that show clearly which species gains electrons and which loses them. For example, Equation 14–1 is made up of the two half-reactions

$$Ce^{4+} + e^- \rightleftarrows Ce^{3+} \qquad \text{(reduction of } Ce^{4+})$$
$$Fe^{2+} \rightleftarrows Fe^{3+} + e^- \qquad \text{(oxidation of } Fe^{2+})$$

The rules for balancing half-reactions are the same as those for balancing ordinary reactions; that is, the number of atoms of each element as well as the net charge must be the same on the two sides of the equation.

Oxidation/reduction reactions are sometimes called redox reactions.

A reducing agent is an electron donor. An oxidizing agent is an electron acceptor.

It is important to understand that a half-reaction is a theoretical concept. A half-reaction *cannot* occur alone in a solution. It is always accompanied by a second half-reaction.

311

Thus, for the oxidation of $Fe^{2+}$ by $MnO_4^-$, the half-reactions are

$$MnO_4^- + 5\ e^- + 8\ H^+ \rightleftharpoons Mn^{2+} + 4\ H_2O$$

$$5\ Fe^{2+} \rightleftharpoons 5\ Fe^{3+} + 5\ e^-$$

In the first half-reaction, the net charge on the left side is $(-1 - 5 + 8) = +2$, which is the same as that on the right. Note also that we have multiplied the $Fe^{2+}/Fe^{3+}$ half-reaction by 5 so that the number of electrons gained by $MnO_4^-$ equals the number lost by $Fe^{2+}$. A balanced net ionic equation for the overall reaction is then obtained by adding the two half-reactions:

$$MnO_4^- + 5\ Fe^{2+} + 8\ H^+ \rightleftharpoons Mn^{2+} + 5\ Fe^{3+} + 4\ H_2O$$

### 14A–1 Comparison of Oxidation/Reduction Reactions with Acid/Base Reactions

Oxidation/reduction reactions can be viewed in a way that is analogous to the Brønsted-Lowry concept of acid/base reactions (Section 6A–2). Both involve the transfer of one or more charged particles from a donor to an acceptor—the particles being electrons in oxidation/reduction and protons in neutralization. When an acid donates a proton, that acid then becomes a conjugate base capable of accepting a proton. By analogy, when a reducing agent donates an electron, that reducing agent becomes an oxidizing agent that can accept an electron. This product could be called a conjugate oxidant (but seldom, if ever, is). We may then write a generalized equation for a redox reaction as

$$Red_1 + Ox_2 \rightleftharpoons Ox_1 + Red_2 \qquad (14-2)$$

Recall that in the Brønsted-Lowry concept an acid/base reaction is described by the equation

$$acid_1 + base_2 \rightleftharpoons base_1 + acid_2$$

Here, oxidant $Ox_2$ accepts electrons from $Red_1$ to form the new reductant, $Red_2$. At the same time, reductant $Red_1$, having given up electrons, becomes an oxidizing agent, $Ox_1$. If we know from chemical evidence that the equilibrium in Equation 14–2 lies to the right, we can state that $Ox_2$ is a better electron acceptor (stronger oxidant) than $Ox_1$. Furthermore, $Red_1$ is a more effective electron donor (better reductant) than $Red_2$.

---

**Feature 14–1**
BALANCING REDOX EQUATIONS

A knowledge of how to balance oxidation/reduction reactions is essential to understanding all the concepts covered in this chapter. Although you probably remember this technique from your general chemistry course, we present a quick review here to remind you of how the process works. For practice, let us complete and balance the following equation after adding $H^+$, $OH^-$, or $H_2O$ as needed:

$$MnO_4^- + NO_2^- \rightleftharpoons Mn^{2+} + NO_3^-$$

First we must write and balance the two half-reactions involved. For $MnO_4^-$ we write

$$MnO_4^- \rightleftharpoons Mn^{2+}$$

To take up the four oxygens on the left, we add eight $H^+$ to give

$$MnO_4^- + 8\ H^+ \rightleftharpoons Mn^{2+} + 4\ H_2O$$

To balance the charge, we need to add five electrons to the left side:

$$MnO_4^- + 8\ H^+ + 5\ e^- \rightleftharpoons Mn^{2+} + 4\ H_2O$$

For the nitrite half-reaction, we add one $H_2O$ to the left side to supply the needed oxygen:

$$NO_2^- + H_2O \rightleftharpoons NO_3^- + 2\ H^+$$

Then we add two electrons to the right side to balance the charge:

$$NO_2^- + H_2O \rightleftharpoons NO_3^- + 2\ H^+ + 2\ e^-$$

Before combining the two equations, we must multiply the first by 2 and the second by 5 so that the electrons will cancel. We then add the two equations and obtain

$$2\ MnO_4^- + 16\ H^+ + \cancel{10\ e^-} + 5\ NO_2^- + 5\ H_2O \rightleftharpoons$$
$$2\ Mn^{2+} + 8\ H_2O + 5\ NO_3^- + 10\ H^+ + \cancel{10\ e^-}$$

which rearranges to the balanced equation

$$2\ MnO_4^- + 6\ H^+ + 5\ NO_2^- \rightleftharpoons 2\ Mn^{2+} + 5\ NO_3^- + 3\ H_2O$$

---

Example 14-1

As written, the following reactions proceed to the right:

$$Sn^{4+} + H_2(g) \rightleftharpoons Sn^{2+} + 2\ H^+$$
$$2\ H^+ + Sn(s) \rightleftharpoons H_2(g) + Sn^{2+}$$
$$2\ Fe^{3+} + Sn^{2+} \rightleftharpoons 2\ Fe^{2+} + Sn^{4+}$$

What can we deduce regarding the strengths of $Sn^{4+}$, $Sn^{2+}$, $H^+$, and $Fe^{3+}$ as electron acceptors or oxidizing agents?

The first reaction shows that $Sn^{4+}$ is a more effective electron acceptor than $H^+$ because it accepts electrons from $H_2$ to give $H^+$. The second reaction shows that $H^+$ is more effective than $Sn^{2+}$ in accepting electrons,

Photograph of the "silver tree" that is formed when a copper wire is immersed in a silver nitrate solution.

When the $CuSO_4$ and $AgNO_3$ solutions are 1 M, the cell develops a potential of 0.462 V as shown in Figure 14–1a.

The equilibrium-constant expression for the reaction shown in Figure 14–1 is

$$K_{eq} = \frac{[Cu^{2+}]}{[Ag^+]^2} = 4.1 \times 10^{15} \quad (14\text{–}4)$$

This expression applies regardless of whether the reaction occurs directly between reactants or indirectly within an electrochemical cell.

and the third reaction reveals that $Fe^{3+}$ is more effective than $Sn^{4+}$. Thus, the order of oxidizing strength is $Fe^{3+} > Sn^{4+} > H^+ > Sn^{2+}$.

### 14A–2 Oxidation/Reduction Reactions in Electrochemical Cells

Many oxidation/reduction reactions can be carried out in two ways that are physically quite different. For example, if we immerse a strip of copper in a solution containing silver nitrate, silver ions migrate to the metal and are reduced:

$$Ag^+ + e^- \rightarrow Ag(s)$$

At the same time, an equivalent quantity of copper is oxidized:

$$Cu(s) \rightleftarrows Cu^{2+} + 2\ e^-$$

Multiplication of the silver half-reaction by 2 followed by addition of the half-reactions yields a net ionic equation for the overall process:

$$2\ Ag^+ + Cu(s) \rightleftarrows 2\ Ag(s) + Cu^{2+} \quad (14\text{–}3)$$

A unique aspect of oxidation/reduction reactions is that the transfer of electrons—and thus the identical net reaction—can often be brought about in an *electrochemical cell* in which the oxidizing agent and the reducing agent are physically separated from one another. Figure 14–1a shows such an arrangement. Note that a *salt bridge* isolates the reactants but maintains electrical contact between the two halves of the cell. The bridge is necessary to prevent $Ag^+$ from reacting directly with the copper metal. An external metallic conductor connects the two metal electrodes. In this cell, metallic copper is oxidized, silver ions are reduced, and electrons flow through the external circuit to the silver electrode. The voltmeter measures the potential difference between the two metals at any instant and is a measure of the tendency of the cell reaction to proceed toward equilibrium. As the reaction proceeds, this tendency, and thus the potential, decreases continuously and approaches zero as the equilibrium state for the overall reaction is approached.

When zero voltage is reached, the concentrations of Cu(II) and Ag(I) ions will have values that satisfy the equilibrium-constant expression in Equation 14–4 (see note in margin). At this point, no further net flow of electrons will occur. *It is important to realize that the overall reaction and its equilibrium position are totally independent of the way the reaction is carried out,* whether by direct reaction in a solution or by indirect action in an electrochemical cell.

## 14B    ELECTROCHEMICAL CELLS

Any oxidation/reduction equilibrium is conveniently studied by measuring the potential of an electrochemical cell in which the two half-reactions

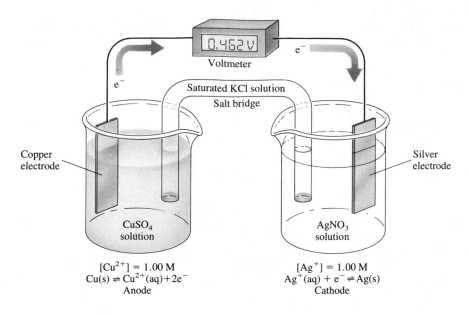

Figure 14–1a
A galvanic cell.

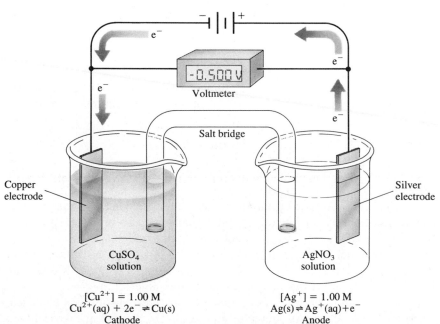

Figure 14–1b
An electrolytic cell.

making up the equilibrium are participants. For this reason, we need to consider some of the characteristics of cells.

An electrochemical cell consists of two conductors called *electrodes,* each of which is immersed in an electrolyte solution. In most of the cells of interest to us, the solutions surrounding the two electrodes are different and must be separated to prevent direct reaction between the reactants. The most common way of avoiding mixing is to insert a salt bridge, such as that shown in Figure 14–1a, between the solutions. Conduction of electricity from one electrolyte solution to the other then occurs by migration of potassium ions from the bridge in one direction and chloride ions in

A salt bridge is a conducting solution that is used to separate the two compartments of an electrochemical cell.

In some cells, known as cells without liquid junction, the electrodes share a common electrolyte.

Reduction occurs at the cathode. Oxidation occurs at the anode.

The reaction

$$2 \, H^+ + 2 \, e^- \rightleftarrows H_2(g)$$

occurs at a cathode when a solution contains no more easily reducible species.

The reaction

$$2 \, H_2O \rightleftarrows O_2(g) + 4 \, H^+ + 4 \, e^-$$

occurs at an anode when a solution contains no more easily oxidizible species.

The $Fe^{2+}$ half-reaction may seem somewhat unusual because a cation rather than an anion migrates to the electrode and gives up an electron. As we shall see, oxidation of a cation at an anode or reduction of an anion at a cathode is not uncommon.

Galvanic cells store electricity; electrolytic cells consume electricity.

For both galvanic and electrolytic cells, remember that

1. Reduction always takes place at the cathode.
2. Oxidation always takes place at the anode.

In a reversible cell, reversing the current reverses the cell reaction. In an irreversible cell, reversing the current causes a different half-reaction to occur at one or both electrodes.

the other. In this way, direct contact between copper metal and silver ions is prevented while electrical continuity is maintained.

## 14B–1 Cathodes and Anodes

The *cathode* in an electrochemical cell is the electrode at which a reduction reaction occurs. The *anode* is the electrode at which an oxidation takes place.

Examples of typical cathodic reactions are

$$Ag^+ + e^- \rightleftarrows Ag(s)$$
$$Fe^{3+} + e^- \rightleftarrows Fe^{2+}$$
$$NO_3^- + 10 \, H^+ + 8 \, e^- \rightleftarrows NH_4^+ + 3 \, H_2O$$

The first reaction takes place at the surface of a silver electrode. The other two occur at an inert electrode, such as platinum.

Typical anodic reactions are

$$Cu(s) \rightleftarrows Cu^{2+} + 2 \, e^-$$
$$2 \, Cl^- \rightleftarrows Cl_2(g) + 2 \, e^-$$
$$Fe^{2+} \rightleftarrows Fe^{3+} + e^-$$

The $Cu/Cu^{2+}$ reaction requires a copper anode, but the other two reactions can be carried out at the surface of an inert platinum electrode.

## 14B–2 Types of Electrochemical Cells

Electrochemical cells are either galvanic or electrolytic. They can also be classified as reversible or irreversible.

*Galvanic,* or *voltaic,* cells are batteries that store electrical energy. The reactions at the two electrodes in such cells tend to proceed spontaneously and produce a flow of electrons from anode to cathode via an external conductor. The cell shown in Figure 14–1a is a galvanic cell that develops a potential of about 0.46 V when electrons move through the external circuit from the copper anode to the silver cathode.

An *electrolytic* cell, in contrast, requires an external source of electrical energy for operation. As shown in Figure 14–1b, the cell in Figure 14–1a can be operated as an electrolytic cell by inserting a 0.5 V dc source into the circuit. The silver electrode is connected to the positive terminal of the source and becomes the anode of the cell. The copper electrode is connected to the negative terminal and is now the cathode. The cell reaction of the electrolytic cell is the reverse of that of the galvanic cell. That is,

$$2 \, Ag(s) + Cu^{2+} \rightleftarrows 2 \, Ag^+ + Cu(s) \tag{14–5}$$

The cell in Figure 14–1 is an example of a *reversible cell,* that is, one in which the direction of the electrochemical reaction is reversed when the direction of electron flow is changed. In an *irreversible* cell, changing the direction of current causes entirely different half-reactions to occur at one or both electrodes.

Allessandro Volta (1745–1827), Italian physicist, was the inventor of the first battery, the so-called voltaic pile. It consisted of alternating disks of copper and zinc separated by disks of cardboard soaked with salt solution. In honor of his many contributions to electrical science, the unit of potential difference, the volt, is named for Volta.

**Feature 14–2**
**THE DANIELL GRAVITY CELL**

The Daniell gravity cell was one of the earliest galvanic cells to find widespread practical application. It was used in the mid-1800s as a battery to power telegraphic communication systems. As shown in Figure 14–2, the cathode was a piece of copper immersed in a saturated solution of copper sulfate. A much less dense solution of dilute zinc sulfate was layered on top of the copper sulfate, and a massive zinc electrode was located in this solution. The two half-cells were separated only by the boundary between the two solutions of different densities. The electrode reactions are

$$Zn(s) \rightleftarrows Zn^{2+} + 2\ e^- \qquad \text{anode}$$
$$Cu^{2+} + 2\ e^- \rightleftarrows Cu(s) \qquad \text{cathode}$$

This cell develops an initial voltage of 1.18 V. In common with all galvanic cells, this potential gradually decreases with use.

## 14B–3 Currents in Electrochemical Cells

As shown in Figure 14–3, electricity is transported through an electrochemical cell by three mechanisms:

1. Electrons carry electricity within the electrodes as well as through the external conductor.

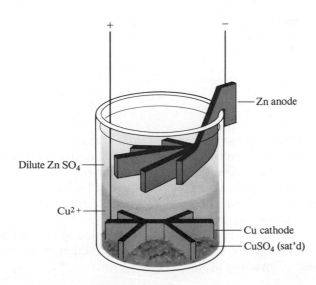

Figure 14–2
A Daniell gravity cell.

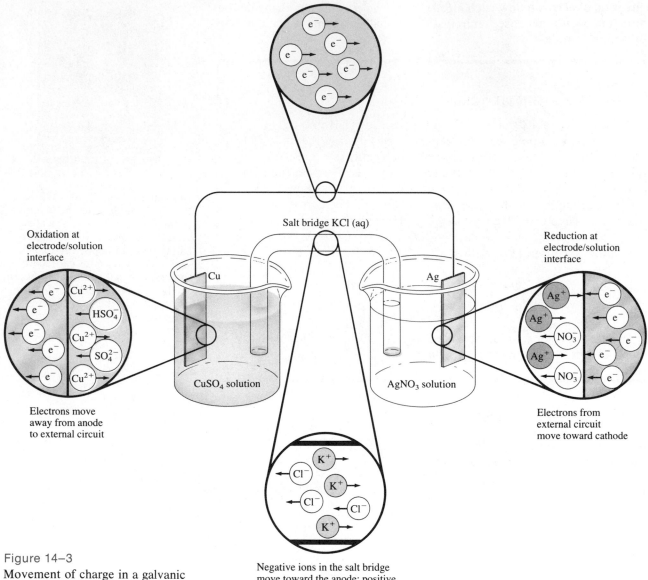

Salt bridge KCl (aq)

Oxidation at
electrode/solution
interface

Electrons move
away from anode
to external circuit

Reduction at
electrode/solution
interface

Electrons from
external circuit
move toward cathode

Cu

CuSO₄ solution

Ag

AgNO₃ solution

Negative ions in the salt bridge
move toward the anode; positive
ions move toward the cathode

Figure 14–3
Movement of charge in a galvanic
cell.

In a cell, electricity is carried by movement of anions toward the anode and
cations toward the cathode.

The phase boundary between an
electrode and its solution is called
an interface.

2. Anions and cations carry electricity within the cell. As shown in Figure
   14–3, copper and silver ions move toward the silver cathode, whereas
   anions, such as sulfate, nitrate, and hydrogen sulfate ions, are attracted to the copper anode. Within the salt bridge, chloride ions migrate toward and into the copper compartment, whereas potassium
   ions move in the opposite direction.
3. The ionic conduction of the solution is coupled to the electronic conduction in the electrodes by the reduction reaction at the cathode and
   the oxidation reaction at the anode.

## 14C    ELECTRODE POTENTIALS

The potential difference that develops between the cathode and the anode of the cell in Figure 14–1a is a measure of the tendency for the reaction

$$2\ Ag^+ + Cu(s) \rightleftarrows 2\ Ag(s) + Cu^{2+}$$

to proceed from a nonequilibrium condition to an equilibrium state. Thus, as shown in Figure 14–4a, when the copper and silver ion activities in the cell are 1.00 M, a potential of 0.462 V develops, which shows that this

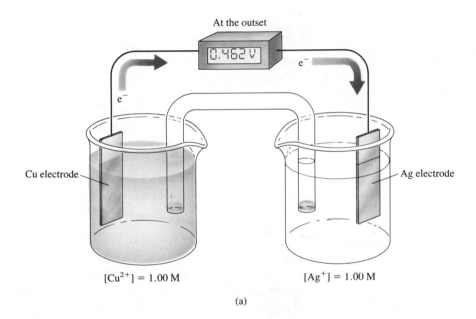

At the outset

0.462 V

e⁻ ... e⁻

Cu electrode

Ag electrode

$[Cu^{2+}] = 1.00\ M$          $[Ag^+] = 1.00\ M$

(a)

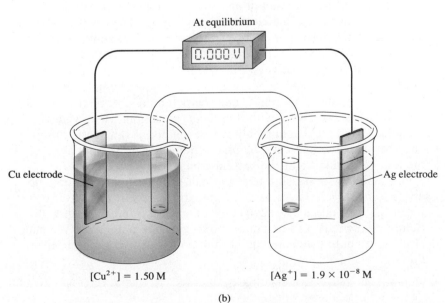

At equilibrium

0.000 V

Cu electrode

Ag electrode

$[Cu^{2+}] = 1.50\ M$          $[Ag^+] = 1.9 \times 10^{-8}\ M$

(b)

Figure 14–4

Change in cell potential after passage of current.

To be able to calculate the cell potential exactly, it is necessary to know the *activities* of the reactants rather than the molar concentrations.

reaction is far from equilibrium. As the reaction proceeds, this potential becomes smaller and smaller until at equilibrium the meter reads 0.000 V. As shown in Figure 14–4b, the copper ion equilibrium concentration is then just slightly less than 1.500 M and the silver ion concentration is $1.9 \times 10^{-8}$ M.

The potential of cells such as those shown in Figure 14–4 is the difference between two *half-cell, or single-electrode, potentials,* one associated with the half-reaction at the silver cathode ($E_{cathode}$) and the other associated with the half-reaction at the copper anode ($E_{anode}$).

Although we cannot determine absolute potentials of electrodes such as these (see Feature 14–3), we can readily determine *relative* electrode potentials. For example, if we replace the copper anode in Figure 14–1a with a cadmium electrode immersed in a cadmium sulfate solution, the voltmeter reads about 0.7 V greater than with the original cell. Since the cathode compartment remains unaltered, we conclude that the half-cell potential for the oxidation of cadmium is about 0.7 V greater than that for copper (that is, cadmium is a stronger reductant than is copper). The tendency for other ions to donate electrons can also be compared by substituting other anode half-cells while keeping the cathode half-cell unchanged.

**Feature 14–3**
WHY ABSOLUTE ELECTRODE POTENTIALS CANNOT BE MEASURED

Although it is not difficult to measure *relative* half-cell potentials, it is impossible to determine absolute half-cell potentials because all voltage-measuring devices measure only *differences* in potential.

To measure the potential of an electrode, one contact of a voltmeter is connected to the electrode in question. The other contact from the meter must then be brought into electrical contact, via another conductor, with the solution in the electrode compartment. This second contact, however, inevitably involves a solid/solution interface that acts as a second half-cell at which chemical change *must occur* if charge is to flow and the potential is to be measured. A potential is associated with this second reaction. Thus what is obtained is not an absolute half-cell potential but rather the difference between the half-cell potential of interest and the potential of the half-cell made up of the second contact and the solution.

Our inability to measure absolute half-cell potentials presents no obstacle because relative half-cell potentials are just as useful, provided they are all measured against a common reference half-cell. Relative potentials can be combined to give cell potentials. We can also use them to calculate equilibrium constants and generate titration curves.

## 14C–1  The Standard Hydrogen Reference Electrode

For relative electrode potential data to be widely applicable and useful, we must have a generally agreed-upon reference half-cell against which all others are compared. Such an electrode must be easy to construct, be reversible, and be highly reproducible in its behavior. The *standard hydrogen electrode* (SHE) meets these specifications and has been used throughout the world for many years as a universal reference electrode. It is a typical *gas electrode*.

Figure 14–5 shows how a hydrogen electrode is constructed. The metal conductor is a piece of platinum that has been coated, or *platinized,* with finely divided platinum (*platinum black*) to increase its specific surface area. This electrode is immersed in an aqueous acidic solution of known, constant hydrogen ion activity. The solution is kept saturated with hydrogen by bubbling the gas at constant pressure over the surface of the electrode. The platinum does not take part in the electrochemical reaction and serves only as the site where electrons are transferred. The half-reaction responsible for the potential that develops at this electrode is

$$2 \text{ H}^+(\text{aq}) + 2 \text{ e}^- \rightleftarrows \text{H}_2(\text{g}) \tag{14–6}$$

The hydrogen electrode is reversible and acts either as an anode or as a cathode, depending upon the half-cell with which it is coupled. Hydrogen is oxidized to hydrogen ion when this electrode serves as the anode. Hydrogen ion is reduced to molecular hydrogen when the hydrogen electrode acts as a cathode.

The potential of a hydrogen electrode depends upon temperature and upon the activities of hydrogen ion and molecular hydrogen in the solution. The latter activity, in turn, is proportional to the pressure of the gas used to keep the solution saturated with hydrogen. For the *standard hydrogen electrode*, the activity of hydrogen ions is specified as unity and the partial pressure of the gas is specified as one atmosphere. *By conven-*

The standard hydrogen electrode is sometimes referred to as the *normal hydrogen electrode*.

Platinum black catalyzes the reaction shown in Equation 14–6. Remember that catalysts do not change the position of equilibrium but simply shorten the time needed to reach equilibrium.

The reaction shown as Equation 14–6 involves two equilibria:

$$2 \text{ H}^+ + 2 \text{ e}^- \rightleftarrows \text{H}_2(\text{aq})$$
$$\text{H}_2(\text{aq}) \rightleftarrows \text{H}_2(\text{g})$$

The continuous stream of gas at constant pressure provides the solution with a constant molecular hydrogen concentration.

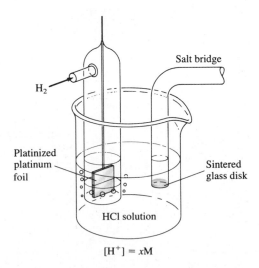

H₂

Salt bridge

Platinized platinum foil

Sintered glass disk

HCl solution

$[\text{H}^+] = x\text{M}$

Figure 14–5
A hydrogen gas electrode.

At $p_{H_2} = 1.00$ atm and $a_{H^+} = 1.00$, the potential of the hydrogen electrode is 0.000 V at all temperatures.

*tion, the potential of the standard hydrogen electrode is assigned a value of zero at all temperatures.* As a consequence of this definition, any potential developed in a galvanic cell consisting of a standard hydrogen electrode and some other electrode is attributed entirely to the other electrode.

Several other reference electrodes that are more convenient for routine measurements have been developed. Some of these are described in Section 17B.

### 14C–2  The Definition of Electrode Potential

An *electrode potential* is defined as the potential of a cell consisting of the electrode in question *acting as a cathode* and the standard hydrogen electrode *acting as an anode*. It should be emphasized that, despite its name, *an electrode potential is in fact the potential of an electrochemical cell involving a carefully defined reference electrode.* It could be more properly called a "relative electrode potential" (but seldom is).

An electrode potential is the potential of a cell in which a standard hydrogen electrode is the anode.

The cell in Figure 14–6 illustrates the definition of electrode potential for the half-reaction

$$Ag^+ + e^- \rightleftarrows Ag(s)$$

Here, the half-cell on the right consists of a strip of pure silver in contact with a solution that has a silver ion activity of 1.00; the electrode on the left is the standard hydrogen electrode. This galvanic cell develops a potential of 0.799 V with the silver electrode functioning *as the cathode;* that is, the spontaneous cell reaction is

$$2 Ag^+ + H_2(g) \rightleftarrows 2 Ag(s) + 2 H^+$$

Because the silver electrode serves as the cathode, the measured poten-

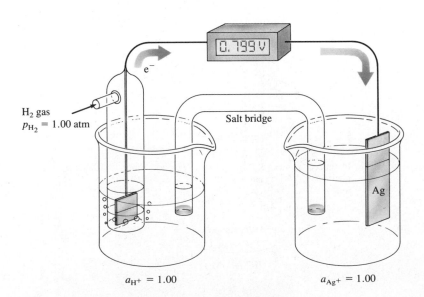

**Figure 14–6**

Definition of the standard electrode potential for $Ag^+ + e^- \rightarrow Ag(s)$.

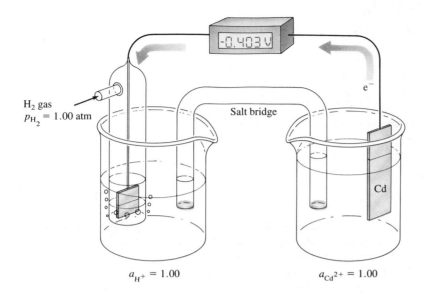

$a_{H^+} = 1.00$   $a_{Cd^{2+}} = 1.00$

**Figure 14–7**
Definition of the standard electrode potential for $Cd^{2+} + 2\ e^- \rightarrow Cd(s)$.

tial is, *by definition,* the electrode potential for the silver half-reaction (or the silver *couple*). Note that the silver electrode is positive with respect to the hydrogen anode (that is, electrons flow from the negative hydrogen anode to the silver cathode). Therefore, the electrode potential is given a positive sign, and we write

> A half-cell is sometimes called a couple.

$$Ag^+ + e^- \rightleftarrows Ag(s) \qquad E_{Ag^+} = +0.799 \text{ V}$$

Figure 14–7 illustrates the definition of the electrode potential for the half-reaction

$$Cd^{2+} + 2\ e^- \rightleftarrows Cd(s)$$

In contrast to the silver electrode, the cadmium electrode acts as the anode of the galvanic cell. That is, the spontaneous cell reaction is

$$Cd(s) + 2\ H^+ \rightleftarrows H_2(g) + Cd^{2+}$$

Here, the cadmium electrode bears a negative charge with respect to the standard hydrogen electrode. In order to reverse this reaction so that the cadmium electrode acts as a cathode, a potential more negative than $-0.403$ V must be applied to the cell. Consequently, the electrode potential of the $Cd/Cd^{2+}$ couple is *by convention* given a negative sign and is equal to $-0.403$ V.

A zinc electrode immersed in a solution with a zinc ion activity of unity develops a potential of $-0.763$ V when paired with a standard hydrogen electrode. Because the zinc electrode also behaves as an anode in the galvanic cell, its electrode potential is also negative.

The electrode potentials for the four half-cells just described can be arranged in the order

| Half-Reaction | Electrode Potential, V |
|---|---|
| $Ag^+ + e^- \rightleftarrows Ag(s)$ | +0.799 |
| $2\ H^+ + 2\ e^- \rightleftarrows H_2(g)$ | 0.000 |
| $Cd^{2+} + 2\ e^- \rightleftarrows Cd(s)$ | −0.403 |
| $Zn^{2+} + 2\ e^- \rightleftarrows Zn(s)$ | −0.763 |

The magnitudes of these electrodc potentials indicate the relative strengths of thc four ionic species as electron acceptors (oxidizing agents); that is, in decreasing strength, $Ag^+ > H^+ > Cd^{2+} > Zn^{2+}$.

## 14C–3 The Sign Convention for Electrode Potentials

Historically, electrochemists have not always used the sign convention just described. Indeed, disagreements regarding the conventions to be used in specifying signs for half-cell processes caused much controversy and confusion in the development of electrochemistry. The International Union of Pure and Applied Chemistry (IUPAC) addressed itself to this problem at its 1953 meeting in Stockholm. The usages adopted at that meeting are collectively referred to as either the *Stockholm Convention* or the *IUPAC Convention* and are now generally accepted. The sign convention described in the previous section and in the paragraphs that follow is based upon the IUPAC recommendations.

Any sign convention must be based upon expressing half-cell processes in a single way—that is, either as oxidations or as reductions. According to the IUPAC convention, the term "electrode potential" (or, more exactly, "relative electrode potential") *is reserved exclusively to describe half-reactions written as reductions.* There is no objection to the use of the term "oxidation potential" to indicate a process written in the opposite sense, but it is not proper to refer to such a potential as an electrode potential.

The sign of an electrode potential is determined by the sign of the half-cell in question when it is coupled to a standard hydrogen electrode. When the half-cell of interest behaves spontaneously as a cathode, it is the positive electrode of the galvanic cell. Thus, its electrode potential is positive. When the half-cell of interest behaves as an anode, the electrode is negative and so is its electrode potential.

It is important to emphasize that the electrode potential refers to a half-cell process written *as a reduction.* For the zinc and cadmium electrodes we have been considering, the spontaneous reactions are oxidations. *It is evident, then, that the sign of an electrode potential indicates whether the reduction is spontaneous with respect to the standard hydrogen electrode.* The positive sign associated with the electrode potential for silver indicates that the process

$$2\ Ag^+ + H_2(g) \rightleftarrows 2\ Ag(s) + 2\ H^+$$

favors the products under ordinary conditions. Similarly, the negative sign of the electrode potential for zinc means that the analogous reaction

$$Zn^{2+} + H_2(g) \rightleftarrows Zn(s) + 2\ H^+$$

An electrode potential is, by definition, a reduction potential. An oxidation potential is the potential for the half-reaction written in the opposite way. The sign of an oxidation potential is therefore opposite that for a reduction potential, but the magnitude is the same.

The IUPAC sign convention is based upon the sign of the half-cell of interest when it is connected with the standard hydrogen electrode.

Zinc reacts with acid, but hydrogen gas will not react with $Zn^{2+}$.

is not spontaneous. That is, the reaction tends to go in the other direction,

$$Zn(s) + 2\ H^+ \rightleftarrows Zn^{2+} + H_2(g)$$

Reference works, particularly those published before 1953, often contain tabulations of electrode potentials that are not in accord with the IUPAC recommendations. For example, in a classic source of oxidation-potential data compiled by Latimer,[1] one finds

$$Zn(s) \rightleftarrows Zn^{2+} + 2\ e^-\qquad E = +0.76\ V$$
$$Cu(s) \rightleftarrows Cu^{2+} + 2\ e^-\qquad E = -0.34\ V$$

In converting these oxidation potentials to electrode potentials as defined by the IUPAC convention, one must mentally (1) express the half-reactions as reductions and (2) change the signs of the potentials.

The sign convention used in a tabulation of electrode potentials may not be explicitly stated. This information can be readily deduced, however, by noting the direction and sign of the potential for a half-reaction with which one is familiar. If the sign agrees with the IUPAC convention, the table can be used as is; if not, the signs of all of the data must be reversed. For example, the reaction

$$O_2(g) + 4\ H^+ + 4\ e^- \rightleftarrows 2\ H_2O\qquad E = +1.229\ V$$

occurs spontaneously with respect to the standard hydrogen electrode and thus carries a positive sign. If the potential for this half-reaction is negative in a tabulation, it and all the other potentials should be multiplied by $-1$.

## 14C–4 The Effect of Concentration on Electrode Potentials: The Nernst Equation

An electrode potential is a measure of the extent to which the existing concentrations in a half-cell differ from their equilibrium values. Thus, for example, there is a greater tendency for the process

$$Ag^+ + e^- \rightleftarrows Ag(s)$$

to occur in a concentrated solution of silver(I) than in a dilute solution of that ion. It follows that the magnitude of the electrode potential for this process must also become larger (more positive) as the silver ion concentration of the solution is increased. We now examine the quantitative relationship between concentration and electrode potential.

Consider the reversible half-reaction

$$aA + bB + ne^- \rightleftarrows cC + dD$$

---

[1] W. M. Latimer, *The Oxidation States of the Elements and Their Potentials in Aqueous Solutions*, 2nd ed. Englewood Cliffs, NJ: Prentice-Hall, 1952.

where the capital letters represent formulas for the participating species (atoms, molecules, or ions), $n$ represents the number of moles of electrons, and the lowercase italic letters indicate the number of moles of each species appearing in the half-reaction as it has been written. The electrode potential for this process is described by the equation

$$E = E^0 - \frac{RT}{nF} \ln \frac{[C]^c [D]^d \cdot \cdot \cdot}{[A]^a [B]^b \cdot \cdot \cdot} \qquad (14\text{-}7)$$

where

$E^0$ = a constant called the *standard electrode potential,* which is characteristic for each half-reaction

$R$ = the gas constant, 8.314 J K$^{-1}$ mol$^{-1}$

$T$ = temperature in K (kelvins)

$n$ = number of moles of electrons that appear in the half-reaction for the electrode process as it has been written

$F$ = the faraday = 96,485 C (coulombs)

ln = the natural logarithm = 2.303 × log

Substituting numerical values for the constants, converting to base 10 logarithms, and specifying 25°C for the temperature give

$$E = E^0 - \frac{0.0592}{n} \log \frac{[C]^c [D]^d \cdot \cdot \cdot}{[A]^a [B]^b \cdot \cdot \cdot} \qquad (14\text{-}8)$$

The meanings of the bracketed terms in Equations 14–7 and 14–8 are

for a solute A, [A] = molar concentration

for a gas B, [B] = $p_B$ = partial pressure in atmospheres

for a pure solid or pure liquid in excess C, [C] = 1.00

for a solvent D, [D] = 1.00

The letters in brackets strictly represent activities, but we shall ordinarily follow our practice of substituting molar concentrations for activities in most calculations. Thus, if some participating species A is a solute, [A] is the concentration of A in moles per liter. If A is a gas, [A] in Equation 14–8 is replaced by $p_A$, the partial pressure of A in atmospheres. If A is a pure liquid or a pure solid as a second phase, [A] has a value of 1.00. Similarly, [A] is assigned a value of unity if A is the solvent. The rationale for these assumptions is the same as that described in Section 6B, which deals with equilibrium-constant expressions.

Equation 14–8 is known as the *Nernst equation* in honor of the German chemist responsible for its development.

Example 14–2

Typical half-cell reactions and their corresponding Nernst expressions follow.

(1)    $Zn^{2+} + 2\,e^- \rightleftarrows Zn(s)$    $E = E^0 - \dfrac{0.0592}{2} \log \dfrac{1}{[Zn^{2+}]}$

The activity of elemental zinc is unity, by definition, because it is a pure second phase. Thus, the electrode potential varies linearly with the logarithm of the reciprocal of the zinc ion concentration.

$$(2) \quad Fe^{3+} + e^- \rightleftharpoons Fe^{2+} \qquad E = E^0 - \frac{0.0592}{1} \log \frac{[Fe^{2+}]}{[Fe^{3+}]}$$

The potential for this couple can be measured with an inert metallic electrode immersed in a solution containing both iron species. The potential depends upon the logarithm of the ratio of the molar concentrations of these ions.

$$(3) \quad 2\,H^+ + 2\,e^- \rightleftharpoons H_2(g) \qquad E = E^0 - \frac{0.0592}{2} \log \frac{p_{H_2}}{[H^+]^2}$$

In this example, $p_{H_2}$ is the partial pressure of hydrogen (in atmospheres) at the surface of the electrode. Ordinarily, its value will be the same as the atmospheric pressure.

$$(4) \quad MnO_4^- + 5\,e^- + 8\,H^+ \rightleftharpoons Mn^{2+} + 4\,H_2O$$

$$E = E^0 - \frac{0.0592}{5} \log \frac{[Mn^{2+}]}{[MnO_4^-][H^+]^8}$$

Here, the potential depends not only upon the concentration of manganese species but also on the pH of the solution.

$$(5) \quad AgCl(s) + e^- \rightleftharpoons Ag(s) + Cl^- \qquad E = E^0 - \frac{0.0592}{1} \log [Cl^-]$$

This half-reaction describes the behavior of a silver electrode immersed in a chloride solution that is *saturated* with AgCl. To ensure this condition, an excess of the solid must always be present. Note that this electrode reaction is the sum of two reactions, namely,

$$AgCl(s) \rightleftharpoons Ag^+ + Cl^-$$

$$Ag^+ + e^- \rightleftharpoons Ag(s)$$

The activities of metallic Ag and of AgCl are equal to unity as long as some of each is in contact with the solution. Therefore, the logarithmic term in the Nernst equation contains the $Cl^-$ concentration only.

Walther Hermann Nernst (1864–1941) was a German physical chemist who made many contributions to our understanding of electrochemistry. He is probably most famous for the equation that bears his name (Equation 14–7), but he was also known for discoveries and inventions in other fields. He invented the Nernst glower, a source of infrared radiation that is shown on the stamp. Nernst received the Nobel Prize in Chemistry in 1920 for his numerous contributions to the field of chemical thermodynamics.

The Nernst expression in part 5 of Example 14–2 applies only if an excess of solid silver chloride is present *so that the solution is saturated* with AgCl at all times.

---

**Feature 14–4**
**THE ACTIVITY QUOTIENT IN THE NERNST EQUATION IS UNITLESS**

Although it is not obvious from our presentation, the quotient in the logarithmic term of the Nernst equation is unitless. Each individual term is a ratio of the activity of the species to its activity in the standard state, which has been arbitrarily assigned a value of unity (see Section 6B–2, p. 124). Thus, for the equation shown in part 3 of Example 14–2,

$$E = E^0 - \frac{0.0592}{2} \log \frac{p_{H_2}/(p_{H_2})_0}{[H^+]^2/([H^+]_0)^2}$$

$$= E^0 - \frac{0.0592}{2} \log \frac{p_{H_2}(\text{atm})/1.00(\text{atm})}{[H^+]^2(\text{mol/L})^2/1.00^2(\text{mol/L})^2}$$

Hydrogen in its standard state $(p_{H_2})_0$ is 1.00 atm. Similarly, $[H^+]_0$ represents the concentration of hydrogen ion in its standard state, 1.00 mol/L. Thus, the units cancel.

### 14C–5 The Standard Electrode Potential, $E^0$

The standard electrode potential, $E^0$, is defined as the electrode potential when all reactants and products that appear in a half-reaction have unit activity.

Examination of Equations 14–7 and 14–8 reveals that the constant $E^0$ is the electrode potential whenever the concentration quotient has a value of unity. This constant is called the *standard electrode potential* for the half-reaction. Note that the quotient is always unity when the activities of the reactants and products of a half-reaction are unity.

The standard electrode potential is an important physical constant that provides quantitative information regarding the driving force for a half-cell reaction.[2] The important characteristics of this constant are

1. The standard electrode potential is a relative quantity in the sense that it is the potential of an electrochemical cell in which the anode is the standard hydrogen electrode, whose potential has been arbitrarily set at 0.000 V.
2. The standard electrode potential for a half-reaction refers exclusively to a reduction reaction; that is, it is a relative reduction potential.
3. The standard electrode potential measures the relative force tending to drive the half-reaction from a state in which all reactants and products are at unit activity to a state in which the reactants and products are at their equilibrium activities relative to the standard hydrogen electrode.
4. The standard electrode potential is independent of number of moles of reactant and products shown in the balanced half-reaction. Thus, the standard electrode potential for the half-reaction

$$Fe^{3+} + e^- \rightleftarrows Fe^{2+} \qquad E^0 = +0.771$$

does not change if we choose to write the reaction as

$$5\,Fe^{3+} + 5\,e^- \rightleftarrows 5\,Fe^{2+} \qquad E^0 = +0.771$$

[2]For further reading on standard electrode potentials, see R. G. Bates, in *Treatise on Analytical Chemistry,* 2nd ed., I. M. Kolthoff and P. J. Elving, Eds., Part I, Vol. 1, Chapter 13. New York: Wiley, 1978.

Note, however, that the Nernst equation must be consistent with the half-reaction as written. For the first case, it is

$$E = 0.771 - \frac{0.0592}{1} \log \frac{[Fe^{2+}]}{[Fe^{3+}]}$$

and for the second

$$E = 0.771 - \frac{0.0592}{5} \log \frac{[Fe^{2+}]^5}{[Fe^{3+}]^5}$$

Note that the two logarithmic terms have identical values:

$$\frac{0.0592}{1} \log \frac{[Fe^{2+}]}{[Fe^{3+}]} = \frac{0.0592}{5} \log \frac{[Fe^{2+}]^5}{[Fe^{3+}]^5}$$

5. A positive electrode potential indicates that the half-reaction in question occurs spontaneously with respect to the half-reaction for the standard hydrogen electrode. That is, the oxidant in the half-reaction is a stronger oxidant than is hydrogen ion. A negative sign indicates just the opposite.
6. The standard electrode potential for a half-reaction is temperature-dependent.

Standard-electrode-potential data are available for an enormous number of half-reactions. Many have been determined directly from electrochemical measurements. Others have been computed from equilibrium studies of oxidation/reduction systems and from thermochemical data associated with such reactions. Table 14–1 contains standard-electrode data for several half-reactions we will be considering in the pages that follow. A more extensive listing is found in Appendix 6.[3]

Table 14–1 and Appendix 6 illustrate the two common ways for tabulating standard potential data. In Table 14–1, potentials are listed in decreasing numerical order. As a consequence, the species in the upper left part (that is, on the left side of the first few half-reactions) are the most effective electron acceptors, as evidenced by their large positive $E^0$ values. They are therefore the strongest *oxidizing agents*. As we proceed down the left side of such a table, each succeeding species is a less effective acceptor of electrons than the one above it. The half-cell reactions at the bottom of the table have little tendency to take place as written. On the other hand, they do tend to occur in the opposite sense. The most effective *reducing agents,* then, are those species on the right side of the last few half-reactions shown in the table.

Compilations of electrode-potential data, such as that shown in Table 14–1, provide the user with qualitative insights into the extent and direction of electron-transfer reactions. For example, the standard potential for silver(I) (+0.799 V) is more positive than that for copper(II) (+0.337 V). We therefore conclude that a piece of copper immersed in a silver(I)

Based upon the $E^0$ values in Table 14–1 for $Fe^{3+}$ and $I_3^-$, which species would you expect to predominate in a solution produced by mixing iron(III) and iodide ions? See the color plates in the middle of the book.

---

[3]Comprehensive sources for standard electrode potentials include *Standard Electrode Potentials in Aqueous Solutions,* A. J. Bard, R. Parsons, and J. Jordan, Eds. New York: Marcel Dekker, 1985; G. Milazzo and S. Caroli, *Tables of Standard Electrode Potentials.* New York: Wiley-Interscience, 1977; M. S. Antelman and F. J. Harris, *Chemical Electrode Potentials.* New York: Plenum Press, 1982. Some compilations are arranged alphabetically by element; others are tabulated according to the value of $E^0$.

Table 14–1
STANDARD ELECTRODE POTENTIALS*

| Reaction | $E^0$ at 25°C, V |
|---|---|
| $Cl_2(g) + 2\,e^- \rightleftharpoons 2\,Cl^-$ | +1.359 |
| $O_2(g) + 4\,H^+ + 4\,e^- \rightleftharpoons 2\,H_2O$ | +1.229 |
| $Br_2(aq) + 2\,e^- \rightleftharpoons 2\,Br^-$ | +1.087 |
| $Br_2(l) + 2\,e^- \rightleftharpoons 2\,Br^-$ | +1.065 |
| $Ag^+ + e^- \rightleftharpoons Ag(s)$ | +0.799 |
| $Fe^{3+} + e^- \rightleftharpoons Fe^{2+}$ | +0.771 |
| $I_3^- + 2\,e^- \rightleftharpoons 3\,I^-$ | +0.536 |
| $Cu^{2+} + 2\,e^- \rightleftharpoons Cu(s)$ | +0.337 |
| $UO_2^{2+} + 4\,H^+ + 2\,e^- \rightleftharpoons U^{4+} + 2\,H_2O$ | +0.334 |
| $Hg_2Cl_2(s) + 2\,e^- \rightleftharpoons 2\,Hg(l) + 2\,Cl^-$ | +0.268 |
| $AgCl(s) + e^- \rightleftharpoons Ag(s) + Cl^-$ | +0.222 |
| $Ag(S_2O_3)_2^{3-} + e^- \rightleftharpoons Ag(s) + 2\,S_2O_3^{2-}$ | +0.017 |
| $2\,H^+ + 2\,e^- \rightleftharpoons H_2(g)$ | 0.000 |
| $AgI(s) + e^- \rightleftharpoons Ag(s) + I^-$ | −0.151 |
| $PbSO_4(s) + 2\,e^- \rightleftharpoons Pb(s) + SO_4^{2-}$ | −0.350 |
| $Cd^{2+} + 2\,e^- \rightleftharpoons Cd(s)$ | −0.403 |
| $Zn^{2+} + 2\,e^- \rightleftharpoons Zn(s)$ | −0.763 |

*See Appendix 6 for a more extensive list.

solution will cause the reduction of that ion and the oxidation of the copper. On the other hand, we would expect no reaction if we place a piece of silver in a copper(II) solution.

Feature 14–5
STANDARD ELECTRODE POTENTIALS FOR SYSTEMS INVOLVING PRECIPITATES OR COMPLEX IONS

In Table 14–1, we find several entries involving Ag(I), including

$$Ag^+ + e^- \rightleftharpoons Ag(s) \qquad\qquad E^0_{Ag^+} = +0.799\text{ V}$$
$$AgCl(s) + e^- \rightleftharpoons Ag(s) + Cl^- \qquad E^0_{AgCl} = +0.222\text{ V}$$
$$Ag(S_2O_3)_2^{3-} + e^- \rightleftharpoons Ag(s) + 2\,S_2O_3^{2-} \qquad E^0_{Ag(S_2O_3)_2^{3-}} = +0.017\text{ V}$$

Each gives the potential of a silver electrode in a different environment. Let us see how the three potentials are related.

The Nernst expression for the first half-reaction is

$$E = E^0_{Ag^+} - \frac{0.0592}{1} \log \frac{1}{[Ag^+]}$$

If we replace $[Ag^+]$ with $K_{sp}/[Cl^-]$, we obtain

$$E = E^0_{Ag^+} - \frac{0.0592}{1} \log \frac{[Cl^-]}{K_{sp}}$$
$$= E^0_{Ag^+} + 0.0592 \log K_{sp} - 0.0592 \log [Cl^-] \qquad (1)$$

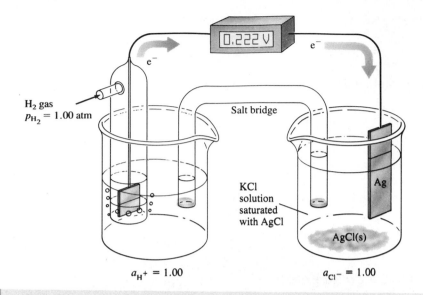

**Figure 14–8**

Definition of the standard electrode potential for an Ag/AgCl electrode.

The standard potentials for $Ag^+$ and AgCl are related by the expression

$$E^0_{Ag^+} = E^0_{AgCl} - 0.0592 \log K_{sp}$$

where $K_{sp}$ is the solubility-product constant for AgCl.

By definition, the standard potential for the Ag/AgCl half-reaction is the potential where $[Cl^-] = 1.00$. That is, when $[Cl^-] = 1.00$, $E = E^0_{AgCl}$. Taking the $K_{sp}$ value for AgCl from Appendix 2 and substituting in Equation 1, we get

$$E^0_{AgCl} = E^0_{Ag^+} + 0.0592 \log 1.82 \times 10^{-10} - 0.0592 \log (1.00)$$
$$= 0.799 + (-0.577) - 0.000 = 0.222 \text{ V}$$

Figure 14–8 illustrates the definition of the standard electrode potential for the Ag/AgCl electrode.

If we proceed in the same way, we obtain for the third equilibrium

$$E^0_{Ag(S_2O_3)_2^{3-}} = E^0_{Ag^+} - 0.0592 \log K_f$$

where $K_f$ is the formation constant for the complex, that is,

$$K_f = \frac{[Ag(S_2O_3)_2^{3-}]}{[Ag^+][S_2O_3^{2-}]^2}$$

In contrast to the data in Table 14–1, Appendix 6 is arranged alphabetically by element to make it easier to locate data for a given electrode reaction.

---

**Example 14–3**

Calculate the electrode potential of a silver electrode immersed in a 0.0500 M solution of NaCl using (a) $E^0_{Ag^+} = 0.799$ V and (b) $E^0_{AgCl} = 0.222$ V.

(a) $Ag^+ + e^- \rightleftarrows Ag(s)$    $E^0 = 0.799$ V

The $Ag^+$ concentration of this solution is given by

$$[Ag^+] = K_{sp}/[Cl^-] = 1.82 \times 10^{-10}/0.0500 = 3.64 \times 10^{-9} \text{ M}$$

Substituting into the Nernst expression gives

$$E = 0.799 - 0.0592 \log \frac{1}{3.64 \times 10^{-9}} = 0.299 \text{ V}$$

(b) Here we may write

$$E = 0.222 - 0.0592 \log [Cl^-] = 0.222 - 0.0592 \log 0.0500$$
$$= 0.299 \text{ V}$$

## 14C–6 Formal Potentials

Electrode-potential calculations for many oxidation/reduction systems are complicated by the presence of other equilibria (such as dissociation, association, complex formation, and solvolysis) that involve one or more of the species involved in the redox half-reaction. The effect of these other equilibria on calculated potentials can be accounted for if equilibrium-constant data are available. Frequently they are not.

Swift[4] proposed substitution of *formal potentials* (also called *conditional potentials*) in place of standard electrode potentials to compensate for activity effects as well as errors due to the existence of competing equilibria. The formal potential of a system is the potential of the half-cell (with respect to the standard hydrogen electrode) when the *concentration* of each solute participating in the half-reaction is exactly 1 M and the concentrations of all other solutes are carefully specified.

> A formal potential is the electrode potential when the *concentrations* of reactants and products of a half-reaction are exactly 1 M and the concentrations of any other solutes are specified.

Formal potentials for many half-reactions are listed in Appendix 6. Note that large differences exist between the formal and standard potentials for some half-reactions. For example, the standard electrode potential for the reduction of iron(III) to iron(II) is 0.771 V. In 1 M perchloric acid, the formal potential for the same half-reaction is 0.732 V. This difference is attributable to the fact that the activity coefficient of iron(III) is considerably smaller than that of iron(II) at the high ionic strength of the 1 M perchloric acid medium. As a consequence, the ratio of activities of the two species is less than unity, a condition that leads to a decrease in the electrode potential. In 1 M hydrochloric acid, the formal potential for this couple is only 0.700 V. Here, the iron(III)/iron(II) activity ratio is even smaller because the chloro complexes of the former are more stable than those of iron(II); a change in potential larger than that seen in perchloric acid is the consequence.

---

[4]E. H. Swift, *A System of Chemical Analysis*, p. 50. San Francisco: Freeman, 1939.

Substitution of formal potentials for standard electrode potentials in the Nernst equation yields better agreement between calculated and experimental results—provided, of course, that the electrolyte concentration of the solution approximates that for which the formal potential is applicable. Not surprisingly, attempts to apply formal potentials to systems that differ substantially in type and in electrolyte concentration can result in errors that are larger than those associated with the use of standard electrode potentials. Henceforth we use whichever is the more appropriate.

---

**Feature 14–6**
**WHY THERE ARE TWO ELECTRODE POTENTIALS FOR $Br_2$ IN TABLE 14–1**

In Table 14–1, we find the following data for $Br_2$:

$$Br_2(aq) + 2\ e^- \rightleftarrows 2\ Br^- \qquad E^0 = +1.087\ V$$
$$Br_2(l) + 2\ e^- \rightleftarrows 2\ Br^- \qquad E^0 = +1.065\ V$$

The second standard potential applies *only to a solution that is saturated with $Br_2$* and not to undersaturated solutions of the element. You should therefore use 1.065 V to calculate the electrode potential of a 0.0100 M solution of KBr that is saturated with $Br_2$ and in contact with an excess of the liquid element. Here,

$$E = 1.065 - \frac{0.0592}{2} \log \frac{[Br^-]^2}{1.00} = 1.065 - \frac{0.0592}{2} \log (0.0100)^2$$

$$= 1.065 - \frac{0.0592}{2} (-4.00) = 1.183\ V$$

In this calculation, the activity of $Br_2$ in the liquid state is constant and assigned a value of 1.00.

The standard electrode potential shown in the entry for $Br_2(aq)$ is *hypothetical* because the solubility of $Br_2$ at 25°C is only about 0.18 M. Thus, the recorded value of 1.087 V is based upon a system that—in terms of our definition of $E^0$—cannot be realized experimentally. Nevertheless, the hypothetical potential does permit us to calculate electrode potentials for solutions that are undersaturated in $Br_2$. For example, if we wish to calculate the electrode potential for a solution that is 0.0100 M in KBr and 0.00100 M in $Br_2$, we would write

$$E = 1.087 - \frac{0.0592}{2} \log \frac{[Br^-]^2}{[Br_2(aq)]} = 1.087 - \frac{0.0592}{2} \log \frac{(0.0100)^2}{0.00100}$$

$$= 1.087 - \frac{0.0592}{2} \log 0.100 = 1.117\ V$$

## 14D   THE POTENTIAL OF ELECTROCHEMICAL CELLS

We can use standard electrode potentials and the Nernst equation to calculate the potential obtainable from a galvanic cell or the potential required to operate an electrolytic cell. The calculated potentials (sometimes called *thermodynamic potentials*) are theoretical in the sense that they refer to cells in which there is no current. Additional factors must be taken into account if a current is involved.

### 14D–1 The Schematic Representation of Cells

Chemists' frequently use a shorthand notation to describe electrochemical cells. The cell in Figure 14–1a, for example, is described by

$$Cu|CuSO_4(1\ M)\|AgNO_3(1\ M)|Ag$$

where the molar concentrations of $CuSO_4$ and $AgNO_3$ are exactly 1 M. An alternative way of writing the cell is

$$Cu|Cu^{2+}(1\ M)\|Ag^+(1\ M)|Ag \qquad (14-9)$$

In this representation, which is more commonly encountered, only the active participants in the half-cell half processes are indicated.

*By convention,* the anodic process is *always* displayed on the left in these representations. A single vertical line indicates a phase boundary, or interface, at which a potential develops. For example, in the first representation, the first vertical line indicates that a potential develops at the phase boundary between the copper anode and the copper sulfate solution. The double vertical line represents two phase boundaries, one at each end of the salt bridge. A *liquid-junction potential* develops at each of the two salt-bridge interfaces. This potential results from differences in the rates at which the ions in the cell compartments and the salt bridge migrate across the interfaces. A liquid-junction potential can amount to as much as several hundredths of a volt, but the junction potentials at the two ends of a salt bridge tend to cancel each other. The net effect on the overall potential of a cell is thus only a few millivolts or less. For our purposes, we will neglect the contribution of liquid-junction potentials to the total potential of the cell.

The cell shown in Figure 14–6 can be represented as

$$Pt,H_2(p = 1.00\ atm)|H^+(a_{H^+} = 1.00\ M)\|Ag^+(a_{Ag^+} = 1.00)|Ag$$

or alternatively as

$$SHE\|Ag^+(a_{Ag^+} = 1.00)|Ag$$

The cell in Figure 14–7 can be represented by

$$Cd|Cd^{2+}(a_{Cd^{2+}} = 1.00)\|H^+(a_{H^+} = 1.00)|H_2(p = 1.00\ atm),Pt$$

> A liquid-junction potential is a potential that develops across the interface between two solutions with different electrolyte composition.

Note that the standard hydrogen electrode is the anode in the first cell and the cathode in the second.

## 14D–2  Cell Potential Calculations

The potential of an electrochemical cell $E_{cell}$ is the difference between the *electrode potential* of the cathode and the *electrode potential* of the anode. That is,

$$E_{cell} = E_{cathode} - E_{anode} \qquad (14\text{–}10)$$

where $E_{cathode}$ and $E_{anode}$ are the half-cell potentials of the cathode and anode, respectively.

Gustav Robert Kirchhoff (1824–1877) was a German physicist who made many important contributions to physics and chemistry. In addition to his work in spectroscopy, he is known for Kirchhoff's laws of electrical circuits. These laws are shown in the form of equations on this stamp.

---

### Example 14–4

Calculate the thermodynamic potential of the following cell

$$Cu | Cu^{2+}(0.100 \text{ M}) \| Ag^+(0.200 \text{ M}) | Ag$$

Note that this cell is similar to the galvanic cell shown in Figure 14–1a.
  The two half-reactions and standard potentials are

$$Ag^+ + 2\,e^- \rightleftarrows Ag(s) \qquad E^0 = 0.799 \text{ V}$$
$$Cu^{2+} + 2\,e^- \rightleftarrows Cu(s) \qquad E^0 = 0.337 \text{ V}$$

The electrode potentials are

$$E_{Ag^+} = 0.799 - 0.0592 \log \frac{1}{0.200} = 0.758 \text{ V}$$

$$E_{Cu^{2+}} = 0.337 - \frac{0.0592}{2} \log \frac{1}{0.100} = 0.307 \text{ V}$$

We see from the cell diagram that the silver electrode is the cathode and the copper electrode is the anode. Therefore, applying Equation 14–10 gives

$$E_{cell} = E_{Ag^+} - E_{Cu^{2+}} = 0.758 - 0.307 = +0.451 \text{ V}$$

---

### Example 14–5

Calculate the thermodynamic potential for the cell

$$Ag | Ag^+(0.200 \text{ M}) \| Cu^{2+}(0.100 \text{ M}) | Cu$$

Note that this cell is similar to the electrolytic cell shown in Figure 14–1b.
  The electrode potentials for the two half-reactions are identical to the electrode potentials calculated in Example 14–4. That is,

$$E_{Ag^+} = 0.758 \text{ V} \qquad \text{and} \qquad E_{Cu^{2+}} = 0.307 \text{ V}$$

In contrast to the previous example, however, the silver electrode is the anode and the copper electrode the cathode. Substituting these electrode potentials into Equation 14–10 gives

A negative cell potential means the cell is electrolytic; a positive potential means the cell is galvanic.

$$E_{cell} = E_{Cu^{2+}} - E_{Ag^+} = 0.307 - 0.758 = -0.451 \text{ V}$$

Examples 14–4 and 14–5 illustrate an important fact. The cell potential computed from Equation 14–10 is positive for a galvanic cell and negative for an electrolytic cell.

Example 14–6

Calculate the potential of the following cell and indicate whether it is galvanic or electrolytic.

$$Pt|UO_2^{2+}(0.0150 \text{ M}), U^{4+}(0.200 \text{ M}), H^+(0.0300 \text{ M})\|$$
$$Fe^{2+}(0.0100 \text{ M}), Fe^{3+}(0.0250 \text{ M})|Pt$$

We see in Table 14–1 that the two half-reactions and their standard potentials are

$$Fe^{3+} + e^- \rightleftarrows Fe^{2+} \qquad\qquad E^0 = +0.771 \text{ V}$$
$$UO_2^{2+} + 4 \text{ H}^+ + 2 \text{ e}^- \rightleftarrows U^{4+} + 2 \text{ H}_2O \qquad E^0 = +0.334 \text{ V}$$

The electrode potential for the cathode is

$$E_{cathode} = 0.771 - 0.0592 \log \frac{[Fe^{2+}]}{[Fe^{3+}]}$$

$$= 0.771 - 0.0592 \log \frac{0.0100}{0.0250} = 0.799 - (-0.0236)$$

$$= 0.7946 \text{ V}$$

The electrode potential for the anode is

$$E_{anode} = 0.334 - \frac{0.0592}{2} \log \frac{[U^{4+}]}{[UO_2^{2+}][H^+]^4}$$

$$= 0.334 - \frac{0.0592}{2} \log \frac{0.200}{(0.0150)(0.0300)^4}$$

$$= 0.334 - 0.2136 = 0.1204 \text{ V}$$

and

$$E_{cell} = E_{cathode} - E_{anode} = 0.7946 - 0.1204 = +0.674 \text{ V}$$

The positive sign means that the cell is galvanic.

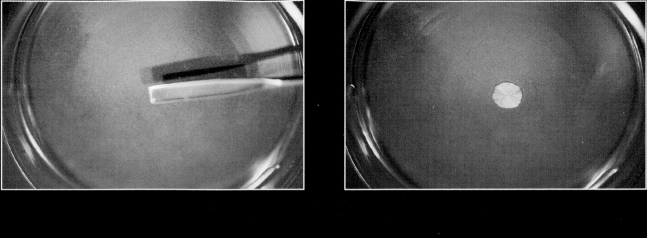

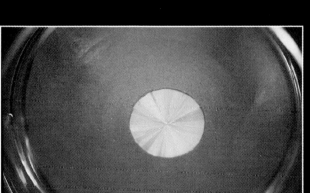

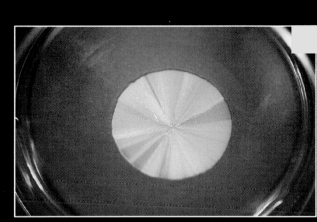

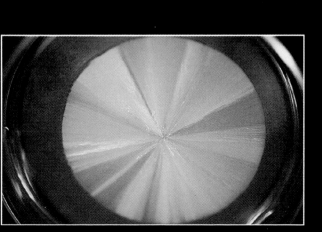

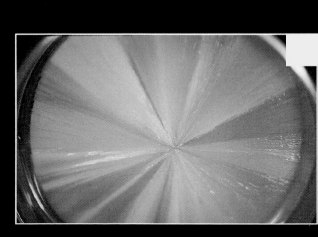

Crystallization of sodium acetate from a
supersaturated solution (Section 4C–1).

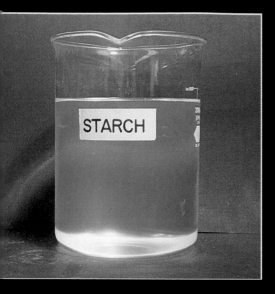 

Tyndall effect (see page 72).

Precipitation of 0.1041 g of Fe(III) with $NH_3$ (left), and homogeneously with urea (right) (Section 4C–5).

Chemical Equilibrium 1: Reaction between iodine and arsenic(III) at pH 1 (Section 6B–1).

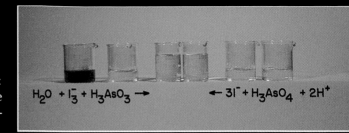

$$H_2O + I_3^- + H_3AsO_3 \rightarrow \quad \leftarrow 3I^- + H_3AsO_4 + 2H^+$$

Chemical Equilibrium 2: Reaction between iodine and arsenic(III) at pH 7 (Section 6B–1).

$$H_2O + I_3^- + H_2AsO_3^- \rightarrow \quad \leftarrow 3I^- + H_2AsO_4^- + 2H^+$$

Chemical Equilibrium 3: Reaction between iodine and ferrocyanide (Section 6B–1).

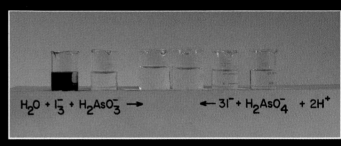

$$I_3^- + 2Fe(CN)_6^{4-} \rightarrow \quad \leftarrow 3I^- + 2Fe(CN)_6^{3-}$$

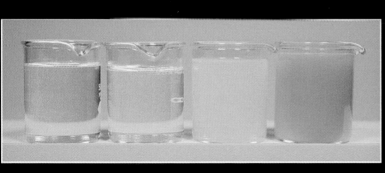

(a)      (b)      (c)      (d)

Argentometric determination of chloride: Fajans method (Section 9B–3). (a) aqueous 2′, 7′-dichlorofluorescein; (b) same, plus 1 mL 0.10 M $Ag^+$. Note the absence of a precipitate; (c) same, plus AgCl and an excess of $Cl^-$; (d) same, plus AgCl and the first slight excess of $Ag^+$.

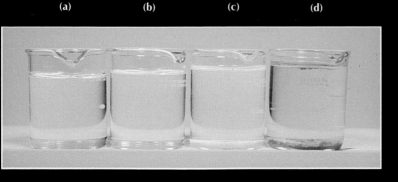

(a)      (b)      (c)      (d)

In (c) and (d) the AgCl has coagulated. Note that in (d) the dye has been carried down on the precipitate and that the supernatant solution is clear while in (c) the dye remains in solution.

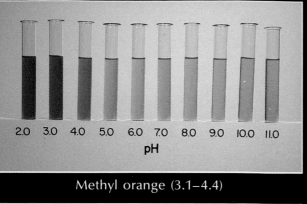

Methyl orange (3.1–4.4)

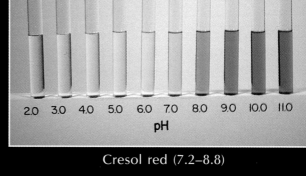

Bromocresol green (3.8–5.4)

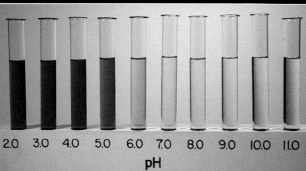

Methyl red (4.2–6.3)

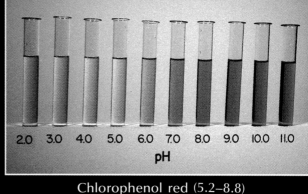

Chlorophenol red (5.2–8.8)

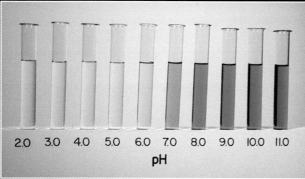

Bromothymol blue (6.0–7.6)

Cresol red (7.2–8.8)

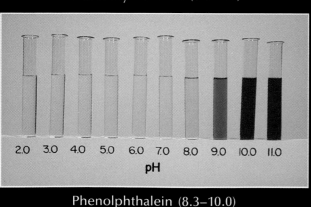

Phenolphthalein (8.3–10.0)

ACID-BASE INDICATORS AND
THEIR TRANSITION pH RANGES
(SECTION 10A–3).

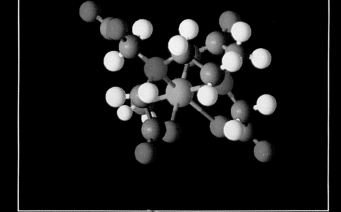

Model of an EDTA/cation chelate (Section 13B–2).

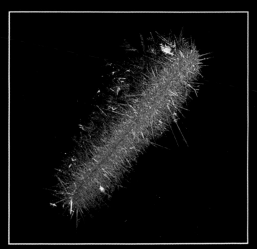

Reduction of silver(I) by direct reaction with copper (Section 14A–2).

Reduction of silver(I) by copper in an electrochemical cell (Section 14A–2).

Reaction between Iron(III) and Iodide (margin note page 329).

(a)    (b)    (c)    (d)

Starch/iodine end point (Sections 15C–1 and 16B–2): (a) iodine solution; (b) same, within a few drops of the equivalence point; (c) same as (b), with starch added; (d) same as (c), at equivalence

Reaction between permanganate and oxalate (Section 16C–1).

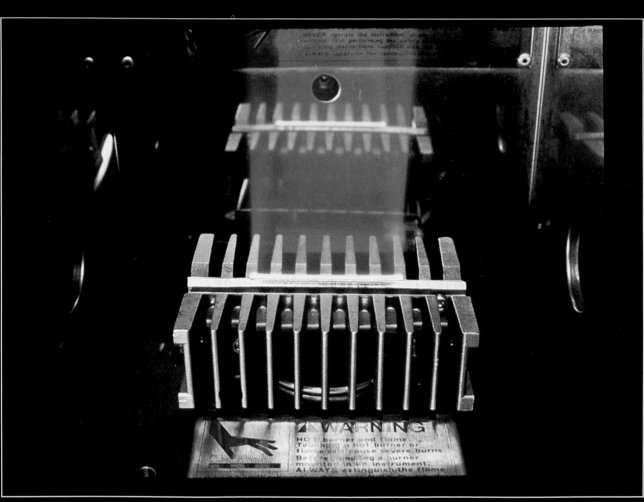

Laminar flow burner for atomic absorption spectroscopy (Section 24B–1). Courtesy of Varian Instruments, Sunnyvale, CA.

Extraction of iodine with chloroform (Section 32C). Left to right: iodine solution; after single 40-mL extraction; after two 20-mL extractions; after four 10-mL extractions.

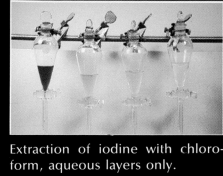

Extraction of iodine with chloroform, aqueous layers only.

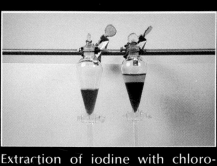

Extraction of iodine with chloroform (Section 32C). Original solution (left) after a single 40-mL extraction (right).

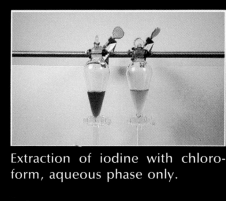

Extraction of iodine with chloroform, aqueous phase only.

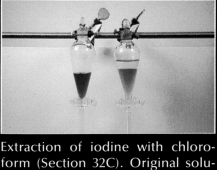

Extraction of iodine with chloroform (Section 32C). Original solution (left); after two 20-mL extractions (right).

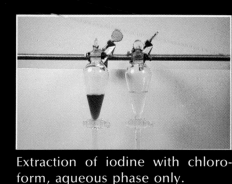

Extraction of iodine with chloroform, aqueous phase only.

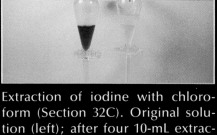

Extraction of iodine with chloroform (Section 32C). Original solution (left); after four 10-mL extractions (right).

Extraction of iodine with chloroform, aqueous phase only.

  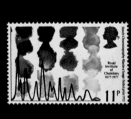 

Stamps from various countries celebrating scientists and events of importance to analytical chemistry. The numbers following each citation refer to the Scott Standard Postage Stamp Catalogue. Stamps commemorating (1) the Centennial of the Metric Convention of 1875, France, #1475; (2) T.W. Richards, Sweden, #1104; (3) Svante Arrhenius, Sweden, #549; (4) Cato Guldberg and Peter Waage, Norway, #453; (5) Walther Nernst, Sweden, #655; (6) Allessandro Volta, Italy, #527; (7) André Marie Ampère, Monaco, #1001; (8) Gustav Robert Kirchhoff, Germany, #9N345; (9) spectroscopic analysis, Vatican, #655; (10) Jöns Jacob Berzelius, Sweden, #1293. (11) Stamp honoring Archer J.P. Martin and Richard L.M. Synge for their work in chromatography, Great Britain, #808. (12) Stamp depicting the visible spectrum and atomic lines, Canada, #613. (13) Stamp announcing an international spectroscopy colloquium in Madrid, Spain, #1570. These stamps are from the private collection of Professor C.M. Lang.

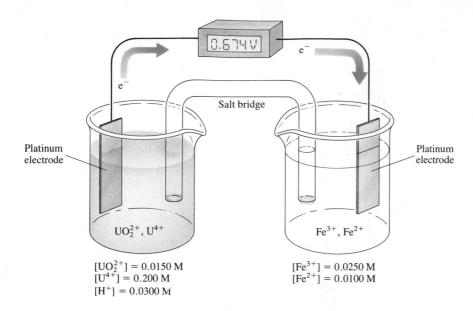

[UO$_2^{2+}$] = 0.0150 M
[U$^{4+}$] = 0.200 M
[H$^+$] = 0.0300 M

[Fe$^{3+}$] = 0.0250 M
[Fe$^{2+}$] = 0.0100 M

Figure 14–9
Cell for Example 14–6.

### Example 14–7

Calculate the theoretical potential for the cell

$$Ag \,|\, AgCl(sat'd), HCl(0.0200 \text{ M}) \,|\, H_2(0.800 \text{ atm}), Pt$$

Note that this cell does not require two compartments (or a salt bridge) because molecular $H_2$ has little tendency to react directly with the low concentration of $Ag^+$ in the electrolyte solution. This cell, shown in Figure 14–10, is an example of a cell *without liquid junction*.

The two half-reactions and their corresponding standard electrode potentials are (Table 14–1)

$$AgCl(s) + e^- \rightleftarrows Ag(s) + Cl^- \qquad E^0 = 0.222 \text{ V}$$
$$2\,H^+ + 2\,e^- \rightleftarrows H_2(g) \qquad\qquad E^0 = 0.000 \text{ V}$$

The two electrode potentials are

$$E_{cathode} = 0.000 - \frac{0.0592}{2} \log \frac{0.800}{(0.0200)^2} = -0.0977 \text{ V}$$

$$E_{anode} = 0.222 - \frac{0.0592}{1} \log 0.0200 = 0.3226 \text{ V}$$

The cell diagram specifies the silver electrode as the anode and the hydrogen electrode as the cathode. Thus,

$$E_{cell} = -0.0977 - 0.3226 = -0.420 \text{ V}$$

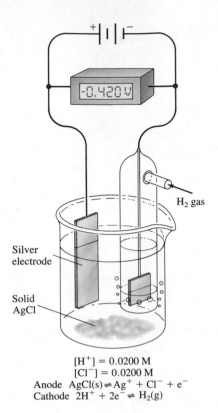

Figure 14–10
Cell for Example 14–7.

$[H^+] = 0.0200$ M
$[Cl^-] = 0.0200$ M
Anode  $AgCl(s) \rightleftharpoons Ag^+ + Cl^- + e^-$
Cathode  $2H^+ + 2e^- \rightleftharpoons H_2(g)$

The negative sign indicates that the cell reaction

$$2\ H^+ + 2\ Ag(s) + 2\ Cl^- \rightleftharpoons H_2(g) + 2\ AgCl(s)$$

is nonspontaneous, and thus the cell is electrolytic. It would require an external power source for operation in this way (Figure 14–10).

---

Example 14–8

Calculate the potential for the following cell employing (a) concentrations and (b) activities:

$$Zn|ZnSO_4(x\text{M}),PbSO_4(\text{sat'd})|Pb$$

where $x = 5.00 \times 10^{-4}, 2.00 \times 10^{-3}, 1.00 \times 10^{-2}, 2.00 \times 10^{-2}$, and $5.00 \times 10^{-2}$.

(a) In a neutral solution, little $HSO_4^-$ is formed and we can assume that

$$[SO_4^{2-}] = c_{ZnSO_4} = x = 5.00 \times 10^{-4}$$

The half-reactions and standard potentials are

$$PbSO_4(s) + 2\ e^- \rightleftharpoons Pb(s) + SO_4^{2-} \qquad E^0 = -0.350 \text{ V}$$
$$Zn^{2+} + 2\ e^- \rightleftharpoons Zn \qquad\qquad\qquad E^0 = -0.763 \text{ V}$$

The potential of the lead electrode is

$$E_{PbSO_4} = -0.350 - \frac{0.0592}{2} \log (5.00 \times 10^{-4}) = -0.252 \text{ V}$$

For the zinc half-reaction.

$$[Zn^{2+}] = 5.00 \times 10^{-4}$$

$$E_{Zn^{2+}} = -0.763 - \frac{0.0592}{2} \log \frac{1}{5.00 \times 10^{-4}} = -0.860 \text{ V}$$

Since the lead electrode is specified as the cathode,

$$E_{cell} = -0.252 - (-0.860) = 0.608 \text{ V}$$

Cell potentials at the other concentrations can be derived in the same way. Their values are given in Table 14–2.

(b) To obtain activity coefficients for $Zn^{2+}$ and $SO_4^{2-}$, we must first calculate the ionic strength with the aid of Equation 7–1:

$$\mu = [5.00 \times 10^{-4} \times (2)^2 + 5.00 \times 10^{-4} \times (2)^2]/2 = 2.00 \times 10^{-3}$$

In Table 7–1, we find $a_A = 4.0$ for $SO_4^{2-}$ and $\alpha_A = 6.0$ for $Zn^{2+}$. Substituting these values into Equation 7–5 gives for sulfate ion

$$-\log f_{SO_4^{2-}} = \frac{0.51 \times (2)^2 \times \sqrt{2.00 \times 10^{-3}}}{1 + 0.33 \times 4.0 \sqrt{2.00 \times 10^{-3}}} = 8.59 \times 10^{-2}$$

$$f_{SO_4^{2-}} = 0.820$$

Repeating the calculations for $Zn^{2+}$, for which $\alpha_A = 6.0$, yields

$$f_{Zn^{2+}} = 0.825$$

The Nernst equation for the lead electrode now becomes

$$E_{PbSO_4} = -0.350 - \frac{0.0592}{2} \log (0.820 \times 5.00 \times 10^{-4}) = -0.250$$

Similarly, for the zinc electrode

$$E_{Zn^{2+}} = -0.763 - \frac{0.0592}{2} \log \frac{1}{0.825 \times 5.00 \times 10^{-4}} = -0.863$$

Thus,

$$E_{cell} = -0.250 - (-0.863) = 0.613 \text{ V}$$

Values for other concentrations are found in Table 14–2 as well as experimentally determined potentials for the cell.

It is evident from the data in Table 14–2 that neglect of activity coefficients causes significant errors in cell potentials. It is also evident from the data in the fifth column of this table that potentials computed with activities agree reasonably well with experiment.

Example 14–9

Calculate the potential required to initiate deposition of copper from a solution that is 0.010 M in $CuSO_4$ and contains sufficient $H_2SO_4$ to give an $H^+$ concentration of $1.00 \times 10^{-4}$ M.

The deposition of copper necessarily occurs at the cathode. Since there is no more easily oxidizable species than water in the system, evolution of $O_2$ occurs at the anode. The required standard electrode potentials are (Table 14–1)

$$O_2(g) + 4\ H^+ + 4\ e^- \rightleftharpoons 2\ H_2O \qquad E^0 = +1.229\ V$$

$$Cu^{2+} + 2\ e^- \rightleftharpoons Cu(s) \qquad E^0 = +0.337\ V$$

The potential for the Cu electrode is given by

$$E = +0.337 - \frac{0.0592}{2} \log \frac{1}{0.010} = +0.278\ V$$

If $O_2$ is evolved at 1.00 atm, the potential for the oxygen electrode is

$$E = +1.229 - \frac{0.0592}{4} \log \frac{1}{(1.00)(1.00 \times 10^{-4})^4} = +0.992\ V$$

and the cell potential is

$$E_{cell} = +0.278 - 0.992 = -0.714\ V$$

Thus, initiation of the reaction

$$2\ Cu^{2+} + 2\ H_2O \rightleftharpoons O_2(g) + 4\ H^+ + 2\ Cu(s)$$

requires the application of a potential greater than $-0.714$ V.

## 14D–3 The Experimental Determination of Standard Potentials

Although standard electrode potentials for hundreds of half-reactions are found in compilations of electrochemical data, it is noteworthy that neither the standard hydrogen electrode nor any other electrode in its standard state can be realized in the laboratory. That is, the standard hydro-

**Table 14-2**
**EFFECT OF IONIC STRENGTH ON THE POTENTIAL OF A GALVANIC CELL***

| Concentration, $ZnSO_4$ | Ionic Strength, $\mu$ | (a) $E$, Based on Concentrations | (b) $E$, Based on Activities | $E$, Experimental Values[†] |
|---|---|---|---|---|
| $5.00 \times 10^{-4}$ | $2.00 \times 10^{-3}$ | 0.608 | 0.613 | 0.611 |
| $2.00 \times 10^{-3}$ | $8.00 \times 10^{-3}$ | 0.573 | 0.582 | 0.583 |
| $1.00 \times 10^{-2}$ | $4.00 \times 10^{-2}$ | 0.531 | 0.550 | 0.553 |
| $2.00 \times 10^{-2}$ | $8.00 \times 10^{-2}$ | 0.513 | 0.537 | 0.542 |
| $5.00 \times 10^{-2}$ | $2.00 \times 10^{-1}$ | 0.490 | 0.521 | 0.529 |

*Cell described in Example 14-8.

[†]Experimental data from I. A. Cowperthwaite and V. K. LaMer, *J. Amer. Chem. Soc.*, **1931**, *53*, 4333.

gen electrode is a *hypothetical electrode*, as is any other electrode system in which the reactants and products are at unit activity or pressure. The reason such electrode systems cannot be prepared experimentally is that chemists lack the knowledge to produce solutions that have ionic activities of exactly unity. That is, no adequate theory exists that permits calculation of the *concentration* of a compound needed to give a solution of unit ionic activity. At such high ionic strengths, the Debye-Hückel relationship (Section 7B-2) is not valid, and no independent experimental method exists for the determination of activity coefficients in such solutions. Thus, for example, the *concentration* of HCl or other acids required to give the unit hydrogen ion activity specified in the standard hydrogen electrode *cannot be calculated or determined experimentally*. Nonetheless, data taken in solutions of low ionic strengths can be extrapolated to give valid measures of standard electrode potentials as theoretically defined. An example of how hypothetical standard electrode potentials might be obtained from experimental data follows.

Example 14-10

D. A. MacInnes[5] found that a cell similar to that shown in Figure 14-10 developed a potential of 0.52053 V. The cell is described by

$$Pt, H_2(1.00\ atm)|HCl(3.215 \times 10^{-3}\ M), AgCl(sat'd)|Ag$$

Calculate the standard electrode potential for the half-reaction

$$AgCl(s) + e^- \rightleftharpoons Ag(s) + Cl^-$$

Here, the electrode potential for the cathode is

$$E_{cathode} = E^0_{AgCl} - 0.0592 \log c_{HCl}f_{Cl^-}$$

[5]D. A. MacInnes, *The Principles of Electrochemistry*, p. 187. New York: Reinhold, 1939.

where $f_{Cl^-}$ is the activity coefficient of $Cl^-$. The second half-cell reaction is

$$H^+ + e^- \rightleftharpoons \tfrac{1}{2} H_2(g)$$

and

$$E_{anode} = E^0_{H_2} = \frac{0.0592}{1} \log \frac{p_{H_2}^{1/2}}{c_{HCl} f_{H^+}}$$

The measured potential is the difference between these potentials (Equation 14–10):

$$E_{cell} = (E^0_{AgCl} - 0.0592 \log c_{HCl} f_{Cl^-}) - \left(0.000 - 0.0592 \log \frac{p_{H_2}^{1/2}}{c_{HCl} f_{H^+}}\right)$$

Combining the two logarithmic terms gives

$$E_{cell} = E^0_{AgCl} - 0.0592 \log \frac{c_{HCl}^2 f_{H^+} f_{Cl^-}}{p_{H_2}^{1/2}}$$

The activity coefficients for $H^+$ and $Cl^-$ can be calculated from Equation 7–5 employing $3.215 \times 10^{-3}$ for the ionic strength $\mu$; these values are 0.945 and 0.939, respectively. Substitution of these activity coefficients and the experimental data into the foregoing equation gives, upon rearrangement,

$$E^0_{AgCl} = 0.52053 + 0.0592 \log \frac{(3.215 \times 10^{-3})^2 (0.945)(0.939)}{1.00^{1/2}}$$

$$= 0.2223 \simeq 0.222 \text{ V}$$

(The mean for this and similar measurements at other concentrations was 0.222 V.)

---

## 14E   QUESTIONS AND PROBLEMS

14–1. Briefly describe or define
  *(a) oxidation.
  (b) liquid junction.
  *(c) standard hydrogen electrode.
  (d) salt bridge.
  *(e) Nernst equation.

14–2. Briefly distinguish between
  *(a) an oxidizing agent and a reducing agent.
  (b) an electrolytic cell and a galvanic cell.
  *(c) a reversible and an irreversible electrochemical cell.
  (d) the anode and the cathode of an electrochemical cell.
  *(e) the standard electrode potential and the formal potential for a half-cell reaction.

14–3. Complete and balance the following equations. Supply $H^+$ and/or $H_2O$, if necessary, to achieve balance:
  *(a) $Br_2 + Sn^{2+} \rightleftharpoons Br^- + Sn^{4+}$
  (b) $Ti^{3+} + Fe^{3+} \rightleftharpoons TiO^{2+} + Fe^{2+}$
  *(c) $IO_3^- + I^- \rightleftharpoons I_2 + H_2O$
  (d) $Cu(s) + NO_3^- \rightleftharpoons Cu^{2+} + HNO_2$

14–4. Write a balanced net-ionic equation for the following. Supply $H^+$ and/or $H_2O$, if necessary, to achieve balance.
  *(a) Metallic iron dissolves in nitric acid (products: $NO_2(g)$, $Fe^{3+}$).
  (b) Manganese(II) is oxidized by periodate (products, $MnO_4^-$, $IO_3^-$).

**\*(c)** Potassium permanganate oxidizes sodium oxalate (products: $CO_2(g)$, $Mn^{2+}$).

**(d)** $V(OH)_4^+$ oxidizes $V^{2+}$ (product: $VO^{2+}$).

**\*(e)** $V(OH)_4^+$ oxidizes $V^{2+}$ (product, $V^{3+}$).

**\*14–5.** The following reactions proceed to essential completion as written:

(1) $Zn(s) + 2 Cr^{3+} \rightleftharpoons Zn^{2+} + 2 Cr^{2+}$

(2) $Sn^{4+} + 2 Cr^{2+} \rightleftharpoons Sn^{2+} + 2 Cr^{3+}$

(3) $I_2 + Sn^{2+} \rightleftharpoons 2 I^- + Sn^{4+}$

(4) $2 HNO_2 + 2 I^- + 2 H^+ \rightleftharpoons I_2 + 2 NO(g) + 2 H_2O$

**(a)** In each equation, identify the reactant that acts as oxidizing agent and the reactant that acts as reducing agent.

**(b)** Write each overall reaction in terms of two balanced half-reactions.

**(c)** Write each half-reaction in (b) as a reduction.

**(d)** Arrange the half-reactions in (c) in order of decreasing effectiveness as electron acceptors.

**14–6.** The following reactions proceed to essential completion as written:

(1) $Fe(CN)_6^{4-} + Fe^{3+} \rightleftharpoons Fe(CN)_6^{3-} + Fe^{2+}$

(2) $2 Fe^{3+} + Sn^{2+} \rightleftharpoons 2 Fe^{2+} + Sn^{4+}$

(3) $V(OH)_4^+ + Fe^{2+} + 2 H^+ \rightleftharpoons VO^+ + Fe^{3+} + 3 H_2O$

(4) $2 Fe(CN)_6^{3-} + Sn^{2+} \rightleftharpoons 2 Fe(CN)_6^{4-} + Sn^{4+}$

**(a)** In each equation, identify the reactant that acts as oxidizing agent and the reactant that acts as reducing agent.

**(b)** Write each overall reaction in terms of two balanced half-reactions.

**(c)** Write each half-reaction in (b) as a reduction.

**(d)** Arrange the half-reactions in (c) in order of decreasing effectiveness as electron acceptors.

**\*14–7.** Based upon the information in Problem 14–5, predict whether you would expect a reaction between

**(a)** $Zn^{2+}$ and $Sn^{2+}$.

**(b)** $Sn^{4+}$ and $Zn(s)$.

**(c)** $I_2$ and $Cr^{2+}$.

**(d)** $I^-$ and $Sn^{4+}$.

**14–8.** Based upon the information in Problem 14–6, predict whether you would expect a reaction between

**(a)** $Fe(CN)_6^{3-}$ and $Fe^{2+}$.

**(b)** $V(OH)_4^+$ and $Sn^{2+}$.

**(c)** $Sn^{4+}$ and $VO^{2+}$.

**\*14–9.** Calculate the potential of a silver electrode in contact with a solution that is

**(a)** 0.0750 M in $AgNO_3$.

**(b)** $6.00 \times 10^{-3}$ M in KI and saturated with AgI.

**(c)** 0.0100 M in $MgBr_2$ and saturated with AgBr.

**(d)** $8.00 \times 10^{-3}$ M in $Ag(S_2O_3)_2^{3-}$ and 0.0400 M in $Na_2S_2O_3$.

**(e)** prepared by mixing 30.0 mL of 0.100 M $AgNO_3$ with 20.0 mL of 0.100 M NaCl.

**(f)** prepared by mixing 30.0 mL of 0.100 M $AgNO_3$ with 20.0 mL of 0.100 M $MgCl_2$.

**14–10.** Calculate the potential of a nickel electrode immersed in a solution that is

**(a)** 0.0600 M in $NiCl_2$.

**(b)** saturated with $Ni(OH)_2$ and has a pH of 8.00. For $Ni(OH)_2$, $K_{sp} = 6.5 \times 10^{-18}$.

**(c)** 0.100 M in $Na_2CO_3$, saturated with $NiCO_3$, and is buffered to a pH of 8.00. For $NiCO_3$, $K_{sp} = 6.6 \times 10^{-9}$.

**(d)** $4.00 \times 10^{-3}$ M in $Ni(NH_3)_6^{2+}$, and 0.100 M in $NH_3$. For $Ni(NH_3)_6^{2+}$, $K_f = 5.0 \times 10^8$.

**(e)** prepared by mixing 25.0 mL of 0.0500 M $NiCl_2$ with an equal volume of 0.0500 M NaOH.

**(f)** prepared by mixing 25.0 mL of 0.0500 M $NiCl_2$ with an equal volume of 0.0500 M $Ba(OH)_2$.

**\*14–11.** Calculate the potential of a platinum electrode in contact with a solution that is

**(a)** 0.0500 M in $Cr^{3+}$ and 0.0150 M in $Cr^{2+}$.

**(b)** 0.0700 M in $K_4Fe(CN)_6$ and 0.0200 M in $K_3Fe(CN)_6$.

**(c)** 0.0700 M in $FeSO_4$ and 0.0200 M in $Fe_2(SO_3)_3$.

**(d)** 0.0200 M in KI and saturated with $PbI_2$. *no reaction*

**14–12.** Calculate the potential of a platinum electrode immersed in a solution that is

**(a)** 0.100 M in $Sn^{2+}$ and 0.0250 M in $Sn^{4+}$.

**(b)** 0.0500 M in $I_3^-$ and 0.100 M in KI.

**(c)** 0.0600 M in $VOSO_4$, 0.0100 M in $V_2(SO_4)_3$, and has a pH of 2.00.

**(d)** 0.0500 M in $Fe^{2+}$ and 0.0360 M in $Fe^{3+}$.

**\*14–13.** Calculate the potential of a platinum electrode in a solution that has been prepared by

**(a)** mixing 30.0 mL of 0.0500 M $FeCl_3$ with 40.0 mL of 0.0800 M $SnCl_2$.

**(b)** mixing 30.0 mL of 0.0500 M $SnCl_2$ with 40.0 mL of 0.0800 M $FeCl_3$.

**(c)** saturating a 0.0300 M KBr solution with $Br_2$.

**(d)** saturating a 0.0200 M $MnCl_2$ solution with $MnO_2$ and adjusting the pH to 7.00.

**(e)** mixing 25.0 mL of 0.0800 M $VCl_2$ with an equal volume of 0.0500 M $VOSO_4$ and adjusting the pH to 2.00.

**(f)** mixing 25.0 mL of 0.0800 M $VOSO_4$ with an equal volume of 0.0500 M $VCl_2$ and adjusting the pH to 2.00.

**14–14.** Calculate the potential of a platinum electrode in a solution that has been produced by

**(a)** mixing 30.0 mL of 0.0100 M $K_2Cr_2O_7$ with 50.0 mL of 0.0500 M $FeCl_2$. The pH of the resulting solution is 1.00.

**(b)** mixing 30.0 mL of 0.0500 M $FeCl_2$ with 50.0 mL of 0.0100 M $K_2Cr_2O_7$. The pH of the resulting solution is 1.00.

**(c)** saturating a 0.0600 M KI solution with $I_2$.

**(d)** saturating a 0.0200 M $SbO^+$ solution with $Sb_2S_5$ and adjusting the pH to 3.00.

**(e)** mixing 50.0 mL of 0.100 M $V(OH)_4^+$ with 25.0 mL of 0.0800 M $V_2(SO_4)_3$ and adjusting the pH to 2.00.

**(f)** mixing 50.0 mL of 0.100 M $V_2(SO_4)_3$ with 25.0 mL of 0.0800 M $V(OH)_4^+$ and adjusting the pH to 2.00.

**14–15.** Calculate the electrode potential for the following half-cells. Indicate whether the half-cell will act as anode or as cathode when coupled with a standard hydrogen electrode in a galvanic cell.
(a) $Pb|Pb^{2+}$ (0.0600 M)
(b) $Pt|Sn^{4+}$ (0.0830 M), $Sn^{2+}$ (0.0178 M)
(c) $Cu|CuI$ (sat'd), $I^-$ (0.100 M)
(d) $Pt, H_2$ (780 torr)$|HCl$ (0.137 M)
(e) $Pt, O_2$ (0.975 atm)$|HCl$ (0.180 M)
(f) $Zn|ZnY^{2-}$ (5.00 $\times$ $10^{-3}$ M), $HY^{3-}$ (9.00 $\times$ $10^{-2}$ M), $H^+$ (1.00 $\times$ $10^{-8}$ M)
(g) $Cu|Cu(NH_3)_4^{2+}$ (0.0100 M), $NH_3$ (0.250 M)

(For $Cu(NH_3)_4^{2+}$, $K_f = 2 \times 10^{13}$.)

**14–16.** Calculate the electrode potential for the following half-cells. Indicate whether the half-cell will act as anode or as cathode when coupled with a standard hydrogen electrode in a galvanic cell.
(a) $Ni|NiSO_4$ (0.0360 M)
(b) $Pt|V^{2+}$ (0.0413 M), $V^{3+}$ (0.0798 M)
(c) $Cd|Cd(OH)_2$ (sat'd), $H^+$ (1.00 $\times$ $10^{-11}$ M)
(d) $Pt, H_2$ (1.05 atm), $HClO_4$ (0.0918 M)
(e) $Pt, O_2$ (765 torr)$|HClO_4$ (0.0411 M)
(f) $Hg|HgY^{2-}$ (2.64 $\times$ $10^{-3}$ M), $H_2Y^{2-}$ (4.55 $\times$ $10^{-2}$ M), $H^+$ (1.00 $\times$ $10^{-3}$ M)
(g) $Hg|HgI_4^{2-}$ (5.75 $\times$ $10^{-3}$ M), $I^-$ (0.0100 M)

(For $HgI_4^{2-}$, $K_f = 6.3 \times 10^{29}$.)

**14–17.** Calculate the theoretical potential for the following cells. Indicate whether the cell, as written, is galvanic or electrolytic.
(a) $Pb|Pb^{2+}$ (0.0400 M)$||Hg^{2+}$ (0.0800 M)$|Hg$
(b) $Ni|Ni^{2+}$ (0.100 M)$||Fe^{2+}$ (0.0200 M)$|Fe$
(c) $Cd|Cd^{2+}$ (0.0250 M)$||H^+$ (1.00 $\times$ $10^{-6}$ M)$|H_2$ (0.980 atm), $Pt$

(d) $Pt, H_2$ (760 torr)$|HOAc$ (0.100 M)$||SHE$
(e) $Ag|AgI$ (sat'd), $I^-$ (2.50 $\times$ $10^{-4}$ M)$||Fe^{3+}$ (0.100 M), $Fe^{2+}$ 0.0500 M$|Pt$
(f) $Bi|BiO^+$ (0.0175 M), $H^+$ (1.00 $\times$ $10^{-3}$ M)$||Cl^-$ (0.0380 M), $AgCl$ (sat'd)$|Ag$
(g) $Cu|Cu^+$ (1.08 $\times$ $10^{-3}$ M)$||I^-$ (3.40 $\times$ $10^{-2}$ M), $CuI$ (sat'd)$|Cu$
(h) $Sn|Sn^{2+}$ (0.0895 M)$||Ti^{3+}$ (4.64 $\times$ $10^{-4}$ M), $Ti^{2+}$ (0.0764 M)$|Pt$
(i) $Pt|MnO_4^-$ (3.15 $\times$ $10^{-2}$ M), $MnO_4^{2-}$ (7.98 $\times$ $10^{-4}$ M)$||Ag^+$ (3.66 $\times$ $10^{-2}$ M)$|Ag$
(j) $Pt|I_3^-$ (0.0497 M), $I^-$ (9.15 $\times$ $10^{-3}$ M)$||Co^{2+}$ (0.0534 M)$|Co$

**14–18.** Calculate the theoretical potential for the following cells. Indicate whether the cell, as written, is galvanic or electrolytic.
(a) $Pd|Pd^{2+}$ (0.0300 M)$||Ag^+$ (0.120 M)$|Ag$
(b) $Zn|Zn^{2+}$ (5.50 $\times$ $10^{-4}$ M)$||Co^{2+}$ (0.0295 M)$|Co$
(c) $Pt, O_2$ (1.04 atm)$|H^+$ (1.00 $\times$ $10^{-4}$ M)$||Pb^{2+}$ (0.0625 M)$|Pb$
(d) $Pt, H_2$ (760 torr)$|HOCl$ (0.1050 M), $NaOCl$ (0.0573 M)$||SHE$
(e) $Cu|CuI$ (sat'd), $KI$ (4.00 $\times$ $10^{-3}$ M)$||Cu^{2+}$ (0.0860 M)$|Cu$
(f) $Pb|PbSO_4$ (sat'd), $SO_4^{2-}$ (6.15 $\times$ $10^{-2}$ M)$||I^-$ (7.44 $\times$ $10^{-2}$ M), $AgI$ (sat'd)$|Ag$
(g) $Ag|Ag^+$ (6.15 $\times$ $10^{-3}$ M)$||Br^-$ (0.0348 M), $AgBr$ (sat'd)$|Ag$
(h) $Pt|V^{3+}$ (8.00 $\times$ $10^{-4}$ M), $VO^{2+}$ (0.0763 M), $H^+$ (0.0132 M)$||V(OH)_4^+$ (0.100 M), $VO^{2+}$ (1.00 $\times$ $10^{-3}$ M), $H^+$ (0.0132 M)$|Pt$
(i) $Pt|Fe(CN)_6^{3-}$ (0.0845 M), $Fe(CN)_6^{4-}$ (2.67 $\times$ $10^{-4}$ M)$||Fe^{3+}$ (1.00 $\times$ $10^{-3}$ M), $Fe^{2+}$ (1.00 $\times$ $10^{-2}$ M)$|Pt$
(j) $Pt|Br_2$ (sat'd), $Br^-$ (7.15 $\times$ $10^{-2}$ M)$||Cl^-$ (2.00 $\times$ $10^{-1}$ M), $Cl_2$ (1.05 atm)$|Pt$

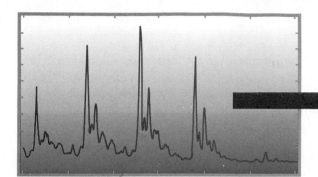

**CHAPTER** 15

# THEORY OF OXIDATION/REDUCTION TITRATIONS

$T$his chapter is concerned with titration curves and indicators for oxidation/reduction titrations. Before describing how such curves are derived, we need to know how equilibrium constants for oxidation/reduction reactions are computed from standard electrode potentials.

## 15A EQUILIBRIUM CONSTANTS FOR OXIDATION/REDUCTION REACTIONS

Let us again consider the equilibrium established when a piece of copper is immersed in a solution containing a dilute solution of silver nitrate:

$$Cu(s) + 2\ Ag^+ \rightleftarrows Cu^{2+} + 2\ Ag(s) \qquad (15-1)$$

The equilibrium constant for this reaction is

$$K_{eq} = \frac{[Cu^{2+}]}{[Ag^+]^2} \qquad (15-2)$$

Reaction 15-1 can also be carried out in the galvanic cell

$$Cu|Cu^{2+}(x\ M)\|Ag^+(y\ M)|Ag$$

A sketch of this cell is shown in Figure 14-1. Its cell potential at any instant is given by

$$E_{cell} = E_{cathode} - E_{anode} = E_{Ag^+} - E_{Cu^{2+}}$$

As the reaction proceeds, the concentration of Cu(II) ions increases and the concentration of Ag(I) ions decreases. These changes make the potential of the copper electrode more positive and that of the silver electrode less positive. As shown in Figure 15-1, the net effect of these changes is a

345

Figure 15–1

Potential of a galvanic cell as it discharges.

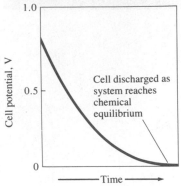

continuous decrease in the potential of the cell as it discharges. Ultimately the concentrations of Cu(II) and Ag(I) attain their equilibrium values as determined by Equation 15–2, and the current ceases. Under these equilibrium conditions, *the potential of the cell becomes zero. Thus, at chemical equilibrium,* we may write

$$E_{cell} = 0 = E_{cathode} - E_{anode} = E_{Ag^+} - E_{Cu^{2+}}$$

or,

$$E_{cathode} = E_{anode} = E_{Ag^+} = E_{Cu^{2+}} \qquad (15\text{–}3)$$

### 15A–1  Electrode Potentials in Equilibrum Systems

We can generalize Equation 15–3 by stating that *at equilibrium, the electrode potentials for all half-reactions in an oxidation/reduction system are equal.* This generalization applies regardless of the number of half-reactions present in the system because interactions among *all systems* must take place until their electrode potentials are identical. For example, if we have four oxidation/reduction systems in a solution, interaction among all four takes place until

$$E_{Ox_1} = E_{Ox_2} = E_{Ox_3} = E_{Ox_4} \qquad (15\text{–}4)$$

where $E_{Ox_1}$, $E_{Ox_2}$, $E_{Ox_3}$, and $E_{Ox_4}$ are the electrode potentials for the four half-reactions.

### 15A–2  The Calculation of Equilibrium Constants

Returning to the reaction shown in Equation 15–1, let us substitute Nernst expressions for the two electrode potentials in Equation 15–3.

$$E^0_{Ag^+} - \frac{0.0592}{2} \log \frac{1}{[Ag^+]^2} = E^0_{Cu^{2+}} - \frac{0.0592}{2} \log \frac{1}{[Cu^{2+}]} \quad (15\text{–}5)$$

It is important to note that the Nernst equation is applied to the silver half-reaction as it appears in the balanced equation (Equation 15–1):

$$2\,Ag^+ + 2\,e^- \rightleftharpoons 2\,Ag(s) \qquad E^0 = 0.799 \text{ V}$$

Rearrangement of Equation 15–5 gives

$$E^0_{Ag^+} - E^0_{Cu^{2+}} = \frac{0.0592}{2} \log \frac{1}{[Ag^+]^2} - \frac{0.0592}{2} \log \frac{1}{[Cu^{2+}]}$$

$$= \frac{0.0592}{2} \log \frac{1}{[Ag^+]^2} + \frac{0.0592}{2} \log \frac{[Cu^{2+}]}{1}$$

Finally, then,

$$\frac{2(E^0_{Ag^+} - E^0_{Cu^{2+}})}{0.0592} = \log \frac{[Cu^{2+}]}{[Ag^+]^2} = \log K_{eq} \qquad (15\text{-}6)$$

The concentration terms in Equation 15–6 are *equilibrium concentrations; the ratio $[Cu^{2+}]/[Ag^+]^2$ in the logarithmic term is therefore the equilibrium constant for the reaction.*

---

Example 15–1

Calculate the equilibrium constant for the reaction shown in Equation 15–1.

Substituting numerical values into Equation 15–6 yields

$$\log K_{eq} = \log \frac{[Cu^{2+}]}{[Ag^+]^2} = \frac{2(0.799 - 0.337)}{0.0592}$$

$$= 15.6$$

$$K_{eq} = \text{antilog } 15.6 = 4.1 \times 10^{15} = 4 \times 10^{15}$$

In making calculations of this sort, you should follow the rounding rule for antilogarithms given on page 29.

---

Example 15–2

Calculate the equilibrium constant for the reaction

$$2\ Fe^{3+} + 3\ I^- \rightleftarrows 2\ Fe^{2+} + I_3^-$$

In Appendix 6 we find

$$2\ Fe^{3+} + 2\ e^- \rightleftarrows 2\ Fe^{2+} \qquad E^0 = 0.771\ V$$
$$I_3^- + 2\ e^- \rightleftarrows 3\ I^- \qquad E^0 = 0.536\ V$$

We have multiplied the first half-reaction by 2 so that the number of moles of $Fe^{3+}$ and $Fe^{2+}$ are the same as in the balanced overall equation. The Nernst equation for $Fe^{3+}$ is thus based upon the half-reaction for a two-electron transfer:

$$E_{Fe^{3+}} = E^0_{Fe^{3+}} - \frac{0.0592}{2} \log \frac{[Fe^{2+}]^2}{[Fe^{3+}]^2}$$

and

$$E_{I_3^-} = E^0_{I_3^-} - \frac{0.0592}{2} \log \frac{[I^-]^3}{[I_3^-]}$$

At equilibrium, the electrode potentials are equal; thus

$$E_{Fe^{3+}} = E_{I_3^-}$$

$$E^0_{Fe^{3+}} - \frac{0.0592}{2} \log \frac{[Fe^{2+}]^2}{[Fe^{3+}]^2} = E^0_{I_3^-} - \frac{0.0592}{2} \log \frac{[I^-]^3}{[I_3^-]}$$

This equation rearranges to

$$E^0_{Fe^{3+}} - E^0_{I_3^-} = \frac{0.0592}{2} \log \frac{[Fe^{2+}]^2}{[Fe^{3+}]^2} + \frac{0.0592}{2} \log \frac{[I_3^-]}{[I^-]^3}$$

Notice that we have changed the sign of the second logarithmic term by inverting the fraction. Further rearrangement gives

$$\log \frac{[Fe^{2+}]^2[I_3^-]}{[Fe^{3+}]^2[I^-]^3} = \frac{2(E^0_{Fe^{3+}} - E^0_{I_3^-})}{0.0592}$$

Recall, however, that the concentration terms here are *equilibrium concentrations* and

$$\log K_{eq} = \frac{(E^0_{Fe^{3+}} - E^0_{I_3^-})2}{0.0592} = \frac{2(0.771 - 0.536)}{0.0592} = 7.94$$

$$K_{eq} = \text{antilog } 7.94 = 8.7 \times 10^7$$

We round the answer to two figures because $\log K_{eq}$ contains only two significant figures (the two to the right of the decimal point).

---

Example 15–3

Calculate the equilibrium constant for the reaction

$$2\ MnO_4^- + 3\ Mn^{2+} + 2\ H_2O \rightleftarrows 5\ MnO_2(s) + 4\ H^+$$

In Appendix 6 we find

$$2\ MnO_4^- + 8\ H^+ + 6\ e^- \rightleftarrows 2\ MnO_2(s) + 4\ H_2O \qquad E^0 = +1.695\ V$$

$$3\ MnO_2(s) + 12\ H^+ + 6\ e^- \rightleftarrows 3\ Mn^{2+} + 6\ H_2O \qquad E^0 = +1.23\ V$$

Again we have multiplied both equations by integers so that the number of electrons are equal. When this system is at equilibrium,

$$E_{MnO_4^-} = E_{MnO_2}$$

$$1.695 - \frac{0.0592}{6} \log \frac{1}{[MnO_4^-]^2[H^+]^8} = 1.23 - \frac{0.0592}{6} \log \frac{[Mn^{2+}]^3}{[H^+]^{12}}$$

Inverting the log term on the right and rearranging lead to

$$\frac{6(1.695 - 1.23)}{0.0592} = \log \frac{[H^+]^{12}}{[MnO_4^-]^2[Mn^{2+}]^3[H^+]^8}$$

$$47.1 = \log \frac{[H^+]^4}{[MnO_4^-]^2[Mn^{2+}]^3} = \log K_{eq}$$

$$K_{eq} = \text{antilog } 47.1 = 1.3 \times 10^{47} = 1 \times 10^{47}$$

Note that the final result has only one significant figure.

### Example 15–4

Calculate the equilibrium constant for the reaction

$$MnO_4^- + 5\ Fe^{2+} + 8\ H^+ \rightleftharpoons Mn^{2+} + 5\ Fe^{3+} + 4\ H_2O$$

The standard electrode potentials are (Appendix 6)

$$MnO_4^- + 5\ e^- + 8\ H^+ \rightleftharpoons Mn^{2+} + 4\ H_2O \qquad E_{MnO_4^-}^0 = +1.51\ V$$
$$Fe^{3+} + e^- \rightleftharpoons Fe^{2+} \qquad\qquad\qquad\qquad\quad E_{Fe^{3+}}^0 = +0.771\ V$$

The balanced net-ionic equation for this reaction involves 5 mol of Fe for 1 mol of Mn; it is therefore necessary to write the half-reaction for the $Fe^{3+}/Fe^{2+}$ couple as

$$5\ Fe^{3+} + 5\ e^- \rightleftharpoons 5\ Fe^{2+} \qquad E_{Fe^{3+}}^0 = +0.771\ V$$

Note that multiplication of this half-reaction by 5 does not alter the value of $E^0$ (page 328).

When the system is in equilibrium,

$$E_{Fe^{3+}} = E_{MnO_4^-}$$

or

$$E_{Fe^{3+}}^0 - \frac{0.0592}{5}\log\frac{[Fe^{2+}]^5}{[Fe^{3+}]^5} = E_{MnO_4^-}^0 - \frac{0.0592}{5}\log\frac{[Mn^{2+}]}{[MnO_4^-][H^+]^8}$$

This equality can be rearranged to provide the logarithm of the equilibrium constant:

$$\frac{0.0592}{5}\log\frac{[Mn^{2+}][Fe^{3+}]^5}{[MnO_4^-][Fe^{2+}]^5[H^+]^8} = \frac{0.0592}{5}\log K_{eq} = E_{MnO_4^-}^0 - E_{Fe^{3+}}^0$$

Finally, then,

$$\log K_{eq} = \frac{5(1.51 - 0.771)}{0.0592} = 62.52 \simeq 62.5$$
$$K_{eq} = 10^{0.52} \times 10^{62} = 3 \times 10^{62}$$

### Feature 15–1
### A GENERAL EXPRESSION FOR CALCULATING EQUILIBRIUM CONSTANTS FROM STANDARD POTENTIALS

In order to derive a general relationship for computing equilibrium constants from standard-potential data, let us consider a reaction in which a species $A_{red}$ reacts with a species $B_{ox}$ to yield $A_{ox}$ and $B_{red}$. The two electrode reactions are

$$A_{ox} + ae^- \rightleftharpoons A_{red}$$
$$B_{ox} + be^- \rightleftharpoons B_{red}$$

In order to obtain a balanced equation for the desired reaction, we must multiply the first equation by $b$ and the second by $a$ to give

$$bA_{ox} + bae^- \rightleftharpoons bA_{red}$$
$$aB_{ox} + bae^- \rightleftharpoons aB_{red}$$

Subtraction of the first equation from the second yields

$$bA_{red} + aB_{ox} \rightleftharpoons bA_{ox} + aB_{red}$$

When this system is in equilibrium, the two electrode potentials $E_A$ and $E_B$ are numerically identical, that is,

$$E_A = E_B$$

Note that the product $ab$ is the total number of electrons gained in the reduction (and lost in the oxidation) for the balanced redox equation. Thus, if $a = b$, it is not necessary to multiply the half reactions by $a$ and $b$. If $a = b = n$, the equilibrium constant is determined from

$$\log K_{eq} = \frac{n(E_B^0 - E_A^0)}{0.0592}$$

Substitution of the Nernst expressions into this equation reveals that, *at equilibrium,*

$$E_A^0 - \frac{0.0592}{ab} \log \frac{[A_{red}]^b}{[A_{ox}]^b} = E_B^0 - \frac{0.0592}{ab} \log \frac{[B_{red}]^a}{[B_{ox}]^a}$$

which rearranges to

$$E_B^0 - E_A^0 = \frac{0.0592}{ab} \log \frac{[A_{ox}]^b[B_{red}]^a}{[A_{red}]^b[B_{ox}]^a} = \frac{0.0592}{ab} \log K_{eq}$$

Finally, then,

$$\log K_{eq} = \frac{ab(E_B^0 - E_A^0)}{0.0592} \tag{15-7}$$

## 15B    REDOX TITRATION CURVES

Because most redox indicators respond to changes in electrode potential, the vertical axis in oxidation/reduction titration curves is an electrode potential instead of the p-functions that were used for precipitation, complex-formation, and neutralization titration curves. The logarithmic relationship between electrode potential and analyte or titrant concentration causes redox titration curves to look much like those for the other types of titrations.

## 15B–1 Electrode Potentials for Redox Titration Systems

In order to demonstrate the nature of the electrode potential used in deriving redox titration curves, let us consider the titration of iron(II) with a standard solution of cerium(IV). The titration is described by the equation

$$Fe^{2+} + Ce^{4+} \rightleftarrows Fe^{3+} + Ce^{3+}$$

This reaction is rapid and reversible, so that the system is at equilibrium at all times throughout the titration. Consequently, the electrode potentials for the two half-reactions are always identical (Section 15A); that is,

$$E_{Ce^{4+}} = E_{Fe^{3+}} = E_{system}$$

where we have termed $E_{system}$ as *the potential of the system*. If a redox indicator is present in this solution, the ratio of the concentrations of its oxidized and reduced forms must adjust so that the electrode potential for the indicator is also equal to the system potential; that is,

$$E_{In} = E_{Ce^{4+}} = E_{Fe^{3+}} = E_{system}$$

Remember that *when redox systems are at equilibrium, the electrode potentials of all systems are identical.* This generality applies whether the reactions take place directly in solution or indirectly in a galvanic cell.

The electrode potential of a system is readily derived from standard-potential data. Thus, for the reaction under consideration, the titration mixture is treated as if it were part of the hypothetical cell

$$SHE \| Ce^{4+}, \ Ce^{3+}, \ Fe^{3+}, \ Fe^{2+} | Pt$$

where SHE symbolizes the standard hydrogen electrode. The potential of the platinum electrode with respect to the standard hydrogen electrode is determined by the tendencies of iron(III) and cerium(IV) to accept electrons—that is, by the tendencies of the following half-reactions to occur:

$$Fe^{3+} + e^- \rightleftarrows Fe^{2+}$$
$$Ce^{4+} + e^- \rightleftarrows Ce^{3+}$$

At equilibrium, the concentration ratios of the oxidized and reduced forms of the two species are such that their attraction for electrons (and thus their electrode potentials) are identical. Note that these concentration ratios vary continuously throughout the titration, as must $E_{system}$. End points are determined from the characteristic variation in $E_{system}$ that occurs during the titration.

Because $E_{system} = E_{Fe^{3+}} = E_{Ce^{4+}}$, data for a titration curve can be obtained by applying the Nernst equation for *either* the cerium(IV) half-reaction or the iron(III) half-reaction. It turns out, however, that one or the other is more convenient, depending upon the stage of the titration. For example, the iron(III) potential is easier to compute in the region short of the equivalence point because here the concentrations of iron(II) and iron(III) are appreciable and are equal to their analytical concentrations. In contrast, the concentration of cerium(IV), which is negligible

Most end points in oxidation/reduction titrations take advantage of rapid changes in $E_{system}$ that occur at or near chemical equivalence.

Before equivalence, $E_{system}$ calculations are easier with the Nernst equation for the analyte. Beyond equivalence, the Nernst equation for the reagent is easier to use.

prior to equivalence because of the large excess of iron(II), can be obtained at this stage only by calculations based upon the equilibrium constant for the reaction. Beyond the equivalence point, the concentrations of cerium(IV) and cerium(III) are readily computed directly from the volumetric data, but that for iron(II) is not. In this region, then, the electrode potential for the cerium(IV) couple is the easier to use. Equivalence-point potentials are derived by the method shown in the next section.

## 15B–2 Equivalence-Point Potentials

At the equivalence point, the concentrations of cerium(IV) and iron(II) are minute and cannot be obtained from the stoichiometry of the reaction. Fortunately, equivalence-point potentials are readily obtained by taking advantage of the fact that the two reactant species and the two product species have known concentration ratios at chemical equivalence.

At the equivalence point in the titration of iron(II) with cerium(IV), the potential of the system $E_{eq}$ is given by both

$$E_{eq} = E^0_{Ce^{4+}} - \frac{0.0592}{1} \log \frac{[Ce^{3+}]}{[Ce^{4+}]}$$

and

$$E_{eq} = E^0_{Fe^{3+}} - \frac{0.0592}{1} \log \frac{[Fe^{2+}]}{[Fe^{3+}]}$$

Adding these two expressions gives

The concentration quotient in Equation 15–8 is *not* the usual ratio of product concentrations and reactant concentrations that appears in equilibirum-constant expressions.

$$2E_{eq} = E^0_{Ce^{4+}} + E^0_{Fe^{3+}} - \frac{0.0592}{1} \log \frac{[Ce^{3+}][Fe^{2+}]}{[Ce^{4+}][Fe^{3+}]} \qquad (15\text{–}8)$$

The definition of equivalence point requires that

$$[Fe^{3+}] = [Ce^{3+}]$$
$$[Fe^{2+}] = [Ce^{4+}]$$

Substitution of these equalities into Equation 15–8 results in the concentration quotient becoming unity and the logarithmic term becoming zero:

$$2E_{eq} = E^0_{Ce^{4+}} + E^0_{Fe^{3+}} - \frac{0.0592}{1} \log \frac{[Ce^{3+}][Ce^{4+}]}{[Ce^{4+}][Ce^{3+}]} = E^0_{Ce^{4+}} + E^0_{Fe^{3+}}$$

$$E_{eq} = \frac{E^0_{Ce^{4+}} + E^0_{Fe^{3+}}}{2} \qquad (15\text{–}9)$$

Example 15–5 illustrates how the equivalence-point potential is derived for a more complex reaction.

Example 15–5

Derive an expression for the equivalence-point potential in the titration of 0.0500 M $U^{4+}$ with 0.1000 M $Ce^{4+}$. Assume both solutions are 1.0 M in $H_2SO_4$.

$$U^{4+} + 2\ Ce^{4+} + 2\ H_2O \rightleftharpoons UO_2^{2+} + 2\ Ce^{3+} + 4\ H^+$$

In Appendix 6 we find

$$UO_2^{2+} + 4\ H^+ + 2\ e^- \rightleftharpoons U^{4+} + 2\ H_2O \qquad E^0 = 0.334\ V$$
$$Ce^{4+} + e^- \rightleftharpoons Ce^{3+} \qquad\qquad\qquad E^f = 1.44\ V$$

Here we use the formal potential for $Ce^{4+}$ in 1.0 M $H_2SO_4$.
  Proceeding as before, we write

$$E_{eq} = E^0_{UO_2^{2+}} - \frac{0.0592}{2} \log \frac{[U^{4+}]}{[UO_2^{2+}][H^+]^4}$$

$$E_{eq} = E^f_{Ce^{4+}} - 0.0592 \log \frac{[Ce^{3+}]}{[Ce^{4+}]}$$

In order to combine the logarithmic terms and eliminate the concentration terms, we must multiply the first equation by 2 to give

$$2E_{eq} = 2E^0_{UO_2^{2+}} - 0.0592 \log \frac{[U^{4+}]}{[UO_2^{2+}][H^+]^4}$$

Adding this to the cerium(IV) equation leads to

$$3E_{eq} = 2E^0_{UO_2^{2+}} + E^f_{Ce^{4+}} - 0.0592 \log \frac{[U^{4+}][Ce^{3+}]}{[UO_2^{2+}][Ce^{4+}][H^+]^4}$$

But at equivalence

$$[U^{4+}] = [Ce^{4+}]/2$$

and

$$[UO_2^{2+}] = [Ce^{3+}]/2$$

Substituting these equations gives, upon rearranging,

$$E_{eq} = \frac{2E^0_{UO_2^{2+}} + E^f_{Ce^{4+}}}{3} - \frac{0.0592}{3} \log \frac{2[Ce^{4+}][Ce^{3+}]}{2[Ce^{3+}][Ce^{4+}][H^+]^4}$$

$$= \frac{2E^0_{UO_2^{2+}} + E^f_{Ce^{4+}}}{3} - \frac{0.0592}{3} \log \frac{1}{[H^+]^4}$$

We see that the equivalence-point potential in this titration is pH-dependent.

---

## 15B–3 The Derivation of Titration Curves

We now show how data for titration curves are computed from standard-potential data.

Consider the titration of 50.00 mL of 0.05000 M $Fe^{2+}$ with 0.1000 M $Ce^{4+}$ in a medium that is 1.0 M in $H_2SO_4$ at all times. Formal potential data for both half-cell processes are available in Appendix 6 and are used for these calculations:

$$Ce^{4+} + e^- \rightleftharpoons Ce^{3+} \qquad E^f = 1.44 \text{ V} \quad (1 \text{ M } H_2SO_4)$$
$$Fe^{3+} + e^- \rightleftharpoons Fe^{2+} \qquad E^f = 0.68 \text{ V} \quad (1 \text{ M } H_2SO_4)$$

### Initial Potential

The solution contains no cerium species at the outset. In all likelihood, a small but unknown amount of $Fe^{3+}$ is present due to air oxidation of $Fe^{2+}$. In any event, we lack sufficient information to calculate an initial potential.

### Potential After Addition of 5.00 mL of Cerium(IV)

Remember: The equation for this reaction is

$$Fe^{2+} + Ce^{4+} \rightleftharpoons Fe^{3+} + Ce^{3+}$$

The quantity $[Ce^{4+}]$ is equal to the concentration of $[Fe^{2+}]$ that *does not react* with $Ce^{4+}$. Thus $[Fe^{3+}]$ must be decreased by this amount, and $[Fe^{2+}]$ must be increased by this amount. $[Ce^{4+}]$ is so small that we can neglect it in both cases.

When oxidant is added, $Ce^{3+}$ and $Fe^{3+}$ are formed, and the solution now contains appreciable and readily calculated concentrations of three of the participants; that of the fourth, $Ce^{4+}$, is vanishingly small. Therefore, it is more convenient to use the concentrations of the two iron species to calculate the electrode potential of the system.

The concentration of Fe(III) is equal to its molar concentration less the equilibrium concentration of the unreacted Ce(IV):

$$[Fe^{3+}] = \frac{5.00 \times 0.1000}{50.00 + 5.00} - [Ce^{4+}] \approx \frac{0.500}{55.00}$$

Similarly, the $Fe^{2+}$ concentration is given by its molarity plus $[Ce^{4+}]$:

$$[Fe^{2+}] = \frac{50.00 \times 0.05000 - 5.00 \times 0.1000}{55.00} + [Ce^{4+}] \approx \frac{2.000}{55.00}$$

The validity of the indicated assumptions is readily shown by computing the equilibrium constant for the reaction between Fe(II) and Ce(IV) using the technique described in Section 15A–2. The large value for this constant ($7 \times 10^{12}$) indicates clearly that the concentration of Ce(IV) must indeed be inconsequential relative to the concentrations of the two iron species.

Substitution for $[Fe^{2+}]$ and $[Fe^{3+}]$ in the Nernst equation gives

$$E_{system} = +0.68 - \frac{0.0592}{1} \log \frac{2.00/\cancel{55.00}}{0.500/\cancel{55.00}} = 0.64 \text{ V}$$

Note that the volumes in the numerator and denominator cancel, which indicates that the potential is independent of dilution. This independence persists until the solution becomes so dilute that the two assumptions made in the calculation become invalid.

It is worth emphasizing again that the use of the Nernst equation for the Ce(IV)/Ce(III) system would yield the same value for $E_{system}$, but to do so would require computing $[Ce^{4+}]$ by means of the equilibrium constant for the reaction.

Additional potentials needed to define the titration curve short of the equivalence point can be obtained similarly. Such data are given in Table 15–1. You may want to confirm one or two of these values.

### Equivalence-Point Potential

Substitution of the two formal potentials into Equation 15–9 yields

$$E_{eq} = \frac{E^f_{Ce^{4+}} + E^f_{Fe^{3+}}}{2} = \frac{1.44 + 0.68}{2} = 1.06 \text{ V}$$

### Potential After Addition of 25.10 mL of Cerium(IV)

The molar concentrations of Ce(III), Ce(IV), and Fe(III) are readily computed at this point, but that for Fe(II) is not. Therefore, calculation of

**Table 15–1**
**ELECTRODE POTENTIAL VERSUS SHE IN TITRATIONS WITH 0.1000 M $Ce^{4+}$**

| Reagent Volume, mL | Potential, V vs. SHE | |
| --- | --- | --- |
| | 50.00 mL of 0.05000 M $Fe^{2+}$ | 50.00 mL of 0.02500 M $U^{4+}$ * |
| 5.00 | 0.64 | 0.316 |
| 15.00 | 0.69 | 0.339 |
| 20.00 | 0.72 | 0.352 |
| 24.00 | 0.76 | 0.375 |
| 24.90 | 0.82 | 0.405 |
| 25.00 | 1.06 ← Equivalence point → | 0.703 |
| 25.10 | 1.30 | 1.30 |
| 26.00 | 1.36 | 1.36 |
| 30.00 | 1.40 | 1.40 |

*$H_2SO_4$ concentration is such that $[H^+] = 1.0$ throughout.

$E_{system}$ based on the cerium half-reaction is convenient. The concentrations of the two cerium ion species are

Again, $[Fe^{2+}]$ is equal to the amount of $Ce^{4+}$ that is left unreacted, so it is added to $c_{Ce^{4+}}$ calculated from the volumes of the two solutions and subtracted from $c_{Ce^{3+}}$.

$$[Ce^{3+}] = \frac{25.00 \times 0.1000}{75.10} + [Fe^{2+}] \approx \frac{2.500}{75.10}$$

$$[Ce^{4+}] = \frac{25.10 \times 0.1000 - 50.00 \times 0.05000}{75.10} + [Fe^{2+}] \approx \frac{0.010}{75.10}$$

The indicated approximations should be reasonable in view of the favorable equilibrium constant. Substitution of these concentrations into the Nernst equation for the cerium couple gives

$$E = +1.44 - \frac{0.0592}{1} \log \frac{[Ce^{3+}]}{[Ce^{4+}]} = +1.44 - \frac{0.0592}{1} \log \frac{2.500/75.10}{0.010/75.10}$$

$$= +1.30 \text{ V}$$

The additional postequivalence potentials in Table 15–1 were derived in a similar fashion.

The titration curve of iron(II) with cerium(IV) appears as $A$ in Figure 15–2. This plot resembles closely the curves encountered in neutralization, precipitation, and complex-formation titrations, with the equivalence point being signaled by a rapid change in the ordinate function. A titration involving 0.00500 M iron(II) and 0.01000 M cerium(IV) yields a curve that is, for all practical purposes, identical to the one derived here, since the electrode potential of the system is independent of dilution.

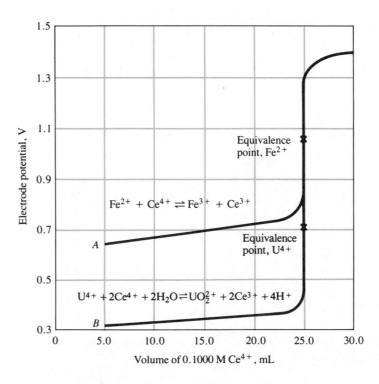

Figure 15–2
Titration curves for 0.1000 M $Ce^{4+}$ titration. ($A$) Titration of 50.00 mL of 0.05000 M $Fe^{2+}$. ($B$) Titration of 50.00 mL of 0.02500 M $U^{4+}$.

Example 15–6

Derive a curve for the titration of 50.00 mL of 0.02500 M $U^{4+}$ with 0.1000 M $Ce^{4+}$. Assume that the solution is 1.0 M in $H_2SO_4$ throughout the titration ([$H^+$] for such a solution will be about 1.0 M).

The analytical reaction is

$$U^{4+} + 2\,H_2O + 2\,Ce^{4+} \rightleftharpoons UO_2^{2+} + 2\,Ce^{3+} + 4\,H^+$$

Why is it impossible to calculate the potential before titrant is added?

and in Appendix 6 we find

$$Ce^{4+} + e^- \rightleftharpoons Ce^{3+} \qquad E^f = +1.44\ V$$
$$UO_2^{2+} + 4\,H^+ + 2\,e^- \rightleftharpoons U^{4+} + 2\,H_2O \qquad E^0 = +0.334\ V$$

Potential After Adding 5.00 mL of $Ce^{4+}$

$$\text{original amount } U^{4+} = 50.00\ \text{mL } U^{4+} \times 0.02500\ \frac{\text{mmol } U^{4+}}{\text{mL } U^{4+}}$$
$$= 1.250\ \text{mmol } U^{4+}$$

$$\text{amount } Ce^{4+} \text{ added} = 5.00\ \text{mL } Ce^{4+} \times 0.1000\ \frac{\text{mmol } Ce^{4+}}{\text{mL } Ce^{4+}}$$
$$= 0.5000\ \text{mmol } Ce^{4+}$$

$$\text{amount } UO_2^{2+} \text{ formed} = 0.5000\ \text{mmol } Ce^{4+} \times \frac{1\ \text{mmol } UO_2^{2+}}{2\ \text{mmol } Ce^{4+}}$$
$$= 0.2500\ \text{mmol } UO_2^{2+}$$

$$\text{amount } U^{4+} \text{ remaining} = 1.250\ \text{mmol } U^{4+} - 0.2500\ \text{mmol } UO_2^{2+}$$
$$\times \frac{1\ \text{mmol } U^{4+}}{\text{mmol } UO_2^{2+}} = 1.000\ \text{mmol } U^{4+}$$

$$\text{total volume of solution} = (50.00 + 5.00)\ \text{mL} = 55.00\ \text{mL}$$

Applying the Nernst equation for $UO_2^{2+}$, we obtain

$$E = 0.334 - \frac{0.0592}{2} \log \frac{[U^{4+}]}{[UO_2^{2+}][H^+]^4}$$
$$= 0.334 - \frac{0.0592}{2} \log \frac{[U^{4+}]}{[UO_2^{2+}](1.00)^4}$$

Substituting concentrations of the two uranium species gives

$$E = 0.334 - \frac{0.0592}{2} \log \frac{1.000\ \text{mmol } U^{4+}/55.00\ \text{mL}}{0.2500\ \text{mmol } UO_2^{2+}/55.00\ \text{mL}} = 0.316\ V$$

Other preequivalence-point data, calculated in the same way, are given in Table 15–1.

Equivalence-Point Potential

Following the procedure shown in Example 15–1, we obtain

$$E_{eq} = \frac{E^f_{Ce^{4+}} + 2E^0_{UO_2^{2+}}}{3} - \frac{0.0592}{3} \log \frac{1}{[H^+]^4}$$

Substituting gives

$$E_{eq} = \frac{1.44 + (2 \times 0.334)}{3} + \frac{0.0592}{3} \log (1.00)^4 = 0.782 \text{ V}$$

Potential After Adding 25.10 mL of $Ce^{4+}$

$$\text{original amount } U^{4+} = 50.00 \text{ mL } U^{4+} \times 0.02500 \frac{\text{mmol } U^{4+}}{\text{mL } U^{4+}}$$

$$= 1.250 \text{ mmol } U^{4+}$$

$$\text{amount } Ce^{4+} \text{ added} = 25.10 \text{ mL } Ce^{4+} \times 0.1000 \frac{\text{mmol } Ce^{4+}}{\text{mL } Ce^{4+}}$$

$$= 2.510 \text{ mmol } Ce^{4+}$$

$$\text{amount } Ce^{3+} \text{ formed} = 1.250 \text{ mmol } U^{4+} \times \frac{2 \text{ mmol } Ce^{3+}}{\text{mmol } U^{4+}}$$

$$= 2.500 \text{ mmol } Ce^{3+}$$

$$\text{amount } Ce^{4+} \text{ remaining} = 2.510 \text{ mmol } Ce^{4+} - 2.500 \text{ mmol } Ce^{3+}$$

$$\times \frac{1 \text{ mmol } Ce^{4+}}{\text{mmol } Ce^{3+}} = 0.010 \text{ mmol } Ce^{4+}$$

$$\text{volume of solution} = 75.10 \text{ mL}$$

$$[Ce^{3+}] = 2.500 \text{ mmol } Ce^{3+}/75.10 \text{ mL}$$

$$[Ce^{4+}] = 0.010 \text{ mmol } Ce^{4+}/75.10 \text{ mL}$$

Substituting into the Nernst expression for the $Ce^{4+}$ formal potential gives

$$E = 1.44 - 0.0592 \log \frac{2.500/75.10}{0.010/75.10} = 1.30 \text{ V}$$

Table 15–1 contains other postequivalence-point data obtained in this same way.

---

The data in the third column of Table 15–1 are plotted as curve *B* in Figure 15–2. The two curves in the figure are identical for volumes greater than 25.00 mL because the concentrations of the two cerium species are identical in this region. It is also interesting that the curve for iron(II) is symmetric around the equivalence point, while that for uranium(IV) is not. In general, symmetric curves are obtained when the analyte and titrant react in a 1 : 1 molar ratio.

Redox titration curves are symmetric when the reactants combine in a 1 : 1 ratio. Otherwise, they are asymmetric.

## 15B–4  The Effect of System Variables on Redox Titration Curves

In earlier chapters, we considered how reactant concentrations and reaction completeness affect titration curves. Here, we describe the effects of these variables on oxidation/reduction titration curves.

### Reactant Concentration

As we have just seen, $E_{system}$ for an oxidation/reduction titration is ordinarily independent of dilution. Consequently, titration curves for oxidation/reduction reactions are usually independent of analyte and reagent concentrations. This behavior is in distinct contrast to that observed in the other types of titration curves we have encountered.

**Feature 15–2**
### WHEN REDOX TITRATION CURVES BECOME CONCENTRATION-DEPENDENT

Electrode potentials become dependent upon dilution when the number of moles of the reactant and product of a half-reaction differ. An example is the reaction

$$I_3^- + 2\,e^- \rightleftharpoons 3\,I^-$$

Application of the Nernst equation gives

$$E = E^0 - \frac{0.0592}{2} \log \frac{[I^-]^3}{[I_3^-]}$$

Here, the concentration term in the numerator bears units of $(mol/L)^3$, but the units in the denominator are mol/L. Consequently, the ratio of the two terms is related to the square of concentration. The result is that, in a titration in which the standard reagent is triiodide ion, potentials at and beyond the equivalence point depend upon dilution.

As mentioned earlier, electrode potentials also become concentration-dependent when we cannot assume that the molar concentrations of the various species are equal to the concentrations derived from stoichiometry.

### Completeness of the Reaction

The change in $E_{system}$ in the equivalence-point region of an oxidation/reduction titration becomes larger as the reaction becomes more complete. This effect is demonstrated in Figure 15–3, which shows curves for the titration of a hypothetical reductant having a standard electrode po-

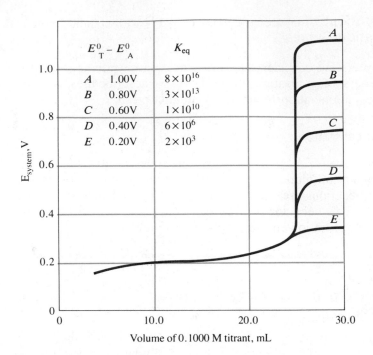

| $E^0_T - E^0_A$ | $K_{eq}$ |
|---|---|
| A    1.00V | $8 \times 10^{16}$ |
| B    0.80V | $3 \times 10^{13}$ |
| C    0.60V | $1 \times 10^{10}$ |
| D    0.40V | $6 \times 10^6$ |
| E    0.20V | $2 \times 10^3$ |

Volume of 0.1000 M titrant, mL

Figure 15–3
Effect of titrant electrode potential upon reaction completeness. The standard electrode potential for the analyte ($E^0_A$) is 0.200 V; starting with curve A, standard electrode potentials for the titrant ($E^0_T$) are 1.20, 1.00, 0.80, 0.60, and 0.40 V, respectively. Both analyte and titrant undergo a one-electron change.

## Feature 15–3
### REACTION RATES AND ELECTRODE POTENTIALS

Standard potentials reveal whether or not a reaction proceeds far enough toward completion to be useful in a particular analytical problem, but they provide no information about the rate at which the equilibrium state is approached. Consequently, a reaction that appears extremely favorable from equilibrium considerations may be totally unacceptable from a kinetic standpoint. The oxidation of arsenic(III) with cerium(IV) in dilute sulfuric acid is a typical example. The reaction is

$$H_3AsO_3 + 2\ Ce^{4+} + H_2O \rightleftharpoons H_3AsO_4 + 2\ Ce^{3+} + 2\ H^+$$

The formal potentials $E^f$ for these two systems are

$$Ce^{4+} + e^- \rightleftharpoons Ce^{3+} \qquad\qquad E^f = +1.4\ V$$
$$H_3AsO_4 + 2\ H^+ + 2\ e^- \rightleftharpoons H_3AsO_3 + H_2O \qquad E^f = +0.56\ V$$

and an equilibrium constant of about $10^{28}$ can be deduced from these data. Even though this reaction is highly favorable from an equilibrium standpoint, titration of arsenic(III) with cerium(IV) is impossible without a catalyst because several hours are needed for the attainment of equilibrium. Fortunately, several substances catalyze the reaction and thus make the titration feasible.

tential of 0.20 V with several hypothetical oxidants with standard potentials ranging from 0.40 to 1.20 V; the corresponding equilibrium constants lie between about $2 \times 10^3$ and $8 \times 10^{16}$. Clearly, the greatest change in potential of the system is associated with the reaction that is most nearly complete. Thus, in this respect, oxidation/reduction titration curves are similar to those involving other types of reactions.

## 15B–5 The Titration of Mixtures

Solutions containing two oxidizing agents or two reducing agents yield titration curves that contain two inflection points, provided the standard potentials involved are sufficiently different from each other. If this difference is greater than about 0.2 V, the end points are usually distinct enough to permit determination of each component. This situation is quite comparable to the titration of two acids having different dissociation constants or to that of two ions forming precipitates that have different solubilities.

In addition, the behavior of a few redox systems is analogous to that of polyprotic acids. For example, consider the two half-reactions

$$VO^{2+} + 2\,H^+ + e^- \rightleftharpoons V^{3+} + H_2O \qquad E^0 = +0.359\text{ V}$$
$$V(OH)_4^+ + 2\,H^+ + e^- \rightleftharpoons VO^{2+} + 3\,H_2O \qquad E^0 = +1.00\text{ V}$$

The curve for the titration of $V^{3+}$ with a strong oxidizing agent, such as permanganate, has two inflection points, the first corresponding to the oxidation of $V^{3+}$ to $VO^{2+}$ and the second to the oxidation of $VO^{2+}$ to $V(OH)_4^+$. The stepwise oxidation of molybdenum(III), first to the +5 oxidation state and subsequently to the +6 state, is another common example. Here again, satisfactory inflections occur in the curves because the difference in the standard potentials of the pertinent half-reactions is about 0.4 V.

The derivation of titration curves for either of these types of reactions is not difficult if the difference in standard potentials is sufficiently great. An example is the titration of a solution containing iron(II) and titanium(III) ions with potassium permanganate. The standard potentials are

$$TiO^{2+} + 2\,H^+ + e^- \rightleftharpoons Ti^{3+} + H_2O \qquad E^0 = +0.099\text{ V}$$
$$Fe^{3+} + e^- \rightleftharpoons Fe^{2+} \qquad E^0 = +0.77\text{ V}$$

The first additions of permanganate are used up by the more readily oxidized titanium(III) ion. As long as an appreciable concentration of this species remains in solution, the potential of the system cannot become high enough to change the concentration of iron(II) ions appreciably. Thus, points defining the first part of the titration curve can be obtained by substituting the stoichiometric concentrations of titanium(III) and titanium(IV) ion into the equation

$$E = +0.099 - 0.0592 \log \frac{[Ti^{3+}]}{[TiO^{2+}][H^+]^2}$$

For all practical purposes then, the first part of this curve is identical to the titration curve for titanium(III) ion by itself. Beyond the first equivalence point, the solution contains both iron(II) and iron(III) ions in significant concentrations and so points on the curve can be most conveniently obtained from the relationship

$$E = E^0_{Fe^{3+}} - 0.0592 \log \frac{[Fe^{2+}]}{[Fe^{3+}]}$$

Throughout this region and beyond the second equivalence point, the curve is essentially identical to that for the titration of iron(II) ion alone. Such calculations leave undefined only the potential at the first equivalence point. A convenient way of estimating its value is to add the Nernst equations for the iron(II) and titanium(III) potentials. Since the electrode potentials for the two systems are identical at equilibrium, we can write

$$2E = +0.099 + 0.771 - 0.0592 \log \frac{[Ti^{3+}][Fe^{2+}]}{[TiO^{2+}][Fe^{3+}][H^+]^2}$$

Iron(III) and titanium(III) ions exist in small and equal amounts as a result of the equilibrium

$$2\,H^+ + TiO^{2+} + Fe^{2+} \rightleftharpoons Fe^{3+} + Ti^{3+} + H_2O$$

Thus, we can write

$$[Fe^{3+}] = [Ti^{3+}]$$

Substitution of this equality into the previous equation for the potential yields

$$E = \frac{+0.87}{2} - \frac{0.0592}{2} \log \frac{[Fe^{2+}]}{[TiO^{2+}][H^+]^2}$$

Finally, if $[TiO^{2+}]$ and $[Fe^{2+}]$ are assumed to be essentially identical to their analytical concentrations, we can compute the equivalence-point potential.

A curve for a mixture of iron(II) and titanium(III) ions titrated with a standard permanganate solution is shown in Figure 15–4.

## 15C    OXIDATION/REDUCTION INDICATORS

Two types of indicators are used for obtaining end points for oxidation/reduction titrations: *general redox indicators* and *specific indicators*.

### 15C–1 General Redox Indicators

General oxidation/reduction indicators are substances that change color upon being oxidized or reduced. In contrast to specific indicators, the

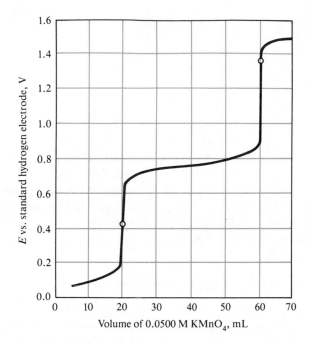

Figure 15–4

Curve for the titration of 50.0 mL of a solution that is 0.0500 M in $Ti^{3+}$ and 0.200 M in $Fe^{2+}$ with 0.0500 M $KMnO_4$. The concentration of $H^+$ is 1.0 M throughout.

color changes of true redox indicators are largely independent of the chemical nature of the analyte and titrant and depend instead upon the changes in the electrode potential of the system that occur as the titration progresses.

*Color changes for general redox indicators depend only upon the potential of the system.*

The half-reaction responsible for color change in a typical general oxidation/reduction indicator can be written as

$$In_{ox} + ne \rightleftharpoons In_{red}$$

If the indicator reaction is reversible, we can write

$$E = E^0_{In} - \frac{0.0592}{n} \log \frac{[In_{red}]}{[In_{ox}]} \qquad (15\text{--}10)$$

Typically, a change from the color of the oxidized form of the indicator to the color of the reduced form requires a change of about 100 in the ratio of reactant concentrations; that is, a color change is seen when

$$\frac{[In_{red}]}{[In_{ox}]} \leq \frac{1}{10}$$

changes to

$$\frac{[In_{red}]}{[In_{ox}]} \geq 10$$

The potential change required to produce the full color change of a typical general indicator can be found by substituting these two values into Equation 15–10, which gives

$$E = E_{In}^0 \pm \frac{0.0592}{n}$$

This equation shows that a typical general indicator exhibits a detectable color change when a titrant causes the system potential to shift from $E_{In}^0 + 0.0592/n$ to $E_{In}^0 - 0.0592/n$, or about $(0.118/n)$ V. For many indicators, $n = 2$ and a change of 0.059 V is thus sufficient.

Table 15–2 lists transition potentials for several redox indicators. Note that indicators functioning in any desired potential range up to about +1.25 V are available. Structures for and reactions of a few of the indicators listed in the table are considered in the paragraphs that follow.

Protons are involved in the reduction of many indicators, and so the range of potentials over which a color change occurs (the *transition potential*) is often pH-dependent.

### Iron(II) Complexes of Orthophenanthrolines

A class of organic compounds known as 1,10-phenanthrolines (or orthophenanthrolines) form stable complexes with iron(II) and certain other ions. The parent compound has a pair of nitrogen atoms located in such positions that each can form a covalent bond with the iron(II) ion.

**Table 15–2**
**SELECTED GENERAL OXIDATION/REDUCTION INDICATORS***

| Indicator | Color Oxidized | Color Reduced | Transition Potential, V | Conditions |
|---|---|---|---|---|
| 5-Nitro-1,10-phenanthroline iron(II) complex | Pale blue | Red-violet | +1.25 | 1 M $H_2SO_4$ |
| 2,3'-Diphenylamine dicarboxylic acid | Blue-violet | Colorless | +1.12 | 7–10 M $H_2SO_4$ |
| 1,10-Phenanthroline iron(II) complex | Pale blue | Red | +1.11 | 1 M $H_2SO_4$ |
| Erioglaucin A | Blue-red | Yellow-green | +0.98 | 0.5 M $H_2SO_4$ |
| Diphenylamine sulfonic acid | Red-violet | Colorless | +0.85 | Dilute acid |
| Diphenylamine | Violet | Colorless | +0.76 | Dilute acid |
| p-Ethoxychrysoidine | Yellow | Red | +0.76 | Dilute acid |
| Methylene blue | Blue | Colorless | +0.53 | 1 M acid |
| Indigo tetrasulfonate | Blue | Colorless | +0.36 | 1 M acid |
| Phenosafranine | Red | Colorless | +0.28 | 1 M acid |

*Data taken in part from I. M. Kolthoff and V. A. Stenger, *Volumetric Analysis*, 2d ed., Vol. 1, p. 140. New York: Interscience, 1942.

Three orthophenanthroline molecules combine with each iron ion to yield a complex with the structure

This complex, which is sometimes called "ferroin," is conveniently written as $(Phen)_3Fe^{2+}$.

The complexed iron in the ferroin undergoes a reversible oxidation/reduction reaction that can be written

$$(Phen)_3Fe^{3+} + e^- \rightleftharpoons (Phen)_3Fe^{2+} \qquad E^0 = +1.06 \text{ V}$$
$$\text{Pale Blue} \qquad\qquad\qquad \text{Red}$$

In practice, the color of the oxidized form is so slight as to go undetected, and the color change associated with this reduction appears to be from nearly colorless to red. Because of the difference in color intensity, the end point is usually taken when only about 10% of the indicator is in the iron(II) form. The transition potential is thus approximately +1.11 V in 1 M sulfuric acid.

Of all the oxidation/reduction indicators, ferroin approaches most closely the ideal substance. It reacts rapidly and reversibly, the color change is pronounced, and solutions of it are stable and readily prepared. In contrast to many indicators, the oxidized form of ferroin is remarkably inert toward strong oxidizing agents. At temperatures above 60°C, ferroin decomposes.

5-Nitro-1,10-phenanthroline

A number of substituted phenanthrolines have been investigated for their indicator properties, and some have proved to be as useful as the parent compound. Among these, the 5-nitro and 5-methyl derivatives are noteworthy, with transition potentials of +1.25 V and +1.02 V, respectively.

5-Methyl-1,10-phenanthroline

## Diphenylamine and Its Derivatives

Diphenylamine, $C_{12}H_{11}N$, which was one of the first redox indicators discovered, was used by Knop in 1924 for the titration of iron(II) with potassium dichromate. In the presence of a strong oxidizing agent, diphenylamine is believed to undergo the reactions

The first reaction is irreversible, but the second can be reversed and constitutes the actual indicator reaction.

The reduction potential for the second reaction is about +0.76 V. Despite the fact that hydrogen ions appear in the equation, variations in acidity have little effect upon the magnitude of this potential, perhaps because the colored product is a protonated species.

Diphenylamine is not very soluble in water, and indicator solutions must be prepared in rather concentrated sulfuric acid. In addition, the indicator is not applicable to solutions containing tungstate ion, which precipitates the violet reaction product. Mercury(II) ions also interfere by inhibiting the indicator reaction.

The sulfonic acid derivative of diphenylamine, which has the structure does not suffer from these disadvantages. The barium or sodium salt of this acid is used to prepare aqueous indicator solutions, which behave in essentially the same manner as the parent substance. The color change is somewhat sharper, however, passing from colorless through green to deep violet. The transition potential is about $+0.8$ V and again is independent of the acid concentration. The sulfonic acid derivative is now widely used in redox titrations.

Diphenylbenzidine, the intermediate in the oxidation of diphenylamine, should behave in redox reactions just as diphenylamine does but should consume less oxidizing agent. Unfortunately, the low solubility of diphenylbenzidine in water and in sulfuric acid has precluded its widespread use. As might be expected, the sulfonic acid derivative of diphenylbenzidine is a satisfactory indicator.

### Starch/Iodine Solutions

Starch, which forms a blue complex with triiodide ion, is a widely used specific indicator in oxidation/reduction reactions involving iodine as an oxidant or iodide ion as a reductant. A starch solution containing a little triiodide or iodide ion can also function as a true redox indicator, however. In the presence of excess oxidizing agent, the concentration ratio of iodine to iodide is high, giving a blue color to the solution. With excess reducing agent, on the other hand, iodide ion predominates, and the blue color is absent. Thus, the indicator system changes from colorless to blue in the titration of many reducing agents with various oxidizing agents. This color change is quite independent of the chemical composition of the reactants, depending only upon the potential of the system at the equivalence point.

### The Choice of Redox Indicator

It is apparent from Figure 15–3 that all the indicators in Table 15–2 except for the first and the last could be employed with reagent A. In contrast, with reagent D, only indigo tetrasulfonate could be employed. The change in potential with reagent E is too small to be satisfactorily detected by an indicator.

### 15C–2  Specific Indicators

Perhaps the best-known specific indicator is starch, which forms a dark blue complex with triiodide ion. This complex signals the end point in titrations in which iodine is either produced or consumed.

Another specific indicator is potassium thiocyanate, which may be employed, for example, in the titration of iron(III) with solutions of titanium(III) sulfate. The end point involves the disappearance of the red color of the iron(III)/thiocyanate complex as a result of the marked decrease in the iron(III) concentration at the equivalence point.

## 15D POTENTIOMETRIC END POINTS

End points for many oxidation/reduction titrations are readily observed by making the solution of the analyte part of the cell:

reference electrode‖analyte solution|Pt

By measuring the potential of this cell during a titration, data for curves analogous to those shown in Figures 15–1 and 15–2 can be generated. End points are readily estimated from such curves. Potentiometric end points are considered in detail in Chapter 17.

## 15E QUESTIONS AND PROBLEMS

*15–1. Make a clear distinction between
  (a) equilibrium and equivalence.
  (b) a specific and a general oxidation/reduction indicator.

15–2. What is meant by the term *electrode potential of the system*?

*15–3. Briefly explain why the electrode potential of the system can be computed with the Nernst expression for either of the participating species.

15–4. How does the stoichiometry among the participants differ at the equivalence point from other points of an oxidation/reduction titration?

*15–5. Under what circumstances are curves asymmetric about the equivalence point in an oxidation/reduction titration?

15–6. Under what circumstances will the equivalence-point potential of an oxidation/reduction titration depend upon
  (a) the concentration of a reactant or product?
  (b) the pH of the environment?

*15–7. Explain why $E_{system}$ is easier to calculate for the analyte than for the titrant prior to the equivalence point.

15–8. Generate equilibrium-constant expressions for the following reactions. Calculate numerical values for $K_{eq}$.
  *(a) $Pb^{2+} + Cd(s) \rightleftharpoons Pb(s) + Cd^{2+}$
  (b) $2 Fe^{3+} + Zn(s) \rightleftharpoons 2 Fe^{2+} + Zn^{2+}$
  *(c) $Sn^{4+} + 2 Cr^{2+} \rightleftharpoons Sn^{2+} + 2 Cr^{3+}$
  (d) $Pb(s) + 2 H^+ \rightleftharpoons Pb^{2+} + 3 H_2(g)$
  *(e) $2 BiO^+ + 4 H^+ + 3 Pb(s) \rightleftharpoons 2 Bi(s) + 3 Pb^{2+} + 2 H_2O$
  (f) $Tl^+ + Ti^{3+} + H_2O \rightleftharpoons Tl(s) + TiO^{2+} + 2H^+$
  *(g) $VO^{2+} + V^{2+} + 2 H^+ \rightleftharpoons 2 V^{3+} + H_2O$
  (h) $BiCl_4^- + 3 Ag(s) \rightleftharpoons Bi(s) + 3 AgCl + Cl^-$
  *(i) $I_3^- + H_2S(g) \rightleftharpoons 3 I^- + S(s) + 2 H^+$
  (j) $PtCl_6^{2-} + 2 Ag(s) \rightleftharpoons PtCl_4^{2-} + 2 AgCl$

*15–9. Calculate the equivalence-point potential for the following titrations. The oxidizing agent is the titrant, the reducing agent the analyte. Where necessary, use 0.0200 M for the initial analyte and titrant concentrations and a constant pH of 1.00 throughout the titration.
  (a) $Fe^{3+} + V^{2+} \rightleftharpoons Fe^{2+} + V^{3+}$
  (b) $2 Ce^{4+} + Sn^{2+} \rightleftharpoons 2 Ce^{3+} + Sn^{4+}$ (1.00 M $HClO_4$, pH = 0.00)
  (c) $Tl^{3+} + 2 Ti^{3+} + 2 H_2O \rightleftharpoons Tl^+ + 2 TiO^{2+} + 4 H^+$
  (d) $PtCl_6^{2-} + U^{4+} + 2 H_2O \rightleftharpoons PtCl_4^{2-} + UO_2^{2+} + 2 Cl^- + 4 H^+$
  (e) $V(OH)_4^+ + V^{3+} \rightleftharpoons 2 VO^{2+} + 2 H_2O$
  (f) $Cr_2O_7^{2-} + 6 Fe^{2+} + 14 H^+ \rightleftharpoons 2 Cr^{3+} + 6 Fe^{3+} + 7H_2O$

15–10. Calculate the equivalence-point potential for the following titrations. The oxidizing agent is the titrant, the reducing agent the analyte. Where necessary, use 0.0200 M for the initial analyte and titrant concentrations and a constant pH of 1.00 throughout the titration.
  (a) $Fe(CN)_6^{3-} + Cr^{2+} \rightleftharpoons Fe(CN)_6^{4-} + Cr^{3+}$
  (b) $Tl^{3+} + 2 Fe^{2+} \rightleftharpoons Tl^+ + 2 Fe^{3+}$
  (c) $VO^{2+} + Ti^{2+} + 2 H^+ \rightleftharpoons V^{3+} + Ti^{3+} + H_2O$
  (d) $PtCl_6^{2-} + H_2SO_3 + H_2O \rightleftharpoons SO_4^{2-} + PtCl_4^{2-} + 4 H^+ + 2 Cl^-$
  (e) $VO^{2+} + V^{2+} + 2 H^+ \rightleftharpoons 2 V^{3+} + H_2O$
  (f) $I_3^- + 2 Ti^{3+} + 2 H_2O \rightleftharpoons 3 I^- + 2 TiO^{2+} + 4 H^+$

*15–11. Select an indicator from Table 15–2 (write NONE, if appropriate) that might be suitable for each titration in Problem 15–9.

15–12. Select an indicator from Table 15–2 (write NONE, if appropriate) that might be suitable for each titration in Problem 15–10.

*15–13. Calculate equilibrium constants for the reactions in Problem 15–9.

15–14. Calculate equilibrium constants for the reactions in Problem 15–10.

*15–15. Calculate the equivalence-point concentration of each species in Problem 15–9.

15–16. Calculate the equivalence-point concentration of each species in Problem 15–10.

15–17. Generate curves for the following titrations. Calculate potentials after the addition of 10.00, 25.00, 40.00, 49.00, 49.90, 50.00, 50.10, 51.00, and 60.00

mL of titrant. Where necessary, assume that the pH is 1.00 throughout.

*(a) 50.00 mL of 0.1000 M $Cr^{2+}$ with 0.1000 M $Fe^{3+}$

(b) 50.00 mL of 0.1000 M $V^{2+}$ with 0.0500 M $Sn^{4+}$

*(c) 50.00 mL of 0.1000 M $Fe(CN)_6^{4-}$ with 0.05000 M $Tl^{3+}$

(d) 50.00 mL of 0.05000 M $U^{4+}$ with 0.1000 M $V(OH)_4^+$

*(e) 50.00 mL of 0.1000 M $V^{3+}$ with 0.1000 M $V(OH)_4^+$

(f) 50.00 mL of 0.1000 M $Ti^{3+}$ with 0.0200 M $MnO_4^-$

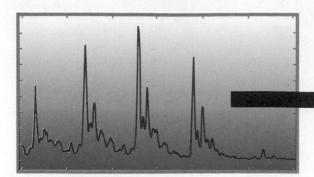

## CHAPTER 16

# APPLICATIONS OF OXIDATION/REDUCTION TITRATIONS

This chapter is concerned with the preparation of standard solutions of oxidants and reductants and with their applications in analytical chemistry. In addition, auxiliary reagents that convert an analyte to a single oxidation state are described.[1]

## 16A  AUXILIARY OXIDIZING AND REDUCING REAGENTS

The analyte in an oxidation/reduction titration must be in a single oxidation state at the outset. Often, however, the steps that precede the titration (dissolution of the sample and separation of interferences) convert the analyte to a mixture of oxidation states. For example, the solution produced when an iron-containing sample is dissolved usually contains a mixture of iron(II) and iron(III). If we choose to use a standard oxidant for the determination of iron, we must first treat the sample solution with an auxiliary reducing agent; if, on the other hand, we plan to titrate with a standard reductant, pretreatment with an auxiliary oxidizing reagent will be needed.[2]

To be useful as a preoxidant or a prereductant, a reagent must react quantitatively with the analyte. In addition, excesses of the reagent must be removeable because such excesses will ordinarily interfere by consuming standard solution.

---

[1]For further reading on redox titrimetry, see J. A. Goldman and V. A. Stenger, in *Treatise on Analytical Chemistry,* I. M. Kolthoff and P. J. Elving, Eds., Part I, Vol. 11, Chapter 119. New York: Wiley, 1975; and I. M. Kolthoff and R. Belcher, *Volumetric Analysis,* Vol. 3. New York: Interscience, 1957.

[2]For a brief summary of auxiliary reagents, see J. A. Goldman and V. A. Stenger, in *Treatise on Analytical Chemistry,* I. M. Kolthoff and P. J. Elving, Eds., Part I, Vol. 11, pp. 7204–7206. New York: Wiley, 1975.

## 16A–1 Auxiliary Reducing Reagents

A number of metals are good reducing agents and have been used for the prereduction of analytes. Included among these are zinc, aluminum, cadmium, lead, nickel, copper, and silver (in the presence of chloride ion). Sometimes, sticks or coils of the metal are immersed directly in the analyte solution. After reduction is judged complete, the solid is removed manually or by filtration.

An alternative to filtration is the use of a *reductor*, such as that shown in Figure 16–1.[3] Here, a finely divided metal is held in a vertical glass tube through which the analyte solution is drawn under a mild vacuum. The metal in a reductor is ordinarily sufficient for hundreds of reductions.

A typical *Jones reductor* has a diameter of about 2 cm and is packed with a 40- to 50-cm column of amalgamated zinc. Amalgamation is accomplished by allowing zinc granules to stand briefly in a solution of mercury(II) chloride:

$$2 \, Zn(s) + Hg^{2+} \rightarrow Zn^{2+} + Zn(Hg)(s)$$

Zinc amalgam is nearly as effective a reducing agent as the pure metal and has the important virtue of inhibiting the reduction of hydrogen ions, a parasitic reaction that not only needlessly uses up the reducing agent but also heavily contaminates the sample solution with zinc(II) ions. Solutions that are quite acidic can be passed through a Jones reductor without significant hydrogen formation.

Table 16–1 lists the principal applications of the Jones reductor. Also listed in Table 16–1 are reductions that can be accomplished with a *Walden reductor,* in which the reductant is granular metallic silver held in a narrow glass column. Silver is not a good reducing agent unless chloride or some other ion that forms a silver salt of low solubility is present. For this reason, prereductions with a Walden reductor are generally carried out from hydrochloric acid solutions of the analyte. The coating of silver chloride produced on the metal is removed periodically by dipping a zinc rod into the solution that covers the packing.

Table 16–1 suggests that the Walden reductor is somewhat more selective in its action than is the Jones reductor.

## 16A–2 Auxiliary Oxidizing Reagents

### Sodium Bismuthate

Sodium bismuthate is an extremely powerful oxidizing agent capable, for example, of converting manganese(II) quantitatively to permanganate ion. This bismuth salt is a sparingly soluble solid with a formula that is usually written as $NaBiO_3$, although its exact composition is somewhat uncertain. Oxidations are performed by suspending the bismuthate in the

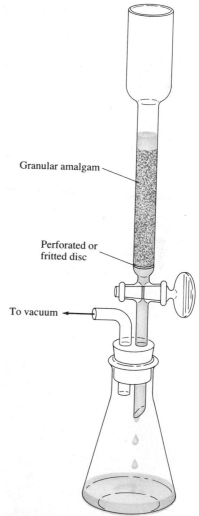

Granular amalgam

Perforated or fritted disc

To vacuum

Figure 16–1
A Jones reductor.

[3]For a discussion of reductors, see F. Hecht, in *Treatise on Analytical Chemistry*, I. M. Kolthoff and P. J. Elving, Eds., Part I, Vol. 11, pp. 6703–6707. New York: Wiley, 1975.

Table 16–1
USES OF THE WALDEN REDUCTOR AND THE JONES REDUCTOR*

| Walden $Ag(s) + Cl^- \rightarrow AgCl(s) + e^-$ | Jones $Zn(Hg)(s) \rightarrow Zn^{2+} + Hg + 2\ e^-$ |
|---|---|
| $Fe^{3+} + e^- \rightarrow Fe^{2+}$ | $Fe^{3+} + e^- \rightarrow Fe^{2+}$ |
| $Cu^{2+} + e^- \rightarrow Cu^+$ | $Cu^{2+} + 2\ e^- \rightarrow Cu(s)$ |
| $H_2MoO_4 + 2\ H^+ + e^- \rightarrow MoO_2^+ + 2\ H_2O$ | $H_2MoO_4 + 6\ H^+ + 3\ e^- \rightarrow Mo^{3+} + 4\ H_2O$ |
| $UO_2^{2+} + 4\ H^+ + 2\ e^- \rightarrow U^{4+} + 2\ H_2O$ | $UO_2^{2+} + 4\ H^+ + 2\ e^- \rightarrow U^{4+} + 2\ H_2O$ |
| | $UO_2^{2+} + 4\ H^+ + 3\ e^- \rightarrow U^{3+} + 2\ H_2O$† |
| $V(OH)_4^+ + 2\ H^+ + e^- \rightarrow VO^{2+} + 3\ H_2O$ | $V(OH)_4^+ + 4\ H^+ + 3\ e^- \rightarrow V^{2+} + 4\ H_2O$ |
| $TiO^{2+}$ not reduced | $TiO^{2+} + 2\ H^+ + e^- \rightarrow Ti^{3+} + H_2O$ |
| $Cr^{3+}$ not reduced | $Cr^{3+} + e^- \rightarrow Cr^{2+}$ |

*From I. M. Kolthoff and R. Belcher, *Volumetric Analysis,* Vol. 3, p. 12. New York: Interscience, 1957. With permission.

†A mixture of oxidation states is obtained. The Jones reductor may still be used for the analysis of uranium, however, because any $U^{3+}$ formed can be converted to $U^{4+}$ by shaking the solution with air for a few minutes.

analyte solution and boiling for a brief period. The unused reagent is then removed by filtration. The half-reaction for the reduction of sodium bismuthate can be written as

$$NaBiO_3(s) + 4\ H^+ + 2\ e^- \rightleftarrows BiO^+ + Na^+ + 2\ H_2O$$

### Ammonium Peroxydisulfate

Ammonium peroxydisulfate, $(NH_4)_2S_2O_8$, is also a powerful oxidizing agent. In acidic solution, it converts chromium(III) to dichromate, cerium(III) to its tetravalent state, and manganese(II) to permanganate. The half-reaction is

$$S_2O_8^{2-} + 2\ e^- \rightleftarrows 2\ SO_4^{2-} \qquad E^0 = 2.01\ V$$

The oxidations are catalyzed by traces of silver ion. The excess reagent is readily decomposed by a brief period of boiling:

$$2\ S_2O_8^{2-} + 2\ H_2O \rightarrow 4\ SO_4^{2-} + O_2(g) + 4\ H^+$$

### Sodium Peroxide and Hydrogen Peroxide

Peroxide is a convenient oxidizing agent either as the solid sodium salt or as a dilute solution of the acid. The half-reaction for hydrogen peroxide in acidic solution is

$$H_2O_2 + 2\ H^+ + 2\ e^- \rightleftarrows 2\ H_2O \qquad E^0 = 1.78\ V$$

After oxidation is complete, the solution is freed of excess reagent by boiling:

$$2\ H_2O_2 \rightarrow 2\ H_2O + O_2(g)$$

## 16B  APPLICATIONS OF STANDARD REDUCTANTS

Standard solutions of most reducing agents tend to react with atmospheric oxygen. For this reason, reductants are seldom used for the direct titration of oxidizing analytes; indirect methods are used instead. The two most common indirect methods, discussed in the paragraphs that follow, are based upon standard solutions of iron(II) and of sodium thiosulfate.

### 16B-1  Iron(II) Solutions

Solutions of iron(II) are readily prepared from iron(II) ammonium sulfate, $Fe(NH_4)_2(SO_4)_2 \cdot 6\ H_2O$ (Mohr's salt), or from the closely related iron(II) ethylenediamine sulfate, $FeC_2H_4(NH_3)_2(SO_4)_2 \cdot 4\ H_2O$ (Oesper's salt). Air-oxidation of iron(II) takes place rapidly in neutral solutions but is inhibited in the presence of acids, with the most stable preparations being about 0.5 M in $H_2SO_4$. Such solutions are stable for no longer than one day, if that long.

Numerous oxidizing agents are conveniently determined by treatment of the sample solution with a measured excess of standard iron(II) followed by immediate titration of the excess with a standard solution of potassium dichromate or cerium(IV) (Sections 16C-1 and 16C-2). Just before or after the samples are titrated, the concentration of the iron(II) solution is established by titrating two or three aliquots of the reagent by itself.

This procedure has been applied to the determination of organic peroxides; chromium(VI); cerium(IV); molybdenum(VI); nitrate, chlorate, and perchlorate ions; and numerous other oxidants.

### 16B-2  Sodium Thiosulfate

Thiosulfate ion is a moderately strong reducing agent that has been widely used to determine oxidizing agents by an indirect procedure that involves iodine as an intermediate. With iodine, thiosulfate ion is oxidized quantitatively to tetrathionate ion, the half-reaction being

$$2\ S_2O_3^{2-} \rightleftharpoons S_4O_6^{2-} + 2\ e^-$$

In its reaction with iodine, each thiosulfate ion gains one electron.

The quantitative aspect of the reaction with iodine is unique. Other oxidants oxidize the tetrathionate ion, wholly or in part, to sulfate ion.

The scheme used for determining oxidizing agents involves adding an unmeasured excess of potassium iodide to a slightly acidic solution of the analyte. Reduction of the analyte produces an amount of iodine that is stoichiometrically related to the number of moles of analyte. The liberated iodine is then titrated with a standard solution of sodium thiosulfate, $Na_2S_2O_3$, one of the few reducing agents that is stable toward air-oxidation. An example of this procedure is the determination of sodium hypochlorite in bleaches. The reactions are

Sodium thiosulfate is one of the few reducing agents that are not oxidized by air.

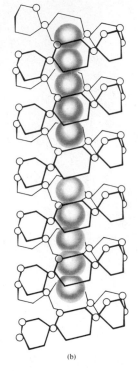

**Figure 16–2**
Thousands of glucose molecules polymerize to form huge molecules of β-amylose as shown schematically in (a) above. Molecules of β-amylose tend to assume a helical structure. The iodine species $I_5^-$ as shown in (b) is incorporated into the amylose helix. For further details see R. C. Teitelbaum, S. L. Ruby, and T. J. Marks, *J. Amer. Chem. Soc.,* **1980,** *102,* 3322.

When titrating iodine solutions with thiosulfate ion, addition of the starch must be delayed until most of the iodine has been consumed in order to avoid decomposition of the starch by the high concentration of iodine.

$$OCl^- + 2\ I^- + 2\ H^+ \rightarrow Cl^- + I_2 + H_2O \qquad \text{(excess KI)}$$
$$I_2 + 2\ S_2O_3^{2-} \rightarrow 2\ I^- + S_4O_6^{2-} \qquad\qquad (16\text{–}1)$$

The quantitative conversion of thiosulfate ion to tetrathionate ion shown in Equation 16–1 requires a pH somewhat lower than 7. If strongly acidic solutions must be titrated, air-oxidation of the excess iodide must be prevented by blanketing the solution with an inert gas, such as carbon dioxide or nitrogen.

### End Points in Iodine/Thiosulfate Titrations

The color of iodine can be discerned in a solution that is about $5 \times 10^{-6}$ M in $I_2$, a concentration that corresponds to less than one drop of a 0.05 M solution in 100 mL. Thus, provided the analyte solution is colorless, the reagent can serve as its own indicator.

More commonly, titrations involving iodine are performed with a suspension of starch as an indicator. The deep blue color of starch solutions containing iodine is believed to arise from the absorption of iodine into the helical chain of β-amylose, a macromolecular component of most starches (Figure 16–2). The closely related α-amylose forms a red adduct with iodine. This reaction is not readily reversible and is thus undesirable. So-called *soluble starch,* which is available from commercial sources, consists principally of β-amylose, the alpha fraction having been removed; indicator solutions are readily prepared from this product.

Aqueous starch suspensions decompose within a few days, primarily because of bacterial action. The decomposition products tend to interfere with the solution's indicator properties and may also be oxidized by iodine. The rate of decomposition can be inhibited by preparing and storing the indicator under sterile conditions and by adding mercury(II) iodide or chloroform as a bacteriostat. Perhaps the simplest alternative is to prepare a fresh suspension of the indicator, which requires only a few minutes, on the day it is to be used.

Starch decomposes irreversibly in solutions containing large concentrations of iodine. Therefore, in titrating solutions of iodine with sodium thiosulfate, as in the indirect determination of oxidants, addition of the indicator is delayed until the titration is nearly complete (indicated by a change in color from deep red to faint yellow). When thiosulfate solutions are being titrated directly with iodine, the indicator can be introduced at the outset.

### The Stability of Sodium Thiosulfate Solutions

Although sodium thiosulfate solutions are resistant to air-oxidation, they do tend to decompose to give sulfur and hydrogen sulfite ion:

$$S_2O_3^{2-} + H^+ \rightleftharpoons HSO_3^- + S(s)$$

Variables that influence the rate of this reaction include pH, presence of microorganisms, solution concentration, presence of copper(II) ions, and exposure to sunlight. These variables may cause the concentration of a thiosulfate solution to change by several percent over a period of a few weeks. On the other hand, proper attention to detail will yield solutions that need only occasional restandardization.

The rate of the decomposition reaction increases markedly with increasing acidity.

The most important single cause for the instability of neutral or slightly basic thiosulfate solutions can be traced to bacteria that metabolize thiosulfate ion to sulfite and sulfate ions as well as elemental sulfur. To minimize this problem, standard solutions of the reagent are prepared under reasonably sterile conditions. Bacterial activity appears to be at a minimum at a pH between 9 and 10, which accounts, at least in part, for the reagent's greater stability in slightly basic solutions. The presence of a bactericide, such as chloroform, sodium benzoate, or mercury(II) iodide, also slows decomposition.

When sodium thiosulfate is added to a strongly acidic medium, a cloudiness develops almost immediately as elemental sulfur forms. Even in neutral solution, this reaction proceeds at such a rate that standard sodium thiosulfate must be restandardized periodically.

## The Standardization of Thiosulfate Solutions

Potassium iodate is an excellent primary standard for thiosulfate solutions. In this application, weighed amounts of primary-standard-grade reagent are dissolved in water containing an excess of potassium iodide. When this mixture is acidified with a strong acid, the reaction

$$IO_3^- + 5 I^- + 6 H^+ \rightleftarrows 3 I_2 + 3 H_2O$$

occurs instantly. The liberated iodine is then titrated with the thiosulfate solution. The overall stoichiometry of the process is

$$1 \text{ mol } IO_3^- \equiv 3 \text{ mol } I_2 \equiv 6 \text{ mol } S_2O_3^{2-}$$

---

Example 16–1

A solution of sodium thiosulfate was standardized by dissolving 0.1210 g of $KIO_3$ (fw = 214.00 g) in water, adding a large excess of KI, and acidifying with HCl. The liberated iodine required 41.64 mL of the thiosulfate solution to decolorize the blue starch/iodine complex. Calculate the molarity of the $Na_2S_2O_3$.

$$\text{amount } NaS_2O_3 = 0.1210 \text{ g } KIO_3 \times \frac{1 \text{ mmol } KIO_3}{0.21400 \text{ g } KIO_3} \times \frac{6 \text{ mmol } Na_2S_2O_3}{\text{mmol } KIO_3}$$

$$= 3.3925 \text{ mmol } Na_2S_2O_3$$

$$c_{Na_2S_2O_3} = \frac{3.3925 \text{ mmol } Na_2S_2O_3}{41.64 \text{ mL } Na_2S_2O_3} = 0.08147 \text{ M}$$

---

Other primary standards for sodium thiosulfate are potassium dichromate, potassium bromate, potassium hydrogen iodate, potassium ferricyanide, and metallic copper. All these standards liberate stoichiometric amounts of iodine when treated with excess potassium iodide.

## Applications of Sodium Thiosulfate Solutions

Numerous substances can be determined by the indirect method involving titration with sodium thiosulfate; typical applications are summarized in Table 16–2.

## 16C   APPLICATIONS OF STANDARD OXIDANTS

Table 16–3 summarizes the properties of five of the most widely used volumetric oxidizing reagents. Note that the standard potentials for these reagents vary from 0.5 to 1.5 V. The choice among them depends upon analyte strength as a reducing agent, rate of reaction between oxidant and analyte, stability of the standard oxidant solutions, cost, and availability of a satisfactory indicator.

### 16C–1   The Strong Oxidants: Potassium Permanganate and Cerium(IV)

Solutions of permanganate ion and cerium(IV) ion are strong oxidizing reagents whose applications closely parallel one another. Half-reactions for the two are

$$MnO_4^- + 8\ H^+ + 5\ e^- \rightleftarrows Mn^{2+} + 4\ H_2O \qquad E^0 = 1.51\ V$$

$$Ce^{4+} + e^- \rightleftarrows Ce^{3+} \qquad\qquad\qquad E^f = 1.44\ V\ (1\ M\ H_2SO_4)$$

The formal potential shown for the reduction of cerium(IV) is for solutions that are 1 M in sulfuric acid. In 1 M perchloric acid and 1 M nitric acid, the potentials are 1.70 and 1.61 V, respectively. Solutions of cerium(IV) in the latter two acids are not very stable and thus find limited application.

### Table 16–2
### SOME APPLICATIONS OF SODIUM THIOSULFATE AS A REDUCTANT

| Substance | Half-Reaction | Special Conditions |
|---|---|---|
| $IO_4^-$ | $IO_4^- + 8\ H^+ + 7\ e^- \rightarrow \frac{1}{2} I_2 + 4\ H_2O$ | Acidic solution |
| | $IO_4^- + 2\ H^+ + 2\ e^- \rightarrow IO_3^- + H_2O$ | Neutral solution |
| $IO_3^-$ | $IO_3^- + 6\ H^+ + 5\ e^- \rightarrow \frac{1}{2} I_2 + 3\ H_2O$ | Strong acid |
| $BrO_3^-, ClO_3^-$ | $XO_3^- + 6\ H^+ + 6\ e^- \rightarrow X^- + 3\ H_2O$ | Strong acid |
| $Br_2, Cl_2$ | $X_2 + 2\ I^- \rightarrow I_2 + 2\ X^-$ | |
| $NO_2^-$ | $HNO_2 + H^+ + e^- \rightarrow NO(g) + H_2O$ | |
| $Cu^{2+}$ | $Cu^{2+} + I^- + e^- \rightarrow CuI(s)$ | |
| $O_2$ | $O_2 + 4\ Mn(OH)_2(s) + 2\ H_2O \rightarrow 4\ Mn(OH)_3(s)$ | Basic solution |
| | $Mn(OH)_3(s) + 3\ H^+ + e^- \rightarrow Mn^{2+} + 3\ H_2O$ | Acidic solution |
| $O_3$ | $O_3(g) + 2\ H^+ + 2\ e^- \rightarrow O_2(g) + H_2O$ | |
| Organic peroxide | $ROOH + 2\ H^+ + 2\ e^- \rightarrow ROH + H_2O$ | |

**Table 16-3**
**SOME COMMON OXIDANTS USED AS STANDARD SOLUTIONS**

| Reagent and Formula | Reduction Product | Standard Potential, V | Standardized with | Indicator* | Stability† |
|---|---|---|---|---|---|
| Potassium permanganate, $KMnO_4$ | $Mn^{2+}$ | 1.51 | $Na_2C_2O_4$, Fe, $As_2O_3$ | $MnO_4^-$ | (b) |
| Potassium bromate, $KBrO_3$ | $Br^-$ | 1.44 | $KBrO_3$ | (1) | (a) |
| Cerium(IV), $Ce^{4+}$ | $Ce^{3+}$ | 1.44‡ | $Na_2C_2O_4$, Fe, $As_2O_3$ | (2) | (a) |
| Potassium dichromate, $K_2Cr_2O_7$ | $Cr^{3+}$ | 1.33 | $K_2Cr_2O_7$, Fe | (3) | (a) |
| Iodine, $I_2$ | $I^-$ | 0.536 | $BaS_2O_3 \cdot H_2O$, $Na_2S_2O_3$ | starch | (c) |

*(1) $\alpha$-Naphthoflavone; (2) 1,10-phenanthroline iron(II) complex (ferroin); (3) diphenylamine sulfonic acid.

†(a) Indefinitely stable; (b) moderately stable, requires periodic standardization; (c) somewhat unstable, requires frequent standardization.

‡$E^f$ in 1 M $H_2SO_4$.

The half-reaction shown for permanganate ion occurs only in solutions that are 0.1 M or greater in strong acid. The product may be Mn(III), Mn(IV), or Mn(VI) in less acidic media.

## Comparison of the Two Reagents

For all practical purposes, the oxidizing strengths of permanganate and cerium(IV) solutions are comparable. Solutions of cerium(IV) in sulfuric acid, however, are stable indefinitely, whereas permanganate solutions decompose slowly and thus require occasional restandardization. Furthermore, cerium(IV) solutions in sulfuric acid do not oxidize chloride ion and thus can be used to titrate hydrochloric acid solutions of analytes. In contrast, permanganate ion cannot be used with hydrochloric acid solutions unless special precautions are taken to prevent the slow oxidation of chloride ion that leads to overconsumption of the standard reagent. A further advantage of cerium(IV) is that a primary-standard-grade salt of the reagent is available, thus making possible the direct preparation of standard solutions.

Despite these several advantages of cerium solutions over permanganate solutions, the latter are more widely used. One reason for the popularity of permanganate solutions is their modest cost. A liter of a given concentration of permanganate may cost as little as 2% of the same volume and the same strength of cerium(IV). Another disadvantage of cerium(IV) solutions is their tendency to form precipitates of basic cerium(IV) salts in solutions that are less than 0.1 M in strong acid.

## End Points

A useful property of a potassium permanganate solution is its intense purple color, which is sufficient to serve as an indicator for most titrations. As little as 0.01 mL of a 0.02 M solution imparts a perceptible color to 100 mL of water. If the permanganate solution is very dilute, diphenylamine sulfonic acid or the 1,10-phenanthroline complex of iron(II) (Table 15–2) provides a sharper end point.

The permanganate end point is not permanent because excess permanganate ions react slowly with the relatively large concentration of manganese(II) ions present at the end point:

$$2 \; MnO_4^- + 3 \; Mn^{2+} + 2 \; H_2O \rightarrow 5 \; MnO_2(s) + 4 \; H^+$$

The equilibrium constant for this reaction is about $10^{47}$, which indicates that the *equilibrium* concentration of permanganate ion is vanishingly small even in highly acidic media. Fortunately, the rate at which this equilibrium is approached is so low that the end point fades only gradually over a period of perhaps 30 s.

Solutions of cerium(IV) are yellow-orange, but the color is not intense enough to act as an indicator in titrations. The most widely used indicator for titrations with cerium(IV) is the iron(II) complex of 1,10-phenanthroline or one of its substituted derivatives (Table 15–2).

## The Preparation and Stability of Standard Solutions

**Potassium Permanganate Solutions.** Aqueous solutions of permanganate are not entirely stable because the ion tends to oxidize water:

$$4 \; MnO_4^- + 2 \; H_2O \rightarrow 4 \; MnO_2(s) + 3 \; O_2(g) + 4 \; OH^-$$

Although the equilibrium constant for this reaction indicates that the products are favored, permanganate solutions, when properly prepared, are reasonably stable because the decomposition reaction is slow.

Decomposition is catalyzed by light, heat, acids, bases, manganese(II), and manganese dioxide. Moderately stable solutions of permanganate ion can be prepared if the effects of these catalysts, particularly manganese dioxide, are minimized. Manganese dioxide is a contaminant in even the best grade of solid potassium permanganate. Furthermore, this compound forms in freshly prepared solutions of the reagent as a consequence of the reaction of permanganate ion with organic matter and dust in the water used to prepare the solution. Removal of manganese dioxide by filtration before standardization markedly enhances the stability of permanganate solutions. Before filtration, the solution is allowed to stand for about 24 hr or is heated for a brief period to hasten oxidation of any organic species in distilled and deionized water. A filtering crucible must be used for filtering because permanganate ion reacts with paper to form additional manganese dioxide.

Standardized permanganate solutions should be stored in the dark. Filtration and restandardization are required if any solid is detected in the

Permanganate solutions are moderately stable provided they are freed of manganese dioxide and stored in a dark container.

solution or on the walls of the storage bottle. In any event, restandardization every one or two weeks is a good precautionary measure.

Solutions containing excess standard permanganate should never be heated because they decompose by oxidizing water. This decomposition cannot be compensated for with a blank. On the other hand, it is possible to titrate hot, acidic solutions of reductants with permanganate without error, provided the reagent is added slowly enough so that large excesses do not accumulate.

---

Example 16–2

Describe how you would prepare 2.0 L of an approximately 0.010 M solution of $KMnO_4$ (fw = 158.04 g).

$$\text{wt } KMnO_4 \text{ needed} = 2.0\,\cancel{L} \times 0.010\,\frac{\cancel{\text{mol } KMnO_4}}{\cancel{L}} \times \frac{158.04 \text{ g } KMnO_4}{\cancel{\text{mol } KMnO_4}}$$

$$= 3.16 \text{ g}$$

Dissolve about 3.2 g of $KMnO_4$ in a little water. After solution is complete, add water to bring the volume to about 2.0 L. Heat the solution to boiling for a brief period, and let stand until cool. Filter through a glass filtering crucible and store in a clean dark bottle.

---

**Cerium(IV) Solutions.** The most widely used compounds for preparing solutions of cerium(IV) are listed in Table 16–4. Primary-standard-grade cerium ammonium nitrate is available commercially and can be used to prepare standard solutions of the cation directly by weight. More commonly, less expensive reagent-grade cerium(IV) ammonium nitrate or cerium(IV) hydroxide is used to give a solution that is subsequently standardized. In either case, the reagent is dissolved in a solution that is at least 0.1 M in sulfuric acid to prevent the precipitation of basic salts.

Sulfuric acid solutions of tetravalent cerium are remarkably stable and can be stored for months or heated at 100°C for prolonged periods without change in concentration.

Primary Standards

Several excellent primary standards are available for the standardization of solutions of potassium permanganate and tetravalent cerium.

Table 16–4
ANALYTICALLY USEFUL CERIUM(IV) COMPOUNDS

| Name | Formula | Formula Weight |
|---|---|---|
| Cerium(IV) ammonium nitrate | $Ce(NO_3)_4 \cdot 2 \, NH_4NO_3$ | 548.2 |
| Cerium(IV) ammonium sulfate | $Ce(SO_4)_2 \cdot 2(NH_4)_2SO_4 \cdot 2 \, H_2O$ | 632.6 |
| Cerium(IV) hydroxide | $Ce(OH)_4$ | 208.1 |
| Cerium(IV) hydrogen sulfate | $Ce(HSO_4)_4$ | 528.4 |

**Sodium Oxalate.** Sodium oxalate is widely used for standardizing permanganate and cerium(IV) solutions. In acidic solutions, the oxalate ion is converted to the undissociated acid. Thus, its reaction with the permanganate ion can be depicted as

$$2\ MnO_4^- + 5\ H_2C_2O_4 + 6\ H^+ \rightleftarrows 2\ Mn^{2+} + 10\ CO_2(g) + 8\ H_2O$$

The reaction between $Ce^{4+}$ and $H_2C_2O_4$ is

$$2\ Ce^{4+} + H_2C_2O_4 \rightleftarrows \\ 2\ Ce^{3+} + 2\ H^+ + 2\ CO_2$$

Carbon dioxide is also the product with cerium(IV).

The reaction between permanganate ion and oxalic acid is complex and proceeds slowly even at elevated temperature unless manganese(II) is present as a catalyst. Thus, when the first few milliliters of standard permanganate are added to a hot solution of oxalic acid, several seconds elapse before the color of the permanganate ion disappears. As the concentration of manganese(II) builds up, however, the reaction proceeds more and more rapidly as a result of autocatalysis.

It has been found that when solutions of sodium oxalate are titrated at 60 to 90°C, the consumption of permanganate is from 0.1 to 0.4% less than theoretical, probably due to the air-oxidation of a fraction of the oxalic acid. This small error can be avoided by adding 90 to 95% of the required permanganate to a cool solution of the oxalate. After the added permanganate is completely used up, as indicated by the disappearance of color, the solution is heated to about 60°C and titrated to a pink color that persists for about 30 s. The disadvantage of this procedure is that it requires a knowledge of the approximate concentration of the permanganate solutions so that a proper initial volume of it can be added. For most purposes, the direct titration of the hot oxalic acid solution provides perfectly adequate data (usually 0.2 to 0.3% high).

Solutions of $KMnO_4$ and $Ce^{4+}$ can also be standardized with electrolytic iron wire and potassium iodide.

Tetravalent cerium standardizations against sodium oxalate are usually performed at 50°C in a hydrochloric acid solution containing iodine monochloride as a catalyst.

---

Example 16–3

You wish to standardize the solution in Example 16–2 against standard $Na_2C_2O_4$ (fw = 134.00 g). If you want to use between 30 and 45 mL of the reagent for the standardization, what range of weights of the primary standard should you take?

For a 30-mL titration:

$$amount\ KMnO_4 = 30\ mL\ KMnO_4 \times 0.010\ \frac{mmol\ KMnO_4}{mL\ KMnO_4}$$

$$= 0.30\ mmol\ KMnO_4$$

$$wt\ Na_2C_2O_4 = 0.30\ mmol\ KMnO_4 \times \frac{5\ mmol\ Na_2C_2O_4}{2\ mmol\ KMnO_4}$$

$$\times 0.134\ \frac{g\ Na_2C_2O_4}{mmol\ Na_2C_2O_4} = 0.101\ g\ Na_2C_2O_4$$

Proceeding in the same way, we find for a 45-mL titration:

$$\text{wt } Na_2C_2O_4 = 45 \times 0.010 \times \frac{5}{2} \times 0.134 = 0.151 \text{ g } Na_2C_2O_4$$

Thus, your samples should weigh between 0.10 and 0.15 g.

---

Example 16–4

Exactly 33.31 mL of the solution in Example 16–2 was required to titrate a 0.1278-g sample of primary-standard $Na_2C_2O_4$. What is the molarity of the $KMnO_4$ reagent?

$$\text{amount } Na_2C_2O_4 = 0.1278 \text{ g } Na_2C_2O_4 \times \frac{1 \text{ mmol } Na_2C_2O_4}{0.13400 \text{ g } Na_2C_2O_4}$$

$$= 0.95373 \text{ mmol } Na_2C_2O_4$$

$$c_{KMnO_4} = 0.95373 \text{ mmol } Na_2C_2O_4 \times \frac{2 \text{ mmol } KMnO_4}{5 \text{ mmol } Na_2C_2O_4}$$

$$\times \frac{1}{33.31 \text{ mL } KMnO_4} = 0.01145 \text{ M } KMnO_4$$

---

Table 16–5
SOME APPLICATIONS OF POTASSIUM PERMANGANATE AND CERIUM(IV) IN ACID SOLUTION

| Substance Analyzed | Half-Reaction | Special Conditions |
|---|---|---|
| Sn | $Sn^{2+} \rightleftharpoons Sn^{4+} + 2 e^-$ | Prereduction with Zn |
| $H_2O_2$ | $H_2O_2 \rightleftharpoons O_2(g) + 2 H^+ + 2 e^-$ | |
| Fe | $Fe^{2+} \rightleftharpoons Fe^{3+} + e^-$ | Prereduction with $SnCl_2$ or with Jones or Walden reductor |
| $Fe(CN)_6^{4-}$ | $Fe(CN)_6^{4-} \rightleftharpoons Fe(CN)_6^{3-} + e^-$ | |
| V | $VO^{2+} + 3 H_2O \rightleftharpoons V(OH)_4^+ + 2 H^+ + e^-$ | Prereduction with Bi amalgam or $SO_2$ |
| Mo | $Mo^{3+} + 4 H_2O \rightleftharpoons MoO_4^{2-} + 8 H^+ + 3 e^-$ | Prereduction with Jones reductor |
| W | $W^{3+} + 4 H_2O \rightleftharpoons WO_4^{2-} + 8 H^+ + 3 e^-$ | Prereduction with Zn or Cd |
| U | $U^{4+} + 2 H_2O \rightleftharpoons UO_2^{2+} + 4 H^+ + 2 e^-$ | Prereduction with Jones reductor |
| Ti | $Ti^{3+} + H_2O \rightleftharpoons TiO^{2+} + 2 H^+ + e^-$ | Prereduction with Jones reductor |
| $H_2C_2O_4$ | $H_2C_2O_4 \rightleftharpoons 2 CO_2 + 2 H^+ + 2 e^-$ | |
| Mg, Ca, Zn, Co, Pb, Ag | $H_2C_2O_4 \rightleftharpoons 2 CO_2 + 2 H^+ + 2 e^-$ | Sparingly soluble metal oxalates filtered, washed, and dissolved in acid; liberated oxalic acid titrated |
| $HNO_2$ | $HNO_2 + H_2O \rightleftharpoons NO_3^- + 3 H^+ + 2 e^-$ | 15-min reaction time; excess $KMnO_4$ back-titrated |
| K | $K_2NaCo(NO_2)_6 + 6 H_2O \rightleftharpoons$ $Co^{2+} + 6 NO_3^- + 12 H^+ + 2 K^+ + Na^+ + 11 e^-$ | Precipitated as $K_2NaCo(NO_2)_6$; filtered and dissolved in $KMnO_4$; excess $KMnO_4$ back-titrated |
| Na | $U^{4+} + 2 H_2O \rightleftharpoons UO_2^{2+} + 4 H^+ + 2 e^-$ | Precipitated as $NaZn(UO_2)_3(OAc)_9$; filtered, washed, dissolved; U determined as above |

## Applications of Potassium Permanganate and Cerium(IV) Solutions

Table 16–5 lists some of the many applications of permanganate and cerium(IV) solutions to the volumetric determination of inorganic species. Both reagents have also been applied to the determination of organic compounds with oxidizable functional groups.

---

Example 16–5

Aqueous solutions containing approximately 3% (w/w) $H_2O_2$ are sold in drugstores as a disinfectant. Propose a method for determining the peroxide content of one of these preparations using the standard solution described in Example 16–4 and between 35 and 45 mL of the reagent per titration. The reaction is

$$5\ H_2O_2 + 2\ MnO_4^- + 6\ H^+ \rightarrow 5\ O_2 + 2\ Mn^{2+} + 8\ H_2O$$

The number of millimoles of $KMnO_4$ in 35 to 45 mL of the reagent is between

$$\text{amount } KMnO_4 = 35\ \text{mL} \times 0.01145\ \frac{\text{mmol } KMnO_4}{\text{mL } KMnO_4}$$
$$= 0.401\ \text{mmol } KMnO_4$$

and

$$\text{amount } KMnO_4 = 45 \times 0.01145 = 0.515\ \text{mmol } KMnO_4$$

The number of millimoles of $H_2O_2$ consumed by these amounts of $KMnO_4$ is

$$\text{amount } H_2O_2 = 0.401\ \text{mmol } KMnO_4^- \times \frac{5\ \text{mmol } H_2O_2}{2\ \text{mmol } KMnO_4}$$
$$= 1.00\ \text{mmol } H_2O_2$$

and

$$\text{amount } H_2O_2 = 0.515 \times \frac{5}{2} = 1.29\ \text{mmol } H_2O_2$$

Thus we need to take samples that contain from 1.0 to 1.3 mmol $H_2O_2$.

$$\text{wt sample} = 1.0\ \text{mmol } H_2O_2 \times 0.03401\ \frac{\text{g } H_2O_2}{\text{mmol } H_2O_2} \times \frac{100\ \text{g sample}}{3\ \text{g } H_2O_2}$$
$$= 1.1\ \text{g sample}$$

to

$$\text{wt sample} = 1.3 \times 0.03401 \times \frac{100}{3} = 1.5 \text{ g sample}$$

Thus our samples should weigh between 1.1 and 1.5 g. They should be diluted to perhaps 75 to 100 mL with water and made slightly acidic with dilute $H_2SO_4$ before titration.

---

## 16C–2 Potassium Dichromate

In its analytical applications, dichromate ion is reduced to green chromium(III) ion:

$$Cr_2O_7^{2-} + 14\ H^+ + 6\ e^- \rightleftarrows 2\ Cr^{3+} + 7\ H_2O \qquad E^0 = 1.33 \text{ V}$$

Dichromate titrations are generally carried out in solutions that are about 1 M in hydrochloric or sulfuric acid. In these media, the formal potential for the half-reaction is 1.0 to 1.1 V.

Potassium dichromate solutions are indefinitely stable, can be boiled without decomposition, and do not react with hydrochloric acid. Moreover, primary-standard reagent is available commercially at a modest cost. The disadvantages of potassium dichromate compared with cerium(IV) and permanganate ion are its lower electrode potential and the slowness of its reaction with certain reducing agents.

### The Preparation, Properties, and Uses of Dichromate Solutions

For most purposes, reagent-grade potassium dichromate is sufficiently pure to permit the direct preparation of standard solutions; the solid is simply dried at 150 to 200°C before being weighed.

The orange color of a dichromate solution is not intense enough for use in end-point detection. However, diphenylamine sulfonic acid (Table 15–2) is an excellent indicator for titrations with this reagent. The oxidized form of the indicator is red-violet, and its reduced form is essentially colorless; thus, the color change observed in a direct titration is from the green of chromium(III) to violet.

### Applications of Potassium Dichromate Solutions

The principal use of dichromate is for the volumetric titration of iron(II) based upon the reaction

$$Cr_2O_7^{2-} + 6\ Fe^{2+} + 14\ H^+ \rightarrow 2\ Cr^{3+} + 6\ Fe^{3+} + 7\ H_2O$$

Often, this titration is performed in the presence of moderate concentrations of hydrochloric acid.

The reaction of dichromate with iron(II) has been widely used for the indirect determination of a variety of oxidizing agents. In these applications, a measured excess of a solution of iron(II) is added to an acidic solution of the analyte. The excess iron(II) is then back-titrated with standard potassium dichromate (Section 16B–1). Standardization of the iron(II) solution by titration with the dichromate is performed concurrently because solutions of iron(II) tend to be air-oxidized. This method

---

Example 16–6

A 5.00-mL sample of brandy was diluted to 1.000 L in a volumetric flask. The ethanol ($C_2H_5OH$) in a 25.00-mL aliquot of the diluted solution was distilled into 50.00 mL of 0.02000 M $K_2Cr_2O_7$, where oxidation to acetic acid occurred with heating:

$$3\ C_2H_5OH + 2\ Cr_2O_7^{2-} + 16\ H^+ \rightarrow 4\ Cr^{3+} + 3\ CH_3COOH + 11\ H_2O$$

After cooling, 20.00 mL of 0.1253 M $Fe^{2+}$ was pipetted into the flask. The excess $Fe^{2+}$ was then titrated with 7.46 mL of the standard $K_2Cr_2O_7$ to a diphenylamine sulfonic acid end point. Calculate the weight/volume percent of $C_2H_5OH$ (fw = 46.07 g) in the brandy.

$$\text{total volume of } K_2Cr_2O_7 = 50.00\ \text{mL} + 7.46\ \text{mL} = 57.46\ \text{mL}$$

$$\text{total amount } K_2Cr_2O_7 = 57.46\ \cancel{\text{mL } K_2Cr_2O_7} \times 0.02000\ \frac{\text{mmol } K_2Cr_2O_7}{\cancel{\text{mL } K_2Cr_2O_7}}$$

$$= 1.1492\ \text{mmol } K_2Cr_2O_7$$

$$\text{amount } Fe^{2+} = 20.00\ \text{mL } Fe^{2+} \times 0.1253\ \frac{\text{mmol } Fe^{2+}}{\text{mL } Fe^{2+}}$$

$$= 2.5060\ \text{mmol } Fe^{2+}$$

$$\text{amount } K_2Cr_2O_7 \text{ consumed by } Fe^{2+}$$

$$= 2.5060\ \cancel{\text{mmol } Fe^{2+}} \times \frac{1\ \text{mmol } K_2Cr_2O_7}{6\ \cancel{\text{mmol } Fe^{2+}}}$$

$$= 0.41767\ \text{mmol } K_2Cr_2O_7$$

$$\text{amount } K_2Cr_2O_7 \text{ consumed by } C_2H_5OH$$

$$= (1.1492 - 0.41767)\ \text{mmol } K_2C_6O_7$$

$$= 0.73153\ \text{mmol } K_2Cr_2O_7$$

$$\text{wt } C_2H_5OH = 0.73153\ \cancel{\text{mmol } K_2Cr_2O_7} \times \frac{3\ \cancel{\text{mmol } C_2H_5OH}}{2\ \cancel{\text{mmol } K_2Cr_2O_7}}$$

$$\times\ 0.04607\ \frac{\text{g } C_2H_5OH}{\cancel{\text{mmol } C_2H_5OH}}$$

$$= 0.050552\ \text{g } C_2H_5OH$$

$$\% C_2H_5OH\ (w/v) = \frac{0.050552\ \text{g } C_2H_5OH}{5.00\ \text{mL sample} \times 25.00\ \cancel{\text{mL}}/1000\ \cancel{\text{mL}}} \times 100\%$$

$$= 40.44\% = 40.4\%\ C_2H_5OH\ (w/v)$$

has been applied to the determination of nitrate, chlorate, permanganate, and dichromate ions as well as organic peroxides and several other oxidizing agents.

## 16C–3 Iodine

Solutions of iodine are weak oxidizing agents that are used for the determination of strong reductants. The most accurate description of the half-reaction for iodine in these applications is

$$I_3^- + 2\,e^- \rightleftarrows 3\,I^- \qquad E^0 = 0.536\ V$$

The applications of standard iodine solutions are more limited than those of the other oxidants we have described because of the significantly lower electrode potential. Occasionally, however, this low potential is advantageous because it imparts a degree of selectivity that makes possible the determination of strong reducing agents in the presence of weak ones. An important advantage of iodine is the availability of a sensitive and reversible indicator for the titrations. Unfortunately, iodine solutions lack stability and must be restandardized regularly.

### The Preparation and Properties of Iodine Solutions

Iodine is not very soluble in water (≈0.001 M). In order to prepare a solution with an analytically useful concentration of the element, iodine is ordinarily dissolved in moderately concentrated potassium iodide. In this medium, iodine is reasonably soluble as a consequence of the reaction

$$I_2(s) + I^- \rightleftarrows I_3^- \qquad K = 7.1 \times 10^2$$

Solutions prepared by dissolving iodine in a concentrated solution of potassium iodide are properly called triiodide solutions. In practice, however, they are usually termed iodine solutions because this terminology accounts for the stoichiometric behavior $(I_2 + 2\,e^- \rightarrow 2\,I^-)$.

Iodine dissolves only slowly in solutions of potassium iodide, particularly when the iodide concentration is low. To ensure complete solution, the solid is always dissolved in a small volume of concentrated potassium iodide, care being taken to avoid dilution of the concentrated solution until the last trace of solid iodine has disappeared. Otherwise, the concentration of iodine in the diluted solution gradually increases with time. This problem can be avoided by filtering the solution through a sintered glass crucible before standardization.

Iodine solutions lack stability for several reasons, one being the volatility of the solute. Losses of iodine from an open vessel occur in a relatively short time even in the presence of an excess of iodide ion. In addition, iodine slowly attacks most organic materials. Consequently, cork or rubber stoppers are never used to close containers of the reagent, and precautions must be taken to protect standard solutions from contact with organic dusts and fumes.

Air-oxidation of iodide ion also causes changes in the molarity of an iodine solution:

$$4\,I^- + O_2(g) + 4\,H^+ \rightarrow 2\,I_2 + 2\,H_2O$$

In contrast to the other effects, this reaction causes the molarity of the iodine to increase. Air-oxidation is promoted by acids, heat, and light.

### The Standardization of Iodine Solutions

Arsenic(III) oxide, for many years the preferred primary standard for iodine solutions, has been identified as a carcinogen; its use is now subject to strict regulation.

Iodine solutions can be standardized against anhydrous sodium thiosulfate, barium thiosulfate monohydrate, or potassium antimony(III) tartrate.

Sodium thiosulfate pentahydrate can be rendered anhydrous by refluxing in methanol for a short time.[4] The product is readily soluble in water, takes up moisture from the atmosphere only slowly, and is an excellent primary standard for triiodide solutions. Anhydrous sodium thiosulfate is available commercially.

Barium thiosulfate monohydrate is also a satisfactory primary standard for iodine solutions.[5] Although the salt is soluble to the extent of only about 0.01 M, the solid reacts so rapidly with iodine solutions that a direct titration is entirely feasible. The reaction is

$$I_2 + Ba_2S_2O_3 \cdot H_2O(s) \rightarrow S_4O_6^{2-} + Ba^{2+} + 2\ I^-$$

It is reported that $BaS_2O_3 \cdot H_2O$ with a purity of 99.85% is readily prepared and stable at room temperature. Barium thiosulfate monohydrate begins to lose water above 50°C; the anhydrous salt, however, is unsuitable for standardization, owing to its low solubility. Therefore, the monohydrate should be used without drying.

The reaction between thiosulfate and iodine is discussed in Section 16B–2.

### Applications of Standard Iodine Solutions

Iodine titrations are generally performed in neutral or acidic media. Basic solutions are avoided because hypoiodite ion forms according to the equation

$$I_2 + OH^- \rightleftharpoons IO^- + I^- + H^+$$

The hypoiodite may subsequently disproportionate to iodide and iodate:

$$3\ IO^- \rightleftharpoons IO_3^- + 2\ I^-$$

The occurrence of these reactions can cause serious errors by disturbing the stoichiometry of the reaction between the standardized reagent and the analyte. Thus, solutions to be titrated with standardized iodine cannot have pH values higher than 9. Occasionally, a pH greater than 7.0 is detrimental.

---

[4]A. A. Woolf, *Anal. Chem.*, **1982**, *54*, 2134.

[5]W. M. MacNevin and O. H. Kriege, *Anal. Chem.*, **1953**, *25*, 767.

| Table 16–6 | |
|---|---|
| **SOME APPLICATIONS OF IODINE SOLUTIONS** | |
| Substance Analyzed | Half-Reaction |
| As | $H_3AsO_3 + H_2O \leftrightarrows H_3AsO_4 + 2\,H^+ + 2\,e^-$ |
| Sb | $H_3SbO_3 + H_2O \leftrightarrows H_3SbO_4 + 2\,H^+ + 2\,e^-$ |
| Sn | $Sn^{2+} \leftrightarrows Sn^{4+} + 2\,e^-$ |
| $H_2S$ | $H_2S \leftrightarrows S(s) + 2\,H^+ + 2\,e^-$ |
| $SO_2$ | $SO_3^{2-} + H_2O \leftrightarrows SO_4^{2-} + 2\,H^+ + 2\,e^-$ |
| $S_2O_3^{2-}$ | $2\,S_2O_3^{2-} \leftrightarrows S_4O_6^{2-} + 2\,e^-$ |
| $N_2H_4$ | $N_2H_4 \leftrightarrows N_2(g) + 4\,H^+ + 4\,e^-$ |
| Ascorbic acid* | $C_6H_8O_6 \rightarrow C_6H_6O_6 + 2\,H^+ + 2\,e^-$ |

*For the structure of ascorbic acid, see Section 34H–4.

Table 16–6 summarizes methods that use iodine as an oxidizing agent. Directions for the preparation, standardization, and application of standard iodine solutions are given in Section 34F.

## 16D   SOME SPECIALIZED OXIDANTS

In this section, we describe three oxidizing agents that are used primarily for determining certain special groups of compounds. Potassium bromate is used for the determination of organic compounds that contain olefinic and certain types of aromatic functional groups; periodic acid reacts selectively with organic compounds having hydroxyl, carbonyl, or amine groups on adjacent carbon atoms; and Karl Fischer reagent is widely employed for the determination of water in a variety of organic and inorganic samples.

### 16D–1 Potassium Bromate as a Source of Bromine

Primary-standard potassium bromate is available from commercial sources and can be used directly to prepare standard solutions that are stable indefinitely. Direct titrations with potassium bromate are relatively few. Instead, the reagent is a convenient and widely used stable source of bromine.[6] In this application, an unmeasured excess of potassium bromide is added to an acidic solution of the analyte. Upon introduction of a measured volume of standard potassium bromate, a stoichiometric quantity of bromine is produced.

$$\underset{\substack{\text{Standard}\\\text{Solution}}}{BrO_3^-} + \underset{\text{Excess}}{5\,Br^-} + 6\,H^+ \rightarrow 3\,Br_2 + 3\,H_2O$$

1 mol $BrO_3^- \equiv$ 3 mol $Br_2$

This indirect generation circumvents the problems associated with the use of standard bromine solutions, which lack stability.

---

[6]For a discussion of bromate solutions and their applications, see M. R. F. Ashworth, *Titrimetric Organic Analysis,* Part I, pp. 118–130. New York: Interscience, 1964.

The primary use of standard potassium bromate is for the determination of organic compounds that react with bromine. Few of these reactions are rapid enough to make direct titration feasible. Instead, a measured excess of standard bromate is added to the solution that contains the sample plus an excess of potassium bromide. After acidification, the mixture is allowed to stand in a glass-stoppered vessel until the bromine/analyte reaction is judged complete. To determine the excess bromine, an excess of potassium iodide is introduced so that the following reaction occurs:

1 mol $BrO_3^-$ ≡ 6 mol e⁻

$$2\ I^- + Br_2 \rightarrow I_2 + 2\ Br^-$$

The liberated iodine is then titrated with standard sodium thiosulfate (Equation 16–1).

Bromine is incorporated into an organic molecule by either substitution or addition.

### Substitution Reactions

Halogen substitution involves the replacement of hydrogen in an aromatic ring by a halogen. Substitution methods have been successfully applied to the determination of aromatic compounds that contain strong ortho-para-directing groups, particularly amines and phenols.

Example 16–7

A 0.2891-g sample of an antibiotic powder was dissolved in HCl and the solution diluted to 100.0 mL. A 20.00-mL aliquot was transferred to a flask, and followed by 25.00 mL of 0.01767 M $KBrO_3$. An excess of KBr was added to form $Br_2$, and the flask was stoppered. After 10 min, during which time the $Br_2$ brominated the sulfanilamide, an excess of KI was added. The liberated iodine was then titrated with 12.92 mL of 0.1215 M sodium thiosulfate. The reactions are

$$BrO_3^- + 5\ Br^- + 6\ H^+ \rightarrow 3\ Br_2 + 3\ H_2O$$

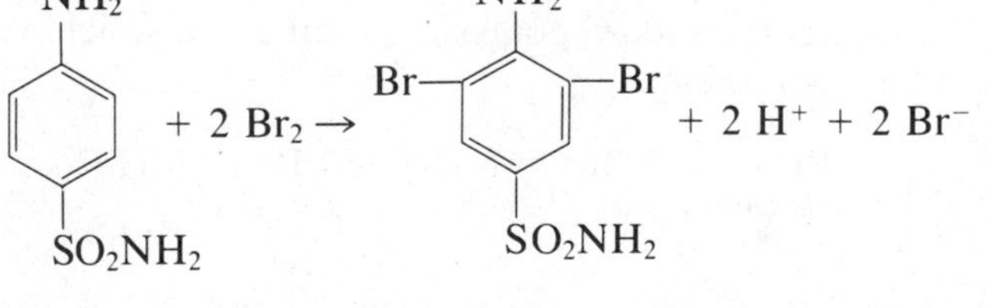

$$Br_2 + 2\ I^- \rightarrow 2\ Br^- + I_2 \quad (\text{excess KI})$$
$$I_2 + 2\ S_2O_3^{2-} \rightarrow S_4O_6^{2-} + 2\ I^-$$

Calculate the percent $NH_2C_6H_4SO_2NH_2$ (fw = 172.21 g) in the powder.

First we determine the amount of $Br_2$:

$$\text{amount } Br_2 = 25.00 \; \cancel{mL \; KBrO_3^-} \times 0.01767 \; \frac{mmol \; \cancel{KBrO_3^-}}{mL \; \cancel{KBrO_3^-}}$$

$$\times \frac{3 \; mmol \; Br_2}{\cancel{mmol \; KBrO_3^-}}$$

$$= 1.32525 \; mmol \; Br_2$$

We next calculate how much $Br_2$ was in excess over that required to brominate the analyte:

$$\text{amount excess } Br_2 = \text{amount } I_2$$

$$= 12.92 \; \cancel{mL \; Na_2S_2O_3} \times 0.1215 \; \frac{mmol \; \cancel{Na_2S_2O_3^-}}{mL \; \cancel{Na_2S_2O_3^-}}$$

$$\times \frac{1 \; mmol \; I_2}{2 \; \cancel{mmol \; Na_2S_2O_3^-}}$$

$$= 0.78489 \; mmol \; Br_2$$

The amount of $Br_2$ consumed by the sample is

$$\text{amount } Br_2 = 1.32525 - 0.78489 = 0.54036 \; mmol \; Br_2$$

$$\text{wt analyte} = 0.54036 \; \cancel{mmol \; Br_2} \times \frac{1 \; \cancel{mmol \; analyte}}{2 \; \cancel{mmol \; Br_2}}$$

$$\times 0.17221 \; \frac{g \; analyte}{\cancel{mmol \; analyte}}$$

$$= 0.046528 \; g \; analyte$$

$$\% \text{ analyte} = \frac{0.046528 \; g \; analyte}{0.2891 \; g \; sample \times 20.00 \; \cancel{mL}/100 \; \cancel{mL}} \times 100\%$$

$$= 80.47\% \; \text{sulfanilamide}$$

---

An important example of the use of a bromine substitution reaction is the determination of 8-hydroxyquinoline:

In contrast to most bromine substitutions, this reaction takes place rapidly enough in hydrochloric acid solution to make direct titration feasible. The titration of 8-hydroxyquinoline with bromine is particularly significant because the former is an excellent precipitating reagent for cations (Section 4E–3). For example, aluminum can be determined according to the sequence

$$Al^{3+} + 3\ HOC_9H_6N \xrightarrow{\text{pH 4-9}} Al(OC_9H_6N)_3(s) + 3\ H^+$$

$$Al(OC_9H_6N)_3(s) + 3\ H^+ \xrightarrow{\text{hot 4 M HCl}} 3\ HOC_9H_6N + Al^{3+}$$

$$3\ HOC_9H_6N + 6\ Br_2 \rightarrow 3\ HOC_9H_4NBr_2 + 6\ HBr$$

The stoichiometric relationships in this case are

$$1\ mol\ Al^{3+} \equiv 3\ mol\ HOC_9H_6N \equiv 6\ mol\ Br_2 \equiv 2\ mol\ KBrO_3$$

### Addition Reactions

Addition reactions involve the opening of an olefinic double bond. For example, 1 mol of ethylene reacts with 1 mol of bromine in the reaction

$$H-\overset{\overset{\displaystyle H}{|}}{C}=\overset{\overset{\displaystyle H}{|}}{C}-H + Br_2 \rightarrow H-\overset{\overset{\displaystyle H}{|}}{\underset{\underset{\displaystyle Br}{|}}{C}}-\overset{\overset{\displaystyle H}{|}}{\underset{\underset{\displaystyle Br}{|}}{C}}-H$$

The literature contains numerous references to the use of bromine for the estimation of olefinic unsaturation in fats, oils, and petroleum products.

A procedure for the determination of phenol in waste water by bromine substitution is found in Section 34H–3, and a direct-oxidation method for the determination of ascorbic acid in vitamin C tablets is given in Section 34H–4.

## 16D–2   Periodic Acid

Aqueous solutions of iodine in the +7 oxidation state are highly complex. In the presence of strong acid, paraperiodic acid, $H_5IO_6$, and its conjugate base predominate, although the metaperiodates $HIO_4$ and $IO_4^-$ are undoubtedly present as well. The reduction of periodic acid to iodate ion is best described by the half-reaction

$$H_5IO_6 + H^+ + 2\ e^- \rightleftharpoons IO_3^- + 3\ H_2O \qquad E^0 = 1.6\ V$$

### The Preparation and Properties of Periodic Acid Solutions

Several periodates are available for the preparation of standard solutions. Among these is paraperiodic acid itself, a crystalline, readily soluble, hygroscopic solid. An even more useful compound is sodium metaperiodate, $NaIO_4$, which is soluble in water to the extent of 0.06 M at 25°C. Sodium paraperiodate, $Na_5IO_6$, is not sufficiently soluble for the preparation of standard solutions; it is, however, readily converted to the more soluble metaperiodate by recrystallization from hot concentrated nitric acid. Potassium metaperiodate can be used as a primary standard for the preparation of periodate solutions. Although its solubility is only about 5 g/L at room temperature, it readily dissolves at elevated temperatures and can be subsequently converted to a more soluble form by the addition of base.

Periodate solutions vary considerably in stability, depending on their mode of preparation and storage. A solution prepared by dissolving sodium metaperiodate in water decomposes at the rate of several percent per week. By way of contrast, a solution of potassium metaperiodate in excess alkali was found to change no more than 0.3 to 0.4% in 100 days. The most stable periodate solutions appear to be those containing an excess of sulfuric acid; such solutions decrease in molarity by less than 0.1% in four months.

## The Standardization of Periodate Solutions

Periodate solutions are most conveniently standardized by buffering aliquots of the reagent with solid borax or sodium hydrogen carbonate to ensure that they remain slightly alkaline. An excess of potassium iodide is then introduced, which results in the formation of 1 mol of iodine for each mole of periodate:

$$H_4IO_6^- + 2\ I^- \longrightarrow IO_3^- + I_2 + 2\ OH^- + H_2O$$

As long as the solution is kept slightly alkaline, further reduction of iodate does not occur, and the liberated iodine can be titrated directly with a standard solution of sodium thiosulfate or sodium arsenite.

## Applications of Periodic Acid

The reason periodic acid is widely used is that it reacts remarkably selectively with organic compounds containing certain combinations of functional groups.[7] Ordinarily, these oxidations are performed at room temperature in the presence of a measured excess of the periodate; most reactions are complete in 0.5 to 1 hr. After oxidation, the excess periodate is determined by the method described for standardization. Alternatively, a reaction product such as ammonia, formaldehyde, or a carboxylic acid may be determined. In this application, the exact quantity of periodate used need not be known.

Periodate oxidations are usually carried out in aqueous solution, although solvents such as methanol, ethanol, or dioxane may be added to enhance the solubility of the sample.

## Compounds Attacked by Periodate

At room temperature, organic compounds containing aldehyde, ketone, or alcohol groups *on adjacent carbon atoms* are rapidly oxidized by peri-

---

[7]The use of periodic acid for this purpose was first reported by L. Malaprade, *Compt. rend.*, **1928,** *186,* 382. See also M. R. F. Ashworth, *Titrimetric Organic Analysis,* Part II, pp. 724–744. New York: Interscience, 1965.

odic acid. Primary and secondary $\alpha$-hydroxyamines are also readily attacked; $\alpha$-diamines are not. With few exceptions, other organic compounds do not react at a significant rate. Thus, compounds containing isolated aldehydes, ketone, alcohol, or amine groups are not affected by periodic acid; nor are compounds with a carboxylic acid group either isolated from or adjacent to any of the reactive groups. At elevated temperatures, the extraordinary selectivity of periodic acid tends to disappear.

Periodate oxidations of organic compounds follow a regular and predictable set of rules:

1. Attack of adjacent functional groups always results in rupture of the carbon-to-carbon bond between the groups.
2. A carbon atom bonded to a hydroxyl group is oxidized to an aldehyde or ketone.
3. A carbonyl group is converted to a carboxylic-acid group.
4. A carbon atom bonded to an amine group loses ammonia (or a substituted amine) and is itself converted to an aldehyde.

The following half-reactions illustrate these rules:

$$\underset{\text{Propylene Glycol}}{CH_3 - \overset{\overset{OH}{|}}{\underset{\underset{H}{|}}{C}} - \overset{\overset{OH}{|}}{\underset{\underset{H}{|}}{C}} - H} \rightarrow CH_3 - \overset{\overset{O}{\|}}{C} - H + H - \overset{\overset{O}{\|}}{C} - H + 2\ H^+ + 2\ e^-$$

$$\underset{\text{Glycerol}}{H - \overset{\overset{OH}{|}}{\underset{\underset{H}{|}}{C}} - \overset{\overset{OH}{|}}{\underset{\underset{H}{|}}{C}} - \overset{\overset{OH}{|}}{\underset{\underset{H}{|}}{C}} - H} + H_2O \rightarrow 2\ H - \overset{\overset{O}{\|}}{C} - H + H - \overset{\overset{O}{\|}}{C} - OH + 4\ H^+ + 4\ e^-$$

For predicting the reaction products of the oxidation of glycerol, the first step can be thought of as producing 1 mol of formaldehyde and 1 mol of an $\alpha$-hydroxyaldehyde (glycolic aldehyde). The latter is then further oxidized and by the second rule produces a second mole of formaldehyde plus 1 mol of formic acid.

Additional examples of the four rules include:

$$\underset{\text{Biacetyl}}{CH_3 - \overset{\overset{O}{\|}}{C} - \overset{\overset{O}{\|}}{C} - CH_3} + 2\ H_2O \rightarrow 2\ CH_3 - \overset{\overset{O}{\|}}{C} - OH + 2\ H^+ + 2\ e^-$$

$$\underset{\text{Acetoin}}{CH_3 - \overset{\overset{O}{\|}}{C} - \overset{\overset{OH}{|}}{C} - CH_3} + H_2O \rightarrow CH_3 - \overset{\overset{O}{\|}}{C} - OH + CH_3 - \overset{\overset{O}{\|}}{C} - H + H^+ + e^-$$

$$\underset{\text{Ethanolamine}}{\overset{\displaystyle \overset{\text{OH}}{|}\ \overset{\text{NH}_2}{|}}{\text{H}-\underset{\overset{|}{\text{H}}}{\text{C}}-\underset{\overset{|}{\text{H}}}{\text{C}}-\text{H}}} + \text{H}_2\text{O} \longrightarrow 2\ \text{H}-\overset{\overset{\displaystyle \text{O}}{\|}}{\text{C}}-\text{H} + \text{NH}_3 + 2\ \text{H}^+ + 2\ e^-$$

**The Determination of α-Hydroxyamines.** As mentioned earlier, the periodate oxidation of compounds having hydroxyl and amino groups on adjacent carbon atoms results in the formation of aldehydes and the liberation of ammonia (rule 4). The latter is readily distilled from the alkaline oxidation mixture and determined by a neutralization titration. This procedure is particularly useful in the analysis of mixtures of the various amino acids that occur in proteins. Because only serine, threonine, β-hydroxyglutamic acid, and hydroxylysine have the requisite structure for the liberation of ammonia, the method is selective for these species.

## 16D–3 Karl Fischer Reagent for Water Determination

A number of chemical methods for the determination of water in solids and organic solvents have been devised (Section 30C). Unquestionably, the most important of these involves the use of Karl Fischer reagent, which is relatively specific for water.[8]

### The Reaction and Stoichiometry

Karl Fischer reagent is composed of iodine, sulfur dioxide, pyridine, and methanol. This mixture reacts with water according to the equation

$$C_5H_5N \cdot I_2 + C_5H_5N \cdot SO_2 + C_5H_5N + H_2O \longrightarrow$$
$$2\ C_5H_5N \cdot HI + C_5H_5N \cdot SO_3 \quad (16\text{–}2)$$
$$C_5H_5N \cdot SO_3 + CH_3OH \longrightarrow C_5H_5N(H)SO_4CH_3 \quad (16\text{–}3)$$

Note that only the first step, which involves the oxidation of sulfur dioxide by iodine to give sulfur trioxide and hydrogen iodide, consumes water. In the presence of a large amount of pyridine, $C_5H_5N$, all reactants and products exist as complexes, as indicated in the equations. The second step, which occurs when an excess of methanol is present, is important to the success of the titration because the pyridine/sulfur trioxide complex is also capable of consuming water:

$$C_5H_5N \cdot SO_3 + H_2O \longrightarrow C_5H_5NHSO_4H \quad (16\text{–}4)$$

[8]For a monograph on this subject, see J. Mitchell and D. M. Smith, *Aquametry,* 2nd ed. New York: Interscience, 1977.

This last reaction is undesirable because it is not as specific for water as the reaction shown in Equation 16–2; it can be prevented completely by having a large excess of methanol present.

## Properties of the Reagent

Equation 16–2 indicates that the stoichiometry of the Karl Fischer titration involves the consumption of 1 mol of iodine, 1 mol of sulfur dioxide, and 3 mol of pyridine for each mole of water. In practice, excesses of both sulfur dioxide and pyridine are employed so that the reagent's combining capacity for water is determined by its iodine content. For typical applications, between 2 and 5 mg of water react with each milliliter of reagent; a twofold excess of sulfur dioxide and a threefold to fourfold excess of pyridine are provided.

Karl Fischer reagent decomposes on standing. Because decomposition is particularly rapid immediately after preparation, it is common practice to prepare the reagent a day or two before it is to be used. Ordinarily, its strength must be established at least daily against a standard solution of water in methanol. A proprietary commercial Karl Fischer reagent reported to require only occasional restandardization is now available.

It is obvious that great care must be exercised to keep atmospheric moisture from contaminating the Karl Fischer reagent and the sample. All glassware must be carefully dried before use, and the standard solution must be stored out of contact with air. It is also necessary to minimize contact between the atmosphere and the solution during the titration.

## End-Point Detection

The end point in a Karl Fischer titration is signaled by the appearance of the first excess of the pyridine/iodine complex when all water has been consumed. The color of the reagent is intense enough for a visual end point; the change is from the yellow of the reaction products to the brown of the excess reagent. With some practice, and in the absence of other colored materials, the end point can be established with a reasonable degree of certainty (that is, to perhaps ±0.2 mL).

Various electrometric end points are also employed for Karl Fischer titrations, the most widely used being the amperometric technique with twin microelectrodes discussed on page 490. Several instrument manufacturers offer fully automated instruments for performing such titrations.

## Applications

Karl Fischer reagent has been applied to the determination of water in numerous types of samples.[9] There are several variations of the basic technique, depending upon the solubility of the material, the state in which the water is retained, and the physical state of the sample. If the

---

[9]For a complete discussion of the applications of the reagent, see J. Mitchell and D. M. Smith, *Aquametry,* 2nd ed. New York: Interscience, 1977.

sample can be dissolved completely in methanol, a direct and rapid titration is usually feasible. This method has been applied to the determination of water in many organic acids, alcohols, esters, ethers, anhydrides, and halides. The hydrated salts of most organic acids, as well as the hydrates of a number of inorganic salts that are soluble in methanol, can also be determined by direct titration.

Direct titration of samples that are only partially dissolved in the reagent usually leads to incomplete recovery of the water. Satisfactory results with this type of sample are often obtained, however, by the addition of an excess of reagent and back-titration with a standard solution of water in methanol after a suitable reaction time. An effective alternative is to extract the water from the sample by refluxing with anhydrous methanol or other organic solvents. The resulting solution is then titrated directly with the Karl Fischer solution.

Difficulty is also encountered in the analysis of sorbed moisture and tightly bound hydrate water. For these, the extraction techniques just described are frequently effective.

Certain substances interfere with the Karl Fischer method. Among these are compounds that react with one of the components of the reagent to produce water. For example, carbonyl compounds combine with methanol to give acetals:

$$RCHO + 2\ CH_3OH \longrightarrow R-CH\underset{OCH_3}{\overset{OCH_3}{}} + H_2O$$

The result is a fading end point. Many metal oxides react with the hydrogen iodide formed in the titration to give water:

$$MO + 2\ HI \rightleftharpoons MI_2 + H_2O$$

Again, erroneous data result. Preliminary treatment of the sample can sometimes prevent these interferences.

Oxidizing or reducing substances frequently interfere with the Karl Fischer titration by reoxidizing the iodide produced or reducing the iodine in the reagent.

## 16E   QUESTIONS AND PROBLEMS

*16-1. Write a balanced net-ionic equation for

(a) the reduction of Mo(VI) in a Walden reductor (see Table 16-1).

(b) the reduction of $V(OH)_4^+$ in a Jones reductor.

(c) the oxidation of chromium(III) with peroxodisulfate.

(d) the reaction between permanganate and hydrazine in a faintly alkaline medium (products: $MnO_2(s)$, $N_2$).

(e) the reaction between potassium iodide and potassium iodate in dilute acid (product: $I_2$).

(f) the oxidation of phosphorous acid by triiodide ion.

(g) the oxidation of cyanide by permanganate in a faintly alkaline medium (products: $MnO_2$, $CNO^-$).

(h) the oxidation of iodide ion by permanganate in a solution containing cyanide (products: $Mn^{2+}$, ICN).

16–2. Write a balanced net-ionic equation for
(a) the reduction of Mo(VI) in a Jones reductor (see Table 16–1).
(b) the reduction of $V(OH)_4^+$ in a Walden reductor.
(c) the oxidation of manganese(II) with peroxodisulfate.
(d) the reaction between permanganate and sulfide in a faintly alkaline medium (products: $MnO_2(s)$, $SO_4^{2-}$).
(e) the reaction between potassium iodide and potassium iodate in strong hydrochloric acid (product: $ICl_2^-$).
(f) the oxidation of thiosulfate ion by triiodide ion.
(g) the oxidation of cyanide by permanganate in a strongly alkaline solution that also contains $Ba^{2+}$ (products: $BaMnO_4(s)$, $CNO^-$).
(h) the oxidation of iodide by permanganate in a strongly alkaline solution containing $Ba^{2+}$ (products: $BaMnO_4(s)$, $IO_4^-$).

*16–3. Write balanced net-ionic equations that demonstrate how potassium dichromate can serve as a primary standard for an iron(II) solution.

16–4. Write balanced net-ionic equations that demonstrate how potassium iodate can serve as a primary standard for a thiosulfate solution.

*16–5. Write balanced net-ionic equations that demonstrate how potassium bromate can serve indirectly as a primary standard for a sodium thiosulfate solution.

16–6. Write balanced net-ionic equations that demonstrate how potassium dichromate can serve indirectly as a primary standard for a sodium thiosulfate solution.

*16–7. What weight of zinc(II) is introduced to the sample during the reduction of 25.00 mL of 0.0500 M $V(OH)_4^+$ in a Jones reductor?

16–8. What weight of silver(I) is needed to reduce 25.00 mL of 0.0500 M $V(OH)_4^+$ in a Walden reductor?

*16–9. What weight of ammonium peroxodisulfate (fw = 228.2) will be needed to oxidize the chromium in 20.00 mL of 0.0320 M Cr(III) to dichromate and provide a 25% excess in addition?

16–10. What weight of $KIO_4$ (fw = 230.0) will be needed to oxidize the manganese in 20.00 mL of $8.00 \times 10^{-3}$ M $MnCl_2$ (fw = 125.8) and provide a 15% excess in addition? Reaction:

$$2\ Mn^{2+} + 5\ IO_4^- + 3\ H_2O \rightarrow$$
$$2\ MnO_4^- + 5\ IO_3^- + 6\ H^+$$

*16–11. Based on the half-reaction

$$Fe^{2+} \rightarrow Fe^{3+} + e^-$$

what weight of iron will react with 25.00 mL of a solution that is 0.01750 M in
(a) $V(OH)_4^+$ (product: $VO^{2+}$)?
(b) $KMnO_4$?
(c) $KIO_3$ (product: $ICl_2^-$)?
(d) $K_2Cr_2O_7$?

16–12. Based on the half-reaction

$$2\ I^- \rightarrow I_2 + 2\ e^-$$

calculate the volume of 0.0500 M KI that will react with 40.00 mL of a solution that is 0.01888 M in
(a) $K_2Cr_2O_7$.
(b) $KIO_3$ (product: $I_2$).
(c) $KIO_3$ (product: $ICl_2^-$).
(d) $K_4Fe(CN)_6$.

*16–13. Calculate the volume of 0.02441 M $KMnO_4$ that will react with 0.100 g of
(a) $K_4Fe(CN)_6$.
(b) $HNO_2$.
(c) $H_2O_2$.
(d) $V_2(SO_4)_3$ (product: $VO_2^+$).
The reactions take place in an acidic solution; see Table 16–5 for additional information.

16–14. Calculate the volume of 0.04667 M $I_2$ needed to react with 20.00 mL of 0.0888 M
(a) $H_2S$.
(b) $Sb_2O_3$.
(c) $H_3PO_3$.
(d) $N_2H_4$.
See Table 16–6 for additional information.

*16–15. What weight of $KMnO_4$ is needed to prepare about 1 liter of 0.020 M solution?

16–16. Describe how you would prepare a liter of approximately 0.050 M Ce(IV), starting with $Ce(HSO_4)_2$ (fw = 528).

*16–17. What weight of primary standard $Na_2C_2O_4$ should be taken to require a 35-mL titration with the $KMnO_4$ solution in Problem 15?

16–18. The Ce(IV) solution in Problem 16 is to be standardized against samples of electrolytic iron wire. What weight of iron will require a titration of about 40 mL?

*16–19. The calcium(II) in a 0.2437-g sample was precipitated as $CaC_2O_4$. The solid was filtered, washed free of excess oxalate, and then redissolved in dilute $H_2SO_4$. The liberated $H_2C_2O_4$ required a 31.44-mL titration with 0.02065 M $KMnO_4$. Express the results of this analysis in terms of percent CaO.

16–20. The iron in a 0.4022-g ore sample was converted to the +2 state in a Jones reductor and titrated with 37.98 mL of 0.01944 M $KMnO_4$. Express the results of this analysis as percent
(a) iron.
(b) magnetite, $Fe_3O_4$.

*16–21. A 7.50-g sample of an ant-control preparation was decomposed to eliminate organic matter. The arsenic in the residue was reduced to the +3 state and subsequently titrated with 28.68 mL of 0.01833 M $I_2$ in a slightly alkaline medium. Calculate the percentage of $As_2O_3$ in the original sample.

16–22. The antimony(III) in a 0.9974-g ore sample required a 38.22-mL titration with 0.04135 M $I_2$. Express the results of this titration in terms of percent
(a) antimony.
(b) stibnite, $Sb_2S_3$.

*16–23. A 25.00-mL sample of household bleach was diluted to 500 mL in a volumetric flask. An unmeasured excess of potassium iodide was added to a 20.00-mL aliquot of the diluted sample; the iodine liberated in the reaction

$$OCl^- + 2 I^- + 2 H^+ \rightarrow I_2 + Cl^- + H_2O$$

required 34.50 mL of 0.0409 M $Na_2S_2O_3$. Calculate the weight-volume percentage of NaOCl in the sample.

16–24. A 0.6465-g sample of bleaching powder was dissolved in dilute acid and treated with an unmeasured excess of KI (see Problem 23 for reaction). The liberated iodine required a 36.92-mL titration with 0.06608 M $Na_2S_2O_3$. Express the results of this analysis in terms of percent $Ca(OCl)_2$ (fw = 143.0).

*16–25. The chromium in a 0.4001-g sample of chromite $(FeO \cdot Cr_2O_3)$ was oxidized to the +6 state with peroxodisulfate. The excess peroxodisulfate was eliminated by boiling, following which the solution was cooled and treated with 50.00 mL of 0.1296 M $Fe^{2+}$. A 3.36-mL back-titration with 0.08771 M dichromate was needed to react with the excess iron(II). Calculate the results of this analysis in terms of percent
(a) Cr.
(b) $FeO \cdot Cr_2O_3$ (fw = 223.9).

16–26. Acetone in aqueous solution is readily determined by making the solution alkaline and adding an excess of standard iodine. The iodine is converted to hypoiodite, which reacts with the acetone to give iodoform. Net reaction:

$$CH_3COCH_3 + 3 IO^- \rightarrow$$
$$CHI_3(s) + CH_3COO^- + 2 OH^-$$

After the reaction is complete, the solution is acidified and the excess $I_2$ is titrated with standard thiosulfate. A 25.00-mL sample was treated with 50.00 mL of 0.0620 M $I_2$; back-titration of the excess iodine required 13.09 mL of 0.0860 M sodium thiosulfate. Calculate the milligrams of acetone (fw = 58.08) in each milliliter of sample.

*16–27. A 5.00-mL portion of table wine was diluted to 100.0 mL in a volumetric flask. The ethanol in a 20.00-mL aliquot was distilled into 100.00 mL of 0.05151 M $K_2Cr_2O_7$. Heating completed oxidation of the alcohol to acetic acid:

$$CH_3CH_2OH + 2 Cr_2O_7^{2-} + 16 H^+ \rightarrow$$
$$3 CH_3COOH + 4 Cr^{3+} + 11 H_2O$$

following which the excess dichromate was titrated with 14.42 mL of 0.02497 M Fe(II). Calculate the weight-volume percentage of ethanol (fw = 46.07) in the wine.

16–28. Dissolved oxygen in water will oxidize an alkaline suspension of $Mn(OH)_2$ to $Mn(OH)_3$:

$$O_2(g) + 4 Mn(OH)_2(s) + 2 H_2O \rightarrow 4 Mn(OH)_3(s)$$

Upon being acidified, the manganese(III) hydroxide produced will oxidize iodide to iodine, which, in turn, is titrated with standard thiosulfate:

$$2 Mn(OH)_3(s) + 2 I^- + 6 H^+ \rightarrow$$
$$2 Mn^{2+} + I_2 + 6 H_2O$$
$$I_2 + 2 S_2O_3^{2-} \rightarrow 2 I^- + S_4O_6^{2-}$$

The iodine produced by dissolved oxygen in a 100.0-mL sample required a titration with 18.73 mL of 0.01147 M thiosulfate. Calculate the milligrams of $O_2$ in each milliliter of sample.

*16–29. Survival of trout requires water in which the oxygen concentration is greater than 5 ppm. A 100-mL sample of lake water was analyzed according to the method outlined in Problem 28, with the liberated iodine requiring a 9.72-mL titration with 0.01235 M thiosulfate. Is the oxygen concentration sufficient to warrant the stocking of trout in this lake?

16–30. Federal regulations mandate that carbon monoxide levels cannot exceed 50 ppm in a working environment. Air in such a locale was passed over heated $I_2O_5$ at a rate of 4.21 L/min for a period of 6.50 min. Reaction:

$$I_2O_5(s) + 5 CO(g) \xrightarrow{150°C} I_2(g) + 5 CO_2(g)$$

The $I_2$ distilled at this temperature and was collected in a KI solution; titration of the $I_3^-$ produced required 10.34 mL of 0.00245 M $Na_2S_2O_3$. Determine whether the air is in compliance; use $1.20 \times 10^{-3}$ g/mL for the density of air.

*16–31. Air in the vicinity of a paper mill was analyzed for its sulfur dioxide content by drawing a quantity through 50.0 mL of 0.01081 M $Ce(SO_4)_2$ at the rate of 3.20 L/min. Reaction:

$$SO_2(g) + 2 Ce^{4+} + 2 H_2O \rightarrow$$
$$SO_4^{2-} + 2 Ce^{3+} + 4 H^+$$

Upon completion of a 75.00-min sampling period, the excess Ce(IV) was titrated with 13.95 mL of 0.03764 M Fe(II). Does the air meet the federal standard of 2 ppm (or less) $SO_2$? Use $1.20 \times 10^{-3}$ g/mL for the density of air.

16–32. The thioglycolic acid in a 17.31-g sample of a hair-setting preparation was titrated with 26.18 mL of 0.03950 M $I_2$. Reaction:

$$2 HS-CH_2COOH + I_2 \rightarrow$$
$$HOOCCH_2-S-S-CH_2COOH + 2 H^+ + 2 I^-$$

Calculate the percentage of thioglycolic acid (fw = 92.1) in the sample.

16–33. *p*-Hydroxyacetanilide is the active ingredient in a headache remedy. A five-tablet sample was dissolved and diluted to 500.0 mL in a volumetric flask. Treatment of 50.00-mL aliquots of this solution with an identical volume of 0.01750 M $KBrO_3$, excess KBr, and acidification caused replacement of two hydrogens with bromine

$$C_8H_9NO_2 + 2\ Br_2 \rightarrow C_8H_7Br_2NO_2 + 2\ HBr$$

Potassium iodide was added, following which the liberated iodine required an average titration of 14.77 mL with 0.06521 M sodium thiosulfate. Calculate the average weight (mg) of *p*-hydroxyacetanilide (fw = 151.6) in each of these tablets.

16–34. The magnesium in a 0.2515-g milk of magnesia tablet was precipitated as $Mg(C_9H_6NO)_2$ with 8-hydroxyquinoline. The solid was filtered, washed free of excess precipitant, and redissolved in acid. Addition of 50.00 mL of 0.07994 M $KBrO_3$ and an unmeasured excess of KBr resulted in bromination of the liberated 8-hydroxyquinoline. Reaction between unused bromate and potassium iodide generated iodine that required a 5.61-mL titration with 0.05005 M $Na_2S_2O_3$. Calculate the weight (in mg) of $MgCO_3$ in the tablet.

*16–35. The $I_2$ in 25.00 mL of an $I_2$/KI solution required 21.89 mL of 0.05268 M $Na_2S_2O_3$. Titration of both the $I_2$ and $I^-$ in 36.18 mL of sample required 30.08 mL of 0.04792 M $KIO_3$ (product: $ICl_2^-$). Calculate the weight-volume percentages of $I_2$ and KI in the sample.

16–36. A square of photographic film 2.0 cm on a side was suspended in a 5% solution of $Na_2S_2O_3$, which caused the silver halides to dissolve. After the film had been removed and washed, the solution and washings were treated with $Br_2$ to oxidize the iodide to iodate and the thiosulfate to sulfate. The solution was then boiled to eliminate the excess bromine, following which an unmeasured excess of KI was introduced. The liberated iodine required 13.70 mL of 0.03520 M thiosulfate.

(a) Write balanced net-ionic equations for the reactions involved in this method.
(b) Calculate the milligrams of AgI on the surface of each square centimeter of film.

*16–37. A 0.5349-g sample of an iron-chromium alloy was dissolved in acid, following which the solution was diluted to 250.0 mL in a volumetric flask. A 50.00-mL aliquot required 34.73 mL of 0.02759 M $KMnO_4$ following passage through a Walden reductor (see Table 16–1). A second 50.00-mL aliquot was passed through a Jones reductor and collected in an unmeasured excess of iron(III) to prevent air oxidation of the chromium(II):

$$Cr^{2+} + Fe^{3+} \rightarrow Cr^{3+} + Fe^{2+}$$

This titration required 44.81 mL of the permanganate solution.
(a) Write balanced net-ionic equations for the reactions involved in this method.
(b) Calculate the percentages of iron and chromium in the alloy.

16–38. A 0.6712-g sample of ferrovanadium was dissolved in acid, and the resulting solution was diluted to 250.0 mL in a volumetric flask. One 50.00-mL aliquot was passed through a Walden reductor (see Table 16–1) and subsequently titrated with 26.10 mL of 0.01909 M $KMnO_4$. A second aliquot was passed through a Jones reductor and collected in an unmeasured excess of iron(III) to avoid air oxidation of vanadium(II):

$$V^{2+} + 2\ Fe^{3+} + H_2O \rightarrow$$
$$VO^{2+} + 2\ Fe^{2+} + 2\ H^+$$

Titration of this solution used 47.09 mL of the permanganate solution.
(a) Write balanced net-ionic equations for the reactions involved in this method.
(b) Calculate the percentages of iron and vanadium in the sample.

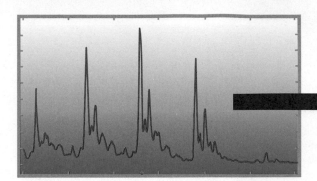

## CHAPTER 17

# POTENTIOMETRIC METHODS

$I$n Chapter 14, we showed that the potential of an electrode relative to the standard hydrogen electrode is determined by the concentration of one or more of the species in the solution in which the electrode is immersed. The present chapter deals with the measurement of electrode potentials and how these data are used to determine the concentration of analytes.[1] Analytical methods based upon potential measurements are termed *potentiometric methods*.

## 17A GENERAL PRINCIPLES

In Feature 14–3, we showed that absolute values for individual half-cell potentials cannot be determined in the laboratory. That is, only *cell* potentials can be obtained experimentally. Figure 17–1 shows a typical cell for a potentiometric analysis. This cell can be depicted as

$$\underbrace{\text{reference electrode}}_{E_{ref}}|\underbrace{\text{salt bridge}}_{E_j}|\text{analyte solution}|\underbrace{\text{indicator electrode}}_{E_{ind}}$$

The *reference electrode* in this diagram is a half-cell with an accurately known electrode potential $E_{ref}$ that is independent of the analyte concentration or of any other ions in the solution under study. By convention, the reference electrode is always treated as the anode in potentiometric measurements.

The *indicator electrode*, which is immersed in the analyte solution, develops a potential $E_{ind}$ that depends upon analyte activity. Most indicator electrodes used in potentiometry are highly selective in their responses.

The third component of a potentiometric cell is a salt bridge, which prevents the components of the analyte solution from mixing with those of the reference electrode. As noted in Chapter 14, a potential $E_j$ develops across the two liquid junctions making up the salt bridge.

Reference electrodes are *always* treated as anodes in this text.

---

[1]For further reading on potentiometric methods, see E. P. Serjeant, *Potentiometry and Potentiometric Titrations*. New York: Wiley, 1984.

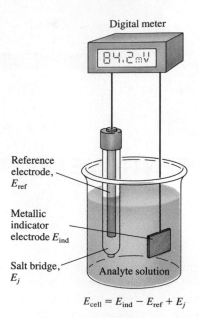

Digital meter

84.2 mV

Reference
electrode,
$E_{ref}$

Metallic
indicator
electrode $E_{ind}$

Salt bridge,
$E_j$

Analyte solution

$$E_{cell} = E_{ind} - E_{ref} + E_j$$

Figure 17–1
A cell for potentiometric analysis.

The potential of the cell we have just considered is given by the equation

$$E_{cell} = E_{ind} - E_{ref} + E_j \qquad (17\text{–}1)$$

The first term on the right contains the information we seek about the analyte concentration. A potentiometric analysis, then, involves measuring a cell potential, correcting this potential for the reference and junction potentials, and computing the analyte concentration from the indicator electrode potential.

In the sections that follow, we shall consider the sources of the three potentials that appear on the right side of Equation 17–1.

## 17B   REFERENCE ELECTRODES

The ideal reference electrode has a potential that is accurately known, constant, and completely insensitive to the composition of the analyte solution. In addition, a reference electrode should be rugged and easy to assemble and should maintain a constant potential while passing small currents.

The standard hydrogen electrode is the universal reference electrode for reporting relative half-cell potentials. The stream of gas needed for the operation of a hydrogen electrode is somewhat hazardous, and preparation and maintenance of the platinized surface are troublesome. As a consequence, more convenient secondary reference electrodes are often substituted for the standard hydrogen electrode. The potentials of these secondary reference electrodes relative to the standard hydrogen elec-

A hydrogen electrode is seldom used as a reference electrode for day-to-day potentiometric measurements because it is somewhat inconvenient and is also a fire hazard.

trode have been carefully determined so that data collected with such electrodes can be converted to a standard hydrogen electrode basis.

## 17B–1 Calomel Electrodes

A calomel electrode can be represented schematically as

$$Hg|Hg_2Cl_2(\text{sat'd}),KCl(x \text{ M})|$$

where $x$ represents the molar concentration of potassium chloride in the solution. Three concentrations of potassium chloride are common: 0.1 M, 1 M, and saturated (about 4.6 M). The saturated calomel electrode (SCE) is the most widely used because it is so easily prepared. Its main disadvantage is its somewhat large temperature coefficient, which is important only in those rare circumstances where substantial temperature changes occur during a measurement. The *electrode* potential of the saturated calomel electrode is 0.2444 V at 25°C.

The electrode reaction in calomel half-cells is

$$Hg_2Cl_2(s) + 2 e^- \rightleftarrows 2 Hg(l) + 2 Cl^-$$

Table 17–1 lists the composition and electrode potential for the three most common calomel electrodes. Note that these half-cells differ only in their potassium chloride concentrations; all are saturated with mercury(I) chloride.

The saturated calomel electrode shown in Figure 17–2 is a typical commercial electrode. It consists of an outer tube that is 5 to 15 cm in length and 0.5 to 1.0 cm in diameter. A mercury/mercury(I) chloride paste in saturated potassium chloride is contained in an inner tube and connected to the saturated potassium chloride solution in the outer tube through a small opening. Contact with the analyte solution is made through a fritted

The "saturated" in a saturated calomel electrode refers to the KCl concentration. All calomel electrodes are saturated with $Hg_2Cl_2$ (calomel).

Figure 17–2

Diagram of a typical commercial saturated calomel electrode.

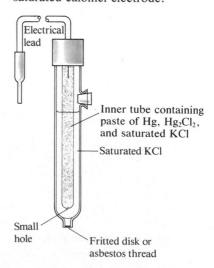

### Table 17–1
### ELECTRODE POTENTIALS FOR REFERENCE ELECTRODES AS A FUNCTION OF COMPOSITION AND TEMPERATURE

| Temperature, °C | Potential (vs. SHE), V | | | | |
|---|---|---|---|---|---|
| | 0.1 M Calomel* | 3.5 M Calomel† | Sat'd Calomel* | 3.5 M Ag/AgCl† | Sat'd Ag/AgCl† |
| 12 | 0.3362 | | 0.2528 | | |
| 15 | 0.3362 | 0.254 | 0.2511 | 0.212 | 0.209 |
| 20 | 0.3359 | 0.252 | 0.2479 | 0.208 | 0.204 |
| 25 | 0.3356 | 0.250 | 0.2444 | 0.205 | 0.199 |
| 30 | 0.3351 | 0.248 | 0.2411 | 0.201 | 0.194 |
| 35 | 0.3344 | 0.246 | 0.2376 | 0.197 | 0.189 |

*From R. G. Bates, in *Treatise on Analytical Chemistry*, 2nd ed., I. M. Kolthoff and P. J. Elving, Eds., Part I, Vol. 1, p. 793. New York: Wiley, 1978.

†From D. T. Sawyer and J. L. Roberts Jr., *Experimental Electrochemistry for Chemists*, p. 42. New York: Wiley, 1974.

disk, a porous fiber, or a piece of porous Vycor ("thirsty glass") sealed in the end of the outer tube.

Figure 17–3 shows a saturated calomel electrode you can construct from materials available in your laboratory. A salt bridge (Section 14A–2) provides electrical contact with the analyte solution.

### 17B–2  Silver/Silver Chloride Electrodes

A system analogous to the saturated calomel electrode consists of a silver electrode immersed in a solution that is saturated in both potassium chloride and silver chloride:

$$\text{Ag} \mid \text{AgCl(sat'd), KCl(sat'd)} \mid$$

The half-reaction is

$$\text{AgCl(s)} + \text{e}^- \rightleftharpoons \text{Ag(s)} + \text{Cl}^-$$

The potential of this electrode is 0.199 V at 25°C.

Silver/silver chloride electrodes of various sizes, shapes, and chloride concentration are on the market (Table 17–1). A simple and easily constructed electrode of this type is shown in Figure 17–4.

### 17C  LIQUID-JUNCTION POTENTIALS

A liquid-junction potential develops across the boundary between two electrolyte solutions that have different compositions. Figure 17–5 shows a very simple liquid junction consisting of a 1 M hydrochloric acid solution in contact with a solution that is 0.01 M in that acid. An inert porous barrier, such as a fritted glass plate, prevents the two solutions from mixing. Both hydrogen ions and chloride ions tend to diffuse across this

A salt bridge is readily constructed by filling a U-tube with a conducting gel prepared by heating about 5 g of agar in 100 mL of water containing about 35 g of potassium chloride. When the liquid cools, it sets up into a gel that is a good electrical conductor.

At 25°C, the potential of the saturated calomel electrode versus the standard hydrogen electrode is 0.244 V; for the saturated silver/silver chloride electrode, this potential is 0.199 V.

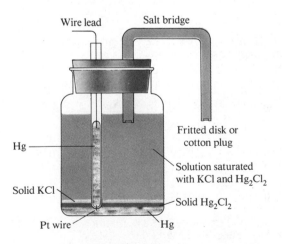

Figure 17–3
Diagram of an easily constructed saturated calomel electrode.

Wire lead   Salt bridge

Hg

Solid KCl

Pt wire

Fritted disk or cotton plug

Solution saturated with KCl and $Hg_2Cl_2$

Solid $Hg_2Cl_2$

Hg

Half-reaction:
$$\text{Hg}_2\text{Cl}_2(\text{s}) + 2\,\text{e}^- \rightleftharpoons 2\,\text{Hg} + 2\,\text{Cl}^-$$

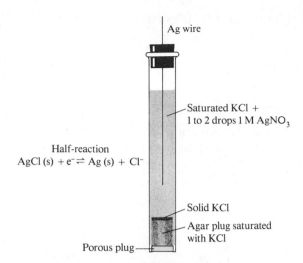

Ag wire

Saturated KCl +
1 to 2 drops 1 M AgNO$_3$

Half-reaction
AgCl (s) + e$^-$ ⇌ Ag (s) + Cl$^-$

Solid KCl

Agar plug saturated
with KCl

Porous plug

**Figure 17–4**
Diagram of a silver/silver chloride electrode.

boundary from the more concentrated to the more dilute solution. The driving force for each ion is proportional to the concentration difference between the two solutions. Hydrogen ions are substantially more mobile than chloride ions. Thus, in this example hydrogen ions diffuse more rapidly than chloride ions, and, as shown in Figure 17–5, a separation of charge results. The more dilute side of the boundary becomes positively charged because of the more rapid diffusion of hydrogen ions. The concentrated side therefore acquires a negative charge from the excess of slower-moving chloride ions. The charge developed tends to counteract the differences in diffusion rates of the two ions so that a condition of equilibrium is attained rapidly. The potential difference resulting from this charge separation may be several hundredths of a volt.

The magnitude of the liquid-junction potential can be minimized by placing a salt bridge between the two solutions. The salt bridge is most effective if the mobilities of the negative and positive ions in the device are nearly equal to each other and if their concentrations are large. A saturated solution of potassium chloride is good from both standpoints. The net junction potential with such a bridge is typically a few millivolts.

All cells used for potentiometric analyses contain a salt bridge that connects the reference electrode to the analyte solution. As we shall see, uncertainties in the magnitude of the junction potential across this bridge place a fundamental limit on the accuracy of potentiometric methods of analysis.

The net junction potential across a typical salt bridge is a few millivolts.

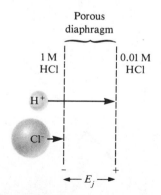

Porous
diaphragm

1 M
HCl

0.01 M
HCl

H$^+$

Cl$^-$

$E_j$

**Figure 17–5**
Schematic representation of a liquid junction, showing the source of the junction potential $E_j$. The length of the arrows corresponds to the relative mobility of the two ions.

## 17D INDICATOR ELECTRODES

An ideal indicator electrode responds rapidly and reproducibly to changes in the concentration of an analyte ion (or group of ions). Although no indicator electrode is absolutely specific in its response, a few are now available that are remarkably selective. There are two types of indicator electrodes: metallic and membrane.

A plot of Equation 17–3 for an electrode of the first kind.

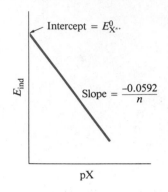

## 17D–1 Metallic Indicator Electrodes

It is convenient to classify metallic indicator electrodes as *electrodes of the first kind, electrodes of the second kind,* and inert *redox electrodes.*

### Electrodes of the First Kind

An electrode of the first kind is a piece of pure metal that is in direct equilibrium with the cation of the metal. A single reaction is involved. For example, the equilibrium between a metal X and its cation $X^{n+}$ is

$$X^{n+} + ne^- \rightarrow X(s)$$

for which

$$E_{\text{ind}} = E^0_{X^{n+}} - \frac{0.0592}{n} \log \frac{1}{a_{X^{n+}}} = E^0_{X^{n+}} + \frac{0.0592}{n} \log a_{X^{n+}} \quad (17\text{–}2)$$

where $E_{\text{ind}}$ is the electrode potential of the metal electrode and $a_{X^{n+}}$ is the activity of the ion (or approximately its molar concentration, $[X^{n+}]$).

We often express the electrode potential of the indicator electrode in terms of the p-function of the cation. Thus, substituting the definition of pX into Equation 17–2 gives

$$E_{\text{ind}} = E^0_{X^{n+}} - \frac{0.0592}{n} \text{pX} \quad (17\text{–}3)$$

Equation 17–3 accurately describes the behavior of a number of common metals that are used as indicator electrodes of the first kind. In contrast, certain harder metals—notably iron, chromium, tungsten, cobalt, and nickel—do not provide reproducible potentials. Moreover, plots of the electrode potential of a metal of this latter kind as a function of pX often yield slopes that differ significantly from the theoretical $(-0.0592/n)$. The nonideal behavior of this type of electrode can be attributed to strains and deformations in the crystal structure of the metal or to the presence of an oxide film on the surface.

### Electrodes of the Second Kind

Metals not only serve as indicator electrodes for their own cations but also respond to the concentration of anions that form sparingly soluble precipitates or stable complexes with such cations. The potential of a silver electrode, for example, correlates reproducibly with the concentration of chloride ion in a solution saturated with silver chloride. Here, the electrode reaction can be written as

$$AgCl(s) + e^- \rightleftharpoons Ag(s) + Cl^- \qquad E^0_{AgCl} = 0.222 \text{ V}$$

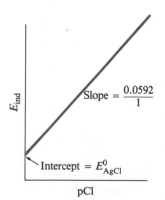

A plot of Equation 17–4 for an electrode of the second kind for $Cl^-$

The Nernst expression for this process is

$$E_{ind} = E^0_{AgCl} - 0.0592 \log [Cl^-] = 0.222 + 0.0592 \, pCl \quad (17\text{--}4)$$

Equation 17–4 shows that the potential of a silver electrode is proportional to pCl, the negative logarithm of the chloride ion concentration. Thus, in a solution saturated with silver chloride, a silver electrode can serve as an indicator electrode of the second kind for chloride ion. Note that the sign of the logarithmic term for an electrode of this type is opposite that for an electrode of the first kind (see Equation 17–3).

Mercury serves as an indicator electrode of the second kind for the EDTA anion $Y^{4-}$. For example, when a small amount of $HgY^{2-}$ is added to a solution containing $Y^{4-}$, the half-reaction at a mercury cathode is

$$HgY^{2-} + 2 \, e^- \rightleftharpoons Hg(l) + Y^{4-} \qquad E^0 = 0.21 \text{ V}$$

for which

$$E_{ind} = 0.21 - \frac{0.0592}{2} \log \frac{[Y^{4-}]}{[HgY^{2-}]}$$

The formation constant for $HgY^{2-}$ is very large ($6.3 \times 10^{21}$), and so the concentration of the complex remains essentially constant over a large range of $Y^{4-}$ concentrations. The Nernst equation for the process can therefore be written as

$$E = K - \frac{0.0592}{2} \log [Y^{4-}] = K + \frac{0.0592}{2} \, pY \qquad (17\text{--}5)$$

where

$$K = 0.21 - \frac{0.0592}{2} \log \frac{1}{[HgY^{2-}]}$$

The mercury electrode is thus a valuable electrode of the second kind for EDTA titrations.

### Inert Metallic Electrodes for Redox Systems

As noted in Chapter 14, an inert metal—such as platinum, gold, palladium, or carbon—responds to the potential of redox systems with which it is in contact. For example, the potential of a platinum electrode immersed in a solution containing cerium(III) and cerium(IV) is

$$E_{ind} = E^0_{Ce(IV)} - 0.0592 \log \frac{[Ce^{3+}]}{[Ce^{4+}]}$$

A platinum electrode is thus a convenient indicator electrode for titrations involving standard cerium(IV) solutions.

The first practical glass electrode of Haber and Klemensiewicz, *Z. Phys. Chem.,* **1909,** *65,* 385.

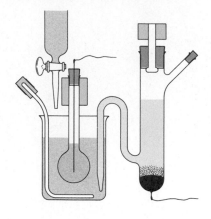

Metal electrodes function by the transfer of electrons across the interface between electrode and solution. Membrane electrodes function by the transfer of ions from one side of the membrane to the other.

## 17D-2  Membrane Electrodes[2]

For many years, the most convenient method for determining pH has involved measurement of the potential that develops across a thin glass membrane that separates two solutions with different hydrogen ion concentrations. The phenomenon upon which the measurement is based was first reported in 1906 and by now has been extensively studied by many investigators. As a result, the sensitivity and selectivity of glass membranes toward hydrogen ions are reasonably well understood. Furthermore, this understanding has led to the development of other types of membranes that respond selectively to more than two dozen other ions.

Membrane electrodes are sometimes called *p-ion electrodes* because the data obtained from them are usually presented as p-functions, such as pH, pCa, or $pNO_3$. The membranes used in constructing these electrodes are classified as *crystalline* and *noncrystalline*. Noncrystalline membranes can be further subdivided into *glass, liquid,* and *immobilized liquid*. In this section, we consider all of these p-ion membranes. In addition, a gas-sensing probe based on a noncrystalline membrane is described.

It is important to note at the outset of this discussion that membrane electrodes are *fundamentally different* from metal electrodes both in design and in principle. We shall use the glass electrode for pH measurements to illustrate these differences.

## 17D-3  The Glass Electrode for pH Measurements

Figure 17–6 shows a typical *cell* for measuring pH. The cell consists of a glass indicator electrode and a saturated calomel reference electrode, both immersed in the solution whose pH is to be determined. The indicator electrode consists of a thin, pH-sensitive glass membrane sealed onto one end of a heavy-walled glass or plastic tube. A small volume of dilute hydrochloric acid saturated with silver chloride is contained in the tube (in some electrodes this solution is a buffer containing chloride ion). A silver wire in this solution forms a silver/silver chloride reference electrode, which is connected to one of the terminals of a potential-measuring device. The calomel electrode is connected to the other terminal.

Figure 17–6 and the schematic representation of this cell in Figure 17–7 show that the system contains *two* reference electrodes: (1) the *external* calomel electrode and (2) the *internal* silver/silver chloride electrode. Although the internal reference electrode is part of the glass electrode, *it is not the pH-sensing element.* Instead, *it is the thin glass membrane at the tip of the electrode that responds to pH.*

---

[2]Some suggested sources for additional information on this topic are *Ion-Selective Electrodes in Analytical Chemistry,* H. Freiser, Ed. New York: Plenum Press, 1978, 1980; A. Evans, *Potentiometry and Ion-Selective Electrodes.* New York: Wiley, 1987; J. Koryta and K. Stulik, *Ion-Selective Electrodes,* 2nd ed. Cambridge: Cambridge University Press, 1983. *Ion-Selective Methodology,* A. K. Covington, Ed. Boca Raton, FL: CRC Press, 1979; R. P. Buck, in *Comprehensive Treatise of Electrochemistry,* J. D. M. Bockris, B. C. Conway, H. E. Yeager, Eds., Vol. 8, Chapter 3. New York: Plenum Press, 1984.

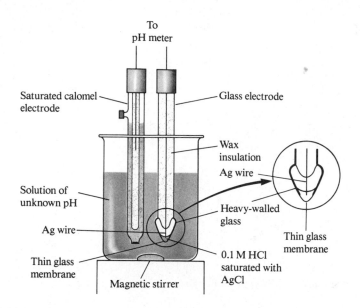

**Figure 17–6**

Typical electrode system for measuring pH.

## The Composition of Glass Membranes

Much systematic investigation has been devoted to how glass composition affects the sensitivity of membranes to protons and other cations, and a number of formulations are now used for the manufacture of electrodes. Corning 015 glass, which has been widely used for membranes, consists of approximately 22% $Na_2O$, 6% $CaO$, and 72% $SiO_2$. This membrane shows excellent specificity toward hydrogen ions up to a pH of about 9. At higher pH values, however, the glass becomes somewhat responsive to sodium as well as to other singly charged cations. Other glass formulations are now in use in which sodium and calcium ions are replaced to various degrees by lithium and barium ions. These membranes have superior selectivity and lifetime.

## The Structure of Membrane Glasses

As shown in Figure 17–8, a silicate glass used for membranes consists of an infinite three-dimensional network of $SiO_4^{4-}$ groups in which each silicon is bonded to four oxygens and each oxygen is shared by two silicons. Within the interstices of this structure are sufficient cations to balance the negative charge of the silicate groups. Singly charged cations, such as

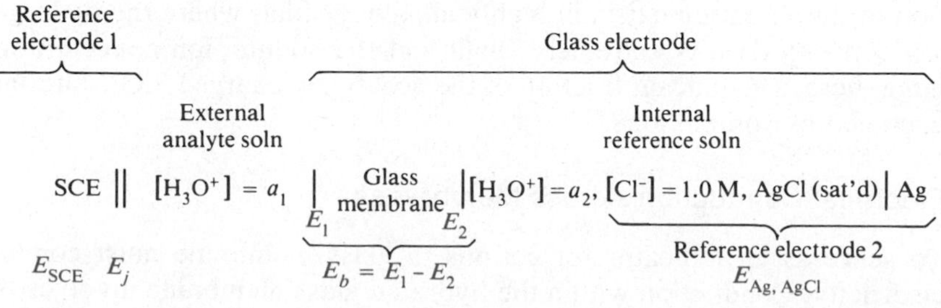

Reference
electrode 1

Glass electrode

External
analyte soln

Internal
reference soln

$$\text{SCE} \; \| \; [H_3O^+] = a_1 \; \left| \; \underset{E_1}{\underset{\text{membrane}}{\text{Glass}}} \; \right| \underset{E_2}{[H_3O^+] = a_2}, [Cl^-] = 1.0 \text{ M, AgCl (sat'd)} \; \bigg| \; Ag$$

$E_{SCE} \quad E_j$

Reference electrode 2

$E_b = E_1 - E_2$

$E_{Ag, AgCl}$

**Figure 17–7**

Diagram of a glass/calomel cell for measurement of pH.

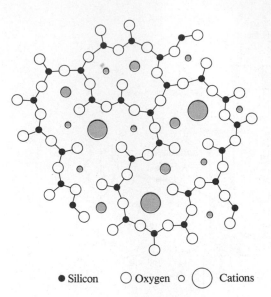

Figure 17–8

Cross-sectional view of a silicate glass structure. In addition to the three Si—O bonds shown, each silicon is bonded to an additional oxygen atom, either above or below the plane of the paper. (Adapted with permission from G. A. Perley, *Anal. Chem.*, **1949,** *21,* 395. Copyright 1949 American Chemical Society.)

● Silicon  ○ Oxygen  ○ ◯ Cations

Glasses that absorb water are said to be hygroscopic.

sodium and lithium, are mobile in the lattice and are responsible for electrical conduction within the membrane.

## The Hygroscopicity of Glass Membranes

The surface of a glass membrane must be hydrated before it will function as a pH electrode. The amount of water involved is approximately 50 mg per cubic centimeter of glass. Nonhygroscopic glasses show no pH function. Even hygroscopic glasses lose their pH sensitivity after dehydration by storage over a desiccant. The effect is reversible, however, and response can be restored by soaking the membrane in water.

The hydration of a pH-sensitive glass membrane involves an ion-exchange reaction between singly charged cations in the glass lattice and protons from the solution in which the electrode is immersed. The process involves univalent cations exclusively because di- and trivalent cations are too strongly held within the silicate structure to exchange with ions in the solution. Typically, then, the ion-exchange reaction can be written as

$$\underset{\text{Soln}}{H^+} + \underset{\text{Glass}}{Na^+Gl^-} \rightleftarrows \underset{\text{Soln}}{Na^+} + \underset{\text{Glass}}{H^+Gl^-} \qquad (17\text{–}6)$$

This process is shown in the schematic of Figure 17–9. The equilibrium constant for this process is so large that the surface of a hydrated glass membrane ordinarily consists entirely of silicic acid ($H^+Gl^-$). An exception to this situation exists in highly alkaline media, where the hydrogen ion concentration is extremely small and the sodium ion concentration large; here, a significant fraction of the negatively charged $SiO_4^{4-}$ sites are occupied by sodium ions.

## Electrical Conduction Across Membranes

To serve as an indicator for cations, a glass membrane must conduct electricity. Conduction within the hydrated glass membrane involves the

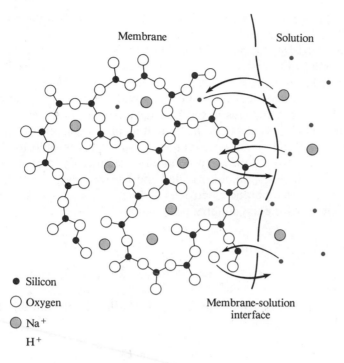

- ● Silicon
- ○ Oxygen
- ⬤ Na$^+$
-    H$^+$

**Figure 17-9**
Ion exchange at the membrane/solution interface. Protons exchange with sodium ions in the silicate structure of the glass membrane.

movement of sodium and hydrogen ions. Sodium ions are the charge carriers in the dry interior of the membrane, and the protons are mobile in the gel layer. Conduction across the solution/gel interfaces occurs by the reactions

$$H^+ + Gl^- \rightleftharpoons H^+Gl^- \qquad (17\text{-}7)$$
$$\text{Soln}_1 \quad \text{Glass}_1 \qquad \text{Glass}_1$$

$$H^+Gl^- \rightleftharpoons H^+ + Gl^- \qquad (17\text{-}8)$$
$$\text{Glass}_2 \qquad \text{Soln}_2 \quad \text{Glass}_2$$

where subscript 1 refers to the interface between glass and analyte solution and subscript 2 refers to the interface between internal solution and glass. The positions of these two equilibria are determined by the hydrogen ion concentrations in the solutions on the two sides of the membrane. Where these positions differ from each other, the surface at which the greater dissociation has occurred is negative with respect to the other surface. A boundary potential $E_b$ thus develops across the membrane. The magnitude of the boundary potential depends upon the ratio of the hydrogen ion concentrations of the two solutions. It is this potential difference that serves as the analytical parameter in a potentiometric pH measurement.

Membrane electrodes function by the exchange of ions at the surfaces of the membrane.

The membrane of a typical glass electrode (with a thickness of 0.03 to 0.1 mm) has an electrical resistance of 50 to 500 megaohms.

## Membrane Potentials

The lower part of Figure 17-7 shows four potentials that develop in a cell when pH is determined with a glass electrode. Two of these, $E_{\text{Ag/AgCl}}$ and $E_{\text{SCE}}$, are reference electrode potentials. There is a third potential across the salt bridge that separates the calomel electrode from the analyte solution. This interface and its associated *junction potential, $E_j$*, are found in all cells used for the potentiometric measurement of ion concentration. The fourth, and most important, potential shown in Figure 17–7 is the *boundary potential, $E_b$, which varies with the pH of the analyte solution.* The two reference electrodes simply provide electrical contacts with the solutions so that changes in the boundary potential can be measured.

Figure 17–7 reveals that the potential of a glass electrode has two components: the fixed potential of a silver/silver chloride electrode $E_{Ag/AgCl}$ and the pH-dependent boundary potential $E_b$. Not shown in Figure 17–7 is a fifth potential, called the *asymmetry potential,* which is found in most membrane electrodes and which changes slowly with time. The sources of the asymmetry potential are obscure.

### The Boundary Potential

As shown in Figure 17–7, the boundary potential consists of two potentials, $E_1$ and $E_2$, each of which is associated with one of the two gel/solution interfaces. The boundary potential is simply the difference between these potentials:

$$E_b = E_1 - E_2 \tag{17-9}$$

The significance of these potentials is shown in the potential profiles of Figure 17–10. The profiles are plotted across the membrane from the analyte solution on the left through the glass membrane to the internal reference solution on the right. The potential $E_1$ is determined by the ratio of the hydrogen ion activity $a_1$ in the analyte solution to the hydrogen ion activity in the gel layer and can be considered a measure of the driving force for the reaction shown in Equation 17–7. Similarly, $E_2$ is related to the ratio of the hydrogen ion activities in the internal reference solution and in the corresponding gel layer and is related to the driving force for the reaction shown in Equation 17–8.

The relationship between the boundary potential and the two hydrogen ion activities is

$$E_b = E_1 - E_2 = 0.0592 \log \frac{a_1}{a_2} \tag{17-10}$$

As shown in Figure 17–10a, if the hydrogen ion activity of the analyte is ten times the activity in the reference solution,

$$E_b = 0.0592 \log \frac{10 a_2}{a_2} = 0.0592 \text{ V}$$

If the two activities are equal, then $E_b = 0$, as illustrated in Figure 17–10b. Finally, Figure 17–10c shows that when $10 a_1 = a_2$, $E_b = -0.0592$ V.

To summarize, the boundary potential depends only upon the hydrogen ion activities of the solutions on either side of the membrane. For a glass

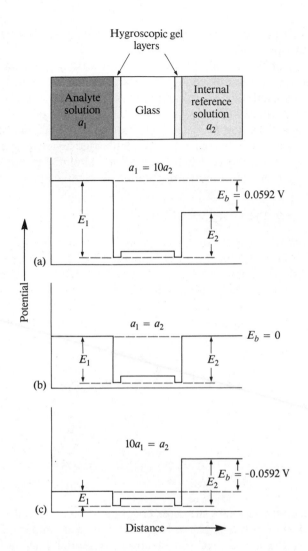

Figure 17–10
Potential profile across a glass membrane from the analyte solution to the internal reference solution. The reference electrode potentials are not shown.

pH electrode, the hydrogen ion activity of the internal solution $a_2$ is held constant, so that Equation 17–10 simplifies to

$$E_b = L' + 0.0592 \log a_1 = L' - 0.0592 \, pH \qquad (17\text{–}11)$$

where

$$L' = -0.0592 \log a_2$$

The boundary potential is then a measure of the hydrogen ion activity of the external solution.

## The Asymmetry Potential

When identical solutions and reference electrodes are placed on the two sides of a glass membrane, the boundary potential should be zero. In fact, however, a small *asymmetry potential* that changes gradually with time is frequently encountered.

The sources of the asymmetry potential are obscure but undoubtedly include such causes as differences in strain on the two surfaces of the membrane imparted during manufacture, mechanical abrasion on the outer surface during use, and chemical etching of the outer surface. In order to eliminate the determinate errors caused by the asymmetry potential, all membrane electrodes must be calibrated against one or more standard analyte solutions. Such calibrations should be carried out at least daily and more often when the electrode receives heavy use.

### The Potential of the Glass Electrode

As noted earlier, the potential of a glass indicator electrode has three components: (1) the boundary potential, given by Equation 17–11, (2) the potential of the internal Ag/AgCl reference electrode, and (3) a small asymmetry potential, $E_{asy}$. In equation form,

$$E_{ind} = E_b + E_{Ag/AgCl} + E_{asy}$$

Substitution of Equation 17–11 for $E_b$ gives

$$E_{ind} = L' + 0.0592 \log a_1 + E_{Ag/AgCl} + E_{asy}$$

or

$$E_{ind} = L + 0.0592 \log a_1 = L - 0.0592 \text{ pH} \qquad (17\text{–}12)$$

where $L$ is a combination of the three constant terms. Compare Equations 17–12 and 17–3. Although these two equations are similar in form, remember that the sources of the electrode potential they describe *are totally different.*

### The Alkaline Error

In basic solutions, glass electrodes respond to the concentration of both hydrogen ion and alkali metal ions. The magnitude of this *alkaline error* for four glass membranes is shown in Figure 17–11 (curves $C$ to $F$). These curves refer to solutions in which the sodium ion concentration was held constant at 1 M while the pH was varied. Note that the error is negative (that is, the measured pH values are lower than the true values), which suggests that the electrode is responding to sodium ions as well as to protons. This observation is confirmed by data obtained for solutions containing different sodium ion concentrations. Thus at pH 12, the electrode with a Corning 015 membrane (curve $C$ in Figure 17–11) registered a pH of 11.3 when immersed in a solution with a sodium ion concentration of 1 M but 11.7 in a solution that was 0.1 M in this ion. All singly charged

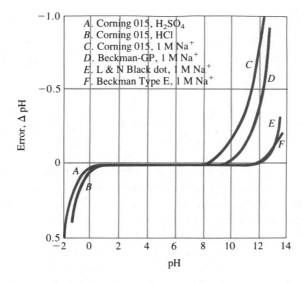

Figure 17-11
Acid and alkaline error for selected glass electrodes at 25°C. (From R. G. Bates, *Determination of pH*, 2nd ed., p. 365. New York: Wiley, 1973. With permission.)

cations induce an alkaline error whose magnitude depends upon both the cation in question and the composition of the glass membrane.

The alkaline error can be satisfactorily explained by assuming an exchange equilibrium between the hydrogen ions on the glass surface and the cations in solution. This process is simply the reverse of that shown in Equation 17-6:

$$H^+Gl^- + B^+ \rightleftarrows B^+Gl^- + H^+$$
$$\text{Glass} \quad \text{Soln} \quad \text{Glass} \quad \text{Soln}$$

where $B^+$ represents some singly charged cation, such as sodium ion.

The equilibrium constant for this reaction is

$$K_{ex} = \frac{a_1 b_1'}{a_1' b_1} \qquad (17-13)$$

where $a_1$ and $b_1$ represent the activities of $H^+$ and $B^+$ in solution and $a_1'$ and $b_1'$ are the activities of these ions on the gel surface. Equation 17-13 can be rearranged to give the ratio of the activities $B^+$ to $H^+$ on the glass surface:

$$\frac{b_1'}{a_1'} = \frac{b_1}{a_1} K_{ex}$$

For the glasses used for pH electrodes, $K_{ex}$ is so small that the activity ratio $b_1'/a_1'$ is ordinarily minuscule. The situation differs in strongly alkaline media, however. For example, $b_1'/a_1'$ for an electrode immersed in a pH 11 solution that is 1 M in sodium ions (Figure 17-11) is $10^{11} K_{ex}$. Here, the activity of the sodium ions relative to that of the hydrogen ions becomes so large that the electrode responds to both species.

### Selectivity Coefficients

The effect of an alkali metal ion on the potential across a membrane can be accounted for by inserting an additional term in Equation 17–11 to give

In Equation 17–14, $b_1$ represents the activity of some singly charged cation such as $Na^-$ or $K^+$.

$$E_b = L' + 0.0592 \log (a_1 + k_{H,B}b_1) \qquad (17\text{–}14)$$

The selectivity coefficient is a measure of the response of an ion-selective electrode to other ions.

where $k_{H,B}$ is the *selectivity coefficient* for the electrode. Equation 17–14 applies not only to glass indicator electrodes for hydrogen ion but also to all other types of membrane electrodes. Selectivity coefficients range from zero (no interference) to values greater than unity. Thus, if an electrode for ion A responds 20 times more strongly to ion B than to ion A, $k_{A,B}$ has a value of 20. If the response of the electrode to ion C is 0.001 of its response to A (a much more desirable situation), $k_{A,C}$ is 0.001.

The product $k_{H,B}b_1$ for a glass pH electrode is ordinarily small relative to $a_1$ provided the pH is less than 9; under these conditions, Equation 17–14 simplifies to Equation 17–11. At high pH values and at high concentrations of a singly charged ion, however, the second term in Equation 17–14 assumes a more important role in determining $E_b$, and an alkaline error is encountered. For electrodes specifically designed for work in highly alkaline media (curve E in Figure 17–11), the magnitude of $k_{H,B}b_1$ is appreciably smaller than for ordinary glass electrodes.

### The Acid Error

As shown in Figure 17–11, the typical glass electrode exhibits a *positive* error when the pH is less than about 0.5; pH readings tend to be too high in this region. The magnitude of the error depends upon a variety of factors and is not very reproducible. The causes of the acid error are not well understood.

### 17D–4  Glass Electrodes for Cations Other Than Protons

The alkaline error in early glass electrodes led to investigations concerning the effect of glass composition upon the magnitude of this error. One consequence has been the development of glasses for which the alkaline error is negligible below about pH 12. Other studies have led to glass compositions that permit the determination of cations other than hydrogen. This latter application requires that the hydrogen ion activity $a_1$ in Equation 17–14 be negligible relative to $k_{H,B}b_1$; under these circumstances, the potential is independent of pH and is a function of pB instead. Incorporation of $Al_2O_3$ or $B_2O_3$ in the glass has the desired effect. Glass electrodes that permit the direct potentiometric measurement of such singly charged species as $Na^+$, $K^+$, $NH_4^+$, $Rb^+$, $Cs^+$, $Li^+$, and $Ag^+$ have been developed. Some of these glasses are reasonably selective toward particular singly charged cations. Glass electrodes for $Na^+$, $Li^+$, $NH_4^+$, and total concentration of univalent cations are now available from commercial sources.

### 17D–5 Liquid-Membrane Electrodes

The potential of a liquid-membrane electrode develops across the interface between the solution containing the analyte and a liquid-ion exchanger that selectively bonds with the analyte ion. These electrodes have been developed for the direct potentiometric measurement of numerous polyvalent cations as well as certain anions.

Figure 17–12 is a schematic of a liquid-membrane electrode for calcium. It consists of a conducting membrane that selectively bonds calcium ions, an internal solution containing a fixed concentration of calcium chloride, and a silver electrode coated with silver chloride to form an internal reference electrode. The active membrane ingredient is a calcium dialkyl phosphate ion exchanger that is nearly insoluble in water. In the electrode shown in Figure 17–12, the ion exchanger is dissolved in an immiscible organic liquid that is forced by gravity into the pores of a hydrophobic porous disk. This disk then serves as the membrane that separates the internal solution from the analyte solution. The ion exchanger in a more recent design is immobilized in a tough polyvinyl chloride gel cemented to the end of a tube that holds the internal solution and reference electrode. Regardless of design, the dissociation equilibrium established at each membrane interface is analogous to Equation 17–8.

$$\underset{\text{Organic}}{[(RO)_2POO]_2Ca} \rightleftharpoons \underset{\text{Organic}}{2\ (RO)_2POO^-} + \underset{\text{Aqueous}}{Ca^{2+}}$$

where R is a high-molecular-weight aliphatic group. As with the glass electrode, a potential develops across the membrane when the extent of dissociation at one surface differs from that at the other surface. This

*Hydrophobic* means water-hating. The hydrophobic disk is porous to organic liquids but repels water.

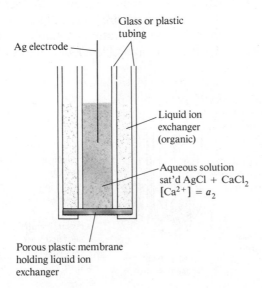

Figure 17–12
Diagram of a liquid-membrane electrode for $Ca^{2+}$.

Glass or plastic tubing

Ag electrode

Liquid ion exchanger (organic)

Aqueous solution sat'd AgCl + CaCl$_2$ [Ca$^{2+}$] = $a_2$

Porous plastic membrane holding liquid ion exchanger

Photograph of a potassium liquid-ion exchanger microelectrode with 125 $\mu$m of ion exchanger inside the tip. The magnification of the original photo was 400×. (From J. L. Walker, *Anal. Chem.*, **1971**, *43(3)N*, 91A. Reproduced by permission of the American Chemical Society.)

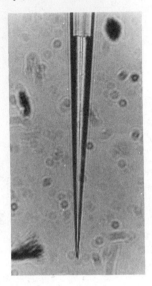

Ion-selective microelectrodes can be used to measure ion activities within a living organism.

potential is a result of differences in the calcium ion activity of the internal and external solutions. The relationship between the membrane potential and the calcium ion activities is given by an equation that is similar to Equation 17–10:

$$E_b = E_1 - E_2 = \frac{0.0592}{2} \log \frac{a_1}{a_2} \qquad (17\text{--}15)$$

where $a_1$ and $a_2$ are the activities of calcium ion in the external and internal solutions, respectively. Since the calcium ion activity of the internal solution is constant,

$$E_b = N + \frac{0.0592}{2} \log a_1 = N - \frac{0.0592}{2} \text{pCa} \qquad (17\text{--}16)$$

where $N$ is a constant (compare Equations 17–16 and 17–11). Note that, because calcium is divalent, a 2 appears in the denominator of the coefficient of the logarithmic term.

Figure 17–13 compares the structural features of a glass-membrane electrode and a commercially available liquid-membrane electrode for calcium ion. The sensitivity of the latter to calcium ion is reported to be 50 times greater than to magnesium ion and 1000 times greater than to sodium or potassium ions. Calcium ion activities as low as $5 \times 10^{-7}$ M can be measured. Electrode performance is independent of pH in the range between 5.5 and 11. At lower pH levels, hydrogen ions undoubtedly replace some of the calcium ions on the exchanger; the electrode then becomes sensitive to pH as well as to pCa.

The calcium ion liquid-membrane electrode is a valuable tool for physiological investigations because this ion plays important roles in such processes as nerve conduction, bone formation, muscle contraction, cardiac expansion and contraction, renal tubular function, and perhaps hypertension. At least some of these processes are more influenced by calcium ion activity than by calcium ion concentration; activity, of course, is the parameter measured by the membrane electrode.

Figure 17–13
Comparison of a liquid-membrane calcium ion electrode with a glass electrode. (Courtesy of Orion Research, Boston, MA.)

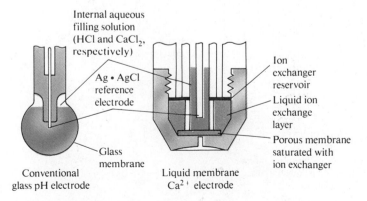

A liquid-membrane electrode specific for potassium ion is also of great value to physiologists because the transport of neural signals appears to involve movement of this ion across nerve membranes. Investigation of this process requires an electrode that can detect small concentrations of potassium ion in media with much larger concentrations of sodium ion. Several liquid-membrane electrodes show promise in meeting this requirement. One is based upon the antibiotic valinomycin, a cyclic ether that has a strong affinity for potassium ion. Of equal importance is the observation that a liquid membrane consisting of valinomycin in diphenyl ether is about $10^4$ times as responsive to potassium ion as to sodium ion.[3]

Table 17-2 lists some liquid-membrane electrodes available from commercial sources. The anion-sensitive electrodes shown make use of a solution containing an anion-exchange resin in an organic solvent. Liquid-membrane electrodes in which the exchange liquid is held in a polyvinyl chloride gel have been developed for $Ca^{2+}$, $K^+$, $NO_3^-$, and $BF_4^-$. These electrodes look like crystalline electrodes, which are considered in the following section.

Table 17-2
CHARACTERISTICS OF LIQUID-MEMBRANE ELECTRODES*

| Analyte Ion | Concentration Range, M | Major Interferences |
|---|---|---|
| $Ca^{2+}$ | $10^0$ to $5 \times 10^{-7}$ | $Pb^{2+}$, $Fe^{2+}$, $Ni^{2+}$, $Hg^{2+}$, $Sr^{2+}$ |
| $Cl^-$ | $10^0$ to $5 \times 10^{-6}$ | $I^-$, $OH^-$, $SO_4^{2-}$ |
| $NO_3^-$ | $10^0$ to $7 \times 10^{-6}$ | $ClO_4^-$, $I^-$, $ClO_3^-$, $CN^-$, $Br^-$ |
| $ClO_4^-$ | $10^0$ to $7 \times 10^{-6}$ | $I^-$, $ClO_3^-$, $CN^-$, $Br^-$ |
| $K^+$ | $10^0$ to $1 \times 10^{-6}$ | $Cs^+$, $NH_4^+$, $Tl^+$ |
| Water hardness ($Ca^{2+}$ + $Mg^{2+}$) | $10^0$ to $6 \times 10^{-6}$ | $Cu^{2+}$, $Zn^{2+}$, $Ni^{2+}$, $Sr^{2+}$, $Fe^{2+}$, $Ba^{2+}$ |

*From *Orion Guide to Ion Analysis*. Boston, MA: Orion Research, 1983. With permission.

[3]M. S. Frant and J. W. Ross Jr., *Science*, **1970**, *167*, 987.

**Feature 17-1**
AN EASILY CONSTRUCTED LIQUID-MEMBRANE
ION-SELECTIVE ELECTRODE

You can make a liquid-membrane ion-selective electrode with glassware and chemicals available in most laboratories.[4] All you need are a pH meter, a pair of reference electrodes, a fritted-glass filter crucible or tube, trimethylchlorosilane, and a liquid ion exchanger.

First, cut the filter crucible (or alternatively, a fritted tube) as shown in Figure 17-14, page 418. Carefully clean and dry the crucible, and then draw a small amount of trimethylchlorosilane into

[4]T. K. Christopoulos and E. P. Diamandis, *J. Chem. Educ.*, **1988**, *65*, 648.

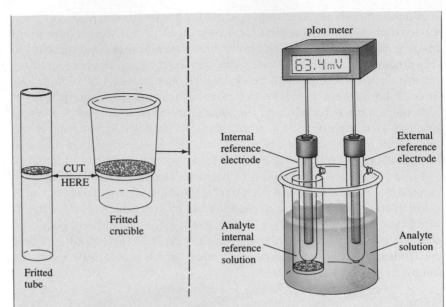

Figure 17–14
A homemade liquid-membrane
electrode.

the frit. This coating makes the glass in the frit hydrophobic. Rinse
the frit with water, dry, and apply a commercial liquid ion exchanger to it. After a minute, remove the excess exchanger. Add a
few milliliters of a $10^{-2}$ M solution of the ion of interest to the crucible, insert a reference electrode into the solution, and voilà, you
have a very nice ion-selective electrode. The exact details of washing, drying, and preparing the electrode are provided in the original
article.

Connect the ion-selective electrode and the second reference
electrode to the pH meter as shown in Figure 17–14. Prepare a
series of standard solutions of the ion of interest, measure the cell
potential for each concentration, plot a working curve of $E_{cell}$ versus log $c$, and perform a least-squares analysis on the data. Compare the slope of the line with the theoretical slope of $(0.0592 \text{ V})/n$.
Measure the potential for an unknown solution of the ion and calculate the concentration from the least-squares parameters.

## 17D–6 Crystalline-Membrane Electrodes

Considerable work has been devoted to the development of solid membranes that are selective toward anions in the same way that some glasses
respond to cations. We have seen that anionic sites on a glass surface
account for the selectivity of a membrane toward certain cations. By
analogy, a membrane with cationic sites might be expected to respond
selectively toward anions.

Membranes prepared from cast pellets of silver halides have been used
successfully in electrodes for the selective determination of chloride, bromide, and iodide ions. In addition, an electrode based upon a polycrystal-
line $Ag_2S$ membrane is offered by one manufacturer for the determination
of sulfide ion. In both types of membranes, silver ions are sufficiently

mobile to conduct electricity through the solid medium. Mixtures of PbS, CdS, and CuS with $Ag_2S$ provide membranes that are selective for $Pb^{2+}$, $Cd^{2+}$, and $Cu^{2+}$, respectively. Silver ion must be present in these membranes to conduct electricity because divalent ions are immobile in crystals. The potential that develops across crystalline solid-state electrodes is described by a relationship similar to Equation 17–16.

A crystalline electrode for fluoride ion has been developed. The membrane consists of a slice of a single crystal of lanthanum fluoride that has been doped with europium(II) fluoride to improve its conductivity. The membrane, supported between a reference solution and the solution to be measured, shows a theoretical response to changes in fluoride ion activity in the range from $10^0$ to $10^{-6}$ M. The electrode is selective for fluoride ion over other common anions by several orders of magnitude; only hydroxide ion appears to offer serious interference.

Some solid-state electrodes available from commercial sources are listed in Table 17–3. Several types of crystalline electrodes are shown in Figure 17–15.

## 17D–7 Gas-Sensing Probes

Figure 17–16 illustrates the essential features of a potentiometric gas-sensing probe, which consists of a tube containing a reference electrode, a specific-ion electrode, and an electrolyte solution. A thin, replaceable, gas-permeable membrane attached to one end of the tube serves as a barrier between the internal and analyte solutions. As can be seen from Figure 17–16, this device is a complete electrochemical cell and is more properly referred to as a probe than as an electrode.

> A gas-sensing probe is a galvanic *cell* whose potential is related to the concentration of a gas in a solution.

### Membrane Composition

A *microporous membrane* is fabricated from a hydrophobic polymer. As the name implies, the membrane is highly porous (the average pore size is

> A hydrophobic substance repels water and is not wetted by aqueous solutions.

Table 17–3
### CHARACTERISTICS OF SOLID–STATE CRYSTALLINE ELECTRODES*

| Analyte Ion | Concentration Range, M | Major Interferences |
|---|---|---|
| $Br^-$ | $10^0$ to $5 \times 10^{-6}$ | $CN^-$, $I^-$, $S^{2-}$ |
| $Cd^{2+}$ | $10^{-1}$ to $1 \times 10^{-7}$ | $Fe^{2+}$, $Pb^{2+}$, $Hg^{2+}$, $Ag^+$, $Cu^{2+}$ |
| $Cl^-$ | $10^0$ to $5 \times 10^{-5}$ | $CN^-$, $I^-$, $Br^-$, $S^{2-}$ |
| $Cu^{2+}$ | $10^{-1}$ to $1 \times 10^{-8}$ | $Hg^{2+}$, $Ag^+$, $Cd^{2+}$ |
| $CN^-$ | $10^{-2}$ to $1 \times 10^{-6}$ | $S^{2-}$ |
| $F^-$ | Sat'd to $1 \times 10^{-6}$ | $OH^-$ |
| $I^-$ | $10^0$ to $5 \times 10^{-8}$ | |
| $Pb^{2+}$ | $10^{-1}$ to $1 \times 10^{-6}$ | $Hg^{2+}$, $Ag^+$, $Cu^{2+}$ |
| $Ag^+/S^{2-}$ | $Ag^+$: $10^0$ to $1 \times 10^{-7}$ | $Hg^{2+}$ |
| | $S^{2-}$: $10^0$ to $1 \times 10^{-7}$ | |
| $SCN^-$ | $10^0$ to $5 \times 10^{-6}$ | $I^-$, $Br^-$, $CN^-$, $S^{2-}$ |

*From *Orion Guide to Ion Analysis*. Cambridge, MA: Orion Research, 1983. With permission.

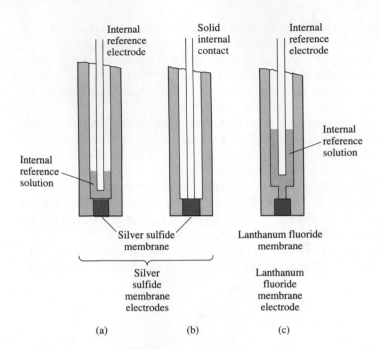

**Figure 17–15**

Types of crystalline ion-selective electrodes. (a) Solid-membrane electrode sensitive to $Ag^+$ and $S^{2-}$; (b) all solid-state solid-membrane electrode sensitive to $Ag^+$ and $S^{2-}$; (c) membrane-configuration ion-sensing electrode for fluoride ion.

less than 1 $\mu$m) and allows the free passage of gases; at the same time, the water-repellent polymer prevents water and solute ions from entering the pores. The thickness of the membrane is about 0.1 mm.

## The Mechanism of Response

Using carbon dioxide as an example, we can represent the transfer of gas to the internal solution by the following set of equations:

$$CO_2(aq) \rightleftarrows CO_2(g)$$

Analyte Solution        Membrane Pores

$$CO_2(g) \rightleftarrows CO_2(aq)$$

Membrane Pores        Internal Solution

$$CO_2(aq) + 2\,H_2O \rightleftarrows HCO_3^- + H_3O^+$$

Internal Solution        Internal Solution

The last equilibrium causes the pH of the internal surface film to change. This change is then detected by the internal glass/calomel electrode system. A description of the overall process is obtained by adding the equations for the three individual equilibria to give

$$CO_2(aq) + 2\,H_2O \rightleftarrows H_3O^+ + HCO_3^-$$

Analyte Solution        Internal Solution

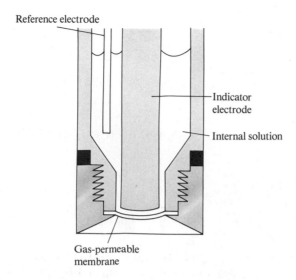

Reference electrode

Indicator electrode

Internal solution

Gas-permeable membrane

Figure 17–16
Diagram of a gas-sensing probe.

The net equilibrium constant is

$$K = \frac{[H_3O^+][HCO_3^-]}{[CO_2(aq)]_{ext}}$$

where $[CO_2(aq)]_{ext}$ is the concentration of the gas in the analyte solution. In order for the measured cell potential to vary linearly with the logarithm of the carbon dioxide concentration of the external solution, the hydrogen carbonate concentration of the internal solution must be sufficiently large so that it is not altered significantly by the carbon dioxide entering from the external solution. Assuming then that $[HCO_3^-]$ is constant, we can rearrange the previous equation to

$$\frac{[H_3O^+]}{[CO_2(aq)]_{ext}} = \frac{K}{[HCO_3^-]} = K_g$$

Letting $a_1$ be the hydrogen ion activity of the internal solution, we can write

$$a_1 = [H_3O^+] = K_g[CO_2(aq)]_{ext} \qquad (17\text{–}17)$$

Substitution of Equation 17–17 for $a_1$ in Equation 17–12 yields

$$E_{ind} = L + 0.0592 \log K_g[CO_2(aq)]_{ext}$$

or

$$E_{ind} = L' + 0.0592 \log [CO_2(aq)]_{ext} \qquad (17\text{–}18)$$

Finally, since

$$E_{cell} = E_{ind} - E_{ref}$$

and

$$E_{cell} = L' + 0.0592 \log [CO_2(aq)]_{ext} \qquad (17-19)$$

where

$$L' = L + 0.0592 \log K_g - E_{ref}$$

Thus, the potential between the glass electrode and the reference electrode in the internal solution is determined by the $CO_2$ concentration in the external solution. *Note that no electrode comes in direct contact with the analyte solution.* Therefore, these devices are gas-sensing *cells*, or *probes*, rather than gas-sensing electrodes. Nevertheless, they continue to be called electrodes in some literature and many advertising brochures.

The only species that interfere are other dissolved gases that permeate the membrane and then affect the pH of the internal solution. The selectivity of gas probes depends upon the selectivity of the internal indicator electrode. Gas-sensing probes for $CO_2$, $NO_2$, $H_2S$, $SO_2$, HF, HCN, and $NH_3$ are now available from commercial sources.

Although sold as gas-sensing electrodes, these devices are complete electrochemical cells that contain *two* electrodes and should be called gas-sensing probes.

## 17E   INSTRUMENTS FOR MEASURING CELL POTENTIALS

Most cells containing an ion electrode have very high electrical resistance (as much as $10^8$ ohms or more). In order to measure potentials of such high-resistance circuits accurately, it is necessary that the voltmeter have an electrical resistance that is several orders of magnitude greater than the resistance of the cell being measured. If the meter resistance is too low, current is drawn from the cell, which has the effect of lowering its output potential, thus creating a negative error. This effect is shown in Figure 17-17, which is a plot of the relative error in potential reading as a function of the ratio of the resistance of a meter to the resistance of a cell. When the meter and the cell have the same resistance, a relative error of $-50\%$ results. When this ratio of meter resistance to cell resistance is 10, the error is about $-9\%$. When it is 1000, the error is less than 0.1% relative.

Numerous high-resistance, direct-reading meters with internal resistances of $10^{11}$ to $10^{12}$ ohms are now on the market. These devices are commonly called *pH meters* but could more properly be referred to as *pIon meters* or *ion meters* since they are frequently used for the measurement of concentrations of other ions as well. Feature 17-2 describes a typical design of an ion meter.

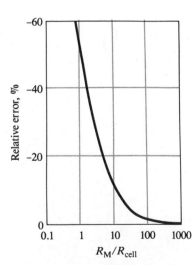

Figure 17-17

Relative error in a cell-potential measurement as a function of the ratio of the electrical resistance of the meter $R_M$ to the resistance of the cell $R_{cell}$.

**Feature 17–2**
## OPERATIONAL-AMPLIFIER VOLTAGE MEASUREMENTS

One of the most important developments in chemical instrumentation over the last several years has been the advent of compact, inexpensive, versatile integrated-circuit amplifiers (op amps).[5] These devices allow us to make potential measurements on high-resistance cells, such as those that contain a glass electrode, without drawing appreciable current. Even a small current ($10^{-7}$ to $10^{-10}$ A) in a glass electrode results in a large error in the measured voltage. One of the most important uses for operational amplifiers is to isolate voltage sources from their measurement circuits. The basic *voltage follower*, which allows this type of measurement, is shown in Figure 17–18a. This circuit has two important characteristics: the output voltage $E_{out}$ is equal to the input voltage $E_{in}$, and the input current $I_{in}$ is essentially zero ($10^{-9}$ to $10^{-11}$ A).

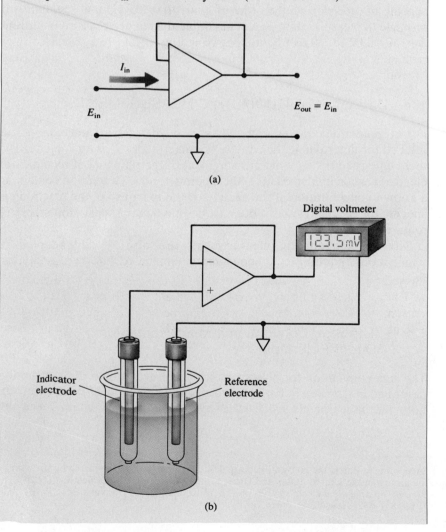

(a)

(b)

Figure 17–18
(a) A voltage-follower operational amplifier. (b) Typical arrangement for potentiometric measurements with a membrane electrode.

A practical application of this circuit is in measuring cell potentials. The cell is connected to the op-amp input as shown in Figure 17–18b, and the output of the op amp is connected to a digital voltmeter to measure the voltage. Modern op amps are nearly ideal voltage-measurement devices and are incorporated in most ion meters and pH meters to monitor high-resistance indicator electrodes with little error.

---

[5]For a detailed description of op-amp circuits, see H. V. Malmstadt, C. G. Enke, and S. R. Crouch, *Electronics and Instrumentation for Scientists*, Chapter 5. Menlo Park, CA: Benjamin-Cummings, 1981.

The readout of ion meters is either digital or analog (in the latter, a needle sweeps a range on a scale from 0 to 14 pH units). Some meters are capable of precision on the order of 0.001 to 0.005 pH unit. Seldom is it possible to measure pH with a comparable degree of *accuracy*. Inaccuracies of $\pm 0.02$ to $\pm 0.03$ pH unit are typical.

Ordinarily, the determinate error in a pH measurement is 0.01 to 0.02 pH unit.

## 17F    DIRECT POTENTIOMETRIC MEASUREMENTS

Direct potentiometric measurements provide a rapid and convenient method for determining the activity of numerous cations and anions. The technique requires only the measurement of the potential of an indicator electrode when immersed in (1) the unknown and (2) a solution containing a known concentration of the analyte. If the response of the electrode is specific for the analyte, as it often is, no preliminary separation steps are required.

Direct potentiometric measurements are also readily adapted to applications requiring continuous and automatic recording of analytical data.

### 17F–1  The Sign Convention and Equations for Direct Potentiometry

The sign convention for potentiometry is consistent with the convention described in Chapter 14 for standard electrode potentials.[6] In this convention, the indicator electrode is *always* treated as the *cathode* and the

---

[6]According to Bates, the convention being described here has been endorsed by standardizing groups in the United States and Great Britain as well as IUPAC. See R. G. Bates, in *Treatise on Analytical Chemistry*, 2nd ed., I. M. Kolthoff and P. J. Elving, Eds., Part I, Vol. 1, pp. 831–832. New York: Wiley, 1978.

reference electrode as the *anode*.[7] For direct potentiometric measurements, the potential of a cell can then be expressed in terms of the potentials developed by the indicator electrode, the reference electrode, and a junction potential:

$$E_{cell} = E_{ind} - E_{ref} + E_j \qquad (17-20)$$

In Section 17D, which deals with indicator electrodes, we describe the response of various types of indicator electrodes to analyte activities. For the cation $X^{n+}$ at 25°C, the electrode response takes the general *Nernstian* form

$$E_{ind} = L - \frac{0.0592}{n}\, pX = L + \frac{0.0592}{n}\, \log a_X \qquad (17-21)$$

where $L$ is a constant and $a_X$ is the activity of the cation. For metallic indicator electrodes, $L$ is ordinarily the standard electrode potential; for membrane electrodes, $L$ is the summation of several constants, including the time-dependent asymmetry potential of uncertain magnitude.

Substitution of Equation 17–21 into Equation 17–20 yields, with rearrangement,

$$pX = -\log a_X = -\frac{E_{cell} - (E_j - E_{ref} + L)}{0.0592/n}$$

The constant terms in parentheses can be combined to give a new constant $K$:

$$pX = -\log a_X = -\frac{E_{cell} - K}{0.0592/n} \qquad (17-22)$$

For an anion $A^{n-}$, the sign of Equation 17–22 is reversed:

$$pA = \frac{E_{cell} - K}{0.0592/n} \qquad (17-23)$$

All direct potentiometric methods are based upon Equation 17–22 or 17–23. The difference in sign in the two equations has a subtle but important consequence in the way that ion-selective electrodes are connected

---

[7]In effect, the sign convention for electrode potentials described in Section 15C–3 also designates the indicator electrode as the cathode by stipulating that half-reactions always be written as reductions; the standard hydrogen electrode, which is the reference electrode in this case, is then the anode.

A plot of Equation 17–24 for cationic electrodes.

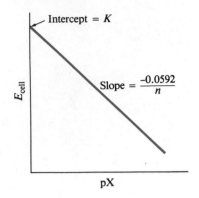

to pH meters and pIon meters. When the two equations are solved for $E_{cell}$, we find that for cations

$$E_{cell} = K - \frac{0.0592}{n} \, pX \qquad (17\text{–}24)$$

and for anions

$$E_{cell} = K + \frac{0.0592}{n} \, pA \qquad (17\text{–}25)$$

Equation 17–24 shows that an increase in pX results in a *decrease* in $E_{cell}$ with a cation-selective electrode. Thus, when a high-resistance voltmeter is connected to the cell in the usual way, with the indicator electrode attached to the positive terminal, the meter reading decreases as pX increases. To eliminate this problem, instrument manufacturers generally reverse the leads so that cation-sensitive electrodes are connected to the *negative* terminal of the voltage-measuring device. Meter readings then increase with increases in pX. Anion-selective electrodes, on the other hand, are connected to the *positive* terminal of the meter so that increases in pA also yield larger readings.

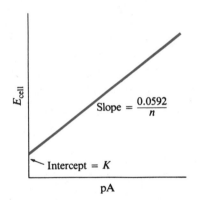

A plot of Equation 17–25 for anionic electrodes.

### 17F–2 The Electrode-Calibration Method

As we saw in Section 17D, the constant $K$ in Equations 17–22 and 17–23 is made up of several constants, at least one of which, the junction potential, cannot be computed from theory or measured directly. So before these equations can be used for the determination of pX or pA, $K$ must be evaluated *experimentally* with a standard solution of the analyte.

In the electrode-calibration method, $K$ in Equations 17–22 and 17–23 is determined by measuring $E_{cell}$ for one or more standard solutions of known pX or pA. The assumption is then made that $K$ is unchanged when the standard is replaced by the analyte solution. The calibration is ordinarily performed at the time pX or pA for the unknown is determined. With membrane electrodes, recalibration may be required if measurements extend over several hours because of changes in the asymmetry potential.

The electrode-calibration method offers the advantages of simplicity, speed, and applicability to the continuous monitoring of pX or pA. It suffers, however, from a somewhat limited accuracy because of uncertainties in junction potentials.

#### Inherent Error in the Electrode-Calibration Procedure

A serious disadvantage of the electrode-calibration method is the inherent error that results from the assumption that $K$ in Equations 17–22 and 17–23 remains constant after calibration. This assumption can seldom, if ever, be exactly true because the electrolyte composition of the unknown almost inevitably differs from that of the solution employed for calibration. The junction-potential term contained in $K$ varies slightly as a consequence, even when a salt bridge is used. This error is frequently on the order of 1 mV or more. Unfortunately, because of the nature of the potential/activity relationship, such an uncertainty has an amplified effect on the inherent accuracy of the analysis.

The magnitude of the error in analyte concentration can be estimated by differentiating Equation 17–22 while holding $E_{cell}$ constant:

$$-\log_{10} e \, \frac{da_1}{a_1} = -0.434 \, \frac{da_1}{a_1} = -\frac{dK}{0.0592/n}$$

$$\frac{da_1}{a_1} = \frac{ndK}{0.0257}$$

Upon replacing $da_1$ and $dK$ with finite increments and multiplying both sides of the equation by 100%, we obtain

$$\text{percent relative error} = \frac{\Delta a_1}{a_1} \times 100\%$$

$$= 3.89 \times 10^3 n\Delta K\% \approx 4000n\Delta K\%$$

The quantity $\Delta a_1/a_1$ is the relative error in $a_1$ associated with an absolute uncertainty $\Delta K$ in $K$. If, for example, $\Delta K$ is $\pm0.001$ V, a relative error in activity of about $\pm 4n\%$ can be expected. *It is important to appreciate that this error is characteristic of all measurements involving cells that contain a salt bridge and cannot be eliminated by even the most careful measurements of cell potentials or the most sensitive and precise measuring devices.*

### Activity Versus Concentration

Electrode response is related to analyte activity rather than analyte concentration. We are usually interested in concentration, however, and the determination of this quantity from a potentiometric measurement requires activity-coefficient data. Activity coefficients are not often available because the ionic strength of the solution is either unknown or else is so large that the Debye-Hückel equation is not applicable.

The difference between activity and concentration is illustrated by Figure 17–19, in which the response of a calcium ion electrode is plotted against a logarithmic function of calcium chloride *concentration*. The nonlinearity is due to the increase in ionic strength—and the consequent decrease in calcium ion activity—with increasing electrolyte concentration. The upper curve is obtained when these concentrations are converted to activities. This straight line has the theoretical slope of 0.0296 (0.0592/2).

Activity coefficients for singly charged species are less affected by changes in ionic strength than are the coefficients for ions with multiple

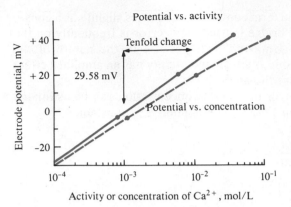

Many chemical reactions of physiological importance depend upon the activity of metal ions rather than their concentration.

charges. Thus, the effect shown in Figure 17–19 is less pronounced for electrodes that respond to $H^+$, $Na^+$, and other univalent ions.

In potentiometric pH measurements, the pH of the standard buffer used for calibration is generally based on the activity of hydrogen ions. Thus, the results are also on an activity scale. If the unknown sample has a high ionic strength, the hydrogen ion *concentration* will differ appreciably from the activity measured.

An obvious way to convert potentiometric measurements from activity to concentration is to make use of an empirical calibration curve, such as the lower plot in Figure 17–19. For this approach to be successful, it is necessary to make the ionic composition of the standards essentially the same as that of the analyte solution. Matching the ionic strength of standards to that of samples is often difficult, particularly for samples that are chemically complex.

Where electrolyte concentrations are not too great, it is often useful to swamp both samples and standards with a measured excess of an inert electrolyte. The added effect of the electrolyte from the sample matrix becomes negligible under these circumstances, and the empirical calibration curve yields results in terms of concentration. This approach has been used, for example, in the potentiometric determination of fluoride ion in drinking water. Both samples and standards are diluted with a solution that contains sodium chloride, an acetate buffer, and a citrate buffer; the diluent is sufficiently concentrated so that the samples and standards have essentially identical ionic strengths. This method provides a rapid means for measuring fluoride concentrations in the part-per-million range with an accuracy of about 5% relative.

TISAB (total ionic strength adjustment buffer) is used to control the ionic strength and the pH of samples and standards in ion-selective electrode measurements for fluoride.

### 17F–3 The Standard-Addition Method

The standard-addition method involves determining the potential of the electrode system before and after a measured volume of a standard has been added to a known volume of the analyte solution. Often an excess of an electrolyte is incorporated into the analyte solution at the outset to prevent any major shift in ionic strength that might accompany the addition of standard. It is also necessary to assume that the junction potential remains constant during the two measurements.

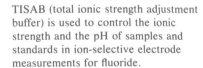

Example 17–1

A cell consisting of a saturated calomel electrode and a lead ion electrode developed a potential of $-0.4706$ V when immersed in 50.00 mL of a sample. A 5.00-mL addition of standard 0.02000 M lead solution caused the potential to shift to $-0.4490$ V. Calculate the molar concentration of lead in the sample.

We shall assume that the activity of $Pb^{2+}$ is approximately equal to $[Pb^{2+}]$ and apply Equation 17–22. Thus,

$$pPb = -\log[Pb^{2+}] = -\frac{E'_{cell} - K}{0.0592/2}$$

where $E'_{cell}$ is the initial measured potential ($-0.4706$ V).

After the standard solution is added, the potential becomes $E''_{cell}$ ($-0.4490$ V) and

$$-\log \frac{50.00 \times [Pb^{2+}] + 5.00 \times 0.0200}{50.00 + 5.00} = -\frac{E''_{cell} - K}{0.0592/2}$$

$$-\log (0.9091[Pb^{2+}] + 1.818 \times 10^{-3}) = -\frac{E''_{cell} - K}{0.0592/2}$$

Subtracting this equation from the first leads to

$$-\log \frac{[Pb^{2+}]}{0.9091[Pb^{2+}] + 1.818 \times 10^{-3}} = \frac{2(E''_{cell} - E'_{cell})}{0.0592}$$

$$= \frac{2[-0.4490 - (-0.4706)]}{0.0592} = 0.7297$$

Taking the antilog of both sides of this equation gives

$$\frac{[Pb^{2+}]}{0.9091[Pb^{2+}] + 1.818 \times 10^{-3}} = 0.1863$$

$$[Pb^{2+}] = 4.08 \times 10^{-4} \text{ M}$$

## 17F–4 Potentiometric pH Measurements with a Glass Electrode[8]

The glass electrode is unquestionably the most important indicator electrode for hydrogen ion. It is convenient to use and subject to few of the interferences that affect other pH-sensing electrodes.

The glass/calomel electrode system is a remarkably versatile tool for measuring pH under many conditions. It can be used without interference

The most common instrumental technique in science is the measurement of pH.

[8]For a detailed discussion of potentiometric pH measurements, see R. G. Bates, *Determination of pH*, 2nd ed. New York: Wiley, 1973.

in solutions containing strong oxidants, strong reductants, proteins, and gases; the pH of viscous or even semisolid fluids can be determined. Electrodes for special applications are available. Included among these are small electrodes for pH measurements in one drop (or less) of solution, in a tooth cavity, or in sweat on the skin; microelectrodes that permit the measurement of pH inside a living cell; rugged electrodes for insertion in a flowing liquid stream to provide a continuous monitoring of pH; and small electrodes that can be swallowed to measure the acidity of the stomach contents (the calomel electrode is kept in the mouth).

## Errors That Affect pH Measurements with the Glass Electrode

The ubiquity of the pH meter and the general applicability of the glass electrode tend to lull the chemist into the attitude that any measurement obtained with such equipment is surely correct. The reader must be alert to the fact that there are distinct limitations to the electrode, some of which were discussed in earlier sections:

1. *The alkaline error.* The ordinary glass electrode becomes somewhat sensitive to alkali metal ions and gives low readings at pH values greater than 9.
2. *The acid error.* Values registered by the glass electrode tend to be somewhat high when the pH is less than about 0.5.
3. *Dehydration.* Dehydration may cause erratic electrode performance.
4. *Error in unbuffered neutral solutions.* Because equilibrium between the bulk of the solution and the layer of solution at the surface of a membrane is achieved only slowly in poorly buffered, approximately neutral solutions, time must be allowed for this equilibrium to be established. Before being used to determine the pH of such solutions, the glass electrode should be thoroughly rinsed with water. Then both electrodes should be immersed in successive portions of the unknown until a constant pH reading is obtained.
5. *Variation in junction potential.* A fundamental source of uncertainty for which a correction cannot be applied is the junction-potential variation resulting from differences in the composition of the standard and the unknown solution.
6. *Error in the pH of the standard buffer.* Any inaccuracies in the preparation of the buffer used for calibration or any changes in its composition during storage cause an error in subsequent pH measurements. The action of bacteria on organic buffer components is a common cause for deterioration.

## The Operational Definition of pH

The utility of pH as a measure of the acidity or alkalinity of aqueous media, the wide availability of commercial glass electrodes, and the relatively recent proliferation of inexpensive solid-state pH meters have made the potentiometric measurement of pH perhaps the most common analytical technique in all of science. It is thus extremely important that pH be defined in a manner that is easily duplicated at various times and in various laboratories throughout the world. To meet this requirement, it is necessary to define pH in operational terms—that is, by the way the measurement is made. Only then will the pH measured by one worker be the same as that measured by another.[9]

The operational definition of pH endorsed by the National Institute of Standards and Technology (NIST), similar organizations in other countries, and the IUPAC is based upon the direct calibration of the meter with carefully prescribed standard buffers followed by potentiometric determination of the pH of unknown solutions.

Consider, for example, the glass/calomel system in Figures 17–6 and 17–7. When these electrodes are immersed in a standard buffer, Equation 17–22 applies and we can write

$$pH_S = -\frac{E_S - K}{0.0592}$$

where $E_S$ is the cell potential when the electrodes are immersed in the standard buffer. Similarly, if the cell potential is $E_U$ when the electrodes are immersed in a solution of unknown pH, we have

$$pH_U = -\frac{E_U - K}{0.0592}$$

By subtracting the first equation from the second and solving for $pH_U$, we find

$$pH_U = pH_S - \frac{(E_U - E_S)}{0.0592} \qquad (17\text{--}26)$$

Equation 17–26 has been adopted throughout the world as the *operational definition of pH*.

Workers at the National Institute of Standards and Technology and elsewhere have used cells without liquid junctions to study primary-standard buffers extensively. Some of the properties of these buffers are presented in Table 17–4 and are discussed in detail elsewhere.[9] Note that the NIST buffers are described by their *molal* concentrations (mol solute/kg solvent) for accuracy and precision of preparation. For general use, the buffers can be prepared from relatively inexpensive laboratory reagents; for careful work, however, certified buffers can be purchased from the NIST. As Table 17–4 shows, the buffers form an internally consistent series over the pH range from 3.5 to 10.3, and their pH values are presented as a function of temperature. The buffer capacity listed for each solution indicates its pH stability with the addition of acid or base.

It should be emphasized that the strength of the operational definition of pH is that it provides a coherent scale for the determination of acidity

By definition, pH is what you measure with a glass electrode and a pH meter.

An operational definition of a quantity defines the quantity in terms of how it is measured.

[9]For a recent discussion of pH scales, see H. B. Kristensen, A. Salomon, and G. Kokholm, *Anal. Chem.*, **1991,** *63,* 885A.

[10]R. G. Bates, *Determination of pH,* 2nd ed., Chapter 4. New York: Wiley, 1973.

Table 17-4
VALUES OF NIST PRIMARY-STANDARD pH SOLUTIONS FROM 0 to 60°C*

| Temperature, °C | Sat'd (25°C) KH Tartrate | 0.05 m KH₂ Citrate† | 0.05 m KH phthalate | 0.025 m KH₂PO₄/ 0.025 m Na₂HPO₄ | 0.008695 m KH₂PO₄/ 0.03043 m Na₂HPO₄ | 0.01 m Na₂B₄O₇ | 0.025 m NaHCO₃/ 0.025 m Na₂CO₃ |
|---|---|---|---|---|---|---|---|
| 0 | — | 3.863 | 4.003 | 6.984 | 7.534 | 9.464 | 10.317 |
| 5 | — | 3.840 | 3.999 | 6.951 | 7.500 | 9.395 | 10.245 |
| 10 | — | 3.820 | 3.998 | 6.923 | 7.472 | 9.332 | 10.179 |
| 15 | — | 3.802 | 3.999 | 6.900 | 7.448 | 9.276 | 10.118 |
| 20 | — | 3.788 | 4.002 | 6.881 | 7.429 | 9.225 | 10.062 |
| 25 | 3.557 | 3.776 | 4.008 | 6.865 | 7.413 | 9.180 | 10.012 |
| 30 | 3.552 | 3.766 | 4.015 | 6.853 | 7.400 | 9.139 | 9.966 |
| 35 | 3.549 | 3.759 | 4.024 | 6.844 | 7.389 | 9.102 | 9.925 |
| 40 | 3.547 | 3.753 | 4.035 | 6.838 | 7.380 | 9.068 | 9.889 |
| 45 | 3.547 | 3.750 | 4.047 | 6.834 | 7.373 | 9.038 | 9.856 |
| 50 | 3.549 | 3.749 | 4.060 | 6.833 | 7.367 | 9.011 | 9.828 |
| 55 | 3.554 | — | 4.075 | 6.834 | — | 8.985 | — |
| 60 | 3.560 | — | 4.091 | 6.836 | — | 8.962 | — |
| Buffer capacity‡ (mol/pH unit) | 0.027 | 0.034 | 0.016 | 0.029 | 0.016 | 0.020 | 0.029 |
| $\Delta pH_{1/2}$ for 1:1 dilution§ | 0.049 | 0.024 | 0.052 | 0.080 | 0.07 | 0.01 | 0.079 |

*Adapted from R. G. Bates, *Determination of pH*, 2nd ed., p. 73. New York: Wiley, 1973.

†$m$ = molality (mol solute/kg H₂O).

‡See page 224.

§Change in pH that occurs when one volume of buffer is diluted with one volume of H₂O.

or alkalinity. However, measured pH values cannot be expected to yield a detailed picture of solution composition that is entirely consistent with solution theory. This uncertainty stems from our fundamental inability to measure single ion activities. That is, the operational definition of pH does not yield the exact pH as defined by the equation

$$pH = - \log [H^+]f_{H^+}$$

## 17G   POTENTIOMETRIC TITRATIONS

A *potentiometric titration* involves measuring the potential of a suitable indicator electrode as a function of titrant volume. The information provided by a potentiometric titration is not the same as that obtained from a direct potentiometric measurement. For example, the direct measurement of 0.100 M solutions of hydrochloric and acetic acids would yield two substantially different hydrogen ion concentrations because the latter is only partially dissociated. In contrast, the potentiometric titration of equal volumes of these two acids would require the same amount of standard base because both solutes have the same number of titratable protons.

Potentiometric titrations provide data that are more reliable than data from titrations that use chemical indicators. They are particularly useful with colored or turbid solutions and for detecting the presence of unsuspected species. They suffer from the disadvantage of being more time-consuming than those involving indicators; on the other hand, they are readily automated.

Figure 17–20 illustrates a typical apparatus for performing a manual potentiometric titration. Its use involves measuring and recording the cell potential (in units of millivolts or pH, as appropriate) after each addition of reagent. The titrant is added in large increments at the outset and in smaller and smaller increments as the end point is approached (as indicated by larger changes in response per unit volume).

*Automatic titrators for carrying out potentiometric titrations are available from several manufacturers.*

### 17G–1   End-Point Detection

Several methods can be used to determine the end point of a potentiometric titration. The most straightforward involves a direct plot of potential as a function of reagent volume, as in Figure 17–21a; the midpoint in the steeply rising portion of the curve is estimated visually and taken as the end point. Various graphical methods have been proposed to aid in the establishment of the midpoint, but it is doubtful that these procedures significantly improve its determination.

A second approach to end-point detection is to calculate the change in potential per unit volume of titrant (that is, $\Delta E/\Delta V$), as in column 3 of Table 17–5. A plot of these data as a function of the average volume $V$ produces a curve with a maximum that corresponds to the point of inflection (Figure 17–21b). Alternatively, this ratio can be evaluated during the titration and recorded in lieu of the potential. Inspection of column 3 of

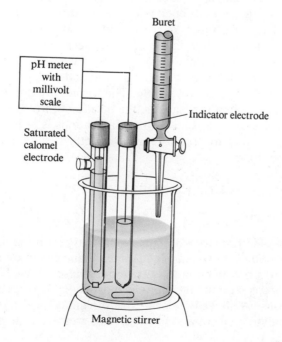

Figure 17–20

Apparatus for a potentiometric titration.

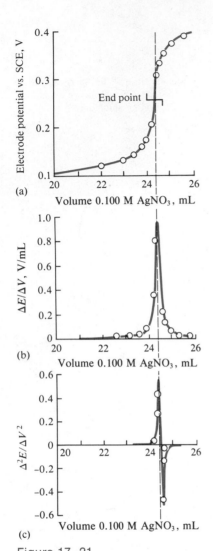

(a)
Volume 0.100 M $AgNO_3$, mL

(b)
Volume 0.100 M $AgNO_3$, mL

(c)
Volume 0.100 M $AgNO_3$, mL

**Figure 17–21**
Titration of 2.433 mmol of chloride ion with 0.1000 M silver nitrate. (a) Titration curve. (b) First-derivative curve. (c) Second-derivative curve.

**Table 17–5**
**POTENTIOMETRIC TITRATION DATA FOR 2.433 MMOL OF CHLORIDE WITH 0.1000 M SILVER NITRATE**

| Volume $AgNO_3$, mL | $E$ vs. SCE, V | $\Delta E/\Delta V$, V/mL | $\Delta^2 E/\Delta V^2$, $V^2/mL^2$ |
|---|---|---|---|
| 5.0 | 0.062 | | |
| 15.0 | 0.085 | 0.002 | |
| 20.0 | 0.107 | 0.004 | |
| 22.0 | 0.123 | 0.008 | |
| 23.0 | 0.138 | 0.015 | |
| 23.50 | 0.146 | 0.016 | |
| 23.80 | 0.161 | 0.050 | |
| 24.00 | 0.174 | 0.065 | |
| 24.10 | 0.183 | 0.09 | |
| 24.20 | 0.194 | 0.11 | 2.8 |
| 24.30 | 0.233 | 0.39 | 4.4 |
| 24.40 | 0.316 | 0.83 | -5.9 |
| 24.50 | 0.340 | 0.24 | -1.3 |
| 24.60 | 0.351 | 0.11 | -0.4 |
| 24.70 | 0.358 | 0.07 | |
| 25.00 | 0.373 | 0.050 | |
| 25.5 | 0.385 | 0.024 | |
| 26.0 | 0.396 | 0.022 | |
| 28.0 | 0.426 | 0.015 | |

Table 17–5 reveals that the maximum is located between 24.30 and 24.40 mL; selection of 24.35 mL would be adequate for most purposes.

Column 4 of Table 17–5 and Figure 17–21c show that the second derivative for the data changes sign at the point of inflection. This change is used as the analytical signal in some automatic titrators.

All the foregoing methods of end-point evaluation are predicated on the assumption that the titration curve is symmetric about the equivalence point and that the inflection in the curve corresponds to this point. This assumption is perfectly valid, provided the participants in the titration react with one another in an equimolar ratio and also provided the electrode reaction is perfectly reversible. The former condition is lacking in many oxidation/reduction titrations; the titration of iron(II) with permanganate is an example. The curve for such titrations is ordinarily so steep, however, that failure to account for asymmetry results in a vanishingly small titration error.

## 17G–2 Potentiometric Precipitation Titrations

### Electrode Systems

The indicator electrode for a precipitation titration is often the metal from which the reacting cation is derived. A membrane electrode responsive to this cation or to the anion in the titration can also be used.

Silver nitrate is without question the most versatile reagent for precipitation titrations. A silver wire serves as the indicator electrode. For reagent and analyte concentrations of 0.1 M or greater, a calomel reference electrode can be located directly in the titration vessel without serious error from the slight leakage of chloride ions from the salt bridge. This leakage can be a source of significant error in titrations that involve very dilute solutions or require high precision, however. The difficulty is eliminated by immersing the calomel electrode in a potassium nitrate solution that is connected to the analyte solution by a salt bridge containing potassium nitrate. Reference electrodes with bridges of this type can be purchased from laboratory supply houses.

## Titration Curves

A theoretical curve for a potentiometric titration is readily derived. For example, the potential of a silver electrode in the argentometric titration of chloride can be described by

$$E_{Ag} = E^0_{AgCl} - 0.0592 \log [Cl^-] = 0.222 - 0.0592 \log [Cl^-]$$

where $E^0_{AgCl}$ is the standard potential for the reduction of AgCl to silver. Alternatively, the standard potential for the reduction of silver ion can be used:

$$E_{Ag} = E^0_{Ag^+} - 0.0592 \log \frac{1}{[Ag^+]} = 0.799 - 0.0592 \log \frac{1}{[Ag^+]}$$

The former potential is more convenient for calculating the potential of the silver electrode when an excess of chloride exists, whereas the latter is preferable for solutions containing an excess of silver ion.

Potentiometric measurements are particularly useful for titrations of mixtures of anions with standard silver nitrate. For example, Figure 9–5 shows a theoretical curve for the titration of a chloride/iodide mixture. An experimental curve has the same general appearance, although the ordinate units are different.

## 17G–3 Complex-Formation Titrations

Both metallic and membrane electrodes have been used to detect end points in potentiometric titrations involving complex formation. The mercury electrode (Figure 17–22) is particularly useful for EDTA titrations of cations that form complexes that are less stable than $HgY^{2-}$ (see page 405 for the half-reactions involved).

These electrodes can be obtained from Kontes Manufacturing Corp., Vineland, NJ 08360.

## 17G–4 Neutralization Titrations

Experimental neutralization curves closely approximate the theoretical curves described in Chapter 10. Ordinarily, however, the experimental curves are somewhat displaced from the theoretical along the pH axis because concentrations rather than activities are used in their derivation; this displacement is of no consequence in locating end points. Potentio-

Figure 17–22
A typical mercury electrode.

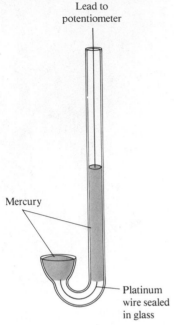

Lead to
potentiometer

Mercury

Platinum
wire sealed
in glass

metric neutralization titrations are particularly valuable for the analysis of mixtures of acids or polyprotic acids. The same considerations apply to bases.

### The Determination of Dissociation Constants

An approximate numerical value for the dissociation constant of a weak acid or base can be estimated from potentiometric titration curves. This quantity can be computed from the pH at any point along the curve, but as a practical matter, the pH at half-neutralization is most convenient. Here,

$$[HA] \approx [A^-]$$

Therefore,

$$K_a = \frac{[H_3O^+][A^-]}{[HA]} = [H_3O^+]$$

$$pK_a = pH$$

It is important to note that the use of concentrations instead of activities may cause the value for $K_a$ to differ from its published value by a factor of 2 or more. A more correct form of the dissociation constant for HA is

$$K_a = \frac{a_{H_3O^+}a_{A^-}}{a_{HA}} = \frac{a_{H_3O^+}[A^-]f_{A^-}}{[HA]f_{HA}}$$

(17–27)

$$K_a = \frac{a_{H_3O^+}f_{A^-}}{f_{HA}}$$

Since the glass electrode provides a good approximation of $a_{H_3O^+}$, the value of $K_a$ measured differs from the thermodynamic value by the ratio of the two activity coefficients. The activity coefficient in the denominator of Equation 17–27 does not change significantly as ionic strength increases because HA is a neutral species. The activity coefficient for $A^-$, on the other hand, decreases as the electrolyte concentration increases. Therefore, the observed hydrogen ion activity is numerically larger than the thermodynamic dissociation constant.

---

### Example 17–2

In order to determine $K_1$ and $K_2$ for $H_3PO_4$ from titration data, careful pH measurements are made after 0.5 and 1.5 mole of base are added for each mole of acid. It is then assumed that the hydrogen ion activities computed from these data are identical to the desired dissociation constants. Calculate the relative error incurred by this assumption if the ionic strength is 0.1 at the time of each measurement. (From Appendix 3, $K_1$ and $K_2$ for $H_3PO_4$ are $7.11 \times 10^{-3}$ and $6.34 \times 10^{-8}$.)

Rearrangement of Equation 17–27 gives

$$K_a(\text{exptl}) = a_{H_3O^+} = K_a(f_{HA}/f_{A^-})$$

The activity coefficient for $H_3PO_4$ is approximately unity since this species is uncharged. In Table 7–1, we find that the activity coefficient for $H_2PO_4^-$ is 0.78 and that for $HPO_4^{2-}$ is 0.36. Substituting these values into the equations for $K_1$ and $K_2$ gives

$$K_1(\text{exptl}) = 7.11 \times 10^{-3} \times 1.0/0.78 = 9.1 \times 10^{-3}$$

$$\text{Error} = \frac{9.1 \times 10^{-3} - 7.11 \times 10^{-3}}{7.11 \times 10^{-3}} \times 100\% = 28\%$$

$$K_2(\text{exptl}) = 6.34 \times 10^{-8} \times 0.78/0.36 = 1.37 \times 10^{-7}$$

$$\text{Error} = \frac{1.37 \times 10^{-7} - 6.34 \times 10^{-8}}{6.34 \times 10^{-8}} \times 100\% = 116\%$$

A single titration of a pure acid can provide sufficient data (equivalent weight and dissociation constant) to make identification of the acid possible.

## 17G–5   Oxidation/Reduction Titrations

An inert indicator electrode constructed of platinum is ordinarily used to detect end points in oxidation/reduction titrations. Occasionally, other inert metals, such as silver, palladium, gold, and mercury, are used instead. Titration curves similar to those derived in Chapter 15 are ordinarily obtained, although they may be displaced along the ordinate axis as a consequence of the effects of high ionic strengths. End points are determined by the methods described earlier in this chapter.

## 17H   THE DETERMINATION OF EQUILIBRIUM CONSTANTS FOR ACID/BASE, PRECIPITATION, AND COMPLEX-FORMATION REACTIONS FROM ELECTRODE POTENTIAL MEASUREMENTS

Numerical values for solubility-product constants, dissociation constants, and formation constants are conveniently evaluated through the measurement of cell potentials. One important virtue of this technique is that the measurement can be made without appreciably affecting any equilibria that may exist in the solution. For example, the potential of a silver electrode in a solution containing silver ion, cyanide ion, and the complex formed between them depends upon the activities of the three species. It is possible to measure this potential with negligible current. Since the activities of the participants are not sensibly altered during the measurement, the position of the equilibrium.

$$Ag^+ + 2\,CN^- \rightleftharpoons Ag(CN)_2^-$$

is likewise undisturbed.

Example 17-3

Calculate the formation constant $K_f$ for $Ag(CN)_2^-$:

$$Ag^+ + 2\ CN^- \rightleftharpoons Ag(CN)_2^-$$

if the cell

$$SCE\|Ag(CN)_2^-(7.50 \times 10^{-3}\ M),CN^-(0.0250\ M)|Ag$$

develops a potential of $-0.625$ V.

Proceeding as in the earlier examples, we have

$$Ag^+ + e^- \rightleftharpoons Ag(s) \qquad E^0 = +0.799\ V$$
$$-0.625 = E_{cathode} - E_{anode} = E_{Ag^+} - 0.244$$
$$E_{Ag^+} = 0.625 - +0.244 = -0.381\ V$$

Application of the Nernst equation for the silver electrode gives

$$-0.381 = 0.799 - \frac{0.0592}{1} \log \frac{1}{[Ag^+]}$$
$$\log [Ag^+] = \frac{-0.381 - 0.799}{0.0592} = -20.0$$
$$[Ag^+] = 1 \times 10^{-20}$$

The result must be rounded to one significant figure:

$$K_f = \frac{[Ag(CN)_2^-]}{[Ag^+][CN^-]^2} = \frac{7.50 \times 10^{-3}}{(1 \times 10^{-20})(2.50 \times 10^{-2})^2} = 1.2 \times 10^{21} \simeq 1 \times 10^{21}$$

In theory, any electrode system in which hydrogen ions are participants can be used to evaluate dissociation constants for acids and bases.

Example 17-4

Calculate the dissociation constant for the weak acid HP if the cell

$$SCE\|HP(0.010\ M),NaP(0.040\ M)|Pt,H_2(1.00\ atm)$$

develops a potential of $-0.591$ V.

The diagram for this cell indicates that the saturated calomel electrode is the cathode. Thus,

$$E_{cell} = E_{cathode} - 0.244$$
$$E_{cathode} = -0.591 + 0.244 = -0.347\ V$$

Application of the Nernst equation for the hydrogen electrode gives

$$-0.347 = 0.000 - \frac{0.0592}{2} \log \frac{1.00}{[H_3O^+]^2} = 0.000 + \frac{2 \times 0.0592}{2} \log [H_3O^+]$$

$$\log [H_3O^+] = \frac{-0.347 - 0.000}{0.0592} = -5.87$$

$$[H_3O^+] = 1.34 \times 10^{-6}$$

Challenge: Calculate the thermodynamic dissociation constant for HP if $f_{P^-} = 0.81$ and $f_{H_3O^+} = 0.86$.

Substitution of this value for $[H^+]$ as well as the concentrations of the weak acid and its conjugate base into the dissociation-constant expression gives

$$K_{HP} = \frac{[H_3O^+][P^-]}{[HP]} = \frac{(1.34 \times 10^{-6})(0.040)}{0.010} = 5.4 \times 10^{-6}$$

## 17I  QUESTIONS AND PROBLEMS

*17–1. Differentiate between an electrode of the first kind and an electrode of the second kind.

17–2. What occurs when a newly manufactured glass electrode is immersed in water?

17–3. What is the source of
  (a) the asymmetry potential in a membrane electrode?
  *(b) the boundary potential in a membrane electrode?
  (c) a junction potential in a glass/calomel electrode system?
  *(d) the potential of a crystalline membrane electrode used to determine the concentration of $F^-$?

*17–4. What is the alkaline error in pH measurement with a glass electrode?

17–5. List the advantages and disadvantages of a potentiometric titration relative to a titration with chemical indicators.

*17–6. A solution of ethylamine is titrated with HCl using a glass/calomel electrode system. Show how $K_b$ can be obtained from the pH at the point of half-neutralization.

*17–7. (a) Calculate the standard potential for the reaction

$$CuSCN(s) + e^- \rightleftharpoons Cu(s) + SCN^-$$

  (b) Sketch a cell having a copper indicator electrode as a cathode and a saturated calomel electrode as an anode that could be used for the determination of $SCN^-$.
  (c) Derive an equation relating the measured potential of the cell in (b) to pSCN. (Assume the junction potential is zero.)

  (d) Calculate the pSCN of a thiocyanate-containing solution that is saturated with CuSCN and employed in conjunction with a copper electrode in the cell sketched in (b) if the resulting potential is $-0.076$ V.

17–8. (a) Calculate the standard potential for the reaction

$$Ag_2S(s) + 2\,e^- \rightleftharpoons 2\,Ag(s) + S^{2-}$$

  (b) Sketch a cell having a silver indicator electrode as the cathode and a saturated calomel electrode as an anode that could be used for determining $S^{2-}$.
  (c) Derive an equation relating the measured potential of the cell in (b) to pS. (Assume the junction potential is zero.)
  (d) Calculate the pS of a solution that is saturated with $Ag_2S$ and then employed with the cell sketched in (b) if the resulting potential is $-0.538$ V.

*17–9. Give a schematic representation of each of the following cells. Derive an equation relating cell potential to p-function. Assume the junction potential is negligible; treat the indicator electrode as the cathode; and specify any necessary concentrations as $1.00 \times 10^{-4}$ M.
  (a) A cell with a mercury indicator electrode for the determination of pCl.
  (b) A cell with a silver indicator electrode for the determination of $pCO_3$.
  (c) A cell with a platinum electrode for the determination of pSn(IV).

17–10. Give a schematic representation of each of the following cells. Derive an equation relating cell poten-

tial and p-function. Assume the junction potential is negligible; treat the indicator electrode as the cathode; and specify any necessary concentrations as $1.00 \times 10^{-4}$ M.
(a) A cell with a lead electrode for the determination of $pCrO_4$.
(b) A cell with a silver indicator electrode for the determination of $pAsO_4$.
(c) A cell with a platinum electrode for the determination of pT1(III).

*17-11. The cell

$$SCE\|Ag_2CrO_4(sat'd),CrO_4^{2-}(x\,M)|Ag$$

is employed for the determination of $pCrO_4$. Calculate $pCrO_4$ when the cell potential is 0.402 V.

17-12. Calculate the potential of the cell

$$SCE\|\text{aqueous solution}|Hg$$

when the aqueous solution is
(a) $7.40 \times 10^{-3}$ M $Hg^{2+}$.
(b) $7.40 \times 10^{-3}$ M $Hg_2^{2+}$.
(c) $Hg_2SO_4(sat'd),SO_4^{2-}(0.0250\,M)$.
(d) $Hg^{2+}(2.00 \times 10^{-3}),OAc^-(0.100\,M)$.

$$Hg^{2+} + 2\,OAc^- \rightleftharpoons Hg(OAc)_2(aq)$$
$$K_f = 2.7 \times 10^8$$

*17-13. The standard potential for the reduction of the EDTA complex of Hg(II) is

$$HgY^{2-} + 2\,e^- \rightleftharpoons Hg(l) + Y^{4-} \qquad E^0 = 0.210\,V$$

Calculate the potential of the cell

$$SCE\|HgY^{2-}(2.00 \times 10^{-4}),Z|Hg$$

when Z is
(a) $H^+(1.00 \times 10^{-4}\,M),EDTA(0.0200\,M)$.
(b) $H^+(1.00 \times 10^{-8}\,M),EDTA(0.0200\,M)$.
(c) $H^+(1.00 \times 10^{-10}\,M),CaCl_2(0.0200\,M),$ $EDTA(0.0200\,M)$.

17-14. Calculate the potential of the cell described in Problem 17-13 when Z is
(a) $H^+(1.00 \times 10^{-6}\,M),EDTA(0.0100\,M)$.
(b) $H^+(1.00 \times 10^{-10}\,M),EDTA(0.0100\,M)$.
(c) $H^+(1.00 \times 10^{-7}\,M),Zn(NO_3)_2(0.0100\,M),$ $EDTA(0.0100\,M)$.

*17-15. The cell

$$SCE\|H^+(a = x)|\text{glass electrode}$$

has a potential of 0.2094 V when the solution in the right-hand compartment is a buffer of pH 4.006. The following potentials are obtained when the buffer is replaced with unknowns: (a) $-0.3011$ V and (b) $+0.1163$ V. Calculate the pH and the hydrogen ion activity of each unknown. (c) Assuming an uncertainty of $\pm0.002$ V in the junction potential,

what is the range of hydrogen ion activities within which the true value might be expected to lie?

17-16. The following cell

$$SCE\|MgA_2(a_{Mg^{2+}} = 9.62 \times 10^{-3})|$$
$$\text{membrane electrode for } Mg^{2+}$$

has a potential of 0.367 V.
(a) When the solution of known magnesium activity is replaced with an unknown solution, the potential is $+0.544$ V. What is the pMg of this unknown solution?
(b) Assuming an uncertainty of $\pm0.002$ V in the junction potential, what is the range of $Mg^{2+}$ activities within which the true value might be expected?

*17-17. The cell

$$SCE\|CdA_2(sat'd),A^-(0.0250\,M)|Cd$$

has a potential of $-0.721$ V. Calculate the solubility product of $CdA_2$, neglecting the junction potential.

17-18. The cell

$$SCE\|HA(0.250\,M),NaA(0.180\,M)|H_2(1.00\,atm),Pt$$

has a potential of $-0.797$ V. Calculate the dissociation constant of HA, neglecting the junction potential.

17-19. A 40.00-mL aliquot of 0.05000 M $HNO_2$ is diluted to 75.00 mL and titrated with 0.08000 M $Ce^{4+}$. The pH of the solution is maintained at 1.00 throughout the titration, and the formal potential of the cerium system is 1.44 V.
*(a) Calculate the potential of the indicator electrode with respect to a saturated calomel reference electrode after the addition of 5.00, 10.00, 15.00, 25.00, 40.00, 49.00, 50.00, 51.00, 55.00, and 60.00 mL of cerium(IV).
(b) Draw a titration curve for these data.

17-20. Calculate the potential of a silver cathode versus the standard calomel electrode after the addition of 5.00, 15.00, 25.00, 30.00, 35.00, 39.00, 40.00, 41.00, 45.00, and 50.00 mL of 0.1000 M $AgNO_3$ to 50.00 mL of 0.0800 M KSeCN. Construct a titration curve from these data. ($K_{sp}$ for AgSeCN = $4.20 \times 10^{-16}$.)

*17-21. Quinhydrone is an equimolar mixture of quinone (Q) and hydroquinone ($H_2Q$). These two compounds react reversibly at a platinum electrode:

$$Q + 2\,H^+ + 2\,e^- \rightleftharpoons H_2Q \qquad E^0 = 0.699\,V$$

The pH of a solution can be determined by saturating it with quinhydrone and making it a part of the cell

$$SCE\|\text{quinhydrone}(sat'd),H^+(x\,M)|Pt$$

If such a cell has a potential of 0.313 V, what is the pH of the solution, assuming the junction potential is zero?

## CHAPTER 18

# ELECTROGRAVIMETRIC AND COULOMETRIC METHODS

In this chapter we describe two related electroanalytical methods: *electrogravimetry* and *coulometry*. In both methods, an electrolysis is carried out long enough to ensure that the analyte is completely oxidized or reduced to a single product of known composition. In electrogravimetric methods, the product is weighed as a deposit on one of the electrodes (the *working electrode*). In coulometric procedures, the quantity of electrons needed to complete the electrolysis is measured.[1]

Electrogravimetry and coulometry are moderately sensitive, rapid, and among the most accurate and precise techniques available to the chemist. Like gravimetry, these methods require no preliminary calibration against standards because the functional relationship between the quantity measured and the analyte concentration can be derived from theory and atomic-weight data.

Electrogravimetry and coulometry differ from potentiometry in that they require a significant current throughout the analytical process. In contrast, potentiometric measurements are performed under conditions of essentially zero current. When there is a current in an electrochemical cell, the cell potential is no longer simply the difference between the electrode potentials of the cathode and the anode. Additional phenomena require application of potentials greater than theoretical for an electrolytic cell and development of potentials smaller than theoretical in a galvanic cell. Before proceeding, we must examine these phenomena in detail.

## 18A THE EFFECT OF CURRENT ON CELL POTENTIALS

Consider the electrolytic cell shown in Figure 18–1. A voltage $E_{applied}$ is applied to the cell in such a way that $E_{cell}$ and $E_{applied}$ have opposing polarities. A current meter is placed in the circuit to indicate the magnitude and direction of any current in the cell. There will be a current $I$ in

Electrogravimetry and coulometry usually have relative errors of a few parts per thousand.

By definition, an electric current is a flow of charge. In metallic conductors the charge is electrons. In electrolytes, the current involves the flow of positive ions in one direction and negative ions in the other. A convention for indicating the direction of currents is needed because charges of opposite sign move in opposite directions under the influence of a voltage. You should appreciate that a positive charge moving in one direction in a circuit is equivalent to a negative charge moving in the other. For simplicity and algebraic consistency, scientists assume that *all charge carriers are positive*. Thus, the current $I$ in Figure 18–2 is given a positive sign and treated as if it were a flow of positive charge from the positive electrode of the battery to the negative electrode even though the actual current is made up of negative electrons.

---

[1]For further information concerning the methods in this chapter, see J. A. Plambeck, *Electroanalytical Chemistry*. New York: Wiley, 1982; *Laboratory Techniques in Electroanalytical Chemistry*, P. T. Kissinger and W. R. Heineman, Eds. New York: Marcel Dekker, 1984.

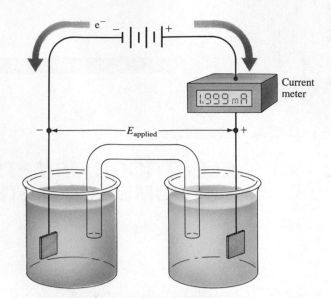

Figure 18–1
An electrolytic cell with a meter to measure the current during electrolysis.

If there is to be a current in an electrolytic cell, it is necessary that $E_{applied} > E_{cell}$.

When a galvanic cell produces a current, $E_{output} < E_{cell}$.

the circuit in the indicated direction only when $E_{applied} > E_{cell}$. When the two potentials are equal, that is when $E_{applied} = E_{cell}$, $I = 0$.

When there is a current, the potential of the cell is less than the thermodynamic potential because one or more of three phenomena are operating: *IR drop, concentration polarization,* and *kinetic polarization.*

The potential of a galvanic cell also changes when a current is present. Here, *IR* drop, concentration polarization, and/or kinetic polarization make the output potential smaller than the thermodynamic potential.

### 18A–1 Ohmic Potential: *IR* Drop

Electrochemical cells, like metallic conductors, resist the flow of charge. In both types of conduction, Ohm's law describes the effect of this resistance. The product of the resistance $R$ of a cell in ohms ($\Omega$) and the current $I$ in amperes (A) is called the *ohmic potential* or the *IR drop* of the cell. In Figure 18–2 we have used a resistor $R$ to represent the cell resistance in the circuit. If we begin with $E_{applied} = E_{cell}$ ($I = 0$) and gradually increase $E_{applied}$, a small current appears in the circuit. The current through the cell resistance $R$ results in a potential drop of $-IR$ volts. In other words, the applied voltage must be greater than the theoretical cell potential by $-IR$ volts. Similarly, when a current of $I$ amperes is produced by a galvanic cell with a resistance of $R$ ohms, the output potential of the cell is decreased by $-IR$ volts. Thus, in the presence of a current, Equation 14–10 for a cell potential must be modified by the addition of the term $-IR$:

$$E_{cell} = E_{cathode} - E_{anode} - IR \qquad (18\text{–}1)$$

Ohm's law: $E = IR$    or    $I = \left(\dfrac{1}{R}\right) E$

where $E_{cathode}$ and $E_{anode}$ are electrode potentials computed with the Nernst equation discussed in Section 14C–4.

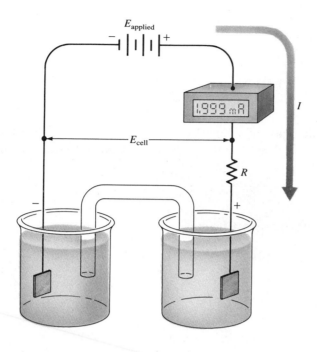

**Figure 18–2**
An electrolytic cell with a resistor $R$ to represent the cell resistance.

---

### Example 18–1

Consider a cell consisting of a copper electrode in contact with 1.00 M $Cu^{2+}$, a cadmium electrode in contact with 1.00 M $Cd^{2+}$, and a connecting salt bridge. The cell has a resistance of 4.00 Ω.

(a) Calculate the potential needed to develop a current of 0.0200 A in the electrolytic cell

$$Cu|Cu^{2+}(1.00\ M)\|Cd^{2+}(1.00\ M)|Cd$$

Since both cation concentrations are 1.00 M, the electrode potentials and the standard electrode potentials are numerically equal. Substituting in Equation 18–1 gives

$$E_{cell} = E_{Cd}^0 - E_{Cu}^0 - IR$$
$$= -0.403\ V - 0.337\ V - 0.0200\ A \times 4.00\ \Omega$$
$$= -0.740\ V - 0.080\ V = -0.820\ V$$

Thus a potential 0.08 V more negative than the theoretical value is needed to deposit cadmium at the rate required to maintain a current of 0.0200 A.

(b) Calculate the cell potential when there is a current of 0.0200 A in the galvanic cell

$$Cd|Cd^{2+}(1.00\ M)\|Cu^{2+}(1.00\ M)|Cu$$

Throughout this chapter, we assume that junction potentials are negligible compared with the other potentials in a cell.

André Marie Ampère (1775–1836), French mathematician and physicist, was the first to apply mathematics to the study of electrical current. Consistent with Benjamin Franklin's definitions of positive and negative charge, Ampère defined a positive current to be the direction of flow of positive charge. Although we now know that negative electrons carry current in metals, Ampère's definition has survived to the present. The unit of current, the ampere, is named in his honor.

Illustration of Equation 18–2.

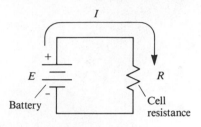

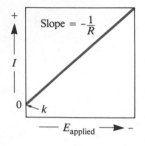

Here,

$$E_{cell} = E^0_{Cu} - E^0_{Cd} - IR$$
$$= 0.337 \text{ V} - (-0.403 \text{ V}) - 0.0200 \text{ A} \times 4.00 \text{ }\Omega$$
$$= 0.740 \text{ V} - 0.080 \text{ V} = 0.660 \text{ V}$$

The production of a current in this cell causes the output potential to be markedly less than the theoretical value because of the $IR$ drop in the cell.

## 18A–2 Polarization Effects

Equation 18–1 can be rearranged to give

$$I = -\frac{1}{R} E_{cell} + \frac{1}{R}(E_{cathode} - E_{anode})$$

For small currents and brief periods of time, $E_{cathode}$ and $E_{anode}$ remain relatively constant during an electrolysis. The cell behavior can then be approximated by the relationship

$$I = -\frac{1}{R} E_{cell} + k \tag{18–2}$$

where $k$ is a constant.

According to Equation 18–2, a plot of current in an electrolytic cell as a function of applied potential should be a straight line with a slope equal to the negative reciprocal of the resistance. As shown in Figure 18–3a, the plot is indeed linear with small currents. As the applied voltage increases, the current deviates significantly from linearity. Figure 18–3b shows that galvanic cells behave in an analogous way. Just as before, there is a linear relationship where small currents are involved, but significant departures from the straight line occur as the current becomes greater.

Cells that exhibit nonlinear behavior are said to be *polarized,* and the degree of polarization is given by an *overvoltage,* or *overpotential,* which is symbolized by $\Pi$ in the figure. Note that polarization requires the application of a potential greater than the theoretical value to give a current of the expected magnitude. Thus, the overpotential required to achieve a current of 0.06 A in the electrolytic cell in Figure 18–3a is about $-0.04$ V. For the galvanic cell in Figure 18–3b, the polarization resulting from a 0.06 A current causes the output cell potential to decrease by about 0.03 V—that is, the overvoltage is $-0.03$ V. Note that in each case the overvoltage is negative. Thus, for a cell affected by overvoltage, Equation 18–1 becomes

$$E_{cell} = E_{cathode} - E_{anode} - IR - \Pi \tag{18–3}$$

*Overvoltage* is the potential difference between the theoretical cell potential and the actual cell potential at a given level of current.

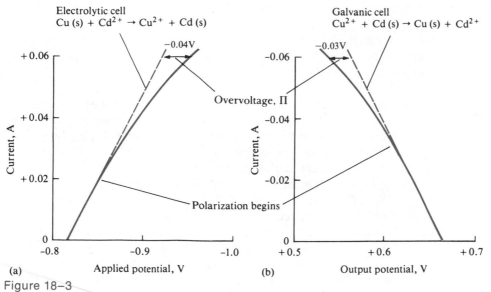

Figure 18–3

Current/voltage curves for (a) an electrolytic and (b) a galvanic cell. In both cells, the overvoltage at 0.06 A is symbolized by Π. For cell (a), Π = −0.040 V; for cell (b), Π = −0.03 V.

Polarization is an electrode phenomenon that may affect either or both of the electrodes in a cell. The degree of polarization of an electrode varies widely. In some instances it approaches zero, but in others it can be so large that the current in the cell becomes independent of potential. Under this circumstance, polarization is said to be complete. Polarization phenomena are conveniently divided into two categories: *concentration polarization* and *kinetic polarization*.

Factors that influence polarization: (1) electrode size, shape, and composition; (2) composition of the electrolyte solution; (3) temperature and stirring rate; (4) current level; (5) physical state of species involved in the cell reaction.

## Concentration Polarization

Electron transfer between a reactive species in a solution and an electrode can take place only from a thin film of solution located immediately adjacent to the electrode surface; this film is only a few angstroms thick and contains a limited number of reactive ions or molecules. In order for there to be a steady current in a cell, this film must be continuously replenished with reactant from the bulk of the solution. That is, as reactant ions or molecules are transformed by the electrochemical reaction, more must be transported into the surface film at a rate that is sufficient to maintain the current. For example, for a current of 0.01 A in the cell described in Example 18–1a, it is necessary to transport cadmium ions to the cathode surface at a rate of about $5 \times 10^{-8}$ mol/s, which is equal to $3 \times 10^{16}$ cadmium ions per second. (Similarly, copper ions must be removed from the surface film of the anode at this same rate.)

Concentration polarization occurs when reactant species do not arrive at the cathode surface or product species do not leave the anode surface fast enough to maintain the desired current. When this happens, the current is limited to values less than those predicted by Equation 18–2.

Reactants are transported to an electrode by (1) diffusion, (2) migration, and (3) convection.

Reactants are transported to an electrode surface by three mechanisms: (1) *diffusion*, (2) *migration*, and (3) *convection*. Products are removed from an electrode surface in the same ways. We will focus on mass-transport processes to the cathode, but our discussion applies equally to anodes.

**Diffusion.** When there is a concentration difference between two regions of a solution, ions or molecules move from the more concentrated region to the more dilute. This process is called *diffusion* and ultimately leads to a disappearance of the concentration difference. The rate of diffusion is directly proportional to the concentration difference. For example, when cadmium ions are deposited at a cathode by a current as illustrated in Figure 18–4, the concentration of $Cd^{2+}$ at the electrode surface $[Cd^{2+}]_0$ is very small. The difference between $[Cd^{2+}]_0$ at the surface and $[Cd^{2+}]$ in the bulk solution creates a concentration gradient that causes cadmium ions to diffuse toward the surface film. The rate of diffusion is given by

Diffusion is a process in which ions or molecules move from a more concentrated part of a solution to a more dilute one.

$$\text{rate of diffusion to cathode surface} = k'([Cd^{2+}] - [Cd^{2+}]_0) \quad (18\text{–}4)$$

where $[Cd^{2+}]$ is the reactant concentration in the bulk of the solution, $[Cd^{2+}]_0$ is its equilibrium concentration at the surface of the cathode, and $k'$ is a proportionality constant. The value of $[Cd^{2+}]_0$ at any instant is fixed by the *potential of the electrode* and can be calculated from the Nernst equation. In this example, the surface cadmium ion concentration is given by

$$E_{\text{cathode}} = E^0_{Cd^{2+}} - \frac{0.0592}{2} \log \frac{1}{[Cd^{2+}]_0}$$

where $E_{\text{cathode}}$ is the potential applied to the cathode. As the applied potential becomes more and more negative, $[Cd^{2+}]_0$ becomes smaller and smaller. The result is that the rate of diffusion and the current become correspondingly larger.

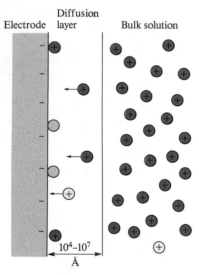

Diffusion
Electrode    layer    Bulk solution

$10^4$–$10^7$
Å

− Electrons
● Freshly deposited Cd atoms
⊕ Cd ions

**Figure 18–4**

Concentration changes at the surface of a cathode. As $Cd^{2+}$ ions are reduced to Cd atoms at the electrode surface, the concentration of $Cd^{2+}$ at the surface becomes very small. Ions then diffuse from the bulk of the solution to the surface as a result of the concentration gradient.

**Migration.** The process by which ions move under the influence of an electric field is called *migration*. This process, shown schematically in Figure 18–5, is the primary cause of mass transfer in the bulk of the solution in a cell. The rate at which ions migrate to or away from an electrode surface generally increases as the electrode potential increases. This charge movement constitutes a current, which also increases with potential. The larger the number of different kinds of ions in a given solution, the smaller is the fraction of the total charge that is carried by a particular species. For example, if a solution contains only cadmium ions (and their corresponding counter ions) the total current in the solution is carried by these ions. If a second salt such as sodium chloride is added to the solution, the fraction of the total current carried by the cadmium ions decreases. When the concentration of sodium ions exceeds that of cadmium ions by a factor of 50 to 100, the fraction of the total current carried by cadmium approaches zero. As a result, the rate of migration of cadmium toward the cathode becomes essentially independent of applied potential.

**Convection.** Reactants can also be transferred to or from an electrode by mechanical means. Thus, forced *convection,* such as stirring or agitation, will tend to decrease the thickness of the diffuse layer at an electrode surface and thus decrease concentration polarization. Natural convection resulting from temperature or density differences also contributes to the transport of species to and from an electrode.

**The Importance of Concentration Polarization.** As noted earlier, concentration polarization sets in when the effects of diffusion, migration, and convection are insufficient to transport a reactant to or from an electrode surface at a rate that produces a current of the magnitude given by Equation 18–2. Concentration polarization requires applied potentials that are larger than theoretical to maintain a given current in an electrolytic cell (Figure 18–3a). Similarly, the phenomenon causes a galvanic cell potential to be smaller than the value predicted from theory and the *IR* drop (Figure 18–3b).

Concentration polarization is important in several electroanalytical methods. In some applications, its effects are deleterious, and steps are taken to eliminate it. In others, it is essential to the analytical method, and every effort is made to promote its occurrence.

## Kinetic Polarization

In kinetic polarization, the magnitude of the current is limited by the rate of one or both electrode reactions—that is, the rate of electron transfer between reactants and electrodes. In order to offset kinetic polarization, an additional potential, or overvoltage, is required to overcome the energy barrier to the half-reaction.

Kinetic polarization is most pronounced for electrode processes that yield gaseous products and is often negligible for reactions that involve the deposition or solution of a metal. Kinetic effects usually decrease with increasing temperature and decreasing current density. These effects also depend upon the composition of the electrode and are most pronounced with softer metals, such as lead, zinc, and particularly mercury. The magnitude of overvoltage effects cannot be predicted from present theory and can only be estimated from empirical information in the literature.[2] In common with *IR* drop, overvoltage effects cause the potential of a galvanic cell to be smaller than theoretical and to require potentials greater than theoretical to operate an electrolytic cell at a desired current.

The overvoltages associated with the formation of hydrogen and oxygen are often 1 V or more and are of considerable importance because these molecules are frequently produced by electrochemical reactions. Of

Figure 18–5
Migration is the movement of ions through a solution as a result of electrostatic attraction between the ions and the electrodes.

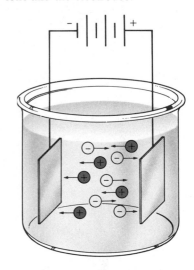

Convection is the mechanical transport of ions or molecules through a solution as a result of stirring, vibration, or temperature and/or density gradients.

Experimental variables that influence degree of concentration polarization: (1) reactant concentration, (2) electrolyte concentration, (3) mechanical agitation, (4) electrode size.

The current in a kinetically polarized cell is governed by the rate of electron transfer rather than the rate of mass transfer.

Current density is defined as amperes per square centimeter ($A/cm^2$) of electrode surface.

[2]Overvoltage data for various gaseous species on different electrode surfaces have been compiled by J. A. Page, in *Handbook of Analytical Chemistry,* L. Meites, Ed., p. 5–184– 5–186. New York: McGraw-Hill, 1963.

particular interest is the high overvoltage of hydrogen on such metals as copper, zinc, and mercury. These metals and several others can therefore be deposited without interference from hydrogen evolution. In theory, it is not possible to deposit zinc from a neutral aqueous solution because hydrogen forms at a potential that is considerably less than that required for zinc deposition. In fact, the metal can be deposited on a copper electrode with no significant hydrogen formation because the rate at which the gas forms on both zinc and copper is negligible, as shown by the high hydrogen overvoltage associated with these metals.

## 18B  THE POTENTIAL SELECTIVITY OF ELECTROLYTIC METHODS

In principle, electrolytic methods offer a reasonably selective means for separating and determining a number of ions. The feasibility of and theoretical conditions for accomplishing a given separation can be readily derived from the standard electrode potentials for the species of interest.

---

Example 18–2

Is a quantitative separation of $Cu^{2+}$ and $Pb^{2+}$ by electrolytic deposition feasible in principle? If so, what range of cathode potentials can be used? Assume that the sample solution is initially 0.1000 M in each ion and that quantitative removal of an ion is realized when only 1 part in 10,000 remains undeposited.

In Appendix 6, we find

$$Cu^{2+} + 2\,e^- \rightleftharpoons Cu(s) \qquad E^0 = 0.337 \text{ V}$$
$$Pb^{2+} + 2\,e^- \rightleftharpoons Pb(s) \qquad E^0 = -0.126 \text{ V}$$

It is apparent that copper will begin to deposit before lead. Let us first calculate the potential required to reduce the $Cu^{2+}$ concentration to $10^{-4}$ of its original concentration (that is, to $1.00 \times 10^{-5}$ M). Substituting into the Nernst equation, we obtain

$$E = 0.337 - \frac{0.0592}{2} \log \frac{1}{1.00 \times 10^{-5}} = 0.189 \text{ V}$$

Similarly, we can derive the cathode potential at which lead begins to deposit:

$$E = -0.126 - \frac{0.0592}{2} \log \frac{1}{0.100} = -0.156 \text{ V}$$

Therefore, if the cathode potential is maintained between 0.189 and −0.156 V, a quantitative separation should in theory occur.

---

Calculations such as the foregoing make it possible to compute the differences in standard electrode potentials theoretically needed for determining one ion without interference from another; these differences range from about 0.04 V for triply charged ions to about 0.24 V for singly charged ions.

These theoretical separation limits can be approached only by maintaining the potential of the *working electrode* (usually the cathode, at which metal deposition occurs) at a level required by theory. The potential of this electrode can *be controlled only by variation of the potential applied to the cell,* however. From Equation 18–5, it is evident that variations in $E_{applied}$ affect not only the cathode potential but also the anode potential, the *IR* drop, and any overvoltages associated with the electrode processes. All these potentials vary continuously as the electrolysis proceeds, and it is not feasible to calculate the fraction of the applied potential attributable to the working electrode. As a consequence, the only *practical* way of achieving the separation of species whose electrode potentials differ by only a few tenths of a volt is to *measure* the cathode potential continuously against a reference electrode whose potential is known; the cell potential can then be adjusted to maintain the potential of the working electrode within the desired range. An analysis performed in this way is called a *controlled-cathode-potential electrolysis* or a *controlled-anode-potential electrolysis*.

> Controlled potential electrolysis is discussed in Section 18C–2.

## 18C   ELECTROGRAVIMETRIC METHODS OF ANALYSIS

Electrolytic precipitation has been used for over a century for the gravimetric determination of metals. In most applications, the metal is deposited on a weighed platinum cathode, and the increase in weight is determined. Important exceptions include the anodic deposition of lead as lead dioxide on platinum and of chloride as silver chloride on silver.

There are two types of electrogravimetric methods. In one, the potential of the working electrode is not controlled and the applied cell potential is held at a more or less constant level that provides a large enough current to complete the electrolysis in a reasonable length of time. The second type is the *potentiostatic method*. This procedure is also called the *controlled-cathode-potential* or the *controlled-anode-potential method*, depending upon whether the working electrode is a cathode or an anode.

> A *working electrode* is the electrode at which the analytical reaction occurs.

> A *potentiostatic method* is an electrolytic procedure in which the potential of the working electrode is maintained at a constant level versus a reference electrode, such as an SCE.

### 18C–1   Electrogravimetry Without Potential Control of the Working Electrode

Electrolytic procedures in which no effort is made to control the potential of the working electrode make use of simple and inexpensive equipment and require little operator attention. Here, the potential applied across the cell is maintained at a more or less constant level throughout the electrolysis.

Figure 18–6
Apparatus for the electrodeposition of metals without cathode-potential control.

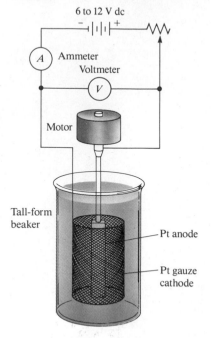

Figure 18–6
Apparatus for the electrodeposition of metals without cathode-potential control.

6 to 12 V dc

Ammeter

Voltmeter

Motor

Tall-form beaker

Pt anode

Pt gauze cathode

## Instrumentation

As shown in Figure 18–6, the apparatus for an analytical electrodeposition without cathode-potential control consists of a suitable cell and a direct-current power supply. The dc power supply usually consists of an ac rectifier, although a 6-V storage battery can also be used. The voltage applied to the cell is controlled by a rheostat. An ammeter and a voltmeter indicate the approximate current and applied voltage. When an analytical electrolysis is performed with this apparatus, the applied voltage is adjusted with the rheostat to give a current of several tenths of an ampere. The voltage is then maintained at about the initial level until the deposition is judged complete.

## Electrolysis Cells

Figure 18–6 shows a typical cell for the deposition of a metal on a solid electrode. The cathode is typically a cylinder of platinum gauze. The anode, located inside of and concentric with the cathode, is held in the chuck of a stirring motor.

## The Physical Properties of Electrolytic Precipitates

Ideally, an electrolytically deposited metal should be strongly adherent, dense, and smooth so that it can be washed, dried, and weighed without mechanical loss or reaction with the atmosphere. Good metallic deposits are fine-grained and have a metallic luster; spongy, powdery, or flaky precipitates are likely to be less pure and less adherent.

The principal factors that influence the physical characteristics of deposits are current density, temperature, and the presence of complexing agents. Ordinarily, the best deposits are formed at current densities that are less than $0.1$ A/cm$^2$. Stirring generally improves the quality of a deposit. The effects of temperature are unpredictable and must be determined empirically.

Many metals form smoother and more adherent films when deposited from solutions in which their ions exist primarily as complexes. Cyanide and ammonia complexes often provide the best deposits. The reasons for this effect are not obvious.

## Applications

In practice, electrolysis at a constant cell potential is limited to the separation of an easily reduced cation from cations that are more difficult to reduce than hydrogen ion or some other easily reduced species, such as nitrate ion. The reason for this limitation is illustrated in Figure 18–7, which shows the changes in current, *IR* drop, and cathode potential during an electrolysis in the cell in Figure 18–6. The analyte here is copper(II) in a solution containing an excess of acid. Initially, the rheostat is adjusted so that the potential applied to the cell is about $-2.5$ V, which, as shown in Figure 18–7a, leads to a current of about 1.5A. The electrolytic deposition of copper is then completed at this applied potential.

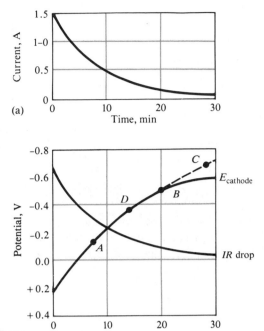

(a)

(b)

Figure 18–7
(a) Current, (b) *IR* drop, and cath-ode-potential change during the electrolytic deposition of copper at a constant applied cell potential.

As shown in Figure 18–7b, the *IR* drop decreases continually as the reaction proceeds. The reason for this decrease is primarily concentration polarization at the cathode, which limits the rate at which copper ions are brought to the electrode surface and thus limits the current. From Equation 18–1, it is apparent that the decrease in *IR* must be offset by an increase in the cathode potential since the cell potential is constant.

Ultimately, the decrease in current and increase in cathode potential are slowed at point *B* by the reduction of hydrogen ions. Because the solution contains a large excess of acid, the current is now no longer limited by concentration polarization, and codeposition of copper and hydrogen goes on simultaneously until the remainder of the copper ions are deposited. Under these conditions, the cathode is said to be *depolarized* by hydrogen ions.

Consider now the fate of a metal ion such as lead(II), which begins to deposit at point *A* on the cathode-potential curve. Clearly this ion would codeposit well before copper deposition was complete, and an interference would result. In contrast, a metal ion such as cobalt(II), which reacts at a cathode potential corresponding to point *C* on the curve, would not interfere because depolarization by hydrogen formation prevents the cathode from reaching this potential.

Codeposition of hydrogen during electrolysis often leads to the formation of analyte deposits that do not adhere to the electrode, a situation that is unsatisfactory for analytical purposes. For this reason, cathode depolarizers that are reduced at less negative cathode potentials than hydrogen ion are introduced. Nitrate ion functions in this manner, its reaction beginning at a potential less negative than that at point *C* in

A depolarizer is a chemical that is easily oxidized (or reduced) and stabilizes the potential of a working electrode by minimizing concentration polarization.

Figure 18–7b. The cathodic reduction of nitrate ion is described by the equation.

$$NO_3^- + 10\ H^+ + 8\ e^- \rightleftharpoons NH_4^+ + 3\ H_2O \qquad (18\text{--}5)$$

Electrolytic methods performed without electrode-potential control, while limited by their lack of selectivity, do have several applications of practical importance. Table 18–1 lists the common elements that can be determined by this procedure.

## 18C–2 Constant-Cathode-Potential Gravimetry

In the discussion that follows, the working electrode is assumed to be a cathode, at which an analyte is deposited as a metal. The remarks are readily extended to an anodic working electrode, however, or to products that do not form as metallic deposits.

### Instrumentation

In order to separate species with electrode potentials that differ by only a few tenths of a volt, it is necessary to use a more elaborate technique than the one just described. These more elaborate techniques are required because concentration polarization at the cathode, if unchecked, causes the potential of that electrode to become so negative that codeposition of the other species begins before the analyte is completely deposited (Figure 18–7b). A large negative drift in the cathode potential can be avoided by employing a three-electrode system, such as that shown in Figure 18–8.

The controlled-potential apparatus shown in Figure 18–8 is made up of two independent electrical circuits that share a common electrode, the *working electrode* at which the analyte is deposited. The *electrolysis circuit* consists of a dc source, a potential divider (*ACB*) that permits continuous variation in the potential applied across the working electrode, a *counter electrode,* and a current meter. The *reference,* or *control, circuit* is made up of a reference electrode (often an SCE), a high-resistance digital voltmeter, and the working electrode. The electrical resistance of

A controlled potential apparatus such as that shown in Figure 18–8 is called a potentiostat. Most potentiostats are automated.

A counter electrode has no effect on the reaction at the working electrode. It simply serves to feed electrons to the working cathode.

Table 18–1
TYPICAL APPLICATIONS OF ELECTROGRAVIMETRIC
METHODS WITHOUT POTENTIAL CONTROL

| Analyte | Weighed as | Cathode | Anode | Conditions |
|---|---|---|---|---|
| $Ag^+$ | Ag | Pt | Pt | Alkaline $CN^-$ solution |
| $Br^-$ | AgBr (on anode) | Pt | Ag | |
| $Cd^{2+}$ | Cd | Cu on Pt | Pt | Alkaline $CN^-$ solution |
| $Cu^{2+}$ | Cu | Pt | Pt | $H_2SO_4/HNO_3$ solution |
| $Mn^{2+}$ | $MnO_2$ (on anode) | Pt | Pt dish | HCOOH/HCOONa solution |
| $Ni^{2+}$ | Ni | Cu on Pt | Pt | Ammoniacal solution |
| $Pb^{2+}$ | $PbO_2$ (on anode) | Pt | Pt | Strong $HNO_3$ solution |
| $Zn^{2+}$ | Zn | Cu on Pt | Pt | Acidic citrate solution |

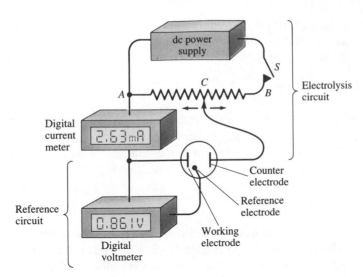

Figure 18–8

Apparatus for controlled-potential electrolysis. Contact *C* is adjusted as necessary to maintain the working electrode (cathode in this example) at a constant potential. The current in the reference-electrode circuit is essentially zero at all times.

the reference circuit is so large that the electrolysis circuit supplies essentially all the current for the deposition.

The purpose of the reference circuit is to monitor continuously the potential between the working electrode and the reference electrode. When this potential reaches a level at which codeposition of an interfering species is about to begin, the potential across the working and counter electrodes is decreased by moving contact *C* to the left. Since the potential of the counter electrode remains constant during this change, the cathode potential becomes smaller, thus preventing interference from codeposition.

The current and applied, or cell, voltage changes that occur in a typical constant-cathode-potential electrolysis are depicted in Figure 18–9. Note that the applied cell potential has to be decreased continuously throughout the electrolysis. As a consequence, the analyst would need to monitor the apparatus throughout the analysis. To avoid such waste of operator

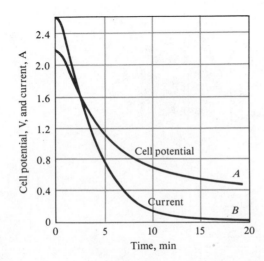

Figure 18–9

Changes in cell potential (*A*) and current (*B*) during a controlled-cathode-potential deposition of copper. The cathode is maintained at −0.36 V versus SCE throughout the experiment. (Data from J. J. Lingane, *Anal Chem. Acta,* **1948,** *2,* 590. With permission.)

time, controlled-cathode-potential electrolyses are generally performed with automated instruments called *potentiostats,* which maintain a constant cathode potential electronically.

### Cells

Figure 18–6 shows a typical cell for the deposition of a metal on a solid electrode. Tall-form beakers are ordinarily employed, and mechanical stirring is provided to minimize concentration polarization; frequently, the anode serves as a mechanical stirrer.

### Electrodes

Ordinarily, the working electrode in electrogravimetry is a metallic gauze cylinder as shown in Figure 18–6. Electrodes are usually constructed of platinum, although copper, brass, and other metals find occasional use. Platinum electrodes have the advantage of being relatively nonreactive and can be ignited to remove any grease, organic matter, or gases that could have a deleterious effect on the physical properties of the deposit.

Certain metals (notably bismuth, zinc, and gallium) cannot be deposited directly onto platinum without causing permanent damage to the electrode. Consequently, a protective coating of copper is always deposited on a platinum electrode before the electrolysis of these metals is undertaken.

### The Mercury Cathode

A mercury cathode, such as that shown in Figure 18–10, is particularly useful for removing easily reduced elements as a preliminary step in an analysis. For example, copper, nickel, cobalt, silver, and cadmium are readily separated at this electrode from such ions as aluminum, titanium,

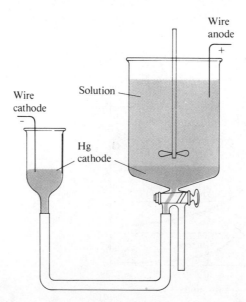

Figure 18–10

A mercury cathode for the electrolytic removal of metal ions from solution.

the alkali metals, sulfates, and phosphates. The precipitated elements dissolve in the mercury with little hydrogen evolution because even at high applied potentials, formation of the gas is prevented by the high overvoltage associated with its evolution on mercury. Ordinarily, the deposited metals are not determined after electrolysis, the goal being simply the removal of interfering ions from a solution of an analyte.

## Applications

The controlled-cathode-potential method is a potent tool for separating and determining metallic species having standard potentials that differ by only a few tenths of a volt. An example illustrating the power of the controlled-cathode-potential method involves determining copper, bismuth, lead, cadmium, zinc, and tin in mixtures by successive deposition of the metals on a platinum cathode that is weighed after each deposition. The first three elements are deposited from a nearly neutral solution containing tartrate ion to complex the tin(IV) and prevent its deposition. First copper is reduced quantitatively by maintaining the cathode potential at $-0.2$ V with respect to a saturated calomel electrode. After being weighed, the copper-plated cathode is returned to the solution, and bismuth is removed at a potential of $-0.4$ V. After the electrode is again weighed, lead is deposited quantitatively by increasing the cathode potential to $-0.6$ V. When lead deposition is complete, the solution is made strongly ammoniacal, and cadmium and zinc are deposited successively at $-1.2$ and $-1.5$ V. Finally, the solution is acidified in order to decompose the tin/tartrate complex by the formation of undissociated tartaric acid. Tin is then deposited at a cathode potential of $-0.65$ V. A fresh cathode must be used here because the zinc redissolves under these conditions.

A procedure such as this is particularly attractive for use with a potentiostat because little operator time is required for the complete analysis.

Table 18–2 lists typical separations performed by the controlled-cathode-potential method.

## 18D   COULOMETRIC METHODS OF ANALYSIS

Coulometric methods are performed by measuring the quantity of electrical charge (electrons) required to convert an analyte quantitatively to a

| Table 18–2 SOME APPLICATIONS OF CONTROLLED-CATHODE-POTENTIAL ELECTROLYSIS | |
| --- | --- |
| Element Determined | Other Elements That Can Be Present |
| Ag | Cu and heavy metals |
| Cu | Bi, Sb, Pb, Sn, Ni, Cd, Zn |
| Bi | Cu, Pb, Zn, Sb, Cd, Sn |
| Sb | Pb, Sn |
| Sn | Cd, Zn, Mn, Fe |
| Pb | Cd, Sn, Ni, Zn, Mn, Al, Fe |
| Cd | Zn |
| Ni | Zn, Al, Fe |

different oxidation state. Coulometric and gravimetric methods share the common advantage that the proportionality constant between the quantity measured and the analyte weight is derived from accurately known physical constants, thus eliminating the need for calibration standards. In contrast to gravimetric methods, coulometric procedures are usually rapid and do not require that the product of the electrochemical reaction be a weighable solid. Coulometric methods are as accurate as their conventional gravimetric and volumetric counterparts and in addition are readily automated.[3]

## 18D-1  The Quantity of Electrical Charge

Units for the quantity of charge include the coulomb (C) and the faraday (F). *The coulomb is the quantity of electrical charge transported by a constant current of one ampere in one second.* Thus, the number of coulombs ($Q$) resulting from a constant current of $I$ amperes operated for $t$ seconds is

1 coulomb = 1 ampere · second

$$Q = It \qquad (18-6)$$

For a variable current $i$, the number of coulombs is given by the integral

$$Q = \int_0^t i \, dt \qquad (18-7)$$

A faraday of charge is equal to one mole of electrons, or $6.022 \times 10^{23}$ electrons.

The equivalent weight of a substance involved in an oxidation/reduction reaction is the formula weight of the substance divided by the number of electrons it donates or accepts.

*The faraday is the quantity of charge that produces one equivalent of chemical change at an electrode.* Since the equivalent in an oxidation/reduction reaction is that amount of a substance that donates or accepts one mole of electrons, the faraday is equal to $6.022 \times 10^{23}$ electrons. The faraday also equals 96,485 C. As shown in Example 18–3, we can use these definitions to calculate the weight of a chemical species formed at an electrode by a current of known magnitude.

---

### Example 18-3

A constant current of 0.800 A is used to deposit copper at the cathode and oxygen at the anode of an electrolytic cell. Calculate the number of grams of each product formed in 15.2 min, assuming no other redox reaction.

The equivalent weights of copper and oxygen are determined from consideration of the two half-reactions

---

[3]For additional information about coulometric methods, see E. Bishop, in *Comprehensive Analytical Chemistry,* C. L. Wilson and D. W. Wilson, Eds., Vol. 11D. New York: Elsevier Scientific, 1975; J. A. Plambeck, *Electroanalytical Chemistry,* Chapter 12. New York: Wiley, 1982; D. J. Curran, in *Laboratory Technique in Electroanalytical Chemistry,* P. T. Kissinger and W. R. Heineman, Eds., pp. 539–568. New York: Marcel Dekker, 1984.

$$Cu^{2+} + 2\ e^- \rightarrow Cu(s)$$

$$2\ H_2O \rightarrow 4\ e^- + O_2(g) + 4\ H^+$$

In this reaction, 1 mol of copper accepts 2 mol of electrons, and 1 mol of oxygen represents 4 mol of electrons. Therefore the equivalent weight of copper is fw Cu/2 and the equivalent weight of oxygen is fw $O_2$/4.

Substituting into Equation 18–6 yields

$$Q = 0.800\ \text{A} \times 15.2\ \cancel{\text{min}} \times 60\ \text{s}/\cancel{\text{min}} = 729.6\ \text{A} \cdot \text{s} = 729.6\ \text{C}$$

To convert this quantity of electrical charge to moles of electrons, we write

$$\frac{729.6\ \cancel{C}}{96{,}485\ \cancel{C}/F} = 7.56 \times 10^{-3}\ F \equiv 7.562 \times 10^{-3}\ \text{mol of electrons}$$

From the definition of the faraday, $7.562 \times 10^{-3}$ equivalent of copper is deposited on the cathode; the same quantity of oxygen is evolved at the anode. Therefore,

$$\text{wt Cu} = 7.562 \times 10^{-3}\ \cancel{\text{mol e}^-} \times \frac{1\ \cancel{\text{mol Cu}}}{2\ \cancel{\text{mol e}^-}} \times \frac{63.54\ \text{g Cu}}{1\ \cancel{\text{mol Cu}}} = 0.240\ \text{g Cu}$$

$$\text{wt O}_2 = 7.562 \times 10^{-3}\ \cancel{\text{mol e}^-} \times \frac{1\ \cancel{\text{mol O}_2}}{4\ \cancel{\text{mol e}^-}} \times \frac{32.00\ \text{g O}_2}{1\ \cancel{\text{mol O}_2}} = 0.0605\ \text{g O}_2$$

Michael Faraday (1791–1867) was one of the foremost chemists and physicists of his time. Among his most important discoveries were Faraday's laws of electrolysis. Faraday, a simple man who lacked mathematical sophistication, was a superb experimentalist and an inspiring teacher and lecturer. The quantity of charge equal to a mole of electrons is named in his honor.

## 18D–2 Types of Coulometric Methods

Two types of methods have been developed that are based on the quantity of charge: *potentiostatic coulometry* and *amperostatic coulometry*. Potentiostatic methods are performed in much the same way as controlled-potential gravimetric methods, with the potential of the working electrode relative to a reference electrode being maintained constant throughout the electrolysis. Here however, the electrolysis current is recorded as a function of time to give a curve similar to curve *B* in Figure 18–9. The analysis is then completed by integrating the current/time curve to obtain the number of coulombs and thus the number of equivalents of analyte.

Coulometric titrations are similar to other titrimetric methods in that analyses are based on measuring the combining capacity of the analyte with a standard reagent. The reagent in a coulometric titration is electrons and the standard solution is a constant current of known magnitude. Electrons are added to the analyte (in this case via the direct current) or to some species that immediately reacts with the analyte until an end point is reached. At that point, the electrolysis is discontinued. The amount of analyte is determined from the magnitude of the current and the time required to complete the titration. The magnitude of the current in amperes is analogous to the molarity of a standard solution, and the time measurement is analogous to the volume measurement in conventional titrimetry.

Amperostatic coulometry is also called coulometric titrimetry.

Electrons are the reagent in a coulometric titration.

## 18D–3 Current-Efficiency Requirements

A fundamental requirement for all coulometric methods is 100% current efficiency; that is, each faraday of electricity must bring about one equivalent of chemical change in the analyte. Note that 100% current efficiency can be achieved without the analyte participating directly in electron transfer at an electrode. For example, chloride ions are readily determined by either of the two coulometric methods by the generating silver ions at a silver anode. These ions then react with the analyte to form a deposit of silver chloride. The quantity of electricity required to complete the silver chloride formation serves as the analytical parameter. In this instance, 100% current efficiency is realized because the number of moles of electrons is exactly equal to the number of moles of chloride ion in the sample despite the fact that these ions do not react directly at the electrode.

## 18D–4 Controlled-Potential Coulometry

In controlled-potential coulometry, the potential of the working electrode is maintained constant at a level at which only the analyte is responsible for the conduction of charge across the electrode/solution interface. The number of coulombs of electricity required to convert the analyte to its reaction product is then determined by recording and integrating the current/time curve derived from the electrolysis or by means of a chemical coulometer.

### Instrumentation

The instrumentation for potentiostatic coulometry consists of an electrolysis cell, a potentiostat, and a device for determining the number of coulombs of electricity consumed by the analyte.

**Cells.** Figure 18–11 illustrates two types of cells used for potentiostatic coulometry. The first consists of a platinum-gauze working electrode, a platinum-wire counter electrode, and a saturated calomel reference electrode. The counter electrode is separated from the analyte solution by a salt bridge that usually contains the same electrolyte as the solution being analyzed. This bridge prevents the reaction products formed at the counter electrode from diffusing into the analyte solution and interfering. For example, hydrogen gas is a common product at a cathodic counter electrode. Unless physically isolated from the analyte solution by the bridge, the gas will react directly with many analytes determined by oxidation at the working anode.

The second type of cell, shown in Figure 18–11b, makes use of a mercury pool. We have noted that a mercury cathode is particularly useful for separating easily reduced elements as a preliminary step in an analysis. In addition, however, it has found considerable use in the coulometric determination of several metallic cations that form mercury-soluble products (*amalgams*). In these applications, no hydrogen evolution occurs even at

An amalgam is a solution of a metal in mercury.

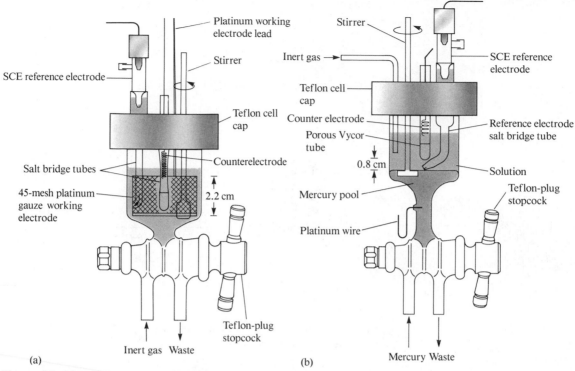

**Figure 18–11**

Electrolysis cells for potentiostatic coulometry. Working electrode: (a) platinum-gauze, (b) mercury-pool. (Reprinted with permission from J. E. Harrar and C. L. Pomernacki, *Anal Chem.*, **1973**, *45*, 57. Copyright 1973 American Chemical Society.)

high applied potentials because of the large overvoltage effects. A coulometric cell such as that shown in Figure 18–11b is also useful for the coulometric determination of certain types of organic compounds.

**Potentiostats and Coulometers.** For controlled-potential coulometry, a potentiostat similar to that shown in Figure 18–8 is required. Generally, however, the potentiostat is automated and is equipped with a recorder that provides a plot of current as a function of time as shown in Figure 18–12. The quantity of analyte is then determined from the area under the current-versus-time curve. This area is usually evaluated with an electronic integrator, although as shown in the following example, a chemical coulometer can be employed.

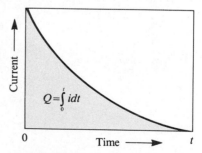

**Figure 18–12**

For a current that varies with time, the quantity of charge $Q$ in a time $t$ is the shaded area under the curve.

$$Q = \int_0^t i\,dt$$

---

Example 18–4

The Fe(III) in a 0.8202-g sample was determined by coulometric reduction to Fe(II) at a platinum cathode. Calculate the percentage of $Fe_2(SO_4)_3$ (fw = 399.88) in the sample if a hydrogen/oxygen coulometer arranged in series with the cell containing the sample evolved 19.37 mL of gas ($H_2$ + $O_2$) at 23°C and 765 torr (after correction for water vapor).

Converting the gas volume to standard conditions gives

$$V = 19.37 \text{ mL} \times \frac{765 \text{ torr}}{760 \text{ torr}} \times \frac{273 \text{ K}}{296 \text{ K}} = 17.982 \text{ mL (STP)}$$

The reactions in the coulometer are

$$4 \text{ H}^+ + 4 \text{ e}^- \rightarrow 2 \text{ H}_2(g)$$

$$2 \text{ H}_2\text{O} \rightarrow \text{O}_2(g) + 4 \text{ H}^+ + 4 \text{ e}^-$$

Thus 3 mol of gas is produced by the passage of 4 mol of electrons, which is equivalent to 0.750 mol of gas per faraday.

$$\text{No. eq Fe}_2(\text{SO}_4)_3 = \frac{17.982 \text{ mL gas}}{22,400 \text{ mL gas/mol gas}} \times \frac{1 \text{ F}}{0.750 \text{ mol gas}}$$

$$\times \frac{1 \text{ eq Fe}_2(\text{SO}_4)_3}{\text{F}}$$

$$= 1.0704 \times 10^{-3}$$

Since 1 mol of $\text{Fe}_2(\text{SO}_4)_3$ consumes 2 mol of electrons, there are 2 eq of $\text{Fe}_2(\text{SO}_4)_3$ in each mole and

$$\text{Wt Fe}_2(\text{SO}_4)_3 = 1.0704 \times 10^{-3} \text{ eq Fe}_2(\text{SO}_4)_3 \times \frac{1 \text{ mol Fe}_2(\text{SO}_4)_3}{2 \text{ eq Fe}_2(\text{SO}_4)_3}$$

$$\times \frac{399.88 \text{ g Fe}_2(\text{SO}_4)_3}{\text{mol Fe}_2(\text{SO}_4)_3}$$

$$= 0.21401 \text{ g Fe}_2(\text{SO}_4)_3$$

$$\% \text{ Fe}_2(\text{SO}_4)_3 = \frac{0.21401 \text{ g Fe}_2(\text{SO}_4)_3}{0.8202 \text{ g sample}} \times 100\% = 26.09\%$$

---

## Applications

Controlled-potential coulometric methods have been applied to the determination of some 55 elements in inorganic compounds.[4] Mercury appears to be favored as the cathode, and methods for the deposition of two dozen or more metals at this electrode have been described. The method has found widespread use in the nuclear-energy field for the relatively interference-free determination of uranium and plutonium.

The controlled-potential coulometric procedure also offers possibilities for the electrolytic determination (and synthesis) of organic compounds. For example, trichloroacetic acid and picric acid are quantitatively re-

---

[4]For a summary of the applications, see J. E. Harrar, in *Electroanalytical Chemistry,* A. J. Bard, Ed., Vol. 8. New York: Marcel Dekker, 1975; E. Bishop, in *Comprehensive Analytical Chemistry,* C. L. Wilson and D. W. Wilson, Eds., Vol. 11D, Chapter XV. New York: Elsevier, 1975.

duced at a mercury cathode whose potential is suitably controlled:

$$Cl_3CCOO^- + H^+ + 2\ e^- \rightarrow Cl_2HCCOO^- + Cl^-$$

Coulometric measurements permit the determination of these compounds with a relative error of a few tenths of a percent.

## 18D–5 Coulometric Titrations[5]

Coulometric titrations are carried out with a constant-current source called an *amperostat,* which senses decreases in cell current and responds by increasing the potential applied to the cell until the current is restored to its original level. Because of the effects of concentration polarization, 100% current efficiency with respect to the analyte can be maintained in only one way: there must be a large excess of an auxiliary reagent that is oxidized or reduced at the electrode to give a product that reacts with the analyte. As an example, consider the coulometric titration of iron(II) at a platinum anode. At the beginning of the titration, the primary anodic reaction is

$$Fe^{2+} \rightarrow Fe^{3+} + e^-$$

As the concentration of iron(II) decreases, however, the requirement of a constant current results in an increase in the applied cell potential. Because of concentration polarization, this increase in potential causes the anode potential to increase to the point where the decomposition of water becomes a competing process:

$$2\ H_2O \rightarrow O_2(g) + 4\ H^+ + 4\ e^-$$

The quantity of electricity required to complete the oxidation of iron(II) then exceeds that demanded by theory, and the current efficiency is less than 100%. The lowered current efficiency is avoided, however, by introducing at the outset an unmeasured quantity of cerium(III), which is oxidized at a lower potential than is water:

$$Ce^{3+} \rightarrow Ce^{4+} + e^-$$

Constant-current generators are called amperostats or galvanostats.

Auxiliary reagents are essential in coulometric titrations.

[5]For further details on this technique, see D. J. Curran, in *Laboratory Techniques in Electroanalytical Chemistry,* P. T. Kissinger and W. R. Heineman, Eds., Chapter 20. New York: Marcel Dekker, 1984.

With stirring, the cerium(IV) produced is rapidly transported from the surface of the electrode to the bulk of the solution, where it oxidizes an equivalent amount of iron(II):

$$Ce^{4+} + Fe^{2+} \rightarrow Ce^{3+} + Fe^{3+}$$

The net effect is an electrochemical oxidation of iron(II) with 100% current efficiency, even though only a fraction of that species is directly oxidized at the electrode surface.

### End Points in Coulometric Titrations

Coulometric titrations, like their volumetric counterparts, require a means for determining when the reaction between analyte and reagent is complete. Generally, the end points described in the chapters on volumetric methods are applicable to coulometric titrations as well. Thus, for the titration of iron(II) just described, an oxidation/reduction indicator, such as 1,10-phenanthroline, can be used; as an alternative, the end point can be established potentiometrically. Similarly, an adsorption indicator or a potentiometric end point can be employed in the coulometric titration of chloride ion by the silver ions generated at a silver anode.

### Instrumentation

As shown in Figure 18–13, the equipment required for a coulometric titration includes a source of constant current that has a range of one to several hundred milliamperes, a titration vessel, a switch, an electric timer, and a device for monitoring current. Movement of the switch to position 1 simultaneously starts the timer and initiates a current in the

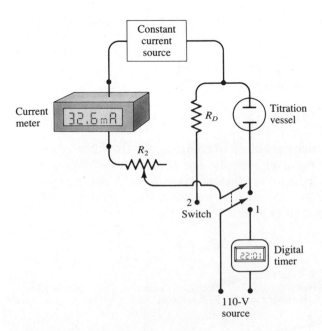

Figure 18–13
Diagram of a coulometric titration apparatus.

titration cell. When the switch is moved to position 2, the electrolysis and the timing are discontinued. With the switch in this position, however, electricity continues to be drawn from the source and passes through a dummy resistor $R_D$ that has about the same electrical resistance as the cell. This arrangement ensures continuous operation of the source, which aids in maintaining the current at a constant level.

The constant-current source for a coulometric titration is often an amperostat, an electronic device capable of maintaining a current of 200 mA or more that is constant to a few hundredths percent of the current. Amperostats are available from several instrument manufacturers. A much less expensive source that produces reasonably constant currents can be constructed from several heavy-duty batteries connected in series to give an output potential of 100 to 300 V. This output is applied to the titration cell, which is in series with a resistor whose resistance is large relative to the cell resistance. Small changes in the cell conductance then have a negligible effect on the current in the resistor and that in the cell.

An ordinary motor-driven electric clock is unsatisfactory for the measurement of the electrolysis time because the rotor of such a device tends to coast when stopped and lag when started. Modern digital electronic timers eliminate this problem.

**Cells for Coulometric Titrations.** Figure 18–14 shows a typical coulometric titration cell consisting of a generator electrode at which the reagent is produced and a counter electrode to complete the circuit. The generator electrode—ordinarily a platinum rectangle, a coil of wire, or a gauze cylinder—has a relatively large surface area to minimize polarization effects. In most instances, the counter electrode is isolated from the reaction medium by a sintered disk or some other porous medium in order to prevent interference by the reaction products from this electrode. For example, hydrogen is often evolved at the cathode as an oxidizing agent is being generated at an anode. Hydrogen reacts rapidly with most oxidizing

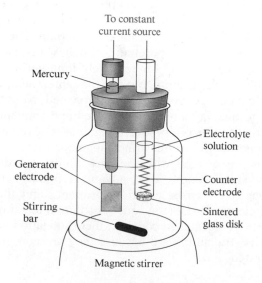

Figure 18–14
A typical coulometric titration cell.

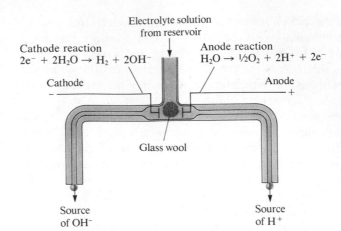

**Figure 18–15**
A cell for the external coulometric generation of acid and base.

agents which leads to a positive determinate error unless the gas is generated in a separate compartment.

An alternative to isolation of the auxiliary electrode is a device in which the reagent is generated externally, such as that shown in Figure 18–15. The apparatus is so arranged that electrolyte flow continues briefly after the current is discontinued, thus flushing the residual reagent into the titration vessel. Note that the apparatus shown in Figure 18–15 provides either hydrogen or hydroxide ions, depending upon which arm is used. The apparatus has also been used for the generation of other reagents. An example is the iodine produced by oxidation of iodide at the anode.

### A Comparison of Coulometric and Conventional Titrations

The various components of the titrator in Figure 18–13 have their counterparts in the reagents and apparatus required for a volumetric titration. The constant-current source of known magnitude serves the same function as the standard solution in a volumetric method. The electronic timer and switch correspond to the buret and stopcock, respectively. Electricity is passed through the cell for relatively long periods of time at the outset of a coulometric titration, but the time intervals are made smaller and smaller as chemical equivalence is approached. Note that these steps are analogous to the way a buret is operated in a conventional titration.

A coulometric titration offers several significant advantages over a conventional volumetric procedure. Principal among these is the elimination of the problems associated with the preparation, standardization, and storage of standard solutions. This advantage is particularly significant with labile reagents such as chlorine, bromine, and titanium(III) ion, which are sufficiently unstable in aqueous solution to seriously limit their use as volumetric reagents. In contrast, their utilization in a coulometric determination is straightforward because they are consumed as soon as they are generated.

Coulometric methods also excel when small amounts of analyte have to be titrated because tiny quantities of reagent are generated with ease and accuracy through the proper choice of current. In contrast, the use of very dilute solutions and the accurate measurement of small volumes are inconvenient at best.

A further advantage of the coulometric procedure is that a single constant-current source provides reagents for precipitation, complex formation, neutralization, or oxidation/reduction titrations. Finally, coulometric titrations are more readily automated since current control is more readily accomplished than is control of liquid flow.

The current/time measurements required for a coulometric titration are inherently as accurate as or more accurate than the comparable volume/molarity measurements of a conventional volumetric method, particularly where small quantities of reagent are involved. When the accuracy of a titration is limited by end-point sensitivity, the two titration methods have comparable accuracies.

> Coulometric methods are as accurate and precise as comparable volumetric methods.

## Applications

Coulometric titrations have been developed for all types of volumetric reactions.[6] Selected applications are described in this section.

**Neutralization Titrations.** Hydroxide ion can be generated at the surface of a platinum cathode immersed in a solution containing the analyte acid:

$$2 H_2O + 2 e^- \rightarrow 2 OH^- + H_2(g)$$

The platinum anode must be isolated by some sort of diaphragm to eliminate potential interference from the hydrogen ions simultaneously produced by the anodic oxidation of water. As a convenient alternative, a silver wire can be substituted for the platinum anode, provided chloride or bromide ions are added to the analyte solution. The anode reaction then becomes

$$Ag(s) + Br^- \rightarrow AgBr(s) + e^-$$

Silver bromide does not interfere with the neutralization reaction.

Coulometric titrations of acids are much less susceptible to the carbonate error encountered in volumetric methods (Section 12A–3). The only measure required to avoid this type of error is to remove the carbon dioxide from the solvent by boiling it or by bubbling an inert gas, such as nitrogen, through the solution for a brief period.

Hydrogen ions generated at the surface of a platinum anode can be used for the coulometric titration of strong as well as weak bases:

$$2 H_2O \rightarrow O_2 + 4 H^+ + 4 e^-$$

---

[6]For additional information, see E. Bishop, in *Comprehensive Analytical Chemistry*, C. L. Wilson and D. W. Wilson, Eds., Vol. 11D, Chapters XVIII to XXIV. New York: Elsevier, 1975; J. T. Stock, *Anal. Chem.*, **1984**, *56*, 1R, and **1980**, *52*, 1R.

Here, the cathode must be isolated from the analyte solution to prevent interference from hydroxide ion.

**Precipitation and Complex-Formation Reactions.** Coulometric titrations with EDTA are carried out by reduction of the ammine mercury(II) EDTA chelate at a mercury cathode:

$$HgNH_3Y^{2-} + NH_4^+ + 2\ e^- \rightarrow Hg(l) + 2\ NH_3 + HY^{3-}\qquad (18\text{--}8)$$

Because the mercury chelate is more stable than the corresponding complexes of such cations as calcium, zinc, lead, and copper, complexation of these ions occurs only after the ligand has been freed by the electrode process.

As shown in Table 18–3, several precipitating reagents can be generated coulometrically. The most widely used of these is silver ion, which is generated at a silver anode.

**Oxidation/Reduction Titrations.** Table 18–4 reveals that a variety of redox reagents can be generated coulometrically. Of particular interest is bromine, the coulometric generation of which forms the basis for a large number of methods. Of interest as well are reagents not ordinarily encountered in conventional volumetric analysis owing to the instability of their solutions; silver(II), manganese(III), and the chloride complex of copper(I) are examples.

## Automatic Coulometric Titrators

A number of instrument manufacturers offer automatic coulometric titrators, most of which employ a potentiometric end point. Some of these instruments are multipurpose and can be used for the determination of a variety of species. Others are designed for a single type of analysis. Examples of the latter are chloride titrators, in which silver ion is generated coulometrically; sulfur dioxide monitors, where anodically generated bromine oxidizes the analyte to sulfate ions; and carbon dioxide monitors, in which the gas, absorbed in monoethanolamine, is titrated with coulometrically generated base.

**Table 18-3**

**SUMMARY OF COULOMETRIC TITRATIONS INVOLVING NEUTRALIZATION, PRECIPITATION, AND COMPLEX-FORMATION REACTIONS**

| Species Determined | Generator Electrode Reaction | Secondary Analytical Reaction |
|---|---|---|
| Acids | $2\ H_2O + 2\ e^- \rightleftharpoons 2\ OH^- + H_2$ | $OH^- + H^+ \rightleftharpoons H_2O$ |
| Bases | $H_2O \rightleftharpoons 2\ H^+ + \frac{1}{2}\ O_2 + 2\ e^-$ | $H^+ + OH^- \rightleftharpoons H_2O$ |
| $Cl^-,\ Br^-,\ I^-$ | $Ag \rightleftharpoons Ag^+ + e^-$ | $Ag^+ + Cl^- \rightleftharpoons AgCl(s)$, etc. |
| Mercaptans | $Ag \rightleftharpoons Ag^+ + e^-$ | $Ag^+ + RSH \rightleftharpoons AgSR(s) + H^+$ |
| $Cl^-,\ Br^-,\ I^-$ | $2\ Hg \rightleftharpoons Hg_2^{2+} + 2\ e^-$ | $Hg_2^{2+} + 2\ Cl^- \rightleftharpoons Hg_2Cl_2(s)$, etc. |
| $Zn^{2+}$ | $Fe(CN)_6^{3-} + e^- \rightleftharpoons Fe(CN)_6^{4-}$ | $3\ Zn^{2+} + 2\ K^+ + 2\ Fe(CN)_6^{4-} \rightleftharpoons$ $K_2Zn_3[Fe(CN)_6]_2(s)$ |
| $Ca^{2+},\ Cu^{2+},\ Zn^{2+},\ Pb^{2+}$ | See Equation 18–8. | $HY^{3-} + Ca^{2+} \rightleftharpoons CaY^{2-} + H^+$, etc. |

Table 18–4
### SUMMARY OF COULOMETRIC TITRATIONS INVOLVING OXIDATION/REDUCTION REACTIONS

| Reagent | Generator Electrode Reaction | Substance Determined |
|---|---|---|
| $Br_2$ | $2\,Br^- \rightleftharpoons Br_2 + 2\,e^-$ | As(III), Sb(III), U(IV), Tl(I), $I^-$, $SCN^-$, $NH_3$, $N_2H_4$, $NH_2OH$, phenol, aniline, mustard gas, mercaptans, 8-hydroxyquinoline, olefins |
| $Cl_2$ | $2\,Cl^- \rightleftharpoons Cl_2 + 2\,e^-$ | As(III), $I^-$, styrene, fatty acids |
| $I_2$ | $2\,I^- \rightleftharpoons I_2 + 2\,e^-$ | As(III), Sb(III), $S_2O_3^{2-}$, $H_2S$, ascorbic acid |
| $Ce^{4+}$ | $Ce^{3+} \rightleftharpoons Ce^{4+} + e^-$ | Fe(II), Ti(III), U(IV), As(III), $I^-$, $Fe(CN)_6^{4-}$ |
| $Mn^{3+}$ | $Mn^{2+} \rightleftharpoons Mn^{3+} + e^-$ | $H_2C_2O_4$, Fe(II), As(III) |
| $Ag^{2+}$ | $Ag^+ \rightleftharpoons Ag^{2+} + e^-$ | Ce(III), V(IV), $H_2C_2O_4$, As(III) |
| $Fe^{2+}$ | $Fe^{3+} + e^- \rightleftharpoons Fe^{2+}$ | Cr(VI), Mn(VII), V(V), Ce(IV) |
| $Ti^{3+}$ | $TiO^{2+} + 2\,H^+ + e^- \rightleftharpoons Ti^{3+} + H_2O$ | Fe(III), V(V), Ce(IV), U(VI) |
| $CuCl_3^{2-}$ | $Cu^{2+} + 3\,Cl^- + e^- \rightleftharpoons CuCl_3^{2-}$ | V(V), Cr(VI), $IO_3^-$ |
| $U^{4+}$ | $UO_2^{2+} + 4\,H^+ + 2\,e^- \rightleftharpoons U^{4+} + 2\,H_2O$ | Cr(VI), Ce(IV) |

## 18E   QUESTIONS AND PROBLEMS

**18–1.** Define
 *(a) coulometric titration.
  (b) kinetic polarization.
 *(c) concentration polarization.
  (d) potentiostat.
 *(e) faraday.
  (f) overvoltage.
 *(g) controlled-cathode-potential electrolysis.
  (h) ohmic potential.
 *(i) cathode depolarizer.
  (j) coulomb.
 *(k) working electrode.
  (l) diffusion.

*18–2. Differentiate between amperostatic coulometry and potentiostatic coulometry.

18–3. Differentiate between kinetic polarization and concentration polarization.

*18–4. Describe three phenomena that cause ions to migrate to an electrode surface from the bulk of a solution.

18–5. Describe variables that tend to decrease concentration polarization.

*18–6. Under what circumstances is kinetic polarization likely to occur?

18–7. List variables that affect the physical properties of electrolytic deposits.

*18–8. Compare a coulometric and a volumetric titration.

18–9. Why is an auxiliary reagent always required in a coulometric titration?

*18–10. How is a constant-cathode-potential electrolysis performed?

*18–11. Nickel is to be deposited from a solution that is 0.200 M in $Ni^{2+}$ and buffered to pH 2.00. Oxygen is evolved at a partial pressure of 1.00 atm at a platinum anode. The cell has a resistance of 3.15 Ω; the temperature is 25°C. Calculate

(a) the thermodynamic potential needed to initiate the deposition of nickel.

(b) the IR drop for a current of 1.10 A.

(c) the initial applied potential, given that the oxygen overvoltage is 0.85 V.

(d) the applied potential needed when $[Ni^{2+}]$ is 0.00020, assuming that all other variables remain unchanged.

18–12. Bismuth is to be deposited from a solution in which the analytical concentration of $BiCl_4^-$ is 0.0800 M and that of KCl is 0.400 M. Oxygen is evolved at the anode at a partial pressure of 765 torr. The cell has a resistance of 1.80 Ω, the temperature is 25°C, and the solution is buffered to pH 1.50. Calculate

(a) the thermodynamic potential needed to initiate the deposition of bismuth.

(b) the IR drop for a current of 0.42 A.

(c) the initial applied potential, given that the oxygen overvoltage is 0.72 V.

(d) the applied potential when the analytical concentration of the undeposited bismuth is $1.00 \times 10^{-5}$ M.

18–13. Calculate the minimum difference in standard electrode potentials needed to lower the concentration of the metal $M_1$ to $1.00 \times 10^{-4}$ M in a solution that is 0.200 M in the less reducible metal $M_2$, where
 *(a) $M_2$ is univalent and $M_1$ is divalent.
  (b) $M_1$ and $M_2$ are both divalent.
 *(c) $M_2$ is trivalent and $M_1$ is univalent.
  (d) $M_2$ is divalent and $M_1$ is univalent.
  (e) $M_2$ is divalent and $M_1$ is trivalent.

*18–14. A solution is 0.150 M in $Co^{2+}$ and 0.0750 M in $Cd^{2+}$. Calculate

(a) the $Co^{2+}$ concentration in the solution as the first cadmium starts to deposit.

**(b)** the cathode potential needed to lower the $Co^{2+}$ concentration to $1 \times 10^{-5}$ M.

**18–15.** A solution is 0.0500 M in $BiO^+$ and 0.0400 M in $Co^{2+}$ and has a pH of 2.50.
  **(a)** What is the concentration of the more readily reduced cation at the onset of deposition of the less reducible one?
  **(b)** What is the potential of the cathode when the concentration of the more easily reduced species is $1.00 \times 10^{-6}$ M?

**\*18–16.** Electrodeposition is to be used to separate the cations in a solution that is buffered to pH 4.00 and is $5.00 \times 10^{-2}$ M in $Cu^{2+}$ and $8.00 \times 10^{-3}$ M in $Ag^+$. Oxygen is evolved at a platinum anode at a pressure of 0.80 atm, the oxygen overvoltage is 0.80 V, and the cell has a resistance of 2.40 $\Omega$.
  **(a)** Which cation deposits first?
  **(b)** Estimate the initial potential that must be applied in order to operate the cell at 0.50 A.
  **(c)** Taking $1.00 \times 10^{-6}$ M as a reasonable estimate for quantitative removal, calculate the range (versus SCE) within which is necessary to maintain the cathode potential.

**18–17.** Electrodeposition is proposed as the means for separating the cations in a solution that is buffered to pH 3.00 and is $1.00 \times 10^{-2}$ M in $Sn^{2+}$ and $5.00 \times 10^{-2}$ M in $Tl^+$. Oxygen is evolved at a platinum anode at a pressure of 750 torr. The cell has a resistance of 1.50 $\Omega$; the oxygen overvoltage is 0.75 V.
  **(a)** Estimate the initial potential that must be applied in order to operate the cell at 0.20 A.
  **(b)** Within what range (versus SCE) should the cathode potential be maintained in order to lower the $Sn^{2+}$ concentration to at least $1.00 \times 10^{-6}$ M without interference from $Tl^+$?

**\*18–18.** Electrogravimetric analysis involving control of the cathode potential is proposed as a means for separating $Bi^{3+}$ and $Sn^{2+}$ in a solution that is 0.200 M in each ion and buffered to pH 1.50.
  **(a)** Calculate the theoretical cathode potential at the start of deposition of the more readily reduced ion.
  **(b)** Calculate the residual concentration of the more readily reduced species at the outset of the deposition of the less readily reduced species.
  **(c)** Propose a range (versus SCE), if such exists, within which the cathode potential should be maintained; consider a residual concentration less than $10^{-6}$ M as constituting quantitative removal.

**\*18–19.** Halide ions can be deposited on a silver anode via the reaction

$$Ag(s) + X^- \rightarrow AgX(s) + e^-$$

  **(a)** If $1.00 \times 10^{-5}$ M is used as the criterion for quantitative removal, is it theoretically feasible to separate $Br^-$ from $I^-$ through control of the

anode potential in a solution that is initially 0.250 M in each ion?
  **(b)** Is a separation of $Cl^-$ and $I^-$ theoretically feasible in a solution that is initially 0.250 M in each ion?
  **(c)** If a separation is feasible in either (a) or (b), what range of anode potentials (versus SCE) should be used?

**\*18–20.** What cathode potential (versus SCE) is required to lower the total Hg(II) concentration of the following solutions to $1.00 \times 10^{-6}$ M (assume the reaction product in each case is elemental mercury):
  **(a)** an aqueous solution of $Hg^{2+}$?
  **(b)** a solution with an equilibrium $SCN^-$ concentration of 0.100 M?

$$Hg^{2+} + 2\,SCN^- \rightleftharpoons Hg(SCN)_2(aq)$$
$$K = 1.8 \times 10^7$$

  **(c)** a solution with an equilibrium $Br^-$ concentration of 0.250 M?

$$HgBr_4^{2-} + 2\,e^- \rightleftharpoons Hg(l) + 4\,Br^-$$
$$E^0 = 0.223\ V$$

**18–21.** What cathode potential (versus SCE) is needed to lower the analytical concentration of a Ni(II)-containing species to $1.00 \times 10^{-6}$ M in a solution that is
  **(a)** 0.0010 M in HCl (assume no chloride complex forms)?
  **(b)** 0.212 M in $CN^-$?

$$Ni^{2+} + 4\,CN^- \rightarrow Ni(CN)_4^{2-} \quad \beta_4 = 1.0 \times 10^{22}$$

  **(c)** 0.215 M in $NH_3$ (analytical concentration) and has a pH of 9.0?

$$Ni(NH_3)_4^{2-} + 2\,e^- \rightarrow Ni(s) + 4\,NH_3$$
$$E^0 = -0.530\ V$$

  **(d)** 0.0612 M in EDTA and buffered to pH 5.00?

**\*18–22.** Calculate the time needed for a constant current of 0.961 A to deposit 0.500 g of Co(II) as
  **(a)** elemental cobalt on the surface of a cathode.
  **(b)** $Co_3O_4$ on an anode.

**18–23.** Calculate the time needed for a constant current of 1.20 A to deposit 0.500 g of
  **(a)** Tl(III) as the element on a cathode.
  **(b)** Tl(I) as $Tl_2O_3$ on an anode.
  **(c)** Tl(I) as the element on a cathode.

**\*18–24.** The cadmium and zinc in a 1.06-g sample were dissolved and subsequently deposited from an ammoniacal solution with a mercury cathode. When the cathode potential was maintained at $-0.95$ V (versus SCE), only the cadmium deposited. When the current ceased at this potential, a hydrogen/oxygen coulometer in series with the cell had evolved 44.6 mL of gas (corrected for water vapor) at 21.0°C and a barometric pressure of 773 mm Hg.

The potential was raised to about $-1.3$ V, whereupon $Zn^{2+}$ ion was reduced. Upon completion of this electrolysis, an additional 31.3 mL of gas was produced under the same conditions. Calculate the percentage of cadmium and zinc in the ore.

**18–25.** A 1.74-g sample of a solid containing $BaBr_2$, KI, and inert species was dissolved, made ammoniacal, and placed in a cell equipped with a silver anode. When the potential was maintained at $-0.06$ V (versus SCE), $I^-$ was quantitatively precipitated as AgI without interference from $Br^-$. The volume of $H_2$ and $O_2$ formed in a gas coulometer in series with the cell was 39.7 mL (corrected for water vapor) at 21.7°C and 748 mm Hg. After precipitation of $I^-$ was complete, the solution was acidified, and the $Br^-$ was removed from solution as AgBr at a potential of 0.016 V. The volume of gas formed under the same conditions was 23.4 mL. Calculate the percentage of $BaBr_2$ and KI in the sample.

**\*18–26.** An excess of $HgNH_3Y^{2-}$ was introduced to 25.00 mL of well water. Express the hardness of the water in terms of ppm $CaCO_3$ if the EDTA needed for the titration was generated at a mercury cathode (Equation 18–8) in 2.02 min by a constant current of 31.6 mA.

**18–27.** A 0.1516-g sample of a purified organic acid was neutralized by the hydroxide ion produced in 5 min and 24 s by a constant current of 0.401 A. Calculate the equivalent weight of the acid.

**\*18–28.** The nitrobenzene in 210 mg of an organic mixture was reduced to phenylhydroxylamine at a constant potential of $-0.96$ V (versus SCE) applied to a mercury cathode:

$$C_6H_5NO_2 + 4 H^+ + 4 e^- \rightarrow C_6H_5NHOH + H_2O$$

The sample was dissolved in 100 mL of methanol; after electrolysis for 30 min, the reaction was judged complete. An electronic coulometer in series with the cell indicated that the reduction required 26.74 C. Calculate the percentage of $C_6H_5NO_2$ in the sample.

**18–29.** Electrolytically generated $I_2$ was used to determine the amount of $H_2S$ in 100.0 mL of brackish water. Following addition of excess KI, a titration required a constant current of 36.32 mA for 10.12 min. The reaction was

$$H_2S + I_2 \rightarrow S(s) + 2 H^+ + 2 I^-$$

Express the results of the analysis in terms of ppm $H_2S$.

**\*18–30.** At a potential of $-1.0$ V (versus SCE), $CCl_4$ in methanol is reduced to $CHCl_3$ at a mercury cathode:

$$2 CCl_4 + 2 H^+ + 2 e^- + 2 Hg(l) \rightarrow$$
$$2 CHCl_3 + Hg_2Cl_2(s)$$

At $-1.80$ V, the $CHCl_3$ further reacts to give $CH_4$:

$$2 CHCl_3 + 6 H^+ + 6 e^- + 6 Hg(l) \rightarrow$$
$$2 CH_4(g) + 3 Hg_2Cl_2(s)$$

A 0.750-g sample containing $CCl_4$, $CHCl_3$, and inert organic species was dissolved in methanol and electrolyzed at $-1.0$ V until the current approached zero. A coulometer indicated that 11.63 C was required to complete the reaction. The potential of the cathode was adjusted to $-1.8$ V. Completion of the titration at this potential required an additional 68.6 C. Calculate the percentage of $CCl_4$ and $CHCl_3$ in the mixture.

**18–31.** A 0.1309-g sample containing only $CHCl_3$ and $CH_2Cl_2$ was dissolved in methanol and electrolyzed in a cell containing a mercury cathode; the potential of the cathode was held constant at $-1.80$ V (versus SCE). Both compounds were reduced to $CH_4$ (see Problem 18–30 for the reaction type). Calculate the percentage of $CHCl_3$ and $CH_2Cl_2$ if 306.7 C was required to complete the reduction.

**\*18–32.** The iron in 0.854-g of ore was converted to the $+2$ state by suitable treatment and then oxidized quantitatively at a platinum anode maintained at $-1.0$ V (versus SCE). The quantity of electricity to complete the oxidation was determined with a chemical coulometer equipped with a platinum anode immersed in an excess of $I^-$ ions. The $I_2$ liberated by the current through the cell required 26.3 mL of 0.0197 M $Na_2S_2O_3$ to reach a starch end point. What was the percentage of $Fe_3O_4$ in the sample?

**18–33.** The phenol content of water downstream from a coking furnace was determined by coulometric analysis. A 100-mL sample was rendered slightly acidic, and an excess of KBr was introduced. To produce $Br_2$ for the reaction

$$C_6H_5OH + 3 Br_2 \rightarrow Br_3C_6H_2OH(s) + 3 HBr$$

a steady current of 0.0313 A for 7 min and 33 s was required. Express the results of this analysis in terms of parts of $C_6H_5OH$ per million parts of water. (Assume that the density of water is 1.00 g/mL.)

**\*18–34.** The $CN^-$ concentration of 10.0 mL of a plating solution was determined by titration with electrogenerated hydrogen ion to a methyl orange end point. A color change occurred after 3 min and 22 s with a current of 43.4 mA. Calculate the number of grams of NaCN per liter of solution.

**18–35.** Traces of $C_6H_5NH_2$ can be determined by reaction with an excess of electrolytically generated $Br_2$:

The polarity of the working electrode is then reversed, and the excess $Br_2$ is determined by a coulometric titration involving the generation of Cu(I):

$$Br_2 + 2\ Cu^+ \rightarrow 2\ Br^- + 2\ Cu^{2+}$$

Suitable quantities of KBr and $CuSO_4$ were added to a 25.0-mL sample containing aniline. Calculate the number of micrograms of $C_6H_5NH_2$ in the sample from the data:

| Working Electrode Functioning as | Generation Time with a Constant Current of 1.51 mA, min |
|---|---|
| Anode | 3.76 |
| Cathode | 0.270 |

**18–36.** Quinone can be reduced to hydroquinone with an excess of electrolytically generated Sn(II):

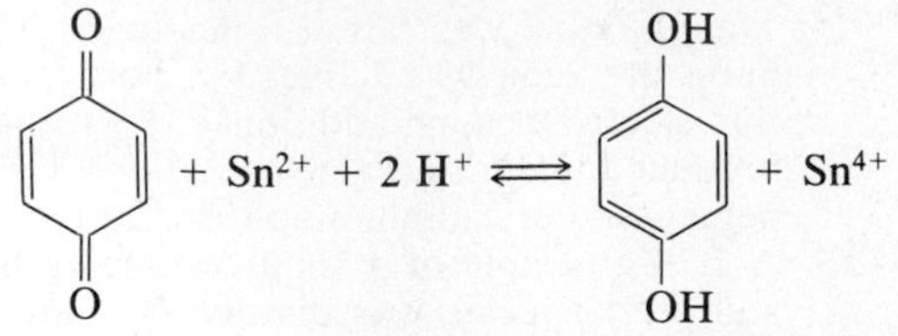

The polarity of the working electrode is then reversed, and the excess Sn(II) is oxidized with $Br_2$ generated in a coulometric titration:

$$Sn^{2+} + Br_2 \rightleftharpoons Sn^{4+} + 2\ Br^-$$

Appropriate quantities of $SnCl_4$ and KBr were added to a 50.0-mL sample. Calculate the weight of $C_6H_4O_2$ in the sample from the data:

| Working Electrode Functioning as | Generation Time with a Constant Current of 1.062 mA, min |
|---|---|
| Cathode | 8.34 |
| Anode | 0.691 |

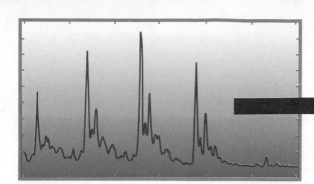

## CHAPTER 19

# VOLTAMMETRY

$V$oltammetry comprises a group of electroanalytical methods in which information about the analyte is derived from the measurement of current as a function of applied potential under conditions that encourage polarization of an indicator, or working, electrode. To bring about polarization, a voltammetric working electrode has a surface area in the range between a few square millimeters and a few square micrometers. Electrodes with such small areas are called *microelectrodes*.

At the outset, it is worthwhile to point out the basic differences between voltammetry and the two types of electrochemical methods that we have discussed in earlier chapters. Voltammetry is based upon the measurement of a current that develops in an electrochemical cell under conditions of complete concentration polarization. In contrast, potentiometric measurements are made at currents that approach zero and where polarization is absent. Voltammetry differs from coulometry in that with the latter, steps are taken to minimize or compensate for the effects of concentration polarization. Furthermore, in voltammetry a minimal consumption of analyte takes place, whereas in coulometry essentially all of the analyte is converted to another state.

Historically, the field of voltammetry developed from *polarography,* which was discovered by the Czechoslovakian chemist Jaroslav Heyrovsky in the early 1920s.[1] Polarography differs from other types of voltammetry in that the working microelectrode takes the form of a *dropping mercury electrode* (DME). The construction and unique properties of this electrode are discussed in a later section.

Voltammetry is widely used by inorganic, physical, and biological chemists for fundamental studies of oxidation and reduction processes in various media, adsorption processes on surfaces, and electron transfer mechanisms at chemically modified electrode surfaces. At one time, voltammetry (particularly classical polarography) was an important tool used by chemists for the determination of inorganic ions and certain organic species in aqueous solutions. In the late 1950s and early 1960s, however, these analytical applications were largely supplanted by various

Working electrodes with surface areas smaller than a few square millimeters are called microelectrodes.

Polarography is voltammetry performed with a dropping mercury electrode.

Heyrovsky was awarded the 1959 Nobel Prize in chemistry for his discovery and development of polarography.

Voltammetry is used as a detection technique for HPLC.

---

[1] J. Heyrovsky, *Chem. Listy,* **1922,** *16,* 256.

spectroscopic methods, and voltammetry ceased to be important in analysis except for certain special applications, such as the determination of molecular oxygen in solutions.

In the mid-1960s, several major modifications of classical voltammetric techniques were developed that enhance significantly the sensitivity and selectivity of the method. At about the same time, the advent of low-cost operational amplifiers made possible the commercial development of relatively inexpensive instruments that incorporated many of these modifications and made them available to chemists. The result has been a recent resurgence of interest in applying voltammetric methods to the determination of a host of species, particularly those of pharmaceutical interest.[2] Furthermore, voltammetry coupled with high-performance liquid chromatography (HPLC, Chapter 28) has become a powerful tool for the analysis of complex mixtures of various kinds. Modern voltammetry also continues to be a potent tool used by various kinds of chemists interested in studying oxidation and reduction processes as well as adsorption processes.

## 19A    EXCITATION SIGNALS IN VOLTAMMETRY

In voltammetry, a variable potential *excitation signal* is impressed upon an electrochemical cell containing a microelectrode. This excitation signal elicits a characteristic current response upon which the method is based. The waveforms of three of the most common excitation signals used in voltammetry are shown in Figure 19–1. The classical voltammetric excitation signal is the linear scan shown in Figure 19–1a, in which the dc potential applied to the cell increases linearly (usually over a 2- to 3-V range) as a function of time. The current that develops in the cell is then recorded as a function of time (and thus as a function of the applied potential).

Two pulse-type excitation signals are shown in Figures 19–1b and c. Currents are measured at various times during the lifetime of these pulses. The last column in Figure 19–1 lists the types of voltammetry that employ the various excitation signals. These techniques are discussed in the sections that follow.

## 19B    LINEAR-SCAN VOLTAMMETRY

A voltammogram is a plot of current versus voltage.

The earliest and simplest voltammetric methods were of the linear-scan type, in which the potential of the working electrode is increased or decreased at a typical rate of 2 to 5 mV/s. The current, usually in microamperes, is then recorded to give a *voltammogram,* which is a plot of current as a function of potential applied to the working electrode. Linear-scan voltammetry is of two types, *hydrodynamic voltammetry* and *polarography.*

---

[2]For a brief summary of several of these modified voltammetric techniques, see J. B. Flato, *Anal. Chem,* **1972,** *44,* 75A.

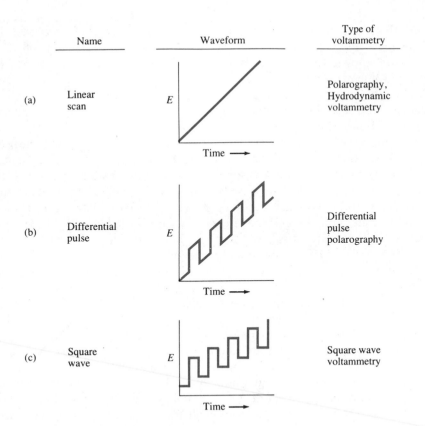

| Name | Waveform | Type of voltammetry |
|---|---|---|
| (a) Linear scan | $E$ vs Time | Polarography, Hydrodynamic voltammetry |
| (b) Differential pulse | $E$ vs Time | Differential pulse polarography |
| (c) Square wave | $E$ vs Time | Square wave voltammetry |

**Figure 19–1**
Potential excitation signals used in voltammetry.

## 19B–1  Voltammetric Systems

Figure 19–2 is a schematic showing the components of an apparatus for carrying out voltammetric measurements. The apparatus is similar to that shown in Figure 18–8. The cell is made up of three electrodes immersed in a solution containing the analyte and also an excess of a nonreactive

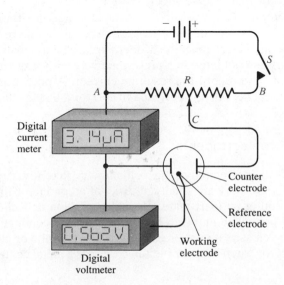

**Figure 19–2**
A manual potentiostat for voltammetry.

electrolyte called a *supporting electrolyte.* One of the three electrodes is the microelectrode, or *working electrode,* whose potential is controlled at a fixed known value or, as in the case of linear-scan voltammetry, is varied linearly with time. Its dimensions are kept small in order to enhance its tendency to become polarized. The second electrode is a reference electrode whose potential remains constant throughout the experiment. The third electrode is a *counter electrode,* which is often a coil of platinum wire or a pool of mercury that simply serves to conduct current from the source through the solution to the microelectrode. The signal source is a variable dc power supply consisting of a battery in series with a variable resistor $R$. The desired potential is selected by moving the contact $C$ to the proper position on the resistor. The electrical resistance of the circuit containing the reference electrode is so large ($> 10^{11}$ $\Omega$) that essentially no current is present in it. Thus, the entire current from the source is carried from the counter electrode to the microelectrode. A voltammogram is recorded by moving the contact $C$ in Figure 19–2 and recording the resulting current as a function of the potential between the working electrode and the reference electrode.

In principle, the manual potentiostat of Figure 19–2 could be used to generate a linear-scan voltammogram. In such an experiment, contact $C$ would be moved at a constant rate from $A$ to $B$ to produce the excitation signal shown in Figure 19–1a. The current and voltage would then be recorded at consecutive equal time intervals during the voltage (or time) scan. Unfortunately, it is difficult to maintain a constant rate of motion of the sliding contact and equally difficult to record the current manually. Thus, in practice we would record a manual voltammogram by moving contact $C$ in small increments indicated by the digital voltmeter in Figure 19–2. For each increment of voltage, the resulting current would be recorded to produce a voltammogram point by point. It is important to emphasize that the independent variable in this experiment is the potential of the *microelectrode* versus the reference electrode and not the potential between the microelectrode and the counter electrode.

The circuit of Figure 19–2 functions quite nicely for recording current at a fixed potential or for generating manual voltammograms. However, complex excitation signals such as those shown in Figure 19–1b and c must be generated electronically. Modern voltammetric instruments automatically vary the potential in a prescribed way with respect to the reference electrode and record the resulting current. A potentiostat based on operational amplifiers that is designed to carry out this task is described in Feature 19–1.

### 19B–2 Microelectrodes

The microelectrodes employed in voltammetry take a variety of shapes and forms. Often, they are small flat disks of a conductor that are press-fitted into a rod of an inert material, such as Teflon or Kel-F, that has a wire contact imbedded in it (see Figure 19–5a). The conductor may be an inert metal, such as platinum or gold; pyrolytic graphite or glassy carbon; a semiconductor, such as tin or indium oxide; or a metal coated with a film

**Feature 19-1**
## VOLTAMMETRIC INSTRUMENTS BASED ON OPERATIONAL AMPLIFIERS

In Feature 17-2 we describe the use of operational amplifiers to measure the potential of electrochemical cells. Op amps also can be used to measure currents and carry out a variety of other control and measurement tasks. Let us consider the measurement of current as illustrated in Figure 19-3.

In this circuit a voltage source is attached to one electrode of an electrochemical cell, which produces a current $I$ in the cell. Because of the very high input resistance of the op amp, essentially all of the current passes through the resistor $R$ to the output of the op amp. The voltage at the output of the op amp is given by Ohm's law, $E_{out} = -IR$. By solving this equation for $I$, we have

$$I = -\frac{E_{out}}{R}$$

In other words, the current in the electrochemical cell is proportional to the voltage output of the op amp. The value of the current then can be calculated from the measured values of $E_{out}$ and the resistance $R$. The circuit is called a *current-to-voltage converter*.

Op amps can be used to construct an automatic three-electrode potentiostat as illustrated in Figure 19-4. Notice that the current-measuring circuit of Figure 19-3 is connected to the working electrode of the cell (op amp C). The reference electrode is attached to a voltage follower (op amp B). As discussed in Feature 17-2, the voltage follower monitors the potential of the reference electrode without drawing any current from the cell. The output of op amp B, which is the reference electrode potential, is fed back to the input of op amp A to complete the circuit. The functions of op amp A are (1) to provide the current in the electrochemical cell between the counter electrode and the working electrode and (2) to maintain the potential difference between the reference electrode and the working electrode at a value provided by the linear-scan voltage generator.

In operation, the linear-scan voltage generator sweeps the potential between the reference and working electrodes, and the current in the cell is monitored by op amp C. The output voltage $E_{out}$ of op amp C, which is proportional to the current $I$ in the cell, is recorded on a stripchart or by a computer for data analysis and presentation.[3]

---

[3]For a complete discussion of op amp three-electrode potentiostats, see P. T. Kissinger and W. R. Heineman, Eds., *Laboratory Technique in Electroanalytical Chemistry*, pp. 171–186. New York: Marcel Dekker, 1984.

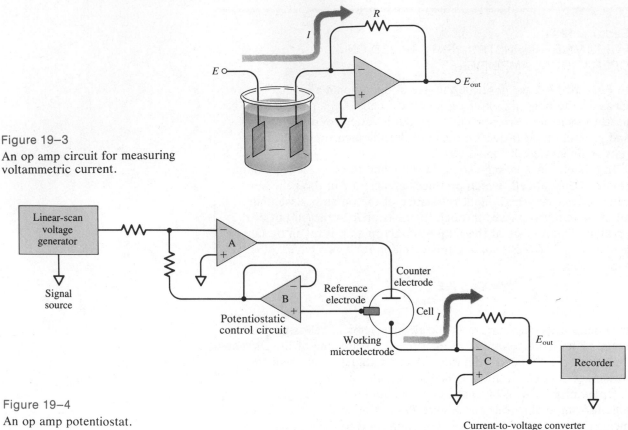

**Figure 19–3**
An op amp circuit for measuring voltammetric current.

**Figure 19–4**
An op amp potentiostat.

of mercury. As shown in Figure 19–6, the range of potentials that can be used with these electrodes in aqueous solutions varies and depends not only upon electrode material but also upon the composition of the solution in which it is immersed. Generally, the positive potential limitations are caused by the large currents that develop due to the oxidation of water to give molecular oxygen. The negative limits arise from the reduction of water to give hydrogen. Note that relatively large negative potentials can be tolerated with mercury electrodes owing to the high overvoltage of hydrogen on this metal.

Mercury microelectrodes have been widely employed in voltammetry for several reasons. One is the relatively large negative potential range just described. Furthermore, a fresh metallic surface is readily formed by simply producing a new drop. In addition, many metal ions are reversibly reduced to amalgams at the surface of a mercury electrode, which simplifies the chemistry. Mercury microelectrodes take several forms. The simplest of these is a mercury film electrode formed by electrodeposition of the metal onto a disk electrode, such as that shown in Figure 19–5a. Figure 19–5b illustrates a *hanging mercury drop electrode* (HMDE). The electrode, which is available from commercial sources, consists of a very fine capillary tube connected to a mercury-containing reservoir. The metal is forced out of the capillary by a piston arrangement driven by a

Large negative potentials can be used with mercury electrodes.

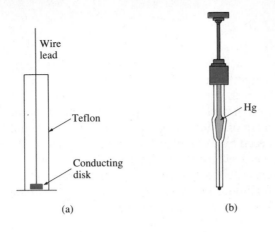

(a)

(b)

Wire
lead

Teflon

Conducting
disk

Hg

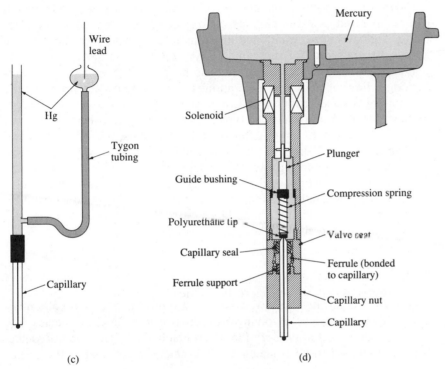

Wire
lead

Hg

Tygon
tubing

Capillary

(c)

Mercury

Solenoid

Plunger

Guide bushing

Compression spring

Polyurethane tip

Valve seat

Capillary seal

Ferrule (bonded
to capillary)

Ferrule support

Capillary nut

Capillary

(d)

Figure 19–5

Some common types of microelectrodes: (a) a disk electrode; (b) a mercury hanging-drop electrode; (c) a dropping mercury electrode; (d) a static mercury dropping electrode.

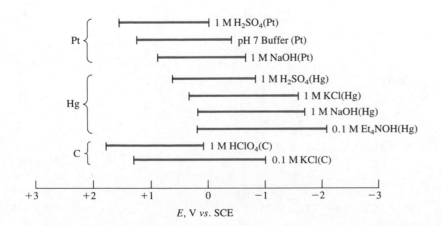

Pt
  1 M H$_2$SO$_4$(Pt)
  pH 7 Buffer (Pt)
  1 M NaOH(Pt)

Hg
  1 M H$_2$SO$_4$(Hg)
  1 M KCl(Hg)
  1 M NaOH(Hg)
  0.1 M Et$_4$NOH(Hg)

C
  1 M HClO$_4$(C)
  0.1 M KCl(C)

+3   +2   +1   0   −1   −2   −3

$E$, V $vs.$ SCE

Figure 19–6

Potential ranges for three types of electrodes in various supporting electrolytes. (Adapted from A. J. Bard and L. R. Faulkner, *Electrochemical Methods,* back cover. New York: Wiley, 1980. With permission.)

Marine barometer tubing was used
formerly for capillaries in polarography.

micrometer screw. The micrometer permits formation of drops having surface areas that are reproducible to 5% or better.

Figure 19–5c shows a typical *dropping mercury electrode* (DME), which was used in nearly all early polarographic experiments. It consists of roughly 10 cm of a fine capillary tubing (i.d. ~0.05 mm) through which mercury is forced by a column of mercury of perhaps 50 cm in height. The diameter of the capillary is such that a new drop forms and breaks off the capillary every 2 to 6 s. The diameter of the drop is 0.5 to 1 mm and is highly reproducible. In some applications, the drop time is controlled by a mechanical knocker that dislodges the drop at a fixed time after it begins to form.

Figure 19–5d shows a commercially available mercury electrode, which can be operated as a dropping mercury electrode or a hanging drop electrode. The mercury is contained in a plastic-lined reservoir about 15 cm above the upper end of the capillary. A compression spring forces the polyurethane-tipped plunger against the head of the capillary, thus preventing a flow of mercury. This plunger is lifted upon activation of the solenoid by a signal from the control system. The capillary is much larger in diameter (0.15 mm) than the typical one. As a result, the formation of the drop is extremely rapid. After 50, 100, or 200 ms, the valve is closed, leaving a full-sized drop in place until it is dislodged by a mechanical knocker that is built into the electrode support block. This system has the advantage that the full-sized drop forms quickly and current measurements can be delayed until the surface area is stable and constant. This procedure largely eliminates the large current fluctuations that are encountered with the classical dropping electrode.

## 19B–3 Voltammograms

Figure 19–7 illustrates the appearance of a typical linear-scan voltammogram for an electrolysis involving the reduction of an analyte species A to give a product P at a mercury film microelectrode. Here, the microelectrode is assumed to be connected to the negative terminal of the linear-

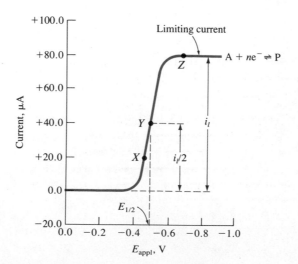

**Figure 19–7**

Linear-scan voltammogram for the reduction of a hypothetical species A to give a product P.

scan generator so that the applied potentials are given a negative sign as shown. By convention, cathodic currents are always treated as being positive, whereas anodic currents are given a negative sign. In this hypothetical experiment, the solution is about $10^{-4}$ M in A, 0.0 M in P, and 0.1 M in KCl, which serves as the supporting electrolyte. The half-reaction at the microelectrode is the reversible reaction

$$A + ne^- \rightleftarrows P \qquad E^0 = -0.26 \text{ V} \qquad (19\text{--}1)$$

For convenience, we have neglected the charges on A and P and also assumed that the standard potential for the half-reaction is $-0.26$ V.

Linear-scan voltammograms generally assume a sigmoidal curve called a *voltammetric wave*. The constant current beyond the steep rise is called the *limiting current* $i_l$ because it arises from a limitation in the rate at which the reactant can be brought to the surface of the electrode by mass-transport processes. Limiting currents are generally directly proportional to reactant concentration. Thus, we may write

$$i_l = kc_A$$

where $c_A$ is the analyte concentration and $k$ is a constant. Quantitative linear-scan voltammetry is based upon this relationship.

The potential at which the current is equal to one half the limiting current is called the *half-wave potential* and given the symbol $E_{1/2}$. The half-wave potential is closely related to the standard potential for the half-reaction but is usually not identical to it. Half-wave potentials are sometimes useful for identification of the components of a solution.

> The half-wave potential occurs when the current is equal to one half of the limiting value.

In order to obtain reproducible limiting currents rapidly, it is necessary that either (1) the solution or the microelectrode be in continuous and reproducible motion or (2) that a dropping electrode, such as a dropping mercury electrode, be employed. Linear-scan voltammetry in which the solution or the electrode is kept in motion is called *hydrodynamic voltammetry*. Voltammetry that employs dropping electrodes is called *polarography*. We shall consider both types.

## 19B-4 Hydrodynamic Voltammetry

Hydrodynamic voltammetry is performed in several ways. One method involves stirring the solution vigorously while it is in contact with a fixed microelectrode. Alternatively, the microelectrode is rotated at a constant high speed in the solution, thus providing the stirring action. Still another way of carrying out hydrodynamic voltammetry involves causing the analyte solution to flow through a tube in which the microelectrode is mounted. The last technique is becoming widely used for detecting oxidizable or reducible analytes as they exit from a liquid chromatographic column (Section 28C-6).

As described in Section 18A-2, during an electrolysis, reactant is carried to the surface of an electrode by three mechanisms: migration under the influence of an electric field, convection resulting from stirring or

> Mass-transport processes include diffusion, migration, and convection.

vibration, and diffusion due to concentration differences between the film of liquid at the electrode surface and the bulk of the solution. In voltammetry, every effort is made to minimize the effect of migration by introducing an excess of an inactive supporting electrolyte. When the concentration of supporting electrolyte exceeds that of the analyte by 50- to 100-fold, the fraction of the total current carried by the analyte approaches zero. As a result, the rate of movement of the analyte toward the electrode of opposite charge becomes essentially independent of applied potential.

### Concentration Profiles at Microelectrode Surfaces During Electrolysis

Throughout this discussion we will consider that the electrode reaction shown in Equation 19–1 takes place at a mercury-coated microelectrode in a solution of A that also contains an excess of a supporting electrolyte. We will assume that the initial concentration of A is $c_A$, while that of P is zero and that P is not soluble in the mercury. We also assume that the reduction reaction is rapid and reversible so that the concentrations of A and P in the film of solution immediately adjacent to the electrode is given at any instant by the Nernst equation. That is,

$$E_{appl} = E_A^0 - \frac{0.0592}{n} \log \frac{c_P^0}{c_A^0} - E_{ref} \qquad (19-2)$$

where $E_{appl}$ is the potential between the microelectrode and the reference electrode and $c_P^0$ and $c_A^0$ are the concentrations of P and A *in a thin layer of solution at the electrode surface only*. Finally we assume that because the electrode is so very small, the electrolysis, over short periods of time, does not alter the bulk concentration of the solution appreciably. That is, the concentration of A in the bulk of the solution $c_A$ is unchanged by the electrolysis and the concentration of P in the bulk of the solution $c_P$ continues to be, for all practical purposes, zero ($c_P \cong 0$).

Electrolysis at a microelectrode does not alter the bulk concentration of the analyte solution during the course of a voltammetry experiment.

**Profiles for Microelectrodes in Stirred Solutions.** Let us consider concentration/distance profiles when a potential is applied to a planar microelectrode immersed in a solution that is stirred vigorously. In order to understand the effect of stirring, it is necessary to develop a picture of liquid flow patterns in such a solution. Three types of flow can be identified as shown in Figure 19–8. (1) *Turbulent flow,* in which liquid motion has no regular pattern, occurs in the bulk of the solution away from the electrode. (2) As the surface is approached, a transition to *laminar flow* takes place. In laminar flow, layers of liquid slide by one another in a direction parallel to the electrode surface. (3) At δ cm from the surface of the electrode, the rate of laminar flow approaches zero as a result of friction between the liquid and the electrode, giving a thin layer of stagnant solution, which is called the *Nernst diffusion layer*. It is only within this stagnant layer that the concentrations of reactant and product vary as a function of distance from the electrode surface. That is, throughout the laminar flow and turbulent flow regions, convection maintains the concen-

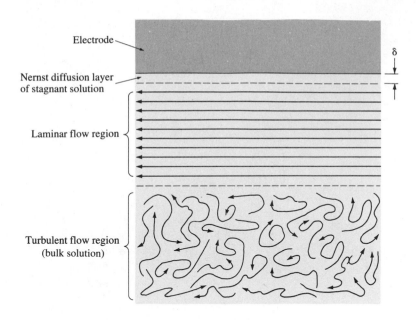

Electrode

Nernst diffusion layer
of stagnant solution

Laminar flow region

Turbulent flow region
(bulk solution)

$\delta$

Figure 19–8
Flow patterns and regions of interest near the working electrode in hydrodynamic voltammetry.

tration of A at its original value and the concentration of P at a level that is vanishingly small.

Figure 19–9a shows two sets of concentration profiles for A and P at three potentials shown as $X$, $Y$, and $Z$ in Figure 19–7. In Figure 19–9, the solution is divided into two regions. One makes up the bulk of the solution where mass transport takes place by mechanical convection brought about by the stirrer. The concentration of A throughout this region is $c_A$. The second region is the Nernst diffusion layer, which is immediately adjacent to the electrode surface and has a thickness of $\delta$ cm. Typically, $\delta$ ranges from 0.01 to 0.001 cm, depending upon the efficiency of the stirring and the viscosity of the liquid. Within the diffusion layer mass transport takes place by diffusion alone just as is the case with an unstirred solution. With the stirred solution, however, diffusion is limited to a narrow layer of liquid and over time cannot extend indefinitely into the solution. As a consequence, steady, diffusion-controlled currents are realized shortly after application of a voltage.

Figure 19–9b gives concentration profiles for P at the three potentials $X$, $Y$, and $Z$. In the Nernst diffusion region the concentration of P decreases linearly with distance from the electrode surface and approaches zero at $\delta$.

Note in the figures that at potential $X$, the equilibrium concentration of A at the electrode surface has been reduced to about 80% of its original value, while the equilibrium concentration P has increased by an equivalent amount. (That is, $c_P^0 = c_A - c_A^0$.) At potential $Y$, which is the half-wave potential, the equilibrium concentrations of the two species at the surface are the same and equal to $c_A/2$. Finally, at potential $Z$ and beyond, the surface concentration of A approaches zero and that of P approaches the original concentration of A, $c_A$. At potentials more negative than $Z$, essentially all A ions entering the surface layer are instantaneously reduced to P. The P formed in this way rapidly diffuses into the bulk of the

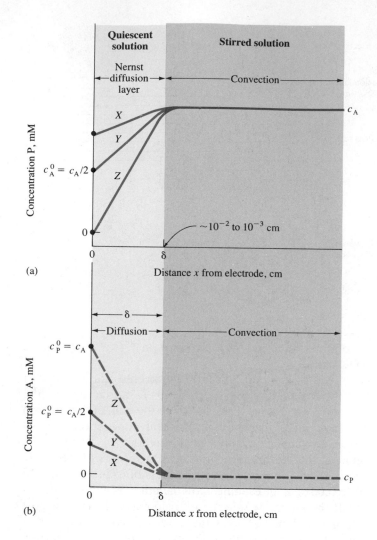

Figure 19–9
Concentration profiles at an electrode/solution interface during the electrolysis $A + ne^- \rightarrow P$ from a stirred solution of A. See Figure 23–5 for potentials corresponding to curves $X$, $Y$, and $Z$.

solution so that the concentration of P in the surface layer remains constant at $c_A$.

## Voltammetric Currents

The current at any point in an electrolysis is determined by a combination of (1) the rate of mass transport of A to the edge of the diffusion layer by convection and (2) the rate of transport of A from the outer edge of the diffusion layer to the electrode surface. Because the product of the electrolysis P diffuses away from the surface and is ultimately swept away by convection, a continuous current is required to maintain the surface concentrations demanded by the Nernst equation. Convection, however, maintains a constant supply of A at the outer edge of the diffusion layer. Thus, a steady-state current results that is determined by the applied potential.

The current in the electrolysis experiment we have been considering is a quantitative measure of how fast A is being brought to the surface of the electrode, and this rate is proportional to the *concentration gradient* $\partial c_A/\partial x$, where $x$ is the distance in centimeters from the electrode surface. For a planar electrode, it can be shown that the current is given by the expression

$$i = nFAD_A \left(\frac{\partial c_A}{\partial x}\right) \tag{19-3}$$

where $i$ is the current in amperes, $n$ is the number of moles of electrons per mole of analyte, $F$ is the faraday, $A$ is the electrode surface area in $cm^2$, $D_A$ is the diffusion coefficient for A in $cm^2 s^{-1}$, and $c_A$ is the concentration of A in $mol/cm^3$. Note that $\partial c_A/\partial x$ is the slope of the initial part of the concentration profiles shown in Figure 19–9a, and these slopes can be approximated by $(c_A - c_A^0)/\delta$. Therefore, Equation 19–4 reduces to

$$i = \frac{nFAD_A}{\delta}(c_A - c_A^0) = k_A(c_A - c_A^0) \tag{19-4}$$

where the constant $k_A$ is equal to $nFAD_A/\delta$.

Equation 19–4 shows that as $c_A^0$ becomes smaller as a result of a larger negative applied potential, the current increases until the surface concentration approaches zero, at which point the current becomes constant and independent of the applied potential. Thus, when $c_A^0 \to 0$, the current becomes the limiting current $i_l$ and

$$i_l = \frac{nFAD_A}{\delta}c_A = k_A c_A \tag{19-5}$$

This derivation is based upon an oversimplified picture of the diffusion layer in that the interface between the moving and stationary layers is viewed as a sharply defined edge where transport by convection ceases and transport by diffusion begins. Nevertheless, this simplified model provides a reasonable approximation of the relationship between current and the variables that affect the current.

### Current/Voltage Relationships for Reversible Reactions

In order to develop an equation for the sigmoidal curve shown in Figure 19–7, we subtract Equation 19–4 from Equation 19–3 and rearrange to obtain

$$c_A^0 = \frac{i_l - i}{k_A} \tag{19-6}$$

The surface concentration of P can also be expressed in terms of the current by employing a relationship similar to Equation 19–4. That is,

$$i = -\frac{nFAD_P}{\delta}(c_P - c_P^0) \tag{19-7}$$

**Challenge:** Show that the units of Equation 19–5 are amperes if the units of the quantities in the equation are as follows:

| Quantity | Units |
|---|---|
| $n$ | mol electrons/mol analyte |
| $F$ | coulomb/faraday = coulomb/mol electrons |
| $A$ | centimeters$^2$ |
| $D_A$ | centimeters$^2$/second |
| $c_A$ | mol analyte/centimeter$^3$ |
| $\delta$ | cm |

Although our model is somewhat oversimplified, it provides a reasonably accurate picture of the processes that occur at the electrode-solution interface.

where the minus sign results from the negative slope of the concentration profile for P. Note that $D_P$ is now the diffusion coefficient of P. But we have said earlier that throughout the electrolysis the concentration of P approaches zero in the bulk of the solution and, therefore, when $c_P \approx 0.0$

$$i = nFAD_Pc_P^0 = k_Pc_P^0$$

where $k_P = nAD_P/\delta$. Rearranging gives

$$c_P^0 = i/k_P \qquad (19\text{--}8)$$

Substituting Equations 19–6 and 19–8 into Equation 19–2 yields, after rearrangement,

$$E_{appl} = E_A^0 - \frac{0.0592}{n} \log \frac{k_A}{k_P} - \frac{0.0592}{n} \log \frac{i}{i_l - i} - E_{ref} \qquad (19\text{--}9)$$

When $i = i_l/2$, the third term on the right side of this equation becomes equal to zero, and, by definition, $E_{appl}$ is the half-wave potential. That is,

The half-wave potential, $E_{1/2}$.

$$E_{appl} = E_{1/2} = E_A^0 - \frac{0.0592}{n} \log \frac{k_A}{k_P} - E_{ref} \qquad (19\text{--}10)$$

Substituting this expression into Equation 19–9 gives an expression for the voltammogram in Figure 19–7. That is,

$$E_{appl} = E_{1/2} - \frac{0.0592}{n} \log \frac{i}{i_l - i} \qquad (19\text{--}11)$$

Often, the ratio $k_A/k_P$ in Equation 19–10 is nearly unity, so that we may write for the species A

$$E_{1/2} \cong E_A^0 - E_{ref} \qquad (19\text{--}12)$$

## Current/Voltage Relationships for Irreversible Reactions

Many voltammetric electrode processes, particularly those associated with organic systems, are irreversible, which leads to drawn-out and less well-defined waves. The quantitative description of such waves requires an additional term (involving the activation energy of the reaction) in Equation 19–11 to account for the kinetics of the electrode process. Although half-wave potentials for irreversible reactions ordinarily show some dependence upon concentration, diffusion currents remain linearly related to concentration; such processes are, therefore, readily adapted to quantitative analysis.

## Voltammograms for Mixtures of Reactants

Ordinarily, the reactants of a mixture will behave independently of one another at a microelectrode; a voltammogram for a mixture is thus simply

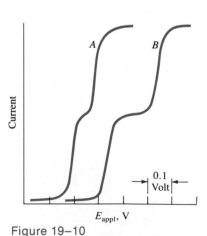

Figure 19–10
Voltammograms for two-component mixtures. Half-wave potentials differ by 0.1 V in curve *A* and by 0.2 V in curve *B*.

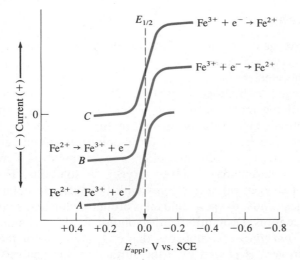

**Figure 19–11**

Voltammetric behavior of iron(II) and iron(III) in a citrate medium. Curve $A$: anodic wave for a solution in which $[Fe^{2+}] = 1 \times 10^{-4}$. Curve $B$: anodic/cathodic wave for a solution in which $[Fe^{2+}] = [Fe^{3+}] = 0.5 \times 10^{-4}$. Curve $C$: cathodic wave for a solution in which $[Fe^{3+}] = 1 \times 10^{-4}$.

the summation of the waves for the individual components. Figure 19–10 shows the voltammograms for a pair of two-component mixtures. The half-wave potentials of the two reactants differ by about 0.1 V in curve $A$ and by about 0.2 V in curve $B$. Clearly, a single voltammogram may permit the quantitative determination of two or more species. Success depends upon the existence of a sufficient difference between succeeding half-wave potentials to permit evaluation of individual diffusion currents. Approximately 0.2 V is required if the more reducible species undergoes a two-electron reduction; a minimum of about 0.3 V is needed if the first reduction is a one-electron process.

Voltammetric half-wave potentials must be separated by 0.2 to 0.3 V in order to resolve mixtures of reactants.

## Anodic and Mixed Anodic/Cathodic Voltammograms

Anodic waves as well as cathodic waves are encountered in voltammetry. An example of an anodic wave is illustrated in curve $A$ of Figure 19–11, where the electrode reaction involves the oxidation of iron(II) to iron(III) in the presence of citrate ion. A limiting current is obtained at about +0.1 V, which is due to the half-reaction

$$Fe^{2+} \rightleftarrows Fe^{3+} + e^-$$

As the potential is made more negative, a decrease in the anodic current occurs; at about −0.02 V, the current becomes zero because the oxidation of iron(II) ion has ceased.

Curve $C$ represents the voltammogram for a solution of iron(III) in the same medium. Here, a cathodic wave results from reduction of the iron(III) to the divalent state. The half-wave potential is identical with that for the anodic wave, indicating that the oxidation and reduction of the two iron species are perfectly reversible at the microelectrode.

Curve $B$ is the voltammogram of an equimolar mixture of iron(II) and iron(III). The portion of the curve below the zero-current line corresponds to the oxidation of the iron(II); this reaction ceases at an applied potential equal to the half-wave potential. The upper portion of the curve is due to the reduction of iron(III).

## Oxygen Waves

Dissolved oxygen is readily reduced at the dropping mercury electrode; an aqueous solution saturated with air exhibits two distinct waves attributable to this element (see Figure 19–12). The first results from the reduction of oxygen to hydrogen peroxide. The second corresponds to the reduction of the molecule to give water. As would be expected from stoichiometric considerations, the wave for the reduction to water is twice as large as the wave for the formation of hydrogen peroxide.

Voltammetric measurements offer a convenient and widely used method for determining dissolved oxygen in solutions. However, the presence of oxygen often interferes with the accurate determination of other species. Thus, oxygen removal is ordinarily the first step in amperometric procedures. Deaeration of the solution for several minutes with an inert gas (*sparging*) accomplishes this end; a stream of the same gas, usually nitrogen, is passed over the surface during analysis to prevent reabsorption of oxygen.

Remove oxygen from analyte solutions by sparging with nitrogen.

## Applications of Hydrodynamic Voltammetry

Currently, the most important uses of hydrodynamic voltammetry include: (1) detection and determination of chemical species as they exit from chromatographic columns or flow-injection apparatus; (2) routine determination of oxygen and certain species of biochemical interest, such as glucose, lactose, and sucrose; (3) detection of end points in coulometric and volumetric titrations; (4) fundamental studies of electrochemical processes.

Flow injection analysis is discussed in Section 22C–3.

**Voltammetric Detectors in Chromatography and Flow-Injection Analysis.** Hydrodynamic voltammetry is becoming widely used for detection and determination of oxidizable or reducible compounds or ions that have been separated by high performance liquid chromatography (Chapter 28) or by flow-injection methods (Section 22C). In these applications a thin

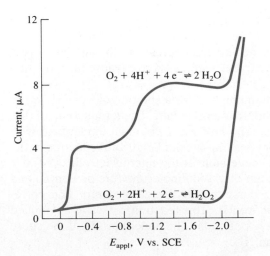

**Figure 19–12**

Voltammogram for the reduction of oxygen in an air-saturated 0.1 M KCl solution. The lower curve is for oxygen-free 0.1 M KCl.

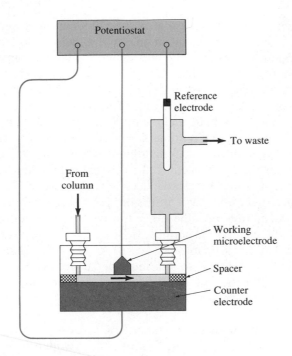

Figure 19–13
A voltammetric system for detecting electroactive species as they exit from a column.

layer cell such as that shown in Figure 19–13 is used. The working electrode in these cells is typically imbedded in the wall of an insulating block that is separated from a counter electrode by a thin spacer as shown. The volume of such a cell is typically 0.1 to 1 $\mu$L. A potential corresponding to the limiting current region for analytes is applied between the metal or glassy carbon working electrode and a silver/silver chloride reference electrode that is located downstream from the detector. In this type of application, detection limits as low as $10^{-9}$ to $10^{-10}$ M of analyte are obtained. This application of hydrodynamic voltammetry is considered further in Section 28C–6.

**Voltammetric Sensors.**  A number of voltammetric systems are produced commercially for the determination of specific species that are of interest in industry and research. These devices are sometimes called electrodes but are, in fact, complete voltammetric cells and are better referred to as sensors. Two of these devices are described here.

The determination of dissolved oxygen in a variety of aqueous environments, such as sea water, blood, sewage, effluents from chemical plants, and soils, is of tremendous importance to industry, biomedical and environmental research, and clinical medicine. One of the most common and convenient methods for making such measurements is with the *Clark oxygen sensor*, which was patented by L. C. Clark Jr. in 1956.[4] A schematic of the Clark oxygen sensor is shown in Figure 19–14. The cell consists of a platinum disk cathodic working electrode imbedded in a

The Clark oxygen sensor is widely used in clinical laboratories for the determination of $O_2$ in blood and other bodily fluids.

[4]For a detailed discussion of the Clark oxygen sensor, see M. L. Hitchman, *Measurement of Dissolved Oxygen*, Chapters 3–5. New York: Wiley, 1978.

Figure 19–14
The Clark voltammetric oxygen sensor. Cathode reaction:
$O_2 + 4 H^+ + 4 e^- \rightleftarrows H_2O$.
Anodic reaction:
$Ag + Cl^- \rightarrow AgCl(s) + e^-$.

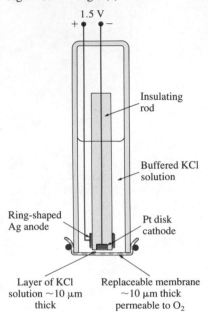

1.5 V
+        −

Insulating rod

Buffered KCl solution

Ring-shaped Ag anode

Pt disk cathode

Layer of KCl solution ∼10 μm thick

Replaceable membrane ∼10 μm thick permeable to $O_2$

centrally located cylindrical insulator. Surrounding the lower end of this insulator is a ring-shaped silver anode. The tubular insulator and electrodes are mounted inside a second cylinder that contains a buffered solution of potassium chloride. A thin (∼20 μm), replaceable, oxygen-permeable membrane of Teflon or polyethylene is held in place at the bottom end of the tube by an O-ring. The thickness of the electrolyte solution between the cathode and the membrane is approximately 10 μm.

When the oxygen sensor is immersed in a flowing or stirred solution of the analyte, oxygen diffuses through the membrane into the thin layer of electrolyte immediately adjacent to the disk cathode, where it diffuses to the electrode and is immediately reduced to water. In contrast to a normal hydrodynamic electrode, two diffusion processes are involved—one through the membrane and the other through the solution between the membrane and the electrode surface. In order for a steady-state condition to be reached in a reasonable period (10 to 20 s), the thickness of the membrane and the electrolyte film must be 20 μm or less. Under these conditions, it is the rate of equilibration of oxygen transfer across the membrane that determines the rate at which steady-state currents are achieved.

A number of enzyme-based voltammetric sensors are offered commercially. An example is a glucose sensor that is widely used in clinical laboratories for the routine determination of glucose in blood serum samples. This device is similar in construction to the oxygen sensor shown in Figure 19–14. The membrane in this case is more complex and consists of three layers. The outer layer is a polycarbonate film that is permeable to glucose but impermeable to proteins and other constituents of blood. The middle layer is an immobilized enzyme—here, glucose oxidase. The inner layer is cellulose acetate membrane, which is permeable to small molecules, such as hydrogen peroxide. When this device is immersed in a glucose-containing solution, glucose diffuses through the outer membrane into the immobilized enzyme, where the following catalytic reaction occurs:

$$\text{glucose} + O_2 \rightarrow H_2O_2 + \text{gluconic acid}$$

The hydrogen peroxide then diffuses through the inner layer of membrane and to the electrode surface, where it is oxidized to give oxygen. That is,

$$H_2O_2 + 2 OH^- \rightleftarrows O_2 + H_2O + 2 e^-$$

The resulting current is directly proportional to the glucose content of the analyte solution.

Several other sensors are available that are based upon the voltammetric measurement of hydrogen peroxide produced by enzymatic oxidations of other species of clinical interest. These analytes include sucrose, lactose, ethanol, and L-lactate. Of course, a different enzyme is required for each species.

**Amperometric Titrations.** Hydrodynamic voltammetry can be employed to estimate the equivalence point of titrations, provided at least one of the participants or products of the reaction involved is oxidized or reduced at a microelectrode. Here, the current at some fixed potential in the limiting current region is measured as a function of the reagent volume (or of time if the reagent is generated by a constant-current coulometric process). Plots of the data on either side of the equivalence point are straight lines with differing slopes; the end point is established by extrapolation to their intersection.

Amperometric titration curves typically take one of the forms shown in Figure 19–15. The curve in Figure 19–15a represents a titration in which the analyte reacts at the electrode while the reagent does not. Figure 19–15b is typical of a titration in which the reagent reacts at the micro-electrode and the analyte does not. Figure 19–15c corresponds to a titra-tion in which both the analyte and the titrant react at the microelectrode.

Two types of amperometric electrode systems are encountered. One employs a single polarizable microelectrode coupled to a reference; the other uses a pair of identical solid state microelectrodes immersed in a stirred solution. For the first type, the microelectrode is often a rotating platinum electrode constructed by sealing a platinum wire into the side of a glass tube that is connected to a stirring motor. A dropping mercury electrode may also be used, in which case the solution is not stirred.

Amperometric titrations with one indicator electrode have, with one notable exception, been confined to titrations in which a precipitate or a stable complex is the product. Precipitating reagents include silver nitrate for halide ions, lead nitrate for sulfate ion, and several organic reagents, such as 8-hydroxyquinoline, dimethylglyoxime, and cupferron, for vari-ous metallic ions that are reducible at microelectrodes. Several metal ions have also been determined by titration with standard solutions of EDTA. The exception just noted involves titrations of certain phenols, aromatic amines, and olefins; hydrazine; and arsenic(III) and antimony(III) with bromine. The bromine is often generated coulometrically; it has also been formed by adding a standard solution of potassium bromate to an acidic

**Figure 19–15**

Typical amperometric titration curves: (a) analyte is reduced, reagent is not; (b) reagent is re-duced, analyte is not; (c) both reagent and analyte are reduced.

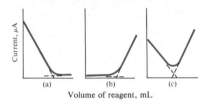

**Figure 19–16**

Typical cell arrangement for am-perometric titrations with a rotat-ing platinum electrode.

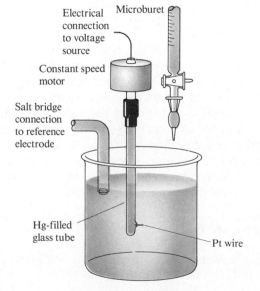

solution of the analyte that also contains an excess of potassium bromide. Bromine is formed in the acidic medium by the reaction

$$BrO_3^- + 5\ Br^- + 6\ H^+ \rightarrow 3\ Br_2 + 3\ H_2O$$

This type of titration has been carried out with a rotating platinum electrode or twin platinum microelectrodes. No current is observed prior to the equivalence point; after chemical equivalence, a rapid increase in current takes place due to electrochemical reduction of the excess bromine.

The use of a pair of identical metallic microelectrodes to establish the equivalence point in amperometric titrations offers the advantage of simplicity of equipment and the avoidance of having to prepare and maintain a reference electrode. This type of system has been incorporated into equipment designed for the routine automatic determination of a single species, usually with a coulometrically generated reagent. An example is a commercially available instrument for the automatic determination of chloride in samples of serum, sweat, tissue extracts, pesticides, and food products. Here, the reagent is silver ion coulometrically generated from a silver anode. The indicator system consists of a pair of twin silver microelectrodes that are maintained at a potential of perhaps 0.1 V. Short of the equivalence point in the titration of chloride ion, there is essentially no current because no easily reduced species is present in the solution. Consequently, electron transfer at the cathode is precluded, and that electrode is completely polarized. Note that the anode is not polarized because the reaction

$$Ag \rightleftarrows Ag^+ + e^-$$

can occur in the presence of a suitable cathodic reactant or depolarizer.

After the equivalence point has been passed, the cathode becomes depolarized, owing to the presence of a significant amount of silver ions, which can react to give silver. That is,

$$Ag^+ + e^- \rightleftarrows Ag$$

A current develops as a result of this half-reaction and the corresponding oxidation of silver at the anode. The magnitude of the current is, as in other amperometric methods, directly proportional to the concentration of the excess reagent. Thus, the titration curve is similar to that shown in Figure 19–15b. In the automatic titrator just mentioned, the amperometric current signal causes the coulometric generator current to cease; the chloride concentration is then computed from the magnitude of the current and the generation time. The instrument is said to have a range of 1 to 999.9 mM $Cl^-$ per liter, a precision of 0.1% relative, and an accuracy of 0.5% relative. Typical titration times are 20 s.

## 19B–5 Polarography

Linear-scan polarography was the first type of voltammetry to be discovered and used. It differs from hydrodynamic voltammetry in two regards.

First, convection is avoided, and second, a dropping mercury electrode, such as that shown in Figure 19–5c, is used as the working electrode. A consequence of the first difference is that polarographic limiting currents are controlled by diffusion alone rather than by both diffusion and convection. Because convection is absent, polarographic limiting currents are generally one or more orders of magnitude smaller than hydrodynamic limiting currents.

Polarographic currents are controlled by diffusion alone, not by convection.

### Polarographic Currents

The current in a cell containing a dropping electrode undergoes periodic fluctuations corresponding in frequency to the drop rate. As a drop dislodges from the capillary, the current falls to zero (see Figure 19–17); it then increases rapidly as the electrode area grows because of the greater surface to which diffusion can occur. The *average current* is the hypothetical *constant* current, which in the drop time $t$ would produce the same quantity of charge as the fluctuating current does during this same period. In order to determine the average current, it is necessary to reduce the large fluctuations in the current by an electronic filter or by sampling the current near the end of each drop, where the change in current with time is relatively small. As shown in Figure 19–18, curve $A$, low-pass filtering limits the oscillations to a reasonable magnitude; the average current (or, alternatively, the maximum current) is then readily determined, provided the drop rate $t$ is reproducible. Note the effect of irregular drops in the upper part of curve $A$, probably caused by vibration of the apparatus.

### Polarograms

Figure 19–18 shows two polarograms, one for a solution that is 1.0 M in hydrochloric acid and $5 \times 10^{-4}$ M in cadmium ion (curve $A$) and a second for the acid in the absence of cadmium ion (curve $B$). The polarographic wave in curve $A$ arises from the reaction

$$Cd^{2+} + 2\,e^- + Hg \rightleftharpoons Cd(Hg) \qquad (19–13)$$

where Cd(Hg) represents elemental cadmium dissolved in mercury giving an amalgam. The sharp increase in current at about $-1$ V in both polarograms is caused by the reduction of hydrogen ions to give hydrogen.

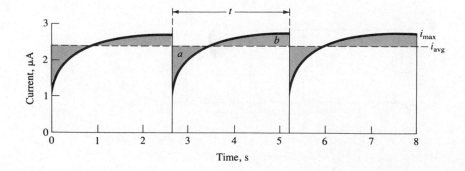

**Figure 19–17**
Effect of drop growth on polarographic current. For $i_{avg}$, area $a =$ area $b$ and $i_{avg} \simeq \frac{6}{7} i_{max}$.

Examination of the polarogram for the supporting electrolyte alone reveals that a small current, called the *residual current*, is present in the cell even in the absence of cadmium ions.

As in hydrodynamic voltammetry, limiting currents are observed when the magnitude of the current is limited by the rate at which analyte can be brought to the electrode surface. In polarography, however, the only mechanism of mass transport is diffusion, and for this reason, polarographic limiting currents are usually termed *diffusion currents* and given the symbol $i_d$. As shown in Figure 19–18, the diffusion current is the difference between the limiting and the residual current. The diffusion current is directly proportional to analyte concentration.

> Diffusion current is proportional to concentration of analyte.

### Diffusion Current at Dropping Electrodes

In deriving an equation for polarographic diffusion currents, it is necessary to take into account the rate of growth of the spherical electrode, which is related to the drop time in seconds $t$ and the rate of flow of mercury through the capillary $m$ in mg/s. These variables are taken into account in the *Ilkovic equation*:

$$(i_d)_{max} = 706 \, nD^{1/2}m^{2/3}t^{1/6}c$$

where $(i_d)_{max}$ is the maximum current in amperes, and $c$ is the analyte concentration in moles per cubic centimeter. To obtain an expression for the average current rather than the maximum, the constant in the foregoing equation becomes 607 rather than 706. That is,

> In polarography, currents are often recorded in microamperes. Concentrations calculated when $\mu A$ are used in Equation 19–14 bear units of millimoles per liter.

$$(i_d)_{ave} = 607 \, nD^{1/2}m^{2/3}t^{1/6}c \qquad (19\text{–}14)$$

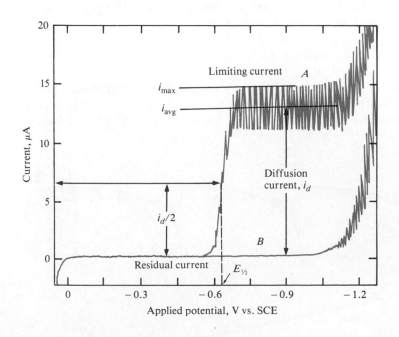

Figure 19–18

Polarograms for (A) a 1 M solution of HCl that is $5 \times 10^{-4}$ M in $Cd^{2+}$ and (B) a 1 M solution of HCl. (From D. T. Sawyer and J. L. Roberts, Jr., *Experimental Electrochemistry for Chemists*. New York: Wiley, 1974. Reprinted by permission of John Wiley & Sons, Inc.)

Note that either the average or the maximum current can be used in quantitative polarography.

The product $m^{2/3}t^{1/6}$ in the Ilkovic equation, called the *capillary constant*, describes the influence of dropping electrode characteristics upon the diffusion current; both $m$ and $t$ are readily evaluated experimentally; comparison of diffusion currents from different capillaries is thus possible.

### Residual Currents

Figure 19–19 shows a residual current curve (obtained at high sensitivity) for a 0.1 M solution of hydrogen chloride. This current has two sources. The first is the reduction of trace impurities that are almost inevitably present in the blank solution; contributors here include small amounts of dissolved oxygen, heavy metal ions from the distilled water, and impurities present in the salt used as the supporting electrolyte.

A second component of the residual current is the so-called *charging* or *condenser current* resulting from a flow of electrons that charge the mercury droplets with respect to the solution; this current may be either negative or positive. At potentials more negative than about $-0.4$ V, an excess of electrons from the dc source provides the surface of each droplet with a negative charge. These excess electrons are carried down with the drop as it breaks; since each new drop is charged as it forms, a small but continuous current results. At applied potentials smaller than about $-0.4$ V, the mercury tends to be positive with respect to the solution; thus, as each drop is formed, electrons are repelled from the surface toward the bulk of mercury, and a negative current is the result. At about $-0.4$ V, the mercury surface is uncharged, and the charging current is zero. The charging current is a type of *nonfaradaic current* in the sense that charge is carried across an electrode/solution interface without an accompanying oxidation/reduction process.

Ultimately, the accuracy and sensitivity of the polarographic method depend upon the magnitude of the nonfaradaic residual current and the accuracy with which a correction for its effect can be determined.

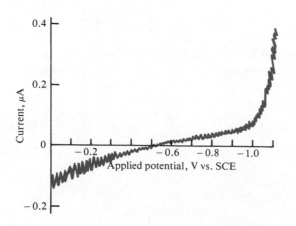

Figure 19–19
Residual current curve for a 0.1 M solution of HCl.

### Comparison of Currents from Dropping and Stationary Planar Electrodes

Constant currents are not obtained with a planar electrode in an unstirred solution because concentration gradients between the electrode surface and the bulk solution are constantly changing with time. In contrast, the dropping electrode exhibits constant reproducible currents essentially instantaneously after an applied voltage adjustment. This behavior represents an advantage of the dropping mercury electrode that accounts for its widespread use in the early years of voltammetry.

The rapid achievement of constant currents results from the highly reproducible nature of the drop formation process and, equally important, the fact that the solution in the electrode area becomes homogenized each time a drop breaks from the capillary. Thus, a concentration gradient is developed only during the brief lifetime of the drop. As we have noted, current changes due to an increase in surface area occur during each lifetime. Changes in the diffusion gradient $dc/dx$ also occur during this period. Nevertheless, these changes are entirely reproducible and lead to currents that are also highly reproducible.

### Effect of Complex Formation on Polarographic Waves

We have already seen (Feature 14–5) that the potential for the oxidation or reduction of a metallic ion is greatly affected by the presence of species that form complexes with that ion. It is not surprising, therefore, that similar effects are observed with polarographic half-wave potentials. The data in Table 19–1 show clearly that the half-wave potential for the reduction of a metal complex is generally more negative than that for reduction of the corresponding simple metal ion. In fact, this negative shift in potential permits the elucidation of the composition of the complex ion and the determination of its formation constant, *provided that the electrode reaction is reversible*. Thus, for the reactions

$$M^{n+} + Hg + ne^- \rightleftarrows M(Hg)$$

and

$$M^{n+} + xA^- \rightleftarrows MA_x^{(n-x)+}$$

Lingane[5] derived the following relationship between the molar concentrations of the ligand $c_L$ and the shift in half-wave potential brought about by its presence

You can evaluate the formula of a complex by this method only if the electrode reaction is reversible.

$$(E_{1/2})_c - E_{1/2} = -\frac{0.0592}{n} \log K_f - \frac{0.0592x}{n} \log c_L \qquad (19\text{--}15)$$

---

[5] J. J. Lingane, *Chem. Rev.*, **1941**, *29*, 1.

| | Table 19-1 | | | |
| --- | --- | --- | --- | --- |

**EFFECT OF COMPLEXING AGENTS ON POLAROGRAPHIC HALF-WAVE POTENTIALS AT THE DROPPING MERCURY ELECTRODE**

| Ion | Noncomplexing Media | 1 M KCN | 1 M KCl | 1 M NH$_3$, 1 M NH$_4$Cl |
| --- | --- | --- | --- | --- |
| Cd$^{2+}$ | $-0.59$ | $-1.18$ | $-0.64$ | $-0.81$ |
| Zn$^{2+}$ | $-1.00$ | NR* | $-1.00$ | $-1.35$ |
| Pb$^{2+}$ | $-0.40$ | $-0.72$ | $-0.44$ | $-0.67$ |
| Ni$^{2+}$ | $-1.01$ | $-1.36$ | $-1.20$ | $-1.10$ |
| Co$^{2+}$ | — | $-1.45$ | $-1.20$ | $-1.29$ |
| Cu$^{2+}$ | $+0.02$ | NR* | $+0.04$ and $-0.22$† | $-0.24$ and $-0.51$† |

*No reduction occurs before involvement of the supporting electrolyte.

†Reduction occurs in two steps having different electrode potentials:

$$Cu^{2+} + 2\,Cl^- + e^- \rightleftarrows CuCl_2^-$$
$$CuCl_2^- + Hg + e^- \rightleftarrows Cu(Hg) + 2\,Cl^-$$

where $(E_{1/2})_c$ and $E_{1/2}$ are the half-wave potentials for the complexed and uncomplexed cations, respectively, $K_f$ is the formation constant for the complex, and $x$ is the molar combining ratio of complexing agent to cation.

Equation 19–15 makes it possible to evaluate the formula for the complex. Thus, a plot of the half-wave potential against log $c_L$ for several ligand concentrations gives a straight line, the slope of which is $0.0592x/n$. If $n$ is known, the combining ratio of ligand to metal ion $x$ is readily calculated. Equation 19–15 can then be employed to calculate $K_f$.

## Effect of pH on Polarograms

Organic electrode processes ordinarily involve hydrogen ions, the typical reaction being represented as

$$R + nH^+ + ne^- \rightleftarrows RH_n$$

where R and RH$_n$ are the oxidized and reduced forms of the organic molecule. Half-wave potentials for organic compounds are therefore markedly pH-dependent. Furthermore, alteration of the pH may result in a change in the reaction product. For example, when benzaldehyde is reduced in a basic solution, a wave is obtained at about $-1.4$ V that is attributable to the formation of benzyl alcohol

$$C_6H_5CHO + 2\,H^+ + 2\,e^- \rightleftarrows C_6H_5CH_2OH$$

If the pH is less than 2, however, a wave occurs at about $-1.0$ V that is just half the size of the foregoing one; here, the reaction involves the production of hydrobenzoin

$$2\,C_6H_5CHO + 2\,H^+ + 2\,e^- \rightleftarrows C_6H_5CHOHCHOHC_6H_5$$

At intermediate pH values, two waves are observed, indicating the occurrence of both reactions.

It should be emphasized that an electrode process that consumes or produces hydrogen ions will alter the pH of the solution *at the electrode surface*, often drastically, unless the solution is well-buffered. These changes affect the reduction potential of the reaction and cause drawn-out, poorly defined waves. Moreover, where the electrode process is altered by pH, as in the case of benzaldehyde, nonlinearity in the diffusion current/concentration relationship will also be encountered. Thus, good buffering is generally vital for the generation of reproducible half-wave potentials and diffusion currents for organic polarography.

### Advantages and Disadvantages of the Dropping Mercury Electrode

In the past, the dropping mercury electrode was the most widely used microelectrode for voltammetry because of its several unique features. The first is the unusually high overvoltage associated with the reduction of hydrogen ions. As a consequence, metal ions such as zinc and cadmium can be deposited from acidic solution even though their thermodynamic potentials suggest that deposition of these metals without hydrogen formation is impossible. A second advantage is that a new metal surface is generated continuously; thus the behavior of the electrode is independent of its past history. In contrast, solid metal electrodes are notorious for their irregular behavior, which is related to adsorbed or deposited impurities. A third unusual feature of the dropping electrode, which has already been described, is that reproducible average currents are *immediately* realized at any given potential regardless of whether this potential is approached from lower or higher settings.

One serious limitation of the dropping electrode is the ease with which mercury is oxidized; this property severely limits the use of the electrode as an anode. At potentials greater than about $+0.4$ V, formation of mercury(I) occurs, giving a wave that masks the curves of other oxidizable species. In the presence of ions that form precipitates or complexes with mercury(I), this behavior occurs at even lower potentials. For example, at the lower left in Figure 19–18, the beginning of an anodic wave can be seen at 0 V due to the reaction:

$$2\,Hg + 2\,Cl^- \rightarrow Hg_2Cl_2(s) + 2\,e^-$$

Incidentally, this anodic wave can be used for the determination of chloride ion.

Another important disadvantage of the dropping mercury electrode is the nonfaradaic residual or charging current, which limits the sensitivity of the classical method to concentrations of about $10^{-5}$ M. At lower concentrations, the residual current is likely to be greater than the diffusion current—a situation that prohibits accurate measurement of the latter. As will be shown later, methods are now available for enhancing detection limits by one to two orders of magnitude.

The detection limit for classical polarography is about $10^{-5}$ M.

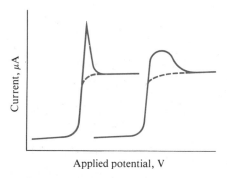

Figure 19–20
Typical current maxima with a
dropping mercury electrode.

Finally, the dropping mercury electrode is cumbersome to use and
tends to malfunction as a result of clogging. An annoying aspect of this
electrode is that it often produces current maxima such as those shown in
Figure 19–20. Although the cause or causes of maxima are not fully
understood, there is considerable empirical knowledge of methods for
eliminating them. Generally, the addition of traces of such high-molecular-weight substances as gelatin, Triton X-100 (a commercial surface-
active agent, or surfactant), methyl red, or other dyes will cause a maximum to disappear. Care must be taken to avoid large amounts of these
reagents, however, because the excess may reduce the magnitude of the
diffusion current. The proper amount of suppressor must be determined
by trial and error; the amount required varies widely from analyte to
analyte.

## Current-Sampled (Tast) Polarography

A simple modification of the classical polarographic technique, and one
that is incorporated into most modern voltammetric instruments, involves
measurement of current only for a period near the end of the lifetime of
each drop. Here, a mechanical knocker is generally used to detach the
drop after a highly reproducible time interval (usually 0.5 to 5 s). Because
of this feature, the method is sometimes called *tast* polarography (from
German *tasten*, to touch). The term *current-sampled polarography* is
more descriptive than tast polarography and thus is preferred inasmuch as
devices other than the mechanical knocker are now used for drop detach-
ment (see Figure 19–5d, for example).

Figure 19–21 contrasts the appearances of classical and current-sampled polarograms. In obtaining the latter type, the potential was scanned
linearly at 5 mV/s. Rather than recording the current continuously, however, it was instead sampled for a 5-ms period just before termination of
each drop. Between sampling periods, the recorder was maintained at its
last current level by means of a sample-and-hold circuit. As is apparent
from the figure, a major advantage of current sampling is that it substantially reduces the large current fluctuations due to the continuous growth
and fall of drops that occur with the dropping electrode. Note in Figure
19–17 that the current near the end of the life of a drop is nearly constant;

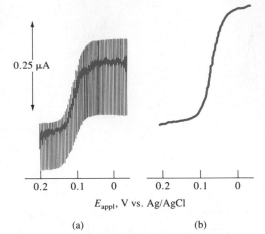

**Figure 19-21**

Comparison of (a) classical and (b) current-sampled polarographic waves for a $1 \times 10^{-4}$ M solution of $Cu^{2+}$ in 1 M $NaNO_3$. (From A. M. Bond and D. R. Canterford, *Anal. Chem.*, **1972**, *44*, 721. With permission. Copyright 1972 American Chemical Society.)

$E_{appl}$, V vs. Ag/AgCl

(a)                    (b)

The detection limit for tast polarography is only about a factor of 3 lower than for classical polarography.

it is this current only that is recorded in the current-sampled technique. The result is a smoothed curve consisting of a series of steps, which are significantly smaller than the current fluctuations encountered in conventional polarography. The improvements in precision and detection limit provided by current sampling alone are not, unfortunately, as great as might be hoped. For example, Bond and Canterford[6] have shown that the detection limits for copper could be lowered from about $3 \times 10^{-6}$ M for conventional polarography to $1 \times 10^{-6}$ M with the current-sampled method, a marginal change at best.

## 19C PULSE POLAROGRAPHIC AND VOLTAMMETRIC METHODS

By the 1960s, linear-scan polarography had ceased to be an important analytical tool in most laboratories. The reason for the decline in use of this once popular technique was not only the appearance of several more convenient spectroscopic methods but also the inherent disadvantages of the method, including slow speed, inconvenient apparatus, and, particularly, poor detection limits. These limitations were largely overcome by pulse methods and the development of electrodes such as those shown in Figure 19-5d. We shall discuss the two most important pulse techniques, *differential pulse polarography* and *square wave polarography*. Both methods have also been applied with electrodes other than the dropping mercury electrode, in which case the procedures are termed differential and square wave voltammetry.

### 19C-1 Differential Pulse Polarography

Figure 19-22 shows the two most common excitation signals that are employed in commercial instruments for differential pulse polarography.

---

[6]A. M. Bond and D. R. Canterford, *Anal. Chem.*, **1972**, *44*, 721.

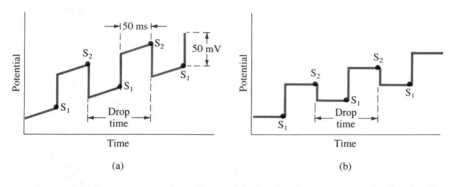

(a)

(b)

Figure 19–22
Excitation signals for differential pulse polarography.

The first (Figure 19–22a), which is used in analog instruments, is obtained by superimposing a periodic pulse on a linear scan. The second (Figure 19–22b), which is ordinarily used in digital instruments, involves combination of a pulse output with a staircase signal. In either case, a 50-mV pulse is applied during the last 50 ms of the lifetime of the mercury drop. Here again, to synchronize the pulse with the drop, the latter is detached at an appropriate time by mechanical means.

As shown in Figure 19–22, two current measurements are made alternately—one at $S_1$, which is 16.7 ms prior to the dc pulse, and one at $S_2$, for 16.7 ms at the end of the pulse. The *difference in current per pulse* ($\Delta i$) is recorded as a function of the linearly increasing voltage. The resulting differential curve consists of a peak (see Figure 19–23) whose height is directly proportional to concentration. For a reversible reaction the peak potential is approximately equal to the standard potential for the half-reaction.

One advantage of the derivative-type polarogram is that individual peak maxima can be observed for substances with half-wave potentials differing by as little as 0.04 to 0.05 V; in contrast, classical and normal pulse polarography require a potential difference of at least 0.2 V for resolution of waves. More importantly, however, differential pulse polarography increases the sensitivity of the polarographic method significantly. This enhancement is illustrated in Figure 19–24. Note that a classical polarogram for a solution containing 180 ppm of the antibiotic tetracycline gives two barely discernible waves; differential pulse polarography, in contrast, provides well-defined peaks at a concentration level that is $2 \times 10^{-3}$ that for the classic wave, or 0.36 ppm. Note also that the current scale for $\Delta i$ is in nA (nanoamperes), or $10^{-3}$ $\mu$A. Generally, detection limits with differential pulse polarography are two to three orders of magnitude lower than those for classical polarography and lie in the range of $10^{-7}$ to $10^{-8}$ M.

The greater sensitivity of differential pulse polarography can be attributed to two sources. The first is an enhancement of the faradaic current, while the second is a decrease in the nonfaradaic charging current. To account for the former, let us consider the events that must occur in the surface layer around an electrode as the potential is suddenly increased by 50 mV. If a reactive species is present in this layer, there will be a surge of current that lowers the reactant concentration to that demanded by the new potential. As the equilibrium concentration for that potential is approached, however, the current decays to a level just sufficient to counteract diffusion, that is, to the diffusion-controlled current. In classical

Derivative polarograms yield peaks that are convenient for qualitative identification of analytes.

Detection limits for differential pulse polarography are two to three orders of magnitude lower than for classical polarography.

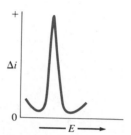

Figure 19–23
Voltammogram for a differential pulse polarography experiment. Here $\Delta i = E_{S_2} - E_{S_1}$. (See Figure 19–22.)

Figure 19–24

(a) Differential pulse polarogram. 0.36 ppm tetracycline · ·HCl in 0.1 M acetate buffer, pH 4, PAR Model 174 polarographic analyzer, dropping mercury electrode, 50 mV pulse amplitude, 1 s drop. (b) DC polarogram. 180 ppm tetracycline · HCl in 0.1 M acetate buffer, pH 4, similar conditions. (Reprinted with permission from J. B. Flato, *Anal. Chem.*, **1972**, *44* (11), 75A. Copyright 1972 American Chemical Society.)

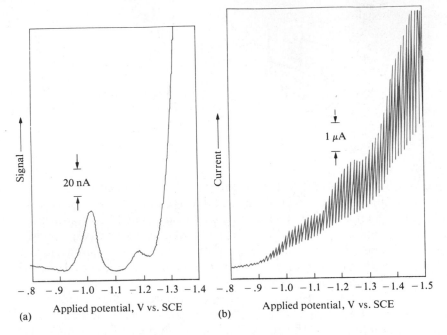

polarography, the initial surge of current is not observed because the time scale of the measurement is long relative to the lifetime of the momentary current. On the other hand, in pulse polarography, the current measurement is made before the surge has completely decayed. Thus, the current measured contains both a diffusion-controlled component and a component that has to do with reducing the surface layer to the concentration demanded by the Nernst expression; the total current is typically several times larger than the diffusion current. It should be noted that when the drop is detached, the solution again becomes homogeneous with respect to the analyte. Thus, at any given voltage, an identical current surge accompanies each voltage pulse.

When the potential pulse is first applied to the electrode, a surge in the nonfaradaic current also occurs as the charge on the drop increases (page 493). This current, however, decays exponentially with time and approaches zero near the end of the life of a drop when its surface area is changing only slightly (see Figure 19–17). Thus, by measuring currents at this time only, the nonfaradaic residual current is greatly reduced, and the signal-to-noise ratio is larger. Enhanced sensitivity results.

Reliable instruments for differential pulse polarography are now commercially available at reasonable cost. The method has thus become the most widely used analytical polarographic procedure.

### 19C–2 Square Wave Polarography and Voltammetry[7]

Square wave polarography is a type of pulse polarography that offers the advantage of great speed and high sensitivity. An entire voltammogram

---

[7]For further information on square wave voltammetry, see J. G. Osteryoung and R. A. Osteryoung, *Anal. Chem.*, **1985**, *57*, 101A.

can be obtained in a few seconds. With a dropping mercury electrode, the scan is performed during the last half of the life of a drop and normalized to compensate for the growth of the drop during the measurement process. Square wave voltammetry has also been used with hanging drop electrodes and with chromatographic detectors.

Figure 19–25c shows the excitation signal in square wave voltammetry, which is obtained by superimposing the pulse train shown in Figure 19–25b onto the staircase signal in 19–25a. The length of each step of the staircase and the period of the pulses ($\tau$) are identical and usually about 5 ms. The potential step of the staircase $\Delta E_s$ is typically 10 mV. The magnitude of the pulse $2E_{sw}$ is often 50 mV. Operating under these conditions, which correspond to a pulse frequency of 200 Hz, a 1-V scan requires 0.5 s. For a reversible reduction reaction, the size of a pulse is great enough that oxidation of the product formed on the forward pulse occurs during the reverse pulse. Thus, as shown in Figure 19–26, the forward pulse produces a cathodic current $i_1$, whereas the reverse pulse gives an anodic current $i_2$. Usually the difference in these currents $\Delta i$ is plotted to give voltammograms. This difference is directly proportional to concentration; the potential of the peak corresponds to the polarographic half-wave potential. Because of the speed of the measurement, it is possible and practical to increase the precision of analyses by signal-averaging data from several voltammetric scans. Detection limits for square wave voltammetry are reported to be $10^{-7}$ to $10^{-8}$ M.

Commercial instruments for square wave voltammetry have recently become available from two manufacturers, and as a consequence, it seems likely that this technique will gain considerable use for analysis of inorganic and organic species. It has also been suggested that square wave voltammetry can be used in detectors for HPLC.

## 19C–3 Applications of Pulse Polarography

In the past, linear scan polarography was used for the quantitative determination of a wide variety of inorganic and organic species, including molecules of biological and biochemical interest. Currently, pulse methods have supplanted the classical method almost completely because of their greater sensitivity, convenience, and selectivity. Generally, quantitative applications are based upon calibration curves in which peak heights are plotted as a function of analyte concentration. In some instances the standard addition method is employed in lieu of calibration curves. In either case, it is essential that the composition of standards resemble as closely as possible the composition of the sample both as to electrolyte concentrations and pH. When this is done, relative precisions and accuracies in the 1 to 3% range can often be realized for concentrations of $10^{-7}$ M and greater.

### Inorganic Applications

The polarographic method is widely applicable to the analysis of inorganic substances. Most metallic cations, for example, are reduced at the drop-

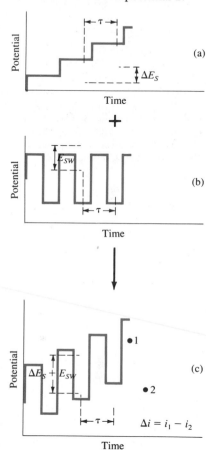

**Figure 19–25**

Generation of a square wave voltammetry excitation signal. The staircase signal in (a) is added to the pulse train in (b) to give the square wave excitation signal in (c). The current response $\Delta i$ is equal to the current at potential 1 minus the current at potential 2.

Multiple scans from multiple drops can be summed to improve the signal-to-noise ratio of a square wave voltammogram.

Detection limits for both differential pulse polarography and square wave voltammetry are $10^{-7}$ to $10^{-8}$ M.

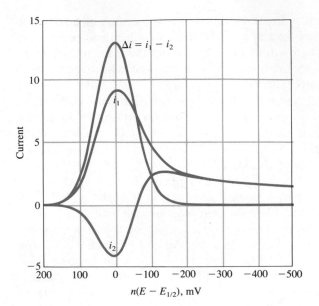

**Figure 19–26**

Current response for a reversible reaction to excitation signal in Figure 19–25c. $i_1$, forward current; $i_2$, reverse current; $i_1 - i_2$, current difference. (From J. J. O'Dea, J. Osteryoung, and R. A. Osteryoung, *Anal. Chem.*, **1981**, *53*, 695. With permission. Copyright 1981 American Chemical Society.)

ping electrode. Even the alkali and alkaline-earth metals are reducible, provided the supporting electrolyte does not react at the high potentials required; here, the tetraalkyl ammonium halides are useful electrolytes because of their high reduction potentials.

The successful polarographic determination of cations frequently depends upon the supporting electrolyte that is used. To aid in this selection, tabular compilations of half-wave potential data are available.[8] The judicious choice of anion often enhances the selectivity of the method. For example, with potassium chloride as a supporting electrolyte, the waves for iron(III) and copper(II) interfere with one another; in a fluoride medium, however, the half-wave potential of the former is shifted by about $-0.5$ V, while that for the latter is altered by only a few hundredths of a volt. The presence of fluoride thus results in the appearance of well-separated waves for the two ions.

The polarographic method is also applicable to the analysis of such inorganic anions as bromate, iodate, dichromate, vanadate, selenite, and nitrite. In general, polarograms for these substances are affected by the pH of the solution because the hydrogen ion is a participant in their reduction. As a consequence, strong buffering to some fixed pH is needed to obtain reproducible data (see the discussion on page 495).

## Organic Applications

Almost from its inception, polarography has been used for the study and determination of organic species, with many papers being devoted to this

---

[8]An extensive source is *Handbook of Analytical Chemistry,* L. Meites, Ed. New York: McGraw-Hill, 1963.

subject. Several common functional groups are oxidized or reduced at working electrodes, making possible the determination of a wide variety of organic compounds.[9]

In general, the reactions of organic compounds at a microelectrode are slower and more complex than those for inorganic species. Consequently, theoretical interpretation of the data is more difficult and often impossible; moreover, a much stricter adherence to detail is required for quantitative work. Despite these handicaps, organic polarography has proved fruitful for the determination of structure, the quantitative analysis of mixtures, and occasionally the qualitative identification of compounds.

## 19D  STRIPPING METHODS

Stripping methods encompass a variety of electrochemical procedures having a common, characteristic initial step.[10] In all of these procedures, the analyte is first deposited on a microelectrode, usually from a stirred solution. After an accurately measured period, the electrolysis is discontinued, the stirring is stopped, and the deposited analyte is determined by one of the voltammetric procedures that have been described in the previous section. During this second step in the analysis, the analyte is redissolved or stripped from the microelectrode; hence the name for these methods. In *anodic stripping,* the microelectrode behaves as a cathode during the deposition step and an anode during the stripping step, with the analyte being oxidized back to its original form. In *cathodic stripping,* the microelectrode behaves as an anode during the deposition step and a cathode during the stripping. The deposition step amounts to an electrochemical preconcentration of the analyte; that is, the concentration of the analyte in the surface of the microelectrode is far greater than it is in the bulk solution.

Figure 19–27a illustrates the voltage excitation program that is followed in an anodic stripping method for determining cadmium and copper in an aqueous solution of these ions. A linear scan voltammetric method is used to complete the analysis. Initially, a constant cathodic potential of about −1 V is applied to the microelectrode, which causes both cadmium and copper ions to be reduced and deposited as metals. The electrode is maintained at this potential for several minutes until a significant amount of the two metals has accumulated at the electrode. The stirring is then stopped for perhaps 30 s while the electrode is maintained at −1 V. The potential of the electrode is then decreased linearly to less negative values, while the current in the cell is recorded as a function of time, or

The following organic functional groups produce one or more polarographic waves.

1. Carbonyl groups
2. Certain carboxylic acids
3. Most peroxides and epoxides
4. Nitro, nitroso, amine oxide, and azo groups
5. Most organic halogen groups
6. Carbon/carbon double bonds
7. Hydroquinones and mercaptans

---

[9]For a detailed discussion of organic voltammetry, see P. Zuman, *Organic Polarographic Analysis.* Oxford: Pergamon Press, 1964; *Polarography of Molecules of Biological Significance,* W. F. Smyth, Ed. New York: Academic Press, 1979; *Topics in Organic Polarography,* P. Zuman, Ed. New York: Plenum Press, 1970; R. N. Adams, *Electrochemistry at Solid Electrodes.* New York: Marcel Dekker, 1969.

[10]For detailed discussions of stripping methods, see J. Wang, *Stripping Analysis.* Deerfield Beach, FL: VCH Publishers, 1985; A. M. Bond, *Modern Polarographic Methods in Analytical Chemistry,* Chapter 9. New York: Marcel Dekker, 1980.

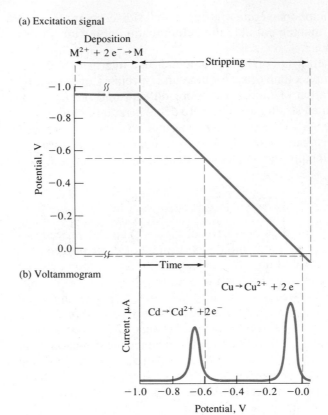

Figure 19–27
(a) Excitation signal for stripping determination of $Cd^{2+}$ and $Cu^{2+}$.
(b) Voltammogram.

potential. Figure 19–27b shows the resulting voltammogram. At a potential somewhat more negative than −0.6 V, cadmium starts to be oxidized, causing a sharp increase in the current. As the deposited cadmium is consumed, the current peaks and then decreases to its original level. A second peak for oxidation of the copper is then observed when the potential has decreased to approximately −0.1 V. The heights of the two peaks are proportional to the masses of deposited metals.

Stripping methods are of prime importance in trace work because the electrochemical preconcentration step of the electrolysis permits the determination of minute amounts of an analyte with reasonable accuracy. Thus, the analysis of solutions in the $10^{-6}$ to $10^{-9}$ M range becomes feasible by methods that are both simple and rapid.

### 19D–1 Electrodeposition Step

Ordinarily, only a fraction of the analyte is deposited during the electrodeposition step; hence, quantitative results depend not only upon control of electrode potential but also upon such factors as electrode size, time of deposition, and stirring rate for both the sample and standard solutions employed for calibration.

Microelectrodes for stripping methods have been formed from a variety of materials, including mercury, gold, silver, platinum, and carbon in

A major advantage of stripping analysis is its capability for electrochemically preconcentrating the analyte prior to the measurement step.

various forms. The most popular electrode is the *hanging mercury drop electrode* (HMDE), which consists of a single drop of mercury in contact with a platinum wire. These electrodes often consist of a microsyringe with a micrometer for exact control of drop size. The drop is then formed at the tip of a capillary by displacement of the mercury in the syringe-controlled delivery system (see Figure 19–5b). The system shown in Figure 19–4d is also capable of producing a hanging drop electrode.

To carry out the determination of a metal ion by anodic stripping, a fresh hanging drop is formed, stirring is begun, and a potential is applied that is a few tenths of a volt more negative than the half-wave potential for the ion of interest. Deposition is allowed to occur for a carefully measured period that may be as short as 60 s or as long as 30 min depending upon the analyte concentration. It should be emphasized that these times seldom result in complete removal of the ion. The electrolysis period is determined by the sensitivity of the method ultimately employed for completion of the analysis.

## 19D–2 Voltammetric Completion of the Analysis

The analyte collected in the hanging drop electrode can be determined by any of several voltammetric procedures. For example, in a linear anodic scan procedure, such as that described at the beginning of Section 19D, stirring is discontinued for perhaps 30 s after termination of the deposition. The voltage is then decreased at a linear fixed rate from its original cathodic value, and the resulting anodic current is recorded as a function of the applied voltage. This linear scan produces a curve of the type shown in Figure 19–27b. Analyses of this type are generally based on calibration with standard solutions of the cations of interest. With reasonable care, analytical precisions of about 2% relative can be obtained.

Most of the other voltammetric procedures described in Section 19C also have been applied to the stripping step. The most widely used of these appears to be an anodic differential pulse technique. The narrow peaks produced by this procedure are desirable for analyzing mixtures. Another method of obtaining narrower peaks is to use a mercury film electrode. Here, a thin mercury film is electrodeposited on an inert microelectrode such as glassy carbon. Ordinarily, the mercury deposition is carried out simultaneously with the analyte deposition. Because the average diffusion path length from the film to the solution interface is much shorter than that in a drop of mercury, escape of the analyte is hastened; the consequence is narrower and larger voltammetric peaks, which lead to greater sensitivity and better resolution of mixtures. On the other hand, the hanging drop electrode appears to give more reproducible results, especially at higher analyte concentrations. Thus, for most applications the hanging drop electrode is employed. Figure 19–28 is a differential-pulse anodic stripping voltammogram for a mixture of cations present at concentrations of 25 ppb, showing good resolution and adequate sensitivity for many purposes.

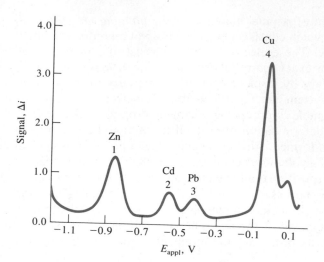

Figure 19–28
Differential-pulse anodic stripping
voltammogram of 25 ppm zinc,
cadmium, lead, and copper. (Re-
printed with permission from
W. M. Peterson and R. V. Wong,
*Amer. Lab.*, **1981**, *13*(11), 116.
Copyright 1981 by International
Scientific Communications, Inc.)

Many other variations of the stripping technique have been developed. For example, a number of cations have been determined by electrodeposition on a platinum cathode. The quantity of electricity required to remove the deposit is then measured coulometrically. Here again, the method is particularly advantageous for trace analyses. Cathodic stripping methods for the halides have also been developed. Here, the halide ions are first deposited as mercury(I) salts on a mercury anode. Stripping is then performed by a cathodic current.

## 19E    QUESTIONS AND PROBLEMS

19–1. Distinguish between
   *(a) voltammetry and polarography.
   (b) linear-scan polarography and pulse polarography.
   *(c) differential pulse polarography and square wave polarography.
   (d) a hanging drop mercury electrode and a dropping mercury electrode.
   *(e) a limiting current and a residual current.
   (f) a limiting current and a diffusion current.
   *(g) laminar flow and turbulent flow.
   (h) the standard electrode potential and the half-wave potential for a reversible reaction at a microelectrode.
19–2. Define
   *(a) voltammograms.
   (b) hydrodynamic voltammetry.
   *(c) Nernst diffusion layer.
   (d) a mercury film electrode.
   *(e) half-wave potential.
19–3. Why is it necessary to buffer solutions in organic voltammetry?
*19–4. List the advantages and disadvantages of the dropping mercury electrode compared with platinum or carbon microelectrodes.

19–5. Suggest how Equation 19–10 could be employed to determine the number of electrons $n$ involved in a reversible reaction at a microelectrode.
*19–6. Quinone undergoes a reversible reduction at a dropping mercury electrode. The reaction is

$$Q + 2\,H^+ + 2\,e^- \rightleftarrows H_2Q \quad E^0 = 0.599\ V$$

   (a) Assume that the diffusion coefficients for quinone and hydroquinone are approximately the same and calculate the approximate half-wave potential (vs. SCE) for the reduction of hydroquinone at a rotating disk electrode from a solution having a pH of 7.0.
   (b) Repeat the calculation in (a) for a solution having a pH of 5.0.
   (c) Sketch a voltammogram for
      (1) a $1.0 \times 10^{-5}$ M solution of quinone.
      (2) a $1.0 \times 10^{-5}$ M solution of hydroquinone.
      (3) a solution that is $5.0 \times 10^{-6}$ M in quinone and $2.5 \times 10^{-6}$ M in hydroquinone.
19–7. Estimate the limiting current for the reduction of $O_2$ to water at a very thin platinum disk-shaped electrode having a diameter of 1.10 mm. Assume

that the thickness of the Nernst diffusion layer is 0.015 cm, that the concentration of $O_2$ is 0.25 mM, and that the solution is stirred vigorously that the diffusion coefficient for $O_2$ is $2.1 \times 10^{-5}$ cm$^2$/s.

*19-8. What are the sources of the residual current in linear-scan polarography? Why are residual currents smaller with differential pulse polarography?

19-9. The polarogram for 20.0 mL of solution that was $3.65 \times 10^{-3}$ M in $Cd^{2+}$ gave a wave for that ion with a diffusion current of 31.3 μA. Calculate the percentage change in concentration of the solution if the current in the limiting current region were allowed to continue for (a) 5 min; (b) 10 min; (c) 30 min.

*19-10. Calculate the milligrams of cadmium in each milliliter of sample, based upon the following data (corrected for residual current):

| | | | Volumes Used, mL | | |
| Solution | Sample | 0.400 M KCl | $2.00 \times 10^{-3}$ M Cd$^{2+}$ | H$_2$O | Current, μA |
| --- | --- | --- | --- | --- | --- |
| (a) | 15.0 | 20.0 | 0.00 | 15.0 | 79.7 |
| | 15.0 | 20.0 | 5.00 | 10.0 | 95.9 |
| (b) | 10.0 | 20.0 | 0.00 | 20.0 | 49.9 |
| | 10.0 | 20.0 | 10.0 | 10.0 | 82.3 |
| (c) | 20.0 | 20.0 | 0.00 | 10.0 | 41.4 |
| | 20.0 | 20.0 | 5.00 | 5.00 | 57.6 |
| (d) | 15.0 | 20.0 | 0.00 | 15.0 | 67.9 |
| | 15.0 | 20.0 | 10.0 | 5.00 | 100.3 |

19-11. The following polarographic data were obtained for the reduction of $Pb^{2+}$ to its amalgam from solutions that were $2.00 \times 10^{-3}$ M in $Pb^{2+}$, 0.100 M in $KNO_3$, and that also had the following concentrations of the anion $A^-$. From the half-wave potentials, derive the formula of the complex as well as its formation constant.

| Concn. A$^-$, M | $E_{1/2}$ vs. SCE, V |
| --- | --- |
| 0.0000 | $-0.405$ |
| 0.0200 | $-0.473$ |
| 0.0600 | $-0.507$ |
| 0.1007 | $-0.516$ |
| 0.300 | $-0.547$ |
| 0.500 | $-0.558$ |

*19-12. Shown in the next column is the polarogram for a solution that was $1.0 \times 10^{-4}$ M in KBr and 0.1 M in $KNO_3$. Offer an explanation of the wave that occurs at +0.12 V and the rapid change in current that starts at about +0.48 V. Would the wave at 0.12 V have any analytical applications? Explain.

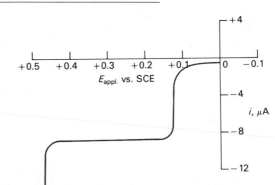

19-13. The following reaction is reversible and has a half-wave potential of $-0.349$ V when carried out at a dropping mercury electrode from a solution buffered to pH 2.5.

$$Ox + 4\,H^+ + 4\,e^- \rightleftarrows R$$

Predict the half-wave potential at pH: (a) 1.0; (b) 3.5; (c) 7.0.

19-14. Why are stripping methods more sensitive than other voltammetric procedures?

19-15. What is the purpose of the electrodeposition step in stripping analysis?

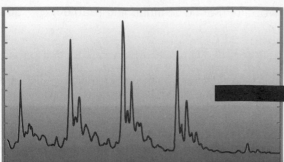

**CHAPTER 20**

# AN INTRODUCTION TO SPECTROSCOPIC METHODS OF ANALYSIS

Historically, the term *spectroscopy* referred to a branch of science in which light (that is, visible radiation) was resolved into its component wavelengths to produce *spectra*. Spectra, of course, played a vital role in the development of modern atomic theory. In addition, spectroscopy has proved to be a powerful tool for qualitative and quantitative analysis.

With the passage of time, the meaning of spectroscopy has become broadened to include studies not only with light but also with other types of electromagnetic radiation, such as X-ray, ultraviolet, infrared, microwave, and radio-frequency radiation. Indeed, current usage extends the meaning of spectroscopic methods still further to include techniques that do not involve electromagnetic radiation. Examples of the last include acoustic, mass, and electron spectroscopy.

The spectroscopic methods described in the next three chapters are largely based upon ultraviolet and visible radiation, although brief mention is made of techniques involving other parts of the electromagnetic spectrum.[1]

## 20A PROPERTIES OF ELECTROMAGNETIC RADIATION

Electromagnetic radiation is a type of energy that is transmitted through space at enormous velocities. Many of the properties of electromagnetic radiation are conveniently described by means of a classical wave model that employs such parameters as wavelength, frequency, velocity, and amplitude. In contrast to other wave phenomena, such as sound, electromagnetic radiation requires no supporting medium for its transmission and readily passes through a vacuum.

---

[1]For further study, see E. J. Meehan, in *Treatise on Analytical Chemistry*, P. J. Elving, E. J. Meehan, and I. M. Kolthoff, Eds., 2nd ed., Part I, Vol. 7, Chapters 1–3. New York: Wiley, 1981; J. D. Ingle Jr. and S. R. Crouch, *Analytical Spectroscopy*. Englewood Cliffs, N.J.: Prentice-Hall, 1988; J. E. Crooks, *The Spectrum in Chemistry*. New York: Academic Press, 1978.

The wave model fails to account for phenomena associated with the absorption or emission of radiant energy. To treat these properties adequately, electromagnetic radiation must be viewed as a stream of discrete wave packets that behave as though they are distinct particles called *photons,* with the energy of the photon being proportional to the frequency of the radiation. These dual views of radiation as particles and waves are not mutually exclusive but, rather, complementary. Indeed, the duality is found to apply to the behavior of streams of electrons and other elementary particles as well and is completely rationalized by wave mechanics.

## 20A–1  Wave Properties

For many purposes, electromagnetic radiation is conveniently pictured as an electric field that undergoes sinusoidal oscillations as it moves through space. Figure 20–1 is a two-dimensional representation of a beam of monochromatic (that is, single-wavelength), plane-polarized radiation. The term *plane-polarized* implies that the oscillations of the electric field are in a single plane. The electric field is represented as a vector whose length is proportional to the field strength. This variation in electric field strength is similar to the time-varying electric field produced by a sinusoidal ac voltage connected between two electrodes in a vacuum. The abscissa in this plot is either time as the radiation passes a fixed point in space or distance at a fixed time. Note that the direction in which the field oscillates is perpendicular to the direction in which the radiation is being propagated.

A wave in which the direction of displacement is perpendicular to the direction of propagation, such as the wave in Figure 20–1, is called a *transverse wave.*

### Wave Parameters

In Figure 20–1, the *amplitude A* of the sinusoidal wave is defined as the length of the electrical vector at the maximum in the wave. The time required for the passage of successive maxima (or minima) through a fixed

The unit of frequency is the hertz (Hz), which corresponds to one oscillation per second.

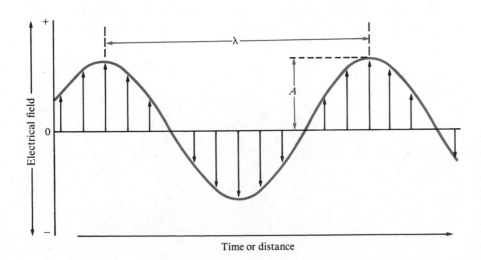

**Figure 20–1**
Representation of a beam of monochromatic radiation of wavelength $\lambda$ and amplitude $A$. The arrows represent the electrical vector of the radiation.

1 Hz = 1 s$^{-1}$

The frequency of a beam of electromagnetic radiation never changes.

| Wavelength Units for Various Spectral Regions | | |
|---|---|---|
| Region | Unit | Definition |
| X-Ray | Angstrom, Å | 10$^{-10}$ m |
| Ultraviolet/ visible | Nanometer, nm | 10$^{-9}$ m |
| Infrared | Micrometer, $\mu$m | 10$^{-6}$ m |

To three significant figures, Equation 20–2 is equally applicable in air and in vacuum.

In contrast to frequency, radiation velocity and wavelength both become smaller as the radiation passes from a vacuum or from air to a denser medium.

The refractive index $\eta$ of a medium is a measure of its interaction with radiation and is defined by $\eta = c/v_i$. For example, the refractive index of water at room temperature is 1.33, which means that radiation passes through water at a rate of $v = c/1.33$, or $2.26 \times 10^{10}$ cm/s.

point in space is called the *period p* of the radiation. The *frequency $\nu$* is the number of oscillations of the field per second and is equal to $1/p$.

It is important to realize that *frequency* is determined by the source and remains *invariant* regardless of the medium traversed by the radiation. In contrast, the *velocity* of propagation $v_i$, the rate at which the wave front moves through a medium is *dependent* upon both the medium and the frequency; the subscript $i$ is employed to indicate this frequency dependence.

Another parameter of interest is the *wavelength $\lambda_i$*, which is the linear distance between successive maxima or minima of a wave. Multiplication of the frequency in waves per second by the wavelength in centimeters gives the velocity of propagation in centimeters per second

$$v_i = \nu\lambda_i \qquad (20\text{–}1)$$

In a vacuum, the velocity of radiation propagation becomes independent of wavelength and is at its maximum. This velocity, which is given the symbol $c$, has been determined to be $2.99792 \times 10^{10}$ cm/s. The velocity of radiation in air differs only slightly from $c$ (it is about 0.03% less). Thus, to three significant figures, Equation 20–2 is equally applicable in air or vacuum

$$c = \nu\lambda = 3.00 \times 10^{10} \text{ cm/s} \qquad (20\text{–}2)$$

The rate of propagation of radiation $v$ is less than $c$ in a medium containing matter because the electromagnetic field of the radiation interacts with the electrons in the atoms or molecules of the medium and is slowed as a consequence. Since the radiant frequency is invariant and fixed by the source, the *wavelength of radiation must decrease* as it passes from a vacuum to a medium containing matter (Equation 20–1). This effect is illustrated in Figure 20–2 for a beam of visible radiation. Note that the wavelength shortens nearly 200 nm, or more than 30%, as the radiation passes from air into glass; a reverse change occurs when the radiation again enters air.

The wavenumber $\bar{\nu}$, defined as the reciprocal of the wavelength in centimeters ($1/\lambda$), is yet another way of describing electromagnetic radia-

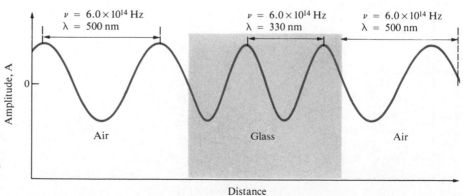

Figure 20–2
Change in wavelength as radiation passes from air into a dense glass and back to air.

tion. Ordinarily, the unit for $\bar{\nu}$ is cm$^{-1}$, which is the number of waves of a particular frequency in a distance of 1 cm.

### Radiant Power and Intensity

The *power P* of radiation is the energy of the beam that reaches a given area per second; the *intensity I* is the power per unit solid angle. These quantities are related to the square of the amplitude $A$ (Figure 20–1). Although it is not strictly correct to do so, power and intensity are often used synonymously.

The wavenumber $\bar{\nu}$ in cm$^{-1}$ is generally used to describe infrared radiation. The most useful part of the infrared spectrum for the detection and determination of organic species is from 2.5 to 15 $\mu$m in wavelength, which corresponds to a wavenumber range of 4000 to 667 cm$^{-1}$.

## 20A–2  The Particle Properties of Radiation

### The Energy of Electromagnetic Radiation

To understand many of the interactions between radiation and matter, it is necessary to postulate that electromagnetic radiation is made up of packets of energy called *photons* (or *quanta*). The energy of a photon depends upon the frequency of the radiation and is given by

$$E = h\nu \qquad (20\text{–}3)$$

where $h$ is Planck's constant ($6.63 \times 10^{-34}$ J · s). In terms of wavelength and wavenumber,

$$E = \frac{hc}{\lambda} = hc\bar{\nu} \qquad (20\text{–}4)$$

Note that wavenumber, like frequency, is directly proportional to energy.

A photon is a particle of electromagnetic radiation having zero mass and an energy of $h\nu$.

Both frequency and wavenumber are proportional to the energy of a photon.

On occasion we speak of "a mole of photons," meaning $6.02 \times 10^{23}$ packets of radiation.

## 20B   THE ELECTROMAGNETIC SPECTRUM

The electromagnetic spectrum encompasses an enormous range of wavelengths and energies. For example, an X-ray photon ($\lambda \cong 10^{-10}$ m) is approximately 10,000 times more energetic than a photon emitted by an incandescent tungsten wire ($\lambda \cong 10^{-6}$ m) and $10^{11}$ times more energetic than a photon in the radio-frequency range.

The major divisions of the electromagnetic spectrum are depicted in the color plate located inside the front cover of this book. Note that both the frequency and the wavelength scales are logarithmic; note also that the region to which the human eye is perceptive (the *visible spectrum*) is but a minute part of the whole spectrum. Feature 20–1 describes the relationship between the colors and wavelengths of radiation. Such diverse radiations as gamma rays or radio waves differ from visible light only in the matter of frequency and, hence, energy.

Figure 20–3 indicates the regions of the spectrum that are useful for analytical spectroscopy and the molecular or atomic transitions responsible for absorption or emission of radiation in each region.

Recall the order of the colors in the spectrum by the mnemonic
**ROY G BIV**.
**R**ed
**O**range
**Y**ellow
**G**reen
**B**lue
**I**ndigo
**V**iolet

The visible region of the spectrum extends from about 380 nm to 780 nm.

## 20C  ABSORPTION OF RADIATION

*Attenuate* means to weaken.

In spectroscopic nomenclature, *absorption* is a process in which a chemical species in a transparent medium selectively *attenuates* (decreases the intensity of) certain frequencies of electromagnetic radiation. According to quantum theory, every elementary particle (atom, ion, or molecule) has a unique set of energy states, the lowest of which is the *ground state*; at room temperature, most elementary particles are in their ground state. When a photon of radiation passes near an elementary particle, absorption becomes probable if (and only if) the energy of the photon matches *exactly* the energy difference between the ground state and one of the higher energy states of the particle. Under these circumstances, the energy of the photon is transferred to the atom, ion, or molecule, converting it to the higher energy state, which is termed an *excited state*. Excitation of a species M to its excited state M* can be depicted by the equation

The lowest energy state of an atom or molecule is called its *ground state*.

*Excitation* is a process in which a chemical species absorbs thermal, electrical, or radiant energy and is promoted to a higher energy state.

$$M + h\nu \rightarrow M^*$$

After a brief period ($10^{-6}$ to $10^{-9}$ s), the excited species *relaxes* to its original, or ground, state, transferring its excess energy to other atoms or molecules in the medium. This process, which causes a small rise in temperature of the surroundings, is described by the equation

*Relaxation* is a process in which an excited species gives up its excess energy and returns to a lower energy state.

$$M^* \rightarrow M + \text{heat}$$

Relaxation may also occur by *photochemical decomposition* of M* to form new species or by the *fluorescent* or *phosphorescent* reemission of radiation. It is important to note that the lifetime of M* is so very short that its concentration at any instant is ordinarily negligible. Furthermore, the amount of thermal energy released during relaxation is usually so

**Figure 20–3**
The regions of the electromagnetic spectrum. (From C. N. Banwell, *Fundamentals of Molecular Spectroscopy,* 3rd ed., p. 7. London: McGraw-Hill, 1983.)

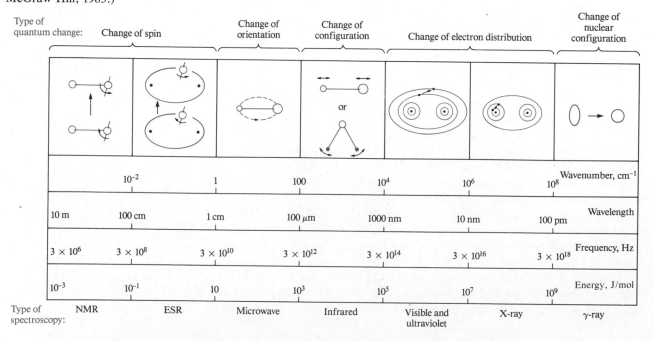

| Type of quantum change: | Change of spin | | Change of orientation | Change of configuration | Change of electron distribution | | Change of nuclear configuration |
|---|---|---|---|---|---|---|---|

| | | | | | | | | Wavenumber, cm$^{-1}$ |
|---|---|---|---|---|---|---|---|---|
| | $10^{-2}$ | | 1 | 100 | $10^4$ | $10^6$ | $10^8$ | |
| | | | | | | | | Wavelength |
| 10 m | 100 cm | | 1 cm | 100 $\mu$m | 1000 nm | 10 nm | 100 pm | |
| | | | | | | | | Frequency, Hz |
| $3 \times 10^6$ | $3 \times 10^8$ | | $3 \times 10^{10}$ | $3 \times 10^{12}$ | $3 \times 10^{14}$ | $3 \times 10^{16}$ | $3 \times 10^{18}$ | |
| | | | | | | | | Energy, J/mol |
| $10^{-3}$ | $10^{-1}$ | | 10 | $10^3$ | $10^5$ | $10^7$ | $10^9$ | |

| Type of spectroscopy: | NMR | | ESR | Microwave | Infrared | Visible and ultraviolet | X-ray | $\gamma$-ray |
|---|---|---|---|---|---|---|---|---|

small as to be undetectable. Thus, absorption measurements have the advantage of creating minimal disturbance of the system under study.

The absorbing characteristics of a species are conveniently described by means of an *absorption spectrum*, which is a plot of some function of the attenuation of a beam of radiation versus wavelength, frequency, or wavenumber. Ultraviolet and visible absorption spectra are usually obtained on a gaseous sample of the analyte or on a dilute solution of the analyte in a transparent solvent. As shown in Figure 20–4, the vertical axis of such plots may be percent transmittance, absorbance, or log absorbance. A plot with log $A$ as ordinate leads to loss of spectral detail but is convenient for comparing samples of different concentrations, since the curves are displaced equally along the vertical axis.

As we shall see, our instruments measure the quantity of light *not absorbed*.

The *transmittance T* of a solution is the fraction of the incident electromagnetic radiation that is transmitted by a sample. The *absorbance* $A = \log(1/T)$ (Section 20C–3). Note that the absorbance increases with decreasing transmittance.

## 20C–1 Atomic Absorption

When a beam of polychromatic ultraviolet or visible radiation passes through a medium containing gaseous atoms, only a few frequencies are attenuated by absorption, and the spectrum consists of a number of very narrow (about 0.005-nm) *absorption lines*. Figure 20–5a is an ultraviolet/visible absorption spectrum for gaseous sodium atoms. The ordinate is *absorbance,* which is a measure of the degree of attenuation of the beam (Section 20C–4).

At a higher sensitivity on the wavelength axis, a handful of additional lines would appear. In addition, the three lines shown are in fact split into pairs called *doublets* that differ in wavelength by only a few tenths of a nanometer. The cause of this splitting is beyond the scope of this text.

**Feature 20–1**
**WHY IS A RED SOLUTION RED?**

A solution such as $Fe(SCN)^{2+}$ is red not because the complex adds red radiation to the solvent. Instead, this complex absorbs the green component from the incoming white radiation and transmits the red component. Thus in a colorimetric analysis of iron based upon its thiocyanate complex, the maximum change in absorbance with respect to concentration occurs with green radiation; the absorbance change with red radiation is negligible. In general, then, the radiation used for a colorimetric analysis should be the complementary color of the analyte solution. The table below shows this relationship for various parts of the visible spectrum.

### THE VISIBLE SPECTRUM

| Wavelength Region, nm | Color | Complementary Color |
|---|---|---|
| 400–435 | Violet | Yellow–green |
| 435–480 | Blue | Yellow |
| 480–490 | Blue–green | Orange |
| 490–500 | Green–blue | Red |
| 500–560 | Green | Purple |
| 560–580 | Yellow–green | Violet |
| 580–595 | Yellow | Blue |
| 595–650 | Orange | Blue–green |
| 650–750 | Red | Green–blue |

The electron volt eV is a unit of energy.

$$1\ eV = 1.60 \times 10^{-19}\ J$$
$$= 3.83 \times 10^{-20}\ calories$$
$$= 1.58 \times 10^{-21}\ L\ atm$$

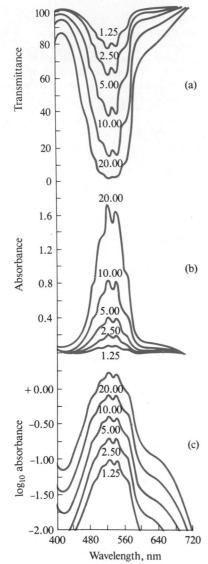

Figure 20–4

Methods for plotting spectral data. The numbers for the curves indicate ppm of $KMnO_4$ in the solution; $b = 2.00$ cm. (From M. G. Mellon, *Analytical Absorption Spectroscopy,* pp. 104–106. New York: Wiley, 1950. With permission.)

Figure 20–5b is a partial energy-level diagram for sodium that shows the transitions responsible for the three absorption lines in (a). The transitions involve excitation of the single outer electron of sodium from its room temperature or ground state $3s$ orbital to the $3p$, $4p$, and $5p$ orbitals. These excitations are brought on by absorption of photons of radiation whose energies *exactly* match the differences in energies between the excited states and the $3s$ ground state.

---

Example 20–1

The energy difference between the $3p$ and the $3s$ orbitals in Figure 20–5b is 2.107 eV. Calculate the wavelength of radiation that would be absorbed in exciting the $3s$ electron to the $3p$ state (1 eV = $1.60 \times 10^{-19}$ J).

Rearranging Equation 20–4 gives

$$\lambda = hc/E$$
$$= \frac{6.63 \times 10^{-34}\ J\ s \times 3.00 \times 10^{10}\ cm\ s^{-1} \times 10^{7}\ nm\ cm^{-1}}{2.107\ eV \times 1.60 \times 10^{-9}\ J/eV} = 590\ nm$$

---

Absorption spectra for the alkali metal atoms are much simpler than those of elements with additional outer electrons. Atomic spectra of the transition metals are particularly rich in lines, with some elements exhibiting several thousand lines.

## 20C–2 Molecular Absorption

Molecules undergo three types of quantized transitions when excited by ultraviolet, visible, and infrared radiation. For ultraviolet and visible radiation, excitation involves promoting an electron residing in a low-energy molecular or atomic orbital to a higher-energy orbital. We have noted that the energy $h\nu$ of the photon must be exactly the same as the energy difference between the two orbital energies. The transition of an electron between two orbitals is called an *electronic transition* and the absorption process is called *electronic absorption*.

In addition to electronic transitions, molecules exhibit two other types of radiation-induced transitions: *vibrational transitions* and *rotational transitions*. Vibrational transitions come about because a molecule has a multitude of quantized energy levels (or *vibrational states*) associated with the bonds that hold the molecule together.

Figure 20–6a is a partial energy-level diagram that depicts some of the processes that occur when a polyatomic species absorbs infrared, visible, and ultraviolet radiation. The energies $E_1$ and $E_2$, two of the several electronically excited states of a molecule, are shown relative to the energy of its ground state $E_0$. In addition, the relative energies of a few of the many vibrational states associated with each electronic state are indicated by the lighter horizontal lines.

An idea of the nature of vibrational states can be gained by picturing a bond in a molecule as a vibrating spring with atoms attached to both ends.

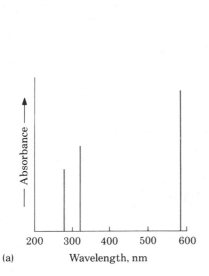

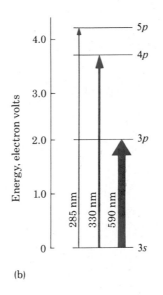

Figure 20–5

(a) Absorption spectrum for sodium vapor. (b) Partial energy-level diagram for sodium, showing the transitions resulting from absorption at 590, 330, and 285 nm.

In Figure 20–7a, two types of stretching vibration are shown. With each vibration, atoms first approach and then move away from one another. The potential energy of such a system at any instant depends upon the extent to which the spring is stretched or compressed. For an ordinary spring, the energy of the system varies continuously and reaches a maximum when the spring is fully stretched or fully compressed. In contrast,

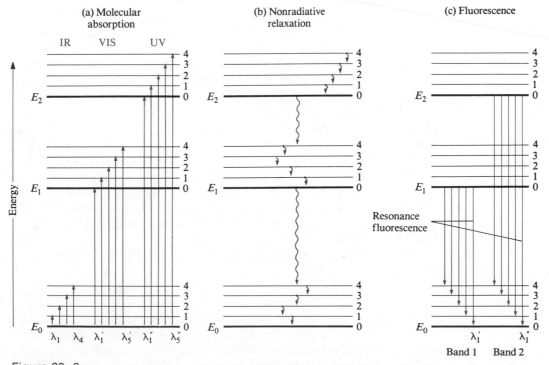

Figure 20–6

Energy-level diagram showing some of the energy changes that occur during absorption, nonradiative relaxation, and fluorescence by a molecular species.

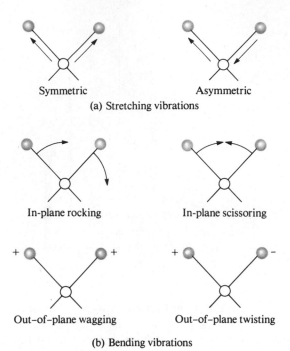

Symmetric          Asymmetric
(a) Stretching vibrations

In-plane rocking          In-plane scissoring

Out-of-plane wagging          Out-of-plane twisting
(b) Bending vibrations

Figure 20-7
Types of molecular vibrations. The plus sign indicates motion from the page toward the reader; the minus sign indicates motion away from the reader.

the energy of a spring system of atomic dimensions can assume only certain discrete energies called vibrational energy levels.

Figure 20–7b shows other types of molecular vibrations. The energies associated with these vibrational states usually differ from one another and from the energies associated with stretching vibrations.

Some of the vibrational energy levels associated with one of the vibrations associated with each of the electronic states of a molecule are depicted by the lines labeled 1, 2, 3, and 4 in Figure 20–6a (the lowest vibrational levels are labeled 0). Note that the differences in energy among the vibrational states are significantly smaller than among energy levels of the electronic states (typically, an order of magnitude smaller).

Although they are not shown, a molecule has a host of quantized rotational states that are associated with the rotational motion of a molecule around its center of gravity. These rotational energy states are superimposed on each of the vibrational states shown in the energy diagram. The energy differences among these states are smaller than those among vibrational states by an order of magnitude.

The overall energy $E$ associated with a molecule is then given by

$$E = E_{\text{electronic}} + E_{\text{vibrational}} + E_{\text{rotational}} + E_{\text{translational}}$$

$$\Delta E_{\text{electronic}} \approx 10\ \Delta E_{\text{vibrational}}$$
$$\approx 100\ \Delta E_{\text{rotational}}$$
$$\gg \Delta E_{\text{translational}}$$

where $E_{\text{electronic}}$ is the energy associated with the electrons in the various outer orbitals of the molecule and $E_{\text{vibrational}}$ is the energy of the molecule as a whole due to interatomic vibrations, and $E_{\text{rotational}}$ accounts for the energy associated with rotation of the molecule about its center of gravity. The last term represents the translational energy of the molecule.

## Infrared Absorption

Infrared radiation generally is not sufficiently energetic to cause electronic transitions but can induce transitions in the vibrational and rotational states associated with *the ground electronic state* of the molecule. Four of these transitions are depicted in the lower left part of Figure 20–6a. For absorption to occur, the source has to emit radiation of frequencies corresponding exactly to the energies indicated by the lengths of the four arrows.

Infrared radiation is not sufficiently energetic to cause electronic transitions.

## Absorption of Ultraviolet and Visible Radiation

The center arrows in Figure 20–6a suggest that the molecules under consideration absorb visible radiation of five wavelengths, thereby promoting electrons to the five vibrational levels of the excited electronic level $E_1$. Ultraviolet photons that are more energetic are required to produce the absorption indicated by the five arrows to the right.

As suggested by Figure 20–6a, molecular absorption in the ultraviolet and visible regions consists of absorption *bands* made up of closely spaced lines. (A real molecule has many more energy levels than shown here; thus the typical absorption band consists of a multitude of lines.) In a solution, the absorbing species are surrounded by solvent, and the band nature of molecular absorption often becomes blurred because collisions tend to spread the energies of the quantum states, thus giving smooth and continuous absorption peaks.

## Relaxation Processes

As noted earlier, the lifetime of an excited species is brief because several mechanisms exist whereby an excited atom or molecule can give up its excess energy and relax to its ground state. Two of the most important of these mechanisms, nonradiative relaxation and fluorescent relaxation, are illustrated in Figures 20–6b and 6c.

Two types of nonradiative relaxation are shown in Figure 20–6b. *Vibrational deactivation, or relaxation,* depicted by the short wavy arrows between vibrational energy levels, takes place during collisions between excited molecules and molecules of the solvent. During the collisions, the excess vibrational energy is transferred to solvent molecules in a series of steps as indicated in the figure. The gain in vibrational energy of the solvent is reflected in a tiny increase in the temperature of the medium. Vibrational relaxation is such an efficient process that the average lifetime of an excited *vibrational* state is only about $10^{-15}$ s.

Nonradiative relaxation between the lowest vibrational level of an excited electronic state and the upper vibrational level of another electronic state can also occur. This type of relaxation, depicted by the two longer wavy arrows in Figure 20–6b, is much less efficient than vibrational relaxation, so that the average lifetime of an electronic excited state is between $10^{-6}$ and $10^{-9}$ s. The mechanisms by which this type of relaxation occurs are not fully understood, but the net effect is again a rise in the temperature of the medium.

Figure 20–6c depicts another relaxation process: fluorescence. Molecular fluorescence is discussed in Chapter 23.

## 20C–3 Terms Employed in Absorption Spectroscopy

Table 20–1 lists the common terms and symbols used in absorption spectroscopy. This nomenclature is recommended by the American Society for Testing Materials as well as the American Chemical Society. Column 3 contains alternative symbols encountered in the older literature. Because a standard nomenclature is highly desirable in order to avoid ambiguities, the reader is urged to learn and use the recommended terms and symbols and avoid those in column 3.

### Transmittance

Figure 20–8 depicts a beam of parallel radiation before and after it has passed through a layer of solution with a thickness $b$ cm and a concentration $c$ of an absorbing species. As a consequence of interactions between the photons and absorbing particles, the power of the beam is attenuated from $P_0$ to $P$. The *transmittance T* of the solution is defined as the fraction of incident radiation transmitted by the solution

> Transmittance is a measure of the quantity of light not absorbed.

$$T = P/P_0 \qquad (20\text{–}5)$$

Transmittance is often expressed as a percentage.

**Table 20–1**
**IMPORTANT TERMS AND SYMBOLS EMPLOYED IN ABSORPTION MEASUREMENT**

| Term and Symbol* | Definition | Alternative Name and Symbol |
|---|---|---|
| Radiant power $P$, $P_0$ | Energy of radiation (in ergs) impinging on a 1-cm$^2$ area of a detector per second | Radiation intensity $I$, $I_0$ |
| Absorbance $A$ | $\log \dfrac{P_0}{P}$ | Optical density $D$; extinction $E$ |
| Transmittance $T$ | $\dfrac{P}{P_0}$ | Transmission $T$ |
| Path length of radiation† $b$ | — | $l$, $d$ |
| Absorptivity† $a$ | $\dfrac{A}{bc}$ | Extinction coefficient $k$ |
| Molar absorptivity‡ $\varepsilon$ | $\dfrac{A}{bc}$ | Molar extinction coefficient |

*Terminology recommended by the American Chemical Society (*Anal. Chem.*, **1990**, *62, 91.*)

†$c$ may be expressed in g/L or in other specified concentration units; $b$ may be expressed in cm or in other units of length.

‡$c$ is expressed in mol/L; $b$ is expressed in cm.

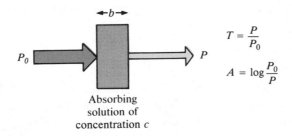

$$T = \frac{P}{P_0}$$

$$A = \log \frac{P_0}{P}$$

Absorbing
solution of
concentration $c$

**Figure 20–8**
Attenuation of a beam of radiation
by an absorbing solution.

### Absorbance

The absorbance $A$ of a solution is defined by the equation

$$A = -\log_{10} T = \log \frac{P_0}{P} \qquad (20\text{–}6)$$

Note that, in contrast to transmittance, the absorbance of a solution increases as the attenuation of the beam becomes greater. If the readout of a spectrophotometer is calibrated in both transmittance and absorbance, Equation 20–6 requires that the absorbance scale be logarithmic. Figure 20–9 shows a typical meter scale.

### 20C–4  The Relationship Between Absorbance and Concentration: Beer's Law

The functional relationship between the quantity measured in an absorption method ($A$) and the quantity sought (the analyte concentration $c$) is known as *Beer's law* and can be written

$$A = \log (P_0/P) = abc \qquad (20\text{–}7)$$

Beer's law is sometimes paraphrased:

The deeper the cup
The darker the blend
The smaller the amount
  of light at the end.

where $a$ is a proportionality constant called the *absorptivity* and $b$ is the path length of the radiation through the absorbing medium. Since absorbance is a unitless quantity, the absorptivity has units that cancel the units of $b$ and $c$.

When the concentration in Equation 20–7 is expressed in moles per liter and $b$ is in centimeters, the proportionality constant is called the *molar absorptivity* and is given the special symbol $\varepsilon$. Thus,

$$A = \varepsilon bc \qquad (20\text{–}8)$$

where $\varepsilon$ has the units of L cm$^{-1}$ mol$^{-1}$.

### 20C–5  The Experimental Measurement of Transmittance and Absorbance

Transmittance and absorbance, as defined by Equations 20–5 and 20–6, cannot be measured in the laboratory because the solution to be studied must be held in some sort of container. Interaction between the radiation

Figure 20–9

A spectrophotometer readout. The readout scales on some spectrophotometers are linear in percent transmittance. As shown below, the absorbance scales must then be logarithmic.

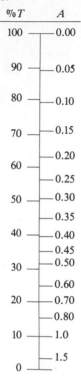

| %T | A |
|---|---|
| 100 | 0.00 |
| 90 | 0.05 |
| 80 | 0.10 |
| 70 | 0.15 |
| | 0.20 |
| 60 | 0.25 |
| 50 | 0.30 |
| | 0.35 |
| 40 | 0.40 |
| | 0.45 |
| 30 | 0.50 |
| | 0.60 |
| 20 | 0.70 |
| | 0.80 |
| 10 | 1.0 |
| | 1.5 |
| 0 | |

## Feature 20–2
## A DERIVATION OF BEER'S LAW

Beer's law can be rationalized by considering the fate of a beam of parallel monochromatic radiation with power $P_0$ after it enters a layer of absorbing matter (solid, liquid, or gas) at a normal angle, as shown in Figure 20–10.[2] After passing through a length $b$ of the material, which contains $n$ absorbing particles (atoms, ions, or molecules), the power of the radiation is decreased to $P$ as a result of absorption. Now consider a cross section of the layer that has an area $S$ and an infinitesimal thickness $dx$ and that contains $dn$ absorbing particles. Associated with each particle, we can imagine a surface at which photon capture occurs. That is, if a photon reaches one of these areas by chance, absorption follows immediately. The total projected area of these capture surfaces within the section of thickness $dx$ is designated $dS$. The ratio of the capture area to the total area is then $dS/S$. On a statistical average, this ratio is equal to the probability for photon capture within the section.

The power of the beam entering the section $P_x$ is proportional to the number of photons per square centimeter per second, and $dP_x$ is the quantity removed per second within the section. The fraction absorbed is then $-dP_x/P_x$, and this ratio also equals the average probability for capture. The term is given a minus sign to indicate that $P$ is decreased. Thus,

$$-\frac{dP_x}{P_x} = \frac{dS}{S} \qquad (20\text{–}9)$$

Recall that $dS$ is the sum of the capture areas for absorbing particles within the section, which must be proportional to the number of particles, or

$$dS = a\,dn \qquad (20\text{–}10)$$

where $dn$ is the number of particles and $a$ is a proportionality constant that is called the *capture cross section*. Combining Equations 20–9 and 20–10 and integrating over the interval between zero and $n$, we obtain

$$-\int_{P_0}^{P} \frac{dP_x}{P_x} = \int_{0}^{n} \frac{a\,dn}{S}$$

When the integrals are evaluated, we find

$$-\ln \frac{P_0}{P} = \frac{an}{S}$$

[2]This discussion is based on a paper by F. C. Strong (*Anal. Chem.*, **1952,** *24*, 338). For a rigorous derivation of the law, see D. J. Swinehart, *J. Chem. Educ.*, **1972,** *39*, 333.

Upon converting to base-10 logarithms and inverting the fraction to change the sign, we obtain

$$\log \frac{P_0}{P} = \frac{an}{2.303\ S} \qquad (20\text{--}11)$$

where $n$ is the total number of particles within the block shown in Figure 20–10. The cross-sectional area $S$ can be expressed in terms of the volume of the block $V$ and its length $b$

$$S = \frac{V}{b}\ cm^2$$

Substitution of this quantity into Equation 20–11 yields

$$\log \frac{P_0}{P} = \frac{anb}{2.303\ V} \qquad (20\text{--}12)$$

Note that $n/V$ has the units of concentration (that is, number of particles per cubic centimeter), which is readily converted to moles per liter

$$c = \frac{n\ \text{particles}}{6.02 \times 19^{23}\ \text{particles/mol}} \times \frac{1000\ cm^3/L}{V\ cm^3} = \frac{1000\ n}{6.02 \times 10^{23}\ V}\ \text{mol/L}$$

Combining this relationship with Equation 20–12 yields

$$\log \frac{P_0}{P} = \frac{(6.02 \times 10^{23})abc}{2.303 \times 1000}$$

Finally, the constants in this equation can be collected into a single factor $\varepsilon$ to give

$$\log \frac{P_0}{P} = \varepsilon bc = A$$

which is a statement of Beer's law.

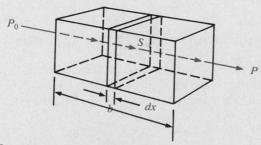

Figure 20–10

Attenuation of radiation with initial power $P_0$ by a solution containing $c$ mol/L of absorbing solute and a path length of $b$ cm ($P < P_0$).

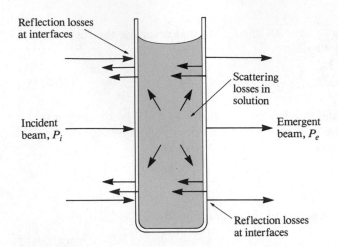

Figure 20–11
Reflection and scattering losses.

and the container walls is inevitable, with losses in power occurring at each interface as a result of reflection and possibly absorption (Figure 20–11). Reflection losses are substantial. For example, it can be shown that about 8.5% of a beam of yellow light is lost by reflection when it passes vertically through a glass beaker filled with water.[3] In addition to reflective losses, scattering by large molecules or inhomogeneities in the solvent may cause a decrease in the power of the beam as it passes through the solution.

In order to compensate for these effects, the power of the beam transmitted through a cell containing an absorbing solution is generally compared with that of a beam that passes through an identical cell containing only solvent. An experimental absorbance is then defined by the equation

$$A = \log \frac{P_{\text{solvent}}}{P_{\text{solution}}} = \log \frac{P_0}{P} \qquad (20\text{–}13)$$

Such experimental absorbances obey Beer's law and are presumably good approximations of true absorbances. Henceforth, the term *absorbance* will refer to the ratio defined by Equation 20–13, $P_0$ being the power of radiation after passage through a cell containing only the solvent and $P$ being the power after passage through an identical cell containing a solution of the analyte.

### 20C–6 The Application of Beer's Law to Mixtures

Beer's law also applies to solutions containing more than one kind of absorbing substance. Provided no interaction occurs among the various species, the total absorbance for a multicomponent system is

$$A_{\text{total}} = A_1 + A_2 + \cdots + A_n = \varepsilon_1 b c_1 + \varepsilon_2 b c_2 + \cdots + \varepsilon_n b c_n \qquad (20\text{–}14)$$

Absorbances are additive.

where the subscripts refer to absorbing components 1, 2, . . . , $n$.

---

[3] D. A. Skoog and J. J. Leary, *Principles of Instrumental Analysis*, 4th ed., p. 68. Philadelphia: Saunders College Publishing, 1992.

## 20C–7 Limitations to the Applicability of Beer's Law

The linear relationship between absorbance and path length at a fixed concentration of an absorbing substance is a generalization for which no exceptions are known. Deviations from the direct proportionality between absorbance and concentration at constant $b$ are frequently encountered, however. Some of these deviations are fundamental and represent *real* limitations to the law. Others occur as a consequence of the manner in which the absorbance measurements are made (*instrumental deviations*) or as a result of chemical changes associated with concentration changes (*chemical deviations*).

### Real Limitations

Beer's law is successful in describing the absorption behavior of dilute solutions only and in this sense is a *limiting law*. At high concentrations (usually $> 0.01$ M), the average distances between particles of the absorbing species are diminished to the point where each particle affects the charge distribution of its neighbors. This interaction can alter the ability of the particles to absorb a given wavelength of radiation. Because the extent of interaction depends upon concentration, the occurrence of this phenomenon causes deviations from the linear relationship between absorbance and concentration. A similar effect is sometimes encountered in dilute solutions of absorbers that contain high concentrations of other species, particularly electrolytes. The close proximity of ions to the absorber alters the molar absorptivity of the latter by electrostatic interactions, which leads to departures from Beer's law.

> Real limitations to Beer's law are encountered only in relatively concentrated solutions of the analyte or in concentrated electrolyte solutions.

While the effect of molecular interactions is ordinarily not significant at concentrations below 0.01 M, some exceptions are encountered among certain large organic ions or molecules. For example, the molar absorptivity at 436 nm for the cation of methylene blue is reported to increase by 88% as the dye concentration is increased from $10^{-5}$ to $10^{-2}$ M; even below $10^{-6}$ M, strict adherence to Beer's law is not observed.

Deviations from Beer's law also arise because molar absorptivity is dependent upon the refractive index of the solution.[4] Thus, if concentration changes cause significant alterations in the refractive index of a solution, departures from Beer's law are observed. In general, this effect is small and rarely significant at concentrations less than 0.01 M.

### Chemical Deviations

Apparent deviations from Beer's law appear when the analyte undergoes association, dissociation, or reaction with the solvent to give products with absorbing characteristics that differ from those of the analyte. As shown in Example 20–2, the extent of such departures can be predicted from the molar absorptivities of the absorbing species and the equilibrium constants for the reactions involved.

> Chemical deviations from Beer's law are encountered when the absorbing species participates in a concentration-dependent equilibrium such as a dissociation or association reaction.

---

[4]G. Kortum and M. Seiler, *Angew. Chem.,* **1939**, *52*, 687.

Example 20-2

A series of solutions containing various concentrations of the acidic indicator HIn ($K_a$ = 1.42 × 10⁻⁵) was prepared in 0.1 M HCl and 0.1 M NaOH. In both media, a linear relationship between absorbance and concentration was observed at 430 and 570 nm. From the magnitude of the acid dissociation constant, it is apparent that, for all practical purposes, the indicator is entirely in the undissociated form (HIn) in the HCl solution and completely dissociated as In⁻ in NaOH. The molar absorptivities at the two wavelengths were found to be

|                   | $\varepsilon_{430}$ | $\varepsilon_{570}$ |
|-------------------|---------------------|---------------------|
| HIn (HCl solution) | 6.30 × 10²          | 7.12 × 10³          |
| In⁻ (NaOH solution) | 2.06 × 10⁴         | 9.60 × 10²          |

Derive absorbance data (1.00-cm cell) at the two wavelengths for unbuffered solutions with indicator concentrations ranging from 2 × 10⁻⁵ to 16 × 10⁻⁵ M. Plot the data.

Let us calculate the concentration of HIn and In⁻ in an unbuffered 2.00 × 10⁻⁵ M solution of the indicator. From the equation for the dissociation reaction, it is apparent that

$$[H^+] = [In^-]$$

Furthermore,

$$[In^-] + [HIn] = 2.00 \times 10^{-5} \text{ M}$$

Substitution of these relationships into the expression for $K_a$ gives

$$\frac{[In^-]^2}{2.00 \times 10^{-5} - [In^-]} = 1.42 \times 10^{-5}$$

Rearrangement yields the quadratic expression

$$[In^-]^2 + (1.42 \times 10^{-5})[In^-] - 2.84 \times 10^{-10} = 0$$

which can be solved to give

$$[In^-] = 1.12 \times 10^{-5}$$
$$[HIn] = 2.00 \times 10^{-5} - 1.12 \times 10^{-5} = 0.88 \times 10^{-5}$$

The absorbance at the two wavelengths are found by substituting into Equation 20-14

$$A_{430} = 6.30 \times 10^2 \times 1.00 \times 0.88 \times 10^{-5} + 2.06 \times 10^4 \times 1.00 \times 1.12 \times 10^{-5}$$
$$= 0.236$$

$$A_{570} = 7.12 \times 10^3 \times 1.00 \times 0.88 \times 10^{-5} + 9.60 \times 10^2 \times 1.00 \times 1.12 \times 10^{-5}$$

$$= 0.073$$

The following data were derived in a similar way and are plotted in Figure 20–12.

| $c_{HIn}$, M | [HIn] | [In$^-$] | $A_{430}$ | $A_{570}$ |
|---|---|---|---|---|
| $2.00 \times 10^{-5}$ | $0.88 \times 10^{-5}$ | $1.12 \times 10^{-5}$ | 0.236 | 0.073 |
| $4.00 \times 10^{-5}$ | $2.22 \times 10^{-5}$ | $1.78 \times 10^{-5}$ | 0.381 | 0.175 |
| $8.00 \times 10^{-5}$ | $5.27 \times 10^{-5}$ | $2.73 \times 10^{-5}$ | 0.596 | 0.401 |
| $12.00 \times 10^{-5}$ | $8.52 \times 10^{-5}$ | $3.48 \times 10^{-5}$ | 0.771 | 0.640 |
| $16.00 \times 10^{-5}$ | $11.9 \times 10^{-5}$ | $4.11 \times 10^{-5}$ | 0.922 | 0.887 |

Figure 20–12 is a plot of the data derived in the foregoing example and illustrates the types of deviations from Beer's law that occur when the absorber is a participant in an association or dissociation equilibrium. Note that the direction of curvature is opposite at the two wavelengths.

### Instrumental Deviations With Polychromatic Radiation

Beer's law is also a limiting law in the sense that it applies only when absorbance is measured with monochromatic radiation. Truly monochromatic sources, such as lasers, are not practical for routine analytical instruments, however. Instead, a polychromatic continuous source is employed in conjunction with a grating or a filter that isolates a more or less symmetric band of wavelengths around the desired one (page 543). The

Deviations from Beer's law often occur when polychromatic radiation is used to measure absorbance.

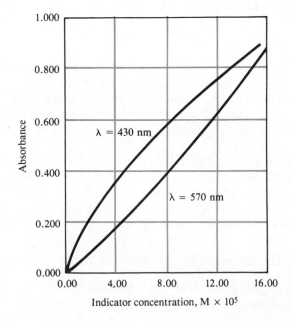

Figure 20–12

Chemical deviations from Beer's law for unbuffered solutions of the indicator HIn. For data, see Example 20–2.

following derivation illustrates how such a source may lead to deviations from Beer's law.

Consider a beam made up of just two wavelengths $\lambda'$ and $\lambda''$, and assume that Beer's law applies strictly to each. With this assumption, we can write for radiation $\lambda'$

$$A' = \log \frac{P_0'}{P'} = \varepsilon' bc$$

$$\frac{P_0'}{P'} = 10^{\varepsilon' bc} \quad \text{and} \quad P' = P_0' \, 10^{-\varepsilon' bc}$$

Similarly, for $\lambda''$

$$\frac{P_0''}{P''} = 10^{\varepsilon'' bc} \quad \text{and} \quad P'' = P_0'' \, 10^{-\varepsilon'' bc}$$

When an absorbance measurement is made with radiation composed of both wavelengths, the power of the beam emerging from the solution is given by $P' + P''$ and that of the beam emerging from the solvent by $P_0' + P_0''$. Therefore, the measured absorbance is

$$A_m = \log \frac{P_0' + P_0''}{P' + P''}$$

which can be rewritten as

$$A_m = \log \frac{P_0' + P_0''}{P_0' \, 10^{-\varepsilon' bc} + P_0'' \, 10^{-\varepsilon'' bc}}$$

$$= \log (P_0' + P_0'') - \log (P_0' \, 10^{-\varepsilon' bc} + P_0'' \, 10^{-\varepsilon'' bc})$$

Now, when $\varepsilon' = \varepsilon''$, this equation simplifies to

$$A_m = \varepsilon' bc$$

and Beer's law is followed. As shown in Figure 20–13, however, the relationship between $A_m$ and concentration is no longer linear when the molar absorptivities differ. Moreover, departures from linearity become greater as the difference between $\varepsilon'$ and $\varepsilon''$ increases. When this treatment is expanded to include additional wavelengths, the effect remains the same.

It is an experimental fact that deviations from Beer's law resulting from the use of a polychromatic beam are not appreciable, provided the radiation used does not encompass a spectral region in which the absorber exhibits large changes in absorbance as a function of wavelength. This observation is illustrated in Figure 20–14.

Generally, the better the instrument, the less it suffers from non-linearity due to polychromatic radiation.

### Instrumental Deviations in the Presence of Stray Radiation

The radiation employed for absorbance measurements is usually contaminated with small amounts of *stray* radiation due to instrumental imperfec-

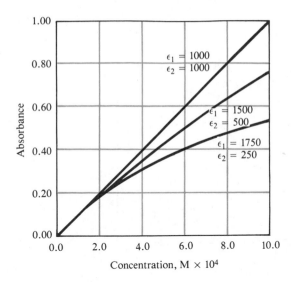

Figure 20–13

Deviations from Beer's law with polychromatic light. The absorber has the indicated molar absorptivities at the two wavelengths $\lambda'$ and $\lambda''$.

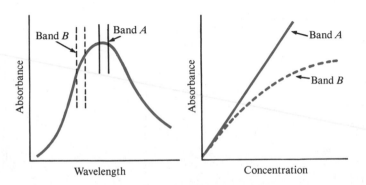

Figure 20–14

The effect of polychromatic radiation upon Beer's law. Band $A$ shows little deviation because $\varepsilon$ does not change greatly throughout the band. Band $B$ shows marked deviation because $\varepsilon$ undergoes significant changes in this region.

tions. Stray radiation is the result of scattering phenomena off the surfaces of prisms, lenses, filters, and windows (page 546). It often differs greatly in wavelength from the principal radiation and, in addition, may not have passed through the sample or solvent.

When measurements are made in the presence of stray radiation, the observed absorbance is given by

$$A' = \log \frac{P_0 + P_s}{P + P_s}$$

where $P_s$ is the power of the stray radiation. Figure 20–15 shows a plot of $A'$ versus concentration for various levels of $P_s$ relative to $P_0$.

Note that the instrumental deviations illustrated in Figures 20–14 and 20–15 result in absorbances that are smaller than theoretical. It can be shown that instrumental deviations always lead to negative absorbance errors.[5]

[5]E. J. Meehan, in *Treatise on Analytical Chemistry,* 2nd ed., P. J. Elving, E. J. Meehan, and I. M. Kolthoff, Eds., Part I, Vol. 7, pp. 71–19. New York: Interscience, 1981.

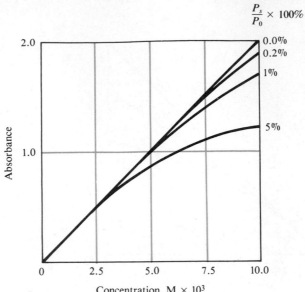

Figure 20-15

Apparent deviation from Beer's law caused by various amounts of stray radiation.

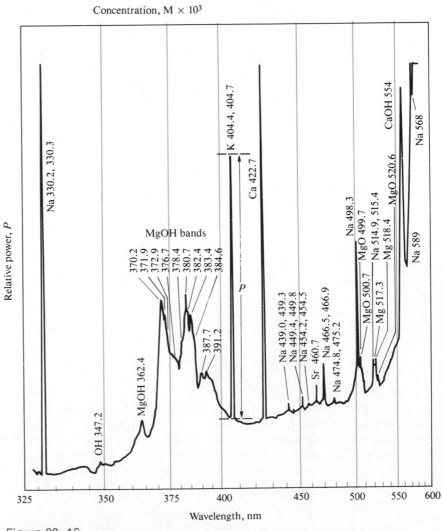

Figure 20-16

Emission spectrum or a brine obtained with an oxyhydrogen flame. (R. Hermann and C. T. J. Alkemade, *Chemical Analysis by Flame Photometry,* 2nd ed., p. 484. New York: Interscience, 1963. With permission.)

## 20D  EMISSION OF ELECTROMAGNETIC RADIATION

Atoms, ions, and molecules can be excited to one or more higher energy levels by any of several processes, including bombardment with electrons or other elementary particles, exposure to a high-voltage ac spark, heat treatment in a flame or arc, or exposure to a source of electromagnetic radiation. The lifetime of an excited species is generally transitory ($10^{-6}$ to $10^{-9}$ s), and relaxation to a lower energy level or the ground state takes place with a release of the excess energy in the form of electromagnetic radiation, heat, or perhaps both.

### 20D–1 Emission Spectra

Radiation from a source is conveniently characterized by means of an *emission spectrum*, which usually takes the form of a plot of the relative power of the emitted radiation as a function of wavelength or frequency. Figure 20–16 illustrates a typical emission spectrum, which was obtained by aspirating a brine solution into an oxyhydrogen flame. Three types of spectra are evident in the figure: *line, band,* and *continuous*. The line spectrum is made up of a series of sharp, well-defined peaks caused by an excitation of individual atoms. The band spectrum consists of several groups of lines so closely spaced that they are not completely resolved. The source of the bands is small molecules or radicals. Finally, the continuous spectrum is responsible for the increase in the background that becomes evident above about 350 nm. The line and band spectra are superimposed on this continuum. The source of the continuum is described on page 531.

### Line Spectra

Line spectra are encountered when the radiating species are individual atomic particles that are well separated, as in a gas. The individual particles in a gaseous medium behave independently of one another, and the spectrum consists of a series of sharp lines with widths of about $10^{-4}$ Å. In Figure 20–16, lines for sodium, potassium, strontium, and calcium are identified.

The energy-level diagram in Figure 20–17a shows the source of two of the lines in a typical emission spectrum of an element. The horizontal line labeled $E_0$ corresponds to the lowest, or ground-state, energy of the atom. The horizontal lines labeled $E_1$ and $E_2$ are two higher energy electronic levels of the species. For example, the single outer electron in the ground state $E_0$ for a sodium atom is located in the $3s$ orbital. Energy level $E_1$ then represents the energy of the atom when this electron has been promoted to the $3p$ state by absorption of thermal, electrical, or radiant energy (see also Figure 20–5). The promotion is depicted by the shorter wavy arrow on the left in Figure 20–17a. After perhaps $10^{-8}$ s, the atom returns to the ground state, emitting a photon whose frequency and wavelength are given by Equations 20–3 and 20–4

$$\nu_1 = (E_1 - E_0)/h$$
$$\lambda_1 = hc/(E_1 - E_0)$$

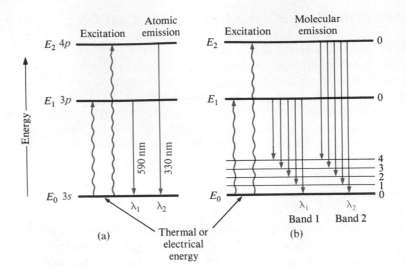

**Figure 20–17**

Energy-level diagrams for a sodium atom and a simple molecule, showing the source of (a) a line spectrum and (b) a band spectrum.

This emission process is illustrated by the shorter straight arrow on the right in Figure 20–17a.

For the sodium atom, $E_2$ in the figure corresponds to the more energetic $4p$ state; the resulting radiation $\lambda_2$ would then appear at a shorter wavelength. The line appearing at about 330 nm in Figure 20–16 results from this transition; the $3p$-to-$3s$ transition provides a line at about 590 nm. It is important to note that the emitted wavelengths *are identical to the wavelengths of the absorption peaks* for sodium (Figure 20–5) because the transitions involved are between the same two states.

### Band Spectra

Band spectra are often encountered in spectral sources because of the presence of gaseous radicals or small molecules. For example, in Figure 20–16 bands for OH, MgOH, and MgO are labeled and consist of a series of closely spaced lines that are not fully resolved by the instrument used to obtain the spectrum. Bands arise from the numerous quantized vibrational levels that are superimposed on the ground-state electronic energy level of a molecule.

Figure 20–17b is a partial energy-level diagram for a molecule showing its ground state $E_0$ and two of its several excited electronic states, $E_1$ and $E_2$. A few of the many vibrational levels associated with the ground state are also shown. Vibrational levels associated with the two excited states have been omitted because the lifetime of an excited vibrational state is brief compared with that of an electronically excited state (about $10^{-15}$ s versus $10^{-8}$ s). A consequence of this tremendous difference in lifetimes is that when an electron is excited to one of the higher vibrational levels of an electronic state, relaxation to the lowest vibrational level of that state occurs before an electronic transition to the ground state can occur. Therefore, the radiation produced by the electrical or thermal excitation of polyatomic species nearly always involves a transition from the *lowest vibrational level of an excited electronic state* to any of the several vibrational levels of the ground state.

The mechanism by which a vibrationally excited species relaxes to the nearest electronic state involves a transfer of its excess energy to other atoms in the system through a series of collisions. As noted, this process takes place at an enormous speed. Relaxation from one electronic state to another can also occur by collisional transfers of energy, but the rate of this process is slow enough that relaxation by photon release is favored.

The energy-level diagram in Figure 20–17b illustrates the mechanism by which two radiation bands consisting of five closely spaced lines are emitted by a molecule excited by thermal or electrical energy. For a real molecule, the number of individual lines is much larger because the ground state contains many more vibrational levels than are shown. In addition, a multitude of rotational states would be superimposed on each of the vibrational levels. The differences in energy among the rotational levels is perhaps an order of magnitude smaller than that for vibrational states. Thus, a real molecular band would be made up of many more lines than shown in Figure 20–17b, and these lines would be much more closely spaced.

## Continuous Spectra

As shown in Figure 20–18, truly continuous radiation is produced when solids are heated to incandescence. Thermal radiation of this kind, which is called *blackbody radiation,* is more characteristic of the temperature of the emitting surface than of the material of which that surface is composed. Blackbody radiation is produced by the innumerable atomic and molecular oscillations excited in the condensed solid by the thermal energy. Note that the energy peaks in Figure 20–18 shift to shorter wavelengths with increasing temperature. It is clear that very high temperatures are needed to cause a thermally excited source to emit a substantial fraction of its energy as ultraviolet radiation.

As noted earlier, part of the continuous background radiation exhibited in the flame spectrum shown in Figure 20–16 is probably thermal emission from incandescent particles in the flame. Note that this background decreases rapidly as the ultraviolet region is approached.

Heated solids are important sources of infrared, visible, and longer-wavelength ultraviolet radiation for analytical instruments.

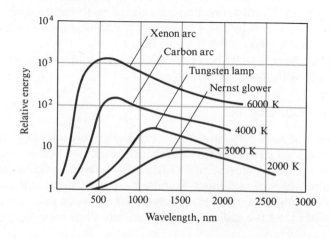

Figure 20–18
Blackbody radiation curves.

**The Effect of Concentration on Line and Band Spectra.** The radiant power $P$ of a line or a band depends directly upon the number of excited atoms or molecules, which in turn is proportional to the total concentration $c$ of the species present in the source. Thus we can write

$$P = kc \qquad (20-15)$$

where $k$ is a proportionality constant. This relationship is the basis of quantitative emission spectroscopy.

## 20D-2 Emission by Fluorescence and Phosphorescence

Fluorescence and phosphorescence are analytically important emission processes in which atoms or molecules are excited by the absorption of a beam of electromagnetic radiation. The excited species then relax to the ground state, giving up their excess energy as photons. Fluorescence takes place much more rapidly than phosphorescence and is generally complete in about $10^{-5}$ s (or less) from the time of excitation. Phosphorescence emission may extend for minutes or even hours after irradiation has ceased.

Fluorescence is considerably more important than phosphorescence in analytical chemistry. Thus, our discussions focus largely on the former.

### Atomic Fluorescence

Gaseous atoms fluoresce when they are exposed to radiation that has a wavelength that exactly matches that of one of the absorption (or emission) lines of the element in question. For example, gaseous sodium atoms are promoted to the excited energy state $E_1$ shown in Figure 20–17b through absorption of 590-nm radiation. Relaxation may then take place by fluorescent reemission of radiation of the identical wavelength. Fluorescence in which the excitation and emission wavelengths are the same is termed *resonance fluorescence*. Sodium atoms could also exhibit resonance fluorescence when exposed to 330-nm radiation. In addition, however, the element could also produce nonresonance fluorescence by first relaxing to energy level $E_1$ through a series of nonradiative collisions with other species in the medium. Further relaxation to the ground state can then take place either by the emission of a 590-nm photon or by further collisional deactivation.

Resonance fluorescence is commonly encountered with atoms and to a lesser extent with molecular species.

### Molecular Fluorescence

The number of molecules that fluoresce is relatively small because fluorescence requires structural features that slow the rate of the nonradiative relaxation processes illustrated in Figure 20–6b and enhance the rate of fluorescence relaxation shown in Figure 20–6c. Most molecules lack these features and undergo nonradiative relaxation at a rate that is significantly greater than the radiative relaxation rate; thus fluorescence is precluded.

As shown in Figure 20–6c, bands of radiation are produced when molecules fluoresce. Like molecular absorption bands, molecular fluorescence bands are made up of a multitude of closely spaced lines that are often difficult to resolve. Note that the lines that terminate the two fluorescence bands on the short-wavelength, or high-energy, side ($\lambda_1'$ and $\lambda_1''$) are resonance lines. That is, molecular fluorescence bands consist largely of lines that are longer in wavelength than the band of absorbed radiation responsible for their excitation. This shift in wavelength is sometimes called the *Stokes shift*. A more detailed discussion of the theory and applications of molecular fluorescence is given in Chapter 23.

## 20E QUESTIONS AND PROBLEMS

20–1. What kind of transitions are responsible for the
*(a) absorption of ultraviolet radiation?
(b) absorption of infrared radiation?
*(c) emission of band spectra?
(d) fluorescence of atoms?
*(e) fluorescence of molecules?
(f) emission of line spectra?

20–2. Define
*(a) ground state.
(b) excited electronic state.
*(c) photon.
(d) band spectra.
*(e) continuous spectra.
(f) line spectra.
*(g) resonance fluorescence.
(h) phosphorescence.
*(i) percent transmittance.
(j) absorbance.
*(k) molar absorptivity.
(l) absorptivity.
*(m) stray radiation.
(n) wavenumber.
*(o) relaxation.
(p) Stokes shift.

20–3. Calculate the frequency in hertz of
*(a) an X-ray beam with a wavelength of 2.65 Å.
(b) an emission line for copper at 211.0 nm.
*(c) the line at 694.3 nm produced by a ruby laser.
(d) the output of a $CO_2$ gas laser at 10.6 $\mu$m.
*(e) an infrared absorption peak at 19.6 $\mu$m.
(f) a microwave beam at 1.86 cm.

20–4. Calculate the wavelength in centimeters of
*(a) an airport tower transmitting at 118.6 MHz.
(b) a VOR (radio navigation aid) transmitting at 114.10 kHz.
*(c) an NMR signal at 105 MHz.
(d) an infrared absorption peak having a wavenumber of 1210.

*20–5. A typical simple infrared spectrophotometer covers a wavelength range from 3 to 15 $\mu$m. Express its range (a) in wavenumbers and (b) in hertz.

20–6. A sophisticated ultraviolet/visible/near-IR instrument has a wavelength range of 185 to 3000 nm. What are its wavenumber and frequency ranges?

20–7. Express the following absorbances in terms of percent transmittance:
*(a) 0.064.    (d) 0.209.
*(b) 0.765.    (e) 0.437.
*(c) 0.318.    (f) 0.413.

20–8. Convert the accompanying transmittance data to absorbances:
*(a) 19.4%.    (d) 4.51%.
*(b) 0.863.    (e) 0.100.
*(c) 27.2%.    (f) 79.8%.

*20–9. Calculate the percent transmittance of solutions having twice the absorbance of the solutions in Problem 20–7.

20–10. Calculate the absorbances of solutions having half the percent transmittance of those in Problem 20–8.

20–11. Use the data in Table 20–2 to evaluate the missing

| | A | % T | $\varepsilon$ | b, cm | c, M | c, ppm | a, $cm^{-1} ppm^{-1}$ |
|---|---|---|---|---|---|---|---|
| *(a) | 0.416 | | | 1.40 | $1.25 \times 10^{-4}$ | | |
| (b) | | 45.5 | | 2.10 | $8.15 \times 10^{-3}$ | | |
| *(c) | 1.424 | | | 0.996 | | | 0.137 |
| (d) | | 19.6 | $5.42 \times 10^3$ | | $2.50 \times 10^{-4}$ | | |
| *(e) | | | $3.46 \times 10^3$ | 2.50 | | 3.33 | |
| (f) | | | $1.214 \times 10^4$ | 1.25 | $7.77 \times 10^{-4}$ | | |
| *(g) | | 48.3 | | 0.250 | | 6.72 | |
| (h) | 0.842 | | $7.73 \times 10^3$ | 2.00 | | | |
| *(i) | | 76.3 | | 1.10 | | | 0.0631 |
| (j) | | 6.54 | $9.82 \times 10^2$ | | $8.64 \times 10^{-3}$ | | |

Table 20–2

quantities. Wherever necessary, assume that the molecular weight of the analyte is 250.

*20–12. A solution containing 4.48 ppm $KMnO_4$ has a transmittance of 0.309 in a 1.00-cm cell at 520 nm. Calculate the molar absorptivity of $KMnO_4$.

20–13. A solution containing 3.75 mg/100 mL of A (fw = 220) has a transmittance of 39.6% in a 1.50-cm cell at 480 nm. Calculate the molar absorptivity of A.

*20–14. A solution containing the complex formed between Bi(III) and thiourea has a molar absorptivity of $9.32 \times 10^3$ L cm$^{-1}$ mol$^{-1}$ at 470 nm.
  (a) What is the absorbance of a $6.24 \times 10^{-5}$ M solution of the complex at 470 nm in a 1.00-cm cell?
  (b) What is the percent transmittance of the solution described in (a)?
  (c) What is the molar concentration of the complex in a solution that has the absorbance described in (a) when measured at 470 nm in a 5.00-cm cell?

20–15. At 580 nm, which is the wavelength of its maximum absorption, the complex $FeSCN^{2+}$ has a molar absorptivity of $7.00 \times 10^3$ L cm$^{-1}$ mol$^{-1}$. Calculate
  (a) the absorbance of a $2.50 \times 10^{-5}$ M solution of the complex at 580 nm in a 1.00-cm cell.
  (b) the absorbance of a solution in which the concentration of the complex is twice that in (a).
  (c) the transmittance of the solutions described in (a) and (b).
  (d) the absorbance of a solution that has half the transmittance of that described in (a).

*20–16. A 2.50-mL aliquot of a solution that contains 3.8 ppm iron(III) is treated with an appropriate excess of KSCN and diluted to 50.0 mL. What is the absorbance of the resulting solution at 580 nm in a 2.50-cm cell? See Problem 20–15 for absorptivity data.

20–17. Zinc(II) and the ligand L form a product that absorbs strongly at 600 nm. As long as the molar concentration of L exceeds that of zinc(II) by a factor of 5, the absorbance is dependent only on the cation concentration. Neither zinc(II) nor L absorbs at 600 nm. A solution that is $1.60 \times 10^{-4}$ M in zinc(II) and $1.00 \times 10^{-3}$ M in L has an absorbance

of 0.464 in a 1.00-cm cell at 600 nm. Calculate
  (a) the percent transmittance of this solution.
  (b) the percent transmittance of this solution in a 2.50-cm cell.
  (c) the molar absorptivity of the complex.

*20–18. The equilibrium constant for the conjugate acid/base pair

$$HIn + H_2O \rightleftarrows H_3O^+ + In^-$$

is $8.00 \times 10^{-5}$. From the additional information

| Species | Absorption Maximum, nm | Molar Absorptivity | |
|---|---|---|---|
| | | **430 nm** | **600 nm** |
| HIn | 430 | $8.04 \times 10^3$ | $1.23 \times 10^3$ |
| In$^-$ | 600 | $0.775 \times 10^3$ | $6.96 \times 10^3$ |

  (a) calculate the absorbance at 430 nm and 600 nm for the following indicator concentrations: $3.00 \times 10^{-4}$ M, $2.00 \times 10^{-4}$ M, $1.00 \times 10^{-4}$ M, $0.500 \times 10^{-4}$, and $0.250 \times 10^{-4}$.
  (b) plot absorbance as a function of indicator concentration.

20–19. The equilibrium constant for the reaction

$$2 CrO_4^{2-} + 2 H^+ \rightleftarrows Cr_2O_7^{2-} + H_2O$$

is $4.2 \times 10^{14}$. The molar absorptivities for the two principal species in a solution of $K_2Cr_2O_7$ are

| $\lambda$ | $\varepsilon_1(CrO_4^{2-})$ | $\varepsilon_2(Cr_2O_7^{2-})$ |
|---|---|---|
| 345 | $1.84 \times 10^3$ | $10.7 \times 10^2$ |
| 370 | $4.81 \times 10^3$ | $7.28 \times 10^2$ |
| 400 | $1.88 \times 10^3$ | $1.89 \times 10^2$ |

Four solutions were prepared by dissolving $4.00 \times 10^{-4}$, $3.00 \times 10^{-4}$, $2.00 \times 10^{-4}$, and $1.00 \times 10^{-4}$ moles of $K_2Cr_2O_7$ in water and diluting to 1.00 L with a pH 5.60 buffer. Derive theoretical absorbance values (1.00-cm cells) for each solution and plot the data for (a) 345 nm; (b) 370 nm; (c) 400 nm.

## CHAPTER 21

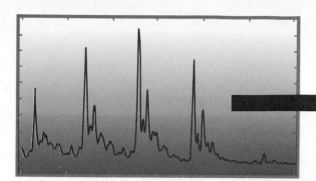

# INSTRUMENTS FOR OPTICAL SPECTROSCOPY

The basic components of analytical instruments for emission, absorption, and fluorescence spectroscopy are remarkably alike in function and general performance requirements regardless of whether the instruments are designed for ultraviolet, visible, or infrared radiation. Because of their similarities, such instruments are frequently referred to as *optical instruments* even when they are applied to spectral regions to which the eye is insensitive. In this chapter, we first examine the characteristics of the components common to all optical instruments and point out those features that are independent of the wavelength region being employed as well as those that are not. We then consider the general design characteristics of typical instruments, particularly those used for absorption spectroscopy.

## 21A  INSTRUMENT COMPONENTS

Most spectroscopic instruments are made up of five components: (1) a stable source of radiant energy, (2) a wavelength selector that permits the isolation of a restricted wavelength region, (3) one or more sample containers, (4) a radiation detector, or transducer, which converts radiant energy to a measurable signal (usually electrical), and (5) a signal processor and readout. Figure 21–1 shows how these components are assembled in instruments for emission, absorption, and fluorescence spectroscopy. Note that the configuration of components 4 and 5 is the same for the three types of instruments.

Emission instruments differ from the other two types in that components 1 and 3 are combined. That is, the sample container is an arc, a spark, a heated surface, or a flame that both holds the sample and causes it to emit characteristic radiation. In contrast, absorption and fluorescence spectroscopy require an external source of radiant energy and a cell to hold the sample. In absorption measurements, the beam from the source passes through the sample after leaving the wavelength selector (in some instruments, the positions of the sample and selector are reversed).

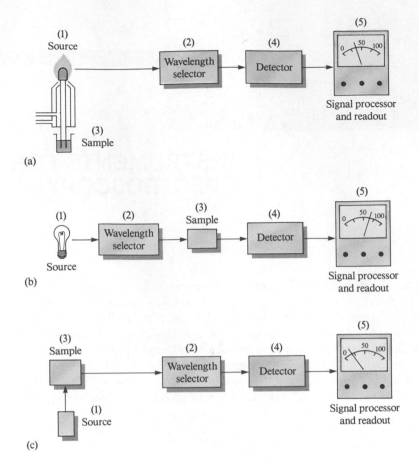

Figure 21–1

Components of various types of instruments for optical spectroscopy: (a) emission spectroscopy; (b) absorption spectroscopy; (c) fluorescence and scattering spectroscopy.

In fluorescence, the source induces the sample to emit characteristic radiation, which is usually measured at a 90-deg angle with respect to the beam from the source.

## 21A–1 The Transmittance of Various Construction Materials

Figure 21–2 shows the spectral-transparency regions for various materials used to construct the windows, lenses, sample containers, and prisms of spectroscopic instruments. Ordinary silicate glass is widely used for instruments designed for use in the visible region only. Fused silica and quartz extend the range of spectroscopic instruments down to 180 to 200 nm in the ultraviolet. Infrared spectroscopy requires the use of windows and cells fashioned from such materials as polished sodium chloride, potassium chloride, or silver chloride. All infrared-transparent materials tend to become fogged as a result of the absorption of moisture and must therefore be polished regularly until clear again.

## 21A–2 Spectroscopic Sources

Absorption and fluorescence spectroscopy require an external radiation source whose output is constant and intense enough to make detection

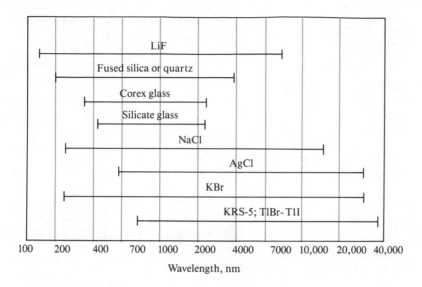

**Figure 21–2**
Transmittance range for various
construction materials.

and measurement easy. Typically, the radiant power of a source varies exponentially with the voltage of its electrical supply. For this reason, voltage regulators are often used to power spectroscopic sources.

The problem of source stability is sometimes circumvented by splitting the output of a source into a reference beam, which passes through the solvent, and a sample beam, which passes through a solution of the analyte. The detector is then illuminated alternately with the two beams, or the beams are monitored by matched detectors. The ratio of the intensities of the two beams then provides an analytical parameter that is largely independent of source fluctuations.

Table 21–1 lists common sources used in the various types of spectroscopy. Note that both continuous and line sources find application. Details concerning line sources are found in Section 24B–6. In this section, we describe the most common continuous sources of ultraviolet, visible, and infrared radiation.

### Hydrogen and Deuterium Lamps

A truly continuous spectrum in the ultraviolet region is produced by the electrical excitation of deuterium or hydrogen at low pressure. The mechanism by which a continuum is produced involves the formation of an excited molecule ($D_2^*$ or $H_2^*$) by absorption of electrical energy. This species then dissociates to give two hydrogen or deuterium atoms plus an ultraviolet photon. The reactions for hydrogen are

$$H_2 + E_e \rightarrow H_2^* \rightarrow H' + H'' + h\nu$$

where $E_e$ is the electrical energy absorbed by the molecule. The energy for the overall process is

$$E_e = E_{H_2^*} = E_{H'} + E_{H''} + h\nu$$

> A continuous source emits radiation of all wavelengths within a spectral region.

Figure 21–3

A deuterium lamp.

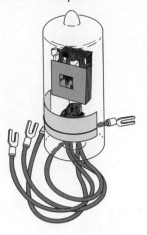

### Table 21–1
### SOURCES FOR SPECTROSCOPY

| Source | Wavelength Region, nm | Type of Spectroscopy |
|---|---|---|
| **Continuous Sources** | | |
| Xenon lamp | 250–600 | Molecular fluorescence; Raman |
| $H_2$ and $D_2$ lamps | 160–380 | UV molecular absorption |
| Tungsten/halogen lamp | 240–2500 | UV/vis/near-IR molecular absorption |
| Tungsten lamp | 350–2200 | Vis/near-IR molecular absorption |
| Nernst glower | 400–20,000 | IR molecular absorption |
| Nichrome wire | 750–20,000 | IR molecular absorption |
| Globar | 1200–40,000 | IR molecular absorption |
| **Line Sources** | | |
| Hollow cathode lamp | UV/vis | Atomic absorption; atomic fluorescence |
| Electrodeless discharge lamp | UV/vis | Atomic absorption; atomic fluorescence |
| Metal vapor lamp | UV/vis | Atomic absorption; molecular fluorescence; Raman |
| Laser | UV/vis/IR | Raman; molecular absorption; molecular fluorescence |

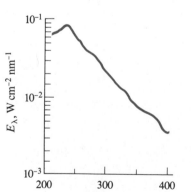

Figure 21–4

Output from a deuterium lamp.

where $E_{H_2^*}$ is the *fixed quantized energy* of $H_2^*$, and $E_{H'}$ and $E_{H''}$ are the *kinetic energies* of the two hydrogen atoms. The sum of the latter two energies can vary from zero to $E_{H_2^*}$. Thus, the energy and the frequency of the photon can also vary within this range of energies. That is, when the two kinetic energies are by chance small, $h\nu$ is large, and when the two energies are large, $h\nu$ is small. The consequence is a truly continuous spectrum from about 160 nm to the beginning of the visible region.

Most modern lamps for generating ultraviolet radiation contain deuterium and are of a low voltage type in which an arc is formed between a heated, oxide-coated filament and a metal electrode (see Figure 21–3). The heated filament provides electrons to maintain a direct current at a potential of about 40 V; a regulated power supply is required for constant intensities.

Both deuterium and hydrogen lamps provide a useful continuous spectrum in the region from 160 to 375 nm as shown in Figure 21–4. The intensity of the deuterium lamp is greater than that of the hydrogen lamp, however, which accounts for the more widespread use of the former. At longer wavelengths (>360 nm), the lamps generate emission lines, which are superimposed on the continuum. For many applications, these lines are a nuisance. They are useful for wavelength calibration of absorption instruments, however.

## Tungsten-Filament Lamps

The most common source of visible and near-infrared radiation is the tungsten-filament lamp, shown in Figure 21–5. The energy distribution of this source approximates that of a blackbody and is thus temperature-dependent. Figure 20–18 illustrates the output of the tungsten-filament lamp at 3000 K. In most absorption instruments, the operating filament temperature is about 2900 K; the bulk of the energy is thus emitted in the infrared region. A tungsten-filament lamp is useful for the wavelength region between 320 and 2500 nm. The lower limit is imposed by absorption by the glass envelope that houses the filament.

The energy output of a tungsten lamp in the visible region varies approximately as the fourth power of the operating voltage, thus making close voltage control essential. For this reason, constant-voltage transformers or electronic voltage regulators are employed. As an alternative, the lamp can be operated from a 6-V storage battery, which provides a remarkably stable voltage if it is maintained in good condition.

Tungsten/halogen lamps contain a small quantity of iodine within a quartz envelope that houses the filament. Quartz allows the filament to be operated at a temperature of about 3500 K, which leads to higher intensities and extends the range of the lamp well into the ultraviolet. The lifetime of a tungsten/halogen lamp is more than double that of an ordinary tungsten lamp because the life of the latter is limited by sublimation of the tungsten from the filament. In the presence of iodine, the sublimed tungsten reacts to give gaseous $WI_2$ molecules, which then diffuse back to the hot filament, where they decompose and redeposit as tungsten atoms. Tungsten/halogen lamps are finding ever-increasing use in modern spectroscopic instruments because of their extended wavelength range, greater intensity, and longer life.

## Infrared Sources

Continuous infrared radiation is obtained from hot inert solids. A *Globar* source consists of a 5 by 50 mm silicon carbide rod. Radiation in the region from 1 to 40 $\mu$m is emitted when the Globar is heated to about 1500°C by the passage of electricity.

A *Nernst glower* is a cylinder of zirconium and yttrium oxides having typical dimensions of 2 by 20 mm; it emits infrared radiation when heated to a high temperature by an electric current. Electrically heated spirals of nichrome wire also serve as infrared sources.

## 21A–3 Wavelength Selectors

Spectroscopic instruments are ordinarily equipped with a device that restricts the radiation being measured to a narrow band that is absorbed or emitted by the analyte. Such devices greatly enhance both the selectivity and the sensitivity of an instrument. In addition, for absorption measurements, narrow bandwidths increase the likelihood of adherence to Beer's law.

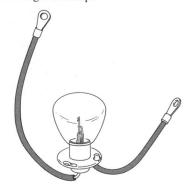

Figure 21–5
A tungsten lamp.

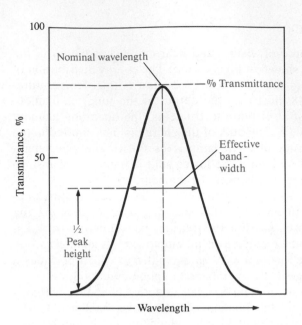

**Figure 21–6**
Output of a typical wavelength selector.

Note that the term *band* in this context has a somewhat different meaning from that used in describing types of spectra.

At the outset, it must be understood that no selector is capable of producing radiation of a single wavelength. Instead, the output of such a device is a range of contiguous wavelengths called a band; these wavelengths are distributed more or less symmetrically about a central *nominal wavelength*. As shown in Figure 21–6, the *effective bandwidth*, or *bandwidth*, of a selector is defined as the width of the band in wavelength units at half-peak height. Note that the ordinate in this plot is the percentage of incident radiation of a given wavelength that is transmitted. Bandwidths vary enormously from one wavelength selector to another. For example, a high-quality monochromator for the visible region may have an effective bandwidth of a few tenths of a nanometer or less, whereas an absorption filter in this same region may possess a bandwidth of 200 nm or more.

As shown in Table 21–2, two general types of wavelength selectors, *filters* and *monochromators,* are used to provide narrow bands of radiation. Monochromators have the advantage that the output wavelength can be varied continuously over a considerable spectral range.

**Table 21–2**
**WAVELENGTH SELECTORS FOR SPECTROSCOPY**

| Type | Wavelength Range, nm | Note |
|---|---|---|
| **Continuously variable** | | |
| Grating | 100–40,000 | 3000 lines/mm for vacuum UV; 50 lines/mm for far IR |
| Prism | 120–30,000 | See Figure 21–2 for construction materials. |
| **Discontinuous** | | |
| Interference filter | 200–14,000 | |
| Absorption filter | 380–750 | |

## Filters

Filters operate by absorbing all but a restricted band of radiation from a continuous source. As shown in Figure 21–6, a filter is generally characterized by its *nominal wavelength,* its maximum percent transmittance, and its effective bandwidth.

**Interference Filters.**  Interference filters find use with ultraviolet and visible radiation, as well as with wavelengths up to about 14 $\mu$m in the infrared region. As the name implies, an interference filter relies on optical interference to provide a relatively narrow band of radiation.

An interference filter consists of a very thin layer of a transparent material (frequently calcium fluoride or magnesium fluoride) coated on both sides with a film of metal that is thin enough so that it transmits approximately half of the radiation striking it and reflects the other half. This array is sandwiched between two glass plates that protect it from the atmosphere. When radiation strikes the central array at a 90-deg angle, approximately half is transmitted by the first metallic layer and the other half reflected. The transmitted radiation undergoes a similar partition when it reaches the second layer of metal. If the reflected portion from the second layer is of the proper wavelength, it is partially reflected from the inner portion of the first layer in phase with the incoming light of the same wavelength. The result is constructive interference of the radiation of this wavelength and destructive removal of most other wavelengths. It is readily shown that the nominal wavelength for an interference filter $\lambda_{max}$ is given by the equation[1]

$$\lambda_{max} = 2tn/\mathbf{n} \tag{21-1}$$

where $t$ is the thickness of the central fluoride layer, $n$ is its refractive index, and $\mathbf{n}$ is an integer called the *interference order*. The glass layers of the filter are often selected to absorb all but one of the wavelengths transmitted by the central layer, thus restricting the transmission of the filter to a single order.

Figure 21–7 illustrates the performance characteristics of a typical interference filter. Most filters of this type have bandwidths of better than 1.5% of the nominal wavelength, although this figure is lowered to 0.15% in some narrow-band filters; the latter have a maximum transmittance of about 10%.

**Absorption Filters.**  Absorption filters, which are generally less expensive and more rugged than interference filters, are limited in application to the visible region. This type of filter usually consists of a colored glass plate that removes part of the incident radiation by absorption. Absorption filters have effective bandwidths that range from perhaps 30 to 250 nm. Filters that provide the narrowest bandwidths also absorb a significant fraction of the desired radiation and may have a transmittance of 1% or

---

[1]For example, see D. A. Skoog and J. J. Leary, *Principles of Instrumental Analysis,* 4th ed., pp. 88–89. Philadelphia: Saunders College Publishing, 1992.

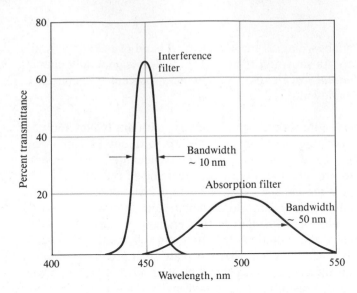

**Figure 21–7**
Bandwidths for two types of filters.

less at their band peaks. Figure 21–7 contrasts the performance characteristics of a typical absorption filter with its interference counterpart.

Glass filters with transmittance maxima throughout the entire visible region are available from commercial sources. While their performance characteristics are distinctly inferior to those of interference filters, their cost is appreciably less, and they are perfectly adequate for many routine applications.

## Monochromators

Monochromators for ultraviolet, visible, and infrared radiation are all similar in construction in the sense that they employ slits, lenses, mirrors, windows, and dispersing devices. To be sure, the materials from which these components are fabricated depend upon the wavelength region of intended use (Figure 21–2).

**The Components of a Monochromator.** Figure 21–8 illustrates the optical elements found in all monochromators: (1) an entrance slit, (2) a collimating lens or mirror to produce a parallel beam, (3) a prism or grating to disperse the radiation into its component wavelengths, and (4) a focusing element that projects a series of rectangular images of the entrance slit upon a planar surface called the *focal plane*. In addition, most monochromators have entrance and exit *windows* to protect the components from dust and corrosive laboratory fumes.

As shown in Figure 21–8, two types of dispersing devices are found in monochromators: reflection gratings and prisms. For purposes of illustration, a beam made up of just two wavelengths, $\lambda_1$ and $\lambda_2$ ($\lambda_1 > \lambda_2$), is shown. The beam enters the monochromator via a narrow rectangular opening (the *slit*), is collimated, and then strikes the surface of the dispersing element at an angle. In the grating monochromator, angular dispersion of the beam into its individual wavelengths results from diffraction at the reflective surface. For the prism instrument, bending, or

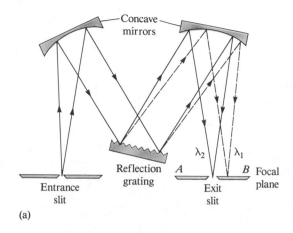

(a)

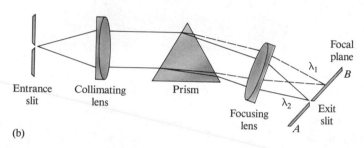

(b)

Figure 21–8
Two types of monochromators: (a) Czerny-Turner grating monochromator; (b) Bunsen prism monochromator. (In both instances, $\lambda_1 > \lambda_2$.)

refraction, of the radiation at the two surfaces leads to dispersion. In either case, the dispersed radiation is focused on the focal plane $AB$, where it appears as two images of the entrance slit (one for each wavelength). These images can be focused on the exit slit by rotating the dispersing element.

If a detector is located at the exit slit of the monochromator shown in Figure 21–8a and the grating is rotated so that one of the lines shown (say, $\lambda_1$) is *scanned* across the slit from $\lambda_1 - \Delta\lambda$ to $\lambda_1 + \Delta\lambda$, where $\Delta\lambda$ is a small wavelength difference, the output of the detector takes the Gaussian shape shown in Figure 21–9. The effective bandwidth of the monochromator, which is defined in the figure, depends upon the size and quality of the dispersing element, the slit widths, and the focal length of the monochromator. A high-quality monochromator will exhibit an effective bandwidth of a few tenths of a nanometer or less in the ultraviolet and visible regions. The effective bandwidth of a monochromator that is satisfactory for most quantitative applications ranges from about 1 to 20 nm.

Reflection gratings serve as the dispersing element in most modern spectroscopic instruments. Thus, the discussion that follows is restricted to grating monochromators exclusively.

**Reflection Gratings.** Most reflection gratings are *replica gratings* prepared from a *master grating,* which consists of a large number of parallel and closely spaced grooves (or blazes) ruled on a hard, polished surface with a suitably shaped diamond tool. A magnified cross-sectional view of a few typical grooves is shown in Figure 21–10. A grating for the ultraviolet and visible regions contains from 300 to 2000 grooves/mm, with 1200

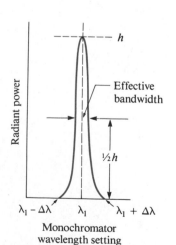

Figure 21–9
Output of an exit slit as monochromator is scanned from $\lambda_1 - \Delta\lambda$ to $\lambda_1 + \Delta\lambda$.

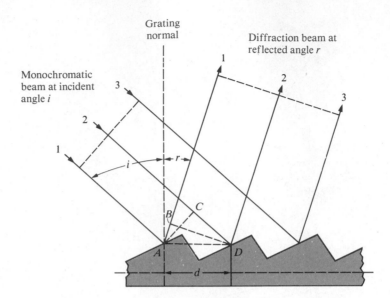

**Figure 21–10**
The mechanism of diffraction from an echellette-type grating.

to 1400 being most common. For the infrared region, 10 to 200 grooves/mm is common.

Replica gratings are formed by evaporating a film of aluminum onto a master grating after it has been coated with a parting agent that permits ready separation of the aluminum from the master. A glass plate is then cemented to the aluminum, and the grooved film is lifted from the master mold to give a finished grating. In recent times, *holographic gratings* have begun to appear in monochromators. This type of grating is manufactured by sophisticated lithographic techniques based upon the use of a pair of laser beams. The performance of these gratings is far superior to that of replica gratings.

**Dispersion by a Reflection Grating.** The grating shown in Figure 21–10 is an *echellette* grating, which is grooved, or *blazed*, to have relatively broad faces from which reflection occurs and narrow unused faces. This geometry provides highly efficient diffraction. Each broad face can be treated as a point source of radiation, giving reflected beams 1, 2, and 3, which interfere with one another. In order for the interference to be constructive, it is necessary that the path lengths differ by an integral multiple **n** of the wavelength of the incident beam.

In Figure 21–10, parallel beams of monochromatic radiation 1 and 2 strike the grating at an incident angle *i* to the *grating normal*. Maximum constructive interference is shown as occurring at the reflected angle *r*. It is evident that beam 2 travels a greater distance than beam 1 and that this difference is equal to $\overline{CD} - \overline{AB}$. For constructive interference to occur, this difference must equal **n**λ:

$$\mathbf{n}\lambda = \overline{CD} - \overline{AB}$$

where **n**, a small whole number, is called the diffraction *order*. Note, however, that angle *CAD* is equal to angle *i* and that angle *BDA* is identical to angle $r$. Therefore, from trigonometry,

$$\overline{CD} = d \sin i$$

where $d$ is the spacing between the reflecting surfaces. It is also seen that

$$\overline{AB} = -d \sin r$$

The minus sign by convention indicates that the angle of reflection $r$ lies on the opposite side of the grating normal from the incident angle $i$ (as in Figure 21–10); angle $r$ is positive when it is on the same side as $i$. Substitution of the last two expressions into the first gives the condition for constructive interference:

$$\mathbf{n}\lambda = d(\sin i + \sin r) \qquad (21\text{–}2)$$

   Equation 21–2 suggests that several values of $\lambda$ exist for a given diffraction angle $r$. Thus, if a first-order line ($\mathbf{n} = 1$) of 900 nm is found at $r$, second-order (450 nm) and third-order (300 nm) lines also appear at this angle. Ordinarily, the first-order line is the most intense; indeed, it is possible to design gratings that concentrate as much as 90% of the incident intensity in this order. The higher-order lines can generally be removed by filters. For example, glass, which absorbs radiation below 350 nm, eliminates the high-order spectra associated with first-order radiation in most of the visible region. The example that follows illustrates these points.

---

Example 21–1

An echellette grating containing 1450 blazes per millimeter was irradiated with a polychromatic beam at an incident angle 48 deg to the grating normal. Calculate the wavelengths of radiation that appear at an angle of reflection +20, +10, and −10 deg.
   To obtain $d$ in Equation 21–2, we write

$$d = \frac{1 \text{ mm}}{1450 \text{ blazes}} \times 10^6 \frac{\text{nm}}{\text{mm}} = 689.7 \frac{\text{nm}}{\text{blaze}}$$

When $r$ equals +20 deg,

$$\lambda = \frac{689.7}{\mathbf{n}} (\sin 48 + \sin 20) = \frac{748.4}{\mathbf{n}}$$

and the wavelengths for the first-, second-, and third-order reflections are 748, 374, and 249 nm, respectively.
   When angle $r$ is −10 deg,

$$\lambda = \frac{689.7}{\mathbf{n}} [\sin 48 + \sin (-10)] = 392.8 \text{ nm}$$

Further calculations of a similar kind yield the following data:

| r, deg | Wavelength, nm | | |
|---|---|---|---|
| | n = 1 | n = 2 | n = 3 |
| +20 | 748 | 374 | 249 |
| +10 | 632 | 316 | 211 |
| 0 | 513 | 256 | 171 |
| −10 | 393 | 196 | 131 |

In contrast to a prism, a grating disperses radiation linearly along the focal plane of the monochromator, which greatly simplifies monochromator design. A second advantage is that replica gratings are significantly less costly than prisms.

**Monochromator Slits.** The slits of a monochromator play an important role in determining the quality of the instrument. Slit jaws are formed by carefully machining two pieces of metal to give sharp edges. Care must be taken to ensure that these edges are parallel to one another and in the same plane.

The effective bandwidth of a monochromator depends upon the dispersion of the prism or grating as well as on the width of the entrance and/or exit slit. Most monochromators are equipped with variable slits so that the effective bandwidth can be changed. Narrow slits, and thus narrow bandwidths, lead to higher instrument *resolution*, which is desirable because greater spectral detail is revealed under such conditions (for example, see Figure 22–5). Because the power of the beam exiting from a monochromator falls off rapidly as the slit widths are narrowed, the resolution at which a monochromator can be operated is generally determined by the sensitivity of its radiation detector.

The shape of the entrance slit determines the shape of the image at the exit slit. Most slits are rectangular.

**Stray Radiation in Monochromators.** The exit beam of a monochromator is usually contaminated with small amounts of radiation having wavelengths far removed from that of the instrument setting. Sources of this unwanted radiation include reflection from various surfaces within the monochromator and scattering by dust particles in the atmosphere or on the surfaces of optical parts. Generally, the effects of spurious radiation are minimized by introducing baffles at appropriate spots in the monochromator and by coating interior surfaces with flat black paint. In addition, the monochromator is sealed with windows over the slits to prevent entrance of dust and fumes. Despite these precautions, however, some spurious radiation still reaches the exit slit of even the best monochromators. The effects of such radiation on absorption measurements are described in Sections 20C–7 and 22A–3.

## 21A–4 Radiation Detectors and Transducers

A *detector* is a device that indicates the existence of some physical phenomenon. Familiar examples of detectors include photographic film for indicating the presence of electromagnetic or radioactive radiation, the pointer of a balance for detecting mass differences, and the mercury level in a thermometer for detecting temperature changes. The human eye is also a detector; it converts visible radiation into an electrical signal that is passed to the brain via the chain of neurons in the optic nerve.

A *transducer* is a special type of detector that converts signals, such as light intensity, pH, mass, and temperature into *electrical* signals that can be subsequently amplified, manipulated, and finally converted into numbers representing the magnitude of the original signal.

> A *transducer* is a type of detector that converts various types of chemical and physical quantities to voltage, charge, or current.

## Properties of Transducers

The ideal electromagnetic radiation transducer responds rapidly to low levels of radiant energy over a broad wavelength range. In addition, it produces an electrical signal that is easily amplified and has a relatively low noise level. Finally, it is essential that the electrical signal produced by the transducer be directly proportional to the beam power $P$:

$$G = KP + K'  \qquad (21-3)$$

where $G$ is the electrical response of the detector in units of current, resistance, or potential. The proportionality constant $K$ measures the sensitivity of the detector in terms of electrical response per unit of radiant power. Many detectors exhibit a small constant response, known as a *dark current* $K'$, even when no radiation impinges upon their surfaces. Instruments with detectors that have a significant dark-current response are ordinarily equipped with a compensating circuit that permits subtraction of a signal proportional to the dark current to reduce $K'$ to zero. Thus, under ordinary circumstances, we can write

$$G = KP  \qquad (21-4)$$

> Generally, the output from analytical instruments fluctuates in a random way as a consequence of the operation of a large number of uncontrolled variables. These fluctuations, which limit the sensitivity of an instrument, are called *noise*. The terminology is derived from radio engineering, where the presence of unwanted signal fluctuations was recognizable to the ear as static, or noise.

> Common causes for noise include vibration, pickup from 60-Hz lines, temperature variations, and frequency or voltage fluctuations in the power supply.

> A dark current is a current produced by a photoelectric detector in the absence of light.

## Types of Transducers

As shown in Table 21-3, two general types of transducers are encountered: One responds to photons, the other to heat. All photon detectors

### Table 21-3
#### DETECTORS FOR SPECTROSCOPY

| Type | Wavelength Range, nm |
|---|---|
| **Photon Detectors** | |
| Phototubes | 150–1000 |
| Photomultiplier tubes | 150–1000 |
| Silicon diodes | 350–1100 |
| Photoconductors | 750–3000 |
| Photovoltaic cells | 380–780 |
| **Heat Detectors** | |
| Thermocouples | 600–20,000 |
| Bolometers | 600–20,000 |
| Pneumatic cells | 600–40,000 |
| Pyroelectric cells | 1000–20,000 |

are based upon the interaction of radiation with a reactive surface to produce electrons (photoemission) or to promote electrons to energy states in which they can conduct electricity (photoconduction). Only ultraviolet, visible, and near-infrared radiation have sufficient energy to cause these processes to occur; thus photon detectors are limited to wavelengths shorter than about 2 $\mu$m.

Generally, infrared radiation is detected by measuring the temperature rise of a blackened material located in the path of the beam. Because the temperature changes resulting from the absorption of the infrared energy are minute, close control of the ambient temperature is required if large errors are to be avoided. It is usually the detector system that limits the sensitivity and precision of an infrared instrument.

### Photon Detectors

Four widely used types of photon detectors are described in the paragraphs that follow: (1) phototubes, (2) photomultiplier tubes, (3) silicon photodiodes, and (4) photovoltaic cells.

**Phototubes.** As shown in Figure 21–11, a phototube consists of a semicylindrical cathode and a wire anode sealed inside an evacuated transparent envelope. The concave surface of the cathode supports a layer of photoemissive material, such as an alkali metal or metal oxide, that tends to emit electrons upon being irradiated. When a potential is applied across the electrodes, the emitted *photoelectrons* flow to the wire anode, producing a current that is readily amplified and displayed or recorded.

The number of electrons ejected from a photoemissive surface is directly proportional to the radiant power of the beam striking that surface. With an applied potential of about 90 V, all these electrons reach the anode to give a current that is also proportional to radiant power. Phototubes frequently produce a small dark current in the absence of radiation (Equation 21–3) that results from thermally induced electron emission.

Photoelectrons are electrons that are ejected from a photosensitive surface as a result of absorption of electromagnetic radiation.

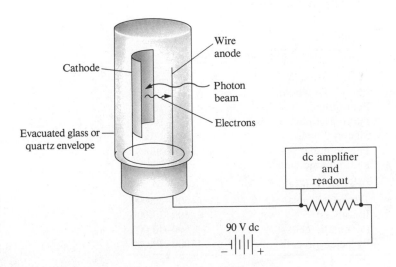

**Figure 21–11**

A phototube and accessory circuit. The photocurrent induced by the radiation causes a potential drop in the resistor, which is then amplified to drive a meter or recorder.

**Photomultiplier Tubes.** The *photomultiplier tube*, shown schematically in Figure 21–12, is similar in construction to the phototube just described but is significantly more sensitive. Its cathode surface is similar in composition to that of a phototube, with electrons being emitted upon exposure to radiation. The emitted electrons are accelerated toward a *dynode* (labeled 1 in the figure) maintained at a potential 90 V more positive than the cathode. Upon striking the dynode surface, each accelerated photoelectron produces several additional electrons, all of which are then accelerated to dynode 2, which is 90 V more positive than dynode 1. Here again, electron amplification occurs. By the time this process has been repeated at each of the remaining dynodes, $10^6$ to $10^7$ electrons have been produced for each photon; this cascade is finally collected at the anode. The resulting current is then further amplified electronically and measured.

**Silicon Photodiodes.** Crystalline silicon is a *semiconductor*—that is, a material whose electrical conductivity is less than that of a metal but greater than that of an electrical insulator. Silicon is a Group IV element and thus has four valence electrons. In a silicon crystal, each of these electrons is combined with electrons from four other silicon atoms to form four covalent bonds. At room temperature, sufficient thermal agitation occurs in this structure to liberate an occasional electron from its bonded state leaving it free to move throughout the crystal. Thermal excitation of an electron leaves behind a positively charged region termed a *hole*, which, like the electron, is also mobile. The mechanism of hole movement is stepwise, with a bound electron from a neighboring silicon atom jumping into the electron-deficient region (the hole) and thereby creating an-

One of the major advantages of photomultipliers is their automatic internal amplification. About $10^6$ to $10^7$ electrons are produced at the anode for each photon that strikes the photocathode of a photomultiplier tube (PMT).

With modern electronic instrumentation, it is possible to detect the arrival of individual photons at the photocathode of a PMT. The photons are counted, and the accumulated count is a measure of the intensity of the electromagnetic radiation impinging on the PMT. Photon counting is advantageous when light intensity, or the frequency of arrival of photons at the photocathode, is low.

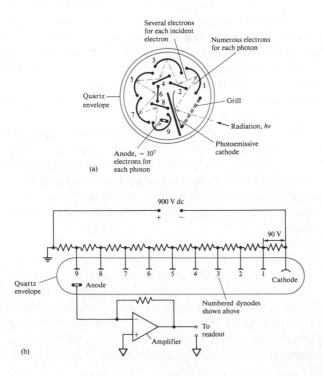

**Figure 21–12**

Photomultiplier tube: (a) cross section; (b) electrical circuit.

Figure 21–13

Two-dimensional representation of *n*-type silicon showing "impurity" atom.

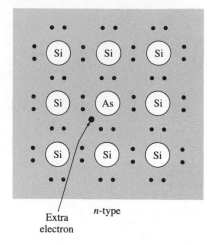

Extra electron          *n*-type

In electronics, a bias is a dc voltage that is inserted in series with a circuit element.

A silicon photodiode is a reverse-biased silicon diode that is used for measuring radiant power.

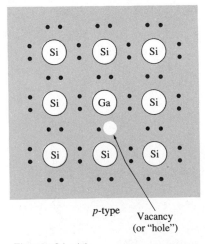

*p*-type      Vacancy
(or "hole")

Figure 21–14

Two-dimensional representation of *p*-type silicon showing "impurity" atom.

other positive hole in its wake. Conduction in a semiconductor involves the movement of electrons and holes in opposite directions.

The conductivity of silicon can be greatly enhanced by *doping*, a process whereby a tiny, controlled amount (approximately 1 ppm) of a Group V or Group III element is distributed homogenously throughout a silicon crystal. For example, when a crystal is doped with a Group V element, such as arsenic, four out of five of the valence electrons of the dopant form covalent bonds with four silicon atoms leaving one electron free to contribute to the conductivity of the crystal, as shown in Figure 21–13. In contrast, when the silicon is doped with a Group III element, such as gallium, which has but three valence electrons, an excess of holes develops, which also enhances conductivity (see Figure 21–14). A semiconductor containing unbonded electrons (*negative* charges) is termed an *n*-type semiconductor, and one containing an excess of holes (*positive* charges) is a *p*-type. In an *n*-type semiconductor, electrons are the *majority carrier*; in a *p*-type, holes are the majority carrier.

Present silicon technology makes it possible to fabricate what is called a *pn junction* or a *pn diode*, which is conductive in one direction and not in the other. Figure 21–15a is a schematic diagram of a silicon diode. The *pn* junction is shown as a dashed line through the middle of the crystal. Electrical wires are attached to both ends of the device. Figure 21–15b shows the junction in its conduction mode, wherein the positive terminal of a dc source is connected to the *p* region and the negative terminal to the *n* region (the diode is said to be *forward-biased* under these conditions). The excess electrons in the *n* region and the positive holes in the *p* region move toward the junction, where they combine and annihilate each other. The negative terminal of the source injects new electrons into the *n* region, which can continue the conduction process. The positive terminal extracts electrons from the *p* region, thus creating new holes that are free to migrate toward the *pn* junction.

Figure 21–15c illustrates the behavior of a silicon diode under *reverse biasing*. Here, the majority carriers are drawn away from the junction, leaving a nonconductive *depletion layer*. The conductance under reverse bias is only about $10^{-6}$ to $10^{-8}$ of that under forward biasing; thus, a silicon diode is a current rectifier.

A reverse-biased silicon diode can serve as a radiation detector because ultraviolet and visible photons are sufficiently energetic to create additional electrons and holes when they strike the depletion layer of a *pn* junction. The resulting increase in conductivity is readily measured and is directly proportional to radiant power. A silicon-diode detector is more sensitive than a simple vacuum phototube but less sensitive than a photomultiplier tube.

**Diode-Array Detectors.** Silicon photodiodes have become of notable importance recently because 1000 or more can be fabricated side by side on a single small silicon chip (the width of individual diodes is about 0.02 mm). With one or two of these *diode-array detectors* placed along the length of the focal plane of a monochromator, all wavelengths can be monitored simultaneously, thus making high-speed spectroscopy possible. Multichannel instruments based upon diode arrays are discussed in Section 21B–2.

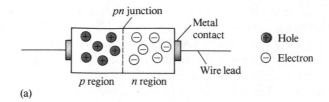

(a)

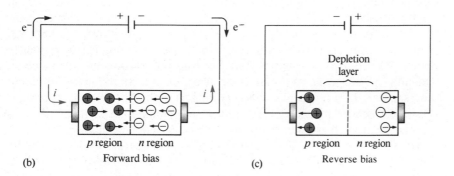

(b)  Forward bias                    (c)        Reverse bias

Figure 21–15
(a) Schematic of a silicon diode.
(b) Flow of electricity under forward bias. (c) Formation of depletion layer that prevents flow of
electricity under reverse bias.

**Photovoltaic Cells.** A photovoltaic cell (or photocell), the simplest of all
radiation transducers, consists of a flat copper or iron electrode upon
which is deposited a layer of a semiconducting material, such as selenium
or copper(I) oxide. The outer surface of the semiconductor is coated with
a thin, transparent film of gold, silver, or lead, which serves as the second, or collector, electrode. When radiation is absorbed on the surface of
the semiconductor, electrons and holes are formed and migrate in opposite directions, thus creating a current. If the two electrodes are connected through a low-resistance external circuit, the current produced is
directly proportional to the power of the incident beam. The currents are
large enough (10 to 100 $\mu$A) to be measured with a simple microammeter
without amplification.

The typical photovoltaic cell has maximum sensitivity at about 550 nm,
with the response falling off to perhaps 10% of maximum at 350 and 750
nm. The photocell constitutes a rugged, low-cost detector of visible radiation that has the advantage of not requiring an external power source. It is
not, however, as sensitive as other detectors and in addition suffers from
*fatigue*, which causes its current output to decrease gradually with continued illumination. Despite these disadvantages, photovoltaic cells are
quite useful for simple, portable, low-cost filter instruments.

### Heat Detectors

The convenient photon detectors discussed in the previous section cannot
be used to measure infrared radiation because photons of these frequencies lack the energy to cause photoemission of electrons; as a consequence, thermal detectors must be used. Unfortunately, the performance
characteristics of thermal detectors are much inferior to those of phototubes, photomultiplier tubes, silicon diodes, or photovoltaic cells.

A thermal detector consists of a tiny blackened surface that absorbs infrared radiation and increases in temperature as a consequence. The temperature rise is converted to an electrical signal that is amplified and measured. Under the best of circumstances, the temperature changes involved are minuscule, amounting to a few thousandths of a degree Celsius. The difficulty of measurement is compounded by thermal radiation from the surroundings, which is always a potential source of uncertainty. To minimize the effects of this background radiation, or noise, thermal detectors are housed in a vacuum and are carefully shielded from their surroundings. To further minimize the effects of this external noise, the beam from the source is chopped by a rotating disk inserted between source and detector. Chopping produces a beam that fluctuates regularly from zero intensity to a maximum. The transducer converts this periodic radiation signal to an alternating electrical current that can be amplified and separated from the dc signal resulting from the background radiation. Despite all these measures, infrared measurements are significantly less precise than measurements of ultraviolet and visible radiation.

As shown in Table 21–3, four types of heat detectors are used for infrared spectroscopy. The most widely used is a tiny thermocouple or a group of thermocouples called a *thermopile*. These devices consist of one or more pairs of dissimilar metal junctions that develop a potential difference when their temperatures differ. The magnitude of the potential depends upon the temperature difference.

A *bolometer* consists of a conducting element whose electrical resistance changes as a function of temperature. Bolometers are fabricated from thin strips of metals, such as nickel or platinum, or from semiconductors consisting of oxides of nickel or cobalt; the latter are called *thermistors*.

A *pneumatic detector* consists of a small cylindrical chamber that is filled with xenon and contains a blackened membrane to absorb infrared radiation and heat the gas. One end of the cylinder is sealed with a window that is transparent to infrared radiation; and the other is sealed with a flexible diaphragm that moves in and out as the gas pressure changes with cooling or heating. The temperature is determined from the position of the diaphragm.

*Pyroelectric detectors* are manufactured from crystals of a pyroelectric material, such as barium titanate or triglycine sulfate. When a crystal of either of these compounds is sandwiched between a pair of electrodes (one of which is transparent to infrared radiation), a temperature-dependent voltage develops that can be amplified and measured.

### Signal Processors and Readouts

A signal processor is ordinarily an electronic device that amplifies the electrical signal from the detector; in addition, it may alter the signal from dc to ac (or the reverse), change the phase of the signal, and filter it to remove unwanted components. The signal processor may also be called upon to perform such mathematical operations on the signal as differentiation, integration, or conversion to a logarithm.

**Feature 21–1**
MEASURING PHOTOCURRENTS WITH
OPERATIONAL AMPLIFIERS

The current produced by a reverse-biased silicon photodiode is typically 0.1 $\mu$A to 100 $\mu$A. These currents are so small that they must be converted to a voltage that is large enough to be measured with a digital voltmeter or other voltage-measuring device. We can perform such a conversion with the op amp circuit shown below. Light striking the reverse-biased photodiode causes a current $I$ in the circuit. Because the op amp has a very large input resistance, essentially no current enters the op amp input designated by the minus sign. Thus, current in the photodiode must also pass through the resistor $R$. The current is conveniently calculated from Ohm's law: $E_{out} = -IR$. Since the current is proportional to the radiant power of the light striking the photodiode, $I = kP$, where $k$ is a constant, and $E_{out} = -IR = -kPR = k'P$. A voltmeter is connected to the output of the op amp to give a direct readout that is proportional to the radiant power of the light. This same circuit can also be used with vacuum photodiodes or photomultipliers.

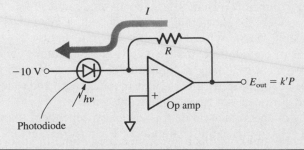

Several types of readout devices are found in modern instruments. Digital meters, scales of potentiometers, recorders, cathode-ray tubes, and monitors of microcomputers are some examples.

## 21A–5 Sample Containers

Sample containers, which are usually called *cells* or *cuvettes*, must have windows fabricated from a material that is transparent in the spectral region of interest. Thus, as shown in Figure 21–2, quartz or fused silica is required for the ultraviolet region (below 350 nm) and may be used in the visible region and to about 3000 nm in the infrared. Because of its lower cost, silicate glass is ordinarily used for the region between 375 and 2000 nm. Plastic containers have also found application in the visible region. The most common window material for infrared studies is crystalline sodium chloride.

Figure 21–16
Typical commercially available
cells.

The best cells have windows that are normal to the direction of the beam to minimize reflection losses. The most common cell length for studies in the ultraviolet and visible regions is 1 cm; matched, calibrated cells of this size are available from several commercial sources. Other path lengths, from shorter than 0.1 cm to 10 cm, can also be purchased. Transparent spacers for shortening the path length of 1-cm cells to 0.1 cm are also available. Some typical cells are shown in Figure 21–16.

For reasons of economy, cylindrical cells are sometimes encountered. Particular care must be taken to duplicate the position of such cells with respect to the beam; otherwise variations in path length and reflection loss at the curved surfaces can cause significant error.

The quality of spectroscopic data is critically dependent upon the way the matched cells are used and maintained. Fingerprints, grease, or other deposits on the walls markedly alter the transmission characteristics of a cell. Thus, thorough cleaning before and after use is imperative, and care must be taken to avoid touching the windows after cleaning is complete. Matched cells should never be dried by heating in an oven or over a flame because this may cause physical damage or a change in path length. Matched cells should be calibrated against each other regularly with an absorbing solution.

Avoid touching or scratching the windows of cuvettes.

## 21B   SPECTROSCOPIC INSTRUMENTS

The components discussed in the previous section have been assembled in various ways to produce dozens of designs for instruments to be used for spectroscopic measurements. These designs run the gamut from remarkably simple to highly sophisticated. Costs also vary widely, from a few hundred dollars to a hundred times that amount or more. No single instrument is best for all purposes, and selection must be determined by the type of work for which the instrument is intended and by the economics of its applications.

### 21B–1  Types of Spectroscopic Instruments

A *spectroscope* is an instrument for identifying the elements in a sample that have been excited in a flame or other hot medium. It consists of a modified monochromator, such as that shown in Figure 21–8b, in which the focal plane containing the exit slit is replaced by a movable eyepiece that permits visual detection of the emission lines (see the diagram in the margin on p. 630). The wavelength of a line is determined from the angle between the incident beam and the path of the line to the eyepiece.

Strictly speaking, a *colorimeter* is an instrument for absorption measurements in which the human eye serves as the detector. One or more comparison standards are required each time the instrument is used.

A *photometer* is a simple instrument that can be used for absorption, emission, or fluorescence measurements with ultraviolet, visible, or infrared radiation. A photometer is distinguished by its use of absorption or interference filters for wavelength selection and a photoelectric device for measuring radiant power. Instruments used for absorption measurements with visible radiation are sometimes called *photoelectric colorimeters* or even simply *colorimeters*. Use of the latter term can lead to ambiguity. A photometer that is to be employed for fluorescence measurements exclusively is often termed a *fluorometer*.

A *spectrograph* records spectra on a photographic plate or film located along the focal plane of a monochromator. Thus, the monochromators in Figure 21–8 could be converted to spectrographs by replacing the focal plane *AB* by a plate or film holder. The spectra would then appear as a series of black images of the entrance slit. Spectrographs are used primarily for qualitative elemental analysis.

A *spectrometer* is a monochromator equipped with a fixed slit at the focal plane. The two monochromators shown in Figure 21–8 are examples of spectrometers. A spectrometer equipped with a phototransducer at the exit slit is called a *spectrophotometer*. Spectrometers can be used for absorption, emission, and fluorescence measurements. Fluorescence spectrometers are often called *spectrofluorometers*.

### 21B–2  Instrument Designs

In this section, we consider four general types of spectroscopic instruments: (1) single-beam, (2) double-beam in space, (3) double-beam in time, and (4) multichannel.

#### Single-Beam Instruments

Figure 21–17a is a schematic of a single-beam instrument for absorption measurements. It consists of one of the radiation sources shown in Table 21–1, a filter or a monochromator for wavelength selection (Table 21–2), matched cells that can be interposed alternately in the radiation beam, one of the detectors listed in Table 21–3, an amplifier, and a readout device.

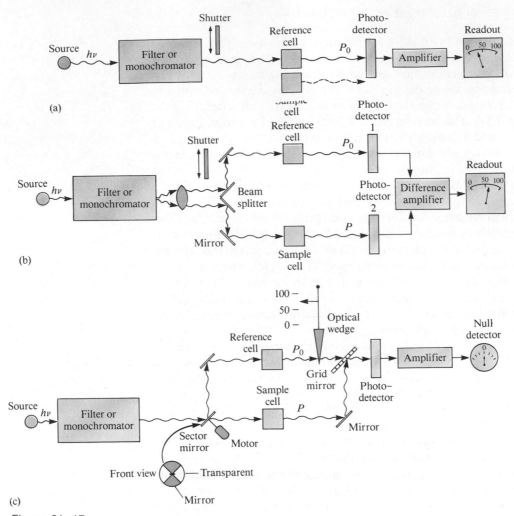

Figure 21-17
Instrument designs for photometers and spectrophotometers: (a) single-beam instrument; (b) double-beam instrument with beams separated in space; (c) double-beam instrument with beams separated in time.

This calibration procedure should be performed before each absorbance measurement.

The measurement of percent transmittance with a manual single-beam instrument involves three steps: (1) the *0% T adjustment*, (2) the *100% T adjustment*, and (3) the *determination of % T for the analyte*. The 0% T adjustment is carried out with the shutter imposed between the source and the photodetector. The meter is mechanically or electrically adjusted until it reads 0. Step 2 is then carried out by placing the solvent in the light path, opening the shutter, and varying the intensity of the radiation or the amplification of the electrical signal from the detector until the meter reads 100 (100% T). The beam intensity can be varied in several ways, including adjusting the electrical power to the source. Alternatively, the beam can be attenuated by a diaphragm, an optical wedge, or an optical comb that physically blocks a fraction of the radiation (the amount removed can be varied). In step 3, the solvent cell is replaced by one containing the analyte, and the percent transmittance is read from the scale. Because the transduced signal from a photodetector is linear with respect to the power of the radiation it receives, the scale reading with the sample in the light path is the percent transmittance (that is, the percent of full scale). Clearly, a logarithmic scale can be substituted to give the absorbance of the solution directly (see Figure 20–9, page 520).

Normally, a single-beam instrument requires a stabilized voltage supply to avoid errors resulting from changes in the beam intensity during the time required to make the 100% $T$ adjustment and determine $\% T$ for the analyte.

Single-beam instruments vary widely in their complexity and performance characteristics. The simplest and least expensive consists of a battery-operated tungsten bulb as the source, a set of glass filters for wavelength selection, test tubes for sample holders, a photovoltaic cell as the detector, and a small microammeter as the readout device. At the other extreme are sophisticated, computer-controlled instruments with a range of 200 to 1000 nm or more. These spectrophotometers have interchangeable tungsten/deuterium lamp sources, use rectangular silica cells, and are equipped with a high resolution grating with variable slits. Photomultiplier tubes are used as detectors, and the output is often digitized and stored so that it can be printed out in several forms.

Many modern Fourier transform infrared spectrometers are single-beam instruments.

## Double-Beam Instruments

Many modern photometers and spectrophotometers are based upon a double-beam design. Figure 21–17b illustrates a double-beam-in-space instrument in which two beams are formed in space by a V-shaped mirror called a beam splitter. One beam passes through the reference solution to a photodetector, and the second simultaneously traverses the sample to a second, matched photodetector. The two outputs are amplified, and their ratio (or the log of their ratio) is determined electronically and displayed by the readout device. With manual instruments, the measurement is a two-step operation involving first the zero adjustment with a shutter in place between selector and beam splitter. In the second step, the shutter is opened and the transmittance or absorbance is read directly from the meter.

The second type of double-beam instrument is illustrated in Figure 21–17c. Here the beams are separated in time by a rotating sector mirror that directs the entire beam from the monochromator first through the reference cell and then through the sample cell. The pulses of radiation are recombined by another sector mirror, which transmits one pulse and reflects the other to the detector. As shown by the insert labeled "front view" in Figure 21–17c, the motor-driven sector mirror is made up of pie-shaped segments, half of which are mirrored and half of which are transparent. The mirrored sections are held in place by blackened metal frames that periodically interrupt the beam and prevent its reaching the detector. The detector circuit is programmed to use these periods to perform the dark-current adjustment.

The instrument shown in Figure 21–17c is a null type, in which the beam passing through the solvent is attenuated until its intensity just matches that of the beam passing through the sample. Attenuation is accomplished in this design with an optical wedge, the transmission of which decreases linearly along its length. Thus, the null point is reached by moving the wedge in the beam until the two electrical pulses are identical as indicated by the null detector. The transmittance (or absorbance) is then read directly from the pointer attached to the wedge.

Double-beam instruments offer the advantage that they compensate for all but the most short-term fluctuations in the radiant output of the source as well as for drift in the detector and amplifier. They also compensate for wide variations in source intensity with wavelength (see Figure 21–4). Furthermore, the double-beam design lends itself well to the continuous recording of transmittance or absorbance spectra. Consequently, most modern ultraviolet and visible recording instruments are double-beam (usually in time). Many infrared spectrophotometers are based on this design.

## Multichannel Instruments

During the last decade, a number of multichannel spectrophotometers have become available. Figure 21–18 is a simplified schematic drawing showing the optical design of a type of multichannel spectrometer called a *diode-array spectrometer*. Radiation from a tungsten or deuterium lamp is focused upon the sample or solvent container and then passes into a monochromator with a fixed grating. The dispersed radiation falls on a photodiode-array detector, which, as mentioned earlier, consists of a linear array of several hundred photodiodes that have been formed along the length of a silicon chip. Typically, the chips are 1 to 6 cm in length and the widths of the individual diodes are 15 to 50 $\mu$m (see Figure 21–19). The chip also contains a capacitor and an electronic switch for each diode. A computer-driven shift register sequentially closes each switch momentarily, which causes each capacitor to be charged to $-5$ V. Radiation imping-

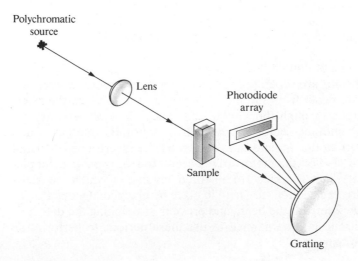

**Figure 21–18**

Diagram of a multichannel spectrophotometer based upon a grating and photodiode detector.

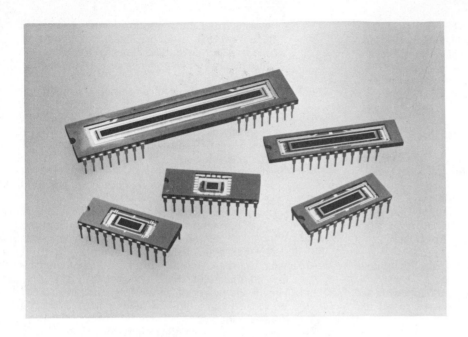

Figure 21–19
Diode arrays of various size.

ing on any diode surface causes partial discharge of its capacitor. This lost charge is replaced during the next switching cycle. The resulting charging currents, which are proportional to the radiant power, are amplified, digitized, and stored in computer memory. The entire cycle is completed in a few milliseconds.

The monochromator slit width of a diode-array instrument is usually made identical to the width of one of the silicon diodes. Thus, the output of each diode corresponds to the radiation of a different wavelength, and a spectrum is obtained by scanning these outputs sequentially. Since the electronic scanning process is remarkably rapid, data for an entire spectrum are accumulated in 1 s or less.

A diode-array instrument is a powerful tool for studying transient intermediates in moderately fast reactions, for kinetic studies, and for the qualitative and quantitative determination of the components exiting from a chromatographic column. The disadvantages of this type of instrument are its somewhat limited resolution (usually 1 to 2 nm) and its moderately high cost.

## 21C   QUESTIONS AND PROBLEMS

**21–1.** Define the term *effective bandwidth of a filter*.

*****21–2.** How many lines per millimeter are required in a grating if the first-order diffraction line at 500 nm is to be observed at a reflection angle of −40 deg when the angle of incidence is 60 deg?

**21–3.** An infrared grating has 72.0 lines/mm. Calculate the wavelengths of the first- and second-order diffraction spectra at reflection angles of (a) −15 deg, (b) 0 deg, and (c) +15 deg. Assume the incident angle is 50 deg.

*****21–4.** Describe how a spectroscope, a spectrograph, and a spectrophotometer differ from each other.

**21–5.** Why do quantitative and qualitative analyses often require different monochromator slit widths?

*****21–6.** The Wien displacement law states that the wavelength maximum in micrometers for blackbody radiation is

$$\lambda_{max} T = 2.90 \times 10^3$$

where $T$ is the temperature in kelvins. Calculate the wavelength maximum for a blackbody that has been heated to *(a) 4000 K, (b) 3000 K, *(c) 2000 K, and (d) 1000 K.

**21–7.** Stefan's law states that the total energy $E_t$ emitted by a blackbody per unit time and per unit area is

$$E_t = \alpha T^4$$

where $\alpha$ is $5.69 \times 10^{-8}$ W/m$^2 \cdot$ K$^4$. Calculate the total energy output in W/m$^2$ for the blackbodies described in Problem 21–6.

**\*21–8.** The relationships described in Problems 21–6 and 21–7 may be of help in solving the following.

  (a) Calculate the wavelength of maximum emission of a tungsten-filament bulb operated at 2870 K and at 3000 K.

  (b) Calculate the total energy output of the bulb in W/cm$^2$.

**21–9.** Describe the differences between the following and list any particular advantages possessed by one over the other:

  *(a) hydrogen- and deuterium-discharge lamps as sources for ultraviolet radiation.

  (b) filters and monochromators as wavelength selectors.

  *(c) photovoltaic cells and phototubes as detectors for electromagnetic radiation.

  (d) phototubes and photomultiplier tubes.

  *(e) photometers and colorimeters.

  (f) spectrophotometers and photometers.

  *(g) single-beam and double-beam instruments for absorbance measurements.

  (h) conventional and diode-array spectrophotometers.

**\*21–10.** A portable photometer with a linear response to radiation registered 73.6 $\mu$A with a blank solution in the light path. Replacement of the blank with an absorbing solution yielded a response of 24.9 $\mu$A. Calculate

  (a) the percent transmittance of the sample solution.

  (b) the absorbance of the sample solution.

  (c) the transmittance to be expected for a solution in which the concentration of the absorber is one third that of the original sample solution.

  (d) the transmittance to be expected for a solution that has twice the concentration of the sample solution.

**21–11.** A photometer with a linear response to radiation gave a reading of 685 mV with a blank in the light path and 179 mV when the blank was replaced by an absorbing solution. Calculate

  (a) the percent transmittance and absorbance of the absorbing solution.

  (b) the expected transmittance if the concentration of absorber is one half that of the original solution.

  (c) the transmittance to be expected if the light path through the original solution is doubled.

**\*21–12.** Why does a deuterium lamp produce a continuous rather than a line spectrum in the ultraviolet?

**21–13.** What are the differences between a photon detector and a heat detector?

**\*21–14.** How is the power of infrared radiation measured?

**21–15.** Why can photomultiplier tubes not be used with infrared radiation?

**\*21–16.** Why is iodine sometimes introduced into a tungsten lamp?

**21–17.** Describe how an absorption photometer and a fluorescence photometer differ from each other.

**\*21–18.** Describe the basic design difference between a spectrometer for absorption measurements and one for emission studies.

**21–19.** What data are needed to describe the performance characteristics of an interference filter?

**21–20.** Define

  *(a) dark current.

  (b) transducer.

  *(c) scattered radiation (in a monochromator).

  (d) $n$-type semiconductor.

  *(e) majority carrier.

  (f) depletion layer.

**\*21–21.** An interference filter is to be constructed for isolation of the CS$_2$ absorption band at 4.54 $\mu$m.

  (a) If the determination is to be based upon first-order interference, how thick should the dielectric layer be (refractive index 1.34)?

  (b) What other wavelengths will be transmitted?

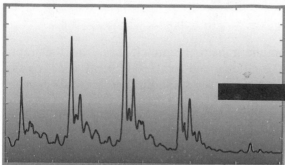

# MOLECULAR ABSORPTION SPECTROSCOPY

Molecular spectroscopy based upon ultraviolet, visible, and infrared radiation is widely used for the identification and determination of myriad inorganic and organic species.[1] Infrared absorption spectroscopy, for example, is one of the most powerful tools available to the chemist for determining the structure of both inorganic and organic compounds. In addition, it is now assuming an important role in quantitative analysis, particularly for determining environmental pollutants.

Molecular ultraviolet/visible absorption spectroscopy is employed primarily for quantitative analysis and is probably more widely used in chemical and clinical laboratories throughout the world than any other single procedure.

Molecular fluorescence methods, while less generally applicable than absorption methods, are of considerable importance because of their high selectivity and extraordinary sensitivity. These methods have proved particularly useful for the quantitative determination of molecules of biological and biochemical interest.

## 22A  ULTRAVIOLET AND VISIBLE ABSORPTION SPECTROSCOPY

In this section, we consider the types of molecular species that absorb ultraviolet or visible radiation and can thus be determined by absorption spectroscopy.

### 22A–1 Absorbing Species

As noted in Section 20C–2, absorption of ultraviolet and visible radiation by molecules generally occurs in one or more electronic absorption bands, each of which is made up of numerous closely packed but discrete

A band consists of a large number of closely spaced vibrational and rotational lines. The energies associated with the lines differ little from one another.

---

[1]For further reading, see E. J. Meehan, in *Treatise on Analytical Chemistry,* 2nd ed., P. J. Elving, E. J. Meehan, and I. M. Kolthoff, Eds., Part I, Vol. 7, Chapters 1–3. New York: Wiley, 1981; R. P. Bauman, *Absorption Spectroscopy.* New York: Wiley, 1962; G. F. Lothian, *Absorption Spectrophotometry,* 3rd ed. London: Adam Hilger, 1969.

lines. Each line arises from the transition of an electron from the ground state to one of the many vibrational and rotational energy states associated with each excited electronic energy state. Because there are so many of these vibrational and rotational states and because their energies differ only slightly, the number of lines contained in the typical band is large and their displacement from one another minute.

Figure 22–1a, which is part of the visible absorption spectrum for 1,2,4,5-tetrazine vapor, shows the fine structure that is due to the numerous rotational and vibrational levels associated with the excited electronic states of this aromatic molecule. In the gaseous state, the individual tetrazine molecules are sufficiently separated from one another to vibrate and rotate freely, and the many individual absorption lines resulting from the multitude of vibrational/rotational energy states are clearly evident. In the condensed state or in solution, however, freedom to rotate is largely lost, and lines due to differences in rotational energy levels are obliterated. Furthermore, in the presence of solvent molecules, energies of the various vibrational levels are modified in an irregular way. Thus, the energy of a given state in an assemblage of molecules takes on a Gaussian distribution; line broadening is the result. This effect is more pronounced in polar solvents, such as water, than in nonpolar hydrocarbon media. This solvent effect is illustrated in Figures 22–1b and c.

## Absorption by Organic Compounds

Two types of electrons are responsible for the absorption of ultraviolet and visible radiation by organic molecules: (1) shared electrons that par-

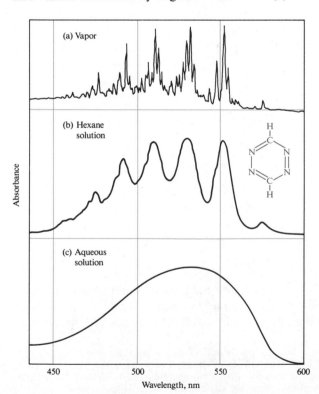

**Figure 22–1**
Typical ultraviolet absorption spectra. The compound is 1, 2, 4, 5-tetrazine. From S. F. Mason, *J. Chem. Soc.* **1959,** 1265. With permission.

ticipate directly in bond formation and are thus associated with more than one atom and (2) unshared outer electrons that are largely localized about such atoms as oxygen, the halogens, sulfur, and nitrogen.

The wavelengths at which an organic molecule absorbs depend upon how tightly its various electrons are bound. Thus, the shared electrons in single bonds such as carbon/carbon or carbon/hydrogen are so firmly held that their excitation requires energies corresponding to wavelengths in the vacuum ultraviolet region (below 180 nm). Single-bond spectra have not been widely exploited for analytical purposes because of the experimental difficulties of working in this region. These difficulties are attributable to the fact that both quartz and atmospheric components absorb radiation below 180 nm, a circumstance that requires the use of evacuated spectrophotometers equipped with lithium fluoride optics.

Organic compounds containing double or triple bonds generally exhibit useful absorption peaks in the readily accessible ultraviolet region because the electrons in unsaturated bonds are relatively loosely held and thus easily excited. Unsaturated organic functional groups that absorb in the ultraviolet and visible regions are termed *chromophores*. Table 22–1 lists common chromophores and the approximate location of their absorption maxima; a few sample spectra are presented in the margin.

> Chromophores are functional groups that are responsible for absorption.

Table 22–1

ASBORPTION CHARACTERISTICS OF SOME COMMON ORGANIC CHROMOPHORES

| Chromophore | Example | Solvent | $\lambda_{max}$, nm | $\varepsilon_{max}$ |
|---|---|---|---|---|
| Alkene | $C_6H_{13}CH{=}CH_2$ | n-Heptane | 177 | 13,000 |
| Conjugated alkene | $CH_2{=}CHCH{=}CH_2$ | n-Heptane | 217 | 21,000 |
| Alkyne | $C_5H_{11}C{\equiv}C{-}CH_3$ | n-Heptane | 178 | 10,000 |
| | | | 196 | 2000 |
| | | | 225 | 160 |
| Carbonyl | $CH_3\overset{O}{\overset{\|}{C}}CH_3$ | n-Hexane | 186 | 1000 |
| | | | 280 | 16 |
| | $CH_3\overset{O}{\overset{\|}{C}}H$ | n-Hexane | 180 | Large |
| | | | 293 | 12 |
| Carboxyl | $CH_3\overset{O}{\overset{\|}{C}}OH$ | Ethanol | 204 | 41 |
| Amido | $CH_3\overset{O}{\overset{\|}{C}}NH_2$ | Water | 214 | 60 |
| Azo | $CH_3N{=}NCH_3$ | Ethanol | 339 | 5 |
| Nitro | $CH_3NO_2$ | Isooctane | 280 | 22 |
| Nitroso | $C_4H_9NO$ | Ethyl ether | 300 | 100 |
| | | | 665 | 20 |
| Nitrate | $C_2H_5ONO_2$ | Dioxane | 270 | 12 |
| Aromatic | Benzene | n-Hexane | 204 | 7900 |
| | | | 256 | 200 |

Spectra for typical organic compounds.

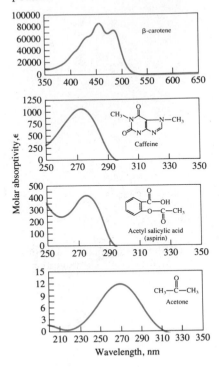

Organic compounds containing sulfur, bromine, and iodine also absorb in the ultraviolet region because these elements contain loosely bound unshared electrons that are more easily excited than the shared electrons of a saturated bond and are thus more readily excited by the absorption of photons.

## Absorption by Inorganic Species

The spectra for most absorbing inorganic complex ions and molecules resemble those for organic compounds (Figures 22–1b and c and spectra shown in the margin), with broad absorption maxima and little fine structure. The spectra for ions of the lanthanide and actinide series represent an important exception. The electrons responsible for absorption by these elements ($4f$ and $5f$, respectively) are shielded from external influences by electrons that occupy orbitals with larger principal quantum numbers. As a consequence, the absorption bands are narrow and relatively unaffected by the nature of the species bonded by the outer electrons.

With few exceptions, the ions and complexes of the 18 elements in the first two transition series are colored in one if not all of their oxidation states. Absorption of visible radiation by these species involves transitions of electrons between filled and unfilled $d$ orbitals that differ in energy as a consequence of ligands bonded to the metal ions. The energy differences between $d$ orbitals (and thus the position of the corresponding absorption peak) depend upon the oxidation state of the element, its position in the periodic table, and the kind of ligand bonded to its ion.

## Charge-Transfer Absorption

For quantitative purposes, *charge-transfer absorption* is particularly important because molar absorptivities are unusually large ($\varepsilon_{max} > 10,000$), a circumstance that leads to high sensitivity. Many inorganic and organic complexes exhibit this type of absorption and are therefore called *charge-transfer complexes*.

A charge-transfer complex consists of an electron-donor group bonded to an electron acceptor. When this product absorbs radiation, an electron from the donor is transferred to an orbital that is largely associated with the acceptor. The excited state is thus the product of a kind of internal oxidation/reduction process. This behavior differs from that of an organic chromophore in which the excited electron is in a *molecular* orbital that is shared by two or more atoms.

Familiar examples of charge-transfer complexes include the phenolic complex of iron(III), the 1,10-phenanthroline complex of iron(II), the iodide complex of molecular iodine, and the ferro/ferricyanide complex responsible for the color of Prussian blue. The red color of the iron(III)/thiocyanate complex is a further example of charge-transfer absorption. Absorption of a photon results in the transfer of an electron from the thiocyanate ion to an orbital that is largely associated with the iron(III) ion. The product is an excited species involving predominantly iron(II) and the thiocyanate radical SCN. As with other types of electronic excitation, the electron in this complex ordinarily returns to its original state

Charge-transfer spectra result when the absorption of a photon promotes an electron from ligand to metal or from metal to ligand in transition metal complexes.

after a brief period. Occasionally, however, an excited complex may dissociate and produce photochemical oxidation/reduction products.

In most charge-transfer complexes involving a metal ion, the metal serves as the electron acceptor. Exceptions are the 1,10-phenanthroline complexes of iron(II) (Section 15C–1) and copper(I), where the ligand is the acceptor and the metal ion the donor. A few other examples of this type of complex are known.

Charge-transfer spectra.

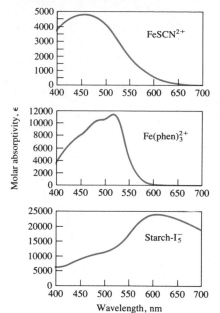

## 22A–2 Instruments for Ultraviolet and Visible Absorption Spectroscopy

### Photometers

Photometers for absorption methods offer the advantages of low cost, simplicity, ruggedness, portability, and ease of maintenance. Moreover, where high spectral purity is not important (and often it is not), the accuracy and precision of measurements made with a photometer can approach those made with a spectrophotometer. The disadvantages of photometers are their lesser versatility, their inability to generate entire spectra, and their generally wider effective bandwidths.

Figure 22–2 is a diagram of a simple single-beam photometer used for quantitative measurements in the visible region. Photometers of this kind are generally supplied with several filters, each of which transmits a different portion of the visible spectrum. Generally, a suitable filter is one whose color is the complement of the color of the analyte solution because it is this complementary color that is absorbed by the solution (see Feature 20–1). If several filters possessing the same general hue are available, the one that provides the greatest absorbance (or least transmittance) for a solution of the analyte should be used.

Ultraviolet photometers have become important detectors in high-performance liquid chromatography. In this application, a mercury-vapor lamp serves as the source and the emission line at 254 or 280 nm is isolated by an interference filter. This type of photometer is described briefly in Section 28A–5.

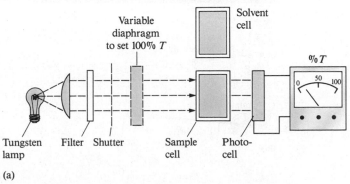

(a)

Figure 22–2
Single-beam photometer for absorption measurements in the visible region.

### Spectrophotometers

Several dozen models of spectrophotometers for the visible or the ultraviolet/visible region are now marketed by various instrument manufacturers.

**Instruments for the Visible Region.** Visible-region spectrophotometers are generally single-beam, grating instruments that are relatively inexpensive, rugged, and readily portable. At least one is battery-operated and small enough to be held in the hand. The most common application of these instruments is for fixed-wavelength quantitative analysis, although several produce surprisingly good absorption spectra that are useful for qualitative analysis also.

Figure 22–3 shows a simple and inexpensive spectrophotometer, the Spectronic 20. The original version of this instrument first appeared on the market in the mid-1950s, and the modified version shown in the figure is still being manufactured and widely sold. Undoubtedly, more of these instruments are currently in use throughout the world than any other single spectrophotometer model.

The Spectronic 20 employs a tungsten-filament light source operated by a stabilized power supply that provides radiation of constant intensity for

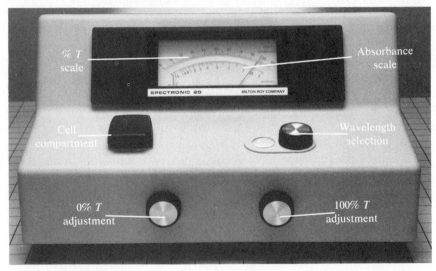

(a)

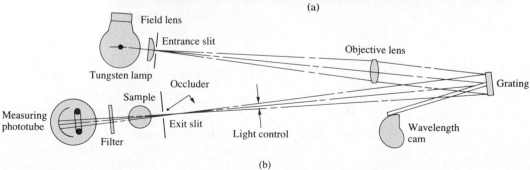

(b)

**Figure 22–3**
(a) The Spectronic 20 spectrophotometer. (b) Its optical diagram. (Courtesy of Milton Roy Company, Analytical Products Division, Rochester, NY.)

sufficient time to provide good reproducibility for absorbance readings. The radiation from the source passes through a fixed slit to the surface of a diffraction grating. A portion of the diffracted radiation then passes through an exit slit to the sample or reference cell and thence to a photo-tube. The amplified electrical signal from the detector powers a meter with a $5\frac{1}{2}$-in. scale that is linear in percent transmittance and logarithmic in absorbance.

The Spectronic 20 is equipped with an occluder, which is a vane that automatically falls between the beam and the detector whenever the cu-vette is removed from its holder. The light-control device, located be-tween the grating and the occluder, is a V-shaped aperture that is moved in and out of the beam in order to control the intensity of the beam falling on the photocell.

With the Spectronic 20, the 0% *T calibration* is made with the cell compartment empty, so that the occluder prevents radiation from reach-ing the detector. The 100% *T adjustment* involves placing a cell containing the blank in the light path and adjusting the light control. Finally, the sample is placed in the cell compartment, and the percent transmittance or the absorbance is read directly off the meter scale.

This calibration procedure should be performed before each absorbance measurement.

The spectral range of the Spectronic 20 is 340 to 625 nm (an accessory phototube extends the range to 950 nm). Other specifications for the instrument include an effective bandwidth of 20 nm and a wavelength accuracy of ±2.5 nm.

Single-beam instruments, such as the Spectronic 20, are well suited for quantitative absorption measurements at a single wavelength. Here, sim-plicity of instrumentation, low cost, and case of maintenance offer dis-tinct advantages.

Several instrument manufacturers offer single-beam instruments for both ultraviolet and visible measurements. The lower wavelength ex-tremes for these instruments range from 190 to 210 nm and the upper from 800 to 1000 nm. All are equipped with interchangeable tungsten and deu-terium or hydrogen lamps. Most employ photomultiplier tubes for detec-tors and gratings for dispersing radiation. Many are equipped with digital readout devices; others are equipped with large meters.

**Recording Spectrophotometers for the Ultraviolet and Visible Regions.** Figure 22–4 shows the optics of a typical double-beam (in time) spectro-photometer designed to operate over a range from about 190 to 750 nm. The instrument is equipped with interchangeable deuterium/tungsten sources, a reflection grating monochromator, and a photomultiplier detec-tor. The beam splitter is a motor-driven circular disk or chopper that is divided into three segments, one of which is transparent, the second reflecting, and the third opaque. With each rotation, the detector receives three signals, the first corresponding to $P_0$, the second to $P$, and the third to the dark current. The resulting electrical signals are then processed electronically to give the transmittance or absorbance on a readout de-vice.

An instrument of the kind shown in Figure 22–4 is usually provided with a motor-driven grating that is synchronized with the paper drive of a recorder so that automatic scanning and recording of an entire spectrum becomes possible.

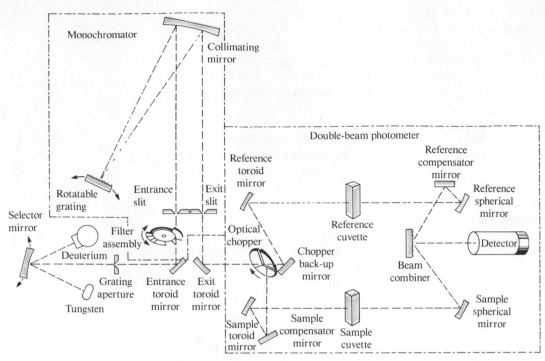

**Figure 22-4**

A double-beam recording spectrophotometer for the ultraviolet and visible regions; the Perkin-Elmer Series. (Courtesy of Coleman Instruments Division, Oak Brook, IL 50421.)

## 22A-3 Qualitative Applications of Ultraviolet Spectrophotometry

Spectrophotometric measurements with ultraviolet radiation are useful for detecting chromophoric groups, such as those shown in Table 22-1.[2] Because large parts of even the most complex organic molecules are transparent to radiation longer than 180 nm, the appearance of one or more peaks in the region from 200 to 400 nm is clear indication of the presence of unsaturated groups or of atoms such as sulfur or halogens. Often, an idea as to the identity of the absorbing groups can be gained by comparing the spectrum of an analyte with those of simple molecules containing various chromophoric groups.[3] Ordinarily, however, ultraviolet spectra lack sufficient fine structure to allow unambiguous identification of an analyte. Thus, ultraviolet qualitative data must be supplemented with other physical or chemical evidence such as infrared, nuclear

[2]For a detailed discussion of ultraviolet absorption spectroscopy in the identification of organic functional groups, see R. M. Silverstein, G. C. Bassler, and T. C. Morrill, *Spectrometric Identification of Organic Compounds,* 5th ed., Chapter 7. New York: Wiley, 1991.

[3]Several organizations publish catalogs of spectra, including American Petroleum Institute, *Ultraviolet Spectral Data, A.P.I. Research Project 44.* Pittsburgh: Carnegie Institute of Technology; *Sadtler Ultraviolet Spectra.* Philadelphia: Sadtler Research Laboratories; and American Society for Testing Materials, Committee E-13, Philadelphia.

Table 22–2
SOLVENTS FOR THE ULTRAVIOLET AND VISIBLE REGIONS

| Solvent | Lower Wavelength Limit, nm | Solvent | Lower Wavelength Limit, nm |
|---------|---------------------------|---------|---------------------------|
| Water | 180 | Carbon tetrachloride | 260 |
| Ethanol | 220 | Diethyl ether | 210 |
| Hexane | 200 | Acetone | 330 |
| Cyclohexane | 200 | Dioxane | 320 |
| | | Cellosolve | 320 |

magnetic resonance, and mass spectra as well as solubility and melting- and boiling-point information.

### Solvents

Ultraviolet spectra for qualitative analysis are most commonly derived for dilute solutions of the analyte. For volatile compounds, however, more useful spectra often result when the sample is examined as a gas (for example, compare Figure 22–1a and 22–1b). Such gas-phase spectra can often be obtained by allowing a drop or two of the pure compound to equilibrate with the atmosphere in a stoppered cuvette.

A solvent for ultraviolet/visible spectroscopy must be transparent throughout this region and should dissolve a sufficient quantity of the sample to give well-defined peaks. Moreover, consideration must be given to possible interactions with the absorbing species. For example, polar solvents, such as water, alcohols, esters, and ketones, tend to obliterate vibrational spectra and should thus be avoided when spectral detail is desired. Nonpolar solvents, such as cyclohexane, often provide spectra that more closely approach that of a gas (compare, for example, the three spectra in Figure 22–1). In addition, the polarity of the solvent often influences the position of absorption maxima. Consequently a common solvent must be employed when comparing spectra for the purpose of identification.

Table 22–2 lists common solvents for studies in the ultraviolet and visible regions and their approximate lower wavelength limits. These limits are strongly dependent upon the purity of the solvent. For example, ethanol and the hydrocarbon solvents are frequently contaminated with benzene, which absorbs below 280 nm.[4]

### The Effect of Slit Width

The effect of variation in slit width, and hence effective bandwidth, is illustrated by the spectra in Figure 22–5. Clearly, peak heights and sepa-

---

[4]Most major suppliers of reagent chemicals in the United States offer spectrochemical grades of solvents. Spectral-grade solvents have been treated so as to remove absorbing impurities and meet or exceed the requirements set forth in *Reagent Chemicals, American Chemical Society Specifications*, 7th ed. Washington, D.C.: American Chemical Society, 1986.

Use minimum slit width for qualitative studies to preserve maximum detail in spectra.

### Figure 22-5

Spectra for reduced cytochrome $c$ obtained with four spectral bandwidths: (1) 20 nm, (2) 10 nm, (3) 5 nm, and (4) 1 nm. At bandwidths <1 nm, peak heights were the same but instrument noise became pronounced. (Courtesy of Varian Instrument Division, Palo Alto, CA.)

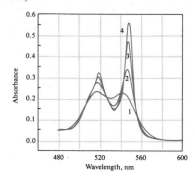

ration are distorted at wider bandwidths. For this reason, spectra for qualitative applications are obtained at minimum slit widths.

### The Effect of Scattered Radiation at the Wavelength Extremes of a Spectrophotometer

Earlier we demonstrated that scattered radiation may lead to instrumental deviations from Beer's law (page 525). Another undesirable effect of this type of radiation is that it occasionally causes false peaks to appear when a spectrophotometer is being operated at its wavelength extremes. Figure 22-6 shows an example of such behavior. Curve $B$ is the true spectrum for a solution of cerium(IV) produced with a research-quality spectrophotometer responsive down to 200 nm or less. Curve $A$ was obtained for the same solution with an inexpensive instrument operated with a tungsten source designed for work in the visible region only. The false peak at about 360 nm is directly attributable to scattered radiation, which was not absorbed because it was made up of wavelengths longer than 400 nm. Under most circumstances, such stray radiation has a negligible effect because its power is only a tiny fraction of the total power of the beam exiting from the monochromator. At wavelength settings below 380 nm, however, radiation from the monochromator is greatly attenuated as a result of absorption by glass optical components and cuvettes. In addition, both the output of the source and the photocell sensitivity fall off dramatically below 380 nm. These factors combine to cause a substantial fraction of the measured absorbance to be due to the scattered radiation of wavelengths to which cerium(IV) is transparent. A false peak results.

This same effect is sometimes observed with ultraviolet/visible instruments when attempts are made to measure absorbances at wavelengths lower than about 190 nm.

### 22A-4 Quantitative Ultraviolet and Visible Photometry and Spectrophotometry

Absorption spectroscopy based upon ultraviolet and visible radiation is one of the most useful tools available to the chemist for quantitative analysis.[5] The important characteristics of spectrophotometric and photometric methods are

1. *Wide applicability*. Enormous numbers of inorganic, organic, and biochemical species absorb ultraviolet or visible radiation and are thus amenable to direct quantitative determination. Many nonabsorbing species can also be determined after chemical conversion to absorbing derivatives. It has been estimated that over 90% of the analyses performed in clinical laboratories are based upon ultraviolet and visible absorption spectroscopy.

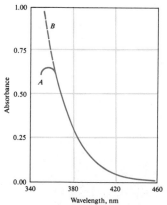

### Figure 22-6

Spectra of cerium(IV) obtained with a spectrophotometer having glass optics ($A$) and quartz optics ($B$). The false peak in $A$ arises from the transmission of stray radiation of longer wavelengths.

---

[5]For a wealth of detailed, practical information on spectrophotometric practices, see *Techniques in Visible and Ultraviolet Spectrometry*, Vol. I, *Standards in Absorption Spectroscopy*, C. Burgess and A. Knowles, Eds. London: Chapman and Hall, 1981; and J. R. Edisbury, *Practical Hints on Absorption Spectrometry*. New York: Plenum Press, 1968.

2. *High sensitivity*. Typical detection limits for absorption spectroscopy range from $10^{-4}$ to $10^{-5}$ M. This range can often be extended to $10^{-6}$ or even $10^{-7}$ M with certain procedural modifications.

3. *Moderate to high selectivity*. Often a wavelength can be found at which the analyte alone absorbs, thus making preliminary separations unnecessary. Furthermore, where overlapping absorption bands do occur, corrections based upon additional measurements at other wavelengths sometimes eliminate the need for a separation step.

4. *Good accuracy*. The relative errors in concentration encountered with a typical spectrophotometric or photometric procedure employing ultraviolet and visible radiation lie in the range from 1 to 5%. Such errors can often be decreased to a few tenths of a percent with special precautions.

5. *Ease and convenience*. Spectrophotometric and photometric measurements are easily and rapidly performed with modern instruments. In addition, the methods readily lend themselves to automation.

## Scope

The applications of quantitative absorption methods not only are numerous but also touch upon every field in which quantitative chemical information is required. You can gain a notion of the scope of spectrophotometry by consulting a series of review articles published biennially in *Analytical Chemistry*[6] and from monographs on the subject.[7]

## Applications to Absorbing Species

The spectrophotometric determination of any organic compound containing one or more of the chromophoric groups listed in Table 22–1 is potentially feasible. In addition, numerous inorganic species absorb ultraviolet or visible radiation and can thus be determined by direct photometric or spectrophotometric procedures. Among these are nitrite and nitrate ions, the oxides of nitrogen, the elemental halogens, ozone, and most of the transition metals in one oxidation state or another.

## Applications to Nonabsorbing Species

Many nonabsorbing inorganic and organic analytes can be determined spectrophotometrically by causing them to react with a chromophoric reagent to yield a product that absorbs in the ultraviolet or visible region. The successful application of such reagents usually requires that the reaction between analyte and reagent be forced to near completion.

---

[6]J. A. Howell and L. G. Hargis, *Anal. Chem.*, **1978**, *50*, 243R; **1980**, *52*, 306R; **1982**, *54*, 171R; **1984**, *56*, 225R; **1986**, *58*, 108R; **1988**, *60*, 131R; **1990**, *62*, 155R.

[7]See, for example, E. B. Sandell and H. Onishi, *Colorimetric Determination of Traces of Metals*, 4th ed. New York: Interscience, 1978; *Colorimetric Determination of Nonmetals*, 2nd ed., D. F. Boltz and J. A. Howell, Eds. New York: Wiley, 1986; Z. Marczenko, *Separation and Spectrophotometric Determination of Elements*. New York: Halsted Press, 1975; and M. Pisez and J. Bartos, *Colorimetric and Fluorometric Analysis of Organic Compounds and Drugs*. New York: Marcel Dekker, 1974.

Typical inorganic color-forming reagents are thiocyanate ion for iron, cobalt, and molybdenum; the anion of hydrogen peroxide for titanium, vanadium, and chromium; and iodide ion for bismuth, palladium, and tellurium. Of even greater importance are organic chelating agents that form stable colored complexes with cations. Examples are diethyldithiocarbamate for the determination of copper, diphenylthiocarbazone for lead, 1,10-phenanthrolene for iron, and dimethylglyoxime for nickel. Figure 22–7 shows the color-forming reactions for the first two of these reagents. The structure of the 1,10-phenanthroline complex of iron(II) is shown on page 365, and the reaction of nickel with dimethylglyoxime to form a red precipitate is described on page 88. In the application of the last reaction to the photometric determination of nickel, an aqueous solution of the cation is extracted with a solution of the chelating agent in an immiscible organic liquid. The absorbance of the resulting bright red organic layer serves as a measure of the concentration of the metal.

## Procedural Details

In developing a photometric or spectrophotometric procedure, conditions that yield a reproducible relationship (preferably linear) between concentration and absorbance must be established at the outset.

**Selection of Wavelength.** Maximum sensitivity is realized when the wavelength selected for photometric measurements is the same as an absorption peak for the analyte, because it is here that the change in absorbance per unit concentration is the greatest. In addition, the absorption curve is often relatively flat at its maximum, which leads to good linearity (Figure 20–14, page 527) and less possibility for error from failure to reproduce precisely the wavelength setting of the instrument. For

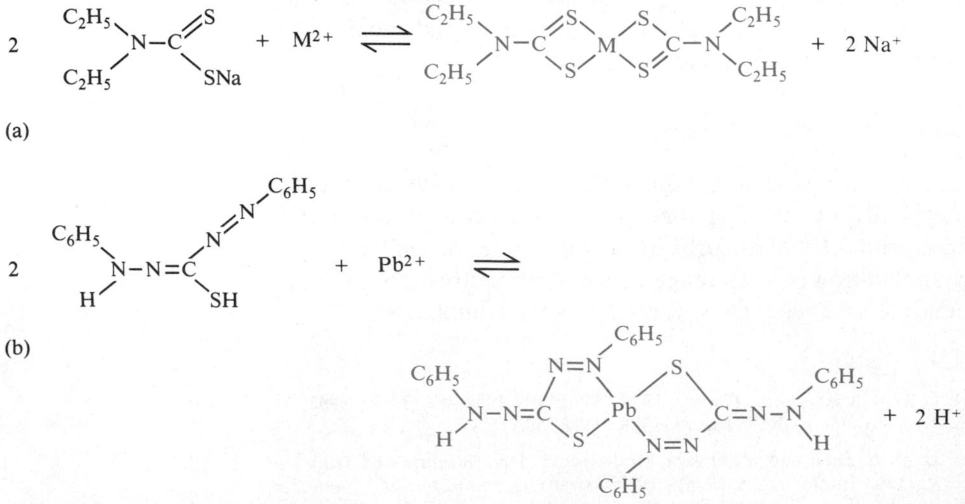

Figure 22–7

Typical chelating reagents for absorption. (a) Diethyldithiocarbamate. (b) Diphenylthiocarbazone.

measurements with a photometer, a filter that has a color that is complementary to that of the analyte solution is chosen. Such a choice leads to enhanced sensitivity as well as greater probability of obtaining a linear calibration curve.

**Variables That Influence Absorbance.** The absorbance of a solution is often influenced by such variables as the nature of the solvent, pH, temperature, electrolyte concentration, reaction time, and presence of interfering substances. The effects of these variables must be studied in order to establish a set of conditions that yield reproducible analytical data.

**The Cleaning and Handling of Cells.** Accurate spectrophotometric measurements require the use of good-quality matched cells that have been regularly calibrated against one another to detect differences arising from scratches, etching, and wear. Cell cleaning techniques, which are described in Section 34M–1, are equally important.[8]

**Determination of the Relationship Between Absorbance and Concentration.** The standard solutions for calibration should approximate the overall composition of the samples as closely as possible and should encompass a reasonable range of analyte concentrations. Seldom, if ever, is it safe to assume adherence to Beer's law and use only a single standard to determine the molar absorptivity. It is even less prudent to base an analysis on a literature value for the molar absorptivity because measured molar absorptivities often vary from instrument to instrument even when the same models are being employed.

**The Standard-Addition Method.** The difficulties that attend the production of a set of standards whose overall composition closely resembles that of the sample can be formidable if not insurmountable. Under these circumstances, a *standard-addition* approach may prove useful. In this procedure, a known quantity of standard is added to an aliquot of the sample, with the absorbance being measured before and after the addition. (See Section 24B–6, page 628.) Provided Beer's law is obeyed (and this must be confirmed experimentally), the analyte concentration can be derived from the two absorbances, the two volumes, and the concentration of the standard.

---

Example 22–1

A 2.00-mL urine specimen was treated with reagents that react with phosphate to produce a color, following which the sample was diluted to 100 mL. Photometric measurement on a 25.0-mL aliquot yielded an absorbance of 0.428. Addition of 1.00 mL of a solution containing 0.0500 mg of phosphate to a second 25.0-mL aliquot resulted in an absorbance of 0.517. Use these data to calculate the number of milligrams of phosphate in each milliliter of the specimen.

---

[8]See J. O. Erickson and T. Surles, *American Laboratory,* **1976,** *8*(6), 50.

The absorbance of the second measurement must be corrected for dilution:

$$\text{corrected absorbance} = 0.517 \times \frac{26.0}{25.0} = 0.538$$

$$\text{absorbance caused by 0.0500 mg phosphate} = 0.538 - 0.428 = 0.110$$

$$\text{weight of phosphate in } \frac{25.0}{100} \text{ of specimen} = \frac{0.428}{0.110} \times 0.0500 = 0.195 \text{ mg}$$

Finally, then,

$$\text{mg phosphate/mL of specimen} = \frac{100}{25.0} \times 0.195 \times \frac{1}{2.00} = 0.390$$

**The Analysis of Mixtures.** The total absorbance of a solution at any given wavelength is equal to the sum of the absorbances of the individual components in the solution (Equation 20–14). This relationship makes it possible, in principle, to determine the concentrations of the individual components of a mixture even if their spectra totally overlap. For example, Figure 22–8 shows three spectra, one for a solution of species A, the second for a solution of B, and the third for a mixture of the two. It is evident that both components contribute to the absorbance of the mixture at every wavelength. To analyze this mixture, molar absorptivities for A and B are first determined at wavelengths $\lambda_1$ and $\lambda_2$ with enough standards to be sure that Beer's law is obeyed over an absorbance range that encompasses the absorbance of the sample. Note that the wavelengths selected are ones at which the two spectra differ significantly. To complete the analysis, the absorbance of the mixture is determined at the same two wavelengths. Example 22–2 demonstrates how the composition of the mixture is derived from data of this kind.

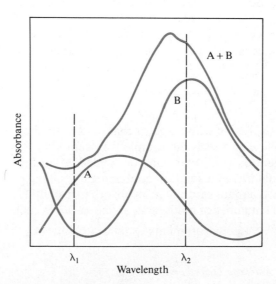

Figure 22–8

Absorption spectrum for a two-component mixture.

Example 22–2

Palladium(II) and gold(III) can be determined simultaneously by complexing the two ions with methiomeprazine ($C_{19}H_{24}N_2S_2$). The absorption maximum for the palladium complex occurs at 480 nm, and that for the gold complex is at 635 nm. Molar absorptivity data at these wavelengths are

| | Molar Absorptivity | |
| --- | --- | --- |
| | **480 nm** | **635 nm** |
| Pd complex | $3.55 \times 10^3$ | $5.64 \times 10^2$ |
| Au complex | $2.96 \times 10^3$ | $1.45 \times 10^4$ |

A 25.0-mL sample was treated with an excess of methiomeprazine and subsequently diluted to 50.0 mL. Calculate the molar concentrations of Pd(II) and Au(III) if the diluted solution had an absorbance of 0.533 at 480 nm and 0.590 at 635 nm when measured in a 1.00-cm cell.

Letting $c_{Pd}$ and $c_{Au}$ be the molar concentrations of the two ions, we can write for 480 nm (see Equation 20–14)

$$0.533 = 3.55 \times 10^3 \times 1.00 \times c_{Pd} + 2.96 \times 10^3 \times 1.00 \times c_{Au}$$

$$c_{Pd} = \frac{0.533 - 2.96 \times 10^3 \times c_{Au}}{3.55 \times 10^3}$$

At 635 nm

$$0.590 = 5.64 \times 10^2 \times 1.00 \times c_{Pd} + 1.45 \times 10^4 \times 1.00 \times c_{Au}$$

Substitution for $c_{Pd}$ in this expression gives

$$0.590 = \frac{5.64 \times 10^2(0.533 - 2.96 \times 10^3 \times c_{Au})}{3.55 \times 10^3} + 1.45 \times 10^4 \times c_{Au}$$

$$= 0.0847 - 4.70 \times 10^2 \times c_{Au} + 1.45 \times 10^4 \times c_{Au}$$

$$c_{Au} = \frac{0.590 - 0.0847}{1.403 \times 10^4} = 3.60 \times 10^{-5} \text{ M}$$

$$c_{Pd} = \frac{0.533 - (2.96 \times 10^3)(3.60 \times 10^{-5})}{3.55 \times 10^3} = 1.20 \times 10^{-4} \text{ M}$$

Since the analysis involved a twofold dilution, the concentrations of Pd(II) and Au(III) in the original sample were $7.20 \times 10^{-5}$ and $2.40 \times 10^{-4}$ M, respectively.

Mixtures containing more than two absorbing species can be analyzed, in principle at least, if one additional absorbance measurement is made for each added component. The uncertainties in the resulting data become greater, however, as the number of measurements increases.[9] Some of the

---

[9] G. Dado and J. Rosenthal, *J. Chem. Educ.*, **1990**, *67*, 797.

newer computerized spectrophotometers are capable of minimizing these uncertainties by overdetermining the system; that is, these instruments record many more data points than unknowns and effectively match the entire spectrum of the unknown as closely as possible by deriving synthetic spectra for various concentrations of the components. The derived spectra are then compared with that of the analyte until a close match is found. Spectra for standard solutions of each component are required, of course.

### The Effect of Instrumental Uncertainties[10]

In the context of this discussion, *noise* refers to random variations in the instrument output due not only to electrical fluctuations but also to such other variables as the way the operator reads the meter, the position of the cell in the light beam, the temperature of the solution, and the output of the source.

The accuracy and precision of spectrophotometric analyses are often limited by the indeterminate error, or *noise*, associated with the instrument. As pointed out in Section 21B–2, a spectrophotometric absorbance measurement entails three steps: a 0% $T$ adjustment, a 100% $T$ adjustment, and a measurement of % $T$. The indeterminate errors associated with each of these steps combine to give a net indeterminate error for the final value obtained for $T$. The relationship between the noise encountered in the measurement of $T$ and the resulting *concentration uncertainty* can be derived by writing Beer's law in the form

$$c = -\frac{1}{\varepsilon b} \log T = \frac{-0.434}{\varepsilon b} \ln T$$

Taking the partial derivative of this equation while holding $\varepsilon b$ constant leads to the expression

$$\partial c = \frac{-0.434}{\varepsilon b T} \partial T$$

where $\partial c$ can be interpreted as the uncertainty in $c$ that results from the noise (or uncertainty) $\partial T$ in $T$. Dividing this equation by the previous one gives

$$\frac{\partial c}{c} = \frac{0.434}{\log T} \times \frac{\partial T}{T} \tag{22–1}$$

where $\partial T/T$ is the *relative* indeterminate error in $T$ attributable to the noise in the three measurement steps, and $\partial c/c$ is the resulting relative indeterminate concentration error.

The best and most useful measure of the indeterminate error $\partial T$ is the standard deviation $\sigma_T$, which is easily measured for a given instrument by making 20 or more replicate transmittance measurements of an absorbing

---

[10]For further reading, see J. D. Ingle Jr. and S. R. Crouch, *Analytical Spectroscopy*, Chapter 5. Englewood Cliffs, NJ: Prentice Hall, 1988.

solution. Substituting $\sigma_T$ and $\sigma_c$ for the corresponding differential quantities in Equation 22–1 leads to

$$\frac{\sigma_c}{c} = \frac{0.434}{\log T} \times \frac{\sigma_T}{T} \qquad (22-2)$$

where $\sigma_c/c$ and $\sigma_T/T$ are relative standard deviations.

It is clear from an examination of Equation 22–2 that the uncertainty in a photometric concentration measurement varies in a complex way with the magnitude of the transmittance. The situation is even more complicated than suggested by the equation, however, because the uncertainty $\sigma_T$ is, under many circumstances, also *dependent upon T*.

In a detailed theoretical and experimental study, Rothman, Crouch, and Ingle[11] described several sources of instrumental indeterminate errors and showed the net effect of these errors on the precision of concentration measurements. The errors fall into three categories: those for which the magnitude of $\sigma_T$ is (1) independent of $T$, (2) proportional to $\sqrt{T^2 + T}$, and (3) proportional to $T$. Table 22–3 summarizes information about these sources of uncertainty. When the three relationships for $\sigma_T$ in the first column are substituted into Equation 22–2, three equations for the relative standard deviation in the concentration are obtained; these are shown in the third column.

**Table 22–3**
**CATEGORIES OF INSTRUMENTAL INDETERMINATE ERRORS IN TRANSMITTANCE MEASUREMENTS**

| Category | Sources | Effect of $T$ on Relative Standard Deviation of Concentration | |
|---|---|---|---|
| $\sigma_T = k_1$ | Readout resolution; thermal detector noise; dark current and amplifier noise | $\dfrac{\sigma_c}{c} = \dfrac{0.434}{\log T} \times \dfrac{k_1}{T}$ | (22–3) |
| $\sigma_T = k_2\sqrt{T^2 + T}$ | Photon detector shot noise | $\dfrac{\sigma_c}{c} = \dfrac{0.434}{\log T} \times k_2 \sqrt{1 + \dfrac{1}{T}}$ | (22–4) |
| $\sigma_T = k_3 T$ | Cell positioning uncertainty; fluctuation in source intensity | $\dfrac{\sigma_c}{c} = \dfrac{0.434}{\log T} \times k_3$ | (22–5) |

*Note:* $\sigma_T$ is the standard deviation of the transmittance measurements; $\sigma_c/c$ is the relative standard deviation of the concentration measurements; $T$ is transmittance; and $k_1$, $k_2$, and $k_3$ are constants for a given instrument.

**Concentration Errors When $\sigma_T = k_1$.** For many photometers and spectrophotometers, the standard deviation in the measurement of $T$ is constant and independent of the magnitude of $T$. This type of indeterminate error is

---

[11]L. D. Rothman, S. R. Crouch, and J. D. Ingle Jr., *Anal. Chem.*, **1975**, *47*, 1226.

often encountered with direct-reading instruments and has its origin in the somewhat limited resolution of the meter scale. The size of a typical scale is such that a reading cannot be reproduced to better than a few tenths of a percent of the full-scale reading, and the magnitude of this uncertainty is the same from one end of the scale to the other. For typical inexpensive instruments, standard deviations of about $0.003T$ ($\sigma_T = \pm0.003T$) are observed.

---

Example 22–3

A spectrophotometric analysis was performed with a manual instrument that exhibited an absolute standard deviation of $\pm0.003T$ throughout its transmittance scale. Calculate the relative standard deviation in concentration that results from this uncertainty when the analyte solution has an absorbance of (a) 1.000 and (b) 2.000.

(a) To convert absorbance to transmittance, we write

$$\log T = -A = -1.00$$
$$T = \text{antilog}\,(-1.00) = 0.100$$

For this instrument, $\sigma_T = k_1 = \pm0.003$ (see first entry in Table 22–3). Substituting this value and $T = 0.100$ into Equation 22–3 yields

$$\frac{\sigma_c}{c} = \frac{0.434}{\log 0.100} \times \frac{\pm0.003}{0.1000} = \pm0.013 \text{ or } \pm1.3\%$$

(b) At $A = 2.000$, $T = \text{antilog}\,(-2.000) = 0.010$

$$\frac{\sigma_c}{c} = \frac{0.434}{\log 0.010} \times \frac{\pm0.003}{0.010} = \pm0.065 \text{ or } \pm6.5\%$$

---

The data plotted as curve $A$ in Figure 22–9 were derived from calculations similar to those in Example 22–3. Note that the relative standard deviation in the concentration passes through a minimum at an absorbance of about 0.5 and rises rapidly when the absorbance is less than about 0.1 or greater than approximately 1.5.

Figure 22–10a is a plot of the relative standard deviation for experimentally determined concentrations as a function of absorbance. It was obtained with the inexpensive spectrophotometer shown in Figure 22–3. The striking similarity between this curve and curve $A$ in Figure 22–9 indicates that the instrument studied is affected by an absolute indeterminate error of about $\pm0.003T$ and that this error is independent of transmittance. It is probable that the source of this uncertainty lies in the limited resolution of the transmittance scale.

Infrared spectrophotometers also exhibit an indeterminate error that is independent of transmittance. With such instruments, the source of this error lies in the thermal detector. Fluctuations in the output of this type of

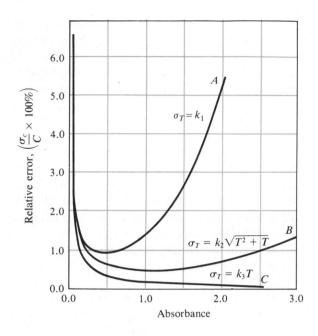

Figure 22–9

Error curves for various categories of instrumental uncertainties.

transducer are independent of the output; indeed, fluctuations are observed even in the absence of radiation. An experimental plot of data from an infrared spectrophotometer is similar in appearance to Figure 22–10a. The curve is displaced upward, however, because of the greater standard deviation associated with infrared measurements.

**Concentration Errors When $\sigma_T = k_2\sqrt{T^2 + T}$.** This type of indeterminate uncertainty is characteristic of the highest-quality spectrophotometers. It has its origin in the so-called *shot noise* that causes the output of photomultipliers and phototubes to fluctuate randomly about a mean value. Equation 22–4 in Table 22–3 describes the effect of shot noise on the relative standard deviation of concentration measurements. A plot of this relationship appears as curve B in Figure 22–9. In obtaining these data, $k_2$ was assumed to be ±0.003, a typical value for high-quality spectrophotometers.

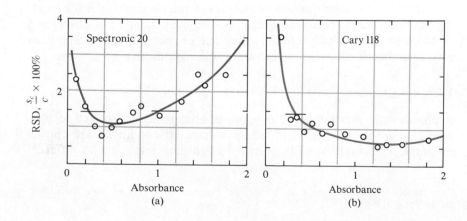

(a)    (b)

Figure 22–10

Experimental curves relating relative concentration uncertainties to absorbance for two spectrophotometers. Data obtained with (a) a Spectronic 20, a low-cost instrument (Figure 22–2), and (b) a Cary 118, a research-quality instrument. (From W. E. Harris and B. Kratochvil, *An Introduction to Chemical Analysis,* p. 384. Philadelphia: Saunders College Publishing, 1981. With permission.)

Figure 22–10b shows an analogous plot of experimental data obtained with an ultraviolet/visible spectrophotometer having performance characteristics similar to those of the instrument shown in Figure 22–4. Note that, in contrast to the less expensive instrument, absorbances of 2.0 or greater can be measured here without serious deterioration in the quality of the data.

**Concentration Errors When $\sigma_T = k_3 T$.** Substitution of $\sigma_T = k_3 T$ into Equation 22–2 reveals that the relative standard deviation in concentration from this type of uncertainty is inversely proportional to the logarithm of the transmittance (Equation 22–5 in Table 22–3). Curve $C$ in Figure 22–9, which is a plot of Equation 22–5, reveals that this type of uncertainty is important at low absorbances (high transmittances) but approaches zero at high absorbances.

At low absorbances, the precision obtained with high-quality double-beam instruments is often described by Equation 22–5. The source of this behavior is failure to position cells reproducibly with respect to the beam during replicate measurements. This position dependence probably is the result of small imperfections in the cell windows, which cause reflective losses and transparency to differ from one area of the window to another.

Evaluation of $k_3$ in Equation 22–5 is possible by comparing the precision of absorbance measurements made in the usual way with measurements in which the cells are left undisturbed at all times with replicate solutions being introduced with a syringe. Experiments of this kind with a high-quality spectrophotometer yielded a value of 0.013 for $k_3$ (footnote 11 page 577). Curve $C$ in Figure 22–9 was obtained by substituting this numerical value into Equation 22–5. Cell positioning errors affect all types of spectrophotometric measurements in which cells are repositioned between measurements.

Fluctuations in source intensity also yield standard deviations that are described by Equation 22–5. This type of behavior is sometimes encountered in inexpensive single-beam instruments that have unstable power supplies and in infrared instruments.

## 22A–5 Photometric and Spectrophotometric Titrations

Photometric and spectrophotometric measurements are useful for locating the equivalence points of titrations.[12] This application of absorption measurements obviously requires that one or more of the reactants or products absorb radiation or that an absorbing indicator be present.

### Titration Curves

A photometric titration curve is a plot of absorbance (corrected for volume change) as a function of titrant volume. If conditions are chosen properly, the curve consists of two straight-line regions with different

---

[12]For further information, see J. B. Headridge, *Photometric Titrations*. New York: Pergamon Press, 1961.

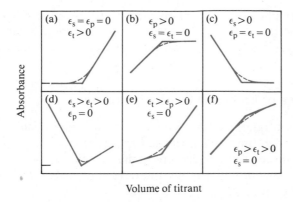

Figure 22–11

Typical photometric titration curves. Molar absorptivities of the substance titrated, the product, and the titrant are $\varepsilon_s$, $\varepsilon_p$, $\varepsilon_t$.

slopes, one occurring at the outset of the titration and the other located well beyond the equivalence-point region; the end point is taken as the intersection of extrapolated linear portions of the two lines.

Figure 22–11 shows typical photometric titration curves. Figure 22–11a is the curve for the titration of a nonabsorbing species with an absorbing titrant that is decolorized by the reaction. An example is the titration of thiosulfate ion with triiodide ion. The titration curve for the formation of an absorbing product from colorless reactants is shown in Figure 22–11b; an example is the titration of iodide ion with a standard solution of iodate ion to form triiodide. The remaining figures illustrate the curves obtained with various combinations of absorbing analytes, titrants, and products.

In order to obtain titration curves with linear portions that can be extrapolated, the absorbing system(s) must obey Beer's law. Furthermore, absorbances must be corrected for volume changes by multiplying the observed absorbance by $(V + v)/V$, where $V$ is the original volume of the solution and $v$ is the volume of added titrant.

## Instrumentation

Photometric titrations are ordinarily performed with a spectrophotometer or a photometer that has been modified so that the titration vessel is held in the light path.[13] After the instrument is set to a suitable wavelength (or an appropriate filter is inserted), the 0% $T$ adjustment is made in the usual way. With radiation passing through the analyte solution to the detector, the instrument is then adjusted to a convenient absorbance reading by varying the source intensity or the detector sensitivity. Ordinarily, no attempt is made to measure the true absorbance since relative values are perfectly adequate for end-point detection. Titration data are then collected without alteration of the instrument settings. The power of the radiation source and the response of the detector must remain constant during a photometric titration. Cylindrical containers are ordinarily used, and care must be taken to avoid any movement of the vessel that might alter the length of the radiation path.

---

[13]Titration flasks and cells for use in the instrument shown in Figure 22–3 are available from the Kontes Manufacturing Corp., Vineland, NJ 08360.

Both filter photometers and spectrophotometers have been employed for photometric titrations. The latter are preferred, however, because their narrower bandwidths enhance the probability of adherence to Beer's law.

### Applications of Photometric Titrations

Photometric titrations often provide more accurate results than a direct photometric determination because the data from several measurements are pooled in determining the end point. Furthermore, the presence of other absorbing species may not interfere since only a change in absorbance is being measured.

One advantage of a photometric end point is that the experimental data are taken well away from the equivalence-point region. Consequently, the equilibrium constants of the reactions need not be as favorable as those required for a titration that depends upon observations near the equivalence point (for example, potentiometric or indicator end points). For the same reason, more dilute solutions may be titrated.

The photometric end point has been applied to all types of reactions.[14] For example, most standard oxidizing agents have characteristic absorption spectra and thus produce photometrically detectable end points. Although standard acids or bases do not absorb, the introduction of acid/base indicators permits photometric neutralization titrations. The photometric end point has also been used to great advantage in titrations with EDTA and other complexing agents. Figure 22–12 illustrates the application of this technique to the successive titration of bismuth(III) and copper(II). At 745 nm, the cations, the reagent, and the bismuth complex formed in the first part of the titration do not absorb but the copper complex does. Thus, the solution exhibits no absorbance until essentially all the bismuth has been titrated. With the first formation of the copper complex, an increase in absorbance occurs. The increase continues until the copper equivalence point is reached. Further reagent additions cause no further absorbance change. Clearly, two well-defined end points result.

The photometric end point has also been adapted to precipitation titrations. The suspended solid product diminishes the radiant power by scattering; titrations are carried to a condition of constant turbidity.

### 22A–6 Spectrophotometric Studies of Complex Ions

Spectrophotometry is a valuable tool for elucidating the composition of complex ions in solution and for determining their formation constants. The power of the technique lies in the fact that quantitative absorption measurements can be performed without disturbing the equilibria under consideration. Although most spectrophotometric studies of complexes involve systems in which a reactant or a product absorbs, nonabsorbing

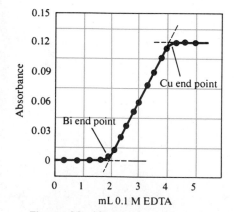

Figure 22–12

Photometric titration curve at 745 nm for 100 mL of a solution that was $2.0 \times 10^{-3}$ M in $Bi^{3+}$ and $Cu^{2+}$. (A. L. Underwood, *Anal. Chem.*, **1954**, *26*, 1322. With permission of the American Chemical Society.)

[14]See, for example, the review by A. L. Underwood in *Advances in Analytical Chemistry and Instrumentation*, C. N. Reilley, Ed., Vol. 3, pp. 31–104. New York: Interscience, 1964.

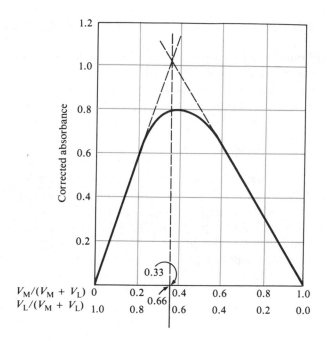

Figure 22–13
Continuous-variation plot for the
$1:2$ complex $ML_2$.

systems can also be investigated successfully. For example, the composition and formation constant for a complex of iron(II) and a nonabsorbing ligand could probably be determined by measuring the color decreases that occur when solutions of the absorbing iron(II) complex of 1,10-phenanthroline are mixed with various amounts of the ligand. For this approach to be successful, the formation constant and the composition of the 1,10-phenanthroline complex have to be known.

The three most common techniques employed for complex-ion studies are (1) the method of continuous variations, (2) the mole-ratio method, and (3) the slope-ratio method.

This section shows how you can determine the composition of a complex in solution without actually isolating the complex as a pure compound.

### The Method of Continuous Variations[15]

In the method of continuous variations, cation and ligand solutions with identical analytical concentrations are mixed in such a way that the total volume and the total moles of reactants in each mixutre is constant but the mole ratio of reactants varies systematically (for example, $1:9$, $8:2$, $7:3$, and so forth). The absorbance of each solution is then measured at a suitable wavelength and corrected for any absorbance the mixture might exhibit if no reaction had occurred. The corrected absorbance is plotted against the volume fraction of one reactant, that is, $V_M/(V_M + V_L)$, where $V_M$ is the volume of the cation solution and $V_L$ that of the ligand. A typical plot is shown in Figure 22–13. A maximum (or minimum if the complex absorbs less than the reactants) occurs at a volume ratio $V_M/V_L$ corresponding to the combining ratio of cation and ligand in the complex. In Figure 22–13, $V_M/(V_M + V_L)$ is 0.33 and $V_L/(V_M + V_L)$ is 0.66; thus,

---

[15]See W. C. Vosburgh and G. R. Cooper, *J. Amer. Chem. Soc.*, **1941**, *63*, 437.

$V_M/V_L$ is 0.33/0.66, which suggests that the complex has the formula $ML_2$.

The curvature of the experimental lines in Figure 22–13 is the result of incompleteness of the complex-formation reaction. A formation constant for the complex can be evaluated from measurements of the deviations from the theoretical straight lines.

### The Mole-Ratio Method

In the mole-ratio method, a series of solutions is prepared in which the analytical concentration of one reactant (usually the cation) is held constant while that of the other is varied. A plot of absorbance versus mole ratio of the reactants is then prepared. If the formation constant is reasonably favorable, two straight lines of different slopes that intersect at a mole ratio that corresponds to the combining ratio in the complex are obtained. Typical mole-ratio plots are shown in Figure 22–14. Note that the ligand of the 1:2 complex absorbs at the wavelength selected so that the slope beyond the equivalence point is greater than zero. We deduce that the uncomplexed cation involved in the 1:1 complex absorbs, because the initial point has an absorbance greater than zero.

Formation constants can be evaluated from the data in the curved portion of mole-ratio plots.

---

### Example 22–4

Derive sufficient equations to permit calculation of the equilibrium concentrations of all the species involved in the 1:2 complex-formation reaction illustrated in Figure 22–14.

Two mass-balance expressions can be written that are based upon the preparatory data. Thus, for the reaction

$$M + 2L \rightleftharpoons ML_2$$

we can write

$$c_M = [M] + [ML_2]$$
$$c_L = [L] + 2[ML_2]$$

where $c_M$ and $c_L$ are the molar concentrations of M and L before reaction occurs. For 1-cm cells, the absorbance of the solution is

$$A = \varepsilon_M[M] + \varepsilon_L[L] + \varepsilon_{ML_2}[ML_2]$$

From the mole-ratio plot, we see that $\varepsilon_M = 0$. Values for $\varepsilon_L$ and $\varepsilon_{ML_2}$ can be obtained from the two straight-line portions of the curve. With one or more measurements of $A$ in the curved region of the plot, sufficient data are available to calculate the three equilibrium concentrations and thus the formation constant.

---

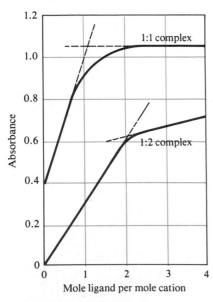

**Figure 22–14**
Mole-ratio plots for a 1:1 and a 1:2 complex. The 1:2 complex is the more stable, as indicated by less curvature near the stoichiometric ratio.

A mole-ratio plot may reveal the stepwise formation of two or more complexes as successive slope changes, provided the complexes have different molar absorptivities and provided the formation constants are sufficiently different from each other.

### The Slope-Ratio Method

This approach is particularly useful for weak complexes but is applicable only to systems in which a single complex is formed. The method assumes (1) that the complex-formation reaction can be forced to completion by a large excess of either reactant and (2) that Beer's law is followed under these circumstances.

Let us consider the reaction in which the complex $M_xL_y$ is formed by the reaction of $x$ moles of the cation M with $y$ moles of a ligand L:

$$x M + y L \rightleftharpoons M_xL_y$$

Mass-balance expressions for this system are

$$c_M = [M] + x[M_xL_y]$$
$$c_L = [L] + y[M_xL_y]$$

where $c_M$ and $c_L$ are the molar analytical concentrations of the two reactants. We now assume that at very high analytical concentrations of L, the equilibrium is shifted far to the right and $[M] \ll x[M_xL_y]$. Under this circumstance, the first mass-balance expression simplifies to

$$c_M = x[M_xL_y]$$

If Beer's law obtains,

$$A_1 = \varepsilon b[M_xL_y] = \varepsilon b c_M/x$$

A plot of absorbance as a function of $c_M$ becomes linear whenever sufficient L is present to satisfy the assumption that $[M] \ll x[M_xL_y]$. The slope of this plot is $\varepsilon b/x$.

When $c_M$ is made very large, we assume that $[L] \ll y[M_xL_y]$, whereupon the second mass-balance equation reduces to

$$c_L = y[M_xL_y]$$

and

$$A_2 = \varepsilon b[M_xL_y] = \varepsilon b c_L/y$$

Again, if our assumptions are valid, a linear plot of $A_2$ versus $c_L$ is observed at high concentrations of M. The slope of this line is $\varepsilon b/y$.

The ratio of the slopes of the two straight lines gives the combining ratio between M and L:

$$\frac{\varepsilon b/x}{\varepsilon b/y} = \frac{y}{x}$$

## 22B    INFRARED ABSORPTION SPECTROSCOPY

Infrared spectrophotometry is one of the most powerful tools available to the chemist for identifying pure organic and inorganic compounds because, with the exception of a few homonuclear molecules, such as $O_2$, $N_2$, and $Cl_2$, all molecular species absorb infrared radiation. Furthermore, with the exception of chiral molecules in the crystalline state, each molecular species has a unique infrared absorption spectrum. Thus, an exact match between the spectrum of a compound of known structure and that of an analyte unambiguously identifies the latter.

Infrared spectroscopy is a less satisfactory tool for quantitative analyses than its ultraviolet and visible counterparts because the narrow peaks that characterize infrared absorption usually lead to deviations from Beer's law. Furthermore, infrared absorbance measurements are considerably less precise. Nevertheless, where modest precision suffices, the unique nature of infrared spectra provides a degree of selectivity in a quantitative measurement that may offset these undesirable characteristics.[16]

### 22B–1    Infrared Absorption Spectra

Vibrational absorption occurs in the infrared region, where the energy of radiation is insufficient to excite electronic transitions. As shown in Figure 22–15, infrared spectra exhibit narrow, closely spaced absorption peaks resulting from transitions among the various vibrational quantum levels. Variations in rotational levels may also give rise to a series of peaks for each vibrational state; with liquid or solid samples, however, rotation is often hindered or prevented, and the effects of these small energy differences are not detected. Thus, a typical infrared spectrum for a liquid, such as that in Figure 22–15, consists of a series of vibrational peaks.

The number of ways a molecule can vibrate is related to the number of atoms, and thus the number of bonds, it contains. For even a simple molecule, the number of possible vibrations is large. For example, *n*-butanal ($CH_3CH_2CH_2CHO$) has 33 vibrational modes, most differing from each other in energy. Not all of these vibrations produce infrared peaks; nevertheless, as shown in Figure 22–15, the spectrum for *n*-butanal is relatively complex.

---

[16]For further reading, see A. L. Smith, in *Treatise on Analytical Chemistry,* 2nd ed., P. J. Elving, E. J. Meehan, and I. M. Kolthoff, Eds., Part I, Vol. 7, Chapter 5. New York: Wiley, 1981.

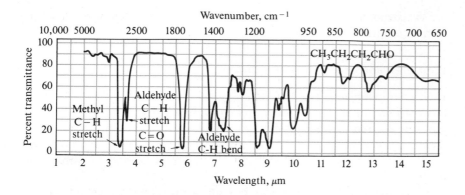

Figure 22–15
Infrared spectrum for *n*-butanal (*n*-butyraldehyde). Note that transmittance rather than absorbance is plotted. [*Catalog of Selected Infrared Spectral Data,* Serial No. 225, Thermodynamics Research Center Data Project Thermodynamics Research Center, Texas A&M University, College Station, TX (loose-leaf data sheets extant, 1964).]

Infrared absorption occurs not only with organic molecules but also with covalently bonded metal complexes, which are generally active in the longer-wavelength infrared region. Infrared spectrophotometric studies have thus provided much useful information about complex metal ions.

## 22B–2 Instruments for Infrared Spectroscopy

Three types of infrared instruments are found in modern laboratories: dispersive spectrometers (or spectrophotometers), Fourier-transform (FTIR) spectrometers, and filter photometers. The first two are used for obtaining complete spectra for qualitative identification, whereas filter photometers are designed for quantitative work. Fourier-transform and filter instruments are nondispersive in the sense that neither employs a grating or prism to disperse radiation into its component wavelengths.

### Dispersive Instruments

With one difference, dispersive infrared instruments are similar in general design to the double-beam (in time) spectrophotometers shown in Figures 21–17 and 22–4. The difference lies in the location of the cell compartment with respect to the monochromator. In ultraviolet/visible instruments, cells are always located between the monochromator and the detector in order to avoid photochemical decomposition, which may occur if samples are exposed to the full power of an ultraviolet or visible source. Infrared radiation, in contrast, is not sufficiently energetic to bring about photodecomposition; thus the cell compartment can be located between the source and the monochromator. This arrangement is advantageous because any scattered radiation generated in the cell compartment is largely removed by the monochromator.

As shown in Section 21A, the components of infrared instruments differ considerably in detail from those in ultraviolet and visible instruments. Thus, infrared sources are heated solids rather than deuterium or tungsten lamps, infrared gratings are much coarser than those required for ultraviolet/visible radiation, and infrared detectors respond to heat rather than photons. Furthermore, the optical components of infrared instruments are constructed from polished solids, such as sodium chloride or potassium bromide.

## Fourier-Transform Spectrometers

Fourier-transform infrared spectrometers offer the advantages of unusually high sensitivity, resolution, and speed of data acquisition (data for an entire spectrum can be obtained in 1 s or less). Offsetting these advantages are the complexity of the instruments and their high cost, because a moderately sophisticated dedicated computer is needed to decode the output data.

Fourier-transform instruments contain no dispersing element, and all wavelengths are detected and measured simultaneously. In order to separate wavelengths, it is necessary to modulate the source signal in such a way that it can subsequently be decoded by a Fourier transformation, a mathematical operation that requires a high-speed computer. The theory of Fourier-transform measurements is beyond the scope of this book.[17]

## Filter Photometers

Infrared photometers designed to monitor the concentration of air pollutants, such as carbon monoxide, nitrobenzene, vinyl chloride, hydrogen cyanide, and pyridine, are now being marketed and used to ensure compliance with regulations established by the Occupational Safety and Health Administration (OSHA). Interference filters, each designed for the determination of a specific pollutant, are available. These transmit narrow bands of radiation in the range of 3 to 14 $\mu$m.

## 22B-3 Qualitative Applications of Infrared Spectrophotometry

An infrared absorption spectrum, even one for a relatively simple compound, often contains a bewildering array of sharp peaks and minima. Peaks useful for the identification of functional groups are located in the shorter-wavelength region of the infrared (from about 2.5 to 8.5 $\mu$m), where the positions of the maxima are only slightly affected by the carbon skeleton to which the groups are attached. Investigation of this region of the spectrum thus provides considerable information regarding the overall constitution of the molecule under investigation. Table 22-4 gives the positions of characteristic maxima for some common functional groups.[18]

Identification of the functional groups in a molecule is seldom sufficient to permit positive identification of the compound, and the entire spectrum

---

[17]For an elementary discussion of the principles of Fourier-transform spectroscopy, see D. A. Skoog and J. J. Leary, *Principles of Instrumental Analysis,* 4th ed., pp. 113–120 and 266–270. Philadelphia: Saunders College Publishing, 1992.

[18]For more detailed information, see N. B. Colthup, *J. Opt. Soc. Amer.,* **1950,** *40,* 397; R. M. Silverstein, G. W. Bassler, and T. C. Morrill, *Spectrometric Identification of Organic Compounds,* 5th ed. New York: Wiley, 1991.

Table 22-4
SOME CHARACTERISTIC INFRARED ABSORPTION PEAKS

| Functional Group | | Absorption Peaks | |
| --- | --- | --- | --- |
| | | Wavenumber, $cm^{-1}$ | Wavelength, $\mu m$ |
| O—H | Aliphatic and aromatic | 3600–3000 | 2.8–3.3 |
| $NH_2$ | Also secondary and tertiary | 3600–3100 | 2.8–3.2 |
| C—H | Aromatic | 3150–3000 | 3.2–3.3 |
| C—H | Aliphatic | 3000–2850 | 3.3–3.5 |
| C≡N | Nitrile | 2400–2200 | 4.2–4.6 |
| C≡C— | Alkyne | 2260–2100 | 4.4–4.8 |
| COOR | Ester | 1750–1700 | 5.7–5.9 |
| COOH | Carboxylic acid | 1740–1670 | 5.7–6.0 |
| C=O | Aldehydes and ketones | 1740–1660 | 5.7–6.0 |
| $CONH_2$ | Amides | 1720–1640 | 5.8–6.1 |
| C=C— | Alkene | 1670–1610 | 6.0–6.2 |
| $\phi$—O—R | Aromatic | 1300–1180 | 7.7–8.5 |
| R—O—R | Aliphatic | 1160–1060 | 8.6–9.4 |

from 2.5 to 15 $\mu m$ must be compared with that of known compounds. Collections of spectra are available for this purpose.[19]

## 22B–4 Quantitative Infrared Photometry and Spectrophotometry

Quantitative infrared absorption methods differ somewhat from their ultraviolet and visible counterparts because of the greater complexity of the spectra, the narrowness of the absorption bands, and the capabilities of the instruments available for measurements in this spectral region.[20]

### Absorbance Measurements

The use of matched cuvettes for solvent and analyte is seldom practical for infrared measurements because of the difficulty in obtaining cells with identical transmission characteristics. Part of this difficulty results from degradation of the transparency of infrared cell windows (typically polished sodium chloride) with use due to attack by traces of moisture in the atmosphere and in samples. Furthermore, path lengths are hard to reproduce because infrared cells are often less than 1 mm thick. Such narrow cells are required to permit the transmission of measurable intensities of infrared radiation through pure samples or through very concentrated solutions of the analyte. Measurements of dilute analyte solutions, as is

---

[19]American Petroleum Institute, *Infrared Spectral Data, A.P.I. Research Project 44.* Pittsburgh: Carnegie Institute of Technology; *Sadtler Standard Spectra.* Philadelphia: Sadtler Research Laboratories.

[20]For an extensive discussion of quantitative infrared analysis, see A. L. Smith, in *Treatise on Analytical Chemistry,* 2nd ed., P. J. Elving, E. J. Meehan, and I. M. Kolthoff, Eds., Part I, Vol. 7, pp. 415–456. New York: Wiley, 1981.

done in ultraviolet or visible spectroscopy, are frequently precluded by the lack of good solvents that transmit over appreciable regions of the infrared spectrum.

For these reasons, a reference absorber is often dispensed with entirely in qualitative infrared work, and the intensity of the radiation passing through the sample is simply compared with that of the unobstructed beam; alternatively, a salt plate may be placed in the reference beam. Either way, the resulting transmittance is ordinarily less than 100%, even in regions of the spectrum where the sample is totally transparent. This ·effect is readily seen by examining the spectrum in Figure 22–15.

For quantitative work, two methods are employed to correct for the scattering and absorption by the solvent and the cell. In the *cell in/cell out* procedure, a single cell is used to obtain successive spectra of the solvent and sample with respect to the unobstructed reference beam. The transmittance of each solution versus the reference beam is then determined at an analyte absorption maximum. These transmittances can be written as

$$T_0 = P_0/P_r$$
$$T_s = P/P_r$$

where $P_r$ is the power of the reference beam and $T_0$ and $T_s$ are the transmittances of the solvent and the sample, respectively, against this reference. If $P_r$ remains constant during the two measurements, the transmittance of the sample with respect to the solvent can be obtained by division of one equation by the other:

$$T = T_s/T_0 = P/P_0$$

An alternative way of obtaining $P_0$ and $T$ is the *base-line* method, in which the solvent transmittance is assumed to be constant or at least to change linearly between the shoulders of the absorption peak. This technique is demonstrated in Figure 22–16.

### Deviations from Beer's Law

Instrumental deviations from Beer's law are common in infrared spectroscopy because the source intensity and detector sensitivity of most infra-

These calculations can be carried out quickly and accurately with a computerized Fourier transform IR spectrophotometer.

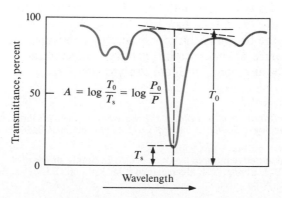

Figure 22–16
Base-line method for determination of absorbance.

$$A = \log \frac{T_0}{T_s} = \log \frac{P_0}{P}$$

red instruments are so low that relatively wide slits are required to produce an output signal that can be measured accurately. As a consequence of these limitations, the bandwidths of radiation from a monochromator are frequently of the same order of magnitude as the widths of the typically narrow infrared absorption peak. As noted on page 527, this combination of circumstances usually leads to a nonlinear relationship between absorbance and concentration. Calibration curves with significant curvature are therefore common in quantitative infrared spectroscopy.

## Applications of Quantitative Infrared Spectroscopy

Infrared spectrophotometry offers the potential for determining an unusually large number of substances because nearly all molecular species absorb in the infrared region. Moreover, the uniqueness of an infrared spectrum provides a degree of specificity that is matched or exceeded by relatively few other analytical methods. This specificity has particular application to the analysis of mixtures of closely related organic compounds.

The recent proliferation of government regulations on atmospheric contaminants has demanded the development of sensitive, rapid, and highly specific methods for a variety of chemical compounds. Infrared absorption procedures appear to meet this need better than any other single analytical tool.

Table 22–5 illustrates the variety of atmospheric pollutants that can be determined with a simple, portable filter photometer equipped with a separate interference filter for each analyte species. Of the more than 400 chemicals for which maximum tolerable limits have been set by OSHA, half or more have absorption characteristics that make them amenable to determination by infrared photometry or spectrophotometry. Obviously, peak overlaps are to be expected with so many compounds absorbing; nevertheless, the method does provide a moderately high degree of selectivity.

**Table 22–5**
**EXAMPLES OF INFRARED VAPOR ANALYSIS FOR OSHA COMPLIANCE**

| Compound | Allowable Exposure, ppm* | Wavelength, $\mu$m | Minimum Detectable Concentration, ppm† |
|---|---|---|---|
| Carbon disulfide | 10 | 4.54 | 0.5 |
| Chloroprene | 10 | 11.4 | 4 |
| Diborane | 0.1 | 3.9 | 0.05 |
| Ethylenediamine | 10 | 13.0 | 0.4 |
| Hydrogen cyanide | 10 | 3.04 | 0.4 |
| Methyl mercaptan | 0.5 | 3.38 | 0.4 |
| Nitrobenzene | 1 | 11.8 | 0.2 |
| Pyridine | 5 | 14.2 | 0.2 |
| Sulfur dioxide | 2 | 8.6 | 0.5 |
| Vinyl chloride | 5 | 10.9 | 0.3 |

Courtesy of The Foxboro Company, Foxboro, MA 02035.

*1986 OSHA exposure limits for 8-hr weighted average.

†For 20.25-m cell.

## 22C   AUTOMATION OF PHOTOMETRIC AND SPECTROPHOTOMETRIC METHODS

The first fully automated instrument for chemical analysis (the Technicon AutoAnalyzer®) appeared on the market in 1957. This instrument was designed to fulfill the needs of clinical laboratories, where blood and urine samples are routinely analyzed for a dozen or more chemical species. The number of such analyses demanded by modern medicine is enormous[21]; the need to keep their cost at a reasonable level is obvious. These two considerations motivated the development of analytical systems that perform several analyses simultaneously with a minimum input of human labor. The use of automatic instruments has spread from clinical laboratories to laboratories for the control of industrial processes and the routine determination of a wide spectrum of species in air, water, soils, and pharmaceutical and agricultural products.[22] In the majority of these applications, the analyses are completed by a photometric or fluorometric measurement.

### 22C–1  Types of Automatic Analytical Systems

Automatic analytical instruments are of two general types, *discrete* and *continuous*; combinations of the two are sometimes encountered. In a discrete instrument, individual samples are maintained as separate entities and kept in separate vessels throughout such unit operations as sampling, definition of the sample (measurement of its weight or volume), dilution, reagent addition, mixing, centrifugation, and transportation to the measuring device. Discrete systems frequently require the use of robots.

In continuous systems, the sample becomes a part of a flowing stream in which the several unit operations of the analysis take place as the sample is carried from the injection point to a flow-through measuring unit and finally to waste.

This section is devoted to continuous flow methods, which most commonly employ photometric and spectrophotometric measurements. Two types of continuous procedures are encountered: *segmented-flow methods*, in which the analytical stream is divided into individual segments by the periodic injection of air bubbles, and *nonsegmented-flow procedures*, in which the analytical stream is unbroken. A nonsegmented-flow analysis is generally termed a *flow-injection analysis*.

### 22C–2  Segmented-Flow Methods

Figure 22–17 is a diagram of a single-channel AutoAnalyzer® used for the analysis of one of the constituents of blood. Ordinarily, several of these

---

[21]For example, in 1976, about 100 million samples were analyzed in domestic clinical laboratories. See Snyder *et al., Anal. Chem.,* **1976,** *48,* 1942A.

[22]For an extensive treatment of automatic analysis, see J. K. Foreman and P. B. Stockwell, *Automatic Chemical Analysis.* New York: Wiley, 1975; V. Cerda and G. Ramis, *An Introduction to Laboratory Automation.* New York: Wiley, 1990.

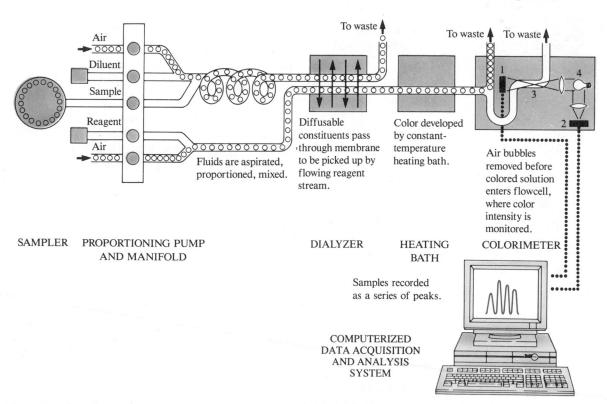

1. Sample photocell
2. Reference photocell
3. Flowcell
4. Light source

To waste

To waste    To waste

Air

Diluent

Sample

Reagent

Air

Fluids are aspirated, proportioned, mixed.

Diffusable constituents pass through membrane to be picked up by flowing reagent stream.

Color developed by constant-temperature heating bath.

Air bubbles removed before colored solution enters flowcell, where color intensity is monitored.

SAMPLER    PROPORTIONING PUMP AND MANIFOLD

DIALYZER    HEATING BATH

COLORIMETER

Samples recorded as a series of peaks.

COMPUTERIZED DATA ACQUISITION AND ANALYSIS SYSTEM

**Figure 22–17**

A single-channel Technicon AutoAnalyzer® system. (Reproduced with permission of Technicon Instruments Corporation. Technicon, SMA, and AutoAnalyzer are registered trademarks of Technicon Instruments Corporation.)

channels are arranged in parallel, each being dedicated to a single type of determination. Multichannel instruments have a sampler module that dilutes the sample and partitions it into aliquots for introduction in each channel.

## The Sample and Reagent Transport System

The heart of a continuous-flow instrument, be it segmented or nonsegmented, is the *peristaltic proportionating pump system*. A peristaltic pump is a device in which a fluid (liquid or gas) is squeezed through plastic tubing by metal rollers mounted on a pair of parallel continuous chains or on a spindle. A spring-loaded platen pinches the tubing against one of the rollers at all times, thus forcing a continuous flow of fluid through the tubing. Generally, peristaltic pumps are driven by a constant-speed motor and are capable of delivering remarkably reproducible volumes. The volume is controlled by the inside diameter of the tubing and by the pumping time. A wide variety of tube sizes permit flow rates as low as 0.015 mL/min and as high as 3.9 mL/min.

Ordinarily, the rollers in continuous-flow instruments are long enough to accommodate several tubes (as many as 28) simultaneously.

### Segmentation

An important feature of the apparatus shown in Figure 22–17 is that provision is made to pump regularly spaced air bubbles into each stream. The spacing of the bubbles is such that every sample is carried through the system in several successive fluid segments. Before final measurement, the stream passes through a debubbler, which removes the bubbles and recombines the segments. Segmenting tends to maintain a sharp concentration profile at the leading and following edges of each sample. In the absence of bubbles, tailing along the tube walls occurs and enhances the probability of contaminating the following sample.

A second purpose of the bubbles is to promote mixing of sample and reagent. As shown in Figure 22–17, each segment is inverted as it rises and falls through the turns of a mixing coil; maximum mixing efficiency occurs when the length of each segment is less than half the coil diameter.

### Separations in Continuous-Flow Analyzers

The instrument shown in Figure 22–17 contains a dialysis module in which the small analyte ions and molecules diffuse through a membrane into a segmented stream of a colorimetric reagent. Larger molecules remain in the original stream and are carried to waste. The membrane is supported between two Lucite or Kel-F plates in which congruent channels have been cut to accommodate the two flows. The transfer of smaller species through this membrane is quite incomplete (less than 50%). Thus, successful quantitative analysis requires close control of temperature and flow rates for both samples and standards.

Another common separation technique used in continuous-flow methods is extraction. Two immiscible liquids are brought together in a mixing coil and then passed into a separator, which is a horizontal glass tube containing an inner tube that partially removes either the lighter or the heavier part of the stream and carries it to the detector module. Separation is again quite incomplete, but the lack of completeness is of no consequence because unknowns and standards are treated in an identical way.

It is important to appreciate that the timing sequences in automatic instruments are sufficiently reproducible so that loss of precision does not accompany incomplete separation or incomplete reactions, as is often the case with manual operations.

### Detectors

Various types of detectors are used with continuous-flow analyzers, including atomic absorption and emission instruments, fluorometers, elec-

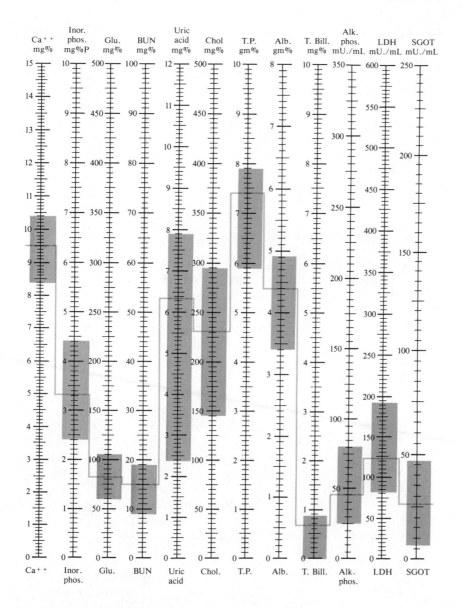

Figure 22–18
Readout from a 12-channel Technicon SMA® system. (Reproduced
with permission of Technicon
Instruments Corporation.)

trochemical systems, refractometers, spectrophotometers, and photometers, the last being by far the most common. The detector shown in Figure
22–17 is a twin-beam photometer in which part of the radiation passes
through a tubular flow cell to a photoelectric detector; after collimation,
the second beam falls on a reference photocell.

Figure 22–18 shows the computer-generated readout from a 12-channel
AutoAnalyzer® employed for the routine analysis of blood. The shaded
areas show the range of concentrations considered normal for the population. A modern AutoAnalyzer® for routine blood screening is capable of
analyzing 200 serum samples per day for 20 constituents. That is, the
instrument performs 4000 separate analyses each day. Interspersed with
the 200 samples are typically 100 standards for calibration.

Blood serum is the clear fluid remaining
after clotting has occurred.

### Applications of Segmented-Flow Analyzers

As noted earlier, a major area of application of segmented-flow analyzers has been in clinical laboratories, where these instruments are the work-horses for routine clinical tests. In addition, however, there now exists a vast literature having to do with the applications of segmented-flow instruments in environmental, pharmaceutical, food, agricultural, and metallurgy areas, for both laboratory and process-monitoring uses. Thus, it has been estimated that continuous-flow automated methods are being used in industry for the determination of more than 300 species in more than 1000 types of sample matrices.[23]

### 22C–3 Flow-Injection Methods

In flow-injection analysis (FIA), the newest continuous-flow method, highly precise sample volumes are introduced into an unsegmented stream. The first description of an unsegmented stream as a medium in which to perform analyses appeared more than a decade ago. By now, it has become apparent that flow injection is an important new way of carrying out automatic wet chemical analyses rapidly and efficiently.[24] Equipment for flow-injection analysis is relatively simple and is marketed by several manufacturers.

Figure 22–19a shows the details of a typical flow-injection instrument. As in segmented-flow analyzers, sample and reagents are transported through flexible plastic tubing by the action of a peristaltic pump. The tubing used in flow-injection methods is typically 0.5 mm in diameter, which is narrower than that used for segmented-flow procedures. In some flow-injection instruments, depulsed positive-displacement pumps are used in place of peristaltic models to reduce detector noise.

All of the various flow-type detectors of segmented-flow analyzers are equally applicable to flow-injection measurements. In addition, separations by extraction or dialysis are possible.

Many modern clinical analyzers are based on the principles of flow-injection analysis.

### Injectors

Sample sizes for flow-injection analysis range from 5 to 200 $\mu$L, with 10 to 30 $\mu$L being typical for most applications. For a successful analysis, it is vital that the sample solution be injected rapidly as a pulse, or plug, of liquid; in addition, the injections must not disturb the flow of the carrier stream. The requirements here are significantly more rigorous than is the case with air-segmented flow because the sample is more likely to spread in the absence of bubbles. To date, the most satisfactory injector systems are based upon sampling loops similar to those used in chromatography (see, for example, Figure 28–4). The method of operation of a sampling

---

[23]A. Coneta, W. T. Dorsheimer, and M. J. F. DuCros, *Amer. Lab.*, **1981,** *13,* 116.

[24]For a monograph dealing with flow-injection analysis, see J. Ruzicka and E. H. Hansen, *Flow Injection Analysis,* 2nd ed. New York: Wiley, 1988. For brief reviews of the method, see J. Ruzicka, *Anal. Chem.*, **1983,** *55,* 1040A; K. K. Stewart, *Anal. Chem.*, **1983,** *55,* 931A; C. B. Ranger, *Anal. Chem.*, **1981,** *53,* 20A; and D. Betteridge, *Anal. Chem.*, **1978,** *50,* 832A.

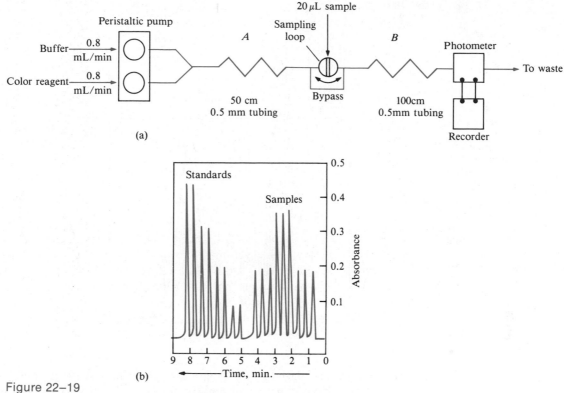

**Figure 22–19**

(a) Flow-injection apparatus for determining calcium in water by formation of a colored complex with *o*-cresolphthalein complexone at pH 10. All tubing had an inside diameter of 0.5 mm. *A* and *B* are reaction coils having the indicated lengths. (b) Recorder output. The four sets of curves at left are for duplicate injections of standards containing 5, 10, 15, and 20 ppm calcium. (From E. H. Hansen, J. Ruzicka, and A. K. Ghose, *Anal. Chim. Acta*, **1978**, *100*, 151. With permission.)

loop can be seen by reference to Figure 22–19. With the valve of the loop in the position shown, reagents flow through the bypass. When a sample has been injected into the loop and the value turned 90 deg, the sample enters the flow as a single, well-defined zone. For all practical purposes, flow through the bypass ceases with the valve in this position because the diameter of the sample loop is significantly greater than that of the bypass tubing.

## The Principles of Flow-Injection Analysis

Immediately after injection, the sample zone in a flow-injection apparatus has a rectangular concentration profile. As the sample moves through the tubing, band broadening, or *dispersion*, takes place and the zone profile then has the shape shown in Figure 22–20c. The width of the zone depends upon the distance between the injector and the detector as well as on the pumping rate.

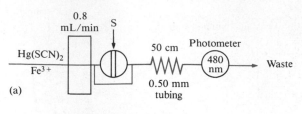

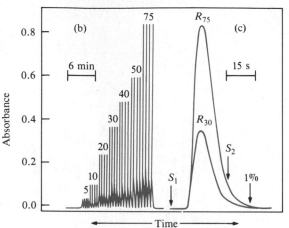

**Figure 22–20**
Flow-injection determination of chloride: (a) flow diagram; (b) recorder readout for quadruplicate runs on standards containing 5 to 75 ppm chloride ion; (c) fast scan of two of the standards to demonstrate the low analyte carryover (less than 1%) from run to run. Note that the point marked 1% corresponds to where the response would just begin for a sample injected at time $S_2$. (From J. Ruzicka and E. H. Hansen, *Flow Injection Methods,* p. 8. New York: Wiley, 1981. Reprinted by permission.)

Figure 22–20a is a flow diagram of the simplest of all flow-injection systems. A colorimetric reagent for chloride ion that consists of a mixture of mercury(II) thiocyanate and iron(III) ion is pumped directly into the sampling valve and thence through a 50-cm reactor coil, where the reagent diffuses into the sample plug and produces a colored product by the sequence of reactions

$$Hg(SCN)_2(aq) + 2\ Cl^- \rightleftharpoons HgCl_2(aq) + 2\ SCN^-$$
$$Fe^{3+} + SCN^- \rightleftharpoons \underset{\text{Red}}{Fe(SCN)^{2+}}$$

The recorder output for a series of standards containing from 5 to 75 ppm of chloride are shown in Figure 22–20b. Note that four injections of each standard were made to demonstrate the reproducibility of the system. This process took 23 min, which corresponds to a sampling rate of 130 samples/hr.

The two curves in Figure 22–20c are scans of samples containing 30 ($R_{30}$) and 75 ($R_{75}$) ppm of chloride. These curves, in contrast to those in Figure 22–20b, were obtained with a high-speed recorder so that their shapes could be studied. The two plots demonstrate that less than 1% of the analyte is present in the flow cell after 28 s, the time of the next injection ($S_2$). This system has been successfully used for the routine determination of chloride ion in brackish and waste waters as well as in serum samples.

Figure 22–19 illustrates a more complicated system for the colorimetric determination of calcium in serum, milk, and drinking water. A borax buffer and a color-forming reagent are combined in a 50-cm mixing coil

(A) prior to sample injection. The recorder output for three samples in triplicate and four standards in duplicate is shown in Figure 22–19b.

### Advantages of Flow-Injection Measurements

Flow-injection equipment differs markedly from its segmented-flow counterpart in that no air-bubble segmentation is used. Before the advent of flow-injection methods, it was believed that air bubbles were vital to the success of continuous-flow techniques, serving to prevent excess sample dispersion, to promote turbulent mixing, and to scrub the walls of the conduit and thus prevent cross-contamination between samples. In fact, however, excess dispersion or dilution and cross-contamination are nearly completely absent in a properly designed system without air bubbles. In addition, the mixing of reagents and sample occurs rapidly, albeit by a different mechanism.

Flow-injection measurements have several important advantages over segmented-flow procedures, including (1) higher analysis rates (typically 100 to 300 samples/hr), (2) enhanced response times (often less than 1 min between sample injection and recorder response), (3) lower solvent costs (because of the narrower tubing), (4) much more rapid startup and shutdown times (less than 5 min for each), and (5) except for the injection system, simpler and more flexible equipment. The last two advantages are of particular importance because they make it feasible and economic to apply automated measurements to a relatively few nonroutine samples. In other words, continuous-flow methods are no longer restricted to situations where the number of samples is large and the analytical method highly routine. A final advantage that is common to both flow-injection and segmented-flow methods relative to classical bulk techniques is the very small amounts of waste reagents that they generate.

## 22D   QUESTIONS AND PROBLEMS

**22–1.** Define
  *(a) chromophore.
  (b) standard-addition method.
  *(c) noise in a spectrophotometric measurement.
  (d) photometric titration.
  *(e) charge-transfer absorption.
  (f) the method of continuous variation.
*22–2. Explain the difference between a segmented-flow and a flow-injection method.
22–3. Explain the difference between a discrete and a continuous system for automatic analyses.
22–4. Identify the characteristics of infrared absorbance measurements that tend to cause departure from Beer's law.
*22–5. Explain how the composition of an absorbing complex is determined by the mole-ratio method.
22–6. Explain how the composition of an absorbing complex is determined by the method of continuous variation.

*22–7. Sketch a photometric titration curve for Fe(III) with $SCN^-$ ion when a photometer with a green filter is used to collect data. Why is a green filter used?
22–8. Sketch a photometric titration curve for the titration of $Sn^{2+}$ with $MnO_4^-$. What color filter should be used for this titration? Explain.
22–9. Why cannot the excellent detector systems available for ultraviolet and visible radiation be used with infrared radiation?
*22–10. Why is the sample compartment located between the monochromator and the detector in an ultraviolet spectrophotometer but between the source and the monochromator in an infrared spectrophotometer?
22–11. A 4.97-g petroleum specimen was decomposed by wet-ashing and subsequently diluted to 500 mL in a volumetric flask. Cobalt was determined by treating 25.00-mL aliquots of this diluted solution as

follows:

| Reagent Volume | | | |
|---|---|---|---|
| **Co(II), 3.00 ppm** | **Ligand** | **H₂O** | Absorbance |
| 0.00 | 20.00 | 5.00 | 0.398 |
| 5.00 | 20.00 | 0.00 | 0.510 |

Assume that the Co(II)/ligand chelate obeys Beer's law, and calculate the percentage of cobalt in the original sample.

*22–12. A two-tablet sample of vitamin/mineral supplement weighing 6.08 g was wet-ashed to eliminate organic matter and then diluted to 1.00 L. Two 10.00-mL aliquots were then analyzed. Calculate the average weight of iron in each tablet, based upon the following information:

| Reagent Volume | | | |
|---|---|---|---|
| **Fe(III), 1.00 ppm** | **Ligand** | **H₂O** | Absorbance |
| 0.00 | 25.00 | 15.00 | 0.492 |
| 15.00 | 25.00 | 0.00 | 0.571 |

22–13. The standard deviation in transmittance for a particular instrument is 0.006, regardless of the magnitude of the transmittance. Calculate the relative standard deviation in concentration due to this source when

*(a) $T = 0.015$      (d) $A = 0.921$
*(b) $A = 0.334$      (e) $T = 0.804$
*(c) $\% T = 64.8$    (f) $\% T = 50.0$

*22–14. The meter of an inexpensive spectrophotometer has a 5-in. scale scribed in linear units from 0 to 100% $T$. The scale, which limits the precision of the instrument, can be read to about ±0.5% $T$. Calculate the relative precision of concentration determinations for an absorbance of (a) 0.020, (b) 0.050, (c) 0.100, (d) 0.400, (e) 0.800, (f) 1.200, (g) 2.000.

22–15. Ethylenediaminetetraacetic acid abstracts bismuth(III) from its thiourea complex:

$$Bi(tu)_6^{3+} + H_2Y^{2-} \rightarrow BiY^- + 6\ tu + 2\ H^+$$

Predict the shape of a photometric titration curve based on this process, given that the Bi(III)/thiourea complex is the only species in the system that absorbs at 465 nm, the wavelength selected for the analysis.

*22–16. The accompanying data (1.00-cm cells) were obtained for the spectrophotometric titration of 10.00 mL of Pd(II) with $2.44 \times 10^{-4}$ M Nitroso R (O. W. Rollins and M. M. Oldham, *Anal. Chem.*, **1971**, *43*, 262):

| Volume of Nitroso R, mL | $A_{500}$ |
|---|---|
| 0 | 0 |
| 1.00 | 0.147 |
| 2.00 | 0.271 |
| 3.00 | 0.375 |
| 4.00 | 0.371 |
| 5.00 | 0.347 |
| 6.00 | 0.325 |
| 7.00 | 0.306 |
| 8.00 | 0.289 |

Calculate the concentration of the Pd(II) solution, given that the ligand-to-cation ratio in the colored product is 2 : 1.

22–17. Solutions containing species A and B obey Beer's law over an extensive concentration range. Molar absorptivity data for each are as follows:

| $\lambda$, nm | $\varepsilon_A$ | $\varepsilon_B$ |
|---|---|---|
| 400 | 893 | 0.00 |
| 420 | 940 | 0.00 |
| 440 | 955 | 0.00 |
| 460 | 936 | 24.6 |
| 480 | 874 | 102 |
| 500 | 795 | 185 |
| 520 | 691 | 289 |
| 540 | 574 | 428 |
| 560 | 440 | 622 |
| 580 | 297 | 980 |
| 600 | 167 | 1178 |
| 620 | 39.7 | 1692 |
| 640 | 3.45 | 1742 |
| 660 | 0.00 | 1806 |
| 680 | 0.00 | 1809 |
| 700 | 0.00 | 1757 |

*(a) A solution containing both solutes has an absorbance of 0.360 in a 1.00-cm cell at 540 nm. What is the concentration of B in this solution if $[A] = 5.00 \times 10^{-4}$ M?

*(b) A solution containing both species has an absorbance of 0.510 both at 440 nm and at 600 nm when measured in a 1.00-cm cell. Calculate the concentrations of A and B.

(c) Construct an absorption spectrum for a $5.80 \times 10^{-4}$ M solution of A in a 1.00-cm cell.

(d) Construct an absorption spectrum for a $3.25 \times 10^{-4}$ M solution of B in a 1.00-cm cell.

(e) Construct an absorption spectrum for a solution that is $5.80 \times 10^{-4}$ M in A and $3.25 \times 10^{-4}$ M in B in a 1.00-cm cell.

*22-18. A. J. Mukhedkar and N. V. Deshpande (*Anal. Chem.*, **1963**, *35*, 47) report on a simultaneous determination for cobalt and nickel based upon absorption by their 8-quinolinol complexes. Molar absorptivities are $\varepsilon_{Co} = 3529$ and $\varepsilon_{Ni} = 3228$ at 365 nm and $\varepsilon_{Co} = 428.9$ and $\varepsilon_{Ni} = 0$ at 700 nm. Calculate the concentration of nickel and cobalt in each of the following solutions (1.00-cm cells):

| Solution | $A_{365}$ | $A_{700}$ |
|---|---|---|
| *1 | 0.724 | 0.0710 |
| 2 | 0.614 | 0.0744 |
| 3 | 0.693 | 0.0460 |

22-19. Solutions of P and of Q individually obey Beer's law over a large concentration range. Spectral data for these species in 1.00-cm cells are

| | Absorbance | |
|---|---|---|
| $\lambda$, nm | $8.55 \times 10^{-5}$ M P | $2.37 \times 10^{-4}$ M Q |
| 400 | 0.078 | 0.550 |
| 420 | 0.087 | 0.592 |
| 440 | 0.096 | 0.599 |
| 460 | 0.102 | 0.590 |
| 480 | 0.106 | 0.564 |
| 500 | 0.110 | 0.515 |
| 520 | 0.113 | 0.433 |
| 540 | 0.116 | 0.343 |
| 560 | 0.126 | 0.255 |
| 580 | 0.170 | 0.170 |
| 600 | 0.264 | 0.100 |
| 620 | 0.326 | 0.055 |
| 640 | 0.359 | 0.030 |
| 660 | 0.373 | 0.030 |
| 680 | 0.370 | 0.035 |
| 700 | 0.346 | 0.063 |

(a) Plot an absorption spectrum for a solution that is $8.55 \times 10^{-5}$ M in P and $2.37 \times 10^{-4}$ M in Q.
(b) Calculate the absorbance (1.00-cm cells) at 440 nm of a solution that is $4.00 \times 10^{-5}$ M in P and $3.60 \times 10^{-4}$ M in Q.
(c) Calculate the absorbance (1.00-cm cells) at 620 nm for a solution that is $1.61 \times 10^{-4}$ M in P and $7.35 \times 10^{-4}$ M in Q.

*22-20. Use the data in the previous problem to calculate the molar concentration of P and Q in each of the following solutions:

| | $A_{440}$ | $A_{620}$ | | $A_{440}$ | $A_{620}$ |
|---|---|---|---|---|---|
| *(a) | 0.357 | 0.803 | (d) | 0.910 | 0.338 |
| (b) | 0.830 | 0.448 | *(e) | 0.480 | 0.825 |
| *(c) | 0.248 | 0.333 | (f) | 0.194 | 0.315 |

22-21. The indicator HIn has an acid dissociation constant of $4.80 \times 10^{-6}$ at ordinary temperatures. The accompanying absorbance data are for $8.00 \times 10^{-5}$ M solutions of the indicator measured in 1.00-cm cells in strongly acidic and strongly alkaline media.

| | Absorbance | |
|---|---|---|
| $\lambda$, nm | pH 1.00 | pH 13.00 |
| 420 | 0.535 | 0.050 |
| 445 | 0.657 | 0.068 |
| 450 | 0.658 | 0.076 |
| 455 | 0.656 | 0.085 |
| 470 | 0.614 | 0.116 |
| 510 | 0.353 | 0.223 |
| 550 | 0.119 | 0.324 |
| 570 | 0.068 | 0.352 |
| 585 | 0.044 | 0.360 |
| 595 | 0.032 | 0.361 |
| 610 | 0.019 | 0.355 |
| 650 | 0.014 | 0.284 |

Estimate the wavelength at which absorption by the indicator becomes independent of pH (that is, the isosbestic point).

*22-22. Calculate the absorbance (1.00-cm cells) at 450 nm of a solution in which the total molar concentration of the indicator described in Problem 22-21 is $8.00 \times 10^{-5}$ and the pH is *(a) 4.92, (b) 5.46, *(c) 5.93, (d) 6.16.

22-23. What is the absorbance at 595 nm (1.00-cm cells) of a solution that is $1.25 \times 10^{-4}$ M in the indicator of Problem 22-21 and has a pH of *(a) 5.30, (b) 5.70, *(c) 6.10?

*22-24. Several buffer solutions were made $1.00 \times 10^{-4}$ M in the indicator of Problem 22-21. Absorbance data (1.00-cm cells) are

| Solution | $A_{450}$ | $A_{595}$ |
|---|---|---|
| *A | 0.344 | 0.310 |
| B | 0.508 | 0.212 |
| *C | 0.653 | 0.136 |
| D | 0.220 | 0.380 |

Calculate the pH of each solution.

22-25. Construct an absorption spectrum for an $8.00 \times 10^{-5}$ M solution of the indicator of Problem 22-21 when measurements are made with 1.00-cm cells and

*(a) $\dfrac{[HIn]}{[In^-]} = 3$  (b) $\dfrac{[HIn]}{[In^-]} = 1$  (c) $\dfrac{[HIn]}{[In^-]} = \dfrac{1}{3}$

*22-26. Molar absorptivity data for the cobalt and nickel complexes with 2,3-quinoxalinedithiol are $\varepsilon_{Co} = $

36,400 and $\varepsilon_{Ni} = 5520$ at 510 nm and $\varepsilon_{Co} = 1240$ and $\varepsilon_{Ni} = 17,500$ at 656 nm.

A 0.425-g sample was dissolved and diluted to 50.0 mL. A 25.0-mL aliquot was treated to eliminate interferences; after addition of 2,3-quinoxalinedithiol, the volume was adjusted to 50.0 mL. This solution had an absorbance of 0.446 at 510 nm and 0.326 at 656 nm in a 1.00-cm cell. Calculate the parts per million of cobalt and nickel in the sample.

22–27. The chelate $ZnQ_2^{2-}$ exhibits a maximum absorption at 480 nm. When the chelating agent is present in at least a fivefold excess, the absorbance is dependent only upon the molar concentration of Zn(II) and obeys Beer's law over a large range. Neither $Zn^{2+}$ nor $Q^{2-}$ absorbs at 480 nm. A solution that is $2.30 \times 10^{-4}$ M in $Zn^{2+}$ and $8.60 \times 10^{-3}$ M in $Q^{2-}$ has an absorbance of 0.690 in a 1.00-cm cell at 480 nm. Under the same conditions, a solution that is $2.30 \times 10^{-4}$ M in $Zn^{2+}$ and $5.00 \times 10^{-4}$ M in $Q^{2-}$ has an absorbance of 0.540. Calculate the numerical value of $K_f$ for the process

$$Zn^{2+} + 2 Q^{2-} \rightleftharpoons ZnQ_2^{2-}$$

*22–28. The sodium salt of 2-quinizarinsulfonic acid (NaQ) forms a complex with aluminum(III) that absorbs strongly at 560 nm (E. G. Owens and J. H. Yoe, *Anal. Chem.*, **1959**, *31*, 385). Use the accompanying data to establish the combining ratio between cation and ligand. In all solutions, $c_{Al} = 3.7 \times 10^{-5}$ M, and all measurements were made in 1.00-cm cells.

| Molar Ligand Concentration | Absorbance |
|---|---|
| $1.00 \times 10^{-5}$ | 0.131 |
| $2.00 \times 10^{-5}$ | 0.265 |
| $3.00 \times 10^{-5}$ | 0.396 |
| $4.00 \times 10^{-5}$ | 0.468 |
| $5.00 \times 10^{-5}$ | 0.487 |
| $6.00 \times 10^{-5}$ | 0.498 |
| $8.00 \times 10^{-5}$ | 0.499 |
| $1.00 \times 10^{-4}$ | 0.500 |

22–29. The accompanying data (1.00-cm cells) were obtained in a slope-ratio investigation of the product formed between Ni(II) and 1-cyclopentene-1-dithiocarboxylic acid (CDA):

| $c_{CDA} = 1.00 \times 10^{-3}$ M | | $c_{Ni} = 1.00 \times 10^{-3}$ M | |
|---|---|---|---|
| $c_{Ni}$, M | $A_{530}$ | $c_{CDA}$, M | $A_{530}$ |
| $5.00 \times 10^{-6}$ | 0.051 | $9.00 \times 10^{-6}$ | 0.031 |
| $1.20 \times 10^{-5}$ | 0.123 | $1.50 \times 10^{-5}$ | 0.051 |
| $3.50 \times 10^{-5}$ | 0.359 | $2.70 \times 10^{-5}$ | 0.092 |
| $5.00 \times 10^{-5}$ | 0.514 | $4.00 \times 10^{-5}$ | 0.137 |
| $6.00 \times 10^{-5}$ | 0.616 | $6.00 \times 10^{-5}$ | 0.205 |
| $7.00 \times 10^{-5}$ | 0.719 | $7.00 \times 10^{-5}$ | 0.240 |

Calculate the ligand-to-cation ratio in this complex.

*22–30. Iron(II) forms a chelate with the ligand P. Evaluate the composition of $FeP_n$ from the accompanying data (1.00-cm cells):

| Fe(II) = $2.00 \times 10^{-3}$ M | | P = $2.00 \times 10^{-3}$ M | |
|---|---|---|---|
| Molar Concentration of P | A | Molar Concentration of Fe | A |
| $4.00 \times 10^{-6}$ | 0.025 | $6.00 \times 10^{-6}$ | 0.113 |
| $1.50 \times 10^{-5}$ | 0.094 | $1.00 \times 10^{-5}$ | 0.189 |
| $3.00 \times 10^{-5}$ | 0.189 | $1.75 \times 10^{-5}$ | 0.330 |
| $5.00 \times 10^{-5}$ | 0.314 | $3.00 \times 10^{-5}$ | 0.566 |
| $7.00 \times 10^{-5}$ | 0.440 | $4.00 \times 10^{-5}$ | 0.754 |

22–31. The accompanying absorbance data (1.00-cm cells) were recorded for a continuous-variation study of the colored product formed between cadmium(II) and the complexing reagent R:

| | Reactant Volume, mL | | |
|---|---|---|---|
| Solution | $1.25 \times 10^{-4}$ M $Cd^{2+}$ | $1.25 \times 10^{-4}$ M R | $A_{390}$ |
| 0 | 10.00 | 0.00 | 0.000 |
| 1 | 9.00 | 1.00 | 0.174 |
| 2 | 8.00 | 2.00 | 0.353 |
| 3 | 7.00 | 3.00 | 0.530 |
| 4 | 6.00 | 4.00 | 0.672 |
| 5 | 5.00 | 5.00 | 0.723 |
| 6 | 4.00 | 6.00 | 0.673 |
| 7 | 3.00 | 7.00 | 0.537 |
| 8 | 2.00 | 8.00 | 0.358 |
| 9 | 1.00 | 9.00 | 0.180 |
| 10 | 0.00 | 10.00 | 0.000 |

(a) Establish the ligand-to-cation ratio in the product.

(b) Determine an average value for the molar absorptivity of the complex; assume that the species in lesser amount is completely incorporated into the complex in the linear portions of the plot.

(c) Evaluate $K_f$ for the complex, using the stoichiometric relationships that exist under conditions of maximum absorbance.

*22–32. (a) Use the following absorbance data (1.00-cm cells) to evaluate the ligand-to-cation ratio in the complex formed between copper(II) and the ligand $H_2B$:

| Solution | Reactant Volume, mL | | $A_{475}$ |
| --- | --- | --- | --- |
| | $8.00 \times 10^{-5}$ M Cu | $8.00 \times 10^{-5}$ M H$_2$B | |
| 0 | 10.00 | 0.00 | 0.000 |
| 1 | 9.00 | 1.00 | 0.104 |
| 2 | 8.00 | 2.00 | 0.210 |
| 3 | 7.00 | 3.00 | 0.314 |
| 4 | 6.00 | 4.00 | 0.419 |
| 5 | 5.00 | 5.00 | 0.507 |
| 6 | 4.00 | 6.00 | 0.571 |
| 7 | 3.00 | 7.00 | 0.574 |
| 8 | 2.00 | 8.00 | 0.423 |
| 9 | 1.00 | 9.00 | 0.211 |
| 10 | 0.00 | 10.00 | 0.000 |

(b) Evaluate an average molar absorptivity for the complex; assume that the species in lesser amount is completely incorporated into the complex in the linear portions of the plot.

(c) Evaluate $K_f$ for the complex, using the stoichiometric relationships that exist under conditions of maximum absorption.

**22-33.** (a) Use the following absorbance data (1.00-cm cells) to evaluate the ligand-to-cation ratio in the complex formed between Co(II) and the bidentate ligand Q:

| Solution | Reactant Volume, mL | | $A_{560}$ |
| --- | --- | --- | --- |
| | $9.50 \times 10^{-5}$ M Co(II) | $9.50 \times 10^{-5}$ M Q | |
| 0 | 10.00 | 0.00 | 0.000 |
| 1 | 9.00 | 1.00 | 0.094 |
| 2 | 8.00 | 2.00 | 0.193 |
| 3 | 7.00 | 3.00 | 0.291 |
| 4 | 6.00 | 4.00 | 0.387 |
| 5 | 5.00 | 5.00 | 0.484 |
| 6 | 4.00 | 6.00 | 0.570 |
| 7 | 3.00 | 7.00 | 0.646 |
| 8 | 2.00 | 8.00 | 0.585 |
| 9 | 1.00 | 9.00 | 0.295 |
| 10 | 0.00 | 10.00 | 0.000 |

(b) Determine an average molar absorptivity for the complex; assume that the species in lesser amount is completely incorporated into the complex in the linear portions of the plot.

(c) Evaluate $K_f$ for the complex, using the stoichiometric relationships that exist under conditions of maximum absorbance.

**\*22-34.** The logarithm of the molar absorptivity for acetone in ethanol is 2.75 at 366 nm. Calculate the range of acetone concentrations that can be used if the percent transmittance is to be greater than 10% and less than 90% with a 1.50-cm cell.

**22-35.** The logarithm of the molar absorptivity of phenol in aqueous solution is 3.812 at 211 nm. Calculate the range of phenol concentrations that can be used if the absorbance is to be greater than 0.100 and less than 2.000 with a 1.25-cm cell.

**\*22-36.** A standard solution was put through appropriate dilutions to give the concentrations of iron shown below. The iron(II)-1,10-phenanthroline complex was then developed in 25.0-mL aliquots of these solutions, following which each was diluted to 50.0 mL. The following absorbances (1.00-cm cells) were recorded at 510 nm:

| Fe(II) Concentration in Original Solutions, ppm | $A_{510}$ |
| --- | --- |
| 4.00 | 0.160 |
| 10.0 | 0.390 |
| 16.0 | 0.630 |
| 24.0 | 0.950 |
| 32.0 | 1.260 |
| 40.0 | 1.580 |

(a) Sketch a calibration curve from these data.

(b) Use the method of least squares to derive an equation relating absorbance and the concentration of iron(II).

(c) Calculate the standard deviation about regression.

(d) Calculate the standard deviation of the slope.

**22-37.** The method developed in Problem 22-36 was used for the routine determination of iron in 25.0-mL aliquots of ground water. Express the concentration (as ppm Fe) in samples that yielded the accompanying absorbance data (1.00-cm cell). Calculate the relative standard deviation of the result. Repeat the calculation assuming the absorbance data are means of three measurements.

| | | |
| --- | --- | --- |
| \*(a) 0.143 | \*(c) 0.068 | \*(e) 1.512 |
| (b) 0.675 | (d) 1.009 | (f) 0.546 |

# MOLECULAR-FLUORESCENCE SPECTROSCOPY

$\mathbf{F}$luorescence is an analytically important emission process in which atoms or molecules are excited by the absorption of a beam of electromagnetic radiation. The excited species then relax to the ground state, giving up their excess energy as photons. In this chapter we consider molecular fluorescence and its applications.

Fluorescence emission is over in $10^{-5}$ s or less. In contrast, phosphorescence may go on for several minutes or even hours. Fluorescence is much more widely used for analyses than phosphorescence.

## 23A  THEORY OF MOLECULAR FLUORESCENCE

Figure 23–1 is the partial energy diagram for a hypothetical molecular species. Three electronic energy states are shown, $E_0$, $E_1$, and $E_2$, where $E_0$ is the ground state and $E_1$ and $E_2$ are excited states. Each electronic state is shown as having four excited vibrational states.

Figure 23–1a shows that when molecules of this species are exposed to a band of visible radiation, $\lambda_1'$ through $\lambda_5'$, all of the five vibrational states of the lower energy electronic state $E_1$ are momentarily populated. Similarly, when the molecules are irradiated with a band of ultraviolet radiation made up of shorter wavelengths $\lambda_1''$ through $\lambda_5''$, the five vibrational levels of the higher energy electronic state $E_2$ become briefly populated.

### 23A–1 Relaxation Processes

As noted in Section 20C, the lifetime of an excited species is brief because there are several ways an excited atom or molecule can give up its excess energy and relax to its ground state. Two of the most important of these mechanisms, nonradiative relaxation and fluorescent relaxation, are illustrated in Figures 23–1b and c.

Vibrational relaxation takes $10^{-15}$ s or less.

Two types of nonradiative relaxation are shown in Figure 23–1b. *Vibrational deactivation,* or *relaxation,* depicted by the short wavy arrows between vibrational energy levels, takes place during collisions between excited molecules and molecules of the solvent. During the collisions, the excess vibrational energy is transferred to solvent molecules in a series of steps, as indicated in the figure. The gain in vibrational energy of the

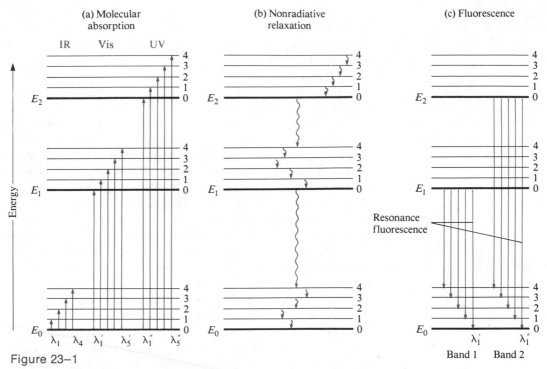

(a) Molecular absorption

(b) Nonradiative relaxation

(c) Fluorescence

Figure 23–1

Energy-level diagram showing some of the energy changes that occur during (a) absorption, (b) nonradiative relaxation, and (c) fluorescence by a molecular species.

solvent results in a slight increase in the temperature of the medium. Vibrational relaxation is such an efficient process that the average lifetime of an excited *vibrational* state is only about $10^{-15}$ s.

Nonradiative relaxation between the lowest vibrational level of an excited electronic state and the upper vibrational level of another electronic state can also occur. This type of relaxation, which is sometimes called *internal conversion,* is depicted by the two longer wavy arrows in Figure 23–1b. Internal conversion is much less efficient than vibrational relaxation, and so the average lifetime of an electronic excited state is between $10^{-6}$ and $10^{-9}$ s. The mechanisms by which this type of relaxation occurs are not fully understood, but the net effect is again a tiny rise in the temperature of the medium.

Internal conversion takes from $10^{-6}$ to $10^{-9}$ s.

Figure 23–1c depicts another relaxation process: fluorescence. Note that bands of radiation are produced when molecules fluoresce because the electronically excited molecules can relax to any of the several vibrational states of the ground electronic state. Like molecular-absorption bands, molecular-fluorescence bands are made up of a multitude of closely spaced lines that are often difficult to resolve.

## Resonance Lines and the Stokes Shift

Note that the lines that terminate the two fluorescence bands on the short-wavelength, or high-energy, side ($\lambda_1'$ and $\lambda_1''$) are identical in energy to the

The wavelength of resonance fluorescence is identical to the wavelength of radiation that caused the fluorescence.

The wavelength of Stokes-shifted fluorescence is longer than that of the radiation that caused the fluorescence.

Quantum yield $= \dfrac{r_f}{r_f + r_r}$, where $r_f$ is the rate of fluorescence relaxation and $r_r$ is the rate of radiationless relaxation.

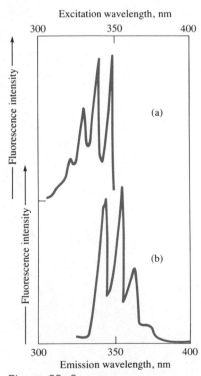

Figure 23–2
Fluorescence spectra for 1 ppm anthracene in alcohol: (a) excitation spectrum; (b) emission spectrum.

two lines labeled $\lambda_1'$ and $\lambda_1''$ in the absorption diagram (Figure 23–1a). These lines are termed *resonance lines* because the fluorescence and absorption wavelengths are identical. Note also that molecular-fluorescence bands are made up largely of lines of longer wavelength, or lower energy, than the band of absorbed radiation responsible for their excitation. This shift to longer wavelengths is sometimes called the *Stokes shift*.

To develop a better understanding of Stokes shifts, let us consider what occurs when the molecule under consideration is irradiated by a single wavelength $\lambda_5''$. As shown in Figure 23–1a, absorption of this radiation promotes an electron into vibrational level 4 of the second excited electronic state $E_2$. In $10^{-15}$ s or less, vibrational relaxation to the zero vibrational level of $E_2$ occurs (Figure 23–1b). At this point, further relaxation can follow either the nonradiative route depicted in Figure 23–1b or the radiative route shown in Figure 23–1c. If the radiative route is followed, relaxation to any of the several vibrational levels of the ground state takes place, giving a band (band 2) of emitted wavelengths, as shown. All of these lines have lower energy, or longer wavelength, than the excitation line $\lambda_5''$.

Let us now turn to those molecules in excited state $E_2$ that undergo internal conversion to electronic state $E_1$. As before, further relaxation can take a nonradiative or a radiative route to the ground state. In the latter case, band 1 of fluorescence is produced. Note that the Stokes shift is from ultraviolet radiation to visible. Note also that band 1 can be produced not only by the mechanism just described but also by the absorption of visible radiation of wavelengths $\lambda_1'$ through $\lambda_5'$ (Figure 23–2a).

## Relationship Between Excitation Spectra and Fluorescence Spectra

Because the energy differences between vibrational states is about the same for both ground and excited states, the absorption spectrum, or, excitation spectrum, and the fluorescence spectrum for a compound often appear as approximate mirror images of one another with overlap occurring at the resonance line. This effect is demonstrated by the spectra in Figure 23–2.

## 23A–2 Fluorescent Species

As shown in Figure 23–1, fluorescence is one of several mechanisms by which a molecule returns to the ground state after it has been excited by absorption of radiation. Thus all absorbing molecules have the potential to fluoresce. Most do not, however, because their structure provides radiationless pathways by which relaxation can occur faster than fluorescent emission.

The *quantum yield* of molecular fluorescence is simply the ratio of the number of molecules that fluoresce to the total number of excited molecules (or the ratio of photons emitted to photons absorbed). Highly fluorescent molecules, such as fluorescein, have quantum efficiencies that approach unity under some conditions. Nonfluorescent species have efficiencies that are essentially zero.

## Fluorescence and Structure

Compounds containing aromatic rings give the most intense and most useful molecular fluorescent emission. While certain aliphatic and alicyclic carbonyl compounds as well as highly conjugated double-bonded structures also fluoresce, their numbers are small in comparison with the number of fluorescent compounds that incorporate aromatic systems.

Most unsubstituted aromatic hydrocarbons fluoresce in solution, with the quantum efficiency increasing with the number of rings and their degree of condensation. The simplest heterocyclics, such as pyridine, furan, thiophene, and pyrrole, do not exhibit molecular fluorescence, but fused-ring structures containing these rings often do.

Substitution on an aromatic ring causes shifts in the wavelength of absorption maxima and corresponding changes in the fluorescence peaks. In addition, substitution frequently affects the fluorescence efficiency. These effects are demonstrated by the data in Table 23–1.

*Many aromatic compounds fluoresce.*

## The Effect of Structural Rigidity

It is found experimentally that fluorescence is particularly favored in rigid molecules. For example, under similar conditions of measurement, the quantum efficiency of fluorene is nearly 1.0, whereas that of biphenyl is about 0.2:

fluorene                    biphenyl

The difference in behavior appears to be largely a result of the increased rigidity furnished by the bridging methylene group in fluorene. This rigidity lowers the rate of nonradiative relaxation to the point where relaxation by fluorescence has time to occur. Many similar examples can be cited. In addition, enhanced emission frequently results when fluorescing dyes are adsorbed on a solid surface; here again, the added rigidity provided by the solid may account for the observed effect.

The influence of rigidity has also been invoked to account for the increase in fluorescence of certain organic chelating agents when they are complexed with a metal ion. For example, the fluorescence intensity of 8-hydroxyquinoline is much less than that of the zinc complex:

Table 23–1

### EFFECT OF SUBSTITUTION ON THE FLUORESCENCE OF BENZENE DERIVATIVES*

| Compound | Relative Intensity of Fluorescence |
|---|---|
| Benzene | 10 |
| Toluene | 17 |
| Propylbenzene | 17 |
| Fluorobenzene | 10 |
| Chlorobenzene | 7 |
| Bromobenzene | 5 |
| Iodobenzene | 0 |
| Phenol | 18 |
| Phenolate ion | 10 |
| Anisole | 20 |
| Aniline | 20 |
| Anilinium ion | 0 |
| Benzoic acid | 3 |
| Benzonitrile | 20 |
| Nitrobenzene | 0 |

*In ethanol solution. Taken from W. West, *Chemical Applications of Spectroscopy* (*Techniques of Organic Chemistry,* Vol. IX, p. 730). New York: Interscience, 1956. Reprinted by permission of John Wiley & Sons.

## Temperature and Solvent Effects

In most molecules, the quantum efficiency of fluorescence decreases with increasing temperature because the increased frequency of collision at elevated temperatures improves the probability of collisional relaxation. A decrease in solvent viscosity leads to the same result.

Rigid molecules and complexes tend to fluoresce.

## 23B    THE EFFECT OF CONCENTRATION ON FLUORESCENCE INTENSITY

The power of fluorescent radiation $F$ is proportional to the radiant power of the excitation beam absorbed by the system:

$$F = K'(P_0 - P) \qquad (23-1)$$

where $P_0$ is the power of the beam incident on the solution and $P$ is its power after it traverses a length $b$ of the medium. The constant $K'$ depends upon the quantum efficiency of the fluorescence. In order to relate $F$ to the concentration $c$ of the fluorescing particle, we write Beer's law in the form

$$\frac{P}{P_0} = 10^{-\varepsilon bc} \qquad (23-2)$$

where $\varepsilon$ is the molar absorptivity of the fluorescing species and $\varepsilon bc$ is the absorbance $A$. By substituting Equation 23-2 into Equation 23-1, we obtain

$$F = K'P_0(1 - 10^{-\varepsilon bc}) \qquad (23-3)$$

Expansion of the exponential term in Equation 23-3 leads to

$$F = K'P_0 \left[ 2.3\varepsilon bc - \frac{(-2.3\varepsilon bc)^2}{2!} - \frac{(-2.3\varepsilon bc)^3}{3!} - \cdots \right] \qquad (23-4)$$

Provided $\varepsilon bc = A < 0.05$, all the subsequent terms in the brackets are small with respect to the first, and so we can write

$$F = 2.3K'\varepsilon bcP_0 \qquad (23-5)$$

or, at constant $P_0$,

$$F = Kc \qquad (23-6)$$

where $K$ is a new constant that is equal to $2.3K'\varepsilon bP_0$. Thus, a plot of the fluorescence power of a solution versus the concentration of the emitting species is linear at low concentrations, as shown in Figure 23-3.

If $c$ becomes large enough that the absorbance is greater than about 0.05 (or the transmittance is smaller than about 90%), linearity is lost and $F$ lies

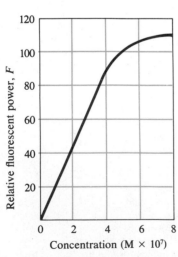

**Figure 23-3**
Calibration curve for the spectro-fluorometric determination of tryptophan in soluble proteins from the lens of a mammalian eye.

below an extrapolation of the straight-line plot. This effect is a result of *self-quenching,* a phenomenon in which analyte molecules absorb the fluorescence produced by other analyte molecules. Indeed, at very high concentrations, $F$ reaches a maximum and then begins to decrease with increasing concentration.

## 23C   FLUORESCENCE INSTRUMENTS

Figure 23–4 shows a typical configuration for the components of *fluorometers* and *spectrofluorometers.* These components are identical to the ones described in Section 21A for ultraviolet and visible spectroscopy. A fluorometer, like a photometer, employs filters for wavelength selection. Most spectrofluorometers, in contrast, employ a filter to limit the excitation radiation and a grating monochromator to disperse the fluorescence from the sample. A few spectrofluorometers have two monochromator systems, one for the excitation radiation and one for the fluorescence.

As shown in Figure 23–4, fluorescence instruments are usually incorporate double-beams to compensate for fluctuations in source power. The beam to the sample first passes through a primary filter or a primary monochromator, which transmits radiation that causes fluorescence but excludes or limits radiation that corresponds to the fluorescence wavelengths. Fluorescence radiation is propagated from the sample in all directions but is most conveniently observed at right angles to the excitation beam; at other angles, increased scattering from the solution and the cell walls may cause large errors in the intensity measurement. The emitted radiation reaches a photoelectric detector after passing through the secondary filter or monochromator, which isolates a fluorescence peak for measurement.

The reference beam passes through an attenuator to decrease its power to approximately that of the fluorescence radiation (the power reduction

Spectrofluorometers with two monochromators are used to obtain fluorescence emission and fluorescence excitation spectra. To obtain emission spectra, the excitation monochromator is set to an absorption peak and the spectrum of the emitted radiation is obtained with the other monochromator. An excitation spectrum is derived by setting the emission monochromator at a fluorescence peak and scanning the sample with the excitation monochromator.

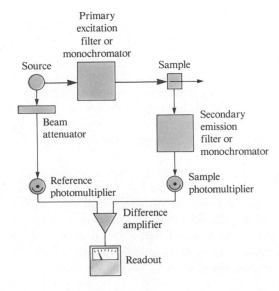

**Figure 23–4**

Components of a fluorometer or a spectrofluorometer.

is usually by a factor of 100 or more). The signals from the reference and sample phototubes are then processed by a difference amplifier whose output is displayed on a meter or recorder. Many fluorescence instruments are of the null type, this state being achieved by optical or electrical attenuators.

The sophistication, performance characteristics, and cost of fluorometers and spectrofluorometers differ as widely as do those of the corresponding instruments for absorption measurements. In many regards, filter-type instruments are better suited for quantitative analytical work than are the more elaborate instruments based on monochromators. Generally, fluorometers are more sensitive than spectrofluorometers because filters have a higher radiation throughput than do monochromators. In addition, source and detector can be positioned closer to the sample in the simpler instrument, a factor that enhances sensitivity.

## 23D APPLICATIONS OF FLUORESCENCE METHODS

Fluorescence methods are generally one to three orders of magnitude more sensitive than methods based upon absorption because the sensitivity of the former can be enhanced either by increasing the power of the excitation beam (Equation 23–5) or amplifying the detector signal. Neither of these options improves the sensitivity of methods based upon absorption, however, because the concentration-related parameter in this case is a ratio:

*Fluorescence methods are 10 to 1000 times more sensitive than absorption methods.*

*Beer's law again.*

$$c = \frac{\log(P_0/P)}{ab}$$

Increasing $P_0$ increases $P$ proportionately and thus has no effect on sensitivity. Similarly, increasing the amplification of the detector signal affects the two measured quantities in an identical way and thus provides no improvement.

### 23D–1 Methods for Inorganic Species

Inorganic fluorometric methods are of two types. Direct methods are based upon the reaction of the analyte with a chelating agent to form a complex that fluoresces. In contrast, indirect methods depend upon the decrease, or *quenching,* of fluorescence of a reagent as a result of its reaction with the analyte. Quenching is used primarily for the determination of anions.

The most successful fluorometric reagents for the determination of cations are aromatic compounds with two or more donor functional groups that permit chelate formation with the metal ion. A typical example is 8-hydroxyquinoline, the structure of which is given on page 88. A few other fluorometric reagents and their applications are listed in Table 23–2. With most of these reagents, the cation is extracted into a solution of the reagent in an immiscible organic solvent, such as chloroform. The fluores-

Table 23–2
SELECTED FLUOROMETRIC METHODS FOR INORGANIC SPECIES*

| Ion | Reagent | Wavelength, nm | | Sensitivity, $\mu$g/mL | Interference |
| | | **Absorption** | **Fluorescence** | | |
|---|---|---|---|---|---|
| $Al^{3+}$ | Alizarin garnet R | 470 | 500 | 0.007 | Be, Co, Cr, Cu, $F^-$, $NO_3^-$, Ni, $PO_4^{3-}$, Th, Zr |
| $F^-$ | Al complex of Alizarin garnet R (quenching) | 470 | 500 | 0.001 | Be, Co, Cr, Cu, Fe, Ni, $PO_4^{3-}$, Th, Zr |
| $B_4O_7^{2-}$ | Benzoin | 370 | 450 | 0.04 | Be, Sb |
| $Cd^{2+}$ | 2-(o-Hydroxyphenyl)-benzoxazole | 365 | Blue | 2 | $NH_3$ |
| $Li^+$ | 8-Hydroxyquinoline | 370 | 580 | 0.2 | Mg |
| $Sn^{4+}$ | Flavanol | 400 | 470 | 0.1 | $F^-$, $PO_4^{3-}$, Zr |
| $Zn^{2+}$ | Benzoin | — | Green | 10 | B, Be, Sb, colored ions |

*From *Handbook of Analytical Chemistry*, L. Meites, Ed., pp. **6**-178 to **6**-181. New York: McGraw-Hill, 1963.

cence of the organic solvent is then measured. For a more complete summary of fluorescent chelating reagents, see the handbook by Meites.[1]

Nonradiative relaxation of transition-metal chelates is so efficient that fluorescence of these species is seldom encountered. It is noteworthy that most transition metals absorb in the ultraviolet or visible region, whereas nontransition-metal ions do not. For this reason, fluorometry often complements spectrophotometry as a method for the determination of cations.

## 23D–2 Methods for Organic and Biochemical Species

The number of applications of fluorometric methods to organic problems is impressive. Weissler and White have summarized the most important of these in several tables.[2] More than 100 entries are found under the heading *Organic and General Biochemical Substances,* including such diverse compounds as adenine, anthranilic acid, aromatic polycyclic hydrocarbons, cysteine, guanidine, indole, naphthols, certain nerve gases, proteins, salicylic acid, skatole, tryptophan, uric acid, and warfarin. Some 50 medicinal agents that can be determined fluorometrically are listed. Included among these are adrenaline, alkylmorphine, chloroquin, digitalis principles, lysergic acid diethylamide (LSD), penicillin, phenobarbital, procaine, and reserpine. Methods for the analysis of ten steroids

Numerous physiologically important compounds fluoresce.

[1]L. Meites, *Handbook of Analytical Chemistry,* pp. **6**-178 to **6**-181. New York: McGraw-Hill, 1963.

[2]A. Weissler and C. E. White, in *Handbook of Analytical Chemistry,* L. Meites, Ed., pp. **6**-182 to **6**-196. New York: McGraw-Hill, 1963.

and an equal number of enzymes and coenzymes are also listed in these tables. Some of the plant products listed are chlorophyll, ergot alkaloids, rauwolfia serpentian alkaloids, flavonoids, and rotenone.

Without question, the most important application of fluorometry is in the analysis of food products, pharmaceuticals, clinical samples, and natural products. The sensitivity and selectivity of the method make it a particularly valuable tool in these fields.

## 23E  QUESTIONS AND PROBLEMS

**23-1.** Define the following terms:
  *(a) resonance fluorescence.
   (b) vibrational relaxation.
  *(c) internal conversion.
   (d) quantum yield.
  *(e) Stokes shift.
   (f) self-quenching.
*23-2. Why is spectrofluorometry potentially more sensitive than spectrophotometry?
 23-3. Which compound below is expected to have a greater fluorescence quantum yield? Explain.

phenolphthalein

fluorescein

*23-4. Why do some absorbing compounds fluoresce and others not?
 23-5. Describe the characteristics of organic compounds that fluoresce.
*23-6. Explain why molecular fluorescence often occurs at a wavelength that is longer than the wavelength of the exciting radiation.
 23-7. Describe the components of a fluorometer.
*23-8. Why are most fluorescence instruments double-beam in design?

 23-9. Why are fluorometers often more useful than spectrofluorometers for quantitative analysis?
*23-10. The reduced form of nicotinamide adenine dinucleotide (NADH) is an important and highly fluorescent coenzyme. It has an absorption maximum at 340 nm and an emission maximum at 465 nm. Standard solutions of NADH gave the following fluorescence intensities:

| Concn NADH, $\mu$mol/L | Relative Intensity |
|---|---|
| 0.100 | 2.24 |
| 0.200 | 4.52 |
| 0.300 | 6.63 |
| 0.400 | 9.01 |
| 0.500 | 10.94 |
| 0.600 | 13.71 |
| 0.700 | 15.49 |
| 0.800 | 17.91 |

  (a) Construct a calibration curve for NADH.
  (b) Derive a least-squares equation for the plot in part (a).
  (c) Calculate the standard deviation of the slope, the standard deviation of the intercept, and the standard deviation about regression for the curve.
  (d) An unknown exhibits a relative fluorescence of 12.16. Calculate the concentration of NADH.
  (e) Calculate the relative standard deviation for the result in part (d).
  (f) Calculate the relative standard deviation for the result in part (d) if the reading of 12.16 was the mean of three measurements.
23-11. The following volumes of a solution containing 1.10 ppm of $Zn^{2+}$ were pipetted into five separatory funnels, each of which contained 5.00 mL of an unknown zinc solution: 0.00, 1.00, 4.00, 7.00, and 11.00 mL. Each solution was then extracted with three 5-mL aliquots of $CCl_4$ containing an excess of 8-hydroxyquinoline. The extracts were then diluted to 25.0 mL and their fluorescence measured. The results were

| mL of Std Zn$^{2+}$ | Fluorometer Reading |
|---|---|
| 0.000 | 6.12 |
| 4.00 | 11.16 |
| 8.00 | 15.68 |
| 12.00 | 20.64 |

(a) Plot the data.
(b) Derive by least squares an equation for the plot.
(c) Calculate the standard deviation of the slope, the standard deviation of the intercept, and the standard deviation about regression.
(d) Calculate the concentration of zinc in the sample.
(e) Calculate a standard deviation for the result in part (d).

*23–12. Fluoride ion quenches the fluorescence of the Al-acid Alizarin Garnet R complex, and the reduction in fluorescence intensity provides a measure of aluminum concentration. To four 10.0-mL aliquots of a water sample were added 0.00, 1.00, 2.00, and 3.00 mL of a standard NaF solution containing 10.0 ppb F$^-$. Exactly 5.00-mL portions of a solution containing an excess of the aluminum complex were added to each, and the solutions were then diluted to 50.0 mL. The fluorescent intensity of the four solutions plus a blank were:

| mL Sample | mL of Std F$^-$ | Meter Reading |
|---|---|---|
| 5.00 | 0.00 | 68.2 |
| 5.00 | 1.00 | 55.3 |
| 5.00 | 2.00 | 41.3 |
| 5.00 | 3.00 | 28.8 |

(a) Plot the data.
(b) By least squares, derive an equation relating the decrease in fluorescence to the volume of standard reagent.
(c) Calculate the standard deviation of the slope, the standard deviation of the intercept, and the standard deviation about regression.
(d) Calculate the ppb F$^-$ in the sample.

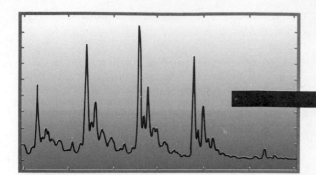

# ATOMIC SPECTROSCOPY BASED ON ULTRAVIOLET AND VISIBLE RADIATION

Throughout this chapter, the word "atomic" will refer not only to neutral atoms but also to elementary ions, such as $K^+$, $Ba^+$, and $Al^+$.

Atomization is a process in which a sample is converted to gaseous atoms.

$A$tomic spectroscopy is used for the qualitative and quantitative determination of perhaps 70 elements. Sensitivities of atomic methods lie typically in the parts-per-million to parts-per-billion range. Additional virtues of these methods are speed, convenience, unusually high selectivity, and moderate instrument costs.

Spectroscopic studies of atoms (or of elementary ions, such as $Fe^+$, $Mg^+$, or $Al^+$) with ultraviolet and visible radiation can be performed only in a gaseous medium in which the individual atoms or ions are well separated from one another. Consequently, the first step in all atomic spectroscopic procedures is *atomization*, a process in which a sample is volatilized and decomposed in such a way as to produce an atomic gas. The efficiency and reproducibility of the atomization step in large measure determine the method's sensitivity, precision, and accuracy; that is, atomization is by far the most critical step in atomic spectroscopy.

As shown in Table 24–1, atomic spectroscopic methods are conveniently categorized on the basis of how the sample is atomized. Note that the atomization temperature varies considerably among the several methods. Note also that atomic methods are based upon absorption, fluorescence, and emission phenomena.[1]

In this chapter we consider spectroscopic methods that are based on the four most common methods of atomization: (1) flame, (2) electrothermal, (3) inductively coupled plasma, and (4) direct-current plasma. Before undertaking this discussion, however, it is worthwhile pointing out the similarities and the differences between atomic spectroscopic methods and the molecular methods described in Chapter 22.

---

[1]References that deal with the theory and applications of atomic spectroscopy include R. D. Sacks, A. Syty, and J. W. Robinson, in *Treatise on Analytical Chemistry*, 2nd ed., P. J. Elving, E. J. Meehan, and I. M. Kolthoff, Eds., Part I, Vol. 7, Chapters 6, 7, and 8. New York: Wiley, 1981; C. Th. J. Alkemade *et al.*, *Metal Vapors in Flames*. Elmsford, NY: Pergamon Press, 1982; B. Magyar, *Guidelines to Planning Atomic Spectrometric Analysis*. New York: Elsevier, 1982; J. D. Ingle Jr. and S. R. Crouch, *Spectrochemical Analysis*. Englewood Cliffs, NJ: Prentice-Hall, 1988.

| Table 24–1 CLASSIFICATION OF ATOMIC SPECTRAL METHODS | | | |
|---|---|---|---|
| Atomization Method | Typical Atomization Temperature, °C | Basis for Method | Common Name and Abbreviation of Method |
| Flame | 1700–3150 | Absorption | Atomic absorption spectroscopy, AAS |
| | | Emission | Atomic emission spectroscopy, AES |
| | | Fluorescence | Atomic fluorescence spectroscopy, AFS |
| Electrothermal | 1200–3000 | Absorption | Electrothermal atomic absorption spectroscopy |
| | | Fluorescence | Electrothermal atomic fluorescence spectroscopy |
| Inductively coupled argon plasma | 6000–8000 | Emission | Inductively coupled plasma spectroscopy, ICP |
| | | Fluorescence | Inductively coupled plasma fluorescence spectroscopy |
| Direct-current argon plasma | 6000–10,000 | Emission | DC plasma spectroscopy, DCP |
| Electric arc | 4000–5000 | Emission | Arc-source emission spectroscopy |
| Electric spark | 40,000(?) | Emission | Spark-source emission spectroscopy |

## 24A   A COMPARISON OF ATOMIC AND MOLECULAR SPECTROSCOPIC METHODS

Atomic spectroscopy is based on absorption, fluorescence, and emission phenomena, whereas only absorption and fluorescence are generally applicable for molecular spectroscopy. The reason that thermal-emission methods are little used for determining molecular species is that most molecules decompose at the temperatures required to excite emission.

There are few conceptual differences between atomic and molecular spectroscopy; the primary difference lies in the types of species studied. Atomic spectroscopy gives information about the identity and concentration of *atoms* in a sample regardless of how these atoms are combined. In contrast, molecular spectroscopy gives qualitative and quantitative information about the *molecules* in a sample.

Atomic spectroscopy is limited to ultraviolet, visible, and X-ray frequencies because only such radiation is energetic enough to cause electronic transitions. Molecules, on the other hand, have vibrational and rotational as well as electronic energy states. As a consequence, molecular spectroscopy is based upon ultraviolet, visible, infrared, microwave, and radio-frequency radiation.

Atomic and molecular spectroscopic instruments have a number of common features. For example, both use monochromators or filters for wavelength selection and photon detectors to determine radiation intensity. Instruments for both atomic and molecular *absorption* studies have a reservoir to contain the sample. In molecular instruments, the reservoir is a cell, or cuvette, that contains a liquid or gaseous solution of the sample. The reservoir for atomic spectroscopy is a flame, a plasma, an arc, or a spark containing the gaseous atomic sample. The atom reservoir serves

Absorption and fluorescence are common to atomic and molecular spectroscopy. Emission is restricted to atomic methods.

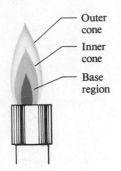

Figure 24–1
Regions in a flame.

Nebulize means to reduce to a fine spray.

Modern atomic absorption instruments use laminar flow burners almost exclusively.

two purposes: It is both the vehicle for atomization and the container of the atomic vapor.

Atomic absorption and fluorescence instruments also have in common a source of radiation. The precise nature of this source is, as we shall see, a major difference between atomic and molecular spectroscopy.

## 24B    ATOMIC SPECTROSCOPY BASED ON FLAME ATOMIZATION

As shown in Table 24–1, three types of atomic methods are based upon flame atomization: (1) atomic absorption spectroscopy (AAS), (2) atomic emission spectroscopy (AES), and (3) atomic fluorescence spectroscopy (AFS).

In flame atomization, an aqueous solution of the sample is dispersed (or *nebulized*) as a fine spray and then mixed with gaseous fuel and oxidant that carry it into a burner. The solvent evaporates in the *base region* of the flame, which is located just above the burner (Figure 24–1). The resulting finely divided solid particles are carried to a region in the center of the flame called the *inner cone*. Here, in this hottest part of the flame, gaseous atoms and elementary ions are formed from the solid particles. Excitation of atomic emission spectra also takes place in this region. Finally, the atoms and ions are carried to the outer edge, or *outer cone*, where oxidation may occur before the atomization products disperse into the atmosphere. Because the velocity of the fuel/oxidant mixture through the flame is high, only a fraction of the sample undergoes all these processes; indeed, a flame is not a very efficient atomizer.

### 24B–1    Flame Atomizers

Two types of burners are used in flame spectroscopy: *turbulent-flow burners* and *laminar-flow burners*. Figure 24–2a is a diagram of a commercially available turbulent-flow, or *total-consumption*, burner in which the nebulizer and burner are combined in a single unit. The sample is drawn up the capillary and nebulized by the Venturi action caused by the flow of gases around the capillary tip. Typical sample flow rates are 1 to 3 mL/min.

Turbulent-flow burners offer the advantage of introducing a relatively large and representative sample into the flame. Disadvantages include a relatively short path length through the flame and problems with clogging of the tip. In addition, these burners are noisy from both the electronic standpoint and the auditory standpoint. Although sometimes used for emission and fluorescence, turbulent-flow burners find little use in present-day absorption instruments.

Figure 24–2b is a diagram of a typical commercial laminar-flow, or *premix*, burner. The sample is nebulized by the flow of oxidant past a capillary tip. The resulting aerosol is then mixed with fuel and flows past a series of baffles that remove all but the finest droplets. As a result, much of the sample collects in the bottom of the mixing chamber, where it

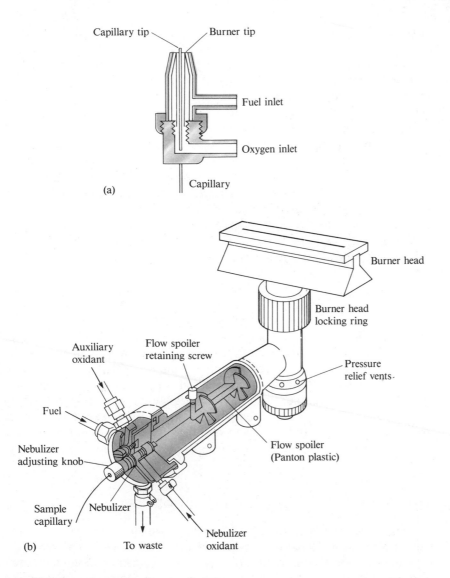

**Figure 24–2**

Diagrams of burners for atomic flame spectroscopy. (a) Turbulent-flow, or total-consumption, burner. (b) Laminar (premix) burner. (a, Courtesy of Beckman Instruments, Fullerton, CA; b, courtesy of Perkin-Elmer Corp., Norwalk, CT.)

drains into a waste container. The aerosol, oxidant, and fuel are then fed into a slotted burner that provides a flame that is usually 5 to 10 cm in length.

Laminar-flow burners provide a relatively quiet flame and a significantly longer path length. These properties tend to enhance sensitivity and reproducibility. Furthermore, clogging is seldom a problem. Disadvantages include a lower rate of sample introduction (which may offset the longer path length advantage), as well as the possibility that mixed solvents will undergo selective evaporation in the mixing chamber, which can lead to analytical uncertainties. Furthermore, the mixing chamber contains a potentially explosive mixture that can be ignited by a flashback. Note that the burner in Figure 24–2b is equipped with pressure-relief vents for this reason. In addition, the burner head is sometimes held in place by stainless steel cables.

## 24B-2 Emission and Absorption Spectra in Flames

Emission and absorption spectra for both atoms and elementary ions are obtained from flames. As described in Section 20D–1, atomic emission spectra are produced when an atom or ion excited by the absorption of energy from a hot source relaxes to its ground state by giving off a photon of radiation. In contrast, atomic absorption takes place when a gaseous atom or ion absorbs a photon of radiation from an external source and is thus excited. It is important to appreciate that, for the same electronic transition, the energy of an emitted photon is *identical* to that of an absorbed photon. Thus, the wavelength of the emitted radiation is the same as the wavelength of the absorbed radiation.

All elements ionize to some degree in a flame, which leads to a mixture of atoms, ions, and electrons in the hot medium. For example, when a sample containing barium is atomized, the equilibrium

$$Ba \rightleftharpoons Ba^+ + e^-$$

is established in the inner cone of the flame. The position of this equilibrium depends on the temperature of the flame and total concentration of barium as well as on the concentration of the electrons produced from the ionization of *all elements* present in the sample. At the temperatures of the hottest flames (>3000 K), nearly half of the barium may be present in ionic form. The spectra of Ba and $Ba^+$ are, however, totally different from one another. Thus, in a high-temperature flame, two spectra for barium are excited, one for the atom and one for its ion. For this reason (and others), control of flame temperatures is important in flame spectroscopy.

## 24B-3 Methods for Obtaining Flame Spectra

In *flame emission spectroscopy*, the excited analyte ions serve as the source of radiation. That is, in contrast to all other types of spectroscopy we have thus far encountered, no *external* source of radiation is required in flame emission spectroscopy. Spectra are recorded by locating the inner cone of the flame in front of the entrance slit of a monochromator, such as one of those shown in Figure 21–8. The output from the exit slit is then monitored as the spectrum is scanned by rotating the grating or prism. A flame emission spectrum obtained in this way is shown in Figure 20–16.

In *atomic absorption spectroscopy*, the radiation from a special type of external source is passed through the inner cone of the flame, through a monochromator (or sometimes an interference filter), and to the surface of a radiation detector. In contrast to molecular absorption methods, atomic absorption methods *do not employ a continuous source of radiation* but instead use sources that emit *lines* of radiation that have the same wavelength as that of an absorption peak of the analyte. The need for and properties of this type of source are considered in Section 24B–6. A consequence of the use of line sources is that complete spectra seldom appear in the literature dealing with atomic absorption spectroscopy.

*For a given transition process, photon energies for absorption and emission are identical.*

*Ionization of an atomic species in a flame is an equilibrium process that can be treated by the law of mass action.*

*The spectrum of an atom is entirely different from that of its ions.*

*The analyte itself is the source of radiation in atomic emission spectroscopy.*

*A line source is required for atomic absorption spectroscopy.*

For *atomic fluorescence spectroscopy*, fluorescence is excited by a beam of radiation that enters the flame at right angles to the path to the entrance slit of the wavelength selector.

## 24B–4  Types of Flames

Several combinations of fuel and oxidant are employed in flame spectroscopy. Low-temperature flames (1750 to 1850°C), which are obtained with propane or natural gas as the fuel and air as the oxidant, have sufficient energy to provide satisfactory spectra for the alkali metals but are cool enough to preclude significant ionization and loss of sensitivity from this cause. Furthermore, few other elements are excited in this type of flame, and so spectra are simple and analyte lines readily isolated even with inexpensive glass filters.

Air/acetylene flames, which have temperatures in the 2200 to 2400°C range, are useful for many atomic absorption methods. This mixture is not satisfactory with elements such as aluminum, silicon, the alkaline earths, and vanadium, which form refractory oxides that are incompletely atomized at these temperatures.

To obtain emission spectra for most of the elements, acetylene is employed as the fuel with oxygen or nitrous oxide as oxidant; these mixtures produce flames having temperatures of 2950 to 3050°C.

Air/hydrogen (2100°C) and oxygen/hydrogen (2700°C) flames are useful for observing lines in the shorter-wavelength ultraviolet because they are transparent in this region, whereas acetylene and other hydrocarbon flames are not.

*Low-temperature flames are preferred for spectroscopy involving the alkali metals. Why?*

## 24B–5  The Effects of Flame Temperature

Both emission and absorption spectra are affected in a complex way by variations in flame temperature. One effect common to the two methods is that increases in atomization efficiency and thus in the total atom population of the flame accompany temperature increases; enhanced sensitivity is the consequence. As noted earlier, however, this increase in atom population can be more than offset by the number of atoms lost to ionization with certain elements.

Another effect of elevated temperatures is line broadening, which is accompanied by a decrease in peak height.

Flame temperature also determines the relative number of excited and unexcited atoms in a flame. In an air/acetylene flame, for example, the ratio of excited to unexcited magnesium atoms can be computed to be about $10^{-8}$, whereas in an oxygen/acetylene flame, which is about 700°C hotter, this ratio is larger by a factor of about 70. Control of temperature is thus of prime importance in flame emission methods. For example, with a 2500°C flame, a temperature increase of 10°C causes the number of sodium atoms in the excited $3p$ state to increase by about 3%. In contrast, the corresponding *decrease* in the much larger number of ground-state atoms is only about 0.002%. Therefore, emission methods, based as they are on the population of *excited atoms*, require much closer control of

*Atomic emission methods require closer control of flame temperature than do atomic absorption methods.*

**Table 24–2**

**COMPARISON OF DETECTION LIMITS FOR VARIOUS ELEMENTS BY FLAME ABSORPTION AND FLAME EMISSION METHODS**

| Flame Emission More Sensitive | Sensitivity About the Same | Flame Absorption More Sensitive |
|---|---|---|
| Al, Ba, Ca, Eu, Ga, Ho, In, K, La, Li, Lu, Na, Nd, Pr, Rb, Re, Ru, Sm, Sr, Tb, Tl, Tm, W, Yb | Cr, Cu, Dy, Er, Gd, Ge, Mn, Mo, Nb, Pd, Rh, Sc, Ta, Ti, V, Y, Zr | Ag, As, Au, B, Be, Bi, Cd, Co, Fe, Hg, Ir, Mg, Ni, Pb, Pt, Sb, Se, Si, Sn, Te, Zn |

Adapted with permission from E. E. Pickett and S. R. Koirtyohann, *Anal. Chem.,* **1969,** *41* (14), 42A. Copyright 1969 American Chemical Society.

flame temperature than do absorption procedures, in which the analytical signal depends upon the number of *unexcited atoms*.

The number of unexcited atoms in a typical flame exceeds the number of excited ones by a factor of $10^3$ to $10^{10}$ or more. This suggests that absorption methods should be significantly more sensitive than emission methods. In fact, however, several other variables also influence sensitivity, and the two methods tend to complement each other in this regard. Table 24–2 illustrates this point.

### 24B–6 Atomic Absorption Spectroscopy Based on Flames

Flame atomic absorption spectroscopy is currently the most widely used of all the atomic methods listed in Table 24–1 because of its simplicity, effectiveness, and relatively low cost.

#### Atomic Absorption Spectra

Typically, the absorption of radiation from an external source by an atomic species in a flame takes the form of a series of narrow peaks (lines) that are the result of transitions of an electron from the ground state to one of several higher energy levels (Section 20C–1). It is important to emphasize that the wavelength of an absorbed line *is identical to* the wavelength of the line that is emitted when the electron returns from the excited state to the ground state. That is, atomic absorption involves *resonance lines* (page 532).

#### Atomic Absorption Line Widths

The natural width of an AA or AE line is about $10^{-5}$ nm.

The natural width of an atomic absorption or an atomic emission line is on the order of $10^{-5}$ nm. Two effects, however, cause observed line widths to be broadened by a factor of 100 (or more). As will become apparent shortly, line broadening is an important consideration in the design of atomic absorption instruments.

**Doppler Broadening.** Doppler broadening results from the rapid motion of atoms as they emit or absorb radiation. Atoms moving toward the detector emit wavelengths that are slightly shorter than the wavelengths emitted by atoms moving at right angles to the detector (Figure 24–3). This difference is a manifestation of the well-known Doppler shift; the effect is reversed for atoms moving away from the detector. The net effect is an increase in the width of the emission line. For precisely the same reason, the Doppler effect also causes broadening of absorption lines. This type of broadening becomes more pronounced as the flame temperature increases because of the consequent increased rate of motion of the atoms.

**Pressure Broadening.** Pressure broadening arises from collisions among atoms that result in slight variations in their ground-state energies and thus slight energy differences between ground and excited states. Like Doppler broadening, pressure broadening becomes greater with increases in temperature. Therefore, broader absorption and emission peaks are always encountered at elevated temperatures.

Both Doppler broadening and pressure broadening are temperature dependent.

### The Effect of Narrow Line Widths on Absorbance Measurements

Since transition energies for atomic absorption lines are unique for each element, analytical methods based upon atomic absorption have the potential of being highly specific. On the other hand, the narrow lines create a problem in quantitative analysis that is not encountered in molecular absorption. No ordinary monochromator is capable of yielding a band of radiation as narrow as the peak width of an atomic absorption line (0.002 to 0.005 nm). Consequently, the use of radiation that has been isolated from a continuous source by a monochromator inevitably causes instru-

The widths of atomic absorption peaks are much less than the effective bandwidths of most monochromators.

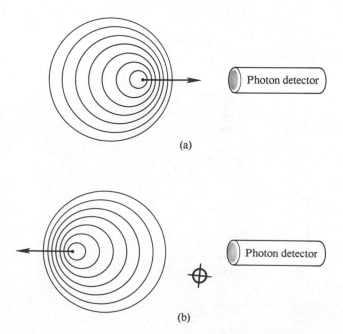

Figure 24–3
Cause of Doppler broadening.
(a) When an atom moves toward a photon detector and emits radiation, the detector sees wave crests more often and detects radiation of higher frequency. (b) When an atom moves away from a photon detector and emits radiation, the detector sees crests less frequently and thus detects radiation of lower frequency.

mental departures from Beer's law (see the discussion of instrument deviations from Beer's law in Section 20C–7). In addition, since the fraction of radiation absorbed from such a beam is small, the detector receives a signal that is only slightly attenuated (that is, $P \rightarrow P_0$) and the sensitivity of the measurement is low.

The problem created by narrow absorption peaks has been surmounted by using radiation from a source that emits not only a *line of the same wavelength* as the one selected for absorption measurements but also one that is *narrower*. For example, a mercury vapor lamp is selected as the external radiation source for the determination of mercury. Gaseous mercury atoms electrically excited in such a lamp return to the ground state by *emitting* radiation with wavelengths that are identical to the wavelengths *absorbed* by the analyte mercury atoms in the flame. Since the lamp is operated at a temperature lower than that of the flame, the Doppler and pressure broadening of the mercury emission lines from the lamp is less than the corresponding broadening of the analyte absorption peaks in the hot flame that holds the sample. As a consequence, the effective bandwidths of the lines emitted by the lamp are significantly less than the corresponding bandwidths of the absorption peaks for the analyte in the flame.

Figure 24–4 illustrates the strategy generally used in measuring absorbances in atomic absorption methods. Figure 24–4a depicts four narrow

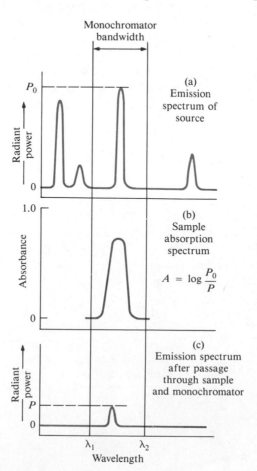

Figure 24–4
Atomic absorption of a resonance line.

*emission* lines from a typical atomic lamp source. Also shown is how one of these lines is isolated by a filter or monochromator. Figure 24–4b shows the flame *absorption spectrum* for the analyte between the wavelengths $\lambda_1$ and $\lambda_2$; note that the width of the absorption peak in the flame is significantly greater than the width of the emission line from the lamp. As shown in Figure 24–3c, the intensity of the incident beam $P_0$ has been decreased to $P$ by passage through the sample. Since the bandwidth of the emission line from the lamp is now significantly less than the bandwidth of the absorption peak in the flame, $\log P_0/P$ is likely to be linearly related to concentration.

### Line Sources for Atomic Absorption Spectroscopy

Two types of lamps are used in atomic absorption instruments: *hollow-cathode lamps* and *electrodeless discharge lamps*.

**Hollow-Cathode Lamps.** The most useful radiation source for atomic absorption spectroscopy is the *hollow-cathode lamp*, shown schematically in Figure 24–5. It consists of a tungsten anode and a cylindrical cathode sealed in a glass tube containing an inert gas, such as argon, at a pressure of 1 to 5 torr. The cathode is either fabricated from the analyte metal or else serves as a support for a coating of that metal.

The application of a potential of about 300 V across the electrodes causes ionization of the argon and generation of a current of 5 to 10 mA as the argon cations migrate toward the cathode and electrons migrate toward the anode. If the potential is sufficiently large, the argon cations strike the cathode with sufficient energy to dislodge some of the metal atoms and thereby produce an atomic cloud; this process is called *sputtering*. Some of the sputtered metal atoms are in excited states and emit their characteristic wavelengths as they return to the ground state. It is important to recall that the atoms producing emission lines in the lamp are at a significantly lower temperature than the analyte atoms in the flame. Thus the emission lines from the lamp are broadened less than the *absorption* peaks in the flame.

The sputtered metal atoms in a lamp eventually diffuse back to the cathode surface (or to the walls of the lamp) and are deposited.

Hollow-cathode lamps for about 40 elements are available from commercial sources. Some are fitted with a cathode containing more than one element; such lamps provide spectral lines for the determination of several species. The development of the hollow-cathode lamp is widely regarded as the single most important event in the evolution of *atomic absorption* spectroscopy.

Sputtering is a process in which atoms or ions are ejected from a surface by a beam of charged particles.

Hollow cathode lamps made atomic absorption spectroscopy practical.

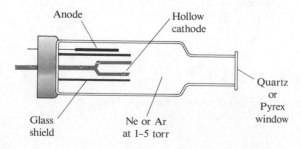

Figure 24–5
Diagram of a hollow-cathode lamp.

**Electrodeless Discharge Lamps.** Electrodeless discharge lamps are useful sources of atomic line spectra and provide radiant intensities that are usually one to two orders of magnitude greater than their hollow-cathode counterparts. A typical lamp is constructed from a sealed quartz tube containing an inert gas, such as argon, at a pressure of a few torr and a small quantity of the analyte metal (or its salt). The lamp contains no electrode but instead is energized by an intense field of radio-frequency or microwave radiation. The argon ionizes in this field, and the ions are accelerated by the high-frequency component of the field until they gain sufficient energy to excite (by collision) the atoms of the metal whose spectrum is sought.

Electrodeless discharge lamps are available commercially for several elements. Their performance does not appear to be as reliable as that of the hollow-cathode lamp.

## Source Modulation

In an atomic absorption measurement, it is necessary to discriminate between radiation from the hollow-cathode or electrodeless discharge lamp and radiation from the flame. Much of the latter is eliminated by the monochromator, which is always located between the flame and the detector. The thermal excitation of a fraction of the analyte atoms in the flame, however, produces radiation of the wavelength at which the monochromator is set. Since such radiation is not removed, it acts as a potential source of interference.

The effect of analyte emission is overcome by *modulating* the output from the hollow-cathode lamp so that its intensity fluctuates at a constant frequency. The detector thus receives an alternating signal from the hollow-cathode lamp and a continuous signal from the flame and converts these signals into the corresponding types of electric current. A relatively simple electronic system then eliminates the unmodulated dc signal produced by the flame and passes the ac signal from the source to an amplifier and finally to the readout device.

Modulation is most often accomplished by interposing a motor-driven circular chopper *between the source and the flame* (see Figure 24–6). Segments of the metal chopper have been removed so that radiation passes through the device half of the time and is reflected the other half. Rotation of the chopper at a constant speed causes the beam reaching the flame to vary periodically from zero intensity to some maximum intensity and then back to zero. As an alternative, the power supply for the source can be designed for intermittent (or ac) operation.

Modulation is defined as changing some property of signal such as frequency, amplitude, or wavelength. In AAS, the frequency of the source is modulated from continuous (dc) to intermittent (ac).

Beam choppers are widely used in spectroscopy to produce pulsed beams of radiation.

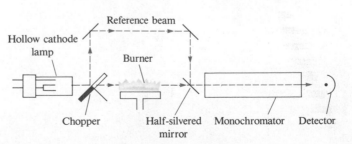

Figure 24–6

Optical paths in a double-beam atomic absorption spectrophotometer.

## Instruments

An atomic absorption instrument contains the same basic components as an instrument designed for molecular absorption measurements: a source, a sample container (here, a flame reservoir), a wavelength selector, and a detector/readout system. Both single- and double-beam instruments are offered by numerous manufacturers. The range of sophistication and the cost are both substantial.

**Photometers.** As a minimum, an instrument for atomic absorption spectroscopy must be capable of providing a sufficiently narrow bandwidth to isolate the line chosen for a measurement from other lines that may interfere with or diminish the sensitivity of the method. A photometer equipped with a hollow-cathode source and glass filters is satisfactory for measuring concentrations of the alkali metals, which have only a few widely spaced resonance lines in the visible region. A more versatile photometer is sold with readily interchangeable interference filters and lamps. A separate filter and lamp combination is needed for each element. Satisfactory results for the analysis of 22 metals are claimed.

**Spectrophotometers.** Most atomic absorption measurements are made with instruments equipped with an ultraviolet/visible grating monochromator. Figure 24–6 is a schematic of a typical double-beam instrument. Radiation from the hollow-cathode lamp is chopped and mechanically split into two beams, one of which passes through the flame and the other around the flame. A half-silvered mirror returns both beams to a single path, by which they pass alternately through the monochromator and to the detector. The signal processor then separates the ac signal generated by the chopped light source from the dc signal produced by the flame. The logarithm of the ratio of the reference and sample components of the ac signal is then computed and sent to the readout device for display as absorbance.

## Interferences

Two types of interference are encountered in atomic absorption methods. *Spectral interferences* occur when particulate matter from the atomization scatters the incident radiation from the source or when the absorption or emission of an interfering species either overlaps or is so close to the analyte wavelength that resolution by the monochromator becomes impossible. *Chemical interferences* result from various chemical processes that occur during atomization and alter the absorption characteristics of the analyte.

**Spectral Interferences.** Interference due to overlapping lines is rare because the emission lines of hollow-cathode sources are so very narrow. Nevertheless, such an interference can occur if the separation between two lines is on the order of 0.01 nm. For example, a vanadium line at 308.211 nm interferes in an analysis based upon the aluminum absorption line at 308.215 nm. The interference is readily avoided, however, by selecting a different aluminum line (309.27 nm, for example).

Spectral interferences are encountered in AAS due to scattering by potential products in the flame and absorption by molecular species in the flame.

Spectral interferences also result from the presence of either *molecular* combustion products that exhibit broad-band absorption or particulate products that scatter radiation. Both diminish the power of the transmitted beam and lead to positive analytical errors. Where the source of these products is the fuel/oxidant mixture alone, corrections are readily obtained from absorbance measurements made with a blank aspirated into the flame.

A much more troublesome problem is encountered when the source of absorption or scattering originates in the sample matrix. In this type of interference, the power of the transmitted beam $P$ is attenuated by the matrix components, but the incident beam power $P_0$ is not; a positive error in absorbance and thus concentration results. A potential matrix interference due to absorption occurs in the determination of barium in alkaline earth mixtures, for example. The wavelength of the barium line used for atomic absorption analysis appears in the center of a broad absorption *band* for molecular CaOH; interference by calcium in a barium analysis results. The net effect is readily eliminated by substituting nitrous oxide for air as the oxidant; the higher temperature decomposes the CaOH and eliminates the absorption band.

Spectral interference due to scattering by products of atomization often occurs when concentrated solutions containing elements such as titanium, zirconium, and tungsten—which form stable oxides—are aspirated into the flame. Metal oxide particles with diameters greater than the wavelength of light appear to be formed and cause scattering of the incident beam.

Fortunately, spectral interferences by matrix products are not widely encountered with flame atomization and usually can be avoided by variations in such analytical parameters as temperature and fuel-to-oxidant ratio. Alternatively, if the source of interference is known, an excess of the interfering substance can be added to both sample and standards, provided the excess is large with respect to the concentration from the sample matrix, the contribution of the latter will become insignificant. The added substance is sometimes called a *radiation buffer*.

The matrix-interference problem is greatly exacerbated with electrothermal atomization and is one of the major causes of the lower accuracy associated with nonflame methods.[2]

**Chemical Interferences.** Chemical interferences can frequently be minimized by a suitable choice of operating conditions. Perhaps the most common type of chemical interference is by anions that form compounds of low volatility with the analyte and thus decrease the rate at which it is atomized. Low results are the consequence. An example is the decrease in calcium absorbance observed with increasing concentrations of sulfate or phosphate ions, which form nonvolatile compounds with calcium ion.

Interferences due to the formation of species of low volatility can often be eliminated or moderated by use of higher temperatures. Alternatively,

A radiation buffer is a substance that is added in large excess to both samples and standards in atomic spectroscopy to swamp the effect of matrix species and thus minimize interference.

Chemical interferences in AAS are caused by reactions between the analyte and the interference that reduce the analyte concentration in the flame.

[2]For a discussion of methods for overcoming matrix interference problems, see D. A. Skoog and J. J. Leary, *Principles of Instrumental Analysis*, 4th ed., pp. 216–218. Philadelphia: Saunders College Publishing, 1992.

*releasing agents*, which are cations that react preferentially with the interference and prevent its interaction with the analyte, can be introduced. For example, the addition of excess strontium or lanthanum ion minimizes interference by phosphate in the determination of calcium. Here, the strontium or lanthanum replaces the analyte in the nonvolatile compound formed with the interfering species.

*Protective agents* prevent interference by preferentially forming stable but volatile species with the analyte. Three common reagents for this purpose are EDTA, 8-hydroxyquinoline, and APDC (the ammonium salt of 1-pyrrolidine-carbodithioic acid). For example, the presence of EDTA has been shown to eliminate interference by silicon, phosphate, and sulfate in the determination of calcium.

The ionization of atoms and molecules is usually inconsequential in combustion mixtures that involve air as the oxidant. In high-temperature flames, however, where oxygen or nitrous oxide serves as the oxidant, ionization becomes appreciable, and a significant concentration of free electrons exists as a consequence of the equilibrium

$$M \rightleftharpoons M^+ + e^-$$

where M represents a neutral atom or molecule and $M^+$ is its ion. Ordinarily, the spectrum of $M^+$ is quite different from that of M, so that ionization of the analyte ions leads to low results. It is important to appreciate that treating the ionization process as an equilibrium—with free electrons as one of the products—implies that the degree of ionization of an analyte atom is strongly influenced by the presence of other ionizable metals in the flame. Thus, if the medium contains not only species M but species B as well, and if B ionizes according to the equation

$$B \rightleftharpoons B^+ + e^-$$

then the degree of ionization of M is decreased by the mass-action effect of the electrons formed from B. The errors caused by analyte ionization can frequently be eliminated by addition of an *ionization suppressor*, which provides a relatively high concentration of electrons to the flame; suppression of analyte ionization results. Potassium salts are frequently used as ionization suppressors because of the low ionization energy of this element.

## Applications of Atomic Absorption Spectroscopy

Atomic absorption spectroscopy provides a sensitive means of determining more than 60 elements. The method is well suited for routine measurements by relatively unskilled operators.

## Region of Flame for Quantitative Measurements

Figure 24–7 shows the absorbance of three elements as a function of distance above the burner tip. For magnesium and silver, the initial rise in absorbance is a consequence of the longer exposure to the heat, which

Figure 24–7
Flame-absorbance profile for three elements.

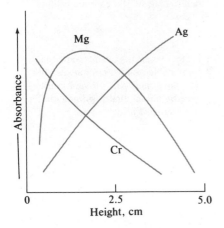

In AAS, it is common practice to run at least one standard every time an analysis is performed.

leads to a greater concentration of atoms in the radiation path. The absorbance for magnesium, however, reaches a maximum near the center of the flame and then falls off as oxidation to magnesium oxide takes place. The effect is not seen with silver because this element is much more resistant to oxidation. For chromium, which forms very stable oxides, maximum absorbance lies immediately above the burner tip. For this element, oxide formation begins as soon as chromium atoms are formed.

It is clear from Figure 24–7 that the part of a flame to be used in an analysis differs from element to element and furthermore that the position of the flame with respect to the source must be reproduced closely during calibration and analysis. Generally, the flame position is adjusted to yield a maximum absorbance reading.

## Calibration

Quantitative atomic absorption methods are usually based on calibration curves, which in principle are linear. Departures from linearity occur, however, and analyses should *never* be based on the measurement of a single standard with the assumption that Beer's law is being followed. In addition, the production of an atomic vapor involves a sufficient number of uncontrollable variables to warrant measuring the absorbance of at least one standard solution each time an analysis is performed. Any deviation of the standard from its original calibration value can then be applied as a correction to the analytical results.

## Standard-Addition Method

The standard-addition method is extensively used in atomic absorption spectroscopy. In this procedure, two or more aliquots of the sample are transferred to volumetric flasks. One is diluted to volume directly, and a known amount of analyte is introduced into the other before dilution to the same volume. The absorbance of each is measured (several different standard additions are recommended if the method is unfamiliar). If a linear relationship exists between absorbance and concentration (and this must be verified experimentally), the following relationships apply:

$$A_x = \frac{kV_x c_x}{V_T}$$

$$A_T = \frac{kV_x c_x}{V_T} + \frac{kV_s c_s}{V_T} \qquad (24\text{–}1)$$

where $V_x$ and $c_x$ are the volume and concentration of the analyte solution, $V_s$ and $c_s$ are the volume and concentration of the standard, $V_T$ is the total volume, $k$ is a proportionality constant, and $A_x$ and $A_T$ are the absorbances of the sample alone and the sample plus standard, respectively. These two equations can be combined to give

$$c_x = \frac{A_x}{A_T - A_x} \times \frac{c_s V_s}{V_x}$$

If several standard additions have been made, $A_T$ in Equation 24–1 can be plotted against $c_s$ as in Figure 24–8. The resulting straight line can then be extrapolated to $A_T = 0$. Substituting this relationship into Equation 24–1 gives, upon rearranging,

$$c_x = \frac{-c_s(V_s)_0}{V_x}$$

where $(V_s)_0$ is the extrapolated volume of standard. Use of the standard-addition method tends to compensate for variations caused by spectral and chemical interferences in the analyte solution.

The standard-addition method compensates for interferences in the sample matrix.

**Sensitivity and Accuracy.** Columns 2 and 3 of Table 24–3 provide information on detection limits for a number of common elements by flame atomic absorption and also by electrothermal methods, which are discussed in Section 24C. For comparison purposes, limits for some of the other atomic procedures are also included. Small differences among the quoted values are not significant. An order of magnitude is probably meaningful, but a factor of 2 or 3 certainly is not.

Table 24–3

DETECTION LIMITS OF ATOMIC SPECTROSCOPY METHODS FOR SELECTED ELEMENTS§

| Element | Absorption, Flame* | Absorption, Electrothermal† | Emission, Flame* | Emission, ICP*‡ |
|---|---|---|---|---|
| Al | 30 | 0.005 | 5 | 2 |
| As | 100 | 0.02 | 0.0005 | 40 |
| Ca | 1 | 0.02 | 0.1 | 0.02 |
| Cd | 1 | 0.0001 | 800 | 2 |
| Cr | 3 | 0.01 | 4 | 0.3 |
| Cu | 2 | 0.002 | 10 | 0.1 |
| Fe | 5 | 0.005 | 30 | 0.3 |
| Hg | 500 | 0.1 | 0.0004 | 1 |
| Mg | 0.1 | 0.00002 | 5 | 0.05 |
| Mn | 2 | 0.0002 | 5 | 0.06 |
| Mo | 30 | 0.005 | 100 | 0.2 |
| Na | 2 | 0.0002 | 0.1 | 0.2 |
| Ni | 5 | 0.02 | 20 | 0.4 |
| Pb | 10 | 0.002 | 100 | 2 |
| Sn | 20 | 0.1 | 300 | 30 |
| V | 20 | 0.1 | 10 | 0.2 |
| Zn | 2 | 0.00005 | 0.0005 | 2 |

*From V. A. Fassel and R. N. Kniseley, *Anal. Chem.*, **1974**, *46*, 1111A. With permission. Copyright 1974 American Chemical Society.

†From C. W. Fuller, *Electrothermal Atomization for Atomic Absorption Spectroscopy*, pp. 65–83. London: The Chemical Society, 1977. With permission of The Royal Society of Chemistry.

‡ICP = inductively coupled plasma.

§All values in nanograms/milliliter = $10^{-3}$ μg/mL = $10^{-3}$ ppm.

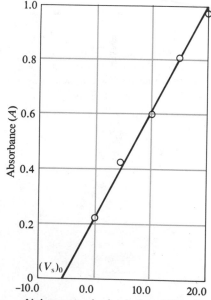

Figure 24–8

A plot of Equation 24–1 for standard-addition data.

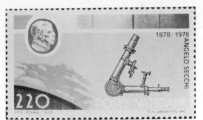

Gustav Robert Kirchhoff (1824–1877) was a German physicist who, along with his colleague, chemist Robert Wilhelm Bunsen (1811–1899), discovered spectroscopic analysis. Bunsen's burner was used to atomize samples of the elements, and Kirchhoff's prism spectroscope was used to analyze the light given off by the incandescent samples. Using this technique, they were able to identify several new elements and demonstrate the presence of a number of elements in the sun by analyzing sunlight. This discovery is depicted on the Vatican stamp shown above. Shown on the stamp are the sun's corona, one of Kirchhoff's spectroscopes, and the absorption spectrum of hydrogen.

Atomic emission spectroscopy can provide both qualitative and quantitative information concerning the analyte.

Atomic-emission and atomic-absorption instruments are similar except no lamp source is required for emission measurements.

One of Kirchhoff's early prism spectroscopes. From his paper in *Abhandl. Berlin Akad.*, **1862**, 227.

Under usual conditions, the relative error associated with a flame absorption analysis is of the order of 1 to 2%. With special precautions, this figure can be lowered to a few tenths of one percent. Errors encountered with nonflame atomization usually exceed those for flame atomization by a factor of 5 to 10.

## 24B–7 Flame Emission Spectroscopy

Atomic emission spectroscopy employing a flame (also called flame emission spectroscopy or flame photometry) has found widespread application in elemental analysis. Its most important uses are in the determination of sodium, potassium, lithium, and calcium, particularly in biological fluids and tissues. Because of its convenience, speed, and relative freedom from interferences, flame emission spectroscopy has become the method of choice for these elements, which are otherwise difficult to determine. The method has also been applied, with various degrees of success, to the determination of perhaps half the elements in the periodic table.

In addition to its quantitative applications, flame emission spectroscopy is also useful for qualitative analysis. Complete spectra are readily recorded; identification of the elements present is then based upon the peak wavelengths, which are unique for each element. In this respect, flame emission has a clear advantage over flame absorption, which does not provide complete absorption spectra because of the discontinuous nature of the radiation sources that must be used.

### Instrumentation

Instruments for flame emission work are similar in design to flame absorption instruments except that in the former the flame acts as the radiation source; a hollow-cathode lamp and chopper are therefore unnecessary. Some instruments can be configured for either emission or absorption measurements. Much of the early work in atomic emission analyses was accomplished with turbulent-flow burners. Laminar-flow burners, however, are becoming more and more widely used.

**Photometers and Spectrophotometers.** For nonroutine analysis, a recording ultraviolet/visible spectrophotometer with a resolution of perhaps 0.05 nm is desirable. Such an instrument can provide complete emission spectra that are useful for the identification of the elements present in a sample. Figure 20–16 is an example of a typical flame emission spectrum excited by an oxyhydrogen flame. The sample was a brine, and spectral lines and bands for several elements are identified. Figure 24–9 illustrates how background corrections are made with a recorded spectrum.

Simple filter photometers often suffice for routine determinations of the alkali and alkaline earth metals. A low-temperature flame is employed to prevent excitation of most other metals. As a consequence, the spectra are simple, and interference filters can be used to isolate the desired emission line.

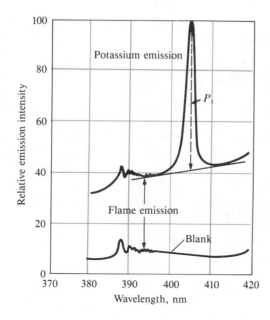

Figure 24–9
A portion of the emission spectrum for potassium in a hydrogen/oxygen flame. The potassium emission is superimposed on the background emission from the flame. For clarity, the spectrum for the blank has been displaced downward.

Several instrument manufacturers supply flame photometers designed specifically for the analysis of sodium, potassium, and lithium in blood serum and other biological samples. In these instruments, the radiation from the flame is split into three beams of approximately equal power. Each beam then passes into a separate photometric system consisting of an interference filter (which transmits an emission line of one of the elements while absorbing those of the other two), a phototube, and an amplifier. The outputs can be measured separately if desired. Ordinarily, however, lithium serves as an *internal standard* for the analysis. For this purpose, a fixed amount of lithium is introduced into each standard and sample. The output ratio between the sodium transducer and the lithium transducer and the potassium transducer and the lithium transducer then serve as analytical parameters. This system provides improved accuracy because the intensities of the three lines are affected in the same way by most analytical variables, such as flame temperature, fuel flow rate, and background radiation. Clearly, lithium must be absent from the sample.

An internal standard is a pure substance that is introduced in known amount into each standard and sample of the analyte. The ratio of the analyte signal to the internal standard signal is then used to determine the analyte concentration.

## Automated Flame Photometers

Fully automated photometers based on the segmented-flow system described in Section 22C–2 are used in clinical laboratories for the determination of sodium and potassium. In these instruments, samples are withdrawn sequentially from a sample turntable, dialyzed to remove protein and particulates, diluted with a lithium internal standard, and aspirated into a flame. Calibration is performed automatically after every nine samples.

## Interferences

The interferences encountered in flame emission spectroscopy have the same sources as those encountered in atomic absorption methods (Section 24B–6); the severity of any given interference often differs from one to the other, however.

### Analytical Techniques

The analytical techniques for flame emission spectroscopy are similar to those described for atomic absorption spectroscopy. Both calibration curves and the standard-addition method are employed. In addition, internal standards can be used to compensate for flame variables.

## 24C ATOMIC ABSORPTION METHODS WITH ELECTROTHERMAL ATOMIZERS

Electrothermal AAS tends to be substantially more sensitive than flame methods.

Flame atomization is not very efficient for two reasons. First, a large portion of the sample flows down the drain (laminar burner) or is not completely atomized (turbulent burner). Second, the residence time of individual atoms in the optical path of the flame is brief ($\cong 10^{-4}$ s). Electrothermal atomizers, which first appeared on the market in the early 1970s, generally provide enhanced sensitivity because the entire sample is atomized in a short period, and the average residence time of the atoms in the optical path is 1 s or more.[3]

Other than the atomizer, the equipment needed for an electrothermal analysis is identical to that used in flame absorption methods. Most modern instruments are designed so that the change from one type of atomizer to the other is a relatively simple matter.

In an electrothermal atomizer, a few microliters of sample are first evaporated at a low temperature and then ashed at a somewhat higher temperature in an electrically heated tube or cup (usually fabricated from carbon). After ashing, the current is rapidly increased to several hundred amperes, which causes the temperature to soar to perhaps 2000 to 3000°C; atomization of the sample occurs in a period of a few milliseconds to seconds. The absorbance of the atomized particles is then determined in the region immediately above the heated conductor.

Figure 24–10a is a cross-sectional view of a commercially available electrothermal atomizer that fits in front of the entrance slit of a monochromator. Atomization occurs in a cylindrical graphite tube that is open at both ends and has a central hole for introduction of sample by means of a micropipette. The tube is about 5 cm long and has an internal diameter of somewhat less than 1 cm. The interchangeable graphite tube fits snugly into a pair of cylindrical graphite electrical contacts located at each end of the tube. These contacts are held in a water-cooled metal housing. Two inert gas streams are provided. The external stream prevents the entrance of outside air and a consequent incineration of the tube. The internal stream flows into the two ends of the tube and out the central sample port.

---

[3]See R. E. Sturgeon, *Anal. Chem.*, **1977**, *49*, 1255A; S. R. Koirtyohann and M. L. Kaiser, *Anal. Chem.*, **1982**, *54*, 1515A; C. W. Fuller, *Electrothermal Atomization for Atomic Absorption Spectroscopy.* London: The Chemical Society, 1978; and A. Varma, *CRC Handbook of Furnace Atomic Absorption Spectroscopy.* Boca Raton: CRC Press, 1989.

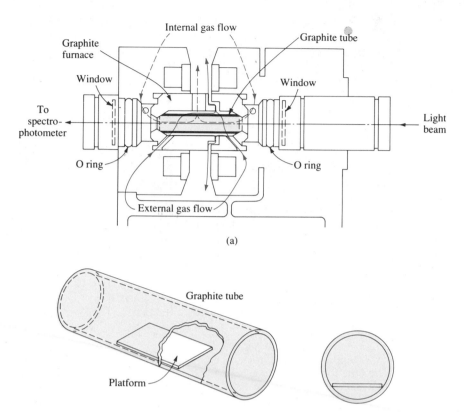

(a)

(b)

Figure 24–10
(a) Cross-sectional view of a graphite furnace. (b) The L'vov platform and its position in the graphite furnace. (a, Courtesy of the Perkin-Elmer Corp., Norwalk, CT; b, Reprinted with permission from: W. Slavin, *Anal. Chem.*, **1982,** *54,* 689A. Copyright 1982 American Chemical Society.)

This stream not only excludes air but also serves to carry away vapors generated from the sample matrix during the first two heating stages.

Figure 24–10b illustrates the so-called L'vov platform, which is often used in graphite furnaces such as that shown in 24–10a. The platform is also graphite and is located beneath the sample entrance port. The sample is evaporated and ashed on this platform in the usual way. When the tube temperature is raised rapidly, however, atomization is delayed, since the sample is no longer directly on the furnace wall. As a consequence, atomization occurs in an environment in which the temperature does not change rapidly. Reproducible peaks are obtained as a result.

With the monochromator set at a wavelength at which absorption occurs, the detector output during electrothermal atomization rises to a maximum after a few seconds of ignition and then rapidly decays back to zero as the atomization products escape into the surroundings. The change is rapid enough (often < 1 s) to require a high-speed recorder or a computerized data-acquisition system. Quantitative analyses are usually based on peak height, although peak area is also useful.

Figure 24–11 shows the output signals (as a function of time) from an atomic absorption spectrophotometer equipped with an electrothermal atomizer. The four peaks on the right were recorded at the wavelength of a lead peak as a 2-$\mu$L sample of canned orange juice was dried, ashed, and atomized. The peaks produced during drying and ashing were caused by particulate ignition products. Comparison of the lead peak on the far right

Electrothermal atomization produces absorption peaks of brief duration.

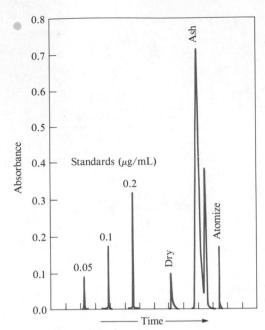

**Figure 24–11**
Typical output from a spectrophotometer equipped with an electrothermal atomizer. The time for drying and ashing are 20 and 60 s, respectively. (Courtesy of Varian Instrument Division, Palo Alto, CA.)

Electrothermal atomizers accommodate small volumes of sample.

with peaks recorded for standards indicates a lead concentration of about 0.1 $\mu$g/mL of juice.

Electrothermal atomizers offer the advantage of unusually high sensitivity for small volumes of sample. Typically, sample volumes are between 0.5 and 10 $\mu$L; under these circumstances, absolute detection limits typically lie in the range from $10^{-10}$ to $10^{-13}$ g of analyte. The relative precision of electrothermal methods is generally in the range of 5 to 10%, compared with the 1 to 2% that can be expected for flame atomization. Furthermore, the interference problems discussed in Section 24B–6 tend to be more severe with electrothermal than with flame atomizers.

## 24D ATOMIC EMISSION METHODS BASED ON ATOMIZATION IN PLASMAS

Plasma atomizers, which became available commercially in the mid-1970s, offer several advantages over flame atomizers.[4] Plasma atomiza-

[4]For a detailed discussion of the various plasma sources, see P. Tschopel, in *Comprehensive Analytical Chemistry*, G. Svehla, Ed., Vol. IX, Chapter 3. New York: Elsevier, 1979; R. D. Sacks, in *Treatise on Analytical Chemistry*, 2nd ed., P. J. Elving, E. J. Meehan, and I. M. Kolthoff, Eds., Part I, Vol. 7, pp. 516–526. New York: Wiley, 1981; J. D. Ingle Jr., and S. R. Crouch, *Spectrochemical Analysis*. Englewood Cliffs, N.J.: Prentice-Hall, 1988.

tion has been used for both thermal emission and fluorescence spectroscopy. It has not been widely used as an atomizer for atomic absorption methods.

By definition, a plasma is a conducting gaseous mixture containing a significant concentration of cations and electrons. In the argon plasma employed for emission analyses, argon ions and electrons are the principal conducting species, although cations from the sample also contribute. Argon ions, once formed in a plasma, are capable of absorbing sufficient power from an external source to maintain the temperature at a level at which further ionization sustains the plasma indefinitely; temperatures as great as 10,000 K are encountered. Three power sources have been employed in argon plasma spectroscopy. One is a dc electrical source capable of maintaining a current of several amperes between electrodes immersed in the argon plasma. The second and third are powerful radio-frequency and microwave-frequency generators through which the argon flows. Of the three, the radio-frequency, or *inductively coupled plasma* (ICP), source appears to offer the greatest advantage in terms of sensitivity and freedom from interference. On the other hand, the *dc plasma source* (DCP) has the virtues of simplicity and lower cost.

> A *plasma* is a gas containing relatively high concentrations of cations and electrons.

## 24D-1  The Inductively Coupled Plasma Source[5]

Figure 24–12a is a schematic drawing of an inductively coupled plasma source. It consists of three concentric quartz tubes through which streams of argon flow at a total rate of between 11 and 17 L/min. The diameter of the largest tube is about 2.5 cm. Surrounding the top of this tube is a water-cooled induction coil powered by a radio-frequency generator capable of producing 2 kW of energy at about 27 MHz. Ionization of the flowing argon is initiated by a spark from a Tesla coil. The resulting ions, and their associated electrons, then interact with the fluctuating magnetic field (labeled H in Figure 24–12) produced by the induction coil I. This interaction causes the ions and electrons within the coil to flow in the closed annular paths depicted in the figure; ohmic heating is the consequence of their resistance to this movement.

During the 1980s, low-flow, low-power torches appeared on the market. Currently, one manufacturer offers only this type of torch, whereas other companies offer them as an option. Typically, these torches require a total argon flow of lower than 10 L/min and require less than 800 W of radio-frequency power.[6]

---

[5]For a more complete discussion, see V. A. Fassel, *Science*, **1978**, *202*, 183; V. A. Fassel, *Anal. Chem.*, **1979**, *51*, 1290A; M. Thompson and J. N. Walsh, *Inductively Coupled Plasma Spectrometry*. London: Blackie, 1983; J. D. Ingle Jr. and S. R. Crouch, *Spectrochemical Analysis*, Chapter 8. Englewood, NJ: Prentice-Hall, 1988; J. W. Olesik, *Anal. Chem.*, **1991**, *63*, 12A; and *Handbook of Inductively Coupled Plasma Spectroscopy*, M. Thompson and J. N. Walsh, Eds. New York: Chapman and Hall, 1989.

[6]See R. N. Savage and G. M. Heiftje, *Anal. Chem.*, **1980**, *52*, 1267; J. J. Urh, *Amer. Lab.*, **1986**, (3), 105.

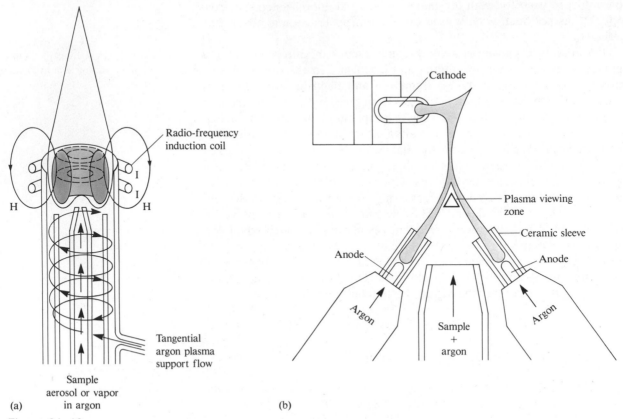

Figure 24–12

Plasma sources. (a) Inductively coupled plasma source. (b) A three-electrode dc plasma jet. (a, From V. A. Fassel, *Science,* **1978,** *202,* 185. With permission. Copyright 1978 by the American Association for the Advancement of Science; b, Courtesy of Applied Research Laboratories, Inc., Sunland, CA.)

The temperature of this plasma is high enough to require that it be thermally isolated from the outer quartz cylinder. Isolation is achieved by flowing argon tangentially around the walls of the tube, as indicated by the arrows in Figure 24–12a. The tangential flow cools the inside walls of the central tube and centers the plasma radially.

### Sample Injection

The sample is carried into the hot plasma at the head of the tubes by argon flowing at about 1 L/min through the central quartz tube. The sample can be an aerosol, a thermally generated vapor, or a fine powder.

### Plasma Appearance and Spectra

The typical plasma has a very intense, brilliant white, nontransparent core topped by a flame-like tail. The core, which extends a few millimeters above the tube, produces a spectral continuum upon which is superimposed the atomic spectrum for argon. The continuum apparently results when argon and other ions recombine with electrons. In the region

10 to 30 mm above the core, the continuum fades and the plasma is optically transparent. Spectral observations are generally made 15 to 20 mm above the induction coil. Here, the background radiation is remarkably free of argon lines and is well suited to spectral measurements. Many of the most sensitive analyte lines in this region of the plasma are from ions such as $Ca^+$, $Cd^+$, $Cr^+$, and $Mn^+$.

### Analyte Atomization and Ionization

By the time the sample atoms reach the observation point in the plasma, they have had a residence time of about 2 ms at temperatures ranging from 6000 to 8000 K. These times and temperatures are two to three times as great as those attainable in the hottest combustion flames (acetylene/nitrous oxide). As a consequence, atomization is more nearly complete, and fewer chemical interferences are encountered. Surprisingly, ionization interference effects are small or nonexistent, perhaps because the large concentration of electrons from the ionization of the argon maintains a more or less constant electron concentration in the plasma.

Several other advantages are associated with the plasma source. First, atomization occurs in a chemically inert environment, which should also enhance the lifetime of the analyte. In addition, and in contrast to flame sources, the temperature cross section of the plasma is relatively uniform. As a consequence, calibration curves tend to remain linear over several orders of magnitude of concentration.

*Plasma temperatures are substantially higher than those in a flame.*

### 24D–2 The Direct-Current Argon Plasma Source

Direct-current plasma jets were first described in the 1920s and were systematically investigated as sources for emission spectroscopy for more than two decades. It was not until recently, however, that a source of this type was designed that provides data reproducible enough to successfully compete with flame and inductively coupled plasma sources.[7]

Figure 24–12b is a diagram of a commercially available dc plasma source that is well suited to the excitation of emission spectra. This plasma-jet source consists of three electrodes arranged in an inverted Y configuration. A graphite anode is located in each arm of the Y, and a tungsten cathode is located at the inverted base. Argon flows from the two anode blocks toward the cathode. The plasma jet is formed when the cathode is momentarily brought into contact with the anodes. Ionization of the argon occurs, and the current that develops ($\simeq$ 14 A) generates additional ions to sustain itself indefinitely. The temperature is perhaps 10,000 K in the arc core and 5000 K in the viewing region. The sample is aspirated into the area between the two arms of the Y, where it is atomized, excited, and its spectrum viewed.

Spectra produced by the plasma jet tend to have fewer lines than those produced by the inductively coupled plasma, and the lines formed in the former are largely from atoms rather than ions. Sensitivities achieved

---

[7]For additional details, see G. W. Johnson, H. E. Taylor, and R. K. Skogerboe, *Anal. Chem.*, **1979**, *51*, 2403; *Spectrochim. Acta, Part B*, **1979**, *34*, 197; J. Reednick, *Amer. Lab.*, **1979**, *11*(3), 53.

with the dc plasma jet appear to range from an order of magnitude lower to about the same as those obtainable with the inductively coupled plasma. The reproducibilities of the two systems are similar. Significantly less argon is required for the dc plasma, and the auxiliary power supply is simpler and less expensive. On the other hand, the graphite electrodes must be replaced every few hours, whereas the inductively coupled plasma source requires little or no maintenance.

## 24D–3 Instruments for Plasma Spectroscopy

Several manufacturers offer instruments for plasma emission spectroscopy. In general, these consist of a high-quality grating spectrophotometer for the ultraviolet and visible regions and a photomultiplier detector. Many are automated, so that an entire spectrum can be scanned sequentially. Others have several photomultiplier tubes located in the focal plane, so that the lines for several elements (two dozen or more) can be monitored simultaneously. Such instruments are very expensive.

## 24D–4 Quantitative Applications of Plasma Sources

Unquestionably, inductively coupled and dc plasma sources yield significantly better quantitative analytical data than other emission sources. The excellence of these results stems from the high stability, low noise, low background, and freedom from interferences of the sources when operated under appropriate experimental conditions. The performance of the inductively coupled plasma source is somewhat better than that of the dc plasma source in terms of detection limits. The latter, however, is less expensive to purchase and operate and is entirely adequate for many applications.

In general, the detection limits with the inductively coupled plasma source appear comparable to or better than those of other atomic spectral procedures. Table 24–3 compares the sensitivity of several of these methods.

---

## 24E   QUESTIONS AND PROBLEMS

*24–1. Describe the basic differences between atomic emission and atomic fluorescence spectroscopy.

24–2. Define *(a) atomization, (b) pressure broadening, *(c) Doppler broadening, (d) turbulent-flow nebulizer, *(e) laminar-flow nebulizer, (f) hollow-cathode lamp, *(g) sputtering, (h) ionization suppressor, *(i) spectral interference, (j) chemical interference, *(k) radiation buffer, (l) releasing agent, *(m) protective agent.

*24–3. Why is atomic emission more sensitive to flame instability than atomic absorption or fluorescence?

24–4. Why is a nonflame atomizer more sensitive than a flame atomizer?

*24–5. Why is source modulation employed in atomic absorption spectroscopy?

24–6. In a hydrogen/oxygen flame, an atomic absorption peak for iron decreased in the presence of large concentrations of sulfate ion.
(a) Suggest an explanation for this observation.
(b) Suggest three possible methods for overcoming the potential interference of sulfate in a quantitative determination of iron.

*24–7. Why are the lines from a hollow-cathode lamp generally narrower than the lines emitted by atoms in a flame?

24–8. Assume that the peaks shown in Figure 24–7 were

obtained for 2-$\mu$L aliquots of standards and sample. Calculate the parts per million of lead in the sample of canned orange juice.

*24–9. Suggest sources of the two peaks in Figure 24–7 that appear during drying and ashing.

24–10. In the concentration range from 500 to 2000 ppm of U, a linear relationship is found between absorbance at 351.5 nm and concentration. At lower concentrations, the relationship becomes nonlinear unless about 2000 ppm of an alkali metal salt is introduced. Explain.

*24–11. What is the purpose of an internal standard in flame emission methods?

24–12. A 5.00-mL sample of blood was treated with trichloroacetic acid to precipitate proteins. After centrifugation, the resulting solution was brought to pH 3 and extracted with two 5-mL portions of methyl isobutyl ketone containing the organic lead-complexing agent APCD. The extract was aspirated directly into an air/acetylene flame and yielded an absorbance of 0.502 at 283.3 nm. Five-milliliter aliquots of standard solutions containing 0.400 and 0.600 ppm of lead were treated in the same way and yielded absorbances of 0.396 and 0.599. Calculate the parts per million of lead in the sample, assuming that Beer's law is followed.

*24–13. The sodium in a series of cement samples was determined by flame emission spectroscopy. The flame photometer was calibrated with a series of standards containing 0, 20.0, 40.0, 60.0, and 80.0 $\mu$g $Na_2O$ per milliliter. The instrument readings for these solutions were 3.1, 21.5, 40.9, 57.1, and 77.3.
(a) Plot the data.
(b) Derive a least-squares line for the data.
(c) Calculate standard deviations for the slope and about regression for the line in (b).
(d) The following data were obtained for replicate 1.000-g samples of cement dissolved in HCl and diluted to 100.0 mL after neutralization.

| | Emission Reading | | | |
| --- | --- | --- | --- | --- |
| | Blank | Sample A | Sample B | Sample C |
| Replicate 1 | 5.1 | 28.6 | 40.7 | 73.1 |
| Replicate 2 | 4.8 | 28.2 | 41.2 | 72.1 |
| Replicate 3 | 4.9 | 28.9 | 40.2 | Spilled |

Calculate the percent $Na_2O$ in each sample. What are the absolute and relative standard deviations for the average of each determination?

24–14. The chromium in an aqueous sample was determined by pipetting 10.0 mL of the unknown into each of five 50.0-mL volumetric flasks. Various volumes of a standard containing 12.2 ppm Cr were added to the flasks, and the solutions were then diluted to volume.

| Unknown, mL | Standard, mL | Absorbance |
| --- | --- | --- |
| 10.0 | 0.0 | 0.201 |
| 10.0 | 10.0 | 0.292 |
| 10.0 | 20.0 | 0.378 |
| 10.0 | 30.0 | 0.467 |
| 10.0 | 40.0 | 0.554 |

(a) Plot absorbance as a function of volume of standard $V_s$.
(b) Derive an expression relating absorbance to the concentrations of standard and unknown ($c_s$ and $c_x$) and the volumes of the standards and unknown ($V_s$ and $V_x$) as well as the volume to which the solutions were diluted ($V_t$).
(c) Derive expressions for the slope and the intercept of the straight line obtained in (a) in terms of the variables listed in (b).
(d) Show that the concentration of the analyte is given by the relationship $c_x = ac_s/bV_x$, where $a$ and $b$ are the slope and the intercept of the straight line in (a).
(e) Determine values for $a$ and $b$ by the method of least squares.
(f) Calculate the standard deviation for the slope and about regression in (e).
(g) Calculate the ppm Cr in the sample using the relationship given in (d).
(h) The relative variance of a result obtained in this way is approximately equal to the sum of the relative variances of the slope and the intercept. Calculate the absolute standard deviation for the result obtained in (g).

# KINETIC METHODS
# OF ANALYSIS

$K$inetic methods of analysis differ in a fundamental way from the equilibrium, or thermodynamic, methods we have dealt with so far. In kinetic methods, measurements are made under *dynamic* conditions in which the concentrations of reactants and products are changing continuously. Here, the rate of appearance of products or disappearance of reactants serves as the analytical parameter. The measurements in thermodynamic methods, in contrast, are performed upon systems that have come to equilibrium so that concentrations are *static*.

The distinction between the two types of methods is illustrated in Figure 25–1, which shows the progress over time of the reaction

$$A + R \rightleftharpoons P \qquad (25-1)$$

where A represents the analyte, R the reagent, and P the product. Thermodynamic methods operate in the region beyond time $t_e$ where the bulk concentrations of reactants and product have become constant and the chemical system is at equilibrium. In contrast, kinetic methods are carried out during the time interval from 0 to $t_e$ where reactant and product concentrations are changing continuously, and the rate of their disappearance or appearance is conveniently measured.

Selectivity in kinetic methods is achieved by choosing reagents and conditions that amplify differences in the *rates* at which the analyte and potential interferences react. Selectivity in thermodynamic methods is realized by choosing reagents and conditions that amplify differences in *equilibrium constants*.

Kinetic methods greatly extend the number of chemical reactions that can be used for analytical purposes because they permit the use of reactions that are too slow or too incomplete for thermodynamics-based procedures.

Two types of kinetic methods are based upon catalyzed reactions. In the first, an analyte is the catalyst and is determined from its catalytic effect upon an *indicator reaction* involving reactants or products that are easily measured. Such methods are among the most sensitive in the chem-

In kinetic methods, we measure *changes* in concentration per unit time. In equilibrium methods, we measure absolute concentrations.

640

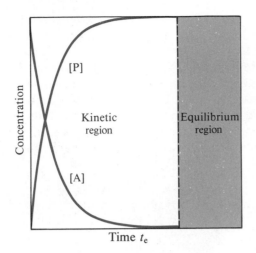

Figure 25–1

Change in concentration of analyte [A] and product [P] as a function of time.

ist's repertoire. In the second type, a catalyst is introduced to hasten the reaction between analyte and reagent. This approach is often highly selective, or even specific, particularly when an enzyme serves as the catalyst.

Undoubtedly, the most widespread use of kinetic methods is in biochemical and clinical laboratories, where the number of analyses based upon kinetics exceeds those based upon thermodynamics.[1]

## 25A   RATES OF CHEMICAL REACTIONS; RATE LAWS

This section provides a brief introduction to the theory of chemical kinetics, which is needed to understand the basis for kinetic methods of analysis.

### 25A–1 Some Terms and Symbols Used in Chemical Kinetics

The *mechanism* for a chemical reaction consists of a series of chemical equations describing the individual elementary steps by which products are formed from reactants. Much of what chemists know about mechanisms has been gained from studies in which the rate at which reactants are consumed or products formed is measured as a function of such variables as reactant and product concentration, temperature, pressure, pH, and ionic strength. Such studies lead to an empirical *rate law* that relates the reaction rate to the concentrations of reactants, products, and intermediates at any instant. Mechanisms are derived by postulating a series of elementary steps that are chemically reasonable and consistent with the empirical rate law.

The rate law for a reaction is determined experimentally.

#### Concentration Terms in Rate Laws

Rate laws are algebraic expressions consisting of concentration terms and constants, which often look somewhat like an equilibrium constant ex-

---

[1]H. O. Mottola, *Kinetic Aspects of Analytical Chemistry*, New York: Wiley, 1988.

Molar concentrations are still symbolized with square brackets. In the context of kinetic methods, however, their numerical values change with time.

pression (see Equation 25–2). You should realize, however, that the square-bracketed terms in a rate expression represent molar concentrations *at a particular instant* rather than equilibrium molar concentrations (as in equilibrium constant expressions). This meaning is frequently emphasized by adding a subscript to show the time to which the concentration refers. Thus, $[A]_t$, $[A]_0$, and $[A]_\infty$ indicate the concentration of A at time $t$, time zero, and infinite time, respectively. Infinite time is regarded as any time greater than that required for the reaction under consideration to come to equilibrium. That is, $t_\infty > t_e$ in Figure 25–1.

## Reaction Order

Let us assume that the empirical rate law for the general reaction shown as Equation 25–1 is found by experiment to take the form

$$\text{rate} = -\frac{d[A]}{dt} = -\frac{d[R]}{dt} = \frac{d[P]}{dt} = k[A]^m[R]^n \qquad (25\text{–}2)$$

Because A and R are being used up, the rates of change of [A] and [R] with respect to time are negative.

where the rate is the derivative of the concentration of A, R, or P with respect to time. Note that the first two rates carry a negative sign because the concentrations of A and R decrease as the reaction proceeds. In this rate expression, $k$ is the *rate constant*, $m$ is the *order of the reaction* with respect to A, and $n$ is the order of the reaction with respect to R. The *overall order* of the reaction is $p = m + n$. Thus, if $m = 1$ and $n = 2$, the reaction is said to be first order in A, second order in R, and third order overall.

## Units for Rate Constants

Since reaction rates are always expressed in terms of concentration per unit time, the units of the rate constant are determined by the overall order $p$ of the reaction according to the relation

The units of $k$ for a first-order reaction are $s^{-1}$.

$$\frac{\text{concentration}}{\text{time}} = (\text{units of } k)(\text{concentration})^p$$

where $p = m + n$. Rearranging leads to

$$\text{units of } k = (\text{concentration})^{1-p} \text{ time}^{-1}$$

Thus, the units for a first-order rate constant are $s^{-1}$, and the units for a second-order rate constant are $M^{-1}s^{-1}$.

## 25A–2 The Rate Law for First-Order Reactions

Radioactive decay is an example of a spontaneous decomposition.

The simplest case in the mathematical analysis of reaction kinetics is that of a spontaneous irreversible decomposition of a species A:

$$A \xrightarrow{k} P \qquad (25\text{–}3)$$

The reaction is first order in A, and the rate is

$$\text{rate} = \frac{-d[A]}{dt} = k[A] \qquad (25\text{-}4)$$

## Pseudo-First-Order Reactions

A first-order decomposition reaction per se is generally of no use in analytical chemistry because an analysis is ordinarily based upon reactions involving at least two species, an analyte and a reagent. Usually, however, the rate law for a reaction involving two species is sufficiently complex as to be of no use for analytical purposes. In fact, the only kinetic methods that are useful are those that can be performed under conditions that permit the chemist to simplify complex rate laws to a form analogous to Equation 25-4. A higher-order reaction that is carried out so that such a simplification is feasible is termed a *pseudo-first-order reaction*. Methods for converting higher-order reactions to pseudo-first-order are dealt with in later sections.

Radioactive decay is an exception to this statement. The technique of neutron activation analysis is based on the measurement of the spontaneous decay of radionuclides created by irradiation of a sample in a nuclear reactor.

## Mathematical Expressions Describing First-Order Behavior

Because essentially all kinetic analyses are performed under pseudo-first-order conditions, it is worthwhile to examine in detail some of the characteristics of reactions having rate laws that approximate Equation 25-4.

By rearranging Equation 25-4, we obtain

$$\frac{d[A]}{[A]} = -k\,dt \qquad (25\text{-}5)$$

The integral of this equation from time zero, when $[A] = [A]_0$, to time $t$, when $[A] = [A]_t$, is

$$\int_{[A]_0}^{[A]_t} \frac{d[A]}{[A]} = -k \int_0^t dt$$

Evaluation of the integrals gives

$$\ln \frac{[A]_t}{[A]_0} = -kt \qquad (25\text{-}6)$$

Finally, by taking the exponential of both sides of Equation 25-6, we obtain

$$\frac{[A]_t}{[A]_0} = e^{-kt} \qquad \text{or} \qquad [A]_t = [A]_0\, e^{-kt} \qquad (25\text{-}7)$$

This integrated form of the rate law gives the concentration of A as a function of the initial concentration $[A]_0$, the rate constant $k$, and the time $t$. A plot of this relationship is depicted in Figure 25-1.

Example 25–1

A reaction is first-order with $k = 0.0370$ s$^{-1}$. Calculate the concentration of reactant remaining 18.2 s after initiation of the reaction if its initial concentration is 0.0100 M.

Substituting into Equation 25–7 gives

$$[A]_{18.2} = (0.0100 \text{ M})e^{-(0.0370 \text{ s}^{-1})(18.2 \text{ s})} = 0.00510 \text{ M}$$

When the rate of a reaction is being followed by the rate of appearance of a product P rather than the rate of disappearance of analyte A, it is useful to modify Equation 25–7 to relate the concentration of P at time $t$ to the initial analyte concentration $[A]_0$. The concentration of A at any time is equal to its original concentration minus the concentration of product (when 1 mol of product forms for 1 mol of analyte). Thus,

$$[A]_t = [A]_0 - [P]_t \qquad (25\text{–}8)$$

Substitution of this expression for $[A]_t$ into Equation 25–7 and rearrangement give

$$[P]_t = [A]_0(1 - e^{-kt}) \qquad (25\text{–}9)$$

A plot of this relationship is also shown in Figure 25–1.

The form of Equations 25–7 and 25–9 is that of a pure exponential, which appears widely in science and engineering. A pure exponential in this case has the useful characteristic that equal elapsed times give equal fractional decreases in reactant concentration or increases in product concentration. As an example, consider a time interval $t = \tau = 1/k$, which upon substitution into Equation 25–7 gives

$$[A]_\tau = [A]_0 e^{-k\tau} = [A]_0 e^{-k/k} = (1/e)[A]_0$$

and likewise for a period of $t = 2\tau = 2/k$

$$[A]_{2\tau} = (1/e)^2[A]_0$$

and so on for successive periods, as shown in Figure 25–2.

The period $\tau = 1/k$ is sometimes referred to as the *natural lifetime* of species A. During time $\tau$, the concentration of A decreases to $1/e$ of its original value. A second period, from $t = \tau$ to $t = 2\tau$, produces an equivalent fractional decrease in concentration to $1/e$ of the value at the beginning of the second interval, which is $(1/e)^2$ of $[A]_0$. A more familiar example of this property of exponentials is found in the half-life $t_{1/2}$ of radionuclides. During a period $t_{1/2}$, half of the atoms in a sample of a radioactive element decay to products; a second period of $t_{1/2}$ reduces the amount of the element to one quarter of its original number, and so on for

The fraction of reactant used (or product formed) in a first-order reaction is the same for any given period of time.

Challenge: Derive an expression for $t_{1/2}$ in terms of $\tau$.

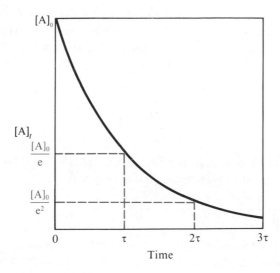

Figure 25–2
Rate curve for a first-order reaction showing that equal elapsed times produce equal fractional decreases in analyte concentration.

succeeding periods. Regardless of the time interval chosen, equal elapsed times produce equal fractional decreases in reactant concentration for a first-order process.

---

**Example 25–2**

Calculate the time required for a first-order reaction with $k = 0.0500 \text{ s}^{-1}$ to proceed to 99.0% completion.

For 99.0% completion, $[A]_t/[A]_0 = (100 - 99)/100 = 0.010$; substitution into Equation 25–6 then gives

$$\ln 0.010 = -kt = -(0.0500 \text{ s}^{-1})t$$

$$t = -\frac{\ln 0.010}{0.0500 \text{ s}^{-1}} = 92 \text{ s}$$

---

## 25A–3  Rate Laws for Second-Order and Pseudo-First-Order Reactions

Let us consider a typical analytical reaction in which 1 mol of analyte A reacts with 1 mol of reagent B to give a single product P. For the present, we assume the reaction is irreversible and write

$$A + R \xrightarrow{k} P \qquad (25\text{–}10)$$

If the reaction occurs in a single elementary step, the rate is proportional to the concentration of each of the reactants, and the rate law is

$$-\frac{d[A]}{dt} = k[A][R] \qquad (25\text{–}11)$$

The reaction is first order in each of the reactants and second order overall. If the concentration of R is chosen such that $[R] \gg [A]$, the concentration of R changes very little during the course of the reaction and we can write $k[R] = \text{constant} = k'$. Equation 25–11 is then rewritten as

$$-\frac{d[A]}{dt} = k'[A] \qquad (25–12)$$

Second-order or other higher-order reactions can be performed under pseudo-first-order conditions through control of experimental conditions.

which is identical in form to the first-order case of Equation 25–4. Hence the reaction is said to be *pseudo first order* in A.

---

Example 25–3

For a pseudo-first-order reaction in which the reagent is present in 100-fold excess, find the relative error resulting from the assumption that $k[R] = k'$ is constant when the reaction is 40% complete.

The initial concentration of the reagent can be expressed as

$$[R]_0 = 100[A]_0$$

At 40% reaction, 60% of A remains. Thus,

$$[A]_{40\%} = 0.60[A]_0$$
$$[R]_{40\%} = [R]_0 - 0.40[A]_0 = 100[A]_0 - 0.40[A]_0 = 99.6[A]_0$$

Assuming pseudo-first-order behavior, the rate at 40% reaction is

$$-\frac{d[A]_{40\%}}{dt} = k[R]_0[A]_{40\%} = k(100[A]_0)(0.60[A]_0)$$

The true rate at 40% reaction is $k(99.6[A]_0)(0.60[A]_0)$. Thus, the relative error is

$$\frac{k(100[A]_0)(0.60[A]_0) - k(99.6[A]_0)(0.60[A]_0)}{k(99.6[A]_0)(0.60[A]_0)} = 0.004 \text{ (or 0.4\%)}$$

---

As Example 25–3 clearly illustrates, the error associated with the determination of the rate of a pseudo-first-order reaction with a 100-fold excess of reagent is negligible. A 50-fold reagent excess leads to a 1% error, an error of this magnitude usually deemed acceptable in kinetic methods. Moreover, the error is even less significant at times when the reaction is less than 40% complete.

It may be necessary to account for the rate of the reverse reaction.

Reactions are seldom completely irreversible, and a rigorous description of the kinetics of a second-order reaction that occurs in a single step must take into account the reverse reaction. The rate of the reaction is the difference between the forward rate and the reverse rate:

$$-\frac{d[A]}{dt} = k_1[A][R] - k_{-1}[P]$$

where $k_1$ is the second-order rate constant for the forward reaction and $k_{-1}$ is the first-order rate constant for the reverse reaction. In deriving this equation we have assumed for simplicity that a single product is formed, but more complex cases can be described as well.[2] As long as conditions are maintained such that $k_{-1}$ and/or [P] are relatively small, the rate of the reverse reaction is negligible, and little error is introduced by the assumption of pseudo-first-order behavior.

## 25A–4  Catalyzed Reactions

Catalyzed reactions, particularly those in which enzymes serve as catalysts, are widely used for the determination of a variety of biological and biochemical species as well as a number of inorganic cations and anions. We shall therefore use enzyme-catalyzed reactions to illustrate catalytic rate laws and to show how these rate laws can be reduced to relatively simple algebraic relationships, such as the pseudo-first-order equation shown as Equation 25–12. These simplified relationships can then be used for analytical purposes.

### Enzyme-Catalyzed Reactions

Enzymes are high-molecular-weight molecules that catalyze reactions of biological and biomedical importance. Enzymes are particularly useful as analytical reagents owing to their high degree of selectivity. Consequently, they are widely used in the determination of molecules with which they combine when acting as catalysts. Such molecules are usually designated as *substrates*. In addition to the determination of substrates, enzyme-catalyzed reactions are employed for the determination of activators, inhibitors, and, of course, enzymes.

> Enzymes serve as highly selective analytical reagents.

> The species acted upon by an enzyme is called a *substrate*.

The behavior of a large number of enzymes is consistent with the general mechanism

$$E + S \underset{k_{-1}}{\overset{k_1}{\rightleftharpoons}} ES \overset{k_2}{\rightarrow} P + E \qquad (25–13)$$

In this equation, the enzyme E is shown as reacting reversibly with the substrate S to form an enzyme-substrate complex ES; this complex then decomposes irreversibly to form the product(s) and the regenerated enzyme. The rate law for this mechanism assumes one of two forms, depending upon the relative rates of the two steps. If the second step is considerably slower than the first (case 1), the reactants and ES essen-

---

[2]See J. H. Espenson, *Chemical Kinetics and Reaction Mechanisms,* pp. 42–48. New York: McGraw-Hill, 1981.

tially achieve equilibrium. In this case, the second step determines the overall rate and is thus termed *the rate-determining step*. When, however, the rates of the two steps are comparable in magnitude (case 2), ES decomposes as rapidly as it is formed, and its concentration can be assumed to be small and relatively constant throughout much of the reaction. In the two sections that follow, we show that in both cases, the reaction conditions can be arranged to yield simple relationships between rate and analyte concentration.

### Case 1: Rate Determined by the Rate of Complex Decomposition

A rate law consistent with the mechanism of Equation 25–13 when the second step is rate-determining is derived in the following way. The concentrations of S and E at time $t$ are given by

$$[\text{S}]_t = [\text{S}]_0 - [\text{ES}]_t \quad \text{and} \quad [\text{E}]_t = [\text{E}]_0 - [\text{ES}]_t \quad (25\text{–}14)$$

Since the equilibrium for the first step is achieved rapidly relative to the rate of the second step, we can equate the forward and reverse rates of the first step and obtain

$$-\frac{d[\text{E}]}{dt} = -\frac{d[\text{S}]}{dt} \simeq \frac{d[\text{ES}]}{dt}$$

At any time $t$, the rate of the forward reaction is proportional to the product of the concentration of the two reactants ($[\text{E}]_t[\text{S}]_t$); the rate of the reverse reaction is then proportional to the instantaneous concentration of the complex ($[\text{ES}]_t$). Thus,

$$k_1[\text{E}]_t[\text{S}]_t = k_{-1}[\text{ES}]_t$$

Rearranging this equation leads to

$$\frac{k_1}{k_{-1}} = \frac{[\text{ES}]_t}{[\text{E}]_t[\text{S}]_t} = K \qquad (25\text{–}15)$$

When the rates of the forward and reverse reactions are equal, the system is *at chemical equilibrium*. Thus the concentration terms in Equation 25–15 are *equilibrium concentrations*, and $K$ in this equation is the *equilibrium constant* for the first step in the reaction. Substitution of $[\text{S}]_t$ and $[\text{E}]_t$ from Equation 25–14 into Equation 25–15 and rearrangement give

$$[\text{ES}]_t = K([\text{S}]_0 - [\text{ES}]_t)([\text{E}]_0 - [\text{ES}]_t) \qquad (25\text{–}16)$$

In theory, either the substrate or the enzyme catalyst can be present in considerable excess; however, enzyme reagents are expensive and are thus nearly always maintained at low levels when a substrate molecule is being determined. Furthermore, natural levels of enzymes are very low so that we can generally assume that $[\text{S}]_0 \gg [\text{ES}]_t$. Equation 25–16 then becomes

$$[ES]_t = K[S]_0([E]_0 - [ES]_t) \qquad (25\text{-}17)$$

When Equation 25–17 is solved for $[ES]_t$, we have

$$[ES]_t = \frac{K[S]_0[E]_0}{1 + K[S]_0} \qquad (25\text{-}18)$$

and the rate of the reaction is

$$\frac{d[P]}{dt} = k_2[ES]_t = \frac{k_2K[S]_0[E]_0}{1 + K[S]_0} = \frac{k_2[S]_0[E]_0}{(1/K) + [S]_0} \qquad (25\text{-}19)$$

It is usually possible to adjust conditions such that Equation 25–19 simplifies to a relationship that can be used for the determination of either the catalyst or the substrate. For enzyme determinations, the concentration of the substrate is made large enough so that $[S]_0 \gg 1/K$; the denominator of Equation 25–19 is then approximately equal to $[S]_0$, and the rate is

To determine an enzyme under case 1 conditions, $[S]_0 \gg 1/K$.

$$\frac{d[P]}{dt} = k_2[E]_0 \qquad (25\text{-}20)$$

Thus, the rate of the reaction is directly proportional to the initial concentration of the enzyme catalyst and can be used for determining that species.

If, however, the substrate is the species to be determined, its initial concentration is made small enough so that $[S]_0 \ll 1/K$. Equation 25–19 then reduces to

To determine a substrate under case 1 conditions, $[S]_0 \ll 1/K$.

$$\frac{d[P]}{dt} = k_2K[S]_0[E]_0$$

In order to determine substrate concentrations, measurements are performed at constant enzyme concentration so that

$$\frac{d[P]}{dt} = k'[S]_0 \qquad (25\text{-}21)$$

where $k' = k_2K[E]_0$. Under these conditions, the rate of the reaction is clearly proportional to substrate concentration. Note that the catalyst is being regenerated continuously (Equation 25–13). Thus $[E]_t \simeq [E]_0$.

## Case 2: Rate Determined by the Steady-State Condition

In the event that the reaction rates of the two steps in the mechanism shown by Equation 25–13 are comparable, it is necessary to use what is called the *steady-state approximation* to arrive at a rate law. In this treatment, which is described in many sources[3] and therefore is not repro-

---

[3]For example, see J. H. Espenson, *Chemical Kinetics and Reaction Mechanisms*, pp. 72–80. New York: McGraw-Hill, 1981.

duced here, the rate of change in the concentration of the ES complex is presumed to be negligible. This assumption leads to the rate law

Equation 25–22 is called the *Michaelis-Menten* equation

$$\frac{d[P]}{dt} = \frac{k_2[E]_0[S]_t}{[S]_t + (k_{-1} + k_2)/k_1} = \frac{k_2[E]_0[S]_t}{[S]_t + K_m} \qquad (25\text{–}22)$$

The *Michaelis constant*, $K_m = (k_{-1} + k_2)/k_1$, is essentially an "equilibrium-like" constant analogous to $1/K$ in case 1; it expresses the ratio of the sum of the rates of the two reactions tending to destroy ES to the rate of the reaction favoring ES formation. The constant is usually expressed in units of millimoles/liter (mM) and ranges from 0.01 mM to 100 mM for common enzymes.

As in case 1, the rate equation can be simplified so that the reaction rate is proportional to either enzyme or substrate concentration. For example, if the concentration of substrate is made large enough so that $[S] \gg K_m$, Equation 25–22 reduces to

$$\frac{d[P]}{dt} = k_2[E]_0 \qquad (25\text{–}23)$$

Under these conditions, when the rate is independent of substrate concentration, the reaction is said to be *pseudo zero order* in substrate, and the rate is directly proportional to the concentration of enzyme.

When conditions are such that the concentration of S is small or when $K_m$ is relatively large, then $[S]_t \ll K_m$, and Equation 25–22 simplifies to

$$\frac{d[P]}{dt} = \frac{k_2}{K_m} [E]_0[S]_t = k'[S]_t$$

Table 25–1

$K_m$ FOR SOME ENZYMES

| Enzyme and Substrate | $K_m$, mM |
|---|---|
| Catalase | |
| H₂O₂ | 25 |
| Hexokinase | |
| Glucose | 0.15 |
| Fructose | 1.5 |
| Chymotrypsin | |
| n-Benzoyltyrosinamide | 2.5 |
| n-Formyltyrosinamide | 12.0 |
| n-Acetyltyrosinamide | 32 |
| Glycyltyrosinamide | 122 |
| Carbonic anhydrase | |
| HCO₃⁻ | 9.0 |
| Glucose oxidase | |
| Glucose (saturated with O₂) | 0.013 |
| Urease | |
| Urea | 2.0 |

where $k' = k_2[E]_0/K_m$. Hence, the kinetics are first order in substrate. In order to use this equation for determining analyte concentrations, it is necessary to measure $d[P]/dt$ at the beginning of the reaction, where $[S]_t \simeq [S]_0$, so that

$$\frac{d[P]}{dt} \simeq k'[S]_0 \qquad (25\text{–}24)$$

To determine enzymes under case 2 conditions, $[S] \gg K_m$.

To determine substrates under case 2 conditions, $[S] \ll K_m$.

The regions where Equations 25–23 and 25–24 are applicable are illustrated in Figure 25–3, in which the initial rate of an enzyme-catalyzed reaction is plotted as a function of substrate concentration. When substrate concentration is low, Equation 25–24, which is linear in substrate concentration, governs the shape of the curve, and it is this region that is used to determine the amount of substrate present.

If it is desired to determine the amount of enzyme, the region of high substrate concentration is employed, where Equation 25–23 applies, and the rate is independent of substrate concentration. The limiting rate of the reaction at large values of [S] is often referred to as $v_{max}$, as indicated in the figure. It can be shown that the value of the substrate concentration at exactly $v_{max}/2$ is equal to the Michaelis constant $K_m$.

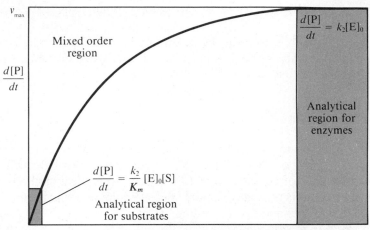

**Figure 25–3**
Change in rate of product formation as a function of substrate concentration, showing the parts of the curve useful for determination of substrate and enzyme.

---

**Example 25–4**

The enzyme urease, which catalyzes the hydrolysis of urea, is widely used for determining urea in blood. Details of this application are given on page 661. The Michaelis constant for urease at room temperature is 2.0 mM, and $k_2 = 2.5 \times 10^4$ s$^{-1}$ at pH 7.5. (a) Calculate the initial rate of the reaction when the urea concentration is 0.030 mM and the urease concentration is 5.0 $\mu$M, and (b) find $v_{max}$.

(a) From Equation 25–22,

$$\frac{d[P]}{dt} = \frac{k_2[E]_0[S]_t}{[S]_t + K_m}$$

At the beginning of the reaction, $[S]_t = [S]_0$ and

$$\frac{d[P]}{dt} = \frac{(2.5 \times 10^4 \text{ s}^{-1})(5.0 \times 10^{-6} \text{ M})(0.030 \times 10^{-3} \text{ M})}{0.030 \times 10^{-3} \text{ M} + 2.0 \times 10^{-3} \text{ M}}$$
$$= 1.8 \times 10^{-3} \text{ M s}^{-1}$$

(b) Figure 25–3 reveals that $d[P]/dt = v_{max}$ when the concentration of substrate is large and Equation 25–23 applies. Thus,

$$d[P]/dt = v_{max} = k_2[E]_0 = (2.5 \times 10^4 \text{ s}^{-1})(5.0 \times 10^{-6} \text{ M}) = 0.125 \text{ M s}^{-1}$$

---

Although our discussion thus far has been concerned with enzymatic methods, an analogous treatment for ordinary catalysis gives rate laws that are similar in form to those for enzymes. These expressions often reduce to the first-order case for ease of data treatment, and many examples of kinetic-catalytic methods are found in the literature.[4]

---

[4]See K. B. Yatsimirskii, *Kinetic Methods of Analysis*. Oxford: Pergamon Press, 1966; H. B. Mark, G. A. Rechnitz, and R. A. Greinke, *Kinetics in Analytical Chemistry*. New York: Wiley-Interscience, 1968.

# 25B    THE DETERMINATION OF REACTION RATES

Several methods are used for the determination of reaction rates. In this section, we describe some of these methods and when in the course of a reaction they are used.

## 25B–1 Experimental Methods

A *fast reaction* is 50% complete in 10 s or less.

The way reaction rates are measured depends upon whether the reaction of interest is fast or slow. A reaction is generally regarded as fast if it proceeds to 50% of completion in 10 s or less. Analytical methods based upon fast reactions generally require special equipment that permits rapid mixing of reagents and fast recording of data.

If a reaction is sufficiently slow, conventional equilibrium methods of analysis can be used to determine the concentration of a reactant or product as a function of time. Often, however, the reaction of interest is too rapid for equilibrium measurements—that is, concentrations change appreciably during the measurement process. Under these circumstances, either the reaction must be stopped (*quenched*) while the measurement is made or an instrumental technique that records concentrations continuously as the reaction proceeds must be employed. In the former case, an aliquot is removed from the reaction mixture and rapidly quenched by mixing it with a reagent that combines with one of the reactants to stop the reaction. Alternatively, quenching is accomplished by lowering the temperature rapidly to slow the reaction to an acceptable level for the measurement step. Unfortunately, quenching techniques tend to be laborious and often time-consuming and are thus not widely used for analytical purposes.

The most convenient approach for obtaining kinetic data is to monitor the progress of the reaction continuously by spectrophotometry, conductometry, potentiometry, or some other instrumental technique. With the advent of inexpensive microcomputer technology, instrumental readings proportional to concentrations of reactants and/or products are often recorded directly as a function of time, stored in the computer's memory, and retrieved later for data processing.

In the following sections, we explore some strategies used to determine rates of reactions and thus concentrations from plots of analyte or product concentration as a function of time.

**Feature 25–1**
**FAST REACTIONS AND STOPPED-FLOW MIXING**

One of the most popular and most reliable methods for carrying out rapid reactions is stopped-flow mixing. In this technique, streams of reagent and sample are mixed rapidly, and the flow of mixed solution is stopped suddenly. The reaction progress is then monitored at a position slightly downstream from the mixing point. The apparatus shown in the figure below is designed to perform stopped-flow mixing.

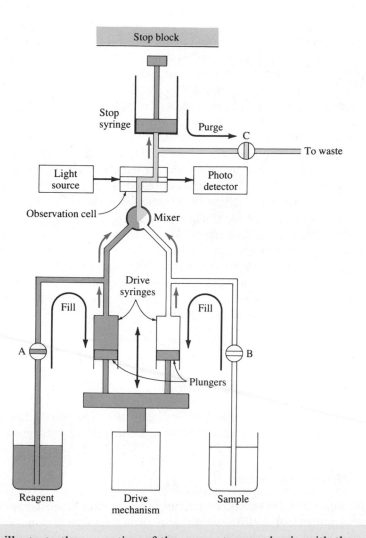

**Stop block**

Stop syringe

Purge

C

To waste

Light source

Photo detector

Observation cell

Mixer

Drive syringes

Fill

Fill

A

B

Plungers

Reagent

Drive mechanism

Sample

A stopped-flow mixing apparatus for rapid reactions.

   To illustrate the operation of the apparatus, we begin with the drive syringes filled with reagent and sample and with valves A, B, and C closed. The stop syringe is empty. The drive mechanism is then activated to move the drive syringe plungers forward rapidly. The reagent and sample pass into the mixer, where they are mixed, and immediately into the observation cell, as indicated by the colored arrows in the figure. The reaction mixture then passes into the stop syringe. Eventually the stop syringe fills and the stop syringe plunger strikes the stop block. This event causes the flow to cease almost instantly with a recently mixed plug of solution in the observation cell. In this example, the observation cell is transparent so that a light beam can be passed through to make absorption measurements. In this way, the progress of the reaction can be recorded. All that is required is that the *dead time*, or the time between the mixing of reagents and the arrival of the sample in the observation cell, be short relative to the time required for the reaction to proceed to completion. For well-designed systems in which the turbulent flow of the mixer provides very rapid and efficient mixing, the dead time is on the order of 2 to 4 ms. Thus, first-order or pseudo-first-order reactions with $\tau \approx 25$ ms ($k \approx 40$ s$^{-1}$) can be examined using the stopped-flow technique.

When the reaction is complete, valve C is opened, and the stop syringe plunger is then pushed down to purge the stop syringe of its contents (gray arrow). Valve C is then closed, valves A and B are opened, and the drive mechanism is moved down to fill the drive syringes with solution (black arrows). At this point, the apparatus is ready for another rapid mixing experiment. The entire apparatus can be placed under the control of a microcomputer, which can also collect and analyze the reaction rate data. Stopped-flow mixing has been used for fundamental studies of rapid reactions and for routine kinetic determinations of analytes involved in fast reactions. The principles of fluid dynamics that make stopped-flow mixing possible and the solution-handling capabilities of this and other similar devices are utilized in many contexts for automatically mixing solutions and measuring analyte concentrations in numerous industrial and clinical laboratories.

## 25B–2 Types of Kinetic Methods

Kinetic methods are classified according to the type of relationship that exists between the measured variable and the analyte concentration.

### The Differential Method

In the *differential method*, concentrations are computed from reaction rates by means of a differential form of a rate expression. Rates are determined by measuring the slope of a curve relating analyte or product concentration to reaction time. To illustrate, let us substitute $[A]_t$ from Equation 25–7 for [A] in Equation 25–4:

$$\text{rate} = -\left(\frac{d[A]}{dt}\right)_t = k[A]_t = k[A]_0 e^{-kt} \qquad (25\text{–}25)$$

As an alternative, the rate can be expressed in terms of the product concentration. That is,

$$\text{rate} = \left(\frac{d[P]}{dt}\right)_t = k[A]_0 e^{-kt} \qquad (25\text{–}26)$$

Equations 25–25 and 25–26 give the dependence of the rate upon $k$, $t$, and, most important, $[A]_0$, the initial concentration of the analyte. At any fixed time $t$, the factor $ke^{-kt}$ is a constant, and the rate is directly proportional to the initial analyte concentration.

Example 25–5

The rate constant for a pseudo-first-order reaction is 0.156 s$^{-1}$. Find the initial concentration of the reactant if its rate of disappearance 10.00 s after the initiation of the reaction is $2.79 \times 10^{-4}$ M s$^{-1}$.

The proportionality constant $ke^{-kt}$ is

$$ke^{-kt} = (0.156 \text{ s}^{-1})e^{-(0.156 \text{ s}^{-1})(10.00 \text{ s})} = 3.28 \times 10^{-2} \text{ s}^{-1}$$

Rearranging Equation 25–25 and substituting numerical values, we have

$$[A]_0 = \text{rate}/ke^{-kt}$$
$$= (2.79 \times 10^{-4} \text{ M s}^{-1})/(3.28 \times 10^{-2} \text{ s}^{-1})$$
$$= 8.51 \times 10^{-3} \text{ M}$$

The choice of the time at which a reaction rate is measured is often based upon such factors as convenience, the existence of interfering side reactions, and the inherent precision of making the measurement at a particular time. It is often advantageous to make the measurement near $t = 0$ because this portion of the exponential curve is nearly linear (see, for example, the initial parts of the curves in Figure 25–1) and the slope is readily estimated from the tangent to the curve. Moreover, if the reaction is pseudo first order, such a small amount of excess reagent is consumed that no error arises from changes in $k$ resulting from changes in reagent concentration. Finally, the relative error in determining the slope is minimal at the beginning of the reaction because the slope is a maximum in this region.

Figure 25–4 illustrates how the differential method is used to determine the concentration of an analyte $[A]_0$ from experimental rate measurements for the reaction shown as Equation 25–1. The solid curves in Figure 25–4a are plots of the experimentally measured product concentration

> Kinetic methods in which data are collected near the beginning of the reaction are called *initial rate methods*.

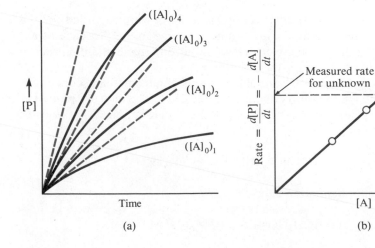

(a)

(b)

Figure 25–4

A plot of data for the determination of A by the differential method. (a) Solid lines are the experimental plots of product concentration as a function of time for four initial concentrations of A. Dashed lines are tangents to the curve at $t \rightarrow 0$. (b) A plot of the slopes obtained from the tangents in (a) as a function of analyte concentration.

[P] as a function of reaction time for four standard solutions of A. These curves are then used to prepare the differential calibration plot shown in Figure 25–4b. To obtain the rates, tangents are drawn to each of the curves in 25–4a at a time near zero (dashed lines in part a). The slopes of the tangents are then plotted as a function of [A], giving the straight line shown in 25–4b. Unknowns are treated in the same way, and analyte concentrations are determined from the calibration curve.

Of course, it is not necessary to record the entire rate curve, as has been done in Figure 25–4a, since only a small portion of the plot is utilized for measuring the slope. As long as sufficient data points are collected to determine the initial slope precisely, time is saved and the entire procedure is simplified. More sophisticated data-handling procedures and numerical analysis of the data make possible high-precision rate measurements at later times as well; under certain circumstances such measurements are more accurate and precise than those made near $t = 0$.

### Integral Methods

> Integral methods use integral forms of the first-order rate equation such as $[A] = [A]_0 e^{-kt}$.

In contrast to the differential method, *integral methods* take advantage of integrated forms of rate laws, such as those shown by Equations 25–6, 25–7, and 25–9.

**Graphical Methods.** Equation 25–6 may be rearranged to give

$$\ln [A]_t = -kt + \ln [A]_0 \qquad (25\text{–}27)$$

Thus a plot of the natural logarithm of experimentally measured concentrations of A (or P) as a function of time should yield a straight line with a slope of $-k$ and a $y$ intercept of $\ln [A]_0$. Use of this procedure for the determination of nitromethane is illustrated in Example 25–6.

---

Example 25–6

The data in the first two columns of Table 25–2 were recorded for the pseudo-first-order decomposition of nitromethane in the presence of excess base. Find the initial concentration of nitromethane and the pseudo-first-order rate constant for the reaction.

Computed values for the natural logarithms of nitromethane concentrations are shown in the third column of Table 25–1. The data are plotted in Figure 25–5. A least-squares analysis of the data (Section 3B–3) leads to an intercept $a$ of

$$a = \ln [CH_3NO_2]_0 = -5.129$$

which upon exponentiation gives

$$[CH_3NO_2]_0 = 5.92 \times 10^{-3} \text{ M}$$

| Table 25-2 DATA FOR THE DECOMPOSITION OF NITROMETHANE | | |
|---|---|---|
| Time, s | $[CH_3NO_2]$, M | $\ln[CH_3NO_2]$ |
| 0.25 | $3.86 \times 10^{-3}$ | $-5.557$ |
| 0.50 | $2.59 \times 10^{-3}$ | $-5.956$ |
| 0.75 | $1.84 \times 10^{-3}$ | $-6.298$ |
| 1.00 | $1.21 \times 10^{-3}$ | $-6.717$ |
| 1.25 | $0.742 \times 10^{-3}$ | $-7.206$ |

The least-squares analysis also gives the slope of the line $b$, which in this case is

$$b = -1.62 = -k$$

and thus

$$k = 1.62 \text{ s}^{-1}$$

**Fixed-Time Methods.** Fixed-time methods are based upon Equation 25–7 or 25–9. The former can be rearranged to

$$[A]_0 = \frac{[A]_t}{e^{-kt}} \qquad (25\text{–}28)$$

The simplest way of employing this relationship is to perform a calibration experiment with a standard solution that has a known concentration $[A]_0$. After a carefully measured reaction time $t$, $[A]_t$ is determined and used to evaluate $e^{-kt}$ by means of Equation 25–28. Unknowns are then analyzed

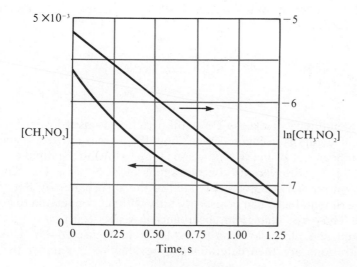

Figure 25–5

Plots of nitromethane concentration and the natural logarithm of nitromethane concentration as a function of time. The data are from Example 25–6.

by measuring $[A]_t$ after exactly the same reaction time and employing the calculated value for $e^{-kt}$ to compute the analyte concentrations.

Equation 25–28 is easily modified for the situation where $[P]$ is measured experimentally rather than $[A]_t$. Equation 25–9 may be rearranged to solve for $[A]_0$. That is,

$$[A]_0 = \frac{[P]_t}{1 - e^{-kt}} \tag{25-29}$$

A more desirable approach to the uses of Equation 25–28 or 25–29 is to measure $[A]$ or $[P]$ at two times $t_1$ and $t_2$. For example, if the product concentration is determined, we can write

$$[P]_{t_1} = [A]_0(1 - e^{-kt_1})$$
$$[P]_{t_2} = [A]_0(1 - e^{-kt_2})$$

Subtracting the first equation from the second yields, upon rearrangement,

$$[A]_0 = \frac{[P]_{t_2} - [P]_{t_1}}{e^{-kt_1} - e^{-kt_2}} = C([P]_{t_2} - [P]_{t_1}) \tag{25-30}$$

The reciprocal of the denominator is constant for constant $t_1$ and $t_2$ and is assigned the symbol $C$.

The use of Equation 25–30 has the fundamental advantage common to most kinetic methods that the absolute determination of concentration or of a variable proportional to concentration is unnecessary. It is the *difference* between two concentrations that is proportional to the initial concentration of the analyte. This means that factors influencing the long-term stability of the measuring instrument are subtracted out.

An important example of uncatalyzed methods is the fixed-time method for the determination of thiocyanate ion based upon spectrophotometric measurements of the red iron(III) thiocyanate complex. The reaction in this application is

$$Fe^{3+} + SCN^- \underset{k_{-1}}{\overset{k_1}{\rightleftharpoons}} Fe(SCN)^{2+}$$
$$\text{Red}$$

Under conditions of excess $Fe^{3+}$, the reaction is pseudo first order in $SCN^-$. The curves in Figure 25–6a show the increase in absorbance due to the appearance of $Fe(SCN)^{2+}$ versus time following the rapid mixing of 0.100 M $Fe^{3+}$ with various concentrations of $SCN^-$ at pH 2. Since the concentration of $Fe(SCN)^{2+}$ is related to the absorbance by Beer's law, the experimental data can be used directly without conversion to concentration. Thus, the change in absorbance $\Delta A$ between times $t_1$ and $t_2$ is computed and plotted versus $[SCN^-]_0$, as in Figure 25–6b. Unknown concentrations are then determined by evaluating $\Delta A$ under the same

A major advantage of kinetic methods is their immunity to errors resulting from long-term drift of the measurement system.

The stopped-flow mixing technique described in Feature 25–1 is often used to carry out the $Fe^{3-}/SCN^-$ reaction.

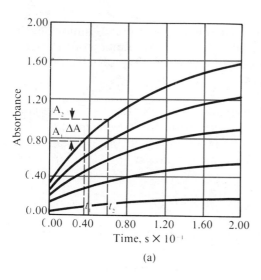

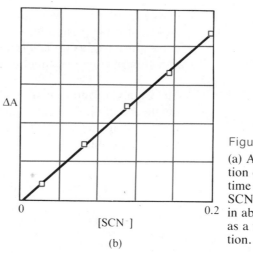

Figure 25–6
(a) Absorbance due to the formation of FeSCN$^{2+}$ as a function of time for five concentrations of SCN$^-$. (b) A plot of the difference in absorbance $\Delta A$ at times $t_2$ and $t_1$ as a function of SCN$^-$ concentration.

experimental conditions and reading the concentration of thiocyanate ion from the linear working curve.

Fixed-time methods are advantageous because the measured quantity is directly proportional to the analyte concentration and because measurements can be made *at any time* during the progress of first-order reactions. When instrumental methods are used for monitoring reactions by fixed-time procedures, the precision of the analytical results approaches the precision of the instrument used.

## 25C   APPLICATIONS OF KINETIC METHODS

Kinetic methods, which are used for the determination of both organic and inorganic species, fall into two categories, catalyzed and uncatalyzed. As noted earlier, uncatalyzed reactions are not nearly as widely used as catalyzed reactions because the selectivity and sensitivity of the former are often quite inferior to those of the latter. Uncatalyzed reactions are used to advantage when high-speed, automated measurements are required, however, or when the sensitivity of the detection method is great.[5]

### 25C–1 Catalytic Methods

#### The Determination of Inorganic Species

Many inorganic cations and anions catalyze *indicator reactions*—that is, reactions whose rates are readily measured by instrumental methods, such as spectrophotometric, fluorometric, or electroanalytical. Condi-

---

[5]For reviews of applications of modern kinetic methods, see M. Kopanica and V. Stara, in *Comprehensive Analytical Chemistry*, G. Svehla, Ed., Vol. XVIII, pp. 11–227. New York: Elsevier, 1983; and G. G. Guilbault, in *Treatise on Analytical Chemistry*, 2nd ed., I. M. Kolthoff and P. J. Elving, Eds., Part I, Vol. 1, Chapter 11. New York: Wiley, 1978.

tions are then employed such that the rate is proportional to the concentration of catalyst, and from the rate data, the concentration of catalyst is determined. An example of this type of method involves the kinetic determination of traces of mercury(II).

**The Catalytic Determination of Mercury.** The catalytic effect of mercury on the reaction of hexacyanoferrate(II) with nitrosobenzene has been employed to determine trace levels of the metal in the +2 state. The uncatalyzed indicator reaction proceeds as follows:

$$Fe(CN)_6^{4-} + H_2O \overset{Slow}{\rightleftharpoons} Fe(CN)_5H_2O^{3-} + CN^- \tag{25-31}$$

$$Fe(CN)_5H_2O^{3-} + C_6H_5NO \overset{Fast}{\rightleftharpoons} Fe(CN)_5C_6H_5NO^{3-} + H_2O \tag{25-32}$$

The presence of $Hg^{2+}$ accelerates the first step, apparently by removing cyanide to form an intermediate complex ion $Hg(CN)^+$, thus forcing the equilibrium to the right. Mercury(II) is then regenerated by reaction of the $Hg(CN)^+$ complex with hydrogen ions in the solution:

$$Hg^{2+} + CN^- \rightarrow Hg(CN)^+$$

$$Hg(CN)^+ + H^+ \rightarrow HCN + Hg^{2+}$$

*Kinetic methods are particularly advantageous when reactions are so slow that it is impractical to wait until equilibrium is achieved.*

Reaction 25–32 is relatively fast and serves as the indicator reaction for monitoring the formation of the product by spectrophotometric measurements at 528 nm. The formation of $Fe(CN)_5C_6H_5NO^{3-}$ is consistent with a general catalyst mechanism similar to that presented in Section 25A–4, so that the rate of the reaction is given by

$$\frac{d[Fe(CN)_5C_6H_5NO^{3-}]}{dt} = k_{cat}[Hg^{2+}]$$

where $k_{cat}$ is a collection of constants.

The reaction is slow enough that a fixed-time measurement of absorbance at $t = 1800$ s permits the determination of $Hg^{2+}$ at levels as low as 3 $\times 10^{-7}$ M. The method has been used for the determination of mercury in biological materials, in industrial by-products, and in water.

**The Scope of Catalyzed Methods in Inorganic Analysis.** Kinetic methods based on catalysis by inorganic analytes are widely applicable. For example, Kopanica and Stara[6] list 40 cations and 15 anions that have been determined by a variety of indicator reactions.

### The Determination of Organic Species

Without question, the most important applications of catalyzed kinetic methods to organic analyses involve the use of enzymes as catalysts.

---

[6]M. Kopanica and V. Stara, in *Comprehensive Analytical Chemistry*, G. Svehla, Ed., Vol. XVIII, pp. 192–209. New York: Elsevier, 1983.

These methods have been used for the determination of both enzymes and substrates and serve as the basis for many of the routine and automated screening tests performed by the thousands in clinical laboratories throughout the world. One of these tests is for determining the quantity of urea in blood and is called the blood urea nitrogen (BUN) test (Figure 22–18). A brief description of this test follows.

**The Enzymatic Determination of Urea.** The determination of urea in blood and urine is frequently carried out by measuring the rate of hydrolysis of urea in the presence of the enzyme urease. The equation for the reaction is

$$CO(NH_2)_2 + 2\ H_2O + H^+ \xrightarrow{\text{Urease}} 2\ NH_4^+ + HCO_3^-$$

As suggested in Example 25–4, urea can be determined by measuring the initial rate of production of the products of this reaction. The high selectivity of the enzyme permits the use of nonselective detection methods, such as electrical conductance, for initial rate measurements. Commercial instruments operate on this principle. The sample is mixed with a small amount of an enzyme-buffer solution in a conductivity cell. The maximum rate of increase in conductance is measured within 10 s of mixing, and the concentration of urea is determined from a calibration curve consisting of a plot of maximum initial rate as a function of urea concentration. The precision of the instrument is on the order of 2 to 5% relative for concentrations in the physiological range of 2 to 10 mM.

Another method for following the rate of urea hydrolysis is based on a specific-ion electrode for ammonium ions (Section 17D–4).

**The Scope of Methods Based on Enzyme-Catalyzed Reactions.** Enzyme-catalyzed reactions are used for the kinetic determination of a large and diverse group of substrate species. These methods have the twin virtues of remarkable selectivity and good sensitivity. In addition, they are generally readily automated.

A number of inorganic species can be determined by enzyme-catalyzed reactions, including ammonia, hydrogen peroxide, carbon dioxide, hydroxylamine, as well as nitrate, phosphate, and pyrophosphate ions.

Methods have also been developed for the determination of individual components in mixtures of closely related organic compounds. For example, a procedure for the determination of the individual components in a mixture of 21 organic acids without preliminary separation has been described (several enzymes are used). Similarly, methods have been developed for the determination of components in mixtures of alcohols and hydroxy compounds, sugars, amines, and steroid alcohols.

Enzymatic methods have been described for the quantitative determination of several hundred enzymes. In addition, some two dozen inorganic cations and anions are known to decrease the rates of certain enzyme-catalyzed indicator reactions. These *inhibitors* can thus be determined from the decrease in rate brought about by their presence.

We may use enzymes for the determination of both inhibitors and activators.

*Enzyme activators* are substances, often inorganic ions, that are required for certain enzymes to become active as catalysts. Activators can thus be determined by their effect on the rates of enzyme-catalyzed reactions. For example, it has been reported that magnesium at concentrations as low as 10 ppb can be determined in blood plasma based on activation by this ion of the enzyme isocitric dehydrogenase.

### 25C–2 Noncatalytic Reactions

As noted earlier, kinetic methods based on uncatalyzed reactions are not nearly as widely used as those in which a catalyst is involved. We have already described two of these methods (pages 656 and 658).

Generally, uncatalyzed reactions can be rendered useful when selective reagents are employed in conjunction with sensitive detection methods. For example, the selectivity of complexing agents can be controlled by adjusting the pH of the medium in the determination of metal ions, as discussed in Section 13B–8. Sensitivity can be achieved through the use of spectrophotometric detection to monitor reagents that form complexes with large molar absorptivities. The determination of $Cu^{2+}$ presented in Problem 25–13 is an example.

A highly sensitive alternative is to select complexes that fluoresce so that the rate of change of fluorescence is used as a measure of analyte concentration (Problem 25–14).

The precision of both noncatalytic and catalytic kinetic methods depends upon such experimental conditions as pH, ionic strength, and temperature. With careful control of these variables, precisions of 1 to 10% relative standard deviation are typical. Automation of kinetic methods and computerized data analysis can often improve the relative precision to 1% or less.

### 25C–3 The Kinetic Determination of Components in Mixtures

An important application of kinetic methods is in the determination of closely related species in mixtures, such as alkaline earth cations or organic compounds with the same functional groups. For example, suppose two species A and B react with a common excess reagent to form products under pseudo-first-order conditions:

$$A + R \xrightarrow{k_A} P'$$

$$B + R \xrightarrow{k_B} P''$$

Generally, $k_A$ and $k_B$ differ from each other. Thus, if $k_A > k_B$, A is depleted before B. It is possible to show that, if the ratio $k_A/k_B$ is greater than about 500, the consumption of A is approximately 99% complete before 1% of B is used up. Thus, a differential determination of A with no significant interference from B is possible provided the rate is measured shortly after mixing.

When the ratio of the two rate constants is small, determination of both species is still possible by more complex methods of data treatment. These methods are beyond the scope of this text.[7]

---

[7]For examples of the use of kinetic methods for the analysis of multicomponent mixtures, see G. M. Ridder and D. W. Margerum, *Anal. Chem.* **1977,** *49,* 2090.

## 25D  QUESTIONS AND PROBLEMS

**25-1.** Define the following terms as they are used in reaction kinetics:

*(a) order of a reaction    *(e) Michaelis constant
(b) pseudo first order    (f) differential method
*(c) enzyme    *(g) integral method
(d) substrate    (h) indicator reaction

**25-2.** The analysis of multicomponent mixtures by kinetic methods is sometimes referred to as "kinetic separations." Explain the significance of this term.

***25-3.** Explain why pseudo-first-order conditions are utilized in most kinetic methods.

**25-4.** List three advantages of kinetic methods.

***25-5.** Develop an expression for the half-life of the reactant in a first-order process in terms of $k$.

**25-6.** Find the natural lifetime in seconds for first-order reactions corresponding to
*(a) $k = 0.497$ s$^{-1}$.
(b) $k = 6.62$ h$^{-1}$.
*(c) $[A]_0 = 3.16$ M and $[A]_t = 0.496$ M at $t = 3650$ s.
(d) $[P]_\infty = 0.176$ M and $[P]_t = 0.0423$ M at $t = 9.62$ s.
*(e) half-life $t_{1/2} = 32.4$ years.
(f) $t_{1/2} = 0.478$ s.

**25-7.** Find the first-order rate constant for a reaction that is 66.6% complete in
*(a) 0.0100 s.    *(c) 1.00 s.    *(e) 26.8 $\mu$s.
(b) 0.100 s.    (d) 5280 s.    (f) 8.86 ns.

**25-8.** Calculate the number of lifetimes $\tau$ required for a pseudo-first-order reaction to achieve the following levels of completion:
*(a) 10%    *(c) 90%    *(e) 99.9%
(b) 50%    (d) 99%    (f) 99.99%

**25-9.** Find the number of half-lives $t_{1/2}$ required to reach the levels of completion listed in Problem 25-8.

**25-10.** Find the relative error associated with the assumption that $k'$ is invariant during the course of a pseudo-first-order reaction under the following conditions.

| | Extent of Reaction, % | Excess of Reagent |
|---|---|---|
| *(a) | 1 | 5X |
| (b) | 1 | 10X |
| *(c) | 1 | 50X |
| (d) | 1 | 100X |
| *(e) | 5 | 5X |
| (f) | 5 | 10X |
| *(g) | 5 | 100X |
| (h) | 63.2 | 5X |
| *(i) | 63.2 | 10X |
| (j) | 63.2 | 50X |
| *(k) | 63.2 | 100X |

***25-11.** Show that for an enzyme reaction obeying Equation 25-22 the substrate concentration for which the rate equals $v_{max}/2$ is equal to $K_m$.

**25-12.** Equation 25-22 can be rearranged to produce the equation

$$\frac{1}{d[P]/dt} = \frac{K_m}{v_{max}[S]} + \frac{1}{v_{max}}$$

where $v_{max} = k_2[E]_0$ when [S] is large.
(a) Suggest a way to employ this equation in the construction of a working curve for the enzymatic determination of substrate.
(b) Describe how the resulting working curve can be used to find $K_m$ and $v_{max}$.

***25-13.** Copper(II) forms a 1 : 1 complex with the organic complexing agent R in acidic medium. The formation of the complex can be monitored by spectrophotometry at 480 nm. Use the following data collected under pseudo-first-order conditions to construct a calibration curve of rate versus concentration of R. Find the concentration of copper(II) in an unknown whose rate under the same conditions was $6.2 \times 10^{-3}$ absorbance $\cdot$ s$^{-1}$.

| $c_{Cu^{2+}}$, ppm | Rate, absorbances $\cdot$ s$^{-1}$ |
|---|---|
| 3.0 | $3.6 \times 10^{-3}$ |
| 5.0 | $5.4 \times 10^{-3}$ |
| 7.0 | $7.9 \times 10^{-3}$ |
| 9.0 | $1.03 \times 10^{-2}$ |

**25-14.** Aluminum forms a 1 : 1 complex with 2-hydroxy-1-naphthaldehyde *p*-methoxybenzoylhydrazonal that exhibits fluorescence emission at 475 nm. Under pseudo-first-order conditions, a plot of the initial rate of the reaction (emission units/s) versus the concentration of aluminum (in $\mu$M) yields a straight line described by the equation

$$\text{rate} = 1.74c_{Al} - 0.225$$

Find the concentration of aluminum in a solution that exhibits a rate of 0.76 emission units/s under the same experimental conditions.

*25–15. The enzyme amine oxidase catalyzes the oxidation of amines to aldehydes. For tryptamine, $K_m$ for the enzyme is $4.0 \times 10^{-4}$ M and $v_{max} = k_2[E]_0 = 1.6 \times 10^{-3}$ $\mu$M/min at pH 8. Find the concentration of a solution of tryptamine that reacts at a rate of 0.18 $\mu$M/min in the presence of amine oxidase under the above conditions. Assume that [tryptamine] $\ll$ $K_m$.

## CHAPTER 26

# AN INTRODUCTION TO CHROMATOGRAPHIC METHODS

Chromatography is widely used for the separation, identification, and determination of the chemical components in complex mixtures. No other separation method is as powerful and as generally applicable as is chromatography.[1]

## 26A A GENERAL DESCRIPTION OF CHROMATOGRAPHY

The word "chromatography" is difficult to define rigorously because the term has been applied to such a variety of systems and techniques. All of these methods, however, have in common the use of a *stationary phase* and a *mobile phase*. Components of a mixture are carried through the stationary phase by the flow of a gaseous or a liquid mobile phase, separations being based on differences in migration rates among the sample components.

## 26A–1 Classification of Chromatographic Methods

Chromatographic methods are of two types. In *column chromatography,* the stationary phase is held in a narrow tube, and the mobile phase is forced through the tube under pressure or by gravity. In *planar chromatography,* the stationary phase is supported on a flat plate or in the pores of a paper. Here the mobile phase moves through the stationary phase by capillary action or under the influence of gravity. We will emphasize column chromatography.

Chromatography was invented by the Russian botanist Mikhail Tswett shortly after the turn of the century. He used the technique to separate various plant pigments, such as chlorophylls and xanthophylls, by passing solutions of them through glass columns packed with finely divided calcium carbonate. The separated species appeared as colored bands on the column, which accounts for the name he chose for the method (Greek *chroma,* meaning "color," and *graphein,* meaning "to write").

Chromatography is a technique in which the components of a mixture are separated based upon the rates at which they are carried through a stationary phase by a gaseous or liquid mobile phase.

Planar and column chromatography are based upon the same types of equilibria.

---

[1]General references on chromatography include *Chromatography: Fundamentals and Applications of Chromatography and Electrophotometric Methods, Part A: Fundamentals, Part B: Applications,* E. Heftmann, Ed. New York: Elsevier, 1983; P. Sewell and B. Clarke, *Chromatographic Separations.* New York: Wiley, 1988; *Chromatographic Theory and Basic Principles,* J. A. Jonsson, Ed. New York: Marcel Dekker, 1987; R. M. Smith, *Gas and Liquid Chromatography in Analytical Chemistry.* New York: Wiley, 1988; J. C. Giddings, *Unified Separation Science.* New York: Wiley, 1991.

Table 26–1

CLASSIFICATION OF COLUMN CHROMATOGRAPHIC METHODS

| General Classification | Specific Method | Stationary Phase | Type of Equilibrium |
|---|---|---|---|
| Liquid chromatography (LC) (mobile phase: liquid) | Liquid-liquid, or partition | Liquid adsorbed on a solid | Partition between immiscible liquids |
| | Liquid-bonded phase | Organic species bonded to a solid surface | Partition between liquid and bonded surface |
| | Liquid-solid, or adsorption | Solid | Adsorption |
| | Ion exchange | Ion-exchange resin | Ion exchange |
| | Size exclusion | Liquid in interstices of a polymeric solid | Partition/sieving |
| Gas chromatography (GC) (mobile phase: gas) | Gas-liquid | Liquid adsorbed on a solid | Partition between gas and liquid |
| | Gas-bonded phase | Organic species bonded to a solid surface | Partition between liquid and bonded surface |
| | Gas-solid | Solid | Adsorption |
| Supercritical-fluid chromatography (SFC) (mobile phase: supercritical fluid) | | Organic species bonded to a solid surface | Partition between supercritical fluid and bonded surface |

Liquid chromatography can be performed in columns and on planar surfaces, but gas chromatography and supercritical fluid chromatography are restricted to column procedures.

As shown in the first column of Table 26–1, chromatographic methods fall into three categories based upon the nature of the mobile phase. The three types of phases include liquids, gases, and supercritical fluids. The second column of the table reveals that there are five types of liquid chromatography and three types of gas chromatography that differ in the nature of the stationary phase and the types of equilibria between phases.

### 26A–2 Elution Chromatography

Elution is a process in which solutes are washed through a stationary phase by the movement of a mobile phase.

Figure 26–1 shows schematically how two components A and B are resolved on a column by *elution chromatography*. Elution involves washing a solute through a column by additions of fresh solvent. A single portion of the sample dissolved in the mobile phase is introduced at the head of the column (at time $t_0$ in Figure 26–1), whereupon components A and B distribute themselves between the two phases. Introduction of additional mobile phase (the *eluent*) forces the dissolved portion of the sample down the column, where further partition between the mobile phase and fresh portions of the stationary phase occurs (time $t_1$).

An eluent is a solvent used to carry the components of a mixture through a stationary phase.

Further additions of solvent carry solute molecules down the column in a continuous series of transfers between the two phases. Because solute movement can occur only in the mobile phase, the average *rate* at which a solute migrates *depends upon the fraction of time it spends in that phase.* This fraction is small for solutes that are strongly retained by the stationary phase (component B in Figure 26–1, for example) and large where retention in the mobile phase is more likely (component A). Ideally, the resulting differences in rates cause the components in a mixture to separate into *bands,* or *zones,* along the length of the column (see Figure

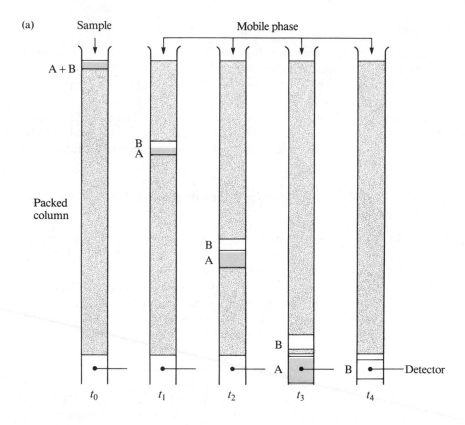

(a)

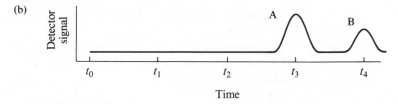

(b)

**Figure 26–1**
(a) Diagram showing the separation of a mixture of components A and B by column elution chromatography. (b) The output of the signal detector at the various stages of elution shown in (a).

26–2). Isolation of the separated species is then accomplished by passing a sufficient quantity of mobile phase through the column to cause the individual bands to pass out the end (to be *eluted* from the column), where they can be collected (times $t_3$ and $t_4$ in Figure 26–1).

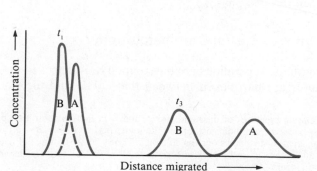

**Figure 26–2**
Concentration profiles of solute bands A and B at two different times in their migration down the column in Figure 26–1. The times $t_1$ and $t_3$ are indicated in Figure 26–1.

## Chromatograms

A chromatogram is a plot of some function of solute concentration versus elution time or elution volume.

If a detector that responds to solute concentration is placed at the end of the column and its signal is plotted as a function of time (or of volume of added mobile phase), a series of symmetric peaks is obtained, as shown in the lower part of Figure 26–1. Such a plot, called a *chromatogram,* is useful for both qualitative and quantitative analysis. The positions of the peaks on the time axis can be used to identify the components of the sample; the areas under the peaks provide a quantitative measure of the amount of each species.

## The Effects of Relative Migration Rates and Band Broadening on Resolution

Relative elution rates and band broadening determine the effectiveness of a chromatographic separation.

Figure 26–2 shows concentration profiles for the bands containing solutes A and B on the column in Figure 26–1 at time $t_1$ and at a later time $t_3$.[2] Because B is more strongly retained by the stationary phase than is A, B lags during the migration. Clearly, the distance between the two increases as they move down the column. At the same time, however, broadening of both bands takes place, which lowers the efficiency of the column as a separating device. While band broadening is inevitable, conditions can often be found where it occurs more slowly than band separation. Thus, as shown in Figure 26–2, a clean resolution of species is possible provided the column is sufficiently long.

Several chemical and physical variables influence the rates of band separation and band broadening. Improved separations can often be realized by the control of those variables that either increase the rate of band separation or decrease the rate of band spreading. These alternatives are illustrated in Figure 26–3.

The variables that influence the relative rates at which solutes migrate through a stationary phase are described in the next section. Following this discussion, we shall turn to those factors that play a part in zone broadening.

## 26B MIGRATION RATES OF SOLUTES

The effectiveness of a chromatographic column in separating two solutes depends in part upon the relative rates at which the two species are eluted. These rates are in turn determined by the partition ratios of the solutes between the two phases.

### 26B–1 Partition Ratios in Chromatography

All chromatographic separations are based upon differences in the extent to which solutes are partitioned between the mobile and the stationary

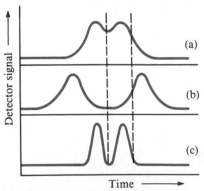

**Figure 26–3**

Two-component chromatographs illustrating two methods of improving separation: (a) original chromatogram with overlapping peaks; improvement brought about by (b) an increase in band separation and (c) a decrease in bandwidth.

---

[2]Note that the relative positions of the bands for A and B in the concentration profile in Figure 26–2 are reversed from their positions in the lower part of Figure 26–1. The difference is that the abscissa is distance along the column in Figure 26–2 but time in Figure 26–1. Thus, in Figure 26–1, the *front* of a peak lies to the left and the *tail* to the right; in Figure 26–2, the reverse is true.

phase. For solute species A, the equilibrium involved is described by the equation

$$A_{\text{mobile}} \rightleftarrows A_{\text{stationary}}$$

The equilibrium constant $K$ for this reaction is called a *partition ratio,* or *partition coefficient,* and is defined as

$$K = \frac{c_S}{c_M} \qquad (26\text{-}1)$$

where $c_S$ is the molar analytical concentration of the solute in the stationary phase and $c_M$ is its analytical concentration in the mobile phase. Ideally, the partition ratio is constant over a wide range of solute concentrations; that is, $c_S$ is directly proportional to $c_M$. At high solute concentrations, however, marked departures from linearity are encountered. Fortunately, most chromatography is performed under conditions in which Equation 26-1 does apply, which greatly simplifies the derivation of expressions that describe the separation process. Chromatography carried out under conditions in which $K$ is more or less constant is termed *linear chromatography.* The discussions that follow deal exclusively with separations of this type.

## 26B-2 Retention Time

Figure 26-4 is a typical chromatogram for a sample containing a single analyte. The time it takes after sample injection for the analyte peak to reach the detector is called the *retention time* and is given the symbol $t_R$. The small peak on the left is for a species that is *not* retained by the column. The sample or the mobile phase will ordinarily contain an unretained species. When they do not, such a species may be added to aid in peak identification. The time $t_M$ for the unretained species to reach the detector is sometimes called the *dead time.* The rate of migration of the unretained species is the same as the average rate of motion of the mobile phase molecules.

The average linear rate of solute migration $\bar{v}$ is

$$\bar{v} = \frac{L}{t_R} \qquad (26\text{-}2)$$

Retention time is the time between sample injection and the appearance of a solute peak at the detector of a chromatographic column.

The dead time is the time it takes for an unretained species to pass through a column.

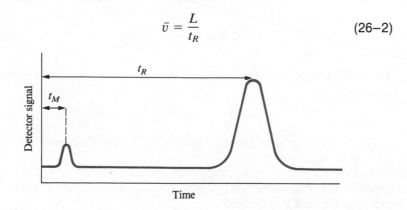

Figure 26-4
A typical chromatogram for a two-component mixture. The small peak on the left represents a solute that is not retained on the column and so reaches the detector almost immediately after elution is started. Thus its retention time $t_M$ is approximately equal to the time required for a molecule of the mobile phase to pass through the column.

where $L$ is the length of the column packing. Similarly, the average linear rate of movement $u$ of the molecules of the mobile phase is

$$u = \frac{L}{t_M} \qquad (26\text{-}3)$$

where $t_M$ is the time required for a molecule of the mobile phase to pass through the column.

## 26B-3 The Relationship Between Retention Time and Partition Ratio

In order to relate the retention time of a solute to its partition ratio, we express its migration rate as a fraction of the velocity of the mobile phase:

$$\bar{v} = u \times \text{fraction of time solute spends in mobile phase}$$

This fraction, however, equals the average number of moles of solute in the mobile phase at any instant divided by the total number of moles of solute in the column:

$$\bar{v} = u \times \frac{\text{moles of solute in mobile phase}}{\text{total moles of solute}}$$

The total number of moles of solute in the mobile phase is equal to the molar concentration $c_M$ of the solute in that phase multiplied by its volume $V_M$. Similarly, the number of moles of solute in the stationary phase is given by the product of the concentration $c_S$ of the solute in the stationary phase and its volume $V_S$. Therefore,

$$\bar{v} = u \times \frac{c_M V_M}{c_M V_M + c_S V_S} = u \times \frac{1}{1 + c_S V_S / c_M V_M}$$

Substitution of Equation 26-1 into this equation gives an expression for the rate of solute migration as a function of its partition ratio and as a function of the volumes of the stationary and mobile phases:

$$\bar{v} = u \times \frac{1}{1 + K V_S / V_M} \qquad (26\text{-}4)$$

The two volumes can be estimated from the method used to prepare the column.

## 26B-4 The Rate of Solute Migration: The Capacity Factor

The *capacity factor* is an important parameter that is widely used to describe the migration rates of solutes on columns. For a solute A, the capacity factor $k'_A$ is defined as

$$k'_A = \frac{K_A V_S}{V_M} \qquad (26\text{--}5)$$

where $K_A$ is the partition ratio for the species A. Substitution of Equation 26–5 into 26–4 yields

$$\bar{v} = u \times \frac{1}{1 + k'_A} \qquad (26\text{--}6)$$

In order to show how $k'_A$ can be derived from a chromatogram, we substitute Equations 26–2 and 26–3 into Equation 26–6:

$$\frac{L}{t_R} = \frac{L}{t_M} \times \frac{1}{1 + k'_A} \qquad (26\text{--}7)$$

This equation rearranges to

$$k'_A = \frac{t_R - t_M}{t_M} \qquad (26\text{--}8)$$

As shown in Figure 26–4, $t_R$ and $t_M$ are readily obtained from a chromatogram. When the capacity factor for a solute is much less than unity, elution occurs so rapidly that accurate determination of its retention time is difficult. When the capacity factor is larger than perhaps 20 to 30, elution times become inordinately long. Ideally, separations are performed under conditions in which the capacity factors for the solutes in a mixture lie in the range between 1 and 5.

Ideally, the capacity factor for analytes in a sample is between 1 and 5.

The capacity factors in gas chromatography can be varied by changing the temperature and the column packing. In liquid chromatography, capacity factors can often be manipulated to give better separations by varying the composition of the mobile phase and the stationary phase.

## 26B–5 Differential Migration Rates: The Selectivity Factor

The *selectivity factor* $\alpha$ of a column for the two species A and B is defined as

The selectivity factor for two analytes in a column provides a measure of how well the column will separate the two.

$$\alpha = \frac{K_B}{K_A} \qquad (26\text{--}9)$$

where $K_B$ is the partition ratio for the more strongly retained species B and $K_A$ is the partition ratio for the less strongly held, or more rapidly eluted, species A. By this definition, $\alpha$ is *always greater than unity*.

Substitution of Equation 26–5 and the analogous equation for solute B into Equation 26–9 provides, after rearrangement, a relationship between the selectivity factor for two solutes and their capacity factors:

$$\alpha = \frac{k'_B}{k'_A} \qquad (26\text{--}10)$$

where $k_B'$ and $k_A'$ are the capacity factors for B and A, respectively. Substitution of Equation 26–8 for the two solutes in Equation 26–10 gives an expression that permits the determination of $\alpha$ from an experimental chromatogram:

$$\alpha = \frac{(t_R)_B - t_M}{(t_R)_A - t_M} \tag{26–11}$$

In Section 26D–1 we show how to use the selectivity factor to compute the resolving power of a column.

## 26C  THE EFFICIENCY OF CHROMATOGRAPHIC COLUMNS

The efficiency of a chromatographic column refers to the amount of band broadening that occurs when a compound passes through the column. Before defining column efficiency in more quantitative terms, let us examine the reasons that bands become broader as they move down a column.

### 26C–1  The Rate Theory of Chromatography

The *rate theory* of chromatography describes the shapes and breadths of elution peaks in quantitative terms, based on a random-walk mechanism for the migration of molecules through a column. A detailed discussion of rate theory is beyond the scope of this text. We can, however, give a qualitative picture of why bands broaden and what variables improve column efficiency by limiting the extent of band broadening.

If you examine the chromatograms shown in this and the next chapter, you will see that the elution peaks look very much like the Gaussian error curves that you encountered in Chapters 2 and 3. As shown in Section 2D, normal error curves are rationalized by assuming that the uncertainty associated with any single measurement is the summation of a much larger number of small, individually undetectable and random uncertainties, each of which has an equal probability of being positive or negative. In a similar way, the typical Gaussian shape of a chromatographic band can be attributed to the additive combination of the random motions of the myriad molecules in the band as it moves down the column.

It is instructive to consider a single solute molecule as it undergoes many thousands of transfers between the stationary and mobile phases during elution. Residence time in either phase is highly irregular. Transfer from one phase to the other requires energy, and the molecule must acquire this energy from its surroundings. Thus, the residence time in a given phase may be transitory for some molecules and relatively long for others. Recall that movement down the column can occur *only while the molecule is in the mobile phase*. As a consequence, certain molecules travel rapidly by virtue of their accidental inclusion in the mobile phase for a majority of the time, whereas others lag because they happen to be incorporated in the stationary phase for a greater-than-average period. The result of these random individual processes is a symmetric spread of velocities around the mean value, which represents the behavior of the average analyte molecule.

The breadth of a band increases as the band moves down the column because more time is allowed for spreading to occur as a result of these various mechanisms. Thus, zone breadth is directly related to residence time in the column and inversely related to the velocity of the mobile phase.

## 26C–2 A Quantitative Definition of Column Efficiency

Two related terms are widely used as quantitative measures of chromatographic column efficiency: (1) *plate height H* and (2) *number of theoretical plates N*. The two are related by the equation

$$N = L/H \qquad (26\text{–}12)$$

where $L$ is the length (usually in centimeters) of the column packing. Feature 26–1 describes how these measures of column efficiency got their names.

The plate height $H$ is also known as the height equivalent of a theoretical plate (HETP).

The efficiency of chromatographic columns increases as the number of plates becomes greater and as the plate height becomes smaller. Enormous differences in efficiencies are encountered in columns as a result of differences in column type and in mobile and stationary phases. Efficiencies in terms of plate numbers can vary from a few hundred to several hundred thousand; plate heights ranging from a few tenths to one thousandth of a centimeter or smaller are not uncommon.

The efficiency of a column is great when $H$ is small and $N$ is large.

### The Definition of Plate Height

In Section 3A–2, we pointed out that the breadth of a Gaussian curve is best described by the standard deviation $\sigma$ and the variance $\sigma^2$. Because chromatographic bands are also Gaussian and because the efficiency of a column is reflected in the breadth of chromatographic peaks, the variance per unit length of column is used by chromatographers as a measure of column efficiency. That is, the plate height $H$ is given by

$$H = \frac{\sigma^2}{L} \qquad (26\text{–}13)$$

This definition of column efficiency is illustrated in Figure 26–5a, which shows a column with a packing $L$ cm in length. Above this schematic is a plot showing the distribution of molecules along the length of the column at the moment the analyte peak reaches the end of the packing (that is, at the retention time $t_R$). The curve is Gaussian, and the locations of $L + 1\sigma$ and $L - 1\sigma$ are indicated as broken vertical lines. Note that $L$ carries units of centimeters and $\sigma^2$ units of centimeters squared; thus $H$ represents a linear distance in centimeters (Equation 26–13). In fact, the plate height

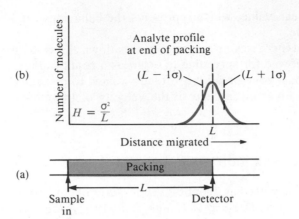

Figure 26–5
Definition of plate height $H = \sigma^2/L$.

can be thought of as the length of column that contains a fraction of the analyte that lies between $L$ and $L - \sigma$. Because the area under a normal error curve bounded by $\pm\sigma$ is about 68% of the total area (page 37), the plate height, as defined, contains 34% of the analyte.

### The Experimental Evaluation of *H* and *N*

Figure 26–6 is a typical chromatogram with time as the abscissa. The variance of the solute peak, which can be obtained by a simple graphical procedure, has units of seconds squared and is usually designated as $\tau^2$ to distinguish it from $\sigma^2$, which has units of centimeters squared. The two standard deviations $\tau$ and $\sigma$ are related by

$$\tau = \frac{\sigma}{L/t_R} \qquad (26\text{–}14)$$

where $L/t_R$ is the average linear velocity of the solute in centimeters per second.

Figure 26–6 illustrates a simple way to approximate $\tau$ and $\sigma$ from an experimental chromatogram. Tangents at the inflection points on the two sides of the chromatographic peak are extended to form a triangle with the baseline of the chromatogram. The area of this triangle can be shown to be approximately 96% of the total area under the peak. In Section 3A–2 it

Figure 26–6
Determination of the standard deviation $\tau$ from a chromatographic peak: $W = 4\tau$.

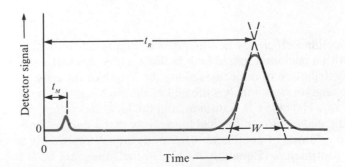

26C   The Efficiency of Chromatographic Columns     675

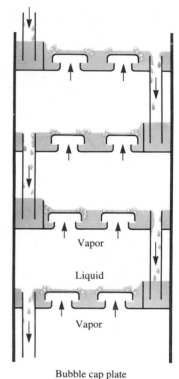

## Feature 26–1
### WHERE DO THE TERMS "PLATE" AND "PLATE HEIGHT" COME FROM?

The 1952 Nobel Prize in Chemistry was awarded to two Englishmen, A. J. P. Martin and R. L. M. Synge, for their work in the development of modern chromatography. In their theoretical studies, they adapted a model that was first developed in the early 1920s to describe separations on fractional distillation columns. Fractionating columns, which were first used in the petroleum industry for separating closely related hydrocarbons, consist of numerous interconnected bubble-cap plates (see Figure 26–7) at which vapor-liquid equilibria are established when a column is operated under reflux conditions. The efficiency of the column as a tool for separation was directly related to the number of plates contained in a unit length of the column.

Martin and Synge treated a chromatographic column as if it were made up of a long series of bubble-cap-like plates, within which equilibrium conditions always prevail. This plate model successfully accounts for the Gaussian shape of chromatographic peaks as well as for factors that influence differences in solute-migration rates. It cannot account for zone broadening, however, because of its basic assumption that equilibrium conditions prevail throughout a column during elution. This assumption can never be valid in the dynamic state that exists in a chromatographic column, where phases are moving past one another at such a pace that sufficient time is not available for equilibration.

Because the plate model is such a poor representation of a chromatographic column, we urge you (1) to avoid attaching any real or imaginary significance to the terms "plate" and "plate height" and (2) to view these terms as designators of column efficiency that are retained for historic reasons only. They have no physical significance. Unfortunately, both terms are so well entrenched in the chromatographic literature that their replacement by more appropriate designations is unlikely.

Bubble cap plate column

Figure 26–7
Plates in a fractionating column.

was shown that about 96% of the area under a Gaussian peak is included within plus or minus two standard deviations ($\pm 2\sigma$) of its maximum. Thus, the intercepts shown in Figure 26–6 occur at approximately $\pm 2\tau$ from the maximum, and $W = 4\tau$, where $W$ is the magnitude of the base of the triangle. Substituting this relationship into Equation 26–14 and rearranging yield

$$\sigma = \frac{LW}{4t_R}$$

Substitution of this equation for $\sigma$ into Equation 26–13 gives

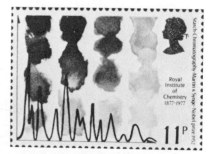

Stamp in honor of biochemists Archer J. P. Martin (1910–) and Richard L. M. Synge (1914–) who were awarded the 1952 Nobel Prize in chemistry for their contributions to the development of modern chromatography.

$$H = \frac{LW^2}{16t_R^2} \qquad (26\text{-}15)$$

To obtain $N$, we substitute into Equation 26–12 and rearrange to get

$$N = 16 \left(\frac{t_R}{W}\right)^2 \qquad (26\text{-}16)$$

Thus, $N$ can be calculated from two time measurements, $t_R$ and $W$; to obtain $H$, the length of the column packing $L$ must also be known.

Another method for approximating $N$, which some workers believe to be more reliable, is to determine $W_{1/2}$, the width of a peak at half its maximum height. The number of theoretical plates is then given by

$$N = 5.54 \left(\frac{t_R}{W_{1/2}}\right)^2 \qquad (26\text{-}17)$$

To compare $N$ and $H$ for two columns, the same compound should be used for the test.

The two parameters $N$ and $H$ are widely used in the literature and by instrument manufacturers as measures of column performance. For these parameters to be meaningful in comparing two columns, it is essential that they be determined with the *same compound*.

## 26C–3 Kinetic Variables Affecting Band Broadening

Band broadening is the consequence of the finite rate at which several mass-transfer processes occur during migration of a solute down a column. Some of these rates can be controlled by the adjustment of experimental variables, thus permitting improvement in separations. Table 26–2 lists the most important of these variables. Their effects on column efficiency, as measured by plate height $H$, are described in the paragraphs that follow.

### The Effect of Mobile-Phase Flow Rate

The magnitude of kinetic effects on column efficiency clearly depends upon the length of time the mobile phase is in contact with the stationary

Table 26–2
**VARIABLES THAT AFFECT COLUMN EFFICIENCY**

| Variable | Symbol | Usual Units |
|---|---|---|
| Linear velocity of mobile phase | $u$ | $\text{cm} \cdot \text{s}^{-1}$ |
| Diffusion coefficient in mobile phase* | $D_M$ | $\text{cm}^2 \cdot \text{s}^{-1}$ |
| Diffusion coefficient in stationary phase* | $D_S$ | $\text{cm}^2 \cdot \text{s}^{-1}$ |
| Capacity factor (Equation 26–8) | $k'$ | Unitless |
| Diameter of packing particle | $d_p$ | cm |
| Thickness of liquid coating on stationary phase | $d_f$ | cm |

*Increases as temperature increases and viscosity decreases.

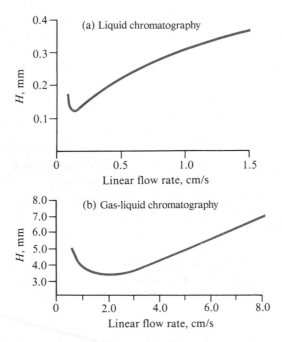

Figure 26–8

Effect of mobile-phase flow rate on plate height for (a) liquid chromatography and (b) gas chromatography.

phase, which in turn depends upon the flow rate of the mobile phase. For this reason, efficiency studies have generally been carried out by determining $H$ (by means of Equation 26–16 or 26–17 and Equation 26–12) as a function of mobile-phase velocity. The data obtained from such studies are typified by the two plots shown in Figure 26–8, the one for liquid chromatography and the other for gas chromatography. While both show a minimum in $H$ (or a maximum in efficiency) at low flow rates, the minimum for liquid chromatography usually occurs at flow rates that are well below those for gas chromatography and are often so low that they are not observed under normal operating conditions.

Generally, liquid chromatograms are obtained at lower flow rates than gas chromatograms. Furthermore, as shown in the figure, plate heights for liquid chromatographic columns are an order of magnitude or more smaller than those encountered with gas chromatographic columns. Offsetting this advantage is the fact that it is impractical to employ liquid columns that are longer than about 25 to 50 cm (because of high pressure drops), whereas gas chromatographic columns may be 50 m or more in length. Consequently, the total number of plates, and thus overall column efficiency, are usually superior with gas chromatographic columns.

## A Theory of Band Broadening

Over the last 30 years, an enormous amount of theoretical and experimental effort has been devoted to developing quantitative relationships describing the effects of experimental variables on plate heights for various types of columns. Perhaps a dozen or more expressions for calculating plate height have been put forward and applied with various degrees of success. It is apparent that none of these is entirely adequate to explain the complex physical interactions and effects that lead to zone broadening

Theoretical studies of zone broadening in the 1950s by Dutch chemical engineers led to the *van Deemter equation*, which can be written in the form

$$H = A + B/u + Cu$$

where the constants $A$, $B$, and $C$ are coefficients of eddy diffusion, longitudinal diffusion, and mass transfer, respectively. This equation is of considerable historic interest; its modernized version takes the form of Equation 26–18 (see S. J. Hawkes, *J. Chem. Educ.*, **1983**, *60*, 393).

and thus lower column efficiencies. Some of the equations, though imperfect, have been of considerable use, however, in pointing the way toward improved column performance. One of these is presented here.

The efficiency of most chromatographic columns can be approximated by the expression

$$H = B/u + C_S u + C_M u \qquad (26\text{--}18)$$

where $H$ is the plate height in centimeters and $u$ is the linear velocity of the mobile phase in centimeters per second. The quantity $B$ is the *longitudinal diffusion coefficient*, while $C_S$ and $C_M$ are *mass transfer coefficients* for the stationary and mobile phases, respectively. The equations given in Table 26–3 reveal the effects of column variables on the three terms in Equation 26–18.

**The Longitudinal Diffusion Term $B/u$.** Diffusion is a process in which species migrate from a more concentrated part of a medium to a more dilute. The rate of migration is proportional to the concentration difference between the regions as well as to the *diffusion coefficient $D_M$* of the species. The latter, which is a measure of the mobility of a substance in a given medium, is a constant equal to the velocity of migration under a unit concentration gradient.

In chromatography, longitudinal diffusion results in the migration of a solute from the concentrated center of a band to the more dilute regions

**Table 26–3**

**KINETIC PROCESSES THAT CONTRIBUTE TO PEAK BROADENING**

| Process | Term in Equation 26–18 | Relationship to Column* and Solute Properties | Equation Number |
|---|---|---|---|
| Longitudinal diffusion | $B/u$ | $\dfrac{B}{u} = \dfrac{2k_D D_M}{u}$ | (26–19) |
| Mass transfer to and from liquid stationary phase† | $C_S u$ | $C_S u = \dfrac{qk'd_f^2 u}{(1 + k')^2 D_S}$ | (26–20) |
| Mass transfer to and from solid stationary phase† | $C_S u$ | $C_S u = \dfrac{2t_d k' u}{(1 + k')^2}$ | (26–21) |
| Mass transfer in mobile phase | $C_M u$ | $C_M u = \dfrac{f(d_p^2, d_c^2, u)}{D_M} u$ | (26–22) |

*$u$, $D_M$, $D_S$, $d_f$, $d_p$, $k'$ are as defined in Table 26–2.

$f$: function of.

$k_D$, $q$: constants.

$t_d$: average desorption time of analyte from surface; $t_d = 1/k_d$, where $k_d$ is first-order rate constant for desorption.

$d_c$: column diameter.

$B$: coefficient of longitudinal diffusion.

$C_S$, $C_M$: coefficients of mass transfer in stationary and mobile phases, respectively.

†Equation 26–20 applies for a liquid stationary phase only. Equation 26–21 applies to a solid stationary phase where adsorption occurs.

on either side (that is, toward and opposed to the direction of flow). Longitudinal diffusion is a common source of band broadening in gas chromatography but is of little significance in liquid chromatography because the rate at which molecules diffuse in a gaseous medium is high. The phenomenon is of little importance in liquid chromatography where diffusion rates are much smaller. The magnitude of the $B$ term in Equation 26–18 is largely determined by the diffusion coefficient $D_M$ of the analyte in the mobile phase and is directly proportional to this constant (Equation 26–19).

As shown by Equation 26–18, the contribution of longitudinal diffusion to plate height is inversely proportional to the linear velocity of the eluent. Such a relationship is not surprising inasmuch as the analyte is in the column for a briefer period when the flow rate is high. Thus, diffusion from the center of the band to the two edges has less time to occur.

The initial decreases in $H$ shown in both curves in Figure 26–8 are a direct consequence of the longitudinal diffusion. Note that the effect is much less pronounced in liquid chromatography because of the much lower diffusion rates in a liquid mobile phase. The striking difference in plate heights shown by the two curves in Figure 26–8 can also be explained by considering the relative rates of longitudinal diffusion in the two mobile phases. That is, diffusion coefficients in gaseous media are orders of magnitude larger than in liquids. Thus band broadening goes on to a much greater extent in gas chromatography than in liquid chromatography.

**The Mass-Transfer Coefficients $C_S$ and $C_M$.** The two mass transfer coefficients $C_S$ and $C_M$ in Equation 26–18 are needed because the equilibrium between the mobile and the stationary phase is established so slowly that a chromatographic column always operates under nonequilibrium conditions. Consequently, analyte molecules at the front of a band are swept ahead before they have time to equilibrate with the stationary phase and thus be retained. Similarly, equilibrium is not reached at the trailing edge of a band, and molecules are left behind in the stationary phase by the fast-moving mobile phase.

Band broadening from mass-transfer effects arises because the many flowing streams of a mobile phase within a column and the layer of immobilized liquid making up the stationary phase both have finite widths. Consequently time is required for solute molecules to diffuse from the interior of these phases to their interface where transfer occurs. This time lag results in the persistence of nonequilibrium conditions along the length of the column. If the rates of mass transfer within the two phases were infinite, broadening of this type would not occur.

Note that the extent of both longitudinal broadening and mass-transfer broadening depend upon the rate of diffusion of analyte molecules but that the direction of diffusion in the two cases is different. Longitudinal broadening originates from the tendency of molecules to move in directions that tend to parallel the flow, whereas mass-transfer broadening occurs from diffusion that tends to be at right angles to the flow. As a consequence, the extent of longitudinal broadening is *inversely* related to flow rate. For mass-transfer broadening, in contrast, the faster the mobile

Solutes are affected by longitudinal diffusion in gas chromatography to a greater extent than in liquid chromatography.

phase moves, the less time there is for equilibrium to be approached. Thus, as shown by the last two terms in Equation 26–18, the mass-transfer effect on plate height is directly proportional to the rate $u$ of movement of the mobile phase.

**The Stationary Phase Mass-Transfer Term $C_S u$.** When the stationary phase is an immobilized liquid, the mass-transfer coefficient is directly proportional to the square of the thickness of the film on the support particles $d_f^2$ and inversely proportional to the diffusion coefficient $D_S$ of the solute in the film (Equation 26–20). These effects can be understood by realizing that both reduce the average frequency at which analyte molecules reach the interface where transfer to the mobile phase can occur. That is, with thick films, molecules must on the average travel farther to reach the surface, and with smaller diffusion coefficients, they travel slower. The consequence is a slower rate of mass transfer and an increase in plate height.

When the stationary phase is a solid surface, the mass-transfer coefficient $C_S$ is directly proportional to the time required for a species to be adsorbed or desorbed, which in turn is inversely proportional to the first-order rate constant for the processes (Equation 26–21).

**The Mobile Phase Mass-Transfer Term $C_M u$.** The mass-transfer processes that occur in the mobile phase are sufficiently complex to defy complete and rigorous analysis, at least to date. On the other hand, a good qualitative understanding of the variables affecting zone broadening from this cause exists, and this understanding has led to vast improvements in all types of chromatographic columns.

The mobile-phase mass-transfer coefficient $C_M$ is known to be inversely proportional to the diffusion coefficient of the analyte in the mobile phase $D_M$ and also to be some function of the square of the particle diameter of the packing $d_p^2$, the square of the column diameter $d_c^2$, and the flow rate (Equation 26–22).

The contribution of mobile-phase mass-transfer to plate height is the product of the mass transfer coefficient $C_M$ (which is a function of solvent velocity) as well as the velocity of the solvent itself. Thus, the net contribution of $C_M u$ to plate height is not linear in $u$ (see the curve labeled $C_M u$ in Figure 26–10) but bears a complex dependency on solvent velocity.

Zone broadening in the mobile phase is due in part to the multitude of pathways by which a molecule (or ion) can find its way through a packed column. As shown in Figure 26–9, the length of these pathways may differ significantly; thus, the residence time in the column for molecules of the same species is also variable. Solute molecules then reach the end of the column over a time interval, which leads to a broadened band. This effect, which is sometimes called *eddy diffusion*, would be independent of solvent velocity if it were not partially offset by ordinary diffusion, which results in molecules being transferred from a stream following one pathway to a stream following another. If the velocity of flow is very low, a large number of these transfers will occur, and each molecule in its movement down the column will sample numerous flow paths, spending a brief time in each. As a consequence, the rate at which each molecule moves

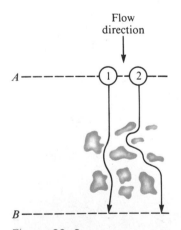

Flow direction

Figure 26–9
Typical pathways of two molecules during elution. Note that the distance traveled by molecule 2 is greater than that traveled by molecule 1. Thus, molecule 2 will arrive at *B* later than molecule 1.

down the column tends to approach that of the average. Thus, at low mobile-phase velocities, the molecules are not significantly dispersed by the multiple-path nature of the packing. At moderate or high velocities, however, sufficient time is not available for diffusion averaging to occur, and band broadening due to the different path lengths is observed. At sufficiently high velocities, the effect of eddy diffusion becomes independent of flow rate.

> Pathways for the mobile phase through the column are numerous and have different lengths.

Superimposed upon the eddy diffusion effect is one that arises from stagnant pools of the mobile phase retained in the stationary phase. Thus, when a solid serves as the stationary phase, its pores are filled with *static* volumes of mobile phase. Solute molecules must then diffuse through these stagnant pools before transfer can occur between the *moving* mobile phase and the stationary phase. This situation applies not only to solid stationary phases but also to liquid stationary phases immobilized on porous solids because the immobilized liquid does not usually fully fill the pores.

> Static pools of solvent contribute to increases in $H$.

The presence of stagnant pools of mobile phase slows the exchange process and results in a contribution to the plate height that is directly proportional to the mobile phase velocity and inversely proportional to the diffusion coefficient for the solute in the mobile phase. An increase in particle diameter $d_p$ also has a significant effect because of the increase in internal volume that accompanies increases in particle size.

**Effect of Mobile Phase Velocity on Terms in Equation 26–18.** Figure 26–10 shows the variation of the three terms in Equation 26–18 as a function of mobile phase velocity. The top curve is the summation of these various effects. Note that an optimum flow rate exists at which the plate height is a minimum and the separation efficiency is a maximum.

**Summary of Methods for Reducing Band Broadening.** Two important controllable variables that affect column efficiency are the diameter of the particles making up the packing and the diameter of the column. The effect of particle diameter is demonstrated by the data shown in Figures 27–1 and 28–1. To take advantage of the effect of column diameter, narrower and narrower columns have been used in recent years.

> Band broadening is minimized by small packing diameter and small column diameter.

With gaseous mobile phases, the rate of longitudinal diffusion can be reduced appreciably by lowering the temperature and thus the diffusion

> The diffusion coefficient $D_M$ has a greater effect on gas chromatography than on liquid chromatography.

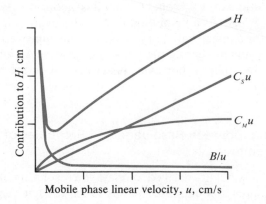

Figure 26–10

Contribution of various mass-transfer coefficients to column plate height. $C_S u$ arises from the rate of mass transfer to and from the stationary phase, $C_M u$ comes from a limitation in the rate of mass transfer in the mobile phase, and $B/u$ is associated with longitudinal diffusion.

coefficient $D_M$. The consequence is significantly smaller plate heights at low temperatures. This effect is usually not noticeable in liquid chromatography because diffusion is slow enough so that the longitudinal diffusion term has little effect on overall plate height.

With liquid stationary phases, the thickness of the layer of adsorbed liquid should be minimized since $C_S$ in Equation 26–18 is proportional to the square of this variable $d_f$ in Equation 26–20.

## 26D    OPTIMIZATION OF COLUMN PERFORMANCE

A chromatographic separation is optimized by varying experimental conditions until the components of a mixture are separated cleanly with a minimum expenditure of time. Optimization experiments are aimed at either (1) reducing zone broadening or (2) altering relative migration rates of the components. As we have shown in Section 26C, zone broadening is increased by those kinetic variables that increase the plate height of a column. Migration rates, on the other hand, are varied by changing those variables that affect the capacity and selectivity factors of the solutes (Section 26B).

### 26D–1  Column Resolution

The *resolution* $R_s$ of a column provides a quantitative measure of its ability to separate two analytes. The significance of this term is illustrated in Figure 26–11, which consists of chromatograms for species A and B on three columns with different resolving powers. The resolution of each column is defined as

$$R_s = \frac{2\Delta Z}{W_A + W_B} = \frac{2[(t_R)_B - (t_R)_A]}{W_A + W_B} \qquad (26\text{–}23)$$

where all of the terms on the right side are as defined in the figure.

It is evident from Figure 26–11 that a resolution of 1.5 gives an essentially complete separation of A and B, whereas a resolution of 0.75 does not. At a resolution of 1.0, zone A contains about 4% B and zone B contains about 4% A. At a resolution of 1.5, the overlap is about 0.3%. The resolution for a given stationary phase can be improved by lengthening the column, thus increasing the number of plates. An adverse consequence of the added plates, however, is an increase in the time required for the resolution.

### The Relationship Between Resolution and Properties of the Column and Solute

A useful equation is readily derived to relate the resolution of a column to the number of plates it contains as well as to the capacity and selectivity

Michael Tswett (1872–1919), a Russian botanist, discovered the basic principles of column chromatography. He separated plant pigments by eluting a mixture of the pigments on a column of powdered alumina. The various pigments separated into colored bands on the column; hence the name chromatography.

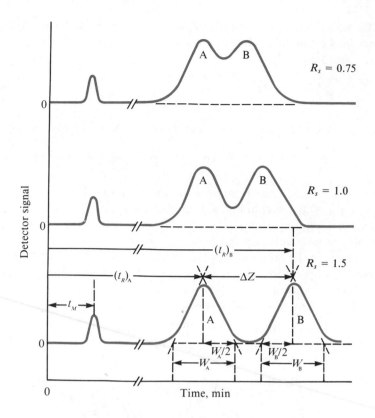

Figure 26–11

Separation at three resolutions: $R_s = 2\Delta Z/(W_A + W_B)$.

factors of a pair of solutes on the column. Thus, it can be shown[3] that for the two solutes A and B in Figure 26–11, the resolution is given by the equation

$$R_s = \frac{\sqrt{N}}{4}\left(\frac{\alpha - 1}{\alpha}\right)\left(\frac{k'_B}{1 + k'_B}\right) \qquad (26-24)$$

where $k'_B$ is the capacity factor of the slower-moving species and $\alpha$ is the selectivity factor. This equation can be rearranged to give the number of plates needed to realize a given resolution:

$$N = 16R_s^2\left(\frac{\alpha}{\alpha - 1}\right)^2\left(\frac{1 + k'_B}{k'_B}\right)^2 \qquad (26-25)$$

## The Relationship Between Resolution and Elution Time

As mentioned earlier, the goal in chromatography is the highest possible resolution in the shortest possible elapsed time. Unfortunately, these goals tend to be incompatible, and a compromise between the two is

---

[3]See D. A. Skoog and J. J. Leary, *Principles of Instrumental Analysis*, 4th ed., pp. 592–593. Philadelphia: Saunders College Publishing, 1992.

usually necessary. The time $(t_R)_B$ required to elute the two species in Figure 26–11 with a resolution of $R_s$ is

$$(t_R)_B = \frac{16 R_s^2 H}{u} \left(\frac{\alpha}{\alpha - 1}\right)^2 \frac{(1 + k_B')^3}{(k_B')^2} \qquad (26\text{–}26)$$

where $u$ is the linear rate of movement of the mobile phase.

---

### Example 26–1

Substances A and B have retention times of 16.40 and 17.63 min, respectively, on a 30.0-cm column. An unretained species passes through the column in 1.30 min. The peak widths (at base) for A and B are 1.11 and 1.21 min, respectively. Calculate (a) column resolution, (b) average number of plates in the column, (c) plate height, (d) length of column required to achieve a resolution of 1.5, and (e) time required to elute substance B on the longer column.

Employing Equation 26–23, we find

(a) $R_s = 2(17.63 - 16.40)/(1.11 + 1.21) = 1.06$

(b) Equation 26–16 permits computation of $N$:

$$N = 16 \left(\frac{16.40}{1.11}\right)^2 = 3493 \quad \text{and} \quad N = 16 \left(\frac{17.63}{1.21}\right)^2 = 3397$$

$$N_{av} = (3493 + 3397)/2 = 3445 = 3.4 \times 10^3$$

(c) $H = L/N = 30.0/3445 = 8.7 \times 10^{-3}$ cm

(d) $k'$ and $\alpha$ do not change with increasing $N$ and $L$. Thus, substituting $N_1$ and $N_2$ into Equation 26–24 and dividing one of the resulting equations by the other yield

$$\frac{(R_s)_1}{(R_s)_2} = \frac{\sqrt{N_1}}{\sqrt{N_2}}$$

where the subscripts 1 and 2 refer to the original and longer columns, respectively. Substituting the appropriate values for $N_1$, $(R_s)_1$, and $(R_s)_2$ gives

$$\frac{1.06}{1.5} = \frac{\sqrt{3445}}{\sqrt{N_2}}$$

$$N_2 = 3445 \left(\frac{1.5}{1.06}\right)^2 = 6.9 \times 10^3$$

But

$$L = NH = 6.9 \times 10^3 \times 8.7 \times 10^{-3} = 60 \text{ cm}$$

(e) Substituting $(R_s)_1$ and $(R_s)_2$ into Equation 26–26 and dividing yield

$$\frac{(t_R)_1}{(t_R)_2} = \frac{(R_s)_1^2}{(R_s)_2^2} = \frac{17.63}{(t_R)_2} = \frac{(1.06)^2}{(1.5)^2}$$

$$(t_R)_2 = 35 \text{ min}$$

Thus, to obtain the improved resolution, the separation time must be doubled.

## 26D–2 Optimization Techniques

Equations 26–24 and 26–26 serve as guides in choosing conditions that lead to a desired degree of resolution with a minimum expenditure of time. An examination of these equations reveals that each is made up of three parts. The first describes the efficiency of the column in terms of $\sqrt{N}$ or $H$. The second, which is the quotient containing $\alpha$, is a selectivity term that depends on the properties of the two solutes. The third component is the capacity term, which is the quotient containing $k'_B$; the term depends on the properties of both the solute and the column.

### Variation in Plate Height

As shown by Equation 26–24, the resolution of a column improves as the square root of the number of plates it contains increases. Example 26–1e reveals, however, that increasing the number of plates is expensive in terms of time unless the increase is achieved by reducing the plate height and not by increasing column length.

Methods for minimizing plate height are discussed in Section 26C and include reducing the particle size of the packing, the diameter of the column, the column temperature (gas chromatography), and the thickness of the liquid film (liquid chromatography). Optimizing the flow rate of the mobile phase is also helpful.

### Variation in the Capacity Factor

Often, a separation can be improved significantly by manipulation of the capacity factor $k'_B$. Increases in $k'_B$ generally enhance resolution (but at the expense of elution time). To determine the optimum range of values for $k'_B$, it is convenient to write Equation 26–24 in the form

$$R_s = Q \frac{k'_B}{1 + k'_B}$$

and Equation 26–26 as

$$(t_R)_B = Q' \frac{(1 + k'_B)^3}{(k'_B)^2}$$

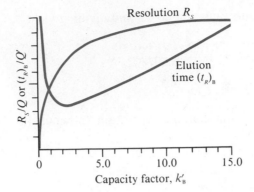

Figure 26–12

Effect of capacity factor $k'_B$ on resolution $R_s$ and elution time $(t_R)_B$. It is assumed that $Q$ and $Q'$ remain constant with variations in $k'_B$.

where $Q$ and $Q'$ contain the rest of the terms in the two equations. Figure 26–12 is a plot of $R_s/Q$ and $(t_R)_B/Q'$ as a function of $k'_B$, assuming $Q$ and $Q'$ remain approximately constant. It is clear that values of $k'_B$ greater than about 10 should be avoided because they provide little increase in resolution but markedly increase the time required for separations. The minimum in the elution-time curve occurs at $k'_B \simeq 2$. Often, then, the optimal value of $k'_B$ lies in the range from 1 to 5.

Usually, the easiest way to improve resolution is by optimizing $k'$. For gaseous mobile phases, $k'$ can often be improved by temperature changes. For liquid mobile phases, changes in the solvent composition often permit manipulation of $k'$ to yield better separations. An example of the dramatic effect that relatively simple solvent changes can bring about is demonstrated in Figure 26–13. Here, modest variations in the methanol/water ratio convert unsatisfactory chromatograms (a and b) to ones with well-separated peaks for each component (c and d). For most purposes, the chromatogram shown in (c) is best since it shows adequate resolution in minimum time.

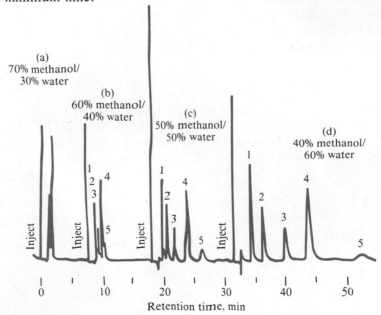

Figure 26–13

Effect of solvent variation on chromatograms. Analytes: (1) 9,10-anthraquinone; (2) 2-methyl-9,10-anthraquinone; (3) 2-ethyl-9,10-anthraquinone; (4) 1,4-dimethyl-9,10-anthraquinone; (5) 2-t-butyl-9,10-anthraquinone. (Courtesy of DuPont Biotechnology Systems, Wilmington, DE.)

## Variation in the Selectivity Factor

Optimizing $k'$ and increasing $N$ are not sufficient to give a satisfactory separation of two solutes in a reasonable time when $\alpha$ approaches unity. Here a means must be sought to increase $\alpha$ while maintaining $k'$ in the range of 1 to 10. Several options are available; in decreasing order of their desirability as determined by promise and convenience, the options are (1) changing the composition of the mobile phase, (2) changing the column temperature, (3) changing the composition of the stationary phase, and (4) using special chemical effects.

An example of the use of option 1 has been reported for the separation of anisole ($C_6H_5OCH_3$) and benzene.[4] With a mobile phase that was a 50% mixture of water and methanol, $k'$ was 4.5 for anisole and 4.7 for benzene, while $\alpha$ was only 1.04. Substitution of an aqueous mobile phase containing 37% tetrahydrofuran gave $k'$ values of 3.9 and 4.7 and an $\alpha$ value of 1.20. Peak overlap was significant with the first solvent system and negligible with the second.

A less convenient but often highly effective method of improving $\alpha$ while maintaining values for $k'$ in their optimal range is to alter the chemical composition of the stationary phase. To take advantage of this option, most laboratories that carry out chromatographic separations frequently maintain several columns that can be interchanged with a minimum of effort.

Increases in temperature usually cause increases in $k'$ but have little effect on $\alpha$ values in liquid-liquid and liquid-solid chromatography. In contrast, with ion-exchange chromatography, temperature effects can be large enough to make exploration of this option worthwhile before resorting to a change in column packing.

A final method for enhancing resolution is to incorporate into the stationary phase a species that complexes or otherwise interacts with one or more components of the sample. A well-known example of the use of this option arises where an adsorbent impregnated with a silver salt improves the separation of olefins as a consequence of the formation of complexes between the silver ions and unsaturated organic compounds.

## 26D-3  The General Elution Problem

In Figure 26–14 are hypothetical chromatograms for a six-component mixture made up of three pairs of components with widely different distribution coefficients and thus widely different capacity factors. In chromatogram (a), conditions have been adjusted so that the capacity factors for components 1 and 2 ($k_1'$ and $k_2'$) are in the optimal range of 1 to 5. The factors for the other components are far larger than the optimum, however. Thus, the peaks for components 5 and 6 appear only after an inordinate length of time has passed; furthermore, these peaks are so broad that they may be difficult to identify unambiguously.

---

[4]L. R. Snyder and J. J. Kirkland, *Introduction to Modern Liquid Chromatography*, 2nd ed., p. 75. New York: Wiley, 1979.

As shown in chromatogram (b), changing conditions to optimize the separation of components 5 and 6 bunches the peaks for the first four components to the point where their resolution is unsatisfactory. Here, however, the total elution time is ideal.

A third set of conditions, in which $k'$ values for components 3 and 4 are optimal, results in chromatogram (c). Again, separation of the other two pairs is not entirely satisfactory.

The phenomenon illustrated in Figure 26–14 is encountered often enough to be given a name—the *general elution problem*. A common solution to this problem is to change conditions that determine the values of $k'$ as the separation proceeds. These changes can be performed in a stepwise manner or continuously. Thus, for the mixture shown in Figure 26–14, conditions at the outset could be those producing chromatogram (a). Immediately after the elution of components 1 and 2, conditions could be changed to those that are optimal for separating components 3 and 4 (as in chromatogram c). With the appearance of peaks for these components, the elution could be completed under the conditions used for producing chromatogram (b). Often such a procedure leads to satisfactory separation of all the components of a mixture in minimal time.

For liquid chromatography, variations in $k'$ are brought about by varying the composition of the mobile phase during elution (*gradient elution* or *solvent programming*). For gas chromatography, temperature increases (*temperature programming*) achieve optimal conditions for separations.

## 26E    A SUMMARY OF IMPORTANT RELATIONSHIPS FOR CHROMATOGRAPHY

The number of quantities, terms, and relationships employed in chromatography is large and often confusing. Tables 26–4 and 26–5 summarize the most important definitions and equations used in this text.

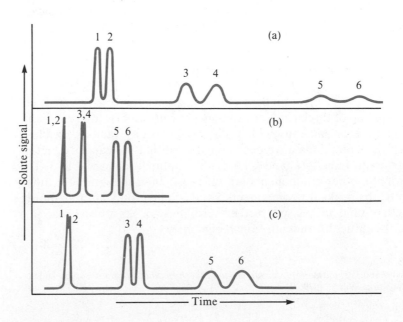

Figure 26–14

The general elution problem in chromatography.

Table 26-4

**IMPORTANT CHROMATOGRAPHIC EXPERIMENTAL QUANTITIES AND RELATIONSHIPS**

| Name | Symbol of Experimental Quantity | Determined From |
|---|---|---|
| Migration time, nonretained species | $t_M$ | Chromatogram (Figure 26-11) |
| Retention times, species A and B | $(t_R)_A, (t_R)_B$ | Chromatogram (Figure 26-11) |
| Adjusted retention time, species A | $(t'_R)_A$ | $(t'_R)_A = (t_R)_A - t_M$ |
| Peak widths, species A and B | $W_A, W_B$ | Chromatogram (Figure 26-11) |
| Length of column packing | $L$ | Direct measurement |
| Flow rate | $F$ | Direct measurement |
| Volume of stationary phase | $V_S$ | Packing preparation data |
| Concentration of solute in mobile and stationary phases | $c_M, c_S$ | Analysis and preparation data |

Table 26-5

**IMPORTANT DERIVED QUANTITIES AND RELATIONSHIPS**

| Name | Calculation of Derived Quantities | Relationship to Other Quantities |
|---|---|---|
| Linear mobile-phase velocity | $u = L/t_M$ | |
| Volume of mobile phase | $V_M = t_M F$ | |
| Capacity factor | $k' = (t_R - t_M)/t_M$ | $k' = \dfrac{K V_S}{V_M}$ |
| Partition coefficient | $K = \dfrac{k' V_M}{V_S}$ | $K = \dfrac{c_S}{c_M}$ |
| Selectivity factor | $\alpha = \dfrac{(t_R)_B - t_M}{(t_R)_A - t_M}$ | $\alpha = \dfrac{k'_B}{k'_A} = \dfrac{K_B}{K_A}$ |
| Resolution | $R_S = \dfrac{2[(t_R)_B - (t_R)_A]}{W_A + W_B}$ | $R_S = \dfrac{\sqrt{N}}{4}\left(\dfrac{\alpha - 1}{\alpha}\right)\left(\dfrac{k'_B}{1 + k'_B}\right)$ |
| Number of plates | $N = 16\left(\dfrac{t_R}{W}\right)^2$ | $N = 16 R_S^2 \left(\dfrac{\alpha}{\alpha - 1}\right)^2 \left(\dfrac{1 + k'_B}{k'_B}\right)^2$ |
| Plate height | $H = L/N$ | |
| Retention time | $(t_R)_B = \dfrac{16 R_S^2 H}{u}\left(\dfrac{\alpha}{\alpha - 1}\right)^2 \dfrac{(1 + k'_B)^3}{(k'_B)^2}$ | |

## 26F  APPLICATIONS OF CHROMATOGRAPHY

Chromatogaphy has grown to be the premiere method for separating closely related chemical species. In addition, it can be employed for the qualitative identification and quantitative determination of separated species.

### 26F-1  Qualitative Analysis

Chromatography is widely used for recognizing the presence or absence of components in mixtures that contain a limited number of possible species whose identities are known. For example, 30 or more amino acids in a protein hydrolysate can be detected with a reasonable degree of certainty by means of a chromatogram. On the other hand, because a chromatogram provides but a single piece of information about each species in a mixture (the retention time), the application of the technique to the qualitative analysis of complex samples of unknown composition is limited. For this reason, it has become common practice to couple chromatographic columns directly with ultraviolet, infrared, and mass spectrometers. The resulting *hyphenated instruments* are powerful tools for identifying the components of complex mixtures.

It is important to note that while a chromatogram may not lead to positive identification of the species in a sample, it often provides sure evidence of the *absence* of species. Thus, failure of a sample to produce a peak at the same retention time as a standard obtained under identical conditions is strong evidence that the compound in question is absent (or present at a concentration below the detection limit of the procedure).

Here are some hyphenated techniques: Gas chromatography–mass spectrometry (GC-MS). High performance liquid chromatography–mass spectrometry (HPLC-MS). Gas chromatography–infrared spectrometry (GC-IR)

### 26F–2 Quantitative Analysis

Chromatography owes its enormous growth in part to its speed, simplicity, relatively low cost, and wide applicability as a separating tool. It is doubtful, however, that its use would have become so widespread had it not been for the fact that it can also provide quantitative information about separated species.

Quantitative chromatography is based upon a comparison of either the height or the area of the analyte peak with that of one or more standards. If conditions are properly controlled, both of these parameters vary linearly with concentration.

#### Analyses Based on Peak Height

The height of a chromatographic peak is obtained by connecting the baselines on the two sides of the peak by a straight line and measuring the perpendicular distance from this line to the peak. This measurement can ordinarily be made with reasonably high precision and yields accurate results, provided variations in column conditions do not alter peak width during the period required to obtain chromatograms for sample and standards. The variables that must be controlled closely are column temperature, eluent flow rate, and rate of sample injection. In addition, care must be taken to avoid overloading the column. The effect of sample-injection rate is particularly critical for the early peaks of a chromatogram. Relative errors of 5 to 10% due to this cause are not unusual with syringe injection.

#### Analyses Based on Peak Area

Peak area is independent of broadening effects caused by the variables mentioned in the previous paragraph. From this standpoint, therefore, area is a more satisfactory analytical parameter than peak height. On the other hand, peak heights are more easily measured and, for narrow peaks, more accurately determined.

Many modern chromatographic instruments are equipped with electronic integrators that provide precise measurements of relative peak areas. If such equipment is not available, a manual estimate must be made. A simple method that works well for symmetric peaks with reasonable widths is to multiply peak height by the width at one-half peak height. Alternatively, a planimeter can be used, or the recorded peak can be cut out and weighed, with its weight then being compared with the weight of a known area of recorder paper. In general, manual integration methods provide areas that are reproducible at the 2 to 5% level; digital integrations are at least an order of magnitude more precise.[5]

### Calibration with Standards

The most straightforward method for quantitative chromatographic analyses involves the preparation of a series of standard solutions that approximate the composition of the unknown. Chromatograms for the standards are then obtained, and peak heights or areas are plotted as a function of concentration. A plot of the data should yield a straight line that passes through the origin; analyses are based upon this plot. Frequent restandardization is necessary for highest accuracy.

The most important source of error in analyses by the method based on calibration standards is usually the uncertainty in the volume of sample; occasionally, the rate of injection is also a factor. Samples are ordinarily small ($\simeq 1\ \mu L$), and the uncertainties associated with the injection of a reproducible volume of this size with a microsyringe can amount to several percent relative. The situation is even worse in gas-liquid chromatography, where the sample must be injected into a heated sample port. Here evaporation from the needle tip may lead to large variations in the volume injected.

Errors in sample volume can be reduced to perhaps 1 to 2% relative by means of a rotary sample valve such as that shown in Figure 26–15. The

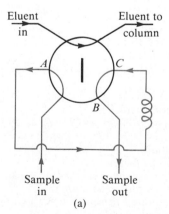

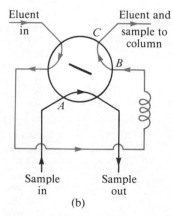

(a)                                    (b)

**Figure 26–15**
A rotary sample valve: (a) valve position for filling sample loop *ACB*; (b) valve position for introduction of sample into column.

[5]See *Chromatographic Integration Methods,* R. M. Smith, Ed. Boca Raton, FL: CRC Press, 1990.

sample loop *ACB* in (a) is filled with liquid; rotation of the valve by 45 deg then introduces a reproducible volume of sample (the volume originally contained in *ACB*) into the mobile-phase stream.

### The Internal-Standard Method

The highest precision for quantitative chromatography is obtained by using internal standards because the uncertainties introduced by sample injection are avoided. In this procedure, a carefully measured quantity of an internal-standard substance is introduced into each standard and sample, and the ratio of analyte peak area (or height) to internal-standard peak area (or height) is the analytical parameter. For this method to be successful, it is necessary that the internal-standard peak be well separated from the peaks of all other components in the sample ($R_s > 1.25$), but it must appear close to the analyte peak. With a suitable internal standard, precisions of 0.5 to 1% relative are reported.

## 26G    QUESTIONS AND PROBLEMS

26-1. Define
    *(a) elution.
    (b) mobile phase.
    *(c) stationary phase.
    (d) partition ratio.
    *(e) retention time.
    (f) capacity factor.
    *(g) selectivity factor.
    (h) plate height.
    *(i) longitudinal diffusion.
    (j) eddy diffusion.
    *(k) column resolution.
    (l) eluent.

*26-2. Describe the general elution problem.

26-3. List the variables that lead to zone broadening.

*26-4. What is the difference between gas-liquid and liquid-liquid chromatography?

26-5. What is the difference between liquid-liquid and liquid-solid chromatography?

*26-6. What variables are likely to affect the $\alpha$ value for a pair of analytes?

26-7. How can the capacity factor for a solute be manipulated?

*26-8. Describe a method for determining the number of plates in a column.

26-9. What are the effects of temperature variation on chromatograms?

*26-10. Why does the minimum in a plot of plate height versus flow rate occur at lower flow rates with liquid chromatography than with gas chromatography?

26-11. What is gradient elution?

*26-12. The following data apply to a column for liquid chromatography:

| | |
|---|---|
| Length of packing | 24.7 cm |
| Flow rate | 0.313 mL/min |
| $V_M$ | 1.37 mL |
| $V_S$ | 0.164 mL |

A chromatogram of a mixture of species A, B, C, and D provided the following data:

| | Retention Time, min | Width of Peak Base ($W$), min |
|---|---|---|
| Nonretained | 3.1 | — |
| A | 5.4 | 0.41 |
| B | 13.3 | 1.07 |
| C | 14.1 | 1.16 |
| D | 21.6 | 1.72 |

Calculate
(a) the number of plates from each peak.
(b) the mean and the standard deviation for $N$.
(c) the plate height for the column.

*26-13. From the data in Problem 26-12, calculate for A, B, C, and D
(a) the capacity factor.
(b) the partition coefficient.

*26-14. From the data in Problem 26-12, calculate for species B and C
(a) the resolution.
(b) the selectivity factor.
(c) the length of column necessary to separate the two species with a resolution of 1.5.
(d) the time required to separate the two species with a resolution of 1.5.

*26-15. From the data in Problem 26-12, calculate for species C and D
(a) the resolution.
(b) the length of column required to separate the two species with a resolution of 1.5.

**26-16.** The following data were obtained by gas-liquid chromatography on a 40-cm packed column:

| Compound | $t_R$, min | $W_{1/2}$, min |
|----------|-----------|----------------|
| Air | 1.9 | — |
| Methylcyclohexane | 10.0 | 0.76 |
| Methylcyclohexene | 10.9 | 0.82 |
| Toluene | 13.4 | 1.06 |

Calculate
(a) an average number of plates from the data.
(b) the standard deviation for the average in (a).
(c) an average plate height for the column.

**26-17.** Referring to Problem 26-16, calculate the resolution for
(a) methylcyclohexene and methylcyclohexane.
(b) methylcyclohexene and toluene.
(c) methylcyclohexane and toluene.

**26-18.** If a resolution of 1.5 is desired in separating methylcyclohexane and methylcyclohexene in Problem 26-16,
(a) how many plates are required?
(b) how long must the column be if the same packing is employed?
(c) what is the retention time for methylcyclohexene on the column in Problem 26-16b?

**26-19.** If $V_S$ and $V_M$ for the column in Problem 26-16 are 19.6 and 62.6 mL, respectively, and a nonretained air peak appears after 1.9 min, calculate the
(a) capacity factor for each compound.
(b) partition coefficient for each compound.
(c) selectivity factor for methylcyclohexane and methylcyclohexene.

**\*26-20.** List the variables that lead to (a) band broadening and (b) band separation.

**\*26-21.** What is the effect on a chromatographic peak of introducing the sample too slowly?

**\*26-22.** From distribution studies, species M and N are known to have water/hexane partition coefficients of 6.01 and 6.20 ($K = [M]_{H_2O}/[M]_{hex}$). The two species are to be separated by elution with hexane in a column packed with silica gel containing adsorbed water. The ratio $V_S/V_M$ for the packing is 0.422.
(a) Calculate the capacity factor for each solute.
(b) Calculate the selectivity factor.
(c) How many plates are needed to provide a resolution of 1.5?
(d) How long a column is needed if the plate height of the packing is $2.2 \times 10^{-3}$ cm?
(e) If a flow rate of 7.10 cm/min is employed, how long will it take to elute the two species?

**26-23.** Repeat the calculations in Problem 26-22, assuming $K_M = 5.81$ and $K_N = 6.20$.

## CHAPTER 27

# GAS-LIQUID CHROMATOGRAPHY

In *gas chromatography* (GC), an inert *carrier gas* serves as the mobile phase that elutes the components of a mixture from a column containing an immobilized stationary phase. In contrast to liquid chromatography, no interaction occurs between the mobile phase and the analyte in gas chromatography; thus, the rate of movement of the analyte is largely independent of the chemical nature of the mobile phase.

As its name implies, the stationary phase in *gas-solid chromatography* is a solid that has a large surface area at which adsorption of the analyte species can take place. Separation occurs because of differences in the positions of adsorption equilibria between the gaseous components of the sample and the solid surface of the stationary phase. Currently, the application of gas-solid chromatography is limited to the separation of low-molecular-weight gaseous species such as carbon monoxide, oxygen, nitrogen, and hydrocarbons. More polar compounds are semipermanently retained on solid surfaces, which precludes the use of solid adsorbents for the gas-chromatographic separation of most species.

*Gas-liquid chromatography* (GLC), the subject of this chapter, is one of the most important and widely used methods for separating and determining the chemical components of complex mixtures. Here the stationary phase is a liquid that is immobilized on the surface of a solid support by adsorption or by chemical bonding. The rate of movement of an analyte through the column is determined by its distribution ratio between the gaseous and the immobilized liquid phase.

Although the concept of gas-liquid chromatography was first enunciated in 1941 by Martin and Synge, its usefulness as a tool for the separation of closely related species was not demonstrated experimentally for more than a decade. Within three years of this demonstration, however, the first commercial apparatus for gas chromatography appeared on the market. Since that time, the growth in applications of this technique has been phenomenal. For example, by 1965 more than 2000 publications on gas chromatography were appearing annually; by 1985, this number had grown to 5000.[1]

> In all types of chromatography, liquid stationary phases are immobilized on solid surfaces by adsorption or by chemical bonding.

> Gas-solid chromatography permits the resolution of low-molecular-weight gas mixtures.

---

[1] For monographs on GLC, see J. A. Perry, *Introduction to Analytical Gas Chromatography*, New York: Marcel Dekker, 1981; *Modern Practice of Gas Chromatography*, 2nd ed., R. L. Grob, Ed. New York: Wiley, 1985; J. Willett, *Gas Chromatography*, New York: Wiley, 1987. See also R. E. Clement, F. L. Onuska, G. A. Eiceman, and H. H. Hill, Jr., *Anal. Chem.*, **1988**, *60*, 279R, **1990**, *62*, 414R.

## 27A   PRINCIPLES OF GAS-LIQUID CHROMATOGRAPHY

The general principles of chromatography, developed in Chapter 26, and the mathematical relationships summarized in Tables 26–4 and 26–5 are applicable to gas chromatography with only a minor modification to correct for the compressibility of gaseous mobile phases.[2]

Equation 26–18 for plate height is applicable to gas chromatography as well as liquid chromatography. The longitudinal diffusion term ($B/u$) in this equation is more important where a gas is the mobile phase, however, because diffusion rates in gases are typically $10^4$ times greater in this medium than in liquids. As a consequence, the minimum in a plot of plate height versus flow rate is considerably broadened in gas chromatography (Figure 26–8b). This effect is also reflected in the curves shown in Figure 27–1, which illustrate the improvement in plate height that accompanies a decrease in the particle size of the column packing.

## 27B   INSTRUMENTS FOR GAS-LIQUID CHROMATOGRAPHY

Figure 27–2 shows the basic components of an instrument for performing gas-liquid chromatographic separations.

### 27B–1  Carrier-Gas Supply

Carrier gases, which must be chemically inert, include helium, argon, nitrogen, and hydrogen, with helium being the most widely used. As is shown later, the choice of gas is often dictated by the type of detector. Associated with the gas supply are pressure regulators, gauges, and flowmeters. In addition, the carrier-gas system often contains a molecular sieve to remove water and other impurities.

Flow rates are controlled by a pressure regulator. Inlet pressures usually range from 10 to 50 psi (lb/in$^2$ above room pressure), which leads to flow rates of 25 to 150 mL/min. As shown in Figure 27–2, flow rates are best established by a simple soap-bubble meter located at the end of the column (see photo in margin). A soap film is formed in the path of the gas by squeezing a bulb containing an aqueous solution of soap or detergent; the time required for this film to move between two graduations on the buret is measured and converted to flow rate.

### 27B–2  Sample-Injection System

In order to avoid band spreading, the sample must be minute and must be introduced rapidly as a "plug" of vapor. Most commonly, a calibrated microsyringe such as those shown in Figure 27–3 is used to inject liquid samples through a rubber or silicone diaphragm or septum into a heated

Helium is the most common mobile phase in gas-liquid chromatography.

A soap-bubble flow meter. (Courtesy of Chrompack Inc., Raritan, NJ.)

[2]See D. A. Skoog and J. J. Leary, *Principles of Instrumental Analysis*, 4th ed., p. 606. Philadelphia: Saunders College Publishing, 1992.

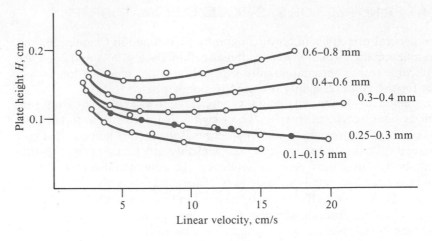

**Figure 27–1**

Effect of particle size on plate height. The numbers to the right are particle diameters. (From J. Boheman and J. H. Purnell, in *Gas Chromatography 1958,* D. H. Desty, Ed. New York: Academic Press, 1958. With permission of Butterworths, Stoneham, MA.)

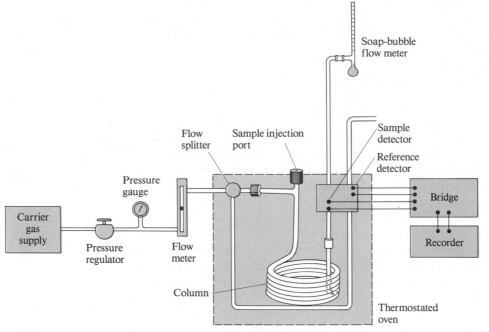

**Figure 27–2**

Diagram of a gas chromatograph.

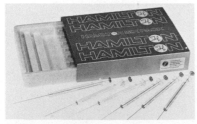

**Figure 27–3**

A set of microsyringes for sample injection. (Courtesy of Chrompack, Inc., Raritan, NJ.)

sample port located at the head of the column. The sample port (Figure 27–4) is ordinarily about 50°C above the boiling point of the least volatile component of the sample so that flash vaporization occurs. For ordinary analytical columns, sample sizes vary from a few tenths of a microliter to 20 $\mu$L. With capillary columns, which require much smaller samples ($\approx 10^{-3} \mu$L), a sample splitter is employed to deliver only a small fraction of the injected sample to the column head, with the remainder going to waste.

Gas samples are best introduced by means of a sample valve similar to that shown in Figure 26–15. Solid samples are commonly introduced as solutions.

## 27B–3  Columns

Historically, all of the pioneering gas-liquid chromatographic studies in the early 1950s were carried out on packed columns in which the stationary phase was a thin film of liquid retained on the surface of a finely divided, inert solid support. From theoretical studies made during this early period, it became apparent, however, that unpacked columns having inside diameters of a few tenths of a millimeter should provide separations that were much superior to packed columns in both speed and column efficiency. In such *capillary columns*, the stationary phase was a uniform film of liquid a few tenths of a micrometer thick that coated the interior of a capillary tubing uniformly. In the late 1950s such *open tubular columns* were constructed, and the predicted performance characteristics were experimentally confirmed in several laboratories with open tubular columns having 300,000 plates or more being described. Despite such spectacular performance characteristics, capillary columns did not gain widespread use until more than two decades after their invention.

The reasons for the delay were several, including: small sample capacities; fragileness of columns; mechanical problems associated with sample introduction and connection of the column to the detector; difficulties in coating the column reproducibly; short lifetimes of poorly prepared columns; tendencies of columns to clog; and patents, which limited commercial development to a single manufacturer (the original patent expired in 1977). By the late 1970s these problems had become manageable and several instrument companies began to offer open tubular columns at a reasonable cost. As a consequence, a major growth in the use of open tubular columns has occurred in the last few years.

### Packed Columns

Present-day packed columns are fabricated from glass, metal (stainless steel, copper, aluminum), or Teflon tubes that typically have lengths of 2 to 3 m and inside diameters of 2 to 4 mm. These tubes are densely packed with a uniform, finely divided packing material, or solid support, that is coated with a thin layer (0.05 to 1 $\mu$m) of the stationary liquid phase. In order to fit in a thermostating oven, the tubes are formed as coils having diameters of roughly 15 cm.

**Solid Support.**  The solid support in a packed column serves to hold the liquid stationary phase in place so that as large a surface area as possible is exposed to the mobile phase. The ideal support consists of small, uniform, spherical particles with good mechanical strength and a specific surface area of at least 1 m$^2$/g. In addition, the material should be inert at elevated temperatures and be uniformly wetted by the liquid phase. No substance that meets all of these criteria perfectly is yet available.

Today, the most widely used support material is prepared from naturally occurring diatomaceous earth, which is made up of the skeletons of thousands of species of single-celled plants that inhabited ancient lakes and seas. Such plants received their nutrients and disposed of their wastes via molecular diffusion through their pores. As a consequence, their re-

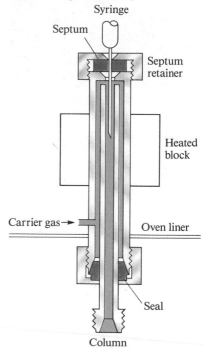

Figure 27–4
A heated sample port. (From H. H. Willard, L. L. Merritt, J. A. Dean, and F. A. Settle, *Instrumental Methods of Analysis,* 7th ed., p. 542. Belmont, CA: Wadsworth, 1988. With permission.)

In 1987, a world record for length of an open tubular column and number of theoretical plates was set, as attested in the *Guinness Book of Records*, by Chrompack International Corporation of the Netherlands. The column was a fused silica column drawn in one piece and having an internal diameter of 0.32 mm and a length of 2100 m or 1.3 miles. The column was coated with a 0.1 $\mu$m film of polydimethyl siloxane. A 1300 m section of this column contained over 2 million plates.

Column efficiency is associated with the particle size of the packing.

Photomicrograph of diatomaceous earth.

mains are well-suited as support materials because gas chromatography is also based upon the same kind of molecular diffusion.

**Particle Size of Supports.** As shown in Figure 27–1, the efficiency of a gas-chromatographic column increases rapidly with decreasing particle diameter of the packing. The pressure difference required to maintain a given flow rate of carrier gas, however, varies inversely as the square of the particle diameter; the latter relationship has placed lower limits on the size of particles employed in gas chromatography because it is not convenient to use pressure differences that are greater than about 50 psi. As a result, the usual support particles are 60 to 80 mesh (250 to 170 $\mu$m) or 80 to 100 mesh (170 to 149 $\mu$m).

## Open Tubular Columns[3]

Open tubular, or capillary, columns are of two basic types, namely, *wall coated open tubular* (WCOT) and *support coated open tubular* (SCOT). Wall coated columns are simply capillary tubes coated with a thin layer of the stationary phase. In support coated open tubular columns, the inner surface of the capillary is lined with a thin film ($\sim$ 30 $\mu$m) of a support material, such as diatomaceous earth. This type of column holds several times as much stationary phase as does a wall coated column and thus has a greater sample capacity. Generally, the efficiency of a SCOT column is less than that of a WCOT column but significantly greater than that of a packed column.

Early WCOT columns were constructed of stainless steel, aluminum, copper, or plastic. Subsequently glass was used. Often the glass was etched with gaseous hydrochloric acid, strong aqueous hydrochloric acid, or potassium hydrogen fluoride to give a rough surface, which bonded the stationary phase more tightly. The newest WCOT columns, which first appeared in 1979, are *fused silica open tubular columns* (FSOT columns) (Figure 27–5). Fused silica capillaries are drawn from specially purified silica that contain minimal amounts of metal oxides. These capillaries have much thinner walls than their glass counterparts. The tubes are given added strength by an outside protective polyimide coating, which is applied as the capillary tubing is being drawn. The resulting columns are quite flexible and can be bent into coils having diameters of a few inches. Silica open tubular columns are available commercially and offer several important advantages such as physical strength, much lower reactivity toward sample components, and flexibility. For most applications, they have replaced the older type WCOT glass columns.

The most widely used silica open tubular columns have inside diameters of 320 and 250 $\mu$m. Higher resolution columns are also sold with diameters of 200 and 150 $\mu$m. Such columns are more troublesome to use and are more demanding upon the injection and detection systems. Thus, a sample splitter must be used to reduce the size of the sample injected onto the column and a more sensitive detector system with a rapid response time is required. Recently, 530 $\mu$m capillaries, sometimes called

Figure 27–5
A 25-m fused-silica capillary column. (Courtesy of Chrompack, Inc., Raritan, NJ.)

---

[3]For a recent monograph on open tubular columns, see M. L. Lee, J. Yang, and K. D. Bartle, *Open Tubular Column Gas Chromatography*. New York: Wiley, 1984.

**Table 27-1**
**PROPERTIES AND CHARACTERISTICS OF TYPICAL GAS CHROMATOGRAPHIC COLUMNS**

|  | Type of Column | | | |
|---|---|---|---|---|
|  | **FSOT*** | **WCOT†** | **SCOT‡** | **Packed** |
| Length, m | 10–100 | 10–100 | 10–100 | 1–6 |
| Inside diameter, mm | 0.1–0.3 | 0.25–0.75 | 0.5 | 2–4 |
| Efficiency, plates/m | 2000–4000 | 1000–4000 | 600–1200 | 500–1000 |
| Sample size, ng | 10–75 | 10–1000 | 10–1000 | $10–10^6$ |
| Relative pressure | Low | Low | Low | High |
| Relative speed | Fast | Fast | Fast | Slow |
| Chemical inertness | Best ————————————————→ Poorest | | | |
| Flexible? | Yes | No | No | No |

*Fused-silica open tubular column.

†Wall-coated open tubular column.

‡Support-coated open tubular column (also called porous-layer open tubular PLOT).

*megabore* columns, have appeared on the market, which will tolerate sample sizes that are similar to those for packed columns. The performance characteristics of megabore open tubular columns are not as good as those of smaller diameter columns but significantly better than those of packed columns.

Table 27-1 compares the performance characteristics of fused silica capillary columns with other types of wall coated columns as well as with support coated and packed columns.

**Adsorption on Column Packings on Capillary Walls.** A problem that has plagued gas chromatography from its inception is the physical adsorption of polar or polarizable analyte species, such as alcohols or aromatic hydrocarbons, on support surfaces. Adsorption results in distorted peaks, which are broadened and often exhibit a tail. It has been established that adsorption is the consequence of silanol groups that form on the surface of diatomaceous silicates by reaction with moisture. Thus, a fully hydrolyzed silicate surface has the structure

$$\begin{array}{cccc} OH & OH & OH & OH \\ | & O & | & O & | & O & | \\ Si & Si & Si & Si \\ | & | & | & | \end{array}$$

The SiOH groups on the support surface have a strong affinity for polar organic molecules and tend to retain them by adsorption.

Support materials can be deactivated by *silanization* with dimethyl-chlorosilane (DMCS). The reaction is

$$—Si—OH + Cl—\underset{\underset{CH_3}{|}}{\overset{\overset{CH_3}{|}}{Si}}—Cl \longrightarrow —Si—O—\underset{\underset{CH_3}{|}}{\overset{\overset{CH_3}{|}}{Si}}—Cl + HCl$$

Adsorption by solute species is minimized by silanization of column packings.

Upon washing with methanol, the second chloride is replaced by a methoxy group:

$$-\underset{\diagdown}{\overset{\diagup}{Si}}-O-\underset{\underset{CH_3}{|}}{\overset{\overset{CH_3}{|}}{C}}-Cl + CH_3OH \longrightarrow -\underset{\diagdown}{\overset{\diagup}{Si}}-O-\underset{\underset{CH_3}{|}}{\overset{\overset{CH_3}{|}}{Si}}-OCH_3 + HCl$$

Silanized supports may still show a residual adsorption, which apparently is the result of mineral impurities in the diatomaceous earth. Acid washing prior to silanization removes these impurities. Packings that have been acid-washed and silanized are sold with the designation —AW—DMCS.

Another silanization reagent is hexamethyldisilazane, whose formula is $(CH_3)_3SiNHSi(CH_3)_3$. Packings treated with this reagent are designated —HMDS.

## 27B–4 Column Thermostating

Temperature programming is a technique in which the temperature of a gas-chromatographic column is increased continuously or in steps during elution.

Column temperature is an important variable that must be controlled to a few tenths of a degree for precise work. Thus, the column is ordinarily housed in a thermostated oven. The optimum column temperature depends upon the boiling point of the sample and the degree of separation required. Roughly, a temperature equal to or slightly above the average boiling point of a sample results in a reasonable elution time (2 to 30 min). For samples with a broad boiling range, it is often desirable to employ temperature programming so that the column temperature is increased either continuously or in steps as the separation proceeds. Figure 27–6c illustrates the effectiveness of temperature programming in improving resolutions.

In general, optimum resolution is associated with minimum temperature; the cost of lowered temperature, however, is an increase in elution time and therefore an increase in the time required to complete an analysis. Figures 27–6a and 6b illustrate this principle.

## 27B–5 Detectors

Detection devices for gas-liquid chromatography must respond rapidly to minute concentrations of solutes as they exit the column. The solute concentration in the carrier gas at any instant is no more than a few parts per thousand and is often smaller than this figure by one or two orders of magnitude. Moreover, the time required for a peak to pass the detector is typically 1 s (or less); thus, the device must be capable of exhibiting its full response during this brief period.

Other desirable properties of a detector are linear response, stability, and uniform response to a wide variety of chemical species or, alternatively, predictable and selective response to one or more classes of solutes. Needless to say, no detector meets all of these desiderata, and it

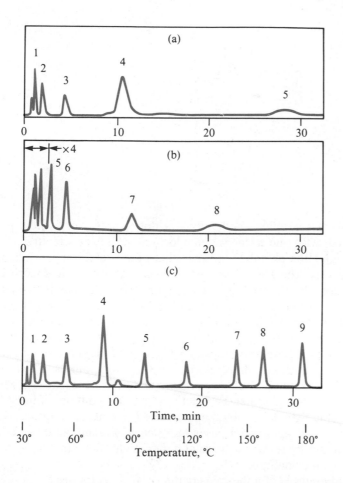

Figure 27–6

Effect of temperature on gas chromatograms. (a) Isothermal at 45°C; (b) isothermal at 145°C; (c) programmed at 30 to 180°C. (From W. E. Harris and H. W. Habgood, *Programmed Temperature Gas Chromatography,* p. 10. New York: Wiley, 1966. Reprinted with permission.)

seems unlikely that such a detector will ever be designed. Three of the most widely used detectors for gas chromatography are described briefly in the paragraphs that follow.

## Thermal Conductivity Detectors

The thermal conductivity detector, or *katharometer*, was one of the earliest detectors employed in gas chromatographic studies. This device, which still finds widespread use, is based upon changes in the thermal conductivity of the gas stream brought about by the presence of analyte molecules. The sensing device in the thermal conductivity detector consists of an electrically heated element whose temperature at constant electrical power depends upon the thermal conductivity of the surrounding gas. The heated element may be a fine platinum, gold, or tungsten wire or a semiconducting thermistor. The resistance of the wire or thermistor is a measure of its temperature, which depends in part upon the rate at which the surrounding gas molecules conduct thermal energy away from the detector element to the walls of a metal block in which it is housed. Figure 27–7 is a schematic drawing of a typical commercial thermal conductivity detector.

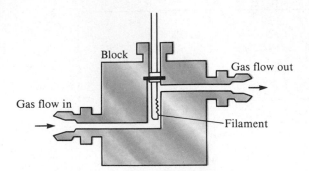

A twin detector system compensates
for the thermal conductivity of the
carrier gas as well as other column
variables.

In chromatographic applications, a double detector system is usually employed, with one element being located in the gas stream *ahead* of the sample-injection chamber and the other immediately beyond the column. Alternatively, the gas stream may be split, as shown in Figure 27–2. In either case, the thermal conductivity of the carrier gas is canceled, and the effects of variation in flow rate, pressure, and electrical power are minimized. The resistances of the twin detectors are usually compared by incorporating them into two arms of a simple Wheatstone bridge circuit.

The thermal conductivities of hydrogen and helium are roughly six to ten times greater than those of most organic compounds. Thus, the presence of even small amounts of organic materials causes a relatively large decrease in the thermal conductivity of the column effluent; consequently, the detector undergoes a marked rise in temperature. The conductivities of other carrier gases more closely resemble those of organic constituents; therefore, a thermal conductivity detector dictates the use of hydrogen or helium.

The advantages of a thermal conductivity detector are its simplicity, its large linear dynamic range ($\simeq 10^5$), its general response to both organic and inorganic species, and its nondestructive character, which permits the collection of solutes after detection. A limitation of the thermal conductivity detector is its relatively low sensitivity ($\simeq 10^{-8}$ g solute/mL of carrier gas). Other detectors exceed this sensitivity by factors as large as $10^4$ to $10^7$. It should be noted that the low sensitivity of thermal detectors precludes their use in conjunction with capillary columns because of the very small samples that can be accommodated by such columns.

### Flame Ionization Detectors

Most organic compounds, when pyrolyzed at the temperature of a hydrogen/air flame, produce ionic intermediates and electrons that provide a mechanism by which electricity can be carried through the plasma. With a burner such as that shown in Figure 27–8, the charged species are attracted to and captured by a collector; the ion current that results is then amplified and recorded. The electrical resistance of a flame plasma is high (perhaps $10^{12}$ $\Omega$), and the resulting currents are therefore minuscule.

The ionization of carbon compounds in a flame is a poorly understood process, although it is observed that the number of ions produced is roughly proportional to the number of *reduced* carbon atoms in the plasma. Functional groups, such as carbonyl, alcohol, halogen, and

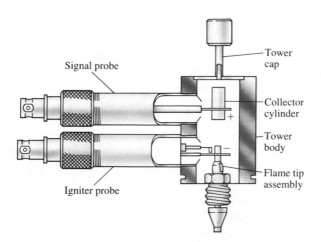

Figure 27–8
A typical flame ionization detector. (Courtesy of Varian Instrument Division, Palo Alto, CA.)

amine, yield fewer ions or none at all. In addition, the detector is insensitive to noncombustible gases, such as $H_2O$, $CO_2$, $SO_2$, and $NO_x$. These properties make the flame ionization detector a useful general detector for most organic samples, including those contaminated with water and the oxides of nitrogen and sulfur.

The flame ionization detector is perhaps the most widely used detector because of its high sensitivity ($\simeq 10^{-13}$ g/mL), large linear response range ($\simeq 10^7$), low noise, ruggedness, and convenience. It is normally the detector used with capillary columns. A disadvantage of the flame ionization detector is that it destroys the sample.

> The high sensitivity of the flame ionization detector is at the expense of the sample, which is destroyed.

### Electron-Capture Detectors

With electron-capture detectors, the effluent from the column is passed over a beta emitter (a radioactive compound that emits electrons) such as nickel-63 or tritium (adsorbed on platinum or titanium foil). An electron from the emitter causes ionization of the carrier gas (often nitrogen) and production of a burst of electrons. In the absence of organic species, a constant standing current between a pair of electrodes results from this ionization. The current decreases, however, in the presence of organic molecules that tend to capture electrons. The response is nonlinear unless the potential across the detector is pulsed.

The electron-capture detector is selective in its response and highly sensitive to electronegative functional groups, such as halogens, peroxides, quinones, and nitro groups. It is insensitive to compounds such as amines, alcohols, and hydrocarbons. An important application of the electron-capture detector has been in the detection and determination of chlorinated pesticides.

Electron-capture detectors are highly sensitive and possess the advantage of not consuming the sample to any significant extent (in contrast to flame detectors).

### Selective Detectors

Gas chromatography is often coupled with the selective techniques of spectroscopy and electrochemistry. The resulting so-called *hyphenated*

> Hyphenated methods couple the separation capabilities of chromatography with the capacity for qualitative and quantitative detection of electrical or spectral methods.

*methods* (for instance, GC-MS and GC-IR) provide the chemist with powerful tools for identifying the components of complex mixtures.[4]

In early hyphenated methods, the eluates from the chromatographic column were collected as separate fractions in a cold trap, a nondestructive, nonselective detector being employed to indicate their appearance. The composition of each fraction was then investigated by nuclear magnetic resonance, infrared, or mass spectroscopy or by electroanalytical measurements. A serious limitation to this approach was the very small (usually micromolar) quantities of solute contained in a fraction; nonetheless, the general procedure proved useful for the qualitative analysis of many multicomponent mixtures.

A second general method, one that now finds widespread use, involves the application of spectroscopic or electroanalytical detectors to monitor the column effluent continuously. Generally, this procedure requires computerized instruments and a large computer memory for storing spectral or electrochemical data for subsequent display as spectra and chromatograms.

## 27C    LIQUID PHASES FOR GAS-LIQUID CHROMATOGRAPHY

Desirable properties for the immobilized liquid phase in a gas-liquid chromatographic column include: (1) *low volatility* (ideally, the boiling point of the liquid should be at 100°C higher than the maximum operating temperature for the column); (2) *thermal stability*; (3) *chemical inertness*; (4) *solvent characteristics* such that $k'$ and $\alpha$ (Sections 26B–4) values for the solutes to be resolved fall within a suitable range.

A myriad of solvents have been proposed as stationary phases in the course of the development of gas-liquid chromatography. By now, only a handful—perhaps a dozen or less—suffice for most applications. The proper choice among these solvents is often critical to the success of a separation. Qualitative guidelines exist for making this choice, but in the end, the best stationary phase can only be determined in the laboratory.

The retention time for a solute on a column depends upon its partition ratio (Equation 26–1), which in turn is related to the chemical nature of the stationary phase. Clearly, to be useful in gas-liquid chromatography, the immobilized liquid must generate different partition ratios for different solutes. In addition, however, these ratios must not be extremely large or extremely small because the former leads to prohibitively long retention times and the latter results in such short retention times that separations are incomplete.

To have a reasonable residence time in the column, a species must show some degree of compatibility (solubility) with the stationary phase. Here, the principle of "like dissolves like" applies, where "like" refers to the polarities of the solute and the immobilized liquid. Polarity is the electrical field effect in the immediate vicinity of a molecule and is measured by the dipole moment of the species. Polar stationary phases con-

---

[4]For a review on hyphenated methods, see T. Hirschfield, *Anal. Chem.*, **1980**, *52*, 297A; C. L. Wilkens, *Science*, **1983**, *222*, 251.

tain functional groups such as —CN, —CO, and —OH. Hydrocarbon-type stationary phases and dialkyl siloxanes are nonpolar, whereas polyester phases are highly polar. Polar solutes include alcohols, acids, and amines; solutes of medium polarity include ethers, ketones, and aldehydes. Saturated hydrocarbons are nonpolar. Generally, the polarity of the stationary phase should match that of the sample components. When the match is good, the order of elution is determined by the boiling point of the eluents.

### Some Widely Used Stationary Phases

Table 27–2 lists the most widely used stationary phases for both packed and open tubular column gas chromatography in order of increasing polarity. These six liquids can probably provide satisfactory separations for 90% or more of the samples encountered by the scientist.

Five of the liquids listed in Table 27–2 are polydimethyl siloxanes that have the general structure

$$
R-\underset{\underset{R}{|}}{\overset{\overset{R}{|}}{Si}}-O\left[\underset{\underset{R}{|}}{\overset{\overset{R}{|}}{Si}}-O\right]\underset{\underset{R}{|}}{\overset{\overset{R}{|}}{Si}}-R
$$

Table 27–2
SOME COMMON STATIONARY PHASES FOR GAS-LIQUID CHROMATOGRAPHY

| Stationary Phase | Common Trade Name | Maximum Temperature, °C | Common Applications |
|---|---|---|---|
| Polydimethyl siloxane | OV-1, SE-30 | 350 | General purpose nonpolar phase; hydrocarbons; polynuclear aromatics; drugs; steroids; PCB's |
| 5% Phenyl-polydimethyl siloxane | OV-3, SE-52 | 350 | Fatty acid methyl esters; alkaloids; drugs; halogenated compounds |
| 50% Phenyl-polydimethyl siloxane | OV-17 | 250 | Drugs; steroids; pesticides; glycols |
| 50% Trifluoropropyl-polydimethyl siloxane | OV-210 | 200 | Chlorinated aromatics; nitroaromatic; alkyl substituted benzenes |
| Polyethylene glycol | Carbowax 20M | 250 | Free acids; alcohols; ethers; essential oils; glycols |
| 50% Cyanopropyl-polydimethyl siloxane | OV-275 | 240 | Polyunsaturated fatty acids; rosin acids; free acids; alcohols |

In the first of these, polydimethyl siloxane, the —R groups are all —$CH_3$ giving a liquid that is relatively nonpolar. In the other polysiloxanes shown in the table, a fraction of the methyl groups are replaced by functional groups such as phenyl (—$C_6H_5$), cyanopropyl (—$C_3H_6CN$), and trifluoropropyl (—$C_3H_6CF_3$). The percentage descriptions in each case gives the amount of substitution of the named group for methyl groups on the polysiloxane backbone. Thus, for example, 5% phenyl polydimethyl-siloxane has a phenyl ring bonded to 5% by number of the silicon atoms in the polymer. These substitutions increase the polarity of the liquids to various degrees.

The fifth entry in Table 27–2 is a polyethylene glycol having the structure

$$HO—CH_2—CH_2—(O—CH_2—CH_2)_n—OH$$

It finds widespread use for separating polar species. Figure 27–9 illustrates applications of the phases listed in Table 27–2 for open tubular columns.

## Bonded and Cross-Linked Stationary Phases

Commercial columns are advertised as having bonded and/or cross-linked stationary phases. The purpose of bonding and cross-linking is to provide a longer lasting stationary phase that can be rinsed with a solvent when the film becomes contaminated. With use, untreated columns slowly

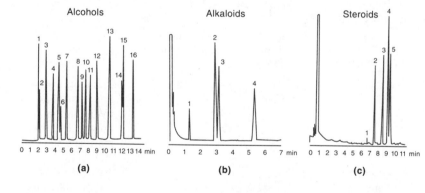

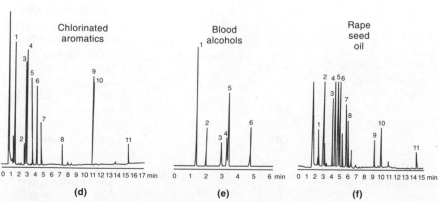

**Figure 27–9**

Typical chromatograms from open tubular columns coated with (a) polydimethyl siloxane, (b) 5% phenyl polydimethyl siloxane, (c) 50% phenyl polydimethyl siloxane, (d) 50% trifluoropropylpolydimethyl siloxane, (e) polyethylene glycol, and (f) 50% cyanopropylpolydimethyl siloxane.

loose their stationary phase due to ''bleeding'' in which a small amount of immobilized liquid is carried out of the column during the elution process. Bleeding is exacerbated when a column must be rinsed with a solvent to remove contaminants. Chemical bonding and cross-linking inhibit bleeding.

Bonding involves attaching a monomolecular layer of the stationary phase to the silica surface of the column by a chemical reaction. For commercial columns, the nature of the reactions is ordinarily proprietary.

Cross-linking is carried out *in situ* after a column is coated with one of the polymers listed in Table 27–2. One way of cross-linking is to incorporate a peroxide into the original liquid. When the film is heated, reaction between the methyl groups in the polymer chains is initiated by a free radical mechanism. The polymer molecules are then cross-linked through carbon to carbon bonds. The resulting films are less extractable and have considerably greater thermal stability than do untreated films. Cross-linking has also been initiated by exposing the coated columns to gamma radiation.

### Film Thickness

Commercial columns are available having stationary phases that vary in thickness from 0.1 to 5 $\mu$m. Film thickness primarily affects the retentive character and the capacity of a column. Thick films are used with highly volatile analytes because such films retain solutes for a longer time thus providing a greater time for separation to take place. Thin films are useful for separating species of low volatility in a reasonable length of time. For most applications with 0.25 or 0.32 mm columns, a film thickness of 0.25 $\mu$m is recommended. With megabore columns, 1 to 1.5 $\mu$m films are often used. Today, columns with 8 $\mu$m films are also marketed.

## 27D   APPLICATIONS OF GAS-LIQUID CHROMATOGRAPHY

In evaluating the importance of gas-liquid chromatography, it is necessary to distinguish between the two roles the method plays. The first is as a tool for performing separations; in this capacity, it is unsurpassed when applied to complex organic, metal-organic, and biochemical systems. The second, and distinctly different, function is analysis. Here, retention times are employed for qualitative identification, and peak heights or areas provide quantitative information. Gas-liquid chromatography is much more limited for qualitative purposes than are most spectroscopic methods. As a consequence, an important trend in the field has been in the direction of combining the remarkable fractionation qualities of chromatography with the superior identification properties of such instruments as mass, infrared, and NMR spectrometers.

### 27D–1   Qualitative Analysis

Gas chromatograms are widely used as criteria of purity for organic compounds. Contaminants are revealed by the appearance of additional peaks. The technique is also useful for evaluating the effectiveness of purification procedures.

In theory, retention times should be useful for the identification of components in mixtures. In fact, however, the applicability of such data is limited by the number of variables that must be controlled in order to obtain reproducible results. Nevertheless, gas chromatography provides an excellent means of confirming the presence or absence of a compound in a mixture, provided an authentic standard sample of the substance is available. The chromatogram of a mixture of the standard and unknown should show no new peaks; enhancement of an existing peak should also be observed. The evidence is particularly convincing if the effect can be duplicated on different columns and at different temperatures.

We have seen (Section 26B–5) that the selectivity factor $\alpha$ for compounds A and B is given by the relationship

$$\alpha = \frac{K_B}{K_A} = \frac{(t_R)_B - t_M}{(t_R)_A - t_M} = \frac{k'_B}{k'_A}$$

If a standard substance is chosen as compound B, then $\alpha$ can provide an index for identification of compound A that is largely independent of column variables other than temperature; that is, numerical tabulations of selectivity factors for pure compounds relative to a common standard can be prepared and then used for the characterization of solutes. The amount of such data available in the literature is presently limited.

## 27D–2 Quantitative Analysis

The detector signal from a gas-liquid chromatographic column is widely used for quantitative and semiquantitative analyses. An accuracy of 1 to 3% relative is attainable under carefully controlled conditions. As with most analytical tools, reliability is directly related to the control of variables. The nature of the sample also plays a part in determining the potential accuracy.

The general discussion of quantitative chromatographic analysis in Section 26F–2 applies to gas chromatography; therefore, no further consideration of this topic is needed.

## 27E    TYPICAL APPLICATIONS OF GAS CHROMATOGRAPHY

Figure 27–10 illustrates some typical applications of gas chromatography to analytical problems. All are based upon gas-liquid equilibria except Figure 27–10a, where two columns are used in series, the first a gas-liquid column and the second a gas-solid. The gas-liquid column retains only the carbon dioxide and passes the remaining gases without retention. While the carbon dioxide is being eluted from the first column, a switch directs the flow around the second column to avoid permanent adsorption of this gas on the solid packing. After the carbon dioxide signal has returned to

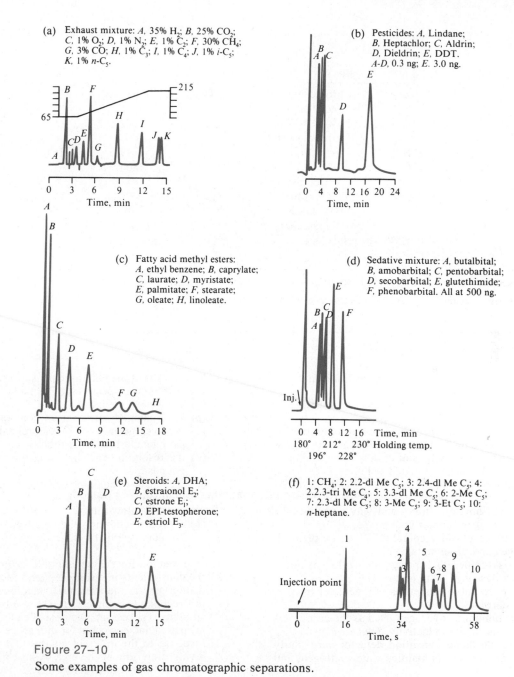

(a) Exhaust mixture: *A*, 35% H$_2$; *B*, 25% CO$_2$; *C*, 1% O$_2$; *D*, 1% N$_2$; *E*, 1% C$_2$; *F*, 30% CH$_4$; *G*, 3% CO; *H*, 1% C$_3$; *I*, 1% C$_4$; *J*, 1% *i*-C$_5$; *K*, 1% *n*-C$_5$.

(b) Pesticides: *A*, Lindane; *B*, Heptachlor; *C*, Aldrin; *D*, Dieldrin; *E*, DDT. *A*-*D*, 0.3 ng; *E*. 3.0 ng.

(c) Fatty acid methyl esters: *A*, ethyl benzene; *B*, caprylate; *C*, laurate; *D*, myristate; *E*, palmitate; *F*, stearate; *G*, oleate; *H*, linoleate.

(d) Sedative mixture: *A*, butalbital; *B*, amobarbital; *C*, pentobarbital; *D*, secobarbital; *E*, glutethimide; *F*, phenobarbital. All at 500 ng.

(e) Steroids: *A*, DHA; *B*, estraionol E$_2$; *C*, estrone E$_1$; *D*, EPI-testopherone; *E*, estriol E$_3$.

(f) 1: CH$_4$; 2: 2.2-dl Me C$_5$; 3: 2.4-dl Me C$_5$; 4: 2.2.3-tri Me C$_4$; 5: 3.3-dl Me C$_5$; 6: 2-Me C$_5$; 7: 2.3-dl Me C$_5$; 8: 3-Me C$_5$; 9: 3-Et C$_5$; 10: *n*-heptane.

Figure 27–10
Some examples of gas chromatographic separations.

zero, the flow is switched back through the second column, thereby permitting separation of the remaining sample components.

Figure 27–10f is particularly noteworthy because it illustrates the power and speed of open tubular columns. Here, methane and nine heptane isomers are separated in just under 1 min.

Pertinent data on column conditions for the chromatograms in Figure 27–10 are found in Table 27–3.

**Table 27-3**

| Chromatogram | Column* | Packing | Detector† | Temperature, °C | Carrier | Flow, mL/min |
|---|---|---|---|---|---|---|
| a | 10′ × $\frac{1}{8}$″ S<br>5′ × $\frac{1}{8}$″ S | Chromosorb 102<br>Molecular sieve 5A | TCD | 65–200 | He | 30 |
| b | 6′ × $\frac{1}{4}$″ G | 1.5% OV-17 on<br>Chromosorb G | ECD | 220 | | |
| c | 6′ × $\frac{1}{8}$″ S | Chromosorb W | TCD | 190 | He | 30 |
| d | 6′ × $\frac{1}{4}$″ G | 1.5% OV-17 on HP<br>Chromosorb G | FID | 180–230 | | |
| e | 6′ × 3.4 mm G | 5% OV-210, 2.5%<br>OV-17 on<br>Supelcoport | ECD | 260 | A | |
| f | 50′ × 0.005″<br>FSOT | Squalene | FID | 20 | $H_2$ | |

*S = packed stainless steel; G = packed glass; FSOT = fused silica open tubular.

†TCD = thermal conductivity; ECD = electron capture; FID = flame ionization.

## 27F QUESTIONS AND PROBLEMS

*27-1. Explain the difference between gas-liquid and gas-solid chromatography.

27-2. Why has gas-solid chromatography been limited in its applications?

*27-3. Describe the differences between an open tubular column and a packed column. What are the advantages and disadvantages of each?

27-4. How do chemically bonded columns differ from conventional gas-liquid columns? What are the advantages and disadvantages of each?

*27-5. What are sample splitters and why must they be used with open tubular columns?

27-6. What are fused-silica columns? What are their advantages and disadvantages?

*27-7. What is temperature programming in gas-liquid chromatography? Why is it used?

27-8. List the desirable properties of a detector for gas-liquid chromatography.

*27-9. Why is helium or hydrogen used as the carrier gas when a thermal conductivity detector is used?

27-10. Why is the flame ionization detector particularly useful for samples containing water, nitrogen, and sulfur?

*27-11. What are hyphenated chromatographic methods?

27-12. A gas-liquid column was labeled —AW—DMCS. What is the meaning of this designation?

*27-13. What is the meaning of silanization when it refers to a column packing?

27-14. Why is it desirable to introduce the sample onto a chromatographic column quickly?

*27-15. The following retention times were found with a 1.10-m gas-liquid chromatographic column: air, 18.0 s; methyl acetate, 1.98 min; methyl propionate, 2.24 min; methyl n-butyrate, 7.93 min. The base widths of the last three peaks were 0.19, 0.23, and 0.79 min, respectively. Calculate

(a) $k'$ for each compound.

(b) $\alpha$ for each adjacent pair of compounds.

(c) the average number of theoretical plates and plate height for the column.

(d) the resolution for each adjacent pair of compounds.

(e) the length of column required to achieve a resolution of 1.5 for the first two peaks, assuming the same flow rate.

(f) the time required to elute methyl n-butyrate for a column with the dimensions of that in part (e).

27-16. One method for the quantitative determination of the constituents in a sample analyzed by gas chromatography is the area-normalization method. In this technique, complete elution of all sample constituents is necessary; the area of each peak is then measured and corrected for differences in detector response to the different eluates. This correction involves division of the area by an empirically determined correction factor. The concentration of the analyte is found from the ratio of its corrected area to the total corrected area of all peaks. For the chromatogram described in Problem 27-15, the areas of the three peaks were 16.4, 45.2, and 30.2 in the order of increasing retention time. Calculate the percentage of each compound if the relative detector responses were 0.60, 0.78, and 0.88, respectively.

*27-17. The stationary liquid in the column described in Problem 27-15 was didecylphthalate, a solvent of intermediate polarity. If a nonpolar solvent such as a silicone oil had been used instead, would the retention times for the three compounds be larger or smaller? Why?

**27-18.** The following retention times were observed with a 1.25-m chromatographic column: air, 0.395 min; isopropylamine, 4.59 min; and n-propylamine, 4.91 min. Peak widths were 0.365 min for isopropylamine and 0.382 min for n-propylamine. Calculate
  (a) $k'$ for each amine.
  (b) an $\alpha$ value.
  (c) an average number of theoretical plates in the column.
  (d) the plate height for the column.
  (e) the resolution for the two amines.
  (f) the length of column required to achieve a resolution of 1.5.
  (g) the time required to achieve a resolution of 1.5, assuming the original linear flow rate.

**27-19.** What effect does each of the following have on the plate height of a column? Explain.
  *(a) increasing the weight of the stationary phase relative to the packing weight
  (b) decreasing the sample-injection rate
  *(c) increasing the injection-port temperature
  (d) increasing the flow rate

*(e) reducing the particle size of the packing
  (f) decreasing the column temperature

**\*27-20.** Peak areas and relative detector responses are to be used to determine the concentration of the five species in a sample. The area-normalization method described in Problem 27-16 is to be used. The relative areas for the five gas chromatographic peaks are given below. Also shown are the relative responses of the detector. Calculate the percentage of each component in the mixture.

| Compound | Relative Peak Area | Relative Detector Response |
|---|---|---|
| A | 32.5 | 0.70 |
| B | 20.7 | 0.72 |
| C | 60.1 | 0.75 |
| D | 30.2 | 0.73 |
| E | 18.3 | 0.78 |

**27-21.** List the variables that lead to (a) band broadening and (b) band separation in gas-liquid chromatography.

# HIGH-PERFORMANCE LIQUID CHROMATOGRAPHY

$M$ost of this chapter deals with four types of column chromatography in which a liquid serves as the mobile phase: *partition*, or *liquid-liquid, chromatography; adsorption*, or *liquid-solid, chromatography; ion-exchange chromatography;* and *size-exclusion*, or *gel-permeation, chromatography.* Also included at the end of the chapter are short sections on planar liquid chromatography and supercritical-fluid chromatography. In the latter method, a supercritical fluid rather than a liquid serves as the mobile phase.

Early liquid chromatography, including Tswett's original work, was carried out in glass columns with diameters of 1 to 5 cm and lengths of 50 to 500 cm. To ensure reasonable flow rates, the diameters of the particles in the solid stationary phase were usually in the 150 to 200 $\mu$m range. Even then, flow rates were only a few tenths of a milliliter per minute. Thus, separation times were long—often several hours. Attempts to speed up the classic procedure by application of vacuum or by pumping were not effective because increases in flow rates increased plate heights beyond the minimum in the typical curve of plate height versus flow rate (Figure 26–8a reproduced here in the margin); decreased efficiencies were the result.

Early in the development of liquid chromatography, it was realized that major increases in column efficiency would accompany decreases in the particle size of packings. This effect is demonstrated by the data plotted in Figure 28–1. Note that in none of these plots is the minimum shown in Figure 26–8a reached because this minimum often occurs only at prohibitively low flow rates in liquid chromatography.

It was not until the late 1960s that the technology for producing and using packings with particle diameters as small as 10 $\mu$m was developed. This technology required sophisticated instruments that contrasted markedly with the simple glass columns of classic liquid chromatography. The name *high-performance liquid chromatography* (HPLC) is often employed to distinguish these newer procedures from the classic methods,

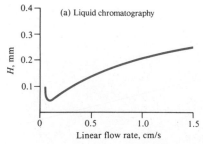

Plate height vs. flow rate for liquid chromatography.

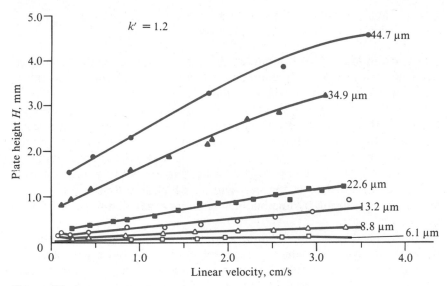

Figure 28–1
Effect of packing particle size and flow rate on plate height in liquid chromatography. Column dimensions: 30 cm × 2.4 mm. Solute: N,N-diethyl-*n*-aminoazobenzene. Mobile phase: mixture of hexane, methylene chloride, isopropyl alcohol. (From R. E. Majors, *J. Chromatogr. Sci.*, **1973**, *11*, 92. With permission.)

which still find considerable use for preparative purposes. The next six sections of this chapter are devoted to HPLC.[1]

## 28A    INSTRUMENTS FOR HIGH-PERFORMANCE LIQUID CHROMATOGRAPHY

Pumping pressures of several hundred atmospheres are required to achieve reasonable flow rates with modern liquid chromatographic packings, which are made up of particles with diameters of 10 $\mu$m or less. As a consequence of these high pressures, the equipment for high-performance liquid chromatography tends to be considerably more elaborate and expensive than that encountered in other types of chromatography. Figure 28–2 shows the important components of a typical high-performance chromatograph.

---

[1]Numerous books on high-performance liquid chromatography are available. Among these are L. R. Snyder and J. J. Kirkland, *Introduction to Modern Liquid Chromatography*, 2nd ed. New York: Wiley, 1979; P. R. Brown and R. A. Hartwick, *High Performance Liquid Chromatography*. New York: Wiley, 1981; V. R. Meyer, *Practical High Performance Liquid Chromatography*. New York: Wiley, 1988. For a recent review dealing with past, current, and future developments in HPLC, see P. R. Brown, *Anal. Chem.*, **1990**, *62*, 995A.

### 28A–1 Mobile-Phase Reservoirs and Solvent-Treatment Systems

As shown in Figure 28–2, a modern high-performance liquid chromatography apparatus is usually equipped with one or more glass or stainless steel reservoirs, each of which contains 500 mL or more of a solvent. Provisions are often included for removing dissolved gases and particulate matter from the liquids. The gases produce band spreading as a result of bubble formation, while both bubbles and particulates interfere with detector performance. Degassers may consist of a vacuum pumping system, a distillation system, a device for heating and stirring, or, as shown in Figure 28–2, a system for *sparging,* in which the dissolved gases are swept out of solution by fine bubbles of an inert gas that is not soluble in the mobile phase.

The simplest way of performing a liquid chromatographic separation is by *isocratic elution,* in which a single solvent is used to carry the analytes through the column. Often, however, a more satisfactory chromatogram can be obtained by a *gradient elution,* in which two (and sometimes more) solvent systems that differ significantly in polarity are employed. The ratio between volumes of the two solvents is varied in a preprogrammed way, sometimes continuously and sometimes in a series of steps. Gradient elution frequently improves separation efficiency, just as temperature programming helps in gas chromatography. Modern HPLC instruments

Sparging is a process in which dissolved gases are swept out of a solvent by bubbles of an inert, insoluble gas.

An isocratic elution in HPLC is one in which the composition of the solvent remains constant.

A gradient elution in HPLC is one in which the composition of the solvent is changed either continuously or in a series of steps.

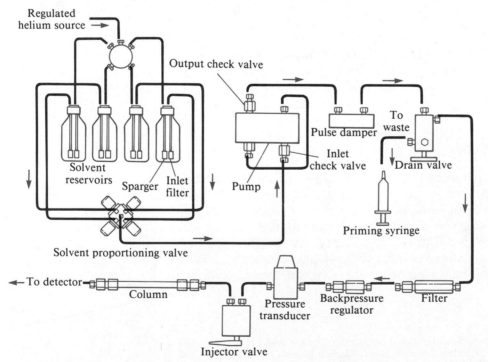

**Figure 28–2**
Schematic of an apparatus for high-performance liquid chromatography. (Courtesy of Perkin-Elmer, Norwalk, CT.)

are often equipped with proportioning valves that introduce liquids from two or more reservoirs at rates that vary continuously (Figure 28-2).

## 28A-2 Pumping Systems

The requirements for liquid chromatographic pumps are severe: (1) generation of pressures up to 6000 psi (pounds per square inch), (2) pulse-free output, (3) flow rates ranging from 0.1 to 10 mL/min, (4) flow reproducibilities of 0.5% relative or better, and (5) resistance to corrosion by a variety of solvents.

Two types of mechanical pumps are employed: a screw-driven syringe type and a reciprocating pump (see Figure 28-3). The former produces a pulse-free delivery whose flow rate is readily controlled; it suffers, however, from a lack of capacity ($\approx$250 mL) and is inconvenient when solvents must be changed. Reciprocating pumps, which are more widely used, usually consist of a small cylindrical chamber that is filled and then emptied by the back-and-forth motion of a piston. The pumping motion produces a pulsed flow that must be subsequently damped. Advantages of reciprocating pumps include small internal volume, high output pressure (up to 10,000 psi), ready adaptability to gradient elution, and constant flow rates, which are largely independent of column back-pressure and solvent viscosity.

Some instruments use a pneumatic pump, which in its simplest form consists of a collapsible solvent container housed in a vessel that can be pressurized by a compressed gas. Pumps of this type are simple, inexpensive, and pulse-free; they suffer from limited capacity and pressure output, however, and from pumping rates that depend upon solvent viscosity. In addition, they are not adaptable to gradient elution.

It should be noted that the high pressures generated by liquid chromatographic pumps do not constitute an explosion hazard because liquids are not very compressible. Thus, rupture of a component results only in solvent leakage, which may, however, constitute a fire hazard.

## 28A-3 Sample-Injection Systems

Syringe injection through an elastomeric septum is often used in liquid chromatography owing to its simplicity. This procedure is not very repro-

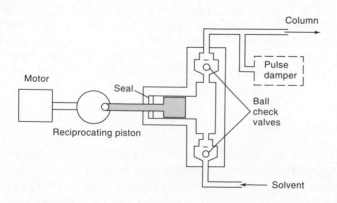

Figure 28-3

A reciprocating pump for HPLC.

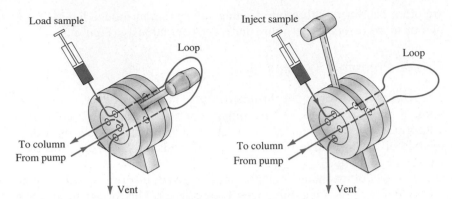

Figure 28–4
A sampling loop for a liquid chromatograph. (Courtesy of Beckman Instruments, Fullerton, CA.)

ducible, however, and is limited to pressures less than about 1500 psi. In *stop-flow* injection, the solvent flow is stopped momentarily, a fitting at the column head is removed, and the sample is injected directly onto the head of the packing by means of a syringe. The most widely used method of sample introduction in liquid chromatography is based upon sampling loops, such as that shown in Figure 28–4. These devices are often an integral part of modern liquid chromatography equipment and have interchangeable loops that provide a choice of sample size ranging from 5 to 500 μL. The precision of injections with a typical sampling loop is a few tenths of a percent relative.

### 28A–4  Columns for High-Performance Liquid Chromatography

High-performance liquid chromatographic columns are usually constructed from stainless steel tubing, although heavy-walled glass tubing is sometimes employed for lower-pressure applications (<600 psi). Most columns range in length from 10 to 30 cm and have an inside diameter of 4 to 10 mm. Column packings typically have particle sizes of 5 or 10 μm. Columns of this type often contain 40,000 to 60,000 plates/m. Recently, high-performance microcolumns with an inside diameter of 1 to 4.6 mm and a length of 3 to 7.5 cm have become available. These columns, which are packed with 3- or 5-μm particles, contain as many as 100,000 plates/m and have the advantages of speed and minimal solvent consumption.

The most common packing for liquid chromatography consists of silica particles, which have been synthesized from submicrometer silica particles under conditions that lead to larger aggregates with highly uniform diameters. The product is often coated with a thin organic film, which is chemically or physically bonded to the surface. Other packing materials include alumina particles, porous-polymer particles, and ion-exchange resins.

### 28A–5  Detectors

No highly sensitive universal detector system, such as those used in gas chromatography, is available for high-performance liquid chromatography. Thus, the system used depends upon the nature of the sample. Table 28–1 lists some of the common detectors and their properties.

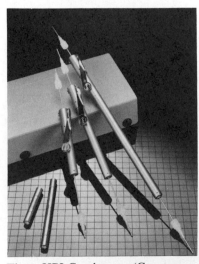

Three HPLC columns. (Courtesy of Chrompack, Inc., Raritan, NJ.)

Table 28–1
PERFORMANCES OF LC DETECTORS

| LC Detector | Commercially Available | Mass LOD (commercial detectors)[a] | Mass LOD (state of the art)[b] |
|---|---|---|---|
| Absorbance | Yes[c] | 100 pg–1 ng | 1 pg |
| Fluorescence | Yes[c] | 1–10 pg | 10 fg |
| Electrochemical | Yes[c] | 10 pg–1 ng | 100 fg |
| Refractive index | Yes | 100 ng–1 $\mu$g | 10 ng |
| Conductivity | Yes | 500 pg–1 ng | 500 ng |
| Mass spectrometry | Yes[d] | 100 pg–1 ng | 1 pg |
| FT–IR | Yes[d] | 1 $\mu$g | 100 ng |
| Light scattering[e] | Yes | 10 $\mu$g | 500 ng |
| Optical activity | No | — | 1 ng |
| Element selective | No | — | 10 ng |
| Photoionization | No | — | 1 pg–1 ng |

[a]Mass LOD (limit of detection) is calculated for injected mass that yields a signal equal to five times the $\sigma$ noise, using a mol wt of 200 g/mol, 10 $\mu$L injected for conventional or 1 $\mu$L injected for microbore LC.

[b]Same definition as $a$, above, but the injected volume is generally smaller.

[c]Commercially available for microbore LC also.

[d]Commercially available, yet costly.

[e]Including low-angle light scattering and nephelometry.

(From E. S. Yeung and R. E. Synovec, *Anal. Chem.*, **1986**, *58*, 1238. With permission. Copyright American Chemical Society)

The most widely used detectors for liquid chromatography are based upon absorption of ultraviolet or visible radiation. Both photometers and spectrophotometers specifically designed for use with chromatographic columns are available from commercial sources. The former often make use of the 254- and 280-nm lines from a mercury source because many organic functional groups absorb in this region (Section 22A–1). Deuterium or tungsten-filament sources with interference filters also provide a simple means of detecting absorbing species. Some modern instruments are equipped with filter wheels. Rotation of the wheel makes it convenient to insert the desired filter in the light path. Spectrophotometric detectors are considerably more versatile than photometers and are also widely used in high-performance instruments.

Another detector that has found considerable application measures the changes in solvent refractive index caused by the presence of analyte molecules. In contrast to most of the other detectors listed in Table 28–1, the refractive-index detector is general rather than selective and responds to all solutes. The disadvantage of this detector is its somewhat limited sensitivity.

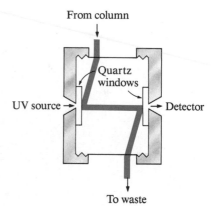

A UV detector for HPLC.

## 28B   HIGH-PERFORMANCE PARTITION CHROMATOGRAPHY

*Partition chromatography*, which has become the most widely used of all liquid chromatographic procedures, is conveniently divided into *liquid-liquid* and *bonded-phase* chromatography. The difference between the two lies in the method by which the stationary phase is held on the

In liquid–bonded-phase partition chromatography, the stationary phase is an organic species that is attached to the surface of the packing particles by covalent bonding.

packing. In liquid-liquid partition packings, retention is by physical adsorption, while in bonded-phase packings, covalent bonds are involved. Early partition chromatography was exclusively liquid-liquid; now, however, bonded-phase methods predominate, with liquid-liquid separations being relegated to certain special applications.

### 28B–1 Bonded-Phase Packings

Most bonded-phase packings are prepared by reaction of an organochlorosilane with the —OH groups formed on the surface of silica particles by hydrolysis in hot dilute hydrochloric acid. The product is an organosiloxane. The reaction for one such SiOH site on the surface of a particle can be written as

$$-Si{\diagup}^{\textstyle CH_3}OH + Cl-Si{\diagup}^{\textstyle CH_3}R \rightarrow -Si{\diagup}^{\textstyle}O-Si{\diagup}^{\textstyle CH_3}R$$

where R is often a straight-chain octyl or octyldecyl group ($C_8H_{17}$— or $C_{18}H_{37}$—). Other organic functional groups that have been bonded to silica surfaces include aliphatic amines, ethers, nitriles, and aromatic hydrocarbons. Thus, a range of polarities for the bonded stationary phase is available.

Bonded-phase packings are significantly more stable than packings in which the stationary phase is retained by physical attraction alone. With the latter, periodic recoating of the solid surfaces is required because the stationary phase is gradually dissolved away in the mobile phase. Furthermore, gradient elution is not practical with liquid-liquid packings, again because of losses of the stationary phase by solubility in the mobile phase. The main disadvantage of bonded-phase packings is their somewhat limited sample capacity.

### 28B–2 Normal- and Reversed-Phase Packings

In normal-phase partition chromatography, the stationary phase is polar and the mobile phase nonpolar. In reversed-phase partition chromatography, the polarity of these phases is reversed.

Two types of partition chromatography are distinguishable based upon the relative polarities of the mobile and stationary phases. Early work in liquid chromatography was based upon highly polar stationary phases, such as triethylene glycol or water; a relatively nonpolar solvent, such as hexane or *i*-propyl ether, served as the mobile phase. For historic reasons, this type of chromatography is now called *normal-phase chromatography.* In *reversed-phase chromatography,* the stationary phase is nonpolar, often a hydrocarbon, and the mobile phase is a relatively polar solvent, such as water, methanol, or acetonitrile.[2] In normal-phase chromatography, the *least* polar component is eluted first; *increasing* the po-

---

[2]For a detailed discussion of reversed-phase HPLC, see A. M. Krstulovic and P. R. Brown, *Reversed-Phase High-Performance Liquid Chromatography.* New York: Wiley, 1982.

larity of the mobile phase *decreases* the elution time. In contrast, in the reversed-phase method, the *most* polar component elutes first, and *increasing* the mobile-phase polarity *increases* the elution time.

It has been estimated that more than three quarters of all high-performance liquid chromatographic separations are currently performed with reversed-phase, bonded octyl or octyldecyl siloxane packings. With such preparations, the long-chain hydrocarbon groups are aligned parallel to one another and perpendicular to the surface of the particle, giving a brush-like, nonpolar hydrocarbon surface. The mobile phase used with these packings is often an aqueous solution containing various concentrations of such solvents as methanol, acetonitrile, and tetrahydrofuran.

In normal-phase chromatography, the least polar analyte is eluted first. In reversed-phase chromatography, the least polar analyte is eluted last.

### 28B–3  Choice of Mobile and Stationary Phases

Successful partition chromatography requires a proper balance of intermolecular forces among the three participants in the separation process—the solute, the mobile phase, and the stationary phase. These intermolecular forces are described qualitatively in terms of the relative polarities of the reactants. In general, the polarities of common organic functional groups in increasing order are: aliphatic hydrocarbons < olefins < aromatic hydrocarbons < halides < sulfides < ethers < nitro compounds < esters ≃ aldehydes ≃ ketones < alcohols ≃ amine < sulfones < sulfoxides < amides < carboxylic acids < water.

Separations by partition chromatography are influenced by the polarities of the solute, the mobile phase, and the stationary phase.

As a rule, most chromatographic separations are achieved by matching the polarity of the analyte to that of the stationary phase; a mobile phase of considerably different polarity is then used. This procedure is generally more successful than one in which the polarities of the analyte and mobile phase are matched but are different from that of the stationary phase. In the latter situation, the stationary phase often cannot compete successfully for the sample components; retention times then become too short for practical application. At the other extreme is the situation where the polarities of the analyte and stationary phase are too much alike, in which case retention times become inordinately long.

### 28B–4  Applications

Figure 28–5 shows typical applications of bonded-phase partition chromatography. Table 28–2 further illustrates the variety of samples to which the technique is applicable.

## 28C   HIGH-PERFORMANCE ADSORPTION CHROMATOGRAPHY

All of the pioneering work in chromatography was based upon liquid-solid adsorption, in which the stationary phase is the surface of a finely divided polar solid. With such a packing, the analyte competes with the mobile phase for sites on the surface of the packing, and retention is the result of adsorption forces.

In adsorption chromatography, analyte species are adsorbed onto the surface of a polar packing.

Figure 28–5
Typical applications of bonded-phase chromatography. (a) Soft-drink additives. Column: 4.6 mm × 250 mm packed with polar (cyano) bonded-phase packing. Isocratic elution: 6% HOAc/94% $H_2O$. Flow rate: 1.0 mL/min (b) Organophosphate insecticides. Column: 4.5 mm × 250 mm, packed with 5-$\mu$m $C_8$ bonded-phase particles. Gradient elution: 67% $CH_3OH$/33% $H_2O$ to 80% $CH_3OH$/20% $H_2O$. Flow rate: 2 mL/min. (a, Courtesy of Du Pont Biotechnology Systems, Wilmington, DE; b, Courtesy of IBM Instruments, Danbury, CT.)

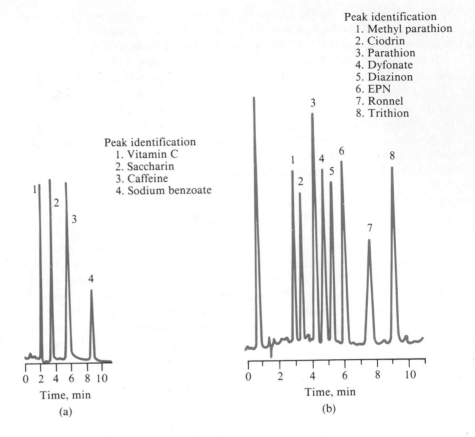

Peak identification
1. Methyl parathion
2. Ciodrin
3. Parathion
4. Dyfonate
5. Diazinon
6. EPN
7. Ronnel
8. Trithion

Peak identification
1. Vitamin C
2. Saccharin
3. Caffeine
4. Sodium benzoate

## 28C–1 Stationary and Mobile Phases

Finely divided silica and alumina are the only stationary phases that find extensive use in adsorption chromatography. Silica is preferred for most (but not all) applications because of its higher sample capacity and its wider range of useful forms. The adsorption characteristics of the two substances parallel one another. For both, retention times become longer as the polarity of the analyte increases.

In adsorption chromatography, the only variable that affects the partition ratios of analytes is the composition of the mobile phase (in contrast

Table 28–2
TYPICAL APPLICATIONS OF HIGH-PERFORMANCE PARTITION CHROMATOGRAPHY

| Field | Typical Mixtures |
| --- | --- |
| Pharmaceuticals | Antibiotics, sedatives, steroids, analgesics |
| Biochemical | Amino acids, proteins, carbohydrates, lipids |
| Food products | Artificial sweeteners, antioxidants, aflatoxins, additives |
| Industrial chemicals | Condensed aromatics, surfactants, propellants, dyes |
| Pollutants | Pesticides, herbicides, phenols, PCBs |
| Forensic chemistry | Drugs, poisons, blood alcohol, narcotics |
| Clinical medicine | Bile acids, drug metabolites, urine extracts, estrogens |

to partition chromatography, where the polarity of the stationary phase is also a factor). Fortunately, enormous variations in capacity factor, selectivity factor, and thus resolution accompany variations in the solvent system; only rarely can a suitable mobile phase not be found.

## 28C–2 Applications of Adsorption Chromatography

Currently, liquid-solid high-performance liquid chromatography is used extensively for separations of relatively nonpolar, water-insoluble organic compounds with molecular weights below about 5000. One particular strength of adsorption chromatography that is not shared by other methods is its ability to resolve isomeric mixtures, such as meta- and para-substituted benzene derivatives.

## 28D   HIGH-PERFORMANCE ION CHROMATOGRAPHY

Ion-exchange resins are used as the stationary phase in *ion chromatography,* a separation procedure in which ions of like charge are separated by elution from a column packed with a finely divided resin.

## 28D–1 Ion-Exchange Resins

Synthetic ion-exchange resins are high-molecular-weight polymeric materials containing many ionic functional groups per molecule. Cation-exchange resins can be either a strong-acid type with sulfonic acid groups ($RSO_3^-H^+$) or a weak-acid type containing carboxylic acid groups (RCOOH); the former have wider application (see Figure 28–6). Anion-exchange resins contain basic amine functional groups attached to the polymer molecule. Strong-base exchangers are quaternary amines [$RN(CH_3)_3^+OH^-$]; weak-base types contain secondary or tertiary amines.

An important property of all ion-exchange resins is that they are essentially insoluble in aqueous media. Thus, when a cation exchanger is immersed in an aqueous solution containing the cation $M^+$, the following exchange equilibrium is quickly established between the solid and solution phases:

$$x\,RSO_3^-H^+ + M^{x+} \rightleftharpoons (RSO_3^-)_x M^{x+} + x\,H^+$$

Solid        Solution        Solid        Solution

where $M^{x+}$ is a cation and R represents *that part of a resin molecule containing one sulfonic acid group.* The analogous process involving a typical anion-exchange resin can be written as

$$x\,RN(CH_3)_3^+OH^- + A^{x-} \rightleftharpoons [RN(CH_3)_3^+]_x A^{x-} + x\,OH^-$$

Solid        Solution        Solid        Solution

where $A^{x-}$ is an anion.

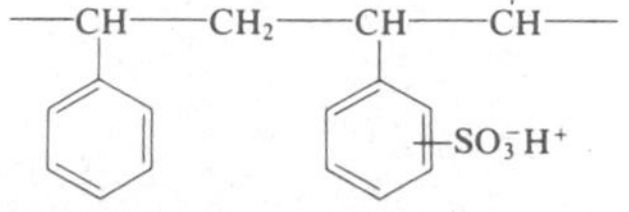

Figure 28–6

Structure of a cross-linked polystyrene ion-exchange resin. Similar resins are used in which the $-SO_3^-H^+$ group is replaced by $-COO^-H^+$, $-NH_3^+OH^-$, and $-N(CH_3)_3^+OH^-$ groups.

### 28D–2 Ion-Exchange Equilibria

Ion-exchange equilibria can be treated by the law of mass action. For example, when a dilute solution of calcium ions is brought into contact with a sulfonic acid resin, the following equilibrium develops:

$$Ca^{2+}(aq) + 2 H^+(res) \rightleftharpoons Ca^{2+}(res) + 2 H^+(aq)$$

Application of the mass law leads to

$$K' = \frac{[Ca^{2+}(res)][H^+(aq)]^2}{[Ca^{2+}(aq)][H^+(res)]^2}$$

where the bracketed terms are molar concentrations (strictly, activities). Note that $[Ca^{2+}(res)]$ and $[H^+(res)]$ are the concentrations of the two ions *in the solid phase*. In contrast to most solids, however, these concentrations can vary from zero to some maximum value at which all the available sites in the resin are occupied by one species only.

Ion-exchange separations are ordinarily performed under conditions in which one of the ions predominates in *both* phases. For example, in the removal of calcium ions from a dilute and somewhat acidic solution, the calcium ion concentration is much smaller than that of hydrogen ions in both the aqueous and the resin phase. Consequently, the concentration of hydrogen ions does not change significantly as a result of the exchange process. Under these circumstances, $[Ca^{2+}(res)] \ll [H^+(res)]$ and $[Ca^{2+}(aq)] \ll [H^+(aq)]$ and the foregoing equilibrium-constant expression can be rearranged to give

$$\frac{[Ca^{2+}(res)]}{[Ca^{2+}(aq)]} = K' \frac{[H^+(res)]^2}{[H^+(aq)]^2} = K \tag{28–1}$$

where $K$ is the distribution ratio as defined by Equation 26–1. Note that $K$ in Equation 28–1 represents the affinity of the resin for calcium ion relative to another ion (here, $H^+$). In general, where $K$ for an ion is large, a strong tendency for the stationary phase to retain that ion exists; where $K$ is small, the opposite is true. Selection of a common reference ion (such as $H^+$) permits a comparison of distribution ratios for various ions on a given type of resin. Such experiments reveal that polyvalent ions are much more strongly retained than singly charged species. Within a given charge group, differences in $K$ appear to be related to the size of the hydrated ion as well as other properties. Thus, for a typical sulfonated cation-exchange resin, values of $K$ for univalent ions decrease in the order $Ag^+ > Cs^+ > Rb^+ > K^+ > NH_4^+ > Na^+ > H^+ > Li^+$. For divalent cations, the order is $Ba^{2+} > Pb^{2+} > Sr^{2+} > Ca^{2+} > Ni^{2+} > Cd^{2+} > Cu^{2+} > Co^{2+} > Zn^{2+} > Mg^{2+} > UO_2^{2+}$.

Polyvalent ions are more strongly held by ion exchange resins than are univalent ions.

### 28D–3 Applications of Ion-Exchange Resins to Chromatography

In ion chromatography, analyte ions are introduced at the head of a column packed with a suitable ion-exchange resin. Elution is then carried out with a solution that contains an ion that competes with the analyte ions for the charged groups on the resin surface. For example, anions such as chloride, thiocyanate, sulfate, and phosphate can be separated on an anion-exchange resin in its basic form. In this application, the sample is first introduced onto the head of the column, where the anions are retained by the reaction

$$x\text{RN}(\text{CH}_3)_3^+\text{OH}^- + \text{A}^{x-} \rightleftharpoons [\text{RN}(\text{CH}_3)_3^+]_x\text{A}^{x-} + x\text{OH}^-$$

Elution is then carried out with a dilute solution of a base, which causes the foregoing reaction to be reversed and the anions released. Since the partition ratios for various anions differ from one another, fractionation occurs during the elution.

One of the attractive aspects of ion chromatography is that conductivity measurements provide a convenient general method for detecting and determining the concentration of the eluted species. Ion chromatography with conductometric detection of ions eluted from the column was first described in 1975 and by now has become a powerful and important technique for the quantitative determination of both inorganic and organic charged species.[3]

Two types of chromatography based on ion-exchange packings are currently in use: *suppressor-based* and *single-column* ion chromatography. They differ in the method used to prevent the conductivity of the eluting electrolyte from interfering with the measurement of the conductivity of the analytes.

## 28D–4  Ion Chromatography Based on Suppressors

Conductivity detectors have many of the properties of the ideal detector alluded to in Section 27B–5. They can be highly sensitive, they are universal for charged species, and as a general rule, they respond in a predictable way to concentration changes. Furthermore, such detectors are simple to operate, inexpensive to construct and maintain, easy to miniaturize, and ordinarily give prolonged, trouble-free service. The only limitation to the use of conductivity detectors, which delayed their general application to ion chromatography until the mid-1970s, was due to the high electrolyte concentrations required to elute most analyte ions in a reasonable time. As a consequence, the conductivity from the mobile-phase components tended to swamp that from the analyte ions, thus greatly reducing the detector sensitivity.

In 1975, the problem created by the high conductance of eluents was solved by the introduction of an *eluent suppressor column* immediately following the ion-exchange column.[4] The suppressor column is packed with a second ion-exchange resin that effectively converts the ions of the eluting solvent to a molecular species of limited ionization without affect-

The conductivity detector is well suited for ion chromatography.

---

[3]For a brief review of ion chromatography, see J. S. Fritz, *Anal. Chem.*, **1987,** *59,* 335A. For a detailed description of the method, see H. Small, *Ion Chromatography.* New York: Plenum Press, 1989. D. T. Gjerde and J. S. Fritz, *Ion Chromatography,* 2nd ed. New York: A. Heuthig, 1987.

[4]H. Small, T. S. Stevens, and W. C. Bauman, *Anal. Chem.*, **1975,** *47,* 1801.

ing the conductivity due to the analyte ions. For example, when cations are being separated and determined, hydrochloric acid is chosen as the eluting reagent, and the suppressor column is an anion-exchange resin in the hydroxide form. The product of the reaction between the eluent and the suppressor is water:

$$H^+(aq) + Cl^-(aq) + resin^+OH^-(s) \rightarrow resin^+Cl^-(s) + H_2O$$

The analyte cations are of course not retained by the anionic groups in this second column.

For anion separations, the suppressor packing is the acid form of a cation-exchange resin and sodium bicarbonate or carbonate is the eluting agent. The reaction in the suppressor is

$$Na^+(aq) + HCO_3^-(aq) + resin^-H^+(s) \rightarrow resin^-Na^+(s) + H_2CO_3(aq)$$

The largely undissociated carbonic acid does not contribute significantly to the conductivity.

An inconvenience associated with the original suppressor columns was the need to regenerate them periodically (typically, every 8 to 10 hr) in order to convert the packing back to the original acid or base form. Recently, however, micromembrane suppressors that operate continuously have become available.[5] For example, where sodium carbonate or bicarbonate is to be removed, the eluent is passed over a series of ultra-thin cation-exchange membranes that separate it from a stream of acidic regenerating solution that flows continuously in the opposite direction. The sodium ions from the eluent exchange with hydrogen ions on the inner surface of the exchanger membrane and then migrate to the other surface for exchange with hydrogen ions from the regenerating reagent. Hydrogen ions from the regeneration solution migrate in the reverse direction, thus preserving electrical neutrality.

Figure 28–7 shows two applications of ion chromatography based upon a suppressor column and conductometric detection. In each, the ions were present in the parts-per-million range; the sample size was 50 $\mu$L in one case and 20 $\mu$L in the other. The method is particularly important for anion analysis because no other rapid and convenient method for handling mixtures of this type now exists.

## 28D–5 Single-Column Ion Chromatography

Recently, equipment has become available commercially for ion chromatography in which no suppressor column is used. This approach depends upon the small differences in conductivity between the eluted sample ions and the prevailing eluent ions. To amplify these differences, low-capacity

---

[5] For a description of this device, see G. O. Franklin, *Amer. Lab.*, **1985** (3), 71.

All anions 10 ppm unless noted

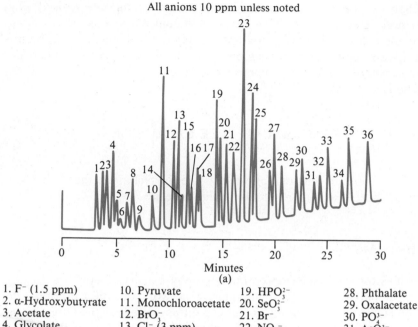

1. F⁻ (1.5 ppm)
2. α-Hydroxybutyrate
3. Acetate
4. Glycolate
5. Butyrate
6. Gluconate
7. α-Hydroxyvalerate
8. Formate (5 ppm)
9. Valerate
10. Pyruvate
11. Monochloroacetate
12. $BrO_3^-$
13. Cl⁻ (3 ppm)
14. Galacturonate
15. $NO_2^-$ (5 ppm)
16. Glucuronate
17. Dichloroacetate
18. Trifluoroacetate
19. $HPO_3^{2-}$
20. $SeO_3^{2-}$
21. Br⁻
22. $NO_3^-$
23. $SO_4^{2-}$
24. Oxalate
25. $SeO_4^{2-}$
26. α-Ketoglutarate
27. Fumarate
28. Phthalate
29. Oxalacetate
30. $PO_4^{3-}$
31. $AsO_4^{3-}$
32. $CrO_4^{2-}$
33. Citrate
34. Isocitrate
35. cis-Aconitate
36. trans-Aconitate

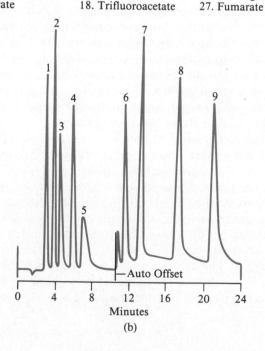

| | ppm |
|---|---|
| 1. $Li^+$ | .5 |
| 2. $Na^+$ | 2 |
| 3. $NH_4^+$ | 3 |
| 4. $K^+$ | 3 |
| 5. Morpholine | 30 |
| 6. Cyclohexylamine | 10 |
| 7. $Mg^{2+}$ | 1 |
| 8. $Ca^{2+}$ | 2 |
| 9. $Sr^{2+}$ | 10 |

**Figure 28–7**

Applications of ion chromatography. (a) Separation of anions on an anion-exchange column. Gradient eluent: 0.00075 to 0.85 M NaOH. Sample size: 20 μL. (b) Separation of cations on a cation-exchange column. Eluent 1 : 0.012 M HCl, 0.00025 M diaminopropionic acid hydrochloride, and 0.00025 M histidine hydrochloride, and 0.004 M histidine hydrochloride. Sample size: 50 μL. Conductivity detector for both samples. (Courtesy of Dionex, Sunnyvale, CA.)

exchangers that permit elution with dilute eluent solutions are used. Furthermore, eluents of low equivalent conductance are chosen.[6]

Single-column ion chromatography offers the advantage of not requiring special equipment for suppression. It is, however, a somewhat less sensitive method for determining anions than suppressor column methods.

## 28E HIGH-PERFORMANCE SIZE-EXCLUSION CHROMATOGRAPHY

Size-exclusion methods extend chromatography to the resolution of high-molecular-weight species.

Size-exclusion chromatography is a powerful technique that is particularly applicable to high-molecular-weight species.[7]

### 28E-1 Packings

Packings for size-exclusion chromatography consist of small ($\approx 10\ \mu$m) silica or polymer particles containing a network of uniform pores into which solute and solvent molecules can diffuse. While in the pores, molecules are effectively trapped and removed from the flow of the mobile phase. The average residence time of analyte molecules depends upon their effective size. Molecules that are significantly larger than the average pore size of the packing are excluded and thus are not retained; that is, they travel through the column at the rate of the mobile phase. Molecules that are appreciably smaller than the pores can penetrate throughout the pore maze and are thus entrapped for the greatest time; they are last to be eluted. Between these two extremes are intermediate-size molecules whose average penetration into the pores of the packing depends upon their diameter. The fractionation that occurs within this group is directly related to molecular size and, to some extent, molecular shape. Note that size-exclusion separations differ from the other chromatographic procedures in the respect that no chemical or physical interactions occur between analyte and stationary phase. Indeed, every effort is made to avoid such interactions because they impair column efficiency.

Numerous size-exclusion packings are on the market. Some are hydrophilic for use with aqueous mobile phases; others are hydrophobic and are used with nonpolar organic solvents. Chromatography based on the former is sometimes called *gel filtration,* while techniques based on the latter are termed *gel permeation.* With both packings, a wide range of pore diameters is available. The average molecular weight suitable for a given packing may be as small as a few hundred or as large as several million. Ordinarily, a given packing will accommodate a 2- to 2.5-decade molecular weight range.

Gel filtration is size-exclusion chromatography with a hydrophilic packing. It is used for separating polar species.

Gel permeation is size-exclusion chromatography with a hydrophobic packing. It is used to separate nonpolar species.

[6]See R. M. Becker, *Anal. Chem.,* **1980,** *52,* 1510; J. R. Benson, *Amer. Lab.,* **1985,** (6), 30; and T. Jupille, *Amer. Lab.,* **1986,** (5), 114.
[7]See *Size-Exclusion Chromatography,* B. J. Hunt and S. R. Holding, Eds. New York: Chapman and Hall, 1989; *Aqueous Size-Exclusion Chromatography,* P. L. Dubin, Ed. Amsterdam: Elsevier, 1988.

## 28E–2 Applications

Figure 28–8a illustrates the application of gel filtration to the determination of three sugars in an aqueous medium. The hydrophilic packing excluded molecular weights greater than 1000. Figure 28–8b is a chromatogram obtained with a hydrophobic packing in which the eluent was tetrahydrofuran. The sample was a commercial epoxy resin in which each monomer unit had a molecular weight of 280.

   Another important application of size-exclusion chromatography is in the rapid determination of the molecular weight or the molecular weight distribution of large polymers or natural products. In this method, the elution volumes of the sample are compared with the elution volumes for a series of standard compounds that have the same chemical characteristics.

## 28F   A COMPARISON OF HIGH-PERFORMANCE LIQUID CHROMATOGRAPHY AND GAS-LIQUID CHROMATOGRAPHY

A comparison of high-performance liquid chromatography and gas-liquid chromatography is presented in Table 28–3. The former is applicable to nonvolatile substances (including inorganic ions) and thermally unstable materials, whereas the latter is not. Yet the resolving power of gas-liquid chromatography, particularly of the open-tubular type, is generally superior to that of high-performance liquid chromatography. When both procedures are applicable to a given separation, gas-liquid chromatography

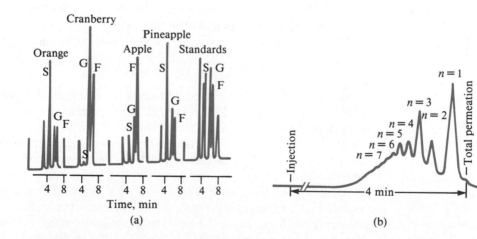

Figure 28–8

Applications of size-exclusion chromatography. (a) Gel-filtration determination of glucose (G), fructose (F), and sucrose (S) in canned juices. A 25-cm column was packed with hydrophilic sulfonated polymer particles with an exclusion limit of 1000. (b) Gel-permeation analysis of a commercial epoxy resin ($n$ = number of monomeric units in the polymer). A porous-silica column, 6.2 mm × 250 mm, was used with a mobile phase of tetrahydrofuran. (Courtesy of Du Pont Biotechnology Systems, Wilmington, DE.)

Table 28–3

COMPARISON OF HIGH-PERFORMANCE LIQUID CHROMATOGRAPHY AND GAS-LIQUID CHROMATOGRAPHY

**Characteristics of Both Methods**
Efficient, highly selective, and widely applicable
Only small sample required
May be nondestructive of sample
Readily adapted to quantitative analysis

**Advantages of HPLC**
Can accommodate nonvolatile and thermally unstable samples
Generally applicable to inorganic ions

**Advantages of GLC**
Simple and inexpensive equipment
Rapid
Unparalleled resolution (with capillary columns)
Easily interfaced with mass spectroscopy

offers the advantage of speed and simplicity of equipment. In other situations the methods tend to be complementary.

## 28G SUPERCRITICAL-FLUID CHROMATOGRAPHY

Supercritical-fluid chromatography (SFC) is a hybrid of gas and liquid chromatography that combines some of the best features of each. While supercritical-fluid chromatography is still in its youth, it appears to be clearly superior to both gas-liquid and high-performance liquid chromatography for certain applications.[8]

### 28G–1 Important Properties of Supercritical Fluids

A *supercritical fluid* is formed whenever a substance is heated above its *critical temperature*. At the critical temperature, a substance can no longer be condensed into its liquid state through the application of pressure. For example, carbon dioxide becomes a supercritical fluid at temperatures above 31°C. In this state, the molecules of carbon dioxide act independently of one another just as they do in a gas.

As shown by the data in Table 28–4, the physical properties of a substance in the supercritical-fluid state can be remarkably different from the same properties in either the liquid or the gaseous state. For example, the density of a supercritical fluid is typically 200 to 400 times greater than that of the corresponding gas and approaches that of the substance in its liquid state. The properties compared in Table 28–4 are those that are of importance in gas, liquid, and supercritical-fluid chromatography.

The density of a supercritical fluid is 200 to 400 times that of its gaseous state, and it is nearly as dense as its liquid state.

[8]M. D. Palmieri, *J. Chem. Educ.*, **1988**, *65*, A254; **1989**, *66*, A141; P. R. Griffiths, *Anal. Chem.*, **1985**, *60*, 593A; R. D. Smith, B. W. Wright, and C. R. Yonker, *Anal. Chem.* **1988**, *60*, 1323A.

Table 28–4
COMPARISON OF PROPERTIES OF SUPERCRITICAL FLUIDS, LIQUIDS,
AND GASES*

|  | Gas (STP) | Supercritical Fluid | Liquid |
|---|---|---|---|
| Density, g/cm$^3$ | $(0.6-2) \times 10^{-3}$ | $0.2-0.5$ | $0.6-1.6$ |
| Diffusion coefficient, cm$^2 \cdot$s$^{-1}$ | $(1-4) \times 10^{-1}$ | $10^{-3}-10^{-4}$ | $(0.2-2) \times 10^{-5}$ |
| Viscosity, g$\cdot$cm$^{-1}\cdot$s$^{-1}$ | $(1-3) \times 10^{-4}$ | $(1-3) \times 10^{-4}$ | $(0.2-3) \times 10^{-2}$ |

*All data order of magnitude only.

An important property of supercritical fluids and one that is related to their high densities (0.2 to 0.5 g/cm$^3$), is their ability to dissolve large nonvolatile molecules. For example, supercritical carbon dioxide readily dissolves n-alkanes containing from 5 to 22 carbon atoms, di-n-alkyl-phthalates in which the alkyl groups contain 4 to 16 carbon atoms, and various polycyclic aromatic hydrocarbons consisting of several rings.[9]

Critical temperatures for fluids used in chromatography vary widely, from about 30°C to above 200°C. Lower critical temperatures are advantageous from several standpoints. For this reason, much of the work to date has focused on such supercritical fluids as carbon dioxide (31°C), ethane (32°C), and nitrous oxide (37°C). Note that these temperatures, and the pressures at these temperatures, are well within the operating conditions of ordinary high-performance liquid chromatography.

*Supercritical fluids tend to dissolve large, nonvolatile molecules.*

## 28G–2 Instrumentation and Operating Variables

Instruments for supercritical-fluid chromatography are similar in design to high-performance liquid chromatographs except that provision is made in the former for controlling and measuring the column pressure. Several manufacturers began to offer apparatus for supercritical-fluid chromatography in the mid-1980s.

### The Effect of Pressure

The density of a supercritical fluid increases rapidly and nonlinearly with pressure increases. Density increases also alter capacity factors ($k'$) and thus elution times. For example, the elution time for hexadecane is reported to decrease from 25 to 5 min as the pressure of carbon dioxide is raised from 70 to 90 atm. Gradient elution can thus be achieved through linear increases in column pressure or through regulation of pressure to obtain linear density increases. An example of the latter is shown in Figure 28–11a, where the density of the carbon dioxide mobile phase was

*Gradient elution can be achieved in SFC by systematically changing the column pressure or the density of the supercritical fluid.*

---

[9]Certain important industrial processes are based upon the high solubility of organic species in supercritical carbon dioxide. For example, this medium has been employed in extracting caffeine from coffee beans to give decaffeinated coffee and in extracting nicotine from cigarette tobacco.

increased linearly from 0.225 to 0.70 g/mL (by increasing the pressure from 73 to 112 atm).

A clear analogy exists between gradient elution through adjustment of pressure or density of a supercritical fluid and by temperature-gradient elution in gas-liquid chromatography and solvent-gradient elution in liquid chromatography.

## Columns

Although supercritical-fluid chromatography has been carried out in both packed and open tubular columns, the latter appear to be favored because their length can be greater, a condition that leads to enhanced column efficiency. Typical columns are similar to the fused-silica open tubular (FSOT) columns described in Table 27–1. Because of the low viscosity of supercritical media, columns can be much longer than those used in liquid chromatography, and column lengths of 10 or 20 m and inside diameters of 50 or 100 $\mu$m are common. For difficult separations, columns 60 m in length and longer have been used.

Many of the column coatings used in liquid chromatography have been applied to supercritical-fluid chromatography as well. Typically, these are polysiloxanes that are chemically bonded to the inner silica wall of the capillary tubing. Film thicknesses are 0.05 to 0.4 $\mu$m.

*Very long columns can be used in SFC because the viscosity of supercritical fluids is so low.*

## Mobile Phases

The most widely used mobile phase for supercritical-fluid chromatography is carbon dioxide. It is an excellent solvent for a variety of organic molecules. In addition, it transmits in the ultraviolet and is odorless, nontoxic, readily available, and remarkably inexpensive relative to other chromatographic solvents. Its critical temperature of 31°C and its pressure of 73 atm at the critical temperature permit a wide selection of temperatures and pressures without exceeding the operating limits of modern high-performance liquid chromatography equipment. In some applications, polar organic modifiers, such as methanol, are introduced in small concentrations ($\simeq 1\%$) to modify $\alpha$ values for analytes.

A number of other substances have served as mobile phases in supercritical chromatography, including ethane, pentane, dichlorodifluoromethane, diethyl ether, and tetrahydrofuran.

## Detectors

A major advantage of supercritical-fluid chromatography is that the sensitive and universal detectors of gas-liquid chromatography are applicable to this technique as well. For example, the convenient flame ionization detector of gas-liquid chromatography can be applied by simply allowing the supercritical carrier to expand through a restrictor and into a hydrogen flame, where ions formed from the analytes cause variations in the electrical conductivity of the medium.

## 28G–3 Supercritical-Fluid Chromatography Versus Other Column Methods

The information in Table 28–4, and other data as well, reveal that several physical properties of supercritical fluids are intermediate between the properties of gases and liquids. As a consequence, this new type of chromatography combines some of the characteristics of both gas and liquid chromatography. Thus, like gas chromatography, supercritical-fluid chromatography is inherently faster than liquid chromatography because of the lower viscosity and higher diffusion rates in the mobile phase. High diffusivity, however, leads to longitudinal band spreading (page 678), which is a significant factor with gas but not with liquid chromatography. Thus, the intermediate diffusivities and viscosities of supercritical fluids result in faster separations than are achieved with liquid chromatography accompanied by less zone spreading than is encountered in gas chromatography.

Figure 28–9 compares the performance characteristics of packed columns when elution is performed with supercritical carbon dioxide and with a conventional liquid mobile phase. Note that the supercritical column yields a plate height of about 0.013 mm at a mobile-phase flow rate of 0.6 cm/s, whereas at this same velocity, the plate height of the conventional column is three times as large, or 0.039 mm. Thus, a decrease in peak width by $\sqrt{3}$ should be realized (Equation 26–16). Alternatively, at a plate height corresponding to the minimum in the HPLC curve (0.012 mm) the mobile-phase velocity in the SFC system is four times that of the HPLC system. Thus, a fourfold decrease in analysis time results. These advantages are reflected in the two chromatograms shown in Figure 28–10.

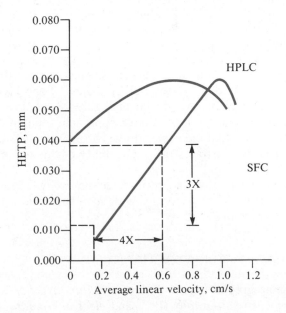

**Figure 28–9**
Performance characteristics of a 5-$\mu$m, particle-size packed column when elution is carried out with a conventional mobile phase (HPLC) and with supercritical carbon dioxide (SFC). (From D. R. Gere, *Application Note 800-3*, Hewlett-Packard, 1983. With permission.)

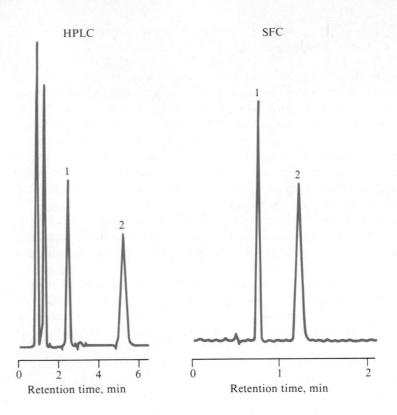

Retention time, min

Retention time, min

**Figure 28–10**

Comparison of chromatograms obtained by conventional partition chromatography (HPLC) and supercritical-fluid chromatography (SFC). Column: 20 cm × 4.6 mm packed with 10-$\mu$m reversed-phase bonded packing. Analytes: (1) biphenyl; (2) terphenyl. For HPLC, mobile phase: 65% $CH_3OH$/35% $H_2O$; flow rate: 4 mL/min; linear velocity: 0.55 cm/s; sample size: 10 $\mu$L. For SFC, mobile phase: $CO_2$; flow rate: 5.4 mL/min; linear velocity: 0.76 cm/s; sample size: 3 $\mu$L. (From D. R. Gere, T. J. Stark, and T. N. Tweeten, *Application Note 800-4*, Hewlett-Packard, 1983. With permission.)

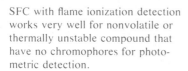

SFC with flame ionization detection works very well for nonvolatile or thermally unstable compound that have no chromophores for photometric detection.

## 28G–4 Applications

Supercritical-fluid chromatography, particularly of the open tubular type, appears to have a potential niche in the spectrum of column-chromatographic methods because it is applicable to a class of compounds that is not readily amenable to either gas-liquid or liquid chromatography. These compounds include species that are nonvolatile or thermally unstable and, in addition, contain no chromophoric groups that can be used for photometric detection. Separation of these compounds is possible with supercritical-fluid chromatography at temperatures below 100°C; furthermore, detection is readily carried out by means of the highly sensitive flame ionization detector. As shown in Figure 28–11, the efficiency of open tubular supercritical columns often approaches that of capillary gas-liquid columns. Note that the supercritical chromatogram was obtained at 40°C, which clearly illustrates the applicability of this technique to thermally labile species.

It is also noteworthy that supercritical columns have the added advantage of being much easier to interface with mass spectrometers than liquid chromatographic columns.

## 28H    PLANAR CHROMATOGRAPHY

Planar chromatographic methods include *thin-layer chromatography* (TLC), *paper chromatography* (PC), and *electrochromatography*. Each

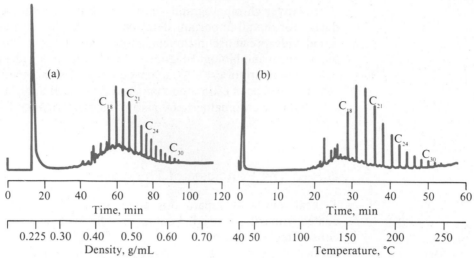

Figure 28-11

Comparison of a supercritical-fluid chromatogram and a capillary gas-liquid chromatogram for an aliphatic fraction from a solvent-refined coal fraction. (a) SFC chromatogram; $CO_2$ mobile phase at 40°C; 34 m × 50 $\mu$m inside-diameter fused capillary column coated with a 0.25-$\mu$m film of cross-linked SE-54; flame ionization detector; linear density program from 0.225 g/mL (73 atm) to 0.70 g/mL (112 atm). (b) Capillary GLC chromatogram; $H_2$ carrier gas; 20 m × 300 $\mu$m inside-diameter fused-silica column coated with a 0.25-$\mu$m film of cross-linked SE-54; flame ionization detector; linear temperature program from 50 to 250°C. (From W. P. Jackson *et al., Ultrahigh Resolution Chromatography,* ACS Symposium Series 250, S. Ahuja, Ed., p. 130. Washington, D.C.: American Chemical Society, 1984. With permission.)

makes use of a flat, relatively thin layer of material that is either self-supporting or is coated on a glass, plastic, or metal surface. The mobile phase moves through the stationary phase by capillary action, sometimes assisted by gravity or an electrical potential. Planar chromatography is sometimes called two-dimensional chromatography, although this description is not strictly correct inasmuch as the stationary phase does have a finite thickness.

Currently, most planar chromatography is based upon the thin-layer technique, which is faster, has better resolution, and is more sensitive than its paper counterpart. This section is devoted to thin-layer methods.

## 28H-1  The Scope of Thin-Layer Chromatography

In terms of theory, the types of stationary and mobile phases, and applications, thin-layer and liquid chromatography are remarkably similar. In fact, thin-layer plates can be profitably used to develop optimal conditions for separations by column liquid chromatography. The advantages of following this procedure are the speed and low cost of the exploratory thin-layer experiments. Some chromatographers take the position that thin-layer experiments should always precede column experiments.

Preliminary thin-layer studies provide valuable clues for successful liquid chromatography.

Thin-layer chromatography has become the workhorse of the drug industry for the all-important determination of product purity. It has also found widespread use in clinical laboratories and is the backbone of many biochemical and biological studies. Finally, it finds widespread use in the industrial laboratories.[10] As a consequence of these many areas of application, it has been estimated that at least as many analyses are performed by thin-layer chromatography as by high-performance liquid chromatography.[11]

## 28H–2 The Principles of Thin-Layer Chromatography

Typical thin-layer separations are performed on a glass plate that is coated with a thin and adherent layer of finely divided particles; this layer constitutes the stationary phase. The particles are similar to those described in the discussion of adsorption, normal- and reversed-phase partition, ion-exchange, and size-exclusion column chromatography. Mobile phases are also similar to those employed in high-performance liquid chromatography.

### The Preparation of Thin-Layer Plates

A thin-layer plate is prepared by spreading an aqueous slurry of the finely ground solid onto the clean surface of a glass or plastic plate or microscope slide. Often a binder is incorporated into the slurry to enhance adhesion of the solid particles to the glass and to one another. The plate is then allowed to stand until the layer has set and adheres tightly to the surface; for some purposes, it may be heated in an oven for several hours. Several chemical supply houses offer precoated plates of various kinds.

### Plate Development

*Plate development* is the process in which a sample is carried through the stationary phase by a mobile phase; it is analogous to elution in liquid chromatography. The most common way of developing a plate is to place a drop of the sample near one edge of the plate (most plates have dimensions of 5 × 20 or 20 × 20 cm) and mark its position with a pencil. After the sample solvent has evaporated, the plate is placed in a closed container saturated with vapors of the developing solvent. One end of the plate is immersed in the developing solvent, with care being taken to avoid direct contact between the sample and the developer (Figure 28–12). After the developer has traversed one half or two thirds of the

[10]Two monographs devoted to the principles and applications of thin-layer chromatography are R. Hamilton and S. Hamilton, *Thin-Layer Chromatography*. New York: Wiley, 1987; and J. C. Touchstone, *Practice of Thin-Layer Chromatography*, 2nd ed. New York: Wiley, 1983. For briefer reviews, see D. C. Fenimore and C. M. Davis, *Anal. Chem.*, **1981**, *53*, 253A; C. F. Poole and S. K. Poole, *Anal. Chem.*, **1989**, *61*, 1257A.

[11]T. H. Mauch II, *Science*, **1982**, *216*, 161.

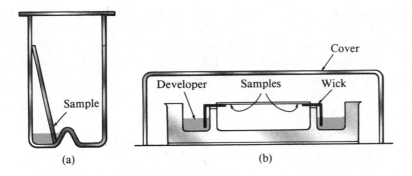

**Figure 28–12**
(a) Ascending-flow developing chamber. (b) Horizontal-flow developing chamber, in which samples are placed on both ends of the plate and developed toward the middle, thus doubling the number of samples that can be accommodated.

length of the plate, the plate is removed from the container and dried. The positions of the components are then determined in any of several ways.

Figure 28–13 illustrates the separation of amino acids in a mixture by development in two directions (*two-dimensional planar chromatography*). The sample was placed in one corner of a square plate, and the plate was developed in the ascending direction with solvent A. This solvent was then removed by evaporation, and the plate was rotated 90 deg, following which ascending development with solvent B was performed. After solvent removal, the positions of the amino acids were determined by spraying with ninhydrin, a reagent that forms a pink to purple product with amino acids. The spots were identified by comparison of their positions with those of standards.

## Locating Analytes on the Plate

Several methods are employed to locate sample components after separation. Two common methods, which can be applied to most organic mixtures, involve spraying with a solution of iodine or sulfuric acid, both of which react with organic compounds to yield dark products. Several specific reagents (such as ninhydrin) are also useful for locating separated species.

The process of locating analytes on a thin layer plate is often termed *visualization.*

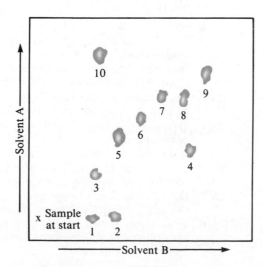

**Figure 28–13**

Two-dimensional thin-layer chromatogram (silica gel) of some amino acids. Solvent A: toluene/2-chloroethanol/pyridine. Solvent B: chloroform/benzyl alcohol/acetic acid. Amino acids: (1) aspartic acid, (2) glutamic acid, (3) serine, (4) $\beta$-alanine, (5) glycine, (6) alanine, (7) methionine, (8) valine, (9) isoleucine, (10) cysteine.

Another method of detection is based upon incorporating a fluorescent material into the stationary phase. After development, the plate is examined under ultraviolet light. The sample components quench the fluorescence of the material so that all of the plate fluoresces except where the nonfluorescing sample components are located.

### Quantitative Analysis

A semiquantitative estimate of the amount of a component present can be obtained by comparing the area of a spot with that of a standard. Better data can be obtained by scraping the spot from the plate, extracting the analyte from the stationary-phase solid, and measuring the analyte by a suitable physical or chemical method. In a third method, a scanning densitometer can be employed to measure the radiation emitted from the spot by fluorescence or reflection.

### 28H–3  Paper Chromatography

Separations by paper chromatography are performed in the same way as those on thin-layer plates. The papers are manufactured from highly purified cellulose with close control over porosity and thickness. Such papers contain sufficient adsorbed water to make the stationary phase aqueous. Other liquids can be made to displace the water, however, thus providing a different type of stationary phase. For example, paper treated with silicone or paraffin oil permits reversed-phase paper chromatography, in which the mobile phase is a polar solvent. Also available commercially are special papers that contain an adsorbent or an ion-exchange resin, thus permitting adsorption and ion-exchange paper chromatography.

## 28I  QUESTIONS AND PROBLEMS

28–1. List the types of substances to which each of the following chromatographic methods is most applicable:
  *(a) gas-liquid
  (b) liquid partition
  *(c) ion-exchange
  (d) liquid adsorption
  *(e) gel permeation
  (f) gel filtration
  *(g) gas-solid
  (h) supercritical fluid

28–2. Define
  *(a) isocratic elution.
  (b) gradient elution.
  *(c) stop-flow injection.
  (d) reversed-phase packing.
  *(e) normal-phase packing.
  (f) ion chromatography.
  *(g) eluent-suppressor column.
  (h) gel filtration.
  *(i) gel permeation.
  (j) critical temperature.
  *(k) FSOT column.
  (l) two-dimensional thin-layer chromatography.
  *(m) supercritical fluid.

28–3. List the differences in properties and roles of the mobile phase in gas, liquid, and supercritical-fluid chromatography. How do these differences influence the characteristics of the three methods?

28–4. For a normal-phase separation, predict the order of elution of
  *(a) n-hexane, n-hexanol, benzene.
  (b) ethyl acetate, diethyl ether, nitrobutane.

28–5. For a reversed-phase separation, predict the order of elution of the solutes in Problem 28–4.

28–6. Describe the physical differences between open tubular and packed columns. What are the advantages and disadvantages of each?

*28–7. Describe the fundamental difference between adsorption and partition chromatography.

28–8. Describe the fundamental difference between ion-exchange and size-exclusion chromatography.

*28–9. What types of species can be separated by high-performance liquid chromatography but not by gas-liquid chromatography?

*28-10. To what types of compounds is supercritical-fluid chromatography particularly applicable?

*28-11. Describe the various kinds of pumps used in high-performance liquid chromatography. What are the advantages and disadvantages of each?

28-12. Describe the difference between single-column and suppressor-column ion chromatography.

*28-13. List the advantages of supercritical fluid chromatography over (a) high-performance liquid chromatography and (b) gas-liquid chromatography.

28-14. Why is there more zone broadening with gas chromatography than with supercritical-fluid chromatography? Why is there less zone broadening with high-performance liquid chromatography than with supercritical-fluid chromatography?

## CHAPTER 25

# THE ANALYSIS OF REAL SAMPLES

$V$ ery early in this text (Section 1C), we pointed out that a quantitative analysis involves a sequence of steps: (1) selecting a method, (2) sampling, (3) preparing a laboratory sample, (4) defining replicate samples (by weight or volume measurements), (5) preparing solutions of the samples, (6) eliminating interferences, (7) completing the analysis by performing measurements that are related in a known way to analyte concentration, and (8) computing the results and estimating their reliability.

Thus far we have focused largely on steps 7 and 8 and to a lesser extent on steps 4 and 6. This emphasis is *not* because the earlier steps are unimportant or easy—quite the opposite. The preliminary steps may be not only more difficult and time-consuming than the two final steps of an analysis but also a greater source of error.

The reasons for postponing a discussion of the preliminary steps to this point are pedagogical. Experience has shown that it is easier to introduce students to analytical techniques by having them first perform measurements on simple materials for which no method selection is required and for which problems with sampling, sample preparation, and sample dissolution are either nonexistent or easily solved. Thus, we have been largely concerned so far with measuring the concentration of analytes in simple aqueous solutions that are uncluttered by interfering species.

The determination of an analyte in a simple solution is usually not particularly difficult once learned because the number of variables that must be controlled is small and the tools available are numerous and easy to use. Furthermore, with simple systems, our theoretical knowledge suffices to allow us to anticipate problems and correct for them. Thus, if chemical analysis involved only determining the concentration of a single species in a simple and readily soluble homogeneous mixture, analytical chemistry could profitably be entrusted to the hands of a skilled technician; certainly, a professional chemist can find more useful and challenging work for mind and hands.

In fact, the materials in which academic and industrial chemists are interested are not as a rule simple. To the contrary, most substances that require analysis are complex, consisting of several species or several tens of species. Such materials are frequently far from ideal in matters of solubility, volatility, stability, and homogeneity, and several steps must precede the final measurement step. Indeed, the final measurement is often anticlimactic in the sense that it is by far easier and less time-consuming than any of the several steps that must precede it.

For example, we showed in earlier chapters that the calcium ion concentration of an aqueous solution is readily determined by titration with a standard EDTA solution or by potential measurements with a specific-ion electrode. Alternatively, the calcium content of a solution can be determined from flame absorption or emission measurement or by the precipitation of calcium oxalate followed by weighing or by titrating with a standard solution of potassium permanganate.

All the aforementioned methods provide a straightforward means of determining the calcium content of a simple salt, such as the carbonate. The chemist is seldom interested in the calcium content of calcium carbonate, however. More likely, what is needed is the percentage of this element in a sample of animal tissue, a silicate rock, or a piece of glass. The analysis thereby acquires new and formidable complexities. For example, none of these materials is soluble in water or dilute aqueous reagents. Before calcium can be determined, therefore, the sample must be decomposed by high-temperature treatment with concentrated reagents. Unless care is taken, this step may cause losses of some of the calcium from the sample or, equally bad, the introduction of that element as a contaminant in the relatively large quantities of reagent usually required to decompose a sample.

Even after the sample has been decomposed to give a solution of calcium ion, the excellent procedures mentioned in the two previous paragraphs cannot ordinarily be applied immediately to complete the analysis, for they are all based upon reactions or properties shared by several elements in addition to calcium. Thus, a sample of animal tissue, silicate rock, or glass almost surely contains one or more components that also react with EDTA, act as a chemical interference in a flame-absorption measurement, or form a precipitate with oxalate ion; furthermore, the high ionic strength resulting from the reagents used for sample decomposition would complicate a direct potentiometric measurement. As a consequence, several additional operations are required before the final measurement to free the solution of interferences.

We have chosen the term *real samples* to describe materials such as those in the preceding illustration. In this context, most of the samples encountered in an elementary quantitative analysis laboratory course definitely are not real, for they are generally homogeneous, usually stable even with rough handling, readily soluble, and, above all, chemically simple. Moreover, well-established and thoroughly tested methods exist for their analysis. There is unquestioned value in introducing analytical techniques with such substances, for they do allow the student to concentrate on the mechanical aspects of an analysis. Once these mechanics have been mastered, however, there is little point in the continued analysis of unreal samples; to do so creates the impression that any chemical

Real samples are far more complex than those that you encounter in the instructional laboratory.

analysis involves nothing more than the slavish adherence to a well-defined and narrow regimen, the end result of which is a number that is accurate to one or two parts per thousand. Unfortunately, all too many chemists retain this view far into their professional lives.

In truth, the pathway leading to knowledge of the composition of real samples is frequently more demanding of intellectual skills and chemical intuition than of mechanical aptitude. Furthermore, a compromise must often be struck between the available time and the accuracy that is believed necessary. The chemist is frequently happy to settle for an accuracy of one or two parts per hundred instead of one or two parts per thousand, knowing that the latter may require several hours or even days of additional effort. In fact, with complex materials, even parts-per-hundred accuracy may be unrealistic.

The difficulties encountered in the analysis of real samples stem from their complexity. As a consequence, the literature may not contain a well-tested analytical route for the kind of sample under consideration, and so an existing procedure must then be modified to take into account compositional differences between the sample in hand and the samples for which the original method was designed. Alternatively, a new analytical method must be developed. In either case, the number of variables that have to be taken into account usually increases exponentially with the number of species contained in the sample.

As an example, let us contrast the problems associated with the flame photometric analysis of calcium carbonate with those for a real calcium-containing sample. In the former, the number of components is small and these variables likely to affect the results are reasonably few. Principal among the variables are the physical losses of analyte due to the evolution of carbon dioxide when the sample is dissolved in acid, the effect of the anion of the acid and of the fuel-to-air ratio on the intensity of the calcium emission line, the position of the inner cone of the flame with respect to the entrance slit to the photometer, and the quality of the standard calcium solutions used for calibration.

The determination of calcium in a sample, such as a silicate rock, which contains a dozen or more other elements, is far more complex. First of all, the sample can be dissolved only by fusing it at a high temperature with a large excess of a reagent such as sodium carbonate. Physical loss of the analyte is possible during this treatment unless suitable precautions are taken. Furthermore, the introduction of calcium from the excess sodium carbonate or the fusion vessel is of real concern. Following fusion, the sample and reagent are dissolved in acid. With this step, all the variables affecting the calcium carbonate sample are operating, but in addition, a host of new variables are introduced because of the dozens of components in the sample matrix. Now, measures must be taken to minimize instrumental and chemical interference brought about by the presence of various anions and cations in the solution being atomized.

The analysis of a real substance is often a challenging problem requiring knowledge, intuition, and experience. The development of a procedure for such materials is not to be taken lightly even by the experienced chemist.

## 29A   CHOICE OF METHOD FOR THE ANALYSIS OF REAL SAMPLES

The choice of a method for the analysis of a complex substance requires good judgment and a sound knowledge of the advantages and limitations of the various analytical tools available. In addition, a familiarity with the literature of analytical chemistry is essential. We cannot be very explicit concerning how an analytical method is selected because there is no single best way that applies under all circumstances. We can, however, suggest a systematic approach to the problem and present some generalities that can aid in making intelligent decisions.

### 29A–1  Definition of the Problem

A first step, which must precede any choice of method, involves a clear definition of the analytical problem. The method of approach selected will be largely governed by the answers to the following questions:

What is the concentration range of the species to be determined?

What degree of accuracy is demanded?

What other components are present in the sample?

What are the physical and chemical properties of the gross sample?

How many samples are to be analyzed?

At the outset, the objectives sought in an analysis must be clearly defined.

The concentration range of the analyte may well limit the number of feasible methods. If, for example, the analyst is interested in an element present to the extent of a few parts per million, gravimetric or volumetric methods can generally be eliminated, and spectrometric, potentiometric, and other more sensitive methods become likely candidates. For components in the parts-per-million range, the chemist must guard against even small losses resulting from coprecipitation or volatility, and contamination from reagents and apparatus becomes a major concern. In contrast, if the analyte is a major component of the sample, these considerations become less important, and a classic analytical method may well be preferable.

The answer to the question of required accuracy is of vital importance in the choice of method and in the way it is performed because the time required to complete an analysis increases, often exponentially, with demands for greater accuracy. Thus, to improve the reliability of analytical results from, say, 2% to 0.2% relative may require the operator time to increase by a factor of 100 or more. Consequently, the chemist should always spend a few minutes before undertaking an analysis in careful consideration of the degree of accuracy really needed. It is usually foolish to produce physical or chemical data having accuracies that are significantly greater than what is demanded by the use to which the data are to be put.

The time required to carry out an analysis increases exponentially with the desired level of accuracy.

The demands of accuracy frequently dictate the procedure chosen for an analysis. For example, if the allowable error in the determination of aluminum is only a few parts per thousand, a gravimetric procedure is probably required. On the other hand, if an error of, say, 50 ppt can be tolerated, a spectroscopic or electroanalytical approach may be preferable.

The way in which an analysis is carried out is also affected by accuracy requirements. If precipitation with ammonia is chosen for the analysis of a sample containing 20% aluminum, the presence of 0.2% iron is of serious concern where accuracy in the parts-per-thousand range is demanded, and a preliminary separation of the two elements is necessary. If an error of 50 ppt is tolerable, however, the separation of iron is not necessary. Furthermore, this tolerance governs other aspects of the method. For example, 1-g samples can be weighed to perhaps 10 mg and certainly no closer than 1 mg. In addition, less care is needed in transferring and washing the precipitate and in other time-consuming operations of the gravimetric method. The intelligent use of shortcuts is not a sign of carelessness but is instead a recognition of realities in matters of time and effort. The question of accuracy, then, must be settled in clear terms at the outset.

In order to choose a method for the determination of one or more species in a sample, it is necessary to know what other elements and/or compounds are present. Lacking such information, a qualitative analysis must be undertaken in order to identify components that are likely to interfere in the various methods under consideration. As we have noted repeatedly, most analytical methods are based on reactions and physical properties that are shared by several elements or compounds. Thus, measurement of the concentration of a given element by a method that is simple and straightforward in the presence of one group of elements or compounds may require many tedious and time-consuming separations in the presence of others. A solvent suitable for one combination of compounds may be totally unsatisfactory when applied for another. Clearly, a knowledge of the qualitative chemical composition of the sample is a prerequisite for selecting a method for the quantitative determination of one or more of its components.

The chemist must also consider the physical state of the sample in order to determine whether it must be homogenized, whether volatility losses are likely, and whether its composition may change under laboratory conditions due to the absorption of water or to efflorescence. It is also important to determine what treatment is sufficient to decompose or dissolve the sample without loss of analyte. Preliminary tests of one sort or another may be needed to provide this type of information.

Finally, the number of samples to be analyzed is an important criterion in selecting a method. If there are many samples, considerable time can be expended in calibrating instruments, preparing reagents, assembling equipment, and investigating shortcuts, since the cost of these operations can be spread over the large number of samples. If, however, a few samples at most are to be analyzed, a longer and more tedious procedure involving a minimum of these preparatory operations may prove to be the wiser choice from the economic standpoint.

Often you can save considerable time by the clever use of shortcuts in an analytical procedure.

Identify the components of a sample before undertaking a quantitative analysis.

Having answered the preliminary questions, the chemist is now in a position to consider possible approaches to the problem. Sometimes, based upon past experience, the route to be followed is obvious. In other instances, it is not, and the chemist must speculate on problems that are likely to be encountered in the analysis and how they can be solved. By now, some methods probably have been eliminated from consideration and others put on the doubtful list. Ordinarily, however, the chemist turns to the analytical literature in order to profit from the experience of others. This, then, is the next logical step in choosing an analytical method.

A little extra time spent in the library can save a tremendous amount of time and effort in the laboratory.

## 29A–2 Investigation of the Literature

A list of reference books and journals concerned with various aspects of analytical chemistry appears in Appendix 1. This list is not an exhaustive catalog but rather one that is adequate for most work. It is divided into several categories. In many instances, the division is arbitrary since some works could be logically placed in more than one category.

The chemist usually begins a search of the literature by referring to one or more of the treatises on analytical chemistry or to those devoted to the analysis of specific types of materials. In addition, it is often helpful to consult a general reference work relating to the compound or element of interest. From this survey, a clearer picture of the problem at hand may develop—what steps are likely to be difficult, what separations must be made, what pitfalls must be avoided. Occasionally, all the answers needed or even a set of specific instructions for the analysis may be found. Alternatively, journal references that lead directly to this information may be discovered. On other occasions, the chemist will acquire only a general notion of how to proceed. Several possible methods may appear suitable; others may be eliminated. At this point, it is often helpful to consider reference works concerned with specific substances or specific techniques. Alternatively, the various analytical journals may be consulted. Monographs on methods for completing the analysis are valuable in deciding among several possible techniques.

The technology for computer-based scientific information retrieval provides an efficient means for surveying the analytical literature.

A major problem in using analytical journals is locating articles pertinent to the problem at hand. The various reference books are useful since most are liberally annotated with references to the original journals. The key to a thorough search of the literature, however, is *Chemical Abstracts* and its *Index Guide*. Such a survey involves the expenditure of a great deal of time, however, and is often made unnecessary by consulting reliable reference works. The advent of computer-aided literature searches promises to minimize the time required for a careful literature survey.

## 29A–3 Choosing or Devising a Method

Having defined the problem and investigated the literature for possible approaches, the chemist must next decide upon the route to be followed in the laboratory. If the choice is simple and obvious, analysis can be undertaken directly. Frequently, however, the decision requires the exercise of considerable judgment and ingenuity; experience, an understanding of chemical principles, and perhaps intuition all come into play.

If the substance to be analyzed occurs widely, the literature survey usually yields several alternative methods for the analysis. Economic considerations may dictate a method that will yield the desired reliability with the least expenditure of time and effort. As mentioned earlier, the number of samples to be analyzed is often a determining factor in this choice.

Investigation of the literature does not invariably reveal a method designed specifically for the type of sample in question. Ordinarily, however, the chemist will encounter procedures for materials that are at least analogous in composition to the one in question; the decision then has to be made as to whether the variables introduced by differences in composition are likely to have any influence on the results. This judgment is often difficult and fraught with uncertainty; recourse to the laboratory may be the only way of obtaining an unequivocal answer.

If it is decided that existing procedures are not applicable, consideration must be given to modifications that may overcome the problems imposed by the variation in composition. Again, it may be possible to propose only tentative alterations, owing to the complexity of the system; whether these modifications will accomplish their purpose without introducing new difficulties can be determined only in the laboratory.

After giving due consideration to existing methods and their modifications, the chemist may decide that none fits the problem and an entirely new procedure must be developed. In doing so, all the facts on the chemical and physical properties of the analyte must be marshalled and given consideration. Several possible ways of performing the desired measurement may become evident from this information. Each possibility must then be examined critically, with consideration given to the influence of the other components in the sample as well as to the reagents that must be used for solution or decomposition. At this point, the chemist must try to anticipate sources of error and possible interferences attributable to interactions among sample components and reagents; it may be necessary to devise strategies to circumvent such problems. In the end, it is to be hoped that one or more tentative methods worth testing will have been located. It is probable that the feasibility of some of the steps in the procedure cannot be determined on the basis of theoretical considerations alone; recourse must be made to preliminary laboratory testing of such steps. Certainly, critical evaluation of the entire procedure can come only from careful laboratory work.

> Preliminary laboratory testing may be needed to validate proposed changes to established methods.

## 29A–4 Testing the Procedure

Once a procedure for an analysis has been selected, a decision must be made as to whether it can be employed directly, without testing, to the problem at hand. The answer to this question is not simple and depends upon a number of considerations. If the method chosen is the subject of a single, or at most a few, literature references, there may be a real point to preliminary laboratory evaluation. With experience, the chemist becomes more and more cautious about accepting claims regarding the accuracy and applicability of a new method. All too often, statements found in the literature tend to be optimistic; a few hours spent in testing the procedure in the laboratory may be enlightening.

Whenever a major modification of a standard procedure is undertaken or an attempt is made to apply it to a type of sample different from that for which it was designed, a preliminary laboratory test is advisable. The effects of such alterations simply cannot be predicted with certainty, and the chemist who dispenses with such precautions is sanguine indeed.

Finally, of course, a newly devised procedure must be extensively tested before it is adapted for general use. We must now consider the means by which a new method or a modification of an existing method can be tested for reliability.

### The Analysis of Standard Samples

Unquestionably, the best technique for evaluating an analytical method involves the analysis of one or more standard samples whose analyte composition is reliably known. For this technique to be of value, however, it is essential that the standards closely resemble the samples to be analyzed with respect to both the analyte concentration range and the overall composition.

Occasionally, standards suitable for method testing can be synthesized by thoroughly homogenizing weighed quantities of pure compounds. Such a procedure is generally inapplicable, however, when the samples to be analyzed are complex, naturally occurring substances, such as rocks, animal tissue, and soils.

The National Institute of Standards and Technology has for sale a variety of standard reference materials (SRM) that have been specifically prepared for validation of analytical methods. Most standard reference materials are substances commonly encountered in commerce or in environmental, pollution, clinical, biological, and forensic studies. The concentration of one or more components in these materials is certified by the Institute based upon measurements using (1) a previously validated reference method, (2) two or more independent, reliable measurement methods, or (3) results from a network of cooperating laboratories, technically competent and thoroughly familiar with the material being tested. More than 800 of these materials are available, including such substances as ferrous and nonferrous metals; ores, ceramics, and cements; environmental gases, liquids, and solids; primary and secondary chemicals; clinical, biological, and botanical samples; fertilizers; and glasses. Several industrial concerns also offer various kinds of standard materials designed for validating analytical procedures.

When standard reference materials are not available (and often they are not), the best the chemist can do is to prepare a solution of known concentration whose composition approximates that of the sample after it has been decomposed and dissolved. Obviously, such a standard gives no information at all concerning the fate of the substance being determined during the important decomposition and solution steps.

The National Institute of Standards and Technology is an important source for standard reference materials.

For literature describing standard reference materials, see the references in footnotes 3 and 4 in Chapter 2 on page 17.

## Analysis by Other Methods

The results of an analytical method can sometimes be evaluated by comparison with data obtained from an entirely different method. Clearly, a second method must exist and in addition should be based on chemical principles that differ as much as possible from the one under examination. Comparable results from the two serve as presumptive evidence that both are yielding satisfactory results, inasmuch as it is unlikely that the same determinate errors would affect each. Such a conclusion does not apply to those aspects of the two methods that are similar.

## Standard Addition to the Sample

Applications of standard-addition methods are presented in Chapters 17, 22, and 24.

When the foregoing approaches are inapplicable, the standard-addition method may prove useful. Here, in addition to being used to analyze the sample, the proposed procedure is tested against portions of the sample to which known amounts of the analyte have been added. The effectiveness of the method can then be established by evaluating the extent of recovery of the added quantity. The standard-addition method may reveal errors arising from the way the sample was treated or from the presence of the other elements and/or compounds in the matrix.

## 29B  THE ACCURACY OBTAINABLE IN THE ANALYSIS OF COMPLEX MATERIALS

To provide a clear idea of the accuracy that can be expected for the analysis of a complex material, data on the determination of four elements in a variety of materials are presented in Tables 29–1 to 29–4. These data were taken from a much larger set of results collected by W. F. Hillebrand and G. E. F. Lundell of the National Bureau of Standards and published in the first edition of their excellent book on inorganic analysis.[1]

The materials analyzed were naturally occurring substances and items of commerce; they were especially prepared to give uniform and homogeneous samples and were distributed among chemists who were, for the most part, actively engaged in the analysis of similar materials. The analysts were allowed to use the methods they considered most reliable and best suited for the problem at hand. In most instances, special precautions were taken, and the results are consequently better than can be expected from the average routine analysis.

The numbers in the second column of Tables 29–1 to 29–4 are best values, obtained by the most painstaking analysis for the measured quantity. Each is considered to be the true value for calculation of the absolute and relative errors shown in the fourth and fifth columns. The fourth column was obtained by discarding extremely divergent results, determining the deviation of the remaining individual data from the best value (second column), and averaging these deviations. The fifth column was

---

[1]W. F. Hillebrand and G. E. F. Lundell, *Applied Inorganic Analysis,* pp. 874–887. New York: Wiley, 1929.

Table 29–1
## DETERMINATION OF IRON IN VARIOUS MATERIALS*

| Materials | Iron, % | Number of Analysts | Average Absolute Error | Average Relative Error, % |
|---|---|---|---|---|
| Soda-lime glass | 0.064 ($Fe_2O_3$) | 13 | 0.01 | 15.6 |
| Cast bronze | 0.12 | 14 | 0.02 | 16.7 |
| Chromel | 0.45 | 6 | 0.03 | 6.7 |
| Refractory | 0.90 ($Fe_2O_3$) | 7 | 0.07 | 7.8 |
| Manganese bronze | 1.13 | 12 | 0.02 | 1.8 |
| Refractory | 2.38 ($Fe_2O_3$) | 7 | 0.07 | 2.9 |
| Bauxite | 5.66 | 5 | 0.06 | 1.1 |
| Chromel | 22.8 | 5 | 0.17 | 0.75 |
| Iron ore | 68.57 | 19 | 0.05 | 0.07 |

*From W. F. Hillebrand and G. E. F. Lundell, *Applied Inorganic Analysis*, p. 878. New York: Wiley, 1929.

Table 29–2
## DETERMINATION OF MANGANESE IN VARIOUS MATERIALS*

| Material | Manganese, % | Number of Analysts | Average Absolute Error | Average Relative Error, % |
|---|---|---|---|---|
| Ferro-chromium | 0.225 | 4 | 0.013 | 5.8 |
| Cast iron | 0.478 | 8 | 0.006 | 1.3 |
|  | 0.897 | 10 | 0.005 | 0.56 |
| Manganese bronze | 1.59 | 12 | 0.02 | 1.3 |
| Ferro-vanadium | 3.57 | 12 | 0.06 | 1.7 |
| Spiegeleisen | 19.93 | 11 | 0.06 | 0.30 |
| Manganese ore | 58.35 | 3 | 0.06 | 0.10 |
| Ferro-manganese | 80.67 | 11 | 0.11 | 0.14 |

*From W. F. Hillebrand and G. E. F. Lundell, *Applied Inorganic Analysis*, p. 880. New York: Wiley, 1929.

Table 29–3
## DETERMINATION OF PHOSPHORUS IN VARIOUS MATERIALS*

| Material | Phosphorus, % | Number of Analysts | Average Absolute Error | Average Relative Error, % |
|---|---|---|---|---|
| Ferro-tungsten | 0.015 | 9 | 0.003 | 20. |
| Iron ore | 0.040 | 31 | 0.001 | 2.5 |
| Refractory | 0.069 ($P_2O_5$) | 5 | 0.011 | 16. |
| Ferro-vanadium | 0.243 | 11 | 0.013 | 5.4 |
| Refractory | 0.45 | 4 | 0.10 | 22. |
| Cast iron | 0.88 | 7 | 0.01 | 1.1 |
| Phosphate rock | 43.77 ($P_2O_5$) | 11 | 0.5 | 1.1 |
| Synthetic mixtures | 52.18 ($P_2O_5$) | 11 | 0.14 | 0.27 |
| Phosphate rock | 77.56 ($Ca_3(PO_4)_2$) | 30 | 0.85 | 1.1 |

*From W. F. Hillebrand and G. E. F. Lundell, *Applied Inorganic Analysis*, p. 882. New York: Wiley, 1929.

Table 29-4
DETERMINATION OF POTASSIUM IN VARIOUS MATERIALS*

| Material | Potassium Oxide, % | Number of Analysts | Average Absolute Error | Average Relative Error, % |
|---|---|---|---|---|
| Soda-lime glass | 0.04 | 8 | 0.02 | 50. |
| Limestone | 1.15 | 15 | 0.11 | 9.6 |
| Refractory | 1.37 | 6 | 0.09 | 6.6 |
| | 2.11 | 6 | 0.04 | 1.9 |
| | 2.83 | 6 | 0.10 | 3.5 |
| Lead-barium glass | 8.38 | 6 | 0.16 | 1.9 |

*From W. F. Hillebrand and G. E. F. Lundell, *Applied Inorganic Analysis*, p. 883. New York: Wiley, 1929.

obtained by dividing the data in the fourth column by the best value (second column) and multiplying by 100%.

The results shown in these tables are typical of the data for 26 elements reported in the original publication. We conclude that analyses reliable to a few tenths of a percent relative are the exception rather than the rule in the analysis of complex mixtures by ordinary methods and that, unless we are willing to invest an inordinate amount of time in the analysis, errors on the order of 1 or 2% must be accepted. If the sample contains less than 1% of the analyte, even larger relative errors are to be expected.

It is clear from these data that the accuracy obtainable in the determination of an element is greatly dependent upon the nature and complexity of the substrate. Thus, the relative error in the determination of phosphorus in two phosphate rocks was 1.1%; in a synthetic mixture, it was only 0.27%. The relative error in an iron determination in a refractory was 7.8%; in a manganese bronze having about the same iron content, it was only 1.8%. Here, the limiting factor in the accuracy is not in the completion step but rather in the dissolution of the samples and the elimination of interferences.

The data in these four tables are more than 60 years old, and it is tempting to think that analyses carried out with more modern tools and

Fundamental sources of uncertainty and error that were with us 60 years ago are still with us today.

Table 29-5
STANDARD DEVIATION OF SILICA RESULTS*

| Year Reported | Sample Type | Number of Results | Standard Deviation (%, absolute) |
|---|---|---|---|
| 1931 | Glass | 5 | 0.28† |
| 1951 | Granite | 34 | 0.37 |
| 1963 | Tonalite | 14 | 0.26 |
| 1970 | Feldspar | 9 | 0.10 |
| 1972 | Granite | 30 | 0.18 |
| 1972 | Syenite | 36 | 1.06 |
| 1974 | Granodiorite | 35 | 0.46 |

*From S. Abbey, *Anal. Chem.*, **1981,** *53* (4), 529A.

†0.09 after eliminating one result.

additional experience are likely to be significantly better in terms of accuracy and precision. A study by S. Abbey suggests that this assumption is not valid, however.[2] For example, the data in Table 29–5, which were taken from his paper, reveal no significant improvement in silicate analyses of standard reference glass and rock samples in the 43-year period from 1931 to 1974. Indeed, the standard deviation among participating laboratories appears to be larger in later years.

From the data in the five tables in this chapter, it is clear that the chemist is well advised to adopt a pessimistic viewpoint regarding the accuracy of an analysis, be it one's own or one performed by someone else.

---

[2]S. Abbey, *Anal. Chem.*, **1981,** *53,* 529A.

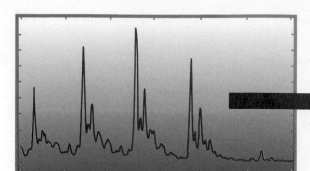

## CHAPTER 30

# PREPARING SAMPLES FOR ANALYSIS

$\text{T}$his chapter deals with two preliminary steps that are important parts of most chemical analyses: (1) sampling and (2) either drying the sample or determining its moisture content. Under some circumstances, neither of these steps is important or necessary, but more commonly, one or both are vital and may limit the accuracy and significance of the analytical data.

## 30A SAMPLING

Generally, a chemical analysis is performed on only a small fraction of the material whose composition is of interest. Clearly, the composition of this fraction must reflect as closely as possible the average composition of the bulk of the material if the results are to have value. The process by which a representative fraction is acquired is termed *sampling*. Often, sampling is the most difficult step in the entire analytical process and the step that limits the accuracy of the procedure. This statement is particularly true when the material to be analyzed is a large and inhomogeneous liquid, such as a lake, or an inhomogeneous solid, such as an ore, a soil, or a piece of animal tissue.

The end product of the sampling step is a quantity of homogeneous material weighing a few grams or, at most, a few hundred grams that may constitute as little as one part in $10^7$ or $10^8$ of the material whose composition is sought. Yet the composition of this minute fraction must, as closely as possible, be identical to the average composition of the total mass. Where, as with the examples just mentioned, the quantity of material to be sampled is large and its composition inherently nonhomogeneous, the task of producing a representative sample is indeed formidable. Clearly, the reliability of the analysis cannot exceed that of the sampling step, and the painstaking analysis of a poor sample is wasted effort.

Sampling is often the most difficult aspect of an analysis.

The composition of the sample must closely resemble the average composition of the total mass of material to be analyzed.

The literature on sampling is extensive;[1] at best we can provide only a brief outline of sampling methods here.

Three steps are involved in sampling bulk materials: (1) identification of the population from which the sample is to be obtained, (2) collection of a *gross sample* that is truly representative of the population being sampled, and (3) reduction of the gross sample to a few hundred grams of a homogeneous *laboratory sample* that is suitable for analysis.

Ordinarily, step 1 is straightforward, with the population being as diverse as a carload of ore, a carton of bottles containing vitamin tablets, a field of wheat, the brain of a rat, or the mud from a stretch of river bottom. Steps 2 and 3 are seldom simple and may require more effort and ingenuity than any other step in the entire analytical scheme.

## 30A–1  The Effects of Sampling Uncertainties

In Chapter 2, we concluded that both determinate and indeterminate errors in analytical data can be traced to instrument, method, and personal causes. Most determinate errors can be eliminated by exercising care, by calibration, and by the proper use of standards, blanks, and reference materials. Indeterminate errors, which are reflected in the precision of data, can generally be kept at an acceptable level by close control of the variables that influence the measurements. Errors due to invalid sampling are unique in the sense that they are not controllable by the use of blanks and standards or by closer control of experimental variables. For this reason, sampling errors are ordinarily treated separately from the other uncertainties associated with an analysis.

For random uncertainties, the overall standard deviation $s_o$ for an analytical measurement is related to the standard deviation of the sampling process $s_s$ and to the standard deviation of the method $s_m$ by the relationship

$$s_o^2 = s_s^2 + s_m^2 \qquad (30-1)$$

In many cases, the method variance will be known from replicate measurements on the product of a single laboratory sample. Under this circumstance, $s_s$ can be computed from measurements of $s_o$ for a series of laboratory samples, each of which is obtained from several gross samples.

Youden has shown that, once the measurement uncertainty has been reduced to one third or less of the sampling uncertainty (that is, $s_m < s_s/3$),

When $s_m < s_s/3$, there is no point in trying to improve the measurement precision.

[1]See, for example, B. W. Woodget and D. Cooper, *Samples and Standards*. London: Wiley, 1987; F. F. Pitard, *Pierre Gy's Sampling Theory and Sampling Practice*. Boca Raton, FL: CRC Press, 1989; F. J. Welcher, Ed., *Standard Methods of Chemical Analysis*, 6th ed., Vol. 2, Part A, pp. 21–55. Princeton, NJ: Van Nostrand, 1963. An extensive bibliography of specific sampling information has been compiled by C. A. Bicking, in *Treatise on Analytical Chemistry*, 2nd ed., I. M. Kolthoff and P. J. Elving, Eds., Part I, Vol. 1, p. 299. New York: Interscience, 1978. For a brief review of sampling problems, see B. Kratochvil and J. K. Taylor, *Anal. Chem.*, **1981**, *53*, 924A.

Identify the population

Collect a gross sample

Reduce the gross sample to a laboratory sample

further improvement in the measurement uncertainty is fruitless.[2] As a consequence, if the sampling uncertainty is large and cannot be improved, it is often wise to switch to a less precise but rapid method of analysis so that more samples can be analyzed in a given length of time, thus improving the precision of the average for the lot.

## 30A–2  The Gross Sample

The gross sample must be representative of the whole not only in the matter of chemical composition but also in its particle size distribution.

Ideally, the gross sample is a miniature replica of the entire mass of material to be analyzed. It corresponds to the whole not only in chemical composition but, *equally important,* in particle-size distribution.

### The Size of the Gross Sample

From the standpoint of convenience and economy, it is desirable that the gross sample weigh no more than absolutely necessary. Basically, gross sample weight is determined by (1) the uncertainty that can be tolerated between the composition of the gross sample and that of the whole, (2) the degree of heterogeneity of the whole, and (3) the level of particle size at which heterogeneity begins.[3]

This last point warrants amplification. A well-mixed, homogeneous solution of a gas or liquid is heterogeneous only on the molecular scale, and the weight of the molecules themselves governs the minimum weight of the gross sample. A particulate solid, such as an ore or a soil, represents the opposite situation. In such materials, the individual pieces of solid differ from each other in composition. Here, heterogeneity develops in particles that may have dimensions on the order of a centimeter or more and may weigh several grams. Intermediate between these extremes are colloidal materials and solidified metals. With the former, heterogeneity is first encountered in the range of $10^{-5}$ cm or less. In an alloy, heterogeneity first occurs in the crystal grains.

The number of particles required in a gross sample ranges from a few particles to $10^{12}$ particles.

In order to obtain a truly representative gross sample, a certain number **n** of the particles referred to in (3) must be taken. The magnitude of this number depends upon (1) and (2) and may involve only a relatively few particles, several millions, or even several millions of millions. The need for large numbers of particles is of no great concern for homogeneous gases and liquids, since heterogeneity among particles first occurs at the molecular level. Thus, even a very small weight of sample will contain more than the requisite number of particles. However, the individual particles of a particulate solid may weigh a gram or more, which sometimes leads to a gross sample that weighs several tons. Sampling of such material is a costly, time-consuming procedure at best; determination of the smallest weight of material required to provide the needed information minimizes this expense.

---

[2]W. J. Youden, *J. Assoc. Off. Anal. Chem.*, **1981,** *50,* 1007.
[3]For a paper on sample weight as a function of particle size, see G. H. Fricke, P. G. Mischler, F. P. Staffieri, and C. L. Housmyer, *Anal. Chem.*, **1987,** *59,* 1213.

The composition of a gross sample removed randomly from a bulk of material is governed by the law of chance; thus, by suitable statistical manipulations, it is possible to predict the probability that a given fraction is similar to the whole. A simple, idealized case serves as an example. A carload of lead ore is made up of just two kinds of particles: galena (lead sulfide) and a gangue containing no lead. All particles have the same size. The car contains, let us say, 100 million particles, and we wish to know how many of these are galena. Since the two components differ in appearance, the composition of the carload could be obtained exactly by counting all the galena particles. This approach would probably involve several lifetimes of work, however. We must therefore settle for the lesser certainty involved in counting some reasonable fraction of the total number of particles. The number of particles contained in this fraction depends, of course, upon the error we are willing to tolerate in the measurement.

The relationship between the allowable error and the number of particles **n** to be counted can be stated as[3,4]

$$\mathbf{n} = \frac{1 - p}{p\sigma_r^2} \qquad (30\text{--}2)$$

where $p$ is the fraction of galena particles, $1 - p$ is the fraction of gangue particles, and $\sigma_r$ is the allowable relative standard deviation in the count of the galena particles. Thus, for example, if 80% of the particles are galena ($p = 0.8$) and the tolerable standard deviation is 1% ($\sigma_r = 0.01$), a random sampling of 2500 particles should be made. A standard deviation of 0.1% requires a sample containing 250,000 particles.

Let us now make the problem more realistic and assume that one of the components in the car contains a higher percentage of lead $P_1$ and the other component contains a lesser amount $P_2$. Furthermore, the average density $d$ of the shipment differs from the densities $d_1$ and $d_2$ of these components. We are now interested in deciding what number of particles and thus what weight we should take to ensure a sample possessing the overall average percent of lead $P$ with a sampling relative standard deviation of $\sigma_r$. Equation 30–2 can be extended to include these stipulations:

$$\mathbf{n} = p(1 - p)\left(\frac{d_1 d_2}{d^2}\right)^2\left(\frac{P_1 - P_2}{\sigma_r P}\right)^2 \qquad (30\text{--}3)$$

From this equation, we see that the demands of accuracy are costly, in terms of the sample size required, because of the inverse-square relationship between the allowable standard deviation and the number of particles taken. Furthermore, a greater number of particles must be taken as the average percentage $P$ of the element of interest becomes smaller.

The degree of heterogeneity as measured by $P_1 - P_2$ has a profound effect on the number of particles required, with the number increasing as

---

[4]For a discussion of the derivation and significance of Equations 30–2 and 30–3, see A. A. Benedetti-Pichler, in *Physical Methods in Chemical Analysis*, W. G. Berl, Ed., Vol. 3, pp. 183–194. New York: Academic Press, 1956; A. A. Benedetti-Pichler, *Essentials of Quantitative Analysis*, Chapter 19. New York: Ronald Press, 1956.

the square of the difference in composition of the two components of the mixture.

The problem of deciding upon the weight of the gross sample for a solid material is ordinarily more difficult than this example because most samples not only contain more than two components but also consist of a range of particle sizes. In most instances, the first of these problems can be met by dividing the sample into an imaginary two-component system. Thus, with an actual lead ore, one component selected might be all the various lead-bearing minerals of the ore and the other all the residual components containing little or no lead. After average densities and percentages of lead are assigned to each part, the system is treated as if it has only two components.

The problem of variable particle size can be handled by calculating the number of particles that would be needed if the sample consisted of particles of a single size. The gross sample weight is then determined by taking into account the particle-size distribution. One approach is to calculate the needed weight by assuming that all particles are the size of the largest. This procedure is not very efficient, however, for it usually calls for removal of a larger weight of material than necessary. Benedetti-Pichler gives alternative methods for computing the weight of gross sample to be chosen.[5]

An interesting conclusion from Equation 30–3 is that the number of particles in the gross sample is *independent* of particle size. The weight of the sample, of course, increases directly as the volume (or as the cube of the particle diameter), so that reduction in the particle size of a given material has a large effect on the weight required in the gross sample.

Clearly, a great deal of information must be known about a substance in order to make use of Equation 30–3. Fortunately, reasonable estimates of the various parameters in the equation can often be made. These estimates can be based upon a qualitative analysis of the substance, visual inspection, and information from the literature on substances of similar origin. Crude measurements of the density of the various sample components may also be necessary.

> To simplify the problem of defining the weight of a gross sample of a multicomponent mixture, assume that the sample is a hypothetical two-component mixture.

> The gross sample of a particulate solid can weigh several tons.

---

Example 30–1

Assume that the average particle in the carload of lead ore just considered is judged to be approximately spherical with a radius of about 5 mm. Roughly 4% of the particles appear to be galena ($\simeq$ 70% Pb), which has a density of 7.6 g/cm$^3$; the remaining particles have a density of about 3.5 g/cm$^3$ and contain little or no lead. How many pounds of ore should the gross sample contain if the sampling uncertainty is to be kept below 0.5% relative?

We first compute values for the average density and percent lead:

$$d = 0.04 \times 7.6 + 0.96 \times 3.5 = 3.7 \text{ g/cm}^3$$

$$P = \frac{(0.04 \times 7.6 \times 0.70) \text{ g Pb/cm}^3}{3.7 \text{ g sample/cm}^3} \times 100\% = 5.8\% \text{ Pb}$$

---

[5]A. A. Benedetti-Pichler, in *Physical Methods in Chemical Analysis*, W. G. Berl, Ed., Vol. 3, p. 192. New York: Academic Press, 1956.

Then, substituting into Equation 30–3 gives

$$\mathbf{n} = 0.04(1 - 0.04) \left[\frac{7.6 \times 3.5}{(3.7)^2}\right]^2 \left(\frac{70 - 0}{0.005 \times 5.8}\right)^2$$

$$= 8.45 \times 10^5 \text{ particles required}$$

$$\text{wt. of sample} = 8.45 \times 10^5 \text{ particles} \times \frac{4}{3}\pi(0.5)^3 \frac{cm^3}{particle} \times \frac{3.7\,g}{cm^3} \times \frac{1}{454\,g/lb}$$

$$= 3.61 \times 10^3 \text{ lb or about 1.8 ton}$$

## Sampling Homogeneous Solutions of Liquids and Gases

For solutions of liquids or gases, the gross sample can be relatively small since ordinarily nonhomogeneity first occurs at the molecular level, and even small volumes of sample will contain many more particles than the number computed from Equation 30–3. Whenever possible, the liquid or gas to be analyzed should be stirred well prior to sampling to make sure that the gross sample is homogeneous. With large volumes of solutions, mixing may be impossible; it is then best to sample several portions of the container with a "sample thief," a bottle that can be opened and filled at any desired location in the solution. This type of sampling is important, for example, in determining the constituents of liquids exposed to the atmosphere. Thus, the oxygen content of lake water may vary by a factor of 1000 or more over a depth difference of a few feet.

Industrial gases or liquids are often sampled continuously as they flow through pipes, with care being taken to ensure that the sample collected represents a constant fraction of the total flow and that all portions of the stream are sampled.

*Solutions of liquids and gases require only a very small sample because they are homogeneous down to the molecular level.*

## Sampling Particulate Solids

Obtaining a random sample from a bulky particulate material is often difficult. It can best be accomplished while the material is being transferred. For example, randomly chosen shovelfuls or wheelbarrow loads may be consigned to a sample pile, or portions of the material may be intermittently removed from a conveyor belt. Alternatively, the material may be forced through a mechanical device called a *riffle* or a series of riffles that continuously isolate a fraction of the stream. Mechanical devices of this sort have been developed for handling coals and ores. Details regarding sampling of these materials are beyond the scope of this book.

## Sampling Metals and Alloys

Samples of metals and alloys are obtained by sawing, milling, or drilling. In general, it is not safe to assume that chips of the metal removed from the surface are representative of the entire bulk, and solid from the interior must be sampled as well. With billets or ingots of metal, a representative sample can be obtained by sawing across the piece at random intervals and collecting the "sawdust" as the sample. Alternatively, the

*Obtain samples of metals and alloys by machining a small sample of chips from a large piece of the material.*

specimen may be drilled, again at various randomly spaced intervals, and the drillings collected as the sample; the drill should pass entirely through the block or halfway through from opposite sides. The drillings can then be broken up and mixed or melted together in a graphite crucible. A granular sample can often then be produced by pouring the melt into distilled water.

## 30A–3  Preparation of a Laboratory Sample

The laboratory sample should have the same number of particles as the gross sample.

For nonhomogeneous solids, the gross sample may weigh several hundred pounds or more, and so reduction of the gross sample to a finely ground and homogeneous laboratory sample, weighing at the most a few pounds, is necessary. This process involves a cycle of operations that includes crushing and grinding, sieving, mixing, and dividing the sample (often into halves) to reduce its weight. During each division, a weight of sample that contains the number of particles computed from Equation 30–3 is retained.

---

### Example 30–2

It is desired to reduce the gross sample in Example 30–1 to a laboratory sample that weighs about one pound. How can this be done?

The laboratory sample should contain the same number of particles as the gross sample, or $8.45 \times 10^5$. Each particle will weigh on the average

$$\text{av. wt. of particle} = 1 \text{ lb} \times 454 \frac{g}{lb} \times \frac{1}{8.45 \times 10^5 \text{ particles}}$$

$$= 5.37 \times 10^{-4} \text{ g/particle}$$

The average weight of a particle is related to its radius by the equation

$$\text{av. wt. of particle} = \frac{4}{3} \pi [r(\text{cm})]^3 \times \frac{3.7 \text{ g}}{\text{cm}^3}$$

Equating these two relationships and solving for $r$ give

$$r = \left( 5.37 \times 10^{-4} \text{ g} \times \frac{3}{4\pi} \times \frac{\text{cm}^3}{3.7 \text{ g}} \right)^{1/3} = 3.3 \times 10^{-2} \text{ cm or 0.3 mm}$$

Thus the sample should be repeatedly ground, mixed, and divided until the particles are about 0.3 mm in diameter.

---

### Crushing and Grinding of Laboratory Samples

Crushing and grinding the sample can sometimes change its composition.

A certain amount of crushing and grinding is ordinarily required to decrease the particle size of solid samples. Because these operations tend to alter the composition of the sample, the particle size should be reduced no more than is required for homogeneity and ready attack by reagents.

Several factors can cause appreciable changes in sample composition as a result of grinding. The heat inevitably generated can cause losses of volatile components. In addition, grinding increases the surface area of the solid and thus increases its susceptibility to reaction with the atmosphere. For example, it has been observed that the iron(II) content of a rock may be decreased by as much as 40% during grinding—apparently a direct result of the iron being oxidized to the +3 state.

Frequently, the water content of a sample is altered substantially during grinding. Increases are observed as a consequence of the increased specific surface area that accompanies a decrease in particle size (page 79). The increased surface area leads to greater amounts of adsorbed water. For example, the water content of a piece of porcelain changed from 0 to 0.6% when it was ground to a fine powder.

In contrast, decreases in the water content of hydrates often take place during grinding as a result of localized frictional heating. For example, the water content of gypsum ($CaSO_4 \cdot 2\,H_2O$) decreased from about 21% to 5% when the compound was ground to a fine powder.

Differences in hardness of the component can also introduce errors during crushing and grinding. Softer materials are ground to fine particles more rapidly than are hard ones and may be lost as dust as the grinding proceeds. In addition, flying fragments tend to contain a higher fraction of the harder components.

Intermittent screening often increases the efficiency of grinding. Screening involves shaking the ground sample on a wire or cloth sieve that will pass particles of a desired size. The residual particles are then returned for further grinding; the operation is repeated until the entire sample passes through the screen. The hardest materials, which often differ in composition from the bulk of the sample, are last to be reduced in particle size and are thus last through the screen. Therefore, grinding must be continued until every particle has been passed if the screened sample is to have the same composition as it had before grinding and screening.

> Crushing and grinding must be continued until the entire sample passes through a screen of the desired mesh size.

A serious error can arise during grinding and crushing as a consequence of sample contamination resulting from the mechanical wear and abrasion of the grinding surfaces. Even though these surfaces are fabricated from hardened steel, agate, or boron carbide, contamination of the sample is nevertheless occasionally encountered. The problem is particularly acute in analyses for minor constituents.

> Mechanical abrasion of the surfaces of the grinding device can contaminate the sample.

A variety of tools are employed for reducing the particle size of solids, including jaw crushers and disk pulverizers for large samples containing large lumps, ball mills for medium-size samples and particles, and various types of mortars for small amounts of material.

The *ball mill* is a useful device for grinding solids that are not too hard. It consists of a porcelain crock with a capacity of perhaps 2-L that can be sealed and rotated mechanically. The container is charged with approximately equal volumes of the sample and flint or porcelain balls with a diameter of 20 to 50 mm. Grinding and crushing occur as the balls tumble in the rotating container. A finely ground and well-mixed powder can be produced in this way.

The *Plattner diamond mortar* (Figure 30–1) is used for crushing hard,

Figure 30–1
A Plattner diamond mortar.

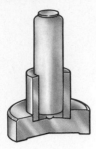

Finely ground materials may segregate after standing for a long time.

brittle materials. It is constructed of hardened tool steel and consists of a base plate, a removable collar, and a pestle. The sample is placed on the base plate inside the collar. The pestle is then fitted into place and struck several blows with a hammer; this reduces the solid to a fine powder that is collected on glazed paper after the apparatus has been disassembled.

### Mixing Solid Laboratory Samples

It is essential that solid materials be thoroughly mixed to ensure random distribution of the components in the analytical samples. A common method for mixing powders involves rolling the sample on a sheet of glazed paper. A pile of the substance is placed in the center and mixed by lifting one corner of the paper enough to roll the particles of the sample to the opposite corner. This operation is repeated many times, with the four corners of the sheet being lifted alternately.

Effective mixing of solids is also accomplished by rotating the sample for some time in a ball mill or a twin-shell V-blender. The latter consists of two connected cylinders that form a V-shaped container for the sample. As the blender is rotated, the sample is split and recombined with each rotation, leading to highly efficient mixing.

It is worthwhile noting that, with long standing, finely ground homogeneous materials may segregate on the basis of particle size and density. For example, analyses of layers of a set of student unknowns that had not been used for several years revealed a regular variation in the analyte concentration from top to bottom of the container. Apparently, segregation occurred as a consequence of vibrations and of density differences in the sample components.

## 30B   MOISTURE IN SAMPLES

Laboratory samples of solids often contain water that is in equilibrium with the atmosphere. As a consequence, unless special precautions are taken, the composition of the sample depends upon the relative humidity and ambient temperature at the time it is analyzed. To cope with this variability in composition, it is common practice to remove moisture from solid samples prior to weighing or, if this is not possible, to bring the water content to some reproducible level that can be duplicated later if necessary. A third alternative involves the determination of the water content at the time the samples are weighed for analysis so that the results can be corrected to a dry basis. In any event, many analyses are preceded by some sort of preliminary treatment designed to take into account the presence of water.

### 30B–1 Forms of Water in Solids

#### Essential Water

*Essential water* forms an integral part of the molecular or crystalline structure of a compound in its solid state. Thus, the *water of crystal-*

*lization* in a stable solid hydrate (for example, $CaC_2O_4 \cdot 2\ H_2O$ and $BaCl_2 \cdot 2\ H_2O$) qualifies as a type of essential water.

*Water of constitution* is a second type of essential water; it is found in compounds that yield stoichiometric amounts of water when heated or otherwise decomposed. Examples of this type of water are found in potassium hydrogen sulfate and calcium hydroxide, which when heated come to equilibrium with the moisture in the atmosphere, as shown by the reactions

$$2\ KHSO_4(s) \rightleftharpoons K_2S_2O_7(s)\ +\ H_2O(g)$$
$$Ca(OH)_2(s) \rightleftharpoons CaO(s)\ +\ H_2O(g)$$

### Nonessential Water

*Nonessential water* is retained by the solid as a consequence of physical forces. It is not necessary for characterization of the chemical constitution of the sample and therefore does not occur in any sort of stoichiometric proportion.

*Adsorbed water* is a type of nonessential water that is retained on the surface of solids. The amount adsorbed is dependent upon humidity, temperature, and the specific surface area of the solid. Adsorption of water occurs to some degree on all solids.

A second type of nonessential water is called *sorbed water* and is encountered with many colloidal substances, such as starch, protein, charcoal, zeolite minerals, and silica gel. In contrast to adsorption, the quantity of sorbed water is often large, amounting to as much as 20% or more of the total weight of the solid. Interestingly enough, solids containing even this amount of water may appear as perfectly dry powders. Sorbed water is held as a condensed phase in the interstices or capillaries of the colloidal solid. The quantity contained in the solid is greatly dependent upon temperature and humidity.

A third type of nonessential moisture is *occluded water,* liquid water entrapped in microscopic pockets spaced irregularly throughout solid crystals. Such cavities often occur in minerals and rocks (and in gravimetric precipitates).

## 30B–2 The Effect of Temperature and Humidity on the Water Content of Solids

In general, the concentration of water in a solid tends to decrease with increasing temperature and decreasing humidity. The magnitude of these effects and the rate at which they manifest themselves differ considerably according to the manner in which the water is retained.

### Compounds Containing Essential Water

The chemical composition of a compound containing essential water is dependent upon temperature and relative humidity. For example, anhydrous barium chloride tends to take up atmospheric moisture to give one

Relative humidity is the ratio of the vapor pressure of water in the atmosphere to its vapor pressure in air that is saturated with moisture. At 25°C, the partial pressure of water in saturated air is 23.76 torr. Thus, when air contains water at a partial pressure of 6 torr, the relative humidity is

$$\frac{6.00}{23.76} = 0.253 \text{ or } 25.3\%$$

The essential water content of a compound depends upon the temperature and relative humidity of its surroundings.

of two stable hydrates, depending upon temperature and relative humidity. The equilibria involved are

$$BaCl_2(s) + H_2O(g) \rightleftharpoons BaCl_2 \cdot H_2O(s)$$
$$BaCl_2 \cdot H_2O(s) + H_2O(g) \rightleftharpoons BaCl_2 \cdot 2 \, H_2O(s)$$

At room temperature and at a relative humidity between 25 and 90%, $BaCl_2 \cdot 2 \, H_2O$ is the stable species. Since the relative humidity in most laboratories is well within these limits, the essential-water content of the dihydrate is ordinarily independent of atmospheric conditions. Exposure of either $BaCl_2$ or $BaCl_2 \cdot H_2O$ to these conditions causes compositional changes that ultimately lead to formation of the dihydrate. On a very dry winter day (relative humidity <25%), however, the situation changes; the dihydrate becomes unstable with respect to the atmosphere, and a molecule of water is lost to form the new stable species $BaCl_2 \cdot H_2O$. At relative humidities less than about 8%, both hydrates lose water and the anhydrous compound is the stable species. From these remarks, it is apparent that the composition of a sample containing essential water is greatly dependent upon the relative humidity of its environment.

Many hydrated compounds can be converted to the anhydrous condition by oven-drying at 100 to 120°C for an hour or two. Such treatment often precedes an analysis of samples containing hydrated compounds.

### Compounds Containing Adsorbed Water

Figure 30–2 shows an *adsorption isotherm*, in which the weight of the water adsorbed on a typical solid is plotted against the partial pressure of water in the surrounding atmosphere. It is apparent from the diagram that the extent of adsorption is particularly sensitive to changes in water-vapor pressure at low partial pressures.

The amount of water adsorbed on a solid decreases as the temperature of the solid increases and generally approaches zero when the solid is heated above 100°C. Adsorption or desorption of moisture usually occurs rapidly, with equilibrium often being reached after 5 or 10 min. The speed of the process is often apparent during the weighing of finely divided anhydrous solids, where a continuous increase in weight will occur unless the solid is contained in a tightly stoppered vessel.

*Adsorbed water* resides on the surface of the particles of a material.

### Compounds Containing Sorbed Water

The quantity of moisture sorbed by a colloidal solid varies tremendously with atmospheric conditions, as shown in Figure 30–2. In contrast to the behavior of adsorbed water, however, equilibrium may require days or even weeks for attainment, particularly at room temperature. Moreover, the amounts of water retained by the two processes are often quite different from each other; typically, adsorbed moisture amounts to a few tenths of a percent of the weight of the solid, whereas sorbed water can amount to 10 or 20%.

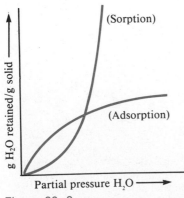

Figure 30–2
Typical adsorption and sorption isotherms.

The amount of water sorbed in a solid also decreases as the solid is heated. Complete removal of this type of moisture at 100°C is by no means a certainty, however, as indicated by the drying curves for an organic compound shown in Figure 30–3. After this material was dried for about 70 min at 105°C, constant weight was apparently reached. It is clear, however, that additional moisture was removed by elevating the temperature. Even at 230°C, dehydration was probably not complete.

*Sorbed water* is contained within the interstices of the molecular structure of a colloidal compound.

## Compounds Containing Occluded Water

Occluded water is not in equilibrium with the atmosphere and is therefore insensitive to changes in humidity. Heating a solid containing occluded water may cause a gradual diffusion of the moisture to the surface, where it evaporates. Frequently, heating is accompanied by *decrepitation*, in which the crystals of the solid are suddenly shattered by the steam pressure created from moisture contained in the internal cavities.

*Occluded water* is trapped in random microscopic pockets of solids, particularly minerals and rocks.

### 30B–3  Drying the Analytical Sample

The methods for dealing with moisture in solid samples depend upon the information desired. When the composition of the material on an as-received basis is needed, the principal concern is that the moisture content not be altered as a consequence of grinding or other preliminary treatment and storage. Where such changes are unavoidable or probable, it is often advantageous to determine the weight loss upon heating at some suitable temperature (say, 105°C) immediately upon receipt of the sample. Then, when the analysis is to be performed, the sample is again dried at this temperature so that the data can be corrected back to the original basis.

Sometimes analytical results are reported on an air-dry basis, where the analysis is performed after the sample has been allowed to equilibrate with the atmosphere. Such a procedure is commonly followed in the

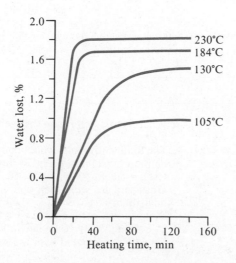

**Figure 30–3**

Removal of sorbed water from an organic compound at various temperatures. (Data from C. O. Willits, *Anal. Chem.*, **1951,** *23,* 1058. With permission of the American Chemical Society. Copyright 1951.)

analysis of metals and alloys as well as particulate matter that does not adsorb moisture strongly.

We have already noted that the moisture content of some substances is markedly changed by variations in humidity and temperature. Colloidal materials containing large amounts of sorbed moisture are particularly susceptible to the effects of these variables. For example, the moisture content of a potato starch has been found to vary from 10 to 21% as a consequence of an increase in relative humidity from 20 to 70%. With substances of this sort, comparable analytical data from one laboratory to another or even within the same laboratory can be achieved only by carefully specifying a procedure for taking the moisture content into consideration. For example, samples are frequently dried to constant weight at 105°C or at some other specified temperature. Analyses are then performed and results reported on this dry basis. While such a procedure may not render the solid completely free of water, it usually lowers the moisture content to a reproducible level.

## 30C   THE DETERMINATION OF WATER IN SAMPLES

Often the only sure way to obtain a result on a dry basis requires a separate determination of moisture in a set of samples taken concurrently with the samples that are to be analyzed. Several methods of determining water in solid samples are available. The simplest involves determining the weight loss after the sample has been heated at 100 to 110°C (or some other specified temperature) until the weight of the dried sample becomes constant. Unfortunately, this simple procedure is not at all specific for water, and large positive determinate errors occur in samples that yield volatile decomposition products (other than water) upon being heated. This method can also yield negative errors when applied to samples containing sorbed moisture (for example, see Figure 30–3).

Several highly selective methods have been developed for the determination of water in solid and liquid samples. One of these, the Karl Fischer method, is described in Section 16D–3. Several others are described in a monograph by Mitchell and Smith.[6]

---

[6] J. J. Mitchell Jr. and D. M. Smith, *Aquametry*, 2nd ed., 3 volumes. New York: Wiley, 1977–80.

---

## 30D   QUESTIONS AND PROBLEMS

*30-1. Describe the steps in a sampling operation.

30-2. What is the object of the sampling step in an analysis?

30-3. Differentiate between
  *(a) sorbed water, adsorbed water, and occluded water.
  (b) water of crystallization and water of constitution.
  *(c) essential water and nonessential water.
  (d) gross sample and laboratory sample.

30-4. What factors determine the weight of a gross sample?

*30-5. The following results were obained for the determination of calcium in a NIST limestone sample: % CaO = 50.38, 50.20, 50.31, 50.22, and 50.41. Five gross samples were then obtained for a carload of limestone. The average % CaO values for the gross samples were found to be 49.53, 50.12, 49.60, 49.87, and 50.49. Calculate the relative standard deviation associated with the sampling step.

| | Number of Particles Counted | Number of Particles Counted That Were | | Percent Relative Standard Deviation of the Count | Absolute Standard Deviation of the Count |
|---|---|---|---|---|---|
| | | Satis-factory | Unsatis-factory | | |
| (a) | 325 | 291 | | | |
| (b) | 325 | | 34 | | |
| (c) | 800 | | 82 | | |
| (d) | 675 | | 68 | | |
| (e) | 1200 | 1025 | | | |

*30-6. The preparation of a heterogeneous catalyst involves coating spherical support particles with a layer of active material. A satisfactory product is entirely coated with a layer of catalyst; the existence of breaks is unacceptable. Complete the accompanying tabulation in the table above.

30-7. A coating that weighs at least 3.00 mg is needed to impart adequate shelf life to a pharmaceutical tablet. A random sampling of 250 tablets revealed that 14 failed to meet this requirement.
(a) Use this information to estimate the relative standard deviation for the measurement.
(b) What is the 90% confidence interval for the number of unsatisfactory tablets?
(c) Assuming that the fraction of rejects remains unchanged, how many tablets should be taken to ensure a relative standard deviation of 10% in this measurement?

30-8. Changes in the method used to coat the tablets lowered the percentage of rejects from 5.6% (Problem 30-7) to 2.0%. How many tablets should be taken for inspection if the permissible relative standard deviation in the measurement is to be
*(a) 25%?      (b) 10%?
*(c) 5%?       (d) 1%?

*30-9. The mishandling of a shipping container loaded with 750 cases of wine caused some of the bottles to break. An insurance adjuster proposed to settle the claim at 20.8% of the value of the shipment, based upon a random 250-bottle sample in which 52 were cracked or broken. Calculate
(a) the relative standard deviation of the adjuster's evaluation.
(b) the absolute standard deviation for the 750 cases (12 bottles/case).
(c) the 90% confidence interval for the total number of bottles.
(d) the size of a random sampling needed for a relative standard deviation of 5.0%, assuming a breakage rate of about 21%.

*30-10. Approximately 15% of the particles in a shipment of silver-bearing ore are judged to be argentite, $Ag_2S$ ($d = 7.3$ g·cm$^{-3}$, 87% Ag); the remainder are siliceous ($d = 2.6$ g·cm$^{-3}$) and contain essentially no silver.
(a) Calculate the number of particles that should be taken for the gross sample if the relative standard deviation due to sampling is to be 1% or less.
(b) Estimate the weight of the gross sample, assuming that the particles are spherical and have an average diameter of 4.0 mm.
(c) The sample taken for analysis is to weigh 0.600 g and contain the same number of particles as the gross sample. To what diameter must the particles be ground to satisfy these criteria?

30-11. The average diameter of the particles in a shipment of copper appears to be 5.0 mm. Approximately 5% of the particles are cuprite (80% Cu) with a density of 6 g·cm$^{-3}$; the remainder is estimated to have a density of 4 g·cm$^{-3}$ and contain 3% Cu.
(a) How many particles of the ore should be sampled if the relative standard deviation due to sampling is to be 4% or less?
(b) What should be the weight of the gross sample be?
(c) To what diameter must the particles be ground in order to yield a sample for analysis that weighs 0.500 g and has the same number of particles as the gross sample?

*30-12. The average particle diameter of an ore sample is 2.0 mm. It is estimated that the stibnite content ($d_{Sb_2S_3} = 4.5$ g·cm$^{-3}$, 71.7% Sb) is approximately 2.0%; the remainder has a density of 3.0 g·cm$^{-3}$ and contains about 1% Sb.
(a) How many particles of the ore should be taken if the relative standard deviation due to sampling is to be 1% or less?
(b) What should be the weight of the gross sample be?
(c) To what diameter must the particles be ground in order to yield a sample for analysis that weighs 0.750 g and has the same number of particles as the gross sample?

30-13. The seller of a mining claim took a random ore sample that weighed approximately 5 lb and had an average particle diameter of 5.0 mm. Inspection revealed that about 1% of the sample was argentite (Problem 30-10), and the remainder had a density of about 2.6 g·cm$^{-3}$ and contained no silver. The prospective buyer insisted upon knowing the silver content of the claim with a relative error no greater than 5%. Establish whether the seller provided a sufficiently large sample to permit such an evaluation.

## CHAPTER 31

# DECOMPOSING AND DISSOLVING THE SAMPLE

$\mathbf{M}$ost analytical measurements are performed on solutions (usually aqueous) of the analyte. While some samples dissolve readily in water or dilute aqueous solutions of the common acids or bases, others require powerful reagents and rigorous treatment. For example, when sulfur or halogens are to be determined in an organic compound, the sample must be subjected to high temperatures and potent reagents in order to rupture the strong bonds between these elements and carbon. Similarly, drastic conditions are usually required to destroy the silicate structure of a siliceous mineral, thus rendering its cations free for analysis.

The proper choice among the various reagents and techniques for decomposing and dissolving analytical samples can be critical to the success of an analysis, particularly where refractory substances are involved or where the analyte is present in trace amounts. This chapter describes some of the more common methods for obtaining aqueous solutions of samples, particularly those that are difficult to decompose or dissolve.[1]

## 31A   SOME GENERAL CONSIDERATIONS

Ideally, the reagent selected should dissolve the entire sample, not just the analyte.

Ideally, a reagent should dissolve the sample completely because attempts to leach an analyte quantitatively from an insoluble residue are usually not successful. In choosing a solvent, consideration must also be given to possible interferences introduced during dissolution or decomposition. When trace amounts of an analyte are being determined, the most important considerations are frequently the purity of the reagent used for sample dissolution and the amounts that must be used.

An important concern when dissolving samples is the possibility that some portion of the analyte may volatilize. For example, carbon dioxide, sulfur dioxide, hydrogen sulfide, hydrogen selenide, and hydrogen telluride are generally volatilized when a sample is dissolved in strong acid, whereas ammonia is commonly lost when a basic reagent is employed.

---

[1]For an extensive discussion of this topic, see R. Bock, *A Handbook of Decomposition Methods in Analytical Chemistry*. New York: Wiley, 1979.

Similarly, hydrofluoric acid reacts with silicates and boron-containing compounds to produce volatile fluorides. Strong oxidizing solvents often cause the evolution of chlorine, bromine, or iodine; reducing solvents may lead to the volatilization of such compounds as arsine, phosphine, and stibine.

A number of elements form volatile chlorides that are partially or completely lost from hot hydrochloric acid solutions. Among these are the chlorides of tin(IV), germanium(IV), antimony(III), arsenic(III), and mercury(II). The oxychlorides of selenium and tellurium also volatilize to some extent from hot hydrochloric acid. The presence of chloride ion in hot concentrated sulfuric or perchloric acid solutions can cause volatilization losses of bismuth, manganese, molybdenum, thallium, vanadium, and chromium.

Boric acid, nitric acid, and the halogen acids are lost from boiling aqueous solutions, and phosphoric acid distills from hot concentrated sulfuric or perchloric acid. Certain volatile oxides can also be lost from hot acidic solutions, including the tetroxides of osmium and ruthenium and the heptoxide of rhenium.

## 31B   AQUEOUS REAGENTS FOR DISSOLVING OR DECOMPOSING SAMPLES

The most common reagents for attacking analytical samples are the mineral acids and, less frequently, aqueous solutions of ammonia and the alkali metal hydroxides.

### 31B–1  Hydrochloric Acid

Concentrated hydrochloric acid is an excellent solvent for many metal oxides as well as for metals more easily oxidized than hydrogen; often, it is a better solvent for oxides than the oxidizing acids. Concentrated hydrochloric acid is about 12 M, but upon heating, hydrogen chloride is lost until a constant-boiling 6 M solution remains (boiling point about 110°C).

### 31B–2  Nitric Acid

Hot concentrated nitric acid dissolves all common metals with the exception of aluminum and chromium, which become passive to this reagent as a consequence of surface oxide formation. When alloys containing tin, tungsten, or antimony are treated with the hot reagent, slightly soluble hydrated oxides, such as $SnO_2 \cdot 4\,H_2O$, form. After coagulation, these colloidal materials can be separated from other metallic species by filtration.

### 31B–3  Sulfuric Acid

Many materials are decomposed and dissolved by hot concentrated sulfuric acid, which owes part of its effectiveness as a solvent to its high boiling point (about 340°C). Most organic compounds are dehydrated and oxidized at this temperature and are thus eliminated from samples as carbon dioxide and water. Most metals and many alloys are attacked by the hot acid.

## 31B–4  Perchloric Acid

Hot concentrated perchloric acid, a potent oxidizing agent, attacks a number of iron alloys and stainless steels that are intractable to other mineral acids. Care must be taken in using the reagent, however, because of *its potentially explosive nature.* The cold concentrated acid is not hazardous, nor are heated dilute solutions; *violent explosions occur, however, when hot concentrated perchloric acid comes into contact with organic materials or easily oxidized inorganic substances.* Because of this property, the concentrated reagent should be heated only in special hoods, which are lined with glass or stainless steel, are seamless, and have a fog system for washing down the walls with water. A perchloric-acid hood should always have its own fan system, one that is independent of all other systems. *With proper precautions,*[2] perchloric acid is a safe and useful reagent.

Perchloric acid is marketed as the 60 to 72% acid. A constant-boiling mixture (72.4% $HClO_4$) is obtained at 203°C.

## 31B–5  Oxidizing Mixtures

More rapid solvent action can sometimes be obtained by the use of mixtures of acids or by the addition of oxidizing agents to a mineral acid. *Aqua regia,* a mixture containing three volumes of concentrated hydrochloric acid and one of nitric acid, is well known. The addition of bromine or hydrogen peroxide to mineral acids often increases their solvent action and hastens the oxidation of organic materials in the sample. Mixtures of nitric and perchloric acid are also useful for this purpose and less dangerous than perchloric acid alone. With this mixture, however, care must be taken to avoid evaporation of all the nitric acid before oxidation of the organic material is complete. *Severe explosions and injuries have resulted from failure to observe this precaution.*

## 31B–6  Hydrofluoric Acid

The primary use of hydrofluoric acid is for the decomposition of silicate rocks and minerals in the determination of species other than silica. In this treatment, silicon is evolved as the tetrafluoride. After decomposition is complete, the excess hydrofluoric acid is driven off by evaporation with sulfuric acid or perchloric acid. Complete removal is often essential to the

---

[2]See A. A. Schilt, *Perchloric Acid and Perchlorates.* Columbus, Ohio: G. Frederick Smith Chemical Company, 1979.

success of an analysis because fluoride ion reacts with several cations to form extraordinarily stable complexes that then interfere with the determination of the cations. For example, precipitation of aluminum (as $Al_2O_3 \cdot x\,H_2O$) with ammonia is quite incomplete if fluoride is present even in small amounts. Frequently, removal of the last traces of fluoride ion from a sample is so difficult and time-consuming as to negate the attractive features of the parent acid as a solvent for silicates.

Hydrofluoric acid finds occasional use in conjunction with other acids in attacking steels that dissolve with difficulty in other solvents.

Because hydrofluoric acid is extremely toxic, dissolution of samples and evaporation to remove excess reagent *should always be carried out in a well-ventilated hood*. Hydrofluoric acid *causes serious damage and painful injury* when brought into contact with the skin. Its effects may not become evident until hours after exposure. If the acid comes into contact with the skin, the affected area should be immediately washed with copious quantities of water. Treatment with a dilute solution of calcium ion, which precipitates fluoride ion, may also be of help.

## 31C  DECOMPOSITION OF SAMPLES BY FLUXES

Many common substances—notably silicates, some mineral oxides, and a few iron alloys—are attacked slowly, if at all, by the aqueous reagents just considered. In such cases, recourse to a fused-salt medium is indicated. Here, the sample is mixed with an alkali metal salt, called the *flux*, and the combination is then fused to form a water-soluble product called the *melt*. Fluxes decompose most substances by virtue of the high temperature required for their use (300 to 1000°C) and the high concentration of reagent brought in contact with the sample.

Where possible, the employment of a flux is avoided, for several dangers and disadvantages attend its use. Among these is the possible contamination of the sample by impurities in the flux. This possibility is exacerbated by the relatively large amount of flux (typically ten times the sample weight) required for a successful fusion. Moreover, the aqueous solution that results when the melt from a fusion is dissolved has a high salt content, which may cause difficulties in the subsequent steps of the analysis. In addition, the high temperatures required for a fusion increase the danger of volatilization losses. Finally, the container in which the fusion is performed is almost inevitably attacked to some extent by the flux; again, contamination of the sample is the result.

While they are very effective solvents, fluxes introduce high concentrations of ionic species to aqueous solutions of the melt.

For a sample containing only a small fraction of material that dissolves with difficulty, it is common practice to employ a liquid reagent first; the undecomposed residue is then isolated by filtration and fused with a relatively small quantity of a suitable flux. After cooling, the melt is dissolved and combined with the major portion of the sample.

### 31C–1 Carrying Out a Fusion

The sample in the form of a very fine powder is mixed intimately with perhaps a tenfold excess of the flux. Mixing is usually carried out in the crucible in which the fusion is to be performed. The time required for fusion can range from a few minutes to hours. The production of a clear melt signals completion of the decomposition, although often this condition is not always obvious.

When the fusion is complete, the mass is allowed to cool slowly; just before solidification, the crucible is rotated to distribute the solid around the walls to produce a thin layer of melt that is easy to dislodge.

## 31C–2 Types of Fluxes

With few exceptions, the common fluxes used in analysis are compounds of the alkali metals. Alkali metal carbonates, hydroxides, peroxides, and borates are basic fluxes employed to attack acidic materials. The acidic fluxes are pyrosulfates, acid fluorides, and boric oxide. If an oxidizing flux is required, sodium peroxide can be used. As an alternative, small quantities of the alkali nitrates or chlorates can be mixed with sodium carbonate.

The properties of the common fluxes are summarized in Table 31–1.

### Sodium Carbonate

Silicates and certain other refractory materials can be decomposed by heating to 1000 to 1200°C with sodium carbonate. This treatment generally converts the cationic constituents of the sample to acid-soluble carbonates or oxides; the nonmetallic constituents are converted to soluble sodium salts.

Carbonate fusions are normally carried out in platinum crucibles.

### Potassium Pyrosulfate

Potassium pyrosulfate is a potent acidic flux that is particularly useful for attacking the more intractable metal oxides. Fusions with this reagent are performed at about 400°C; at this temperature, the slow evolution of the highly acidic sulfur trioxide takes place:

$$K_2S_2O_7 \rightarrow K_2SO_4 + SO_3(g)$$

Potassium pyrosulfate can be prepared by heating potassium hydrogen sulfate:

$$2 KHSO_4 \rightarrow K_2S_2O_7 + H_2O$$

### Other Fluxes

Table 31–1 contains data for several other common fluxes. Noteworthy are boric oxide and the calcium carbonate/ammonium chloride mixture. Both are employed to decompose silicates for the analysis of alkali metals. Boric oxide is removed after solution of the melt by evaporation to dryness with methyl alcohol; methyl borate, $B(OCH_3)_3$, distills.

**Table 31-1**
**COMMON FLUXES**

| Flux | Melting Point, °C | Type of Crucible for Fusion | Type of Substance Decomposed |
|---|---|---|---|
| $Na_2CO_3$ | 851 | Pt | Silicates and silica-containing samples, alumina-containing samples, sparingly soluble phosphates and sulfates |
| $Na_2CO_3$ + an oxidizing agent, such as $KNO_3$, $KClO_3$, or $Na_2O_2$ | — | Pt (not with $Na_2O_2$), Ni | Samples requiring an oxidizing environment; that is, samples containing S, As, Sb, Cr, etc. |
| NaOH or KOH | 318 380 | Au, Ag, Ni | Powerful basic fluxes for silicates, silicon carbide, and certain minerals (main limitation is purity of reagents) |
| $Na_2O_2$ | Decomposes | Fe, Ni | Powerful basic oxidizing flux for sulfides; acid-insoluble alloys of Fe, Ni, Cr, Mo, W, and Li; platinum alloys; Cr, Sn, Zr minerals |
| $K_2S_2O_7$ | 300 | Pt, porcelain | Acidic flux for slightly soluble oxides and oxide-containing samples |
| $B_2O_3$ | 577 | Pt | Acidic flux for silicates and oxides where alkali metals are to be determined |
| $CaCO_3$ + $NH_4Cl$ | — | Ni | Upon heating the flux, a mixture of CaO and $CaCl_2$ is produced; used to decompose silicates for the determination of the alkali metals |

## 31D   DECOMPOSITION OF ORGANIC COMPOUNDS FOR ELEMENTAL ANALYSIS[3]

Determination of the elemental composition of an organic sample generally requires drastic treatment to convert the element of interest to a form susceptible to the common analytical techniques. These treatments are usually oxidative and involve conversion of carbon to carbon dioxide and

---

[3]For a thorough treatment of this topic, see T. S. Ma and R. C. Rittner, *Modern Organic Elemental Analysis*. New York: Marcel Dekker, 1979.

hydrogen to water; occasionally, however, heating the sample with a potent reducing agent is sufficient to rupture the covalent bonds in the compound and free the analyte element from the carbonaceous residue.

Oxidation procedures can be grouped in two categories. *Wet-ashing* (or wet-oxidation) makes use of liquid oxidizing agents, such as sulfuric, nitric, and perchloric acids. *Dry-ashing* usually implies ignition of the organic compound in air or in a stream of oxygen. In addition, oxidations can be carried out in certain fused-salt media; sodium peroxide is the most common flux for this purpose.

In the sections that follow, we consider briefly some of the methods for decomposing organic substances prior to elemental analysis.

## 31D–1 Wet-Ashing Procedures

Solutions of strong oxidizing agents are often used to decompose organic samples for the determination of metallic constituents. The main problem associated with the use of these reagents is possible losses of the elements of interest by volatilization.

We have already encountered an example of wet-ashing in the Kjeldahl method for the determination of nitrogen in organic compounds (page 269), where concentrated sulfuric acid is the oxidizing agent. This reagent is also frequently employed for the decomposition of organic materials in which metallic constituents are to be determined. Nitric acid can be added periodically to the solution to hasten the oxidation rate.[4] The following elements are volatilized (at least partially) by this procedure, particularly if the sample contains chlorine: arsenic, boron, germanium, mercury, antimony, selenium, tin, the halogens, sulfur, and phosphorus.

A reagent even more effective than a sulfuric/nitric acid mixture is perchloric acid mixed with nitric acid. *Great care must be exercised in the use of this reagent,* however, because of the tendency of hot anhydrous perchloric acid to react explosively with organic material. It is essential to start the oxidation with a mixture in which nitric acid predominates; this reagent attacks the easily oxidized components in the early stages. With continued heating, water and nitric acid are lost by decomposition and evaporation, and the solution becomes a progressively stronger oxidant as the perchloric acid concentration increases. If the solution becomes too concentrated in perchloric acid before most of the oxidation is complete, it will darken or turn black. *Should darkening occur, the mixture should be immediately removed from the heat and diluted with water and nitric acid.* The heating can then be continued. As mentioned on page 766, perchloric acid oxidations should be carried out only in a special hood. If properly performed, oxidations with a mixture of nitric and perchloric acids are rapid, and losses of metallic ions are negligible.[5] *It can-*

Hot, concentrated perchloric acid solutions must be treated with the utmost respect.

[4]*Official Methods of Analysis of the AOAC*, 11th ed., p. 400. Washington, D.C.: Association of Official Analytical Chemists, 1970.

[5]T. T. Gorsuch, *Analyst,* **1959,** *84,* 135; G. F. Smith, *Anal. Chim. Acta,* **1953,** *8,* 397; G. F. Smith, *The Wet Chemical Oxidation of Organic Compositions Employing Perchloric Acid.* Columbus, Ohio: G. F. Smith Chemical Company, 1965.

> **Feature 31–1**
> A PERCHLORIC ACID PRECAUTION
>
> Willard and Diehl[6] state, "The very properties of perchloric acid which make it extremely useful in the analytical laboratory are those which make its improper use hazardous. . . . It follows, then, that perchloric acid is dangerous only when hot and concentrated, and in the presence of some easily oxidized material such as organic matter. . . . A progression of colors, in the oxidation of organic material, changing from a light yellow, to a straw, to light brown, to dark brown, generally precedes a perchlorate explosion. If such a color change is observed, dilute the solution immediately or leave the vicinity hurriedly."
>
> ———————
> [6]H. H. Willard and H. Diehl, *Advanced Quantitative Analysis*, pp. 8–9. New York: D. Van Nostrand & Co. Inc., 1943.

*not be too strongly emphasized that hot concentrated perchloric acid reacts explosively with organic materials and other easily oxidizable substances and that extraordinary care is needed in using this reagent.*

## High-Pressure Wet-Ashing

An apparatus has recently been described for automatically wet-ashing various kinds of organic samples preliminary to determination of their metal content by atomic spectroscopy.[7] The oxidation is performed in a closed quartz vessel mounted in an aluminum heating block. The vessel and block are contained in an autoclave that can be pressurized with nitrogen up to 100 atm. Concentrated nitric acid and nitric/hydrochloric acid mixtures are used for the oxidation, and these reagents reach temperatures of 250 to 300°C at the elevated pressures made possible by the autoclave. Organic samples such as foodstuffs, biological materials, petroleum products, and coals are completely oxidized in 15 to 120 min in the apparatus.

## Microwave Wet-Ashing

A recent development in wet-ashing at elevated pressures (and thus elevated temperatures) is based upon the microwave decomposition of samples in nitric and other mineral acids contained in sealed Teflon vessels.[8] Heating is performed in a specially designed microwave oven that is now on the market. The rate of wet-ashing and efficiency of decomposition are

———————

[7]G. Knapp and A. Grillo, *Amer. Lab.*, **1986,** *18* (3), 76.

[8]*Anal. Chem.*, **1986,** *58,* 1424A; P. Aysola, P. Anderson, and C. H. Langford, *Anal. Chem.*, **1987,** *59,* 1582; H. M. Kingston and L. B. Jassie, *Anal. Chem.*, **1986,** *58,* 2534; *Introduction to Microwave Sample Preparation: Theory and Practice,* H. M. Kingston and L. B. Jassie, Eds. Washington, DC: American Chemical Society, 1988.

reported to increase dramatically with this type of equipment. For example, decompositions that take several hours at atmospheric conditions are said to be completed in a few minutes by the microwave technique. A further advantage is that the amount of reagent required for the digestion is much smaller, thus decreasing the magnitude of reagent blanks significantly.

## 31D–2 Dry-Ashing Procedures

The simplest method for decomposing an organic sample prior to determining the cations it contains is to heat the sample over a flame in an open dish or crucible until all carbonaceous material has been oxidized to carbon dioxide. Red heat is often required to complete the oxidation. Analysis of the nonvolatile components follows dissolution of the residual solid. Unfortunately, there is always substantial uncertainty about the completeness of recovery of supposedly nonvolatile elements from a dry-ashed sample. Some losses probably result from the entrainment of finely divided particulate matter in the convection currents around the crucible. In addition, volatile metallic compounds may be lost during the ignition. For example, copper, iron, and vanadium are appreciably volatilized when samples containing porphyrin compounds are ashed.

Although dry-ashing is the simplest method for decomposing organic compounds, it is often the least reliable. It should not be employed unless tests have demonstrated its applicability to a given type of sample.

## 31D–3 Combustion-Tube Methods

Several common and important elemental components of organic compounds are converted to gaseous products as a sample is pyrolyzed in the presence of oxygen. With suitable apparatus, it is possible to trap these volatile compounds quantitatively, thus making them available for the analysis of the element of interest. The heating is commonly performed in a glass or quartz combustion tube through which a stream of carrier gas is passed. The stream transports the volatile products to parts of the apparatus where they are separated and retained for the measurement; the gas may also serve as the oxidizing agent. Elements susceptible to this type of treatment are carbon, hydrogen, oxygen, nitrogen, the halogens, sulfur, and oxygen. Table 31–2 gives details for some of these methods.

Automated combustion-tube analyzers are now available for the determination of either carbon, hydrogen, and nitrogen or carbon, hydrogen, and oxygen in a single sample.[9] The apparatus requires essentially no attention by the operator, and the analysis is complete in less than 15 min. In one such analyzer, the sample is ignited in a stream of helium and oxygen and passes over an oxidation catalyst consisting of a mixture of silver vanadate and silver tungstate. Halogens and sulfur are removed

[9]For a description of these instruments, see Chapters 2, 3, and 4 of the reference in footnote 3.

Table 31–2
COMBUSTION-TUBE METHODS FOR THE ELEMENTAL ANALYSIS OF ORGANIC SUBSTANCES

| Element | Name of Method | Method of Oxidation | Method of Completion of Analysis |
|---|---|---|---|
| Halogens | Pregl | Sample burned in stream of oxygen over red-hot platinum catalyst; halogens converted primarily to HX and $X_2$ | Gas stream passed through carbonate solution containing $SO_3^{2-}$ (to reduce halogens and oxyhalogens to halides); $X^-$ then determined by usual procedures |
| | Grote | Sample burned in stream of air over hot silica catalyst; products are HX and $X_2$ | Same as above |
| Sulfur | Pregl | Similar to halogen determination; combustion products are $SO_2$ and $SO_3$ | Gas stream passed through aqueous $H_2O_2$ to convert sulfur oxides to $H_2SO_4$, which can then be titrated with standard base |
| | Grote | Similar to halogen determination; products are $SO_2$ and $SO_3$ | Similar to above |
| Nitrogen | Dumas | Sample oxidized by hot CuO to $CO_2$, $H_2O$, and $N_2$ | Gas stream passed through concentrated KOH solution, leaving only $N_2$, which is measured volumetrically |
| Carbon and hydrogen | Pregl | Similar to halogen analysis; products are $CO_2$ and $H_2O$ | $H_2O$ adsorbed on desiccant and $CO_2$ on Ascarite; determined gravimetrically |
| Oxygen | Unterzaucher | Sample pyrolyzed over carbon; oxygen converted to CO; $H_2$ used as carrier gas | Gas stream passed over $I_2O_5$ [5 CO(g) + $I_2O_5$(s) → 5 $CO_2$(g) + $I_2$(g)] and liberated $I_2$ titrated |

with a packing of silver salts. A packing of hot copper is located at the end of the combustion train to remove oxygen and convert nitrogen oxides to nitrogen. The exit gas, consisting of a mixture of water, carbon dioxide, nitrogen, and helium, is collected in a glass bulb. The analysis of this mixture is accomplished with three thermal-conductivity measurements (Section 27B–5). The first is made on the intact mixture, the second is made on the mixture after water has been removed by passage of the gas through a dehydrating agent, and the third is made on the mixture after carbon dioxide has been removed by an absorbent. The relationship between thermal conductivity and concentration is linear, and the slope of the curve for each constituent is established by calibration with a pure compound such as acetanilide.

## 31D–4 Combustion with Oxygen in a Sealed Container

A relatively straightforward method for the decomposition of many organic substances involves combustion with oxygen in a sealed container. The reaction products are absorbed in a suitable solvent before the reaction vessel is opened; they are subsequently analyzed by ordinary methods.

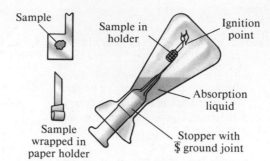

Figure 31-1
Schöniger combustion apparatus.
(Courtesy Thomas Scientific,
Swedesboro, NJ 08085.)

A remarkably simple apparatus for performing such oxidations has been suggested by Schöniger (Figure 31-1).[10] It consists of a heavy-walled flask of 300- to 1000-mL capacity fitted with a ground-glass stopper. Attached to the stopper is a platinum gauze basket that holds from 2 to 200 mg of sample. If the substance to be analyzed is a solid, it is wrapped in a piece of low-ash filter paper cut in the shape shown in Figure 31-1. Liquid samples are weighed into gelatin capsules, which are then wrapped in a similar fashion. The paper tail serves as the ignition point.

A small volume of an absorbing solution (often sodium carbonate) is placed in the flask, and the air in the flask is displaced by oxygen. The tail of the paper is ignited, the stopper is quickly fitted into the flask, and the flask is inverted to prevent the escape of the volatile oxidation products. The reaction ordinarily proceeds rapidly, being catalyzed by the platinum gauze surrounding the sample. During the combustion, the flask is shielded to minimize damage in case of explosion.

After cooling, the flask is shaken thoroughly and disassembled, and the inner surfaces are carefully rinsed. The analysis is then performed on the resulting solution. This procedure has been applied to the determination of halogens, sulfur, phosphorus, fluorine, arsenic, boron, carbon, and various metals in organic compounds.

---

[10]W. Schöniger, *Mikrochim. Acta,* **1955,** 123; **1956,** 869. See also the review articles by A. M. G. MacDonald, in *Advances in Analytical Chemistry and Instrumentation,* C. E. Reilley, Ed., Vol. 4, p. 75. New York: Interscience, 1965; and M. E. McNally and R. L. Grob, *Amer. Lab.,* **1981,** *13* (1), 31.

---

## 31E QUESTIONS AND PROBLEMS

*31-1. Differentiate between wet-ashing and dry-ashing.

31-2. What is a flux? When is its use called for?

*31-3. What fluxes are suitable for the determination of alkali metals in silicates?

31-4. What flux is commonly used for the decomposition of certain refractory oxides?

*31-5. Under what conditions is the use of perchloric acid likely to be dangerous?

31-6. How are organic compounds decomposed for the determination of
*(a) halogens?
 (b) sulfur?
*(c) nitrogen?
 (d) heavy-metal species?

# ELIMINATING INTERFERENCES

Interferences are encountered in chemical analyses whenever the sample matrix contains a species that either produces a signal indistinguishable from that of the analyte or else attenuates the analyte signal. Few analytical measurements are entirely specific and thus free from interference. Consequently, an important step in most analyses is one that eliminates interfering species.

Two general methods are available for coping with interferences. The first makes use of a *masking agent,* which is a substance that reacts with the interference to form a species that no longer contributes to or attenuates the analyte signal. Obviously, the masking agent must not have a significant effect on the behavior of the analyte. An example of a masking agent is fluoride ion in the iodometric determination of copper in ores that contain significant amounts of iron. Fluoride ion does not inhibit the reaction of the analyte with iodide ion to give iodine, which is subsequently titrated. It does, however, prevent iron(III) from undergoing an analogous reaction and thus interfering. The masking effect of the fluoride ion is attributable to the formation of stable iron(III) fluoride complexes that lower the electrode potential of the iron(III) to a point where it is no longer capable of producing iodine from iodide ion.

> A masking agent is a reagent that chemically binds an interference and prevents it from causing errors in an analysis.

The second approach to the interference problem is to physically separate the analyte from potential interferences. In Chapters 26 through 28, we considered the chemist's most powerful tool for performing separations: chromatography. In this chapter, we consider precipitation, solvent extraction, ion exchange, and distillation as ways of eliminating interferences.

> Chemical species are generally separated by converting them to different phases that can then be mechanically isolated.

## 32A THE NATURE OF THE SEPARATION PROCESS

All separation processes have in common the distribution of the components in a mixture between two phases that subsequently can be separated mechanically. If the ratio of the amount of one particular component in each phase (the *distribution ratio*) differs significantly from that of another, a separation of the two is potentially feasible. To be sure, the complexity of the separation depends upon the magnitude of this difference in distribution ratios. Where the difference is large, a single-stage process suffices. For example, a single precipitation with silver ion is adequate for the isolation of chloride from many other anions because the ratio of the amount of chloride ion in the solid phase to that in the aqueous phase is immense while comparable ratios for, say, nitrate or perchlorate ion approach zero.

A more complex situation prevails when the distribution ratio for one component is essentially zero, as in the foregoing example, but the ratio for the other is not very large. Here, a multistage process is required. For example, uranium(VI) can be extracted into ether from an aqueous nitric acid solution. Although the distribution ratio approaches unity for a single extraction, uranium(VI) can nevertheless be isolated quantitatively by repeated, or *exhaustive,* extraction of the aqueous solution with fresh portions of ether.

The most complex procedures are required when the distribution ratios of the species to be separated are both greater than zero and approach one another in magnitude; here, multistage *fractionation* techniques, such as chromatography, are necessary. Fractionation, like its simpler counterparts, is based upon differences in the distribution ratios of solutes. Two factors, however, account for the gain in separation efficiency with fractionation. First, the number of times that partitioning occurs between phases is increased enormously; second, distribution occurs between fresh portions of both phases. An exhaustive extraction differs from a fractionation in that fresh portions of only one phase are involved in the former.

## 32B    SEPARATION BY PRECIPITATION

Separations by precipitation require large solubility differences between analyte and potential interferences. The theoretical feasibility of this type of separation can be determined by solubility calculations such as those shown in Section 8C. Unfortunately, several other factors may preclude the use of precipitation to achieve a separation. For example, the various coprecipitation phenomena described in Section 4C–4 may cause extensive contamination of a precipitate by an unwanted component, even though the solubility product of the contaminant has not been exceeded. Likewise, the rate of an otherwise feasible precipitation may be too slow to be useful for a separation. Finally, when precipitates form as colloidal suspensions, coagulation may be difficult and slow, particularly when the isolation of a small quantity of a solid phase is attempted.

Many precipitating agents have been employed for quantitative inorganic separations. Some of the most generally useful are described in the sections that follow.

### 32B–1 Separations Based on Control of Acidity

Enormous differences exist in the solubilities of the hydroxides, hydrous oxides, and acids of various elements. Moreover, the concentration of hydrogen or hydroxide ions in a solution can be varied by a factor of $10^{15}$

Table 32–1
SEPARATIONS BASED UPON CONTROL OF ACIDITY

| Reagent | Species Forming Precipitates | Species Not Precipitated |
|---|---|---|
| Hot concd HNO$_3$ | Oxides of W(VI), Ta(V), Nb(V), Si(IV), Sn(IV), Sb(V) | Most other metal ions |
| NH$_3$/NH$_4$Cl buffer | Fe(III), Cr(III), Al(III) | Alkali and alkaline earths, Mn(II), Cu(II), Zn(II), Ni(II), Co(II) |
| HOAc/NH$_4$OAc buffer | Fe(III), Cr(III), Al(III) | Common dipositive ions |
| NaOH/Na$_2$O$_2$ | Fe(III), most dipositive ions, rare earths | Zn(II), Al(III), Cr(VI), V(V), U(VI) |

or more and can be readily controlled by the use of buffers. As a consequence, many separations based on pH control are, in theory, available to the chemist. In practice, these separations can be grouped in three categories: (1) those made in relatively concentrated solutions of strong acids, (2) those made in buffered solutions at intermediate pH values, and (3) those made in concentrated solutions of sodium or potassium hydroxide. Table 32–1 lists common separations that can be achieved by control of acidity.

## 32B–2 Sulfide Separations

With the exception of the alkali and alkaline-earth metals, most cations form sparingly soluble sulfides whose solubilities differ greatly from one another. Because it is relatively easy to control the sulfide ion concentration of an aqueous solution by adjustment of pH (Section 8C–2), separations based on the formation of sulfides have found extensive use. Sulfides can be conveniently precipitated from homogeneous solution, with the anion being generated by the hydrolysis of thioacetamide (Table 4–1).

A theoretical treatment of the ionic equilibria influencing the solubility of sulfide precipitates was considered in Section 8C–2. Such treatment may fail to provide realistic conclusions regarding the feasibility of separations, however, because of coprecipitation and the slow rate at which some sulfides form. As a consequence, resort must be made to empirical observations.

Table 32–2 shows some common separations that can be accomplished with hydrogen sulfide through control of pH.

## 32B–3 Other Inorganic Precipitants

No other inorganic ion is as generally useful for separations as hydroxide and sulfide ions. Phosphate, carbonate, and oxalate ions are often employed as precipitants for cations, but their behavior is nonselective; therefore separations must ordinarily precede their use.

Chloride and sulfate are useful because of their highly selective behavior. The former is used to separate silver from most other metals, and the

Recall p. 188 that

$$[S^{2-}] = \frac{6.8 \times 10^{-24}}{[H_3O^+]^2}$$

**Table 32–2**
**PRECIPITATION OF SULFIDES**

| Elements | Conditions for Precipitation* | Conditions for No Precipitation* |
|---|---|---|
| Hg(II), Cu(II), Ag(I) | 1, 2, 3, 4 | |
| As(V), As(III), Sb(V), Sb(III) | 1, 2, 3 | 4 |
| Bi(III), Cd(II), Pb(II), Sn(II) | 2, 3, 4 | 1 |
| Sn(IV) | 2, 3 | 1, 4 |
| Zn(II), Co(II), Ni(II) | 3, 4 | 1, 2 |
| Fe(II), Mn(II) | 4 | 1, 2, 3 |

*1 = 3 M HCl; 2 = 0.3 M HCl; 3 = buffered to pH 6 with acetate; 4 = buffered to pH 9 with $NH_3/(NH_4)_2S$.

latter is frequently employed to isolate a group of metals that includes lead, barium, and strontium.

## 32B–4 Organic Precipitants

Selected organic reagents for the isolation of various inorganic ions were discussed in Section 4E–3. Some of these organic precipitants, such as dimethylglyoxime, are useful because of their remarkable selectivity in forming precipitates with a few ions. Others, such as 8-hydroxyquinoline, yield slightly soluble compounds with a host of cations. The selectivity of this sort of reagent is owing to the wide range of solubility products among its reaction products and also the fact that the precipitating reagent is ordinarily an anion that is the conjugate base of a weak acid. Thus, separations based on pH control can be realized just as with hydrogen sulfide.

## 32B–5 The Separation of Constituents Present in Trace Amounts

A problem often encountered in trace analysis is that of isolating the species of interest, which may be present in microgram quantities, from the major components of the sample. Although such a separation is sometimes based on a precipitation, the techniques required differ from those used when the analyte is present in generous amounts.

Several problems attend the quantitative separation of a trace element by precipitation even when solubility losses are not important. Supersaturation often delays formation of the precipitate, and coagulation of small amounts of a colloidally dispersed substance is often difficult. In addition, it is likely that an appreciable fraction of the solid will be lost during transfer and filtration. To minimize these difficulties, a quantity of some other ion that also forms a precipitate with the reagent is often added to the solution. The precipitate from the added ion is called a *collector* and carries the desired minor species out of solution. For example, in isolating manganese as the sparingly soluble manganese dioxide, a small amount of iron(III) is frequently added to the analyte solution before the introduction of ammonia as the precipitating reagent. The basic iron(III) oxide carries down even the smallest traces of the dioxide. Other examples are basic aluminum oxide as a collector of trace amounts of titanium and copper sulfide for collection of traces of zinc and lead. Many other collectors are described by Sandell and Onishi.[1]

A collector may entrain a trace constituent as a result of similarities in their solubilities. Others involve coprecipitation, in which the minor component is adsorbed on or incorporated into the collector precipitate as the result of mixed-crystal formation. Clearly, the collector must not interfere with the method selected for determining the trace component.

> A collector is used to remove trace constituents from solution.

### 32B–6 Separation by Electrolytic Precipitation

Electrolytic precipitation constitutes a highly useful method for accomplishing separations. In this process, the more easily reduced species, be it the wanted or the unwanted component of the mixture, is isolated as a separate phase. The method becomes particularly effective when the potential of the working electrode is controlled at a predetermined level (Section 18C–2).

The mercury cathode (page 454) has found wide application in the removal of many metal ions prior to the analysis of the residual solution. In general, metals more easily reduced than zinc are conveniently deposited in the mercury, leaving such ions as aluminum, beryllium, the alkaline earths, and the alkali metals in solution. The potential required to decrease the concentration of a metal ion to any desired level is readily calculated from polarographic data.

## 32C    EXTRACTION METHODS

The extent to which solutes, both inorganic and organic, distribute themselves between two immiscible solvents differs enormously from one species to another, and these differences have been exploited for decades to separate them. This section deals with applications of extractions to analytical separations.

### 32C–1 Theory

Two terms are employed to describe the distribution of a solute between two immiscible solvents: *distribution coefficient* and *distribution ratio*. It is important to have a clear understanding of the distinction between the two.

#### The Distribution Coefficient

The distribution coefficient is an equilibrium constant that describes the distribution of a solute between two immiscible solvents. For example, when an aqueous solution of an organic solute A is shaken with an organic

> Be sure you understand the difference between the distribution coefficient and the distribution ratio.

---

[1]E. B. Sandell and H. Onishi, *Colorimetric Determination of Traces of Metals,* 4th ed., pp. 709–721. New York: Interscience, 1978.

solvent, such as hexane, there is quickly established an equilibrium that is described by the equation

$$A(aq) \rightleftharpoons A(org) \tag{32-1}$$

where (aq) and (org) refer to the aqueous and organic phases. Ideally, the ratio of the activities of A in the two phases is constant and independent of the total quantity of A. That is, at any given temperature

$$K_d = \frac{[A(org)]}{[A(aq)]} \tag{32-2}$$

where the equilibrium constant $K_d$ is the distribution coefficient. The terms in brackets are strictly the activities of A in the two solvents, but molar concentrations can frequently be substituted without serious error. Often, $K_d$ is approximately equal to the ratio of the solubility of A in the two solvents.

When the solute exists in different states of aggregation in the two solvents, the equilibrium becomes

$$xA_y(aq) \rightleftharpoons yA_x(org)$$

and the distribution coefficient takes the form

$$K_d = \frac{[A_x(org)]^y}{[A_y(aq)]^x}$$

## The Distribution Ratio

The distribution ratio $D$ for an analyte is defined as the ratio of its *analytical* concentration in two immiscible solvents. For a simple system, such as that described by Equation 32-1, the distribution ratio is identical to the distribution coefficient. For more complex systems, however, the two can be quite different. For example, for the distribution of a fatty acid HA between water and diethyl ether, we can write

$$D = \frac{c_{org}}{c_{aq}} \tag{32-3}$$

where $c_{org}$ and $c_{aq}$ are the molar *analytical* concentrations of HA in the two phases. In the aqueous medium, the analytical concentration of the acid is equal to the sum of the equilibrium concentrations of the weak acid and its conjugate base:

$$c_{aq} = [HA(aq)] + [A^-(aq)]$$

In contrast, no significant dissociation of the acid occurs in the nonpolar organic layer so that the analytical and equilibrium concentrations of HA are identical, and we can write

$$c_{org} = [HA(org)]$$

Substituting the last two relationships into Equation 32–3 gives

$$D = \frac{[HA(org)]}{[HA(aq)] + [A^-(aq)]}$$

In order to relate $D$ to $K_d$ for the *species* HA, we substitute the dissociation-constant expression for HA into this equation:

$$D = \frac{[HA(org)]}{[HA(aq)] + [HA(aq)]\, K_a/[H_3O^+(aq)]}$$

where $K_a$ is the acid dissociation constant for HA. Factoring out [HA(aq)] leads to

$$D = \frac{[HA(org)]}{[HA(aq)]} \times \frac{1}{1 + K_a/[H_3O^+]}$$

Substituting $K_d$ for the ratio of the equilibrium HA concentrations and rearranging gives

$$D = \frac{K_d[H_3O^+]}{K_a + [H_3O^+]} = \frac{c_{org}}{c_{aq}} \qquad (32\text{–}4)$$

This equation can be used to compute the extent of extraction of HA from buffered aqueous solutions. (Note the difference between the distribution coefficient $K_d$ and the distribution ratio $D$.)

---

Example 32–1

The distribution coefficient for a weak acid between diethyl ether and water is found to be 800, and its acid dissociation constant in water is $1.50 \times 10^{-5}$. Calculate the analytical concentration of HA remaining in an aqueous solution after the extraction of 50.0 mL of 0.0500 M HA with 25.0 mL of ether, assuming the aqueous solution is buffered to a pH of (a) 2.00 and (b) 8.00.

(a) Substituting $[H_3O^+] = 1.00 \times 10^{-2}$ and the two equilibrium constants into Equation 32–4 yields

$$D = \frac{c_{org}}{c_{aq}} = \frac{800 \times 1.00 \times 10^{-2}}{1.50 \times 10^{-5} + 1.00 \times 10^{-2}} = 799 \qquad (32\text{–}5)$$

The total number of millimoles of the acid contained in the two solvents is equal to the original number of millimoles in the aqueous solution:

$$\text{total no. mmol HA} = 50.0 \times 0.0500 = 2.50$$

After extraction, the 2.50 mmol of HA is distributed between the two solvents so that

$$50.0 c_{aq} + 25.0 c_{org} = 2.50$$

where $c_{aq}$ and $c_{org}$ are the analytical concentrations in the two solvents. Rearranging Equation 32–5 shows that

$$c_{org} = 799c_{aq}$$

and so we can write

$$50.0c_{aq} + (25.0)(799c_{aq}) = 2.50$$
$$c_{aq} = 1.25 \times 10^{-4} \text{ M}$$

(b) When $[H_3O^+] = 1.00 \times 10^{-8}$,

$$D = \frac{800 \times 1.00 \times 10^{-8}}{1.50 \times 10^{-5} + 1.00 \times 10^{-8}} = 0.533$$

Proceeding as in part (a), we have

$$50.0 \times c_{aq} + (25.0)(0.533c_{aq}) = 2.50$$
$$c_{aq} = 3.95 \times 10^{-2} \text{ M}$$

## The Completeness of Multiple Extractions

Distribution coefficients and distribution ratios are useful because they provide guidance as to the most efficient way to perform extractive separations. For example, consider again the extraction of a species HA from an aqueous solution with a pH of 2.00 where Equation 32–5 applies. Suppose that $V_{aq}$ milliliters of water containing $a_0$ millimoles of HA is extracted with $V_{org}$ milliliters of diethyl ether. At equilibrium, $a_1$ millimoles of HA remain in the aqueous layer and $a_0 - a_1$ millimoles have been transferred to the organic layer. The analytical concentration of HA in each layer is

$$c_{aq1} = \frac{a_1}{V_{aq}}$$

$$c_{org1} = \frac{a_0 - a_1}{V_{org}}$$

Substitution of these quantities into Equation 32–5 and rearrangement gives

$$a_1 = \left(\frac{V_{aq}}{V_{org}D + V_{aq}}\right) a_0 \qquad (32\text{–}6)$$

The number of millimoles $a_2$ remaining in the aqueous layer after a second extraction with an identical volume of solvent is, by the same reasoning,

$$a_2 = \left(\frac{V_{aq}}{V_{org}D + V_{aq}}\right) a_1$$

When this expression is multiplied by Equation 32–6, $a_1$ cancels and we obtain

$$a_2 = \left(\frac{V_{aq}}{V_{org}D + V_{aq}}\right)^2 a_0$$

After $n$ extractions, the number of millimoles of HA remaining in the aqueous layer is

$$a_n = \left(\frac{V_{aq}}{V_{org}D + V_{aq}}\right)^n a_0 \qquad (32–7)$$

Equation 32–7 can be rewritten in terms of the initial and final analytical concentrations of HA in the water by substituting the relationships

$$a_n = (c_{aq})_n V_{aq} \quad \text{and} \quad a_0 = (c_{aq})_0 V_{aq}$$

where $(c_{aq})_n$ is the analytical concentration of HA in the aqueous phase after $n$ extractions. Substitution of these relationships into Equation 32–7 gives

$$(c_{aq})_n = \left(\frac{V_{aq}}{V_{org}D + V_{aq}}\right)^n (c_{aq})_0 \qquad (32–8)$$

As shown in the example that follows, a more efficient extraction is achieved with several small volumes of solvent than a single large one.

---

Example 32–2

The distribution ratio of $I_2$ between $CCl_4$ and $H_2O$ is 85. Calculate the concentration of $I_2$ remaining after 50.0 mL of an aqueous $1.00 \times 10^{-3}$ M solution of $I_2$ has been extracted with (a) one 50.0-mL portion of $CCl_4$, (b) two 25.0-mL portions, and (c) five 10.0-mL portions.
(a) Substituting into Equation 32–8 gives

It is always better to use several small portions of solvent to extract a sample than to extract with one large portion.

$$(c_{aq})_1 = \left(\frac{50.0}{50.0 \times 85 + 50.0}\right)^1 \times 1.00 \times 10^{-3} = 1.16 \times 10^{-5} \text{ M}$$

(b) $$(c_{aq})_2 = \left(\frac{50.0}{25.0 \times 85 + 50.0}\right)^2 \times 1.00 \times 10^{-3} = 5.28 \times 10^{-7} \text{ M}$$

(c) $$(c_{aq})_5 = \left(\frac{50.0}{10.0 \times 85 + 50.0}\right)^5 \times 1.00 \times 10^{-3} = 5.29 \times 10^{-10} \text{ M}$$

---

Figure 32–1 demonstrates that the improved efficiency brought about by multiple extractions falls off rapidly as the number of extractions increases. Clearly, little is gained by dividing the extracting solvent into more than five or six portions.

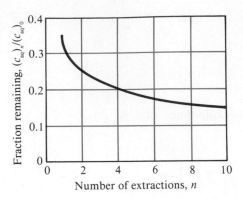

Figure 32–1

Plot of Equation 32–8, assuming $K_d = 2$ and $V_{aq} = 100$. The total volume of the organic solvent was also assumed to be 100, so that $V_{org} = 100/n$.

Exhaustive extractions can be performed with a Soxhlet extractor.

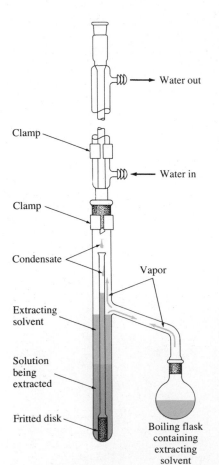

### 32C–2 Types of Extraction Procedures

Three types of separation procedures are based upon distribution equilibria between immiscible solvents: *simple, exhaustive, and countercurrent extractions.*

#### Simple Extractions

When the distribution ratio for one species in a mixture is reasonably favorable (on the order of 5 to 10 or greater) and that for the others is unfavorable (<0.001), an extractive separation can be simple, rapid, and quantitative. The solution containing the analyte is extracted successively with up to six portions of fresh solvent. An ordinary separatory funnel is used with either the original solution or the extract being retained for completing the analysis.

#### Exhaustive Extractions

Exhaustive extraction permits the separation of components of a mixture that have relatively unfavorable distribution ratios (<1) from those having ratios that approach zero. An apparatus is used in which the organic solvent is automatically distilled, condensed, and caused to pass continuously through the aqueous layer. Thus, the equivalent of several hundred extractions with fresh solvent is accomplished in 1 hr or less with equipment that requires no attention.

#### Countercurrent Fractionation

Automated devices that permit hundreds of automatic successive extractions have been developed. With these instruments, fractionation occurs by a *countercurrent* scheme in which distribution between *fresh* portions of the two phases occurs in a series of discrete steps. An exhaustive extraction differs from the countercurrent technique in that fresh portions of only one phase are introduced in the former.

The countercurrent method permits the separation of components with nearly identical partition ratios. For example, Craig[2] has demonstrated

[2]L. C. Craig, *Anal. Chem.,* **1950,** *22,* 1346.

that ten amino acids can be separated by countercurrent extraction even though their partition coefficients differ by less than 0.1.

## 32D    APPLICATIONS OF EXTRACTION PROCEDURES

Extraction is often more attractive than a classic precipitation for separating inorganic species because the equilibration and separation of phases in a separatory funnel are less tedious and time-consuming than precipitation, filtration, and washing. In addition, the problems of coprecipitation and postprecipitation are avoided. Finally, extraction procedures are ideally suited for the isolation of trace quantities of a species.

### 32D-1  The Extractive Separation of Metal Ions as Chelates

Many organic chelating agents are weak acids that react with metal ions to give uncharged complexes that are highly soluble in such organic solvents as ethers, hydrocarbons, ketones, and chlorinated species (including chloroform and carbon tetrachloride). Distribution ratios for such reagents vary widely among cations and can be controlled by changes in pH and reagent concentration, thus making possible many useful extractive separations.

After being chelated, many cations can be extracted from aqueous media into organic solvents.

#### The Effect of pH and Reagent Concentration on Distribution Ratios

Several chelating agents have proved useful for separations based upon the selective extraction of metal ions from a buffered aqueous solution into a nonaqueous solvent containing these agents. As shown in Figure 32–2, such a process involves several equilibria and several species. Among the latter are the undissociated ligand HL, its conjugate base $L^-$, the metal-ligand complex $ML_n$, and metal and hydronium ions. The important equilibria are

$$HL(aq) \rightleftharpoons HL(org) \qquad\qquad K_{d1} = \frac{[HL(org)]}{[HL(aq)]} \qquad (32\text{–}9)$$

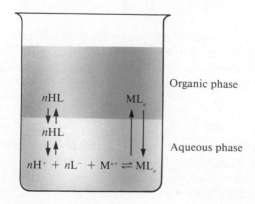

Organic phase

Aqueous phase

Figure 32–2

Equilibria in the extraction of the aqueous cation $M^{n+}$ into an immiscible organic phase containing the organic chelating agent HL.

$$HL(aq) + H_2O \rightleftharpoons H_3O^+(aq) + L^-(aq) \qquad K_a = \frac{[H_3O^+(aq)][L^-(aq)]}{[HL(aq)]}$$

$$(32\text{--}10)$$

$$M^{n+}(aq) + nL^-(aq) \rightleftharpoons ML_n(aq) \qquad K_f = \frac{[ML_n(aq)]}{[M^{n+}(aq)][L^-(aq)]^n}$$

$$(32\text{--}11)$$

$$ML_n(aq) \rightleftharpoons ML_n(org) \qquad K_{d2} = \frac{[ML_n(org)]}{[ML_n(aq)]} \qquad (32\text{--}12)$$

Organic chelating agents as well as neutral metal chelates are usually highly soluble in organic liquids so that the distribution coefficients $K_{d1}$ and $K_{d2}$ are generally large numerically. Furthermore, the concentration of $M^{n+}$ in the nonpolar organic layer ordinarily approaches zero in most cases. The selectivity of the reagent is determined by the relative magnitudes of the formation constants $K_f$ for various cations. As shown by Equation 32–10, the concentration of the active reagent $L^-$ is pH-dependent. Thus, by controlling pH, one can control the concentration of $L^-$ and thus which cations are extracted and which are not.

In order to derive an expression relating the amount of a cation extracted to pH and concentration of chelating agent, we employ the distribution ratio, which for the system shown in Figure 32–2 takes the form

$$D = \frac{c_{org}}{c_{aq}} = \frac{[ML_n(org)]}{[M^{n+}(aq)] + [ML_n(aq)]} \simeq \frac{[ML_n(org)]}{[M^{n+}(aq)]} \qquad (32\text{--}13)$$

where $c_{org}$ and $c_{aq}$ are the molar analytical concentrations of $M^{n+}$ in the organic and aqueous phases. Ordinarily, the assumption that $[ML_n(aq)]$ $\ll [M^{n+}(aq)]$ is valid because (1) the chelate is generally not very soluble in water and (2) that which is in solution is largely dissociated. As shown by the following derivation, $D$ is independent of the total amount of metal in the two phases but dependent upon both the concentration of HL in the organic layer and the hydronium ion concentration in the aqueous solution.

If $c_L$ is the original molar concentration of HL in the organic phase, mass balance requires that

$$c_L = [HL(org)] + [HL(aq)] + [L^-(aq)] + n[ML_n(aq)] + n[ML_n(org)]$$

Ordinarily, extractions are carried out with such a large excess of chelating agent that the concentration of the *species* HL in the organic layer far exceeds the concentration of all other species containing L. Thus, the foregoing mass-balance expression simplifies to

$$c_L \simeq [HL(org)] \qquad (32\text{--}14)$$

In order to arrive at an expression that relates $D$ for this system to the original concentration of the chelating agent in the organic solution and to the pH of the aqueous solution, let us multiply Equation 32–11 by 32–12 and rearrange, which leads to

$$[ML_n(org)] = K_f K_{d2} [M^{n+}(aq)][L^-(aq)]^n$$

Substituting into Equation 32–13 gives

$$D = \frac{c_{org}}{c_{aq}} = K_f K_{d2}[L^-(aq)]^n \qquad (32\text{–}15)$$

Dividing Equation 32–10 by 32–9 allows us to express $[L^-(aq)]$ in terms of $[H_3O^+]$ and $[HL(org)]$:

$$[L^-(aq)] = \frac{K_a}{K_{d1}} \frac{[HL(org)]}{[H_3O^+(aq)]}$$

Substitution of this equation and Equation 32–14 into Equation 32–15 leads to the desired relationship:

$$D = \frac{c_{org}}{c_{aq}} = \frac{K_f K_{d2} K_a^n}{K_{d1}^n} \times \frac{c_L^n}{[H_3O^+(aq)]^n} \qquad (32\text{–}16)$$

Combining the four equilibrium constants into a single constant $K_{ex}$ yields

$$D = \frac{c_{org}}{c_{aq}} = \frac{K_{ex} c_L^n}{[H_3O^+(aq)]^n} \qquad (32\text{–}17)$$

---

Example 32–3

Lead forms a neutral complex $PbL_2$ with the ligand $L^-$. The constant $K_{ex}$ for the distribution of this complex between water and $CCl_4$ has been found by experiment to be $2.0 \times 10^4$. A 25.0-mL aliquot of an aqueous solution that is $5.00 \times 10^{-4}$ M in $Pb^{2+}$ and 0.500 M in $HClO_4$ is extracted with two 10.0-mL portions of $CCl_4$ that are 0.0250 M in HL. Calculate the percentage of unrecovered $Pb^{2+}$ in the aqueous solution.

Substituting into Equation 32–17 gives

$$D = \frac{(2.0 \times 10^4)(0.0250)^2}{(0.500)^2} = 50.0$$

Substituting into Equation 32–8 for two extractions gives

$$(c_{aq})_2 = \left( \frac{25.0}{10.0 \times 50.0 + 25.0} \right)^2 (5.00 \times 10^{-4}) = 1.13 \times 10^{-6}$$

$$\% \text{ unextracted } Pb^{2+} = \frac{1.13 \times 10^{-6}}{5.0 \times 10^{-4}} \times 100\% = 0.23\%$$

---

Some Separations Based on the Extraction of Metal Chelates

**Extractions with Diphenylthiocarbazone.** Diphenylthiocarbazone, or dithizone, is a useful reagent for separating minute quantities of a dozen or more metal ions. Its reaction with a divalent cation such as $Pb^{2+}$, can be written as

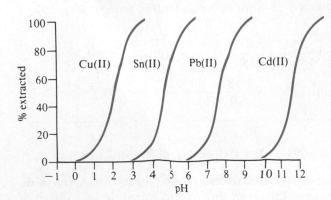

Both dithizone and its metal chelates are essentially insoluble in water but dissolve readily in solvents such as carbon tetrachloride and chloroform. Solutions of the reagent are deep green, whereas solutions of the metal chelates are intensely red, violet, orange, or yellow and provide a sensitive means for the photometric determination of the separated ions.

The distribution ratio of the various metal dithizonates is described by Equations 32–17 and 32–13. Thus, for lead,

$$D = \frac{c_{org}}{c_{aq}} = \frac{[PbDz_2(org)]}{[Pb^{2+}(aq)]} = \frac{K_{ex}c_{Dz}^2}{[H_3O^+(aq)]^2}$$

Figure 32–3 shows how the percentages of lead and other metal ions extracted into a chloroform solution of dithizone change as the pH of the aqueous medium changes. The data for these curves were derived in a way analogous to the calculation shown in Example 32–3. It is apparent from these curves that clean separations of copper, lead, and cadmium can be achieved by pH control. Depending upon its absorption spectrum, tin(II) may interfere slightly in the determination of copper and lead.

**Extractions with 8-Hydroxyquinoline.** Many of the chelates of 8-hydroxyquinoline (Section 3D–3) are readily extracted into various organic solvents. The extraction process can be formulated as

$$2(HQ)_{org} + (M^{2+})_{aq} \rightleftharpoons (MQ_2)_{org} + 2(H^+)_{aq}$$

where HQ symbolizes the chelating agent. The equilibrium is clearly pH-

**Figure 32–3**

Effect of pH on the extraction of cations with $CCl_4$ solutions of dithizone.

Table  32–3
DIETHYL ETHER EXTRACTION OF VARIOUS CHLORIDES
FROM 6 M HYDROCHLORIC ACID*

| Percent Extracted | Elements and Oxidation State |
|---|---|
| 90–100 | Fe(III), 99%; Sb(V), 99%**; Ga(III), 97%; Ti(III), 95%**; Au(III), 95% |
| 50–90 | Mo(VI), 80–90%; As(III), 80%**†; Ge(IV), 40–60% |
| 1–50 | Te(IV), 34%; Sn(II), 15–30%; Sn(IV), 17%; Ir(IV), 5%; Sb(III), 2.5%** |
| <1 >0 | As(V),** Cu(II), In(III), Hg(II), Pt(IV), Se(IV), V(V), V(IV), Zn(II) |
| 0 | Al(III), Bi(III), Cd(II), Cr(III), Co(II), Be(II), Fe(II), Pb(II), Mn(II), Ni(II), Os(VIII), Pd(II), Rh(III), Ag(I), Th(IV), Ti(IV), W(VI), Zr(IV) |

*From E. H. Swift, *Introductory Quantitative Analysis*, p. 431. Englewood Cliffs, NJ: Prentice-Hall, 1950. With permission.

**Isopropyl ether employed rather than diethyl ether.

†8 M HCl rather than 6 M.

dependent, which permits the separation of metals through control of the aqueous-phase pH. The method has proved particularly useful for the separation of trace amounts of metals.

**Extractions with Other Chelating Agents.** Separations that make use of other organic chelating agents are described in several reference works.[3]

## 32D–2  The Extraction of Metal Chlorides

The data in Table 32–3 indicate that a substantial number of metal chlorides can be extracted into diethyl ether from 6 M hydrochloric acid solution; equally important, a large number of metal ions are either unaffected or extracted only slightly under these conditions. Thus, many useful separations are possible. One of the most important is the separation of iron(III) (99% extracted) from a host of other cations. For example, the greater part of the iron from steel or iron ore samples can be removed by extraction prior to analysis for such trace elements as chromium, aluminum, titanium, and nickel. The species extracted has been shown to be the ion pair $H_3O^+FeCl_4^-$. It has also been shown that the percentage of iron transferred to the organic phase depends upon the hydrochloric acid content of the aqueous phase (little is removed from solutions that are below 3 M and above 9 M HCl) and, to some extent, upon the iron content. Unless special precautions are taken, extraction of the last traces of iron is incomplete.

---

[3]G. H. Morrison and H. Freiser, *Solvent Extraction in Analytical Chemistry*. New York: Wiley, 1957; A. K. De, S. M. Khophar, and R. A. Chalmers, *Solvent Extraction of Metals*. New York: Van Nostrand, 1970; E. B. Sandell and H. Onishi, *Colorimetric Determination of Traces of Metals*, 4th ed. New York: Interscience, 1978.

Methyl isobutyl ketone also extracts iron from hydrochloric acid solutions. It is reported to have a somewhat more favorable distribution ratio than diethyl ether and has the added advantage of being less flammable.

### 32D–3  The Extraction of Nitrates

Certain nitrate salts are selectively extracted by diethyl ether as well as other organic solvents. For example, uranium(VI) is conveniently separated from such elements as lead and thorium by ether extraction of an aqueous solution that is saturated with ammonium nitrate and has a nitric acid concentration of about 1.5 M. Bismuth and iron(III) nitrates are also extracted to some extent under these conditions.

## 32E    ION-EXCHANGE SEPARATIONS

Ion-exchange resins have several applications in analytical chemistry. The most important is in high-performance liquid chromatography, described in Section 28D. A brief description of other analytical applications of these useful materials follows.

### 32E–1  The Separation of Interfering Ions
of Opposite Charge

Ion exchange is particularly effective where the charge on the analyte is opposite that of the interfering ion(s).

Ion-exchange resins are useful for the removal of interfering ions, particularly where these ions have a charge opposite that of the analyte. For example, iron(III), aluminum(III), and other cations interfere in the gravimetric determination of sulfate by virtue of their tendency to coprecipitate with barium sulfate. Passage of a solution to be analyzed through a column containing a cation-exchange resin results in the retention of all the cations and the liberation of a corresponding number of protons. The sulfate ion passes freely through such a column and the analysis can then be performed on the effluent. In a similar manner, phosphate ion, which interferes in the determination of barium and calcium ions, can be removed by passing the sample through an anion-exchange resin.

### 32E–2  The Concentration of Traces of an Electrolyte

A useful application of ion exchangers is the concentration of traces of an ion from a very dilute solution. Cation-exchange resins, for example, have been employed to collect traces of metallic elements from large volumes of natural waters. The ions are then liberated by treating the resin with acid; the result is a considerably more concentrated solution for analysis.

### 32E–3  The Conversion of Salts to Acids or Bases

The total salt content of a sample can be determined by titrating the hydrogen ions released as an aliquot of sample is washed through a cation

**Table 32–4**
**SEPARATION OF SOME INORGANIC SPECIES BY DISTILLATION**

| Analyte | Sample Treatment | Volatile Species | Method of Collection |
|---|---|---|---|
| $CO_3^{2-}$ | Acidification | $CO_2$ | $Ba(OH)_2(aq) + CO_2(g) \rightarrow$ $BaCO_3(s) + H_2O$ or on Ascarite |
| $SO_3^{2-}$ | Acidification | $SO_2$ | $SO_2(g) + H_2O_2(aq) \rightarrow$ $H_2SO_4(aq)$ |
| $S^{2-}$ | Acidification | $H_2S$ | $Cd^{2+}(aq) + H_2S(g) \rightarrow$ $CdS(s) + 2 H^+$ |
| $F^-$ | Addition of $SiO_2$ and acidification | $H_2SiF_6$ | Basic solution |
| Si | Addition of HF | $SiF_4$ | Basic solution |
| $H_3BO_3$ | Addition of $H_2SO_4$ and methanol | $B(OCH_3)_3$ | Basic solution |
| $Cr_2O_7^{2-}$ | Addition of concd HCl | $CrO_2Cl_2$ | Basic solution |
| $NH_4^+$ | Addition of NaOH | $NH_3$ | Acidic solution |
| As, Sb | Addition of concd HCl and $H_2SO_4$ | $AsCl_3$ $SbCl_3$ | Water |
| Sn | Addition of HBr | $SnBr_4$ | Water |

exchanger in its acidic form. Similarly, a standard hydrochloric acid solution can be prepared by passing a solution containing a known weight of sodium chloride through a cation-exchange resin in its acid form. The effluent and washings are collected in a volumetric flask and diluted to volume. In an analogous way, standard sodium hydroxide solutions can be prepared by treatment of an anion-exchange resin with a known quantity of sodium chloride.

## 32F   THE SEPARATION OF INORGANIC SPECIES BY DISTILLATION

Distillation permits the separation of components whose solution/vapor-phase distribution ratios differ significantly from one another. If one species has a distribution ratio that is large compared with those of the other components of the mixture, the separation is simple. Table 32–4 lists some inorganic species that can be conveniently separated by simple distillation.

## 32G   QUESTIONS AND PROBLEMS

32–1. What is a masking agent and how does it function?

32–2. Differentiate between
(a) an exhaustive extraction and a countercurrent extraction.
(b) a distribuion coefficient and a distribution ratio.

*32–3. What is a collector and when is it used?

32–4. Suggest a precipitation method for separating
*(a) Fe(III) from Al(III).
(b) Fe(III) from Cu(II).
*(c) Fe(III) from Sn(IV).
(d) As(III) from Cd(II).
*(e) Hg(II) from Mn(II), Zn(II), and Fe(III).

*32–5. The distribution coefficient for X between chloroform and water is 9.6. Calculate the concentration of X remaining in the aqueous phase after 50.0 mL of 0.150 M X is treated by extraction with the following quantities of chloroform: (a) one 40.0-mL portion, (b) two 20.0-mL portions, (c) four 10.0-mL portions, and (d) eight 5.00-mL portions.

**32–6.** The distribution coefficient for Z between $n$-hexane and water is 6.25. Calculate the percent of Z remaining in 25.0 mL of water that was originally 0.0600 M in Z after extraction with the following volumes of $n$-hexane: (a) one 25.0-mL portion, (b) two 12.5-mL portions, (c) five 5.00-mL portions, and (d) ten 2.50-mL portions.

**\*32–7.** What volume of $CHCl_3$ is required to decrease the concentration of X in Problem 32–5 to $1.00 \times 10^{-4}$ M if 25.0 mL of 0.0500 M X is extracted with (a) 25.0-mL portions of $CHCl_3$, (b) 10.0-mL portions of $CHCl_3$, and (c) 2.0-mL portions of $CHCl_3$?

**32–8.** What volume of $n$-hexane is required to decrease the concentration of Z in Problem 32–6 to $1.00 \times 10^{-5}$ M if 40.0 mL of 0.0200 M Z is extracted with (a) 50.0-mL portions of $n$-hexane, (b) 25.0-mL portions, and (c) 10.0-mL portions?

**\*32–9.** What is the minimum distribution coefficient that permits removal of 99% of a solute from 50.0 mL of water with (a) two 25.0-mL extractions with benzene and (b) five 10.0-mL extractions with benzene?

**32–10.** If 30.0 mL of water that is 0.0500 M in Q is to be extracted with four 10.0-mL portions of an immiscible organic solvent, what is the minimum distribution coefficient that allows transfer of all but the following percentages of the solute to the organic layer: \*(a) $1.00 \times 10^{-4}$, (b) $1.00 \times 10^{-2}$, (c) $1.00 \times 10^{-3}$?

**\*32–11.** A 0.150 M aqueous solution of the weak organic acid HA was prepared from the pure compound, and three 50.0-mL aliquots were transferred to 100-mL volumetric flasks. Solution 1 was diluted to 100 mL with 1.0 M $HClO_4$, solution 2 was diluted to the mark with 1.0 M NaOH, and solution 3 was diluted to the mark with water. A 25.0-mL aliquot of each was extracted with 25.0-mL of $n$-hexane. The extract from solution 2 contained no detectable trace of A-containing species, indicating that $A^-$ is not soluble in the organic solvent. The extract from solution 1 contained no $ClO_4^-$ or $HClO_4$ but was found to be 0.0454 M in HA (by extraction with standard NaOH and back-titration with standard HCl). The extract from solution 3 was found to be 0.0225 M in HA. Assume that HA does not associate or dissociate in the organic solvent, and calculate
(a) the distribution ratio for HA between the two solvents.
(b) the concentration of the *species* HA and $A^-$ in aqueous solution 3 after extraction.
(c) the dissociation constant of HA in water.

**32–12.** To determine the equilibrium constant for the reaction

$$I_2 + 2\,SCN^- \rightleftharpoons I(SCN)_2^- + I^-$$

25.0 mL of a 0.0100 M aqueous solution of $I_2$ was extracted with 10.0 mL of $CCl_4$. After extraction, spectrophotometric measurements revealed that the $I_2$ concentration *of the aqueous layer* was $1.12 \times 10^{-4}$ M. An aqueous solution that was 0.0100 M in $I_2$ and 0.100 M in KSCN was then prepared. After extraction of 25.0 mL of this solution with 10.0 mL of $CCl_4$, the concentration of $I_2$ *in the $CCl_4$ layer* was found from spectrophotometric measurement to be $1.02 \times 10^{-3}$ M.
(a) What is the distribution coefficient for $I_2$ between $CCl_4$ and $H_2O$?
(b) What is the formation constant for $I(SCN)_2^-$?

**\*32–13.** The total cation content of natural water is often determined by exchanging the cations for hydrogen ions on a strong-acid ion-exchange resin. A 25.0-mL sample of a natural water was diluted to 100 mL with distilled water, and 2.0 g of a cation-exchange resin was added. After stirring, the mixture was filtered and the solid remaining on the filter paper was washed with three 15.0-mL portions of water. The filtrate and washings required 15.3 mL of 0.0202 M NaOH to give a bromocresol green end point.
(a) Calculate the number of milliequivalents of cation present in exactly 1 L of sample. (Here, the equivalent weight of a cation is its formula weight divided by its charge.)
(b) Report the results in terms of milligrams of $CaCO_3$ per liter.

**32–14.** An organic acid was isolated and purified by recrystallization of its barium salt. To determine the equivalent weight of the acid, a 0.393-g sample of the salt was dissolved in about 100 mL of water. The solution was passed through a strong-acid ion-exchange resin, and the column was then washed with water; the eluate and washings were titrated with 18.1 mL of 0.1006 M NaOH to a phenolphthalein end point.
(a) Calculate the equivalent weight of the organic acid.
(b) A potentiometric titration curve of the solution resulting when a second sample was treated in the same way revealed two end points, one at pH 5 and the other at pH 9. What is the molecular weight of the acid?

**\*32–15.** Describe the preparation of exactly 2 L of 0.1500 M HCl from primary-standard-grade NaCl using a cation-exchange resin.

**32–16.** An aqueous solution containing $MgCl_2$ and HCl was analyzed by first titrating a 25.00-mL aliquot to a bromocresol green end point with 18.96 mL of 0.02762 M NaOH. A 10.00-mL aliquot was then diluted to 50.00 mL with distilled water and passed through a strong-acid ion-exchange resin. The eluate and washings required 36.54 mL of the NaOH solution to reach the same end point. Report the molar concentrations of HCl and $MgCl_2$ in the sample.

*32–17. Copper(II) reacts with the chelating agent HL to give a complex $CuL_2$ that is readily soluble in $CHCl_3$. A spectrophotometric study revealed that when a $1.00 \times 10^{-4}$ aqueous solution of copper(II) was extracted with $CHCl_3$ that was 0.0100 M in HL, the analytical concentrations of copper in the two phases were identical at pH 5.65.

(a) Write equations describing the equilibria in the system, assuming that dissociation of $CuL_2$ in the organic phase is negligible.

(b) Calculate $K_{ex}$.

(c) Calculate the distribution ratio for the system at pH 6.00.

(d) If 50.0 mL of $5.00 \times 10^{-5}$ M $Cu^{2+}$ in a pH 6.00 buffer were to be extracted with 25.0 mL portions of 0.0100 M HL in $CHCl_3$, how many extractions would be required to remove 99% of the copper from the aqueous phase?

(e) Repeat the calculations in part (d) for 99.9% removal.

32–18. Exactly 25.0 mL of a standard solution that was $2.24 \times 10^{-4}$ M in $Ag^+$ and buffered to pH 4.30 was extracted with 10.0 mL of a n-hexane solution that was 0.025 M in the chelating agent $H_2Z$ (the extracted complex has the formula $Ag_2Z$). After the phases were separated, the silver concentration of the aqueous phase was found to be $5.41 \times 10^{-6}$ M.

(a) Calculate the distribution ratio for the system.

(b) Calculate $K_{ex}$ for the system, assuming that the organic phase contained no uncomplexed $Ag^+$.

(c) How many 10.0-mL extractions would be required to extract 99.5% of the original $Ag^+$ at pH 4.30?

(d) Repeat the calculations in part (c) for a solution buffered to pH 3.00 and having a silver ion concentration of $5.15 \times 10^{-4}$ M.

# THE CHEMICALS, APPARATUS, AND UNIT OPERATIONS OF ANALYTICAL CHEMISTRY

$T$his chapter is concerned with the practical aspects of the unit operations encountered in an analytical laboratory as well as with the apparatus and chemicals used in these operations.

## 33A  THE SELECTION AND HANDLING OF REAGENTS AND OTHER CHEMICALS

The purity of reagents has an important bearing upon the accuracy attained in any analysis. It is therefore essential that the quality of a reagent be consistent with the use for which it is intended.

### 33A–1  The Classification of Commercial Chemicals

#### Reagent Grade

Reagent-grade chemicals conform to the minimum standards set forth by the Reagent Chemical Committee of the American Chemical Society[1] and are used wherever possible in analytical work. Some suppliers label their products with the maximum limits of impurity allowed by the ACS specifications, whereas others print actual assays for the various impurities.

#### Primary-Standard Grade

The qualities required of a *primary standard*—in addition to extraordinary purity—are set forth in Section 5A–3. Primary-standard reagents, which are available from commercial sources, have been carefully analyzed, and the assay is printed on the container label. The National Insti-

The National Institute of Standards and Technology (NIST) is the current name of what was formerly the National Bureau of Standards (NBS).

[1]Committee on Analytical Reagents, *Reagent Chemicals,* 7th ed. Washington, D.C.: American Chemical Society, 1986.

tute of Standards and Technology is an excellent source for primary standards. This agency also provides *reference standards,* which are complex substances that have been exhaustively analyzed.[2]

### Special-Purpose Reagent Chemicals

Chemicals that have been prepared for a specific application are also available. Included among these are solvents for spectrophotometry and high-performance liquid chromatography and reagents for nonaqueous spectroscopy and electron microscopy. Information pertinent to the intended use is supplied with these reagents. Data provided with a spectrophotometric solvent, for example, might include its absorbance at selected wavelengths and its ultraviolet cutoff wavelength as well as its assay.

### 33A–2 Rules for Handling Reagents and Solutions

High quality in a chemical analysis requires the availability of reagents and solutions with established purity. A freshly opened bottle of a reagent-grade chemical can ordinarily be used with confidence; whether this same confidence is justified when the bottle is half empty depends entirely on the way it has been handled after being opened. The following rules should be observed to prevent the accidental contamination of reagents and solutions.

1. Select the best grade of chemical available for analytical work. If a choice exists, pick the smallest bottle that will supply the desired quantity.
2. Replace the top of every container *immediately* after the removal of the reagent; do not rely on someone else to do this.
3. Hold the stoppers of reagent bottles between your fingers; never set a stopper on a desk top.
4. *Unless specifically directed to the contrary, never return any excess reagent to a bottle.* The minor financial saving that might accompany the return of an excess is overshadowed by the risk of contaminating the entire bottle.
5. Unless directed otherwise, never insert spatulas, spoons, or knives into a bottle that contains a solid chemical. Instead, shake the capped bottle vigorously or tap it gently against a wooden table to break up any encrustation; then pour out the desired quantity. These measures are occasionally ineffective, and in such cases a clean porcelain spoon should be used.
6. Keep the reagent shelf and the laboratory balance clean and neat. Clean up any spillages immediately, even though someone else is waiting to use the same chemical or reagent.
7. Dispose of solid chemicals and solutions in accordance with local regulations.

---

[2]United States Department of Commerce, *NIST Standard Reference Materials Catalog, 1991–92, NIST Special Publication 260.* Washington, D.C. 20234: Government Printing Office, 1991.

## 33B    THE CLEANING AND MARKING OF LABORATORY WARE

Because a chemical analysis is ordinarily performed in duplicate or triplicate, each vessel that holds a sample must be marked so that its contents can always be identified. Flasks, beakers, and some crucibles have small etched areas on which semipermanent markings can be made with a pencil. Special marking inks are available for porcelain surfaces. The marking is baked permanently into the glaze by heating at a high temperature. A saturated solution of iron(III) chloride, while not as satisfactory as the commercial preparation, can also be used for marking.

Every beaker, flask, or crucible that will contain the sample must be thoroughly cleaned before being used. The apparatus should be washed with a hot detergent solution and then rinsed—initially with copious amounts of tap water and finally with several small portions of distilled or deionized water. A properly cleaned object will be coated with a uniform and unbroken film of water. *It is seldom necessary to dry the interior surface of glassware before use;* drying is ordinarily a waste of time at best and a potential source of contamination at worst.

> Unless you are directed otherwise, do not dry the interior surfaces of glassware or porcelain ware.

## 33C    THE EVAPORATION OF LIQUIDS

It is frequently necessary to decrease the volume of a solution without loss of a nonvolatile solute. Figure 33–1 illustrates how this operation is performed. The ribbed cover glass permits vapors to escape and protects the remaining solution from accidental contamination. Less satisfactory are glass hooks that provide space between the rim of the beaker and a conventional cover glass.

Evaporation is frequently difficult to control, owing to the tendency of some solutions to overheat locally. The bumping that results can be sufficiently vigorous to cause partial loss of the solution. The danger of such loss is minimized through careful and gentle heating. If their use is permissible, glass beads are also helpful in minimizing bumping.

Some unwanted species can be eliminated during evaporation. For example, chloride and nitrate can be removed from a solution by adding sulfuric acid and evaporating until copious white fumes of sulfur trioxide are observed (this operation must be performed in a hood). Nitrate ion and nitrogen oxides can be eliminated from acidic solutions by adding urea, evaporating to dryness, and gently igniting the residue. The removal of ammonium chloride is best accomplished by adding concentrated nitric acid and evaporating the solution to a small volume. Ammonium ion is rapidly oxidized upon heating; the solution is then evaporated to dryness.

Organic constituents can frequently be eliminated from a solution by adding sulfuric acid and heating to the appearance of sulfur trioxide fumes (hood); this process is one type of *wet-ashing*. Nitric acid can be added toward the end of heating to hasten the oxidation of the last traces of organic matter.

> Bumping is sudden, often violent boiling that tends to spatter solution out of its container.

> Wet-ashing is the oxidation of the organic constituents of a sample with oxidizing reagents such as nitric acid, sulfuric acid, hydrogen peroxide, aqueous bromine, or a combination of these reagents.

Figure 33–1

Arrangement for the evaporation of a liquid.

## 33D  THE MEASUREMENT OF MASS

Highly reliable mass data are ordinarily required at one stage or another in the course of a chemical analysis; an *analytical balance* is used to acquire these data. A less precise but more rugged *laboratory balance* suffices for weighings where the demands for reliability are not critical.

### 33D-1  The Distinction Between Mass and Weight

The reader should appreciate the difference between mass and weight. *Mass,* which is the more fundamental quantity, is an invariant measure of the amount of matter in an object. *Weight* is the force of attraction that exists between an object and its surroundings, principally the earth. Because gravitational attraction varies with geographical location, the weight of an object depends upon where it is weighed. For example, the weight of a crucible is less in Denver than in Atlantic City (both cities are at approximately the same latitude) because the attractive force between crucible and earth is smaller at the higher altitude. Similarly, the crucible weighs more in Seattle than in Panama (both cities are at sea level) because the earth is somewhat flattened at the poles, and the force of attraction increases measurably with latitude. The mass of the crucible remains constant regardless of the location at which it is measured.

> Mass is an invariant measure of the amount of matter in an object. Weight is the force of gravitational attraction between that object and the earth.

Weight and mass are related by the familiar expression

$$W = Mg$$

where $W$ is the weight of an object, $M$ is its mass, and $g$ is the acceleration due to gravity.

A chemical analysis is always based upon mass in order to free the results from a dependence upon locality. A balance is used to compare the weight of an object with the weight of a set of standard masses. Because $g$ affects both unknown and known identically, an equality between their weights indicates an equality in mass.

The distinction between mass and weight tends to be lost in common usage; that is, the operation of comparing masses is ordinarily referred to as *weighing,* and the objects of known mass as well as the results of weighing are called *weights*. It should always be borne in mind, however, that analytical data are based upon mass rather than weight.

### 33D-2  Types of Analytical Balances

By definition, an *analytical balance* is a weighing instrument that has a maximum capacity that ranges from 1 g to a few kilograms with a precision of at least 1 part in $10^5$ at maximum capacity. The precision and accuracy of many modern analytical balances exceed one part in $10^6$ at full capacity.

The most commonly encountered analytical balances have a maximum capacity ranging between 160 and 200 g; measurements can be made with

> An analytical balance has a maximum capacity that ranges from 1 g to several kilograms and a precision at maximum capacity of at least one part in $10^5$.

> The most common type of analytical balance, a macrobalance, has a maximum load of 160 to 200 g and a precision of 0.1 mg.

A semimicrobalance has a maximum load of 10 to 30 g and a precision of 0.01 mg.

A microbalance has a maximum load of 1 to 3 g and a precision of 0.001 mg, or 1 $\mu$g.

a standard deviation of $\pm0.1$ mg. *Semimicroanalytical balances* have a maximum loading of 10 to 30 g with a precision of $\pm0.01$ mg. Typical *microanalytical balances* have a capacity of 1 to 3 g and a precision of $\pm0.001$ mg.

The analytical balance has undergone a dramatic evolution over the past several decades. The traditional analytical balance had two pans attached to either end of a light-weight beam that pivoted about a knife edge located in the center of the beam. The object to be weighed was placed on one pan; sufficient standard weights were then added to the other pan to restore the beam to its original position. Weighing with such an *equal-arm* balance was tedious and time consuming.

The first *single-pan analytical balance* appeared on the market in 1946. The speed and convenience of weighing with this balance were vastly superior to what could be realized with the traditional equal-arm balance. Consequently, the single-pan balance rapidly replaced the latter in most laboratories. The single-pan balance is still extensively used. The design and operation of a single-pan balance are discussed in Section 33D–3.

The *hybrid electronic analytical balance* dates from the early 1970's. This balance retained the beam and knife edge of a single-pan balance but employed a solenoid in place of standard weights to restore the beam to its original position. The current in a solenoid is proportional to the electromagnetic force it creates and is thus proportional to the weight of the object on the opposite end of the beam as well.

The brief era of the hybrid balance ended with the appearance of the true *electronic analytical balance,* which has neither a beam nor a knife edge; this type of balance is discussed briefly in Section 33D–4. It seems likely that both mechanical and hybrid balances will be supplanted by electronic balances in the years to come.

### 33D–3 The Single-Pan Mechanical Analytical Balance

#### Components

To avoid damage to the knife edges and bearing surfaces, the arrest system for a mechanical balance should be engaged at all times other than during actual weighing.

Although they differ considerably in appearance and performance characteristics, all mechanical balances—equal-arm as well as single-pan—have several common components. Figure 33–2 is a diagram of a typical single-pan balance. Fundamental to this instrument is a light-weight *beam* that is supported on a planar surface by a prism-shaped knife edge labeled A in the figure. Attached to the left end of the beam is a pan for holding the object to be weighed and a full set of weights held in place by hangers. These weights can be lifted from the beam one at a time by a mechanical arrangement that is controlled by a set of knobs on the exterior of the balance case. The right end of the beam holds a counterweight of such a size as to just balance the pan and weights on the left end of the beam.

A second knife edge (B) is located near the left end of the beam and serves to support a second planar surface, which is located in the inner side of a stirrup that couples the pan to the beam. The two knife edges and their planar surfaces are fabricated from extraordinarily hard materials (agate or synthetic sapphire) and form two bearings that permit motion of

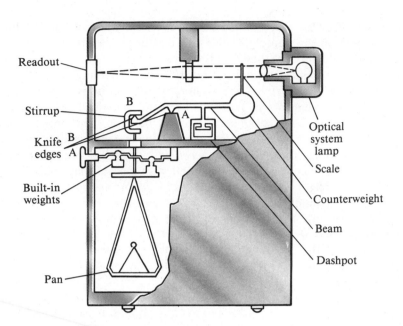

Readout

B

Stirrup

Knife B
edges A

A

Built-in
weights

Pan

Optical
system
lamp

Scale

Counterweight

Beam

Dashpot

**Figure 33–2**
Modern single-pan analytical balance. (From R. M. Schoonover, *Anal. Chem.*, **1982**, *54*, 973A. Published 1982 American Chemical Society.)

the beam and pan with a minimum of friction. The performance of a mechanical balance is critically dependent upon the mechanical perfection of these two bearings.

To protect the bearings from damage and wear when an object is placed on the pan or when the balance is not being used, balances are equipped with a *beam arrest* and a *pan arrest*. The beam arrest is a mechanical device that raises the beam so that the central knife edge no longer touches its bearing surface and simultaneously frees the stirrup from contact with the outer knife edge. The purpose of the arrest mechanism is to prevent damage to the bearings of the balance while objects are being placed upon or removed from the pan. When engaged, the pan arrest supports most of the weight of the pan and its contents and thus prevents oscillation. Both arrests are controlled by a lever mounted on the outside of the balance case and should be engaged whenever the balance is not in use.

An *air damper* (sometimes called a *dashpot*) is mounted near the end of the beam opposite the pan. This device consists of a piston that moves within a concentric cylinder attached to the balance case. Air in the cylinder undergoes expansion and contraction as the beam is set in motion; the beam rapidly comes to rest as a result of this opposition to motion.

Protection from air currents is needed to permit discrimination between small differences in weight (<1 mg). An analytical balance is thus always enclosed in a case equipped with doors to permit the introduction or removal of objects.

## Weighing with a Single-Pan Balance

The beam of a single-pan balance is free to oscillate about knife edge A when restraints are removed. With no object on the pan and all of the weights in place, its rest position will be essentially horizontal. Placement of an object on the pan causes the left end of the beam to be displaced downward. Weights are then removed systematically one by one from the beam until the imbalance is less than 100 mg. The angle of deflection of the beam with respect to its original horizontal position is directly proportional to the milligrams of additional weight that must be removed to restore the beam to its original horizontal position. The optical system shown in the upper part of Figure 33–2 measures this deflection and converts the angle to milligrams. A *reticle*, which is a small transparent screen mounted on the beam, is scribed with a scale that reads 0 to 100 mg. A beam of light passes through the scale to an enlarging lens, which in turn focuses a small part of the enlarged scale onto a frosted glass plate located on the front of the balance. A vernier makes it possible to read this scale to the nearest 0.1 mg.

### Stability

An important property of a balance is *stability,* which causes the beam of the balance to return to its rest position after being displaced by the momentary application of a slight force to one end. Stability requires the center of gravity of the beam (including the stirrup, pan, and any load) to be below the central knife edge so that the weight of the beam acts as a restoring force to offset the displacement.

### Instructions for the Use of a Single-Pan Balance

1. Zero the empty balance. Rotate the arrest control to the full release position. Adjust the vernier so that the scale reading is zero.
2. Arrest the balance again, and place the object to be weighed on the pan. Turn the arrest knob to the partially released position. Rotate the knob controlling the heaviest likely weight for the object until the illuminated scale changes or the instruction "remove weight" appears; then turn the weight knob back one stop. Repeat this procedure with all the other knobs, working systematically through the lighter weights.
3. Turn the arrest control to its fully released position and allow time for the balance to come to equilibrium. The weight of the object is the sum of the weights indicated on the dials and that displayed on the illuminated scale. Use the vernier to establish the weight to the nearest 0.1 mg.

**Note.** Directions for operation differ somewhat with make and model. Consult the instructor for any modifications to this procedure that may be required.

### Summary of Rules for the Use of a Single-Pan Analytical Balance

The single-pan mechanical analytical balance is a delicate instrument. Care must be taken to adhere to the accompanying rules.

### To Minimize Wear or to Avoid Damage

1. Be certain that the arresting mechanisms for the beam are engaged whenever the loading is being changed and whenever the balance is not in use.
2. Center the load on the pan insofar as possible.
3. Protect the balance from corrosion. Objects to be placed on the pan should be limited to nonreactive metals, nonreactive plastics, and vitreous materials.
4. Observe special precautions (Section 33E–6) for the weighing of liquids.
5. Consult with the instructor if the balance appears to need adjustment.
6. Keep the balance and its case scrupulously clean. A camel's-hair brush is useful for the removal of spilled material or dust.

### To Obtain Reliable Mass Data

7. Always allow an object that has been heated to return to room temperature before weighing it.
8. Use tongs or finger pads to prevent the uptake of moisture by dried objects.

Be sure to observe these precautions when you use a single-pan analytical balance.

## 33D–4  The Electronic Analytical Balance[3]

Figure 33–3 is a diagram of an electronic analytical balance. The pan rides above a hollow metal cylinder that is surrounded by a coil and fits over the inner pole of a cylindrical permanent magnet. An electric current in the coil creates a magnetic field that supports or levitates the cylinder, the pan

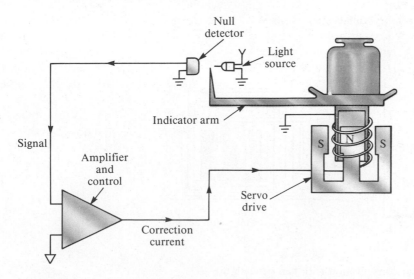

Figure 33–3
Electronic analytical balance. (From R. M. Schoonover, *Anal. Chem.*, **1982,** *54,* 974A. Published 1982 American Chemical Society.)

[3]For a more detailed discussion, see R. M. Schoonover, *Anal. Chem.*, **1982,** *54,* 973A; K. M. Lang, *Amer. Lab.*, **1983,** *15* (3), 72.

and indicator arm, and whatever load is on the pan. The current is adjusted so that the level of the indicator arm is in the null position when the pan is empty. Placing an object on the pan causes the pan and indicator arm to move downward, which increases the amount of light striking the photocell of the null detector. The increased current from the photocell is amplified and fed into the coil, creating a larger magnetic field, which returns the pan to its original null position. A device such as this, in which a small electric current causes a mechanical system to maintain a null position, is called a *servo system*. The current required to keep the pan and object in the null position is directly proportional to the weight of the object and is readily measured, digitized, and displayed. The calibration of an electronic balance involves the use of a standard mass and adjustment of the current so that the mass of the standard is exhibited on the display.

Figure 33–4 shows the configurations for two electronic analytical balances. In each, the pan is tethered to a system of constraints known collectively as a *cell*. The cell incorporates several *flexures* that permit limited movement of the pan and prevent torsional forces (resulting from off-center loading) from disturbing the alignment of the balance mechanism. At null, the beam is parallel to the gravitational horizon and each flexure pivot is in a relaxed position.

Figure 33–4a shows an electronic balance with the pan located below the cell. Higher precision is achieved with this arrangement than with the

> A servo system is a device in which a small electric signal causes a mechanical system to return to a null position.

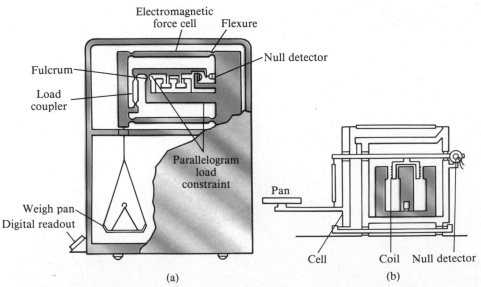

**Figure 33–4**
Electronic analytical balances. (a) Classical configuration with pan beneath the cell. (b) A top-loading design. Note that the mechanism is enclosed in a windowed case. (a, From R. M. Schoonover, *Anal. Chem.*, **1982,** *54,* 976A. Published 1982, American Chemical Society; b, Reprinted with permission from K. M. Lang, *Amer. Lab.*, **1983,** *15*(3), 72. Copyright 1983 by International Scientific Communications, Inc.)

top-loading design shown in Figure 33–4b. Even so, top-loading electronic balances have a precision that equals or exceeds that of the best mechanical balances and additionally provide unencumbered access to the pan.

Electronic balances generally feature an automatic *taring control* that causes the display to read zero with a container (such as a boat or weighing bottle) on the pan. Most balances permit taring to 100% of capacity.

Some electronic balances have dual capacities and dual precisions. These features permit the capacity to be decreased from that of a macrobalance to that of a semimicrobalance (30 g) with a concomitant gain in precision to 0.01 mg. Thus, the chemist has effectively two balances in one.

A modern electronic analytical balance provides unprecedented speed and ease of use. For example, one instrument is controlled by touching a single bar at various positions along its length. One position turns the instrument on or off, another automatically calibrates the balance against a standard mass, and a third zeros the display, either with or without an object on the pan. Reliable weighing data are obtainable with little or no instruction.

A few disadvantages attend the use of an electronic balance. Problems are sometimes encountered in the weighing of ferromagnetic materials. Strong electromagnetic radiation may also interfere with performance. Finally, the precision and accuracy of these instruments appear to be more susceptible to the effects of dust than do the precision and accuracy of their mechanical counterparts.

> A tare is the mass of an empty sample container. Taring is the process of setting a balance to read zero in the presence of the tare.

## 33D–5 Sources of Error in Weighing

### Correction for Buoyancy[4]

A *buoyancy error* will affect weighing data if the density of the object being weighed differs significantly from that of the standard weights. This error has its origin in the difference in the buoyant force exerted by the medium (air) upon the object and upon the weights. Correction for buoyancy is accomplished with the equation

> A buoyancy error is the weighing error that develops when the density of the object weighed is significantly different from that of the standard weights.

$$W_1 = W_2 + W_2 \left( \frac{d_{air}}{d_{obj}} - \frac{d_{air}}{d_{wts}} \right) \qquad (33-1)$$

where $W_1$ is the corrected weight of the object, $W_2$ is the mass of the standard weights, $d_{obj}$ is the density of the object, $d_{wts}$ is the density of the weights, and $d_{air}$ is the density of the air displaced by them; $d_{air}$ has a value of 0.0012 $g \cdot cm^{-3}$.

The consequences of Equation 33–1 are shown in Figure 33–5, in which the relative error due to buoyancy is plotted against the density of objects weighed in air against stainless steel weights. Note that this error is less than 0.1% for objects that have a density of 2 $g \cdot cm^{-3}$ or greater. It is thus

---

[4]For further information, see R. Battino and A. G. Williamson, *J. Chem. Educ.*, **1984**, *64*, 51.

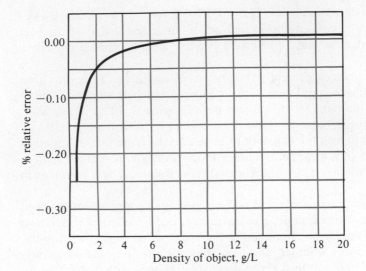

**Figure 33–5**
Effect of buoyancy on weighing data (stainless steel weights). Plot of relative error as a function of the density of the object weighed.

seldom necessary to apply a correction to the weight of most solids. The same cannot be said for low-density solids, liquids, or gases, however; for these, the effects of buoyancy are significant and a correction must be applied.

The density of weights used in single-pan balances or for calibrating electronic balances ranges from 7.8 to 8.4 $g \cdot cm^{-3}$, depending upon the manufacturer. In most cases, use of a density of 8 $g \cdot cm^{-3}$ provides a sufficiently accurate correction. If a greater accuracy is required, the specifications for the balance to be used should be consulted for the necessary density data.

---

**Example 33–1**

A bottle weighed 7.6500 g empty and 9.9700 g after introduction of an organic liquid with a density of 0.92 $g \cdot cm^{-3}$. The balance was equipped with stainless steel weights having a density of 8.0 $g \cdot cm^{-3}$. Correct the weight of the sample for the effects of buoyancy.

The apparent weight of the liquid is 9.9700 − 7.6500 = 2.3200 g. The same buoyant force acts on the container during both weighings; thus, we need to consider only the force that acts on the 2.3200 g of liquid. Substitution of 0.0012 $g \cdot cm^{-3}$ for $d_{air}$, 0.92 $g \cdot cm^{-3}$ for $d_{obj}$, and 8.0 $g \cdot cm^{-3}$ for $d_{wts}$ in Equation 33–1 gives

$$W_1 = 2.3200 + 2.3200 \left( \frac{0.0012}{0.92} - \frac{0.0012}{8.0} \right) = 2.3227 \text{ g}$$

---

## Temperature Effects

Always allow heated objects to return to room temperature before you attempt to weigh them.

Attempts to weigh an object whose temperature is different from that of its surroundings will result in a significant error. Failure to allow sufficient time for a heated object to return to room temperature is the commonest

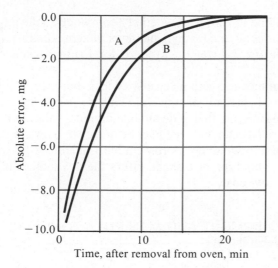

Figure 33–6
Effect of temperature on weighing data. Absolute error in weight as a function of time after the object was removed from a 110°C drying over. A: porcelain filtering crucible. B: weighing bottle containing about 7.5 g of KCl.

source of this problem. Errors due to a difference in temperature have two sources. First, convection currents within the balance case exert a buoyant effect on the pan and object. Second, warm air trapped in a closed container weighs less than the same volume at a lower temperature. Both effects cause the apparent weight of the object to be low. This error can amount to as much as 10 or 15 mg for a typical porcelain filtering crucible or a weighing bottle (Figure 33–6). Heated objects must always be cooled to room temperature before being weighed.

## Other Sources of Error

A porcelain or glass object will occasionally acquire a static charge that is sufficient to cause a balance to perform erratically; this problem is particularly serious when the relative humidity is low. Spontaneous discharge frequently occurs after a short period. A low-level source of radioactivity (such as a photographer's brush) in the balance case will provide sufficient ions to relieve the charge. Alternatively, the object can be wiped with a faintly damp chamois.

The optical scale of a single-pan balance should be checked regularly for accuracy, particularly under loading conditions that require the full scale range. A standard 100-mg weight is used for this check.

## 33D–6 Auxiliary Balances

Balances that are less precise than analytical balances find much use in the analytical laboratory. These offer the advantages of speed, ruggedness, large capacity, and convenience and should be used whenever high sensitivity is not required.

Top-loading auxiliary balances are particularly convenient. A sensitive top-loading balance will accommodate 150 to 200 g with a precision of about 1 mg—an order of magnitude less than a macro analytical balance. Some balances of this type tolerate loads as great as 25,000 g with a precision of ±0.05 g. Most are equipped with a taring device that brings

Use auxiliary laboratory balances for weighings that do not require great accuracy.

the balance reading to zero with an empty container on the pan. Some are fully automatic, require no manual dialing or weight handling, and provide a digital readout of the weight. Modern top-loading balances are electronic.

A triple-beam balance with a sensitivity less than that of a typical top-loading auxiliary balance is also useful. This is a single-pan balance with three decades of weights that slide along individual calibrated scales. The precision of a triple-beam balance may be one or two orders of magnitude less than that of a top-loading instrument but is adequate for many weighing operations. This type of balance offers the advantages of simplicity, durability, and low cost.

## 33E    THE EQUIPMENT AND MANIPULATIONS ASSOCIATED WITH WEIGHING

The weight of many solids changes with humidity, owing to their tendency to absorb weighable amounts of moisture. This effect is especially pronounced when a large surface area is exposed, as with a reagent chemical or a sample that has been ground to a fine powder. The first step in a typical analysis, then, involves drying the sample so that the results will not be affected by the humidity of the surrounding atmosphere.

> Weighing to constant weight is a process in which a solid is cycled through heating, cooling, and weighing steps until its weight becomes constant to within 0.2 to 0.3 mg.

A sample, a precipitate, or a container is brought to *constant weight* by a cycle that involves heating (ordinarily for 1 hr or more) at an appropriate temperature, cooling, and weighing. This cycle is repeated as many times as needed to obtain successive weighings that agree within 0.2 to 0.3 mg of one another. The establishment of constant weight provides some assurance that the chemical or physical processes that occur during the heating (or ignition) are complete.

### 33E-1 Weighing Bottles

Solids are conveniently dried and stored in *weighing bottles,* two common varieties of which are shown in Figure 33–7. The ground-glass portion of the cap-style bottle shown on the left is on the outside and does not come into contact with the contents; this design eliminates the possibility of some of the sample becoming entrained upon and subsequently lost from the ground-glass surface.

Plastic weighing bottles are available; ruggedness is the principal advantage of these bottles over their glass counterparts.

> A desiccator is a device for drying substances or objects.

### 33E-2 Desiccators and Desiccants

Oven-drying is the most common way of removing moisture from solids. To be sure, this approach is not appropriate for substances that decompose or for those from which water is not removed at the ambient temperature of the oven; see Section 30B.

Dried materials are stored in *desiccators* while they cool in order to minimize the uptake of moisture. Figure 33–8 shows the components of a

Figure 33–7
Typical weighing bottles.

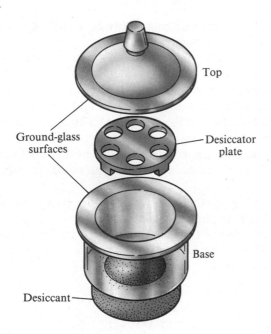

Top

Ground-glass
surfaces

Desiccator
plate

Base

Desiccant

Figure 33–8
Components of a typical desiccator.

typical desiccator. The base section contains a chemical drying agent, such as anhydrous calcium chloride, calcium sulfate (Drierite[5]), anhydrous magnesium perchlorate (Anhydrone[6] or Dehydrite[7]), or phosphorus pentoxide. The ground-glass surfaces are lightly coated with grease.

The lid of a desiccator is removed or replaced by a sliding motion rather than by a vertical one to minimize the likelihood of disturbing the sample. An airtight seal is achieved by slight rotation and downward pressure upon the positioned lid.

When a heated object is placed in a desiccator, the increase in pressure as the enclosed air is warmed may be sufficient to break the seal between lid and base. Conversely, if the seal is not broken, the cooling of heated objects can produce a partial vacuum in the dessicator. Both of these conditions can cause the contents of the desiccator to be physically lost or contaminated. Although it defeats the purpose of the desiccator somewhat, it is advisable to allow some cooling to occur before the lid is seated. It is also helpful to break the seal once or twice during cooling to relieve any excessive vacuum that might otherwise develop. Finally, it is prudent to lock the lid in place with your thumbs while moving the desiccator from place to place.

Very hygroscopic materials should be stored in containers equipped with snug covers, such as weighing bottles; the covers remain in place while in the desiccator. Most other solids can be safely stored uncovered.

---

[5]W. A. Hammond Drierite Co.

[6]J. T. Baker Co.

[7]Thomas Scientific Co.

Figure 33-9
Arrangement for the drying of
samples.

Be sure to provide identifying markings
to all pieces of equipment before placing them in the drying oven.

### 33E-3 The Manipulation of Weighing Bottles

Heating at 105 to 110°C is sufficient to remove the moisture from the
surface of most solids. Figure 33-9 depicts the arrangement recommended for drying a sample. The weighing bottle is contained in a labeled
beaker covered with a ribbed cover glass. This arrangement protects the
sample from accidental contamination and also allows for the free access
of air. Crucibles that contain a precipitate that can be freed of moisture by
simple drying can be treated similarly. The beaker containing the weighing bottle or crucible to be dried must be carefully marked to permit
identification.

The handling of a dried object with your fingers can result in the transfer of a weighable amount of water or oil from your skin to the object.
Weighing bottles should thus be manipulated with tongs, chamois finger
cots, clean cotton gloves, or strips of paper. A weighing bottle being
manipulated with strips of paper is shown in Figure 33-10.

### 33E-4 Weighing by Difference

Weighing by difference is a simple method for determining a series of
sample weights. First the bottle and its contents are weighed. One sample
is then transferred from the bottle to a container; gentle tapping of the
bottle with its top and slight rotation of the bottle provide control over the
amount of sample removed. Following transfer, the bottle and its residual
contents are weighed. The weight of the sample is the difference between
the two weighings. It is essential that all the solid removed from the
weighing bottle be transferred to the container without loss.

### 33E-5 Weighing Hygroscopic Solids

Special precautions are needed for the weighing of hygroscopic solids
because of the rapid rate with which they equilibrate with moisture in the
atmosphere. The approximate amount of each sample is placed in individual weighing bottles. After the contents have been dried, the bottles are
capped and cooled in a desiccator. Their exact weights are then determined by difference as described above, care being taken to replace the
cap of the weighing bottle as quickly as possible after transfer.

### 33E-6 Weighing Liquids

The weight of a liquid is always obtained by difference. Liquids that are
noncorrosive and relatively nonvolatile can be transferred to previously
weighed containers with snugly fitting covers (such as weighing bottles);
the weight of the container is subtracted from the total weight.

A volatile or corrosive liquid should be sealed in a weighed glass ampoule. The ampoule is heated, and the neck is then immersed in the
sample; as cooling occurs, the liquid is drawn into the bulb. The ampoule
is then inverted and the neck sealed off with a small flame. The ampoule

Figure 33-10
Method for quantitative transfer of
a solid sample. Note the use of
paper strips to avoid contact between glass and skin.

and its contents, along with any glass removed during sealing, are cooled to room temperature and weighed. The ampoule is then transferred to an appropriate container and broken. A volume correction for the glass of the ampoule may be needed if the receiving vessel is a volumetric flask.

## 33F   WEIGHT TITRATIONS

Weight (or gravimetric) titrimetry is discussed in Section 5D. Directions for the weight titration of chloride ion by the Mohr method are found in Section 34B–3.

A convenient reagent dispenser for a weight titration is a small polyethylene bottle equipped with a fine delivery tip (Figure 33–11). Weighings are normally performed with a top-loading balance that has a sensitivity of 1 mg or 0.001 mL. It has been found that the weight of such a bottle does not change significantly when it is handled with bare hands.

### 33F–1  Directions for Preparing a Reagent Dispenser

A dispenser is readily fashioned from a 60-mL polyethylene bottle with a screw cap [such as the Nalge 2-oz ($\simeq$ 60 mL) or 4-oz ($\simeq$125 mL) Boston Round Bottle]. With a cork borer, make a hole in the cap that is slightly smaller than the outside diameter of the tip (Note). Carefully force the tip through the hole; apply a bead of epoxy cement to seal the tip to the cap. Use a pressure-sensitive label to identify the contents of the bottle.

**Note.** A tip can be prepared by constricting the opening of an ordinary medicine dropper in a flame. Equally satisfactory are glass tips for conventional Kimax® burets.

### 33F–2  Directions for Performing a Weight Titration

Fill the reagent dispenser with a quantity of the standard titrant, wipe away any liquid spilled on the outside of the container with an absorbing

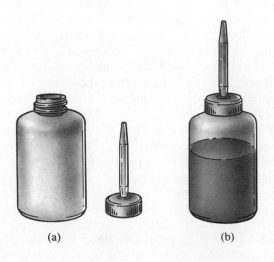

(a)                    (b)

Figure 33–11
Reagent dispenser for weight titrations.

tissue, and tighten the screw cap firmly. Weigh the bottle and its contents to the nearest milligram. Introduce a suitable indicator into the solution of the analyte. Grasp the dispenser in one hand and the titration flask in the other. Tilt the dispenser so that its tip is below the lip of the flask and deliver several increments of the reagent by squeezing the bottle while rotating the flask with the other hand. When it is judged that only a few more drops of reagent are needed, ease the pressure on the bottle so that the flow ceases; then touch the tip to the inside of the flask and further reduce the pressure on the dispenser so that the liquid in the tip is drawn back into the bottle as the tip is removed from the flask. Set the dispenser on a piece of clean, dry glazed paper and rinse down the inner walls of the flask with a stream of distilled or deionized water (Note 1). Add reagent a drop at a time until the end point is reached (Note 2). Weigh the dispenser and record the data.

**Notes**

1. Instead of rinsing the walls, the titration flask can be tilted and rotated so that the bulk of the liquid picks up droplets that adhere to the inner surface.
2. Increments smaller than an ordinary drop can be added by forming a partial drop on the tip and then touching the tip to the wall. This partial drop is then combined with the bulk of the solution by rinsing the walls with wash water or tilting the flask until the droplet is incorporated into the bulk.

## 33G    THE EQUIPMENT AND MANIPULATIONS FOR FILTRATION AND IGNITION

### 33G–1 Apparatus

#### Simple Crucibles

Simple crucibles serve only as containers. The more common types maintain constant weight, within the limits of experimental error, and are used principally to convert precipitates to suitable weighing forms. The solid is first collected on a filter paper. The filter and contents are then transferred to a weighed crucible, and the paper is ignited. Porcelain, aluminum oxide, silica, and platinum are used for the manufacture of these crucibles.

Simple crucibles of nickel, iron, silver, and gold are used as containers for the high-temperature fusion of samples that are not soluble in aqueous reagents. Attack by both the atmosphere and the contents may cause these crucibles to suffer weight changes. Moreover, such attack will contaminate the sample with species derived from the crucible. The chemist selects the crucible whose products will offer the least interference in subsequent steps of the analysis.

#### Filtering Crucibles

Filtering crucibles serve not only as containers but also as filters. A vacuum is used to hasten the filtration; a tight seal between crucible and

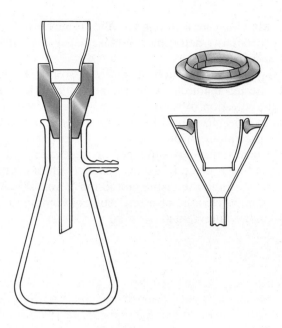

Figure 33–12
Adaptors for filtering crucibles.

filtering flask is accomplished with any of several types of flexible adaptors (Figure 33–12; a complete filtration train is shown in Figure 33–17). Collection of a precipitate with a filtering crucible is frequently less time-consuming than with paper.

*Sintered-glass* (also called *fritted-glass*) crucibles are manufactured in three porosities: fine, medium, and coarse (marked *f, m,* and *c*). The upper temperature limit for a sintered-glass crucible is ordinarily about 200°C. Filtering crucibles made entirely of quartz can tolerate substantially high temperatures without damage. The same is true for crucibles with unglazed porcelain or aluminum oxide frits. The latter are not as costly as quartz.

A *Gooch crucible* has a perforated bottom that supports a fibrous mat. Asbestos was at one time the filtering medium of choice for a Gooch crucible; current regulations concerning this material have virtually eliminated its use. Small circles of glass matting have now replaced asbestos; they are used in pairs to protect against disintegration during the filtration. Glass mats can tolerate temperatures in excess of 500°C and are substantially less hygroscopic than asbestos.

### Filter Paper

Paper is an important filtering medium. Ashless paper is manufactured from cellulose fibers that have been treated with hydrochloric and hydrofluoric acids to remove metallic impurities and silica; ammonia is then used to neutralize the acids. The residual ammonium salts in many filter papers may be sufficient to affect the analysis for nitrogen by the Kjeldahl method (Section 34C–11).

All papers tend to pick up moisture from the atmosphere, and ashless paper is no exception. It is thus necessary to destroy the paper by ignition if the precipitate collected on it is to be weighed. Typically, 9- or 11-cm circles of ashless paper leave a residue that weighs less than 0.1 mg, an amount that is ordinarily negligible. Ashless paper is available in several porosities (Appendix 8).

Gelantinous precipitates, such as hydrous iron(III) oxide, clog the pores of any filtering medium. A coarse-porosity ashless paper is most effective for the filtration of such solids, but even here clogging occurs. This problem can be minimized by mixing a dispersion of ashless filter paper with the precipitate prior to filtration. Filter paper pulp is available in tablet form from chemical suppliers; if necessary, the pulp can be prepared by treating a piece of ashless paper with concentrated hydrochloric acid and washing the disintegrated mass free of acid.

Table 33–1 summarizes the characteristics of common filtering media. None satisfies all requirements.

## Heating Equipment

Many precipitates can be weighed directly after being brought to constant weight in a low-temperature drying oven. Such an oven is electrically heated and capable of maintaining a constant temperature to within 1°C (or better). The maximum attainable temperature ranges from 140 to 260°C, depending upon make and model; for many precipitates, 110°C is a satisfactory drying temperature. The efficiency of a drying oven is greatly increased by the forced circulation of air. The passage of predried air through an oven designed to operate under a partial vacuum represents an additional improvement.

Microwave laboratory ovens are currently appearing on the market. Where applicable, these greatly shorten drying cycles. For example,

### Table 33–1
### Comparison of Filtering Media for Gravimetric Analysis

| Characteristic | Paper | Gooch Crucible — Asbestos Mat* | Gooch Crucible — Glass Mat | Glass Crucible | Porcelain Crucible | Aluminum Oxide Crucible |
|---|---|---|---|---|---|---|
| Speed of filtration | Slow | Rapid | Rapid | Rapid | Rapid | Rapid |
| Convenience and ease of preparation | Troublesome, inconvenient | Troublesome, inconvenient | Convenience | Convenient | Convenient | Convenient |
| Maximum ignition temperature, °C | None | 1200 | >500 | 200–500 | 1100 | 1450 |
| Chemical reactivity | Carbon has reducing properties | Inert | Inert | Inert | Inert | Inert |
| Porosity | Many available | Difficult to control | Several available | Several available | Several available | Several available |
| Convenience with gelatinous precipitates | Satisfactory | Unsuitable; filter tends to clog | Unsuitable; filter tends to clog | Unsuitable; filter tends to clog | Unsuitable; filter tends to clog | Unsuitable; filter tends to clog |
| Cost | Low | Low | Low | High | High | High |

*The use of asbestos, a confirmed carcinogen, is subject to stringent control in many areas.

slurry samples that require 12 to 16 hr for drying in a conventional oven are reported to be dried within 5 to 6 min in a microwave oven.[8]

An ordinary heat lamp can be used to dry a precipitate that has been collected on ashless paper and to char the paper as well. The process is conveniently completed by ignition at an elevated temperature in a muffle furnace.

Burners are convenient sources of intense heat. The maximum attainable temperature depends upon the design of the burner and the combustion properties of the fuel. Of the three common laboratory burners, the Meker provides the highest temperatures, followed by the Tirrill and Bunsen types.

A heavy-duty electric furnace (*muffle furnace*) is capable of maintaining controlled temperatures of 1100°C or higher. Long-handled tongs and heat-resistant gloves are needed for protection when transferring objects to or from such a furnace.

### 33G–2 The Manipulations Associated with Filtration and Ignition

#### Preparation of Crucibles

A crucible used to convert a precipitate to a form suitable for weighing must maintain—within the limits of experimental error—a constant weight throughout the drying or ignition. The crucible is first cleaned thoroughly (filtering crucibles are conveniently cleaned by backwashing on a filtration train) and subjected to the same regimen of heating and cooling as that required for the precipitate. This process is repeated until constant weight (page 806) has been achieved, that is, until consecutive weighings differ by 0.3 mg or less.

Backwashing a filtering crucible is done by turning the crucible upside down in the adaptor (Figure 33–12) and drawing water through the inverted crucible.

#### Filtration and Washing of Precipitates

Three steps are involved in filtering an analytical precipitate: *decantation, washing,* and *transfer.* In decantation, as much supernatant liquid as possible is passed through the filter while the precipitated solid is kept essentially undisturbed in the beaker where it was formed. This procedure speeds the overall filtration rate by delaying the time at which the pores of the filtering medium become clogged with precipitate. A stirring rod is used to direct the flow of decantate (Figure 33–13). When flow ceases, the drop of liquid at the end of the pouring spout is collected with the stirring rod and returned to the beaker. Wash liquid is next added to the beaker and thoroughly mixed with the precipitate. The solid is allowed to settle, following which this liquid is also decanted through the filter. Several such washings may be required, depending upon the precipitate. Most washing should be carried out *before* the solid is transferred; this results in a more thoroughly washed precipitate and a more rapid filtration.

Decantation is the process of pouring a liquid gently so as to not disturb a solid in the bottom of the container.

---

[8]D. G. Kuehn, R. L. Brandvig, D. C. Lundean, and R. H. Jefferson, *Amer. Lab.,* **1986,** *18* (7), 31.

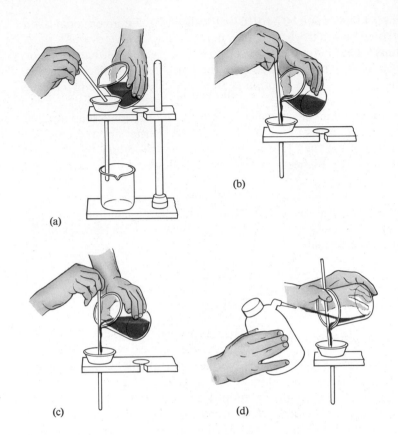

**Figure 33–13**

Steps in filtering. (a and b) Washing by decantation; (c and d) transfer of the precipitate.

The transfer process is illustrated in Figures 33–13c and d. The bulk of the precipitate is moved from beaker to filter by suitably directed streams of wash liquid. As in decantation and washing, a stirring rod provides direction for the flow of material to the filtering medium.

The last traces of precipitate that cling to the inside of the beaker are dislodged with a *rubber policeman,* which is a small section of rubber tubing that has been crimped on one end. The open end of the tubing is fitted onto the end of a stirring rod and is wetted with wash liquid before use. Any solid collected with it is combined with the main portion on the filter. Small pieces of ashless paper can be used to wipe the last traces of hydrous oxide precipitates from the wall of the beaker; these papers are ignited along with the paper that holds the bulk of the precipitate.

Many precipitates possess the exasperating property of *creeping,* or spreading over a wetted surface against the force of gravity. Filters are never filled to more than three quarters of capacity, owing to the possibility that some of the precipitate could be lost as the result of creeping. The addition of a small amount of nonionic detergent, such as Triton-X-100, to the supernatant liquid or to the wash liquid can be helpful in minimizing creeping.

A gelatinous precipitate must be completely washed before it is allowed to dry because drying causes the solid to shrink and develop cracks. Further additions of wash liquid simply pass through these cracks and accomplish little or no washing.

Creeping is a process in which a solid moves up the side of a wetted container or filter paper.

Do not permit a gelatinous precipitate to dry until it has been washed completely.

## 33G–3 Directions for the Filtration and Ignition of Precipitates

### Preparation of a Filter Paper

Figure 33–14 shows the sequence followed in folding a filter paper and seating it in a 58-deg funnel (or 60-deg fluted funnel). The paper is folded exactly in half (a), firmly creased, and folded again (b). A triangular piece from one of the corners is torn off parallel to the second fold (c). The paper is then opened so that the untorn quarter forms a cone (d). The cone is fitted into the funnel, and the second fold is creased (e). Seating is completed by dampening the cone with water from a wash bottle and *gently* patting it with a finger. There is no leakage of air between the funnel and a properly seated cone; in addition, the stem of the funnel will be filled with an unbroken column of liquid.

### The Transfer of Paper and Precipitate to a Crucible

After filtration and washing have been completed, the filter and its contents must be transferred from the funnel to a crucible that has been brought to constant weight. Ashless paper has very low wet strength and must be handled with care during the transfer. The danger of tearing is lessened considerably if the paper is allowed to dry somewhat before it is removed from the funnel.

Figure 33–15 illustrates the transfer process. The triple-thick portion of the filter paper is drawn across the funnel to flatten the cone along its upper edge (a); the corners are next folded inward (b); the top edge is then folded over (c). Finally, the paper and its contents are eased into the crucible (d) so that the bulk of the precipitate is near the bottom.

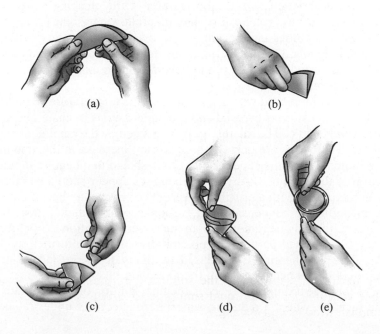

(a)

(b)

(c)

(d)

(e)

Figure 33–14
Folding and seating a filter paper.

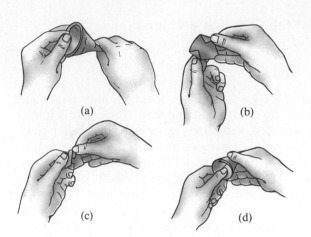

(a)          (b)

(c)          (d)

Figure 33–15

Transferring a filter paper and precipitate to a crucible.

## Ashing of a Filter Paper

If a heat lamp is to be used, the crucible is placed on a clean, nonreactive surface, such as a wire screen covered with aluminum foil. The lamp is then positioned about 1 cm above the rim of the crucible and turned on. Charring takes place without further attention. The process is considerably accelerated if the paper is moistened with no more than one drop of concentrated ammonium nitrate solution. Elimination of the residual carbon is accomplished with a burner, as described in the next paragraph.

Considerably more attention must be paid if a burner is used to ash a filter paper. The burner produces much higher temperatures than a heat lamp. The possibility thus exists for the mechanical loss of precipitate if moisture is expelled too rapidly in the initial stages of heating or if the paper bursts into flame. Also, partial reduction of some precipitates can occur through reaction with the hot carbon of the charring paper; such reduction is a serious problem if reoxidation following ashing is inconvenient. These difficulties can be minimized by positioning the crucible as illustrated in Figure 33–16. The tilted position allows for the ready access of air; a clean crucible cover should be available to extinguish any flame that might develop.

You should have a burner for each crucible. You can tend to the ashing of several filter papers at the same time.

Heating is commenced with a small flame. The temperature is gradually increased as moisture is evolved and the paper begins to char. The intensity of heating that can be tolerated can be gauged by the amount of smoke given off. Thin wisps are normal. A significant increase in the amount of smoke indicates that the paper is about to flash and that heating should be temporarily discontinued. Any flame that does appear should be immediately extinguished with a crucible cover. (The cover may become discolored, owing to the condensation of carbonaceous products; these products must ultimately be removed from the cover by ignition to confirm the absence of entrained particles of precipitate.) When no further smoking can be detected, heating is increased to eliminate the residual carbon. Strong heating, as necessary, can then be undertaken.

This sequence ordinarily precedes the final ignition of a precipitate in a muffle furnace, where a reducing atmosphere is equally undesirable.

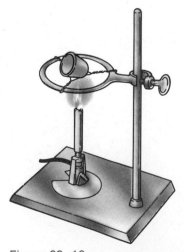

Figure 33–16

Ignition of a precipitate. Proper crucible position for preliminary charring.

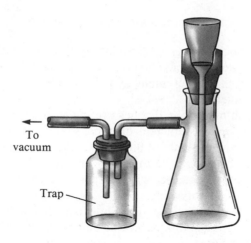

Figure 33–17
Train for vacuum filtration.

### The Use of Filtering Crucibles

A vacuum filtration train (Figure 33–17) is used where a filtering crucible can be used instead of paper. The trap isolates the filter flask from the source of vacuum.

## 33G–4  Rules for the Manipulation of Heated Objects

Careful adherence to the following rules will minimize the possibility of accidental loss of a precipitate.

1. Practice unfamiliar manipulations before putting them to use.
2. *Never* place a heated object on the benchtop; instead, place it on a wire gauze or a heat-resistant ceramic plate.
3. Allow a crucible that has been subjected to the full flame of a burner or to a muffle furnace to cool momentarily (on a wire gauze or ceramic plate) before transferring it to the desiccator.
4. Keep the tongs and forceps used to handle heated objects scrupulously clean. In particular, do not allow the tips to touch the benchtop.

## 33H  THE MEASUREMENT OF VOLUME

The precise measurement of volume is as important to many analytical methods as is the precise measurement of mass.

## 33H–1  Units of Volume

The unit of volume is the *liter* (L), defined as one cubic decimeter. The *milliliter* (mL) is one one-thousandth of a liter and is used where the liter represents an inconveniently large volume unit.

## 33H–2 The Effect of Temperature on Volume Measurements

The volume occupied by a given mass of liquid varies with temperature; to a lesser extent, so also does the volume of the container that holds the liquid. The accurate measurement of volume may require that both of these effects be taken into account.

Most volumetric measuring devices are made of glass, a material that fortunately has a small coefficient of expansion. The volume of a soft-glass container changes by about 0.003%/°C; the volume of a heat-resistant glass changes by about one third this amount. Clearly, variations in the volume of a glass container with temperature need to be considered only for the most exacting work.

The coefficient of expansion for dilute aqueous solutions (approximately 0.025%/°C) is such that a 5°C change has a measurable effect upon the reliability of ordinary volumetric measurements.

---

Example 33–2

A 40.00-mL sample is taken from an aqueous solution at 5°C; what volume does it occupy at 20°C?

$$V_{20°} = V_{5°} + 0.00025(20 - 5)(40.00) = 40.00 + 0.15 = 40.15 \text{ mL}$$

---

Volumetric measurements must be referred to some standard temperature; this reference point is ordinarily 20°C. The ambient temperature of most laboratories is sufficiently close to 20°C to eliminate the need for temperature corrections in volume measurements for aqueous solutions. In contrast, the coefficient of expansion for organic liquids may require corrections for temperature differences of 1°C or less.

---

Example 33–3

The coefficient of expansion for an alcoholic KOH solution is 0.11%/°C. Calculate the relative error in the volume of an aliquot delivered at 25.2°C by a 50.00-mL pipet if no correction for temperature is made.

$$V_{20°} = 50.00 - 0.0011(25.2 - 20.0)(50.00) = 49.71 \text{ mL}$$

$$E_{rel} = \frac{49.71 - 50.00}{50.00} \times 1000 \text{ ppt} = -5.8 \text{ ppt}$$

---

## 33H–3 Apparatus for the Precise Measurement of Volume

Volume is measured reliably with a *pipet,* a *buret*, and a *volumetric flask.*

Volumetric equipment is marked by the manufacturer to indicate not only the manner of calibration (usually TD for "to deliver" or TC for "to contain") but also the temperature at which the calibration strictly applies. Pipets and burets are ordinarily calibrated to deliver specified volumes, whereas volumetric flasks are calibrated on a to-contain basis.

## Pipets

Pipets permit the transfer of accurately known volumes from one container to another. Common types are shown in Figure 33–18; information concerning their use is given in Table 33–2. A *volumetric,* or *transfer,* pipet (Figure 33–18a) delivers a single, fixed volume between 0.5 and 200 mL. Many such pipets are color-coded by volume for convenience in identification and sorting. *Measuring* pipets (Figures 33–18b and c) are calibrated in convenient units to permit delivery of any volume up to a maximum capacity ranging from 0.1 to 25 mL.

Volumetric and measuring pipets are filled to a calibration mark at the outset; the manner in which the transfer is completed depends upon the particular type of pipet. Because an attraction exists between most liquids and glass, a small amount of liquid tends to remain in the tip after the pipet is emptied. This residual liquid is never blown out of a volumetric pipet or from some measuring pipets; it is blown out of other types of pipets (Table 33–2).

| TOLERANCES OF CLASS A TRANSFER PIPETS | |
|---|---|
| Capacity, mL | Tolerances, mL |
| 0.5 | ±0.006 |
| 1 | ±0.006 |
| 2 | ±0.006 |
| 5 | ±0.01 |
| 10 | ±0.02 |
| 20 | ±0.03 |
| 25 | ±0.03 |
| 50 | ±0.05 |
| 100 | ±0.08 |

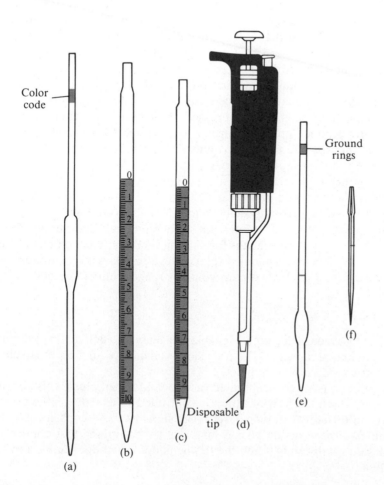

Figure 33–18

Typical pipets: (a) volumetric, (b) Mohr, (c) serological, (d) Eppendorf micropipet, (e) Ostwald-Folin, (f) lambda.

Table 33–2
CHARACTERISTICS OF PIPETS

| Name | Type of Calibration* | Function | Available Capacity, mL | Type of Drainage |
|---|---|---|---|---|
| Volumetric | TD | Delivery of fixed volume | 1–200 | Free |
| Mohr | TD | Delivery of variable volume | 1–25 | To lower calibration line |
| Serological | TD | Delivery of variable volume | 0.1–10 | Blow out last drop† |
| Serological | TD | Delivery of variable volume | 0.1–10 | To lower calibration line |
| Ostwald-Folin | TD | Delivery of fixed volume | 0.5–10 | Blow out last drop† |
| Lambda | TC | Containment of fixed volume | 0.001–2 | Wash out with suitable solvent |
| Lambda | TD | Delivery of fixed volume | 0.001–2 | Blow out last drop† |
| Eppendorf | TD | Delivery of variable or fixed volume | 0.001–1 | Tip emptied by air displacement |

*TD—to deliver; TC—to contain.

†A frosted ring near the top of recently manufactured pipets indicates that the last drop is to be blown out.

RANGE AND PRECISION OF TYPICAL EPPENDORF MICROPIPETS

| Volume Range, μL | Standard Deviation, μL |
|---|---|
| 1–20 | <0.04 @ 2 μL |
| | <0.06 @ 20 μL |
| 10–100 | <0.10 @ 15 μL |
| | <0.15 @ 100 μL |
| 20–200 | <0.15 @ 25 μL |
| | <0.30 @ 200 μL |
| 100–1000 | <0.6  @ 250 μL |
| | <1.3  @ 1000 μL |
| 500–5000 | <3 @ 1.0 mL |
| | <8 @ 5.0 mL |

TOLERANCES, CLASS A BURETS

| Volume, mL | Tolerances, mL |
|---|---|
| 5 | ±0.01 |
| 10 | ±0.02 |
| 25 | ±0.03 |
| 50 | ±0.05 |
| 100 | ±0.20 |

Hand-held Eppendorf micropipets (Figure 33–18d) deliver adjustable microliter volumes of liquid. With these pipets, a known and adjustable volume of air is displaced from the plastic disposable tip by depressing the pushbutton on the top of the pipet to a first stop. This button operates a spring-loaded piston that forces air out of the pipet. The volume of displaced air can be varied by a locking digital micrometer adjustment located on the front of the device. The plastic tip is then inserted into the liquid and the pressure on the button released, causing liquid to be drawn into the tip. The tip is then placed against the wall of the receiving vessel, and the pushbutton is again depressed to the first stop. After one second, the pushbutton is depressed further to a second stop, which completely empties the tip. The range of volumes and precision of typical pipets of this type are shown in the margin.

Numerous *automatic* pipets are available for situations that call for the repeated delivery of a particular volume. In addition, a motorized, computer-controlled microliter pipet is now available (see Figure 33–19). This device is programmed to function as a pipet, a dispenser of multiple volumes, a buret, and a means for diluting samples. The volume desired is entered on a keyboard and is displayed on a panel. A motor-driven piston dispenses the liquid. Maximum volumes range from 10 to 2500 μL.

Burets

Burets, like measuring pipets, enable the analyst to deliver any volume up to a maximum capacity. The precision attainable with a buret is substantially greater than with a pipet.

A buret consists of a calibrated tube to hold titrant plus a valve arrangement by which the flow of titrant is controlled. This valve is the principal source of difference among burets. The simplest valve consists of a close-fitting glass bead inside a short length of rubber tubing that connects the buret and its tip; only when the tubing is deformed does liquid flow past the bead.

A buret equipped with a glass stopcock for a valve relies upon a lubricant between the ground-glass surfaces of stopcock and barrel for a liquid-tight seal. Some solutions, notably bases, cause a glass stopcock to freeze upon long contact; therefore thorough cleaning is needed after each use. Valves made of Teflon are commonly encountered; these are unaffected by most common reagents and require no lubricant.

### Volumetric Flasks

Volumetric flasks are manufactured with capacities ranging from 5 mL to 5 L and are usually calibrated to contain a specified volume when filled to a line etched on the neck. They are used for the preparation of standard solutions and for the dilution of samples to a fixed volume prior to taking aliquots with a pipet. Some are also calibrated on a to-deliver basis; these are readily distinguished by the two reference lines on the neck. If delivery of the stated volume is desired rather than containment, the flask is filled to the upper line.

### 33H–4   General Considerations Concerning the Use of Volumetric Equipment

Volume markings are blazed upon clean volumetric equipment by the manufacturer. An equal degree of cleanliness is needed in the laboratory if these markings are to have their stated meanings. Only clean glass surfaces support a uniform film of liquid. Dirt or oil causes breaks in this film; the existence of breaks is a certain indication of an unclean surface.

### Cleaning

A brief soaking in a warm detergent solution is usually sufficient to remove the grease and dirt responsible for water breaks. Prolonged soaking should be avoided because a rough area or ring is likely to develop at a detergent-air interface. This ring cannot be removed and causes a film break that destroys the usefulness of the equipment.

After being cleaned, the apparatus must be thoroughly rinsed with tap water and then with three or four portions of deionized water. It is seldom necessary to dry volumetric ware. As a general rule, calibrated glass equipment should not be heated.

### Avoiding Parallax

The top surface of a liquid confined in a narrow tube (such as a pipet or buret or the neck of a volumetric flask) exhibits a marked curvature, or *meniscus*. It is common practice to use the bottom of the meniscus as the point of reference in calibrating and using volumetric equipment. This minimum can be established more exactly by holding an opaque card or piece of paper behind the graduations (Figure 33–20).

Your eye must be at the level of the liquid surface to avoid an error due to *parallax*, a condition that causes the volume to appear smaller than its

**Figure 33–19**
A hand-held, battery-operated motorized pipet. (Courtesy of Rainin Instrument Co., Inc., Woburn, MA.)

| TOLERANCES OF CLASS A VOLUMETRIC FLASKS | |
| --- | --- |
| Capacity, mL | Tolerances, mL |
| 5 | ±0.02 |
| 10 | ±0.02 |
| 25 | ±0.03 |
| 50 | ±0.05 |
| 100 | ±0.08 |
| 250 | ±0.12 |
| 500 | ±0.20 |
| 1000 | ±0.30 |
| 2000 | ±0.50 |

A meniscus is the curved surface of a liquid at its interface with the atmosphere.

Parallax is the apparent displacement of a liquid level or of a pointer as an observer changes position. It occurs when the observer's line of vision is not perpendicular to the surface of the calibrated scale being read.

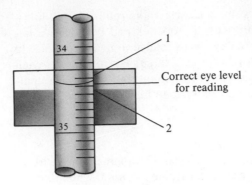

Correct eye level
for reading

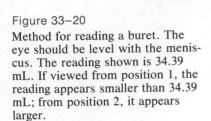

Figure 33–20
Method for reading a buret. The
eye should be level with the meniscus. The reading shown is 34.39
mL. If viewed from position 1, the
reading appears smaller than 34.39
mL; from position 2, it appears
larger.

actual value if the meniscus is viewed from above and larger if the meniscus is viewed from below (Figure 33–20).

## 33H–5 Directions for the Use of a Pipet

The following directions pertain specifically to volumetric pipets but can
be modified for the use of other types as well.

Liquid is drawn into a pipet through the application of a slight vacuum.
*Never use your mouth for suction because you may accidentally ingest
the liquid being pipetted.* Instead, a rubber suction bulb or a rubber tube
connected to a vacuum source should be used (Figure 33–21a).

### Cleaning

Use a rubber bulb to draw detergent solution to a level 2 to 3 cm above the
calibration mark of the pipet. Drain this solution and then rinse the pipet
with several portions of tap water. Inspect for film breaks; repeat this
portion of the cleaning cycle if necessary. Finally, fill the pipet with
distilled water to perhaps one third of its capacity and carefully rotate it so
that the entire interior surface is wetted. Repeat this rinsing step at least
twice.

### Measurement of an Aliquot

Use a rubber bulb to draw a small volume of the liquid to be sampled into
the pipet and thoroughly wet the entire interior surface. Repeat with *at
least* two additional portions. Then carefully fill the pipet to a level somewhat above the graduation mark (Figure 33–21a). Quickly replace the
bulb with your *forefinger* to arrest the outflow of liquid (Figure 33–21b).
Make certain there are no bubbles in the bulk of the liquid or foam at the
surface. Tilt the pipet slightly from the vertical and wipe the exterior free
of adhering liquid (Figure 33–21c). Touch the tip of the pipet to the wall of
a glass vessel (*not* the container into which the aliquot is to be transferred), and slowly allow the liquid level to drop by partially releasing your
forefinger (Note 1). Halt further flow as the bottom of the meniscus coin-
cides exactly with the graduation mark. Then place the pipet tip well
within the receiving vessel, and allow the liquid to drain. When free flow

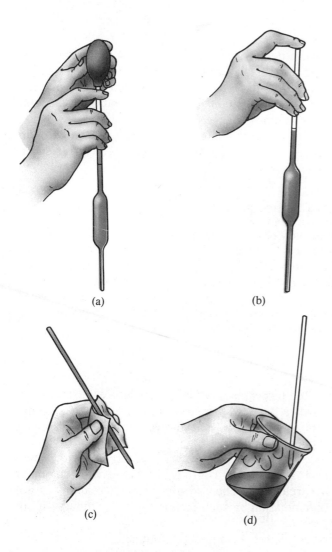

(a)                    (b)

(c)              (d)

Figure 33–21
Steps in dispensing an aliquot.

ceases, rest the tip against the inner wall of the receiver for a full 10 s (Figure 33–21d). Finally, withdraw the pipet with a rotating motion to remove any liquid adhering to the tip. *The small volume remaining inside the tip of a volumetric pipet should not be blown or rinsed into the receiving vessel* (Note 2).

**Notes**

1. The liquid can best be held at a constant level if your forefinger is *faintly* moist. Too much moisture makes control impossible.
2. Rinse the pipet thoroughly after use.

## 33H–6 Directions for the Use of a Buret

Before it is placed in service, a buret must be scrupulously clean; in addition, its valve must be liquid-tight.

## Cleaning

A brief soaking in a warm detergent solution is usually sufficient to remove the grease and dirt responsible for water breaks. Prolonged soaking should be avoided because a rough area or ring is likely to develop at the detergent/air interface. This ring cannot be removed and causes a film break that destroys the usefulness of the equipment.

After being cleaned, the apparatus must be thoroughly rinsed with tap water and then with three or four portions of distilled water. It is seldom necessary to dry volumetric ware.

## Lubrication of a Glass Stopcock

Carefully remove all old grease from a glass stopcock and its barrel with a paper towel and dry both parts completely. Lightly grease the stopcock, taking care to avoid the area adjacent to the hole. Insert the stopcock into the barrel and rotate it vigorously with slight inward pressure. A proper amount of lubricant has been used when (1) the area of contact between stopcock and barrel appears nearly transparent, (2) the seal is liquid-tight, and (3) no grease has worked its way into the tip.

**Notes**

1. Cleaning solution often disperses more stopcock lubricant than it destroys and thus leaves a buret with a heavier grease film than before treatment. This solution should *never* be allowed to come into contact with lubricated stopcock assemblies.
2. Grease films that are unaffected by cleaning solution may yield to an organic solvent such as acetone. Thorough washing with detergent should follow such treatment. The use of silicone lubricants is not recommended; contamination by such preparations is difficult—if not impossible—to remove.
3. So long as the flow of liquid is not impeded, fouling of a buret tip with stopcock grease is not a serious matter. Removal is best accomplished with organic solvents. A stoppage during a titration can be freed by *gentle* warming of the tip with a lighted match.
4. Before a buret is returned to service after reassembly, it is advisable to test for leakage. Simply fill the buret with water and establish that the volume reading does not change with time.

## Filling

Make certain the stopcock is closed. Add 5 to 10 mL of the titrant, and carefully rotate the buret to wet the interior completely. Allow the liquid to drain through the tip. *Repeat this procedure at least two more times.* Then fill the buret well above the zero mark. Free the tip of air bubbles by rapidly rotating the stopcock and permitting small quantities of the titrant to pass. Finally, lower the level of the liquid just to or somewhat below the zero mark. Allow for drainage ($\simeq 1$ min), and then record the initial volume reading, estimating to the nearest 0.01 mL.

## Titration

Figure 33–22 illustrates the preferred method for the manipulation of a stopcock; with your hand held as shown, any tendency for lateral movement by the stopcock will be in the direction of firmer seating. Be sure the tip of the buret is well within the titration vessel (ordinarily a flask). Introduce the titrant in increments of about 1 mL. Swirl (or stir) constantly to ensure thorough mixing. Decrease the size of the increments as the titration progresses; add titrant dropwise in the immediate vicinity of the end point (Note 2). When it is judged that only a few more drops are needed, rinse the walls of the container (Note 3). Allow for drainage (at least 30 s) at the completion of the titration. Then record the final volume, again to the nearest 0.01 mL.

**Figure 33–22**
Recommended method for manipulation of a buret stopcock.

Buret readings should be made to the nearest 0.01 mL.

### Notes

1. When unfamiliar with a particular titration, many chemists prepare an extra sample. No care is lavished on its titration since its functions are to reveal the nature of the end point and to provide a rough estimate of titrant requirements. This deliberate sacrifice of one sample frequently results in an overall saving of time.
2. Increments smaller than one drop can be taken by allowing a small volume of titrant to form on the tip of the buret and then touching the tip to the wall of the flask. This partial drop is then combined with the bulk of the liquid as in Note 3.
3. Instead of being rinsed toward the end of a titration, the flask can be tilted and rotated so that the bulk of the liquid picks up any drops that adhere to the inner surface.

## 33H–7   Directions for the Use of a Volumetric Flask

Before being put into use, volumetric flasks should be washed with detergent and thoroughly rinsed. Only rarely do they need to be dried. If required, however, drying is best accomplished by clamping the flask in an inverted position. Insertion of a glass tube connected to a vacuum line hastens the process.

### Direct Weighing into a Volumetric Flask

The direct preparation of a standard solution requires the introduction of a known weight of solute to a volumetric flask. Determine the weight of solute by difference (p. 808). Use a powder funnel to ensure a quantitative transfer. Rinse the funnel thoroughly; collect the washings in the flask.

   The foregoing procedure is inappropriate if heating is needed to dissolve the solute. Instead, weigh the solid into a beaker or flask, add solvent, heat to dissolve the solute, and allow the solution to cool to room temperature. Transfer this solution quantitatively to the volumetric flask, as described in the next section.

### The Quantitative Transfer of Liquid to a Volumetric Flask

Insert a funnel into the neck of the volumetric flask; use a stirring rod to direct the flow of liquid from the beaker into the funnel. Tip off the last drop of liquid on the spout of the beaker with the stirring rod. Rinse both the stirring rod and the interior of the beaker with distilled water and transfer the washings to the volumetric flask, as before. Repeat the rinsing process *at least* two more times.

### Dilution to the Mark

The solute should be completely dissolved *before* you dilute to the mark.

After the solute has been transferred, fill the flask about half-full and swirl the contents to hasten solution. Add more solvent and again mix well. Bring the liquid level almost to the mark, and allow time for drainage ($\simeq 1$ min); then use a medicine dropper to make such final additions of solvent as are necessary (Note). Firmly stopper the flask, and invert it repeatedly to ensure thorough mixing. Transfer the contents to a storage bottle that either is dry or has been thoroughly rinsed with several small portions of the solution from the flask.

**Note.** If, as sometimes happens, the liquid level accidentally exceeds the calibration mark, the solution can be saved by correcting for the excess volume. Use a gummed label to mark the actual location of the meniscus. After the flask has been emptied, carefully refill to the manufacturer's etched mark with water. Use a buret to determine the additional volume needed to fill the flask so that the meniscus is at the gummed-label mark. This volume must be added to the nominal volume of the flask when calculating the concentration of the solution.

## 33I   THE CALIBRATION OF VOLUMETRIC WARE

The reliability of a volumetric analysis depends upon agreement between the volumes purportedly and actually contained in (or delivered by) the apparatus. Calibration simply verifies this agreement or provides a correction if agreement is lacking. The latter involves the assignment of corrections to the existing volume markings or the striking of new markings that more closely agree with the nominal values.

A calibration consists of determining the mass of a liquid of known density that is contained in (or delivered by) volumetric ware. Although calibration appears to be a straightforward process, a number of important variables must be controlled. Principal among these is temperature, which influences a calibration in two ways. First, and more important, the volume occupied by a given mass of liquid varies with temperature. Second, the volume of the apparatus itself is variable, owing to the tendency of the glass to expand or contract with temperature changes.

We have noted (page 803) that the effect of buoyancy upon weighing data is most pronounced when the density of the object differs significantly from that of the weights. If water is used as the calibration liquid, a correction for buoyancy is generally necessary.

Finally, the liquid selected for calibration requires consideration. Water is the liquid of choice for most work. Mercury is also useful, particularly for small volumes. The volume contained in an apparatus will be identical with that delivered by it because mercury does not wet glass surfaces. A small correction must be applied to account for the convex meniscus of mercury. The magnitude of this correction depends upon the diameter of the apparatus at the graduation mark.

The calculations associated with calibration, while not difficult, are somewhat involved. The raw weighing data are first corrected for buoyancy with Equation 33–1. Next, the volume of the apparatus at the temperature of calibration ($T$) is obtained by dividing the density of the liquid at that temperature into the corrected weight. Finally, this volume is corrected to the standard temperature of 20°C as in Example 33–2.

Table 33–3 is provided to ease the computational burden of calibration. Corrections for buoyancy with respect to stainless steel or brass weights (the density difference between the two is small enough to be neglected) and for the volume change of water and of glass containers have been incorporated into these data. Multiplication by the appropriate factor from Table 33–3 converts the mass of water at temperature $T$ to (1) the corresponding volume at that temperature and (2) the volume at 20°C. ·

| Table 33–3 VOLUME OCCUPIED BY 1.0000 g OF WATER WEIGHED IN AIR AGAINST STAINLESS STEEL WEIGHTS* | | |
|---|---|---|
| | Volume, mL | |
| Temperature, $T$ °C | At $T$ | Corrected to 20°C |
| 10 | 1.0013 | 1.0016 |
| 11 | 1.0014 | 1.0016 |
| 12 | 1.0015 | 1.0017 |
| 13 | 1.0016 | 1.0018 |
| 14 | 1.0018 | 1.0019 |
| 15 | 1.0019 | 1.0020 |
| 16 | 1.0021 | 1.0022 |
| 17 | 1.0022 | 1.0023 |
| 18 | 1.0024 | 1.0025 |
| 19 | 1.0026 | 1.0026 |
| 20 | 1.0028 | 1.0028 |
| 21 | 1.0030 | 1.0030 |
| 22 | 1.0033 | 1.0032 |
| 23 | 1.0035 | 1.0034 |
| 24 | 1.0037 | 1.0036 |
| 25 | 1.0040 | 1.0037 |
| 26 | 1.0043 | 1.0041 |
| 27 | 1.0045 | 1.0043 |
| 28 | 1.0048 | 1.0046 |
| 29 | 1.0051 | 1.0048 |
| 30 | 1.0054 | 1.0052 |

*Corrections for buoyancy (stainless steel weights) and change in container volume have been applied.

Example 33–4

A 25-mL pipet delivers 24.976 g of water weighed against stainless steel weights at 25°C. Use the data in Table 33–3 to calculate the volume delivered by this pipet at 25 and 20°C.

$$\text{At } 25°C\colon \quad V = 24.976 \text{ g} \times 1.0040 \text{ mL/g} = 25.08 \text{ mL}$$

$$\text{At } 20°C\colon \quad V = 24.976 \text{ g} \times 1.0037 \text{ mL/g} = 25.07 \text{ mL}$$

## 33I–1  General Directions for Calibration Work

All volumetric ware should be painstakingly freed of water breaks before being calibrated. Burets and pipets need not be dry; volumetric flasks should be thoroughly drained and dried at room temperature. The water used for calibration should be in thermal equilibrium with its surroundings. This condition is best established by drawing the water well in advance, noting its temperature at frequent intervals, and waiting until no further changes occur.

Although an analytical balance can be used for calibration, weighings to the nearest milligram are perfectly satisfactory except for all but the very smallest volumes. Thus a top-loading balance is more conveniently used. Weighing bottles or small, well-stoppered conical flasks can serve as receivers for the calibration liquid.

### Calibration of a Volumetric Pipet

Determine the empty weight of the stoppered receiver to the nearest milligram. Transfer a portion of temperature-equilibrated water to the receiver with the pipet (Section 33H–5), weigh the receiver and its contents (again, to the nearest milligram), and calculate the weight of water delivered from the difference in these weights. Calculate the volume delivered with the aid of Table 33–3. Repeat the calibration several times; calculate the mean volume delivered and its standard deviation.

### Calibration of a Buret

Fill the buret with temperature-equilibrated water and make sure that no air bubbles are trapped in the tip. Allow about 1 min for drainage; then lower the liquid level to bring the bottom of the meniscus to the 0.00-mL mark. Touch the tip to the wall of a beaker to remove any adhering drop. Wait 10 min and recheck the volume; if the stopcock is tight, there should be no perceptible change. During this interval, weigh (to the nearest milligram) a 125-mL conical flask fitted with a rubber stopper.

Once tightness of the stopcock has been established, slowly transfer (at about 10 mL/min) approximately 10 mL of water to the flask. Touch the tip to the wall of the flask. Wait 1 min, record the volume that was apparently delivered, and refill the buret. Weigh the flask and its contents to the nearest milligram; the difference between this weight and the initial value gives the mass of water delivered. Use Table 33–3 to convert this mass to the true volume. Subtract the apparent volume from the true volume. This difference is the correction that should be applied to the apparent volume to give the true volume. Repeat the calibration until agreement within ±0.02 mL is achieved.

Starting again from the zero mark, repeat the calibration, this time delivering about 20 mL to the receiver. Test the buret at 10-mL intervals over its entire volume. Prepare a plot of the correction to be applied as a function of volume delivered. The correction associated with any interval can be determined from this plot.

### Calibration of a Volumetric Flask with a Single-Pan Balance

Weigh the clean, dry flask to the nearest milligram. Then fill to the mark with equilibrated water and reweigh. Calculate the volume contained with the aid of Table 33–3.

A glass tube that has been drawn out to a tip or a medicine dropper is useful in making final adjustments to the liquid level.

### Calibration of a Volumetric Flask Relative to a Pipet

The calibration of a volumetric flask relative to a pipet provides an excellent method for partitioning a sample into aliquots. These directions pertain to a 50-mL pipet and a 500-mL volumetric flask; other combinations are equally convenient.

Carefully transfer ten 50-mL aliquots from the pipet to a dry 500-mL volumetric flask. Mark the location of the meniscus with a gummed label. Cover with a label varnish to ensure permanence. Once the flask is calibrated in this way, dilution to the label permits the same pipet to deliver precisely a one-tenth aliquot of the solution in the flask. Note that if the pipet is changed, recalibration with the new pipet is necessary.

An aliquot is a measured fraction of the volume of a liquid sample.

## 33J   THE LABORATORY NOTEBOOK

A laboratory notebook is needed to record measurements and observations concerning an analysis. The book should be permanently bound with consecutively numbered pages (if necessary, the pages should be hand-numbered before any entries are made). Most notebooks have more than ample room; there is no need to crowd entries.

The first few pages should be saved for a table of contents that is updated as entries are made.

### 33J–1 Rules for the Maintenance of a Laboratory Notebook

1. *Record all data and observations directly into the notebook in ink.* Neatness is greatly to be desired. Nevertheless, the pursuit of this quality must not include the transcribing of observations from a sheet of paper to the notebook or from one notebook to another. The risk of

Remember that you can discard an experimental measurement *only if you have certain knowledge that you made an experimental error*. Thus, you must carefully record experimental observations in your notebook as soon as they occur.

misplacing—or incorrectly transcribing—crucial data and thereby ruining an experiment is unacceptable.

2. Supply each entry or series of entries with a heading or label. Weight data for a set of empty crucibles should carry the heading "empty crucible weights" (or something similar), for example, and the weight of each crucible should be identified by the same number or letter used to label the crucible. The significance of such an entry is obvious when it is recorded but may become unclear with the passage of time.

3. Date each page of the notebook as it is used.

4. *Never* attempt to erase or obliterate an incorrect entry. Instead, cross it out with a single horizontal line and locate the correct entry as nearby as possible. Do not write over incorrect numbers; with time, it may become impossible to distinguish the correct entry from the incorrect one.

5. Never remove a page from the notebook. Draw diagonal lines across any page that is to be disregarded. Provide a brief rationale for disregarding the page.

An entry in a laboratory notebook should never be erased but should be crossed out instead.

## 33J–2 Format

The instructor should be consulted concerning the format to be used in keeping the laboratory notebook.[10] One convention involves using each page consecutively for the recording of data and observations as they occur. The completed analysis is then summarized on the next available page spread (that is, left and right facing pages). As shown in Figure 33–23, the first of these two facing pages should contain the following entries:

1. The title of the experiment ("The Gravimetric Determination of Chloride").

2. A brief statement of the principles upon which the analysis is based.

3. A complete summary of the weighing, volumetric, and/or instrument response data needed to calculate the results.

4. A report of the best value for the set and a statement of its precision.

The second page should contain the following items:

1. Equations for the principal reactions in the analysis.

2. An equation showing how the results were calculated.

3. A summary of observations that appear to bear upon the validity of a particular result or the analysis as a whole. *Any such entry must have been originally recorded in the notebook at the time the observation was made.*

---

[10]See also Howard M. Kanare, *Writing the Laboratory Notebook*. Washington, D.C. 20036: The American Chemical Society, 1985.

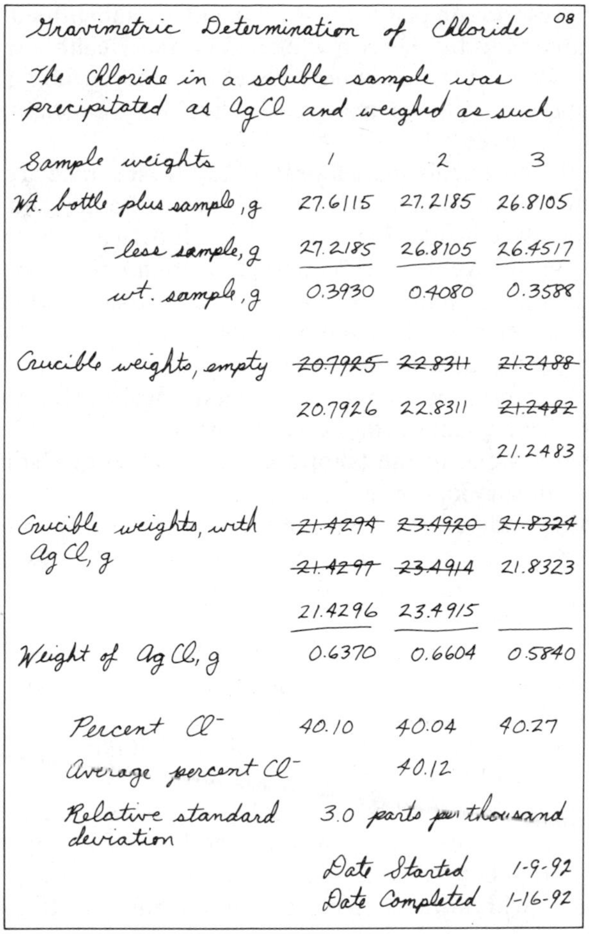

Gravimetric Determination of Chloride 08

The Chloride in a soluble sample was precipitated as AgCl and weighed as such

| Sample weights | 1 | 2 | 3 |
| --- | --- | --- | --- |
| Wt. bottle plus sample, g | 27.6115 | 27.2185 | 26.8105 |
| – less sample, g | 27.2185 | 26.8105 | 26.4517 |
| wt. sample, g | 0.3930 | 0.4080 | 0.3588 |
| Crucible weights, empty | ~~20.7925~~ | ~~22.8311~~ | ~~21.2488~~ |
| | 20.7926 | 22.8311 | ~~21.2482~~ |
| | | | 21.2483 |
| Crucible weights, with AgCl, g | ~~21.4294~~ | ~~23.4920~~ | ~~21.8324~~ |
| | ~~21.4297~~ | ~~23.4914~~ | 21.8323 |
| | 21.4296 | 23.4915 | |
| Weight of AgCl, g | 0.6370 | 0.6604 | 0.5840 |
| Percent $Cl^-$ | 40.10 | 40.04 | 40.27 |
| Average percent $Cl^-$ | | 40.12 | |
| Relative standard deviation | 3.0 parts per thousand | | |

Date Started 1-9-92
Date Completed 1-16-92

Figure 33–23
Summary data page of a laboratory notebook.

## 33K SAFETY IN THE LABORATORY

Work in a chemical laboratory necessarily involves a degree of risk; accidents can and do happen. Strict adherence to the following rules will go far toward preventing (or minimizing the effect of) accidents.

1. At the outset, learn the location of the nearest eye fountain, fire blanket, shower, and fire extinguisher. Learn the proper use of each, and do not hesitate to use this equipment should the need arise.
2. *WEAR EYE PROTECTION AT ALL TIMES*. The potential for serious and perhaps permanent eye injury makes it mandatory that adequate eye protection be worn at all times by students, instructors, and visitors. Eye protection should be donned before entering the laboratory and should be used continuously until it is time to leave. Serious eye injuries have occurred to people performing such innocuous tasks as computing or writing in a laboratory notebook; such incidents are usually the result of someone else's losing control of an experiment.

Regular prescription glasses are not adequate substitutes for eye protection approved by the Office of Safety and Health Administration (OSHA). Contact lenses should never be used in the laboratory because laboratory fumes may react with them and have a harmful effect on the eyes.

3. Most of the chemicals in a laboratory are toxic; some are very toxic, and some—such as concentrated solutions of acids and bases—are highly corrosive in addition. Avoid contact between these liquids and the skin. In the event of such contact, *immediately* flood the affected area with copious quantities of water. If a corrosive solution is spilled on clothing, remove the garment immediately. Time is of the essence; modesty cannot be a matter of concern.

4. *NEVER* perform an unauthorized experiment. Such activity is grounds for disqualification at many institutions.

5. Never work alone in the laboratory; be certain that someone is always within earshot.

6. Never bring food or beverages into the laboratory. Do not drink from laboratory glassware. Do not smoke in the laboratory.

7. Always use a bulb to draw liquids into a pipet; *NEVER* use your mouth to provide suction.

8. Wear adequate foot covering (no sandals). Confine long hair with a net. A laboratory coat or apron will provide some protection and may be required.

9. Be extremely tentative in touching objects that have been heated; hot glass looks just like cold glass.

10. Always fire-polish the ends of freshly cut glass tubing. *NEVER* attempt to force glass tubing through the hole of a stopper. Instead, make sure that both tubing and hole are wet with soapy water. Protect your hands with several layers of towel while inserting glass into a stopper.

11. Use fume hoods whenever toxic or noxious gases are likely to be evolved. Be cautious in testing for odors; use your hand to waft vapors above containers toward your nose.

12. Notify the instructor in the event of an injury.

13. Dispose of solutions and chemicals as instructed. It is illegal to flush solutions containing heavy metal ions or organic liquids down the drain in many localities; alternative arrangements are required for the disposal of such liquids.

## CHAPTER 34

# SELECTED METHODS
# OF ANALYSIS

This chapter contains detailed directions for performing a variety of chemical analyses. The methods have been chosen with the aim of providing experience with common analytical techniques that are widely used by chemists.

The chances of success in the laboratory greatly improve when time is taken at the outset to read carefully and *understand* each step in an analytical method and to develop a plan for how and when each step is to be performed. For greatest efficiency, such study and planning should take place *before entering the laboratory*.

The recommendations that follow regarding laboratory work will not only help provide you with an understanding of the various experiments to be undertaken but will also result in more efficient use of the time available for laboratory work.

## Background Information

Before an analysis is undertaken, you should understand the significance of each step in the procedure in order to avoid the pitfalls and potential sources of error present in all analytical methods. Information about these steps can usually be found in the discussion section that precedes most of the procedures, in sections of earlier chapters that are referred to in the discussion section, and in the "Notes" that follow many of the procedures. If these sources fail to explain the need for certain of the recommended steps, the instructor should be consulted before the laboratory work is begun.

The better you understand the reasons for the steps in an analysis, the greater are your chances for success.

## The Accuracy of Measurements

In studying a procedure, it is wise to distinguish between those measurements that must be made with maximum precision, and thus with maximum care, and those that can be carried out rapidly with little concern for precision. Generally, measurements that appear in the equation used to compute the results must be performed with maximum precision. The remaining measurements can *and should* be made less carefully to conserve time. The words *about* and *approximately* are often used to indicate that a measurement does not have to be done carefully. It is a waste of time and effort to measure, let us say, a volume to ±0.02 mL when an uncertainty of ±0.5 mL or even ±5 mL will have no discernible effect on the results.

In some procedures, a statement such as "weigh three 0.5-g samples to the nearest 0.1 mg . . ." is encountered. Here, samples of perhaps 0.4 to 0.6 g are to be taken and weighed to the nearest 0.1 mg. The number of significant figures in the specification of a volume or a weight is also a guide as to the care that should be taken in making a measurement. For example, the statement "add 10.00 mL of a solution to the beaker" indicates that the volume should be measured carefully with a buret or pipet with the aim of limiting the uncertainty to perhaps ±0.02 mL. In contrast, the instruction "add 10 mL to the beaker" suggests that a graduated cylinder will be adequate to measure the needed volume.

### Time Utilization

The time requirements of the several unit operations involved in an analysis should be studied before work is started. Such study will reveal operations that require considerable elapsed, or clock, time but little or no operator time—for example, when a sample is dried in an oven, cooled in a desiccator, or evaporated on a hot plate. The experienced chemist plans to use such periods of waiting to perform other operations or perhaps to begin a new analysis. Some workers find it worthwhile to prepare a written time schedule for each laboratory period to avoid periods when no work can be done.

Time planning is also needed to identify places where an analysis can be interrupted for overnight or longer, as well as those operations that must be completed without a break.

### Reagents

Directions for the preparation of reagents accompany many of the procedures. Before preparing such reagents, it is wise to check to see if they are already prepared and available for general use.

If a reagent is known to pose a hazard, you should plan *in advance of the laboratory period* what steps you should take to minimize injury or damage. Furthermore, rules for the disposal of waste liquids and solids should be known and followed. These rules vary from one part of the country to another and from one laboratory to another.

### Water

Some laboratories use deionizers to purify water; others employ stills for this purpose. The terms "distilled water" and "deionized water" are used interchangeably in the directions that follow. Either type is satisfactory for analytical work.

Tap water should be employed only for preliminary cleaning of glassware. The cleaned glassware is then rinsed with at least three small portions of distilled or deionized water.

## 34A   GRAVIMETRIC METHODS OF ANALYSIS

General aspects, calculations, and typical applications of gravimetric analysis are discussed in Chapter 4.

### 34A–1   The Gravimetric Determination of Chloride in a Soluble Sample

**Discussion**

The chloride content of a soluble salt can be determined by precipitation of the anion as silver chloride:

$$Ag^+ + Cl^- \rightarrow AgCl(s)$$

The precipitate is collected in a weighed filtering crucible and washed; its weight is determined after it has been dried to constant weight at 110°C.

The solution containing the sample is kept somewhat acidic during the precipitation to eliminate possible interference from anions of weak acids (such as $CO_3^{2-}$) that form sparingly soluble silver salts in a neutral environment. A moderate excess of silver ion is needed to diminish the solubility of silver chloride, but an excess is avoided to minimize coprecipitation of silver nitrate.

Silver chloride forms first as a colloid and is subsequently coagulated with heat. Nitric acid and the small excess of silver nitrate promote coagulation by providing a moderately high electrolyte concentration. Nitric acid in the wash solution maintains the electrolyte concentration and eliminates the possibility of peptization during the washing step; the acid subsequently decomposes to give volatile products when the precipitate is dried. See Section 4C–2 for additional information concerning the properties and treatment of colloidal precipitates.

In common with other silver halides, finely divided silver chloride undergoes photodecomposition:

$$2\ AgCl(s) \rightarrow 2\ Ag(s) + Cl_2(g)$$

The elemental silver produced in this reaction is responsible for the violet color that develops in the precipitate. In principle, this reaction leads to low results for chloride ion. In practice, however, its effect is negligible provided direct and prolonged exposure to sunlight is avoided.

If photodecomposition of silver chloride occurs before filtration, the additional reaction

$$3\ Cl_2(aq) + 3\ H_2O + 5\ Ag^+ \rightarrow 5\ AgCl(s) + ClO_3^- + 6\ H^+$$

tends to cause high results.

Some photodecomposition of silver chloride is inevitable as the analysis is ordinarily performed. It is worthwhile to minimize exposure of the solid to intense sources of light as far as possible.

Iodide, bromide, and thiocyanate, if present, precipitate along with silver chloride and cause high results. Additional interference can be expected from tin and antimony, which are likely to precipitate as oxychlorides under the conditions of the analysis.

Because silver nitrate is expensive, any unused reagent should be collected in a storage container; similarly, precipitated silver chloride should be retained after the analysis is complete.[1]

---

### Procedure

Clean three sintered-glass or porecelain filtering crucibles by allowing about 5 mL of concentrated $HNO_3$ to stand in each for about 5 min. Use a vacuum (Figure 33–17) to draw the acid through the crucible. Rinse each crucible with three portions of tap water, and then discontinue the vacuum. Next, add about 5 mL of 6 M $NH_3$ and wait for about 5 min before drawing it through the filter. Finally, rinse each crucible with six to eight portions of distilled or deionized water. Provide each crucible with an identifying mark. Bring the crucibles to constant weight by heating at 110°C while the other steps in the analysis are being carried out. The first drying should be for at least 1 hr; subsequent heating periods can be somewhat shorter (30 to 40 min).

Transfer the unknown to a weighing bottle and dry it at 110°C (Figure 33–9) for 1 to 2 hr; allow the bottle and contents to cool to room temperature in a desiccator. Weigh (to the nearest 0.1 mg) individual samples by difference (page 524) into 400-mL beakers (Note 1). Dissolve each sample in about 100 mL of distilled water to which 2 to 3 mL of 6 M $HNO_3$ has been added.

*Be sure to label your beakers.*

Slowly, and with good stirring, add 0.2 M $AgNO_3$ to each of the cold sample solutions until AgCl is observed to coagulate (Notes 2, 3); then introduce an additional 3 to 5 mL. Heat almost to boiling, and digest the solids for about 10 min. Add a few drops of $AgNO_3$ to confirm that precipitation is complete. If additional precipitate forms, add about 3 mL of $AgNO_3$, digest, and again test for completeness of precipitation. Pour any unused $AgNO_3$ into a waste container (NOT into the original reagent bottle). Cover each beaker, and store in a dark place for at least 2 hr and preferably until the next laboratory period.

Read the instructions for filtration in Section 33G–2. Decant the supernatant liquids through weighed filtering crucibles. Wash the precipitates several times (while they are still in the beaker) with a wash solution consisting of 2 to 5 mL of 6 M $HNO_3$ per liter of distilled water; decant these washings through the filters. Quantitatively transfer the AgCl from

---

[1]Silver can be recovered from silver chloride and from surplus reagent by reduction with ascorbic acid; see J. W. Hill and L. Bellows, *J. Chem. Educ.,* **1986,** *63* (4), 357; see also J. P. Rawat and S. Iqbal M. Kamoonpuri, *ibid.,* **1986,** *63* (6), 537 for recovery (as $AgNO_3$) based on ion exchange. See also D. D. Perrin, W. L. F. Amrarejo, and D. R. Perrin, *Chemistry International,* **1987,** *9* (1), 3, concerning a potential explosion hazard in the recovery of silver nitrate.

the beakers to the individual crucibles with fine streams of wash solution; use rubber policemen to dislodge any particles that adhere to the walls of the beakers. Continue washing until the filtrates are essentially free of $Ag^+$ ion (Note 4).

Dry the precipitate at 110°C for at least 1 hr. Store the crucibles in a desiccator while they cool. Determine the weight of the crucibles and their contents. Repeat the cycle of heating, cooling, and weighing until consecutive weighings agree to within 0.2 mg. Calculate the percentage of $Cl^-$ in the sample.

Upon completion of the analysis, remove the precipitates by gently tapping the crucibles over a piece of glazed paper. Transfer the collected AgCl to a container for silver wastes. Remove the last traces of AgCl by filling the crucibles with 6 M $NH_3$ and allowing them to stand.

**Notes**
1. Consult with the instructor concerning an appropriate sample size.
2. Determine the approximate amount of $AgNO_3$ needed by calculating the volume that would be required if the unknown were pure NaCl.
3. Use a separate stirring rod for each sample and leave it in its beaker throughout the determination.
4. To test the washings for $Ag^+$, collect a small volume in a test tube and add a few drops of HCl. Washing is judged complete when little or no turbidity develops.

## 34A–2  The Gravimetric Determination of Tin in Brass

### Discussion

Brasses are important alloys. Copper is ordinarily the principal constituent, with lesser amounts of lead, zinc, tin, and possibly other elements as well. Treatment of a brass with nitric acid results in the formation of the sparingly soluble "metastannic acid" $H_2SnO_3 \cdot xH_2O$; all other constituents are dissolved. The solid is filtered, washed, and ignited to $SnO_2$.

The gravimetric determination of tin provides experience in the use of ashless filter paper and is frequently performed in conjunction with a more inclusive analysis of a brass sample.

### Procedure

Provide identifying marks on three porcelain crucibles and their covers. During waiting periods in the experiment, bring these to constant weight by ignition at 900°C in a muffle furnace.

Do not dry the unknown. If so instructed, rinse it with acetone to remove any oil or grease. Weigh (to the nearest 0.1 mg) approximately 1 g samples of the unknown into 250-mL beakers. Cover the beakers with watch glasses. Place the beakers in the HOOD, and cautiously introduce a mixture containing about 15 mL of concentrated $HNO_3$ and 10 mL of $H_2O$. Digest the samples for at least 30 min; add more $HNO_3$ if necessary.

Rinse the watch glasses; then evaporate the solutions to about 5 mL but not to dryness (Note 1).

Add about 5 mL of 3 M $HNO_3$, 25 mL of distilled water, and one quarter of a tablet of filter paper pulp to each sample; heat without boiling for about 45 min. Collect the precipitated $H_2SnO_3 \cdot xH_2O$ on fine-porosity ashless filter papers (Section 33G–3 and Notes 2, 3). Use many small volumes of hot 0.3 M $HNO_3$ to wash the last traces of copper from the precipitate. Test for completeness of washing with a drop of $NH_3$ on the top of the precipitate; wash further if the precipitate turns blue.

Remove the filter paper and its contents from the funnels, fold, and place in crucibles that, with their covers, have been brought to constant weight (Figure 33–15). Ash the filter paper at as low a temperature as possible. There must be free access of air throughout the charring (Section 33G–3 and Figure 33–16). Gradually increase the temperature until all the carbon has been removed. Then bring the covered crucibles and their contents to constant weight in a 900°C furnace (Note 4). Calculate the percentage of tin in the unknown.

**Notes**
1. It is often time-consuming and difficult to redissolve the soluble components of the residue after a sample has been evaporated to dryness.
2. The filtration step can be quite time-consuming and once started cannot be interrupted.
3. If the unknown is to be analyzed electrolytically for its lead and copper content (Section 34J–1), collect the filtrates in tall-form beakers. The final volume should be about 125 mL; evaporate to that volume if necessary. If the analysis is for tin only, the volume of washings is not important.
4. Partial reduction of $SnO_2$ may cause the ignited precipitate to appear gray. In this case, add a drop of nitric acid, cautiously evaporate, and ignite again.

## 34A–3 The Gravimetric Determination of Nickel in Steel

### Discussion

The nickel in a steel sample can be precipitated from a slightly alkaline medium with an alcoholic solution of dimethylglyoxime (page 88). Interference from iron(III) is eliminated by masking with tartaric acid. The product is freed of moisture by drying at 110°C.

The bulky character of nickel dimethylglyoxime limits the weight of nickel that can be accommodated conveniently and thus the sample weight. Care must also be taken to control the excess of alcoholic dimethylglyoxime used. If too much is added, the alcohol concentration becomes great enough to dissolve appreciable amounts of the nickel dimethylglyoxime, which leads to low results. If the alcohol concentration becomes too low, however, some of the reagent may precipitate, giving a positive error.

## Preparation of Solutions

(a) *Dimethylglyoxime, 1% (w/v).* Dissolve 10 g of dimethylglyoxime in 1 L of ethanol. (Sufficient for about 50 precipitations.)
(b) *Tartaric acid, 15% (w/v).* Dissolve 225 g of tartaric acid in sufficient water to give 1500 mL of solution. Filter before use if the solution is not clear. (Sufficient for about 50 precipitations.)

## Procedure

Clean and mark three medium-porosity sintered-glass crucibles (Note 1); bring them to constant weight by drying at 110°C for at least 1 hr.

Weigh (to the nearest 0.1 mg) samples containing between 30 and 35 mg of nickel into individual 400-mL beakers (Note 2). In the HOOD, dissolve each sample in about 50 mL of 6 M HCl with gentle warming. Carefully add approximately 15 mL of 6 M $HNO_3$, and boil gently to expel any oxides of nitrogen that may have been produced. Dilute to about 200 mL and heat to boiling. Introduce about 30 mL of 15% tartaric acid and sufficient concentrated $NH_3$ to produce a faint odor of $NH_3$ in the vapors over the solutions (Note 3); then add another 1 to 2 mL of $NH_3$. If the solutions are not clear at this stage, proceed as directed in Note 4. Make the solutions acidic with HCl (no odor of $NH_3$), heat to 60 to 80°C, and add about 20 mL of the 1% dimethylglyoxime solution. With good stirring, add 6 M $NH_3$ until a slight excess exists (faint odor of $NH_3$) plus an additional 1 to 2 mL. Digest the precipitates for 30 to 60 min, cool for at least 1 hr, and filter.

Wash the solids with water until the washings are free of $Cl^-$ (Note 5). Bring the crucibles and their contents to constant weight at 110°C. Report the percentage of nickel in the sample. The dried precipitate has the composition $Ni(C_4H_7O_2N_2)_2$(fw = 288.93).

## Notes

1. Medium-porosity porcelain filtering crucibles or Gooch crucibles with glass pads can be substituted for sintered-glass crucibles in this determination.
2. Use a separate stirring rod for each sample and leave it in the beaker throughout the analysis.
3. The existence or absence of excess $NH_3$ is readily established by odor; the vapors over the container should be wafted toward the nose with a waving motion of the hand.
4. If $Fe_2O_3 \cdot xH_2O$ forms upon addition of $NH_3$, acidify the solution with HCl, introduce additional tartaric acid, and neutralize again. Alternatively, remove the solid by filtration. Thorough washing with a hot $NH_3/NH_4Cl$ solution is required; the washings are combined with the solution containing the bulk of the sample.
5. Test the washings for $Cl^-$ by collecting a small portion in a test tube, acidifying with $HNO_3$, and adding a drop or two of 0.1 M $AgNO_3$. Washing is judged complete when little or no turbidity develops.

## 34B PRECIPITATION TITRATIONS

As noted in Chapter 9, most precipitation titrations make use of a standard silver nitrate solution as titrant. Directions follow for the volumetric titration of chloride ion using an adsorption indicator and for the weight titration of the same species by the Mohr method, in which chromate ion serves as the indicator.

### 34B–1 Preparation of a Standard Silver Nitrate Solution

#### Procedure

Use a top-loading balance to transfer the approximate weight of $AgNO_3$ to a weighing bottle (Note 1). Dry at 110°C for about 1 hr but not much longer (Note 2), and then cool to room temperature in a desiccator. Weigh the bottle and its contents (to the nearest 0.1 mg). Transfer the bulk of the $AgNO_3$ to a volumetric flask using a powder funnel. Cap the weighing bottle; reweigh it and any solid that remains. Rinse the powder funnel thoroughly. Dissolve the $AgNO_3$, dilute to the mark with water, and mix well (Note 3). Calculate the molar concentration of this solution.

#### Notes

1. Consult with your instructor concerning the volume and concentration of $AgNO_3$ to be prepared. The weight of $AgNO_3$ to be taken is as follows:

| Silver Ion Concentration, M | Approximate Weight (g) of $AgNO_3$ Needed to Prepare | | |
|---|---|---|---|
| | **1000 mL** | **500 mL** | **250 mL** |
| 0.10 | 16.9 | 8.5 | 4.2 |
| 0.05 | 8.5 | 4.2 | 2.1 |
| 0.02 | 3.4 | 1.8 | 1.0 |

2. Prolonged heating causes partial decomposition of $AgNO_3$. Some discoloration may occur, even after only 1 hr at 110°C; the effect of this decomposition on the purity of the reagent is ordinarily imperceptible.
3. Silver nitrate solutions should be stored in a dark place when not in use.

### 34B–2 The Determination of Chloride by Titration with an Adsorption Indicator

#### Discussion

In this titration, the anionic adsorption indicator dichlorofluorescein is used to locate the end point. With the first excess of titrant, the indicator becomes incorporated in the counter-ion layer surrounding the silver chloride and imparts color to the solid (Section 9A–4). In order to obtain a satisfactory color change, it is desirable to maintain the particles of silver chloride in the colloidal state. Dextrin is added to the solution to stabilize the colloid and prevent its coagulation.

## Preparation of Solutions

*Dichlorofluorescein indicator*. Dissolve 0.2 g of dichlorofluorescein in a solution prepared by mixing 75 mL of ethanol and 25 mL of water. (Sufficient for several hundred titrations.)

## Procedure

Dry the unknown at 110°C for about 1 hr; allow it to return to room temperature in a desiccator. Weigh individual samples (to the nearest 0.1 mg) into individual conical flasks, and dissolve them in appropriate volumes of distilled water (Note 1). To each, add about 0.1 g of dextrin and 5 drops of indicator. Titrate (Note 2) with $AgNO_3$ to the first permanent pink color of silver dichlorofluoresceinate. Report the percentage of $Cl^-$ in the unknown.

Read your buret to the nearest 0.01 mL.

### Notes
1. Use 0.25-g samples for 0.1 M $AgNO_3$ and about half that amount for 0.05 M reagent. Dissolve the former in about 200 mL of distilled water and the latter in about 100 mL. If 0.02 M $AgNO_3$ is to be used, a 0.4-g sample should be weighed into a 500-mL volumetric flask, and 50-mL aliquots should be taken for titration.
2. Colloidal AgCl is sensitive to photodecomposition, particularly in the presence of the indicator; attempts to perform the titration in direct sunlight will fail. If photodecomposition appears to be a problem, establish the approximate end point with a rough preliminary titration, and use this information to estimate the volumes of $AgNO_3$ needed for the other samples. For each subsequent sample, add the indicator and dextrin only after most of the $AgNO_3$ has been added, and then complete the titration without delay.

## 34B–3 The Determination of Chloride by a Weight Titration Based on the Mohr Method

### Discussion

The Mohr method uses $CrO_4^{2-}$ ion as an indicator in the titration of chloride ion with silver nitrate. The first excess of titrant results in the formation of a red silver chromate precipitate, which signals the end point.

Instead of a buret, a balance is employed in this procedure to determine the weight of silver nitrate solution needed to reach the end point. The concentration of the silver nitrate is most conveniently determined by standardization against primary-standard sodium chloride, although direct preparation by weight is also feasible. The reagent concentration is expressed as weight molarity (mmol $AgNO_3$/g of solution). See Section 5D for additional details.

## Preparation of Solutions

(a) *Silver nitrate, approximately 0.1 mmol/g of solution.* (Sufficient for about ten titrations.) Dissolve about 4.5 g of $AgNO_3$ in about 500 mL of distilled water. Standardize the solution against weighed quantities of reagent-grade NaCl as directed in *Procedure* (Note). Express the concentration as weight molarity (mmol $AgNO_3$/g of solution). When not in use, store the solution in a dark place.

(b) *Potassium chromate, 5%.* (Sufficient for about ten titrations.) Dissolve about 1.0 g of $K_2CrO_4$ in about 20 mL of distilled water.

**Note.** Alternatively, standard $AgNO_3$ can be prepared directly by weight. To do so, follow the directions in Section 34B–1 for weighing out a known amount of primary-standard $AgNO_3$. Use a powder funnel to transfer the weighed $AgNO_3$ to a 500-mL polyethylene bottle that has been previously weighed to the nearest 10 mg. Add about 500 mL of water and weigh again. Calculate the weight molarity.

## Procedure

Dry the unknown at 110°C for at least 1 hr (Note 1). Cool in a desiccator. Consult with your instructor for a suitable sample size. Weigh (to the nearest 0.1 mg) individual samples into 250-mL conical flasks, and dissolve in about 100 mL of distilled water. Add small quantities of $NaHCO_3$ until effervescence ceases. Introduce about 2 mL of $K_2CrO_4$ solution, and titrate (Note 2) to the first permanent appearance of red $Ag_2CrO_4$.

Determine an indicator blank by suspending a small amount of chloride-free $CaCO_3$ in 100 mL of distilled water containing 2 mL of $K_2CrO_4$.

Correct reagent weights for the blank. Report the percentage of $Cl^-$ in the unknown.

Dispose of AgCl and reagents as directed by the instructor.

### Notes
1. The $AgNO_3$ is conveniently standardized concurrently with the analysis. Dry reagent-grade NaCl for about 1 hr. Cool; then weigh (to the nearest 0.1 mg) 0.25-g portions into conical flasks and titrate as above.
2. Directions for performing a weight titration are given in Section 33F–2.

## 34C  NEUTRALIZATION TITRATIONS

Neutralization titrations are performed with standard solutions of strong acids or bases. While a single solution (of either acid or base) is sufficient for the titration of a given type of analyte, it is convenient to have standard solutions of both acid and base available in the event a back-titration is needed to locate end points more exactly. The concentration of one solution is established by titration against a primary standard; the concentration of the other is then determined from the acid/base ratio (that is, the volume of acid needed to neutralize 1.000 mL of the base).

The directions in this section call for volumetric measurements. The methods are readily adapted to the weight titration technique described in Section 33F, however. In the latter case, weight molarity (mmol reagent/g solution) is computed from standardization data and used to calculate analyte concentrations.

## 34C–1  The Effect of Atmospheric Carbon Dioxide on Neutralization Titrations

Water in equilibrium with the atmosphere is about $1 \times 10^{-5}$ M in carbonic acid as a consequence of the equilibrium

$$CO_2(g) + H_2O \rightleftharpoons H_2CO_3(aq)$$

At this concentration level, the amount of 0.1 M base consumed by the carbonic acid in a typical titration is negligible. With more dilute reagents (<0.05 M), however, the water used as a solvent for the analyte and in the preparation of reagents must be freed of carbonic acid by boiling for a brief period.

Water that has been purified by distillation rather than by deionization is often supersaturated with carbon dioxide and may thus contain sufficient acid to affect the results of an analysis.[2] The instructions that follow are based upon the assumption that the amount of carbon dioxide in the water supply can be neglected without causing serious error. For further discussion on the effects of carbon dioxide in neutralization titrations, see Section 12A–3.

## 34C–2  Preparation of Indicator Solutions for Neutralization Titrations

### Discussion

The theory of acid/base indicators is discussed in Section 10A–2, and the structures of some common indicators are shown in Section 10G–1. An indicator exists for virtually any pH range between 1 and 13.[3] Directions follow for the preparation of indicator solutions suitable for most neutralization titrations.

---

[2]Water that is to be used for neutralization titrations can be tested by adding 5 drops of phenolphthalein to a 500-mL portion. Less than 0.2 to 0.3 mL of 0.1 M $OH^-$ should suffice to produce the first faint pink color of the indicator. If a larger volume is needed, the water should be boiled and cooled before it is used to prepare standard solutions or to dissolve samples.

[3]See, for example, J. Beukenkemp and W. Rieman III, in *Treatise on Analytical Chemistry*, I. M. Kolthoff and P. J. Elving, Eds., Part I, Vol. 11, pp. 6987–7001. New York: Wiley, 1974.

## Procedure

Stock solutions ordinarily contain between 0.5 and 1.0 g of indicator per liter. (One liter of indicator is sufficient for hundreds of titrations.)

(a) *Methyl orange, methyl red, and sulphonthaleins.* Dissolve the sodium salt directly in deionized water. (The common sulphonthaleins include bromocresol green, bromothymol blue, bromophenol blue, thymol blue, cresol red, and phenol red.)

(b) *Phenolphthalein, thymolphthalein.* Dissolve the solid indicator in a solution consisting of 800 mL ethanol and 200 mL of distilled or deionized water.

## 34C–3 Preparation of Dilute Hydrochloric Acid Solutions

### Discussion

The preparation and standardization of acids are considered in Sections 12A–1 and 12A–2.

### Procedure

For a 0.1 M solution, add about 8 mL of concentrated HCl to about 1 L of deionized water (Note). Mix thoroughly, and store in a glass-stoppered bottle.

**Note.** It is advisable to eliminate $CO_2$ from the water by a preliminary boiling if very dilute solutions (<0.05 M) are being prepared.

## 34C–4 Preparation of Carbonate-Free Sodium Hydroxide

### Discussion

See Sections 12A–3 and 12A–4 for information concerning the preparation and standardization of bases.

### Procedure

If so directed by your instructor, prepare a bottle for protected storage (Figure 12–2; Note 1). Transfer 1 L of distilled or deionized water to the storage bottle (see the Note in Section 34C–3). Decant 4 to 5 mL of 50% NaOH into a small container (Note 2), add it to the water, and *mix thoroughly*. USE EXTREME CARE IN HANDLING 50% NaOH, which is highly corrosive. If the reagent comes into contact with your skin, IMMEDIATELY flush the area with *copious* amounts of water.

Protect the solution from unnecessary contact with the atmosphere.

**Notes**

1. A solution of base that will be used up within two weeks can be stored in a tightly capped polyethylene bottle. After each removal of base, squeeze the bottle while tightening the cap to minimize the air space above the reagent. The bottle will become embrittled after extensive use as a container for bases.

2. Be certain that any solid $Na_2CO_3$ in the 50% NaOH has settled to the bottom of the container and that the decanted liquid is absolutely clear. If necessary, filter the base through a glass mat in a Gooch crucible; collect the clear filtrate in a test tube inserted in the filter flask.

## 34C–5 The Determination of the Acid/Base Ratio

### Discussion

When both acid and base solutions have been prepared, it is useful to determine their volumetric combining ratio so that the two solutions can be used to establish end points more accurately in any of the methods given later in this section. In addition, knowledge of this ratio and the concentration of one solution permits calculation of the molarity of the other.

### Procedure

Instructions for placing a buret into service are given in Sections 33H–4 and 33H–6; consult these instructions if necessary. Place a test tube or small beaker over the top of the buret that holds the NaOH solution to minimize contact between the solution and the atmosphere.

Record the initial volumes of acid and base in the burets to the nearest 0.01 mL. Deliver 35 to 40 mL of the acid into a 250-mL conical flask. Touch the tip of the buret to the inside wall of the flask, and rinse down with a little distilled water. Add two drops of phenolphthalein (Note 1) and then sufficient base to render the solution a definite pink. Introduce acid dropwise to discharge the color, and again rinse down the walls of the flask. Carefully add base until the solution acquires a faint pink hue that persists for at least 30 s (Notes 2, 3). Record the final buret volumes (again, to nearest 0.01 mL). Repeat the titration. Calculate the acid/base volume ratio. The ratios for duplicate titrations should agree to within 1 to 2 ppt. Perform additional titrations, if necessary, to achieve this order of precision.

*Read buret volumes to the nearest 0.01 mL.*

**Notes**

1. The volume ratio can also be determined with an indicator that has an acidic transition range, such as bromocresol green (page 215). If the NaOH is contaminated with carbonate, the ratio obtained with this indicator will differ significantly from the value obtained with phenolphthalein. In general, the acid/base ratio should be evaluated with the indicator that is to be used in subsequent titrations.

2. Fractional drops can be formed on the buret tip, touched to the wall of the flask, and then rinsed down with a small amount of water.
3. The phenolphthalein end point fades as $CO_2$ is absorbed from the atmosphere.

### 34C–6 Standardization of Hydrochloric Acid Against Sodium Carbonate

#### Discussion

See Section 12A–2.

#### Procedure

Dry a quantity of primary-standard $Na_2CO_3$ for about 2 hr at 110°C (Figure 33–9), and cool in a desiccator. Weigh individual 0.20- to 0.25-g samples (to the nearest 0.1 mg) into 250-mL conical flasks, and dissolve each in about 50 mL of distilled water. Introduce 3 drops of bromocresol green, and titrate with HCl until the solution just begins to change from blue to green. Boil the solution for 2 to 3 min, cool to room temperature (Note 1), and complete the titration (Note 2).

Determine an indicator correction by titrating approximately 100 mL of 0.05 M NaCl and 3 drops of indicator. Boil briefly, cool, and complete the titration. Subtract any volume needed for the blank from the titration volumes. Calculate the concentration of the HCl solution.

#### Notes

1. The indicator should change from green to blue as $CO_2$ is removed during heating. If no color change occurs, an excess of acid was added originally. This excess can be back-titrated with base, provided the acid/base ratio is known; otherwise, the sample must be discarded.
2. It is permissible to back-titrate with base to establish the end point with greater certainty.

### 34C–7 Standardization of Sodium Hydroxide Against Potassium Hydrogen Phthalate

#### Discussion

See Section 12A–4.

#### Procedure

Dry a quantity of primary-standard potassium hydrogen phthalate (KHP) for about 2 hr at 110°C (Figure 33–9), and cool in a desiccator. Weigh individual 0.7- to 0.8-g samples (to the nearest 0.1 mg) into 250-mL conical flasks, and dissolve each in 50 to 75 mL of distilled or deionized water.

Add 2 drops of phenolphthalein; titrate with base until the pink color of the indicator persists for 30 s (Note). Calculate the concentration of the NaOH solution.

**Note.** It is permissible to back-titrate with acid to establish the end point more precisely. The volume of acid must, of course, be recorded so that a correction to the volume of base used in the titration can be computed from the acid/base ratio.

## 34C–8  The Determination of Potassium Hydrogen Phthalate in an Impure Sample

### Discussion

The unknown is a mixture of KHP and a neutral salt. This analysis is conveniently performed concurrently with the standardization of the base.

### Procedure

Consult with the instructor concerning an appropriate sample size. Then follow the directions in Section 34C–7.

## 34C–9  The Determination of the Acid Content of Vinegars and Wines

### Discussion

The total acid content of a vinegar or a wine is readily determined by titration with a standard base. It is customary to report the acid content of vinegar in terms of acetic acid, the principal acidic constituent, even though other acids are present. Similarly, the acid content of a wine is expressed in terms of percent tartaric acid, notwithstanding the presence of other acids in the sample. Most vinegars contain about 5% acid (w/v) expressed as acetic acid; wines ordinarily contain somewhat under 1% acid (w/v) expressed as tartaric acid.

### Procedure

(a) *If the unknown is a vinegar* (Note 1), pipet 25.00 mL into a 250-mL volumetric flask and dilute to the mark with distilled water. Mix thoroughly, and pipet 50.00-mL aliquots into 250-mL conical flasks. Add about 50 mL of water and 2 drops of phenolphthalein (Note 2) to each, and titrate with standard 0.1 M NaOH to the first permanent (≈30 s) pink color.

   Report the acidity of the vinegar as percent (w/v) $CH_3COOH$ (fw = 60.053).

(b) *If the unknown is a wine,* pipet 50.00-mL aliquots into 250-mL conical flasks, add about 50 mL of distilled water and 2 drops of phenolphthalein to each (Note 2), and titrate to the first permanent ($\approx$30 s) pink color.

Express the acidity of the sample as percent (w/v) tartaric acid $C_2H_4O_2(COOH)_2$ (fw = 150.09).

**Notes**
1. The acidity of bottled vinegar tends to decrease on exposure to air. It is recommended that unknowns be stored in individual vials with snug covers.
2. The amount of indicator used should be increased as necessary to make the color change visible in colored samples.

## 34C–10 The Determination of Sodium Carbonate in an Impure Sample

### Discussion

The titration of sodium carbonate is discussed in Section 12A–2 in connection with its use as a primary standard; the same considerations apply for the determination of carbonate in an unknown that has no interfering contaminants.

### Procedure

Dry the unknown at 110°C for 2 hr, and then cool in a desiccator. Consult with the instructor on an appropriate sample size. Weigh (to the nearest 0.1 mg) individual samples into 250-mL conical flasks. Dissolve each in 50 to 75 mL of distilled water, add 2 drops of bromocresol green, and titrate with standard acid until the indicator just begins to turn green. Boil the solution for 2 to 3 min, cool to room temperature, and complete the titration. If additional acid is not required after boiling, the solution already contains an excess. Either back-titrate this excess with standard base, or discard the sample if a base solution is not available.

Report the percentage of $Na_2CO_3$ in the sample.

## 34C–11 The Determination of Amine Nitrogen by the Kjeldahl Method

### Discussion

These directions are suitable for the determination of protein in materials such as blood meal, wheat flour, pasta products, dry cereals, and pet foods. A simple modification permits the analysis of unknowns that con-

tain more highly oxidized forms of nitrogen.[4] The chemistry of the
Kjeldahl method is described in Section 12B–1.

---

## Procedure

### Preparation of Samples
Consult with your instructor for an appropriate sample size. *If the unknown is powdered* (such as blood meal), weigh samples onto individual
9-cm filter papers (Note 1). Fold the paper around the sample and drop
each into a Kjeldahl flask (the paper keeps the samples from clinging to
the neck of the flask). *If the unknown is not powdered* (such as breakfast
cereals or pasta), the samples can be weighed directly into the Kjeldahl
flasks without the paper.

Add 25 mL of concentrated $H_2SO_4$, 10 g of powdered $K_2SO_4$, and the
catalyst (Note 2) to each flask.

### Digestion
Clamp the flasks in a slanted position in a hood or vented digestion rack.
Heat carefully to boiling. Discontinue heating briefly if foaming becomes
excessive; never allow the foam to reach the neck of the flask. Once
foaming ceases and the acid is boiling vigorously, the samples can be left
unattended; prepare the distillation apparatus during this time. Continue
digestion until the solution becomes colorless or faint yellow; 2 to 3 hr
may be needed for some materials. If necessary, allow the mixtures to
cool somewhat, and *cautiously* replace the acid lost by evaporation.

When digestion is complete, discontinue heating, and allow the flasks to
cool to room temperature; swirl the flasks if the contents show signs of
solidifying. Cautiously add 250 mL of water to each flask and again allow
the solution to cool to room temperature. If mercury was used as the
catalyst, introduce 25 mL of 4% (w/v) $Na_2S$ solution (Note 3).

### Distillation of Ammonia
Arrange a distillation apparatus similar to that shown in Figure 12–3b
(page 271). Pipet 50.00 mL of standard 0.1 M HCl into the receiver flask
(Note 4). Clamp the flask so that the tip of the adapter extends below the
surface of the standard acid. Circulate water through the condenser
jacket.

Hold the Kjeldahl flask at an angle and gently introduce about 60 mL of
50% (w/v) NaOH solution, taking care to minimize mixing with the solution in the flask. *The concentrated caustic solution is highly corrosive and
should be handled with great care* (Note 5). Add several pieces of granulated zinc (Note 6) and a small piece of litmus paper. *Immediately* connect
the Kjeldahl flask to the spray trap. Cautiously mix the contents by gentle
swirling. The litmus paper should be blue after mixing is complete, indicating that the solution is basic.

Bring the solution to a boil, and distill at a steady rate until one half to
one third of the original volume remains. Control the rate of heating to

---

[4]See *Official Methods of Analysis,* 14th ed., p. 16. Washington, D.C.: Association of Official
Analytical Chemists, 1984.

prevent the liquid in the receiver flask from being drawn back into the Kjeldahl flask. After distillation is judged complete, lower the receiver flask to bring the adapter well clear of the liquid. Discontinue heating, disconnect the apparatus, and rinse the inside of the condenser with small portions of distilled water, collecting the washings in the receiver flask. Add 2 drops of bromocresol green to the receiver flask, and titrate the residual HCl with standard 0.1 M NaOH to the color change of the indicator.

Report the percentage of nitrogen and the percentage of protein (Note 7) in the unknown.

**Notes**
1. If filter paper is used to hold the sample, carry a similar piece through the analysis as a blank. Acid-washed filter paper is frequently contaminated with measurable amounts of ammonium ion and should be avoided if possible.
2. Any of the following catalyze the digestion: a drop of mercury, 0.5 g of HgO, a crystal of $CuSO_4$, 0.1 g of selenium, 0.2 g of $CuSeO_3$. The catalyst can be omitted, if desired.
3. Mercury(II) ions must be precipitated as the sulfide to prevent retention of some of the ammonia as a mercury complex.
4. A modification of this procedure uses about 50 mL of 4% boric acid solution in lieu of the standard HCl in the receiver flask (page 271). After distillation is complete, the ammonium borate produced is titrated with standard 0.1 M HCl, with 2 to 3 drops of bromocresol green as indicator.
5. If any sodium hydroxide solution comes into contact with your skin, wash the affected area IMMEDIATELY with copious amounts of water.
6. Granulated zinc (10 to 20 mesh) is added to minimize bumping during the distillation; it reacts slowly with the base to give small bubbles of hydrogen that prevent superheating of the liquid.
7. The percentage of protein in the unknown is calculated by multiplying the % N by an appropriate factor: 5.70 for cereals, 6.25 for meats, and 6.38 for dairy products.

---

## 34D   COMPLEX-FORMATION TITRATIONS WITH EDTA

See Chapter 13 for a discussion of the analytical uses of EDTA as a chelating reagent. Directions follow for a direct titration of magnesium, a displacement titration of calcium with the magnesium/EDTA complex, and a determination of the hardness of a natural water. Although these procedures are written for volumetric titrimetry, they are readily adapted to weight titrations (Section 34F).

### 34D–1 Preparation of Solutions

A pH-10 buffer and an indicator solution are needed for these titrations.
 (a) *Buffer solution pH 10*. (Sufficient for 80 to 100 titrations.) Dilute 57

mL of concentrated $NH_3$ and 7 g of $NH_4Cl$ in sufficient distilled water to give 100 mL of solution.

(b) *Eriochrome Black T indicator.* (Sufficient for about 100 titrations.) Dissolve 100 mg of the solid in a solution containing 15 mL of ethanolamine and 5 mL of absolute ethanol. This solution should be freshly prepared every two weeks; refrigeration slows its deterioration.

(c) *Calmagite indicator.* (Sufficient for about 200 titrations.) Dissolve 0.05 g of the indicator in sufficient distilled water to give 50 mL of solution.

## 34D–2  Preparation of Standard 0.01 M EDTA Solution

### Discussion

See Section 13B–1 for a description of the properties of reagent-grade $Na_2H_2Y \cdot 2 H_2O$ and its use in the direct preparation of standard EDTA solutions.

### Procedure

Dry about 4 g of the purified dihydrate $Na_2H_2Y \cdot 2 H_2O$ (Note 1) at 80°C to remove superficial moisture. Cool to room temperature in a desiccator. Weigh (to the nearest milligram) about 3.8 g into a 1-L volumetric flask (Note 2). Use a powder funnel to ensure quantitative transfer; rinse the funnel well with water before removing it from the flask. Add no more than 600 to 800 mL of water (Note 3) and swirl periodically. Dissolution may take 15 min or longer. When all the solid has dissolved, dilute to the mark with water and mix well (Note 4). In calculating the molarity of the solution, correct the weight of the salt for the 0.3% moisture it ordinarily retains after drying at 80°C.

Be sure to use a drying oven that is set to 80°C to dry the solid.

### Notes
1. Directions for the purification of the disodium salt are described by W. J. Blaedel and H. T. Knight, *Anal. Chem.,* **1954,** *26*(4), 741.
2. The solution can be prepared from the anhydrous disodium salt, if desired. The weight taken should be about 3.6 g.
3. Water used in the preparation of standard EDTA solutions must be totally free of polyvalent cations. If any doubt exists concerning its quality, the water should be passed through a cation-exchange resin before use.
4. As an alternative, an EDTA solution that is approximately 0.01 M can be prepared and standardized by direct titration against a $Mg^{2+}$ solution of known concentration (using the directions in Section 34D–3) or against primary-standard $CaCO_3$ by displacement titration (Section 34D–4).

The solid should be completely dissolved before diluting to the mark.

### 34D-3 The Determination of Magnesium by Direct Titration

**Discussion**

See Section 13B-7.

---

**Procedure**

Transfer the aqueous solution of the unknown to a clean 500-mL volumetric flask, dilute to the mark with water, and mix thoroughly. Transfer 50.00-mL aliquots to 250-mL conical flasks, add 1 to 2 mL of pH-10 buffer and 3 to 4 drops of Erio T or Calmagite indicator to each. Titrate with 0.01 M EDTA until the color changes from red to pure blue (Notes 1, 2).
Express the results as parts per million of $Mg^{2+}$ in the sample.

**Notes**
1. The color change tends to be slow in the vicinity of the end point. Care must be taken to avoid overtitration.
2. Other alkaline earths, if present, are titrated along with the $Mg^{2+}$; removal of $Ca^{2+}$ and $Ba^{2+}$ can be accomplished with $(NH_4)_2CO_3$. Most polyvalent cations are also titrated. Precipitation as hydroxides or the use of a masking reagent may be needed to eliminate this source of interference.

---

### 34D-4 The Determination of Calcium by Displacement Titration

**Discussion**

A solution of the magnesium/EDTA complex is useful for the titration of cations that form stabler complexes than the magnesium complex but for which no indicator is available. Magnesium ions in the complex are displaced by a chemically equivalent quantity of analyte cations. The remaining uncomplexed analyte and the liberated magnesium ions are then titrated; either Eriochrome Black T or Calmagite can serve as indicator. Note that the concentration of the magnesium solution is not important; all that is necessary is that the molar ratio between $Mg^{2+}$ and EDTA in the reagent be exactly unity.

---

**Procedure**

Preparation of the Magnesium/EDTA Complex, 0.1 M
(Sufficient for 90 to 100 titrations.) To 3.72 g of $Na_2H_2Y \cdot 2 H_2O$ in 50 mL of distilled water, add an equivalent quantity (2.46 g) of $MgSO_4 \cdot 7 H_2O$. Add a few drops of phenolphthalein, followed by sufficient 0.1 M NaOH to turn the solution faintly pink. Dilute to about 100 mL with water. The addition of a few drops of Erio T to a portion of this solution buffered to pH 10 should cause development of a dull violet color. Moreover, a single

drop of 0.01 M $Na_2H_2Y$ solution added to the violet solution should cause a color change to blue, and an equal quantity of 0.01 M $Mg^{2+}$ should cause a change to red. The composition of the original solution should be adjusted with additional $Mg^{2+}$ or $H_2Y^{2-}$ until these criteria are met.

Titration

Weigh a sample of the unknown (to the nearest 0.1 mg) into a 500-mL beaker (Note 1). Cover with a watch glass, and carefully add 5 to 10 mL of 6 M HCl. After the sample has dissolved, remove $CO_2$ by adding about 50 mL of deionized water and boiling gently for a few minutes. Cool, add a drop or two of methyl red, and neutralize with 6 M NaOH until the red color is discharged. Quantitatively transfer the solution to a 500-mL volumetric flask, and dilute to the mark. Take 50.00-mL aliquots of the diluted solution for titration, treating each as follows: Add about 2 mL of pH-10 buffer, 1 mL of Mg/EDTA solution, and 3 to 4 drops of Erio T or Calmagite indicator. Titrate (Note 2) with standard 0.01 M $Na_2H_2Y$ to a color change from red to blue.

Report the number of milligrams of CaO in the sample.

**Notes**
1. The sample taken should contain 150 to 160 mg of $Ca^{2+}$.
2. Interferences with this titration are substantially the same as those encountered in the direct titration of $Mg^{2+}$ and are eliminated in the same way.

---

### 34D–5  The Determination of Hardness in Water

Discussion

See Section 13B–9.

---

Procedure

Acidify 100.0-mL aliquots of the sample with a few drops of HCl, and boil gently for a few minutes to eliminate $CO_2$. Cool, add 3 to 4 drops of methyl red, and neutralize with 0.1 M NaOH. Introduce 2 mL of pH-10 buffer, 3 to 4 drops of Erio T or Calmagite, and titrate with standard 0.01 M $Na_2H_2Y$ to a color change from red to pure blue (Note).

Report the results in terms of milligrams of $CaCO_3$ per liter of water.

**Note.** The color change is sluggish if $Mg^{2+}$ is absent. In this event, add 1 to 2 mL of 0.1 M $MgY^{2-}$ (Section 34D–4, Procedure) before starting the titration.

---

### 34E  OXIDATION/REDUCTION TITRATIONS WITH POTASSIUM PERMANGANATE

The properties and uses of potassium permanganate are described in Section 16C–1. Directions follow for the determination of iron in an ore and calcium in a limestone. These directions are readily adapted to weight titrimetry as described in Section 33F.

## 34E–1  Preparation of 0.02 M Potassium Permanganate

### Discussion

See page 378 for a discussion of the precautions needed in the preparation and storage of permanganate solutions.

### Procedure

Dissolve about 3.2 g of $KMnO_4$ in 1 L of distilled water. Keep the solution at a gentle boil for about 1 hr. Cover and let stand overnight. Remove $MnO_2$ by filtration (Note 1) through a fine-porosity sintered-glass crucible (Note 2) or through a Gooch crucible fitted with glass mats. Transfer the solution to a clean glass-stoppered bottle; store in the dark when not in use.

#### Notes
1. The heating and filtering can be omitted if the permanganate solution is standardized and used on the same day.
2. Remove the $MnO_2$ that collects on the fritted plate with 1 M $H_2SO_4$ containing a few milliliters of 3% $H_2O_2$, followed by a rinse with copious quantities of water.

## 34E–2  Standardization of Potassium Permanganate Solutions

### Discussion

See page 380 for a discussion of sodium oxalate and other primary standards for permanganate solutions.

### Procedure

Dry about 1.5 g of primary-standard-grade $Na_2C_2O_4$ at 110°C for at least 1 hr. Cool in a desiccator; weigh (to the nearest 0.1 mg) individual 0.2- to 0.3-g samples into 400-mL beakers. Dissolve each in about 250 mL of 1 M $H_2SO_4$. Heat each solution to 80 to 90°C, and titrate with $KMnO_4$ while stirring with a thermometer. The pink color imparted by one addition should be permitted to disappear before any further titrant is introduced (Notes 1, 2). Reheat if the temperature drops below 60°C. Take the first persistent ($\approx$30 s) pink color as the end point (Notes 3, 4). Determine a blank by titrating an equal volume of the 1 M $H_2SO_4$.

Correct the titration data for the blank, and calculate the concentration of the permanganate solution (Note 5).

**Notes**

1. Promptly wash any $KMnO_4$ that spatters on the walls of the beaker into the bulk of the liquid with a stream of water.

2. Finely divided $MnO_2$ will form along with $Mn^{2+}$ if the $KMnO_4$ is added too rapidly and will cause the solution to acquire a faint brown discoloration. Precipitate formation is not a serious problem so long as sufficient oxalate remains to reduce the $MnO_2$ to $Mn^{2+}$; the titration is simply discontinued until the brown color disappears. The solution must be free of $MnO_2$ at the end point.

3. The surface of the permanganate solution rather than the bottom of the meniscus can be used to measure titrant volumes. Alternatively, backlighting with a flashlight or a match permits reading of the meniscus in the conventional manner.

4. A permanganate solution should not be allowed to stand in a buret any longer than necessary because partial decomposition to $MnO_2$ may occur. Freshly formed $MnO_2$ can be removed from a glass surface with 1 M $H_2SO_4$ containing a small amount of 3% $H_2O_2$.

5. As noted on page 380, this procedure yields molarities that are a few tenths of a percent low. For more accurate results, introduce from a buret sufficient permanganate to react with 90 to 95% of the oxalate (about 40 mL of 0.02 M $KMnO_4$ for a 0.3-g sample). Let the solution stand until the permanganate color disappears. Then warm to about 60°C and complete the titration, taking the first permanent pink ($\simeq 30$ s) as the end point (Notes 3, 4). Determine a blank by titrating an equal volume of the 1 M $H_2SO_4$.

---

## 34E-3  The Determination of Iron in an Ore

### Discussion

The common ores of iron are hematite ($Fe_2O_3$), magnetite ($Fe_3O_4$), and limonite ($2 Fe_2O_3 \cdot 3 H_2O$). Steps in the analysis of these ores are (1) dissolution of the sample, (2) reduction of iron to the divalent state, and (3) titration of iron(II) with a standard oxidant.

**The Decomposition of Iron Ores.**  Iron ores often decompose completely in hot concentrated hydrochloric acid. The rate of attack by this reagent is increased by the presence of a small amount of tin(II) chloride, which probably acts by reducing sparingly soluble iron(III) oxides on the surface of the ore to more soluble iron(II) species. The tendency of iron(II) and iron(III) to form chloro complexes accounts for the effectiveness of hydrochloric acid over nitric or sulfuric acid as a solvent for iron ores.

Many iron ores contain silicates that may not be entirely decomposed by treatment with hydrochloric acid. Incomplete decomposition is indicated by a dark residue that remains after prolonged treatment with the acid. A white residue of hydrated silica, which does not interfere in any way, is indicative of complete decomposition.

**The Prereduction of Iron.** Because part or all of the iron is in the trivalent state after decomposition of the sample, prereduction to iron(II) must precede titration with the oxidant. Any of the methods described in Section 16A–1 can be used. Perhaps the most satisfactory prereductant for iron is tin(II) chloride:

$$2\ Fe^{3+} + Sn^{2+} \rightarrow 2\ Fe^{2+} + Sn^{4+}$$

The only other common species reduced by this reagent are the high-oxidation states of arsenic, copper, mercury, molybdenum, tungsten, and vanadium.

The excess reducing agent is eliminated by the addition of mercury(II) chloride:

$$Sn^{2+} + 2\ HgCl_2 \rightarrow Hg_2Cl_2(s) + Sn^{4+} + 2\ Cl^-$$

The slightly soluble mercury(I) chloride does not reduce permanganate, nor does the excess mercury(II) chloride reoxidize iron(II). Care must be taken, however, to prevent the occurrence of the alternative reaction

$$Sn^{2+} + HgCl_2 \rightarrow Hg(l) + Sn^{4+} + 2\ Cl^-$$

Elemental mercury reacts with permanganate and causes the results of the analysis to be high. The formation of mercury, which is favored by an appreciable excess of tin(II), is prevented by careful control of this excess and by the rapid addition of excess mercury(II) chloride. A proper reduction is indicated by the appearance of a small amount of a silky white precipitate after the addition of mercury(II). Formation of a gray precipitate at this juncture indicates the presence of metallic mercury; the total absence of a precipitate indicates that an insufficient amount of tin(II) chloride was used. In either event, the sample must be discarded.

Discard the sample if (1) the precipitate is gray or (2) no precipitate is formed when mercury(II) is added.

**The Titration of Iron(II).** The reaction of iron(II) with permanganate is smooth and rapid. The presence of iron(II) in the reaction mixture, however, *induces* oxidation of chloride ion by permanganate, a reaction that does not ordinarily proceed rapidly enough to cause serious error. High results are obtained if this parasitic reaction is not controlled. Its effects can be eliminated through removal of the hydrochloric acid by evaporation with sulfuric acid or by introduction of *Zimmermann-Reinhardt reagent*, which contains manganese(II) in a fairly concentrated mixture of sulfuric and phosphoric acids.

The oxidation of chloride ion during a titration is believed to involve a direct reaction between this species and the manganese(III) ions that form as an intermediate in the reduction of permanganate ion by iron(II). The presence of manganese(II) in the Zimmermann-Reinhardt reagent is believed to inhibit the formation of chlorine by decreasing the potential of the manganese(III)/manganese(II) couple. Phosphate ion is believed to exert a similar effect by forming stable manganese(III) complexes. Moreover, phosphate ions react with iron(III) to form nearly colorless complexes so that the yellow color of the iron(II)/chloro complexes does not interfere with the end point.[5]

---

## Preparation of Reagents

The following solutions suffice for about 100 titrations.

(a) *Tin(II) chloride, 0.25 M*. Dissolve 60 g of iron-free $SnCl_2 \cdot 2\ H_2O$ in 100 mL of concentrated HCl; warm if necessary. After the solid has dissolved, dilute to 1 L with distilled water and store in a well-stoppered bottle. Add a few pieces of mossy tin to help preserve the solution.

(b) *Mercury(II) chloride, 5% (w/v)*. Dissolve 50 g of $HgCl_2$ in 1 L of distilled water.

(c) *Zimmermann-Reinhardt reagent*. Dissolve 300 g of $MnSO_4 \cdot 4\ H_2O$ in 1 L of water. Cautiously add 400 mL of concentrated $H_2SO_4$, 400 mL of 85% $H_3PO_4$, and dilute to 3 L.

---

## Procedure

### Sample Preparation

Dry the ore at 110°C for at least 3 hr, and then allow it to cool to room temperature in a desiccator. Consult with your instructor for a sample size that will require from 25 to 40 mL of standard 0.02 M $KMnO_4$. Weigh samples into 500-mL conical flasks. To each, add 10 mL of concentrated HCl and about 3 mL of 0.25 M $SnCl_2$ (Note 1). Cover each flask with a small watch glass or Tuttle flask cover. Heat the flasks in a hood at just below boiling until the samples are decomposed and the undissolved solid—if any—is pure white (Note 2). Use another 1 or 2 mL of $SnCl_2$ to eliminate any yellow color that may develop as the solutions are heated. Heat a blank consisting of 10 mL of HCl and 3 mL of $SnCl_2$ for the same amount of time.

After the ore has been decomposed, remove the excess Sn(II) by the dropwise addition of 0.02 M $KMnO_4$ until the solutions become faintly yellow. Dilute to about 15 mL. Add sufficient $KMnO_4$ solution to impart a faint pink color to the blank; then decolorize with one drop of the $SnCl_2$ solution.

*Take samples and blank individually through subsequent steps to minimize air-oxidation of iron(II).*

### Reduction of Iron

Heat the sample solution nearly to boiling, and make dropwise additions of 0.25 M $SnCl_2$ until the yellow color just disappears; then add two more drops (Note 3). Cool to room temperature, and *rapidly* add 10 mL of 5% $HgCl_2$ solution. A small amount of silky white $Hg_2Cl_2$ should precipitate (Note 4). The blank solution should be treated with the $HgCl_2$ solution.

---

[5]The mechanism by which Zimmermann-Reinhardt reagent acts has been the subject of much study. For a discussion of this work, see H. A. Laitinen and W. E. Harris, *Chemical Analysis*, 2nd ed., pp. 369–372. New York: McGraw-Hill, 1975.

Titration

Following addition of the $HgCl_2$, wait 2 to 3 min. Then add 25 mL of Zimmermann-Reinhardt reagent and 300 mL of water. Titrate *immediately* with standard 0.02 M $KMnO_4$ to the first faint pink that persists for 15 to 20 s. Do not add the $KMnO_4$ rapidly at any time. Correct the titrant volume for the blank.

Report the percentage of $Fe_2O_3$ in the sample.

**Notes**

1. The $SnCl_2$ hastens decomposition of the ore by reducing iron(III) oxides to iron(II). Insufficient $SnCl_2$ is indicated by the appearance of yellow iron(III)/chloride complexes.
2. If dark particles persist after the sample has been heated with acid for several hours, filter the solution through ashless paper, wash the residue with 5 to 10 mL of 6 M HCl, and retain the filtrate and washings. Ignite the paper and its contents in a small platinum crucible. Mix 0.5 to 0.7 g of $Na_2CO_3$ with the residue and heat until a clear melt is obtained. Cool, add 5 mL of water, and then cautiously add a few milliliters of 6 M HCl. Warm the crucible until the melt has dissolved, and combine the contents with the original filtrate. Evaporate the solution to 15 mL and continue the analysis.
3. The solution may not become entirely colorless but instead may acquire a faint yellow-green hue. Further additions of $SnCl_2$ will not alter this color. If too much $SnCl_2$ is added, it can be removed by adding 0.2 M $KMnO_4$ and repeating the reduction.
4. The absence of precipitate indicates that insufficient $SnCl_2$ was used and that the reduction of iron(III) was incomplete. A gray residue indicates the presence of elemental mercury, which reacts with $KMnO_4$. The sample must be discarded in either event.

---

## 34E–4 The Determination of Calcium in a Limestone

### Discussion

In common with a number of other cations, calcium is conveniently determined by precipitation with oxalate ion. The solid calcium oxalate is filtered, washed free of excess precipitating reagent, and dissolved in dilute acid. The oxalic acid liberated in this step is then titrated with standard permanganate or some other oxidizing reagent. This method is applicable to samples that contain magnesium and the alkali metals, but most other cations must be absent since they either precipitate or coprecipitate as oxalates and cause positive errors in the analysis.

**Factors Affecting the Composition of Calcium Oxalate Precipitates.** It is essential that the mole ratio between calcium and oxalate be exactly unity in the precipitate and thus in solution at the time of titration. A number of precautions are needed to ensure this condition. For example, the calcium oxalate formed in a neutral or ammoniacal solution is likely to be contaminated with calcium hydroxide or a basic calcium oxalate, either of which leads to low results. The formation of these compounds is prevented by adding the oxalate to an acidic solution of the sample and slowly forming the precipitate by the dropwise addition of ammonia. The coarsely crystalline calcium oxalate that is produced under these conditions is readily filtered. Losses resulting from the solubility of calcium oxalate are negligible above pH 4, provided washing is limited to freeing the precipitate of excess oxalate.

Coprecipitation of sodium oxalate becomes a source of positive error in the determination of calcium whenever the concentration of sodium in the sample exceeds that of calcium. The error from this source can be eliminated by reprecipitation (Section 4C–4).

Magnesium, if present in high concentration, may precipitate as the oxalate and contaminate the analytical precipitate. An excess of oxalate ion helps prevent this interference through the formation of soluble oxalate complexes of magnesium. Prompt filtration of the calcium oxalate can also help prevent interference because of the pronounced tendency of magnesium oxalate to form supersaturated solutions from which precipitate formation occurs only after an hour or more. For samples containing more magnesium than calcium, these measures do not suffice to give accurate results and reprecipitation of the calcium becomes necessary.

**The Composition of Limestones.**  Limestones are composed principally of calcium carbonate; dolomitic limestones contain large amounts of magnesium carbonate as well. Calcium and magnesium silicates occur in smaller amounts, along with the carbonates and silicates of iron, aluminum, manganese, titanium, sodium, and other metals.

Hydrochloric acid is an effective solvent for most limestones. Only silica, which does not interfere with the analysis, remains undissolved. Some limestones are more readily decomposed after they have been ignited; a few yield only to a carbonate fusion (Section 31C–2).

The method that follows is remarkably effective for determining calcium in most limestones. Iron and aluminum, in amounts equivalent to that of calcium, do not interfere. Small amounts of manganese and titanium can also be tolerated.

---

Procedure

Sample Preparation
Dry the unknown for 1 to 2 hr at 110°C, and cool in a desiccator. If the material is readily decomposed in acid, weigh 0.25- to 0.30-g samples (to the nearest 0.1 mg) into 250-mL beakers. Add 10 mL of water to each sample and cover with a watch glass. Add 10 mL of concentrated HCl dropwise, taking care to avoid losses due to spattering as the acid is introduced. Proceed to the paragraph labeled ''Precipitation of Calcium Oxalate.''

If the limestone is not completely decomposed by acid, weigh the sample into a small porcelain crucible and ignite. Raise the temperature slowly to 800 to 900°C and maintain this temperature for about 30 min. After cooling, place the crucible and its contents in a 250-mL beaker, add 5 mL of water, and cover with a watch glass. Introduce 10 mL of concentrated HCl dropwise, and then heat to boiling. Remove the crucible with a stirring rod and rinse it thoroughly with water; combine the washings with the solution containing the sample.

### Precipitation of Calcium Oxalate

Add 5 drops of saturated bromine water to oxidize any iron in the samples and boil gently (HOOD) for 5 min to remove the excess $Br_2$. Dilute each sample solution to about 50 mL, heat to boiling, and add 100 mL of hot 6% (w/v) $(NH_4)_2C_2O_4$ solution. Add 3 to 4 drops of methyl red, and precipitate $CaC_2O_4$ by slowly adding 6 M $NH_3$. When the indicator just begins to change color, add the $NH_3$ at a rate of one drop every 3 to 4 s. Continue until the solutions turn to the intermediate yellow-orange color of the indicator (pH 4.5 to 5.5). Allow the solutions to stand for no more than 30 min (Note) and filter; medium-porosity filtering crucibles or Gooch crucibles with glass mats are satisfactory. Wash the precipitates with several 10-mL portions of cold water. Rinse the crucibles to remove residual $(NH_4)_2C_2O_4$, and return them to the beakers in which the $CaC_2O_4$ was formed.

### Titration

Add 100 mL of water and 50 mL of 3 M $H_2SO_4$ to each of the beakers containing the precipitated calcium oxalate and the crucible. Heat to 80 to 90°C, and titrate with 0.02 M permanganate. The temperature should be above 60°C throughout the titration; reheat if necessary.

Report the percentage of CaO in the unknown.

**Note.** The period of standing can be longer if the unknown contains no $Mg^{2+}$.

## 34F    IODIMETRIC TITRATIONS

The oxidizing properties of iodine, the composition and stability of triiodide solutions, and the applications of this reagent in volumetric analysis are discussed in Section 16C–3. Starch is ordinarily employed as an indicator for iodimetric titrations. The directions that follow are readily adapted to weight titrimetry, which is discussed in Section 33F.

### 34F–1 Preparation of Reagents

(a) *Iodine, approximately 0.05 M.* Weigh about 40 g of KI into a 100-mL beaker. Add 12.7 g of $I_2$ and 10 mL of water. Stir for several minutes (Note 1). Introduce an additional 20 mL of water, and stir again for several minutes. Carefully decant the bulk of the liquid into a storage bottle containing 1 L of distilled water. Mix well. It is essential that any undissolved iodine remain in the beaker (Note 2).

**Notes**
1. Iodine dissolves slowly in the KI solution. Thorough stirring is needed to hasten the process.

2. Any solid $I_2$ inadvertently transferred to the storage bottle will cause the concentration of the solution to increase gradually. Filtration through a sintered-glass crucible eliminates this potential source of difficulty.

(b) *Starch indicator.* Rub 1 g of soluble starch and 15 mL of water into a paste. Dilute to about 500 mL with boiling water, and heat until the mixture is clear. Cool; store in a tightly stoppered bottle. For most titrations, 3 to 5 mL of the indicator is used. (Sufficient for about 100 titrations.)

   The indicator is readily attacked by airborne organisms and should be freshly prepared every few days.

## 34F–2 Standardization of Iodine Solutions

### Discussion

Arsenic(III) oxide, long a favored primary standard for iodine solutions, is now seldom used because of the elaborate federal regulations governing the use of even small amounts of arsenic-containing compounds. Barium thiosulfate monohydrate and anhydrous sodium thiosulfate have been proposed as alternative standards.[6,7] Perhaps the most convenient method for determining the concentration of an iodine solution is the titration of aliquots with a sodium thiosulfate solution that has been standardized against pure potassium iodate. Instructions for this method follow.

### Preparation of Reagents

(a) *Sodium thiosulfate, 0.1 M.* Follow the directions in Sections 34G–1 and 34G–2 for the preparation and standardization of this solution.
(b) *Starch indicator.* See Section 34F–1.

### Procedure

Transfer 25.00-mL aliquots of the iodine solution to 250-mL conical flasks, and dilute to about 50 mL. Introduce approximately 1 mL of 3 M $H_2SO_4$, and titrate immediately with standard sodium thiosulfate until the solution becomes a faint straw yellow. Add about 5 mL of starch indicator, and complete the titration, taking as the end point the change in color from blue to colorless (Note).

**Note.** The blue color of the starch/iodine complex may reappear after the titration has been completed, owing to the air-oxidation of iodide ion.

[6]W. M. McNevin and O. H. Kriege, *Anal. Chem.*, **1953,** 25 (5), 767.
[7]A. A. Woolf, *Anal. Chem.*, **1982,** 54 (12), 2134.

## 34F–3 The Determination of Antimony in Stibnite

### Discussion

The analysis of stibnite, a common antimony ore, is a typical application of iodimetry and is based upon the oxidation of Sb(III) to Sb(V):

$$SbO_3^{3-} + I_2 + H_2O \rightleftharpoons SbO_4^{3-} + 2\,I^- + 2\,H^+$$

Hydrogen carbonate ion reacts with $H^+$ and thus forces the reaction between $I_2$ and Sb(III) to completion.

The position of this equilibrium is strongly dependent upon the hydrogen ion concentration. In order to force the reaction to the right, it is common practice to carry out the titration in the presence of an excess of sodium hydrogen carbonate, which consumes the hydrogen ions as they form.

Stibnite is an antimony sulfide ore containing silica and other contaminants. Provided the material is free of iron and arsenic, the analysis of stibnite for its antimony content is straightforward. Samples are decomposed in hot concentrated hydrochloric acid to eliminate sulfide as gaseous hydrogen sulfide. In order to avoid the loss of antimony chloride by volatilization, excess potassium chloride is added to the solvent to convert the antimony to nonvolatile chloro complexes, such as $SbCl_4^-$ and $SbCl_6^{3-}$.

Sparingly soluble basic antimony salts, such as SbOCl, often form when the excess hydrochloric acid is neutralized; these react incompletely with iodine and cause low results. The difficulty is overcome by adding tartaric acid, which forms a soluble complex ($SbOC_4H_4O_6^-$) from which antimony is rapidly oxidized by the reagent.

### Procedure

Dry the unknown at 110°C for 1 hr, and allow it to cool in a desiccator. Weigh individual samples (Note 1) into 500-mL conical flasks. Introduce about 0.3 g of KCl and 10 mL of concentrated HCl to each flask. Heat the mixtures (HOOD) just below boiling until only white or slightly gray residues of $SiO_2$ remain.

Add 3 g of tartaric acid to each sample and heat for an additional 10 to 15 min. Then, with good swirling, add water (Note 2) from a pipet or buret until the volume is about 100 mL. If reddish $Sb_2S_3$ forms, discontinue dilution and heat further to eliminate $H_2S$; add more HCl if necessary.

Add 3 drops of phenolphthalein, and neutralize with 6 M NaOH to the first faint pink of the indicator. Discharge the color by the dropwise addition of 6 M HCl, and then add 1 mL in excess. Introduce 4 to 5 g of $NaHCO_3$, taking care to avoid losses of solution by spattering during the addition. Add 5 mL of starch indicator, rinse down the inside of the flask, and titrate with standard 0.05 M $I_2$ to the first blue color that persists for 30 s.

Report the percentage of $Sb_2S_3$ in the unknown.

### Notes

1. Samples should contain between 1.5 and 2 mmol of antimony; consult with your instructor for an appropriate sample size. Weighings to the nearest milligram are adequate for samples larger than 1 g.

2. The slow addition of water, with efficient stirring, is essential to prevent the formation of SbOCl.

## 34G   IODOMETRIC METHODS OF ANALYSIS

Numerous methods are based upon the reducing properties of iodide ion:

$$2\,I^- \rightarrow I_2 + 2\,e^-$$

Iodine, the reaction product, is ordinarily titrated with a standard sodium thiosulfate solution, with starch serving as the indicator:

$$I_2 + 2\,S_2O_3^{2-} \rightarrow 2\,I^- + S_4O_6^{2-}$$

A discussion of iodometric methods is found in Section 16C–3. All the procedures described in this section can also be performed conveniently by weight titrimetry (Section 33F).

### 34G–1  Preparation of 0.1 M Sodium Thiosulfate

#### Procedure

Boil about 1 L of distilled water for 10 to 15 min. Allow the water to cool to room temperature; then add about 25 g of $Na_2S_2O_3 \cdot 5\,H_2O$ and 0.1 g of $Na_2CO_3$. Stir until the solid has dissolved. Transfer the solution to a clean glass or plastic bottle, and store in a dark place.

### 34G–2  Standardization of Sodium Thiosulfate Against Potassium Iodate

#### Discussion

Solutions of sodium thiosulfate are conveniently standardized by titration of the iodine produced when an unmeasured excess of potassium iodide is added to a known volume of an acidified standard potassium iodate solution. The reaction is

$$IO_3^- + 5\,I^- + 6\,H^+ \rightarrow 3\,I_2 + 3\,H_2O$$

Note that each formula weight of iodate results in the production of three formula weights of iodine. The procedure that follows is based upon this reaction.

#### Preparation of Solutions

(a) *Potassium iodate, 0.0100 M.* Dry about 1.2 g of primary-standard $KIO_3$ at 110°C for at least 1 hr and cool in a desiccator. Weigh (to the nearest 0.1 mg) about 1.1 g into a 500-mL volumetric flask; use a powder funnel to ensure quantitative transfer of the solid. Rinse the funnel well, dissolve the $KIO_3$ in about 200 mL of distilled water, dilute to the mark, and mix thoroughly.

(b) *Starch indicator.* See Section 34F–1.

## Procedure

Pipet 50.00-mL aliquots of standard iodate solution into 250-mL conical flasks. *Treat each sample individually from this point to minimize error caused by the air-oxidation of iodide ion.* Introduce 2 g of iodate-free KI, and swirl the flask to hasten solution. Add 2 mL of 6 M HCl, and immediately titrate with thiosulfate until the solution becomes pale yellow. Introduce 5 mL of starch indicator, and titrate with constant stirring to the disappearance of the blue color. Calculate the molarity of the iodine solution.

## 34G–3 Standardization of Sodium Thiosulfate Against Copper

### Discussion

Thiosulfate solutions can also be standardized against pure copper wire or foil. This procedure is advantageous when the solution is to be used for the determination of copper because any determinate error in the method tends to be canceled.

Copper(II) is reduced quantitatively to copper(I) by iodide ion:

$$2\ Cu^{2+} + 4\ I^- \rightarrow 2\ CuI(s) + I_2$$

The importance of CuI formation in forcing this reaction to completion can be seen from the following standard electrode potentials:

$$Cu^{2+} + e^- \rightleftharpoons Cu^+ \qquad E^0 = 0.15\ V$$
$$I_2 + 2\ e^- \rightleftharpoons 2\ I^- \qquad E^0 = 0.54\ V$$
$$Cu^{2+} + I^- + e^- \rightleftharpoons CuI(s) \qquad E^0 = 0.86\ V$$

The first two potentials suggest that iodide should have no tendency to reduce copper(II); the formation of CuI, however, favors the reduction. The solution must contain at least 4% excess iodide to force the reaction to completion. Moreover, the pH must be below 4 to prevent the formation of basic copper species that react slowly and incompletely with iodide ion. The acidity of the solution cannot be greater than about 0.3 M, however, because of the tendency of iodide ion to undergo air-oxidation, a process catalyzed by copper salts. Nitrogen oxides also catalyze the air-oxidation of iodide ion. A common source of these oxides is the nitric acid ordinarily used to dissolve metallic copper and other copper-containing solids. Urea is used to scavenge nitrogen oxides from solutions:

*The pH must be kept between 0.5 and 4.*

$$(NH_2)_2CO + 2\ HNO_2 \rightarrow 2\ N_2(g) + CO_2(g) + 3\ H_2O$$

The titration of iodine by thiosulfate tends to yield slightly low results, owing to the adsorption of small but measurable quantities of iodine upon solid CuI. The adsorbed iodine is released only slowly, even when thiosulfate is present in excess; transient and premature end points result. This difficulty is largely overcome by the addition of thiocyanate ion. The sparingly soluble copper(I) thiocyanate replaces part of the copper iodide at the surface of the solid:

$$CuI(s) + SCN^- \rightarrow CuSCN(s) + I^-$$

Accompanying this reaction is the release of the adsorbed iodine, which thus becomes available for titration. The addition of thiocyanate must be delayed until most of the iodine has been titrated to prevent interference from a slow reaction between the two species, possibly

$$2\ SCN^- + I_2 \rightarrow 2\ I^- + (SCN)_2$$

## Preparation of Solutions

(a) *Urea, 5% (w/v)*. Dissolve about 5 g of urea in sufficient water to give 100 mL of solution. Approximately 10 mL is needed for each titration.
(b) *Starch indicator*. See Section 34F–1.

## Procedure

Use scissors to cut copper wire or foil into 0.20- to 0.25-g portions. Wipe the metal free of dust and grease with a filter paper; do not dry it. The pieces of copper should be handled with paper strips, cotton gloves, or tweezers to prevent contamination by contact with the skin.

Use a weighed watch glass or weighing bottle to obtain the weight of individual copper samples by difference (to the nearest 0.1 mg). Transfer each sample to a 250-mL conical flask. Add 5 mL of 6 M $HNO_3$, cover with a small watch glass, and warm gently (HOOD) until the metal has dissolved. Dilute with about 25 mL of distilled water, add 10 mL of 5% (w/v) urea, and boil briefly to eliminate nitrogen oxides. Rinse the watch glass, collecting the rinsings in the flask. Cool.

Add concentrated $NH_3$ dropwise and with thorough mixing to produce the intensely blue $Cu(NH_3)_4^{2+}$; the solution should smell faintly of ammonia (Note). Make dropwise additions of 3 M $H_2SO_4$ until the color of the complex just disappears, and then add 2.0 mL of 85% $H_3PO_4$. Cool to room temperature.

*Treat each sample individually from this point on to minimize the air-oxidation of iodide ion.* Add 4.0 g of KI to the sample, and titrate immediately with $Na_2S_2O_3$ until the solution becomes pale yellow. Add 5 mL of starch indicator, and continue the titration until the blue color becomes faint. Add 2 g of KSCN; swirl vigorously for 30 s. Complete the titration, using the disappearance of the blue starch/$I_2$ color as the end point.

Calculate the molarity of the $Na_2S_2O_3$ solution.

**Note.** Vapors should not be sniffed directly from the flask but instead should be wafted toward your nose with a waving motion of your hand.

---

### 34G–4  The Determination of Copper in Brass

#### Discussion

The standardization procedure described in Section 34G–3 is readily adapted to the determination of copper in brass, an alloy that also contains appreciable amounts of tin, lead, and zinc (and perhaps minor amounts of nickel and iron). The method is relatively simple and applicable to brasses with less than 2% iron. A weighed sample is treated with nitric acid, which causes the tin to precipitate as a hydrated oxide of uncertain composition (Section 34A–2). Evaporation with sulfuric acid to the appearance of sulfur trioxide eliminates the excess nitrate, redissolves the tin compound, and possibly causes the formation of lead sulfate. The pH is adjusted through the addition of ammonia, followed by acidification with a measured amount of phosphoric acid. An excess of potassium iodide is added, and the liberated iodine is titrated with standard thiosulfate. See Section 34G–3 for additional discussion.

---

#### Procedure

If so directed, free the metal of oils by treatment with an organic solvent; briefly heat in an oven to drive off the solvent. Weigh (to the nearest 0.1 mg) 0.3-g samples into 250-mL conical flasks, and introduce 5 mL of 6 M $HNO_3$ into each; warm (HOOD) until solution is complete. Add 10 mL of concentrated $H_2SO_4$, and evaporate (HOOD) until copious white fumes of $SO_3$ are given off. Allow the mixture to cool. Cautiously add 30 mL of distilled water, boil for 1 to 2 min, and again cool.

Follow the instructions in the third and fourth paragraphs of the *Procedure* in Section 34G–3.

Report the percentage of Cu in the sample.

---

### 34H  TITRATIONS WITH POTASSIUM BROMATE

Applications of standard bromate solutions to the determination of organic functional groups are described in Section 16D–1. Directions follow for the determination of phenol in an aqueous solution and for ascorbic acid in vitamin C tablets.

### 34H–1  Preparation of Solutions

(a) *Potassium bromate, 0.015 M*. Transfer about 1.5 g of reagent-grade potassium bromate to a weighing bottle, and dry at 110°C for at least 1 hr. Cool in a desiccator. Weigh approximately 1.3 g (to the nearest 0.1

mg) into a 500-mL volumetric flask; use a powder funnel to ensure quantitative transfer of the solid. Rinse the funnel well, and dissolve the $KBrO_3$ in about 200 mL of distilled water. Dilute to the mark, and mix thoroughly.

*Solid potassium bromate can cause a fire if it comes into contact with damp organic material (such as paper toweling in a waste container).* Consult with your instructor concerning the disposal of any excess.

(b) *Sodium thiosulfate, 0.05 M.* Follow the directions in Section 34G–1; use about 12.5 g of $Na_2S_2O_3 \cdot 5\ H_2O$ per liter of solution.

(c) *Starch indicator.* See Section 34F–1.

## 34H–2 Standardization of Sodium Thiosulfate Against Potassium Bromate

### Discussion

Iodine is generated by the reaction between a known volume of standard potassium bromate and an unmeasured excess of potassium iodide:

$$BrO_3^- + 6\ I^- + 6\ H^+ \rightarrow Br^- + 3\ I_2 + 3\ H_2O$$

The iodine produced is titrated with the sodium thiosulfate solution.

### Procedure

Pipet 25.00-mL aliquots of the $KBrO_3$ solution into 250-mL conical flasks and rinse the interior wall with distilled water. *Treat each sample individually beyond this point.* Introduce 2 to 3 g of KI and about 5 mL of 3 M $H_2SO_4$. Immediately titrate with 0.05 M $Na_2S_2O_3$ until the solution is pale yellow. Add 5 mL of starch indicator, and titrate to the disappearance of the blue color.

Calculate the concentration of the thiosulfate solution.

## 34H–3 The Determination of Phenol by Bromination

### Discussion

The phenol content of waste waters from manufacturing processes is conveniently determined by mixing the sample with a known excess of standard bromate followed by an unmeasured excess of bromide. The bromine liberated upon acidification reacts with the phenol:

$$BrO_3^- + 5\ Br^- + 6\ H^+ \rightarrow 3\ Br_2 + 3\ H_2O$$
$$C_6H_5OH + 3\ Br_2 \rightarrow C_6H_2Br_3OH + 3\ H^+ + 3\ Br^-$$

After the bromination is complete, the excess bromine is determined by the procedure used for the standardization.

## Procedure

Transfer a sample containing between 1 and 1.5 mmol of phenol to a 250-mL volumetric flask, dilute to the mark with water, and mix well. Pipet 25.00 mL aliquots of the diluted sample into 250-mL conical flasks, and add 25.00 mL aliquots of standard $KBrO_3$ solution. Add about 1 g of KBr and about 5 mL of 3 M $H_2SO_4$ to each flask. *Stopper each flask immediately after acidification* to prevent the loss of $Br_2$. Mix and let stand for about 10 min. Introduce 2 to 3 g of KI to each flask, and immediately restopper. Swirl the solutions until the KI has dissolved. Titrate the liberated iodine with standard 0.05 M $Na_2S_2O_3$ until the solution is pale yellow. Add 5 mL of starch indicator, and complete the titration.

Report the number of milligrams of phenol contained in each milliliter of the unknown.

### 34H–4  The Determination of Ascorbic Acid in Vitamin C Tablets by Titration with Potassium Bromate

## Discussion

Ascorbic acid, $C_6H_8O_6$, is cleanly oxidized to dehydroascorbic acid by bromine:

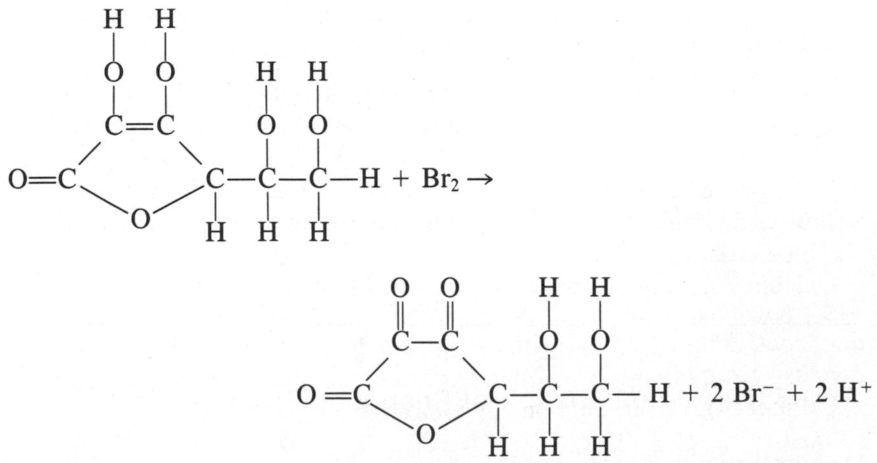

An unmeasured excess of potassium bromide is added to an acidified solution of the sample. The solution is titrated with standard potassium bromate to the first permanent appearance of excess bromine; this excess is then determined iodometrically with standard sodium thiosulfate. The entire titration must be performed without delay to prevent air-oxidation of the ascorbic acid.

## Procedure

Weigh (to the nearest milligram) 3 to 5 vitamin C tablets (Note 1). Pulverize them thoroughly in a mortar, and transfer the powder to a dry weighing bottle. Weigh individual 0.40- to 0.50-g samples (to the nearest 0.1 mg) into dry 250-mL conical flasks. *Treat each sample individually beyond this point.* Dissolve the sample (Note 2) in 50 mL of 1.5 M $H_2SO_4$; then add about 5 g of KBr. Titrate immediately with standard $KBrO_3$ to the first faint yellow due to excess $Br_2$. Record the volume of $KBrO_3$ used. Add 3 g of KI and 5 mL of starch indicator; back-titrate (Note 3) with standard 0.05 M $Na_2S_2O_3$.

Calculate the average weight (in milligrams) of ascorbic acid (fw = 176.13) in each tablet.

**Notes**
1. This method is not applicable to chewable vitamin C tablets.
2. The binder in many vitamin C tablets remains in suspension throughout the analysis. If the binder is starch, the characteristic color of the complex with iodine appears upon the addition of KI.
3. The volume of thiosulfate needed for the back-titration seldom exceeds a few milliliters.

## 34I  POTENTIOMETRIC METHODS

Potentiometric measurements provide a highly selective method for the quantitative determination of numerous cations and anions. A discussion of the principles and applications of potentiometric measurements is found in Chapter 17. Detailed instructions are given in this section on the use of potentiometric measurements to locate end points in volumetric titrations. In addition, a procedure for the direct potentiometric determination of fluoride ion in drinking water and in toothpaste is described.

### 34I–1  General Directions for Performing a Potentiometric Titration

The procedure that follows is applicable to the three titrimetric methods described in this section. With the proper choice of indicator electrode, it can also be applied to most of the volumetric and gravimetric titrimetry experiments given in Sections 34B through 34H.

1. Dissolve the sample in 50 to 250 mL of water. Rinse a suitable pair of electrodes with distilled water, and immerse them in the sample solution. Provide magnetic (or mechanical) stirring. Position the buret so that reagent can be delivered without splashing.
2. Connect the electrodes to the meter, commence stirring, and measure and record the initial buret volume as well as the initial potential (or pH).
3. Measure and record the meter reading and buret volume after each addition of titrant. Introduce fairly large volumes (about 5 mL) at the outset. Withhold a succeeding addition until the meter reading remains constant within 1 to 2 mV (or 0.05 pH unit) for at least 30 s (Note). Judge the volume of reagent to be added by estimating a value for

$\Delta E/\Delta V$ after each addition. In the immediate vicinity of the equivalence point, introduce the reagent in 0.1-mL increments. Continue the titration 2 to 3 mL beyond the equivalence point, increasing the volume increments as $\Delta E/\Delta V$ again becomes smaller.

**Note.** Stirring motors occasionally cause erratic meter readings; it may be advisable to turn off the motor while meter readings are being made.

### 34I–2  The Potentiometric Titration of Chloride and Iodide in a Mixture

### Discussion

The potentiometric titration of halide mixtures is discussed in Section 17G–2. The silver indicator electrode can be a commercial billet type or simply a polished wire. A calomel electrode can be used as reference, although diffusion of chloride ion from the salt bridge may cause the results of the titration to be measurably high. This source of error can be eliminated by placing the calomel electrode in a potassium nitrate solution that is in contact with the analyte solution by means of a $KNO_3$ salt bridge. Alternatively, the analyte solution can be made slightly acidic with several drops of nitric acid; a glass electrode can then serve as the reference electrode because the pH of the solution and thus its potential remain essentially constant throughout the titration.

The titration of $I^-/Cl^-$ mixtures demonstrates how a potentiometric titration can have multiple end points. The potential of the silver electrode is proportional to pAg. Thus, a plot of $E_{Ag}$ against titrant volume yields an experimental curve with the same shape as the theoretical curve shown in Figure 9–5 (the ordinate units will be different, of course).

Experimental curves for the titration of $I^-/Cl^-$ mixtures do not show the sharp discontinuity that occurs at the first equivalence point of the theoretical curve (Figure 9–5). More important, the volume of silver nitrate needed to reach the $I^-$ end point is generally somewhat greater than theoretical; the total volume closely approaches the correct amount, however. This effect is the result of coprecipitation of the more soluble AgCl during formation of the less soluble AgI. An overconsumption of reagent thus occurs in the first part of the titration.

Despite this coprecipitation error, the potentiometric method is useful for the analysis of halide mixtures. With approximately equal quantities of iodide and chloride, relative errors can be kept to 2% or less.

### Preparation of Reagents

(a) *Silver nitrate, 0.05 M.* Follow the instructions in Section 34B–1.
(b) *Potassium nitrate salt bridge.* Bend an 8-mm glass tube into a U-shape with arms that are long enough to extend nearly to the bottom of two 100-mL beakers. Heat 50 mL of water to boiling, and stir in 1.8 g of powdered agar; continue to heat and stir until a uniform suspension is formed. Dissolve 12 g of $KNO_3$ in the hot suspension. Allow the mixture to cool somewhat. Clamp the U-tube with the openings

facing up, and use a medicine dropper to fill it with the warm agar suspension. Cool the tube under a cold-water tap to form the gel. When the bridge is not in use, immerse the ends in 2.5 M $KNO_3$.

## Procedure

Obtain the unknown in a clean 250-mL volumetric flask; dilute to the mark with water, and mix well.

Transfer 50.00 mL of the sample to a clean 100-mL beaker, and add a drop or two of concentrated $HNO_3$. Place about 25 mL of 2.5 M $KNO_3$ in a second 100-mL beaker, and make contact between the two solutions with the salt bridge. Immerse a silver electrode in the analyte solution and a calomel reference electrode in the second beaker. Titrate with $AgNO_3$ as described in Section 34I–1. Use small increments of titrant in the vicinity of the two end points.

Plot the data, and establish end points for the two analyte ions. Plot a theoretical titration curve, assuming the measured concentrations of the two constituents to be correct.

Report the number of milligrams of $I^-$ and $Cl^-$ in the sample or as otherwise instructed.

## 34I–3 The Potentiometric Determination of Solute Species in a Phosphate Mixture

### Discussion

The use of a glass/calomel electrode system to locate end points in neutralization titrations and to estimate dissociation constants is discussed in Section 17G–4. As a preliminary step to the titrations, the electrode system is standardized against a buffer of known pH.

The unknown is issued as an aqueous solution prepared from one or perhaps two adjacent members of the following series: HCl, $H_3PO_4$, $NaH_2PO_4$, $Na_2HPO_4$, $Na_3PO_4$, and NaOH. The object is to determine which of these components were used to prepare the unknown as well as the weight percent of each solute.

Most unknowns require a titration with either standard acid or standard base. A few may require separate titrations, one with acid and the other with base. The initial pH of the unknown provides guidance concerning the appropriate titrant(s); a study of curve $A$ in Figure 11–3 may be helpful in interpreting the data.

### Preparation of Solutions

*Standardized 0.1 M HCl and/or 0.1 M NaOH.* Follow the directions in Sections 34C–3 through 34C–7.

### Procedure

Obtain the unknown in a clean 250-mL volumetric flask. Dilute to the mark and mix well. Transfer a small amount of the diluted unknown to a

Suggestion: perform one titration with a few drops of bromocresol green added and another with phenolphthalein to see whether either would be a useful indicator for your titration.

beaker, and determine its pH. Titrate a 50.00-mL aliquot with standard acid or standard base (or perhaps both). Use the resulting titration curves to select indicator(s) suitable for end-point detection, and perform duplicate titrations with these.

Identify the solute species in the unknown, and report the weight/volume percent of each. Calculate the approximate dissociation constant that can be obtained for any phosphate-containing species from the titration data.

## 34I–4 The Potentiometric Titration of Copper with EDTA

### Discussion

Mercury serves as an electrode of the second kind (Section 17D–1) for the titration of many cations with EDTA. A small volume of the mercury(II)/EDTA complex is added to a solution of the sample. The formation constant of $HgY^{2-}$ is so favorable that $[HgY^{2-}]$ remains essentially constant throughout the titration; the electrode is thus responsive to $[Y^{4-}]$ and indirectly to the concentration of the cation being titrated (in this experiment, $Cu^{2+}$). See Section 17D–1 for additional information.

### Preparation of Solutions

(a) *Copper(II), 0.02 M.* Weigh (to the nearest 0.1 mg) about 0.32 g of copper into a 150-mL beaker. Cover with a watch glass, and dissolve in a minimum volume of dilute $HNO_3$ (HOOD); warm gently, if necessary, to hasten solution. Rinse the underside of the watch glass and the inner wall of the beaker. Transfer the solution quantitatively to a 250-mL volumetric flask, dilute to the mark, and mix well.

Recall that the solid should be dried at 80°C.

(b) *EDTA, about 0.02 M.* Weigh about 1.9 g (to the nearest 0.1 mg) of purified, dried $Na_2H_2Y \cdot 2\ H_2O$ into a 250-mL volumetric flask and dissolve in water, following the instructions given in Section 34D–2.

(c) *Acetate buffer.* (Sufficient for 40 to 50 titrations.) Dilute about 7 mL of glacial acetic acid to about 100 mL. With a pH meter, adjust the pH to 4.6 through the addition of 6 M NaOH. Dilute to about 250 mL with water.

(d) *Mercury(II), 0.02 M.* (Sufficient for several hundred titrations.) Dissolve 0.40 g of Hg in a minimum volume of $HNO_3$ (HOOD). Transfer quantitatively to a 100-mL volumetric flask, dilute to the mark, and mix well.

### Procedure

Adjust the meter to read in millivolts. Connect the mercury electrode and a calomel electrode to the meter. Transfer a 25.00-mL aliquot of the copper solution to a 250-mL beaker. Add about 5 mL of acetate buffer and one drop of a solution prepared by mixing equal volumes of the EDTA and Hg(II) solutions. Add water, if needed, to cover the ends of the electrodes (Note 1). Titrate (Note 2) with EDTA according to Section 34I–1.

Compare the milligrams of copper found with the amount taken. Calculate the relative error in the titration.

**Notes**
1. Sufficient chloride ion may leak from the calomel electrode to interfere with the functioning of the mercury indicator electrode. If necessary, place the reference electrode in a beaker that is connected by a $KNO_3$ bridge to the analyte container; see Section 341–2.
2. The titration must be performed slowly in the vicinity of the end point.

## 341–5 The Direct Potentiometric Determination of Fluoride Ion

### Discussion

The solid-state fluoride electrode (Section 17D–6) has found extensive use in the determination of fluoride in a variety of materials. Directions follow for the determination of this ion in drinking water and in toothpaste. A total ionic strength adjustment buffer (TISAB) is used to adjust all unknowns and standards to essentially the same ionic strength; when this is done, the concentration of fluoride, rather than its activity, is measured. The pH of the buffer is about 5, a level at which $F^-$ is the predominant fluorine-containing species. The buffer also contains cyclohexylaminedinitrilotetraacetic acid, which forms stable chelates with iron(III) and aluminum(III), thus freeing fluoride ion from its complexes with these cations.

Before undertaking these experiments, we suggest that you review Sections 17D and 17F.

### Preparation of Solutions

(a) *Total ionic strength adjustment buffer (TISAB).* This solution is marketed commercially under the trade name TISAB.[8] Sufficient buffer for 15 to 20 determinations can be prepared by mixing (with stirring) 57 mL of glacial acetic acid, 58 g of NaCl, 4 g of cyclohexylaminedinitrilotetraacetic acid, and 500 mL of distilled water in a 1-L beaker. Cool the contents in a water or ice bath, and carefully add 6 M NaOH until a pH of 5.0 to 5.5 is reached. Dilute to 1 L with water, and store in a plastic bottle.

(b) *Standard fluoride solution, 100 ppm.* Dry a quantity of NaF at 110°C for 2 hr. Cool in a desiccator; then weigh (to the nearest milligram) 0.22 g into a 1-L volumetric flask. (CAUTION! NaF IS HIGHLY TOXIC. *Immediately* wash any skin touched by this compound with copious quantities of water.) Dissolve in water, dilute to the mark, mix well, and store in a plastic bottle. Calculate the exact concentration of fluoride in parts per million.

A standard $F^-$ solution can be purchased from commercial sources.

---

[8]Orion Research, Boston, MA.

## Procedure

The apparatus for this experiment consists of a solid-state fluoride electrode, a saturated calomel electrode, and a pH meter. A sleeve-type calomel electrode is needed for the toothpaste determination because the measurement is made on a suspension that tends to clog the liquid junction. The sleeve must be loosened momentarily to renew the interface after each series of measurements.

### Determination of Fluoride in Drinking Water

Transfer 50.00-mL portions of the water to 100-mL volumetric flasks, and dilute to the mark with TISAB solution.

Prepare a 5-ppm F⁻ solution by diluting 25.0 mL of the 100-ppm standard to 500 mL in a volumetric flask. Transfer 5.00-, 10.0-, 25.0-, and 50.0-mL aliquots of the 5-ppm solution to 100-mL volumetric flasks, add 50 mL of TISAB solution, and dilute to the mark. (These solutions correspond to 0.5, 1.0, 2.5, and 5.00 ppm F⁻ in the sample.)

After thorough rinsing and drying with paper tissue, immerse the electrodes in the 0.5-ppm standard. Stir mechanically for 3 min; then measure and record the potential. Repeat with the remaining standards and samples.

Plot the measured potential against the log of the concentration of the standards. Use this plot to determine the concentration in parts per million of fluoride in the unknown.

### Determination of Fluoride in Toothpaste[9]

Weigh (to the nearest milligram) 0.2 g of toothpaste into a 250-mL beaker. Add 50 mL of TISAB solution, and boil for 2 min with good mixing. Cool and then transfer the suspension quantitatively to a 100-mL volumetric flask, dilute to the mark with distilled water, and mix well. Follow the directions for the analysis of drinking water, beginning with the second paragraph.

Report the parts per million of F⁻ in the sample.

## 34J    ELECTROGRAVIMETRIC METHODS

A convenient example of an electrogravimetric method of analysis is the simultaneous determination of copper and lead in a sample of brass. Additional information concerning electrogravimetric methods is found in Section 18C.

### 34J–1  The Electrogravimetric Determination of Copper and Lead in Brass

#### Discussion

This procedure is based upon the deposition of metallic copper on a cathode and of lead as $PbO_2$ on an anode. As a first step, the hydrous

---

[9]From T. S. Light and C. C. Cappuccino, *J. Chem. Educ.*, **1975,** *52,* 247.

oxide of tin ($SnO_2 \cdot xH_2O$) that forms when the sample is treated with nitric acid must be removed by filtration. Lead dioxide is deposited quantitatively at the anode from a solution with a high nitrate ion concentration; copper is only partially deposited on the cathode under these conditions. It is therefore necessary to eliminate the excess nitrate after deposition of the $PbO_2$ is complete. Removal is accomplished through the addition of urea:

$$6 \, NO_3^- + 6 \, H^+ + 5(NH_2)_2CO \rightarrow 8 \, N_2(g) + 5 \, CO_2(g) + 13 \, H_2O$$

Copper then deposits quantitatively from the solution after the nitrate ion concentration has been decreased.

---

## Procedure

### Preparation of Electrodes

Immerse the platinum electrodes in hot 6 M $HNO_3$ for about 5 min (Note 1). Wash them thoroughly with distilled or deionized water, rinse with several small portions of acetone or ethanol, and dry in an oven at 110°C for 2 to 3 min. Cool and weigh both anodes and cathodes to the nearest 0.1 mg.

### Preparation of Samples

It is not necessary to dry the unknown. A rinse with acetone is recommended if there is evidence of oil on the surface of the metal. Weigh (to the nearest 0.1 mg) 1-g samples into 250-mL beakers. Cover the beakers with watch glasses. Cautiously add about 35 mL of 6 M $HNO_3$ (HOOD). Digest for at least 30 min; add more acid if necessary. Evaporate to about 5 mL but never to dryness (Note 2).

To each sample, add 5 mL of 3 M $HNO_3$, 25 mL of water, and one quarter of a tablet of filter paper pulp; digest without boiling for about 45 min. Filter off the $SnO_2 \cdot xH_2O$, using a fine-porosity filter paper (Note 3); collect the filtrates in tall-form electrolysis beakers. Use many small washes with hot 0.3 M $HNO_3$ to remove the last traces of copper; test for completeness with a few drops of $NH_3$. The final volume of filtrate and washings should be between 100 and 125 mL; either add water or evaporate to attain this volume.

### Electrolysis

With the current switch off, attach the cathode to the negative terminal and the anode to the positive terminal of the electrolysis apparatus. Briefly turn on the stirring motor to be sure the electrodes do not touch. Cover the beakers with split watch glasses and commence the electrolysis. Maintain a current of 1.3 A for 35 min.

Rinse the cover glasses and add 10 mL of 3 M $H_2SO_4$ followed by 5 g of urea to each beaker. Maintain a current of 2 A until the solutions are colorless. To test for completeness of the electrolysis, remove one drop of the solution with a medicine dropper, and mix it with a few drops of $NH_3$ in a small test tube. If the mixture turns blue, rinse the contents of the tube back into the beaker and continue the electrolysis for an additional 10 min. Repeat the test until no blue $Cu(NH_3)_4^{2+}$ is produced.

When electrolysis is complete, discontinue stirring but leave the current on. Rinse the electrodes thoroughly with water as they emerge from the solution. After rinsing is complete, turn off the electrolysis apparatus (Note 4), disconnect the electrodes, and dip them in acetone. Dry the cathodes for about 3 min and the anodes for about 15 min at 110°C. Allow the electrodes to cool in air, and then weigh them.

Report the percentages of lead (Note 5) and copper in the brass.

**Notes**

1. Alternatively, grease and organic materials can be removed by heating platinum electrodes to redness in a flame. Electrode surfaces should not be touched with your fingers after cleaning because grease and oil cause nonadherent deposits that can flake off during washing and weighing.

2. Chloride ion must be totally excluded from this determination because it attacks the platinum anode during electrolysis. This reaction is not only destructive but also causes positive errors in the analysis by codepositing platinum with copper on the cathode.

3. If desired, the tin content can be determined by ignition of the $SnO_2 \cdot xH_2O$. See Section 34A–2.

4. It is important to maintain a potential between the electrodes until they have been removed from the solution and washed. Some copper may redissolve if this precaution is not observed.

5. Experience has shown that a small amount of moisture is retained by the $PbO_2$ and that better results are obtained if 0.8643 is used instead of 0.8662, the stoichiometric factor.

---

## 34K  COULOMETRIC TITRATIONS

In a coulometric titration, the "reagent" is a constant direct current of exactly known magnitude. The time required for this current to oxidize or reduce the analyte quantitatively (directly or indirectly) is measured. See Section 18D–5 for a discussion of this electroanalytical method.

### 34K–1 The Coulometric Titration of Cyclohexene

#### Discussion[10]

Many olefins react sufficiently rapidly with bromine to permit their direct titration. The reaction is carried out in a largely nonaqueous environment with mercury(II) as a catalyst. A convenient way of performing this titration is to add excess bromide ion to a solution of the sample and generate the bromine at an anode that is connected to a constant-current source. The electrode processes are

$$2\,Br^- \rightarrow 2\,Br_2 + 2\,e^- \quad \text{Anode}$$
$$2\,H^+ + 2\,e^- \rightarrow H_2(g) \quad \text{Cathode}$$

---

[10]This procedure was described by D. H. Evans in *J. Chem. Educ.*, **1968,** *45* (1), 88.

The hydrogen produced does not react with bromine rapidly enough to interfere. The bromine reacts with an olefin, such as cyclohexene, to give the addition product:

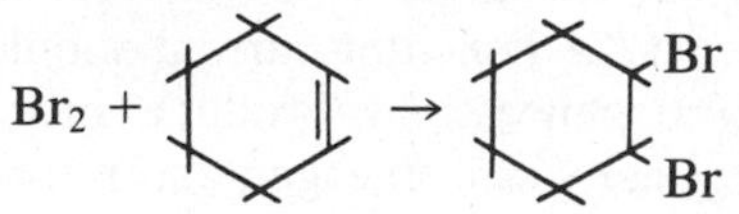

The amperometric method with twin-polarized electrodes (page 490) provides a convenient way to detect the end point in this titration. A potential difference of 0.2 to 0.3 V is maintained between two small electrodes. This potential is not sufficient to cause the generation of hydrogen at the cathode. Thus, short of the end point, the cathode is polarized and no current is observed. At the end point, the first excess of bromine depolarizes the cathode and produces a current. The electrode reactions at the twin indicator electrodes are

$$2\ Br^- \rightarrow Br_2 + 2\ e^- \qquad \text{Anode}$$
$$Br_2 + 2\ e^- \rightarrow 2\ Br^- \qquad \text{Cathode}$$

The current is proportional to the bromine concentration and is readily measured with a microammeter.

A convenient way to perform several analyses is to initially generate sufficient bromine in the solvent to give a readily measured current, say, 20 $\mu$A. An aliquot of the sample is then introduced, whereupon the current immediately decreases and approaches zero. Generation of bromine is again commenced, and the time needed to regain a current of 20 $\mu$A is measured. A second aliquot of the sample is added to the same solution, and the process is repeated. Several samples can thus be analyzed without changing the solvent.

The procedure that follows is for the determination of cyclohexene in a methanol solution. Other olefins can be determined as well.

## Preparation of Solvent

Dissolve about 9 g of KBr and 0.5 g of mercury(II) acetate (Note 1) in a mixture consisting of 300 mL of glacial acetic acid, 130 mL of methanol, and 65 mL of water. (Sufficient for about 35 mmol of $Br_2$.) (CAUTION! Mercury compounds are highly toxic and the solvent is a skin irritant. If inadvertent contact occurs, flood the affected area with copious quantities of water.)

## Procedure

Obtain the unknown in a 100-mL volumetric flask; dilute to the mark with methanol, and mix well. The temperature of the methanol should be between 18 and 20°C (Note 2).

Add sufficient acetic acid/methanol solvent to cover the indicator and generator electrodes in the electrolysis vessel. Apply about 0.2 V to the indicator electrodes. Activate the generator electrode system, and generate bromine until a current of about 20 $\mu$A is indicated on the microammeter. Stop the generation of bromine, record the indicator current to the nearest 0.1 $\mu$A, and set the timer to zero. Transfer 10.00 mL of the unknown to the solvent; the indicator current should decrease to almost zero. Resume bromine generation. Produce bromine in smaller and smaller increments by activating the generator for shorter and shorter periods as the indicator current rises and approaches the previously recorded value. Read and record the time needed to reach the original indicator current. Reset the timer to zero, introduce a second aliquot of sample (make the volume larger if the time needed for the first titration was too short, and conversely), and repeat the process. Titrate several aliquots.

Report the number of milligrams of cyclohexene in the unknown.

**Notes**
1. Mercury(II) ions catalyze the addition of bromine to olefinic double bonds.
2. The coefficient of expansion for methanol is 0.11%/°C; thus, significant volumetric errors result if the temperature is not controlled.

## 34L   VOLTAMMETRY

Various aspects of polarographic and amperometric methods are considered in Chapter 19. Two examples that illustrate these methods are described in this section. Enormous diversity exists in the instrumentation available for these determinations. It will thus be necessary for you to consult the manufacturer's operating instructions concerning the details of operation for the particular instrument you will use.

### 34L–1 The Polarographic Determination of Copper and Zinc in Brass

#### Discussion

The percentage of copper and zinc in a sample of brass can be determined from polarographic measurements. The method is particularly useful for rapid, routine analyses; in return for speed, however, the accuracy is considerably lower than that obtained with volumetric or gravimetric methods.

The sample is dissolved in a minimum amount of nitric acid. It is not necessary to remove the $SnO_2 \cdot xH_2O$ produced. Addition of an ammonia/ammonium chloride buffer causes the precipitation of lead as a basic oxide. A polarogram of the supernatant liquid has two copper waves. The one at about $-0.2$ V (versus SCE) corresponds to the reduction of copper(II) to copper(I), and the one at about $-0.5$ V represents further reduction to the metal. The analysis is based upon the total diffusion current of the two waves. The zinc concentration is determined from its wave at $-1.3$ V. For instruments that permit current offset, the copper waves are measured at the highest feasible sensitivity. These waves are then suppressed by the offset control of the instrument, and the zinc wave is obtained, again at the highest possible sensitivity setting.

## Preparation of Solutions

(a) *Copper(II) solution, $2.5 \times 10^{-2}$ M.* Weigh (to the nearest milligram) 0.4 g of copper wire. Dissolve in 5 mL of concentrated $HNO_3$ (HOOD). Boil briefly to remove oxides of nitrogen; then cool, dilute with water, transfer quantitatively to a 250-mL volumetric flask, dilute to the mark with water, and mix thoroughly.

(b) *Zinc(II) solution, $2.5 \times 10^{-2}$ M.* Dry reagent-grade ZnO for 1 hr at 110°C, cool in a desiccator, and weigh 0.5 g (to the nearest milligram) into a small beaker. Dissolve in a mixture of 25 mL of water and 5 mL of concentrated $HNO_3$. Transfer to a 250-mL volumetric flask, and dilute to the mark with water.

(c) *Gelatin, 0.1%.* Add about 0.1 g of gelatin to 100 mL of boiling water.

(d) *Ammonia/ammonium chloride buffer.* (Sufficient for about 15 polarograms.) Mix 27 g of $NH_4Cl$ and 35 mL of concentrated ammonia in sufficient distilled water to give about 500 mL. This solution is about 1 M in $NH_3$ and 1 M in $NH_4^+$.

## Procedure

### Preparation of Calibration Standards

Use a buret to transfer 0-, 1-, 8-, and 15-mL portions of standard Cu(II) solution to 50-mL volumetric flasks. Add 5 mL of gelatin solution and 30 mL of buffer to each. Dilute to the mark, and mix well. Prepare an identical series of Zn(II) solutions.

Rinse the polarographic cell three times with small portions of a copper(II) solution; then fill the cell. Bubble nitrogen through the solution for 10 to 15 min to remove oxygen. Apply a potential of about $-1.6$ V, and adjust the sensitivity to cause the detector to give a response that is essentially full scale. Obtain a polarogram, scanning from 0 to $-1.5$ V (versus SCE). Measure the limiting current at a potential just beyond the second wave. Obtain a diffusion current by subtracting the current for the blank (Note) at this same potential. Calculate $i_d/c$.

Repeat the foregoing with the other two copper solutions and with the zinc solutions.

### Sample Preparation

Weigh 0.10- to 0.15-g (to the nearest 0.5 mg) samples of brass into 50-mL beakers, and dissolve in 2 mL of concentrated $HNO_3$ (HOOD). Boil briefly to eliminate oxides of nitrogen. Cool, add 10 mL of distilled water, and transfer each quantitatively to a 50-mL volumetric flask; dilute to the mark with water, and mix well.

Transfer 10.00 mL of the diluted sample to another 50.0-mL volumetric flask, add 5 mL of gelatin and 30 mL of buffer, dilute to the mark with water, and mix well.

### Analysis

Follow the directions in the second paragraph of *Preparation of calibration standards*. Evaluate the diffusion currents for copper and zinc.

Calculate the percentage of Cu and Zn in the brass sample.

**Note.** The polarogram for the blank (that is, 0 mL of standard) should be obtained at the sensitivity setting used for the standard with the lowest metal-ion concentration.

## 34L–2  The Amperometric Titration of Lead

### Discussion

Amperometric titrations are discussed in Section 19B–3. In the procedure that follows, the lead concentration of an aqueous solution is determined by titration with a standard potassium dichromate solution. The reaction is

$$Cr_2O_7^{2-} + 2\ Pb^{2+} + H_2O \rightarrow 2\ PbCrO_4(s) + 2\ H^+$$

The titration can be performed with a dropping mercury electrode maintained at either 0 or $-1.0$ V (versus SCE). At 0 V, the current remains near zero short of the end point but rises rapidly immediately thereafter, owing to the reduction of dichromate, which is now in excess. At $-1.0$ V, both lead ion and dichromate ion are reduced. The current thus decreases in the region short of the end point (reflecting the decreases in the lead ion concentration as titrant is added), passes through a minimum at the end point, and rises as dichromate becomes available. The end point at $-1.0$ V should be the easier of the two to locate exactly. Removal of oxygen is unnecessary, however, for the titration at 0 V.

### Preparation of Solutions

(a) *Supporting electrolyte.* Dissolve 10 g of $KNO_3$ and 8.2 g of sodium acetate in about 500 mL of distilled water. Add glacial acetic acid to bring the pH to 4.2 (pH meter); about 20 mL of the acid will be required. (Sufficient for about 20 titrations.)

(b) *Gelatin, 0.1%.* Add 0.1 g of gelatin to 100 mL of boiling water.

(c) *Potassium dichromate, 0.01 M.* Use a powder funnel to weigh about 0.75 g (to the nearest 0.5 mg) of primary-standard $K_2Cr_2O_7$ into a 250-mL volumetric flask. Dilute to the mark with distilled water, and mix well.

(d) *Potassium nitrate salt bridge.* See Section 34I–2, *Preparation of Reagents*.

### Procedure

Obtain the unknown (Note) in a clean 100-mL volumetric flask; dilute to the mark with distilled water, and mix well. Perform the titration in a 100-

mL beaker. Locate a saturated calomel electrode in a second 100-mL beaker, and provide contact between the solutions in the two containers with the $KNO_3$ salt bridge.

Transfer a 10.00-mL aliquot of the unknown to the titration beaker. Add 25 mL of the supporting electrolyte and 5 mL of gelatin. Insert a dropping mercury electrode in the sample solution. Connect both electrodes to the polarograph. Measure the current at an applied potential of 0 V. Add 0.01 M $K_2Cr_2O_7$ in 1-mL increments; record the current and volume after each addition. Continue the titration to about 5 mL beyond the end point. Correct the currents for volume change and plot the data. Evaluate the end-point volume.

Repeat the titrations at $-1.0$ V. Here, you must bubble nitrogen through the solution for 10 to 15 min before the titration and after each addition of reagent. The flow of nitrogen must, of course, be interrupted during the current measurements. Again, correct the currents for volume change, plot the data, and determine the end-point volume.

Report the number of milligrams of Pb in the unknown.

The corrected current is given by:

$$(i_d)_{\text{corr}} = i_d \left( \frac{V + v}{V} \right)$$

where $V$ is the original volume $v$, is the volume of reagent added, and $i_d$ is the uncorrected current.

**Note.** A stock 0.0400 M solution can be prepared by dissolving 13.5 g of reagent-grade $Pb(NO_3)_2$ in 10 mL of 6 M $HNO_3$ and diluting to 1 L with water. Unknowns should contain 15 to 25 mL of this solution.

---

## 34M   METHODS BASED ON THE ABSORPTION OF RADIATION

Molecular absorption methods are discussed in Chapter 22. Directions follow for (1) the use of a calibration curve for the determination of iron in water, (2) the use of a standard-addition procedure for the determination of manganese in steel, and (3) a spectrophotometric determination of the pH of a buffer solution.

### 34M-1   The Cleaning and Handling of Cells

The accuracy of spectrophotometric measurements is critically dependent upon the availability of good-quality matched cells. These should be calibrated against one another at regular intervals to detect differences resulting from scratches, etching, and wear. Equally important is the proper cleaning of the exterior sides (the *windows*) just before the cells are inserted into a photometer or spectrophotometer. The preferred method is to wipe the windows with a lens paper soaked in methanol; the methanol is then allowed to evaporate, leaving the windows free of contaminants. It has been shown that this method is far superior to the usual procedure of wiping the windows with a dry lens paper, which tends to leave a residue of lint and a film on the window.[11]

---

[11]For further information, see J. O. Erickson and T. Surles, *Amer. Lab.*, **1976**, 8 (6), 50.

### 34M–2 The Determination of Iron in a Natural Water

#### Discussion

The red-orange complex that forms between iron(II) and 1,10-phenanthroline (orthophenanthroline) is useful for determining iron in water supplies. The reagent is a weak base that reacts to form phenanthrolinium ions, PhenH$^+$, in acidic media. Complex formation with iron is thus best described by the equation

For the structure of the iron(II) complex of 1,10-phenanthroline, see p. 365.

$$Fe^{2+} + 3\ PhenH^+ \rightleftharpoons Fe(Phen)_3^{2+} + 3\ H^+$$

The formation constant for this equilibrium is $2.5 \times 10^6$ at 25°C. Iron(II) is quantitatively complexed in the pH range between 3 and 9. A pH of about 3.5 is ordinarily recommended to prevent precipitation of iron salts, such as phosphates.

An excess of a reducing reagent, such as hydroxylamine or hydroquinone, is needed to maintain iron in the +2 state. The complex, once formed, is very stable.

This determination can be performed with a spectrophotometer set at 508 nm or with a photometer equipped with a green filter.

---

#### Preparation of Solutions

(a) *Standard iron solution, 0.01 mg/mL.* Weigh (to the nearest 0.2 mg) 0.0702 g of reagent-grade $Fe(NH_4)_2(SO_4)_2 \cdot 6\ H_2O$ into a 1-L volumetric flask. Dissolve in 50 mL of water that contains 1 to 2 mL of concentrated sulfuric acid; dilute to the mark, and mix well.

(b) *Hydroxylamine hydrochloride.* Dissolve 10 g of $H_2NOH \cdot HCl$ in about 100 mL of distilled water. (Sufficient for 80 to 90 measurements.)

(c) *Orthophenanthroline solution.* (Sufficient for 80 to 90 measurements.) Dissolve 1.0 g of orthophenanthroline monohydrate in about 1 L of water. Warm slightly if necessary. Each milliliter is sufficient for no more than about 0.09 mg of Fe. Prepare no more reagent than needed; it darkens on standing and must then be discarded.

(d) *Sodium acetate, 1.2 M.* (Sufficient for 80 to 90 measurements.) Dissolve 166 g of $NaOAc \cdot 3\ H_2O$ in 1 L of distilled water.

---

#### Procedure

Preparation of a Calibration Curve
Transfer 25.00 mL of the standard iron solution to a 100-mL volumetric flask and 25 mL of distilled water to a second 100-mL volumetric flask. Add 1 mL of hydroxylamine, 10 mL of sodium acetate, and 10 mL of orthophenanthroline to each flask. Allow the mixtures to stand for 5 min; dilute to the mark and mix well.

Clean a pair of matched cells for the instrument. Rinse each cell with at least three portions of the solution it is to contain. Determine the absorbance of the standard with respect to the blank.

Repeat the above procedure with at least three other volumes of the standard iron solution; attempt to encompass an absorbance range between 0.1 and 1.0. Plot a calibration curve.

Determination of Iron

Transfer 10.00 mL of the unknown to a 100-mL volumetric flask; treat in exactly the same way as the standards, measuring the absorbance with respect to the blank. Alter the volume of unknown taken to obtain absorbance measurements for replicate samples that are within the range of the calibration curve.

Report the parts per million of iron in the unknown.

## 34M–3 The Determination of Manganese in Steel

Discussion

Small quantities of manganese are readily determined photometrically by the oxidation of Mn(II) to the intensely colored permanganate ion. Potassium periodate is an effective oxidizing reagent for this purpose. The reaction is

$$5\,IO_4^- + 2\,Mn^{2+} + 3\,H_2O \rightarrow 5\,IO_3^- + 2\,MnO_4^- + 6\,H^+$$

Permanganate solutions that contain an excess of periodate are quite stable.

Interferences to the method are few. The presence of most colored ions can be compensated for with a blank. Cerium(III) and chromium(III) are exceptions; these yield oxidation products with periodate that absorb to some extent at the wavelength used for the measurement of permanganate.

The method given here is applicable to steels that do not contain large amounts of chromium. The sample is dissolved in nitric acid. Any carbon in the steel is oxidized with peroxodisulfate. Iron(III) is eliminated as a source of interference by complexation with phosphoric acid. The standard-addition method (page 628) is used to establish the relationship between absorbance and amount of manganese in the sample.

A spectrophotometer set at 525 nm or a photometer with a green filter can be used for the absorbance measurements.

Preparation of Solutions

(a) *Standard manganese(II) solution.* (Sufficient for several hundred analyses.) Weigh 0.1 g (to the nearest 0.1 mg) of manganese into a 50-mL beaker, and dissolve in about 10 mL of 6 M $HNO_3$ (HOOD). Boil gently to eliminate oxides of nitrogen. Cool; then transfer the solution quantitatively to a 1-L volumetric flask. Dilute to the mark with water, and mix thoroughly. The manganese in 1 mL of the standard solution, after being converted to permanganate, causes a volume of 50 mL to increase in absorbance by about 0.09.

Peroxodisulfate and persulfate are synonyms. So also are bisulfite and hydrogen sulfite.

## Procedure

The unknown does not require drying. If there is evidence of oil, rinse with acetone and dry briefly. Weigh (to the nearest 0.1 mg) duplicate samples (Note 1) into 150-mL beakers. Add about 50 mL of 6 M $HNO_3$, and boil gently (HOOD); heating for 5 to 10 min should suffice. Cautiously add about 1 g of ammonium peroxodisulfate, and boil gently for an additional 10 to 15 min. If the solution is pink or has a deposit of $MnO_2$, add 1 mL of $NH_4HSO_3$ (or 0.1 g of $NaHSO_3$) and heat for 5 min. Cool; transfer quantitatively (Note 2) to 250.0-mL volumetric flasks. Dilute to the mark with water, and mix well. Use a 20.00-mL pipet to transfer three aliquots of each sample to individual beakers. Treat as follows:

| Aliquot | Volume of 85% $H_3PO_4$, mL | Volume of Standard Mn, mL | Weight of $KIO_4$, g |
|---|---|---|---|
| 1 | 5 | 0.00 | 0.4 |
| 2 | 5 | 5.00 (Note 3) | 0.4 |
| 3 | 5 | 0.00 | 0.0 |

Boil each solution gently for 5 min, cool, and transfer quantitatively to a 50-mL volumetric flask. Mix well. Measure the absorbance of aliquots 1 and 2 using aliquot 3 as the blank (Note 4).

Report the percentage of manganese in the unknown.

### Notes

1. The sample size depends upon the manganese content of the unknown; consult with your instructor.
2. If there is evidence of turbidity, filter the solutions as they are transferred to the volumetric flasks.
3. The volume of the standard addition may be dictated by the absorbance of the sample. It is useful to obtain a rough estimate by generating permanganate in about 20 mL of sample, diluting to about 50 mL, and measuring the absorbance.
4. A single blank can be used for all measurements, provided the samples weigh within 50 mg of one another.

## 34M–4 The Spectrophotometric Determination of pH

### Discussion

The pH of an unknown buffer is determined by addition of an acid/base indicator and spectrophotometric measurement of the absorbance of the resulting solution. Because overlap exists between the spectra for the acid and base forms of the indicator, it is necessary to evaluate individual molar absorptivities for each form at two wavelengths. See page 574 for further discussion.

The relationship between the two forms of bromocresol green in an aqueous solution is described by the equilibrium

$$HIn + H_2O \rightleftharpoons H_3O^+ + In^-$$

for which

$$K_a = \frac{[H_3O^+][In^-]}{[HIn]} = 1.6 \times 10^{-5}$$

The spectrophotometric evaluation of $[In^-]$ and $[HIn]$ permits the calculation of $[H_3O^+]$.

## Preparation of Solutions

(a) *Bromocresol green, $1.0 \times 10^{-4}$ M.* (Sufficient for about five determinations.) Dissolve 40.0 mg (to the nearest 0.1 mg) of the sodium salt of bromocresol green (fw = 720) in water, and dilute to 500 mL in a volumetric flask.
(b) *HCl, 0.5 M.* Dilute 4 mL of concentrated HCl to approximately 100 mL with water.
(c) *NaOH, 0.4 M.* Dilute 7 mL of 6 M NaOH to about 100 mL with water.

## Procedure

### Determination of Individual Absorption Spectra
Transfer 25.00-mL aliquots of the bromocresol green indicator solution to two 100-mL volumetric flasks. To one add 25 mL of 0.4 M HCl; to the other add 25 mL of 0.4 M NaOH. Dilute to the mark and mix well.

Obtain the absorption spectra for the acid and conjugate-base forms of the indicator between 400 and 600 nm, using water as a blank. Record absorbance values at 10-nm intervals routinely and at closer intervals as needed to define maxima and minima in the two curves. Evaluate the molar absorptivity for HIn and In⁻ at wavelengths corresponding to their absorption maxima.

### Determination of the pH of an Unknown Buffer
Transfer 25.00 mL of the stock bromocresol green indicator to a 100-mL volumetric flask. Add 50.0 mL of the unknown buffer, dilute to the mark, and mix well. Measure the absorbance of the diluted solutions at the wavelengths for which absorptivity data were calculated.

Report the pH of the buffer.

## 34N   MOLECULAR FLUORESCENCE

The phenomenon of fluorescence and its analytical applications are discussed in Chapter 23. Directions follow for the determination of quinine in beverages, in which the quinine concentration is typically between 25 and 60 ppm.

### 34N–1 The Determination of Quinine in Beverages[12]

#### Discussion

Solutions of quinine fluoresce strongly when excited by radiation at 350 nm. The relative intensity of the fluorescent peak at 450 nm provides a sensitive method for the determination of quinine in beverages. Preliminary measurements are needed to define a concentration region in which fluorescent intensity is either linear or nearly so. The unknown is then diluted as necessary to produce readings within this range.

#### Preparation of Reagents

(a) *Sulfuric acid, 0.05 M*. Add about 17 mL of 6 M $H_2SO_4$ to 2 L of distilled water.

(b) *Quinine sulfate standard, 1 ppm*. Weigh (to the nearest 0.5 mg) 0.100 g of quinine sulfate into a 1-L volumetric flask, and dilute to the mark with 0.05 M $H_2SO_4$. (Sufficient for 60 to 70 analyses.) Transfer 10.00 mL of this solution to another 1-L volumetric flask, and again dilute to the mark with 0.05 M $H_2SO_4$. This latter solution contains 1 ppm of quinine; it should be prepared daily and stored in the dark when not in use.

#### Procedure

##### Determination of a Suitable Concentration Range

To find a suitable working range, measure the relative fluorescent intensity of the 1 ppm standard at 450 nm (or with a suitable filter that transmits in this range). Use a graduated cylinder to dilute 10 mL of the 1 ppm solution with 10 mL of 0.05 M $H_2SO_4$; again measure the relative fluorescence. Repeat this dilution and measurement process until the relative intensity approaches that of a blank consisting of 0.05 M $H_2SO_4$. Make a plot of the data, and select a suitable range for the analysis (that is, a region within which the plot is linear).

##### Preparation of a Calibration Curve

Use volumetric glassware to prepare three or four standards that span the linear region; measure the fluorescence intensity for each. Plot the data.

##### Analysis

Obtain an unknown. Make suitable dilutions with 0.05 M $H_2SO_4$ to bring its fluorescence intensity within the calibration range.

Calculate the parts per million of quinine in the unknown.

### 34O ATOMIC SPECTROSCOPY

Several methods of analysis based upon atomic spectroscopy are discussed in Chapter 24. One such application is atomic absorption, which is demonstrated in the experiment that follows.

---

[12]These directions were adapted from J. E. O'Reilly, *J. Chem. Educ.*, **1975,** *52* (9), 610.

## 34O–1 The Determination of Lead in Brass by Atomic Absorption Spectroscopy

### Discussion

Brasses and other copper-based alloys contain from 0 to about 10% lead as well as tin and zinc. Atomic absorption spectroscopy permits the quantitative estimation of these elements. The accuracy of this procedure is not as great as that obtainable with gravimetric or volumetric measurements, but the time needed to acquire the analytical information is considerably less.

A weighed sample is dissolved in a mixture of nitric and hydrochloric acids, the latter being needed to prevent the precipitation of tin as metastannic acid, $SnO_2 \cdot xH_2O$. After suitable dilution, the sample is aspirated into a flame, and the absorption of radiation from a hollow cathode lamp is measured.

---

### Preparation of Solutions

(a) *Standard lead solution, 100 mg/L.* Dry a quantity of reagent-grade $Pb(NO_3)_2$ for about 1 hr at 110°C. Cool; weigh (to the nearest 0.1 mg) 0.17 g into a 1-L volumetric flask. Dissolve in a solution of 5 mL water and 1 to 3 mL of concentrated $HNO_3$. Dilute to the mark with distilled water, and mix well.

---

### Procedure

Weigh duplicate samples of the unknown (Note 1) into 150-mL beakers. Cover with watch glasses, and then dissolve (HOOD) in a mixture consisting of about 4 mL of concentrated $HNO_3$ and 4 mL of concentrated HCl (Note 2). Boil gently to remove oxides of nitrogen. Cool; transfer the solutions quantitatively to 250-mL volumetric flasks, dilute to the mark with water, and mix well.

Use a buret to deliver 0-, 5-, 10-, 15-, and 20-mL portions of the standard lead solution to individual 50-mL volumetric flasks. Add 4 mL of concentrated $HNO_3$ and 4 mL of concentrated HCl to each, and dilute to the mark with water.

Transfer 10.00-mL aliquots of each sample to 50-mL volumetric flasks, add 4 mL of concentrated HCl and 4 mL of concentrated $HNO_3$, and dilute to the mark with water.

Set the monochromator at 283.3 nm, and measure the absorbance for each standard and the sample at that wavelength. Take at least three—and preferably more—readings for each measurement.

Plot the calibration data. Report the percentage of lead in the brass.

### Notes

1. The weight of sample depends on the lead content of the brass and on the sensitivity of the instrument used for the absorption measurements. A sample containing 6 to 10 mg of lead is reasonable. Consult with your instructor.

2. Brasses that contain a large percentage of tin require additional HCl to prevent the formation of metastannic acid. The diluted samples may develop some turbidity on prolonged standing; a slight turbidity has no effect on the determination of lead.

---

### 34O–2 The Determination of Sodium, Potassium, and Calcium in Mineral Waters by Atomic Emission Spectroscopy

### Discussion

A convenient method for the determination of alkali and alkaline earth metals in water and in blood serum is based upon the characteristic spectra these elements emit upon being aspirated into a natural gas/air flame. The accompanying directions are suitable for the analysis of three such elements in water samples. Radiation buffers (Section 24B–6) are used to minimize the effect of each element upon the emission intensity of the others.

---

### Preparation of Solutions

(a) *Standard calcium solution, 500 ppm.* Dry a quantity of $CaCO_3$ for about 1 hr at 110°C. Cool in a desiccator; weigh (to the nearest milligram) 1.25 g into a 600-mL beaker. Add about 200 mL of distilled water and about 10 mL of concentrated HCl; cover the beaker with a watch glass during the addition of the acid to avoid loss due to spattering. After reaction is complete, transfer the solution quantitatively to a 1-L volumetric flask, dilute to the mark, and mix well.

(b) *Standard potassium solution, approximately 500 ppm.* Dry a quantity of KCl for about 1 hr at 110°C. Cool; weigh (to the nearest milligram) about 0.95 g into a 1-L volumetric flask. Dissolve in distilled water, and dilute to the mark.

(c) *Standard sodium solution, approximately 500 ppm.* Proceed as in (b), using 1.25 g (to the nearest milligram) of dried NaCl.

(d) *Radiation buffer for the determination of calcium.* Prepare about 100 mL of a solution that has been saturated with NaCl, KCl, and $MgCl_2$, in that order.

(e) *Radiation buffer for the determination of potassium.* Prepare about 100 mL of a solution that has been saturated with NaCl, $CaCl_2$, and $MgCl_2$, in that order.

(f) *Radiation buffer for the determination of sodium.* Prepare about 100 mL of a solution that has been saturated with $CaCl_2$, KCl, and $MgCl_2$, in that order.

---

### Procedure

### Preparation of Working Curves

Add 5.00 mL of the appropriate radiation buffer to each of a series of 100-mL volumetric flasks. Add volumes of standard that will produce solutions that range from 0 to 10 ppm in the cation to be determined. Dilute to the mark with water, and mix well.

Measure the emission intensity for each solution, taking at least three readings for each. Aspirate distilled water between each set of measurements. Correct the average values for background luminosity, and prepare a working curve from the data.

Repeat for the other two cations.

### Analysis of a Water Sample

Prepare duplicate aliquots of the unknown as directed for preparation of working curves. If necessary, use a standard to calibrate the response of the instrument to the working curve; then measure the emission intensity of the unknown. Correct the data for background. Determine the cation concentration in the unknown by comparison with the working curve.

---

## 34P   SEPARATION OF CATIONS BY ION EXCHANGE

The application of ion-exchange resins to the separation of ionic species of opposite charge is discussed in Section 32E–1. Directions follow for the ion-exchange separation of nickel(II) from zinc(II) based upon converting zinc ions to the negatively charged chloride complexes. After separation, each of the cations is determined by EDTA titration.

### 34P–1  The Separation of Nickel(II) and Zinc(II) by Ion Exchange

#### Discussion

The separation of the two cations is based upon differences in their tendency to form anionic complexes. Stable chlorozincate(II) complexes (such as $ZnCl_3^-$ and $ZnCl_4^{2-}$) are formed in 2 M hydrochloric acid and retained on an anion-exchange resin. In contrast, nickel(II) is not complexed appreciably in this medium and passes rapidly through such a column. After separation is complete, elution with water effectively decomposes the chloro complexes and permits removal of the zinc.

---

#### Preparation of Ion-Exchange Columns

A typical ion-exchange column is a cylinder 25 to 40 cm in length and 1 to 1.5 cm in diameter. A stopcock at the lower end permits adjustment of the flow of liquid through the column. A buret makes a convenient column. It is recommended that two columns be prepared to permit the simultaneous treatment of duplicate samples.

Insert a plug of glass wool to retain the resin particles. Then introduce sufficient strong-base anion-exchange resin (Note) to give a 10- to 15-cm column. Wash the column with about 50 mL of 6 M $NH_3$, followed by 100 mL of water and 100 mL of 2 M HCl. At the end of this cycle, the flow should be stopped so that the liquid level remains about 1 cm above the resin column. *At no time should the liquid level be allowed to drop below the top of the resin.*

---

### Procedure

Obtain the unknown, which should contain between 2 and 4 mmol of $Ni^{2+}$ and $Zn^{2+}$, in a clean 100-mL volumetric flask. Add 16 mL of 12 M HCl, dilute to the mark with distilled water, and mix well. The resulting solution is approximately 2 M in acid. Transfer 10.00 mL of the diluted unknown onto the column. Place a 250-mL conical flask beneath the column, and slowly drain until the liquid level is barely above the resin. Rinse the interior of the column with several 2- to 3-mL portions of the 2 M HCl; lower the liquid level to just above the resin surface after each washing. Elute the nickel with about 50 mL of 2 M HCl at a flow rate of 2 to 3 mL/min. After elution is complete, evaporate the collected liquid just to dryness on a hot plate or steam bath.

Elute the Zn(II) by passing about 100 mL of water through the column, using the same flow rate; collect the liquid in a 500-mL conical flask.

**Note.** Amberlite® CG 400 or its equivalent can be used.

---

### 34P–2 The Titration of Nickel and Zinc with EDTA

### Discussion

Both nickel and zinc are determined by titration with standard EDTA at pH 10. A weight titration is recommended. Eriochrome Black T is the indicator for the zinc titration; bromopyrogallol or murexide is used for the nickel titration.

---

### Preparation of Reagents

(a) *Standard EDTA.* See Section 34D–2. Express the concentration in terms of mmol EDTA/g of solution.
(b) *pH-10 buffer.* See Section 34D–1.
(c) *Eriochrome Black T indicator.* See Section 34D–1.
(d) *Bromopyrogallol indicator.* Dissolve 0.5 g of the solid indicator in 100 mL of 50% (v/v) ethanol. (Sufficient for 100 titrations.)
(e) *Murexide indicator.* The solid is approximately 0.2% indicator by weight in NaCl. Approximately 0.2 g is needed for each titration. The solid preparation is used because solutions of the indicator are quite unstable.

---

### Procedure

(a) *Titration of nickel.* Evaporate the solution containing the nickel to dryness to eliminate excess HCl. Avoid overheating; the residual $NiCl_2$ must not be permitted to decompose to NiO. Dissolve the residue in 100 mL of distilled water, and add 10 to 20 mL of pH-10 buffer. Add 15 drops of bromopyrogallol indicator or 0.2 g of murexide. Titrate to the color change (blue to purple for bromopyrogallol, yellow to purple for murexide).

Calculate the weight in milligrams of nickel in the unknown.

(b) *Titration of zinc.* Add 10 to 20 mL of pH-10 buffer and 1 to 2 drops of Eriochrome Black T to the eluate. Titrate with standard EDTA solution to a color change from red to blue.

Calculate the weight in milligrams of zinc in the unknown.

---

## 34Q   GAS-LIQUID CHROMATOGRAPHY

As noted in Chapter 27, gas-liquid chromatography enables the analyst to resolve the components of complex mixtures. The accompanying directions are for the determination of ethanol in beverages.

### 34Q–1 The Gas Chromatographic Determination of Ethanol in Beverages[13]

Discussion

Ethanol is conveniently determined in aqueous solutions by means of gas chromatography. The method is readily extended to measurement of the *proof* of alcoholic beverages. By definition, the proof of a beverage is two times its volume percent of ethanol at 60°F.

The operating instructions pertain to a $\frac{1}{4}$-in. (o.d.) $\times$ 0.5-m Poropack column containing 80- to 100-mesh packing. A thermal conductivity detector is needed. (Flame ionization is not satisfactory because of its insensitivity to water.)

The analysis is based upon a calibration curve in which the ratio of the area under the ethanol peak to the area under the ethanol-plus-water peak is plotted as a function of the volume percent of ethanol:

$$\text{vol \% EtOH} = \frac{\text{vol EtOH}}{\text{vol soln}}$$

This relationship is not strictly linear. At least two reasons can be cited to account for the curvature. First, the thermal conductivity detector responds linearly to mass ratios rather than volume ratios. Second, at the high concentrations involved, the volumes of ethanol and water are not strictly additive, as would be required for linearity. That is,

$$\text{vol EtOH} + \text{vol H}_2\text{O} \neq \text{vol soln}$$

---

[13]Adapted from J. J. Leary, *J. Chem. Educ.*, **1983**, 60, 675.

## Preparation of Standards

Use a buret to measure 10.00, 20.00, 30.00, and 40.00 mL of absolute ethanol into separate 50-mL volumetric flasks (Note). Dilute to volume with distilled water, and mix well.

**Note.** The coefficient of thermal expansion for ethanol is approximately five times that for water. It is thus necessary to keep the temperature of the solutions used in this experiment to ±1°C during volume measurements.

## Procedure

The following operating conditions have yielded satisfactory chromatograms for this analysis:

| | |
|---|---|
| Column temperature | 100°C |
| Detector temperature | 130°C |
| Injection-port temperature | 120°C |
| Bridge current | 100 mA |
| Flow rate | 60 mL/min |

Inject a 1-$\mu$L sample of the 20% (v/v) standard, and record the chromatogram. Obtain additional chromatograms, adjusting the recorder speed until the water peak has a width of about 2 mm at half-height. Then vary the volume of sample injected and the attenuation until peaks with a height of at least 40 mm are produced. Obtain chromatograms for the remainder of the standards (including pure water and pure ethanol) in the same way. Measure the area under each peak, and plot $area_{EtOH}/(area_{EtOH} + area_{H_2O})$ as a function of the volume percentage of ethanol.

Obtain chromatograms for the unknown. Report the volume percentage of ethanol.

# SELECTED REFERENCES TO THE LITERATURE OF ANALYTICAL CHEMISTRY

## Treatises

As used here, the term *treatise* means a comprehensive presentation of one or more broad areas of analytical chemistry.

N. H. Furman and F. J. Welcher, Eds., *Standard Methods of Chemical Analysis,* 6th ed. New York: Van Nostrand, 1962–66. In five parts; largely devoted to specific applications.

I. M. Kolthoff and P. J. Elving, Eds., *Treatise on Analytical Chemistry.* New York: Wiley, 1959–86. Part I (12 volumes) is devoted to theory; Part II (17 volumes) deals with analytical methods for the elements; Part III (4 volumes) treats industrial analytical chemistry. Early volumes of the second edition of Part I of this monumental work began to appear in 1978.

G. Svehla, C. L. Wilson, and D. W. Wilson, Eds., *Comprehensive Analytical Chemistry.* New York: Elsevier, 1959–1990. To 1990, 25 volumes of this work have appeared.

A. Weissberger, Ed., *Techniques of Chemistry,* Volume I, *Physical Methods of Chemistry,* 4th ed. New York: Interscience, 1971–1986. This work consists of a large number of individually bound books dealing with various instruments employed for chemical measurements.

## Official Methods of Analysis

These publications are often single volumes that provide a useful source of analytical methods for the determination of specific substances in articles of commerce. The methods have been developed by various scientific societies and serve as standards in arbitration as well as in the courts.

*Standard Methods for the Examination of Water and Wastewater,* 17th ed. New York: American Public Health Association, 1989.

*ASTM Book of Standards.* Philadelphia: American Society for Testing Materials. This 48-volume work is revised annually and contains methods for both physical testing and chemical analysis. Volume 3.05, *Chemical Analysis of Metals and Metal-Bearing Ores,* 1990, is one of the useful volumes for the chemist.

N. H. Hanson, *Official, Standardized and Recommended Methods of Analysis,* 2nd ed. London: Society for Analytical Chemistry, 1973.

*Official Methods of Analysis,* 15th ed. Washington, D.C.: Association of Official Analytical Chemists, 1990. This is a very useful source of methods for the analysis of such materials as drugs, food, pesticides, agricultural materials, cosmetics, vitamins, and nutrients. It is revised every five years.

## Review Serials

*Analytical Chemistry, Fundamental Reviews.* These reviews appear biennially as a supplement to the April issue of *Analytical Chemistry* in even-numbered years. Most of the significant developments occurring in the past two years in 30 or more areas of analytical chemistry are covered.

*Analytical Chemistry, Application Reviews.* These reviews appear as part of *Analytical Chemistry* biennially in odd-numbered years. The articles are devoted to recent analytical work in 18 specific areas, such as water analysis, clinical chemistry, petroleum products, and air pollution.

*Critical Reviews in Analytical Chemistry.* This publication appears quarterly and provides in-depth reviews of various aspects of analytical chemistry.

D. Glick, Ed., *Methods of Biochemical Analysis.* This annual publication consists of a series of review articles covering the newest developments in the analysis of biochemical substances.

## Tabular Compilations

E. Hoegfeldt and D. D. Perrin, *Stability Constants of Metal-Ion Complexes.* London: The Chemical Society, 1979 and 1981. Two volumes.

L. Meites, Ed., *Handbook of Analytical Chemistry.* New York: McGraw-Hill, 1963.

G. Milazzo, S. Caroli, and V. K. Sharma, *Tables of Standard Electrode Potentials.* New York: Wiley, 1978.

A. J. Bard, R. Parsons, and T. Jordan, Eds., *Standard Potentials in Aqueous Solution.* New York: Marcel Dekker, 1985.

## Advanced Analytical and Instrumental Textbooks

J. N. Butler, *Ionic Equilibrium: A Mathematical Approach.* Reading, MA: Addison-Wesley, 1964.

G. D. Christian and J. E. O'Reilly, *Instrumental Analysis,* 2nd ed. Boston: Allyn and Bacon, 1986.

H. A. Laitinen and W. E. Harris, *Chemical Analysis,* 2nd ed. New York: McGraw-Hill, 1975.

E. D. Olsen, *Modern Optical Methods of Analysis*. New York: McGraw-Hill, 1975.

D. A. Skoog and J. J. Leary, *Principles of Instrumental Analysis,* 4th ed. Philadelphia: Saunders College Publishing, 1992.

H. Strobel and W. R. Heineman, *Chemical Instrumentation,* 3rd ed. Boston: Addison-Wesley, 1989.

H. H. Willard, L. L. Merritt Jr., J. A. Dean, and F. A. Settle, *Instrumental Methods of Analysis,* 7th ed. New York: Wadsworth, 1988.

## Monographs

Hundreds of monographs devoted to limited areas of analytical chemistry are available. In general, these are authored by experts and are excellent sources of information. Representative monographs in various areas are listed below.

### Gravimetric and Titrimetric Methods

M. R. F. Ashworth, *Titrimetric Organic Analysis*. New York: Interscience, 1965. Two volumes.

L. Erdey, *Gravimetric Analysis*. Oxford: Pergamon, 1965.

J. S. Fritz, *Acid-Base Titrations in Nonaqueous Solvents*. Boston: Allyn and Bacon, 1973.

W. F. Hillebrand, G. E. F. Lundell, H. A. Bright, and J. I. Hoffman, *Applied Inorganic Analysis,* 2nd ed. New York: Wiley, 1953, reissued 1980.

I. M. Kolthoff, V. A. Stenger, and R. Belcher, *Volumetric Analysis*. New York: Interscience, 1942–1957. Three volumes.

T. S. Ma and R. C. Rittner, *Modern Organic Elemental Analysis*. New York: Marcel Dekker, 1979.

W. Wagner and C. J. Hull, *Inorganic Titrimetric Analysis*. New York: Marcel Dekker, 1971.

### Organic Analysis

S. Siggia and J. G. Hanna, *Quantitative Organic Analysis via Functional Groups,* 4th ed. New York: Wiley, 1979.

F. T. Weiss, *Determination of Organic Compounds: Methods and Procedures*. New York: Wiley-Interscience, 1970.

### Spectrometric Methods

D. F. Boltz and J. A. Howell, *Colorimetric Determination of Nonmetals,* 2nd ed. New York: Wiley-Interscience, 1978.

J. A. Dean and T. C. Rains, Eds., *Flame Emission and Atomic Absorption Spectroscopy*. New York: Marcel Dekker, 1969–1975. Three volumes.

J. D. Ingle and S. R. Crouch, *Analytical Spectroscopy*. Englewood Cliffs, NJ: Prentice-Hall, 1988.

W. J. Price, *Spectrochemical Analysis by Atomic Absorption.* London: Heyden, 1979.

E. B. Sandell and H. Onishi, *Colorimetric Determination of Traces of Metals,* 4th ed. New York: Interscience, 1978–1989. Two volumes.

F. D. Snell, *Photometric and Fluorometric Methods of Analysis.* New York: Wiley, 1978–1981. Two volumes.

## Electroanalytical Methods

A. J. Bard and L. R. Faulkner, *Electrochemical Methods.* New York: Wiley, 1980.

P. T. Kissinger and W. R. Heineman, Eds., *Laboratory Techniques in Electroanalytical Chemistry.* New York: Marcel Dekker, 1984.

J. J. Lingane, *Electroanalytical Chemistry,* 2nd ed. New York: Interscience, 1954.

D. T. Sawyer and J. L. Roberts Jr., *Experimental Electrochemistry for Chemists.* New York: Wiley, 1974.

## Analytical Separations

E. Heftmann, Ed., *Chromatography: Fundamentals and Applications of Chromatography and Electrophoretic Methods, Part A: Fundamentals, Part B: Applications.* New York: Elsevier, 1983.

P. Sewell and B. Clarke, *Chromatographic Separations.* New York: Wiley, 1988.

J. A. Jonsson, Ed., *Chromatographic Theory and Basic Principles.* New York: Marcel Dekker, 1987.

R. M. Smith, *Gas and Liquid Chromatography in Analytical Chemistry.* New York: Wiley, 1988.

E. Katz, *Quantitative Analysis Using Chromatographic Techniques.* New York: Wiley, 1987.

J. C. Giddings, *Unified Separation Science.* New York: Wiley, 1991.

## Miscellaneous

R. G. Bates, *Determination of pH: Theory and Practice,* 2nd ed. New York: Wiley, 1973.

R. Bock, *Decomposition Methods in Analytical Chemistry.* New York: Wiley, 1979.

G. H. Morrison, Ed., *Trace Analysis.* New York: Interscience, 1965.

D. D. Perrin, *Masking and Demasking Chemical Reactions.* New York: Wiley, 1970.

M. Pinta, *Modern Methods for Trace Element Analysis.* Ann Arbor, MI: Ann Arbor Science, 1978.

W. Rieman and H. F. Walton, *Ion Exchange in Analytical Chemistry.* Oxford: Pergamon, 1970.

W. J. Williams, *Handbook of Anion Determination.* London: Butterworths, 1979.

## Periodicals

Numerous journals are devoted to analytical chemistry; these are primary sources of information in the field. Some of the best-known titles are listed below. (The boldface portion of the title is the *Chemical Abstracts* abbreviation for the journal.)

*American **Lab**oratory*

***Analyst**, The*

*Analytical **Biochem**istry*

*Analytical **Chem**istry*

*Analytica **Chimica Acta***

*Analytical **Instrum**entation*

*Analytical **Letters***

*Applied **Spectros**copy*

*Clinical **Chem**istry*

*Journal of the **Association** of **Official Analytical Chem**ists*

*Journal of **Chromatog**raphic Science*

*Journal of **Chromatog**raphy*

*Journal of **Electroanal**ytical **Chem**istry and **Interfacial Electrochem**istry*

***Microchemical Journal***

***Mikrochimica Acta***

***Separation Science***

***Spectrochimica Acta***

***Talanta***

*Zeitschrift für **Analytische Chem**ie*

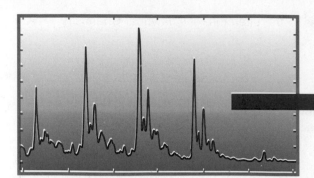

# SOLUBILITY-PRODUCT CONSTANTS

| Substance | Formula | $K_{sp}$ |
|---|---|---|
| Aluminum hydroxide | $Al(OH)_3$ | $2 \times 10^{-32}$ |
| Barium carbonate | $BaCO_3$ | $5.1 \times 10^{-9}$ |
| Barium chromate | $BaCrO_4$ | $1.2 \times 10^{-10}$ |
| Barium iodate | $Ba(IO_3)_2$ | $1.57 \times 10^{-9}$ |
| Barium manganate | $BaMnO_4$ | $2.5 \times 10^{-10}$ |
| Barium oxalate | $BaC_2O_4$ | $2.3 \times 10^{-8}$ |
| Barium sulfate | $BaSO_4$ | $1.3 \times 10^{-10}$ |
| Bismuth oxide chloride | $BiOCl$ | $7 \times 10^{-9}$ |
| Bismuth oxide hydroxide | $BiOOH$ | $4 \times 10^{-10}$ |
| Cadmium carbonate | $CdCO_3$ | $2.5 \times 10^{-14}$ |
| Cadmium hydroxide | $Cd(OH)_2$ | $5.9 \times 10^{-15}$ |
| Cadmium oxalate | $CdC_2O_4$ | $9 \times 10^{-8}$ |
| Cadmium sulfide | $CdS$ | $2 \times 10^{-28}$ |
| Calcium carbonate | $CaCO_3$ | $4.8 \times 10^{-9}$ |
| Calcium fluoride | $CaF_2$ | $4.9 \times 10^{-11}$ |
| Calcium oxalate | $CaC_2O_4$ | $2.3 \times 10^{-9}$ |
| Calcium sulfate | $CaSO_4$ | $2.6 \times 10^{-5}$ |
| Copper(I) bromide | $CuBr$ | $5.2 \times 10^{-9}$ |
| Copper(I) chloride | $CuCl$ | $1.2 \times 10^{-6}$ |
| Copper(I) iodide | $CuI$ | $1.1 \times 10^{-12}$ |
| Copper(I) thiocyanate | $CuSCN$ | $4.8 \times 10^{-15}$ |
| Copper(II) hydroxide | $Cu(OH)_2$ | $1.6 \times 10^{-19}$ |
| Copper(II) sulfide | $CuS$ | $6 \times 10^{-36}$ |
| Iron(II) hydroxide | $Fe(OH)_2$ | $8 \times 10^{-16}$ |
| Iron(II) sulfide | $FeS$ | $6 \times 10^{-18}$ |
| Iron(III) hydroxide | $Fe(OH)_3$ | $4 \times 10^{-38}$ |
| Lanthanum iodate | $La(IO_3)_3$ | $6.2 \times 10^{-12}$ |
| Lead carbonate | $PbCO_3$ | $3.3 \times 10^{-14}$ |
| Lead chloride | $PbCl_2$ | $1.6 \times 10^{-5}$ |
| Lead chromate | $PbCrO_4$ | $1.8 \times 10^{-14}$ |
| Lead hydroxide | $Pb(OH)_2$ | $2.5 \times 10^{-16}$ |
| Lead iodide | $PbI_2$ | $7.1 \times 10^{-9}$ |
| Lead oxalate | $PbC_2O_4$ | $4.8 \times 10^{-10}$ |
| Lead sulfate | $PbSO_4$ | $1.6 \times 10^{-8}$ |
| Lead sulfide | $PbS$ | $7 \times 10^{-28}$ |

| Substance | Formula | $K_{sp}$ |
|---|---|---|
| Magnesium ammonium phosphate | $MgNH_4PO_4$ | $3 \times 10^{-13}$ |
| Magnesium carbonate | $MgCO_3$ | $1 \times 10^{-5}$ |
| Magnesium hydroxide | $Mg(OH)_2$ | $1.8 \times 10^{-11}$ |
| Magnesium oxalate | $MgC_2O_4$ | $8.6 \times 10^{-5}$ |
| Manganese(II) hydroxide | $Mn(OH)_2$ | $1.9 \times 10^{-13}$ |
| Manganese(II) sulfide | $MnS$ | $3 \times 10^{-13}$ |
| Mercury(I) bromide | $Hg_2Br_2$ | $5.8 \times 10^{-23}$ |
| Mercury(I) chloride | $Hg_2Cl_2$ | $1.3 \times 10^{-18}$ |
| Mercury(I) iodide | $Hg_2I_2$ | $4.5 \times 10^{-29}$ |
| Silver arsenate | $Ag_3AsO_4$ | $1 \times 10^{-22}$ |
| Silver bromide | $AgBr$ | $5.2 \times 10^{-13}$ |
| Silver carbonate | $Ag_2CO_3$ | $8.1 \times 10^{-12}$ |
| Silver chloride | $AgCl$ | $1.82 \times 10^{-10}$ |
| Silver chromate | $Ag_2CrO_4$ | $1.1 \times 10^{-12}$ |
| Silver cyanide | $AgCN$ | $7.2 \times 10^{-11}$ |
| Silver iodate | $AgIO_3$ | $3.0 \times 10^{-8}$ |
| Silver iodide | $AgI$ | $8.3 \times 10^{-17}$ |
| Silver oxalate | $Ag_2C_2O_4$ | $3.5 \times 10^{-11}$ |
| Silver sulfide | $Ag_2S$ | $6 \times 10^{-50}$ |
| Silver thiocyanate | $AgSCN$ | $1.1 \times 10^{-12}$ |
| Strontium oxalate | $SrC_2O_4$ | $5.6 \times 10^{-8}$ |
| Strontium sulfate | $SrSO_4$ | $3.2 \times 10^{-7}$ |
| Thallium(I) chloride | $TlCl$ | $1.7 \times 10^{-4}$ |
| Thallium(I) sulfide | $Tl_2S$ | $1 \times 10^{-22}$ |
| Zinc hydroxide | $Zn(OH)_2$ | $1.2 \times 10^{-17}$ |
| Zinc oxalate | $ZnC_2O_4$ | $7.5 \times 10^{-9}$ |
| Zinc sulfide | $ZnS$ | $4.5 \times 10^{-24}$ |

From R. P. Frankenthal, in *Handbook of Analytical Chemistry*, L. Meites, Ed., pp. 1-13–1-19. New York: McGraw-Hill, 1963. With permission.

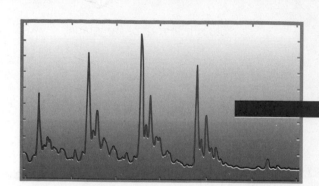

# DISSOCIATION CONSTANTS FOR ACIDS

| Acid | Formula | Dissociation Constant at 25°C | | |
|---|---|---|---|---|
| | | $K_1$ | $K_2$ | $K_3$ |
| Acetic | $CH_3COOH$ | $1.75 \times 10^{-5}$ | | |
| Arsenic | $H_3AsO_4$ | $6.0 \times 10^{-3}$ | $1.05 \times 10^{-7}$ | $3.0 \times 10^{-12}$ |
| Arsenous | $H_3AsO_3$ | $6.0 \times 10^{-10}$ | $3.0 \times 10^{-14}$ | |
| Benzoic | $C_6H_5COOH$ | $6.14 \times 10^{-5}$ | | |
| Boric | $H_3BO_3$ | $5.83 \times 10^{-10}$ | | |
| 1-Butanoic | $CH_3CH_2CH_2COOH$ | $1.51 \times 10^{-5}$ | | |
| Carbonic | $H_2CO_3$ | $4.45 \times 10^{-7}$ | $4.7 \times 10^{-11}$ | |
| Chloroacetic | $ClCH_2COOH$ | $1.36 \times 10^{-3}$ | | |
| Citric | $HOOC(OH)C(CH_2COOH)_2$ | $7.45 \times 10^{-4}$ | $1.73 \times 10^{-5}$ | $4.02 \times 10^{-7}$ |
| Ethylenediamine-tetraacetic | $H_4Y$ | $1.0 \times 10^{-2}$ | $2.1 \times 10^{-3}$ | $6.9 \times 10^{-7}$ |
| | | | $K_4 = 5.5 \times 10^{-11}$ | |
| Formic | $HCOOH$ | $1.77 \times 10^{-4}$ | | |
| Fumaric | $trans\text{-}HOOCCH:CHCOOH$ | $9.6 \times 10^{-4}$ | $4.1 \times 10^{-5}$ | |
| Glycolic | $HOCH_2COOH$ | $1.48 \times 10^{-4}$ | | |
| Hydrazoic | $HN_3$ | $1.9 \times 10^{-5}$ | | |
| Hydrogen cyanide | $HCN$ | $2.1 \times 10^{-9}$ | | |
| Hydrogen fluoride | $H_2F_2$ | $7.2 \times 10^{-4}$ | | |
| Hydrogen peroxide | $H_2O_2$ | $2.7 \times 10^{-12}$ | | |
| Hydrogen sulfide | $H_2S$ | $5.7 \times 10^{-8}$ | $1.2 \times 10^{-15}$ | |
| Hypochlorous | $HOCl$ | $3.0 \times 10^{-8}$ | | |
| Iodic | $HIO_3$ | $1.7 \times 10^{-1}$ | | |
| Lactic | $CH_3CHOHCOOH$ | $1.37 \times 10^{-4}$ | | |
| Maleic | $cis\text{-}HOOCCH:CHCOOH$ | $1.20 \times 10^{-2}$ | $5.96 \times 10^{-7}$ | |
| Malic | $HOOCCHOHCH_2COOH$ | $4.0 \times 10^{-4}$ | $8.9 \times 10^{-6}$ | |
| Malonic | $HOOCCH_2COOH$ | $1.40 \times 10^{-3}$ | $2.01 \times 10^{-6}$ | |
| Mandelic | $C_6H_5CHOHCOOH$ | $3.88 \times 10^{-4}$ | | |
| Nitrous | $HNO_2$ | $5.1 \times 10^{-4}$ | | |
| Oxalic | $HOOCCOOH$ | $5.36 \times 10^{-2}$ | $5.42 \times 10^{-5}$ | |
| Periodic | $H_5IO_6$ | $2.4 \times 10^{-2}$ | $5.0 \times 10^{-9}$ | |
| Phenol | $C_6H_5OH$ | $1.00 \times 10^{-10}$ | | |
| Phosphoric | $H_3PO_4$ | $7.11 \times 10^{-3}$ | $6.34 \times 10^{-8}$ | $4.2 \times 10^{-13}$ |
| Phosphorous | $H_3PO_3$ | $1.00 \times 10^{-2}$ | $2.6 \times 10^{-7}$ | |
| o-Phthalic | $C_6H_4(COOH)_2$ | $1.12 \times 10^{-3}$ | $3.91 \times 10^{-6}$ | |
| Picric | $(NO_2)_3C_6H_2OH$ | $5.1 \times 10^{-1}$ | | |

| Acid | Formula | Dissociation Constant at 25°C | | |
|------|---------|------|------|------|
| | | $K_1$ | $K_2$ | $K_3$ |
| Propanoic | $CH_3CH_2COOH$ | $1.34 \times 10^{-5}$ | | |
| Pyruvic | $CH_3COCOOH$ | $3.24 \times 10^{-3}$ | | |
| Salicylic | $C_6H_4(OH)COOH$ | $1.05 \times 10^{-3}$ | | |
| Sulfamic | $H_2NSO_3H$ | $1.03 \times 10^{-1}$ | | |
| Sulfuric | $H_2SO_4$ | Strong | $1.20 \times 10^{-2}$ | |
| Sulfurous | $H_2SO_3$ | $1.72 \times 10^{-2}$ | $6.43 \times 10^{-8}$ | |
| Succinic | $HOOCCH_2CH_2COOH$ | $6.21 \times 10^{-5}$ | $2.32 \times 10^{-6}$ | |
| Tartaric | $HOOC(CHOH)_2COOH$ | $9.20 \times 10^{-4}$ | $4.31 \times 10^{-5}$ | |
| Trichloroacetic | $Cl_3CCOOH$ | $1.29 \times 10^{-1}$ | | |

From L. Meites, *Handbook of Analytical Chemistry,* p. **1**-21. New York: McGraw-Hill, 1963. With permission.

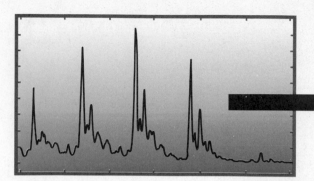

# DISSOCIATION CONSTANTS FOR BASES

| Base | Formula | Dissociation Constant at 25°C |
|---|---|---|
| Ammonia | $NH_3$ | $1.76 \times 10^{-5}$ |
| Aniline | $C_6H_5NH_2$ | $3.94 \times 10^{-10}$ |
| 1-Butylamine | $CH_3(CH_2)_2CH_2NH_2$ | $4.0 \ \times 10^{-4}$ |
| Dimethylamine | $(CH_3)_2NH$ | $5.9 \ \times 10^{-4}$ |
| Ethanolamine | $HOC_2H_4NH_2$ | $3.18 \times 10^{-5}$ |
| Ethylamine | $CH_3CH_2NH_2$ | $4.28 \times 10^{-4}$ |
| Ethylenediamine | $NH_2C_2H_4NH_2$ | $K_1 = 8.5 \ \times 10^{-5}$ |
| | | $K_2 = 7.1 \ \times 10^{-8}$ |
| Hydrazine | $H_2NNH_2$ | $1.3 \ \times 10^{-6}$ |
| Hydroxylamine | $HONH_2$ | $1.07 \times 10^{-8}$ |
| Methylamine | $CH_3NH_2$ | $4.8 \ \times 10^{-4}$ |
| Piperidine | $C_5H_{11}N$ | $1.3 \ \times 10^{-3}$ |
| Pyridine | $C_5H_5N$ | $1.7 \ \times 10^{-9}$ |
| Trimethylamine | $(CH_3)_3N$ | $6.25 \times 10^{-5}$ |

From L. Meites, *Handbook of Analytical Chemistry*, p. **1**-21. New York: McGraw-Hill, 1963. With permission.

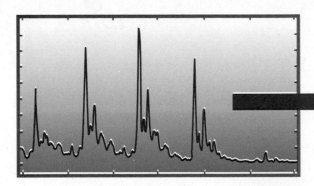

# STEPWISE FORMATION CONSTANTS

| Ligand | Cation | $\log K_1$ | $\log K_2$ | $\log K_3$ | $\log K_4$ | $\log K_5$ | $\log K_6$ |
|---|---|---|---|---|---|---|---|
| $CH_3COO^-$ | $Ag^+$ | 0.4 | $-0.2$ | | | | |
| | $Cd^{2+}$ | 1.3 | 1.0 | 0.1 | $-0.4$ | | |
| | $Cu^{2+}$ | 2.2 | 1.1 | | | | |
| | $Hg^{2+}$ | $\log K_1K_2 = 8.4$ | | | | | |
| | $Pb^{2+}$ | 2.7 | 1.5 | | | | |
| $NH_3$ | $Ag^+$ | 3.3 | 3.8 | | | | |
| | $Cd^{2+}$ | 2.6 | 2.1 | 1.4 | 0.9 | $-0.3$ | $-1.7$ |
| | $Co^{2+}$ | 2.1 | 1.6 | 1.0 | 0.8 | 0.2 | $-0.6$ |
| | $Cu^{2+}$ | 4.3 | 3.7 | 3.0 | 2.3 | $-0.5$ | |
| | $Ni^{2+}$ | 2.8 | 2.2 | 1.7 | 1.2 | 0.8 | 0.0 |
| | $Zn^{2+}$ | 2.4 | 2.4 | 2.5 | 2.1 | | |
| $Br^-$ | $Ag^+$ | $AgBr(s) + Br^- \rightleftharpoons AgBr_2^-$ | | | $\log K_{s2} = -4.7$ | | |
| | | $AgBr_2^- + Br^- \rightleftharpoons AgBr_3^{2-}$ | | | $\log K_3 = 0.7$ | | |
| | $Hg^{2+}$ | 9.0 | 8.3 | 1.4 | 1.3 | | |
| | $Pb^{2+}$ | 1.2 | | | | | |
| $Cl^-$ | $Ag^+$ | $AgCl(s) + Cl^- \rightleftharpoons AgCl_2^-$ | | | $\log K_{s2} = -4.7$ | | |
| | | $AgCl_2^- + Cl^- \rightleftharpoons AgCl_3^{2-}$ | | | $\log K_3 = 0.0$ | | |
| | $Bi^{3+}$ | 2.4 | 2.0 | 1.4 | 0.4 | 0.5 | |
| | $Cd^{2+}$ | 1.5 | 0.4 | 0.4 | | | |
| | $Cu^+$ | $Cu^+ + 2\,Cl^- \rightleftharpoons CuCl_2^-$ | | | $\log K_1K_2 = 4.9$ | | |
| | $Fe^{2+}$ | 0.4 | 0.0 | | | | |
| | $Fe^{3+}$ | 1.5 | 0.6 | $-1.0$ | | | |
| | $Hg^{2+}$ | 6.7 | 6.5 | 0.9 | 1.0 | | |
| | $Pb^{2+}$ | 1.6 | $Pb^{2+} + 3\,Cl^- \rightleftharpoons PbCl_3^-$ | | $\log K_1K_2K_3 = 1.7$ | | |
| | $Sn^{2+}$ | 1.1 | 0.6 | 0.0 | | | |
| $CN^-$ | $Ag^+$ | $Ag^+ + 2\,CN^- \rightleftharpoons Ag(CN)_2^-$ | | | $\log K_1K_2 = 21.1$ | | |
| | $Cd^{2+}$ | 5.5 | 5.1 | 4.6 | 3.6 | | |
| | $Hg^{2+}$ | 18.0 | 16.7 | 3.8 | 3.0 | | |
| | $Ni^{2+}$ | $Ni^{2+} + 4\,CN^- \rightleftharpoons Ni(CN)_4^-$ | | | $\log K_1K_2K_3K_4 = 31$ | | |
| EDTA | See Table 13–1 (page 290) | | | | | | |
| $F^-$ | $Al^{3+}$ | 6.1 | 5.0 | 3.8 | 2.7 | 1.6 | 0.5 |
| | $Fe^{3+}$ | 5.3 | 4.0 | 2.8 | | | |
| $OH^-$ | $Al^{3+}$ | 8.9 | $Al(OH)_3(s) + OH^- \rightleftharpoons Al(OH)_4^-$ | | $\log K_{s4} = 1.0$ | | |
| | $Cd^{2+}$ | 2.3 | | | | | |
| | $Cu^{2+}$ | 6.5 | | | | | |

| Ligand | Cation | $\log K_1$ | $\log K_2$ | $\log K_3$ | $\log K_4$ | $\log K_5$ | $\log K_6$ |
|---|---|---|---|---|---|---|---|
| | $Fe^{2+}$ | 3.9 | | | | | |
| | $Fe^{3+}$ | 11.1 | 10.7 | | | | |
| | $Hg^{2+}$ | 10.3 | | | | | |
| | $Ni^{2+}$ | 4.6 | | | | | |
| | $Pb^{2+}$ | 6.2 | $Pb(OH)_2(s) + OH^- \rightleftharpoons Pb(OH)_3^-$ | | | $\log K_{s3} = -1.3$ | |
| | $Zn^{2+}$ | 4.4 | $Zn(OH)_2(s) + 2\,OH^- \rightleftharpoons Zn(OH)_4^{2-}$ | | | $\log K_{s4} = -0.9$ | |
| $I^-$ | $Cd^{2+}$ | 2.4 | 1.6 | 1.0 | 1.1 | | |
| | $Cu^+$ | | $CuI(s) + I^- \rightleftharpoons CuI_2^-$ | | $\log K_{s2} = -3.1$ | | |
| | $Hg^{2+}$ | 12.9 | 11.0 | 3.8 | 2.3 | | |
| | $Pb^{2+}$ | 1.3 | $PbI_2(s) + I^- \rightleftharpoons PbI_3^-$ | | $\log K_{s3} = -4.7$ | | |
| | | | $PbI_3^- + I^- \rightleftharpoons PbI_4^{2-}$ | | $\log K_4 = -3.8$ | | |
| $C_2O_4^{2-}$ | $Al^{3+}$ | $\log K_1K_2 = 13$ | | 3.8 | | | |
| | $Fe^{3+}$ | 9.4 | 6.8 | 4.0 | | | |
| | $Mg^{2+}$ | 3.4 | 1.0 | | | | |
| | $Mn^{2+}$ | 3.9 | 1.9 | | | | |
| | $Pb^{2+}$ | $Pb^{2+} + 2\,C_2O_4^{2-} \rightleftharpoons Pb(C_2O_4)_2^{2-}$ | | | $\log K_1K_2 = 6.5$ | | |
| $SO_4^{2-}$ | $Al^{3+}$ | 3.2 | 1.9 | | | | |
| | $Cd^{2+}$ | 2.3 | | | | | |
| | $Cu^{2+}$ | 2.4 | | | | | |
| | $Fe^{3+}$ | 3.0 | 1.0 | | | | |
| $SCN^-$ | $Ag^+$ | | $AgSCN(s) + SCN^- \rightleftharpoons Ag(SCN)_2^-$ | | | $\log K_{s2} = -7.2$ | |
| | $Cd^{2+}$ | 1.0 | 0.7 | 0.6 | 1.0 | | |
| | $Co^{2+}$ | 2.3 | 0.7 | $-0.7$ | 0.0 | | |
| | $Cu^{2+}$ | | $CuSCN(s) + SCN^- \rightleftharpoons Cu(SCN)_2^-$ | | | $\log K_{s2} = -3.4$ | |
| | $Fe^{3+}$ | 2.1 | 1.3 | | | | |
| | $Hg^{2+}$ | $\log K_1K_2 = 17.3$ | | 2.7 | 1.8 | | |
| | $Ni^{2+}$ | 1.2 | 0.5 | 0.2 | | | |

From L. Meites, *Handbook of Analytical Chemistry,* p. 1–39, New York: McGraw-Hill, 1963. With permission.

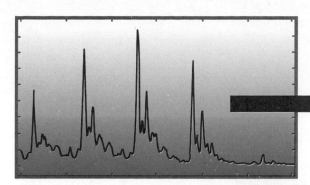

# SOME STANDARD AND FORMAL ELECTRODE POTENTIALS

| Half-Reaction | $E^0$, V* | Formal Potential, V† |
|---|---|---|
| **Aluminum** | | |
| $Al^{3+} + 3\ e^- \rightleftharpoons Al(s)$ | $-1.662$ | |
| **Antimony** | | |
| $Sb_2O_5(s) + 6\ H^+ + 4\ e^- \rightleftharpoons 2\ SbO^+ + 3\ H_2O$ | $+0.581$ | |
| **Arsenic** | | |
| $H_3AsO_4 + 2\ H^+ + 2\ e^- \rightleftharpoons H_3AsO_3 + H_2O$ | $+0.559$ | 0.577 in 1 M HCl, HClO₄ |
| **Barium** | | |
| $Ba^{2+} + 2\ e^- \rightleftharpoons Ba(s)$ | $-2.906$ | |
| **Bismuth** | | |
| $BiO^+ + 2\ H^+ + 3\ e^- \rightleftharpoons Bi(s) + H_2O$ | $+0.320$ | |
| $BiCl_4^- + 3\ e^- \rightleftharpoons Bi(s) + 4\ Cl^-$ | $+0.16$ | |
| **Bromine** | | |
| $Br_2(l) + 2\ e^- \rightleftharpoons 2\ Br^-$ | $+1.065$ | 1.05 in 4 M HCl |
| $Br_2(aq) + 2\ e^- \rightleftharpoons 2\ Br^-$ | $+1.087‡$ | |
| $BrO_3^- + 6\ H^+ + 5\ e^- \rightleftharpoons \frac{1}{2}\ Br_2(l) + 3\ H_2O$ | $+1.52$ | |
| $BrO_3^- + 6\ H^+ + 6\ e^- \rightleftharpoons Br^- + 3\ H_2O$ | $+1.44$ | |
| **Cadmium** | | |
| $Cd^{2+} + 2\ e^- \rightleftharpoons Cd(s)$ | $-0.403$ | |
| **Calcium** | | |
| $Ca^{2+} + 2\ e^- \rightleftharpoons Ca(s)$ | $-2.866$ | |
| **Carbon** | | |
| $C_6H_4O_2$ (quinone) $+ 2\ H^+ + 2\ e^- \rightleftharpoons C_6H_4(OH)_2$ | $+0.699$ | 0.696 in 1 M HCl, HClO₄, H₂SO₄ |
| $2\ CO_2(g) + 2\ H^+ + 2\ e^- \rightleftharpoons H_2C_2O_4$ | $-0.49$ | |
| **Cerium** | | |
| $Ce^{4+} + e^- \rightleftharpoons Ce^{3+}$ | | $+1.70$ in 1 M HClO₄; $+1.61$ in 1 M HNO₃; $+1.44$ in 1 M H₂SO₄ |
| **Chlorine** | | |
| $Cl_2(g) + 2\ e^- \rightleftharpoons 2\ Cl^-$ | $+1.359$ | |
| $HClO + H^+ + e^- \rightleftharpoons \frac{1}{2}\ Cl_2(g) + H_2O$ | $+1.63$ | |
| $ClO_3^- + 6\ H^+ + 5\ e^- \rightleftharpoons \frac{1}{2}\ Cl_2(g) + 3\ H_2O$ | $+1.47$ | |

A.13

| Half-Reaction | $E^0$, V* | Formal Potential, V† |
|---|---|---|
| **Chromium** | | |
| $Cr^{3+} + e^- \rightleftharpoons Cr^{2+}$ | $-0.408$ | |
| $Cr^{3+} + 3\ e^- \rightleftharpoons Cr(s)$ | $-0.744$ | |
| $Cr_2O_7^{2-} + 14\ H^+ + 6\ e^- \rightleftharpoons 2\ Cr^{3+} + 7\ H_2O$ | $+1.33$ | |
| **Cobalt** | | |
| $Co^{2+} + 2\ e^- \rightleftharpoons Co(s)$ | $-0.277$ | |
| $Co^{3+} + e^- \rightleftharpoons Co^{2+}$ | $+1.808$ | |
| **Copper** | | |
| $Cu^{2+} + 2\ e^- \rightleftharpoons Cu(s)$ | $+0.337$ | |
| $Cu^{2+} + e^- \rightleftharpoons Cu^+$ | $+0.153$ | |
| $Cu^+ + e^- \rightleftharpoons Cu(s)$ | $+0.521$ | |
| $Cu^{2+} + I^- + e^- \rightleftharpoons CuI(s)$ | $+0.86$ | |
| $CuI(s) + e^- \rightleftharpoons Cu(s) + I^-$ | $-0.185$ | |
| **Fluorine** | | |
| $F_2(g) + 2\ H^+ + 2\ e^- \rightleftharpoons 2\ HF(aq)$ | $+3.06$ | |
| **Hydrogen** | | |
| $2\ H^+ + 2\ e^- \rightleftharpoons H_2(g)$ | $0.000$ | $-0.005$ in 1 M HCl, $HClO_4$ |
| **Iodine** | | |
| $I_2(s) + 2\ e^- \rightleftharpoons 2\ I^-$ | $+0.5355$ | |
| $I_2(aq) + 2\ e^- \rightleftharpoons 2\ I^-$ | $+0.615$‡ | |
| $I_3^- + 2\ e^- \rightleftharpoons 3\ I^-$ | $+0.536$ | |
| $ICl_2^- + e^- \rightleftharpoons \frac{1}{2}\ I_2(s) + 2\ Cl^-$ | $+1.056$ | |
| $IO_3^- + 6\ H^+ + 5\ e^- \rightleftharpoons \frac{1}{2}\ I_2(s) + 3\ H_2O$ | $+1.196$ | |
| $IO_3^- + 6\ H^+ + 5\ e^- \rightleftharpoons \frac{1}{2}\ I_2(aq) + 3\ H_2O$ | $+1.178$‡ | |
| $IO_3^- + 2\ Cl^- + 6\ H^+ + 4\ e^- \rightleftharpoons ICl_2^- + 3\ H_2O$ | $+1.24$ | |
| $H_5IO_6 + H^+ + 2\ e^- \rightleftharpoons IO_3^- + 3\ H_2O$ | $+1.601$ | |
| **Iron** | | |
| $Fe^{2+} + 2\ e^- \rightleftharpoons Fe(s)$ | $-0.440$ | |
| $Fe^{3+} + e^- \rightleftharpoons Fe^{2+}$ | $+0.771$ | 0.700 in 1 M HCl; 0.732 in 1 M $HClO_4$; 0.68 in 1 M $H_2SO_4$ |
| $Fe(CN)_6^{3-} + e^- \rightleftharpoons Fe(CN)_6^{4-}$ | $+0.36$ | 0.71 in 1 M HCl; 0.72 in 1 M $HClO_4$, $H_2SO_4$ |
| **Lead** | | |
| $Pb^{2+} + 2\ e^- \rightleftharpoons Pb(s)$ | $-0.126$ | $-0.14$ in 1 M $HClO_4$; $-0.29$ in 1 M $H_2SO_4$ |
| $PbO_2(s) + 4\ H^+ + 2\ e^- \rightleftharpoons Pb^{2+} + 2\ H_2O$ | $+1.455$ | |
| $PbSO_4(s) + 2\ e^- \rightleftharpoons Pb(s) + SO_4^{2-}$ | $-0.350$ | |
| **Lithium** | | |
| $Li^+ + e^- \rightleftharpoons Li(s)$ | $-3.045$ | |
| **Magnesium** | | |
| $Mg^{2+} + 2\ e^- \rightleftharpoons Mg(s)$ | $-2.363$ | |
| **Manganese** | | |
| $Mn^{2+} + 2\ e^- \rightleftharpoons Mn(s)$ | $-1.180$ | |
| $Mn^{3+} + e^- \rightleftharpoons Mn^{2+}$ | | 1.51 in 7.5 M $H_2SO_4$ |
| $MnO_2(s) + 4\ H^+ + 2\ e^- \rightleftharpoons Mn^{2+} + 2\ H_2O$ | $+1.23$ | |
| $MnO_4^- + 8\ H^+ + 5\ e^- \rightleftharpoons Mn^{2+} + 4\ H_2O$ | $+1.51$ | |
| $MnO_4^- + 4\ H^+ + 3\ e^- \rightleftharpoons MnO_2(s) + 2\ H_2O$ | $+1.695$ | |
| $MnO_4^- + e^- \rightleftharpoons MnO_4^{2-}$ | $+0.564$ | |

| Half-Reaction | $E^0$, V* | Formal Potential, V† |
|---|---|---|
| **Mercury** | | |
| $Hg_2^{2+} + 2\ e^- \rightleftharpoons Hg(l)$ | +0.788 | 0.274 in 1 M HCl; 0.776 in 1 M HClO$_4$; 0.674 in 1 M H$_2$SO$_4$ |
| $2\ Hg^{2+} + 2\ e^- \rightleftharpoons Hg_2^{2+}$ | +0.920 | 0.907 in 1 M HClO$_4$ |
| $Hg^{2+} + 2\ e^- \rightleftharpoons 2\ Hg(l)$ | +0.854 | |
| $Hg_2Cl_2(s) + 2\ e^- \rightleftharpoons 2\ Hg(l) + 2\ Cl^-$ | +0.268 | 0.244 in sat'd KCl; 0.282 in 1 M KCl; 0.334 in 0.1 M KCl |
| $Hg_2SO_4(s) + 2\ e^- \rightleftharpoons 2\ Hg(l) + SO_4^{2-}$ | +0.615 | |
| **Nickel** | | |
| $Ni^{2+} + 2\ e^- \rightleftharpoons Ni(s)$ | −0.250 | |
| **Nitrogen** | | |
| $N_2(g) + 5\ H^+ + 4\ e^- \rightleftharpoons N_2H_5^+$ | −0.23 | |
| $HNO_2 + H^+ + e^- \rightleftharpoons NO(g) + H_2O$ | +1.00 | |
| $NO_3^- + 3\ H^+ + 2\ e^- \rightleftharpoons HNO_2 + H_2O$ | +0.94 | 0.92 in 1 M HNO$_3$ |
| **Oxygen** | | |
| $H_2O_2 + 2\ H^+ + 2\ e^- \rightleftharpoons 2\ H_2O$ | +1.776 | |
| $HO_2^- + H_2O + 2\ e^- \rightleftharpoons 3\ OH^-$ | +0.88 | |
| $O_2(g) + 4\ H^+ + 4\ e^- \rightleftharpoons 2\ H_2O$ | +1.229 | |
| $O_2(g) + 2\ H^+ + 2\ e^- \rightleftharpoons H_2O_2$ | +0.682 | |
| $O_3(g) + 2\ H^+ + 2\ e^- \rightleftharpoons O_2(g) + H_2O$ | +2.07 | |
| **Palladium** | | |
| $Pd^{2+} + 2\ e^- \rightleftharpoons Pd(s)$ | +0.987 | |
| **Platinum** | | |
| $PtCl_4^{2-} + 2\ e^- \rightleftharpoons Pt(s) + 4\ Cl^-$ | +0.73 | |
| $PtCl_6^{2-} + 2\ e^- \rightleftharpoons PtCl_4^{2-} + 2\ Cl^-$ | +0.68 | |
| **Potassium** | | |
| $K^+ + e^- \rightleftharpoons K(s)$ | −2.925 | |
| **Selenium** | | |
| $H_2SeO_3 + 4\ H^+ + 4\ e^- \rightleftharpoons Se(s) + 3\ H_2O$ | +0.740 | |
| $SeO_4^{2-} + 4\ H^+ + 2\ e^- \rightleftharpoons H_2SeO_3 + H_2O$ | +1.15 | |
| **Silver** | | |
| $Ag^+ + e^- \rightleftharpoons Ag(s)$ | +0.799 | 0.228 in 1 M HCl; 0.792 in 1 M HClO$_4$; 0.77 in 1 M H$_2$SO$_4$ |
| $AgBr(s) + e^- \rightleftharpoons Ag(s) + Br^-$ | +0.073 | |
| $AgCl(s) + e^- \rightleftharpoons Ag(s) + Cl^-$ | +0.222 | 0.228 in 1 M KCl |
| $Ag(CN)_2^- + e^- \rightleftharpoons Ag(s) + 2\ CN^-$ | −0.31 | |
| $Ag_2CrO_4(s) + 2\ e^- \rightleftharpoons 2\ Ag(s) + CrO_4^{2-}$ | +0.446 | |
| $AgI(s) + e^- \rightleftharpoons Ag(s) + I^-$ | −0.151 | |
| $Ag(S_2O_3)_2^{3-} + e^- \rightleftharpoons Ag(s) + 2\ S_2O_3^{2-}$ | +0.017 | |
| **Sodium** | | |
| $Na^+ + e^- \rightleftharpoons Na(s)$ | −2.714 | |
| **Sulfur** | | |
| $S(s) + 2\ H^+ + 2\ e^- \rightleftharpoons H_2S(g)$ | +0.141 | |
| $H_2SO_3 + 4\ H^+ + 4\ e^- \rightleftharpoons S(s) + 3\ H_2O$ | +0.450 | |
| $SO_4^{2-} + 4\ H^+ + 2\ e^- \rightleftharpoons H_2SO_3 + H_2O$ | +0.172 | |
| $S_4O_6^{2-} + 2\ e^- \rightleftharpoons 2\ S_2O_3^{2-}$ | +0.08 | |
| $S_2O_8^{2-} + 2\ e^- \rightleftharpoons 2\ SO_4^{2-}$ | +2.01 | |

| Half-Reaction | $E^0$, V* | Formal Potential, V† |
|---|---|---|
| **Thallium** | | |
| $Tl^+ + e^- \rightleftharpoons Tl(s)$ | −0.336 | −0.551 in 1 M HCl; −0.33 in 1 M HClO$_4$, H$_2$SO$_4$ |
| $Tl^{3+} + 2\ e^- \rightleftharpoons Tl^+$ | +1.25 | 0.77 in 1 M HCl |
| **Tin** | | |
| $Sn^{2+} + 2\ e^- \rightleftharpoons Sn(s)$ | −0.136 | −0.16 in 1 M HClO$_4$ |
| $Sn^{4+} + 2\ e^- \rightleftharpoons Sn^{2+}$ | +0.154 | 0.14 in 1 M HCl |
| **Titanium** | | |
| $Ti^{3+} + e^- \rightleftharpoons Ti^{2+}$ | −0.369 | |
| $TiO^{2+} + 2\ H^+ + e^- \rightleftharpoons Ti^{3+} + H_2O$ | +0.099 | 0.04 in 1 M H$_2$SO$_4$ |
| **Uranium** | | |
| $UO_2^{2+} + 4\ H^+ + 2\ e^- \rightleftharpoons U^{4+} + 2\ H_2O$ | +0.334 | |
| **Vanadium** | | |
| $V^{3+} + e^- \rightleftharpoons V^{2+}$ | −0.256 | −0.21 in 1 M HClO$_4$ |
| $VO^{2+} + 2\ H^+ + e^- \rightleftharpoons V^{3+} + H_2O$ | +0.359 | |
| $V(OH)_4^+ + 2\ H^+ + e^- \rightleftharpoons VO^{2+} + 3\ H_2O$ | +1.00 | 1.02 in 1 M HCl, HClO$_4$ |
| **Zinc** | | |
| $Zn^{2+} + 2\ e^- \rightleftharpoons Zn(s)$ | −0.763 | |

*G. Milazzo, S. Caroli, and V. K. Sharma, *Tables of Standard Electrode Potentials*. London: Wiley, 1978. Reprinted by permission of John Wiley & Sons, Ltd.

†E. H. Swift and E. A. Butler, *Quantitative Measurements and Chemical Equilibria*. New York: Freeman, 1972.

‡These potentials are hypothetical because they correspond to solutions that are 1.00 M in Br$_2$ or I$_2$. The solubilities of these two compounds at 25°C are 0.18 M and 0.0020 M, respectively. In saturated solutions containing an excess of Br$_2$(l) or I$_2$(s), the standard potentials for the half-reaction Br$_2$(l) + 2 e$^-$ $\rightleftharpoons$ 2 Br$^-$ or I$_2$(s) + 2 e$^-$ $\rightleftharpoons$ 2 I$^-$ should be used. In contrast, at Br$_2$ and I$_2$ concentrations less than saturation, these hypothetical electrode potentials should be employed.

# DESIGNATIONS AND POROSITIES FOR FILTERING CRUCIBLES

| Crucible Type | Designation | | | | |
|---|---|---|---|---|---|
| Glass, Pyrex† | C (60)* | M (15) | F (5.5) | | |
| Glass, Kimax‡ | EC (170–220) | C (40–60) | M (10–15) | F (4–4.5) | VF (2–2.5) |
| Porcelain, Coors U.S.A.§ | Medium (40) | Fine (5) | VF (1.2) | | |
| Porcelain, Selas‖ | XF (100) | XFF (40) | #10 (8.8) | #01 (6) | |
| Aluminum oxide, ALUNDUM¶ | Extra Coarse (30) | Coarse (20) | Medium (5) | Fine (0.1) | |

*The numbers in parentheses are nominal maximum pore diameters in micrometers.

†Corning Glass Works, Corning, NY

‡Owens-Illinois, Toledo, OH.

§Coors Porcelain, Golden, CO.

‖Selas Corporation of America, Dresher, PA.

¶Norton, Worcester, MA.

# DESIGNATIONS CARRIED BY ASHLESS FILTER PAPERS

| Manufacturer | Fine Crystals | Moderately Fine Crystals | Coarse Crystals | | Gelatinous Precipitates | |
|---|---|---|---|---|---|---|
| Schleicher and Schuell* | 507, 590 589 blue ribbon | 589 white ribbon | 589 green ribbon | 589 black ribbon | 589 black ribbon | 589-1H |
| Munktell† | OOH | OK OO | OOR | | OOR | |
| Whatman‡ | 42 | 44, 40 | 41 | | 41 | 41H |
| Eaton-Dikeman§ | 90 | 80 | 60 | | 50 | |

Manufacturers' literature should be consulted for more complete specifications. Tabulated are manufacturer designations of papers suitable for filtration of the indicated type of precipitate.

*Schleicher and Schuell, Inc., Keene, NH.

†E. H. Sargent and Company, Chicago, IL, agents.

‡H. Reeve Angel and Company, Inc., Clifton, NJ, agents.

§Eaton-Dikeman Company, Mount Holly Springs, PA.

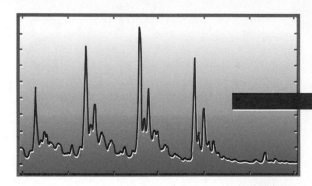

# USE OF EXPONENTIAL NUMBERS

Scientists frequently find it necessary (or convenient) to use exponential notation to express numerical data. A brief review of this notation follows.

## Exponential Notation

An exponent is used to describe the process of repeated multiplication or division. For example, $3^5$ means

$$3 \times 3 \times 3 \times 3 \times 3 = 3^5 = 243$$

The 5 is the exponent of the number (or base) 3; thus, 3 raised to the fifth power is equal to 243.

A negative exponent represents repeated divison. For example, $3^{-5}$ means

$$\frac{1}{3} \times \frac{1}{3} \times \frac{1}{3} \times \frac{1}{3} \times \frac{1}{3} = \frac{1}{3^5} = 3^{-5} = 0.00412$$

Note that changing the sign of the exponent yields the *reciprocal* of the number; that is,

$$3^{-5} = \frac{1}{3^5} = \frac{1}{243} = 0.00412$$

It is important to note that a number raised to the first power is the number itself, and any number raised to the zero power has a value of 1. For example,

$$4^1 = 4$$
$$4^0 = 1$$
$$67^0 = 1$$

## Fractional Exponents

A fractional exponent symbolizes the process of extracting the root of a number. The fifth root of 243 is 3; this process is expressed exponentially as

$$(243)^{1/5} = 3$$

Other examples are

$$25^{1/2} = 5$$

$$25^{-1/2} = \frac{1}{25^{1/2}} = \frac{1}{5}$$

## The Combination of Exponential Numbers in Multiplication and Division

Multiplication and division of exponential numbers having the same base are accomplished by adding and subtracting the exponents. For example,

$$3^3 \times 3^2 = (3 \times 3 \times 3)(3 \times 3) = 3^{(3+2)} = 3^5 = 243$$

$$3^4 \times 3^{-2} \times 3^0 = (3 \times 3 \times 3 \times 3)(\tfrac{1}{3} \times \tfrac{1}{3}) \times 1 = 3^{(4-2+0)} = 3^2 = 9$$

$$\frac{5^4}{5^2} = \frac{5 \times 5 \times 5 \times 5}{5 \times 5} = 5^{(4-2)} = 5^2 = 25$$

$$\frac{2^3}{2^{-1}} = \frac{(2 \times 2 \times 2)}{1/2} = 2^4 = 16$$

Note that in the last equation the exponent is given by the relationship

$$3 - (-1) = 3 + 1 = 4$$

## Extraction of the Root of an Exponential Number

To obtain the root of an exponential number, the exponent is divided by the desired root. Thus,

$$(5^4)^{1/2} = (5 \times 5 \times 5 \times 5)^{1/2} = 5^{(4/2)} = 5^2 = 25$$

$$(10^{-8})^{1/4} = 10^{(-8/4)} = 10^{-4}$$

$$(10^9)^{1/2} = 10^{(9/2)} = 10^{4.5}$$

## The Use of Exponents in Scientific Notation

Scientists and engineers are frequently called upon to use very large or very small numbers for which ordinary decimal notation is either awkward or impossible. For example, to express Avogadro's number in decimal notation would require 21 zeros following the number 602. In scientific notation the number is written as a multiple of two numbers, the one number in decimal notation and the other expressed as a power of 10.

Thus, Avogadro's number is written as $6.02 \times 10^{23}$. Other examples are

$$4.32 \times 10^3 = 4.32 \times 10 \times 10 \times 10 = 4320$$

$$4.32 \times 10^{-3} = 4.32 \times \frac{1}{10} \times \frac{1}{10} \times \frac{1}{10} = 0.00432$$

$$0.002002 = 2.002 \times \frac{1}{10} \times \frac{1}{10} \times \frac{1}{10} = 2.002 \times 10^{-3}$$

$$375 = 3.75 \times 10 \times 10 = 3.75 \times 10^2$$

It should be noted that the scientific notation for a number can be expressed in any of several equivalent forms. Thus,

$$4.32 \times 10^3 = 43.2 \times 10^2 = 432 \times 10^1 = 0.432 \times 10^4 = 0.0432 \times 10^5$$

The number in the exponent is equal to the number of places the decimal must be shifted to convert a number from scientific to purely decimal notation. The shift is to the right if the exponent is positive and to the left if it is negative. The process is reversed when decimal numbers are converted to scientific notation.

## Arithmetic Operations With Scientific Notation

The use of scientific notation is helpful in preventing decimal errors in arithmetic calculations. Some examples follow.

### Multiplication

Here, the decimal parts of the numbers are multiplied and the exponents are added; thus,

$$420{,}000 \times 0.0300 = (4.20 \times 10^5)(3.00 \times 10^{-2})$$
$$= 12.60 \times 10^3 = 1.26 \times 10^4$$
$$0.0060 \times 0.000020 = 6.0 \times 10^{-3} \times 2.0 \times 10^{-5}$$
$$= 12 \times 10^{-8} = 1.2 \times 10^{-7}$$

### Division

Here, the decimal parts of the numbers are divided; the exponent in the denominator is subtracted from that in the numerator. For example,

$$\frac{0.015}{5000} = \frac{15 \times 10^{-3}}{5 \times 10^3} = 3.0 \times 10^{-6}$$

### Addition and Subtraction

Addition or subtraction in scientific notation requires that all numbers be expressed to a common power of 10. The decimal parts are then added or subtracted, as appropriate. Thus,

$$2.00 \times 10^{-11} + 4.00 \times 10^{-12} - 3.00 \times 10^{-10} =$$
$$2.00 \times 10^{-11} + 0.400 \times 10^{-11} - 30.0 \times 10^{-11} = -2.76 \times 10^{-11}$$
$$= -2.76 \times 10^{-10}$$

### Raising to a Power a Number Written in Exponential Notation

Here, each part of the number is raised to the power separately. For example,

$$(2 \times 10^{-3})^4 = (2.0)^4 \times (10^{-3})^4 = 16 \times 10^{-(3\times4)}$$
$$= 16 \times 10^{-12} = 1.6 \times 10^{-11}$$

### Extraction of the Root of a Number Written in Exponential Notation

Here, the number is written in such a way that the exponent of 10 is evenly divisible by the root. Thus,

$$(4.0 \times 10^{-5})^{1/3} = \sqrt[3]{40 \times 10^{-6}} = \sqrt[3]{40} \times \sqrt[3]{10^{-6}}$$
$$= 3.4 \times 10^{-2}$$

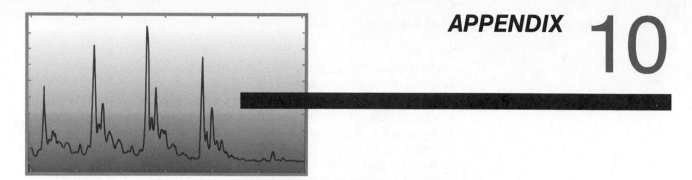

# APPENDIX 10

# VOLUMETRIC CALCULATIONS USING NORMALITY AND EQUIVALENT WEIGHT

$T$he *normality* of a solution expresses the number of equivalents of solute contained in 1 L of solution or the number of milliequivalents in 1 mL. The equivalent and milliequivalent, like the mole and millimole, are units for describing the amount of a chemical species. The former are defined, however, in such a way that it is possible to state that, at the equivalence point in *any* titration,

> no. meq analyte present = no. meq standard reagent added   (A10–1)

or

> no. eq analyte present = no. eq standard reagent added   (A10–2)

As a consequence, stoichiometric ratios such as those described in Section 5C–5 need not be derived every time a volumetric calculation is performed. Instead, the stoichiometry is taken into account by the way equivalent or milliequivalent weight is defined.

## A10–1 The Definition of Equivalent and Milliequivalent

In contrast to the mole, the amount of a substance contained in one equivalent can vary from reaction to reaction. Consequently, the weight of one equivalent of a compound can never be computed *without reference to a chemical reaction* in which that compound is, directly or indirectly, a participant. Similarly, the normality of a solution can never be specified *without knowledge about how the solution will be used*.

### Equivalent Weights in Neutralization Reactions

One equivalent weight of a substance participating in a neutralization reaction is that amount of substance (molecule, ion, or paired ion such as

A.23

NaOH) that either reacts with or supplies 1 mol of hydrogen ions *in that reaction*.[1] A milliequivalent is simply 1/1000 of an equivalent.

The relationship between equivalent weight (eqw) and formula weight is straightforward for strong acids or bases and for other acids or bases that contain a single reactive hydrogen or hydroxide ion. For example, the equivalent weights of potassium hydroxide, hydrochloric acid, and acetic acid are equal to their formula weights because each has but a single reactive hydrogen ion or hydroxide ion. Barium hydroxide, which contains two identical hydroxide ions, reacts with two hydrogen ions in any acid/base reaction, and so its equivalent weight is one half its formula weight:

$$\text{eqw Ba(OH)}_2 = \frac{\text{fw Ba(OH)}_2}{2}$$

The situation becomes more complex for acids or bases that contain two or more reactive hydrogen or hydroxide ions with different tendencies to dissociate. With certain indicators, for example, only the first of the three protons in phosphoric acid is titrated:

$$H_3PO_4 + OH^- \rightarrow H_2PO_4^- + H_2O$$

With certain other indicators, a color change occurs only after two hydrogen ions have reacted:

$$H_3PO_4 + 2\,OH^- \rightarrow HPO_4^{2-} + 2\,H_2O$$

For a titration involving the first reaction, the equivalent weight of phosphoric acid is equal to the formula weight; for the second, the equivalent weight is one half the formula weight. (Because it is not practical to titrate the third proton, an equivalent weight that is one third the formula weight is not generally encountered for $H_3PO_4$.) If it is not known which of these reactions is involved, an unambiguous definition of the equivalent weight for phosphoric acid *cannot be made*.

### Equivalent Weights in Oxidation/Reduction Reactions

The equivalent weight of a participant in an oxidation/reduction reaction is that weight which directly or indirectly produces or consumes 1 mol of electrons. The numerical value for the equivalent weight is conveniently

---

[1]An alternative definition, proposed by the International Union of Pure and Applied Chemistry, is as follows: An equivalent is "that amount of a substance, which, in a specified reaction, releases or replaces that amount of hydrogen that is combined with 3 g of carbon-12 in methane $^{12}CH_4$" (see *Information Bulletin* No. 36, International Union of Pure and Applied Chemistry, August 1974). This definition applies to acids. For other types of reactions and reagents, the amount of hydrogen referred to may be replaced by the equivalent amount of hydroxide ions, electrons, or cations. The reaction to which the definition is applied must be specified.

established by dividing the formula weight of the substance of interest by the change in oxidation number associated with its reaction. As an example, consider the oxidation of oxalate ion by permanganate ion:

$$5\ C_2O_4^{2-} + 2\ MnO_4^- + 16\ H^+ \rightarrow 10\ CO_2 + 2\ Mn^{2+} + 8\ H_2O \quad (A10-3)$$

In this reaction, the change in oxidation number for manganese is 5 because the element passes from the +7 to the +2 state; the equivalent weights for $MnO_4^-$ and $Mn^{2+}$ are therefore one fifth their formula weights. Each carbon atom in the oxalate ion is oxidized from the +3 to the +4 state, leading to the production of two electrons by that species. Therefore, the equivalent weight of sodium oxalate is one half of its formula weight. It is also possible to assign an equivalent weight to the carbon dioxide produced by the reaction. Since this molecule contains but a single carbon atom and since that carbon undergoes a change in oxidation number of 1, the formula weight and equivalent weight of the two are identical.

It is important to note that in evaluating the equivalent weight of a substance, *only* its change in oxidation number during the titration is considered. For example, suppose the manganese content of a sample containing $Mn_2O_3$ is to be determined by a titration based upon the reaction given in Equation A10-3. The fact that each manganese in the $Mn_2O_3$ has an oxidation number of +3 plays no part in determining equivalent weight. That is, we must assume that by suitable treatment, all the manganese is oxidized to the +7 state before the titration is begun. Each manganese from the $Mn_2O_3$ is then reduced from the +7 to the +2 state in the titration step. The equivalent weight is thus the formula weight of $Mn_2O_3$ divided by $2 \times 5 = 10$.

As in neutralization reactions, the equivalent weight for a given oxidizing or reducing agent is not invariant. Potassium permanganate, for example, reacts under some conditions to give $MnO_2$:

$$MnO_4^- + 3\ e^- + 2\ H_2O \rightarrow MnO_2(s) + 4\ OH^-$$

The change in the oxidation state of manganese in this reaction is from +7 to +4, and the equivalent weight of potassium permanganate is now equal to its formula weight divided by 3 (instead of 5 as in the earlier example).

## Equivalent Weights in Precipitation and Complex-Formation Reactions

The equivalent weight of a participant in a precipitation or a complex-formation reaction is that weight which reacts with or provides one mole of the *reacting* cation if it is univalent, one half mole if it is divalent, one third mole if it is trivalent, and so on. It is important to note that the cation referred to in this definition is always *the cation directly involved in the analytical reaction* and not necessarily the cation contained in the compound whose equivalent weight is being defined.

Example A10–1

Define equivalent weights for $AlCl_3$ and $BiOCl$ if the two compounds are determined by a precipitation titration with $AgNO_3$:

$$Ag^+ + Cl^- \rightarrow AgCl(s)$$

In this instance, the equivalent weight is based upon the number of moles of *silver ions* involved in the titration of each compound. Since 1 mol of $Ag^+$ reacts with 1 mol of $Cl^-$ provided by one third mole of $AlCl_3$, we can write

$$\text{eqw } AlCl_3 = \frac{\text{fw } AlCl_3}{3}$$

Because each mole of $BiOCl$ reacts with only 1 $Ag^+$ ion,

$$\text{eqw } BiOCl = \frac{\text{fw } BiOCl}{1}$$

Note that the fact that $Bi^{3+}$ (or $Al^{3+}$) is trivalent has no bearing because the definition is based *upon the cation involved in the titration:* $Ag^+$.

## A10–2  The Definition of Normality

The normality $c_N$ of a solution expresses the number of milliequivalents of solute contained in 1 mL of solution or the number of equivalents contained in 1 L. Thus, a 0.20 N hydrochloric acid solution contains 0.20 meq of HCl in each milliliter of solution or 0.20 eq in each liter.

The normal concentration of a solution is defined by equations analogous to Equation 5–1. Thus, for a solution of the species A, the normality $c_{N(A)}$ is given by the equations

$$c_{N(A)} = \frac{\text{no. meq A}}{\text{no. mL solution}} \tag{A10–4}$$

$$c_{N(A)} = \frac{\text{no. eq A}}{\text{no. L solution}} \tag{A10–5}$$

## A10–3  Some Useful Algebraic Relationships

Two pairs of algebraic equations, analogous to Equations 5–5 and 5–6 as well as 5–7 and 5–8, apply when normal concentrations are used:

$$\text{amount A} = \text{no. meq A} = \frac{\text{wt A (g)}}{\text{meqw A (g/meq)}} \tag{A10–6}$$

$$\text{amount A} = \text{no. eq A} = \frac{\text{wt A (g)}}{\text{eqw A (g/eq)}} \tag{A10–7}$$

$$\text{amount A} = \text{no. meq A} = V \text{ (mL)} \times c_{N(A)} \text{ (meq/mL)} \quad \text{(A10–8)}$$
$$\text{amount A} = \text{no. eq A} = V \text{ (L)} \times c_{N(A)} \text{ (eq/L)} \quad \text{(A10–9)}$$

## A10–4 Calculation of the Normality of Standard Solutions

Example A10–2 shows how the normality of a standard solution is computed from preparatory data. Note the similarity between this example and Example 5–10 (Chapter 5).

---

### Example A10–2

Describe the preparation of 5.000 L of 0.1000 N $Na_2CO_3$ from the primary-standard solid, assuming the solution is to be used for titrations in which the reaction is

$$CO_3^{2-} + 2\,H^+ \rightarrow H_2O + CO_2$$

Applying Equation A10–9 gives

$$\text{amount } Na_2CO_3 = V \text{ soln (L)} \times c_{N(Na_2CO_3)} \text{ (eq/L)}$$
$$= 5.000 \text{ L} \times 0.1000 \text{ eq/L} = 0.5000 \text{ eq } Na_2CO_3$$

Rearranging Equation A10–7 gives

$$\text{wt } Na_2CO_3 = \text{no. eq } Na_2CO_3 \times \text{eqw } Na_2CO_3$$

But 2 eq of $Na_2CO_3$ are contained in each mole of the compound; therefore,

$$\text{wt } Na_2CO_3 = 0.5000 \text{ eq } Na_2CO_3 \times \frac{105.99 \text{ g } Na_2CO_3}{2 \text{ eq } Na_2CO_3} = 26.50 \text{ g}$$

Therefore, dissolve 26.50 g in water and dilute to 5.000 L.

---

It is worth noting that, when the carbonate ion reacts with two protons, the weight of sodium carbonate required to prepare a 0.10 N solution is just one half that required to prepare a 0.10 M solution.

## A10–5 The Treatment of Titration Data With Normalities

### Calculation of Normalities from Titration Data

Examples A10–3 and A10–4 illustrate how normality is computed from standardization data. Note that these examples are similar to Examples 5–13 and 5–14.

Example A10–3

Exactly 50.00 mL of an HCl solution required 29.71 mL of 0.03926 N $Ba(OH)_2$ to give an end point with bromocresol green indicator. Calculate the normality of the HCl.

Note that the molarity of $Ba(OH)_2$ is one half its normality. That is,

$$c_{Ba(OH)_2} = 0.03926 \ \frac{\text{meq}}{\text{mL}} \times \frac{1 \ \text{mmol}}{2 \ \text{meq}} = 0.01963 \ \text{M}$$

Because we are basing our calculations on the milliequivalent, we write

$$\text{no. meq HCl} = \text{no. meq } Ba(OH)_2$$

The number of milliequivalents of standard is obtained by substituting into Equation A10–8:

$$\text{amount } Ba(OH)_2 = 29.71 \ \text{mL } Ba(OH)_2 \times 0.03926 \ \frac{\text{meq } Ba(OH)_2}{\text{mL } Ba(OH)_2}$$

To obtain the number of milliequivalents of HCl, we write

$$\text{amount HCl} = (29.71 \times 0.03926) \ \text{meq } Ba(OH)_2 \times \frac{1 \ \text{meq HCl}}{1 \ \text{meq } Ba(OH)_2}$$

Equating this result to Equation A10–8 yields

$$\begin{aligned} \text{amount HCl} &= 50.00 \ \text{mL HCl} \times c_{N(HCl)} \\ &= (29.71 \times 0.03926 \times 1) \ \text{meq HCl} \end{aligned}$$

$$c_{N(HCl)} = \frac{(29.71 \times 0.03926 \times 1) \ \text{meq HCl}}{50.00 \ \text{mL HCl}} = 0.02333 \ \text{N}$$

Example A10–4

A 0.2121-g sample of pure $Na_2C_2O_4$ (fw = 134.00) was titrated with 43.31 mL of $KMnO_4$. What was the normality of the $KMnO_4$ solution? The chemical reaction is

$$2 \ MnO_4^- + 5 \ C_2O_4^{2-} + 16 \ H^+ \rightarrow 2 \ Mn^{2+} + 10 \ CO_2 + 8 \ H_2O$$

By definition, at the equivalence point in the titration,

$$\text{no. meq } Na_2C_2O_4 = \text{no. meq } KMnO_4$$

Substituting Equations A10–8 and A10–6 into this relationship gives

$$V_{KMnO_4} \times c_{N(KMnO_4)} = \frac{\text{wt } Na_2C_2O_4 \ (\text{g})}{\text{meqw } Na_2C_2O_4 \ (\text{g/meq})}$$

$$43.31 \text{ mL KMnO}_4 \times c_{N(KMnO_4)} = \frac{0.2121 \text{ g Na}_2\text{C}_2\text{O}_4}{0.13400 \text{ g Na}_2\text{C}_2\text{O}_4/2 \text{ meq}}$$

$$c_{N(KMnO_4)} = \frac{0.2121 \text{ g Na}_2\text{C}_2\text{O}_4}{43.31 \text{ mL KMnO}_4 \times 0.1340 \text{ g Na}_2\text{C}_2\text{O}_4/2 \text{ meq}}$$

$$= 0.073093 \text{ meq/mL KMnO}_4 = 0.07309 \text{ N}$$

Note that the normality found here is five times the molarity computed in Example 5–14.

## Calculation of the Quantity of Analyte from Titration Data

The examples that follow illustrate how analyte concentrations are computed when normalities are involved.

### Example A10–5

A 0.8040-g samples of an iron ore was dissolved in acid. The iron was then reduced to $Fe^{2+}$ and titrated with 47.22 mL of 0.1121 N (0.02242 M) $KMnO_4$ solution. Calculate the results of this analysis in terms of (a) percent Fe (fw = 55.847) and (b) percent $Fe_3O_4$ (fw = 231.54). The reaction of the analyte with the reagent is described by the equation

$$\text{MnO}_4^- + 5 \text{ Fe}^{2+} + 8 \text{ H}^+ \rightarrow \text{Mn}^{2+} + 5 \text{ Fe}^{3+} + 4 \text{ H}_2\text{O}$$

(a) At the equivalence point, we know that

$$\text{no. meq KMnO}_4 = \text{no. meq Fe}^{2+} = \text{no. meq Fe}_3\text{O}_4$$

Substituting Equations A10–8 and A10–6 leads to

$$V_{KMnO_4} \text{ (mL)} \times c_{N(KMnO_4)} \text{ (meq/mL)} = \frac{\text{wt Fe}^{2+} \text{ (g)}}{\text{meqw Fe}^{2+} \text{ (g/meq)}}$$

Substituting numerical data into this equation gives, after rearranging,

$$\text{wt Fe}^{2+} = 47.22 \text{ mL KMnO}_4 \times 0.1121 \frac{\text{meq}}{\text{mL KMnO}_4} \times \frac{0.055847 \text{ g}}{1 \text{ meq}}$$

Note that the milliequivalent weight of the $Fe^{2+}$ is equal to its milliformula weight. The percentage of iron is

$$\text{percent Fe}^{2+} = \frac{(47.22 \times 0.1121 \times 0.055847) \text{ g Fe}^{2+}}{0.8040 \text{ g sample}} \times 100\%$$

$$= 36.77\%$$

(b) Here,

$$\text{no. meq KMnO}_4 = \text{no. meq Fe}_3\text{O}_4$$

and

$$V_{KMnO_4} \text{ (mL)} \times c_{N(KMnO_4)} \text{ (meq/mL)} = \frac{\text{wt } Fe_3O_4 \text{ (g)}}{\text{meqw } Fe_3O_4 \text{ (g/meq)}}$$

Substituting numerical data and rearranging give

$$\text{wt } Fe_3O_4 = 47.22 \text{ mL} \times 0.1121 \frac{\text{meq}}{\text{mL}} \times 0.23154 \frac{\text{g } Fe_3O_4}{3 \text{ meq}}$$

Note that the milliequivalent weight of $Fe_3O_4$ is one third its formula weight because each $Fe^{2+}$ undergoes a one-electron change and the compound is converted to 3 $Fe^{2+}$ before titration. The percentage of $Fe_3O_4$ is then

$$\text{percent } Fe_3O_4 = \frac{(47.22 \times 0.1121 \times 0.23154/3) \text{ g } Fe_3O_4}{0.8040 \text{ g sample}} \times 100\%$$
$$= 50.81\%$$

Note that the answers to this example are identical to those in Example 5–15.

---

### Example A10–6

A 0.4755-g sample containing $(NH_4)_2C_2O_4$ and inert compounds was dissolved in water and made alkaline with KOH. The liberated $NH_3$ was distilled into 50.00 mL of 0.1007 N (0.05035 M) $H_2SO_4$. The excess $H_2SO_4$ was back-titrated with 11.13 mL of 0.1214 N NaOH. Calculate the percentage of N (fw = 14.007) and of $(NH_4)_2C_2O_4$ (fw = 124.10) in the sample.

At the equivalence point, the number of milliequivalents of acid and base are equal. In this titration, however, two bases are involved: NaOH and $NH_3$. Thus,

$$\text{no. meq } H_2SO_4 = \text{no. meq } NH_3 + \text{no. meq NaOH}$$

After rearranging,

$$\text{no. meq } NH_3 = \text{no. meq N} = \text{no. meq } H_2SO_4 - \text{no. meq NaOH}$$

Substituting Equations A10–6 and A10–8 for the number of milliequivalents of N and $H_2SO_4$, respectively, yields

$$\frac{\text{wt N (g)}}{\text{meqw N (g/meq)}} = 50.00 \text{ mL } H_2SO_4 \times 0.1007 \frac{\text{meq}}{\text{mL } H_2SO_4}$$

$$-11.13 \text{ mL NaOH} \times 0.1214 \frac{\text{meq}}{\text{mL NaOH}}$$

$$\text{wt N} = (50.00 \times 0.1007 - 11.13 \times 0.1214) \text{ meq} \times 0.014007 \text{ g N/meq}$$

$$\text{percent N} = \frac{(50.00 \times 0.1007 - 11.13 \times 0.1214) \times 0.014007 \text{ g N}}{0.4755 \text{ g sample}} \times 100\%$$

$$= 10.85\%$$

The number of milliequivalents of $(NH_4)_2C_2O_4$ is equal to the number of milliequivalents of $NH_3$ and N, but the milliequivalent weight of the $(NH_4)_2C_2O_4$ is equal to one half its formula weight. Thus,

$$\text{wt } (NH_4)_2C_2O_4 = (50.00 \times 0.1007 - 11.13 \times 0.1214) \text{ meq}$$
$$\times 0.12410 \text{ g/2 meq})$$

$$\text{percent } (NH_4)_2C_2O_4$$

$$= \frac{(50.00 \times 0.1007 - 11.13 \times 0.1214) \times 0.06205 \text{ g } (NH_4)_2C_2O_4}{0.4755 \text{ g sample}}$$

$$\times 100\% = 48.07\%$$

Note that the results obtained here are identical to those obtained in Example 5–17.

## Compounds Recommended for the Preparation of Standard Solutions of Some Common Elements*

| Element | Compound | FW | Solvent† | Notes |
|---------|----------|-----|----------|-------|
| Aluminum | Al metal | 26.98 | Hot dil HCl | a |
| Antimony | $KSbOC_4H_4O_6 \cdot \frac{1}{2} H_2O$ | 333.93 | $H_2O$ | c |
| Arsenic | $As_2O_3$ | 197.84 | dil HCl | i,b,d |
| Barium | $BaCO_3$ | 197.35 | dil HCl | |
| Bismuth | $Bi_2O_3$ | 465.96 | $HNO_3$ | |
| Boron | $H_3BO_3$ | 61.83 | $H_2O$ | d,e |
| Bromine | KBr | 119.01 | $H_2O$ | a |
| Cadmium | CdO | 128.40 | $HNO_3$ | |
| Calcium | $CaCO_3$ | 100.09 | dil HCl | i |
| Cerium | $(NH_4)_2Ce(NO_3)_6$ | 548.23 | $H_2SO_4$ | |
| Chromium | $K_2Cr_2O_7$ | 294.19 | $H_2O$ | i,d |
| Cobalt | Co metal | 58.93 | $HNO_3$ | a |
| Copper | Cu metal | 63.55 | dil $HNO_3$ | a |
| Fluorine | NaF | 41.99 | $H_2O$ | b |
| Iodine | $KIO_3$ | 214.00 | $H_2O$ | i |
| Iron | Fe metal | 55.85 | HCl, hot | a |
| Lanthanum | $La_2O_3$ | 325.82 | HCl, hot | f |
| Lead | $Pb(NO_3)_2$ | 331.20 | $H_2O$ | a |
| Lithium | $Li_2CO_3$ | 73.89 | HCl | a |
| Magnesium | MgO | 40.31 | HCl | |
| Manganese | $MnSO_4 \cdot H_2O$ | 169.01 | $H_2O$ | g |
| Mercury | $HgCl_2$ | 271.50 | $H_2O$ | b |
| Molybdenum | $MoO_3$ | 143.94 | 1 M NaOH | |
| Nickel | Ni metal | 58.70 | $HNO_3$, hot | a |
| Phosphorus | $KH_2PO_4$ | 136.09 | $H_2O$ | |
| Potassium | KCl | 74.56 | $H_2O$ | a |
| | $KHC_8H_4O_4$ | 204.23 | $H_2O$ | i,d |
| | $K_2Cr_2O_7$ | 294.19 | $H_2O$ | i,d |
| Silicon | Si metal | 28.09 | NaOH, concd | |
| | $SiO_2$ | 60.08 | HF | j |
| Silver | $AgNO_3$ | 169.87 | $H_2O$ | a |

| Element | Compound | FW | Solvent† | Notes |
|---|---|---|---|---|
| Sodium | NaCl | 58.44 | $H_2O$ | i |
|  | $Na_2C_2O_4$ | 134.00 | $H_2O$ | i,d |
| Strontium | $SrCO_3$ | 147.63 | HCl | a |
| Sulfur | $K_2SO_4$ | 174.27 | $H_2O$ |  |
| Tin | Sn metal | 118.69 | HCl |  |
| Titanium | Ti metal | 47.90 | $H_2SO_4$, 1:1 | a |
| Tungsten | $Na_2WO_4\cdot2\,H_2O$ | 329.86 | $H_2O$ | h |
| Uranium | $U_3O_8$ | 842.09 | $HNO_3$ | d |
| Vanadium | $V_2O_5$ | 181.88 | HCl, hot |  |
| Zinc | ZnO | 81.37 | HCl | a |

*The data in this table were taken from a more complete list assembled by B. W. Smith and M. L. Parsons, *J. Chem. Educ.*, **1973**, *50*, 679. Unless otherwise specified, compounds should be dried to constant weight at 110°C.

†Unless otherwise specified, acids are concentrated analytical grade.

aConforms well to the criteria listed in Section 5B–1 and approaches primary standard quality.

bHighly toxic.

cLoses ½ $H_2O$ at 110°C. After drying, fw = 324.92. The dried compound should be weighed quickly after removal from the desiccator.

dAvailable as a primary standard from the National Institute of Standards and Technology.

e$H_3BO_3$ should be weighed directly from the bottle. It loses 1 $H_2O$ at 100°C and is difficult to dry to constant weight.

fAbsorbs $CO_2$ and $H_2O$. Should be ignited just before use.

gMay be dried at 110°C without loss of water.

fLoses both waters at 110°C. fw = 293.82. Keep in desiccator after drying.

iPrimary standard.

jHF is highly toxic and dissolves glass.

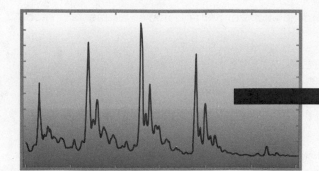

# DERIVATION OF ERROR PROPAGATION EQUATIONS

$I$n this appendix we derive several equations that permit the calculation of the standard deviation for the results from various types of arithmetical computations.

## A12A Propagation of Measurement Uncertainties

The calculated result for a typical analysis ordinarily requires data from several independent experimental measurements, each of which is subject to an indeterminate uncertainty and each of which contributes to the net indeterminate error of the final result. For the purpose of showing how such indeterminate uncertainties affect the outcome of an analysis, let us assume that a result $y$ is dependent upon the experimental variables, $a$, $b$, $c$, . . . , each of which fluctuates in a random and independent way. That is, $y$ is a function of $a$, $b$, $c$, . . . so that we may write

$$y = f(a, b, c, \ldots) \qquad \text{(A12--1)}$$

The uncertainty $dy_i$ is generally given in terms of the deviation from the mean or $(y_i - \bar{y})$, which will depend upon the size and sign of the corresponding uncertainties $da_i$, $db_i$, $dc_i$, . . . . That is,

$$dy_i = (y_i - \bar{y}) = f(da_i, db_i, dc_i, \ldots)$$

The variable in $dy$ as a function of the uncertainties in $a$, $b$, $c$, . . . can be derived by taking the total differential of Equation A12–1. That is,

$$dy = \left(\frac{\partial y}{\partial a}\right)_{b,c,\ldots} da + \left(\frac{\partial y}{\partial b}\right)_{a,c,\ldots} db + \left(\frac{\partial y}{\partial c}\right)_{a,b,\ldots} dc + \cdots \quad \text{(A12--2)}$$

In order to develop a relationship between the standard deviation of $y$ and the standard deviations of $a$, $b$, and $c$ for $N$ replicate measurements, we employ Equation 2–2 (p. 8), which requires that we square Equation A12–2, sum between $i = 0$ and $i = N$, divide by $N - 1$, and take the square root of the result. The square of Equation A12–2 takes the form

$$(dy)^2 = \left[ \left(\frac{\partial y}{\partial a}\right)_{b,c,\ldots} da + \left(\frac{\partial y}{\partial b}\right)_{a,c,\ldots} db + \left(\frac{\partial y}{\partial c}\right)_{a,b,\ldots} dc + \cdots \right]^2 \quad \text{(A12–3)}$$

This equation must then be summed between the limits of $i = 1$ to $i = N$.

In squaring Equation A12–2, two types of terms emerge from the right-hand side of the equation: (1) square terms and (2) cross terms. Square terms take the form

$$\left(\frac{\partial y}{\partial a}\right)^2 da^2, \quad \left(\frac{\partial y}{\partial b}\right)^2 db^2, \quad \left(\frac{\partial y}{\partial c}\right)^2 dc^2, \ldots$$

Square terms are always positive and can, therefore, *never* cancel when summed. In contrast, cross terms may be either positive or negative in sign. Examples are

$$\left(\frac{\partial y}{\partial a}\right)\left(\frac{\partial y}{\partial b}\right) da\,db, \quad \left(\frac{\partial y}{\partial a}\right)\left(\frac{\partial y}{\partial c}\right) da\,dc, \ldots$$

If *da, db,* and *dc* represent *independent* and *random uncertainties,* some of the cross terms will be negative and others positive. Thus, the *sum of all such terms should approach zero,* particularly when $N$ is large.

As a consequence of the tendency of cross terms to cancel, the sum of Equation A12–3 from $i = 1$ to $i = N$ can be assumed to be made up exclusively of square terms. This sum then takes the form

$$\Sigma(dy_i)^2 = \left(\frac{\partial y}{\partial a}\right)^2 \Sigma(da_i)^2 + \left(\frac{\partial y}{\partial b}\right)^2 \Sigma(db_i)^2 + \left(\frac{\partial y}{\partial c}\right)^2 \Sigma(dc_i)^2 + \cdots \quad \text{(A12–4)}$$

Dividing through by $N - 1$ gives

$$\frac{\Sigma(dy_i)^2}{N-1} = \left(\frac{\partial y}{\partial a}\right)^2 \frac{\Sigma(da_i)^2}{N-1} + \left(\frac{\partial y}{\partial b}\right)^2 \frac{\Sigma(db_i)^2}{N-1} + \left(\frac{\partial y}{\partial c}\right)^2 \frac{\Sigma(dc_i)^2}{N-1} + \cdots \quad \text{(A12–5)}$$

From Equation 2–2, however, we see that

$$\frac{\Sigma(dy_i)^2}{N-1} = \frac{\Sigma(y_i - \bar{y})^2}{N-1} = s_y^2$$

where $s_y^2$ is the variance of $y$. Similarly,

$$\frac{\Sigma(da_i)^2}{N-1} = \frac{\Sigma(a_i - \bar{a})^2}{N-1} = s_a^2$$

and so forth. Thus, Equation A12–5 can be written in terms of the variances of the variables; that is,

$$s_y^2 = \left(\frac{\partial y}{\partial a}\right)^2 s_a^2 + \left(\frac{\partial y}{\partial b}\right)^2 s_b^2 + \left(\frac{\partial y}{\partial c}\right)^2 s_c^2 + \cdots \quad \text{(A12–6)}$$

## A12B    The Standard Deviation of Computed Results

In this section, we employ Equation A12–6 to derive relationships that permit calculation of standard deviations for the results produced by five types of arithmetic operations.

### A12B–1 Addition and Subtraction

Consider the case where we wish to compute the quantity $y$ from the three experimental quantities $a$, $b$, and $c$ by means of the equation

$$y = a + b - c$$

We assume that the standard deviations for these quantities are $s_y$, $s_a$, $s_b$, and $s_c$. Applying Equation A12–6 leads to

$$s_y^2 = \left(\frac{\partial y}{\partial a}\right)_{b,c}^2 s_a^2 + \left(\frac{\partial y}{\partial b}\right)_{a,c}^2 s_b^2 + \left(\frac{\partial y}{\partial c}\right)_{a,b}^2 s_c^2$$

The partial derivatives of $y$ with respect to the three experimental quantities are

$$\left(\frac{\partial y}{\partial a}\right)_{b,c} = 1; \qquad \left(\frac{\partial y}{\partial b}\right)_{a,c} = 1; \qquad \left(\frac{\partial y}{\partial c}\right)_{a,b} = -1$$

Therefore, the variance of $y$ is given by

$$s_y^2 = (1)^2 s_a^2 + (1)^2 s_b^2 + (-1)^2 s_c^2 = s_a^2 + s_b^2 + s_c^2$$

or the standard deviation of the result is given by

$$s_y = \sqrt{s_a^2 + s_b^2 + s_c^2} \tag{A12–7}$$

Thus, the *absolute* standard deviation of a sum or difference is equal to the square root of the sum of the squares of the *absolute* standard deviation of the numbers making up the sum or difference.

### A12B–2 Multiplication and Division

Let us now consider the case where

$$y = \frac{ab}{c}$$

The partial derivatives of $y$ with respect to $a$, $b$, and $c$ are

$$\left(\frac{\partial y}{\partial a}\right)_{b,c} = \frac{b}{c}; \qquad \left(\frac{\partial y}{\partial b}\right)_{a,c} = \frac{a}{c}; \qquad \left(\frac{\partial y}{\partial c}\right) = -\frac{ab}{c^2}$$

Substituting into Equation A12–6 gives

$$s_y^2 = \left(\frac{b}{c}\right)^2 s_a^2 + \left(\frac{a}{c}\right)^2 s_b^2 + \left(-\frac{ab}{c^2}\right)^2 s_c^2$$

Dividing this equation by the square of the original equation ($y^2 = a^2 b^2/c^2$) gives

$$\frac{s_y^2}{y^2} = \frac{s_a^2}{a^2} + \frac{s_b^2}{b^2} + \frac{s_c^2}{c^2}$$

or

$$\frac{s_y}{y} = \sqrt{\left(\frac{s_a}{a}\right)^2 + \left(\frac{s_b}{b}\right)^2 + \left(\frac{s_c}{c}\right)^2} \qquad \text{(A12–8)}$$

Thus for products and quotients, the *relative* standard deviation of the result is equal to the sum of the squares of the *relative* standard deviation of the number making up the product or quotient.

### A12B–3 Exponential Calculations

Consider the following computation

$$y = a^x$$

Here, Equation A12–6 takes the form

$$s_y^2 = \left(\frac{\partial a^x}{\partial y}\right)^2 s_a^2$$

or

$$s_y = \frac{\partial a^x}{\partial y} s_a$$

But

$$\frac{\partial a^x}{\partial y} = x a^{(x-1)}$$

Thus

$$s_y = x a^{(x-1)} s_a$$

and dividing by the original equation ($y = a^x$) gives

$$\frac{s_y}{y} = \frac{x a^{(x-1)} s_a}{a^x} = x \frac{s_a}{a} \qquad \text{(A12–9)}$$

Thus the relative error of the result is equal to the relative error of numbers to be exponentiated, multiplied by the exponent.

It is important to note that the error propagated in taking a number to a power is different from the error propagated in multiplication. For example, consider the uncertainty in the square of $4.0(\pm 0.2)$. Here the relative error in the result (16.0) is given by Equation A12–9

$$s_y/y = 2 \times (0.2/4) = 0.1 \qquad \text{or} \qquad 10\%$$

Consider now the case when $y$ is the product of two *independently measured* numbers that by chance happen to have values of $a = 4.0(\pm 0.2)$ and $b = 4.0(\pm 0.2)$. In this case the relative error of the product $ab = 16.0$ is given by Equation A12–8:

$$s_y/y = \sqrt{(0.2/4)^2 + (0.2/4)^2} = 0.07 \qquad \text{or} \qquad 7\%$$

The reason for this apparent anomaly is that in the second case the sign associated with one error can be the same or different from that of the other. If they happen to be the same, the error is identical to that encountered in the first case, where the signs *must* be the same. In contrast, the possibility exists that one sign could be positive and the other negative, in which case the relative errors tend to cancel one another. Thus, the probable error lies between the maximum (10%) and zero.

### A12B–4 Calculation of Logarithms

Consider the computation

$$y = \log_{10} a$$

In this case, we can write Equation A12–6 as

$$s_y^2 = \left(\frac{\partial \log_{10} a}{\partial y}\right)^2 s_a^2$$

But

$$\frac{\partial \log_{10} a}{\partial y} = \frac{0.434}{a}$$

and

$$s_y = 0.434\frac{s_a}{a} \qquad\qquad\qquad \text{(A12–10)}$$

Thus the absolute standard deviation of a logarithm is determined by the *relative* standard deviation of the number.

### A12B–5 Calculation of Antilogarithms

Consider the relationship

$$y = \text{antilog}_{10} a = 10^a$$

$$\left(\frac{\partial y}{\partial a}\right) = 10^a \log_e 10 = 10^a \ln 10 = 2.303 \times 10^a$$

$$s_y^2 = \left(\frac{\partial y}{\partial a}\right)^2 s_a^2$$

or

$$s_y = \frac{\partial y}{\partial a} s_a = 2.303 \times 10^a s_a$$

Dividing by the original relationship gives

$$\frac{s_y}{y} = 2.303 s_a \qquad\qquad (A12–11)$$

Thus, the *relative* standard deviation of the antilog of a number is determined by the absolute standard deviation of the number.

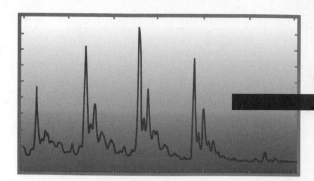

# ANSWERS TO QUESTIONS AND PROBLEMS

## Chapter 2

**2-1. (a)** *Accuracy* is the closeness of a result to its true or accepted value. *Precision* is the closeness of a result to other results measured in exactly the same way.

**(c)** The *mean* of a set of data is obtained by summing the results and dividing the sum by the number of results. The *median* of a set of data is the middle result when the data are arranged in order of size. For an even number of data, the median is obtained by averaging the middle two data.

**(e)** *Constant errors* are the same in magnitude regardless of the size of the sample. *Proportional errors* are proportional in size to the sample size.

**2-2. (a)** The *range* of a set of data is the numerical difference between the largest and the smallest value in the set.

**(c)** A *histogram* is a bar graph that is obtained by dividing a set of data into equal-size cells and plotting the percentage of measurements falling into each cell as a function of the measured quantity.

**(e)** *Personal bias,* or prejudice, is the tendency of an experimenter to read instrument scales so as to cause the results to reflect a consciously or subconsciously preconceived notion of the outcome.

**2-3.** Instrumental error, method error, and personal error.

**2-5.** Constant determinate errors can be detected by varying sample size because the effect of the error becomes smaller as the sample size increases.

**2-7.**

|  | A | C | E |
|---|---|---|---|
| **(a)** mean | 16.34 | 35.65 | 41.37 |
| **(b)** median | 16.35 | 35.62 | 41.37 |
| **(c)** spread | 0.14 | 0.23 | 0.40 |
| **(d)** $s$ | 0.059 | 0.10 | 0.14 |
| **(e)** CV, % | 0.36 | 0.29 | 0.35 |

**2-8.**

|  | A | C | E |
|---|---|---|---|
| **(a)** absolute error | 0.06 | −0.05 | −0.03 |
| **(b)** relative error, ppt | 3 | 1 | 1 |

**2-9. (a)** 0.6%
**(c)** 0.1%

**2-10. (a)** 1%
**(c)** 0.1%

**2-11. (a)** 200 mg
**(b)** 1 g

**2-12.**

|  | $s$ | CV, % | Rounded Result |
|---|---|---|---|
| **(a)** | 0.030 | 5.2 | 0.57($\pm$0.03) |
| **(b)** | 0.089 | 0.42 | 21.3($\pm$0.1) |
| **(c)** | $0.14 \times 10^{-16}$ | 2.0 | $6.9(\pm0.1) \times 10^{-16}$ |
| **(d)** | $1.4 \times 10^3$ | 0.77 | $1.84(\pm0.01) \times 10^5$ |
| **(e)** | $0.51 \times 10^{-2}$ | 8.5 | $6.0(\pm0.5) \times 10^{-2}$ or $7(\pm1) \times 10^{-2}$ |
| **(f)** | $0.11 \times 10^{-3}$ | 1.3 | $8.1(\pm0.1) \times 10^{-3}$ |

**2-13. (a)** 0.238 **(c)** 23.7796 **(e)** $3.4 \times 10^{-4}$ **(g)** 9.8

## Chapter 3

**3-1. (a)** A *chemical sample* is the physical entity that is to be analyzed. A *statistical sample* is the number of replicate measurements that are made on the chemical sample. These measurements are but a small fraction of the infinite number of measurements that could be made in principle.

**3-3.** The *confidence interval* is a range of values defined by the confidence limits around the experimental mean $\bar{x}$ within which the true mean $\mu$ lies with a given level of confidence.

**3-4.**

| Sample | $s$, % K |
|---|---|
| 1 | 0.095 |
| 2 | 0.12 |
| 3 | 0.11 |
| 4 | 0.10 |
| 5 | 0.10 |

$s_{\text{pooled}} = 0.11\%$ $K^+$

**3–6.** $s_{pooled} = 0.29\%$ heroin

**3–7.**

|       | A     | C     | E     |
| ----- | ----- | ----- | ----- |
| $\bar{x}$ | 16.22 | 2.774 | 55.95 |
| $s$   | 0.012 | 0.054 | 0.15  |

**3–9.** (a) 80% CI $= 18.5 \pm 1.29 \times 2.4/\sqrt{1}$
$\qquad\qquad\quad = 18.5 \pm 3.1\ \mu g/mL$
$\qquad$ 95% CI $= 18.5 \pm 1.96 \times 2.4/\sqrt{1}$
$\qquad\qquad\quad = 18.5 \pm 4.7\ \mu g/mL$
$\quad$ (b) 80% CI $= 18.5 \pm 1.29 \times 2.4/\sqrt{2}$
$\qquad\qquad\quad = 18.5 \pm 2.2\ \mu g/mL$
$\qquad$ 95% CI $= 18.5 \pm 1.96 \times 2.4/\sqrt{2}$
$\qquad\qquad\quad = 18.5 \pm 3.3\ \mu g/mL$
$\quad$ (c) 80% CI $= 18.5 \pm 1.29 \times 2.4/\sqrt{4}$
$\qquad\qquad\quad = 18.5 \pm 1.5\ \mu g/mL$
$\qquad$ 95% CI $= 18.5 \pm 1.96 \times 2.4/\sqrt{4}$
$\qquad\qquad\quad = 18.5 \pm 2.4\ \mu g/mL$

**3–11.** 10 and 17 measurements

**3–13.** (a) $7.24 \pm 0.42\%$
$\qquad$ (b) $7.24 \pm 0.26\%$

**3–16.** (a) mean $= 16.57$
$\qquad$ median $= 16.64$
$\quad$ (b) $Q_{expt} = 0.24/0.40 = 0.60$
$\qquad$ $Q_{crit} = 0.642$; therefore retain
$\quad$ (c) The "best" value is probably the median because it is not influenced strongly by the outlier.

**3–17.** (a) $(\bar{x} - \mu)_{actual} = -0.006$
$\qquad$ $(\bar{x} - \mu)_{exp} = ts/\sqrt{N} = (2.36)(0.097 \times 0.090)/$
$\qquad$ $\sqrt{8} = 0.0073 > 0.006$
$\qquad$ No determinate error demonstrated.
$\quad$ (b) $(\bar{x} - \mu)_{actual} = -0.03$
$\qquad$ $(\bar{x} - \mu)_{exp} = (2.78)(0.36 \times 0.055)/\sqrt{5} = 0.025 <$
$\qquad$ $0.03$
$\qquad$ Determinate error is suggested.
$\quad$ (c) $(\bar{x} - \mu)_{actual} = 0.03$
$\qquad$ $(\bar{x} - \mu)_{exp} = (2.45)(0.55 \times 0.076)/\sqrt{7} = 0.039 >$
$\qquad$ $0.03$
$\qquad$ No determinate error demonstrated.

**3–19.** $s_{pooled} = \sqrt{\dfrac{\Sigma(x_{i(top)} - \bar{x}_{(top)})^2 + \Sigma(x_{i(bot)} - \bar{x}_{(bot)})^2}{N_1 + N_2 - 2}}$
$\qquad\qquad = 0.051$
$\quad$ (a) $(\bar{x}_1 - \bar{x}_2)_{actual} = 0.052$

$\qquad$ $(\bar{x}_1 - \bar{x}_2)_{cal} = \pm ts_{pooled}\sqrt{\dfrac{N_1 + N_2}{N_1 N_2}} = 0.10$

$\qquad$ Nonhomogeneity is not demonstrated.

$\quad$ (b) $(\bar{x}_1 - \bar{x}_2)_{calc} = \pm z\sigma\sqrt{\dfrac{N_1 + N_2}{N_1 N_2}} = 0.045$

$\qquad$ Nonhomogeneity is indicated.

**3–21.** (a) $F = 1.44$
$\qquad$ $F_{12,6} = 4.00$
$\qquad$ No difference demonstrated.
$\quad$ (c) $F = 1.96$
$\qquad$ $F_{6,20} = 2.60$
$\qquad$ No difference demonstrated.

**3–22.** (a) 0.34 $\qquad$ (c) 0.58

**3–23.** (a) No improvement demonstrated.
$\qquad$ (c) Improvement suggested.

**3–24.** (a) See broken line in the figure in the next column.

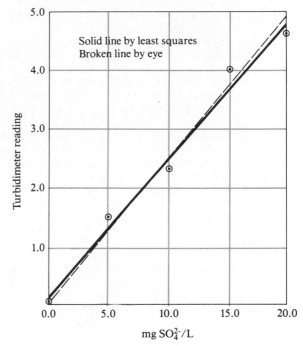

(b) $R = 0.232 c_x + 0.162$
(c) See the solid line in the figure above.
(d) $s_m = 0.017$; $\quad s_b = 0.21$
(e) $x_c = 15.1$ mg $SO_4^{2-}/mL$; $\quad s_c = 1.3$;
$\quad$ CV $= 9.3\%$
(f) $s_c = 0.81$; $\quad$ CV $= 5.4\%$

## Chapter 4

**4–1.** (a) The individual particles of a *colloid* are smaller than about $10^{-5}$ mm in diameter while those of a crystalline precipitate are larger. As a consequence, *crystalline precipitates* settle out of solution relatively rapidly, whereas colloidal particles do not unless they can be caused to agglomerate.

$\quad$ (c) *Precipitation* is the process by which a solid phase forms and is carried out of solution when the solubility product of a species is exceeded. *Coprecipitation* is the process in which a *normally soluble* species is carried out of solution during the formation of a precipitate.

$\quad$ (e) *Occlusion* is a type of coprecipitation in which an impurity is entrapped in a pocket formed by a rapidly growing crystal. *Mixed-crystal formation* is a type of coprecipitation in which a foreign ion is incorporated into a growing crystal in a lattice position that is ordinarily occupied by one of the ions of the precipitate. The contaminant has the same charge and is about the same size as the ion it replaces.

**4–2.** (a) *Digestion* is a process for improving the purity and filterability of a precipitate by heating the solid in contact with the solution from which it is formed (the mother liquor).

(c) In *reprecipitation,* a precipitate is filtered, washed, redissolved, and then re-formed from the new solution. Because the concentration of contaminant is lower in this new solution than in the original, the second precipitate contains less coprecipitated impurity.

(e) The *counter-ion layer* is a layer of solution surrounding a colloidal particle that contains a sufficient excess of ions of opposite charge to just equal the number of primarily adsorbed ions on the particle. ·

(g) Relative supersaturation = $(Q - S)/S$ where $Q$ is the concentration of solute at any time and $S$ is its equilibrium solubility.

**4-3.** *Peptization* is the process in which a coagulated colloid returns to its original dispersed state as a consequence of a reduction in the electrolyte concentration of the solution in contact with the precipitate. Peptization during the washing of a coagulated colloid can be avoided by washing with an electrolyte solution rather than with pure water.

**4-5.** (a) positive charge    (b) adsorbed $Ag^+$
  (c) $NO_3^-$

**4-8.** (a) 0.677 mol $H_2O$
  (c) 0.0700 mol $K_2SO_4$
  (e) 0.149 mol $NaC_2H_3O_2$

**4-9.** (a) 0.333 g FeO
  (c) 0.739 g $Fe_2O_3$
  (e) 1.706 g $K_4Fe(CN)_6$

**4-11.** (a) $6.540 \times 10^{-3}$ mol CaO
  (b) 6.540 mmol CO
  (c) 0.2878 g $CO_2$

**4-13.** (a) 0.1081 g
  (b) 0.3160 g
  (c) 0.3050 g
  (d) 0.5620 g

**4-14.** (a) 0.1268 g AgCl
  (b) 0.2268 g $BaF_2$
  (c) 0.2914 g $BaSO_4$
  (d) 0.4616 g $Ba_3(PO_4)_2$

**4-15.** (a) 0.945 g urea
  (c) 0.268 g thioacetamide
  (e) 0.225 g trichloroacetic acid

**4-16.** (a) $\dfrac{\text{fw } CaCO_3}{\text{fw } CaC_2O_4}$    (c) $\dfrac{2 \text{ fw } NaIO_3}{\text{fw } PbI_2}$

  (e) $\dfrac{\text{fw } Na_2B_4O_7}{2 \text{ fw } B_2O_3}$    (g) $\dfrac{2 \text{ fw } KI}{\text{fw } Ba(IO_3)_2}$

  (i) $\dfrac{2 \text{ fw } HgO}{\text{fw } Hg_2I_2}$

**4-17.** (a) 0.4190 g $PbI_2$
  (c) 0.4807 g $PbI_2$

**4-18.** 0.222 g AgCl

**4-20.** 95.36% KCl

**4-22.** (a) 6970% $NH_4Al(SO_4)_2$

**4-24.** 0.0353% $H_2S$

**4-26.** 38.82% $Hg_2Cl_2$

**4-28.** 3.641 g Ag/L

**4-30.** 0.617 g sample yields 0.825 g $BaSO_4$

**4-33.** 28.72% KI;   19.95% $NH_4Cl$

**4-35.** 51.30% $NH_4NO_3$; 31.90% $(NH_4)_2SO_4$

**4-36.** (a) 80.00% Ag;   20.00% Cu
  (c) 90.00% Ag;   10.00% Cu
  (e) 50.00% Ag;   50.00% Cu

# Chapter 5

**5-1.** (a) The *millimole* is an amount of an elementary species, such as an atom, an ion, a molecule, or an electron. A millimole contains

$$6.02 \times 10^{23} \text{ particles} \times 10^{-3} \frac{\text{mole}}{\text{millimole}}$$
$$= 6.02 \times 10^{20} \text{ particles}$$

  (c) A *stoichiometric factor* is a simple whole number ratio of elementary species given by a balanced chemical equation.

**5-2.** (a) The *equivalence point* in a titration is that point at which sufficient titrant has been added so that stoichiometrically equivalent amounts of analyte and titrant are present. The *end point* in a titration is the point at which an observable physical change occurs that signals the equivalence point.

  (c) The *analytical molarity* of a solution is the total concentration of a solute in all forms and can be taken as a recipe for the preparation of a solution. On the other hand, *species molarity* is the concentration of a given species at equilibrium.

**5-4.** For a dilute solution, 1 L = 1000 mL ≈ 1000 g, so

$$\frac{\text{mg}}{\text{L}} = \frac{10^{-3} \text{ g}}{1000 \text{ g}} = 10^{-6} = 1 \text{ ppm}$$

**5-5.** (a) 0.1453 mol;   145.3 mmol
  (c) 0.01263 mol;   12.63 mmol
  (e) 0.06171 mol;   61.71 mmol

**5-6.** (a) 0.377 g KCl
  (c) 0.652 g $MgCl_2$
  (e) 2.040 g $CuSO_4$

**5-8.** (a) 31.80 mL
  (c) 10.60 mL
  (e) 15.90 mL

**5-11.** (a) 3.063 ppm $NH_4^+$

**5-13.** (a) 6.16 M    (c) 4.67 M

**5-14.** (a) Dilute 80 g ethanol to 500 mL.
  (b) Dilute 63 g ethanol to 500 mL.
  (c) Add 80 g ethanol to 420 g $H_2O$.

**5-16.** (a) pNa = pBr = 2.000
  (c) pBa = 2.46;   pOH = 2.15
  (e) pCa = 2.28;   pBa = 2.44;   pCl = 1.75

**5-17.** (a) pNa = 0.618;   pCl = 0.936;
   pOH = 0.903
  (c) pH = −0.176;   pZn = 0.921;   pCl = 0.241
  (e) pK = 4.249;   pFe(CN)$_6$ = 5.421;   pOH = 4.385

5–18. **(a)** $2.1 \times 10^{-9}$ M
**(c)** $0.925$ M
**(e)** $2.1$ M
**(g)** $0.99$ M

5–19. **(a)** $6.2 \times 10^{-10}$ M
**(c)** $3.5 \times 10^{-1}$ M
**(e)** $4.8 \times 10^{-8}$ M
**(g)** $1.6$ M

5–20. $0.1203$ M HCl

5–22. $0.2217$ M $H_2SO_4$

5–24. $0.02713$ M $KMnO_4$

5–27. $93.75\%$ $Na_2CO_3$

5–29. $135.7$ ppm Fe

5–32. $5.802$ g ethyl acetate/100 mL

5–34. **(a)** $0.02966$ mol $KMnO_4$/kg solution
**(b)** $31.68\%$ $Fe_2O_3$

5–36. **(a)** $9.36 \times 10^{-3}$ M $Ba(OH)_2$
**(b)** $2 \times 10^{-5}$ M
**(c)** $E_{rel} = 3$ ppt;   $E_{abs} = 3 \times 10^{-5}$ M

## Chapter 6

6–1. **(a)** A *salt* is the product (other than water) of an acid/base reaction.
**(c)** The *hydronium ion* is the product of the reaction of a proton with one molecule of water. Its formula is $H_3O^+$.
**(e)** A *base* is a proton acceptor.
**(g)** The *Le Châtelier principle* states that the position of a chemical equilibrium is always shifted in such a direction as to relieve the effect of an applied stress.
**(i)** An *amphiprotic solvent* is one that undergoes self-ionization to form a pair of ionic species. For example,

$$2\ H_2O \rightleftarrows H_3O^+ + OH^-$$
$$2\ C_2H_5OH \rightleftarrows C_2H_5OH_2^+ + C_2H_5O^-$$

6–2. Water is a *leveling solvent* toward mineral acids because all of these acids are completely dissociated in that medium and are thus equal in strength. In ethanol, the degree of dissociation of the various mineral acids differs significantly. Thus, ethanol is termed a *differentiating* solvent because it differentiates among the strengths of the acids.

6–3. The hydroxide concentration in the $K_{sp}$ for $Yb(OH)_3$ is cubed, and $[OH^-]$ is squared in the $K_{sp}$ for $Ni(OH)_2$. Thus, for $Yb(OH)_3$, $K_{sp} = 27S^3$, for $Ni(OH)_2$ $K_{sp} = 4S^2$, and $S$ is approximately equal for both salts.

6–4. **(a)** $Mg(OH)_2(s) \rightleftarrows Mg^{2+}(aq) + 2\ OH^-(aq)$;
$K_{sp} = [Mg^{2+}][OH^-]^2$
**(c)** $OCl^-(aq) + H_2O \rightleftarrows HOCl(aq) + OH^-(aq)$;
$K_b = \dfrac{[HOCl][OH^-]}{[OCl^-]}$
**(e)** $HOCl(aq) + H_2O \rightleftarrows H_3O^+(aq) + OCl^-(aq)$;
$K_a = \dfrac{[H_3O^+][OCl^-]}{[HOCl]}$

6–5. **(a)** $1.8 \times 10^{-11}$
**(c)** $K_b = \dfrac{K_w}{K_a} = \dfrac{1.0 \times 10^{-14}}{3.0 \times 10^{-8}} = 3.3 \times 10^{-7}$
**(e)** $K_a = 3.0 \times 10^{-8}$

6–6. $H_2PO_4^-(aq) + H_2O \rightleftarrows H_3O^+ + HPO_4^{2-}(aq)$
$K_2 = \dfrac{[H_3O^+][HPO_4^{2-}]}{[H_2PO_4^-]} = 6.34 \times 10^{-8}$
$H_2PO_4^-(aq) + H_2O \rightleftarrows H_3PO_4(aq) + OH^-(aq)$
$\dfrac{K_w}{K_1} = \dfrac{[H_3PO_4][OH^-]}{[H_2PO_4^-]} = 1.58 \times 10^{-7}$

6–7. **(a)** $Mg(OH)_2(s) + 2\ H_3O^+(aq) \rightleftarrows Mg^{2+}(aq) + 4\ H_2O$
$K = \dfrac{[Mg^{2+}]}{[H_3O^+]^2}$
**(c)** $H_2S(aq) + I_2(aq) \rightleftarrows S(s) + 2\ H^+(aq) + 2\ I^-(aq)$
$K = \dfrac{[H^+]^2[I^-]^2}{[H_2S][I_2]}$
**(e)** $Ag^+(aq) + 2\ CN^-(aq) \rightleftarrows Ag(CN)_2^-(aq)$
$K_f = \dfrac{[Ag(CN)_2^-]}{[Ag^+][CN^-]^2}$

6–8. **(a)** $4.0 \times 10^{-16}$
**(c)** $2.0 \times 10^{-2}$
**(e)** $3.5 \times 10^{-10}$

6–9. **(b)** $[Sr^{2+}] = [SO_4^{2-}] = 5.7 \times 10^{-4}$;   $S = 105$ mg/L
**(d)** $K_{sp} = 4.3 \times 10^{-29}$;   $S = 2.8 \times 10^{-23}$ mg/L
**(e)** $[La^{3+}] = 6.9 \times 10^{-4}$;   $[IO_3^-] = 2.1 \times 10^{-3}$;   $S = 460$ mg/L
**(g)** $[Fe(CN)_6^{4-}] = 4.7 \times 10^{-6}$;   $S = 1.61$ mg/L

6–11. $PbI_2 > TlI > BiI_3 > AgI$

6–13. **(a)** $[CrO_4^{2-}] = 6.9 \times 10^{-8}$
**(b)** $[CrO_4^{2-}] = 4.4 \times 10^{-2}$

6–15. **(a)** $[Ce^{3+}] = 2.5 \times 10^{-2}$
~~**(b)** $[Ce^{3+}] = 1.7 \times 10^{-2}$~~
**(c)** $[Ce^{3+}] = 1.9 \times 10^{-3}$
**(d)** $[Ce^{3+}] = 7.6 \times 10^{-7}$

6–17. **(a)** No   **(c)** Yes   **(e)** Yes

6–18. **(a)** No   **(c)** No

6–21. **(a)** pH = 4.84   **(c)** pH = 2.88
**(e)** pH = 1.96

6–22. **(a)** pH = 5.34   **(c)** pH = 3.39
**(e)** pH = 2.51

6–23. **(a)** pH = 12.05   **(c)** pH = 10.84
**(e)** pH = 10.56

6–24. **(a)** pH = 11.48   **(c)** pH = 10.33
**(e)** pH = 10.03

## Chapter 7

7–2. The species $CH_3COOH$ is uncharged; therefore its activity coefficient is unity.

7–4. The mean activity coefficient is an experimentally measured quantity that contains a term for each ion in an electrolyte.

$$f_\pm = (f_A^m \cdot f_B^n)^{\frac{1}{m+n}}$$

It is impossible to determine experimentally the individual activity coefficients $f_A$ and $f_B$.

7–5. (a) 0.060    (c) 0.300    (e) 0.180
7–6. (a) 0.20    (c) 0.073
7–7. (a) 0.20    (c) 0.079
7–8. (a) 0.90    (c) 0.75    (e) 0.67
     (g) 0.55    (h) 0.64
7–9. (a) $K'_{sp} = 1.30 \times 10^{-16}$
     (c) $K'_{sp} = 4.5 \times 10^{-12}$
     (e) $K'_{sp} = 3.8 \times 10^{-17}$
7–10. (a) $1.1 \times 10^{-8}$    (c) $1.7 \times 10^{-4}$
     (e) $2.1 \times 10^{-6}$

7–11.

| | Solubility | |
|---|---|---|
| | **Activity (1)** | **Molar Concn (2)** |
| (a) | $1.4 \times 10^{-6}$ | $1.0 \times 10^{-6}$ |
| (b) | $2.0 \times 10^{-3}$ | $1.2 \times 10^{-3}$ |
| (c) | $3.1 \times 10^{-5}$ | $1.1 \times 10^{-5}$ |
| (d) | $3.1 \times 10^{-5}$ | $4.7 \times 10^{-5}$ |

7–13.

| | $[H_3O^+]$ | |
|---|---|---|
| | **Activity (1)** | **Molar Concn (2)** |
| (a) | $3.50 \times 10^{-3}$ | $2.95 \times 10^{-3}$ |
| (c) | $1.25 \times 10^{-11}$ | $1.07 \times 10^{-11}$ |
| (e) | $4.60 \times 10^{-6}$ | $4.77 \times 10^{-6}$ |
| (g) | $1.63 \times 10^{-8}$ | $1.13 \times 10^{-8}$ |

7–14. (b) $2.4 \times 10^{-4}$
7–15. (a) $1.6 \times 10^{-6}$
7–16.

| | $[Hg^+]$ | |
|---|---|---|
| | **Activity (1)** | **Molar Concn (2)** |
| (a) | $5.4 \times 10^{-6}$ | $5.4 \times 10^{-6}$ |
| (c) | $8.5 \times 10^{-13}$ | $2.5 \times 10^{-13}$ |
| (e) | $1.1 \times 10^{-11}$ | $3.1 \times 10^{-12}$ |

## Chapter 8

8–1. In the calculation of the molar solubility of $Fe(OH)_2$, the hydronium ion concentration may be assumed to be negligibly small. This assumption cannot be made in the case of $Fe(OH)_3$.

8–3. The concentrations of ions must be multiplied by the absolute values of their charges because this process gives the number of moles per liter of change that the ion contributes.

8–6. (a) $K_{sp} = [Ag^+][IO_3^-]$
     (b) $K_{sp} = [Ag^+]^2[SO_3^{2-}]$
     (c) $K_{sp} = [Ag^+]^3[AsO_4^{3-}]$
     (d) $K_{sp} = [Pb^{2+}][Cl^-][F^-]$
8–8. (a) $K_{sp} = 1.4 \times 10^{-10}$
     (c) $K_{sp} = 6.19 \times 10^{-12}$
8–9. $K_{sp} = 8.01 \times 10^{-16}$
8–10. (a) $1.3 \times 10^{-4}$    (b) $1.2 \times 10^{-3}$
     (c) $4.0 \times 10^{-6}$    (d) $1.6 \times 10^{-2}$
8–12. (a) $[IO_3^-] = 3.8 \times 10^{-3}$    (b) $[IO_3^-] = 0.19$
8–14. (a) $[OH^-] = 4.20 \times 10^{-5}$, $[Cl^-] = 0.0204$, $[Mg^{2+}] = 0.0102$
     (b) $[H_3O^+] = 6.68 \times 10^{-2}$, $[Cl^-] = 0.204$, $[Mg^{2+}] = 6.86 \times 10^{-2}$
8–15. (a) $[I^-] = 0.0360$, $[K^+] = 0.0240$, $[Mg^{2+}] = 6.0 \times 10^{-3}$, $[OH^-] = 5.5 \times 10^{-5}$
     (b) $[Ba^{2+}] = 0.0300$, $[I^-] = 0.0360$, $[OH^-] = 2.40 \times 10^{-2}$, $[Mg^{2+}] = 3.1 \times 10^{-8}$

     (c) $[Mg^{2+}] = 0.0180$, $[NO_3^-] = 0.0300$, $[I^-] = 6.00 \times 10^{-3}$, $[Ag^+] = 1.4 \times 10^{-14}$
     (d) $[Mg^{2+}] = 0.0180$, $[NO_3^-] = 0.0360$, $[Ag^+] = 9.1 \times 10^{-9}$, $[I^-] = 9.1 \times 10^{-9}$
8–17. (a) $0.100 = [H_3PO_4] + [H_2PO_4^-] + [HPO_4^{2-}] + [PO_4^{3-}]$
     (c) $0.150 = [HNO_2^-] + [NO_2^-]$
        $0.0500 = [Na^+]$
     (e) $[OH^-] = 0.100 + 2S - 2[Zn(OH)_4^{2-}]$
        $[Na^+] = 0.100$
        $[Zn^{2+}] = S - [Zn(OH)_4^{2-}]$
     (g) $2S = [F^-] + [HF]$
        $S = [Ca^{2+}]$
     (i) $S = [Cd^{2+}] + [Cd(NH_3)^{2+}] + [Cd(NH_3)_2^{2+}] + [Cd(NH_3)_3^{2+}] + [Cd(NH_3)_4^{2+}] + [Cd(NH_3)_5^{2+}] + [Cd(NH_3)_6^{2+}]$
        $0.0100 = [NH_3] + [Cd(NH_3)^{2+}] + 2[Cd(NH_3)_2^{2+}] + \cdots$
8–18. (a) $[H_3O^+] = [OH^-] + [H_2PO_4^-] + 2[HPO_4^{2-}] + 3[PO_4^{3-}]$
     (c) $[H_3O^+] + [Na^+] = [OH^-] + [NO_2^-]$
     (e) $[H_3O^+] + [Na^+] + 2[Zn^{2+}] = [OH^-] + [Zn(OH)_3^-] + 2[Zn(OH)_4^{2-}]$
     (g) $[H_3O^+] + 2[Ca^{2+}] = [OH^-] + [F^-]$
     (i) $[H_3O^+] + [NH_4^+] + 2[Cd^{2+}] + 2[Cd(NH_3)^{2+}] + 2[Cd(NH_3)_2^{2+}] + \cdots = [OH^-]$
8–21. (a) $[H_2C_2O_4] = \dfrac{[H_3O^+]^2[C_2O_4^{2-}]}{K_1 K_2}$

        $[HC_2O_4^-] = \dfrac{[H_3O^+]^2[C_2O_4^{2-}]}{K_2}$

     (b) $[Pb^{2+}] = \dfrac{[H_3O^+][C_2O_4^{2-}]}{K_1 K_2}$
                $+ \dfrac{[H_3O^+][C_2O_4^{2-}]}{K_2} + [C_2O_4^{2-}]$

     (c) $\dfrac{[C_2O_4^{2-}]}{S} = \dfrac{K_1 K_2}{[H_3O^+]^2 + K_1[H_3O^+] + K_1 K_2}$
8–25. (a) $S = 5.2 \times 10^{-3}$ M
     (c) $S = 3.6 \times 10^{-4}$ M
8–26. (a) $S = 1.5 \times 10^{-4}$ M
     (c) $S = 7.4 \times 10^{-5}$ M
8–29. (a) $S = 7 \times 10^{-1}$ M
8–30. (a) $S = 1.9 \times 10^{-6}$ M
     (b) $S = 9.3 \times 10^{-6}$ M
8–32. (a) $S = 4 \times 10^{-3}$ M
     (b) $S = 5 \times 10^{-2}$ M
     (c) $S = 1 \times 10^{-4}$ M
8–34. (a) $Cu(OH)_2$
     (b) $[OH^-] = 1.8 \times 10^{-9}$
     (c) $[Cu^{2+}] = 3.4 \times 10^{-8}$
8–36. (a) $[Ag^+] = 8.3 \times 10^{-11}$
     (b) $[Ag^+] = 1.6 \times 10^{-11}$
     (c) $[SCN^-]/[I^-] = 1.3 \times 10^4$
     (d) $[SCN^-]/[I^-] = 1.3 \times 10^4$
8–38. 1.88 g
8–40. (a) $S = 1.0 \times 10^{-2}$, 50% undissociated
     (b) $S = 7.1 \times 10^{-2}$, 70% undissociated

## Chapter 9

9–1. (b) $Ag^+(adsored) + Fl^-(aq) \rightleftarrows Ag^+Fl^-(adsored)$

**9–3.** The range of concentrations during a titration is enormous. The use of p-values compressed the concentration scale to provide a convenient presentation of the data.

**9–5.** In the Volhard method for $CO_3^{2-}$, it is necessary to filter the $Ag_2CO_3$ from the solution before back-titration with KSCN because $Ag_2CO_3$ is more soluble than AgSCN. As a consequence, $SCN^-$ present during the back-titration reacts with the precipitated $Ag_2CO_3$ and gives a fading end point:

$$Ag_2CO_3 + 2\ SCN^- \rightleftarrows 2\ AgSCN(s) + CO_3^{2-}$$

**9–7.** One of the seven chlorine atoms is titrated.

**9–9.** (a) 0.04877 M   (c) 47.76 mL   (e) 2.24 mg $COCl_2$

**9–11.** (a) 0.01930 M   (b) 0.2424 M
(c) 0.02384 M   (d) 0.09292 M

**9–12.** (a) 0.08623 M
(d) 0.09560 M
(e) 0.1077 M
(g) 0.1122 M

**9–13.** (b) 40.8 mL   (d) 38.1 mL

**9–14.** 0.09127 M

**9–15.** 71.07%

**9–17.** (a) 2.57% $As_2O_3$   (b) 15.72 mL

**9–19.** 12.33 mg $C_2H_2$/L

**9–20.** 15.60 mg saccharin/tablet

**9–22.** 2.38% $CHI_3$

**9–24.** 8.65 mg acetone

**9–27.** 18.0 ppm

**9–29.** 55.69% aldrin

**9–31.** 33.8 ppm

**9–33.** $6.0 \times 10^{-3}$ M

**9–35.** 20.54% $NH_4Cl$,   53.82% $(NH_4)_2SO_4$

**9–36.** 10.60% $Cl^-$, 55.65% $ClO_4^-$

**9–38.** 73.21% KCl, 26.79% $K_2SO_4$

**9–39.** (a) $K = 1.65 \times 10^2$

**9–40.** (a) 0.063%

**9–41.**

| Vol $AgNO_3$, mL | $[Ag^+]$ | pAg |
|---|---|---|
| 5.00 | $1.63 \times 10^{-11}$ | 10.79 |
| 40.00 | $7.2 \times 10^{-7}$ | 6.14 |
| 45.00 | $2.6 \times 10^{-3}$ | 2.58 |

**9–42.**

| Vol $AgNO_3$ mL | $[Ag^+]$ (a) | (c) |
|---|---|---|
| 5.00 | $1.30 \times 10^{-15}$ | $2.86 \times 10^{-9}$ |
| 20.00 | $2.90 \times 10^{-15}$ | $6.37 \times 10^{-9}$ |
| 30.00 | $6.64 \times 10^{-15}$ | $1.46 \times 10^{-8}$ |
| 35.00 | $1.41 \times 10^{-14}$ | $3.09 \times 10^{-8}$ |
| 39.00 | $7.39 \times 10^{-14}$ | $1.62 \times 10^{-7}$ |
| 40.00 | $9.11 \times 10^{-9}$ | $1.35 \times 10^{-5}$ |
| 41.00 | $1.10 \times 10^{-3}$ | $1.10 \times 10^{-3}$ |
| 45.00 | $5.26 \times 10^{-3}$ | $5.26 \times 10^{-3}$ |
| 50.00 | $1.00 \times 10^{-2}$ | $1.00 \times 10^{-2}$ |

# Chapter 10

**10–1.** The *equivalence point* in a titration is the theoretical point in the titration when an amount of reagent has been added that is chemically equivalent to the amount of analyte. The *end point* in a titration is that point where an observable physical change occurs that is related to the condition of chemical equivalence.

**10–3.** Nitric acid is seldom used as a standard because it is an oxidizing agent and thus will react with reducible species in titration mixtures.

**10–5.** Temperature, ionic strength, and the presence of organic solvents and colloidal particles

**10–6.** (a) bromocresol green, bromothymol blue, or phenolphthalein
(c) bromothymol blue
(e) phenolphthalein

**10–7.** cresol purple or phenol red

**10–9.** Prior to the equivalence point, the two curves differ because of the buffering effect of the HOCl/$OCl^-$ in the solution. After the equivalence point, the curves are essentially identical because all HOCl has been converted to $OCl^-$. In both curves, the pH changes as though a strong base is being added to a neutral solution.

**10–11.** (a) Because $[H_3O^+]$ and $[OH^-]$ are inversely related, if $[H_3O^+]$ is appreciable, $[OH^-]$ is not, and vice versa.
(b) Both $[H_3O^+]$ and $[OH^-]$ can be neglected when pH $\approx$ 7 or when $c_{HA}$ and $c_{A^-}$ are relatively large.

**10–12.** (a) acidic   (c) basic   (e) basic

**10–14.** (a) $[HOCl]/[OCl^-] = 2.2$
(c) $[HONH_3^+]/[HONH_2] = 7.2 \times 10^{-2}$
(e) $[(CH_3)_3CCOOH]/[(CH_3)_3CCOO^-] = 7.2 \times 10^{-3}$

**10–15.** (a) benzoic acid/benzoate or anilinium/amiline
(c) hydrogen phthalate/phthalate or pyridinium/pyridine or propanic acid/propionate
(e) $NH_4^+/NH_3$   or   $HOC_2H_4NH_3^+/HOC_2H_4NH_2$
(g) $H_2NNH_3^+/H_2NNH_2$   or   $HCN/CN^-$
(i) $HCO_3^-/CO_3^{2-}$   or butylammonium/butylamine or methyl ammonium/methyl amine

**10–16.** KOH > $CH_3NH_2$ > $(CH_3)_3N$ > KCN > HOCl

**10–18.** (a) pH = 1.35
(c) pH = 1.81
(e) pH = 12.60

**10–19.** (a) pH = 1.44
(c) pH = 9.39

**10–20.** (a) pH = 4.26
(b) pH = 4.76
(c) pH = 5.76

**10–21.** (a) pH = 10.26
(b) pH = 9.76
(c) pH = 8.75

**10–24.** (a) pH = 12.03

**10–25.** (b) pH = 2.02

**10–26.** (c) pH = 4.00

**10–27.** (a) pH = 1.30
(b) pH = 1.37

**10–29.** (a) pH = 2.41
(b) pH = 12.35
(c) pH = 8.35
(d) pH = 3.97
(e) pH = 3.85

**10–31.** (a) pH = 3.61

(b) pH = 9.32
(c) pH = 10.92
(d) pH = 6.68
(e) pH = 1.74

10–33. (a) $\Delta$pH = 0
(c) $\Delta$pH = −1.000
(e) $\Delta$pH = 0.500
(g) $\Delta$pH = 0.000

10–35. (a) $\Delta$pH = 5.00
(b) $\Delta$pH = 0.097
(c) $\Delta$pH = 0.079
(d) $\Delta$pH = 1.032
(e) $\Delta$pH = 3.37
(f) $\Delta$pH = 0.176
(g) $\Delta$pH = 0.018

10–37. (a) $\alpha_0$ = 0.88     $\alpha_1$ = 0.12
(b) $\alpha_0$ = 0.19     $\alpha_1$ = 0.81
(c) $\alpha_0$ = 0.0048     $\alpha_1$ = 0.9952

10–39. (a) $\alpha_0$ = $6.7 \times 10^{-4}$     $\alpha_1$ = 1.00
(b) $\alpha_0$ = 0.054     $\alpha_1$ = 0.946
(c) $\alpha_0$ = 0.746     $\alpha_1$ = 0.254

10–42. 152.3 g sodium formate

10–44. 74.6 mL

10–46.

| Vol, mL | pH |
|---|---|
| 0.00 | 13.00 |
| 10.00 | 12.82 |
| 25.00 | 12.52 |
| 40.00 | 12.05 |
| 45.00 | 11.72 |
| 49.00 | 11.00 |
| 50.00 | 7.00 |
| 51.00 | 3.00 |
| 55.00 | 2.32 |
| 60.00 | 2.04 |

10–48. (b)

| Vol, mL | pH |
|---|---|
| 0.00 | 9.51 |
| 5.00 | 6.98 |
| 15.00 | 6.40 |
| 25.00 | 6.03 |
| 40.00 | 5.43 |
| 45.00 | 5.08 |
| 49.00 | 4.34 |
| 50.00 | 3.67 |
| 51.00 | 3.00 |
| 55.00 | 2.32 |
| 60.00 | 2.04 |

# Chapter 11

11–1. (a) $H_2A + H_2O \rightleftarrows H_3O^+HA^-$     $K_1 = 10^{-2}$
$HA^- + H_2O \rightleftarrows H_3O^+A^{2-}$     $K_2 = 10^{-6}$
$c_{H_2A} = [H_2A] + [HA^-] + [A^{2-}]$
We can usually assume that $[A^{2-}] \ll [H_2A]$ + $[HA^-] \approx c_{H_2A}$. If this is correct;

$[HA^-] \approx [H_3O^+]$

Substituting this relationship into the expression for $K_2$ gives

$$K_2 = 10^{-6} = \frac{[H_3O^+][A^-]}{[HA^-]}$$

Thus, the assumption is valid if $c_{H_2A} \geq 1 \times 10^{-4}$.

(b) The data in Table 6–4, page 134, show that for an acid with a dissociation constant of $10^{-2}$, the following errors are incurred by not employing the quadratic equation.

| $c_{HA}$, M | Error, % |
|---|---|
| $1.00 \times 10^{-3}$ | 244 |
| $1.00 \times 10^{-2}$ | 61 |
| $1.00 \times 10^{-1}$ | 17 |

(c) A mixture of $H_2A$ and $Na_2A$ reacts to form $HA^-$:

$$H_2A + A^{2-} \rightleftarrows 2 HA^-   \qquad K = \frac{[HA^-]^2}{[H_2A][A^{2-}]}$$

The equilibrium constant for this reaction is given by $K = K_1/K_2$, which is generally a reasonably large positive number since $K_2 \ll K_1$. Thus little $H_2A$ and $A^{2-}$ exist in a mixture of $H_2A$ and $Na_2A$.

11–3. (a) bromocresol green
(b) methyl yellow
(c) bromocresol green
(d) bromocresol green
(e) methyl yellow
(f) phenolphthalein

11–5.

| | Approximate Method | Quadratic Method |
|---|---|---|
| (a) | 2.28 | 2.33 |
| (b) | 1.73 | 1.93 |
| (c) | 2.32 | 2.38 |
| (d) | 4.03 | 4.03 |

11–7. (a) 4.28
(b) 4.57
(c) 4.18
(d) 8.34

11–9. (a) 9.05
(b) 9.49
(c) 8.90
(d) 11.36   (11.34 by quadratic method)

11–11. (a) 1.59   (1.53 by approximate method)
(b) 4.69
(c) 9.78
(d) 12.73   (12.63 by quadratic method)

11–13. (a) $H_2PO_4^-$
(b) $H_2PO_4^-/H_3PO_4$   (in a ratio of 2 : 3)
(c) $H_2PO_4^-/HPO_4^-$   (in a ratio of 2 : 1)

**(d)** $HPO_4^{2-}/H_2PO_4$   (in a ratio of 2:5)
**(e)** $H_2PO_4^-/HPO_4^{2-}$   (in a ratio of 9:1)

**11–15.** (a) 2.76   (2.82 by quadratic method)
     (b) 4.55
     (c) 7.19
     (d) 9.92

**11–17.** (a) 2.36   (2.18 by approximate method)
     (b) 6.41
     (c) 6.72
     (d) 2.34   (2.22 by approximate method)

**11–19.** (a) 7.10
     (b) 2.41   (2.24 by approximate method)
     (c) 8.10
     (d) 11.90   (12.10 by approximate method)

**11–21.** (a) 2.92   (2.89 by approximate method)
     (b) 5.47

**11–23.** 14.8 g

**11–25.** (a) Mix 438 mL of $NaH_2PO_4$ with 62 mL of $H_3PO_4$.
     (b) Mix 223 mL of NaOH with 207 mL of $H_3PO_4$.
     (c) Mix 13.7 mL of HCl with 486 mL of $NaH_2PO_4$.
     (d) Mix 408 mL of $Na_2HPO_4$ with 92 mL of HCl.

**11–27.** (a) $\alpha_0 = 3.13 \times 10^{-3}$
       $\alpha_1 = 0.989$
       $\alpha_2 = 8.13 \times 10^{-3}$
     (b) $\alpha_0 = 1.74 \times 10^{-2}$
       $\alpha_1 = 0.977$
       $\alpha_2 = 5.78 \times 10^{-3}$
     (c) $\alpha_0 = 7.06 \times 10^{-2}$
       $\alpha_1 = 0.650$
       $\alpha_2 = 0.280$
     (d) $\alpha_0 = 4.02 \times 10^{-4}$
       $\alpha_1 = 0.960$
       $\alpha_2 = 4.01 \times 10^{-2}$
       $\alpha_3 = 4.79 \times 10^{-8}$
     (e) $\alpha_0 = 1.45 \times 10^{-12}$
       $\alpha_1 = 2.75 \times 10^{-4}$
       $\alpha_2 = 0.913$
       $\alpha_3 = 8.66 \times 10^{-2}$

**11–29.**

| mL, Reagent | pH | mL, Reagent | pH |
|---|---|---|---|
| 0.00 | 1.00 | 35.00 | 7.25 |
| 10.00 | 1.30 | 44.00 | 8.80 |
| 20.00 | 1.84 | 45.00 | 10.07 |
| 24.00 | 2.57 | 46.00 | 11.31 |
| 25.00 | 4.40 | 50.00 | 12.00 |
| 26.00 | 6.24 | | |

**11–31.**

| | pH | | | pH | |
|---|---|---|---|---|---|
| mL, Reagent | (a) | (c) | mL, Reagent | (a) | (c) |
| 0.00 | 11.66 | 0.96 | 37.50 | 6.35 | 2.14 |
| 12.50 | 10.33 | 1.26 | 45.00 | 5.75 | 2.65 |
| 20.00 | 9.73 | 1.48 | 49.00 | 4.97 | 3.40 |
| 24.00 | 8.95 | 1.61 | 50.00 | 3.83 | 7.31 |
| 25.00 | 8.34 | 1.64 | 51.00 | 2.70 | 11.30 |
| 26.00 | 7.73 | 1.67 | 60.00 | 1.74 | 12.26 |

## Chapter 12

**12–1.** When carbon dioxide is dissolved in water, it is largely present as $CO_2$ and $H_2CO_3$, which upon heating decomposes to give $CO_2$ (and $H_2O$). The $CO_2$ is not strongly bonded by water molecules, however, and thus is readily volatilized from aqueous media. Gaseous HCl molecules, on the other hand, are fully dissociated into $H_3O^+$ and $Cl^-$ when dissolved in water; neither of these species is volatile.

**12–3.** The solvent autoprotolysis constant, the strength of the solvent as acid or base, and the dielectric constant of the solvent influence the choice of a solvent for nonaqueous acid/base titration.

**12–5.** The *equivalent weight* of an acid is that weight which contains one mole of titratable protons. The stoichiometry of the reaction in question must be known in order to use equivalent weights.

**12–7.** (a) Dilute 750 mL of 0.400 M $HClO_4$ to 2.5 L.
     (b) Dilute about 14 mL of $H_3PO_4$ to 2.5 L.
     (c) Dilute about 26 mL of concentrated $NH_3(aq)$ to 2.5 L.
     (d) Dissolve about 4.3 g $Ba(OH)_2$ in sufficient $H_2O$ to give 2.5 L of solution.

**12–9.** Dilute 43.26 g of HCl solution to 2000 mL.

**12–11.** 0.1402 M HCl

**12–13.** 0.1000 M $HClO_4$;   0.9301 M NaOH

**12–15.** 0.08974 M KOH

**12–17.**

| $c_{KOH}$, M |
|---|
| 0.1083 |
| 0.1097 |
| 0.1079 |
| 0.1085 |

     $s_c = 7.8 \times 10^{-4}$ M

**12–19.** 81.81% $Na_2CO_3$

**12–21.** 40.1 mg HOAc/mL

**12–23.** (a) 0.0544 M
     (b) 33.7 mL
     (c) 0.1152 M
     (d) 31.5 mL

**12–25.** 26.7% $ThSiO_4$

**12–26.** 3353 ppm $CO_2$

**12–28.** 76.0% biguanide

**12–29.** 20.55% formaldehyde

**12–31.** 56.59% dimethyl phthalate

**12–33.** (a) 60.65 mL
     (b) 28.96 mL
     (c) 58.37 mL

**12–35.** (a) $H_3PO_4$   ($V_2 = 2V_1$)
     (b) HCl   ($V_2 = V_1$)
     (c) $H_2PO_4^-$   ($V_2 = 0$)
     (d) HCl + $H_3PO_4$   ($V_2 < 2V_1$)
     (e) $H_3PO_4$ + $HPO_4^-$   ($V_2 > 2V_1$)

**12–37.** 28.50% $NaHCO_3$,   44.19% $Na_2CO_3$

# Chapter 13

**13-1.** (a) A *chelate* is a cyclic complex consisting of a metal ion and a reagent that contains two or more electron-donor groups located in such a position that they can bond with the metal ion to form a heterocyclic ring.

(c) A *ligand* is a species that contains one or more electron-pair donor groups that tend to form bonds with metal ions.

(e) A *conditional formation constant* is a nonthermodynamic equilibrium constant for the reaction between a metal ion and a complexing agent that applies only when the pH and/or the concentration of other complexing ions is carefully specified.

(g) The hardness of water is the concentration of calcium carbonate that is equivalent to the total concentration of all multivalent metal carbonates in the water.

**13-2.** Three general methods for performing EDTA titrations are (1) *direct titration*, (2) *back-titration*, and (3) *displacement titration*. Method 1 is simple and rapid and requires but one standard reagent. Method 2 is advantageous for those metals that react so slowly with EDTA as to make direct titration inconvenient. In addition, this procedure is useful for cations for which a satisfactory indicator is not available. Also, Method 2 is useful for analyzing samples that contain anions that form sparingly soluble precipitates with the analyte under the analytical conditions. Method 3 is particularly useful in situations where no satisfactory indicators are available for direct titration.

**13-4.** Multidentate ligands offer the advantage that they usually form more stable complexes than do unidentate ligands. Furthermore, they often form but a single complex with the cation, which simplifies their titration curves and makes end-point detection easier.

**13-5.** (a) $Ag^+ + S_2O_3^{2-} \rightleftharpoons Ag(S_2O_3)^- \quad K_1 = \dfrac{[Ag(S_2O_3)^-]}{[Ag^+][S_2O_3^{2-}]}$

$Ag(S_2O_3)^- + S_2O_3^{2-} \rightleftharpoons Ag(S_2O_3)_2^{3-}$

$$K_2 = \frac{[Ag(S_2O_3)_2^{3-}]}{[Ag(S_2O_3)^-][S_2O_3^{2-}]}$$

(c) $Cd^{2+} + NH_3 \rightleftharpoons Cd(NH_3)^{2+} \quad K_1 = \dfrac{[CdNH_3^{2+}]}{[Cd^{2+}][NH_3]}$

$Cd(NH_3)^{2+} + NH_3 \rightleftharpoons Cd(NH_3)_2^{2+}$

$$K_2 = \frac{[Cd(NH_3)_2^{2+}]}{[Cd(NH_3)^{2+}][NH_3]}$$

$Cd(NH_3)_2^{2+} + NH_3 \rightleftharpoons Cd(NH_3)_3^{2+}$

$$K_3 = \frac{[Cd(NH_3)_3^{2+}]}{[Cd(NH_3)_2^{2+}][NH_3]}$$

$Cd(NH_3)_3^{2+} + NH_3 \rightleftharpoons Cd(NH_3)_4^{2+}$

$$K_4 = \frac{[Cd(NH_3)_4^{2+}]}{[Cd(NH_3)_3^{2+}][NH_3]}$$

**13-7.** The $MgY^{2-}$ is added to assure a sufficient analytical concentration of $Mg^{2+}$ to provide a sharp end point with Eriochrome Black T indicator.

**13-8.** 0.01032 M

**13-10.** 0.01005 M

**13-12.** 323.9 mg of $Ca^{2+}$ and 256.4 mg of $Mg^{2+}$; both normal.

**13-14.** 94.40%

**13-16.** 31.48% $NaBr$ and 48.58% $NaBrO_3$

**13-18.** 0.08431 M

**13-20.** 7.515% Pb;  4.304% Mg;  9.456% Zn

**13-22.** 8.517% Pb;  24.85% Zn;  64.07% Cu;  2.56% Sn

**13-23.** (a) $1.4 \times 10^9$  (b) $3.3 \times 10^{11}$  (c) $2.2 \times 10^{13}$

**13-25.** (a) $K''_{Cd} = 8.1 \times 10^{12}$
(b) $K''_{Cd} = 1.3 \times 10^{14}$
(c) $K''_{Cd} = 4.9 \times 10^9$
(d) $K''_{Cd} = 8.1 \times 10^9$

**13-27.** $K'_{Sr} = 3.66 \times 10^8$

| Vol, mL | pSr | Vol, mL | pSr |
|---------|------|---------|------|
| 0.00 | 2.00 | 25.00 | 5.37 |
| 10.00 | 2.30 | 25.10 | 6.16 |
| 24.00 | 3.57 | 26.00 | 7.16 |
| 24.90 | 4.57 | 30.00 | 7.86 |

**13-29.** $K''_{Co} = 5.74 \times 10^{13}$

| Vol, mL | pCo | Vol, mL | pCo |
|---------|------|---------|------|
| 0.00 | 2.66 | 20.00 | 8.96 |
| 5.00 | 2.86 | 21.00 | 13.72 |
| 10.00 | 3.10 | 30.00 | 14.72 |
| 19.00 | 4.20 | | |

# Chapter 14

**14-1.** (a) *Oxidation* involves the loss of electrons by a chemical species and the promotion of the species losing the electrons to a higher oxidation state.

(c) The *standard hydrogen electrode* consists of a platinized platinum surface in a solution that has hydrogen-ion activity of 1.00, with hydrogen at 1.00 atm being continuously passed through to keep the solution saturated. The potential of this electrode is assigned a value of 0.000 V at all temperatures.

(e) The *Nernst equation* describes the relation between the potential of a half-cell process and the concentrations (strictly, the activities) of the species involved in the process.

**14-2.** (a) An *oxidizing agent* abstracts electrons from some other species and is reduced in the process. A *reducing agent* is promoted to a higher oxidation state as a result of electron donation.

(c) A change in the direction of current in a *reversible* electrochemical cell simply reverses the direction of the cell reaction. The same alteration in an *irreversible* cell results in an entirely different overall cell reaction.

(e) The *standard electrode potential* is the potential of a half-cell process when all participants have *unit activity* (strictly, it is the potential of an electrochemical cell in which the standard hydrogen electrode, with a potential of 0.000 V by definition, acts as an anode). The *formal potential* of a system is the potential of the system relative to the standard hydrogen electrode when the *concentrations* of the reactants and products are 1M and the concentrations of all other species in the solution are carefully specified.

**14–3.** (a) $Br_2 + Sn^{2+} \rightleftarrows 2\ Br^- + Sn^{4+}$
(c) $IO_3^- + 5\ I^- + 6\ H^+ \rightleftarrows 3\ I_2 + 3\ H_2O$

**14–4.** (a) $Fe(s) + 3\ NO_3^- + 6\ H^+ \rightleftarrows Fe^{3+} + 3\ NO_2(g) + 3\ H_2O$
(c) $2\ MnO_4^- + 5\ H_2C_2O_4 + 6\ H^+ \rightleftarrows 2\ Mn^{2+} + 10\ CO_2(g) + 8\ H^+$
(d) $2\ V(OH)_4^+ + V^{2+} + 2\ H^+ \rightleftarrows 3\ VO^{2+} + 5\ H_2O$
(e) $V(OH)_4^+ + 2\ V^{2+} + 4\ H^+ \rightleftarrows 3\ V^{3+} + 4\ H_2O$

**14–5.** (a)

| | Oxidizing Agent | Reducing Agent |
|---|---|---|
| (1) | $Cr^{3+}$ | $Zn(s)$ |
| (2) | $Sn^{4+}$ | $Cr^{2+}$ |
| (3) | $I_2$ | $Sn^{2+}$ |
| (4) | $HNO_2$ | $I^-$ |

(b,c) (1) $Cr^{3+} + e^- \rightleftarrows Cr^{2+}$
$\quad\quad Zn^{2+} + 2\ e^- \rightleftarrows Zn(s)$
(2) $Sn^{4+} + 2\ e^- \rightleftarrows Sn^{2+}$
$\quad\quad Cr^{3+} + e^- \rightleftarrows Cr^{2+}$
(3) $I_2 + 2\ e^- \rightleftarrows 2\ I^-$
$\quad\quad Sn^{4+} + 2\ e^- \rightleftarrows Sn^{2+}$
(4) $HNO_2 + H^+ + e^- \rightleftarrows NO(g) + H_2O$
$\quad\quad I_2 + 2\ e^- \rightleftarrows 2\ I^-$
(d) $HNO_2 + H^+ + e^- \rightleftarrows NO(g) + H_2O$
$\quad\quad I_2 + 2\ e^- \rightleftarrows 2\ I^-$
$\quad\quad Sn^{4+} + 2\ e^- \rightleftarrows Sn^{2+}$
$\quad\quad Cr^{3+} + e^- \rightleftarrows Cr^{2+}$
$\quad\quad Zn(s) + 2\ e^- \rightleftarrows Zn^{2+}$

**14–7.** (a) no reaction
(b) reaction
(c) reaction
(d) no reaction

**14–9.** (a) +0.732 V
(b) −0.0195 V
(c) +0.172 V
(d) +0.058 V
(e) +0.698 V
(f) +0.323

**14–11.** (a) −0.377 V
(b) +0.33 V

(c) +0.757 V
(d) −0.267 V

**14–13.** (a) +0.169 V
(b) −0.701 V
(c) +1.155 V
(d) +0.451 V
(e) −0.225 V
(f) +0.091 V

**14–15.** (a) −0.162 V (anode)
(b) +0.174 V (cathode)
(c) −0.126 V (anode)
(d) −0.0514 V (anode)
(e) +1.185 V (cathode)
(f) −1.221 V (anode)
(g) −0.045 V (anode)

**14–17.** (a) +0.989 V galvanic
(b) −0.211 V electrolytic
(c) +0.096 V galvanic
(d) +0.170 V galvanic
(e) +0.727 V galvanic
(f) +0.139 V galvanic
(g) −0.443 V electrolytic
(h) −0.333 V electrolytic
(i) +0.055 V galvanic
(j) −0.992 V electrolytic

## Chapter 15

**15–1.** (a) *Equilibrium* is the state characterized by the apparent completion of the reaction in question; that is, all measurable physical and chemical variables are constant. *Equivalence* is the point in a redox reaction when the number of moles of electrons lost by the reductant is equal to the number of moles of electrons gained by the oxidant.

(b) *General oxidation-reduction indicators* respond to the change in electrode potential of the system being titrated; ordinarily, their behavior is independent of the chemical nature of the analyte and the reagent. *Specific indicators*, in contrast, change color as a consequence of their reaction with one of the species involved in the titration.

**15–3.** The system potential $E_{system}$ can be calculated from either $E_{red}$ or $E_{ox}$ because $E_{system} = E_{ox} = E_{red}$.

**15–5.** Titration curves are asymmetric when the numbers of electrons in the two half-reactions are unequal; that is, when $n \neq m$.

**15–7.** $E_{system}$ is easier to calculate from the analyte because the analyte is present in excess prior to the equivalence point. The concentrations of the oxidized and reduced forms of the analyte are readily calculated from the volumes and concentrations of the solutions.

**15–8.** (a) $K_{eq} = \dfrac{[Cd^{2+}]}{[Pb^{2+}]} = 2.3(\pm 0.2) \times 10^9$

(c) $K_{eq} = \dfrac{[Cr^{3+}]^2[Sn^{2+}]}{[Cr^{2+}]^2[Sn^{4+}]} = 9.7(\pm 0.8) \times 10^{18}$

(e) $K_{eq} = \dfrac{[Pb^{2+}]^2}{[BiO^+][H^+]^4} = 1.6(\pm 0.4) \times 10^{45}$

(g) $K_{eq} = \dfrac{[V^{3+}]}{[VO^{2+}][V^{2+}][H^+]^2} = 2.4(\pm 0.1) \times 10^{10}$

(i) $K_{eq} = \dfrac{[I^-][H^+]^2}{P_{H_2S}[I_3^-]} = 2.2(\pm 0.2) \times 10^{13}$

**15–9.** (a) $E_{eq} = 0.258$ V
(b) $E_{eq} = 0.58$ V
(c) $E_{eq} = 0.83$ V
(d) $E_{eq} = 0.50$ V
(e) $E_{eq} = 0.56$ V
(f) $E_{eq} = 1.15$ V

**15–11.** (a) phenosafranine
(b) methylene blue
(c) diphenylamine sulfonic acid
(e) methylene blue
(f) 1,10-phenanthroline iron(II) or diphenylamine dicarboxylic acid

**15–13.** (a) $K_{eq} = 2.2(\pm 0.1) \times 10^{17}$
(b) $K_{eq} = 3(\pm 2) \times 10^{43}$
(c) $K_{eq} = 8(\pm 6) \times 10^{38}$
(d) $K_{eq} = 5(\pm 4) \times 10^{11}$
(e) $K_{eq} = 7(\pm 3) \times 10^{10}$
(f) $K_{eq} = 5 \times 10^{56}$ $(4 \times 10^{55} - 5 \times 10^{57})$

**15–15.** (a) $[Fe^{2+}] = [V^{3+}] = 0.0100$ M, $[V^{2+}] = [Fe^{3+}] = 2 \times 10^{-11}$ M
(b) $[Ce^{3+}] = 0.0133$ M, $[Sn^{4+}] = 0.00666$ M
$[Ce^{4+}] = 4 \times 10^{-17}$ M, $[Sn^{2+}] = 2 \times 10^{-17}$ M
(c) $[Tl^+] = 0.00666$ M, $[TiO^+] = 0.0133$ M
$[Tl^{3+}] = 4 \times 10^{-17}$ M, $[Ti^{3+}] = 8 \times 10^{-17}$ M
(d) $[PtCl_4^{2-}] = [UO_2^{2+}] = 0.0100$ M, $[Cl^-] = 0.0200$ M, $[H^+] = 0.100$ M
$[PtCl_6^{2-}] = [U^{4+}] = 3 \times 10^{-12}$ M
(e) $[VO^{2+}] = 0.0200$ M, $[V(OH)_4^+] = [V^{3+}] = 8 \times 10^{-8}$ M
(f) $[Fe^{3+}] = 1.71 \times 10^{-2}$ M, $[Cr^{3+}] = 5.71 \times 10^{-3}$ M, $[H^+] = 0.100$ M
$[Fe^{2+}] = 7 \times 10^{-9}$ M, $[Cr_2O_7^{2-}] = 1 \times 10^{-9}$ M

**15–17.**

| Vol titrant, mL | (a) $E_{system}$ | (c) $E_{system}$ | (e) $E_{system}$ |
|---|---|---|---|
| 10.00 | −0.444 | 0.32 | 0.223 |
| 25.00 | −0.408 | 0.36 | 0.259 |
| 40.00 | −0.372 | 0.40 | 0.276 |
| 49.00 | −0.308 | 0.46 | 0.341 |
| 49.90 | −0.249 | 0.52 | 0.400 |
| 50.00 | 0.182 | 0.95 | 0.56 |
| 50.10 | 0.611 | 1.17 | 0.70 |
| 51.00 | 0.671 | 1.20 | 0.76 |
| 60.00 | 0.730 | 1.23 | 0.82 |

## Chapter 16

**16–1.** (a) $2 H_2MoO_4 + 12 H^+ + 3 Zn(Hg) \rightarrow 2 Mo^{3+} + 3 Zn^{2+} + 3 Hg + 8 H_2O$
(b) $2 V(OH)_4^+ + 8 H^+ + 3 Zn(Hg) \rightarrow 2 V^{2+} + 3 Zn^{2+} + 3 Hg + 8 H_2O$
(c) $2 Cr^{3+} + 3 S_2O_8^{2-} + 7 H_2O \rightarrow Cr_2O_7^{2-} + 6 SO_4^{2-} + 14 H^+$
(d) $4 MnO_4^- + 3 N_2H_5^+ + H^+ \rightarrow 4 MnO_2 + 3 N_2 + 8 H_2O$
(e) $IO_3^- + 5 I^- + 6 H^+ \rightarrow I_2 + 3 H_2O$
(f) $H_3PO_3 + I_3^- + H_2O \rightarrow H_3PO_4 + 3 I^- + 2 H^+$
(g) $3 CN^- + 2 MnO_4^- + 2 H^+ \rightarrow 2 MnO_2 + 3 CNO^- + H_2O$
(h) $5 I^- + 2 MnO_4^- + 10 CN^- + 16 H^+ \rightarrow 5 I(CN)_2^- + 2 Mn^{2+} + 8 H_2O$

**16–3.** $Cr_2O_7^{2-} + 6 Fe^{2+} + 14 H^+ \rightarrow 2 Cr^{3+} + 6 Fe^{3+} + 7 H_2O$

**16–5.** $BrO_3^- + 6 I^- + 6 H^+ \rightarrow 3 I_2 + Br^- + 3 H_2O$
$3 I_2 + 6 S_2O_3^{2-} \rightarrow 6 I^- + 3 S_4O_6^{2-}$

**16–7.** 0.1226 g Zn(II)

**16–9.** 0.219 g $(NH_4)_2S_2O_8$

**16–11.** (a) 0.02443 g Fe
(b) 0.1222 g Fe
(c) 0.0977 g Fe
(d) 0.1466 g Fe

**16–13.** (a) 11.12 mL $KMnO_4$
(b) 174.3 mL $KMnO_4$
(c) 60.22 mL $KMnO_4$
(d) 21.00 mL $KMnO_4$

**16–15.** 3.2 g $KMnO_4$

**16–17.** 0.234 g $Na_2C_2O_4$

**16–19.** 37.35% CaO

**16–21.** 2.77% $As_2O_3$

**16–23.** 5.25% NaOCl

**16–25.** 20.4% Cr
43.93% $FeO \cdot Cr_2O_3$

**16–27.** 11.72% $C_2H_5OH$

**16–29.** 9.6 ppm $O_2$; yes

**16–31.** 1.72 ppm $SO_2$; yes

**16–33.** 325 mg/tablet

**16–35.** 0.585 g $I_2$/100 mL
0.768 g KI/100 mL

**16–37.** (a) **Walden reductor reactions:**

$Fe^{3+} + Ag(s) + Cl^- \rightarrow Fe^{2+} + AgCl(s)$

$5 Fe^{2+} + MnO_4^- + 8 H^+ \rightarrow 5 Fe^{3+} + Mn^{2+} + 4 H_2O$

**Jones reductor reactions:**

$2 Fe^{3+} + Zn(Hg) \rightarrow 2 Fe^{2+} + Zn^{2+} + Hg(l)$

$2 Cr^{3+} + Zn(Hg) \rightarrow 2 Cr^{2+} + Zn^{2+} + Hg(l)$

$5 Cr^{2+} + MnO_4^- + 8 H^+ \rightarrow 5 Cr^{3+} + Mn^{2+} + 4 H_2O$

(b) 50.02% Fe; 13.52% Cr

## Chapter 17

**17-1.** An *electrode of the first kind* is a metal electrode that is used to determine the concentration of the ion derived from that metal. For example, a copper electrode is an electrode of the first kind for determining the concentration of $Cu^{2+}$ ions in a solution.

An *electrode of the second kind* is a metal electrode that responds to the concentration of an anion that forms a precipitate of limited solubility or a stable complex with the cation derived from that metal. For example, a silver electrode serves as an electrode of the second kind for $Br^-$ ions. In this application, the analyte solution is saturated with AgBr before measurements are made.

**17-3. (b)** The *boundary potential* for a membrane electrode develops when the membrane separates two solutions that have different concentrations of a cation or an anion that the membrane selectively binds. For an aqueous solution the following equilibria develop when the membrane is positioned between two solutions of $A^+$

$$\underset{\text{membrane}_1}{A^+M^-} \rightleftarrows \underset{\text{soln}_1}{A^+} + \underset{\text{membrane}_1}{M^-}$$

$$\underset{\text{membrane}_2}{A^+M^-} \rightleftarrows \underset{\text{soln}_2}{A^+} + \underset{\text{membrane}_2}{M^-}$$

where the subscripts refer to the two sides of the membrane. A potential develops across this membrane if one of these equilibria proceeds further to the right than the other, and this potential is the boundary potential. For example, if the concentration of $A^+$ is greater in solution 1 than in solution 2, the negative charge on side 1 will lie farther to the left. Thus, a greater fraction of the negative charge on side 1 will be neutralized by $A^+$.

**(d)** The membrane in a solid-state electrode for $F^-$ is crystalline $LaF_3$, which when immersed in aqueous solution dissociates to a slight extent according to the equation

$$LaF_3 \rightleftarrows La^{3+} + 3\ F^-$$

Thus, a boundary potential develops across this membrane when it separates two solutions with different $F^-$ ion concentrations. The source of this potential is described in part (b) of this answer.

**17-4.** The *alkaline error* in a glass electrode develops in a solution having a low concentration of $H_3O^+$ and a high concentration of an alkali metal ion. Under these circumstances, the electrode begins to respond to the alkali ion concentration as well as the $H_3O^+$ concentration. A negative pH error results.

**17-6.** The equilibrium involved is

$$C_2H_5NH_2 + H_2O \rightleftarrows C_2H_5NH_3^+ + OH^-$$

$$K_b = \frac{[OH^-][C_2H_5NH_3^+]}{[C_2H_5NH_2]}$$

At half-neutralization,

$$[C_2H_5NH_3^+] = [C_2H_5NH_2]$$

and the base dissociation constant expression simplifies to

$$K_b = [OH^-]$$

Taking the negative logarithm of this equation yields

$$-\log K_b = -\log [OH^-] = pOH = 14.00 - pH$$
$$K_b = \text{antilog (pH} - 14.00)$$

**17-7. (a)** $E^0_{CuSCN} = -0.327$ V
**(b)** $SCE \| CuSCN(\text{sat'd}), SCN^-(x\ M)|Cu$
$pSCN = (E_{cell} + 0.571)/0.0592$
**(c)** $pSCN = 8.36$

**17-9. (a)** $SCE \| Hg_2Cl_2(\text{sat'd}), Cl^-(x\ M)|Hg$

$$pCl = (E_{cell} - 0.024)/0.0592$$

**(b)** $SCE \| Ag_2CrO_4(\text{sat'd}), CO_3^{2-}(x\ M)|Ag$

$$E^0_{Ag_2CO_3} = 0.471\ V$$
$$pCO_3 = (E_{cell} - 0.227)/0.0296$$

**(c)** $SCE \| Sn^{4+}(x\ M), Sn^{2+}(1.00 \times 10^{-4}\ M)|Pt$

$$pSn(IV) = (0.028 - E_{cell})/0.0296$$

**17-11.** $pCrO_4 = 6.76$

**17-13. (a)** 0.157 V   **(b)** −0.026 V   **(c)** 0.047 V

**17-15. (a)** pH = 12.629  and  $a_{H^+} = 2.35 \times 10^{-13}$
**(b)** pH = 5.579  and  $a_{H^+} = 2.64 \times 10^{-6}$
**(c)** For (a), pH = 12.63 (±0.03)
$a_{H^+} = 2.3\ (\pm0.2) \times 10^{-13}$
For (b), pH = 5.58 (± 0.03)
$a_{H^+} = 2.6\ (\pm0.2) \times 10^{-6}$

**17-17.** $K_{sp} = 2.0 \times 10^{-6}$

**17-19. (a)**

| mL, Reagent | E vs. SCE | mL, Reagent | E vs. SCE |
|---|---|---|---|
| 5.00 | 0.58 | 49.00 | 0.66 |
| 10.00 | 0.59 | 50.00 | 0.80 |
| 15.00 | 0.60 | 51.00 | 1.10 |
| 25.00 | 0.61 | 55.00 | 1.14 |
| 40.00 | 0.62 | 60.00 | 1.15 |

**17-21.** pH = 2.40

## Chapter 18

18–1. (a) A coulometric titration is an electroanalytical method in which a constant current of known magnitude is used to generate a reagent that reacts with the analyte (in some instances, the analyte itself may react partially at the working electrode). The time required to complete the reaction between the electrogenerated reagent and the analyte is measured.

(c) Concentration polarization is encountered when the current in an electrochemical cell is limited by the rate at which reactants are brought to or removed from the surface of one or both electrodes. Under this circumstance, the current in the cell is no longer directly proportional to the cell potential.

(e) The faraday is the quantity of charge (electricity) that produces one equivalent of chemical change at an electrode. It consists of 1 mol, or $6.02 \times 10^{23}$, electrons.

(g) A controlled-cathode-potential electrolysis is one in which the cathode potential is monitored continuously against a reference electrode and the cell potential is adjusted to maintain the cathode at a fixed, predetermined level.

(i) A cathode depolarizer is a chemical species introduced into an electrochemical cell in excess to prevent undesired concentration polarization. For example, nitrate ion is often used as a cathode depolarizer to keep the cathode potential low enough to prevent hydrogen evolution.

(k) A working electrode is the electrode in an electrochemical cell at which the analyte reacts directly or indirectly.

18–2. In amperostatic coulometry, the current in the electroanalytical cell is held constant. In potentiostatic coulometry, the potential of the working electrode is maintained constant.

18–4. (1) Migration, which results from electrostatic attraction between the ions and the electrode, (2) diffusion, which results from concentration differences between the film of solution at an electrode surface and the bulk of the solution, and (3) convection, which brings ions to the electrode surface mechanically.

18–6. Kinetic polarization is often encountered when the reaction product is a gas, particularly when the electrode is a softer metal, such as mercury, zinc, or copper. It is likely to occur at low temperatures and high current densities.

18–8. In a coulometric titration, the reagent is an electric current of known strength; in a volumetric titration, the reagent is a standard solution. In the former, time is measured; in the latter, volume is determined. Both require an end point. In a coulometric method, increments of reagent are introduced by means of a switch that connects the current source to an electrochemical cell. In a vol-

umetric titration, the flow of reagent is controlled by the stopcock of a buret.

18–10. A constant-cathode-potential electrolysis is performed by employing a reference electrode to monitor the potential of a working cathode. The potential applied across the working cathode and an anodic counter electrode is then varied in such a way that the potential of the cathode remains constant at a desired level throughout the electrolysis.

18–11. (a) $-1.381$ V
(b) $-3.46$ V
(c) $-5.70$ V
(d) $-5.79$ V

18–13. (a) $E_1^0 - E_2^0 = 0.077$ V
(c) $E_1^0 - E_2^0 = 0.223$ V

18–14. (a) $4.25 \times 10^{-6}$ M
(b) $-0.425$ V

18–16. (a) Ag deposits first.
(b) $-2.316$ V
(c) $E_{cathode} = 0.200$ to $0.054$ V

18–19. (a) not feasible
(b) feasible
(c) $+0.099$ to $-0.014$ V

18–20. (a) $0.432$ V
(b) $0.277$ V
(c) $-0.127$ V

18–22. (a) $28.4$ min
(b) $9.47$ min

18–24. $13.3\%$ Cd,   $5.43\%$ Zn

18–26. $79.5$ ppm

18–28. $4.062\%$

18–30. $2.47\%$ CCl$_4$;   $1.85\%$ CHCl$_3$

18–32. $4.68\%$

18–34. $0.445$

18–36. $2.73 \times 10^{-4}$ g

## Chapter 19

19–1. (a) Voltammetry is an analytical technique that is based upon measuring the current that develops in a microelectrode as the applied potential is varied. Polarography is a particular type of voltammetry in which the microelectrode is a dropping mercury electrode.

(c) As shown in Figures 19–22 (page 499) and 19–24 (page 501), differential pulse polarography and square wave polarography differ in the type of excitation signal used.

(e) A residual current in polarography is a nonfaradaic charging current that arises from charging individual drops of mercury as they form and fall. A limiting current is a constant faradaic current that is limited in magnitude by the rate at which a reactant is brought to the surface of a microelectrode.

(g) Turbulent flow is a type of liquid flow that has no regular pattern. Laminar flow is a type of liquid flow in which layers of liquid move parallel to each other at the same velocity.

**19-2. (a)** Voltammograms are plots of current as a function of voltage applied to a microelectrode.

**(c)** The Nernst diffusion layer is the thin layer of stagnant solution that is immediately adjacent to the surface of an electrode.

**(e)** The half-wave potential is the potential on a voltammetric wave at which the current is one half of the limiting current.

**19-4.** The advantages of the dropping mercury electrode compared with other types of microelectrodes include: (1) its high hydrogen overvoltage, (2) its continuously produced fresh metal surface, and (3) its reproducible average currents that are immediately realized at any given applied potential. Its disadvantages include: (1) its poor anodic potential range, (2) its relatively large residual currents, (3) its tendency to yield current maxima, and (4) its inconvenience.

**19-6. (a)** $-0.060$ V
**(b)** $+0.059$ V
**(c)**

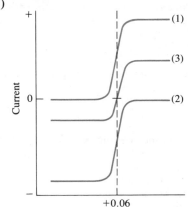

Potential, V vs SCE

**19-8. (a)** $-0.05\%$
**(b)** $-0.14\%$
**(c)** $-0.41\%$

**19-10.** $PbA_2$; $7.2 \times 10^5$

**19-12. (a)** $-0.260$ V
**(b)** $-0.408$ V
**(c)** $-0.615$ V

## Chapter 20

**20-1. (a)** Transitions among various electronic energy levels.

**(c)** Transitions from the lowest vibrational levels of the various excited electronic states to the various vibrational states of the ground state.

**(e)** As in part (c).

**20-2. (a)** The ground state is the lowest energy state of an atom, ion, or molecule and the one occupied by most of the atoms, ions, or molecules of a species at room temperature.

**(c)** A photon is a particle of radiant energy whose magnitude is given by $h\nu$, where $h$ is Planck's constant and $\nu$ is the frequency of a radiation.

**(e)** A continuous spectrum is one produced by a heated body. Even at highest resolution, a continuous spectrum cannot be dispersed into lines.

**(g)** Resonance fluorescence is a type of fluorescence in which the emitted radiation is identical in frequency to the excitation frequency.

**(i)** $\% T = (P/P_0) \times 100\%$, where $T$ is the transmittance, $P_0$ is the power of a beam of radiation incident upon an absorbing medium, and $P$ is the power of the beam after it has passed through a layer of the medium having a thickness of $b$ cm.

**(k)** The molar absorptivity $\varepsilon$ of a species is defined by the equation $\varepsilon = A/bc$, where $A$ is the absorbance of a solution that is $c$ molar in the absorbing species, and $b$ is the path length in centimeters of the radiation used to measure $A$. The absorbance is given by $A = \log(P_0/P) = \log(1/T)$, where the terms used in defining $A$ are found in the answer to part (i).

**(o)** Relaxation is a process whereby an excited species loses energy and returns to a lower energy state.

**20-3. (a)** $1.13 \times 10^{18}$ Hz
**(c)** $4.32 \times 10^{14}$ Hz
**(e)** $1.53 \times 10^{13}$ Hz

**20-4. (a)** 252.8 cm   **(c)** 286 cm

**20-5. (a)** $3 \times 10^3$ to $6.7 \times 10^2$   **(b)** $10 \times 10^{13}$ Hz to $2.0 \times 10^{13}$ Hz

**20-7. (a)** 86.3%   **(b)** 17.2%   **(c)** 48.1%

**20-8. (a)** 0.712   **(b)** 0.064   **(c)** 0.565

**20-9. (a)** 74.5%   **(b)** 3.0%   **(c)** 23.1%

**20-10. (a)** 1.013   **(b)** 0.365   **(c)** 0.866

**20-11. (a)** $\% T = 38.4$; $\varepsilon = 2.38 \times 10^3$; $c = 31.2$ ppm; $a = 0.0133$

**(c)** $\% T = 3.77$; $\varepsilon = 3.42 \times 10^4$; $c = 4.18 \times 10^{-5}$ M; $c = 10.44$ ppm

**(e)** $A = 0.115$; $\% T = 76.7$; $c = 1.33 \times 10^{-5}$ M; $a = 0.0138$

**(g)** $A = 0.316$; $\varepsilon = 4.70 \times 10^4$; $c = 2.69 \times 10^{-5}$ M; $a = 0.188$

**(i)** $A = 0.1175$; $\varepsilon = 1.58 \times 10^4$; $c = 6.76 \times 10^{-6}$ M; $c = 1.69$ ppm

**20-12.** $1.80 \times 10^4$

**20-14. (a)** 0.582   **(b)** 26.2%   **(c)** $1.25 \times 10^{-5}$ M

**20-16.** 0.0595

**20-18. (a)**

| $c_{ind}$, M | $A_{430}$ | $A_{600}$ |
|---|---|---|
| $3.00 \times 10^{-4}$ | 1.54 | 1.06 |
| $2.00 \times 10^{-4}$ | 0.935 | 0.777 |
| $1.00 \times 10^{-4}$ | 0.383 | 0.455 |
| $0.500 \times 10^{-4}$ | 0.149 | 0.261 |
| $0.250 \times 10^{-4}$ | 0.0557 | 0.145 |

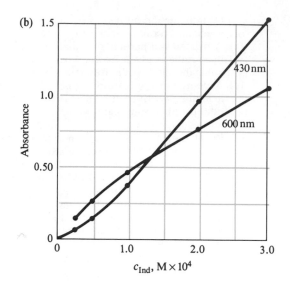

**(b)**

Chapter 21

21–2. **(a)** 446 lines/mm

21–4. A spectroscope consists of a monochromator that has been modified so that the focal plane contains a movable eyepiece that permits visual detection of the various emission lines of the elements. A spectrograph is a monochromator equipped with a photographic film or plate holder located along its focal plane; spectra are recorded photographically. A spectrophotometer is a monochromator equipped with a photoelectric detector that is located behind the exit slit.

21–6. **(a)** 725 nm      **(c)** 1.45 $\mu$m

21–7. **(a)** $1.46 \times 10^7$ W/m²      **(c)** $9.10 \times 10^5$ W/m²

21–8. **(a)** 1.01 $\mu$m,   0.967 $\mu$m
      **(b)** $3.86 \times 10^6$ W/m²,   $4.61 \times 10^6$ W/m²

21–9. **(a)** Hydrogen and deuterium lamps differ only in the gases they contain. The latter produces radiation of somewhat higher intensity.
      **(c)** A phototube is a vacuum tube equipped with a photoemissive cathode. It has a high electrical resistance and requires a potential of 90 V or more to produce a photocurrent. The currents are generally small enough to require considerable amplification before they can be measured. A photovoltaic cell consists of a photosensitive semiconductor sandwiched between two electrodes. A current is generated between the electrodes when radiation is absorbed by the semiconducting layer. The current is generally large enough to be measured directly with a microammeter. The advantages of a phototube are greater sensitivity and wavelength range as well as better reproducibility. The advantages of the photocell are its simplicity, low cost, and general ruggedness. In addition, it does not require an external power supply or elaborate

electronic circuitry. Its use is limited to visible radiation.
      **(e)** A photometer is an instrument for absorption measurements that consists of a source, a filter, and a photoelectric detector. A colorimeter differs from a photometer in that the human eye serves as the detector in the former. The photometer offers the advantage of greater precision and the ability to discriminate between colors, provided they are not too much alike. The main advantages of a colorimeter are simplicity, low cost, and the fact that no power supply is needed.
      **(g)** A single-beam spectrophotometer employs a fixed beam of radiation that irradiates first the solvent and then the analyte solution. In a double-beam instrument, the solvent and solution are irradiated simultaneously or nearly so. The advantages of the double-beam instrument are freedom from problems arising from fluctuations in the source intensity and from drift in electronic circuits; in addition, it is more easily adapted to automatic spectral recording. The single-beam instrument offers the advantages of simplicity and lower cost.

21–10. **(a)** 33.8%      **(b)** 0.471
       **(c)** 0.697      **(d)** 0.114

21–12. In a deuterium lamp, the input energy from the power source produces an excited deuterium *molecule* that dissociates into two atoms in the ground state and a photon of radiation. As the excited deuterium molecule relaxes, its quantized energy is distributed between the energy of the photon and the kinetic energies of the two deuterium atoms. The latter can vary from nearly zero to the original energy of the excited molecule. Therefore, the energy of the radiation, which is the difference between the quantized energy of the excited molecule and the kinetic energies of the atoms, can also vary continuously over the same range. Consequently, the emission spectrum is continuous.

21–14. The power of an infrared beam is measured with a heat detector, which consists of a tiny blackened surface that is warmed as a consequence of radiation absorption. This surface is attached to a transducer, which converts the heat signal to an electrical one. Heat transducers are of three types: (1) a thermopile, which consists of several dissimilar metal junctions that develop a potential that depends upon the difference in temperature of the junctions; (2) a bolometer, which is fashioned from a conductor whose electrical resistance depends upon its temperature; and (3) a pneumatic detector whose internal pressure is temperature-dependent.

21–16. Tungsten/halogen lamps contain a small amount of iodine in the evacuated quartz envelope that holds the tungsten filament. The iodine prolongs the life of the lamp and permits it to operate at a higher temperature. The iodine combines with gaseous tungsten that sublimes from the filament and causes the metal to be redeposited, thus adding to the life of the lamp.

21–18. The basic difference between an absorption spectrometer and an emission spectrometer is that the former requires a separate radiation source and a sample compartment that holds containers for the sample and its solvent. With an emission spectrometer, the sample container is a hot flame, a heated surface, or an electric arc or spark that also serves as the radiation source.

21–20. (a) The dark current is the small current that develops in a radiation transducer in the absence of radiation.

 (c) Scattered radiation in a monochromator is unwanted radiation that reaches the exit slit as a result of reflection and scattering. Its wavelength is usually different from that of the radiation reaching the slit directly from the dispersing element.

 (e) A majority carrier in a semiconductor can be electrons or holes, depending on which is present in the greater amount. The majority carrier, as the name implies, conducts a greater fraction of current than does the minority carrier.

21–21. (a) $1.694 \ \mu$m

 (b) $(4.54/2) \ \mu$m, $(4.54/3) \ \mu$m, and so forth

# Chapter 22

22–1. (a) A chromophore is an unsaturated functional group in an organic molecule that absorbs characteristic wavelengths of ultraviolet or visible radiation.

 (c) Noise in a spectrophotometric measurement refers to the random variations in the output of a spectrophotometer due to uncontrollable fluctuations in electronic circuits, power supplies, sources, and readout devices. Spectrophotometric noise also arises from cell-positioning uncertainties and from uncontrolled variables that affect the chemical behavior of the system under study.

 (e) Charge-transfer absorption arises from the radiation-induced transfer of an electron from a donor group in a complex to an acceptor group. It is characterized by a very large molar absorptivity.

 (g) Quantum efficiency in fluorescence refers to the ratio of the number of quanta absorbed by a sample to the number of photons emitted as fluorescence.

22–2. In segmented-flow methods, solutions are segmented by the introduction of regularly spaced air bubbles into the flowing streams of sample and reagent. In flow-injection analysis, analyte and reagents are introduced into continuously flowing streams of sample and reagent that contain no air bubbles.

22–5. In the mole-ratio method for studying complexes, solutions are prepared in which the concentration of one of the reactants is held constant while the concentration of the other is varied over a considerable range. A plot of absorbance as a function of the mole ratio of the reactants exhibits a change in slope at the ratio that corresponds to the combining ratio of the reactants in the complex.

22–7.

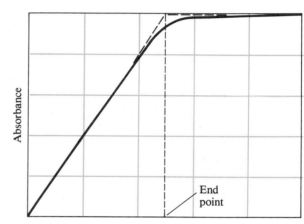

Volume thiocyanate

22–10. The sample compartment in an ultraviolet/visible instrument is located between the monochromator and the detector in order to reduce the possibility of photochemical decomposition of samples that might occur if samples were exposed to the full power of the source. Infrared radition, in contrast, is not energetic enough to cause photochemical problems but suffers instead from a problem caused by the relatively high percentage of radiation scattered by the sample and its container. With the monochromator located between the sample and the detector, much of this scattered radiation is removed and does not reach the detector.

22–12. $4.67 \times 10^{-3}$ g Fe/tablet

22–13. (a) $\pm 9.5\%$ (b) $\pm 1.7\%$ (c) $\pm 2.1\%$

22–14. (a) $1.1\%$ (b) $4.9\%$ (c) $2.7\%$
 (d) $1.4\%$ (e) $1.7\%$
 (f) $2.9\%$ (g) $1.1\%$

22–16. $0.389$ M

22–17. (a) $1.71 \times 10^{-4}$ M
 (b) $c_A = 5.34 \times 10^{-4}$ M; $c_B = 3.57 \times 10^{-4}$ M

22–18. (1) $c_{Co} = 1.65 \times 10^{-4}$ M; $c_{Ni} = 4.33 \times 10^{-5}$ M

22–20.

| | $c_P$, M | $c_Q$, M |
|---|---|---|
| (a) | $2.08 \times 10^{-4}$ | $4.90 \times 10^{-5}$ |
| (c) | $8.36 \times 10^{-5}$ | $6.10 \times 10^{-5}$ |
| (e) | $2.11 \times 10^{-4}$ | $9.64 \times 10^{-5}$ |

22–22. (a) $0.492$ (c) $0.190$

22–23. (a) $0.302$ (c) $0.491$

22–24. A, $5.60$; C, $4.80$

22–25. (a)

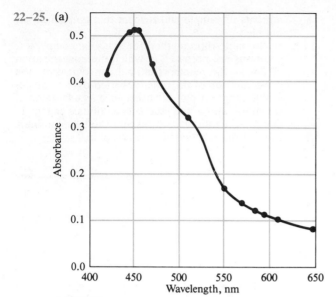

22–27. $1.8 \times 10^8$
22–29. ligand-to-cation ratio = 3
22–31. (a) $CdR^{2+}$      (b) $1.42\ (\pm0.02) \times 10^4$
         (c) $3.8 \times 10^5$
22–33. (a) $CoQ_3^{2+}$      (b) $3.05\ (\pm0.04) \times 10^4$
         (c) $1 \times 10^{17}$
22–34. $5.4 \times 10^{-5}$   to   $1.2 \times 10^{-3}$ M
22–36. (a)

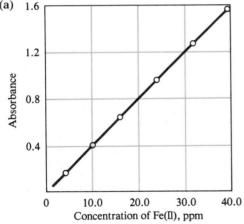

(b) $A = 3.95 \times 10^{-2}c_{Fe} - 1.01 \times 10^{-3}$
(c) $s_r = 3.3 \times 10^{-3}$
(d) $s_b = 1.1 \times 10^{-4}$
22–37. (a) $c_{Fe}$ = 3.65 ppm; $s_r(1)$ = ±2.8%;   $s_r(3)$ = 2.1%
         (c) $c_{Fe}$ = 1.75 ppm; $s_r(1)$ = ±6.1%;   $s_r(3)$ = 4.6%
         (e) $c_{Fe}$ = 38.3 ppm; $s_r(1)$ = ±0.3%;   $s_r(3)$ = 0.2%

## Chapter 23

23–1. (a) *Resonance fluorescence* is observed when ex-
         cited atoms in a vapor emit radiation of the
         same wavelength as that used to excite them.
      (c) *Internal conversion* is the nonradiative relaxa-
         tion of a molecule from the lowest vibrational

level of an excited electronic state to the highest vibrational level of a lower electronic state.
(e) The *Stokes shift* is the difference in wavelength
between the radiation used to excite fluorescence and the wavelength of the emitted radiation. The shift is to longer wavelengths or lower
frequencies.

23–2. For spectrofluorometry, the analytical signal $F$ is
given by $F = K'\varepsilon bcP_0$. The magnitude of $F$, and
thus the sensitivity, can be enhanced by increasing
the source intensity $P_0$ or the transducer sensitivity.
      For spectrophotometry, the analytical signal $A$
is given by $A = \log P/P_0$. Increasing $P_0$ or the detector's response to $P_0$ is accompanied by a corresponding increase in $P$. Thus, the ratio does not
change nor does the analytical signal. Consequently, no improvement accompanies these measurements.

23–4. Compounds that fluoresce have structures that
slow the rate of nonradiative relaxation to the point
where there is time for fluorescence to occur. Compounds that do not fluoresce have structures that
permit rapid relaxation by nonradiative processes.

23–6. Excitation of fluorescence usually involves transfer of an electron to a high vibrational state of an
upper electronic state. Relaxation to a lower vibrational state of this electronic state goes on much
more rapidly than fluorescence relaxation. When
fluorescence relaxation occurs, it is to a high vibrational state of the ground state or to a high vibrational state of an electronic state that is above the
ground state. Such transitions involve less energy
than the excitation energy. Therefore, the emitted
radiation is longer in wavelength than the excitation wavelength.

23–8. Most fluorescence instruments are double-beamed
in order to compensate for fluctuations in the analytical signal arising from fluctuations in the power
supply to the source.

23–10. (a)

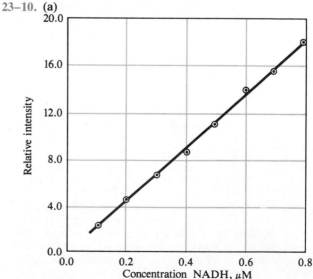

**(b)** $y = 0.000357 + 22.35x$
**(c)** $s_r = 0.17 \quad s_m = 0.27 \quad s_b = 0.136$
**(d)** $x = 0.544 \ \mu M$
**(e)** $s_c = 0.0084$
**(f)** $s_c = 0.0054$

**23–12. (a)**

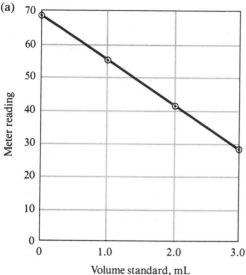

Volume standard, mL

**(b)** $M = 68.23 - 13.22 \ V_s$, where M is the meter reading and $V_s$ is the volume of standard F⁻ solutions.
**(c)** $s_y = 0.435 \quad s_m = 0.19 \quad s_b = 0.36$
**(d)** $c_x = 10.3$ ppb F⁻

## Chapter 24

**24–1.** In atomic emission, the analyte emits radiation as a consequence of excitation in a flame, arc, spark, or plasma. In atomic fluorescence, the analyte emits radiation as a consequence of being excited by a beam of electromagnetic radiation of the wavelength it absorbs. Fluorescence methods require an external source, which is usually located at right angles to the radiation path to the detector. Atomic emission requires no external source of radiation.

**24–3.** In atomic emission spectroscopy, the analytical signal is produced by *excited* atoms or ions, whereas in atomic absorption, the signal results from absorption by *unexcited* species. Typically, the number of unexcited species exceeds the number of excited species by several orders of magnitude. The ratio of unexcited to excited atoms in a hot medium varies exponentially with temperature. Thus a small change in temperature brings about a large relative change in the number of excited atoms. The relative number of unexcited atoms changes very little because they are present in an enormous excess. Therefore, emission spectroscopy is more sensitive to temperature changes than is absorption spectroscopy.

**24–5.** In atomic absorption spectroscopy, the source radiation must be modulated to an ac signal to prevent radiation emitted by the analyte from interfering with the absorption signal. The detector is then made to reject the dc signal from the flame and measure the modulated signal from the source.

**24–7.** The temperature in a hollow cathode lamp is significantly lower than that in a flame. As a consequence, Doppler broadening is less pronounced in the former and narrower lines result.

**24–9.** These peaks are caused by light scattering by particulate matter generated during sample ashing.

**24–11.** When an internal standard is used in emission methods, the ratio of the intensity of the analyte line to the intensity of an internal standard line serves as the analytical parameter. This procedure tends to compensate for indeterminate errors arising from fluctuations in flame temperature.

**24–12.** 0.504 ppm

**24–14. (a)**

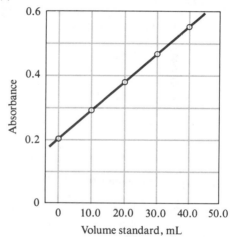

Volume standard, mL

**(b)** If Beer's law is followed, the absorbance is
$$A = \frac{\varepsilon b c_x V_x}{V_T} + \frac{\varepsilon b c_s V_s}{V_T}$$
**(c)** The slope of a plot of $A$ versus $V_s$ is
$$b = \frac{\varepsilon b c_s}{V_T}$$
and the intercept is
$$a = \frac{\varepsilon b c_x V_x}{V_T}$$
**(d)** Dividing the second equation in (c) by the first leads to
$$\frac{a}{b} = \frac{\varepsilon b c_x V_x / V_T}{\varepsilon b c_s / V_T}$$
$$c_x = \frac{a c_s}{b V_x}$$
**(e)** $b = 8.81 \times 10^{-3}$
   $a = 0.202$
**(f)** $s_m = 4.1 \times 10^{-5}$
   $s_r = 1.3 \times 10^{-3}$
**(g)** 28.0 ppm Cr
**(h)** 0.22 ppm Cr

## Chapter 25

25–1. (a) The order of a reaction is the numerical sum of the exponents of the concentration terms in the rate law for the reaction.

(c) Enzymes are high-molecular-weight organic molecules that catalyze reactions of biochemical importance.

(e) The Michaelis constant is an equilibrium-like constant defined by the equation $K_m = (k_{-1} + k_2)/k_1$, where $k_{-1}$ and $k_1$ are the rate constants for the forward and reverse reactions in the formation of an enzyme/substrate complex, which is the intermediate in an enzyme-catalyzed reaction. The term $k_2$ is the rate constant for the decomposition of the complex to give products.

(g) Integral methods use integrated forms of rate equations to compute concentrations from kinetic data.

25–3. Pseudo-first-order conditions are used in kinetic methods because under these conditions the reaction rate is directly proportional to the concentration of the analyte.

25–5. $t_{1/2} = 0.693/k$

25–6. (a) 2.01 s  (c) $1.97 \times 10^3$ s  (e) $1.47 \times 10^9$ s

25–7. (a) 40.5 s$^{-1}$  (c) 0.405 s$^{-1}$  (e) $1.51 \times 10^4$ s$^{-1}$

25–8. (a) 0.105  (c) 2.3  (e) 6.9

25–9. (a) 0.15  (c) 3.3  (e) 10.0

25–10. (a) 0.2%  (c) 0.02%  (e) 1.0%  (g) 0.05%  (i) 6.8%  (k) 0.64%

25–11. $\dfrac{dP}{dt} = \dfrac{k_2[E]_0[S]}{[S] + K_m}$  (Equation 25-22)

At $v_{max}$, $\dfrac{dP}{dt} = k_2[E]_0$. Thus at $v_{max}/2$, we can write

$$\frac{dP}{dt} = \frac{k_2[E]_0}{2} = \frac{k_2[E]_0[S]}{[S] + K_m}$$

$$[S] + K_m = 2[S]$$
$$K_m = [S]$$

25–13.

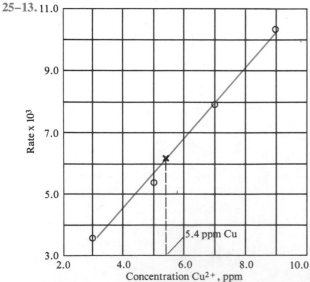

5.4 ppm Cu

Rate x 10$^3$ (y-axis), Concentration Cu$^{2+}$, ppm (x-axis)

25–15. $4.5 \times 10^{-2}$ $\mu$mol/L

## Chapter 26

26–1. (a) Elution is a process in which species are washed through a chromatographic column by additions of fresh solvent.

(c) The stationary phase in a chromatographic column is a solid or liquid that is fixed in place. A mobile phase passes over or through the stationary phase.

(e) The retention time for an analyte is the time interval between its injection onto a column and the appearance of its peak at the other end of the column.

(g) The selectivity factor $\alpha$ of a column toward two species is given by the equation $\alpha = K_B/K_A$, where $K_B$ is the partition ratio of the more strongly held species B and $K_A$ is the corresponding ratio for the less strongly held solute A.

(i) Longitudinal diffusion is a source of band broadening in a column that occurs when a solute diffuses from the concentrated center of a band to the more dilute regions on either side. This movement is thus toward and opposed to the direction of flow of the mobile phase.

(k) The resolution $R_s$ of a column toward two species A and B is given by the equation $R_s = 2\Delta Z/(W_A + W_B)$, where $\Delta Z$ is the distance (in units of time) between the peaks for the two species and $W_A$ and $W_B$ are the widths (also in units of time) of the peaks at their bases.

26–2. The general elution problem arises whenever chromatograms are obtained on samples that contain species with widely different partition ratios. When conditions are such that good separation of the more strongly held species is realized, lack of resolution among the weakly retained species is observed. Conversely, when conditions are chosen to give satisfactory separation of the weakly retained compounds, severe band broadening and long retention times are encountered for the strongly bound species. The general elution problem is often solved in liquid chromatography by gradient elution; temperature programming serves the same purpose in gas chromatography.

26–4. In gas-liquid chromatography, the mobile phase is a gas, whereas in liquid-liquid chromatography it is a liquid.

26–6. Variables that affect $\alpha$ values include composition of the mobile phase, column temperature, composition of the stationary phase, and chemical interaction between the stationary phase and one of the solutes being separated.

26–8. The number of plates in a column can be determined by measuring the retention time $t_R$ and width of a peak at its base $W$. The number of plates $N$ is then given by the equation $N = 16(t_R/W)^2$.

**26-10.** The minima observed in plots of plate height versus flow rate are caused by longitudinal diffusion, which, in contrast to other broadening sources, goes on to a greater extent at low flow rates than at high ones. The rate of longitudinal diffusion is orders of magnitude larger in a gaseous mobile phase than in a liquid, however. Thus, the phenomenon becomes noticeable at higher flow rates in gases than in liquids.

**26-12.** (a) $N_A = 2775$; $N_B = 2472$; $N_C = 2364$; $N_D = 2523$
(b) $N = 2534(\pm174) = 2.5(\pm0.2) \times 10^3$ plates
(c) $H = 0.0097$ cm

**26-13.**

|  | A | B | C | D |
|---|---|---|---|---|
| (a) $k'$ | 0.74 | 3.3 | 3.5 | 6.0 |
| (b) $K$ | 6.2 | 28 | 29 | 50 |

**26-14.** (a) 0.72   (b) 1.1   (c) 108 cm
(d) 62 min

**26-15.** (a) 5.2   (b) 2.1 cm

**26-20.** (a) Variables that lead to band broadening: (1) very low or very high flow rates; (2) high viscosity; (3) low temperature; (4) large particle size for packing; (5) for liquid stationary phases, thick films; (6) long columns; (7) slow introduction of sample; (8) large samples.
(b) Variables that lead to band separation: (1) packings that produce partition coefficients that differ significantly; (2) increased packing length; (3) variation in solvent composition; (4) optimum temperature; (5) change in mobile-phase pH; (6) incorporation of species in the stationary phase that selectively complex certain analytes.

**26-21.** Slow sample introduction leads to band broadening.

**26-22.** (a) $k'_M = 2.54$; $k'_N = 2.62$
(b) $\alpha_{N,M} = 6.20/6.01 = 1.03$
(c) $8.1 \times 10^4$ plates
(d) $1.8 \times 10^2$ cm
(e) 91 min

## Chapter 27

**27-1.** In gas-liquid chromatography, the stationary phase is a liquid immobilized on a solid. Retention of sample constituents involves equilibria between a gaseous and a liquid phase. In gas-solid chromatography, the stationary phase is a solid surface that retains analytes by physical adsorption. Here, separations involve adsorption equilibria.

**27-3.** In open tubular columns, the stationary phase is retained by adsorption or chemical bonding on the inner surface of a capillary tubing. In packed columns, the stationary phase is a liquid immobilized on uniform particles of an inert solid contained in a narrow glass or metal tube. The main advantage of open tubular columns is their remark-

able resolution. Packed columns are easier and more convenient to use, however, and in addition accommodate much larger amounts of sample.

**27-5.** Sample splitters are mechanical devices that deliver a fixed small fraction of a sample to the head of a column; the remainder of the sample goes to waste. Capillary columns require such small samples that reproducible volumes cannot be introduced with a syringe. This problem is overcome by delivering easily measured volumes to the splitter, which then reduces the volume delivered to the column in a precise way.

**27-7.** Temperature programming involves increasing the temperature of a gas-chromatographic column in a programmed way as a separation is being carried out. This technique is particularly useful for samples that contain constituents that boil over a large temperature range. The lower-boiling constituents are separated initially at temperatures that provide good resolution. As the separation proceeds, the column temperature is increased so that the higher-boiling constituents come off the column with good resolution and at reasonable lengths of time.

**27-9.** Hydrogen or helium is used as the carrier gas with thermal-conductivity detectors because the conductivities of these gases are several times greater than those of most organic compounds. Thus, the change in conductivity brought about by the presence of organic species in the effluent from a column is larger than it would be with gases having lower thermal conductivities.

**27-11.** In hyphenated methods, chromatographic columns are coupled with highly selective detectors such as ultraviolet, infrared, and mass spectrometers and various kinds of electroanalytical instruments. These hyphenated methods are powerful tools for identifying the components of complex mixtures.

**27-13.** The label -AW-DMCS stands for acid-washed dimethylchlorosilane column packing.

**27-15.** (a) $k'_1 = 5.60$; $k'_2 = 6.46$; $k'_3 = 25.4$
(b) $\alpha_{2,1} = 1.16$; $\alpha_{3,2} = 3.93$
(c) $1.6 \times 10^3$ plates;   0.068 cm/plate
(d) $(R_s)_{2,1} = 1.2$; $(R_s)_{3,2} = 11$
(e) $2.5 \times 10^3$ plates, or 1.7-m column
(f) 12 min

**27-17.** Retention times would be shorter because the nonpolar stationary phase would have less tendency to retain the polar analyte molecules.

**27-19.** (a) This would lead to a thicker layer of liquid stationary phase, which increases band broadening and thus increases plate heights.
(c) Increasing the port temperature leads to greater band broadening owing to a greater longitudinal diffusion; larger plate heights are the result.
(e) Reducing the particle size of the packing improves efficiency and thus reduces plate height.

**27-20.** %A = 21.1; %B = 13.1; %C = 36.4; %D = 18.7; %E = 10.7

## Chapter 28

**28–1.** (a) Substances that are somewhat volatile and are thermally stable.

(c) Substances that are ionic.

(e) High-molecular-weight compounds that are soluble in nonpolar solvents.

(g) Relatively nonpolar, water-insoluble compounds with molecular weights lower than 5000. Good for separating isomers.

**28–2.** (a) In an isocratic elution, the solvent composition is held constant throughout the elution.

(c) In a stop-flow injection, the flow of solvent is stopped, a fitting at the head of the column is removed, and the sample is injected directly onto the head of the column. The fitting is then replaced and pumping is resumed.

(e) In a normal-phase packing, the stationary phase is quite polar and the mobile phase is relatively nonpolar.

(g) An eluent-suppressor column is located after the ion-exchange column in ion chromatography. It converts the ionized species used to elute analyte ions to largely undissociated molecules that do not interfere with conductometric detection.

(i) Gel-permeation chromatography is a type of size-exclusion chromatography in which the packings are hydrophobic and the eluents are nonaqueous. It is used for separating high-molecular-weight nonpolar species.

(k) FSOT columns are fused silica open tubular columns used in gas chromatography.

(m) A supercritical fluid is a substance that is heated above its critical temperature so that it cannot be condensed into a liquid no matter how great the pressure.

**28–4.** (a) n-hexane, benzene, n-hexanol

**28–5.** (a) n-hexanol, benzene, n-hexane

**28–7.** In adsorption chromatography, the sample components are selectively retained on the surface of a solid stationary phase by adsorption. In partition chromatography, selective retention occurs in a liquid or liquid-like stationary phase.

**28–9.** Nonvolatile and thermally unstable compounds.

**28–11.** The simplest type of pump for liquid chromatography is a pneumatic pump, which consists of a collapsible solvent container housed in a vessel that can be pressurized by a compressed gas. This type of pump is simple, inexpensive, and pulse-free. It has limited capacity and pressure output, it is not adaptable to gradient elution, and its pumping rate depends upon the viscosity of the solvent. A screw-driven syringe pump consists of a large syringe in which the piston is moved in or out by means of a motor-driven screw. It also is pulse-free, and the rate of delivery is easily varied. It suffers from lack of capacity and is inconvenient to use when solvents must be changed. The most versatile and widely used pump is the reciprocating pump, which usually consists of a small cylindrical chamber that is filled and then emptied by the back-and-forth motion of a piston. Advantages of the reciprocating pump include small internal volume, high output pressures, adaptability to gradient elution, and flow rates that are constant and independent of viscosity and back pressure. The main disadvantage is the pulsed output that must subsequently be damped.

**28–13.** (a) The advantages over HPLC are (1) that it can be used with general detectors, such as the flame ionization detector, and (2) that it is inherently faster or gives better resolution for the same speed.

(b) The advantage over GLC is that it can be used for nonvolatile and thermally unstable samples.

## Chapter 30

**30–1.** The steps in sampling are (1) identification of the population from which the sample is to be drawn, (2) collection of a gross sample, and (3) reduction of the gross sample to a small quantity of homogeneous material for analysis.

**30–3.** (a) Sorbed water is that held as a condensed liquid phase in the capillaries of a colloid. Adsorbed water is that retained by adsorption on the surface of a finely ground solid. Occluded water is that held in cavities distributed irregularly throughout a crystalline solid.

(c) Essential water is chemically bound water that occurs as an integral part of the molecular or crystalline structure of a compound in its solid state. Nonessential water is that retained by a solid as a consequence of physical forces.

**30–5.** For NIST sample, mean = 50.30% CaO; $s$ = 0.09% CaO; $s_{rel}$ = 1.8 ppt
For gross samples, mean = 49.92% CaO; $s$ = 0.39% CaO; $s_{rel}$ = 7.8 ppt
Relative variance of sampling = $(7.8)^2 - (1.8)^2$ = 5.8
Relative standard deviation of sampling = 7.6 ppt

**30–6.**

|  | $\sigma_{rel}$, % | $\sigma_{abs}$, **particles** |
|---|---|---|
| (a) | 1.9 | 6 |
| (b) | 16 | 6 |
| (c) | 10 | 80 |
| (d) | 12 | 8 |
| (e) | 1.2 | 13 |

**30–8.** (a) 784 tablets   (c) $2.0 \times 10^4$ tablets

**30–9.** (a) $\sigma_{rel}$ = 12%

(b) $\sigma_{abs}$ = $1.1 \times 10^3$ bottles

(c) $9000 \pm 1800$ bottles

(d) $1.5 \times 10^3$ bottles

**30–10.** (a) $3.5 \times 10^4$ particles

(b) $3.9 \times 10^3$ g (8.5 lb)

(c) 0.22 mm

**30–12.** (a) $2.2 \times 10^5$ particles

(b) $2.8 \times 10^3$ g (6.1 lb)

(c) 0.13 mm

## Chapter 31

**31–1.** Dry-ashing is carried out by igniting the sample in air or sometimes oxygen. Wet-ashing is carried out by heating the sample in an aqueous medium containing such oxidizing agents as $H_2SO_4$, $HClO_4$, $HNO_3$, $H_2O_2$, or some combination of these.

**31–3.** $B_2O_3$ or $CaCO_3/NH_4Cl$

**31–5.** When hot concentrated $HClO_4$ comes in contact with organic materials or other oxidizable species, explosions are highly probable.

**31–6.** (a) Samples for halogen determination can be decomposed in a Schöniger combustion flask, combusted in a tube furnace in a stream of oxygen, or fused in a peroxide bomb.

(c) Samples for nitrogen determination are decomposed in hot concentrated $H_2SO_4$ in a Kjeldahl flask or oxidized by CuO in a tube furnace in the Dumas method.

## Chapter 32

**32–1.** A masking agent is a complexing reagent that reacts selectively with one or more components of a solution to prevent them from interfering in an analysis.

**32–3.** A collector is a species that is added to a sample when traces of an analyte are to be precipitated. The collector forms a precipitate with the reagent used to precipitate the analyte and prevents the small amount of analyte precipitate from becoming physically lost.

**32–4.** (a) NaOH. The basic oxide of iron precipitates, whereas the aluminum remains in solution as the aluminate ion.

(c) Hot concentrated $HNO_3$ precipitates Sn(IV) as $SnO_2 \cdot xH_2O$.

(e) Precipitate $Hg^{2+}$ as the sulfide from 3 M HCl.

**32–5.** (a) $1.73 \times 10^{-2}$ M    (b) $6.40 \times 10^{-3}$ M
(c) $2.06 \times 10^{-3}$ M    (d) $6.89 \times 10^{-4}$ M

**32–7.** (a) 75 mL    (b) 40 mL    (c) 22 mL

**32–9.** (a) 18.0    (b) 7.56

**32–11.** (a) $K_d = 1.53$
(b) [HA] = 0.0147;   $[A^-] = 0.0378$
(c) $K_a = 9.7 \times 10^{-2}$

**32–13.** (a) 12.4 meq $Ca^{2+}$/L
(b) $6.19 \times 10^2$ mg $CaCO_3$/L

**32–15.** Dissolve 17.53 g of NaCl in about 100 mL of water and pass the solution through a column packed with a cation exchange resin in its acid form. Wash the column with several hundred milliliters of water, collecting the liquid from the original solution and the washings in a 2-L volumetric flask. Dilute to the mark and mix well.

**32–17.** (a) $H_2L(aq) \rightleftharpoons H_2L$ (org)
$H_2L(aq) + H_2O \rightleftharpoons H_3O^+ + HL^-(aq)$
$HL^-(aq) + H_2O \rightleftharpoons H_3O^+ + L^{2-}(aq)$
$Cu^{2+}(aq) + L^{2-}(aq) \rightleftharpoons CuL(aq)$
$CuL(aq) \rightleftharpoons CuL(org)$
(b) $K_{ex} = 5.01 \times 10^{-8}$
(c) At pH 6.00, $K = 5.01$.
(d) four extractions
(e) six extractions

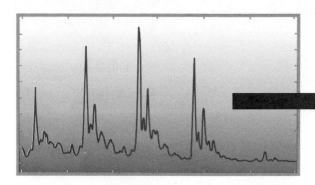

# INDEX

Boldface page numbers refer to specific laboratory directions.
Page numbers followed by *t* refer to tabular entries; page numbers
preceded by A. refer to appendices.

Nitrous oxide
  as oxidant, atomic spectroscopy, 619, 626
  supercritical fluid chromatography with, 729
Noise, instrumental, 547, 576, 577*t*
Nominal wavelength, 540
Nonaqueous solvents
  influence of, on solute behavior, 279
  selection of, 280
  titrations in, 277–282, **876–878**
Noncatalytic kinetic methods, 662
Noncrystalline membrane electrodes, 406
Nonequilibrium mass transfer, linear chromatography, 679
Nonessential water, 759
Nonfaradaic currents, 493, 496, 499, 500
Nonflame atomizers, for atomic spectroscopy, 632
Nonradiative relaxation, of excited molecules, 507, 604
Nonresonance fluorescence, 606
Nonsegmented flow analyzers, 592 (*see also* Flow injection analysis)
Normal error curve, 19, 36–38
Normality, $c_N$, 97, A.23, A.26
Normal phase chromatography, 718
Notebook, rules for keeping, 829
n-type semiconductor, 550
Nucleation, 73
Nuclides, half-life of, 644
Null detector, in optical instruments, 558
Null hypothesis, 48
Number of theoretical plates, $N$, 673, 689*t*
  estimation of, from zone breadth, 676

Occluder, 567
Occlusion
  in crystalline precipitates, 78, 81
  of water in solids, 759, 761
Occupational Safety and Health Administration, OSHA, air-quality standards of, 591*t*
Oesper's salt, 373
Ohmic potential, 442 (*see also IR* drop)
Olefins, coulometric titration of, 467*t* (*see also* Alkene group)
One hundred percent $T$ adjustment, absorption spectroscopy, 556, 567
  effect of uncertainties in, 576
Open tubular columns, GLC, 697, 698

Operational amplifier, 423, 475, 553
Operator time, in chemical analysis, 84, 834
Optical density, $D$, 518*t*
Optical methods of analysis, 508 (*see also* Absorption spectroscopy, listings under specific methods)
Optical wedge, 558
Order
  of chemical reactions, 641
  spectral, **n**, 541, 544
Organic compounds
  absorption by, uv/visible regions, 562–565
  coulometric analysis of, 460, **876–878**
  decomposition of, 769–774 (*see also* Kjeldahl method)
  fluorescence analysis for, 611
Organic functional group analysis
  gravimetric, 89
  kinetic, 660
  neutralization, 275
  with periodate, 391 (*see also* Periodic acid)
  polarographic, 495, 503
  spectrophotometric, 562, 568, 588, 611
Organic halogen groups, polarographic behavior of, 503
Organic peroxides, iodometric determination of, 373, 376*t*, 385
Organic phosphates, gravimetric analysis for, 89*t*
Organic reagents
  for gravimetric analysis, 86–89
  separations involving, 778
  for spectrophotometric analysis, 572
Orthophenanthrolines, 364 (*see also* 1,10-Phenanthroline-iron(II) complex)
OSHA tolerances, for exposure levels, 588, 591*t*
Ostwald-Folin pipets, 820*t*
Outer cone, of flame, 616
Outliers, 13
  criteria for rejection of, 45–48
Oven drying, 806
Overall formation constant, $\beta_n$, 142
Overall order, of a chemical reaction, $p$, 641
Overdetermination, of spectra, 576
Overvoltage, 444 (*see also* Kinetic polarization)
Oxalate ion
  homogeneous generation of, 83*t*
  titration of, 208*t*
Oxalic acid
  as reducing reagent, 87*t*

Oxalic acid (*continued*)
  titration of, 381*t*, 467*t* (*see also* Sodium oxalate)
Oxidant
  titration of, 311 (*see also* Oxidizing agent)
Oxidation, 311
  during grinding of solids, 757
Oxidation potential, 324
Oxidation/reduction indicators, 362–368
Oxidation/reduction reactions, 311–314
  definition of equivalent weight for, A.24
  equilibria in, 125*t*, 349
Oxidation/reduction titrations, 345–368 (*see also* listings for specific reagents)
  amperometric, 489
  coulometric, 466, 467*t*
  curves for, 350–362, 437
  equivalence-point potential for, 352–354, 355, 358
  indicator electrodes for, 405
  of mixtures, 361
  potentiometric, 368, 437
Oxides, interference by, in Karl Fischer titrations, 395
Oxidizing agent, 311, 329
  auxiliary, 370, 371
  as flux, 769*t*
  interference by, in Karl Fischer titrations, 395
  as solvents, for samples, 766
  titration of, 373, 384
  volumetric, 376–393, **853–862**
Oxygen
  combustion-tube analysis for, 773*t*
  iodometric determination of, 376*t*
  overvoltage effects with, 447
  voltammetric behavior of, 486, 487
Ozone, iodometric determination of, 376*t*

Packed columns, GLC, 697, 699*t*
Packings, effect on column efficiency, 676*t*, 681, 712
Palladium electrode, for potentiometric titrations, 405, 437
Pan, of analytical balance, 798
Pan arrests, analytical balance, 799
Paper, as filtering medium, 811 (*see also* Ashless filter paper)
Paper chromatography, 732, 736
Parallax, error due to, 821
Particle growth, 73
Particle size
  and column efficiency, 676*t*, 698
  effect on sampling, 752–754

# Periodic Table of the Elements

Legend:
- Metals
- Metalloids
- Nonmetals

| Group IA | IIA | IIIB | IVB | VB | VIB | VIIB | VIIIB | | | IB | IIB | IIIA | IVA | VA | VIA | VIIA | VIII |
|---|---|---|---|---|---|---|---|---|---|---|---|---|---|---|---|---|---|
| Hydrogen 1 **H** 1.00797 | | | | | | | | | | | | | | | | Hydrogen 1 **H** 1.0079 | Helium 2 **He** 4.0026 |
| Lithium 3 **Li** 6.942 | Beryllium 4 **Be** 9.0122 | | | | | | | | | | | Boron 5 **B** 10.811 | Carbon 6 **C** 12.01115 | Nitrogen 7 **N** 14.0067 | Oxygen 8 **O** 15.9994 | Fluorine 9 **F** 18.9984 | Neon 10 **Ne** 20.183 |
| Sodium 11 **Na** 22.9898 | Magnesium 12 **Mg** 24.312 | | | | | | | | | | | Aluminum 13 **Al** 26.9815 | Silicon 14 **Si** 28.086 | Phosphorus 15 **P** 30.9738 | Sulfur 16 **S** 32.064 | Chlorine 17 **Cl** 35.453 | Argon 18 **Ar** 39.948 |
| Potassium 19 **K** 39.102 | Calcium 20 **Ca** 40.08 | Scandium 21 **Sc** 44.956 | Titanium 22 **Ti** 47.90 | Vanadium 23 **V** 50.942 | Chromium 24 **Cr** 51.996 | Manganese 25 **Mn** 54.9380 | Iron 26 **Fe** 55.847 | Cobalt 27 **Co** 58.9332 | Nickel 28 **Ni** 58.70 | Copper 29 **Cu** 63.54 | Zinc 30 **Zn** 65.37 | Gallium 31 **Ga** 69.72 | Germanium 32 **Ge** 72.59 | Arsenic 33 **As** 74.9216 | Selenium 34 **Se** 78.96 | Bromine 35 **Br** 79.909 | Krypton 36 **Kr** 83.80 |
| Rubidium 37 **Rb** 85.4678 | Strontium 38 **Sr** 87.62 | Yttrium 39 **Y** 88.9059 | Zirconium 40 **Zr** 91.224 | Niobium 41 **Nb** 92.9064 | Molybdenum 42 **Mo** 95.94 | Technetium 43 **Tc** (98) | Ruthenium 44 **Ru** 101.07 | Rhodium 45 **Rh** 102.9055 | Palladium 46 **Pd** 106.4 | Silver 47 **Ag** 107.870 | Cadmium 48 **Cd** 112.40 | Indium 49 **In** 114.82 | Tin 50 **Sn** 118.69 | Antimony 51 **Sb** 121.75 | Tellurium 52 **Te** 127.60 | Iodine 53 **I** 126.9044 | Xenon 54 **Xe** 131.30 |
| Cesium 55 **Cs** 132.905 | Barium 56 **Ba** 137.34 | Lanthanum 57 *****La** 138.91 | Hafnium 72 **Hf** 178.49 | Tantalum 73 **Ta** 180.948 | Tungsten 74 **W** 183.85 | Rhenium 75 **Re** 186.2 | Osmium 76 **Os** 190.2 | Iridium 77 **Ir** 192.2 | Platinum 78 **Pt** 195.09 | Gold 79 **Au** 196.967 | Mercury 80 **Hg** 200.59 | Thallium 81 **Tl** 204.37 | Lead 82 **Pb** 207.19 | Bismuth 83 **Bi** 208.980 | Polonium 84 **Po** (209) | Astatine 85 **At** (210) | Radon 86 **Rn** (222) |
| Francium 87 **Fr** (223) | Radium 88 **Ra** (226) | Actinium 89 ******Ac** (227) | Unniquadium 104 **Unq** (261) | Unnilpentium 105 **Unp** (262) | Unnilhexium 106 **Unh** (263) | Unnilseptium 107 **Uns** (262) | Unniloctium 108 **Uno** (265) | Unnilennium 109 **Une** (266) | | | | | | | | | |

*Lanthanide Series

| Cerium 58 **Ce** 140.12 | Praseodymium 59 **Pr** 140.907 | Neodymium 60 **Nd** 144.24 | Promethium 61 **Pm** (145) | Samarium 62 **Sm** 150.35 | Europium 63 **Eu** 151.96 | Gadolinium 64 **Gd** 157.25 | Terbium 65 **Tb** 158.924 | Dysprosium 66 **Dy** 162.50 | Holmium 67 **Ho** 164.930 | Erbium 68 **Er** 167.26 | Thulium 69 **Tm** 168.934 | Ytterbium 70 **Yb** 173.04 | Lutetium 71 **Lu** 174.97 |
|---|---|---|---|---|---|---|---|---|---|---|---|---|---|

**Actinide Series

| Thorium 90 **Th** 232.038 | Protactinium 91 **Pa** 231.0359 | Uranium 92 **U** 238.03 | Neptunium 93 **Np** (237) | Plutonium 94 **Pu** (244) | Americium 95 **Am** (243) | Curium 96 **Cm** (247) | Berkelium 97 **Bk** (247) | Californium 98 **Cf** (251) | Einsteinium 99 **Es** (252) | Fermium 100 **Fm** (257) | Mendelevium 101 **Md** (258) | Nobelium 102 **No** (259) | Lawrencium 103 **Lr** (260) |
|---|---|---|---|---|---|---|---|---|---|---|---|---|---|

# International Atomic Weights

| Element | Symbol | Atomic number | Atomic weight | Element | Symbol | Atomic number | Atomic weight |
|---------|--------|--------------|--------------|---------|--------|--------------|--------------|
| Actinium | Ac | 89 | (227) | Mercury | Hg | 80 | 200.59 |
| Aluminum | Al | 13 | 26.9815 | Molybdenum | Mo | 42 | 95.94 |
| Americium | Am | 95 | (243) | Neodymium | Nd | 60 | 144.24 |
| Antimony | Sb | 51 | 121.75 | Neon | Ne | 10 | 20.183 |
| Argon | Ar | 18 | 39.948 | Neptunium | Np | 93 | (237) |
| Arsenic | As | 33 | 74.9216 | Nickel | Ni | 28 | 58.70 |
| Astatine | At | 85 | (210) | Niobium | Nb | 41 | 92.906 |
| Barium | Ba | 56 | 137.34 | Nitrogen | N | 7 | 14.0067 |
| Berkelium | Bk | 97 | (247) | Nobelium | No | 102 | (259) |
| Beryllium | Be | 4 | 9.0122 | Osmium | Os | 76 | 190.2 |
| Bismuth | Bi | 83 | 208.980 | Oxygen | O | 8 | 15.9994 |
| Boron | B | 5 | 10.811 | Palladium | Pd | 46 | 106.4 |
| Bromine | Br | 35 | 79.909 | Phosphorus | P | 15 | 30.9738 |
| Cadmium | Cd | 48 | 112.40 | Platinum | Pt | 78 | 195.09 |
| Calcium | Ca | 20 | 40.08 | Plutonium | Pu | 94 | (244) |
| Californium | Cf | 98 | (251) | Polonium | Po | 84 | (209) |
| Carbon | C | 6 | 12.01115 | Potassium | K | 19 | 39.102 |
| Cerium | Ce | 58 | 140.12 | Praseodymium | Pr | 59 | 140.907 |
| Cesium | Cs | 55 | 132.905 | Promethium | Pm | 61 | (145) |
| Chlorine | Cl | 17 | 35.453 | Protactinium | Pa | 91 | 231.0359 |
| Chromium | Cr | 24 | 51.996 | Radium | Ra | 88 | (226) |
| Cobalt | Co | 27 | 58.9332 | Radon | Rn | 86 | (222) |
| Copper | Cu | 29 | 63.54 | Rhenium | Re | 75 | 186.2 |
| Curium | Cm | 96 | (247) | Rhodium | Rh | 45 | 102.905 |
| Dysprosium | Dy | 66 | 162.50 | Rubidium | Rb | 37 | 85.47 |
| Einsteinium | Es | 99 | (252) | Ruthenium | Ru | 44 | 101.07 |
| Erbium | Er | 68 | 167.26 | Samarium | Sm | 62 | 150.35 |
| Europium | Eu | 63 | 151.96 | Scandium | Sc | 21 | 44.956 |
| Fermium | Fm | 100 | (257) | Selenium | Se | 34 | 78.96 |
| Fluorine | F | 9 | 18.9984 | Silicon | Si | 14 | 28.086 |
| Francium | Fr | 87 | (223) | Silver | Ag | 47 | 107.870 |
| Gadolinium | Gd | 64 | 157.25 | Sodium | Na | 11 | 22.9898 |
| Gallium | Ga | 31 | 69.72 | Strontium | Sr | 38 | 87.62 |
| Germanium | Ge | 32 | 72.59 | Sulfur | S | 16 | 32.064 |
| Gold | Au | 79 | 196.967 | Tantalum | Ta | 73 | 180.948 |
| Hafnium | Hf | 72 | 178.49 | Technetium | Tc | 43 | (99) |
| Helium | He | 2 | 4.0026 | Tellurium | Te | 52 | 127.60 |
| Holmium | Ho | 67 | 164.930 | Terbium | Tb | 65 | 158.924 |
| Hydrogen | H | 1 | 1.00797 | Thallium | Tl | 81 | 204.37 |
| Indium | In | 49 | 114.82 | Thorium | Th | 90 | 232.038 |
| Iodine | I | 53 | 126.9044 | Thulium | Tm | 69 | 168.934 |
| Iridium | Ir | 77 | 192.2 | Tin | Sn | 50 | 118.69 |
| Iron | Fe | 26 | 55.847 | Titanium | Ti | 22 | 47.90 |
| Krypton | Kr | 36 | 83.80 | Tungsten | W | 74 | 183.85 |
| Lanthanum | La | 57 | 138.91 | Uranium | U | 92 | 238.03 |
| Lawrencium | Lw | 103 | (260) | Vanadium | V | 23 | 50.942 |
| Lead | Pb | 82 | 207.19 | Xenon | Xe | 54 | 131.30 |
| Lithium | Li | 3 | 6.942 | Ytterbium | Yb | 70 | 173.04 |
| Lutetium | Lu | 71 | 174.97 | Yttrium | Y | 39 | 88.905 |
| Magnesium | Mg | 12 | 24.312 | Zinc | Zn | 30 | 65.37 |
| Manganese | Mn | 25 | 54.9380 | Zirconium | Zr | 40 | 91.22 |
| Mendelevium | Mv | 101 | (258) | | | | |

Numbers in parentheses indicate mass number of most stable known isotope.